Y0-BSD-912

ALLE·ZEIT·WACH
1842

Kenneth R. Lang

Astrophysical Formulae

A Compendium for the Physicist and Astrophysicist

Second Corrected and Enlarged Edition

With 46 Figures

Springer-Verlag Berlin Heidelberg New York 1980

Dr. KENNETH R. LANG

Professor of Astronomy
Department of Physics
Tufts University
Medford, Massachusetts 02155
U.S.A.

ISBN 3-540-55040-2 Springer-Verlag Berlin Heidelberg New York (pbk.)
ISBN 0-387-55040-2 Springer-Verlag New York Heidelberg Berlin (pbk.)
ISBN 3-540-09933-6 Springer-Verlag Berlin Heidelberg New York (hdbk.)
ISBN 0-387-09933-6 Springer-Verlag New York Heidelberg Berlin (hdbk.)

ISBN 3-540-06605-5 1. Auflage Springer-Verlag Berlin Heidelberg New York
ISBN 0-387-06605-5 1st Edition Springer-Verlag New York Heidelberg Berlin

Library of Congress Cataloging in Publication Data. Lang, Kenneth R., Astrophysical formulae. "Springer study edition". Bibliography: p. Includes indexes. 1. Astrophysics-Formulae. I. Title. QB461.L36 1980 523.01'02'12 80-12918

9 8 7 6 5 4

Printed and bound by Edwards Brothers, Inc., Ann Arbor, Michigan.
Printed in the United States of America.

2153/3130-543210

To Marcella

Ebbene, forse voi credete
che l'arco senza fondo della volta spaziale
sia un vuoto vertiginoso di silenzi.
Vi posso dire allora che verso
questa terra sospettabile appena
l'universo giá dilaga di pensieri.

(MARIO SOCRATE, *Favole paraboliche*)

Well, maybe you think
That the endless arch of the space vault
Is a giddy, silent hollowness.
But I can tell you that,
Overflowing with thought,
the universe is approaching
this hardly guessable earth.

(MARIO SOCRATE, *Parabolic Fables*)

Preface to the Second Edition

This second edition contains corrections of misprints and errors found by the author, as well as those suggested during the Russian translation of the first printing. The Russian editors and translators who kindly supplied this material include V. E. CHERTOPRUD, A. G. DOROSHKEVICH, V. L. HOHLOVA, M. Yu. KHLOPOV, D. K. NADIOZHIN, L. M. OZERNOI, I. G. PERSIANTSEV, L. A. POKROVSKII, A. V. ZASOV, and Yu. K. ZEMTSOV. Supplemental references for the period 1974 to 1980 have also been added as appendix where they are included under the headings of general references and specific references for each chapter. Although specialized references come mainly from American journals, references to reviews and books are also included to help guide the reader to other sources. The author encourages suggestions for additions and corrections to possible future editions of this volume.

KENNETH R. LANG

Department of Physics, Tufts University
Medford, Massachusetts

January, 1980

Preface

This book is meant to be a reference source for the fundamental formulae of astrophysics. Wherever possible, the original source of the material being presented is referenced, together with references to more recent modifications and applications. More accessible reprints and translations of the early papers are also referenced. In this way the reader is provided with the often ignored historical context together with an orientation to the more recent literature. Any omission of a reference is, of course, not meant to reflect on the quality of its contents. In order to present a wide variety of concepts in one volume, a concise style is used and derivations are presented for only the simpler formulae. Extensive derivations and explanatory comments may be found in the original references or in the books listed in the selected bibliography which follows. Following the convention in astrophysics, the c.g.s. (centimeter-gram-second) system of units is used unless otherwise noted. To conserve space, the fundamental constants are not always defined, and unless otherwise noted the following symbols have the following meaning and value.

Symbol	Meaning	Value
c	Speed of light in vacuum	$2.997924562(11)\times10^{10}$ cm s^{-1}
$h=2\pi\hbar$	Planck's constant	$6.626196(50)\times10^{-27}$ erg s
$\hbar=h/2\pi$	Rationalized Planck's constant	$1.0545919(80)\times10^{-27}$ erg s
k	Boltzmann's constant	$1.380622(59)\times10^{-16}$ erg $\mathring{K}^{-1}$
e	Elementary charge of an electron	$4.803250(21)\times10^{-10}$ esu
m	Electron rest mass	$9.109558(54)\times10^{-28}$ g
G	Gravitational constant	$6.6732(31)\times10^{-8}$ dyn cm^2 g^{-2}
N_A	Avogadro's number	$6.022169(40)\times10^{23}$ mole^{-1}
a.m.u.$=u$	Atomic mass unit	$1.660531(11)\times10^{-24}$ g
α	Fine-structure constant	$7.297351(11)\times10^{-3}$
e/m	Electron charge to mass ratio	$5.272759(16)\times10^{17}$ esu g^{-1}
R_∞	Rydberg constant	$1.09737312(11)\times10^{5}$ cm^{-1}
a_0	Bohr radius	$5.2917715(81)\times10^{-9}$ cm
$h/(mc)$	Compton wavelength	$2.4263096(74)\times10^{-10}$ cm
$r_0=e^2/(mc^2)$	Classical electron radius	$2.817939(13)\times10^{-13}$ cm
R	Gas constant	$8.31434(35)\times10^{7}$ erg $\mathring{K}^{-1}$ mole^{-1}
σ	Stefan-Boltzmann constant	$5.66961(96)\times10^{-5}$ erg cm^{-2} s^{-1} $\mathring{K}^{-4}$
σ_T	Thomson cross section	$6.652453(62)\times10^{-25}$ cm^2
a.u.	Astronomical unit	$1.49597892(1)\times10^{13}$ cm
pc	Parsec	$3.0856(1)\times10^{18}$ cm
l.y.	One light year	9.4605×10^{17} cm $=6.324\times10^{4}$ a.u.
$M_\odot$	Solar mass	$1.989(2)\times10^{33}$ g
$R_\odot$	Solar radius	$6.9598(7)\times10^{10}$ cm

Symbol	Meaning	Value
$L_{\odot}$	Solar luminosity	$3.826(8) \times 10^{33}$ erg s^{-1}
1 e.V.	One electron volt	
	associated wavelength	$12{,}396.3 \times 10^{-8}$ cm
	associated wavenumber	8067.1 cm^{-1}
	associated frequency	2.41838×10^{14} Hz
	associated energy	1.60184×10^{-12} erg
	associated temperature	11,605.9 °K

Here, as elsewhere in this book, the digits in parenthesis following each quoted value represents the standard deviation error in the final digits of the quoted value. The value for the velocity of light is from K. M. EVENSON *et al.*, Phys. Rev. Lett. **29**, 1346 (1972), whereas the values for the other physical constants are from B. N. TAYLOR, W. H. PARKER, and D. N. LANGENBERG, Rev. Mod. Phys. **41**, 375 (1969), and the five astronomical constants are from Chapter 5 of this volume. As this book is meant to serve as a reference source, the table of contents is more elaborate than that of most books. The tables include the observed parameters and physical characteristics of most of the well known astronomical objects.

A substantial fraction of this book was completed during two summers as a visiting fellow at the Institute of Theoretical Astronomy, Cambridge, and I am especially grateful for the hospitality and courtesy which the members of the Institute have shown me. I am also indebted to the California Institute of Technology for the freedom to complete this book. The staff of the scientific periodicals library of the Cambridge Philosophical Society and the library of the Hale observatories are especially thanked for their aid in supplying and checking references. Those who have read portions of this book and have supplied critical comment and advice include Drs. L. H. ALLER, H. ARP, R. BLANFORD, R. N. BRACEWELL, A. G. W. CAMERON, E. CHURCHWELL, D. CLAYTON, W. A. FOWLER, J. GREENSTEIN, H. GRIEM, J. R. JOKIPII, B. KUCHOWICZ, M. G. LANG, A. G. MICHALITSANOS, R. L. MOORE, E. N. PARKER, W. H. PRESS, M. J. REES, J. A. ROBERTS, J. R. ROY, W. L. W. SARGENT, W. C. SASLAW, M. SCHMIDT, I. SHAPIRO, P. SOLOMON, E. SPIEGEL, and S. E. WOOSLEY.

KENNETH R. LANG

California Institute of Technology, Pasadena, California

and

Tufts University, Medford, Massachusetts

October, 1974

Contents

1. Continuum Radiation

4. Nuclear Astrophysics and High Energy Particles

1. Continuum Radiation

"What is light? Since the time of Young and Fresnel we know that it is wave motion. We know the velocity of the waves, we know their lengths, and we know that they are transverse; in short, our knowledge of the geometrical conditions of the motion is complete. A doubt about these things is no longer possible; a refutation of these views is inconceivable to the physicist. The wave theory of light is, from the point of view of human beings, certainty."

HERTZ, 1889

"It is now, I believe, generally admitted that the light which we receive from the clear sky is due in one way or another to small suspended particles which divert the light from its regular course ... There seems to be no reason why the color of the compound light thus scattered should not agree with that of the sky ... Suppose for distinctness of statement, that the primary ray is vertical, and that the plane of vibration is that of the meridian. The intensity of the light scattered by a small particle is constant, and a maximum for rays which lie in the vertical plane running east and west, while there is no scattered ray along the north and south line. If the primary ray is unpolarized, the light scattered north and south is entirely due to that component which vibrates east and west, and is therefore perfectly polarized."

Lord RAYLEIGH, 1871

1.1. Static Electric Fields

The experimentally determined Coulomb force, $\boldsymbol{F}$, between two static point charges, q_1 and q_2 is (COULOMB, 1785)

$$\boldsymbol{F} = \frac{q_1 q_2}{R^2} \boldsymbol{n}_r , \tag{1-1}$$

where R is the distance between the charges, and $\boldsymbol{n}_r$ is a unit vector directed from one charge to the other. The static electric field, $\boldsymbol{E}$, of a point charge q_1 is defined so that

$$\boldsymbol{F} = q_2 \boldsymbol{E}, \quad \text{where } \boldsymbol{E} = \frac{q_1}{R^2} \boldsymbol{n}_r \tag{1-2}$$

and R is the distance from q_1. Integrating equation (1-2) over a closed spherical surface we obtain Gauss's law

$$\oint_S \boldsymbol{E} \cdot \boldsymbol{n} \, ds = 4\pi q = 4\pi \int_V \rho \, dv , \tag{1-3}$$

where ρ is the charge density, $\oint_s$ denotes the closed surface integral, $\boldsymbol{E}\cdot\boldsymbol{n}$ is the component of $\boldsymbol{E}$ which is normal to the surface element ds, and $\int_v \rho\, dv$ is the amount of charge within the closed surface. Using the equations of vector analysis, Eq. (1–3) may be expressed as Poisson's equation (POISSON, 1813)

$$\nabla\cdot\boldsymbol{E} = -\nabla^2\varphi = 4\pi\rho, \tag{1-4}$$

where

$$\boldsymbol{E} = -\nabla\varphi, \tag{1-5}$$

and φ is called the scalar electric potential. For one static point charge, q, we have

$$\varphi = q/R, \tag{1-6}$$

where R is the radial distance from the charge. For an electrostatic dipole consisting of two charges, $+q$ and $-q$ separated by a distance a along the z axis

$$\varphi = aq\cos\theta/R^2 \tag{1-7}$$

so that

$$\boldsymbol{E}_r = \frac{2d}{R^3}\cos\theta\,\boldsymbol{n}_r \tag{1-8}$$

$$\boldsymbol{E}_\theta = \frac{d\sin\theta}{R^3}\boldsymbol{n}_\theta$$

and

$$\boldsymbol{E}_\varphi = 0,$$

where the dipole moment $d = aq$, the angle θ is the angle between the z axis and the radial direction, R is the radial distance from the dipole, and $\boldsymbol{n}_\theta$ and $\boldsymbol{n}_r$ are unit vectors in the θ and the radial direction.

1.2. Static Magnetic Fields

The experimentally determined magnetic force, $\boldsymbol{F}$, between two current elements $I_1\,d\boldsymbol{l}_1/c$ and $I_2\,d\boldsymbol{l}_2/c$ is (AMPÈRE, 1827)

$$\boldsymbol{F} = \frac{I_1 I_2}{c^2 R^2}\,d\boldsymbol{l}_2\times(d\boldsymbol{l}_1\times\boldsymbol{n}_r), \tag{1-9}$$

where I_1 and I_2 are the two currents, $d\boldsymbol{l}_1$ and $d\boldsymbol{l}_2$ are unit vector elements of length whose directions are the same as that of the current flow, R is the distance between the two elements, and $\boldsymbol{n}_r$ is a unit vector directed from length element $d\boldsymbol{l}_1$ towards element $d\boldsymbol{l}_2$. The attractive force per unit length between two long straight parallel wires is, for example,

$$F = \frac{2I_1 I_2}{c^2 D}, \tag{1-10}$$

where D is the perpendicular distance between the wires. The static magnetic field, $\boldsymbol{B}$, of current element $I_1 d\boldsymbol{l}_1/c$ may be defined so that (BIOT and SAVART, 1820)

$$F = \frac{I_2 dl_2}{c} \times B \quad \text{where } B = \frac{I_1}{c} \frac{dl_1 \times n_r}{R^2}. \tag{1-11}$$

Integrating equation (1–11) over a closed contour we obtain Ampère's law

$$\oint_c B \cdot dl = \frac{4\pi}{c} I = \frac{4\pi}{c} \int_s J \cdot n \, ds, \tag{1-12}$$

where $\oint_c$ denotes a closed contour integral, I is the current passing through the closed contour, $\int_s$ is an open surface integral over any surface bounded by the closed curve, and $J \cdot n$ is the component of the current density, J, which is normal to the surface element ds. Using the equations of vector analysis, Eq. (1-12) may be expressed in the differential form

$$\nabla \times B = \frac{4\pi}{c} J. \tag{1-13}$$

As an example, it follows directly from Eq. (1–12) that the magnetic field of a long wire of current, I, extending along the z axis is

$$B_\varphi = \frac{2I}{cR} n_\varphi, \tag{1-14}$$

where R is the perpendicular distance from a point on the z axis and n_φ is unit vector in the φ direction of a spherical coordinate system whose positive z axis is in the direction of current flow. It follows directly from Eq. (1-11) that we may express the static magnetic field B in terms of a magnetic vector potential A such that

$$B = \nabla \times A \quad \text{where } A = \frac{I \, dl}{cR}, \tag{1-15}$$

and therefore

$$\nabla \cdot B = \nabla \cdot \nabla \times A = 0. \tag{1-16}$$

For each Cartesian coordinate of the vector A, we have

$$\nabla^2 A = \frac{-4\pi}{c} J, \tag{1-17}$$

and

$$A = \frac{1}{c} \int_v \frac{J}{R} dv,$$

where R is the distance to the volume element dv. As an example, consider a closed loop of current, I, and radius, a, placed in the xy plane and centered at the z axis. Using Eq. (1-15) we obtain

$$A = \frac{\pi a^2 I}{cR^2} \sin\theta \, n_\varphi, \tag{1-18}$$

so that

$$\boldsymbol{B}_{\mathrm{r}} = \frac{2m\cos\theta}{R^3}\,\boldsymbol{n}_{\mathrm{r}} \quad \text{for } R \gg a\,, \tag{1-19}$$

$$\boldsymbol{B}_{\theta} = \frac{m\sin\theta}{R^3}\,\boldsymbol{n}_{\theta} \quad \text{for } R \gg a$$

and

$$\boldsymbol{B}_{\varphi} = 0\,,$$

where the magnetic dipole moment $m = \pi a^2 I/c$, the angle θ is the angle between the z axis and the radial direction, R is the radial distance from the center of the loop, and $\boldsymbol{n}_{\varphi}$, $\boldsymbol{n}_{\theta}$, and $\boldsymbol{n}_{r}$ are unit vectors in the directions of the spherical coordinate system centered in the current loop and with the z axis perpendicular to the plane of the loop. For a spherical mass of radius, r, the latitude $\lambda = \pi/2 - \theta$ if often used, and the dipole moment of the static magnetic field is given by

$$m = r^3 B_{\mathrm{p}}/2\,, \tag{1-20}$$

where B_{p} is the polar magnetic field strength at the surface of the sphere. For the earth, $\mathrm{m} \approx 8.1 \times 10^{25}$ gauss cm^3 which corresponds to a polar field strength of 0.62 gauss. Surface magnetic field strengths of other objects are given in Table 4.

1.3. Electromagnetic Fields in Matter—Constitutive Relations

Experiments on the flow of current in metals (Davy, 1821) which are subjected to an applied electric potential were explained by Ohm (1826). Ohm's law states that the current density, $\boldsymbol{J}$, is given by

$$\boldsymbol{J} = \sigma \boldsymbol{E}\,, \tag{1-21}$$

where the constant σ is called the conductivity and $\boldsymbol{E}$ is the electric field strength. Kelvin (1850) suggested that the magnetic induction $\boldsymbol{B}$ induced in matter by a magnetizing force, $\boldsymbol{H}$, is given by

$$\boldsymbol{B} = \mu \boldsymbol{H}\,, \tag{1-22}$$

where μ is called the magnetic permeability. Similarly, Maxwell (1861) suggested that the electric displacement, $\boldsymbol{D}$, caused by the application of an electromotive force to matter is given by

$$\boldsymbol{D} = \varepsilon \boldsymbol{E}\,, \tag{1-23}$$

where ε is called the dielectric constant or permittivity. In most astrophysical situations $\mu = 1$, its free space value. In free space, $\sigma = 0$ and $\varepsilon = 1$. The constants ε and μ define the index of refraction, n, of a medium with zero conductivity and no magnetic field

$$n = \sqrt{\varepsilon\mu}\,. \tag{1-24}$$

The indices of refraction for a conductive medium and for a medium with a magnetic field are discussed in Sects. 1.12 and 1.33, respectively.

The macroscopic constant, ε, was related to the atomic properties of matter by Lorentz (1880) and Lorenz (1881). They argued that the mean or observed field, $\boldsymbol{E}$, coming from a region containing a great number of atoms or molecules will be different from the effective field, $\boldsymbol{E}'$, acting on each individual molecule. It was assumed that under the influence of $\boldsymbol{E}'$ a molecule will polarize itself forming an electric dipole moment

$$\boldsymbol{d} = \alpha \boldsymbol{E}' , \tag{1-25}$$

where the constant, α, is called the polarizability. From Eq. (1-8) the resulting dipole field at the surface of a spherical molecule will be

$$\boldsymbol{E}_{\mathrm{s}} = \frac{\boldsymbol{d}}{a^3} = \frac{\alpha \boldsymbol{E}'}{a^3} , \tag{1-26}$$

where a is the molecular radius. Thus the observed field, $\boldsymbol{E}$, from an isotropic medium with a number density of N molecules will be

$$\boldsymbol{E} = \boldsymbol{E}' - N\left[\frac{4}{3}\pi a^3 \boldsymbol{E}_{\mathrm{s}}\right] = \left[1 - \frac{4\pi}{3} N \alpha\right] \boldsymbol{E}' . \tag{1-27}$$

The total electric polarization of the medium, $\boldsymbol{P}$, is given by

$$\boldsymbol{P} = N \alpha \boldsymbol{E}' = \chi_{\mathrm{e}} \boldsymbol{E} = \left[\frac{N\alpha}{1 - \frac{4}{3}\pi N \alpha}\right] \boldsymbol{E} , \tag{1-28}$$

where $\boldsymbol{P}$ is the volume density of the electric dipole moment, and the constant χ_{e} is called the electric susceptibility. The dielectric constant, ε, is introduced by noting that

$$\boldsymbol{D} = \varepsilon \boldsymbol{E} = \boldsymbol{E} + 4\pi \boldsymbol{P} = (1 + 4\pi \chi_e) \boldsymbol{E} , \tag{1-29}$$

and the polarizability, α, is given by

$$\alpha = \frac{3}{4\pi N} \frac{\varepsilon - 1}{\varepsilon + 2} = \frac{3}{4\pi N} \frac{n^2 - 1}{n^2 + 2} , \tag{1-30}$$

where N is the number density of molecules, and n is the index of refraction. The macroscopic version of Eq. (1-4) which excludes the contribution of electric dipole charges is

$$\nabla \cdot \boldsymbol{D} = 4\pi \rho . \tag{1-31}$$

In a similar way, a magnetic susceptibility, χ_{m}, may be defined so that the magnetic field $\boldsymbol{H}$ is given by

$$\boldsymbol{M} = \chi_{\mathrm{m}} \boldsymbol{H} , \tag{1-32}$$

where the magnetization, $\boldsymbol{M}$, is the volume density of the dipole magnetic moment. In this case we have

$$\boldsymbol{B} = \mu \boldsymbol{H} = \boldsymbol{H} + 4\pi \boldsymbol{M} , \tag{1-33}$$

and the macroscopic version of Eq. (1-13) which excludes the current density of the magnetic dipoles is

$$\nabla \times \boldsymbol{H} = \frac{4\pi \boldsymbol{J}}{c} . \tag{1-34}$$

1.4. Induced Electromagnetic Fields

FARADAY (1843) first showed that an electric potential was induced in a circuit moving through the lines of force of a magnet, and his law for the induced electric potential is

$$\oint_{c} \boldsymbol{E} \cdot d\boldsymbol{l} = -\frac{1}{c}\frac{\partial}{\partial t}\int_{s} \boldsymbol{B} \cdot \boldsymbol{n}\, ds, \tag{1-35}$$

where $\oint_{c}$ denotes the closed contour integral, $\boldsymbol{B}$ is called the magnetic induction, $\int_{s}$ denotes an open surface integral over any surface bounded by the closed contour, and $\boldsymbol{B} \cdot \boldsymbol{n}$ denotes the component of $\boldsymbol{B}$ which is normal to the surface element ds. The line integral in Eq. (1-35) is called the induced electromotive force, and the magnetic flux (or number of lines of force) is defined as

$$\Phi = \int_{s} \boldsymbol{B} \cdot \boldsymbol{n}\, ds.$$

Using the equations of vector analysis, the differential form of Eq. (1–35) is

$$\nabla \times \boldsymbol{E} + \frac{1}{c}\frac{\partial \boldsymbol{B}}{\partial t} = 0. \tag{1-36}$$

Here the electric field vector, $\boldsymbol{E}$, and the magnetic induction, $\boldsymbol{B}$, are defined in the same reference frame.

1.5. Continuity Equation for the Conservation of Charge

The conservation of electric charge is expressed by the continuity equation

$$\frac{d}{dt}\int_{v} \rho\, dv + \oint_{s} \boldsymbol{J} \cdot \boldsymbol{n}\, ds = 0, \tag{1-37}$$

where ρ is the charge density, $\boldsymbol{J} \cdot \boldsymbol{n}$ is the component of current density normal to the surface element ds, $\oint_{s}$ denotes a closed surface integral, and $\int_{v} \rho\, dv$ is a volume integral within the closed surface. This equation states that the current flowing into a closed surface area is equal to the time rate of change of charge within the area, and was first confirmed experimentally by FARADAY (1843). Using the equations of vector analysis, the differential form of Eq. (1-37) is

$$\frac{\partial \rho}{\partial t} + \nabla \cdot \boldsymbol{J} = 0. \tag{1-38}$$

1.6. Maxwell's Equations

MAXWELL (1861, 1873) first noted that the electrostatic equations (1-31) and (1-34) were inconsistent with the continuity equation (1-38) for time varying charges and current. He suggested that a displacement current $(1/4\pi)\partial D/\partial t$ be added so that Eqs. (1-16), (1-31), and (1-34) for the static fields together with Eq. (1-35) for time varying fields define the electromagnetic field equations:

$$\nabla\times\boldsymbol{H}-\frac{1}{c}\frac{\partial\boldsymbol{D}}{\partial t}=\frac{4\pi}{c}\boldsymbol{J} \quad \text{or} \quad \oint_c \boldsymbol{H}\cdot d\boldsymbol{l}-\frac{1}{c}\frac{\partial}{\partial t}\int_s \boldsymbol{D}\cdot\boldsymbol{n}\,ds=\frac{4\pi}{c}\int_s \boldsymbol{J}\cdot\boldsymbol{n}\,ds, \tag{1-39}$$

$$\nabla\times\boldsymbol{E}+\frac{1}{c}\frac{\partial\boldsymbol{B}}{\partial t}=0 \quad \text{or} \quad \oint_c \boldsymbol{E}\cdot d\boldsymbol{l}+\frac{1}{c}\frac{\partial}{\partial t}\int_s \boldsymbol{B}\cdot\boldsymbol{n}\,ds=0, \tag{1-40}$$

$$\nabla\cdot\boldsymbol{D}=4\pi\rho \quad \text{or} \quad \oint_s \boldsymbol{D}\cdot\boldsymbol{n}\,ds=4\pi\int_v \rho\,dv, \tag{1-41}$$

and

$$\nabla\cdot\boldsymbol{B}=0 \quad \text{or} \quad \oint_s \boldsymbol{B}\cdot\boldsymbol{n}\,ds=0. \tag{1-42}$$

Thus five vectors describe the electromagnetic field and its effect on material objects. They are the electric vector, $\boldsymbol{E}$, the magnetic vector, $\boldsymbol{H}$, the electric displacement, $\boldsymbol{D}$, the magnetic induction, $\boldsymbol{B}$, and the electric current density, $\boldsymbol{J}$. The continuity equation (1-38) for the conservation of electric charge is assumed to hold, and the current density $\boldsymbol{J}$ is given by

$$\boldsymbol{J}=\sigma\boldsymbol{E}+\rho\boldsymbol{v}=\boldsymbol{J}_\sigma+\boldsymbol{J}_c, \tag{1-43}$$

where the conduction current $\boldsymbol{J}_\sigma=\sigma\boldsymbol{E}$ for material with conductivity, σ, and the convection current $\boldsymbol{J}_c=\rho\boldsymbol{v}$ for charges of density ρ moving with velocities $\boldsymbol{v}$.

1.7. Boundary Conditions

Maxwell's equations (1-39)—(1-42) apply for material in which the dielectric constant, ε, and the magnetic permeability, μ, are continuous. At surfaces of discontinuity in ε and/or μ, the electromagnetic fields at each side of the discontinuity are related by the following equations which may be obtained directly from the integral form of Maxwell's equations

$$\boldsymbol{n}_{12}\cdot(\boldsymbol{B}_2-\boldsymbol{B}_1)=0, \tag{1-44}$$

$$\boldsymbol{n}_{12}\cdot(\boldsymbol{D}_2-\boldsymbol{D}_1)=4\pi\rho_s, \tag{1-45}$$

$$\boldsymbol{n}_{12}\times(\boldsymbol{E}_2-\boldsymbol{E}_1)=0, \tag{1-46}$$

$$\boldsymbol{n}_{12}\times(\boldsymbol{H}_2-\boldsymbol{H}_1)=4\pi\boldsymbol{J}_s/c, \tag{1-47}$$

where $\boldsymbol{n}_{12}$ is a unit vector normal to the surface of discontinuity and pointing from medium 1 into medium 2. Eqs. (1-44) and (1-45) apply to components of $\boldsymbol{B}$ and $\boldsymbol{D}$ which are normal to the surface, whereas Eqs. (1-46) and (1-47) apply to components of $\boldsymbol{E}$ and $\boldsymbol{H}$ which are tangential to the surface. The ρ_s and $\boldsymbol{J}_s$ denote, respectively, the charge density and current density at the boundary surface.

1.8. Energy Density of the Electromagnetic Field

The total energy density, U, of the electromagnetic field is (KELVIN, 1853; MAXWELL, 1861, 1873)

$$U = U_E + U_H = \frac{\varepsilon}{8\pi} E^2 + \frac{\mu}{8\pi} H^2 = \frac{1}{8\pi} [\boldsymbol{E} \cdot \boldsymbol{D} + \boldsymbol{B} \cdot \boldsymbol{H}], \tag{1-48}$$

where U_E and U_H are, respectively, the electric and magnetic energy densities, ε and μ are, respectively, the dielectric constant and magnetic permeability of the medium, and E and H are, respectively, the magnitudes of the electric and magnetic vectors, $\boldsymbol{E}$ and $\boldsymbol{H}$.

1.9. Poynting Energy Flux

The amount of energy per unit time which crosses a unit area normal to the direction of the electric and magnetic vectors, $\boldsymbol{E}$ and $\boldsymbol{H}$, is given by the magnitude of the Poynting vector (POYNTING, 1884)

$$\boldsymbol{S} = \frac{c}{4\pi} (\boldsymbol{E} \times \boldsymbol{H}), \tag{1-49}$$

for time varying electromagnetic fields.

1.10. Electromagnetic Momentum and Radiation Pressure

The momentum density of the electromagnetic field is (MAXWELL, 1861)

$$\boldsymbol{g} = \frac{\boldsymbol{S}}{c^2} = \frac{(\boldsymbol{E} \times \boldsymbol{H})}{4\pi c}. \tag{1-50}$$

The radiation pressure, $\boldsymbol{P}$, is given by the net rate of momentum transfer per unit area. From Eq. (1-50), then

$$\boldsymbol{P} = \frac{\boldsymbol{S}}{c} = \frac{\boldsymbol{E} \times \boldsymbol{H}}{4\pi}. \tag{1-51}$$

1.11. Lorentz Force Law

The vector force, $\boldsymbol{F}$, exerted by an electromagnetic field on a charge, q, moving at the velocity, v, is (HEAVISIDE, 1889; LORENTZ, 1892)

$$\boldsymbol{F} = q\left(\boldsymbol{E} + \frac{\mu}{c} \boldsymbol{v} \times \boldsymbol{H}\right), \tag{1-52}$$

where the magnitude of $\boldsymbol{v} \times \boldsymbol{H}$ is the component of velocity in the direction perpendicular to the magnetic field when μ is unity.

Let us suppose that a charged particle of charge, q, has initial velocity components $v_\perp$ and $v_\parallel$ perpendicular and parallel to the direction of a uniform constant magnetic field, $\boldsymbol{H}$. It follows from Eq. (1-52) that the particle motion in the presence of a $\boldsymbol{H}$ will be the superposition of a constant velocity, $v_{\|}$, in the direction of $\boldsymbol{H}$ together with a circular motion in the plane perpendicular to $\boldsymbol{H}$. As first pointed out by HEAVISIDE (1904) the circular radian frequency (the gyrofrequency) is given by

$$\omega_H = \frac{q\mu H}{\gamma m c} = \frac{qB}{\gamma m c}, \tag{1-53}$$

and the gyroradius (the Larmor radius) is

$$r_H = \frac{v_\perp}{\omega_H} = \frac{m c \gamma v_\perp}{qB}, \tag{1-54}$$

where m is the mass of the particle, and the particle momentum is $\boldsymbol{p} = \gamma m \boldsymbol{v}$ where $\gamma = [1-(v^2/c^2)]^{-1/2}$. For low velocities, $v \ll c$, we have $\gamma \approx 1$.

In the presence of a uniform constant electric field, $\boldsymbol{E}$, directed along the x axis, a charged particle of charge q, momentum $\gamma m v$, total energy $\gamma m c^2$, and kinetic energy $(\gamma-1)mc^2$ will follow a trajectory specified by Eq. (1-52).

$$\begin{aligned} x &= \frac{\gamma m c^2}{qE} \cosh\left(\frac{eEy}{\gamma m v c}\right) \\ &\approx \frac{qE}{2mv^2} y^2 \quad \text{for } v \ll c, \end{aligned} \tag{1-55}$$

where the initial velocity, v, is in the xy plane.

The average motion of a charged particle in the presence of uniform constant electric and magnetic fields, $\boldsymbol{E}$ and $\boldsymbol{H}$, will be in a direction perpendicular to both fields. The average drift velocity is

$$\boldsymbol{v} = \frac{c\boldsymbol{E} \times \boldsymbol{H}}{H^2}, \tag{1-56}$$

where it has been assumed that $v \ll c$ and $\boldsymbol{E} \ll \boldsymbol{H}$.

1.12. Electromagnetic Plane Waves

Maxwell's equations (1-39)—(1-42) in a charge and current free region (zero conductivity and no magnetic field) assume the form

$$\nabla^2 E - \frac{\varepsilon\mu}{c^2} \frac{\partial^2 E}{\partial t^2} = 0, \tag{1-57}$$

and

$$\nabla^2 H - \frac{\varepsilon\mu}{c^2} \frac{\partial^2 H}{\partial^2 t} = 0,$$

where ε and μ are, respectively, the dielectric constant and the magnetic permeability. Eqs. (1-57) are wave equations for electromagnetic waves which propagate with the velocity

$$v = c/\sqrt{\varepsilon\mu}\,. \tag{1-58}$$

Of special interest is the harmonic solution to Eq. (1–57) for which

$$\boldsymbol{E} = \boldsymbol{E}_0 \exp[i(\boldsymbol{k} \cdot \boldsymbol{r} - \omega t)], \tag{1-59}$$

where t is the time variable and the harmonic frequency, ω, is related to the propagation or wave vector, $\boldsymbol{k}$, by

$$\boldsymbol{k} = \frac{2\pi}{\lambda} \boldsymbol{s} = \frac{\omega}{c} n \boldsymbol{s}, \tag{1-60}$$

where λ is the wavelength, $\boldsymbol{s}$ is the direction of wave propagation which is perpendicular to $\boldsymbol{E}$, and n is the index of refraction. The phase velocity of the wave is

$$v_p = \frac{\omega}{k} = \frac{c}{n} = \frac{c}{\sqrt{\varepsilon\mu}}, \tag{1-61}$$

whereas the group velocity is given by the differential

$$v_g = \frac{d\omega}{dk}. \tag{1-62}$$

It follows directly from Maxwell's equations that when the electric field vector $\boldsymbol{E}$ is given by Eq. (1-59), the magnetic field vector $\boldsymbol{H}$ is given by

$$\boldsymbol{H} = \left(\frac{\varepsilon}{\mu}\right)^{1/2} \boldsymbol{s} \times \boldsymbol{E},$$

where $\boldsymbol{s}$ is a unit vector denoting the direction of propagation of the wave. It follows that the field vectors of a plane wave are perpendicular to each other and perpendicular to the direction of propagation, and hence they are called transverse plane waves.

From Eq. (1-49), the energy flux in the plane wave is

$$S = \frac{c}{4\pi} \sqrt{\frac{\varepsilon}{\mu}} E^2 = \frac{c}{4\pi} \sqrt{\frac{\mu}{\varepsilon}} H^2 \tag{1-63}$$

which is directed along the direction of wave propagation. The energy density, U, in the plane wave is, from Eq. (1–48),

$$U = \frac{\varepsilon E^2}{4\pi} = \frac{\mu H^2}{4\pi}. \tag{1-64}$$

When the medium has a non-zero conductivity, σ, the propagation vector, k, is complex and is given by

$$\begin{aligned} k &= \left[\varepsilon\mu \frac{\omega^2}{c^2}\left(1 + i\frac{4\pi\sigma}{\omega\varepsilon}\right)\right]^{1/2} \\ &\approx \sqrt{\mu\varepsilon}\frac{\omega}{c} + i\frac{2\pi}{c}\sqrt{\frac{\mu}{\varepsilon}}\sigma \quad \text{for } \sigma \ll \omega\varepsilon \\ &\approx (1+i)\frac{\sqrt{2\pi\omega\mu\sigma}}{c} \quad \text{for } \sigma \gg \omega\varepsilon, \end{aligned} \tag{1-65}$$

and the waves are attenuated exponentially with distance. The phase velocity of the wave is $v_p=\omega/\alpha=c/n$ where α is the real part of k and n is the index of refraction. When a magnetic field is present there are additional real and imaginary terms for k, and these terms are given in Sect. 1.33.

1.13. Polarization of Plane Waves—The Stokes Parameters

If the propagation direction of a harmonic plane wave is the z axis of a Cartesian coordinate system, the components of the electric field vector are

$$E_x = a_1 \cos(\omega t - \boldsymbol{k}\cdot\boldsymbol{r} + \delta_1)$$

$$E_y = a_2 \cos(\omega t - \boldsymbol{k}\cdot\boldsymbol{r} + \delta_2) \tag{1-66}$$

$$\text{and} \quad E_z = 0,$$

where a_1 and a_2 are scalar amplitudes, ω is the harmonic frequency, t is the time variable, $\boldsymbol{k}$ is the wave vector, $\boldsymbol{r}$ is the vector radius, and δ_1 and δ_2 are arbitrary phases. It follows directly from Eqs. (1-66) that E_x and E_y are related by

$$\left(\frac{E_x}{a_1}\right)^2 + \left(\frac{E_y}{a_2}\right)^2 - \frac{2E_x E_y}{a_1 a_2}\cos\delta = \sin^2\delta, \tag{1-67}$$

where $\delta=\delta_2-\delta_1$. Eq. (1-67) describes an ellipse whose tilt angle, ψ, with respect to the x axis, and the length of the major and minor axis, $2a$ and $2b$ are related by the equations

$$\tan 2\psi = (\tan 2\alpha)\cos\delta,$$

$$\tan\chi = \pm b/a,$$

$$a^2+b^2 = a_1^2+a_2^2, \tag{1-68}$$

$$\text{and} \quad \sin 2\chi = (\sin 2\alpha)\sin\delta,$$

$$\text{where} \quad \tan\alpha = a_2/a_1.$$

In general, the wave is said to be elliptically polarized. When $\delta=\delta_2-\delta_1=m\pi$ where m is an integer, the ellipse reduces to a straight line and the wave is said to be linearly polarized. When $a_1=a_2=a$ and $\delta=\pm\pi/2+2m\pi$, where m is an integer, the wave is said to be circularly polarized. By definition the wave is right hand or left hand circularly polarized according as the $\pm$ sign is plus or minus.

The Stokes parameters for the wave are (STOKES, 1852)

$$I=a_1^2+a_2^2,$$

$$Q=a_1^2-a_2^2=I\cos 2\chi\cos 2\psi,$$

$$U=2a_1 a_2\cos\delta=I\cos 2\chi\sin 2\psi, \tag{1-69}$$

$$\text{and} \quad V=2a_1 a_2\sin\delta=I\sin 2\chi,$$

$$\text{where} \quad I^2=Q^2+U^2+V^2,$$

and I represents the intensity of the wave. The degree of polarization π, of any electromagnetic wave is defined as

$$\pi = [Q^2 + U^2 + V^2]^{1/2}/I\,, \tag{1-70}$$

when I, Q, U, and V are defined in terms of time averages.

1.14. Reflection and Refraction of Plane Waves

By considering the boundary conditions for the electromagnetic fields (Eqs. (1-44)—(1-47)) the laws of reflection and refraction of the harmonic plane wave may be obtained. If a plane wave falls on a boundary between two homogeneous media, the angle θ_i, between the direction of propagation of the incident wave and the normal to the surface is related to the angle θ_t, between the direction of propagation of the transmitted wave and the surface normal by Snell's law of refraction (SNELL, 1621)

$$\frac{\sin\theta_i}{\sin\theta_t} = \left(\frac{\varepsilon_2\mu_2}{\varepsilon_1\mu_1}\right)^{1/2} = \frac{n_2}{n_1} = \frac{v_1}{v_2}\,, \tag{1-71}$$

where ε is the dielectric constant, μ is the magnetic permeability, $n=\sqrt{\varepsilon\mu}$ is called the index of refraction, v is the phase velocity, and subscripts 1 and 2 denote, respectively, the incident medium and the medium of the transmitted wave. The similar relation for the angles θ_i and θ_r which the incident and reflected waves make with the surface normal is

$$\sin\theta_i = \sin\theta_r\,, \tag{1-72}$$

which follows from Fermat's (1627) principle that light follows that path which brings it to its destination in the shortest possible time.

The ratio of the amount of energy in the reflected wave to that in the incident wave is given by (FRESNEL, 1822)

$$\begin{aligned} R_{\parallel} &= \frac{\tan^2(\theta_i-\theta_t)}{\tan^2(\theta_i+\theta_t)}\,, \\ R_{\perp} &= \frac{\sin^2(\theta_i-\theta_t)}{\sin^2(\theta_i+\theta_t)}\,, \end{aligned} \tag{1-73}$$

where the reflection coefficients $R_{\parallel}$ and $R_{\perp}$ are, respectively, for waves which are linearly polarized in the plane of incidence and normal to the plane of incidence. Using Snell's law, Eq. (1-71), and assuming $\mu=1$, these equations become

$$\begin{aligned} R_{\perp} &= \left[\frac{\cos\theta_i-(\varepsilon-\sin^2\theta_i)^{1/2}}{\cos\theta_i+(\varepsilon-\sin^2\theta_i)^{1/2}}\right]^2, \\ \text{and}\quad R_{\parallel} &= \left[\frac{\varepsilon\cos\theta_i-(\varepsilon-\sin^2\theta_i)^{1/2}}{\varepsilon\cos\theta_i+(\varepsilon-\sin^2\theta_i)^{1/2}}\right]^2, \end{aligned} \tag{1-74}$$

where ε is the dielectric constant of the medium. Use of Eq. (1-74) in determining the dielectric constant of the Moon and Venus is discussed by HEILES and DRAKE

(1963), CLARK and KUZMIN (1965), BERGE and GREISEN (1969), and MUHLEMAN (1969).

The ratio of the energy in the transmitted wave to that in the incident wave is given by the transmission coefficients

$$T_{\|} = 1 - R_{\|} \quad \text{and} \quad T_{\perp} = 1 - R_{\perp} . \tag{1-75}$$

For normal incidence, Eqs. (1-74) and (1-75) reduce to

$$R = \left(\frac{n-1}{n+1}\right)^2 \quad \text{and} \quad T = \frac{4n}{(n+1)^2} , \tag{1-76}$$

where

$$n = \frac{n_2}{n_1} = \left(\frac{\varepsilon_2 \mu_2}{\varepsilon_1 \mu_1}\right)^{1/2} .$$

When a wave from free space is normally incident upon a medium with $\mu = 1$, the reflection coefficient becomes

$$R_0 = \left[\frac{1-\sqrt{\varepsilon}}{1+\sqrt{\varepsilon}}\right]^2 . \tag{1-77}$$

If the medium is partially conducting with conductivity, σ, then the term ε appearing in Eq. (1-77) must be replaced by

$$\frac{1}{\mu}\left[\varepsilon + \frac{i\sigma}{2\pi\nu}\right], \tag{1-78}$$

where ν is the wave frequency. In this case the amplitude of the transmitted wave will fall to $1/e$ of its initial value in the skin depth distance

$$\delta = c/\sqrt{4\pi^2 \mu\nu\sigma} = c/\sqrt{2\pi\mu\omega\sigma} . \tag{1-79}$$

Radar astronomers often measure the radar cross section given as

$$\sigma_r = g R_0 \pi r^2 , \tag{1-80}$$

where r is the radius of the planet, and the coefficient, g, accounts for surface roughness. Radar cross section measurements of Venus, Mercury, Mars and the Moon are reviewed by EVANS (1969). For a surface which has a Gaussian distribution of slopes with an rms value of a, the coefficient $g = 1 + a^2$.

Of special interest is the Brewster angle of incidence for which the electric vector of the reflected radiation is linearly polarized in a plane normal to the plane of incidence. This angle is given by (BREWSTER, 1815)

$$\tan\theta_i = n = \frac{n_2}{n_1} = \left(\frac{\varepsilon_2 \mu_2}{\varepsilon_1 \mu_1}\right)^{1/2} , \tag{1-81}$$

which is $\sqrt{\varepsilon}$ for incidence from free space upon a medium with $\mu = 1$.

Eqs. (1-51), (1-64), (1-72), and (1-73) can be combined to give the net radiation pressure at the surface of reflection.

$$P_{\mathrm{N}} = \frac{E^2}{4\pi}(1+R)\cos^2\theta_i\,, \tag{1-82}$$

and $$P_{\mathrm{T}} = \frac{E^2}{4\pi}(1-R)\sin\theta_i\cos\theta_i\,,$$

where $E^2/4\pi$ is the energy density of the wave, P_{N} and P_{T} denote the normal and tangential components of pressure, R is the reflection coefficient, and θ_i is the angle of incidence.

1.15. Dispersion Relations

An electron of charge e and mass m which is part of an atomic system acts like a harmonic oscillator whose motion in the x direction is given by

$$x(t) = x_0\cos(\omega_0 t)\,,$$

where t is the time variable and ω_0 is the radian frequency of the harmonic oscillation. When this oscillator is subjected to a harmonic plane wave of radian frequency, ω, its equation of motion as specified by the Lorentz force law (Eq. (1-52)) is

$$m\ddot{x} + m\omega_0^2 x = eE(t) = eE_0\exp[i\omega t]\,, \tag{1-83}$$

where E_0 is the amplitude of the electric field vector $\boldsymbol{E}$ which is assumed to have the x direction, and $\ddot{}$ denotes the second derivative with respect to time. Eq. (1-83) has the solution

$$x(t) = \frac{eE(t)}{m(\omega_0^2-\omega^2)}\,,$$

so that the dipole moment is

$$d(t) = ex(t) = \frac{e^2E(t)}{m(\omega_0^2-\omega^2)}\,. \tag{1-84}$$

Using Eqs. (1-25), (1-30), and (1-84), the index of refraction, n, of a density of N such bound electrons is given by the dispersion relation

$$\frac{n^2-1}{n^2+2} = \frac{4\pi Ne^2}{3m(\omega_0^2-\omega^2)}\,. \tag{1-85}$$

For a gas, $n \approx 1$ and we have

$$n \approx \left[1 - \frac{4\pi Ne^2}{m(\omega_0^2-\omega^2)}\right]^{1/2}, \tag{1-86}$$

a relation first derived by MAXWELL (1899). For a system of N atoms, the quantum mechanical form of Eq. (1-86) is (LADENBURG, 1921; KRAMERS, 1924)

$$n \approx \left[1 - \frac{N e^2}{\pi m} \sum_k \frac{f_{1k}}{\nu_{1k}^2 - \nu^2}\right]^{1/2}, \tag{1-87}$$

where f_{1k} and ν_{1k} are, respectively, the oscillator strength and the frequency of the transition from the ground state to the k level. The oscillator strength is defined in terms of the Einstein transition probabilities in Sect. 2.4. For a gas with free electron density, N_e, Eq. (1-86) becomes

$$n \approx \left[1 - \left(\frac{\omega_p}{\omega}\right)^2\right]^{1/2}, \tag{1-88}$$

where the plasma frequency, ω_p, is given by

$$\omega_p = \left(\frac{4\pi N_e e^2}{m}\right)^{1/2}.$$

When absorption must be taken into account the expression for the index of refraction becomes more complex as shown in Sect. 1.33. There is a general relation between refraction and absorption, however, which was first derived by KRAMERS and KRONIG (1928). If the electric susceptibility $\chi = \chi_1 + i\chi_2$, where χ_1 and χ_2 are real, then the Kramers-Kronig relations are

$$\chi_1(\omega_0) = \frac{2}{\pi} \int_0^\infty \frac{\omega \chi_2(\omega)\, d\omega}{\omega^2 - \omega_0^2} \tag{1-89}$$

and

$$\chi_2(\omega_0) = -\frac{2\omega_0}{\pi} \int_0^\infty \frac{\chi_1(\omega)\, d\omega}{\omega^2 - \omega_0^2},$$

where ω_0 is the frequency at which χ_1 or χ_2 are being evaluated. When a medium has a total extinction cross section, σ_e, then Eq. (1-89) becomes

$$\chi_1(\omega_0) = \frac{2c}{\pi V} \int_0^\infty \frac{\sigma_e(\omega)\, d\omega}{\omega^2 - \omega_0^2}, \tag{1-90}$$

where V is the volume of the medium. As first derived by FEENBERG (1932), the extinction cross section for matter with an absorption cross section, σ_a, and scattering cross section, σ_s, is given by $\sigma_e = \sigma_a + \sigma_s$.

1.16. Lorentz Coordinate Transformation

The Lorentz transformation is a means by which the coordinates x, y, z, t of an event in one inertial system (the K system) can be transferred into the coordinates x', y', z', t' of the same event in another inertial system (the K' system).

An inertial system is defined as one in which a freely moving body proceeds with constant velocity, and according to the first postulate of special relativity the laws of nature are invariant with respect to the Lorentz transformation of the coordinates from one inertial system to another. If the K system is moving at a uniform velocity, v, along the x axis and away from the K' system, then the Lorentz transformation (LORENTZ, 1904) is

$$x' = \gamma[x - vt], \quad y' = y, \quad z' = z, \quad \text{and } t' = \gamma[t - (xv)/c^2], \tag{1-91}$$

where

$$\gamma = \left[1 - \frac{v^2}{c^2}\right]^{-1/2}.$$

In his 1904 paper LORENTZ showed that Maxwell's equations are invariant when subjected to his coordinate transformation.

1.17. Lorentz Transformation of the Electromagnetic Field

If the K system has electromagnetic fields, $\boldsymbol{E}$ and $\boldsymbol{H}$, then the fields $\boldsymbol{E}'$ and $\boldsymbol{H}'$ in the K' system moving at the velocity v with respect to the K system may be calculated by using the Lorentz coordinate transformations. The components of E and H parallel to the direction of motion remain unchanged, whereas the components perpendicular to $\boldsymbol{v}$ are given by

$$\boldsymbol{E}' = \gamma[\boldsymbol{E} + (\boldsymbol{v}/c) \times \boldsymbol{H}] \tag{1-92}$$

and

$$\boldsymbol{H}' = \gamma[\boldsymbol{H} - (\boldsymbol{v}/c) \times \boldsymbol{E}],$$

where $\gamma = [1 - (v^2/c^2)]^{-1/2}$, and we assume that $\mu = 1$ as it is for free space. In many astrophysical situations, $\boldsymbol{E} \ll \boldsymbol{H}$ and $v \ll c$ so that Eqs. (1-92) become

$$\boldsymbol{E}' \approx \boldsymbol{E} + (\boldsymbol{v}/c) \times \boldsymbol{H} \tag{1-93}$$

and

$$\boldsymbol{H}' \approx \boldsymbol{H}.$$

1.18. Induced Electric Fields in Moving or Rotating Matter (Unipolar [Homopolar] Induction)

From Eq. (1-93), and the boundary conditions given by Eqs. (1-45) and (1-47), a uniform slab of perfectly conducting matter moving at velocity $\boldsymbol{v}$ through a magnetic field $\boldsymbol{B}$ will have an induced electric field given by

$$\boldsymbol{E} = -\frac{\boldsymbol{v}}{c} \times \boldsymbol{B}, \tag{1-94}$$

whose lines of force terminate in the surface charge density

$$\rho_s = E/4\pi. \tag{1-95}$$

A surface current density

$$J_s = cB/4\pi \tag{1-96}$$

must also be present.

Similarly, the electric field, $\boldsymbol{E}$, induced on the surface of a sphere of radius, $\boldsymbol{R}$, rotating at an angular velocity, Ω, through a magnetic field $\boldsymbol{B}$ is

$$\boldsymbol{E} = -\frac{\boldsymbol{\Omega} \times \boldsymbol{R}}{c} \times \boldsymbol{B}, \tag{1-97}$$

which also follows from the Lorentz force law (Eq. (1-52)). When the magnetic field is a magnetic dipole (Eqs. (1-19)), it follows from Eq. (1-97) that the potential difference induced between latitude λ and the equator is

$$V = \frac{\Omega B_p R^2}{2} (\cos^2 \lambda - 1), \tag{1-98}$$

where R is the sphere's radius, and B_p is the polar magnetic field strength.

1.19. Electromagnetic Field of a Point Charge Moving with a Uniform Velocity

When a point charge, q, is at rest it has no magnetic field and an electric field, E, given by Eq. (1-2)

$$\boldsymbol{E} = \frac{q}{R^2} \boldsymbol{n}_{\mathrm{r}},$$

where R is the radial distance from the charge and $\boldsymbol{n}_{\mathrm{r}}$ is a unit vector in the radial direction. When the charge is moving at the uniform velocity, $\boldsymbol{v}$, the Lorentz transformations give (POINCARÉ, 1905)

$$\boldsymbol{E} = \frac{q}{R^2} \frac{1 - \dfrac{v^2}{c^2}}{\left(1 - \dfrac{v^2}{c^2} \sin^2 \theta\right)^{3/2}} \boldsymbol{n}_{\mathrm{r}}, \tag{1-99}$$

and

$$\boldsymbol{H} = -(\boldsymbol{v} \times \boldsymbol{E})/\mathrm{c},$$

where θ is the angle between the direction of motion and the unit radial vector, $\boldsymbol{n}_{\mathrm{r}}$.

1.20. Vector and Scalar Potentials (The Retarded and Liénard-Wiechert Potentials)

Because the divergence of the curl of any vector is zero, it follows from Eq. (1-42) that the magnetic induction, $\boldsymbol{B}$, may be expressed as

$$\boldsymbol{B} = \nabla \times \boldsymbol{A}, \tag{1-100}$$

where $\boldsymbol{A}$ is called the magnetic vector potential. It then follows from Eq. (1-40) that

$$\boldsymbol{E} = -\frac{1}{c} \frac{\partial \boldsymbol{A}}{\partial t} - \nabla \varphi, \tag{1-101}$$

where φ is called the scalar electric potential. The $\boldsymbol{A}$ and φ may be related by the Lorentz condition (LORENTZ, 1892)

$$\nabla \cdot \boldsymbol{A} + \frac{1}{c}\frac{\partial \varphi}{\partial t} = 0. \tag{1-102}$$

When the Lorentz condition is used, it follows from Maxwell's Eqs. (1-39)—(1-42) that in a vacuum ($\varepsilon = \mu = 1$) the potentials satisfy the wave equations

$$\nabla^2 \boldsymbol{A} - \frac{1}{c^2}\frac{\partial^2 \boldsymbol{A}}{\partial t^2} = -\frac{4\pi}{c}\boldsymbol{J} \tag{1-103}$$

and

$$\nabla^2 \varphi - \frac{1}{c^2}\frac{\partial^2 \varphi}{\partial t^2} = -4\pi\rho,$$

where $\boldsymbol{J}$ is the current density and ρ is the charge density. These equations have the retarded solutions (LORENTZ, 1892)

$$\boldsymbol{A}(R,t) = \frac{1}{c}\int \frac{\boldsymbol{J}\left(t - \frac{R}{c}\right)}{R}\,dv$$

and

$$\varphi(R,t) = \int \frac{\rho\left(t - \frac{R}{c}\right)}{R}\,dv, \tag{1-104}$$

where the integrals are volume integrals of charge and current density at the retarded time $t' = t - R/c$, and R is the radial distance.

If each volume element of charge is moving at the velocity, $\boldsymbol{v}$, the volume integrals in Eqs. (1-104) must be corrected for the changing charge density. Consider a spherical shell of thickness dr at the distance R from the origin of the coordinate system used in the integration. During the time $dt = dr/c$, the amount of charge flowing through the inner surface of the shell is

$$\rho\,\boldsymbol{n}_r \cdot \boldsymbol{v}\,dt = \rho\,\frac{\boldsymbol{n}_r \cdot \boldsymbol{v}}{c}\,dr, \tag{1-105}$$

where $\boldsymbol{n}_r$ is a unit vector in the radial direction and $\boldsymbol{v}$ is the velocity vector. The sum of the charge elements, dq, to be accounted for in the volume element, dv, is

$$dq = \rho\left[1 - \frac{\boldsymbol{n}_r \cdot \boldsymbol{v}}{c}\right]dv. \tag{1-106}$$

For a point charge, q, moving at the velocity, v, Eqs. (1-104) and (1-106) give the Liénard-Wiechert potentials (LIÉNARD, 1898; WIECHERT, 1901)

$$\varphi(R,t) = \left.\frac{q}{R\left[1 - \frac{\boldsymbol{n}_r \cdot \boldsymbol{v}}{c}\right]}\right|_{t' = t - R/c} \tag{1-107}$$

$$\boldsymbol{A}(R,t) = \left.\frac{q\boldsymbol{v}}{cR\left[1 - \frac{\boldsymbol{n}_r \cdot \boldsymbol{v}}{c}\right]}\right|_{t' = t - R/c},$$

where R is the distance from the charge.

1.21. Electromagnetic Radiation from an Accelerated Point Charge

It follows directly from Eqs. (1-101) and (1-107) that the electric field vector $\boldsymbol{E}$ and magnetic induction $\boldsymbol{B}$ of a point charge q moving along the z axis at a velocity $v \ll c$ are given by

$$\boldsymbol{E} = \frac{q\dot{v}}{c^2 R} \sin\theta\, \boldsymbol{n}_\theta \bigg|_{t'=t-R/c} \tag{1-108}$$

$$\boldsymbol{B} = \frac{q\dot{v}}{c^2 R} \sin\theta\, \boldsymbol{n}_\varphi \bigg|_{t'=t-R/c},$$

where denotes the first derivative with respect to time, $\dot{v}$ is the acceleration, θ is the angle between the z axis and the radial vector to the point of observation, $\boldsymbol{n}_\theta$ and $\boldsymbol{n}_\varphi$ are unit vectors in the θ and φ directions defined by the spherical coordinate system, and it is assumed that the radial distance, R, is large. It then follows from Eqs. (1-63) and (1-64) that the total energy radiated per unit time per unit solid angle, $d\Omega$, is

$$\frac{dP}{d\Omega} = \frac{q^2 \dot{v}^2}{4\pi c^3} \sin^2\theta \bigg|_{t'=t-R/c}, \tag{1-109}$$

and the total energy radiated per unit time in all directions is

$$P = \frac{2}{3} \frac{q^2}{c^3} \dot{v}^2 \bigg|_{t'=t-R/c}, \tag{1-110}$$

a result first obtained by LARMOR (1897).

When the velocities are large, the correct results are (LIÉNARD, 1898; JACKSON, 1962)

$$\boldsymbol{E} = \left[q \frac{(\boldsymbol{n}-\boldsymbol{\beta})(1-\beta^2)}{(1-\boldsymbol{n}\cdot\boldsymbol{\beta})^3 R^2} \right] \bigg|_{t'=t-R/c} + \frac{q}{c} \left[\frac{\boldsymbol{n} \times \{(\boldsymbol{n}-\boldsymbol{\beta}) \times \dot{\boldsymbol{\beta}}\}}{(1-\boldsymbol{n}\cdot\boldsymbol{\beta})^3 R} \right] \bigg|_{t'=t-R/c} \tag{1-111}$$

$$\boldsymbol{H} = \boldsymbol{n} \times \boldsymbol{E}$$

$$\frac{dP}{d\Omega} = \frac{q^2}{4\pi c} \frac{|\boldsymbol{n} \times \{(\boldsymbol{n}-\boldsymbol{\beta}) \times \dot{\boldsymbol{\beta}}\}|^2}{(1-\boldsymbol{n}\cdot\boldsymbol{\beta})^5} \bigg|_{t'=t-R/c}$$

and

$$P = \frac{2}{3} \frac{q^2}{c} \gamma^6 [(\dot{\boldsymbol{\beta}})^2 - (\boldsymbol{\beta} \times \dot{\boldsymbol{\beta}})^2] \bigg|_{t'=t-R/c},$$

where $\boldsymbol{\beta} = \boldsymbol{v}/c$, the unit vector in the direction of observation is $\boldsymbol{n}$, and $\gamma = [1-(v/c)^2]^{-1/2}$.

1.22. Electromagnetic Radiation from Electric and Magnetic Dipoles

The electric dipole moment $d(t)$ of a charge q is related to its acceleration $\dot{v}(t)$ by $\dot{v}(t) = \ddot{d}(t)/q$. It therefore follows from Eqs. (1-108) that for an electric dipole oriented along the z axis, the electric and magnetic field vectors are

$$E = \frac{\ddot{d}(t)}{c^2 R} \sin\theta\, \boldsymbol{n}_\theta \Bigg|_{t'=t-R/c} \tag{1-112}$$

$$H = \frac{\ddot{d}(t)}{c^2 R} \sin\theta\, \boldsymbol{n}_\varphi \Bigg|_{t'=t-R/c},$$

where the distance R is large, θ is the angle between the z axis and the radial vector to the point of observation, and $\boldsymbol{n}_\theta$ and $\boldsymbol{n}_\varphi$ are unit vectors in the θ and φ direction. It also follows from Eqs. (1-63) and (1-64) that the total energy radiated per unit time per unit solid angle is

$$\frac{dP}{d\Omega} = \frac{[\ddot{d}(t)]^2}{4\pi c^3} \sin^2\theta \Bigg|_{t'=t-R/c} \tag{1-113}$$

and the total energy radiated per unit time in all directions is

$$P = \frac{2}{3} \frac{[\ddot{d}(t)]^2}{c^3} \Bigg|_{t'=t-R/c}. \tag{1-114}$$

As an example, for an electric dipole rotating in the $x-y$ plane with angular velocity, Ω, the total energy radiated per unit time per unit solid angle is (LANDAU and LIFSHITZ, 1962)

$$\frac{dP}{d\Omega} = \frac{d_0^2 \Omega^4}{8\pi c^3} (1 + \cos^2\theta) \tag{1-115}$$

and the total energy radiated per unit time in all directions is

$$P = \frac{2 d_0^2 \Omega^4}{3 c^3}, \tag{1-116}$$

where d_0 is the maximum value of the dipole moment $d(t)$ and quantities have been averaged over the period of the rotation. As another example, for harmonic oscillation of a charge, q, at frequency, ν_0, the total energy radiated per unit time is

$$P = \frac{64\pi^4 \nu_0^4}{3c^3} \left| \frac{q x_0}{2} \right|^2, \tag{1-117}$$

where $|q x_0/2|$ is the rms value of the dipole moment, and x_0 is the peak value of the harmonic displacement. The corresponding quantities for a magnetic dipole with magnetic dipole moment $m(t)$ are given by Eqs. (1-112)—(1-117) with $d(t)$ replaced with $m(t)$, with the exception that the roles of $\boldsymbol{E}$ and $\boldsymbol{H}$ are reversed. That is, the magnetic dipole formula for $\boldsymbol{H}$ is given by the electric dipole formula for $\boldsymbol{E}$ with $d(t)$ replaced by $m(t)$. The equations for dipoles moving at speed v comparable to c were first obtained by HEAVISIDE (1902).

Eqs. (1-112)—(1-117) are solutions for the far field ($R \gg c\dot{d}/\ddot{d}$ and $R > cd/\dot{d}$). The complete solution for the electromagnetic fields of a dipole was obtained by HERTZ (1889). The components of the fields in the spherical coordinate directions are

$$E_r = 2\left(\frac{[d]}{R^3} + \frac{[\dot{d}]}{cR^2}\right)\cos\theta \tag{1-118}$$

$$E_\theta = \left(\frac{[d]}{R^3} + \frac{[\dot{d}]}{cR^2} + \frac{[\ddot{d}]}{c^2 R}\right)\sin\theta$$

$$H_\varphi = \left(\frac{[\dot{d}]}{cR^2} + \frac{[\ddot{d}]}{c^2 R}\right)\sin\theta,$$

where [ ] denotes an evaluation at the retarded time $t' = t - (R/c)$, the dipole direction is along the z axis, θ is the angle from the z axis to the point of observation, and R is the distance to the point of observation. The fields for a magnetic dipole of magnetic dipole moment, $m(t)$, are given by Eqs. (1-118) with $d(t)$ replaced by $m(t)$, E replaced by H and H replaced by E.

1.23. Thermal Emission from a Black Body

The brightness, $B_\nu(T)$, of the radiation from a black body, a perfect absorber, in thermodynamic equilibrium at the temperature T is (PLANCK, 1901, 1913; WIEN, 1893, 1894; RAYLEIGH, 1900, 1905; JEANS, 1905, 1909; MILNE, 1930)

$$B_\nu(T) = \frac{2h\nu^3}{c^2}\frac{n_\nu^2}{[\exp(h\nu/kT) - 1]} \quad \text{Planck's law} \tag{1-119}$$

$$= \frac{2h\nu^3 n_\nu^2}{c^2}\exp(-h\nu/kT) \quad \text{if } h\nu \gg kT \quad \text{Wien's law} \tag{1-120}$$

$$= \frac{2n_\nu^2 \nu^2 kT}{c^2} \quad \text{if } h\nu \ll kT \quad \text{Rayleigh-Jeans law} \tag{1-121}$$

where n_ν is the index of refraction of the medium at frequency ν, c is the velocity of light, and h and k are, respectively, Planck's and Boltzmann's constants. The Rayleigh-Jeans approximation is used at radio frequencies, $\nu \approx 10^9$ Hz, for the criterion $h\nu \ll kT$, or $\nu < 2 \times 10^{10}\, T$, is valid. Wien's approximation is sometimes used at optical frequencies, where $\nu \approx 10^{15}$ Hz. The temperature, T, is called the equivalent brightness temperature and the units of brightness are erg sec^{-1} cm^{-2} Hz^{-1} rad^{-2}. Plots of $B_\nu(T)$ for various T are shown in Fig. 1.

That frequency, ν_m, for which the brightness is a maximum is given by (WIEN, 1894)

$$\nu_m \approx \frac{3kT}{h} \approx 6 \times 10^{10}\, T \text{ Hz} \quad \text{Wien displacement law} \tag{1-122}$$

or $\lambda_m \approx 0.51\, T^{-1}$,

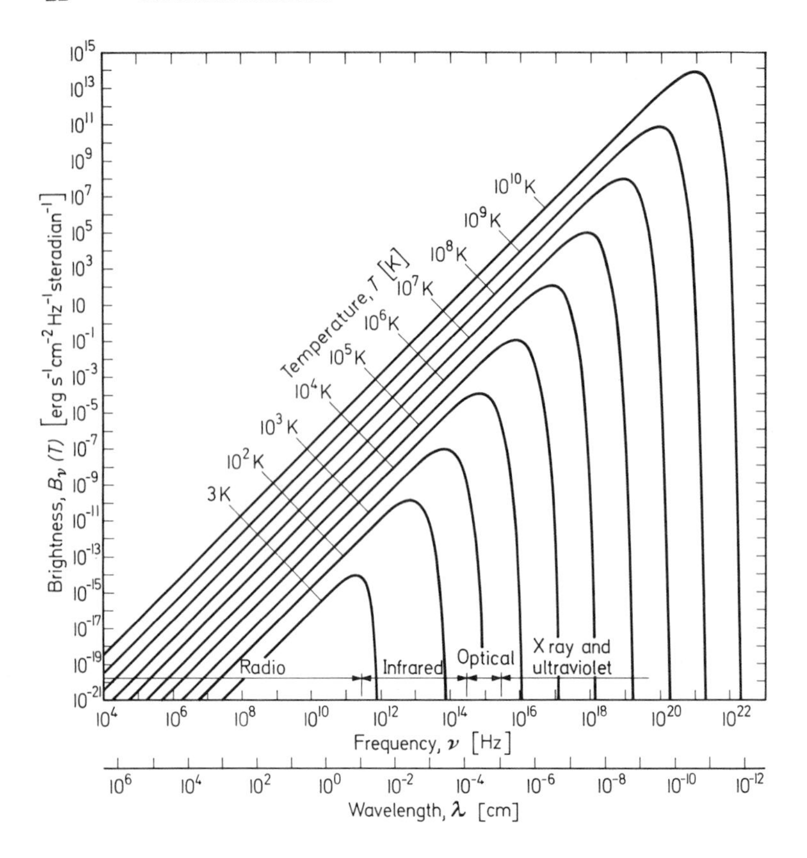

Fig. 1. The brightness, $B_\nu(T)$, of a black-body radiator at frequency, ν, and temperature, T. The Planck function, $B_\nu(T)$, is given by Eq. (1-119)

where λ_m is the wavelength of maximum brightness in centimeters. When the brightness is expressed in terms of a unit wavelength rather than a unit frequency, $\lambda_m \approx 0.29\, T^{-1}$ cm.

When $B_\nu(T)$ is integrated over all frequencies, the total emittance, πB, of the black body is found to be (STEFAN, 1879; BOLTZMANN, 1884; MILNE, 1930)

$$\pi B = \sigma T^4 n^2 \quad \text{Stefan-Boltzmann law,} \tag{1-123}$$

where n is the index of refraction of the medium and the Stefan-Boltzmann constant $\sigma = 2\pi^5 k^4/(15 c^2 h^3) = 5.669 \times 10^{-5}$ erg cm^{-2} sec^{-1} °K^{-4}. The total black body intensity, I, is therefore given by

$$I = \frac{\sigma n^2}{\pi} T^4 = 1.8046 \times 10^{-5}\, T^4 \text{ erg cm}^{-2}\text{ sec}^{-1}\text{ steradian}^{-1}, \tag{1-124}$$

and the radiation energy density, U, is

$$U = a T^4 n^3, \tag{1-125}$$

where the radiation density constant, $a = 4\sigma/c = 7.564 \times 10^{-15}$ erg cm^{-3} °K^{-4}.

For a graybody in the region of frequencies where the Rayleigh Jeans law is valid, the effective temperature, T_{eff}, is given by

$$T_{eff} = [1 - R_0(\lambda)]\, T, \tag{1-126}$$

where $R_0(\lambda)$ is the reflection coefficient at the wavelength, λ, of observation given by Eqs. (1-77) or (1-78), $[1 - R_0(\lambda)]$ is called the emissivity, and T is the black body temperature.

For a planet which has no heat conductivity and no atmosphere the effective temperature is

$$T_{eff} = T_{\odot} \left(\frac{R_{\odot}}{a}\right)^{1/2} (1-A)^{1/4} \approx 392 (1-A)^{1/4} a^{-1/2}, \tag{1-127}$$

where the temperature of the Sun, $T_{\odot} \approx 5778\,^{\circ}$K, the solar radius $R_{\odot} \approx 6.96 \times 10^{10}$ cm, the semi-major axis of the planetary orbit is a and is given in a.u. in the numerical approximation, and the Bond albedo (BOND, 1863), A, of a planet is defined (RUSSELL, 1916) as the ratio of the total amount of radiant energy reflected by the planet in all directions to the amount it receives from the Sun. If the planet rotates slowly and radiates predominantly from its sunlit side, the effective temperature is:

$$T_{eff} = \frac{T_{\odot}}{(2)^{1/4}} \left(\frac{R_{\odot}}{a}\right)^{1/2} (1-A)^{1/4} \approx 330 (1-A)^{1/4} a^{-1/2}, \tag{1-128}$$

and for the rapidly rotating major planets,

$$T_{eff} = \frac{T_{\odot}}{\sqrt{2}} \left(\frac{R_{\odot}}{a}\right)^{1/2} (1-A)^{1/4} \approx 277 (1-A)^{1/4} a^{-1/2}. \tag{1-129}$$

Measured values of planet temperatures are summarized in Table 1. The albedo and temperatures from Eq. (1-127) are given in Table 57.

For a black body whose radius is r and whose distance is D, the flux density $S_\nu(T)$ incident at the Earth is:

$$S_\nu(T) = \Omega_s B_\nu(T) \approx \frac{\pi r^2}{D^2} B_\nu(T), \tag{1-130}$$

where Ω_s is the solid angle subtended by the source and $B_\nu(T)$ is the brightness of the source at frequency ν and temperature T. The flux observed, $S_{0\nu}(T)$, with an antenna of efficiency, η_A, and beam area, Ω_A, is given by

$$S_{0\nu}(T) = \eta_A \Omega_s B_\nu(T) \quad \text{if } \Omega_s < \Omega_A \tag{1-131}$$

$$\text{or } \eta_A \Omega_A B_\nu(T) \quad \text{if } \Omega_s \geq \Omega_A. \tag{1-132}$$

At radio frequencies, the Rayleigh-Jeans approximation is used with Eq. (1-131) to obtain

$$S_{0\nu}(T) = 3.07 \times 10^{-37} \eta_A \Omega_s \nu^2 T \text{ erg sec}^{-1} \text{ cm}^{-2} \text{ Hz}^{-1} \quad \text{if } \Omega_s < \Omega_A. \tag{1-133}$$

Radio astronomers measure flux density in flux units, where one flux unit $= 10^{-23}$ erg sec^{-1} cm^{-2} Hz^{-1} $= 10^{-26}$ watt m^{-2} Hz^{-1}.

Table 1. Brightness temperatures of the moon and the planets at radio wavelengths[1]

Wavelength (cm)	Moon (°K)	Mercury (°K)	Venus (°K)	Mars (°K)	Jupiter (°K)	Saturn (°K)	Uranus (°K)	Neptune (°K)
0.10	229			171±31	155±15	140±15		
0.12					155±80			
0.13	216			184±50				
0.15	265							
0.18	240							
0.225			240±40					
0.32	210				111±15	97±50		
0.33	196	227±53		178±18				
0.34			296±30	198±40		130±15		
0.35							133±23	84±13
0.387								
0.40	230							
0.43					105±15	133±24		
0.80	211					132± 9		
0.815			380±40	173±18				
0.85			350±40		144±23			
0.86	225		425±40	240±42	113±11	96±20		
1.18	220		400±36		123±11			
1.25	215							
1.28					116±10			
1.34					98±11			
1.37	220							
1.43			451±40		106±11			
1.46			440±50					
1.50						141±15		
1.53		450±60		165±33				
1.58			477±57		105±21			
1.65	207			196±16	165±12		201±16	194±24
1.90		288±46			180±27	140±15		
1.94			495±25	168±17				
2.00	190						205±35	155±30
2.70				186±12	201±13		212±16	225±20
3.00					171±30			
3.10			592±40			137±12	158±20	115±36
3.14	195			211±28	145±26			
3.17					173±30			
3.20	216							
3.40						106±21		
3.60		390±100						
3.70						159±16		
3.75			675±20	210±11	226±34			
4.52			653±30					
6.00			700±35	196±27	291±25	176±10	210±17	227±33
7.50			630±30					
9.00						165±25		
9.40	220				658±58	177±30		
9.60	230							
10.0			640±35	184±18	690±62	196±55		
10.2					640±85			
10.3								
10.6			650±40		826	172±20		
11.0	214			169±19		200±30		

Table 1 (continued)

Wavelength (cm)	Moon (°K)	Mercury (°K)	Venus (°K)	Mars (°K)	Jupiter (°K)	Saturn (°K)	Uranus (°K)	Neptune (°K)
11.3		300 ± 15	630 ± 20	170 ± 19	706.	196 ± 20	130 ± 40	
20.8	205				1,312			
21.0	232				3,570	200 ± 30		
21.2			587 ± 25	233 ± 65	2,745	286 ± 37		
21.3			590 ± 20	163 ± 35		303 ± 50		
21.6					3,043			
22.0	270				2,979			
25.0	226							
30.2	227							
31.0					5,447			
31.25	227							
32.3	233							
35.2	223							
40.0	224							
42.9					11,280			
48.4					14,320			
49.0					14,990			
49.1					11,880			
54.0	218.5							
60.24	216.5							
68.0					23,790			
69.8					27,190			
70.16	217				20,000	1690 ± 430		
73.0					28,120			
75.0					28,050			
154.0					138,650			
168.0					135,000			
368.0					579,440			

[1] The data are from Dickel, Degioanni, and Goodman (1970), Epstein (1971), Hagfors (1970), Kellermann (1970), Mayer (1970), Mayer and McCullough (1971), Morrison (1970), Pollack and Morrison (1970), and Troitskii (1970). At wavelengths less than ten centimeters the brigthness temperature of the moon, mercury, venus and mars has been observed to vary with phase. The variable component for the moon increases from 15 °K to 115 °K as the wavelength decreases from 3 cm to 0.1 cm, whereas mercury, venus, and mars have respective variable components of about 90 °K, 50 °K and 40 °K.

At optical frequencies, the Wien approximation may be used with Eq. (1-131) to obtain

$$S_{0\nu}(T) = 1.47 \times 10^{-47} \eta_A \Omega_s \nu^3 \exp\left(-4.8 \times 10^{-11} \frac{\nu}{T}\right) \text{erg sec}^{-1} \text{cm}^{-2} \text{Hz}^{-1}. \quad \text{(1-134)}$$

Optical astronomers measure flux in terms of apparent magnitude, m. For two stars whose apparent magnitudes are m_1 and m_2, the ratio of the measured total flux from the two stars, s_1/s_2, is given by the relation

$$m_1 - m_2 = -2.5 \log(s_1/s_2). \quad \text{(1-135)}$$

One magnitude is equivalent to minus four decibels, and a calibration visual magnitude for the Sun is $m_\odot = -26.73 \pm 0.03$ and $s_\odot = 1.35 \times 10^6 \text{ erg sec}^{-1} \text{ cm}^{-2}$

at the Earth. Further formulae relating flux densities and magnitudes to luminosity, and formulae for magnitude corrections, are given in Chap. 5.

The Lorentz transformations given in Eqs. (1-91) were first used by MOSENGEIL (1907) to calculate the radiation from a black body moving at a velocity, v, away from an observer. At the radiation source, the total flux, s, of the radiation in the frequency range $d\nu$ and solid angle, $d\Omega$, is

$$s = B_\nu(T)\, d\nu\, d\Omega\,, \tag{1-136}$$

where $B_\nu(T)$ is the brightness of the black body at temperature T and frequency, ν. The observed frequency, ν_{obs}, is given by (LORENTZ, 1904)

$$\nu_{obs} = \frac{\nu}{(1-\beta^2)^{1/2}}(1-\beta\cos\theta)\,, \tag{1-137}$$

which accounts for the frequency shift of a moving source which was first observed by DOPPLER (1843). Here $\beta = v/c$ and θ denotes the angle between the velocity vector and the wave vector of the radiating source. The observed solid angle, $d\Omega_{obs}$, is given by

$$d\Omega_{obs} = \frac{1-\beta^2}{(1-\beta\cos\theta)^2}\, d\Omega\,. \tag{1-138}$$

It follows from Eqs. (1-92) and (1-119) that the total flux density seen by the observer, S_{obs}, and the observed brightness, $B_{\nu\,obs}(T_{obs})$, are given by

$$S_{obs} = B_\nu(T)\frac{(1-\beta\cos\theta)^2}{(1-\beta^2)}\, d\nu\, d\Omega = B_{\nu\,obs}(T_{obs})\, d\nu_{obs}\, d\Omega_{obs} \tag{1-139}$$

and

$$B_{\nu\,obs}(T_{obs}) = \frac{B_\nu(T)(1-\beta\cos\theta)^3}{(1-\beta^2)^{3/2}} = \frac{2h\nu_{obs}^3}{c^2\exp\{[h\nu_{obs}/(kT)][(1-\beta^2)^{1/2}/(1-\beta\cos\theta)]\}-1} \tag{1-140}$$

It follows from Eqs. (1-140) that the observed spectrum is that of a black-body radiator at the temperature $T_{obs} = T(1-\beta\cos\theta)(1-\beta^2)^{-1/2}$. For $\beta \ll 1$ and for an expanding background radiation, we have $T_{obs} \approx T(1+\beta\cos\theta)$.

1.24. Radiation Transfer and Observed Brightness

If a beam of radiation of intensity, I_0, passes through an absorbing cloud of thickness, L, the intensity of the radiation when leaving the cloud is given by

$$I_{0\nu} = I_0 \exp(-\tau_\nu)\,, \tag{1-141}$$

where the optical depth $\tau_\nu = \int_0^L \alpha_\nu\, dx$, the absorption coefficient per unit length in the cloud is α_ν, and the subscript ν denotes the frequency dependence of the variables.

An absorber also emits radiation, and the emission coefficient, ε_ν, is defined as the amount of energy a unit volume of material emits per second per unit solid angle in the frequency range ν to $\nu+d\nu$. The total intensity I_ν emitted by a column of gas of unit cross sectional area and length, L, is therefore

$$I_\nu = \int_0^L \varepsilon_\nu \exp[-\alpha_\nu x]\, dx\,. \tag{1-142}$$

For matter in thermodynamic equilibrium at temperature, T, we have (KIRCHHOFF, 1860; MILNE, 1930)

$$\varepsilon_\nu = n_\nu^2\, \alpha_\nu B_\nu(T) \qquad \text{(Kirchhoff's law)}, \tag{1-143}$$

where $B_\nu(T)$ is the vacuum brightness of a black body at temperature, T, and n_ν is the index of refraction of the medium.

It follows from Eqs. (1-142) and (1-143) that

$$\begin{aligned} B_{c\nu}(T) &= B_\nu(T)[1-\exp(-\tau_\nu)] \\ &= B_\nu(T) \qquad \text{if } \tau_\nu \gg 1 \quad \text{(optically thick)} \\ &= \tau_\nu B_\nu(T) \quad \text{if } \tau_\nu \ll 1 \quad \text{(optically thin)}, \end{aligned} \tag{1-144}$$

where $B_{c\nu}(T)$ is called the brightness of the cloud at frequency, ν, and the index of refraction is assumed to be unity.

If a source of brightness, $B_{s\nu}(T)$, at frequency ν, is irradiating the cloud, the total brightness, $B_{0\nu}(T)$ of the cloud will be

$$B_{0\nu}(T) = B_{s\nu}(T)\exp[-\tau_\nu] + B_{c\nu}(T)\,, \tag{1-145}$$

where $B_{c\nu}(T)$ is given by Eq. (1-144). The flux density observed by an antenna when observing a gas cloud will be given by Eqs. (1-131) or (1-132) when $B_{c\nu}(T)$ or $B_{0\nu}(T)$ are substituted for $B_\nu(T)$.

1.25. Magnetobremsstrahlung or Gyroradiation (Gyromagnetic and Synchrotron Radiation) of a Single Electron

When an electron moves linearly at the velocity, v, and with the acceleration, $\dot{v}$, the total power radiated per unit solid angle is (LIÉNARD, 1898)

$$\frac{dP}{d\Omega} = \frac{e^2 \dot{v}^2}{4\pi c^3} \frac{\sin^2\theta}{(1-\beta\cos\theta)^5}\,, \tag{1-146}$$

where $\beta = v/c$ and θ is the angle between the line of sight and the common direction of v and $\dot{v}$ (here we assume the acceleration vector is parallel to the velocity vector). At low velocities this distribution becomes the Larmor distribution (Eq. (1-109)). At high velocities the distribution becomes a narrow cone of half-angle, θ, given by

$$\theta = \gamma^{-1} = [1-\beta^2]^{1/2} = mc^2 E^{-1} = 8.2\times 10^{-7} E^{-1} \text{ radians}, \tag{1-147}$$

where $\beta = v/c$ and E is the total energy of the electron. This directed beam of radiation is typical regardless of the relation between the acceleration and velocity vectors.

When the electron moves about a circular orbit of radius, ρ, the observer sees a pulse of radiation of approximate duration $\rho/(\beta c \gamma)$ in the electron frame, and $\rho/(\beta c \gamma^3)$ in his own frame. Thus each pulse of radiation contains frequencies up to the critical frequency, ν_c, given by

$$2\pi\nu_c = \left(\frac{c\beta}{\rho}\right)\gamma^3 \approx 6\times 10^{28}\,\frac{E^3}{\rho}\ \mathrm{Hz}. \tag{1-148}$$

When the electron motion is assumed to be periodically circular, the observed radiation will consist of harmonics of the rotation frequency $\beta c/\rho$, up to the critical harmonic, γ^3.

The total power radiated, P_r, is given by (JACKSON, 1962)

$$P_{r_\parallel} = \frac{2}{3}\,\frac{e^2}{c^3}\,\dot{v}^2\gamma^6, \tag{1-149}$$

where $\dot{v}$ denotes the first derivative of the velocity, v, with respect to time. The subscript $\parallel$ in Eq. (1-149) denotes the case where the acceleration and velocity vectors are parallel. When they are perpendicular, the total power radiated is

$$P_{r_\perp} = \frac{2}{3}\,\frac{e^2}{c^3}\,\dot{v}^2\gamma^4 \tag{1-150}$$

a result first obtained by LIÉNARD (1898).

When an electron moves at a velocity, v, in a magnetic field of strength H, the frequency of gyration, ω_H, is (HEAVISIDE, 1904)

$$\omega_H = \frac{c\beta}{\rho} = \frac{eH}{\gamma mc}\sin\psi \approx 1.76\times 10^7\,\frac{H}{\gamma}\sin\psi\ \mathrm{Hz} \approx 14.4\,\frac{H}{E}\sin\psi\ \mathrm{Hz}, \tag{1-151}$$

where the pitch angle, ψ, is the angle between the H vector and the velocity vector. This means that the radius of gyration, ρ, is

$$\rho = \frac{v}{\omega_H} \approx 2\times 10^9\,E H^{-1}\ \mathrm{cm} \quad \text{for } v \approx c. \tag{1-152}$$

The resulting gyroradiation is called gyromagnetic when $v \ll c$. When the electron velocity is relativistic, the radiation is called synchrotron radiation. In this case, the radiation is primarily directed in the direction transverse to the magnetic field, and in a narrow beam of width γ^{-1}. The idea that cosmic radio sources might be radiating synchrotron radiation was first suggested by ALFVÉN and HERLOFSON (1950). Subsequently, GINZBURG (1953), SHKLOVSKI (1953) and GORDON (1954) suggested that the optical radiation of the Crab Nebula was synchrotron radiation and would be found to be polarized—as it was at optical wavelengths by DOMBROVSKII (1954) and in the radio wavelength region by OÖRT and WALRAVEN (1956).

As pointed out by EPSTEIN and FELDMAN (1967) the conventional formulae for the synchrotron radiation of a single electron must be corrected for the Doppler shifted gyrofrequency, ω_{DH}, given by

$$\omega_{DH} = \frac{\omega_H}{(1-\beta\cos^2\psi)} \approx \frac{\omega_H}{\sin^2\psi}. \tag{1-153}$$

As these effects cancel out when considering the radiation of an ensemble of electrons (SCHEUER, 1968), the Doppler shift effect will be ignored in what follows.

The frequency near which the synchrotron emission is a maximum is called the critical frequency and is defined as

$$\nu_c = \frac{3}{4\pi} \frac{eH}{mc} \gamma^2 \sin\psi \approx 6.266 \times 10^{18} H E^2 \sin\psi \text{ Hz}. \tag{1-154}$$

The constant in Eq. (1-154) is 16.1 when H is given in microgauss, E is in GeV and ν is in MHz. From Eqs. (1-150) and (1-151) the total radiated power, P_r, is obtained

$$\begin{aligned} P_r = \frac{2}{3} \frac{e^4}{m^2 c^3} \beta^2 \gamma^2 H^2 \sin^2\psi &\approx 2 \times 10^{-3} H^2 E^2 \sin^2\psi \text{ erg sec}^{-1} \\ &\approx 1.6 \times 10^{-15} H^2 \gamma^2 \sin^2\psi \text{ erg sec}^{-1}. \end{aligned} \tag{1-155}$$

O'DELL and SARTORI (1970) note that this equation and similar equations to follow are only valid if $\gamma \sin\psi \gg 1$ or if the radian frequency $\omega \gg 10^7 H/\sin\psi$. From Eq. (1-155) the total power radiated per unit frequency interval ν to $\nu + d\nu$ centered at the maximum frequency is

$$P(\nu_m) d\nu \approx \frac{P_r}{\nu_c} \approx 3 \times 10^{-22} H \text{ erg sec}^{-1} \text{ Hz}^{-1}. \tag{1-156}$$

The lifetime, τ_r, of the electron due to radiation damping is

$$\tau_r \approx \frac{E}{P_r} \approx 500 E^{-1} H^{-2} \text{ sec}. \tag{1-157}$$

Detailed calculations for the radiation from a relativistic particle in circular motion were first obtained by SCHOTT (1912). When these equations are applied to the ultrarelativistic motion ($E \gg mc^2$) of an electron in a magnetic field (VLADIMIRSKII, 1948; SCHWINGER, 1949; WESTFOLD, 1959) the following formulae are obtained.

The average power radiated per unit frequency interval in all directions may be divided into two components $P_1(\nu)$ and $P_2(\nu)$ according as the component is parallel or perpendicular to the projection of the H field line in the plane normal to the direction of observation.

$$P_1(\nu) d\nu = \frac{\sqrt{3} e^3 H}{2\pi m c^2} \frac{\nu}{2\nu_c} \left[\int_{\nu/\nu_c}^{\infty} K_{5/3}(\eta) d\eta - K_{2/3}\left(\frac{\nu}{\nu_c}\right) \right] d\nu, \tag{1-158}$$

$$P_2(\nu) d\nu = \frac{\sqrt{3} e^3 H}{2\pi m c^2} \frac{\nu}{2\nu_c} \left[\int_{\nu/\nu_c}^{\infty} K_{5/3}(\eta) d\eta + K_{2/3}\left(\frac{\nu}{\nu_c}\right) \right] d\nu, \tag{1-159}$$

where it is assumed that the electron has the appropriate pitch angle to radiate towards the observer, and K is the modified Bessel function (a Bessel function of the second kind with imaginary argument).

The total power radiated in frequency interval between ν and $\nu+d\nu$ is therefore given by

$$P(\nu)d\nu = \frac{\sqrt{3}e^3}{mc^2} H_\perp \frac{\nu}{\nu_c} \int_{\nu/\nu_c}^{\infty} K_{5/3}(\eta)d\eta d\nu$$

$$\approx 5.04 \times 10^{-22} H_\perp \left(\frac{\nu}{\nu_c}\right)^{1/3} d\nu \text{ erg sec}^{-1} \text{ Hz}^{-1} \qquad \text{for } \nu \ll \nu_c \qquad (1\text{-}160)$$

$$\approx 2.94 \times 10^{-22} H_\perp \left(\frac{\nu}{\nu_c}\right)^{1/2} \exp\left(\frac{-\nu}{\nu_c}\right) d\nu \text{ erg sec}^{-1} \text{ Hz}^{-1} \quad \text{for } \nu \gg \nu_c,$$

where $H_\perp$ is the component of H which is perpendicular to the velocity vector. The function $P(\nu)d\nu$ has its maximum at $\nu = 0.3\,\nu_c$, and describes the spectral distribution of power shown in Fig. 2.

For ultrarelativistic motion ($E \gg mc^2$), the total angular spectrum can also be divided into two components $P_1(\psi)$ and $P_2(\psi)$ according as the direction of polarization is parallel or perpendicular to the projection of the magnetic field in the plane normal to the direction of observation (WESTFOLD, 1959; GINZBURG and SYROVATSKII, 1965)

$$P_2(\psi) = \frac{3}{4\pi^2 r^2} \frac{e^3 H}{mc^2 \zeta} \left(\frac{\nu}{\nu_c}\right)^2 \left(1 + \frac{\psi^2}{\zeta^2}\right)^2 [K_{2/3}(g_\nu)]^2$$

and

$$P_1(\psi) = \frac{3}{4\pi^2 r^2} \frac{e^3 H}{mc^2 \zeta} \left(\frac{\nu}{\nu_c}\right)^2 \frac{\psi^2}{\zeta^2} \left(1 + \frac{\psi^2}{\zeta^2}\right) [K_{1/3}(g_\nu)]^2, \qquad (1\text{-}161)$$

where

$$g_\nu = \frac{\nu}{2\nu_c} \left(1 + \frac{\psi^2}{\zeta^2}\right)^{3/2}.$$

The r^2 term has been added to denote the dependence of the radiated power on the distance, r, from the source, $\zeta = \gamma^{-1} = mc^2/E$, and here ψ is not the pitch angle but the angle between the direction of observation and the nearest velocity vector of the radiation cone. The $P_1(\psi)$ and $P_2(\psi)$ are illustrated in Fig. 2. In general, then, the polarization is elliptical with the axes parallel and perpendicular to the projection of the magnetic field on the plane transverse to the direction of observation. The direction of the ellipse is right or left hand according as $\psi >$ or < 0. The polarization is seen to be linear only when $\psi = 0$. The degree of polarization of the total power per unit frequency interval at a given frequency ν, is (WESTFOLD, 1959)

$$\pi = \frac{K_{2/3}(\nu/\nu_c)}{\int_{\nu/\nu_c}^{\infty} K_{5/3}(\eta)d\eta}$$

$$\approx \frac{1}{2} \qquad \text{for } \nu \ll \nu_c \qquad (1\text{-}162)$$

$$\approx 1 - \frac{2\nu_c}{3\nu} \qquad \text{for } \nu \gg \nu_c.$$

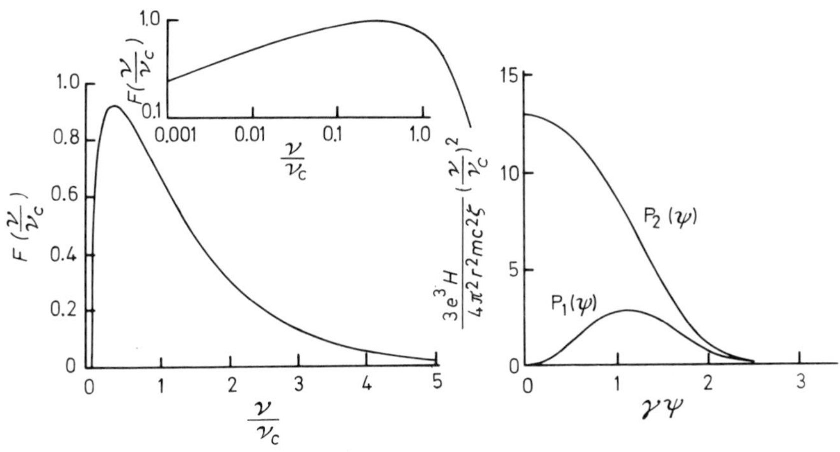

Fig. 2. The function $F(\nu/\nu_c)=(\nu/\nu_c)\int_{\nu/\nu_c}^{\infty} K_{5/3}(\eta)d\eta$, which characterizes the spectral distribution of synchrotron radiation from a single electron, is shown in both linear and logarithmic plots (cf. VLADIMIRSKII, 1948; SCHWINGER, 1949). The total synchrotron power radiated per unit frequency interval is related to $F(\nu/\nu_c)$ by Eq. (1-160), and the critical frequency, ν_c, is given in Eq. (1-154). Also shown is the angular spectrum for the synchrotron radiation of a single electron in directions parallel, $P_1(\psi)$, and perpendicular, $P_2(\psi)$, to the projection of the magnetic field on the plane of the figure (after GINZBURG and SYROVATSKII, 1965, by permission of Annual Reviews, Inc.). The angle, ψ, is the angle between the direction of observation and the nearest velocity vector of the radiation cone, H is the magnetic field intensity, r is the distance from the radiating electron, $\gamma=\zeta^{-1}=[1-(v/c)^2]^{-1/2}$ where v is the velocity of the electron, and $P_1(\psi)$ and $P_2(\psi)$ are given by Eqs. (1-161). The angular spectrum plots are for $\nu/\nu_c=0.29$

Here the degree of polarization, π, is related to the maximum and minimum observable values of intensity, $P_{\max}$ and $P_{\min}$, by $\pi=(P_{\max}-P_{\min})/(P_{\max}+P_{\min})$. It is related to the Stokes parameters by Eq. (1-70). Detailed formulae for the Stokes parameters of the synchrotron radiation from an ultrarelativistic electron are given by LEGG and WESTFOLD (1968).

The rate at which an electron loses its energy, E, by synchrotron radiation is given by

$$\frac{dE}{dt}=-\int_0^{\infty} P(\nu)d\nu=-2.368\times10^{-3}H_\perp^2E^2\,\mathrm{erg\,sec}^{-1}, \qquad (1\text{-}163)$$

where $P(\nu)d\nu$ is given by Eq. (1-160). The electron energy, $E(t)$, as a function of time, t, is given by

$$E(t)=\frac{E_0}{(1+t/t_{1/2})}, \qquad (1\text{-}164)$$

where the time required for the electron to lose half its initial energy, E_0, is

$$t_{1/2}=\frac{4.223\times10^2}{H_\perp^2E_0}\,\mathrm{sec}. \qquad (1\text{-}165)$$

The constant 4.223×10^2 has the more practical units of 8.352×10^9 years (μ gauss)^2 GeV. As pointed out by TAKAKURA (1960), the radiation lifetime may not be so small as the collision lifetime, t_c, when thermal electrons are present. For example, the basic relaxation time, τ_r, for an electron of energy, E, moving through singly charged ions of density, N_i, is given by (TRUBNIKOV, 1965)

$$\tau_r \approx \frac{\sqrt{m}}{15\pi\sqrt{2}\,e^4}\frac{E^{3/2}}{N_i}$$
$$\approx 10^8\,E^{3/2} N_i^{-1} \text{ sec},$$

where the numerical approximation is for E in keV and N_i in cm^{-3}.

Detailed calculations of the angular spectrum and the frequency spectrum of the synchrotron radiation for moderate electron energies, $E \approx mc^2$, include motions parallel to the H field (TRUBNIKOV, 1958; TAKAKURA, 1960; BEKEFI, 1966). The total power radiated per unit solid angle in the nth harmonic at the angle θ to the H field line is

$$\frac{dP_n}{d\Omega} = \frac{e^4 H^2 (1-\beta)^2 \gamma^2}{2\pi m^2 c^3 (1-\beta_{\parallel}\cos\theta)^2}\left\{\left(\frac{\cos\theta - \beta_{\parallel}}{\beta_{\perp}\sin\theta}\right)^2 J_n^2(\gamma\sin\theta) + J_n'^2(\gamma\sin\theta)\right\}, \quad (1\text{-}166)$$

where $\beta^2 = \beta_{\parallel}^2 + \beta_{\perp}^2 = (v_{\parallel}/c)^2 + (v_{\perp}/c)^2 = (v/c)^2$, the $v_{\parallel}$ and $v_{\perp}$ are the instantaneous particle velocities along and perpendicular to the magnetic field,

$$\gamma = \frac{n\beta_{\perp}}{1-\beta_{\parallel}\cos\theta},$$

J_n is a Bessel function of order n, and $'$ denotes the first derivative with respect to the argument. This radiation is at the critical radian frequency

$$\omega = \frac{neH(1-\beta^2)^{1/2}}{mc(1-\beta_{\parallel}\cos\theta)}.$$

The angular distribution of the radiation from a mildly relativistic electron is shown in Fig. 3.

The total power radiated in the nth harmonic is

$$P_n = \frac{2e^4 H^2}{m^2 c^3}\frac{1-\beta_0^2}{\beta_0}\left[n\beta_0^2 J_{2n}'(2n\beta_0) - n^2(1-\beta_0)^2 \int_0^{\beta_0} J_{2n}(2nt)\,dt\right],$$

where $\beta_0 = \beta_{\perp}/(1-\beta_{\parallel}^2)^{1/2}$.

The total power radiated over all harmonics is

$$P_r = \frac{2e^4 H^2}{3m^2 c^3}\left[(1-\beta_{\parallel}^2)(1-\beta^2)^{-1} - 1\right] = 1.59 \times 10^{-15}\frac{\beta_{\perp}^2 H^2}{1-\beta^2} \text{ erg sec}^{-1}. \quad (1\text{-}167)$$

The first and second terms in the brackets { } of Eq. (1-166) respectively represent polarized components of emission with the electric vector parallel, $\parallel$, and perpendicular, $\perp$, to the magnetic field. Analytic approximations for the

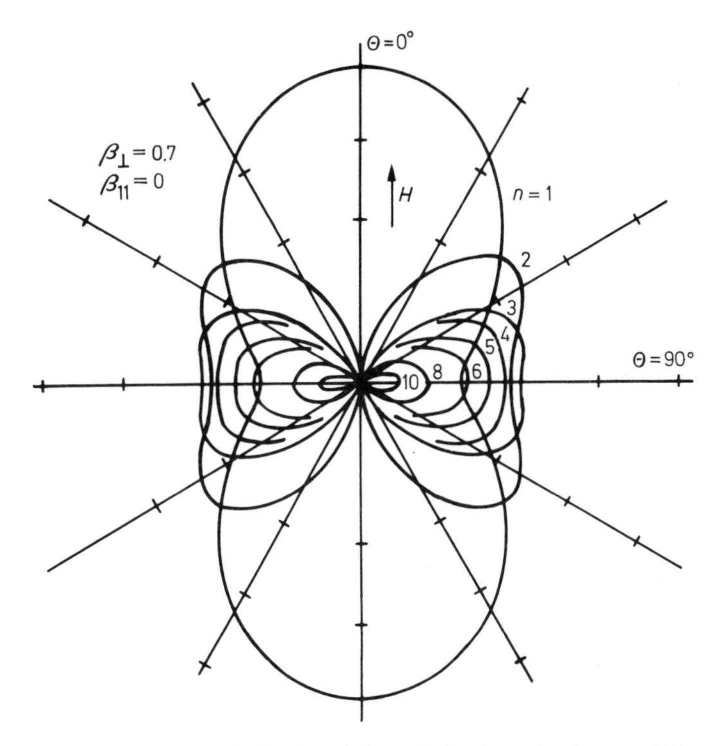

Fig. 3. The angular distribution of the radiation intensity from a mildly relativistic electron (after OSTER, 1960). The intensity distribution is shown for various harmonics, n, as a function of the angle, θ, between the observer and the direction of the magnetic field, H. The figure is for a fixed particle energy corresponding to $\beta_{\perp}=V_{\perp}/c=0.7$ where the electron velocity is V and $\perp$ denotes the component perpendicular to the direction of the magnetic field. The short bars on the radial lines denote equal increments of intensity; and the complete polar surface is obtained by rotating the figure about the vertical axis. The figure illustrates the fact that higher frequencies are emitted in a narrow angular range about the electron orbital plane defined by $\theta=90°$

Bessel function terms of Eq. (1-166) are given by WILD and HILL (1971), and they obtain

$$P_{\|} = A z^2\left(\frac{1.5}{n_c}+\frac{0.5033}{n}\right)^{-1/3} n\exp\left(-\frac{n}{n_0}\right) \tag{1-168}$$

$$P_{\perp}=A\left(\frac{1.5}{n_c}+\frac{1.193}{n}\right)^{1/3}\left(1-\frac{1}{5n^{2/3}}\right)^2 n\exp\left(-\frac{n}{n_0}\right),$$

where P is the power emitted by a single relativistic or sub-relativistic electron per unit solid angle per unit frequency interval in the nth harmonic,

$$z=\frac{(\cos\theta-\beta\cos\varphi)}{(1-\beta\cos\varphi\cos\theta)},$$

$$A=\frac{e^3 H}{2\pi m c^2\sin^2\theta}\,\frac{(1-\beta^2)^{1/2}}{(1-\beta\cos\varphi\cos\theta)},$$

φ is the angle between the electron velocity vector and the magnetic field direction, θ is the angle between the magnetic field direction, and the observers line of sight, and

$$n_c = \tfrac{3}{2}(1-X^2)^{-3/2}$$

and

$$(2n_0)^{-1} = \ln[1+(1-X^2)^{1/2}] - \ln X - (1-X^2)^{1/2},$$

where

$$X = \frac{\beta \sin\varphi \sin\theta}{(1-\beta\cos\varphi\cos\theta)}.$$

The polarization of the radiation from mildly relativistic electrons will be elliptical, with the ratio of the axes of the ellipse given by

$$R_n = -\frac{(\cos\theta - \beta_{\|}) J_n(\gamma \sin\theta)}{\beta_{\perp} \sin\theta J_n'(\gamma\sin\theta)}. \tag{1-169}$$

In the case of gyroradiation where the electron kinetic energy $E \ll mc^2$, Eq. (1-166) for the total power radiated per unit solid angle in the nth harmonic can be simplified to the expression

$$\frac{dP_n}{d\Omega} = \frac{e^4 H^2}{2\pi m^2 c^3} \frac{(n\beta_{\perp}/2)^{2n}}{[(n-1)!]^2} (\sin\theta)^{2n-2}(1+\cos^2\theta), \tag{1-170}$$

where θ is the angle between the direction of wave propagation and the magnetic field, and $\beta_{\perp} = (v_{\perp}/c)$ corresponds to the component of velocity, $v_{\perp}$, perpendicular to the magnetic field. For the case of a Maxwellian distribution of electron kinetic energies, the absorption coefficient per unit length, α_ν, due to gyroradiation is given by (SITENKO and STEPANOV, 1957; WILD, SMERD, and WEISS, 1963)

$$\alpha_\nu = \frac{4\pi^{3/2}}{c} \frac{\nu_p^2}{\nu} \frac{(n/2)^{2n}}{n!} \frac{(\sin\theta)^{2n-2}(1+\cos^2\theta)}{\cos\theta} \beta_0^{2n-3} \exp\left[-\frac{(1-n\nu_H/\nu)^2}{\beta_0^2\cos^2\theta}\right], \tag{1-171}$$

where the plasma frequency $\nu_p = [e^2 N_e/(\pi m)]^{1/2}$, the frequency of the radiation is ν, the $\beta_0^2 = 2kT/(mc^2)$, and the gyrofrequency $\nu_H = eH/(2\pi mc)$. For optical depth calculations, the effective thickness, L, of a layer of thermal electrons is given by $L = 2L_H \beta_0 \cos\theta$ where the scale of the magnetic field $L_H = H/\nabla H$, is the characteristic distance over which a change in ν_H is significant.

When an electron loses a nonnegligible amount of kinetic energy in its orbit in a magnetic field, quantum effects must be taken into account. This condition is fulfilled when (CANUTO and CHIU, 1971)

$$\left(\frac{e^2}{\hbar c}\right)\gamma^2\left(\frac{H}{H_q}\right) \gg 1$$

or

$$\gamma^2 H \gg 6\times 10^{15} \quad \text{gauss}.$$

Here the total energy of the electron is $E = \gamma mc^2$ and $H_q = m^2c^3/(e\hbar) \approx 4.414 \times 10^{13}$ gauss. Quantum effects also become important for $E - mc^2 \approx H/H_q$ or for

$H/H_q \gg 1$ (ERBER, 1966). When the electron orbits become quantized, the energy of the electron in a magnetic field is given by (RABI, 1928)

$$E = mc^2\left[1 + \left(\frac{p_z}{mc}\right)^2 + \frac{2nH}{H_q}\right]^{1/2}, \tag{1-172}$$

where p_z is the component of electron momentum in the direction of the magnetic field, H is the magnetic field intensity, and the principle quantum number $n = 0, 1, 2, \ldots$. The transition between two quantum states n and n' gives rise to a photon of energy $h\nu$ given by (CANUTO and CHIU, 1971)

$$h\nu = \frac{E - p_z c\cos\theta}{\sin^2\theta}\left\{1 - \left[1 - \frac{m^2c^4\sin^2\theta(n-n')}{(E - p_z c\cos\theta)^2}\right]^{1/2}\right\}$$

$$\approx mc^2\left\{(n-n')\frac{H}{H_q} + \theta\left[\left(\frac{h\nu}{mc^2}\right)^2\cos^2\theta\right]\right\} \quad \text{for } H \ll H_q,$$

where θ is the angle of the emitted photon with respect to the axis of the magnetic field. For the $n=1$ to $n'=0$ transition, all of the radiation will have an energy of $1.16 \times 10^{-8} H$ eV and the total power radiated per unit volume, P, is given by (CHIU and FASSIO-CANUTO, 1969)

$$P = \frac{1}{2}\,\frac{e^2 mc}{\hbar^2}\,mc^2 N_e\left(\frac{h\nu}{mc^2}\right)^2\left(\frac{H}{H_q}\right)(1+\cos^2\theta)$$

$$\approx 2.85 \times 10^{-5} N_e H_8^3 (1+\cos^2\theta) \quad \text{erg sec}^{-1}\,\text{cm}^{-3},$$

where N_e is the number density of electrons in quantum state $n=1$, and $H_8 = H/10^8$. The lifetime of the electron in this quantum regime is given by

$$\tau = \left[\frac{2}{3}\,\frac{e^2}{\hbar c}\,\frac{mc^2}{\hbar}\left(\frac{H}{H_q}\right)^2\gamma\right]^{-1} \approx 2.6 \times 10^{-19}\left(\frac{H_q}{H}\right)^{-2}\gamma^{-1}\ \text{sec}.$$

1.26. Synchrotron Radiation from an Ensemble of Particles

The volume emissivity (power per unit frequency interval per unit volume per unit solid angle) of the ultrarelativistic radiation of a group of electrons is

$$\varepsilon(\nu) = \int P(\nu)N(E)\,dE, \tag{1-173}$$

where $P(\nu)$ is the total power radiated per unit frequency interval by one electron (Eq. (1-160)), and $N(E)\,dE$ is the number of electrons per unit volume per unit solid angle along the line of sight which are moving in the direction of the observer and whose energies lie in the range E to $E + dE$.

If all of the ultrarelativistic electrons possess the same energy, E, then Eqs. (1-160) and (1-173) give the total intensity

$$I(\nu) \approx 5.04 \times 10^{-22} H_\perp\left(\frac{\nu}{\nu_c}\right)^{1/3}\int N(E)\,dl \ \text{erg sec}^{-1}\,\text{cm}^{-2}\,\text{Hz}^{-1}\,\text{rad}^{-2} \quad \text{for } \nu \ll \nu_c$$

$$\approx 2.94 \times 10^{-22} H_\perp\left(\frac{\nu}{\nu_c}\right)^{1/2}\exp\left(\frac{-\nu}{\nu_c}\right)\int N(E)\,dl \ \text{erg sec}^{-1}\,\text{cm}^{-2}\,\text{Hz}^{-1}\,\text{rad}^{-2} \quad \text{for } \nu \gg \nu_c,$$

where ν is the frequency, $\nu_c \approx 6 \times 10^{18} H_\perp E^2$ Hz, the component of H along the line of sight is $H_\perp$, and $\int N(E)\,dl$ is the number of electrons per unit solid angle per square centimeter along the line of sight whose velocities are directed towards the observer. The degree of polarization in this case is given by Eq. (1-162)

$$\pi = \frac{1}{2}\left\{1 + \frac{\Gamma(\frac{1}{3})}{2}\left(\frac{\nu}{2\nu_c}\right)^{2/3}\right\} \quad \text{for } \nu \ll \nu_c$$

$$= 1 - \frac{2}{3}\frac{\nu_c}{\nu} \quad \text{for } \nu \gg \nu_c .$$

Because the observed cosmic rays have a power law spectrum, and because the spectra of many observed radio sources are power law, it is often assumed that

$$N(E)\,dE = K E^{-\gamma} dE, \tag{1-174}$$

where K is a constant. The intensity of the ultrarelativistic synchrotron radiation of a homogeneous and isotropic distribution of electrons whose $N(E)$ are given by Eq. (1-174), and which are imbedded in a homogeneous magnetic field of strength H, may be obtained from Eqs. (1-160) and (1-173)

$$\begin{aligned} I(\nu) = l\varepsilon(\nu) = K l \alpha(\gamma) \frac{\sqrt{3}}{8\pi}\frac{e^3}{mc^2}\left[\frac{3e}{4\pi m^3 c^5}\right]^{(\gamma-1)/2} H_\perp^{(\gamma+1)/2}\,\nu^{-(\gamma-1)/2} \\ \approx 0.933 \times 10^{-23} \alpha(\gamma) K l H_\perp^{(\gamma+1)/2}\left(\frac{6.26 \times 10^{18}}{\nu}\right)^{(\gamma-1)/2} \ \text{erg sec}^{-1}\,\text{cm}^{-2}\,\text{Hz}^{-1}\,\text{rad}^{-2}, \end{aligned} \tag{1-175}$$

where ν is the frequency, l is the dimension of the radiating region along the line of sight, and $\alpha(\gamma)$ is a slowly varying function of γ which is of order unity and is given by

$$\alpha(\gamma) = 2^{(\gamma-3)/2}\frac{\gamma + 7/3}{\gamma + 1}\,\Gamma\left(\frac{3\gamma - 1}{12}\right)\Gamma\left(\frac{3\gamma + 7}{12}\right)$$

for $\gamma > \frac{1}{3}$. The degree of polarization is, in this case:

$$\pi = (\gamma + 1)/(\gamma + \tfrac{7}{3}). \tag{1-176}$$

If the magnetic field is randomly distributed and $N(E)$ is given by Eq. (1-174) then $I(\nu)$ is given by Eq. (1-175) with a new constant of order unity substituted for $\alpha(\gamma)$. In this case, however, the degree of polarization is zero.

Measurements of the degree of circular polarization, π_0, of the synchrotron radiation from an ensemble of ultrarelativistic electrons ($E \gg mc^2$) may be used to determine the magnetic field intensity, H, of the radiating source (LEGG and WESTFOLD, 1968; MELROSE, 1971). For a monoenergetic distribution of electrons, the degree of circular polarization is given by

$$\begin{aligned} \pi_0 = \frac{2\sqrt{2}}{\sqrt{3}}\frac{\cot\theta}{F(\nu/\nu_c)}\left(\frac{\nu_H \sin\theta}{\nu}\right)^{1/2}\left\{\left(\frac{\nu}{\nu_c}\right)^{3/2} K_{1/3}\left(\frac{\nu}{\nu_c}\right)\right. \\ \left. + \left[2 + \frac{\varphi'(\theta)}{\varphi(\theta)}\tan\theta\right]\left(\frac{\nu}{\nu_c}\right)^{-1/2}\left[\frac{\nu}{\nu_c}K_{2/3}\left(\frac{\nu}{\nu_c}\right) - \frac{1}{2}F\left(\frac{\nu}{\nu_c}\right)\right]\right\}, \end{aligned} \tag{1-177}$$

where ν is the frequency of radiation, the gyrofrequency $\nu_H = eH/(2\pi mc)$, the critical frequency, ν_c, is given by Eq. (1-154), K is the modified Bessel function, the angle $\theta = \alpha - \psi$ where α is the pitch angle and ψ is the angle between the direction of observation and the velocity vector of the electron, the proportion of electrons having values of θ within the range θ to $\theta + d\theta$ in the solid angle $d\Omega = 2\pi \sin\theta \, d\theta$ is $\varphi(\theta) d\Omega$, the first derivative of $\varphi(\theta)$ with respect to θ is $\varphi'(\theta)$ and is unity for an isotropic distribution, and the function $F(x)$ is given by

$$F(x) = x \int_x^{\infty} K_{5/3}(\eta) \, d\eta \,. \tag{1-178}$$

For an ensemble of electrons with a power law energy distribution of index, γ, as given by Eq. (1-174), the degree of circular polarization is given by

$$\pi_0 = \frac{4}{\sqrt{3}} \frac{\gamma+1}{\gamma(\gamma+\frac{7}{3})} \cot\theta \left(\frac{\nu_H \sin\theta}{\nu}\right)^{1/2} \frac{\Gamma\left(\frac{3\gamma+4}{12}\right)\Gamma\left(\frac{3\gamma+8}{12}\right)}{\Gamma\left(\frac{3\gamma-1}{12}\right)\Gamma\left(\frac{3\gamma+7}{12}\right)} \times \left[\gamma + 2 + \tan\theta \frac{\varphi'(\theta)}{\varphi(\theta)}\right], \tag{1-179}$$

for $\gamma > \frac{1}{3}$. Values of the gamma functions, Γ, are tabulated by LEGG and WESTFOLD (1968) for values of γ in the range 0.4—9.0. The degree of circular polarization given by Eqs. (1-177) and (1-179) is the ratio of the fourth Stokes parameter, V, to the first Stokes parameter, I. MELROSE (1971) has extended these formulae to include the effects of reabsorption and differential Faraday rotation on the degree of circular polarization. Both effects can cause the circular polarization to become slightly smaller and to reverse in sign. The polarization of gyro-synchrotron radiation in a magnetoactive plasma is given by RAMATY (1969).

From Eq. (1-175) we see that the synchrotron radiation from electrons with a power law distribution of index $-\gamma$ has a power law frequency spectrum proportional to ν^α, where the spectral index, α, is given by

$$\alpha = -\frac{(\gamma - 1)}{2} \,. \tag{1-180}$$

Most radio sources are observed to have a power law frequency spectrum, and some examples are shown in Fig. 4. Spectral indices for radio galaxies and quasi-stellar sources with known redshifts are included in Table 31.

It follows from Eqs. (1-164) and (1-174) that the total electron energy, U_e, is given by

$$U_e = 422 \frac{L}{H^2} \frac{3-\gamma}{2-\gamma} \left(\frac{E_2^{2-\gamma} - E_1^{2-\gamma}}{E_2^{3-\gamma} - E_1^{3-\gamma}}\right) \text{ erg}, \tag{1-181}$$

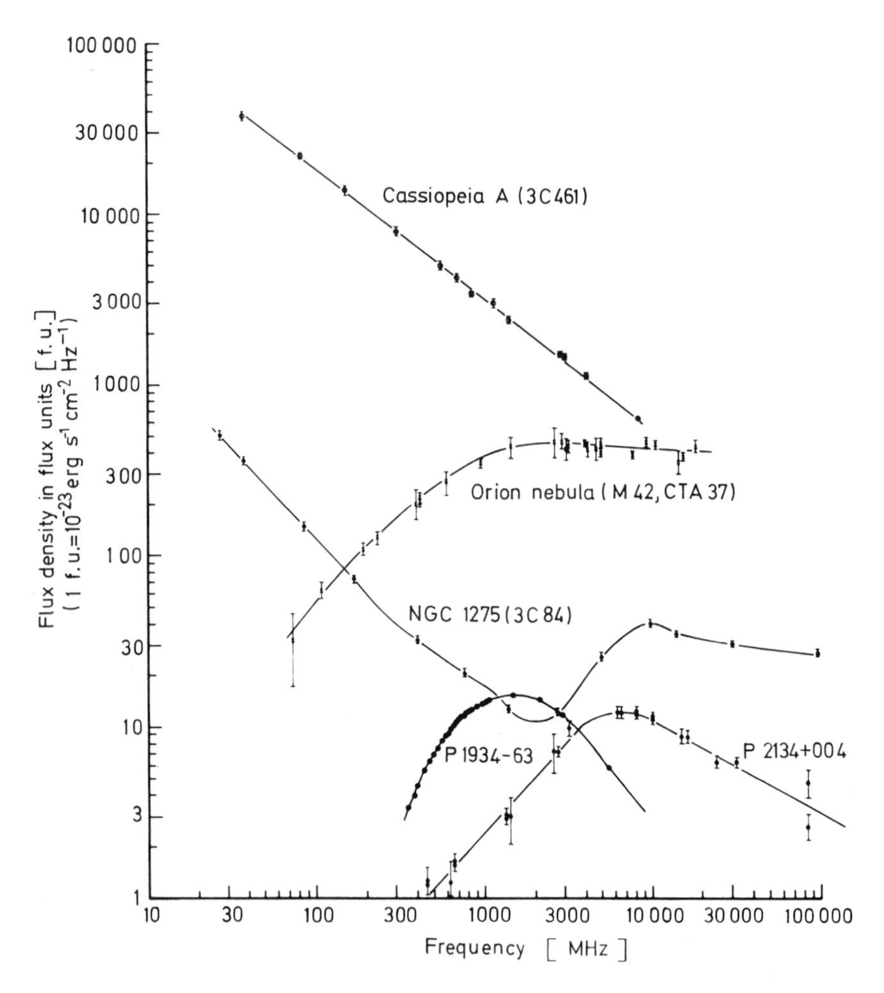

Fig. 4. Radiofrequency spectra of sources exhibiting the power law spectrum of synchrotron radiation (Casseopeia A), the flat spectrum of thermal bremsstrahlung radiation with low frequency self absorption (Orion Nebula), unusual high frequency radiation (NGC 1275), and low frequency absorption processes (P 1934 − 63 and P 2134 + 004). The data for P 2134 + 004 are from E. K. Conklin, and the other data are from Kellermann (1966), Hjellming, and Churchwell (1969), Terzian and Parrish (1970), Kellermann, Pauliny-Toth, and Williams (1969), and Kellermann *et al.* (1971)

where E_1 and E_2 denote, respectively, the lowest and highest electron energies, it is assumed that the electron energy distribution index $\gamma \neq 2$ or 3, and the total luminosity, L, of the source is given by

$$L = 4\pi r^2 \left(1 + \frac{z}{2}\right)^2 \frac{S_{\nu_r}}{\nu_r^{\alpha}} \int_{\nu_1}^{\nu_2} \nu^{\alpha}\, d\nu, \tag{1-182}$$

where $r = cz/H_0$ is the distance to the source, H_0 is the Hubble constant, z is the redshift, S_{ν_r} is the source flux density measured at some reference frequency, ν_r, and ν_1 and ν_2 are the lower and upper cutoff frequencies beyond which the source radiation is negligible. Assuming that each electron radiates only at its critical frequency given by Eq. (1-154), Eq. (1-181) may be written as

$$U_e = 10^{12} \frac{L}{H^{3/2}} \frac{\nu_2^{1/2+\alpha} - \nu_1^{1/2+\alpha}}{\nu_2^{1+\alpha} - \nu_1^{1+\alpha}} \frac{(2\alpha+2)}{(2\alpha+1)} \text{ erg}, \tag{1-183}$$

provided that $\alpha \neq -\frac{1}{2}$ or -1. As discussed in Sect. 1.38, the luminosity of synchrotron radiation is limited by the inverse Compton effect.

The total energy in the source, U_T, is given by (Burbidge, 1959)

$$U_T = U_p + U_m = a\,U_e + \frac{VH^2}{8\pi}, \tag{1-184}$$

where the total particle energy is U_p, the constant $a \approx 100$ denotes the increase in energy of the more energetic baryons above that of the relativistic electrons, the energy stored in the magnetic field is U_m, and V is the volume of the source. The value of $a = 100$ is typical for cosmic rays. Luminosities of radio galaxies and quasars range from 10^{40} to 10^{44} erg sec^{-1}, whereas those for normal galaxies are between 10^{34} and 10^{38} erg sec^{-1}. A typical field of $H \approx 10^{-4}$ gauss gives $U_m \approx U_p \approx 10^{60}$ erg if $a = 100$.

1.27. Synchrotron Radiation in a Plasma

The total power radiated per unit frequency interval by a single relativistic electron in a plasma is given by (Ginzburg and Syrovatskii, 1965)

$$P(\nu) = \sqrt{3}\,\frac{e^3 H \sin\psi}{mc^2}\left[1 + \left(\frac{\nu_p E}{\nu m c^2}\right)^2\right]^{-1/2} \frac{\nu}{\nu_c'} \int_{\nu/\nu_c'}^{\infty} K_{5/3}(\eta)\, d\eta, \tag{1-185}$$

where the magnetic field intensity is H, the pitch angle is ψ, the total energy of the electron is E, the frequency of the synchrotron radiation is ν, the modified Bessel function is K, the critical frequency, ν_c', is given by

$$\nu_c' = \frac{3eH\sin\psi}{4\pi mc} \frac{mc^2}{E}\left[1 - \left(\frac{n_\nu v}{c}\right)^2\right]^{-3/2} \tag{1-186}$$

or

$$\nu_c' \approx \nu_c\left[1 + \left(\frac{\nu_p E}{\nu m c^2}\right)^2\right]^{-3/2} = \nu_c\left[1 + (1 - n_\nu^2)\left(\frac{E}{mc^2}\right)^2\right]^{-3/2}, \tag{1-187}$$

where the critical frequency, ν_c, in a vacuum is given in Eq. (1-154), the electron velocity is v, the index of refraction, n_ν, is assumed to be less than unity, and the plasma frequency, ν_p, is given by

$$\nu_p = \nu(1 - n_\nu^2)^{1/2} = \left(\frac{e^2 N_e}{\pi m}\right)^{1/2} \approx 9 \times 10^3\, N_e^{1/2} \text{ Hz}, \tag{1-188}$$

where N_e is the number density of thermal electrons. The general formulae for the synchrotron radiation emitted in the nth harmonic by an electron in an isotropic plasma were first derived by EIDMAN (1958), slightly modified by LIEMOHN (1965), and numerically computed by RAMATY (1969). WILD and HILL (1971) give these formulae together with an approximate expression for the power radiated in the parallel component by a relativistic or subrelativistic electron.

As first pointed out by TSYTOVITCH (1951) and rigorously derived by EIDMAN (1958) and RAZIN (1960), the synchrotron radiation in a plasma will be appreciably suppressed at frequencies, ν, below the critical frequency, ν_R, given by

$$\nu_R = \frac{4ecN_e}{3H\sin\psi} \approx 20\frac{N_e}{H\sin\psi} \approx \frac{2\nu_p^2}{3\nu_H}. \tag{1-189}$$

The total power radiated, P, by a single relativistic electron in a plasma is obtained by integrating Eq. (1-185) over all frequencies and is given by (SIMON, 1969)

$$P = \pi\frac{3^{3/4}}{\sqrt{2}}\frac{e^3 H\sin\psi}{mc^2}\frac{E}{mc^2}\nu_p \exp\left[\frac{-2\sqrt{3}\pi mc\nu_p}{\gamma eH\sin\psi}\right], \tag{1-190}$$

where $\gamma = E/(mc^2)$. The asymptotic form for the volume emissivity, $\varepsilon(\nu)$, of the synchrotron radiation from a group of electrons with a power law energy distribution is given by (SIMON, 1969)

$$\varepsilon(\nu) = 6.9\,N_0\left(\frac{\nu}{\nu_R}\right)^{3/2}\nu_p\left(\frac{\nu}{\sqrt{2}\nu_p}\right)^{-\gamma}\exp\left(-3.7\frac{\nu_R}{\nu}\right)\ \text{erg sec}^{-1}\,\text{Hz}^{-1}\,\text{cm}^{-3} \quad \text{for } \nu < \nu_R, \tag{1-191}$$

where $\varepsilon(\nu)$ is the power per unit frequency per unit volume, and it is assumed that the number density of electrons per unit solid angle along the line of sight which are moving in the direction of the observer and whose energies lie in the range E to $E+dE$ is given by

$$N(E)dE = N_0\left(\frac{E}{mc^2}\right)^{-\gamma} dE. \tag{1-192}$$

It follows from Eqs. (1-175) and (1-191) that the spectral index $\alpha = -(\gamma - 1)/2$ for frequencies $\nu > \nu_R$ and is roughly $\alpha = \frac{3}{2} - \gamma$ for $\nu < \nu_R$.

Synchrotron radiation will also be suppressed by thermal absorption in a plasma at those frequencies for which the optical depth is greater than unity. The absorption coefficient and optical depth of a plasma are discussed in detail in Sect. 1.30. The formula for the absorption coefficient at radio frequencies $\nu > \nu_p$ leads to the conclusion that radiation will be thermally absorbed at frequencies below the critical frequency, ν_T, given by

$$\nu_T = [l\,T^{-3/2} N_e^2]^{-2.1}, \tag{1-193}$$

where l is the extent of the plasma along the line of sight to the radiating source, and T and N_e are, respectively, the kinetic temperature and the number density of the thermal electrons. When the thermal electrons are mixed with the relativistic electrons the spectral index will be $\alpha_0 = -(\gamma-1)/2$ for $\nu > \nu_T$ and $\alpha = 2.1 + \alpha_0$ for $\nu < \nu_T$. If the synchrotron radiation is created in a vacuum but subsequently passes through a plasma, the intensity of the radiation will fall off as $\exp[-(\nu/\nu_T)^{-2.1}]$ for $\nu < \nu_T$.

1.28. Additional Modifications of the Synchrotron Radiation Spectrum

As first suggested by TWISS (1954) and calculated by LE ROUX (1961) and MCCRAY (1969), synchrotron radiation may become self absorbed with an absorption coefficient per unit length, α_ν, given by

$$\alpha_\nu = -\frac{c^2}{8\pi\nu^2}\int_0^\infty E^2 \frac{d}{dE}\left[\frac{N(E)}{E^2}\right] P(\nu)\, dE, \tag{1-194}$$

where ν is the frequency, $N(E)dE$ is the number density of electrons per unit solid angle along the line of sight which are moving in the direction of the observer and whose kinetic energies lie in the range E to $E+dE$, the total power radiated per unit frequency interval by a single relativistic electron is $P(\nu)$ and is given by Eq. (1-160), and it is assumed that the electrons are distributed isotropically. For electrons which have the power law energy distribution given by Eq. (1-174), Eq. (1-194) becomes (LE ROUX, 1961).

$$\begin{aligned}\alpha_\nu &= g(\gamma)\frac{e^3}{2\pi m}\left(\frac{3e}{2\pi m^3 c^5}\right)^{\gamma/2} K(H\sin\psi)^{(\gamma+2)/2}\nu^{-(\gamma+4)/2} \\ &\approx 0.019(3.5\times 10^9)^\gamma K(H\sin\psi)^{(\gamma+2)/2}\nu^{-(\gamma+4)/2} \quad \mathrm{cm}^{-1},\end{aligned} \tag{1-195}$$

where $g(\gamma)$ is a constant of order unity given by

$$g(\gamma) = \frac{\sqrt{3}}{16}\left(\gamma + \frac{10}{3}\right)\Gamma\left(\frac{3\gamma+2}{12}\right)\Gamma\left(\frac{3\gamma+10}{12}\right).$$

Here γ is the index of the power law energy distribution given in Eq. (1-174), H is the magnetic field intensity, ψ is the average pitch angle, and the constant K which also appears in Eq. (1-174) is a function of the intensity I_ν of the source and is given by

$$K \approx 7.4\times 10^{21}\frac{I_\nu}{lH}\left(\frac{\nu}{6.26\times 10^{18} H}\right)^{(\gamma-1)/2}, \tag{1-196}$$

where l is the extent of the radiating source along the line of sight. The synchrotron radiation will become appreciably suppressed at frequencies, ν, below the critical frequency, ν_s, for which the optical depth $\tau_\nu = \alpha_\nu l$ is unity. For a source of maximum flux density S_{ν_s}, magnetic field intensity, H, angular size, θ, and redshift, z, the critical frequency, ν_s, is given by (SLISH, 1963; WILLIAMS, 1963)

$$\nu_s \approx 34\left(\frac{S_{\nu_s}}{\theta^2}\right)^{2/5} H^{1/5}(1+z)^{1/5}, \tag{1-197}$$

where the constant has been chosen so that ν_s is in MHz when S_{ν_s} is in flux units, θ in seconds of arc, and H in gauss. The spectral index of the radiation intensity will be $\alpha = -(\gamma-1)/2$ for $\nu > \nu_s$ and $\alpha = 2.5$ for $\nu < \nu_s$. Observations of the maximum flux density, critical frequency, and angular size of many of the compact radio sources (KELLERMANN and PAULINY-TOTH, 1969; CLARKE *et al.*, 1969) indicate that $H=10^{-4\pm1}$ gauss if the observed low frequency cutoff in their spectra is due to synchrotron self-absorption.

It follows from Eq. (1-163) that high energy electrons lose their energy by synchrotron radiation faster than low energy electrons and therefore the spectrum of the radiation will become steeper with increasing time. When synchrotron radiation losses predominate, a source which is initially supplied with electrons with a power law energy distribution of index $-\gamma$ will develop a spectral index of $\alpha=-(2\gamma+1)/3$ for frequencies higher than the critical frequency, ν_K, given by (KARDASHEV, 1962)

$$\nu_K = 3.4\times10^8 H^{-3} t^{-2} \quad \text{Hz}, \tag{1-198}$$

where H is the magnetic field intensity in gauss, and t is the age of the radio source in years. Spectral changes are also induced by energy losses due to the expansion of the source, collisions of electrons and heavy particles, the inverse Compton effect, and ionization. The details of these changes are given by KARDASHEV (1962) and DE LA BEAUJARDIÉRE (1966). The emission spectrum can also be altered by energy cutoffs in the electron distribution, but the steepest low energy cutoff that is possible has a spectral index of only one third.

O'DELL and SARTORI (1970) have recently pointed out that there is a natural low frequency cutoff in the spectrum of the synchrotron radiation of a single electron (cf. Fig. 2) which falls off as the one-third power of the frequency for frequencies, ν, less than the critical frequency

$$\nu_c = \frac{eH}{2\pi m c \psi} \approx 3\times10^6 \frac{H}{\psi} \quad \text{Hz}, \tag{1-199}$$

where ψ is the average pitch angle. This cyclotron turnover induces a low frequency spectral index of $\alpha=\gamma$ for a group of electrons with a power law energy distribution of index $-\gamma$, whereas the high frequency spectral index is $\alpha=-(\gamma-1)/2$.

1.29. Bremsstrahlung (Free-Free Radiation) of a Single Electron

When a fast charged particle encounters an atom, molecule, or ion it is acccelerated and emits radiation called bremsstrahlung (braking radiation). Consider a nonrelativistic electron of mass, m, charge, e, and velocity, v. When the electron enters the Coulomb force field of a charge Ze, the angular deflection θ, of the particle is given by (RUTHERFORD, 1911)

$$\tan\frac{\theta}{2} = \frac{Ze^2}{mv^2 b}, \tag{1-200}$$

where the impact parameter, b, designates the perpendicular distance from the Ze charge to the original path of the electron.

It follows from the LARMOR (1897) formula for the total power radiated by an accelerated electron (Eq. (1-110)), and from Eq. (1-200) that the total bremsstrahlung energy radiated in frequency interval between ν and $\nu+d\nu$ by a non-relativistic electron during a collision with a charge Ze is given by

$$W(\nu)d\nu \approx \frac{2e^2}{3\pi c^3}|\Delta v|^2 d\nu \approx \frac{2e^2 v^2}{3\pi c^3}(1-\cos\theta)d\nu \approx \frac{8}{3\pi}\left(\frac{e^2}{mc^2}\right)^2 \frac{Z^2 e^2}{c}\left(\frac{c}{vb}\right)^2 d\nu$$

$$\text{for } \nu < \frac{v}{b} \tag{1-201}$$

$$\approx 0 \quad \text{for } \nu > \frac{v}{b},$$

where Δv is the change in electron velocity caused by the collision. The frequency spectrum is flat out to the critical frequency $\nu_c \approx v/b$, and the total energy radiated during one collision is $[v W(\nu)]/b$.

The total number of encounters per unit time between an electron and a volume density of N_i ions in the parameter range b to $b+db$ is

$$N_i v 2\pi b\, db = N_i v 2\pi \frac{d\sigma_s}{d\Omega}\sin\theta\, d\theta\,, \tag{1-202}$$

where the differential scattering cross section, $d\sigma_s/d\Omega$, is given by

$$\frac{d\sigma_s}{d\Omega} = \frac{b}{\sin\theta}\left|\frac{db}{d\theta}\right| = \frac{Z^2}{4}\left(\frac{e^2}{mv^2}\right)^2 \frac{1}{\sin^4\left(\frac{\theta}{2}\right)}. \tag{1-203}$$

Using Eqs. (1-201) and (1-202), the total bremsstrahlung power, $P_i(v,\nu)d\nu$, radiated in frequency interval between ν and $\nu+d\nu$ in collisions with N_i ions is given by

$$P_i(v,\nu)d\nu = N_i v Q_r(v,\nu)d\nu\,, \tag{1-204}$$

where the radiation cross section, $Q_r(v,\nu)$ is given by

$$Q_r(v,\nu) = \frac{16}{3}\frac{Z^2 e^6}{m^2 c^3 v^2}\int_{b_{min}}^{b_{max}} \frac{db}{b} = \frac{16}{3}\frac{Z^2 e^6}{m^2 c^3 v^2}\ln\left(\frac{b_{max}}{b_{min}}\right), \tag{1-205}$$

which has the dimensions of area energy/frequency. The photon cross section, $\sigma_r(v,\hbar\omega)$, is also often used, and is defined by

$$\sigma_r(v,\hbar\omega) = \frac{Q_r(v,\nu)}{\hbar^2\omega} = \frac{16}{3}Z^2\frac{e^2}{\hbar c}\left(\frac{e^2}{mc^2}\right)^2\frac{c^2}{\hbar\omega v^2}\ln\left(\frac{b_{max}}{b_{min}}\right), \tag{1-206}$$

which has the dimensions of area per unit photon energy. Using Eqs. (1-201) and (1-202), the total bremsstrahlung power, $P_a(v,\nu)d\nu$, radiated in frequency interval between ν and $\nu+d\nu$ in collisions with a volume density of N_a neutral molecules is given by

$$P_a(v,\nu)d\nu = \frac{2}{3}\frac{e^2 v^2}{\pi c^3}\nu_c(v)d\nu\,, \tag{1-207}$$

where the collision frequency for momentum transfer is

$$\nu_c(v) = N_a v \int_0^\pi \frac{d\sigma_s}{d\Omega}(1-\cos\theta)\,2\pi\sin\theta\,d\theta \approx N_a v \pi a^2, \tag{1-208}$$

and πa^2 is the effective differential cross section for elastic collisions by a sphere of radius, a, which may be measured in the laboratory.

The logarithmic Gaunt factor $\ln(b_{max}/b_{min})$ depends on the radian frequency, ω, of the bremsstrahlung radiation and the velocity, v, of the electron according to the formulae (BOHR, 1915; BETHE, 1930)

$$\begin{aligned}
\ln\left(\frac{b_{max}}{b_{min}}\right) &= \ln\left(\frac{\gamma^2 m v^3}{Z e^2 \omega}\right) && \text{for } v < \frac{Ze^2}{\hbar} \text{ and } \omega < \frac{mv^3}{Ze^2} \\
&= 0 && \text{for } v < \frac{Ze^2}{\hbar} \text{ and } \omega > \frac{mv^3}{Ze^2}, \\
\ln\left(\frac{b_{max}}{b_{min}}\right) &= \left[\ln\left(\frac{2\gamma^2 m v^2}{\hbar\omega}\right) - \frac{v^2}{c^2}\right] && \text{for } v > \frac{Ze^2}{\hbar} \text{ and } \omega < \frac{mv^2}{\hbar} \\
&= 0 && \text{for } v > \frac{Ze^2}{\hbar} \text{ and } \omega > \frac{mv^2}{\hbar},
\end{aligned} \tag{1-209}$$

where $\gamma = E/(mc^2)$ is not $\gg 1$, the kinetic energy of the electron is E, and it is assumed that ω is larger than the plasma frequency $\omega_p = (4\pi N_e e^2/m)^{1/2} \approx 5.64 \times 10^4 N_e^{1/2}$, where N_e is the electron density. When $\omega < \omega_p$, Eqs. (1-209) are appropriate with ω_p substituted for ω. As suggested by BETHE and HEITLER (1934), Eqs. (1-209) are best used with the average velocity $v = [E^{1/2} + (E - \hbar\omega)^{1/2}]/\sqrt{2m}$, where E is the initial energy of the electron and $\hbar\omega$ is the photon energy. For the special case of electron-electron scattering at high velocities but in the nonrelativistic case, we have (BETHE, 1930)

$$\begin{aligned}
\ln\left(\frac{b_{max}}{b_{min}}\right) &= \ln\left[\frac{(\gamma-1)(\gamma+1)^{1/2}}{\sqrt{2}}\frac{mc^2}{\hbar\omega}\right] && \text{for } \omega < \frac{mv^2}{\hbar} \\
&\approx 0 && \text{for } \omega > \frac{mv^2}{\hbar}.
\end{aligned} \tag{1-210}$$

When the electron is relativistic, $\gamma \gg 1$, the differential scattering cross section, $d\sigma_s/d\Omega$, is given by the Mott formula (MOTT and MASSEY, 1965)

$$\frac{d\sigma_s}{d\Omega} = \frac{Z^2 e^4}{4m^2 v^4} \frac{1}{\sin^4\left(\frac{\theta}{2}\right)} \frac{1 - \beta^2 \sin^2\left(\frac{\theta}{2}\right)}{\gamma^2}, \tag{1-211}$$

where $\beta = v/c$, the electron velocity is v, and θ is the angular deflection of the electron. Approximate formulae for the total bremsstrahlung power radiated per unit frequency interval, $P_i(v, \nu)$ are given by Eqs. (1-204) and (1-205) when

$$\ln\left(\frac{b_{\max}}{b_{\min}}\right) \approx \ln\left(\frac{\gamma^2 mc^2}{\hbar\omega}\right) \quad \text{for } \gamma \gg 1 \quad \text{and } \omega < \frac{(\gamma-1)mc^2}{\hbar}$$
$$\approx 0 \quad \text{for } \gamma \gg 1 \quad \text{and } \omega > \frac{(\gamma-1)mc^2}{\hbar}. \tag{1-212}$$

The exact formula for the photon cross section, σ_r, of a high energy electron incident on an unshielded static charge Ze is given by (cf. BETHE, 1930; BETHE and HEITLER, 1934; WHEELER and LAMB, 1939; HEITLER, 1954; JAUCH and ROHRLICH, 1955)

$$\sigma_r = 4Z^2 \frac{e^2}{\hbar c}\left(\frac{e^2}{mc^2}\right)^2 \frac{1}{\hbar\omega}\frac{1}{E_i^2}\left(E_i^2 + E_f^2 - \frac{2}{3}E_i E_f\right)\left[\ln\left(\frac{2E_i E_f}{mc^2\hbar\omega}\right) - \frac{1}{2}\right], \tag{1-213}$$

where the initial and final electron energy are, respectively, E_i and E_f, and $E_i - E_f = \hbar\omega$, the photon energy. Eqs. (1-212) and (1-213) are equally valid for the collisions of a relativistic electron with free electrons, free protons, or with hydrogen atoms (cf. BLUMENTHAL and GOULD, 1970). As first pointed out by FERMI (1940), however, the screening effects of atoms limits the maximum impact parameter to the atomic radius, a. For the collision of relativistic electrons with atoms we have the additional condition that

$$\ln\left(\frac{b_{\max}}{b_{\min}}\right) = \ln\left(\frac{amv}{\hbar}\right) = \ln\left(\frac{1.4}{Z^{1/3}}\frac{\hbar c}{e^2}\frac{v}{c}\right) \quad \text{for } \omega < \omega_s, \tag{1-214}$$
$$\text{where } \omega_s = \frac{Z^{1/3}}{1.4}\frac{e^2}{\hbar c}\frac{\gamma^2 mc^2}{\hbar},$$

$b_{\max}$ is the atomic radius $a = 1.4\hbar^2/(Z^{1/3}me^2)$, and the fine structure constant $e^2/\hbar c = 1/137$.

For relativistic electrons, energy losses by radiation predominate over collisional energy losses. It follows from Eqs. (1-204), (1-205), (1-212), and (1-214) that the radiative energy loss, dE_r/dx, of a relativistic electron when traversing a unit distance of matter is given by

$$\frac{dE_r}{dx} \approx \frac{16}{3} N_i Z^2 \left(\frac{e^2}{\hbar c}\right)\left(\frac{e^2}{mc^2}\right)^2 E \ln\gamma \quad \text{for } \omega > \omega_s$$
$$\approx \frac{16}{3} N_i Z^2 \left(\frac{e^2}{\hbar c}\right)\left(\frac{e^2}{mc^2}\right)^2 E \ln\left(\frac{1.4}{Z^{1/3}}\frac{\hbar c}{e^2}\right) \quad \text{for } \omega < \omega_s, \tag{1-215}$$

where the electron energy $E = \gamma mc^2$, the fine structure constant $e^2/\hbar c \approx 1/137$, and the classical electron radius $e^2/mc^2 \approx 2.8 \times 10^{-13}$ cm.

For nonrelativistic electrons, collisional energy loss predominates over radiative energy loss. The total energy lost per unit length, dE_c/dx, in collisions with N atoms with Z electrons is (BOHR, 1915; BETHE, 1930)

$$\frac{dE_c}{dx} = 4\pi NZ \frac{e^4}{mv^2} \ln\left(\frac{b_{\max}}{b_{\min}}\right), \tag{1-216}$$

where $\ln(b_{\max}/b_{\min})$ is given by Eqs. (1-209).

1.30. Bremsstrahlung (Free-Free Radiation) from a Plasma

The total bremsstrahlung power radiated from a plasma per unit volume per unit solid angle per unit frequency interval ν to $\nu+d\nu$ is called the volume emissivity, ε_ν, and is given by

$$\varepsilon_\nu = n_\nu \frac{N_e}{4\pi} \int P_r(v,\nu) f(v)\, dv\,, \tag{1-217}$$

where n_ν is the index of refraction, N_e is the electron density, $f(v)$ is the electron velocity distribution, and $P_r(v,\nu)$ is the total power radiated per unit frequency interval in the collision of an electron of velocity, v, with N_i ions and is given by Eqs. (1-204) to (1-214). The velocity criterion $v \lessgtr Ze^2/\hbar$ may be converted to the temperature criterion $T \lessgtr 3.16 \times 10^5 Z^2$ through the relation

$$v \approx \left(\frac{kT}{m}\right)^{1/2} \approx 3.89 \times 10^5\, T^{1/2} \quad \text{cm sec}^{-1}. \tag{1-218}$$

When $f(v)$ is Maxwellian (Eq. (3-114)), Eq. (1-217) becomes

$$\varepsilon_\nu d\nu = \frac{8}{3}\left(\frac{2\pi}{3}\right)^{1/2} \frac{n_\nu Z^2 e^6}{m^2 c^3} \left(\frac{m}{kT}\right)^{1/2} N_i N_e g(\nu,T) \exp(-h\nu/kT)\, d\nu$$

$$\approx 5.4 \times 10^{-39} n_\nu Z^2 \frac{N_i N_e}{T^{1/2}} g(\nu,T) \exp(-h\nu/kT)\, d\nu \quad \text{erg sec}^{-1}\,\text{cm}^{-3}\,\text{Hz}^{-1}\,\text{rad}^{-2}, \tag{1-219}$$

and the Gaunt factor, $g(\nu,T)$ is given by

$$g(\nu,T) = \frac{\sqrt{3}}{\pi} \ln \Lambda \approx 0.54 \ln \Lambda\,, \tag{1-220}$$

where $\ln \Lambda$ is given in Table 2. Discussions of the appropriate Gaunt factor in various domains are given by GAUNT (1930), SAUTER (1933), SOMMERFELD (1931), ELWERT (1954), SCHEUER (1960), OSTER (1961), BRUSSAARD and VAN DE HULST (1962), MERCIER (1964), and OSTER (1970). Values have been tabulated by KARZAS and LATTER (1961).

Table 2. The logarithm factor, Λ, where the Gaunt factor $g(\nu,T) = (\sqrt{3}/\pi)\ln\Lambda$. The frequency is ν, the constants $\gamma = 1.781$ and $e' = 2.718$, the plasma frequency is $\omega_p \approx 5.64 \times 10^4 N_e^{1/2}$ Hz, the electron density is N_e, the electron temperature is T, and the ion charge is Ze

	$T < 3.6 \times 10^5 Z^2$ °K	$T > 3.6 \times 10^5 Z^2$ °K
$\omega \gg \omega_p$	$\Lambda = \left(\frac{2}{\gamma}\right)^{5/2}\left(\frac{kT}{m}\right)^{1/2}\left(\frac{kT}{2\pi Z e^2 \nu}\right)$ $\approx 5.0 \times 10^7 (T^{3/2}/Z\nu)$	$\Lambda = \frac{4kT}{\gamma h \nu}$ $\approx 4.7 \times 10^{10} (T/\nu)$
$\omega \leq \omega_p$	$\Lambda = \left(\frac{2}{\gamma}\right)^2 \frac{1}{e'^{1/2}} \left(\frac{kT}{m}\right)^{1/2} \frac{kT}{Ze^2\omega_p}$ $\approx 3.1 \times 10^3 (T^{3/2}/Z N_e^{1/2})$	$\Lambda = \left(\frac{8}{e'\gamma}\right)^{1/2}\left(\frac{kT}{m}\right)^{1/2} \frac{(mkT)^{1/2}}{\hbar\omega_p}$ $\approx 3.0 \times 10^6 (T/N_e^{1/2})$

The absorption coefficient per unit length for electron-ion bremsstrahlung is given by (KIRCHHOFF, 1860; MARGENAU, 1946; SMERD and WESTFOLD, 1949)

$$\alpha_\nu = \frac{\varepsilon_\nu}{n_\nu^2} \frac{c^2}{2h\nu^3}\left[\exp\left(\frac{h\nu}{kT}\right) - 1\right] = \frac{\varepsilon_\nu}{n_\nu^2 B_\nu(T)}, \tag{1-221}$$

where $B_\nu(T)$ is the vacuum brightness of a black-body radiator. For frequencies $\nu \ll 10^{10}\,T$, the Rayleigh-Jeans approximation for $B_\nu(T)$ can be used with Eqs. (1-219) and (1-221) to obtain

$$\begin{aligned}\alpha_\nu &= \frac{N_e N_i}{n_\nu (2\pi\nu)^2}\left[\frac{32\pi^2 Z^2 e^6}{3(2\pi)^{1/2} m^3 c}\right]\left(\frac{m}{kT}\right)^{3/2} \ln\Lambda \\ &\approx \frac{9.786\times 10^{-3}}{n_\nu \nu^2} \frac{N_e N_i}{T^{3/2}} \ln[4.954\times 10^7 (T^{3/2}/Z\nu)] \quad \text{cm}^{-1} \quad \text{for } T < 3.16\times 10^5\ {}^\circ\text{K} \\ &\approx \frac{9.786\times 10^{-3}}{n_\nu \nu^2} \frac{N_e N_i}{T^{3/2}} \ln[4.7\times 10^{10} (T/\nu)] \quad \text{cm}^{-1} \quad \text{for } T > 3.16\times 10^5\ {}^\circ\text{K}\end{aligned} \tag{1-222}$$

The optical depth, $\tau_\nu = \int \alpha_\nu\, dl$ is often expressed as the approximation (ALTENHOFF, MEZGER, STRASSL, WENDKER, and WESTERHOUT, 1960)

$$\tau_\nu \approx 8.235\times 10^{-2}\, T_e^{-1.35}\, \nu^{-2.1} \int N_e^2\, dl, \quad \text{for } \nu \ll 10^{10}\,T, \quad \text{and } T < 9\times 10^5\ {}^\circ\text{K}, \tag{1-223}$$

where T_e is the electron temperature in °K, the frequency ν is in GHz or 10^9 Hz, and the emission measure $\int N_e^2\, dl$ is in parsec cm^{-6}, where one parsec $= 3\times 10^{18}$ cm. Eq. (1-223) is the same as the more exact result found by using Eq. (1-222) if the former is multiplied by a factor $a(\nu, T)$ which is very nearly unity, and has been tabulated by MEZGER and HENDERSON (1967).

It follows from Eq. (1-223) that the optical depth, τ_ν, becomes unity at the critical frequency, ν_T, given by

$$\nu_T \approx 0.3 (T_e^{-1.35} N_e^2 l)^{1/2}, \tag{1-224}$$

where ν_T is in GHz and the extent, l, of the plasma along the line of sight is in parsecs. At radio frequencies where $h\nu \ll kT$, or $\nu \ll 10^{10}\,T$, the exponential factor in Eq. (1-219) becomes unity and the radiation spectrum is nearly flat for frequencies $\nu > \nu_T$. For frequencies $\nu < \nu_T$, the radiation is self-absorbed and the radiation spectrum is that of a black-body falling off as ν^2 at lower frequencies. Thermal bremsstrahlung from regions of ionized hydrogen, HII regions, exhibit such a spectrum, and an example is shown in Fig. 4 for the Orion nebula. The energy density, $U(\nu)\,d\nu$, of the radiation of a thermal plasma in the frequency range ν to $\nu + d\nu$ is given by (SMERD and WESTFOLD, 1949)

$$U(\nu)\,d\nu = \frac{4\pi}{c} n_\nu B_\nu(T)\, d\nu, \tag{1-225}$$

where the brightness of a black-body radiator, $B_\nu(T)$ is given by Eq. (1-119) and the index of refraction, n_ν, is related to the plasma frequency and the electron density by Eq. (1-188).

The total bremsstrahlung luminosity, L, of a thermal plasma is given by integrating Eq. (1-219) over all frequencies to obtain

$$L = 4\pi V \int \varepsilon_\nu d\nu = \frac{32\pi}{3}\left(\frac{2\pi}{3}\right)^{1/2} \frac{e^6}{mc^3 h}\left(\frac{kT}{m}\right)^{1/2} N_e N_i V Z^2 g$$
$$\approx 1.43 \times 10^{-27} N_e N_i T^{1/2} V Z^2 g \quad \text{erg sec}^{-1}, \qquad (1\text{-}226)$$

where the volume of the source is V, and it has been assumed that the index of refraction is unity.

In the presence of an intense magnetic field ($H > 10^{13}$ gauss) the electron bremsstrahlung differs from the ordinary bremsstrahlung process in that in a magnetic field the electrons are one dimensional particles free to move in the direction of the field, but bound in the plane perpendicular to the magnetic field in quantized circular orbits with energy (in the nonrelativistic limit) in multiples of $1.19 \times 10^{-8} H$ eV, where H is the magnetic field intensity in gauss. Formulae for the emission rate and absorption coefficients of electron bremsstrahlung in intense magnetic fields are given by CANUTO, CHIU, and FASSIO-CANUTO (1969) for a vacuum, and by CANUTO and CHIU (1970) for a plasma. The emissivity for the zero level quantum state is given by (CHIU, CANUTO, and FASSIO-CANUTO, 1969)

$$\varepsilon = 3.01 \times 10^{-29} N_i N_e H_8 Z^2 E_5^{5/2} \nu_8^{-2} \quad \text{erg sec}^{-1}\ \text{cm}^{-3} (10^8\ \text{Hz})^{-1}, \qquad (1\text{-}227)$$

where N_e and N_i are the electron and ion number densities, respectively, Z is the charge of the nucleus, E_5 is the electron energy measured in $k \cdot 10^5\ ^\circ\text{K} = 8.63$ eV, and ν_8 is the frequency of the radiation in 10^8 Hz.

1.31. Photoionization and Recombination (Free-Bound) Radiation

In the recombination of an electron of mass, m, charge, e, and velocity, v, with an ion of charge eZ to form a bound atom in a state with principal quantum number, n, or, for the converse process of photoionization from the level, n, the frequency, ν, of emission or absorption is given by (EINSTEIN, 1905)

$$h\nu = \tfrac{1}{2} m v^2 + E_i - E_n, \qquad (1\text{-}228)$$

where E_i is the ionization energy, and E_n is the energy of the nth state. For hydrogen-like atoms, Eq. (1-228) becomes

$$h\nu = \tfrac{1}{2} m v^2 + Z^2 E_H / n^2, \qquad (1\text{-}229)$$

where $E_H = 2\pi^2 e^4 m / h^2 = 13.6\ \text{eV} \approx 2.2 \times 10^{-11}$ erg is the ionization energy of hydrogen, and $Z^2 E_H$ is the ionization energy of a hydrogenic ion of charge $e(Z-1)$.

For the photoionization of a hydrogen-like system consisting of one electron in the field of charge, eZ, of one ion, the total absorption cross section, $\sigma_a(\nu, n)$ for light of frequency, ν, is (KRAMERS, 1923)

$$\sigma_a(\nu, n) = \frac{32\pi^2 e^6 R_\infty Z^4}{3^{3/2} h^3 \nu^3 n^5} \approx 2.8 \times 10^{29} \frac{Z^4}{\nu^3 n^5} \quad \text{cm}^2, \tag{1-230}$$

where $R_\infty = 2\pi^2 e^4 m/(h^3 c)$ is the Rydberg constant, and n is the principle quantum number of the initial state. The more exact expression for $\sigma_a(\nu, n)$ is given by (GAUNT, 1930; MENZEL and PEKERIS, 1935; BURGESS, 1958)

$$\sigma_a(\nu, n)\, g_{\text{fb}}(T, \nu), \tag{1-231}$$

where the free-bound Gaunt factor, $g_{\text{fb}}(T, \nu)$, is nearly unity at optical frequencies. Gaunt factors have been tabulated by MENZEL and PEKERIS (1935) and KARZAS and LATTER (1961). Measured values of $\sigma_a(\nu, n)$ are given in Table 3 for certain threshold or maximum wavelengths. The cross section for photoionization in which two k-shell electrons leave any atom is given in Chap. 4.

Table 3. Values of the absorption cross section, σ_T, at the spectral head, and σ_m at the maximum or minimum of the absorption associated with the given ionization potential, V_i [1]

System	V_i (eV)	λ_T (Å)	σ_T	System	V_i (eV)	λ_T (Å)	σ_T	λ_m (Å)	σ_m
H	13.6	912	6.3	Ne	21.6	575	4.0	400	8.0
He	24.6	504	7.4	Na	5.14	2,412	0.12	1,900	0.013
Li	5.39	2,300	2.5	K	4.34	2,860	0.012	2,700	0.006
Be	9.32	1,330	(8.2)	Ca	6.11	2,028	0.45	1,930	<0.1
B	8.30	1,490	19	Kr	14.7	845	35	800	35
C	11.3	1,100	(11)	Rb	4.18	2,970	0.11		<0.004
N	14.6	852	(9)	Cs	3.89	3,185	0.22	2,750	0.08
O	13.6	910	(2.6)	Tl	6.11	2,030	4.5	1,730	<0.1
F	17.4	713	(6)	H_2	15.4	804		780	7.4
Ne	48.5	256	2.5		12.6	1,019			
Mg	7.64	1,620	1.18	O_2	16.1	770		550	22
Al	5.98	2,070		N_2	15.6	792		750	26
Ar	15.8	787	35	CH_4	12.8	967		960	(56)
Ar	15.9	778		H_2O	12.5	985		346	(100)
Ar	32	420	(3)	CO	14.21	868		600	16.5
Ca		1,589	(0.45)		(16.5)	(750)			~ 1
Ga	6.00	2,070	(0.2)		(20)	(620)			~ 1
Kr	14.0	885		CO_2	14.4	860		(800)	18
In	5.79	2,140	(0.3)		(17.7)	(700)		(600)	(15)
Xe	12.1	1,020		NH_3	10.3	1,210		(900)	10
He^+	54.4	228	1.6		(16.5)	(750)		(700)	(25)
C^+	24.4	508	3.7	N_2O	12.9	960		700	35
N^+	29.6	419	(6.4)			?		(357)	(30)
O^+	35.2	353	(8.1)			?		(225)	(30)
F^+	35.0	354	2.5	NO	9.25	1,340		880	20
Ne^+	41.0	302	4.5		(14)	(880)			?
Na^+	47.3	262	7.1		(16)	(730)			?
Mg^+	15.0	825			(18.7)	(660)		(600)	(20)
Si^+	16.3	760		H_2^+	16.3	763		400	0.67
Ca^+	11.9	1,040	0.17						

[1] From DITCHBURN and ÖPIK (1962) by permission of Academic Press. Uncertain values are given in parenthesis.

The recombination cross section, $\sigma_r(\nu,n)$, for the recombination radiation of a single electron-ion encounter may be obtained by using the MILNE (1921) relation

$$\frac{\sigma_a}{\sigma_r} = \frac{m^2 c^2 v^2 g_Z(1)}{\nu^2 h^2 g_{Z-1}(n)}, \tag{1-232}$$

where $g_Z(1)$ is the statistical weight of the ion in the ground state and $g_{Z-1}(n)$ is the statistical weight of the nth level of the atom. For hydrogen-like systems $g_Z(1)=1$ and $g_{Z-1}(n)=2n^2$. Thus Eqs. (1-230) and (1-232) give

$$\sigma_r(\nu,n) = \frac{128\pi^4 e^{10} Z^4 g_{fb}}{3\sqrt{3}\, m c^3 h^4 \nu n^3 v^2} \approx 3\times 10^{10} \frac{Z^4 g_{fb}}{\nu n^3 v^2} \text{ cm}^2. \tag{1-233}$$

For a plasma with electron density, N_e, and ion density, N_i, the total number of recombinations between ground-state ions and electrons per unit volume from the velocity range v to $v+dv$ is

$$N_i N_e f(v) \sigma_r(\nu,n) v\, dv, \tag{1-234}$$

where $f(v)$ is the electron velocity distribution and $\sigma_r(\nu,n)$ is the single electron-ion cross section for recombination to level n (given by Eq. (1-233) for hydrogen-like atoms). When the free electrons are in local thermodynamic equilibrium at temperature, T, the velocity distribution is Maxwellian. In this case, the recombination power, $P_r(n,\nu)\,d\nu$, radiated per unit volume in the frequency interval ν to $\nu+d\nu$ when recombining to level n is, for hydrogen like atoms, given by:

$$P_r(n,\nu)\,d\nu = N_e N_i \frac{Z^4 K}{(kT)^{3/2}} \frac{g_{fb}}{n^3} \exp\left[-\frac{h\nu}{kT} + \frac{Z^2 E_H}{n^2 kT}\right] d\nu, \tag{1-235}$$

where

$$K = \frac{64\pi^{1/2} e^4 h}{3^{3/2} m^2 c^3} (E_H)^{3/2} \approx 3.4\times 10^{-40} (E_H)^{3/2} \text{ erg}^{5/2}\text{ cm}^3$$

and $E_H = 2\pi^2 e^4 m/h^2 \approx 2.2\times 10^{-11}$ erg is the ionization energy of hydrogen. The total recombination power $P_r(\nu)\,d\nu$, radiated per unit volume in the frequency interval ν to $\nu+d\nu$ is obtained by summing $P_r(n,\nu)$ over all permitted quantum levels.

$$P_r(\nu)\,d\nu = N_e N_i \frac{Z^4 K}{(kT)^{3/2}} \exp\left(\frac{-h\nu}{kT}\right) \sum_{n_0}^{\infty} n^{-3} \exp\left(\frac{Z^2 E_H}{n^2 kT}\right) g_{fb}\, d\nu, \tag{1-236}$$

where $n_0 \geq (Z^2 E_H/h\nu)^{1/2}$.

When detailed values of $P_r(\nu)\,d\nu$ are required, the approximation formulae of BRUSSAARD and VAN DE HULST (1962) may be used to evaluate the summation in Eq. (1-236). The total volume emissivity, ε_ν, of both thermal bremsstrahlung and recombination radiation is given by (BRUSSAARD and VAN DE HULST, 1962)

$$\begin{aligned} \varepsilon_\nu\, d\nu &= \frac{8}{3}\left(\frac{2\pi}{3}\right)^{1/2} \frac{Z^2 e^6}{m^2 c^3} \left(\frac{m}{kT}\right)^{1/2} N_i N_e b\, d\nu \\ &\approx 5.4\times 10^{-39} Z^2 \frac{N_i N_e}{T^{1/2}} b\, d\nu \text{ erg sec}^{-1}\text{ cm}^{-3}\text{ Hz}^{-1}\text{ rad}^{-2} \end{aligned} \tag{1-237}$$

where

$$b=[g(\nu,T)+f(\nu,T)]\exp\left(\frac{-h\nu}{kT}\right), \tag{1-238}$$

the free-free Gaunt factor $g(\nu,T)$ is given by Eqs. (1-220) and Table 2, and the free-bound Gaunt factor, $f(\nu,T)$, is given by

$$f(\nu,T)=2\theta\sum_{n=m}^{5}g_n\frac{e^{\theta/n^2}}{n^3}+2\theta g_\infty\sum_{n=6}^{\infty}\frac{e^{\theta/n^2}}{n^3}+50\theta(g_5-g_\infty)\sum_{n=6}^{\infty}\frac{e^{\theta/n^2}}{n^5}, \tag{1-239}$$

where m is the principal quantum number of the lowest level to which emmision at the frequency, ν, can occur, $g_n(\nu)$ is the Gaunt factor for the transition from the free to bound state, $\theta=2\pi^2 me^4Z^2/(h^2kT)$. A plot of the spectral function b for $T=1.58\times10^5$ °K is shown in Fig. 5 together with the spectrum of bremsstrahlung radiation alone. These plots indicate that recombination radiation is not important when compared with bremsstrahlung for $h\nu\ll kT$ or $\nu\ll10^{10}\,T$. At optical frequencies recombination radiation of neutral and ionized hydrogen and helium become important, as does two photon emission of hydrogen. These effects are discussed in detail in Sect. 2.11, and by CHANDRASEKHAR and BREEN (1946) and BROWN and MATHEWS (1970). The continuous spectra arising from free-free and free-bound transitions at X-ray wavelengths (1 Å to 30 Å) are calculated by CULHANE (1969) for temperatures between 0.8×10^6 and 10^8 °K.

A lower limit to the total luminosity, L, of the recombination radiation may be obtained by integrating Eq. (1-236) over all frequencies resulting from capture

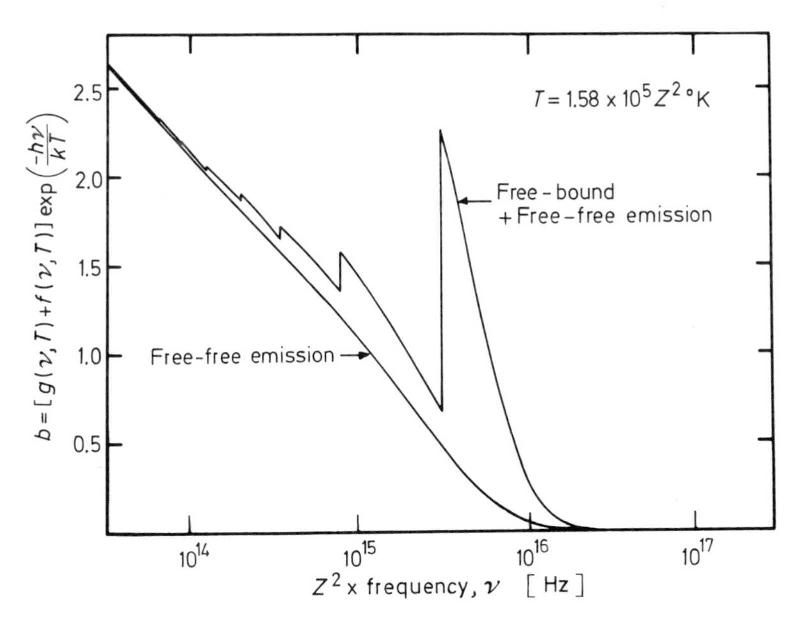

Fig. 5. The emission spectrum of a Maxwellian distribution of electrons at a temperature of $T=2\pi^2me^4Z^2/(h^2k)=1.58\times10^5\,Z^2$ °K with exact Gaunt factors taken into account (after BRUSSAARD and VAN DE HULST, 1962). The lower curve illustrates the spectrum of thermal bremsstrahlung emission, whereas the upper curve illustrates the combined spectrum of thermal bremsstrahlung and recombination radiation. The volume emissivity, ε_ν, is related to b by Eq. (1-237)

into the ground state only ($n=1$, $\nu \geq Z^2 E_H/h$). This term has the value (COOPER, 1966)

$$L_{min} \approx 10^{-21} N_e N_i T^{-1/2} V Z^4 \text{ erg sec}^{-1}, \tag{1-240}$$

where the volume of the source is V. A comparison of Eqs. (1-226) and (1-240) indicates that the total luminosity due to recombination radiation may be greater than that due to thermal bremsstrahlung for temperatures less than 10^6 °K.

1.32. Astrophysical Plasmas

Almost all matter in space can be considered to be a plasma, an ionized gas consisting of electrons (e), ions (i), and neutral atoms (a) or molecules (m) with respective densities, N_e, N_i, N_a, N_m; respective masses, m, M_i, M_a, M_m; and respective charges, e, Ze, and zero for the neutral particles. The plasma is often taken to be in local thermodynamic equilibrium at the temperature, T; and often contains a magnetic field of strength, H. Typical values of N_e, N_a+N_m, T_e, and H are given in Table 4 for well known astrophysical plasmas.

Table 4. Electron densities, N_e, molecular density, N_m, + atomic density, N_a, electron temperature, T_e, and magnetic field strength, H, for astrophysical plasmas.

Plasma	N_e (cm^{-3})	N_m+N_a (cm^{-3})	T_e (°K)	H (Gauss)
Ionosphere:				
$D \sim$ 70 km	10^3	2×10^{15}	2×10^2	$\sim3\times10^{-1}$
$E \sim$ 100 km	10^5 day, 10^3 night	6×10^{12}	2–3×10^2	$\sim3\times10^{-1}$
$F_1 \sim$ 200 km	10^5 day, absent night	10^{10}	10^3	$\sim3\times10^{-1}$
$F_2 \sim$ 300 km	10^6 day, 10^5 night	10^9	1–3×10^3	$\sim3\times10^{-1}$
Interplanetary Space	1–10^4	≈0	10^2–10^3	10^{-6}–10^{-5}
Solar Corona	10^4–10^8	≈0	10^3–10^6	10^{-5}–1
Solar Chromosphere	10^{12}	≈0	10^4	10^3
Stellar Interiors	10^{27}	≈0	$10^{7.5}$	–
Planetary Nebulae	10^3–10^5	≈0	10^3–10^4	10^{-4}–10^{-3}
HII Regions	10^2–10^3	≈0	10^3–10^4	10^{-6}
White Dwarfs	10^{32}	≈0	10^7	10^6 (surface)
Pulsars	10^{42} center, 10^{12} surface	≈0		10^{12} (surface)
Interstellar Space	10^{-3}–10 (av ~0.03)	10^{-2}–10^5 (av ~1)	10^2	10^{-6}
Intergalactic Space	$\lesssim10^{-5}$	≈0	10^5–10^6	$\lesssim10^{-8}$

For a plasma in thermodynamic equilibrium at temperature, T, the electron velocity distribution is Maxwellian (MAXWELL, 1860, Eq. (3-114)), and the rms velocity, v_{th}, and the average kinetic energy, K.E., are given by

$$\begin{aligned} v_{th} &= (3kT/m)^{1/2} \approx 6.7\times10^5\, T^{1/2} \text{ cm sec}^{-1} \\ \text{K.E.} &= 3kT/2 \approx 2.1\times10^{-16}\, T \text{ erg.} \end{aligned} \tag{1-241}$$

For an electron density, N_e, the average separation between the electrons, $\langle r\rangle$, and the average Coulomb energy, C.E., are given by

$$\begin{aligned}\langle r\rangle &= (3N_e/4\pi)^{-1/3} \approx N_e^{-1/3} \text{ cm} \\ \text{C.E.} &= e^2/\langle r\rangle \approx 2.3\times10^{-19} N_e^{1/3} \text{ erg.}\end{aligned} \tag{1-242}$$

It follows from Poisson's equation (POISSON, 1813, Eq. (1-4)) that the electric field intensity, E, due to a linear displacement, $x(t)$, of the electrons is given by

$$E = 4\pi e N_e x(t)\,. \tag{1-243}$$

By using the Lorentz force equation (LORENTZ, 1892, Eq. (1-52)) with Eq. (1-243), the equation of motion for the displaced charge is obtained

$$m\ddot{x}(t) + 4\pi e^2 N_e x(t) = 0\,, \tag{1-244}$$

where $\ddot{}$ denotes the second derivative with respect to time, t. Eq. (1-244) is the equation of motion of a harmonic oscillator whose oscillating frequency is the plasma frequency, ω_p, given by

$$\begin{aligned}\omega_p &= \left(\frac{4\pi e^2 N_e}{m}\right)^{1/2} \approx 5.64\times10^4 N_e^{1/2} \quad \text{rad sec}^{-1} \\ \text{or} \quad \nu_p &= \omega_p/2\pi \approx 8.9\times10^3 N_e^{1/2} \quad \text{Hz,}\end{aligned} \tag{1-245}$$

a result first obtained by TONKS and LANGMUIR (1929). It also follows directly from the Lorentz force equation (1-52) that a charged particle moving in a magnetic field of intensity, H, will gyrate about the field at the cyclotron frequency, ω_H, given by

$$\begin{aligned}\omega_H &= \frac{ZeH}{Mc\gamma}\sin\psi \approx 1.76\times10^7 \frac{ZmH}{M\gamma}\sin\psi \quad \text{rad sec}^{-1} \\ \text{or} \quad \nu_H &\approx 2.8\times10^6 H\gamma^{-1}\sin\psi \quad \text{Hz} \quad \text{for } M=m\end{aligned} \tag{1-246}$$

a result first obtained by HEAVISIDE (1904). Here the particle mass is M, its charge is Ze, the velocity is v, and the $\gamma=[1-(v/c)^2]^{-1/2}=E/mc^2$ where E is the particle energy. The pitch angle, ψ, is the angle between the velocity vector and the magnetic field vector. The radius of gyration, r_H, is given by

$$r_H = \frac{v}{\omega_H} \approx \frac{v_{th}}{\omega_H} = \frac{c(3kTm)^{1/2}M\gamma}{ZeHm} \approx 3.8\times10^{-2}\frac{T^{1/2}M\gamma}{ZHm} \text{ cm.} \tag{1-247}$$

When a test charge, Ze, is placed in a plasma in thermodynamic equilibrium, its electrostatic potential, $\varphi(r)$, will cause the electrons to move into a spatial density distribution $N(r)$ given by Boltzmann's equation (3-4) and Maxwell's distribution (Eq. 3-114)

$$\begin{aligned}N(r) &= N_e \exp\left[-\frac{e\varphi(r)}{kT}\right] \\ &\approx N_e\left[1-\frac{e\varphi(r)}{kT}\right] \quad \text{for } e\varphi(r) \ll kT,\end{aligned} \tag{1-248}$$

where r is the distance from the test charge. Using the approximate form of Eq. (1-248) together with Poisson's equation (POISSON, 1813, Eq. (1-4)) we obtain the equation

$$\varphi(r) = \frac{Ze}{r}\exp\left(-\frac{r}{r_D}\right), \tag{1-249}$$

where the Debye radius, r_D, is given by

$$r_D = 3^{-1/2}\left(\frac{v_{th}}{\omega_p}\right) = \left(\frac{kT}{4\pi e^2 N_e}\right)^{1/2} \approx 6.9\left(\frac{T}{N_e}\right)^{1/2} \text{ cm}. \tag{1-250}$$

Eqs. (1-249) and (1-250) were first obtained by DEBYE and HÜCKEL (1923), and they show that the electrons of a plasma move in such a way as to screen out the Coulomb field of a test charge for distances $r > r_D$. LANDAU (1946) has shown that the wavelength of electron plasma oscillations must be greater than the Debye length in order for the waves to exist.

It follows from Maxwell's equations (MAXWELL, 1860, Eqs. (1-39)—(1-42)) that the magnetic field intensity, H, the electric field intensity, E, and the current density, J, in a vacuum are related by the equation

$$c\nabla \times H = 4\pi J + \frac{\partial E}{\partial t}. \tag{1-251}$$

When electrons of density, N_e, are introduced into the vacuum, $J = N_e e v$, where the velocity, v, is given by $v = eE_0/(i\omega m)$, the latter equation following from the Lorentz force law for an electron subjected to a oscillatory electric field given by $E = E_0 \exp(i\omega t)$. Eq. (1-251) may then be written as

$$c\nabla \times H = i\omega\left[1 - \frac{4\pi e^2 N_e}{m\omega^2}\right]E_0 = i\omega\varepsilon_\nu E_0, \tag{1-252}$$

where ε_ν is the dielectric constant. It then follows that the index of refraction, $n_{\nu\perp}$, for transverse ($\perp$) electromagnetic waves of frequency, ν, is given by

$$n_{\nu\perp} = \sqrt{\varepsilon_\nu} = \left[1 - \left(\frac{\nu_p}{\nu}\right)^2\right]^{1/2} \approx \left[1 - 8.06\times 10^7 \frac{N_e}{\nu^2}\right]^{1/2}, \tag{1-253}$$

an equation first derived by MAXWELL (1899).

Eq. (1-253) is valid if the thermal speed of the electrons is sufficiently low that the mean distance between most of the interacting particles does not change appreciably during the period of an oscillation. When the thermal motion of the electrons is taken into account it can be shown that longitudinal plasma oscillations are possible. If φ is the average potential due to the motion of plasma particles, Poisson's equation (1-4) may be written as

$$\begin{aligned} -\nabla^2\varphi = 4\pi\rho &= 4\pi e N_e\left[1 - \int \frac{f(v)\,dv}{\left(1 + \frac{2e\varphi}{mU^2}\right)^{1/2}}\right] \\ &\approx \frac{4\pi K^2 e^2 N_e \varphi}{m}\int \frac{f(v)\,dv}{(\omega - K\cdot v)^2}, \end{aligned} \tag{1-254}$$

where U is the wave velocity in the wave coordinate system, v is the velocity in the laboratory system, $f(v)$ is the velocity distribution function, ω is the wave frequency, and the wave number $K=2\pi/\lambda$, where λ is the wavelength. Eq. (1-254) was first obtained by VLASOV (1945). Assuming a Maxwellian velocity distribution for a plasma in thermodynamic equilibrium (Eq. (3-114)), expanding the integral, and solving for ω we obtain

$$\omega^2 \approx \frac{4\pi e^2 N_e}{m} + \left(\frac{3kT}{m}\right) K^2 = \omega_p^2 + v_{th}^2 K^2, \tag{1-255}$$

to give an index of refraction $n_{\nu\|}$ for longitudinal waves of

$$n_{\nu\|} = \frac{cK}{\omega} = \left(\frac{c}{v_{th}}\right)\left[1-\left(\frac{\nu_p}{\nu}\right)^2\right]^{1/2} \approx 4.5\times 10^4\, T^{-1/2}\left[1-8.06\times 10^7\,\frac{N_e}{\nu^2}\right]^{1/2}. \tag{1-256}$$

Eqs. (1-255) and (1-256) were first derived by BOHM and GROSS (1949), and have been shown to be valid for waves with frequencies $\nu \approx \nu_p$.

Black-body radiation intensity for a plasma (PLANCK, 1901; MILNE, 1930)

$$\begin{aligned} B_{\nu p}(T) &= n_\nu^2\,\frac{2h\nu^3}{c^2}\left[\exp\left(\frac{h\nu}{kT}\right)-1\right]^{-1} \\ &\approx 2n_\nu^2\,\nu^2 kT/c^2 \quad \text{for } \nu < 10^{10}\,T. \end{aligned} \tag{1-257}$$

Radiation energy density in frequency interval between ν and $\nu+d\nu$ for a plasma (SMERD and WESTFOLD, 1949)

$$\begin{aligned} U(\nu)\,d\nu &= \frac{4\pi}{c}\,B_{\nu p}(T)\left|\frac{\partial \omega n_\nu}{\partial \omega}\right| d\nu \qquad (1\text{-}258) \\ &\approx 8\pi\,\frac{\nu^2 kT}{c^3}\left[1-\left(\frac{\nu_p}{\nu}\right)^2\right]^{1/2} d\nu \qquad \text{for } \nu<10^{10}\,T, \text{ transverse waves} \\ &\approx \frac{8\pi\nu^2 kT}{3\sqrt{3}\,c^3}\left(\frac{mc}{kT}\right)^3\left[1-\left(\frac{\nu_p}{\nu}\right)^2\right]^{1/2} d\nu \quad \text{for } \nu<10^{10}\,T, \text{ longitudinal waves } \nu\approx\nu_p. \end{aligned}$$

Kirchhoff's law for a plasma (KIRCHHOFF, 1860; MILNE, 1930)

$$\varepsilon_\nu = \alpha_\nu B_{\nu p}(T), \tag{1-259}$$

where ε_ν is the emission coefficient (power per unit volume per unit frequency interval per unit solid angle), α_ν is the absorption coefficient per unit length, and $B_{\nu p}(T)$ is the black body radiation intensity for a plasma (n_ν^2 times that for free space).

Absorption coefficient per unit length (KRAMERS, 1923; GAUNT, 1930)

$$\alpha_\nu = \frac{1}{n_\nu}\left(\frac{\nu_p}{\nu}\right)^2 \frac{\nu_{eff}}{c}, \quad \text{for } \nu<10^{10}\,T. \tag{1-260}$$

Effective collision frequency for electrons with ions, i, or molecules, m,

$$\nu_{\text{eff}}(i) = \frac{4}{\sqrt{2}}\left(\frac{\pi}{3}\right)^{3/2} Z^2 \frac{e^4}{m^2}\left(\frac{m}{kT}\right)^{3/2} N_{\text{i}}\, g(\nu, T) \approx 50\frac{N_{\text{i}}}{T^{3/2}} \quad \text{Hz}$$
$$\nu_{\text{eff}}(m) = \frac{4\pi}{3} a^2 v_{\text{th}} N_{\text{m}} \approx 9\times 10^5 \pi a^2 T^{1/2} N_{\text{m}} \quad \text{Hz}, \tag{1-261}$$

where the Gaunt factor $g(\nu, T) = (\sqrt{3}/\pi)\ln\Lambda$. The logarithmic factors are given in Table 2. The effective collision cross section for molecules of radius a is πa^2 which has the value $4.4\times 10^{-16}\ \text{cm}^2$ for air.

Mean free path:

$$r_{\text{m}} = \frac{v_{\text{th}}}{\nu_{\text{eff}}} \approx 10^4 (T^2/N_{\text{i}}) \ \text{cm}. \tag{1-262}$$

1.33. Propagation of Electromagnetic (Transverse) Waves in a Plasma

Following RATCLIFFE (1959) we define the dimensionless parameters

$$X = \left(\frac{\nu_{\text{p}}}{\nu}\right)^2, \quad Y = \frac{\nu_{\text{H}}}{\nu} \quad \text{and} \quad Z = \frac{\nu_{\text{eff}}}{2\pi\nu}, \tag{1-263}$$

where ν is the frequency of the electromagnetic wave, and ν_{p}, ν_{H}, and ν_{eff} are defined in Sect. 1.32. The complex index of refraction, ε_ν, for any arbitrary direction of the magnetic field is then given by

$$\varepsilon_\nu^2 = [n_\nu - i q_\nu]^2 = 1 - \frac{X}{\dfrac{1 - iZ - \frac{1}{2}Y_T^2}{1 - X - iZ} \pm \left[\dfrac{\frac{1}{4}Y_T^4}{(1 - X - iZ)^2} + Y_L^2\right]^{1/2}}, \tag{1-264}$$

where the + and − signs of the ± denote, respectively, the ordinary and extraordinary waves, $Y_T = Y\sin\psi$ and $Y_L = Y\cos\psi$, where the angle ψ is the angle between the propagation of the electromagnetic wave and the magnetic field vector. The index of refraction is n_ν and the absorption coefficient, α_ν, is $4\pi q_\nu \nu/c$.

For the important special case of quasi-longitudinal propagation, where the wave nearly propagates along the magnetic field line, we have the index of refraction

$$n_\nu \approx \{1 - \nu_{\text{p}}^2[\nu(\nu \pm \nu_{\text{H}})]^{-1}\}^{1/2} = \left[1 - \frac{X}{1 \pm Y}\right]^{1/2} \tag{1-265}$$

and the absorption coefficient

$$\alpha_\nu \approx \frac{\nu_{\text{eff}}}{n_\nu c}\left(\frac{\nu_{\text{p}}}{\nu}\right)^2\left[1 \pm \left(\frac{\nu_{\text{H}}}{\nu}\right)\right]^{-2} = \frac{\pi\nu}{n_\nu c}\,\frac{XZ}{(1 \pm Y)^2 + Z^2}. \tag{1-266}$$

The dispersion relation in this case is that of a "cold" plasma given by APPLETON (1932) and HARTREE (1931). The ordinary and extraordinary waves are both nearly completely circularly polarized, the sense of the ordinary wave being left hand. The phase difference, $\Delta\varphi$, between the extraordinary and ordinary waves after traversing the length dl of the plasma is:

$$\Delta\varphi = \frac{\omega}{2c}(n_{\nu+} - n_{\nu-})\,dl \approx \frac{\pi\nu}{c}\left(\frac{\nu_p}{\nu}\right)^2\left(\frac{\nu_H}{\nu}\right)dl = \frac{\pi\nu}{c}XY\,dl, \tag{1-267}$$

where $n_{\nu+}$ and $n_{\nu-}$ are given by Eq. (1-265) with the + and − signs. Thus a linearly polarized electromagnetic wave, which is incident on a plasma, will be rotated through Ω radians, where Ω is given by

$$\Omega = \int_0^L \Delta\varphi\,dl = \frac{e^3}{2\pi m^2 c^2 \nu^2}\int_0^L N_e H\cos\psi\,dl \approx \frac{2.36\times10^4}{\nu^2}\int_0^L N_e H\cos\psi\,dl \text{ radians} \tag{1-268}$$

after traversing a thickness, L, of the plasma. The angle ψ is the angle between the magnetic field and the line of sight. This effect is called Faraday rotation after its discovery by FARADAY (1844). As long as $\nu \gg \nu_H$, the rotation due to components of H which are perpendicular to the line of sight is negligible compared with Ω. The line integral $\int_0^L N_e H\cos\psi\,dl$ is often measured by observing the periodic rise and fall of intensity at closely spaced frequencies received with a linearly polarized feed. The period, P, of this sinusoidal oscillation is given by

$$P = 1.33\times10^{-4}\nu^3\left[\int_0^L N_e H\cos\psi\,dl\right]^{-1} \text{Hz} \approx 10^9\,[\lambda^3\,\text{R.M.}]^{-1}\ \text{Hz}, \tag{1-269}$$

where the rotation measure, R.M., is given by

$$\text{R.M.} = 8.1\times10^5 \int N_e H\cos\psi\,dl \quad \text{rad}\,m^{-2}, \tag{1-270}$$

and the rotation angle $\Omega = \lambda^2\,$R.M. radians when λ is in meters and dl is in parsecs for Eq. (1-270) and cm for Eq. (1-269).

For transverse propagation, the index of refraction is given by

$$n_{\nu 0} = [1-(\nu_p/\nu)^2]^{1/2} = [1-X]^{1/2} \tag{1-271}$$

whereas for quasi-transverse propagation

$$n_{\nu 0} = \{1-\nu_p^2[\nu^2+(\nu^2-\nu_p^2)\cot^2\theta]^{-1}\}^{1/2},$$

where θ represents the angle between the direction of propagation and the magnetic field. For both cases

$$n_{\nu e} = \left\{1-\left(\frac{\nu_p}{\nu}\right)^2\frac{[1-(\nu_p/\nu)^2]}{1-(\nu_p/\nu)^2-(\nu_H/\nu)^2}\right\}^{1/2} = \left\{1-\frac{X(1-X)}{1-X-Y^2}\right\}^{1/2}, \tag{1-272}$$

and the absorption coefficient is given by

$$\alpha_{\nu 0} = \frac{\nu_{\text{eff}}}{n_\nu c}\left(\frac{\nu_p}{\nu}\right)^2 = \frac{\nu_{\text{eff}} X}{n_\nu c}, \tag{1-273}$$

$$\alpha_{\nu e} = \frac{\nu_{\text{eff}}}{n_\nu c}\left(\frac{\nu_p}{\nu}\right)^2\frac{[1+(\nu_H/\nu)^2]}{[1-(\nu_H/\nu)^2]^2} = \frac{\nu_{\text{eff}} X}{n_\nu c}\,\frac{1+Y^2}{(1-Y^2)^2}, \tag{1-274}$$

where the subscripts 0 and e denote, respectively, the ordinary and extraordinary waves. The two waves in this case are nearly completely linearly polarized, the electric vector of the ordinary wave being parallel to the magnetic field.

In the absence of a magnetic field, the dielectric constant, ε_ν, and the conductivity, σ_ν, of a plasma are given by (GINZBURG, 1961)

$$\varepsilon_\nu = 1 - \frac{\nu_p^2}{\nu^2 + (\nu_{eff}/2\pi)^2} \tag{1-275}$$

and

$$\sigma_\nu = \frac{\nu_p^2 \nu_{eff}}{4\pi[\nu^2 + (\nu_{eff}/2\pi)^2]} \approx \frac{\nu_{eff}}{4\pi}\left(\frac{\nu_p}{\nu}\right)^2 . \tag{1-276}$$

The index of refraction, n_ν, and the absorption coefficient α_ν, are given by

$$\begin{aligned} n_\nu &= \left\{\frac{\varepsilon_\nu}{2} + \left[\left(\frac{\varepsilon_\nu}{2}\right)^2 + \left(\frac{\sigma_\nu}{\nu}\right)^2\right]^{1/2}\right\}^{1/2} \\ &\approx \left[1 - \frac{\nu_p^2}{\nu^2 + (\nu_{eff}/2\pi)^2}\right]^{1/2} \quad \text{for } \varepsilon_\nu \gg \frac{\sigma_\nu}{\nu}, \end{aligned} \tag{1-277}$$

and

$$\begin{aligned} \alpha_\nu &= 4\pi\frac{\nu}{c}\left\{-\frac{\varepsilon_\nu}{2} + \left[\left(\frac{\varepsilon_\nu}{2}\right)^2 + \left(\frac{\sigma_\nu}{\nu}\right)^2\right]^{1/2}\right\}^{1/2} \\ &\approx \frac{1}{n_\nu}\left(\frac{\nu_p}{\nu}\right)^2 \frac{\nu_{eff}}{c} \approx \frac{4\pi\sigma_\nu}{n_\nu c} \quad \text{for } \varepsilon_\nu \gg \frac{\sigma_\nu}{\nu}. \end{aligned} \tag{1-278}$$

When a wave of frequency, ν, propagates through a plasma with $\nu > \nu_H$, its group velocity is

$$u = c n_\nu \approx c\left[1 - \left(\frac{\nu_p}{\nu}\right)^2\right]^{1/2} . \tag{1-279}$$

A pulse of electromagnetic radiation will therefore be delayed by the time

$$\tau_D = \int_0^L \frac{dl}{u} \approx \int_0^L \frac{1}{c}\left[1 + \frac{1}{2}\left(\frac{\nu_p}{\nu}\right)^2\right] dl \approx 0.3 \times 10^{-10} L + \frac{1.35 \times 10^{-3}}{\nu^2} \int_0^L N_e \, dl \quad \text{sec}, \tag{1-280}$$

when propagating through a plasma of thickness, L. The differential dispersion will be

$$\Delta\tau_D = \frac{e^2}{2\pi m c}\left[\frac{1}{\nu_1^2} - \frac{1}{\nu_2^2}\right] \int_0^L N_e \, dl \approx 1.35 \times 10^{-3}\left[\frac{1}{\nu_1^2} - \frac{1}{\nu_2^2}\right] \int_0^L N_e \, dl \text{ sec}, \tag{1-281}$$

which is the delay between a pulse received at a high frequency, ν_1, and that received at a low frequency, ν_2, after propagating through a plasma of thickness, L. The parameter $\int_0^L N_e \, dl$ is called the dispersion measure and has c.g.s. units of cm^{-2}. The units of parsec cm^{-3} are also often used for dispersion measure (one parsec $= 3.0857 \times 10^{18}$ cm). In Table 5, dispersion measures are given together with other data for eighty-four pulsars.

Table 5. Positions, periods, P, reference epoch, half width, dispersion measure, $\int N_e dl$, period change, dP/dt, and age, $P/(dP/dt)$ for eighty-four pulsars[1]

Designation	Pulsar	α (1950.0) h m s	δ (1950.0) ° ′ ″	l^{II} (°)	b^{II} (°)
P0031−07	MP 0031	0 31 36	−07 38 26	110.4	−69.8
P0105+65	PSR 0105+65	1 05 00	65 50	124.6	3.3
P0138+59	PSR 0138+59	1 38	59 45	129.1	−2.3
P0153+61	PSR 0153+61	1 53 00	61 55	130.5	0.2
P0254−54	MP 0254	2 54 24	−54	270.9	−54.9
P0301+19	PSR 0301+19	3 01 45	19 23 36	161.1	−33.2
P0329+54	CP 0329	3 29 11	54 24 38	145.0	−1.2
P0355+54	PSR 0355+54	3 55 00	54 13	148.1	0.9
P0450−18	MP 0450	4 50 22	−18 04 14	217.1	−34.1
P0525+21	NP 0525	5 25 52	21 58 18	183.8	−6.9
P0531+21	NP 0531	5 31 31	21 58 55	184.6	−5.8
P0540+23	PSR 0540+23	5 40 10	23 30	184.4	−3.3
P0611+22	PSR 0611+22	6 11 03	22 35	188.7	2.4
P0628−28	PSR 0628−28	6 28 51	−28 34 08	237.0	−16.8
P0736−40	MP 0736	7 36 51	−40 35 18	254.2	−9.2
P0740−28	PSR 0740−28	7 40 48	−28 15 16	243.8	−2.4
P0809+74	CP 0809	8 09 03	74 38 10	140.0	31.6
P0818−13	MP 0818	8 18 06	−13 41 23	235.9	12.6
P0823+26	AP 0823+26	8 23 51	26 47 18	197.0	31.7
P0833−45	PSR 0833−45	8 33 39	−45 00 19	263.6	−2.8
P0834+06	CP 0834	8 34 26	06 20 47	219.7	26.3
P0835−41	MP 0835	8 35 34	−41 24 54	260.9	−0.3
P0904+77	PSR 0904+77	9 04	77 40	135.3	33.7
P0940−56	MP 0940	9 40 40	−56	278.8	−2.5
P0943+10	PP 0943	9 43 20	10 05 33	225.4	43.2
P0950+08	CP 0950	9 50 31	08 09 43	228.9	43.7
P0959−54	MP 0959	9 59 51	−54 37	280.1	0.3
P1055−51	MP 1055−51	10 55 49	−51 40	286.0	7.0
P1112+50	PSR 1112+50	11 12 49	50 40	155.1	60.7
P1133+16	CP 1133	11 33 28	16 07 33	241.9	69.2
P1154−62	MP 1154	11 54 45	−62 08 36	296.7	−0.2
P1237+25	AP 1237+25	12 37 12	25 10 17	252.2	86.5
P1240−64	MP 1240	12 40 20	−64 07 12	302.1	−1.6
P1359−50	MP 1359	13 59 43	−50	314.5	11.0
P1426−66	MP 1426	14 26 34	−66 09 54	312.3	−6.3
P1449−65	MP 1449	14 49 22	−65	315.3	−5.3
P1451−68	PSR 1451−68	14 51 29	−68 32	313.9	−8.6
P1508+55	HP 1508	15 08 04	55 42 56	91.3	52.3
P1530−53	MP 1530	15 30 23	−53 30	325.7	1.9
P1541+09	AP 1541+09	15 41 14	09 38 43	17.8	45.8
P1556−44	MP 1556	15 56 12	−44 31 30	334.5	6.4
P1604−00	MP 1604	16 04 38	−00 24 41	10.7	35.5
P1642−03	PSR 1642−03	16 42 25	−03 12 30	14.1	26.1

[1] These data were kindly provided by Professor Y. TERZIAN, Cornell University.

Table 5 (continued)

Designation	Pulsar	α (1950.0) h m s	δ (1950.0) ° ′ ″	l^{II} (°)	b^{II} (°)
P1700−18	MP 1700−18	17 00 56	−18	4.0	14.0
P1706−16	MP 1706	17 06 33	−16 37 21	5.8	13.7
P1717−29	PSR 1717−29	17 17 10	−29 32	356.5	4.3
P1718−32	PSR 1718−32	17 18 40	−32 05	354.5	2.5
P1727−47	MP 1727	17 27 50	−47 42 18	342.6	−7.6
P1747−46	MP 1747	17 47 56	−46 56 12	345.0	−10.2
P1749−28	PSR 1749−28	17 49 49	−28 06 00	1.5	−1.0
P1818−04	MP 1818	18 18 14	−04 29 03	25.5	4.7
P1819−22	PSR 1819−22	18 19 50	−22 53	9.4	−4.3
P1822−09	PSR 1822−09	18 22 40	−09 36	21.4	1.3
P1826−17	PSR 1826−17	18 26 15	−17 54	14.6	−3.3
P1831−03	PSR 1831−03	18 31 00	−03 40	27.7	2.3
P1845−01	OP 1845−01	18 45	−01 27	31.3	0.2
P1845−04	JP 1845	18 45 10	−04 05 32	28.9	−1.0
P1846−06	PSR 1846−06	18 46 07	−06 43	26.7	−2.4
P1857−26	MP 1857	18 57 44	−26 04 49	10.5	−13.5
P1858+03	JP 1858	18 58 40	03 27 02	37.2	−0.6
P1900−06	PSR 1900−06	19 00 30	−06 35	28.5	−5.6
P1907+02	PSR 1907+02	19 07 20	02 56	37.7	−2.7
P1911−04	MP 1911	19 11 15	−04 45 59	31.3	−7.1
P1915+13	OP 1915+13	19 15 25	13 50 00	48.3	0.6
P1917+00	PSR 1917+00	19 17 15	00 18	36.5	−6.1
P1919+21	CP 1919	19 19 36	21 47 17	55.8	3.5
P1929+10	PSR 1929+10	19 29 52	10 53 03	47.4	−3.9
P1933+16	JP 1933+16	19 33 32	16 09 58	52.4	−2.1
P1944+17	MP 1944	19 44 38	17 58 44	55.3	−3.5
P1946+35	JP 1946	19 46 35	35 28 36	70.6	5.0
P1953+29	JP 1953	19 53 00	29 15 03	66.0	0.7
P2002+30	JP 2002	20 02 35	30	67.7	−0.7
P2016+28	AP 2016+28	20 16 00	28 30 31	68.1	−4.0
P2020+28	PSR 2020+28	20 20 33	28 44 30	68.9	−4.7
P2021+51	JP 2021	20 21 25	51 45 08	87.9	8.4
P2045−16	PSR 2045−16	20 45 47	−16 27 48	30.5	−33.1
P2111+46	JP 2111	21 11 38	46 31 42	89.0	−1.3
P2148+63	PSR 2148+63	21 48 00	63 10	104.1	7.4
P2154+40	PSR 2154+40	21 54 56	40 02 30	90.5	−11.4
P2217+47	PSR 2217+47	22 17 46	47 39 48	98.4	−7.6
P2256+58	PSR 2256+58	22 56 30	58 50	108.8	−0.7
P2303+30	AP 2303+30	23 03 34	30 43 49	97.7	−26.7
P2305+55	PSR 2305+55	23 05 00	55 26	108.6	−4.2
P2319+60	JP 2319	23 19 42	60 00	112.0	−0.6

Table 5 (continued)

Period, P (1 sec)	Epoch (J.D. 24+)	Half width (m sec)	$\int N_e dl$ (pc cm^{-3})	dP/dt (n sec/day)	$P/(dP/dt)$ (years)
0.9429507566	40,690	51.0	10.89	0.0366	7.1E+07
1.283651	41,554	(29)	30		
1.222949	41,544	35	34.5		
2.351615	41,554	(43)	60		
0.4476	(41,347)	10	10		
1.38758356	41,347	27	15.7		
0.71451864244	40,622	6	26.776	0.177	1.1E+07
0.15638004	41,594	3.5	57.0		
0.548935369	41,536	19	39.9		
3.7454934468	40,957	181.0	50.8	3.461	2.9E+06
0.0331296454	41,221	3	56.805	36.526	2.5E+03
0.245964	41,554	(16)	72		
0.33491850	41,603	(8)	96.7	7.377	1.3E+05
1.244414895981	40,242	(50)	34.36	0.217	1.6E+07
0.374918324	40,221	22	100	<1.728	>5.9E+05
0.166750167	41,027	8	80		
1.2922412852	40,689	41.3	5.84	0.0138	2.5E+08
1.23812810726	41,006	22	40.9	0.1820	1.9E+07
0.530659599042	40,242	5.9	19.4	0.144	1.0E+07
0.0892187479	40,307	2	63	10.823	2.3E+04
1.27376349759	40,626	23.7	12.90	0.587	5.9E+06
0.76698	(41,347)	20	120		
1.57905	(40,222)	<80			
0.662	(40,402)	30	145		
1.097707	40,519	50	15.35		
0.25306504317	40,622	8.8	2.965	0.0198	3.5E+07
1.436551	40,553	50	90		
0.1971	(41,347)		<30		
1.65643810	41,594	14	9.2		
1.18791116405	40,622	27.7	4.834	0.323	1.0E+07
0.40052	(41,347)		270		
1.38244857195	40,626	49.9	9.254	0.0825	4.6E+07
0.38850	(41,347)	60	220		
0.690	(40,585)	20	20		
0.7874	(41,347)	10	60		
0.180	(40,282)	(5)	90		
0.263376764	(40,545)	25	8.6	<0.259	>2.8E+06
0.73967787630	40,626	10.3	19.60	0.433	4.7E+06
1.368852	40,553	25	20		
0.7484483	(41,347)	63	35.0		
0.25705	(41,347)				
0.42181607579	41,005	15	10.72	0.0265	4.4E+07
0.38768877965	40,622	3.7	35.71	0.154	6.9E+06
0.802	(41,347)		<40		
0.65305045437	40,622	12	24.99	0.550	3.3E+06
0.620449	41,554	(32)	45		
0.477154	41,554	(35)	120		

Table 5 (continued)

Period, P (1 sec)	Epoch (J.D. 24+)	Half width (m sec)	$\int N_e dl$ (pc cm^{-3})	$dP \to dt$ (n sec→day)	$P \to (dP \to dt)$ (years)
0.829683	40,553	30	121		
0.742349	40,553	20	40		
0.562553168299	40,127	6	50.88	0.705	2.2E+06
0.59807262183	40,622	10.3	84.48	0.545	3.0E+06
1.874398	41,554	(76)	140		
0.768948	41,554	11	19.3		
0.307128	41,554	(59)	207		
0.686675	41,554	(32)	235		
0.659475	(41,192)	(80)	90		
0.59773452	40,998	20	141.9		
1.451323	41,554	(38)	152		
0.612204	41,026	25	35		
0.655444	40,754	(170)	402		
0.431885	41,554	(14)	180		
0.494914	41,554	(11)	190		
0.82593366503	(40,624)	8.0	89.41	0.351	6.5E+06
0.1946255	41,274	15	94		
1.272254	41,554	(27)	85		
1.33730115212	40,690	31.2	12.43	0.116	3.2E+07
0.22651703833	40,625	5.3	3.176	0.100	6.2E+06
0.35873542051	40,690	8.0	158.53	0.519	1.9E+06
0.4406179	(40,618)	30	16.3		
0.717306	40,659	21	129.1		
0.426676	40,754	13	20		
2.111206	40,756	15	233		
0.55795339053	40,689	13.9	14.16	0.0129	1.2E+08
0.343400790	41,348	6.7	24.6		
0.52919531221	40,626	6.6	22.580	0.263	5.5E+06
1.96156682076	40,695	79.0	11.51	0.945	5.7E+06
1.01468444509	41,006	29	141.4	0.0620	4.5E+06
0.380142	41,554	(29)	125		
1.52526334	41,532	52	71.0		
0.53846737844	40,624	7.9	43.52	0.239	6.2E+06
0.368241	41,554	(14)	148		
1.575884410	41,006	26.0	49.9	0.2514	1.7E+07
0.475068	41,554	(27)	45		
2.25648387	41,536	65	96		

1.34. Propagation of Longitudinal (P mode) Waves in a Plasma: Plasma Line Radiation and Cerenkov Radiation

Because a plasma is macroscopically neutral, local perturbations in the electron or ion density set up electron or ion oscillations near the plasma frequency, ω_p. As TONKS and LANGMUIR (1929) pointed out, these oscillations will be propagated as longitudinal plasma waves of frequency, ω_p. For a hot plasma of temperature, T, the correct frequency, ω, for the wave is (BOHM and GROSS, 1949)

$$\omega = [\omega_p^2 + v_{th}^2 K^2]^{1/2}, \tag{1-282}$$

where $v_{th} = [3kT/m]^{1/2}$ and $K = 2\pi/\lambda$, where λ is the wavelength of the wave. The refractive index, $n_{\|}$, of the plasma for the longitudinal wave is given by

$$n_{\|}^2 = \left(\frac{c}{v_{th}}\right)^2 \left[1 - \left(\frac{v_p}{v}\right)^2\right]. \tag{1-283}$$

The wave will only exist for $n_{\|}^2 > 1$, which provides a lower limit to ν near ν_p.

The wave will be seriously damped (Landau damping) for wavelengths, less than the Debye radius, r_D, given by (DEBYE and HÜCKEL, 1923; LANDAU, 1946)

$$r_D = \frac{1}{3^{1/2}}\left(\frac{v_{th}}{\omega_p}\right) \approx 6.9\left(\frac{T}{N_e}\right)^{1/2} \text{cm}, \tag{1-284}$$

or, equivalently, using Eq. (1-282) the wave will be damped for frequencies $\nu \gtrsim \sqrt{2.7}\,\nu_p$ providing an upper limit to ν. The longitudinal plasma waves may be converted into electromagnetic waves when the plasma is nonuniform (cf. ZHELEZNYAKOV, 1970, for details of several such processes). Of special interest is the electromagnetic radiation produced by the scattering of longitudinal plasma waves on the small-scale thermal fluctuations in plasma density, N (GINZBURG and ZHELEZNYAKOV, 1958, 1959; SALPETER, 1960; SMERD, WILD, and SHERIDAN, 1962). Rayleigh scattering by the quasi-neutral random fluctuations due to the thermal motion of the ions produces electromagnetic radiation at the ion frequency $\nu_i = (e^2 N_i/\pi m_i)^{1/2}$ and at the plasma frequency $\nu_p = (e^2 N_e/\pi m_e)^{1/2}$. The total flux of energy, $W(\nu_p)$, scattered at the plasma frequency in a volume, V, is given by (GINZBURG and ZHELEZNYAKOV, 1959)

$$W(\nu_p) \approx n_\nu \frac{e^4 N_e V}{6m^2 c^3} E_0^2, \tag{1-285}$$

where the index of refraction $n_\nu = \sqrt{3}\,v_{th}/v_\varphi$, the phase velocity of the plasma wave is v_φ, and E_0 is the electric field amplitude of the plasma wave. The efficiency parameter, Q, for the change of plasma wave energy into the energy of electromagnetic waves at the plasma frequency is given by

$$Q = \frac{W(\nu_p)}{L^2 S_p} \approx \frac{4\pi e^4 N_e L}{3m^2 c^3 v_{th}}, \tag{1-286}$$

where L is the linear extent of the scattering region, and S_p is the plasma wave energy flux density. If the emission takes place over a frequency range $\Delta\nu$, across an area A, and into a solid angle $\Delta\Omega$, then an effective temperature, T_b, defined by using the Rayleigh-Jeans law for the brightness of a black body, is given by

$$T_b = \frac{c^2 W(\nu_p) e^{-\tau}}{2\nu_p^2 k \Delta\nu \Delta\Omega A}, \tag{1-287}$$

where τ denotes the optical depth between the observer and the source.

Combination scattering by the space-charge fluctuations due to the thermal motion of the electrons produces electromagnetic radiation at the frequency $2\nu_p$ whose total flux of energy, $W(2\nu_p)$ is given by (ZHELEZNYAKOV, 1970)

$$W(2\nu_p) \approx \frac{4\sqrt{3}}{5} \frac{e^4 N_e V}{m^2 c^3} \frac{kT}{mc^2} E_0^2, \tag{1-288}$$

provided that $v_\varphi \ll c/\sqrt{3}$. This emission is beamed strongly in the direction normal to that of the plasma wave, and has an angular spectrum of intensity, $I(\theta)$, given by (SMERD, WILD, and SHERIDAN, 1962)

$$I(\theta) \propto \sin^3\theta[1+3(v_0/c)^2 - 2\sqrt{3}(v_0/c)\cos\theta], \tag{1-289}$$

where θ is the angle from the direction of the plasma wave, and v_0 is the mean velocity of the stream of particles exciting the plasma wave.

For a stream of particles of density $N_s \ll N_e$ and velocity $u \gg v_{\varphi'}$ where v_φ is the phase velocity of the plasma, the amplitude of the steady-state plasma wave is given by (BOHM and GROSS, 1949)

$$\begin{aligned} E_0 &= \frac{m}{2eK^3}\left\{\frac{16}{3} v_\varphi (2\pi\nu_{ps})^2 \frac{\nu_s}{\nu}\left(\frac{df(u)}{du}\right)_{u=v_\varphi}\right\}^2 \\ &\approx \frac{\pi m \nu_p}{e v_0}\left\{\frac{16}{3}\frac{v_0^3 m}{kT_s}\left(\frac{N_s}{N_e}\right)^2\right\}^2, \end{aligned} \tag{1-290}$$

where $K = 2\pi\nu_p/v_\varphi$, the plasma frequency $\nu_p = [e^2 N_e/\pi m]^{1/2}$, the plasma frequency of the stream $\nu_{ps} = [e^2 N_s/\pi m]^{1/2}$, the collision frequency of the stream, ν_s, and the plasma, ν, are related by $\nu_s/\nu \approx N_s/N_e$, and the velocity distribution of the stream, $f(u)$, is assumed to be Maxwellian with a mean value of $v_0 \approx v_\varphi$ and $(df(u)/du)_{u=v_\varphi} \approx v_{th}^{-2}$.

Longitudinal waves may also be generated as a consequence of the accelerated motions of electrons scattered by ions in analogy with thermal bremsstrahlung (cf. BEKEFI, 1966). In this case the frequency of the longitudinal waves are again concentrated near ν_p, but in this frequency range the volume emissivity is greater than that of the thermal bremsstrahlung by the factor $(c/v_{th})^2$.

If a charged particle of velocity, u, is injected into a plasma for which the phase velocity, v_φ, is less than u, the particle will excite longitudinal waves in a manner which is analogous to the optical Cerenkov effect (CERENKOV, 1937). Optical Cerenkov radiation is concentrated into a cone centered on the trajectory of the charged particle and given by

$$\cos\psi = \left[\frac{c}{u n_\nu}\right], \tag{1-291}$$

where n_ν is the refractive index of the medium. The total energy radiated per unit frequency interval per unit length of the medium is, in the optical case, given by

$$I_\nu = \frac{2\pi e^2 \nu}{c^2}\left[1 - \frac{c^2}{n_\nu^2 u^2}\right], \tag{1-292}$$

where it is understood that $u > c/n_\nu$.

The frequency and angular spectrum for the Cerenkov radiation of an electron in a plasma have been calculated by COHEN (1961). The radiation is symmetric about the trajectory of the charged particle. If θ denotes the angle from this trajectory, the angular spectrum of the radiated energy is

$$I_\theta = 6.3 \times 10^5 \frac{e^2 v_p^2}{u^2} \frac{\cos\theta}{\cos^2\theta - (v_{th}/u)^2} \text{ erg rad}^{-2} \text{ cm}^{-1}. \tag{1-293}$$

Here u is the velocity of the incident electron and $v_{th} = (3kT_e/m)^{1/2}$ where T_e is the electron temperature. The total energy radiated per unit frequency interval per unit path of the electron is

$$\begin{aligned} I_\nu &= 2\pi \frac{e^2 \nu_p}{u^2} \frac{(\nu/\nu_p)}{(\nu/\nu_p)^2 - 1} \quad \text{for } \nu > \nu_c \\ &= 0 \quad \text{for } \nu < \nu_c, \end{aligned} \tag{1-294}$$

where the low frequency cutoff is

$$\nu_c = \nu[1 - (v_{th}/u)^2]^{-1/2}. \tag{1-295}$$

Plots of the spectral form of I_ν and the angular form of I_θ are shown in Fig. 6. The high frequency cutoff is assumed to be limited to $\sqrt{2}\,\omega_p$ by Landau damping ($v_\varphi \geq \sqrt{2}\, v_{th}$) and the total radiated energy per unit length of path is

$$\begin{aligned} W &= 2\pi^2 \frac{e^2 \nu_p^2}{u^2} \ln\left[\left(\frac{u}{v_{th}}\right)^2 - 1\right] \\ &\approx 5.1 \times 10^{-39} \nu_p^2 \left(\frac{c}{u}\right)^2 \ln\left[\left(\frac{u}{v_{th}}\right)^2 - 1\right] \text{ erg cm}^{-1}. \end{aligned} \tag{1-296}$$

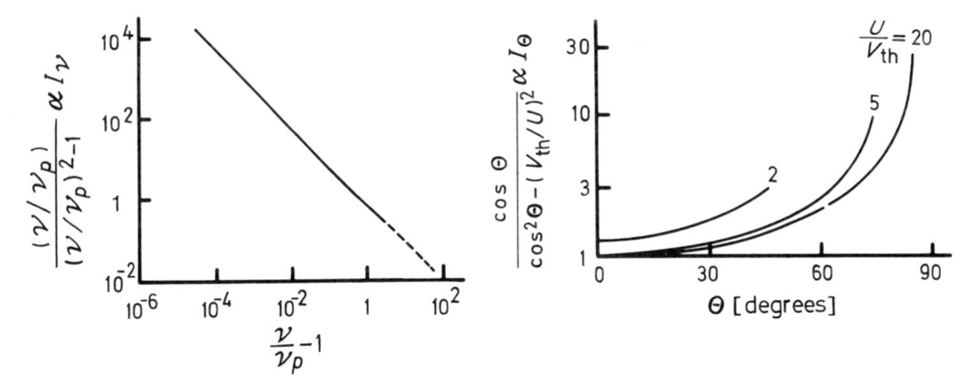

Fig. 6. The angular, I_θ, and frequency, I_ν, power spectrum of the Cerenkov radiation of a single charged particle in a plasma (after COHEN, 1961). The electron velocity is U, the thermal velocity is $V_{th} = (3kT/m)^{1/2}$, and the plasma frequency is $\nu_p = (e^2 N_e/\pi m)^{1/2}$, where the temperature is T and the electron density is N_e. The radiation is symmetrical about the trajectory of the electron, and θ is the angle from this trajectory. The I_θ and I_ν are given, respectively, by Eqs. (1–293) and (1–294). The frequency spectrum has a low frequency cutoff at $\nu_c = \nu_p[1 - (V_{th}/U)^2]^{-1/2}$, and in the dashed region Landau damping inhibits wave propagation. The Cerenkov angular spectrum also exhibits a high angle cutoff which results from Landau damping

1.35. Scattering from a Harmonic Oscillator

The equation of motion of a harmonic oscillator which is driven by a harmonic plane wave is

$$m\ddot{x}+m(2\pi\nu_0)^2 x=eE_0\exp(i2\pi\nu t)-m\gamma_{\text{cl}}\dot{x}, \tag{1-297}$$

where $\dot{}$ and $\ddot{}$ denote, respectively, the first and second derivatives with respect to time, x is the direction of the electric field vector, E_0 and ν are, respectively, the strength and frequency of the electric field, m and e are, respectively, the mass and charge of the oscillator, ν_0 is the resonance frequency of the oscillator, and the classical damping constant is

$$\gamma_{\text{cl}}=\frac{8\pi^2 e^2\nu_0^2}{3mc^3}\approx 2.5\times 10^{-22}\nu_0^2 \ \text{sec}^{-1}. \tag{1-298}$$

This equation has the steady state solution

$$x(t)=\frac{e}{4\pi^2 m}\mathscr{R}\left[\frac{E_0\exp[i2\pi\nu t]}{(\nu^2-\nu_0^2)+i(\gamma_{\text{cl}}\nu/2\pi)}\right], \tag{1-299}$$

where $\mathscr{R}$ denotes the real part of the term in paranthitis.

From Eq. (1-109) the total energy radiated per unit time per unit solid angle, $d\Omega$, is

$$\frac{dP}{d\Omega}=\frac{3}{8\pi}\sigma_{\text{s}}cU\sin^2\theta, \tag{1-300}$$

where $U=E_0^2/4\pi$ is the energy density of the incident field, θ is the angle between the x axis and the direction of observation, and the scattering cross section, σ_{s}, is given by

$$\begin{aligned}\sigma_{\text{s}}&=\frac{8\pi}{3}\left(\frac{e^2}{mc^2}\right)^2\frac{\nu^4}{(\nu^2-\nu_0^2)^2+(\gamma_{\text{cl}}^2\nu^2/4\pi^2)}\\ &\approx 6.65\times 10^{-25}\frac{\nu^4}{(\nu^2-\nu_0^2)^2+(\gamma_{\text{cl}}^2\nu^2/4\pi^2)}\ \text{cm}^2.\end{aligned} \tag{1-301}$$

Similarly, the total energy radiated per unit time in all directions, P, is obtained from Eq. (1-110)

$$P=\sigma_{\text{s}}cU. \tag{1-302}$$

The total scattering cross section obtained by integrating σ_{s} over all frequencies is

$$\sigma_{\text{tot}}=\frac{\pi e^2}{mc}\approx 0.0265\ \text{cm}^2. \tag{1-303}$$

The corresponding result for a quantum mechanical oscillator is (LADENBURG, 1921; KRAMERS, 1924)

$$\sigma_{\text{tot}}=\frac{\pi e^2}{mc}f_{nm}, \tag{1-304}$$

where f_{nm} is the oscillator strength of the $m-n$ transition.

1.36. Rayleigh Scattering by Bound Electrons

When plane wave radiation of frequency, ν, is incident upon atoms or molecules whose characteristic transition frequencies, ν_0, are much larger than ν, the total energy radiated per unit time per unit solid angle, $dP/d\Omega$, and the total energy radiated per unit time in all directions, P, are given by Eqs. (1-300) and (1-302) with

$$\begin{aligned}\sigma_s = \sigma_R &= \frac{8\pi}{3}\left(\frac{e^2}{mc^2}\right)^2 \frac{\nu^4}{(\nu^2-\nu_0^2)^2} \quad \text{for } \nu<\nu_0 \\ &\approx 6.65\times10^{-25}\left(\frac{\nu}{\nu_0}\right)^4 \text{ cm}^2 \quad \text{for } \nu \ll \nu_0 .\end{aligned} \tag{1-305}$$

This result, which was first obtained by RAYLEIGH (1899), follows from Eq. (1-301) when we assume $\nu_0 \gg \gamma_{\rm cl}$ when $\nu \ll \nu_0$.

The corresponding quantum mechanical expression for the Rayleigh scattering cross section is (LADENBURG, 1921; KRAMERS, 1924)

$$\sigma_R \approx 6.65\times10^{-25}\left(\sum_k f_{1k}\left[\frac{\nu_{1k}^2}{\nu^2}-1\right]^{-1}\right)^2 \text{ cm}^2, \tag{1-306}$$

where f_{1k} is the oscillator strength of the $1-k$ transition.

From Eq. (1-86), the index of refraction, n, of a gas of density N is

$$n = \left[1 - \frac{4\pi N e^2}{4\pi^2 m(\nu_0^2-\nu^2)}\right]^{1/2}. \tag{1-307}$$

Substituting Eq. (1-307) into Eq. (1-305) we obtain

$$\sigma_R = \frac{8\pi^3}{3}\left(\frac{\nu}{c}\right)^4\left(\frac{n^2-1}{N}\right)^2 \approx \frac{128\pi^5\alpha^2}{3\lambda^4}, \tag{1-308}$$

where λ is the wavelength of the incident radiation and α is the polarizability.

1.37. Thomson Scattering by a Free Electron

The total energy radiated per unit time per unit solid angle by a free electron is (LARMOR, 1897, Eq. (1-109))

$$\frac{dP}{d\Omega} = \frac{e^2}{4\pi c^3}\dot{v}^2 \sin^2\theta, \tag{1-309}$$

where $\dot{v}$ is the acceleration and θ is the angle between the direction of acceleration and the direction of observation. From the Lorentz force equation (1-52) an incident electric field of strength E_0 gives the electron an acceleration $\dot{v} = eE_0/m$. Substituting into Eq. (1-309) we obtain

$$\frac{dP}{d\Omega} = \frac{3}{8\pi}\sigma_T c U \sin^2\theta, \tag{1-310}$$

where $U = E_0^2/4\pi$ is the energy density of the incident plane wave and the Thomson cross section (THOMSON, 1903) is given by

$$\sigma_T = \frac{8\pi}{3}\left(\frac{e^2}{mc^2}\right)^2 \approx 6.65\times10^{-25}\,\text{cm}^2 \quad \text{for } \nu < 10^{20}\,\text{Hz}. \tag{1-311}$$

This result could have been obtained from Eq. (1-301) with $\nu_0 = 0$. The total energy radiated per unit time in all directions, P, is obtained by integrating Eq. (1-310) over all angles

$$P = \sigma_T c U. \tag{1-312}$$

When the photon energy $h\nu$ is greater than or near mc^2, or $\nu \gtrsim 10^{20}$ Hz, the correct scattering cross section for free electrons is (KLEIN and NISHINA, 1929)

$$\sigma_{KN} = \frac{3}{4}\sigma_T\left\{\frac{1+\alpha}{\alpha^2}\left[\frac{2(1+\alpha)}{1+2\alpha} - \frac{1}{\alpha}\ln(1+2\alpha)\right] + \frac{1}{2\alpha}\ln(1+2\alpha) - \frac{1+3\alpha}{(1+2\alpha)^2}\right\}, \tag{1-313}$$

where $\alpha = h\nu/(mc^2)$. At low and high photon energies this equation becomes

$$\sigma_{KN} = \sigma_T\left\{1 - \frac{2h\nu}{mc^2} + \frac{26}{5}\left(\frac{h\nu}{mc^2}\right)^2 + \cdots\right\} \quad \text{for } \nu \ll 10^{20}\,\text{Hz} \tag{1-314}$$

$$\begin{aligned} &= \frac{3}{8}\sigma_T\left(\frac{mc^2}{h\nu}\right)\left[\ln\left(\frac{2h\nu}{mc^2}\right) + \frac{1}{2}\right] \quad \text{for } \nu \gg 10^{20}\,\text{Hz} \\ &\approx \frac{3\times10^{-5}}{\nu}\left[\ln(1.6\times10^{-20}\nu) + \frac{1}{2}\right]\,\text{cm}^2 \quad \text{for } \nu \gg 10^{20}\,\text{Hz}. \end{aligned} \tag{1-315}$$

The total energy radiated per unit time is, in this case, given by

$$P = \sigma_{KN} c U, \tag{1-316}$$

where U is the energy density of the incident wave. The general formula for the angular distribution of the scattered radiation is, however, given by

$$I = \left(\frac{r_0}{R}\right)^2 \frac{\sin^2\psi\, I_0}{[1+\alpha(1-\cos\theta)]^3}\left\{1 + \frac{\alpha^2(1-\cos\theta)^2}{2\sin^2\psi[1+\alpha(1-\cos\theta)]}\right\}, \tag{1-317}$$

where I and I_0 are, respectively, the intensities of the incident and scattered radiation, $r_0 = e^2/mc^2 \approx 2.8\times10^{-13}$ cm is the classical electron radius, R is the distance from the free electron to the observation point, and ψ and θ are, respectively, the angles that the incident electric vector and the direction of incident wave propagation make with the direction of observation.

In the presence of a magnetic field of intensity, H, the Thomson scattering cross section is reduced below the classical value given by Eq. (1-311) when the photon frequency, ν, becomes smaller than the gyrofrequency, $\nu_H \approx 2\times10^6 H$ Hz. The reduction factor is roughly $(\nu/\nu_H)^2$ for $\nu \ll \nu_H$ and is discussed in detail by CANUTO, LODENQUAI, and RUDERMAN (1971).

1.38. Compton Scattering by Free Electrons and Inverse Compton Radiation

Let a photon of energy, $h\nu_1$, collide with an electron which is at rest. If the incident photon energy is large enough, the frequency of the scattered photon, ν_2, will be smaller than ν_1 by an observable amount. The effect is named after Compton who first observed and explained the effect (COMPTON, 1923). If the photon is deflected by an angle, φ, and if the electron recoils with a velocity, v, at an angle, θ, with respect to the initial trajectory of the photon, the relativistic equations for conservation of energy and momentum are

$$h\nu_1 = h\nu_2 + mc^2\left\{\left(1-\frac{v^2}{c^2}\right)^{-1/2} - 1\right\},$$

$$\frac{h\nu_1}{c} = \frac{h\nu_2}{c}\cos\varphi + \frac{mv\cos\theta}{\left[1-\frac{v^2}{c^2}\right]^{1/2}}, \tag{1-318}$$

and

$$\frac{h\nu_2}{c}\sin\varphi = \frac{mv}{\left[1-\frac{v^2}{c^2}\right]^{1/2}}\sin\theta .$$

It follows from Eqs. (1-318) that

$$\nu_2 = \nu_1\left[1+\frac{h\nu_1}{mc^2}(1-\cos\varphi)\right]^{-1} \tag{1-319}$$

and that the increment in wavelength is

$$\Delta\lambda = \frac{2h}{mc}\sin^2\frac{\varphi}{2} \approx 4.8\times10^{-10}\sin^2\frac{\varphi}{2}\ \text{cm},$$

where φ is the deflection angle of the incident photon.

These formulae assume that the electron is at rest. If α is the angle which the incident photon makes with the trajectory of a moving electron whose total energy $E=\gamma mc^2$, the energy of a photon in the rest frame of the electron is

$$h\nu_1 = \gamma h\nu(1+\beta\cos\alpha),$$

where $h\nu$ is the energy of the photon in its own frame, and $\beta = v/c$ for an electron of velocity, v. When β is large, we have the scattered frequency

$$\nu_2 \approx \gamma^2\nu \quad \text{for } \gamma h\nu \ll mc^2 . \tag{1-320}$$

The scattering cross section for this case is (FEENBERG and PRIMAKOFF, 1948; DONAHUE, 1951).

$$\sigma_s \approx \gamma^2\sigma_T \quad \text{for } \gamma h\nu \ll mc^2, \tag{1-321}$$

where $\sigma_T \approx 6.65\times10^{-25}\ \text{cm}^2$, and the total energy radiated per unit time, P, by an electron passing through radiation of energy density, U, is given by

$$P = \gamma^2\sigma_T c U \quad \text{for } \gamma h\nu \ll mc^2 . \tag{1-322}$$

When the electron velocity is high and $\gamma h\nu \gg mc^2$, all of the electron energy, γmc^2, is transferred into scattered radiation regardless of the incident photon frequency. Hence,

$$\nu_2 \approx \gamma mc^2/h \quad \text{for } \gamma h\nu \gg mc^2 , \tag{1-323}$$

and using the Klien-Nishina cross section

$$\sigma_s \approx \frac{3}{8}\sigma_T \left(\frac{mc^2}{\gamma h\nu}\right)\left[\ln\left(\frac{2\gamma h\nu}{mc^2}\right)\right] \quad \text{for } \gamma h\nu \gg mc^2 , \tag{1-324}$$

so that the total energy radiated per unit time by one electron in passing through radiation of energy density, U, is

$$P = \frac{3}{8}\sigma_T c\, U \left[\frac{mc^2}{\gamma h\nu}\right] \ln\left(\frac{2\gamma h\nu}{mc^2}\right) \quad \text{for } \gamma h\nu \gg mc^2 . \tag{1-325}$$

The Compton scattering of relativistic particles is further discussed in Chap. 4, where it is noted that other processes such as pair production are also important. Applications of scattering formulae to astronomical objects are given by FELTEN and MORRISON (1963, 1966), GOLDREICH and MORRISON (1964), and GOULD and SCHRÉDER (1966). Scattering of high energy particles has been reviewed by BLUMENTHAL and GOULD (1970).

In order to maintain synchrotron radiation against inverse Compton losses, the magnetic field energy density in the region where the electrons are radiating must dominate over the local radiation energy density. For a homogeneous, isotropic source of luminosity, L, and radius, R, this condition means that (FELTEN and MORRISON, 1966; HOYLE, BURBIDGE, and SARGENT, 1966)

$$H^2 > \frac{8L}{R^2 c} , \tag{1-326}$$

where H is the magnetic field intensity, and the luminosity of the source is given by $L \approx 4\pi R^2 \nu S/\theta^2$, where ν is the frequency, S is the flux density, and θ is the angular size of the source. Using Eq. (1-197) in Eq. (1-326), this means that the Compton effect will set in for sources which would otherwise have a frequency, ν_{max}, of maximum flux density given by

$$\nu_{max} \gtrsim 10^{13} H \text{ Hz}. \tag{1-327}$$

For radio sources with an upper cutoff frequency, ν_c, and maximum effective brightness temperature, T_{max}, it follows from Eq. (1-121) that the brightness $B \approx 2\nu_c^2 kT/c^2$. Using Eq. (1-154) with $E \approx kT$, we obtain $H \approx 10^{-19}\nu_c (kT)^{-2}$ and substituting these values of B and H into Eq. (1-326) we obtain (KELLERMANN and PAULINY-TOTH, 1969)

$$10^{-72}\, T_{max}^5\, \nu_c < 1 . \tag{1-328}$$

Probably $\nu_c \approx 10^{11}$ Hz so that $T_{max} < 10^{12}$ °K for an incoherent synchrotron radiator.

It follows from Eqs. (1-327) and (1-197) that the minimum observable angular size, θ_{IC}, for a source undergoing quenching by inverse Compton radiation is given by

$$\theta_{IC} \approx 10^{-3} S^{1/2} \nu^{-1} \quad \text{seconds of arc}, \tag{1-329}$$

where the flux density, S, is in flux units and the frequency, ν, is in GHz. For comparison, the synchrotron self-absorption angle, θ_{SA}, given by Eq. (1-197) is

$$\theta_{SA} \approx 10^{-3} S^{1/2} \nu^{-5/4} \quad \text{seconds of arc}, \tag{1-330}$$

where the magnetic field has been assumed to be 10^{-4} gauss, and the units are the same as those for Eq. (1-329). The angular size deduced from flux variations on time scales of τ hours is (Rees, 1966)

$$\theta \lesssim 10^{-9} \tau \gamma z^{-1} \quad \text{seconds of arc}, \tag{1-331}$$

where $\gamma = [1-(v/c)^2]^{-1/2}$ for a source expanding at the velocity, v, the redshift is z, and the source distance $D = cz/H_0$ where the Hubble constant $H_0 = 75$ km sec^{-1} Mpc^{-1}. These angular sizes may also be compared with that caused by interstellar scattering, θ_{scat}, which follows from Eq. (1-385) together with measurements of Lang (1971) and Harris, Zeissig, and Lovelace (1970).

$$\theta_{scat} \approx 25 D^{1/2} \nu^{-2} \quad \text{seconds of arc}, \tag{1-332}$$

where D is the source distance in parsecs (one parsec $\approx 3 \times 10^{18}$ cm) and ν is the frequency in MHz.

1.39. Rayleigh Scattering from a Small Sphere

Let a harmonic plane wave which is linearly polarized in the x direction and propagating in the z direction be incident upon a dielectric sphere of radius, a. Small spheres for which $a < 0.05\,\lambda_0$, where λ_0 is the wavelength of the incident radiation, will oscillate along the x axis in synchronism with the electric field of the plane wave. The sphere will therefore radiate as an electric dipole whose dipole moment, $d(t)$, is given by Eqs. (1-25) and (1-30).

$$d(t) = a^3 \left(\frac{n^2-1}{n^2+2}\right) E_0(t) \approx a^3 \left[\frac{\varepsilon_2 - \varepsilon_1}{\varepsilon_2 + 2\varepsilon_1}\right] E_0(t), \tag{1-333}$$

where $E_0(t)$ is the strength of the electric field vector of the plane wave, ε_2 and ε_1 are, respectively, the dielectric constants of the sphere and the surrounding medium, and the relative index of refraction is $n = [\varepsilon_2 \mu_2/(\varepsilon_1 \mu_1)]^{1/2}$ where μ is the magnetic permeability. Using a spherical coordinate system centered at the center of the sphere, the total energy radiated per unit time per unit solid angle, $dP/d\Omega$, may be obtained from Eq. (1-113)

$$\frac{dP}{d\Omega} = \frac{4\pi^3 c a^6}{\lambda_0^4} \left(\frac{n^2-1}{n^2+2}\right)^2 \sin^2\theta\, E_0^2(t), \tag{1-334}$$

where θ is the angle between the x axis and the radial vector to the point of observation. From Eq. (1-114), the total energy radiated per unit time in all directions is

$$P = \sigma_s c U, \tag{1-335}$$

where $U = E_0^2(t)/4\pi$ is the energy density of the plane wave, and the scattering cross section, σ_s, is given by (Rayleigh, 1871)

$$\sigma_s = \frac{128\pi^5 a^6}{3\lambda_0^4}\left(\frac{n^2-1}{n^2+2}\right)^2 \quad \text{for } a < 0.05\,\lambda_0. \tag{1-336}$$

The efficiency factor, Q_s, for scattering is defined as the ratio of σ_s to the geometric cross section

$$Q_s = \frac{\sigma_s}{\pi a^2} = \frac{128\pi^4 a^4}{3\lambda_0^4}\left(\frac{n^2-1}{n^2+2}\right)^2. \tag{1-337}$$

From Eq. (1-113) the intensity of the scattered radiation, I, for an incident wave of unit intensity is

$$I = \frac{16\pi^4 a^6}{R^2\lambda_0^4}\left(\frac{n^2-1}{n^2+2}\right)^2 \sin^2\theta, \tag{1-338}$$

where R is the distance from the sphere to the observation point.

When the incident plane wave is linearly polarized, the scattered radiation is also linearly polarized. When the incident wave is unpolarized, both the incident and scattered waves may be regarded as a superposition of two linearly polarized waves. When the plane of observation is defined as the plane that contains both the direction of propagation of the incident wave and the direction of observation, the intensity of the scattered light may be resolved into components $I_\parallel$ and $I_\perp$ which are parallel and perpendicular to this plane. For unit intensity of unpolarized incident radiation, the scattered intensities are

$$I_\perp = \frac{16\pi^4 a^6}{R^2\lambda_0^4}\left(\frac{n^2-1}{n^2+2}\right)^2, \tag{1-339}$$

and

$$I_\parallel = \frac{16\pi^4 a^6}{R^2\lambda_0^4}\left(\frac{n^2-1}{n^2+2}\right)^2 \cos^2\theta, \tag{1-340}$$

and the total scattered intensity is

$$I = \frac{I_\perp + I_\parallel}{2} = \frac{8\pi^4 a^6}{R^2\lambda_0^4}\left(\frac{n^2-1}{n^2+2}\right)^2 (1+\cos^2\theta), \tag{1-341}$$

where θ is the angle between the direction of propagation of the incident wave and the direction of observation.

Even when the incident radiation is unpolarized, the scattered radiation is partly polarized. The degree of polarization, P, is defined as

$$P = \frac{I_\perp - I_\parallel}{I_\perp + I_\parallel} = \frac{\sin^2\theta}{1+\cos^2\theta}, \tag{1-342}$$

where here again θ is the angle between the directions of propagation and observation. What is actually observed at optical frequencies is a magnitude difference, Δm_{p}, given by

$$\Delta m_{\mathrm{p}} = 2.50 \log \frac{I_{\perp}}{I_{\parallel}} \approx 2.172 P \quad \text{for } P \ll 1\,, \tag{1-343}$$

where $I_{\perp}$ and $I_{\parallel}$ are, respectively, the maximum and minimum observed intensities.

When the scattering sphere is not a perfect dielectric, but is partially conducting, part of the incident radiation is absorbed as well as scattered. If σ is the conductivity of a conducting sphere in a dielectric medium, then the index of refraction of the sphere, m_2, is given by

$$m_2 = \sqrt{\varepsilon_2 \mu_2} + i \frac{2\pi\sigma}{\omega} \sqrt{\frac{\mu_2}{\varepsilon_2}} \quad \text{for } \sigma \ll \omega\varepsilon \tag{1-344}$$

and

$$m_2 = \left(\frac{2\pi\mu_2\sigma}{\omega}\right)^{1/2} (1+i) \quad \text{for } \sigma \gg \omega\varepsilon\,, \tag{1-345}$$

where ε_2 and μ_2 are, respectively, the dielectric constant and magnetic permeability of the sphere, and ω is the radian frequency of the harmonic plane wave. When the surrounding space is a perfect dielectric, its refractive index is $m_1 = \sqrt{\varepsilon_1 \mu_1}$ where ε_1 and μ_1 are, respectively, the dielectric constant and magnetic permeability of the medium. The relative dielectric constant for a badly conducting sphere is then given by

$$m = \frac{m_2}{m_1} = \left(\frac{\varepsilon_2}{\varepsilon_1}\right)^{1/2} + i \frac{2\pi\sigma}{\omega\sqrt{\varepsilon_1 \varepsilon_2}} \tag{1-346}$$

or

$$m^2 \approx \frac{\varepsilon_2}{\varepsilon_1} + i \frac{4\pi\sigma}{\omega\varepsilon_1}\,. \tag{1-347}$$

It follows directly from a limiting case of the Mie scattering theory discussed in Sect. 1.41 that the scattering cross section in this case is

$$\sigma_{\mathrm{s}} = \frac{128\pi^5 a^6}{3\lambda_0^4} \mathscr{R}\left(\frac{m^2-1}{m^2+2}\right)^2 \quad \text{for } a < 0.05\,\lambda_0\,, \tag{1-348}$$

where $\mathscr{R}$ denotes the real part of the term within parenthesis.

1.40. Extinction and Reddening of Stars

The total extinction cross section, σ_{e}, of a sphere of radius a is given by (FEENBERG, 1932; VAN DE HULST, 1949, 1957)

$$\sigma_{\mathrm{e}} = \sigma_{\mathrm{s}} + \sigma_{\mathrm{a}} = -\frac{8\pi^2 a^3}{\lambda_0} \mathscr{I}\left(\frac{m^2-1}{m^2+2}\right) \quad \text{for } a < 0.05\,\lambda_0\,, \tag{1-349}$$

where the complex index of refraction is m, the σ_s is given by Eq. (1-348), the $\mathscr{I}$ denotes the imaginary part of the term within the parenthesis, and σ_a is the cross section for the absorption of radiation by the sphere. The extinction cross section is the ratio of the total radiated energy scattered and absorbed per unit time to the incident radiant energy per unit time per unit area. An extinction coefficient is defined as $Q_e = \sigma_e/(\pi a^2)$, and the albedo, A, is defined as $A = Q_s/Q_e$. For a totally reflecting sphere of high conductivity

$$Q_s \approx Q_e \approx \frac{10}{3}\left(\frac{2\pi a}{\lambda_0}\right)^4 \quad \text{for } a < 0.05\,\lambda_0\,, \tag{1-350}$$

whereas for a nonconducting sphere $\sigma_e \approx \sigma_s$ where σ_s is given by Eq. (1-348).

Optical astronomers observe a magnitude difference, Δm, between nearby and distant stars of the same type. This difference is given by

$$\Delta m = 1.086\, N_c \pi a^2 Q_e = 1.086\, N_c \sigma_e \approx 1.086\, N \sigma_e D. \tag{1-351}$$

where N_c is the total column density of scattering spheres, N is the volume density of scatterers, and D is the distance to the object. The existence of interstellar absorption was first discovered by TRUMPLER (1930). He found that stars in open galactic clusters have systematically smaller apparent magnitudes (when distance is taken into account) which may be explained if the average extinction is $\Delta m/D = 0.79$ magnitudes per kiloparsec.

The wavelength dependence of interstellar extinction was first given by WHITFORD (1958). He found that distant stars are apparently reddened as opposed to nearby stars of the same type. An interstellar reddening curve is usually expressed in terms of a normalized color excess given by

$$E(\lambda) = \frac{\Delta m(\lambda) - \Delta m(\lambda_1)}{\Delta m(\lambda_1) - \Delta m(\lambda_2)} = \frac{Q_e(\lambda) - Q_e(\lambda_1)}{Q_e(\lambda_2) - Q_e(\lambda_1)}, \tag{1-352}$$

where λ is the wavelength of observation, λ_1 and λ_2 are reference wavelengths, and $\Delta m(\lambda) = m_1(\lambda) - m_2(\lambda)$ is the magnitude difference between a reddened and unreddened star. Whitford found that $E(\lambda)$ went as λ^{-1} over a large part of the optical spectrum. Observed reddening curves and the reddening correction to obtain actual magnitudes from observed magnitudes are given in Chap. 5.

1.41. Mie Scattering from a Homogeneous Sphere of Arbitrary Size

The formulae for scattering from a sphere of radius a, have been derived from Maxwell's equations by MIE (1908) and DEBYE (1909). When the plane of observation is defined as that plane that contains both the direction of propagation of the incident wave and the direction of observation, the intensity of the scattered light may be resolved into two components, $I_\parallel$ and $I_\perp$, which are parallel and perpendicular to this plane. For unit intensity of unpolarized incident radiation, Mie's results for the scattered intensities are (BORN and WOLF, 1970)

$$I_\perp = \frac{\lambda^2}{4\pi^2 R^2} |S_1|^2 \sin^2\varphi\,,$$

and (1-353)

$$I_\| = \frac{\lambda^2}{4\pi^2 R^2} |S_2|^2 \cos^2\varphi\,,$$

where R is the distance from the sphere, and

$$S_1 = \sum_{n=1}^{\infty} \frac{2n+1}{n(n+1)} \{a_n \pi_n(\cos\theta) + b_n \tau_n(\cos\theta)\}\,,$$

$$S_2 = \sum_{n=1}^{\infty} \frac{2n+1}{n(n+1)} \{a_n \tau_n(\cos\theta) + b_n \pi_n(\cos\theta)\}\,,$$

$$a_n = \frac{\psi_n(\alpha)\psi_n'(\beta) - m\psi_n(\beta)\psi_n'(\alpha)}{\zeta_n(\alpha)\psi_n'(\beta) - m\psi_n(\beta)\zeta_n'(\alpha)}\,,$$

$$b_n = \frac{m\psi_n(\alpha)\psi_n'(\beta) - \psi_n(\beta)\psi_n'(\alpha)}{m\zeta_n(\alpha)\psi_n'(\beta) - \psi_n(\beta)\zeta_n'(\alpha)}\,,$$

$$\alpha = 2\pi m_2 a/\lambda_0\,,$$

$$\beta = 2\pi m_1 a/\lambda_0\,,$$

λ_0 is the wavelength in vacuum, m_2 and m_1 are, respectively, the complex indices of refraction in the sphere and the surrounding medium, $m = m_1/m_2$, and the Riccati-Bessel functions are related to the Bessel, J. Neumann, N, and Hankel, H, functions by

$$\psi_n(Z) = (\pi Z/2)^{1/2} J_{n+1/2}(Z)$$

$$\zeta_n(Z) = (\pi Z/2)^{1/2} H^{(1)}_{n+1/2}(Z) = \psi_n(Z) + i\chi_n(Z),$$

$$\chi_n(Z) = -(\pi Z/2)^{1/2} N_{n+1/2}(Z)\,,$$

and

$$\pi_n(\cos\theta) = \frac{P_n^{(1)}(\cos\theta)}{\sin\theta}\,,$$

$$\tau_n(\cos\theta) = \frac{d}{d\theta} P_n^{(1)}(\cos\theta)\,,$$

where $P_n^{(1)}(\cos\theta)$ is the associated Legendre function of the first kind, θ is the angle between the direction of propagation of the incident wave and the direction of observation, and φ is the other angle of the spherical coordinate system.

The scattering cross section, σ_s, and the extinction cross section, σ_e, are given by

$$\sigma_s = \left(\frac{\lambda^2}{2\pi}\right) \sum_{n=1}^{\infty} (2n+1)\{|a_n|^2 + |b_n|^2\}\,, \tag{1-354}$$

and

$$\sigma_e = \left(\frac{\lambda^2}{2\pi}\right) \sum_{n=1}^{\infty} (2n+1)\{\mathcal{R}(a_n + b_n)\}\,, \tag{1-355}$$

where λ is the wavelength in the medium outside the sphere, and $\mathscr{R}$ ( ) denotes the real part of the term in parenthesis. The absorption cross section, σ_a, is given by

$$\sigma_a = \sigma_e - \sigma_s . \tag{1-356}$$

By definition, the extinction cross section is related to the total energy absorbed and scattered per unit time, P, by

$$P = \sigma_e c U , \tag{1-357}$$

where U is the energy density of the incident radiation and cU is the incident energy per unit time per unit area. The associated extinction efficiency, Q_e, is given by $\sigma_e/(\pi a^2)$ where πa^2 is the geometrical cross section of the sphere.

When the radius a is less than $0.05\,\lambda_0$,

$$a_1 = -\frac{2}{3} i \left(\frac{m^2-1}{m^2+2}\right)\left(\frac{2\pi a}{\lambda_0}\right)^3 , \tag{1-358}$$

where the square of the relative index of refraction, m^2, is given by

$$m^2 = \left(\frac{m_2}{m_1}\right)^2 \approx \frac{\varepsilon_2}{\varepsilon_1} + i\left(\frac{4\pi\sigma}{\omega\varepsilon_1}\right) \tag{1-359}$$

for a badly conducting sphere of conductivity, σ, and dielectric constant ε_2. The medium surrounding the sphere has dielectric constant, ε_1, and the radian frequency of the radiation is ω. In this case, we obtain the formulae for Rayleigh scattering (Rayleigh, 1871) which are given in the previous section.

When the radius, a, is much larger than the wavelength, λ, the effective extinction cross section becomes

$$\sigma_e = 2\pi a^2 . \tag{1-360}$$

In this case, for small angles, θ, between the direction of propagation of the incident wave and the direction of observation, the observed intensity for unit incident intensity is given by

$$I \approx \frac{Q_e}{4}\left[\frac{J_1\left(\frac{2\pi a}{\lambda}\sin\theta\right)}{\sin\theta}\right]^2 ,$$

where J is the Bessel function, and the intensity pattern is the Fraunhofer diffraction pattern of a circular aperature, first derived by Airy (1835).

1.42. Radar Backscatter

The radar cross section, σ_R, is defined to be 4π times the ratio of the reflected power per unit solid angle in the direction of the source to the power per unit area in the incident wave. That is, the radar cross section of a target is the projected area of a perfectly conducting sphere which, if placed in the same position as the real target, would scatter the same amount of energy to the observer.

For a planet at a distance, R, with a radar cross section, σ_R, the echo power, P_R, observed with a radar transmitting a c.w. signal of power, P_T, is given by the radar equation

$$P_R = \frac{GA}{(4\pi R^2)^2}\sigma_R P_T, \tag{1-361}$$

where G is the gain of the transmitting antenna over an isotropic radiator in the direction of the planet, and A is the effective collecting area of the receiving antenna for signals arriving from the direction of the planet. The backscatter cross section, σ_b, is the radar cross section for the case when the observer and transmitter occupy the same point. It follows from the previous sections that the backscatter cross section of a free electron is given by

$$\sigma_b = \frac{3}{2}\sigma_T = 4\pi\left(\frac{e^2}{mc^2}\right)^2, \tag{1-362}$$

where the THOMSON (1903) scattering cross section $\sigma_T \approx 6.65 \times 10^{-25}$ cm^2. For Rayleigh scattering from a poorly conducting sphere of radius, a,

$$\sigma_b = 4\pi a^2\left(\frac{2\pi a}{\lambda_0}\right)^4\left(\frac{m^2-1}{m^2+2}\right)^2 \quad \text{for } a < 0.05\,\lambda_0, \tag{1-363}$$

where m is the relative index of refraction and λ_0 is the wavelength. For a planet of radius, r_0, the cross section is

$$\sigma_b = g R_0 \pi r_0^2, \tag{1-364}$$

where the reflection coefficient for normal incidence, R_0, is given by Eqs. (1-77) and (1-78), and the coefficient g is a measure of the surface roughness. For a smooth surface with a Gaussian distribution of surface slopes, for example, $g = 1 + s^2$ where s is the r.m.s. slope (HAGFORS, 1964).

For spherical targets it is often of interest to measure the backscatter cross section $\sigma_b(\varphi)$ as a function of the angle φ between the direction of illumination and an arbitrary normal to the surface. When a pulse of radiation of duration, τ, is used, it illuminates a surface annulus of linear width $c\tau/(2\tan\varphi)$ at the angle

$$\varphi = \cos^{-1}\left(1 - \frac{ct}{2r_0}\right), \tag{1-365}$$

where the range delay, t, is the time after which the pulse first strikes the surface, and r_0 is the radius of the sphere (cf. Fig. 7). Range delay mapping of planetary surfaces provides a measure of $\sigma_b(\varphi)$ which can then be compared with theory to determine the roughness of the surface. As first pointed out by RAYLEIGH (1879) the image of a scattered wavefront is not seriously affected unless the wavefront deformation is greater than $\lambda/8$, where λ is the wavelength. Radar signals will, therefore, detect surface features whose sizes are larger than $\lambda/8$. HAGFORS (1961, 1964) has assumed that the probability that the height of the true surface will depart from the mean by an amount h is proportional to $\exp[-\frac{1}{2}(h/h_0)^2]$, where h_0 is the r.m.s. height variation. The horizontal structure of the surface is then specified by an autocorrelation function, $\rho(\Delta r)$, of the height, $h(r)$, as a function of the distance, r, from any arbitrary point on the surface. Here

$$\rho(\Delta r) = \langle h(r)\,h(r+\Delta r)\rangle / h_0^2. \tag{1-366}$$

HAGFORS finds that the backscatter cross section normalized to a unit surface area is given by

$$\sigma_b(\varphi) \propto \frac{1}{\cos^4\varphi} \exp\left[\frac{-r_0^2 \tan^2\varphi}{2h_0^2}\right] \quad \text{for } \rho(\Delta r) = \exp\left(\frac{-\Delta r^2}{2r_0^2}\right) \tag{1-367}$$

and

$$\sigma_b(\varphi) \propto \frac{1}{\left[\cos^4\varphi + \left(\frac{r_0\lambda}{4\pi h_0^2}\right)^2 \sin^2\varphi\right]^{3/2}} \quad \text{for } \rho(\Delta r) = \exp\left(\frac{-\Delta r}{r_0}\right).$$

If the surface were white so that its brightness is the same in all directions, independently of the direction from which it is illuminated, then $\sigma_b(\varphi)$ is given by Lambert's law (LAMBERT, 1760)

$$\sigma_b(\varphi) \propto \cos^2(\varphi). \tag{1-368}$$

The sidereal rotational velocity, Ω_s, of a planet can be determined by measuring its Doppler effect in radar echoes. For an annulus defined by the range delay, t, the total bandwidth, B, of the echo is given by (SHAPIRO, 1967)

$$B = \frac{2\nu_0}{c}[\Omega_s + \Omega_a][ct(4r_0 - ct)]^{1/2}[1 - (\boldsymbol{\Omega}\cdot\boldsymbol{r})^2]^{1/2}, \tag{1-369}$$

where ν_0 is the transmitted frequency, Ω_a is the apparent angular rotational velocity of the planet caused by the relative motion of the radar site and the planet, r_0 is the radius of the planet, $\boldsymbol{\Omega}$ is a unit vector in the direction of the total rotation, $\Omega_s + \Omega_a$, $\boldsymbol{r}$ is a unit vector in the direction from the Earth to the planet at the time of reflection, and Ω_s is the intrinsic rotational velocity of the planet.

When radar echoes are viewed in a narrow frequency interval, $\Delta\nu$, centered about a frequency, ν, different from the transmitted frequency, ν_0, the received radiation comes from a narrow strip of surface of linear width $c\Delta\nu/(2\Omega_s\nu)$ which is parallel to the axis of rotation. The combination of both delay and Doppler measurements then define two sections on the planet's surface which are equidistant from its equator, and which have the backscatter cross section, $\sigma_b(t, \nu)$, given by (GREEN, 1968)

$$\sigma_b(t,\nu) = \sigma(t,\nu)\sec\varphi = \frac{c^2 \sec\varphi\,\sigma(\varphi)}{2\nu_0\Omega_s\cos\alpha}\left[\frac{ct}{2r_0}\left(2 - \frac{ct}{2r_0}\right) - \left(\frac{\nu c}{2\nu_0 r_0 \Omega_s \cos\alpha}\right)^2\right]^{-1/2} \tag{1-370}$$

provided that the planet is a uniform sphere of radius, r_0, and rotational velocity, Ω_s, with an angle α between its polar axis and the plane perpendicular to the line of sight. Here $\sigma(t, \nu)$ is the cross section per unit area of the surface, and $\sigma(\varphi)$ is the same cross section at the angle $\varphi = \cos^{-1}[1-(ct/2r_0)]$. The delay—Doppler mapping technique is illustrated in Fig. 7.

For a plasma with plasma frequency $\nu_p = [e^2 N_e/(\pi m)]^{1/2} \approx 8.97 \times 10^3 N_e^{1/2}$ Hz, the backscatter cross section $\sigma_b dv$ per unit volume, dv, is given by (BOOKER and GORDON, 1950; BOOKER, 1956)

$$\sigma_b dv = \frac{4\pi^3}{c^4}\left|\frac{\Delta N_e}{N_e}\right|^2 \nu_p^4 P(2kl, 2km, 2kn), \tag{1-371}$$

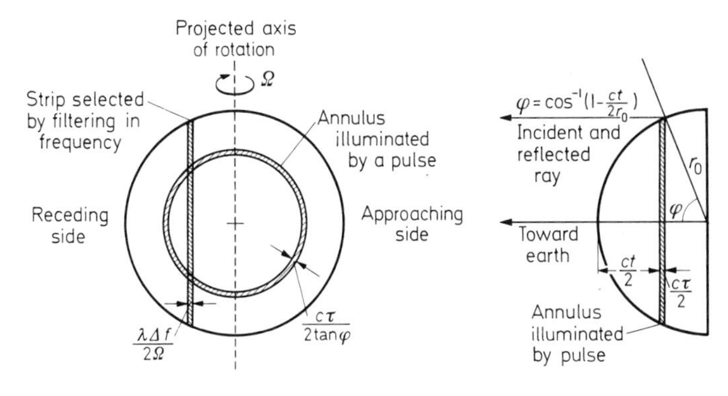

Fig. 7. The disk and near hemisphere of a planet illustrating the annulus of constant range delay, t, for a pulse of duration, τ, and the strip selected by filtering in frequency with a resolution, Δf, at a frequency determined by the wavelength, λ, and the instantaneous apparent angular velocity, Ω, of the planet. The planet radius is r_0 and the angle of incidence and reflection is $\phi = \cos^{-1}[1 - ct/(2r_0)]$. The range delay, t, is the time after the pulse first strikes the surface

where N_e is the electron density, ΔN_e is the fluctuating component of electron density, the frequency of transmission, v, is assumed to be $\gg v_p$, and the three-dimensional wave number spectrum of the density variations, $P(2kl, 2km, 2kn)$ is a function of the propagation constant $k = 2\pi v/c$, and l, m, n is a unit vector in the direction of observation. For spherically symmetric irregularities, $\Delta N_e/N_e$, of scale size, a,

$$P(2kl, 2km, 2kn) = (2\pi)^{3/2} a^3 \exp\left(-\frac{8\pi a^2}{\lambda^2}\right), \tag{1-372}$$

where λ is the wavelength of observation. For a plasma which is in thermal equilibrium, $\sigma_b dv$ is given by (FEJER, 1960)

$$\sigma_b dv = 4\pi \left(\frac{e^2}{mc^2}\right)^2 N_e \approx \frac{3}{2} \sigma_T N_e \quad \text{for } \lambda < r_D, \tag{1-373}$$

where the THOMSON (1903) scattering cross section $\sigma_T \approx 6.65 \times 10^{-25}$ cm^2, and the Debye radius $r_D = [k T_e/(4\pi e^2 N_e)]^{1/2} \approx 6.9 (T_e/N_e)^{1/2}$ cm, where T_e is the electron temperature of the plasma. For wavelengths λ larger than r_D ions begin to contribute to the backscatter cross section and we have the relations (BUNEMAN, 1962; FARLEY, 1966)

$$\sigma_b dv = 4\pi \left(\frac{e^2}{mc^2}\right)^2 N_e \left[\frac{1}{(1+\alpha^2)(2+\alpha^2)}\right] \quad \text{for } T_e = T_i, \tag{1-374}$$

and

$$\sigma_b dv \approx 4\pi \left(\frac{e^2}{mc^2}\right)^2 N_e \left[\frac{\alpha^2}{1+\alpha^2} + \frac{1}{(1+\alpha^2)(1+\alpha^2+T_e/T_i)}\right] \quad \text{for } T_e \lesssim 3\,T_i,$$

where the first and last terms in the brackets of the second expression denote, respectively, the electronic and ionic contributions, and $\alpha = 4\pi r_D/\lambda$. As shown by DOUGHERTY and FARLEY (1960), SALPETER (1960), and HAGFORS (1961), the ratio T_e/T_i may be determined by observing the frequency spectrum of the radar echo signal. The power spectrum $W(\omega)$ is given by

$$W(\omega)=\frac{\left|1+\left(\frac{\lambda}{4\pi}\right)^2\sum_i\left(\frac{1}{r_{Di}}\right)^2F_i(\omega)\right|^2\overline{|N_e^0(\omega)|^2}+\left(\frac{\lambda}{4\pi r_{De}}\right)^4|F_e(\omega)|^2\sum_i\overline{|N_i^0(\omega)|^2}}{\left|1+\left(\frac{\lambda}{4\pi}\right)^2\left\{\left(\frac{1}{r_{De}}\right)^2\cdot F_e(\omega)+\sum_i\left(\frac{1}{r_{Di}}\right)^2F_i(\omega)\right\}\right|^2}, \tag{1-375}$$

where

ω = angular radio wave frequency displacement from the transmitted frequency
λ = radio wavelength
r_{De} = electron Debye length $=[k\,T_e/(4\pi e^2 N_e)]^{1/2}$
r_{Di} = ion Debye length $=[k\,T_i/(4\pi Z^2 e^2 N_i)]^{1/2}$

$$F_e(\omega)=1-\omega\int_0^\infty\exp\left(-\frac{16\pi^2kT_e}{\lambda^2m_e}\tau^2\right)\sin(\omega\tau)\,d\tau$$

$$-i\omega\int_0^\infty\exp\left(-\frac{16\pi^2kT_e}{\lambda^2m_e}\tau^2\right)\cos(\omega\tau)\,d\tau\,,$$

$$F_i(\omega)=1-\omega\int_0^\infty\exp\left(-\frac{16\pi^2kT_i}{\lambda^2m_i}\tau^2\right)\sin(\omega\tau)\,d\tau$$

$$-i\omega\int_0^\infty\exp\left(-\frac{16\pi^2kT_i}{\lambda^2m_i}\tau^2\right)\cos(\omega\tau)\,d\tau\,.$$

The term $\overline{|N_e^0(\omega)|^2}$ is the fluctuation spectrum of the independent electrons. That is,

$$\overline{|N_e^0(\omega)|^2}=2N\int_0^\infty\exp\left(-\frac{16\pi^2kT_e}{\lambda^2m_e}\tau^2\right)\cos(\omega\tau)\,d\tau\,, \tag{1-376}$$

and $\overline{|N_i^0(\omega)|^2}$ is the corresponding spectrum for the ions obtained by replacing N_e by N_i, T_e by T_i, and m_e by m_i. Here N_e is the number density of electrons, T_e is the electron temperature, m_e is the electron mass, and with the subscript i, they represent the corresponding quantities for ions.

1.43. Phase Change and Scattering Angle due to Fluctuations in Electron Density

The electric field vector, $\boldsymbol{E}$, of an electromagnetic plane wave is given by

$$\boldsymbol{E}=E_0\exp[i(\boldsymbol{k}\cdot\boldsymbol{r}-\omega t)]\boldsymbol{\eta}_\perp\,, \tag{1-377}$$

where E_0 is the amplitude of the wave, ω is its radian frequency, $\boldsymbol{\eta}_\perp$ is a unit vector which is perpendicular to the direction of propagation, and the wave vector $\boldsymbol{k}$ is given by

$$\boldsymbol{k} = \frac{\omega}{c} n \boldsymbol{s}, \tag{1-378}$$

where n is the index of refraction of the medium and $\boldsymbol{s}$ is a unit vector in the direction of propagation. When the plane wave is propagating in an ionized gas, the index of refraction is given by Eq. (1-253)

$$n = \left[1 - \frac{4\pi N_e e^2}{m\omega^2}\right]^{1/2}, \tag{1-379}$$

where N_e is the free electron density. A small change, Δn, in the index of refraction will give rise to a small phase change, $\Delta\varphi$, of the electric field vector. From Eqs. (1-377) to (1-379), a change in electron density, ΔN_e, occurring in a turbule of size, a, will give rise to the phase change.

$$\Delta\varphi = \frac{\omega\Delta n a}{c} = \frac{e^2 a \Delta N_e}{m\nu c} = r_0 a \Delta N_e \lambda, \tag{1-380}$$

where $r_0 = e^2/(mc^2) \approx 2.8 \times 10^{-13}$ cm is the classical electron radius, and ν and λ are, respectively, the frequency and wavelength of the plane wave.

Let the spatial distribution of the gas turbules be Gaussian, and describe an isotropic Gaussian phase autocorrelation function, $\rho(r)$, given by

$$\rho(r) = \langle\varphi(x)\varphi(x+r)\rangle/\varphi_0^2 = \exp[-r^2/(2a^2)], \tag{1-381}$$

where $\langle\ \rangle$ denotes a spatial average, $\varphi(r)$ describes the phase distribution across the wave front, and the r.m.s. value of this distribution is $\varphi_0 = \langle[\varphi(x)]^2\rangle^{1/2}$. Then from Eq. (1-380), a thickness, L, of a gas with turbules distributed according to Eq. (1-381) will cause an r.m.s. phase shift, φ_0, given by

$$\begin{aligned}\varphi_0 &= (2\pi)^{1/4}\left(\frac{L}{a}\right)^{1/2}\langle\Delta\varphi^2\rangle^{1/2} = (2\pi)^{1/4}\frac{e^2}{mc\nu}(La)^{1/2}\langle\Delta N_e^2\rangle^{1/2}\\ &\approx 10^{-2}\frac{(La)^{1/2}}{\nu}\langle\Delta N_e^2\rangle^{1/2}\ \text{radians}.\end{aligned} \tag{1-382}$$

When studying the ionosphere, φ_0 is often expressed in the alternate form

$$\varphi_0 \approx \pi(La)^{1/2}\nu_p^2\frac{\lambda}{c^2}\left\langle\left(\frac{\Delta N_e}{N_e}\right)^2\right\rangle^{1/2}, \tag{1-383}$$

where the plasma frequency, ν_p, is defined by the condition that the index of refraction given by Eq. (1-379) becomes zero

$$\nu_p = \left(\frac{e^2 N_e}{\pi m}\right)^{1/2} \approx 8.97 \times 10^3 N_e^{1/2}\ \text{Hz}. \tag{1-384}$$

When a ray of the plane wave radiation has traversed a thickness, L, of a gas, the r.m.s. value, θ_{scat}, of its angular deviation, θ, from a straight line path is (Chandrasekar, 1952; Fejer, 1953)

$$\theta_{\text{scat}} \approx (2\pi)^{-3/4} \left(\frac{L}{a}\right)^{1/2} \langle \Delta n^2 \rangle^{1/2} \approx (2\pi)^{-3/4} \left(\frac{L}{a}\right)^{1/2} \frac{c}{\omega a} \langle \Delta \varphi^2 \rangle^{1/2}$$
$$\approx 10^8 \left(\frac{L}{a}\right)^{1/2} \frac{\langle \Delta N_e^2 \rangle^{1/2}}{\nu^2} \quad \text{radians}, \tag{1-385}$$

where the index of refraction change, Δn, and the phase change, $\Delta \varphi$, are given by Eq. (1-380), and it has been assumed that the turbules are distributed according to Eq. (1-381). The parameter θ_{scat} determines the minimum detectable size of an astronomical source. That is, even if a source is smaller than θ_{scat}, it will appear to have the angular size θ_{scat}. Values of θ_{scat} are given in Table 6 together with other scintillation parameters for the atmosphere, the ionosphere, and the interplanetary and interstellar medium.

Table 6. Representative values of scintillation parameters for the atmosphere, the ionosphere, the interplanetary medium, and the interstellar medium[1]. The rms scattering angle is θ_{scat}, the scale size is a, the wind velocity is v, the decorrelation time scale of the intensity fluctuations is τ_ν, the r.m.s. electron density is $\langle N_e^2 \rangle^{1/2}$, the r.m.s. fluctuating electron density is $\langle \Delta N_e^2 \rangle^{1/2}$, the thickness of the scintillating screen is L, and the effective screen distance is D

Medium	θ_{scat}, (seconds of arc)	a (cm)	v (km/sec)	τ_ν (sec)	$\frac{\langle \Delta N_e^2 \rangle^{1/2}}{\langle N_e^2 \rangle^{1/2}}$	$\langle N_e^2 \rangle^{1/2}$ (cm^{-3})	L (cm)	D (cm)
Atmosphere (at 4000 Å)	0.75	10	10^{-4}	1	10^{-4}	10^{19}	10^4	10^5
Ionosphere (at 45 MHz)	5	10^4–10^5	0.1–0.3	30	10^{-3}	10^5	5×10^6	3×10^7
Interplanetary medium (at 200 MHz)	10^{-3}–10^4	10^7 *	250–700	0.2	10^{-2}	10–10^8	10^{13}	10^{13}
Interstellar medium (at 318 MHz)	10^{-3}	10^{11}	30–200	10^3	10^{-2}	0.03	10^{22}	10^{22}

[1] Values are for the frequencies given in parenthesis and the L and D which are tabulated. The references are Chandrasekhar (1952) for the atmosphere, Hewish (1951) and Aarons, Whitney, and Allen (1971) for the ionosphere, Dennison and Hewish (1967) and Cohen, Gundermann, Hardebeck, and Sharp (1967) for the interplanetary medium, and Lang and Rickett (1970), Rickett (1970) and Lang (1971) for the interstellar medium.

* Spacecraft observations indicate a characteristic scale or correlation length of 10^{11} cm for the turbulence of the interplanetary medium.

1.44. The Scintillation Pattern

Booker, Ratcliffe, and Shinn (1950) first showed that the autocorrelation function of a wave disturbance over any plane parallel to a thin phase changing screen is the same as that over the wave front just after it has emerged from the screen. The observable statistical properties of the variations in wave intensity across the Earth are therefore directly related to the spatial variations of the wave phase across the wave front at the screen. In particular, the spatial autocorrelation function, $\varphi_0^2 \rho(r)$, of the phase, $\varphi(x)$, across the wave front may be directly related to the spatial autocorrelation function, $M(r)$, of the fluctuation $\Delta I(x) = I(x) - \langle I(x) \rangle$ of the wave intensity, $I(x)$, at the Earth. Here

$$M(r) = \langle \Delta I(x) \Delta I(x+r) \rangle , \tag{1-386}$$

where $\langle \ \rangle$ denotes a spatial average. The $M(r)$ is usually specified by computing its Fourier transform, $M(q)$, given by

$$M(q) = \int_{-\infty}^{+\infty} M(r) \exp(-irq) dr . \tag{1-387}$$

The $M(q)$ is called the power spectrum of the intensity fluctuations because it is equal to the square of the Fourier transform of $\Delta I(x)$. That is,

$$M(q) = \left| \int_{-\infty}^{+\infty} \Delta I(x) \exp(-ixq) dx \right|^2 . \tag{1-388}$$

When a source is not a point source, the intensity fluctuations due to different parts of the source will overlap and the net result will be to smear out the intensity fluctuations. That is, intensity fluctuations due to a point source will be deeper than those due to an extended source, and very extended sources will not be observed to scintillate. This condition is illustrated by the formula for the power spectrum, $M_s(q)$, of an extended source (SALPETER, 1967)

$$M_s(q) = M(q) |V(qD)|^2 , \tag{1-389}$$

where D is the distance to the scintillating screen, $M(q)$ is the power spectrum for a point source, and the visibility function, $V(qD)$, is the Fourier transform of the source brightness distribution.

Assuming that the intensity fluctuations in time, $I(t)$, observed by a fixed antenna are caused by the motion of a thin phase changing screen across the observers line of sight, the autocorrelation function, $M(\tau)$, of the intensity fluctuations, $I(t)$, is given by

$$M(\tau) = M\left(\frac{r}{v}\right) = \frac{\langle [I(t) - \langle I(t) \rangle][I(t+\tau) - \langle I(t+\tau) \rangle] \rangle}{\sigma^2} , \tag{1-390}$$

where the screen moves with the velocity, v, perpendicular to the line of sight to the source, the angular brackets denote a time average, and the mean square deviation of $I(t)$ is σ^2.

If measurements of intensity fluctuations are made with an antenna pointing ON a source and then OFF of it, the depth of modulation of the intensity fluctuations may be measured. This modulation or scintillation index is given by (COHEN *et al.*, 1967)

$$m = \{[(\sigma_{ON})^2 - (\sigma_{OFF})^2] / [\langle I_{ON}(t) \rangle - \langle I_{OFF}(t) \rangle]^2\}^{1/2} , \tag{1-391}$$

where $I(t)$ is the observed intensity variation in time, σ^2 denotes the mean square value of the intensity, and the angular brackets denote a time average. As mentioned before, an extended source decreases the modulation index below its value for a point source. When a source is broader than the critical angle, ψ_c, the modulation depth is considerably reduced (SALPETER, 1967)

$$\begin{aligned} \psi_c &\approx a/D \quad \text{for small } m , \\ \psi_c &\approx a/(\varphi_0 D) \quad \text{for } m \approx 1 . \end{aligned} \tag{1-392}$$

Another statistical parameter which may be measured is the probability distribution, $P(I)$, of the intensity. If distributions are measured while pointing ON and OFF a source, they are related by the convolution equation

$$P_{ON}(x) = \int_{-\infty}^{+\infty} P(I) P_{OFF}(x-I)\, dI\,, \tag{1-393}$$

where $P(I)$ is the probability distribution due to the scintillating screen.

A decorrelation time, τ_ν, may be defined as the equivalent width of the autocorrelation function, $M(t)$, given by Eq. (1-390)

$$\tau_\nu = \frac{\int_{-\infty}^{+\infty} M(t)\, dt}{M(t=0)} \approx \frac{a}{\varphi_0 v} \approx \frac{c}{\nu v \theta_{scat}} \approx 10^3 \left(\frac{a}{L}\right)^{1/2} \frac{\nu}{v \langle \Delta N_e^2 \rangle^{1/2}} \text{ sec}, \tag{1-394}$$

where v is the velocity at which the scattering turbules are moving transverse to the line-of-sight. The decorrelation time is roughly the quarter period of the intensity fluctuations, and representative values are given in Table 6. A decorrelation frequency, f_ν, may be similarly defined as the equivalent width of a frequency correlation function. If two radiometers, 1 and 2, record intensities $I_1(t)$ and $I_2(t)$, respectively, at two different frequencies, then the cross-correlation function, $\Gamma_{12}(\tau)$, is given by

$$\Gamma_{12}(\tau) = \frac{1}{\sigma_1 \sigma_2} \langle [I_1(t) - \langle I_1(t) \rangle][I_2(t+\tau) - \langle I_2(t+\tau) \rangle] \rangle\,, \tag{1-395}$$

where the angular brackets denote a time average and σ_1^2 and σ_2^2 are the mean square deviations of $I_1(t)$ and $I_2(t)$, respectively. The f_ν may then be defined as the equivalent width of a plot of $\Gamma_{12}(0)$ versus frequency separation of the radiometers. For strong phase fluctuations where geometric optics applies, the path difference between a direct and scattered ray is $D\theta_{scat}^2/4$ if the screen distance, D, is half the distance to the source. Consequently, the frequency difference, f_ν, for which the two waves interfere is given by

$$f_\nu \approx \frac{4c}{D\theta_{scat}^2} \approx \frac{10^{-5} a \nu^4}{\langle \Delta N_e^2 \rangle D^2} \text{ Hz} \quad \begin{array}{l}\text{for strong scintillations}\\ \text{in an extended medium.}\end{array} \tag{1-396}$$

A pulse of radiation will be broadened in time by f_ν^{-1}. That is, the time profile of a scattered pulse will be the convolution of the emitted pulse with an exponential function whose $1/e$ decay time is f_ν^{-1}. Theoretical formulae for f_ν in different conditions of scattering have been derived together with formulae for $M(q)$, m, and $P(I)$ by Hewish (1951), Fejer (1953), Mercier (1962), Salpeter (1967), Jokipii and Hollweg (1970), and Lovelace *et al.* (1971). A few of these formulae are given in Table 7.

If a stable diffraction pattern is moving with a velocity, v, at an angle θ to the projected baseline, B, between two antennas, then an upper limit to the velocity is given by

$$v = B|\cos\theta|/T \leq B/T\,, \tag{1-397}$$

Table 7. Formulae for the power spectrum, $M(q)$, the modulation index, m, the probability distribution of intensity, $P(I)$, and the decorrelation frequency, f_v, for three regions of scintillation. Here a thin phase scattering screen is assumed to be located at a distance, D, to have turbules of scale size, a, and to cause an r.m.s. phase fluctuation, ϕ_0. For a Gaussian screen, $\phi^2(q) = (2\pi)^{-1/2} a \phi_0^2 \exp(-q^2 a^2/2)$

Region	$M(q)$	m	$P(I)$	f_v
$\phi_0 \ll 1$ and $D \ll 2\pi a^2/\lambda$ or $\phi_0 > 1$ and $D \ll 2\pi a^2/(\lambda \phi_0)$	$4\phi^2(q) \sin^2\left(\frac{q^2 D \lambda}{4\pi}\right)$ $\approx \left(\frac{q^2 D \lambda}{2\pi}\right)^2 \phi^2(q)$	$\frac{\phi_0 D \lambda}{2\pi a^2} \propto \lambda^2$	$\propto \exp\left[\frac{-I^2}{2m^2}\right]$	$\to v$
$\phi_0 \ll 1$ and $D \gg 2\pi a^2/\lambda$	$4\phi^2(q) \sin^2\left(\frac{q^2 D \lambda}{4\pi}\right)$ $\approx 2\phi^2(q)$	$\sqrt{2}\,\phi_0 \propto \lambda$	$\propto \exp\left[\frac{-I^2}{2m^2}\right]$	$\frac{2\pi a^2 c}{\lambda^2 D} \propto \lambda^{-2}$
$\phi_0 \gg 1$ and $D \gg 2\pi a^2/(\lambda \phi_0)$	$\exp\left[-\left(\frac{qa}{2\phi_0}\right)^2\right]$	1	$\propto \exp[-I]$	$\frac{2\pi a^2 c}{\phi_0^2 \lambda^2 D} \propto \lambda^{-4}$

where T is the time displacement between the arrival of the pattern at the two antennas. If the intensity fluctuations seen at the two antennas are highly correlated, then we have the lower limit (LANG and RICKETT, 1970)

$$v \geq B[T^2 + \tau_v^2]^{-1/2}, \tag{1-398}$$

where τ_v is the decorrelation time. If the stable diffraction pattern is observed simultaneously with three antennas, then a unique velocity may be specified (cf. DENNISON and HEWISH, 1967). If, however, the scintillating structure rearranges itself as it moves, this apparent velocity overestimates the true drift velocity (cf. BRIGGS, PHILLIPS, and SHINN, 1950). Finally, LOVELACE *et al.* (1971) have shown that under certain circumstances the Bessel transform of the time autocorrelation function measured by a single antenna may give the drift velocity, v. Values of v for the atmosphere, ionosphere, and the interplanetary and interstellar medium were giben in Table 6.

2. Monochromatic (Line) Radiation

"They (atoms) move in the void and catching each other up jostle together, and some recoil in any direction that may chance, and others become entangled with one another in various degrees according to the symmetry of their shapes and sizes and position and order, and they remain together and thus the coming into being of composit things is effected."

SIMPLICUS (6th Century A. D.)

"I write about molecules with great diffidence, having not yet rid myself of the tradition that atoms are physics, but molecules are chemistry, but the new conclusion that hydrogen is abundant seems to make it likely that the above-mentioned elements H, O, and N will frequently form molecules."

Sir Arthur EDDINGTON, 1937

"Modern improvements in optical methods lend additional interest to an examination of the causes which interfere with the absolute homogeneity of spectrum lines. So far as we know these may be considered under five heads, and it appears probable that the list is exhaustive."

(I) The translatory motion of the radiating particles in the line of sight, operating in accordance with Doppler's principle.
(II) A possible effect of the rotation of the particles.
(III) Disturbance depending on collisions with other particles either of the same or another kind.
(IV) Gradual dying down of the luminous vibrations as energy is radiated away.
(V) Complications arising from the multiplicity of sources in the line of sight. Thus if the light of a flame be observed through a similar one, the increase of illumination near the centre of the spectrum line is not so great as towards the edges and the line is effectively widened."

Lord RAYLEIGH, 1915

2.1. Parameters of the Atom

Classical electron radius. Consider the orbital motion of an electron of mass, m, and charge, e, about a nucleus of charge Ze. Equating the Coulomb force of attraction to the force from centripetal acceleration we obtain

$$\frac{Ze^2}{r^2} = \frac{mv^2}{r}, \tag{2-1}$$

where r is the radius of the orbit, and v is the velocity of the electron. Solving for r,

$$r = r_0\left(\frac{c}{v}\right)^2 Z, \tag{2-2}$$

where the classical electron radius

$$r_0 = e^2/(mc^2) \approx 2.818 \times 10^{-13}\,\text{cm}. \tag{2-3}$$

Radius of the first Bohr orbit. Following BOHR (1913) we may assume that the angular momentum of the electron is quantized so that

$$m v r = \frac{h n}{2\pi}, \tag{2-4}$$

where $h \approx 6.625 \times 10^{-27}$ erg sec is Planck's constant and n is an integer. Using Eqs. (2-1) and (2-4) we obtain the radius r_n of the nth Bohr orbit.

$$r_n = a_0 \frac{n^2}{Z}, \tag{2-5}$$

where the radius of the first Bohr orbit of hydrogen ($Z=1$) is

$$a_0 = \frac{h^2}{4\pi^2 m e^2} \approx 0.529 \times 10^{-8}\,\text{cm}. \tag{2-6}$$

Line frequency. For atomic or molecular radiation resulting from the transition between two levels of energy, E_m and E_n, the frequency of radiation, ν_{mn}, is given by (PLANCK, 1910; BOHR, 1913)

$$\nu_{mn} = |E_m - E_n|/h, \tag{2-7}$$

where $|\,|$ denotes the absolute value, and h is Planck's constant.

Rydberg constant for infinite mass. The total energy, E_n, of an electron whose velocity is v and orbital radius is r, is given by

$$E_n = -\frac{Z e^2}{r} + \frac{1}{2} m v^2 = \frac{-2\pi^2 m e^4 Z^2}{h^2 n^2}, \tag{2-8}$$

where n is an integer and it has been assumed that the angular momentum is quantized according to Eq. (2-4). Using Eq. (2-8) in Eq. (2-7) the frequency corresponding to this energy is

$$\nu_{mn} = c R_\infty Z^2 \left| \frac{1}{n^2} - \frac{1}{m^2} \right|, \tag{2-9}$$

where m and n are integers, the Rydberg constant for infinite mass is (RYDBERG, 1890),

$$R_\infty = \frac{2\pi^2 m_e e^4}{c h^3} \approx 1.097 \times 10^5\,\text{cm}^{-1}, \tag{2-10}$$

and m_e is the electron mass.

Compton frequency and wavelength. The photon energy, $h\nu$, becomes equal to the electron rest mass energy, mc^2, at the Compton frequency

$$\nu_0 = m c^2/h \approx 1.23 \times 10^{20}\ \text{Hz}, \tag{2-11}$$

which corresponds to the Compton wavelength (COMPTON, 1923)

$$\lambda_0 = \frac{h}{2\pi m c} \approx 3.862 \times 10^{-11}\,\text{cm}. \tag{2-12}$$

Zeeman displacement—Larmor frequency of precession. Assume that the moving electron is a harmonic oscillator which in the absence of electromagnetic fields exhibits rectilinear oscillation given by

$$\vec{r}(t) = \vec{r}_0 \cos(2\pi v_0 t), \tag{2-13}$$

where v_0 is the frequency of oscillation, $\vec{r}_0$ is the peak linear displacement, and t is the time variable. The equation of motion of the forced vibration of the oscillator in a magnetic field of strength, $\vec{H}$, directed along the z axis is (LORENTZ, 1897)

$$m\ddot{\vec{r}} + m(2\pi v_0)^2 \vec{r} = \frac{e}{c} \dot{\vec{r}} x \vec{H}, \tag{2-14}$$

where $\dot{}$ and $\ddot{}$ denote, respectively, the first and second derivatives with respect to time, and the magnetic force term is the term on the right side of the equation. Solving Eq. (2-14) for the frequency of oscillation, v, we obtain

$$v = v_0 + o \quad \text{for } o \ll v_0, \tag{2-15}$$

where the Zeeman displacement (ZEEMAN, 1896, 1897; LORENTZ, 1897)

$$o = \frac{eH}{4\pi mc} \approx 1.4 \times 10^6 H \text{ Hz}, \tag{2-16}$$

when the magnetic field intensity, H, is in gauss. The latter constant is also called the Larmor frequency of precession because LARMOR (1897) showed that a superimposed H field leaves the motion of an electron in its orbit alone except for a uniform precession of the orbit about the direction of the lines of force, the precessional velocity being $2\pi o$.

Bohr magneton and nuclear magneton. The magnetic dipole moment of an electric current is equal to the product of the current strength, the area enclosed by the circulating current, and c^{-1}. For an electron of charge, e, and angular velocity, ω, the current strength is $e\omega/2\pi$. Multiplying this strength by the area πr^2 where r is the radius of the orbit, and using Eq. (2-4) for the quantization of angular momentum, we obtain the magnetic moment

$$M = L\mu_B, \tag{2-17}$$

where L is an integer, and the Bohr magneton, μ_B, is given by

$$\mu_B = \frac{eh}{4\pi mc} \approx 9.273 \times 10^{-21} \text{ erg gauss}^{-1}. \tag{2-18}$$

The corresponding magneton for the hydrogen nucleus is the nuclear magneton

$$\mu_K = \mu_B \left(\frac{m}{m_p}\right) \approx 5.05 \times 10^{-24} \text{ erg gauss}^{-1}, \tag{2-19}$$

where m_p is the proton mass.

From Eqs. (2-7), (2-15), and (2-16) we also see that the energy associated with the magnetic field of strength H is

$$E = \mu_B H. \tag{2-20}$$

Fine structure constant. From Eqs. (2-4) and (2-6) we obtain the ratio of the velocity of the first Bohr orbit to the velocity of light. This fine structure constant is

$$\alpha = \frac{v}{c} = \frac{2\pi e^2}{hc} = \frac{2\mu_B}{e a_0} = \frac{r_0}{\lambda_0} = \frac{\lambda_0}{a_0} \approx \frac{1}{137.037} \approx 7.3 \times 10^{-3}, \tag{2-21}$$

where μ_B is the Bohr magnetron, a_0 is the radius of the first Bohr orbit, r_0 is the classical electron radius, and λ_0 is the Compton wavelength.

SOMMERFELD (1916) showed that the total energy of an electron in an elliptical orbit is given by Eq. (2-8) unless relativity corrections are considered. In this case, the energy of level n is given by

$$E_n = \frac{-2\pi^2 m e^4}{h^2} \frac{Z^2}{n^2}\left[1 + \frac{\alpha^2 Z^2}{n}\left(\frac{1}{k} - \frac{3}{4n}\right)\right], \tag{2-22}$$

where α is the fine structure constant, the ratio of major to minor axes of the ellipse is n/k, and the integer k is called the azimuthal quantum number.

2.2. Einstein Probability Coefficients

When considering spontaneous and radiation induced transitions between atomic energy levels, the probability per unit time, P_{mn}, that an atom will undergo a transition from a high state of energy, E_m, to a lower state of energy, E_n, is (EINSTEIN, 1917)

$$P_{mn} = A_{mn} + B_{mn} U_\nu, \tag{2-23}$$

where A_{mn} is the Einstein coefficient for spontaneous transition between the two states (spontaneous emission), and B_{mn} is the Einstein stimulated emission coefficient for a transition induced by radiation of energy density $U_\nu d\nu$ in the frequency range ν to $\nu + d\nu$. The probability per unit time for a radiation induced absorption is

$$P_{nm} = B_{nm} U_\nu. \tag{2-24}$$

From Eq. (1-64), $U_\nu = E_\nu^2/4\pi$ for plane wave radiation with electric field strength, E_ν, at frequency, ν. From Eq. (1-119), the energy density of black body radiation at the temperature, T, in the frequency range ν to $\nu + d\nu$ is (PLANCK, 1901)

$$U_\nu = \frac{8\pi h \nu^3}{c^3}\left[\exp\left(\frac{h\nu}{kT}\right) - 1\right]^{-1} = \frac{4\pi}{c} B_\nu(T), \tag{2-25}$$

where $B_\nu(T)$ is the brightness of the radiator at frequency, ν, and temperature, T. Because the number of downward transitions must equal the number of upperward transitions, and because in thermodynamic equilibrium each state has a population determined by the Boltzmann distribution, Eq. (3-126), it follows that

$$\frac{g_n}{g_m} \exp[-E_n/kT] B_{nm} U_\nu = \exp[-E_m/kT](A_{mn} + B_{mn} U_\nu). \tag{2-26}$$

Choosing $h\nu = E_m - E_n$ in Eq. (2-25), it follows from Eq. (2-26) that

$$A_{mn} = \frac{8\pi h \nu^3}{c^3} B_{mn}, \tag{2-27}$$

and

$$g_m B_{mn} = g_n B_{nm}, \tag{2-28}$$

where g_m is the statistical weight of the mth level. Some authors define P_{nm} as $B_{nm} B_\nu(T)$ where $B_\nu(T)$ is given by Eq. (2-25). In this case, Eq. (2-27) becomes $A_{mn} = 2h\nu^3 B_{mn}/c^2$. Although relations (2-27) and (2-28) were determined under the assumption of thermodynamic equilibrium, the Einstein coefficients must be properties of the atom only and therefore Eqs. (2-27) and (2-28) are true regardless of the nature of the radiation.

When only spontaneous and induced transitions are considered, the total energy emitted per unit time per unit volume, dP/dv, is given by

$$dP/dv = N_m h \nu_{mn} A_{mn}, \tag{2-29}$$

where N_m is the volume density of atoms in level m and $h\nu_{mn} = E_m - E_n$.

2.3. Einstein Probability Coefficient for Spontaneous Emission from an Electric Dipole

The electric dipole moment, $d(t)$, of an electron of charge, e, which is constrained to move harmonically in the x direction is

$$d(t) = e\,x(t) = e\,x_0 \cos(2\pi \nu_0 t), \tag{2-30}$$

where ν_0 is the frequency of the oscillation and x_0 is the peak displacement of the oscillation. Using the LARMOR (1897) relation for the total energy radiated per unit time in all directions by a dipole (Eq. (1-114)), we obtain the average emitted power:

$$P = \frac{2}{3} \frac{\langle |\ddot{d}(t)|^2 \rangle}{c^3} = \frac{64\pi^4 \nu_0^4}{3c^3} \left(\frac{e\,x_0}{2}\right)^2, \tag{2-31}$$

where $\ddot{}$ denotes the second derivative with respect to time, $|\,|$ denotes the absolute value, $\langle\,\rangle$ denotes a time average, and $(e\,x_0/2)$ is the mean dipole moment.

By comparing Eqs. (2-29) and (2-31), the spontaneous emission coefficient, A_{mn}, for electric dipole radiation is obtained.

$$A_{mn} = \frac{64\pi^4 \nu_{mn}^3}{3hc^3} |\mu_{mn}|^2 \approx 1.2 \times 10^{-2} \nu_{mn}^3 |\mu_{mn}|^2 \sec^{-1}, \tag{2-32}$$

where the electric dipole matrix element, μ_{mn}, is often given by

$$\mu_{mn}^2 = S_{mn}/g_m \approx (e^2 x_0^2), \tag{2-33}$$

where S_{mn} is called the strength of the electric dipole and g_m is the statistical weight of the mth level.

For an estimate of the value of the A_{mn} of an electric dipole transition, let

$$\nu_{mn} = c R_\infty \approx 3.3 \times 10^{15} \text{ Hz} \tag{2-34}$$

be the radiation frequency of an electron in the first Bohr orbit, whose squared electric dipole matrix element is

$$|\mu_{mn}|^2 \approx e^2 a_0^2 \approx 6.46 \times 10^{-36} \text{ cm}^2 \text{ e.s.u.}^2, \tag{2-35}$$

where a_0 is the radius of the first Bohr orbit. Substituting Eqs. (2-34) and (2-35) into Eq. (2-32) we obtain

$$A_{mn} \approx 10^9 \text{ sec}^{-1}. \tag{2-36}$$

Electric dipole moments for the simple molecules of the most abundant elements are given in Table 21. The moments are given in Debye units, where one Debye $= 10^{-18}$ e.s.u. cm.

2.4. Relation of the Electric Dipole Emission Coefficient to the Classical Damping Constant and the Oscillator Strength

The kinetic energy, E, of a harmonic oscillator is given by

$$E = \frac{m[\dot{x}(t)]^2}{2} = 2\pi^2 m \nu_0^2 x_0^2 \cos^2(2\pi\nu_0 t), \tag{2-37}$$

where $x(t)$ is given by Eq. (2-30). It follows from Eq. (2-31) for the time rate of change of energy, dE/dt, and from Eq. (2-37), that

$$\frac{dE}{dt} = -\gamma_{\text{cl}} E, \tag{2-38}$$

where the classical damping constant

$$\gamma_{\text{cl}} = \frac{8\pi^2 e^2 \nu_0^2}{3mc^3} \approx 2.5 \times 10^{-22} \nu_0^2 \text{ sec}^{-1}. \tag{2-39}$$

Comparing Eq. (2-38) with Eq. (2-29), we obtain

$$A_{mn} = -3\gamma_{\text{cl}} f_{mn}, \tag{2-40}$$

where the oscillator strength, f_{mn}, is the effective number of electrons per atom. The oscillator strength has the property

$$g_m f_{mn} = -g_n f_{nm}, \tag{2-41}$$

where g_m is the statistical weight of the nth level. When m is larger than n, as we have assumed in the previous discussion, f_{mn} and f_{nm} are called, respectively, the emission and absorption oscillator strengths. The oscillator strength obeys the Reiche-Thomas-Kuhn sum rule (THOMAS, 1925; KUHN, 1925)

$$\sum_{n<m} f_{mn} + \sum_{n>m} f_{mn} = \sum_{n<m} f_{mn} - \sum_{m<n} \frac{g_n}{g_m} f_{nm} = Z, \tag{2-42}$$

for an atom with Z optical electrons. Typical values of the oscillator strength at optical frequencies are on the order of unity.

2.5. Probability Coefficient for Spontaneous Emission from a Magnetic Dipole

Let a magnetic dipole have the oscillatory magnetic dipole moment, $m(t)$, given by

$$m(t) = m_0 \cos(2\pi\, \nu_0\, t)\,, \tag{2-43}$$

where ν_0 is the harmonic frequency. As the time averaged energy radiated per unit time is given by Eq. (2-31) with $d(t)$ replaced by $m(t)$, it follows from Eq. (2-32) that the spontaneous emission coefficient, A_{mn}, for magnetic dipole radiation is

$$A_{mn} = \frac{64\pi^4\, \nu_{mn}^3}{3hc^3}\, |\mu_{mn}|^2 \approx 1.2\times 10^{-2}\, \nu_{mn}^3\, |\mu_{mn}|^2\ \mathrm{sec}^{-1}\,, \tag{2-44}$$

where the magnetic dipole matrix element, μ_{mn}, is given by

$$\mu_{mn}^2 = S_{mn}/g_m\,, \tag{2-45}$$

where S_{mn} is the strength of the magnetic dipole and g_m is the statistical weight of the level m.

For an estimate of the value of A_{mn} of a magnetic dipole transition, let

$$\nu_{mn} = c\, R_\infty \approx 3.3\times 10^{15}\ \mathrm{Hz} \tag{2-46}$$

be the radiation frequency of an electron in the first Bohr orbit, whose magnetic moment is

$$|\mu_{mn}|^2 = \mu_B^2 = \left(\frac{eh}{4\pi mc}\right)^2 \approx 8.6\times 10^{-41}\ \mathrm{erg}^2\ \mathrm{gauss}^{-2}\,, \tag{2-47}$$

where μ_B is the Bohr magneton. Substituting Eqs. (2-46) and (2-47) into Eq. (2-44), we obtain

$$A_{mn} \approx 10^4\ \mathrm{sec}^{-1}\,. \tag{2-48}$$

2.6. Probability Coefficient for Spontaneous Emission from an Electric Quadrupole

The quadrupole moment of a set of charges of magnitude, e, is (LANDAU and LIFSHITZ, 1962)

$$D_{\alpha\beta} = \sum e(3\, x_\alpha\, x_\beta - \delta_{\alpha\beta}\, r^2)\,, \tag{2-49}$$

where x_α and x_β are the x coordinates of the radius vector, r, between two charges α and β, $\delta_{\alpha\beta}$ is the Kronecker delta, and the summation is performed over all the charges e. The electric, E, and magnetic, H, field strengths of the quadrupole radiation at a far distance, R, from the set of charges are given by

$$E = H = \frac{\dddot{\vec{D}}}{6c^3 R}\, \sin\theta\,, \tag{2-50}$$

where $\dddot{}$ denotes the third derivative with respect to time, and θ is the angle between the direction of observation and the vector $\vec{D}$ with the components $D_\alpha = D_{\alpha\beta}\, n_\beta$ where $\vec{n}$ is the unit vector, $\vec{n} = \vec{R}/R$. The total energy emitted per unit time in all directions, P, is obtained by integrating the Poynting flux $cE^2/4\pi$

over all angles and noting that $D_{\alpha\beta}$ is a symmetric tensor with five independent components.

$$P = \frac{\dddot{D}^2_{\alpha\beta}}{180\,c^5}\,. \tag{2-51}$$

If we assume that $D_{\alpha\beta}$ is oscillatory in time and is given by

$$D_{\alpha\beta} = 3e\,x_0^2\cos(2\pi\,\nu_0\,t)\,, \tag{2-52}$$

where ν_0 is the oscillatory frequency, Eq. (2-51) gives a time averaged power

$$P = \frac{32\pi^6\,\nu_0^6}{5c^5}\left(\frac{e\,x_0^2}{2}\right)^2, \tag{2-53}$$

where $e\,x_0^2/2$ is the average electric quadrupole moment.

By comparing Eq. (2-29) with Eq. (2-53), the spontaneous emission coefficient, A_{mn}, for electric quadrupole radiation is obtained

$$A_{mn} = \frac{32\pi^6\,\nu_{mn}^5}{5hc^5}\left(\frac{S_{mn}}{g_m}\right) \approx 3.8\times 10^{-23}\,\nu_{mn}^5\left(\frac{S_{mn}}{g_m}\right)\ \mathrm{sec}^{-1}, \tag{2-54}$$

where S_{mn} is the strength of the electric quadrupole, and g_m is the statistical weight of the level m.

For an estimate of the value of A_{mn} of an electric quadrupole transition, let

$$\nu_{mn} = c\,R_\infty \approx 10^{15}\ \mathrm{Hz}\,, \tag{2-55}$$

be the radiation frequency of an electron in the first Bohr orbit whose electric quadrapole moment is

$$\frac{S_{mn}}{g_m} = a_0^4\,e^2 \approx 1.8\times 10^{-52}\ \mathrm{cm}^4\ \mathrm{e.s.u.}^2\,, \tag{2-56}$$

where a_0 is the radius of the first Bohr orbit. Substituting Eqs. (2-55) and (2-56) into Eq. (2-54), we obtain

$$A_{mn} \approx 10\ \mathrm{sec}^{-1}\,. \tag{2-57}$$

2.7. Radiation Transfer

The equation of radiation transfer for the intensity, $I(\nu, x)$, of radiation at the frequency, ν, and at the point, x, measured along the line of sight is (Kirchhoff, 1860; Woolley, 1947; Smerd and Westfold, 1949)

$$\frac{d}{d\tau_\nu}\left[\frac{I(\nu,x)}{n_\nu^2}\right] = S_\nu - \frac{I(\nu,x)}{n_\nu^2}\,, \tag{2-58}$$

where the source function, S_ν, is given by

$$S_\nu = \frac{\varepsilon_\nu}{\alpha_\nu\,n_\nu^2}\,, \tag{2-59}$$

and the optical depth, τ_ν, is defined by

$$d\tau_\nu = \alpha_\nu(x)dx. \tag{2-60}$$

The volume emissivity, ε_ν, is the power emitted per unit volume per unit frequency interval per unit solid angle, the absorption coefficient per unit length is α_ν, and n_ν is the index of refraction. For a plasma, n_ν will be unity as long as $\nu \gg \nu_p = 8.9 \times 10^3 N_e^{1/2}$ Hz, where ν_p is the plasma frequency and N_e is the free electron density.

For an isotropic and homogeneous plasma of thickness, L, the optical depth may be written

$$\tau_\nu(L) = \int_0^L \alpha_\nu(x)dx = \alpha_\nu L = N\sigma_\nu L, \tag{2-61}$$

where N is the volume density of absorbing atoms and σ_ν is called the absorption cross section.

Under conditions of local thermodynamic equilibrium (LTE) at temperature, T, the source function, S_ν, is given by (PLANCK, 1901; MILNE, 1930)

$$\begin{aligned} S_\nu = \frac{\varepsilon_\nu}{\alpha_\nu n_\nu^2} = B_\nu(T) &= \frac{2h\nu^3}{c^2}\left[\exp\left(\frac{h\nu}{kT}\right) - 1\right]^{-1} \\ &\approx \frac{2h\nu^3}{c^2}\exp\left[-\left(\frac{h\nu}{kT}\right)\right] \quad \text{for } \nu \gg 10^{10}\,T \\ &\approx \frac{2\nu^2 kT}{c^2} \quad \text{for } \nu \ll 10^{10}\,T, \end{aligned} \tag{2-62}$$

which is the expression for the vacuum brightness, $B_\nu(T)$, of a black body radiator at temperature, T, and frequency, ν.

The general solution to Eq. (2-58) for the intensity at some point A on the line of sight in terms of some other point B on the line of sight is

$$\frac{I(\nu, A)}{n_\nu^2(A)} = \frac{I(\nu, B)}{n_\nu^2(B)}\exp[\tau_\nu(B) - \tau_\nu(A)] - \int_{\tau_\nu(A)}^{\tau_\nu(B)} S_\nu(t)\exp[t - \tau_\nu(A)]\,dt, \tag{2-63}$$

where $\tau_\nu(A)$ is the optical depth from $\tau = 0$ to the point A and similarly for $\tau_\nu(B)$. It follows from Eqs. (2-62) and (2-63) that the emergent intensity from an isothermal radiator of thickness L is

$$\begin{aligned} I(\nu, 0) &= n_\nu^2(0)B_\nu(T)\{1 - \exp[-\tau_\nu(L)]\} \\ &\approx n_\nu^2(0)B_\nu(T) \quad \text{for } \tau_\nu(L) \gtrsim 1 \text{ (optically thick)} \\ &\approx n_\nu^2(0)B_\nu(T)\tau_\nu(L) \quad \text{for } \tau_\nu(L) < 1 \text{ (optically thin)}. \end{aligned} \tag{2-64}$$

2.8. Resonance Absorption of Line Radiation

The quantum mechanical absorption coefficient per unit length, α_ν, for the self absorption of the line radiation resulting from a transition from a high energy level m to a lower level n is (EINSTEIN, 1917)

$$\alpha_\nu = \frac{c^2 N_n g_m}{8\pi \nu^2 g_n n_\nu^2} A_{mn} \left[1 - \frac{g_n N_m}{g_m N_n}\right] \varphi_{mn}(\nu), \tag{2-65}$$

where N_n is the population of level n, A_{mn} is the Einstein coefficient for spontaneous transition from level m to level n, the index of refraction of the medium, n_ν, usually does not differ appreciably from unity, the statistical weight $g_n = 2J+1$ where J is the total angular quantum number of the level, and $\varphi_{mn}(\nu)$ is the spectral intensity distribution over the line. Detailed descriptions of the $\varphi_{mn}(\nu)$ for various processes are given in Sect. 2.18 to 2.22.

The absorption coefficient must be normalized to satisfy the criteria (LADENBURG, 1921; KRAMERS, 1924)

$$\int_0^{+\infty} \alpha_\nu d\nu = \frac{c^2 N_n g_m}{8\pi \nu^2 g_n n_\nu^2} A_{mn} \left[1 - \frac{g_n N_m}{g_m N_n}\right]. \tag{2-66}$$

It follows from Eqs. (2-40) and (2-41) that Eq. (2-66) may also be expressed in the form

$$\int_0^{\infty} \alpha_\nu d\nu = \frac{\pi e^2}{mc} f_{nm} N_n \left[1 - \frac{g_n N_m}{g_m N_n}\right], \tag{2-67}$$

where f_{nm} is the absorption oscillator strength for the $m-n$ transition. These criteria are the same as the criterion that

$$\int_0^{\infty} \varphi_{mn}(\nu) d\nu = 1. \tag{2-68}$$

For a Gaussian line profile of full width to half maximum, $\Delta\nu_{\mathrm{D}}$, this criterion means that $\varphi_{mn}(\nu_{mn}) = (\ln 2/\pi)^{1/2}(2/\Delta\nu_{\mathrm{D}})$. For a Lorentz line profile of full width to half maximum, $\Delta\nu_{\mathrm{L}}$, the $\varphi_{mn}(\nu_{mn}) = (2/\pi)\Delta\nu_{\mathrm{L}}^{-1}$. In the following we will pick $\varphi_{mn}(\nu_{mn}) = \Delta\nu_L^{-1}$, noting there must be a small correction depending on the exact line profile.

Under conditions of local thermodynamic equilibrium (LTE) at temperature, T, the populations of levels m and n are given by Boltzmann's equation (3-126). This equation specifies that $\exp[-h\nu_{mn}/(kT)] = g_n N_m/(g_m N_n)$, and that $N_n = g_n N_{\mathrm{tot}} \exp[-\chi_n/(kT)]/U$, where N_{tot} is the total density of particles, U is the partition function (cf. Chap. 3), and χ_n is the excitation energy of level n from the ground state. It follows that at the line frequency, ν_{mn}, Eq. (2-65) becomes, in LTE conditions,

$$\alpha_{\nu_{mn}}(\mathrm{LTE}) = \frac{c^2}{8\pi \nu_{mn}^2} \frac{N_m}{\Delta\nu_{\mathrm{L}}} \left[\exp\left(\frac{h\nu_{mn}}{kT}\right) - 1\right] A_{mn}. \tag{2-69}$$

At radio frequencies the factor $[1 - \exp(-h\nu_{mn}/kT)]$ becomes $h\nu_{mn}/kT$, whereas it is approximately unity at optical frequencies. Thus, for example, Eq. (2-67) becomes

$$\int_0^{\infty} \alpha_\nu d\nu \approx \frac{\pi e^2}{mc} f_{nm} N_n \quad \text{for } h\nu_{mn} \gg kT. \tag{2-70}$$

This result shows that the integrated absorption coefficient is a constant if N_n is constant, irrespective of the physical process responsible for the line broadening. Using Eqs. (2-32), (2-33), and (2-40), the self absorption coefficient, $\alpha_{\nu_{mn}}(\mathrm{LTE})$, at the line frequency, ν_{mn}, under conditions of local thermodynamic equilibrium may also be written in the forms

$$\begin{aligned}\alpha_{\nu_{mn}}(\mathrm{LTE}) &= \frac{8\pi^3 \nu_{mn}}{3hc}|\mu_{mn}|^2 \frac{N_m}{\Delta\nu_\mathrm{L}}\left[\exp\left(\frac{h\nu_{mn}}{kT}\right)-1\right] \\ &= \frac{8\pi^3 \nu_{mn}}{3hc}\left|\frac{S_{mn}}{g_m}\right| \frac{N_m}{\Delta\nu_\mathrm{L}}\left[\exp\left(\frac{h\nu_{mn}}{kT}\right)-1\right] \\ &= \frac{\pi e^2}{mc} f_{nm}\frac{N_n}{\Delta\nu_\mathrm{L}}\left[1-\exp\left(\frac{-h\nu_{mn}}{kT}\right)\right],\end{aligned} \tag{2-71}$$

where μ_{mn} is the dipole matrix element, S_{mn} is the dipole strength, g_m is the statistical weight of the mth level, and we recall that the population density, N_m, of the mth level may also be given by $N_m = g_m N_\mathrm{tot} \exp[-\chi_m/(kT)]/U$, where N_tot is the total density of particles, χ_m is the excitation energy of the mth level above the ground state, and U is the partition function.

2.9. Line Intensities under Conditions of Local Thermodynamic Equilibrium

Under conditions of local thermodynamic equilibrium, the optical depth at the center of a line coming from a homogeneous and isotropic plasma of thickness, L, is

$$\tau_{\nu_{mn}}(L) = L\alpha_{\nu_{mn}}(\mathrm{LTE}), \tag{2-72}$$

where only self absorption of the line has been considered, and $\alpha_{\nu_{mn}}(\mathrm{LTE})$ is given by Eqs. (2-69) or (2-71). Using the solution of the radiation transfer equation (2-64), together with Eqs. (2-71) and (2-72), the intensity $I(\nu_{mn}, 0)$ at the center of the line emerging from an optically thin plasma is obtained.

$$I(\nu_{mn},0) = B_{\nu_{mn}}(T)\frac{\pi e^2 f_{nm}}{mc\Delta\nu_\mathrm{L}}\left[1-\exp\left(\frac{-h\nu_{mn}}{kT}\right)\right]LN_n, \tag{2-73}$$

where the brightness of a black body at temperature, T, is $B_{\nu_{mn}}(T)$ and is given by Eqs. (2-62). Once the oscillator strength, f_{nm}, is calculated, and the intensity, $I(\nu_{mn}, 0)$, and the line width, $\Delta\nu_\mathrm{L}$, are measured, the column density, LN_n, may be obtained using Eq. (2-73).

In actual practice, in addition to the line radiation, an antenna also sees continuum radiation from the cloud itself, c, from discrete sources, s, behind the cloud, and from the general background, B, of the Galaxy and the isotropic 3 °K radiation. If the respective brightness temperatures are T_c, T_s, and T_B, and the solid angles of the two discrete sources are Ω_s and Ω_c, an antenna with efficiency, η_A, and beam solid angle, Ω_A, will observe the antenna temperature

$$T_A = \eta_A T_{0c} + \eta_A T_{0e}[1-\exp(-\tau_{\nu_{mn}})] + \eta_A[T_{0s}+T_B]\exp(-\tau_{\nu_{mn}}), \tag{2-74}$$

where

$$T_{0e} = T_{exc} \quad \text{or} \quad T_{exc}\,\Omega_c/\Omega_A \quad \text{according as } \Omega_c > \text{ or } < \Omega_A$$

$$T_{0c} = T_c \quad \text{or} \quad T_c\,\Omega_c/\Omega_A \quad \text{according as } \Omega_c > \text{ or } < \Omega_A$$

and

$$T_{0s} = T_s \quad \text{or} \quad T_s\,\Omega_s/\Omega_A \quad \text{according as } \Omega_s > \text{ or } < \Omega_A\,,$$

where T_{exc} is the excitation or spin temperature of the line radiation. When the receiver is frequency switched between ν_{mn} and another frequency different from ν_{mn}, the difference in the observed antenna temperature is

$$\begin{aligned} \Delta T_A &= \eta_A[T_{0e} - (T_{0s} + T_B)][1 - \exp(-\tau_{\nu_{mn}})] \\ &\approx \eta_A[T_{0e} - (T_{0s} + T_B)]\,\tau_{\nu_{mn}} \quad \text{for } \tau_{\nu_{mn}} \ll 1\,. \end{aligned} \tag{2-75}$$

When $T_{0e} \gg T_{0s} + T_B$, we see the line in emission and when $T_{0e} \ll T_{0s} + T_B$ we see it in absorption.

For the special case of an optically thin gas ($\tau_{mn} \ll 1$) in local thermodynamic equilibrium, and seen in emission at radio frequencies ($h\nu_{mn} \ll kT$), Eqs. (2-71), (2-72), and (2-75) give

$$\Delta T_A \Delta \nu_L = \eta_A \frac{\pi e^2 h}{mc} \frac{\nu_{mn}}{k} \frac{LN_{tot}}{U} g_n f_{nm}\,, \tag{2-76}$$

where we assume $\Omega_c > \Omega_A$. In this case, the area of the line profile leads to a direct measurement of LN_{tot}, the number of atoms in a cylinder of unit cross sectional area and length L, which are radiating at the frequency ν_{mn}.

For the special case of an optically thin gas ($\tau_{mn} \ll 1$) in local thermodynamic equilibrium, and seen in absorption at optical frequencies ($h\nu_{mn} \gg kT$), Eqs. (2-71), (2-72), and (2-75) give

$$\frac{\Delta T_A \Delta \nu_L}{T_0} = -\eta_A \frac{\pi e^2}{mc} \frac{LN_{tot}}{U} g_n f_{nm} \exp\left(\frac{-\chi_m}{kT}\right), \tag{2-77}$$

where T_0 is the temperature of the background continuum, and we assume $\Omega_s > \Omega_A$. In this case, the normalized area of the line profile (the equivalent width) leads to a measurement of LN_{tot}. Details depend on the exact form of the line broadening function. Nevertheless, comparisons of the equivalent widths of lines from the same object do lead to the relative densities of the elements in the object. Calculations of line intensities in stellar atmospheres must include the effects of both scattering and absorption, and these effects are discussed in detail in Sect. 2.16. Absorption of line and continuum radiation by photoionization, free-free transitions, and scattering is described by Eqs. (2-61) and (2-72) together with the absorption and scattering cross sections given in Sects. 1.30, 1.31, and 1.37.

2.10. Line Intensities when Conditions of Local Thermodynamic Equilibrium do not Apply

Let b_n designate the ratio of the actual population of level n to the population in conditions of local thermodynamic equilibrium (obtainable from the Saha-

Boltzmann equation (3-129)). The absorption coefficient at the frequency ν_{mn} of the $m-n$ transition is (GOLDBERG, 1966)

$$\alpha_{\nu_{mn}} = b_n \beta_{nm} \alpha_{\nu_{mn}}(\mathrm{LTE}), \tag{2-78}$$

where

$$\beta_{nm} = \frac{1-(b_m/b_n)\exp(-h\nu_{mn}/kT)}{1-\exp(-h\nu_{mn}/kT)},$$

and the absorption coefficient, $\alpha_{\nu_{mn}}(\mathrm{LTE})$, in conditions of local thermodynamic equilibrium is given by Eqs. (2-69) or (2-71). The emission coefficient is given by

$$\varepsilon_{\nu_{mn}} = b_m \alpha_{\nu_{mn}}(\mathrm{LTE})\, B_{\nu_{mn}}(T), \tag{2-79}$$

where the black body brightness, $B_{\nu_{mn}}(T)$, at temperature, T, is given by

$$B_{\nu_{mn}}(T) = \frac{2h\nu_{mn}^3}{c^2}\,\frac{1}{[\exp(h\nu_{mn}/kT)-1]}.$$

Using Eq. (2-71) for $\alpha_{\nu_{mn}}(\mathrm{LTE})$ in Eq. (2-78), we obtain

$$\alpha_{\nu_{mn}} = \frac{\pi N_n e^2 f_{nm} b_n}{mc\,\Delta\nu_{\mathrm{L}}}\left[1-\left(\frac{b_m}{b_n}\right)\exp\left(\frac{-h\nu_{nm}}{kT}\right)\right], \tag{2-80}$$

where f_{nm} is the oscillator strength of the $n-m$ transition, $\Delta\nu_{\mathrm{L}}$ is the width of the line, and N_n is the population of the nth level. This equation reduces to

$$\alpha_{\nu_{mn}} \approx \alpha_{\nu_{mn}}(\mathrm{LTE}) \qquad \text{for } \nu \gg 10^{10}\,T$$

and

$$\alpha_{\nu_{mn}} \approx N_n \frac{\pi e^2 f_{nm}}{mc\,\Delta\nu_{\mathrm{L}}}\,\frac{h\nu_{mn}}{kT}\, b_m\left\{1-\left(\frac{kT}{h\nu_{mn}}\right)\left(\frac{b_m-b_n}{b_m}\right)\right\} \qquad \text{for } \nu \ll 10^{10}\,T, \tag{2-81}$$

where

$$\frac{b_m-b_n}{b_n} \approx (m-n)\frac{d\ln b_n}{dn}.$$

The absorption coefficient is very sensitive to small changes in level populations at low frequencies. If $b_m - b_n$ exceeds $b_m h\nu_{mn}/(kT)$, the absorption coefficient becomes negative providing maser amplification of the line.

The population, N_n, of level n may be determined from the equation of statistical equilibrium

$$N_n \sum_{m=1}^{\infty} P_{nm} = N_c P_{cn} + \sum_{m=1}^{\infty} N_m P_{mn}, \tag{2-82}$$

where P_{mn} denotes the probability per unit time of a transition from level m to level n, and the subscript c denotes the continuum. A few of the possible processes which determine level populations are given below together with their probabilities.

a) Spontaneous transition between levels. The probability per unit time of a spontaneous transition from a high level m to a lower level n is A_{mn}, the Einstein coefficient for spontaneous emission, which is given in Sect. 2.12 for hydrogen-like atoms.

b) Radiation induced transitions between levels. The probability per unit time for a radiation induced transition between a high level m and a lower level n is

$$B_{mn} U_\nu \quad \text{or} \quad B_{nm} U_\nu , \tag{2-83}$$

where the B_{mn} and B_{nm} are the Einstein stimulated emission coefficients for a transition induced by radiation of energy density $U_\nu d\nu$ in the frequency range ν to $\nu + d\nu$. The interrelationships between A_{mn}, B_{mn}, and B_{nm} are given by Eqs. (2-27) and (2-28).

c) Radiation induced ionizations. It follows from Eq. (1-230) for the absorption cross section, σ_ν, for photoionization by radiation of frequency, ν, that the probability per unit time of a radiation induced ionization is

$$P_{nc} = \int_0^\infty \frac{c\, U_\nu \sigma_\nu d\nu}{h\nu} \approx 1.3 \times 10^{66} Z^4 \int_0^\infty \frac{U_\nu d\nu}{\nu^4 n^5} \quad \text{sec}^{-1} \tag{2-84}$$

for a hydrogen-like atom. Here the photon energy is $h\nu$, U_ν is the energy density of the radiation, and n is the energy level of the unionized atom.

d) Radiative recombination of ions with electrons. The probability per unit time of a radiative recombination is

$$P_{cn} = N_e \int_0^\infty \sigma_r(v, n) \left[1 + \frac{c^3 U_\nu}{8\pi h \nu^3} \right] v f(v) dv , \tag{2-85}$$

where N_e is the free electron density, U_ν is the radiation energy density at frequency, ν, the distribution of velocity, v, is $f(v)$, and $\sigma_r(v,n)$ is the cross section for the recombination of a free electron of velocity, v, with an ion to form an atom with level, n. Using the Milne (1921) relation,

$$\sigma_r(v, n) = \frac{h^2 g(n) \nu^2}{m^2 c^2 g(1) v^2} \sigma_\nu , \tag{2-86}$$

where $g(n)$ is the statistical weight of level, n, the statistical weight of the ground state of the ion is $g(1)$, and σ_ν is the cross section for photoionization from level n and is given by Eq. (1-230) for hydrogen-like atoms, for which $g(n) = 2n^2$.

e) Collision induced transitions between levels. The probability per unit time of a collision induced transition between levels m and n is

$$P_{mn} = N \gamma_{mn} = N \int_0^\infty \sigma_{mn}(v) v f(v) dv , \tag{2-87}$$

where N is the density of colliding particles, γ_{mn} is called the rate coefficient for collisional excitation, $\sigma_{mn}(v)$ is the collisional cross section for collisions with a particle of velocity, v, and $f(v)$ is the velocity distribution. For example, for a Maxwellian distribution at temperature, T, we have (Eq. (3-114))

$$f(v)dv = \left(\frac{2}{\pi}\right)^{1/2} v^2 \left(\frac{m_r}{kT}\right)^{3/2} \exp\left[-\frac{m_r v^2}{2kT}\right] dv, \tag{2-88}$$

where the reduced mass, $m_r = m_a m_p/(m_a + m_p)$ and m_a and m_p are, respectively, the masses of the atom and the colliding particle. The equation of detailed balancing for collisional excitation is

$$g_m \gamma_{mn} = g_n \gamma_{nm} \exp\left[\frac{-h\nu_{mn}}{kT}\right], \tag{2-89}$$

where g_m is the statistical weight of the mth level, and γ_{mn} is the rate coefficient for collisional deexcitation.

The collision cross section, $\sigma_{mn}(v)$, is often expressed in terms of the collision strength, $\Omega(m,n)$, by the expression (HEBB and MENZEL, 1940; SEATON, 1958)

$$\sigma_{mn}(v) = \frac{\pi}{g_m}\left(\frac{h}{2\pi m_e v}\right)^2 \Omega(m,n), \tag{2-90}$$

where $\Omega(m,n)$ is on the order of unity for most collisions, and v is the velocity of the particle before collision. Another estimate of the collision cross section is, of course, the geometric cross section, πa^2, where a is the atomic radius. For electric dipole moment transitions of oscillator strength f_{mn}, the collision strength $\Omega(m,n)$ is given by (SEATON, 1962)

$$\Omega(m,n) \approx \frac{1.6\pi}{3^{1/2}} \frac{g_n f_{nm} E_1}{h\nu_{mn}}, \tag{2-91}$$

where $h\nu_{mn}$ is the energy of the $m-n$ transition, and E_1 is the binding energy of the neutral hydrogen atom. An approximate expression for P_{nm} of hydrogen-like atoms in the case of a Maxwellian velocity distribution of free electrons is therefore (SEATON, 1962)

$$P_{nm} \approx 4\pi a_0^2 \left(\frac{8kT}{m}\right)^{1/2} \left(\frac{E_1}{kT}\right)^2 \left[\frac{\exp(-\chi_{nm})}{\chi_{nm}} - 1\right] 3 f_{nm} N_e, \tag{2-92}$$

where $\pi a_0^2 = 8.797 \times 10^{-17}\,\mathrm{cm}^2$, $\chi_{nm} = h\nu_{mn}/(kT)$, and P_{nm} and P_{mn} are related by the equation

$$n^2 \exp(\chi_n) P_{nm} = m^2 \exp(\chi_m) P_{mn}. \tag{2-93}$$

When the transitions have no electric dipole moment, $\Omega(m,n)$ is independent of velocity, and Eqs. (2-87) and (2-90) lead to

$$P_{mn} \approx \frac{N_e h^2 \Omega(m,n)}{g_m (2\pi m_e)^{3/2} (kT)^{1/2}} \approx 8.6 \times 10^{-6} T^{-1/2} N_e \frac{\Omega(m,n)}{g_m} \ \mathrm{sec}^{-1} \tag{2-94}$$

for collisions with free electrons of density, N_e, whose velocity distribution is Maxwellian. This rate is appropriate for collisional deexcitation of positive ions, and the appropriate collisional excitation rate is given by Eqs. (2-93) and (2-94). Collision strengths for the well known forbidden transitions of oxygen, nitrogen, neon, sulpher, and argon are given in Table 8 of Sect. 2.11.

f) Collisional ionization by thermal electrons and the inverse process of three body recombination.
The semi-empirical formula for the probability per unit time of a collisional induced ionization of a hydrogen-like atom is (SEATON, 1960; JEFFRIES, 1968)

$$P_{nc} \approx 7.8 \times 10^{-11} T_e^{1/2} n^3 \exp(-\chi_n) N_e \sec^{-1}, \tag{2-95}$$

where n is the principal quantum number of the energy level, χ_n is the ionization potential of the nth level in units of kT_e, and N_e is the free electron density. Here the principle of detailed balance gives

$$P_{cn} = \frac{N_e h^3}{(2\pi m k T_e)^{3/2}} n^2 \exp(\chi_n) P_{nc}, \tag{2-96}$$

where the statistical weight of the nth level is $g_n = 2n^2$.

g) Dielectronic recombination of an ion with an electron to give a doubly excited atom followed by a radiative transition to a singly excited state.
A complex ion with one or more electrons may recombine with an electron to form an atom in level i, nl by way of the intermediate doubly excited level j, nl. In this case the term

$$\sum_j \sum_l N_{j,nl} b(j, nl) A_{ji} \tag{2-97}$$

must be added to the right-hand side of the equation of statistical equilibrium (Eq. (2-82)). Here the factor $b(j,nl)$ by which the population of j,nl differs from its value in thermodynamic equilibrium, $N_{j,nl}$, is given by (BATES and DALGARNO, 1962; DAVIES and SEATON, 1969)

$$b(j, nl) = \frac{A_a}{A_a + A_r}, \tag{2-98}$$

where A_a is the probability coefficient for autoionization from level j, nl and $A_r = A_{ji}$ is the radiative probability coefficient for the transition $j,nl \rightarrow i,nl$. The equilibrium population, $N_{j,nl}$, is given by the Saha-Boltzmann equation (Eq. (3-127))

$$\begin{aligned} N_{j,nl} &= N_i N_e \frac{h^3}{(2\pi m k T)^{3/2}} \frac{g_{j,nl}}{2 g_{i,nl}} \exp\left(\frac{-h\nu_{ij}}{kT}\right) \\ &\approx 2.1 \times 10^{-16} N_i N_e T^{-3/2} \frac{g_{j,nl}}{g_{i,nl}} \exp\left(\frac{-h\nu_{ij}}{kT}\right) \text{cm}^{-3}, \end{aligned} \tag{2-99}$$

where N_i and N_e are, respectively, the ion and electron densities, $g_{i,nl}$ is the statistical weight of the level i, nl, and $h\nu_{ij}$ is the energy of the $i-j$ transition.

The level populations, N_n, or equivalently, b_n, for high temperature (5000 °K < T < 10,000 °K) conditions in ionized hydrogen regions have been calculated by SEJNOWSKI and HJELLMING (1969) and BROCKLEHURST (1970) for electron densities in the range $10\ \text{cm}^{-3} < N_e < 10^4\ \text{cm}^{-3}$. These b_n values, which include effects from processes $(a-f)$, have been used to calculate the ratio of hydrogen line temperatures to continuum temperatures (ANDREWS and HJELLMING, 1969; HJELLMING, ANDREWS, and SEJNOWSKI, 1969). In general, the ratio is enhanced beyond that of the LTE case for values of n larger than 60, and the amount of enhancement depends upon the electron density, the temperature, and the size of the region. The most detailed comparison of observed hydrogen recombination line intensities and non-LTE theory has been carried out for the Orion nebula (HJELLMING and GORDON, 1971), and these observations are consistent with theoretical non-LTE results. Observationally, the ratio, ρ, of the central line intensities of the $n+1 \rightarrow n$ transition and a nearby higher order $m+\Delta m \rightarrow m$ transition is compared with the LTE value

$$\rho_{\text{LTE}} = \frac{n^2 f_{n,n+1}}{m^2 f_{m,m+\Delta m}}, \tag{2-100}$$

where the oscillator strength $f_{m,m+\Delta m}$ is given by (MENZEL, 1969) $f_{m,m+\Delta m}/m \approx 0.194$, 0.0271, 0.00841, 0.00365, and 0.00191 for $\Delta m = 1, 2, 3, 4,$ and 5, respectively. Dielectronic recombination has been employed to explain the temperatures of the solar corona (BURGESS, 1964, 1965; BURGESS and SEATON, 1964), and to explain the anomalous intensity of the radio frequency carbon line (PALMER *et al.*, 1967; GOLDBERG and DUPREE, 1967; DUPREE, 1969).

2.11. Forbidden Lines, Recombination Spectra, the Balmer Decrement, and Planetary Nebulae

Certain very hot stars radiate sufficient energy in the ultraviolet wavelength region to ionize the surrounding gas according to the cross section given in Eq. (1-230). After photoionization, the ejected electrons lose energy by inelastic collisions followed by radiation. As a result a Maxwellian distribution of electron velocities with electron temperature, T_e, is established (BOHM and ALLER, 1947). As first suggested by BOWEN (1928), the forbidden lines observed in the nebulae are excited by electron impact. It follows from Eq. (2-94) that for a Maxwellian distribution of electrons the probability per second, a_{mn}, for collisional deactivation from level m to level n is given by

$$a_{mn} = \frac{8.63 \times 10^{-6}}{g_m T_e^{1/2}} N_e \Omega(n,m)\ \text{sec}^{-1} \quad \text{for } E_n < E_m, \tag{2-101}$$

where g_m is the statistical weight of level m, the electron density is N_e, and the collision strength $\Omega(n,m)$ has been assumed to be independent of energy, as it is for most positive ions. For collisional excitation, the probability per second, b_{mn}, is given by

$$b_{nm} = \frac{g_m}{g_n} a_{mn} \exp\left[-\frac{(E_m - E_n)}{kT_e}\right] \quad \text{for } E_n < E_m . \tag{2-102}$$

Most of the observed forbidden lines are due to transitions within the ground configurations of atoms or ions with order $2p^q$ or $3p^q$ electrons where $q=2,3,4$. These configurations have spectral terms denoted by the numbers 1, 2, and 3 in order of increasing excitation energy.

number	$q=2$ and 4	$q=3$
3	^1S	^2P
2	^1D	^2D
1	^3P	^4S

Values of the collision strengths $\Omega(1,2)$, $\Omega(1,3)$, and $\Omega(2,3)$ for some frequently observed forbidden transitions are given in Table 8. A more complete set of collision strengths is given by CZYZAK *et al.* (1968).

The intensity, I_{mn}, of the $m \to n$ transition is given by

$$I_{mn} = N_m A_{mn} h \nu_{mn} , \tag{2-103}$$

where N_m is the population of level m, the spontaneous transition probability is A_{mn}, and ν_{mn} is the frequency of the $m \to n$ transition. The population N_m is determined by the equation of statistical equilibrium (Eq. (2-82)) with both collision induced and spontaneous transitions being taken into account. The populations N_2 and N_3 are related, for example, by the equation

$$\frac{N_2}{N_3} = \frac{b_{12}(A_{31} + a_{31}) + (b_{12} + b_{13})(A_{32} + a_{32})}{b_{12}(b_{12} + b_{13}) + b_{13}(A_{21} + a_{21})} . \tag{2-104}$$

Spontaneous transition probabilities, A_{mn}, and wavelengths, λ_{mn}, for frequently observed forbidden lines are also given in Table 8. A more complete set of wavelengths and transition probabilities is given by GARSTANG (1968). It follows from Eqs. (2-101) to (2-104) that the ratio of observed line intensities leads to a relation between electron density, N_e, and electron temperature, T_e. Because typical planetary nebulae have $N_e \approx 10^4\,\text{cm}^{-3}$ and $T_e \approx 10^4\,°\text{K}$, it is customary to use the variables x and t defined by the relation

$$x = \frac{0.01\,N_e}{T_e^{1/2}} = \frac{10^{-4}\,N_e}{t^{1/2}} , \tag{2-105}$$

so that $x \approx t \approx 1$ for a typical nebula. Using Eqs. (2-101) to (2-105) together with the constants given in Table 8, the following relations are obtained (SEATON, 1960; ALLER and CZYZAK, 1968)

Table 8. Wavelengths, λ_{ij}, transition probabilities, A_{ij}, and collision strengths, $\Omega(i,j)$, for the forbidden transitions of the most abundant elements[1]

Element	λ_{21} (Å)	A_{21} (sec^{-1})	$\Omega(1,2)$	λ_{31} (Å)	A_{31} (sec^{-1})	$\Omega(1,3)$	λ_{32} (Å)	A_{32} (sec^{-1})	$\Omega(3,2)$
O II	3,728.8	4.8×10^{-5}	1.43	2,470.4	0.060	0.428	7,319.4	0.115	1.70
	+3,726.0	$+1.70\times10^{-4}$		+2,470.3	+0.0238		+7,330.7	+0.061	
							+7,318.6	+0.061	
							+7,329.9	+0.100	
O III	5,006.8 (N_1)	0.021	2.39	2,321.1	0.23	0.335	4,363.2	1.60	0.310
	+4,958.9 (N_2)	+0.0071							
N II	6,583.4	0.003	3.14	3,063.0	0.034	0.342	5,754.6	1.08	0.376
	+6,548.1	+0.00103							
Ne III	3,868.8	0.17	1.27	1,814.8	2.2	0.164	3,342.5	2.8	0.188
	+3,967.5	+0.052							
Ne IV	2,441.3	5.9×10^{-4}	1.04	1,608.8	1.33	0.427	4,714.3	0.40	1.42
	+2,438.6	$+5.6\times10^{-3}$		+1,609.0	+0.53		+4,724.2	+0.44	
							+4,715.6	+0.11	
							+4,725.6	+0.39	
Ne V	3,425.9	0.38	1.38	1,575.2	4.2	0.218	2,972	2.60	0.185
	+3,345.8	+0.138							
S II	6,716.4	4.7×10^{-5}	3.07	4,068.6	0.34	1.28	10,320.6	0.21	6.22
	+6,730.8	$+3.0\times10^{-4}$		+4,076.4	+0.134		+10,287.1	+0.17	
							+10,372.6	+0.087	
							+10,338.8	+0.20	
S III	9,532.1	0.064	4.97	3,721.7	0.85	1.07	6,312.1	2.54	0.961
	+9,069.4	+0.025		+3,796.7	+0.016				
Ar III	7,135.8	0.32	4.75	3,109.0	4.0	0.724	5,191.8	3.1	0.665
	+7,751.0	+0.083		+3,005.1	+0.043				
Ar IV	4,740.2	0.028	1.43	2,854.8	2.55	0.645	7,237.3	0.67	4.92
	+4,711.3	0.0022		+2,869.1	+0.97		+7,170.6	+0.91	
							+7,332.0	+0.122	
							+7,262.8	+0.68	
Ar V	7,005.7	0.51	1.19	2,691.4	6.8	0.141	4,625.5	3.78	0.945
	+6,435.1	+0.22		+2,784.4	+0.081				

[1] After GARSTANG (1968) and CZYZAK *et al.* (1968) by permission of the International Astronomical Union.

For [OIII]:

$$\frac{I(\lambda 4959)+I(\lambda 5007)}{I(\lambda 4363)}=\frac{7.15}{1+0.028x}D\left(\frac{14300}{T_e}\right).$$

For [OII]:

$$\frac{I(\lambda 3729)+I(\lambda 3726)}{I(\lambda 7320)+I(\lambda 7330)}=\frac{5.5}{\varepsilon}\left[\frac{1+0.36\varepsilon+5.3x(1+0.82\varepsilon)}{1+13.8x(1+0.38\varepsilon)+38.4x^2(1+0.78\varepsilon)}\right],$$

$$\frac{I(\lambda 3729)}{I(\lambda 3726)}=1.5\left[\frac{1+0.33\varepsilon+2.30x(1+0.75\varepsilon)}{1+0.40\varepsilon+9.9x(1+0.84\varepsilon)}\right]. \tag{2-106}$$

For [NII]:

$$\frac{I(\lambda 6548)+I(\lambda 6584)}{I(\lambda 5755)}=\frac{8.5}{1+0.29x}D\left(\frac{10800}{T_e}\right),$$

and for [SII]

$$\frac{I(\lambda 4068)+I(\lambda 4076)}{I(\lambda 6716)+I(\lambda 6730)}=0.164\left\{3.8+x\left[1+1.32D\left(-\frac{6000}{T_e}\right)\right]\right\}D\left(-\frac{6000}{T_e}\right),$$

where

$$D(y)=10^y$$

and

$$\varepsilon=\exp\left(-\frac{1.96}{t}\right).$$

These relations are shown in Fig. 8 for [OII], [OIII], [NII], [NeIII], [NeV] and [SII]. Similar relations relating the abundances of ions in various stages of ionization are given by ALLER and CZYZAK (1968). Measured values of N_e and T_e are given together with other parameters for well known planetary nebulae in Table 9.

The electron temperature, T_e, may be approximated by equating the thermal energy, kT_e, to the excitation energy of approximately 2 eV for the forbidden lines to obtain $T_e \approx 10^4$ °K. A more exact method is to measure the relative intensities of the allowed transitions of the emission lines of hydrogen and helium. It follows from Eqs. (2-82), and (2-101) to (2-103) that the observed intensity, I_{mn}, of the $m \to n$ transition is given by

$$I_{mn}=\varepsilon h\nu_{mn}N_e N_+\alpha_{mn}\frac{R}{3\theta^2}\ \text{erg cm}^{-2}\,\text{sec}^{-1}\,\text{rad}^{-2}, \tag{2-107}$$

where the filling factor, ε, denotes the fraction of the observed volume of the nebula which is occupied by line emitting atoms or ions, N_e is the electron density, N_+ is the atom or ion density, ν_{mn} is the frequency of the $m \to n$ transition, R and θ denote, respectively, the radius and angular extent of the nebula, and α_{mn} is the effective recombination coefficient.

The recombination coefficient may be calculated using the equation of statistical equilibrium, Eq. (2-82), together with the recombination cross section $\sigma_r(\nu,n)$ given by MILNE (1921), Eq. (1-233), or the photoionization cross section, $\sigma_a(\nu,n)$ given by KRAMERS (1923), MENZEL and PEKERIS (1935), and BURGESS (1958), Eq. (1-230), and the radiative transition probability, A_{mn}. The effective recombination coefficient, $\alpha_{mn}(T_e)$, for hydrogen-like atoms is given by (BAKER and MENZEL, 1938)

$$\alpha_{mn}(T_e)=\frac{10.394\times 10^{-14}g_{bb}(m,n)}{mn(m^2-n^2)}Z\lambda^{3/2}b_m\exp(x_m)\ \ \text{cm}^3\,\text{sec}^{-1}, \tag{2-108}$$

where the bound-bound Kramers-Gaunt factor, $g_{bb}(m,n)$, and the equilibrium departure coefficients, b_m, are tabulated by BAKER and MENZEL (1938) and GREEN, RUSH, and CHANDLER (1957), the nuclear charge of the atom is Z,

$$\lambda=\frac{Z^2 I_H}{kT_e}=157{,}890\frac{Z^2}{T_e},$$

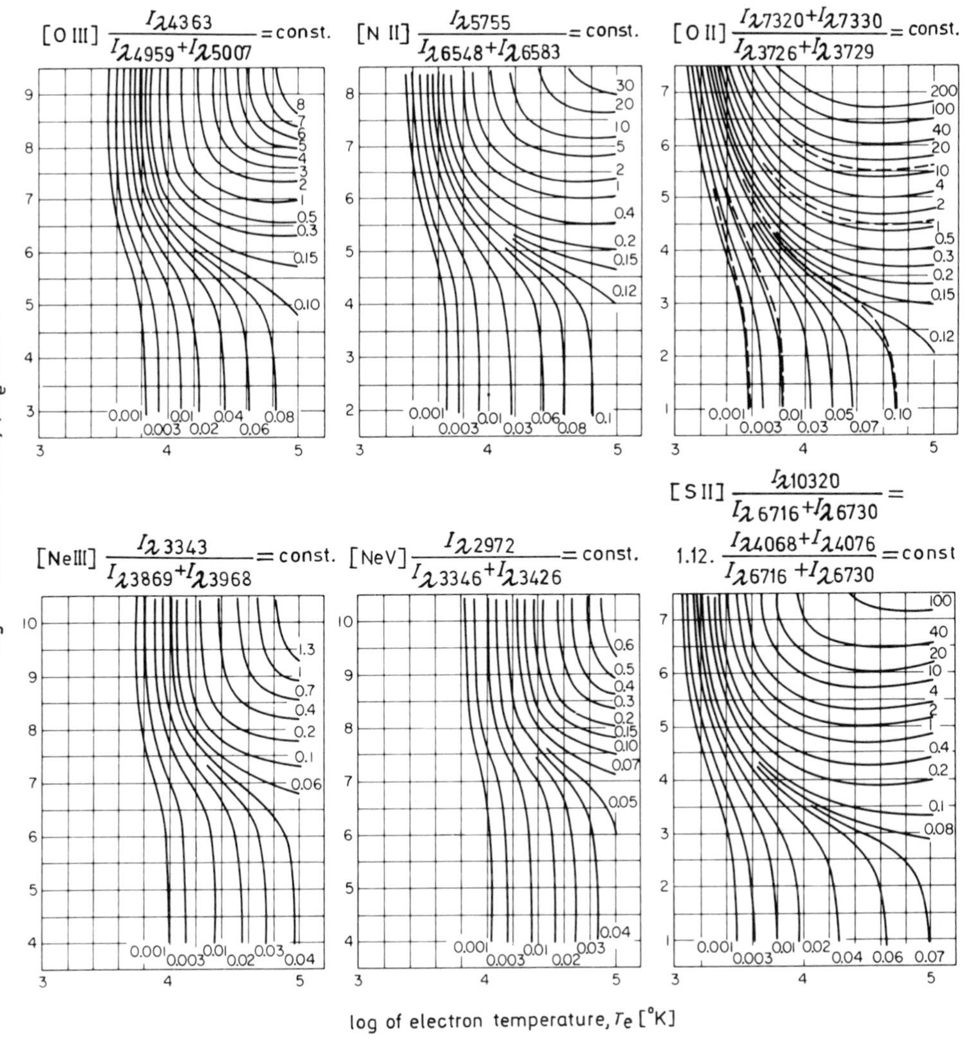

Fig. 8. Line intensity ratios for forbidden transitions as a function of electron density, N_e, and electron temperature, T_e (after Boyarchuk, Gershberg, Godovnikov, and Pronik, 1969)

where the ionization potential of hydrogen, $I_H = 13.60$ eV, and

$$x_m = \frac{I_m}{kT_e} = \frac{\lambda}{m^2},$$

where the threshold ionization energy of level m is $I_m = Z^2 I_H/m^2$. Because α_{mn}/Z depends only on T_e/Z^2, it is clear from Eq. (2-107) that a measurement of the relative intensities of lines of a given atom or ion will lead to a measure of T_e, whereas a measure of the intensity of one line will give the volume integral of $N_e N_+$ assuming a value for T_e.

Table 9. Positions, angular radius, $H\beta$ flux, $F(H\beta)$, received at the earth, extinction, electron density, and the central star temperature, T_c, for ninety planetary nebulae[1]

Nebula	Designation	α(1950.0) h m s	δ (1950.0) ° ′	Radius (arc seconds)	log $F(H\beta)$ (erg cm^{-2} sec^{-1})
NGC 40	120 + 9°1	0 10 17	+72 15	18.2	−10.64
Hu 1–1	119 − 6°1	0 25 30	+55 41	2.5	−11.73
NGC 246	118 −74°1	0 44 30	−12 09	118.0	−10.20
NGC 650–1	130 −10°1	1 39 12	+51 19	32.0	−10.67
IC 1747	130 + 1°1	1 53 58	+63 05	6.4	−11.45
IC 289	138 + 2°1	3 06 16	+61 08	22.5	−11.67
IC 351	159 −15°1	3 44 20	+34 54	3.5	−11.40
IC 2003	161 −14°1	3 53 12	+33 43	2.5	−11.18
NGC 1501	144 + 6°1	4 02 42	+60 47	27.0	−11.26
NGC 1514	165 −15°1	4 06 08	+30 39	50.0	(−10.70)
NGC 1535	206 −40°1	4 11 54	−12 52	9.2	−10.36
J 320	190 −17°1	5 02 49	+10 38	3.2	−11.37
IC 418	215 −24°1	5 25 10	−12 44	6.2	− 9.53
NGC 2022	196 −10°1	5 39 24	+ 9 04	9.8	−11.15
IC 2149	166 +10°1	5 52 36	+46 07	4.3	−10.50
IC 2165	221 −12°1	6 19 23	−12 57	4.0	—10.87
J 900	194 + 2°1	6 23 01	+17 49	4.7	−11.27
NGC 2346	215 + 3°1	7 06 47	− 0 43	27.3	−12.39
NGC 2371–2	189 +19°1	7 22 24	+29 35	27.0	−10.96
NGC 2392	197 +17°1	7 26 11	+21 01	23.0	−10.39
NGC 2438	231 + 4°2	7 39 32	−14 37	34.0	−10.97
NGC 2440	234 + 2°1	7 39 41	−18 05	16.2	−10.45
IC 2448	285 −14°1	9 06 33	−69 44	4.5	−10.83
NGC 2792	265 + 4°1	9 10 35	−42 14	6.5	−11.23
NGC 2867	278 − 5°1	9 20 05	−58 06	6.2	−10.57
NGC 3132	272 +12°1	10 04 54	−40 12	(28)	−10.43
NGC 3211	286 − 4°1	10 16 12	−62 26	6.9	−10.99
NGC 3242	261 +32°1	10 22 22	−18 23	(20)	− 9.77
NGC 3587	148 +57°1	11 11 54	+55 17	101.0	−10.33
NGC 3918	294 + 4°1	11 47 48	−56 54	6.5	− 9.99
NGC 4361	294 +43°1	12 21 54	−18 31	(55)	−10.48
IC 3568	123 +34°1	12 31 47	+82 50	9.0	−10.81
IC 4406	319 +15°1	14 19 18	−43 55	17.0	−10.73
NGC 5882	327 +10°1	15 13 30	−45 27	3.5	−10.40
NGC 6058	64 +48°1	16 02 42	+40 49	13.0	−11.70
NGC 6072	342 +10°1	16 09 41	−36 06	35.0	−11.37
IC 4593	25 +40°1	16 09 54	+12 12	6.0	−10.55
NGC 6153	341 + 5°1	16 28 00	−40 08	11.7	−10.85
NGC 6210	43 +37°1	16 42 24	+23 53	<8.0	−10.06
IC 4634	0 +12°1	16 58 36	−21 44	4.2	−10.92
IC 4642	334 − 9°1	17 07 38	−55 20	7.7	−11.34
NGC 6302	349 + 1°1	17 10 21	−37 02	22.3	−10.53
NGC 6309	9 +14°1	17 11 18	−12 51	7.5	−11.25
NGC 6326	338 − 8°1	17 16 48	−51 42	5.9	−11.06
NGC 6369	2 + 5°1	17 26 18	−23 43	14.3	−11.30
NGC 6445	8 + 3°1	17 46 18	−20 00	16.6	−11.15
NGC 6543	96 +29°1	17 58 36	+66 38	10.0	− 9.60

N_e, electron temperature, T_e, excitation class, apparent visual, V, or photographic, m_p, magnitude,

Extinction	log N_e (*F*) (cm^{-3})	(*Av*) (cm^{-3})	T_e (°K/10^4)	Excitation class	V or m_p (mag)	T_c (°K/10^3)
0.82	3.0[1]		{(1.2) 0.8	2	11.6	34
0.32		4.0	1.14	6	—	
0.00		2.0	(1.0)		11.92	52
0.13		2.75	1.23		17.0*p*:	157
0.94	3.52[1]		1.1	6.2	15*p*:	40
1.1		3.09	1.63	10	16.8*p*:	45
0.57		3.94	1.27	8	15*p*:	75
0.53		3.71	1.07	8	—	
1.10		3.28	(1.0)		14.3	70
0.71		2.92	1.4		9.5	74
0.00		3.95	1.3	7	11.6	41
0.27		3.34	1.5	5	13.5:	40
0.23	4.40[1]	4.30	1.13	3	9.7	42
0.45	3.10[1]		1.3	10	14.9	107
0.10	3.38[1]	3.51	1.25	4	10.5*p*:	44
0.51	3.50[1]	3.60	1.22	9	—	55
0.73		4.25	1.19	7	—	
1.67		3.07	(1.7)	7	10.8*p*:	
0.00		3.05	1.95	9*p*	14.8	95
0.15	3.96[1]	3.75	1.7	8*p*	10.5	65
0.45		2.45	(1.5)		17.5*p*:	140
0.48	3.55[1]	3.13	1.34	9	—	100
0.15		3.37	1.30	6	—	
0.66		3.47	1.8	8	—	
0.44		3.66	1.09	7	—	
0.27		3.23	1.14	6*p*	8.8:	48
0.27		3.47	1.30	8	—	
0.18	3.8[1]		1.15	7	$>$11.3	92
0.00		2.30			14.3*p*:	90
0.36	4.0	4.2	1.32	7	—	
0.00	3.60[1]		2.0	10	12.9	150
0.0	3.81	2.9	1.24	5	11.39	32
0.26		2.79	—	5	17:	
0.46		3.95	—		—	
0?		2.73	2.0		13.7	72
1.05		2.54	—		17.5:	110
0.06		3.29	—	4	10.8	29
0.93		3.84	1.56	6	—	
0.05	3.95	(3.51)	1.06	5	11.13	49
0.53		3.53	1.06	5	—	
0.25		3.29	2.17	10+	—	
1.42	3.85	38	2.1	10+*p*	—	
0.83	3.2–3.7		1.53	8	13:	94
0.24		3.23	1.9	6.5	—	
2.10		3.74			16:	79
0.80	3.04		1.5	8	19:	180
0.17	3.9	3.77	0.82	5	10.8	55

Table 9 (continued)

Nebula	Designation	α(1950.0) h m s	δ (1950.0) °	Radius (arc seconds)	$\log F(H\beta)$ (erg cm^{-2} sec^{-1})
NGC 6537	10 + 0°1	18 02 18	−19 51	3.7	−11.71
NGC 6565	3 − 4°5	18 08 43	−28 11	4.5	−11.23
NGC 6563	358 − 7°1	18 08 44	−33 51	22.6	−10.96
NGC 6572	34 +11°1	18 09 41	+ 6 50	6.2	− 9.75
NGC 6567	11 − 0°2	18 10 48	−19 05	4.4	−10.95
NGC 6578	10 − 1°1	18 13 18	−20 28	4.8	−11.75
NGC 6620	5 − 6°1	18 19 47	−26 51	2.3	−12.31
NGC 6629	9 − 5°1	18 22 40	−23 14	7.5	−10.94
NGC 6644	8 − 7°2	18 29 30	−25 10	1.3	−11.10
IC 4732	10 − 6°1	18 31 02	−22 39	2.0	−11.50
IC 4776	2 −13°1	18 42 36	−33 24	3.5	−10.77
NGC 6720	63 +13°1	18 51 44	+32 58	35.0	−10.06
NGC 6741	33 − 2°1	19 00 00	−00 31	4.0	−11.49
NGC 6751	29 − 5°1	19 03 12	−06 04	10.5	−11.25
NGC 6772	33 − 6°1	19 12 00	− 2 47	32.5	−11.65
IC 4846	27 − 9°1	19 13 00	− 9 05	(1.0)?	−11.33
NGC 6778	34 − 6°1	19 15 48	− 1 40	8.0	−11.26
NGC 6781	41 − 2°1	19 16 02	+ 6 27	53.0	−11.19
NGC 6790	37 − 6°1	19 20 35	+ 1 25	3.7	−10.73
NGC 6803	46 − 4°1	19 28 54	+ 9 57	2.8	−11.15
NGC 6804	45 − 4°1	19 29 12	+ 9 07	16.0	−11.28
NGC 6807	42 − 6°1	19 32 06	+ 5 34	1.0?	−11.62
BD+30° 3639	64 + 5°1	19 32 48	+30 24	1.5	−11.50
NGC 6818	25 −17°1	19 41 06	−14 16	9.2	−10.26
NGC 6826	83 +12°1	19 43 27	+50 24	12.7	− 9.92
NGC 6833	82 +11°1	19 48 21	+48 50	1.0:	−11.22
NGC 6853	60 − 3°1	19 57 24	+22 35	100 (180)	− 9.44
NGC 6884	82 + 7°1	20 08 49	+46 18	3.8	−11.11
NGC 6886	60 − 7°2	20 10 30	+19 50	3.0	−11.50
NGC 6891	54 −12°1	20 12 48	+12 33	6.3	−10.60
NGC 6894	69 − 2°1	20 14 23	+30 25	22.0	−11.46
IC 4997	58 −10°1	20 17 55	+16 35	0.8	−10.49
NGC 6905	61 − 9°1	20 20 12	+19 57	21.5	−10.90
NGC 7008	93 + 5°2	20 59 05	+54 21	38.6	−10.86
NGC 7009	37 −34°1	21 01 30	−11 34	13.4	− 9.78
NGC 7026	89 + 0°1	21 04 35	+47 39	5.6	−10.90
NGC 7027	84 − 3°1	21 05 09	+42 02	7.1	−10.12
IC 5117	89 − 5°1	21 30 37	+44 22	1.0:	−11.71
Anon 21^h31^m	86 − 8°1	21 31 07	+39 25	2.5	−11.20
IC 5217	100 − 5°1	22 21 56	+50 43	3.4	−11.18
NGC 7293	36 −57°1	22 26 54	−21 03	330 (400)	− 9.35
NGC 7354	107 + 2°1	22 38 26	+61 01	10.0	−11.55
NGC 7662	106 −17°1	23 23 29	+42 16	7.6	− 9.98

[1] These data were kindly prepared by Professor L. H. ALLER of the University of California, Los Angeles, from ALLER (1956), ALLER and FAULKER (1964), ALLER and LILLER (1960), BÖHM (1968), CAPRIOTTI (1964), CAPRIOTTI and DAUB (1960), CURTIS (1918), HIGGS (1971), KALER (1970), LILLER (1955), LILLER and ALLER (1955), MILNE and ALLER (to be publ.), MOTTEMANN (1972), O'DELL (1962), O'DELL (1963), PEREK (1971), PEREK and KOHOUTEK (1967), WILSON (1950) and WILSON (1958). Angular radii which are uncertain because of an irregular or fuzzy image are placed in parentheses.

Extinction	$\log N_e$ (*F*) (cm^{-3})	(*Av*) (cm^{-3})	T_e (°K/10^4)	Excitation class	*V* or m_p (mag)	T_c (°K/10^3)
1.77		3.58	2.15		—	
0.31		3.96	1.15	6	—	
0.34		2.79	—		18:	>100
0.36	4.00	4.0	1.1	5	9.06	57
0.68		3.82	1.09	5	—	64
1.72		3.67	—		—	39
0.92					—	36
0.95		3.34	0.92	4	12.5*p*:	44
			—		—	
0.75		4.1:	1.7:	5	12.0*p*:	
0.08			—	6*p*	—	
0.03	2.89	3.00	1.3	6*p*	14.7:	130
1.26	3.50	3.60	1.1	8	—	
0.40		3.29	1.5	7*p*	12.9	72
1.07		3.80	—		18.7*p*	105
0.53		4.6:	1.14	5		
0.34	3.0	3.5	1.0	5	14.8*p*	55
1.26		2.69	—		16.2*p*	65
0.64		4.29	1.34	6	10.5*p*	
0.71		4.13	1.02	6	14*p*:	44
0.85		3.08	—		14.1	72
0.34		4.34	1.6	5	—	
1.72	4.6	4.95	—	1	10.1	38
0.01		3.8	1.8	9	>12	138
0.03	3.32	3.25	1.03	5	> 9.9	42
0.10		4.00:	1.3	4		
0.00		3.47	1.3	7*p*	13.9	130
0.44		3.46	1.11	6		
0.94		3.93	1.2	8		90
0.24	3.4	3.33	1.0	5	11.09*p*:	41
0.56		2.54	—	6	17.5*p*	90
0.00	4.4	4.8	1.8	5	13.7*p*:	
0.09		2.9	1.4	7	13.9*p*:	70
0.61		2.73	1.4	7*p*	13.3	93
0.15	3.85	3.6	1.05	6	>10.9	81
0.80		3.3	1.0	6	15*p*:	97
1.35	4.0–5.6	—	(1.3)	10*p*	—	>200
1.38		4.0	1.22	6		
0.61		3.02	1.9	10		>200
0.78	4.25	3.78	1.2	6	14.6*p*	50
0.00		(2.68)	—		13.5	150
1.87		3.83	—	8	—	70
0.29	4.5	4.0	1.3	8	12.5	100

Electron densities obtained from forbidden line data are recorded in column (F), whereas those obtained from the measured surface brightness are recorded in column (*Av*). In the latter case, the filling factors are from KALER (1970), and the distance estimates are from MOTTEMANN (1972) or by the so-called Shklovsky method. The excitation designation is described in ALLER (1956), and the photographic magnitude is designated by the suffix *p*.

BAKER and MENZEL (1938) considered two cases, case A and case B, which correspond, respectively, to an optically thin and an optically thick nebula, the later case being considered first by ZANASTRA (1927). Case A assumes that the excited states of a hydrogen atom are populated by radiative capture from the continuum, and by cascade from all higher states, and depopulated by cascade to lower levels. Case B assumes that the rate of depopulation of excited states by emission of Lyman lines is exactly equal to the rate of population by absorption of Lyman quanta. BURGESS (1958), SEATON (1960), and PENGALLY (1964) have considered these two cases when orbital degeneracy is taken into account. Of particular interest are the hydrogen (HI) line of $H(\beta)=H(4,2)$ at $\lambda=4861$ Å and the ionized helium (HeII) line of $He^+(4,3)$ at $\lambda=4686$ Å. For these lines we have (SEATON, 1960)

	Case A	*Case B*
$\alpha_{4,2}$(HI)	1.98×10^{-14} cm^3 sec^{-1}	2.99×10^{-14} cm^3 sec^{-1}
$\alpha_{4,3}$(HeII)	$9.0\ \times10^{-14}$ cm^3 sec^{-1}	$20.8\ \times10^{-14}$ cm^3 sec^{-1},

for $T_e=10^4$ °K. Values of $\alpha_{mn}(T_e)$ may be approximated for other values of T_e by assuming that $\alpha_{mn}\propto T_e^{-1}$. PENGALLY (1964) has given values of $h\nu_{mn}\alpha_{mn}(T_e)$ for the λ4861 and the λ4686 lines as a function of T_e for case A and case B. He has also tabulated the relative intensities of the hydrogen Balmer and Paschen series relative to $I_{4,2}(\text{HI})=100$ and the Pickering and Pfund series of ionized helium relative to $I_{4,3}(\text{HeII})=100$, in both cases for case A and case B at $T_e=1$ and 2×10^4 °K. The wavelengths of most of these transitions are given in Table 11.

The observed intensities must be corrected for reddening, or absorption by interstellar dust particles, before Eqs. (2-107) and (2-108) may be used. If $I_c(\lambda)$ and $I_0(\lambda)$ denote, respectively, the corrected and observed intensities at wavelength λ, then we have (BERMAN, 1936; WHITFORD 1948, 1958; BURGESS, 1958)

$$\log I_c(\lambda)=\log I_0(\lambda)+C f(\lambda)\,, \tag{2-109}$$

where the units of $f(\lambda)$ are chosen so that $f(\lambda)=0$ for the $H(\beta)$ line at $\lambda=4861$ Å, and $f(\infty)=-1$. In this way the observed value of the $H(\beta)$ intensity must be multiplied by 10^C to correct for extinction. The mean value of the constant C is 0.19 (PENGALLY, 1964). Observations at radio frequencies do not need to be corrected for reddening, and the bremsstrahlung formula (1-222) leads to the relation

$$E=\int_0^L N_e^2\,ds=\frac{S_\nu T_e^{1/2}c^2}{2k\xi\Omega}\,, \tag{2-110}$$

for an optically thick nebula. Here the emission measure, E, is the line integral of the square of electron density, N_e, across the extent, L, of the nebula, S_ν is the observed flux density at frequency, ν, the apparent solid angle of the source is Ω, and

$$\xi=9.786\times10^{-3}\ln\left(4.9\times10^7\,\frac{T_e^{3/2}}{\nu}\right).$$

Observed flux densities at $\nu = 5000$ MHz are given by TERZIAN (1968). At this frequency, Eq. (2-110) becomes

$$E \approx \frac{S_\nu T_e^{1/2}}{2\theta^2} \text{ cm}^{-6} \, pc,$$

where θ is the angular extent in degrees and S_ν is the flux density in flux units. The mass, M, of the nebula is given by

$$M = \frac{4\pi R^3}{3} \varepsilon N_e m_H,$$

where $m_H = 1.673 \times 10^{-24}$ grams is the mass of the hydrogen atom. It follows from Eqs. (2-107) and (2-109) that the distance, D, to the nebula is given by

$$D^5 = \frac{3}{\varepsilon} \left(\frac{M}{4\pi m_H} \right)^2 \frac{hc}{\lambda \theta^3 F \times 10^C},$$

where F is the flux observed at wavelength, λ, the angular radius of the nebula is θ, and C is the extinction correction.

As it was shown in Sect. 1.31, recombination continuum radiation becomes important at optical frequencies, and the magnitude of its contribution is a sensitive function of frequency, ν, electron density, N_e, and electron temperature, T_e. The total continuum power, $P(\nu) d\nu$, radiated per unit volume per frequency interval between frequencies ν and $\nu + d\nu$ is

$$P(\nu) d\nu = N(\text{HII}) N_e \gamma_{\text{eff}} d\nu, \tag{2-111}$$

where N (HII) is the density of ionized hydrogen, N_e is the electron density, and the effective emission coefficient, γ_{eff}, is given by

$$\gamma_{\text{eff}} = \gamma(\text{HI}) + \gamma(2q) + \frac{N(\text{HeII})}{N(\text{HI})} \gamma(\text{HeI}) + \frac{N(\text{HeIII})}{N(\text{HII})} \gamma(\text{HeII}),$$

where γ(HI), γ(HeI) and γ(HeII) denote, respectively, the emission coefficients of neutral hydrogen, neutral helium, and singly ionized helium, $\gamma(2q)$ is the emission coefficient for two photon emission from the $2^2 S_{1/2}$ level of hydrogen, and N(HeII) and N(HeIII) denote, respectively, the densities of HeII and HeIII. The emission coefficients for γ(HI) and γ(HeI) are given by

$$\gamma = \frac{P_r(\nu)}{N_i N_e} + \frac{4\pi \varepsilon_\nu}{N_i N_e} = \gamma_{\text{fb}} + \gamma_{\text{ff}},$$

where the recombination power density, $P_r(\nu)$, is given in Eq. (1-236), the volume emissivity for bremsstrahlung, ε_ν, is given in Eq. (1-219), and the subscripts ff and fb denote, respectively, free-free and free-bound emission. BROWN and MATHEWS (1970) have tabulated γ(HI) and γ(HeI) as a function of temperature for optical wavelengths. Their results are illustrated in Fig. 9 for a temperature of 10^4 °K. These data indicate that $\gamma(\text{HI}) \approx \gamma(\text{HeI})$, and that two photon emission becomes important near the Balmer discontinuity at $\lambda = 3647$ Å. For normal nebulae, $N(\text{HeII})/N(\text{HII}) \approx 0.1$ and $N(\text{HeIII})/N(\text{HII}) \approx 0.01$, and the contribution of helium to the continuum radiation is of second order.

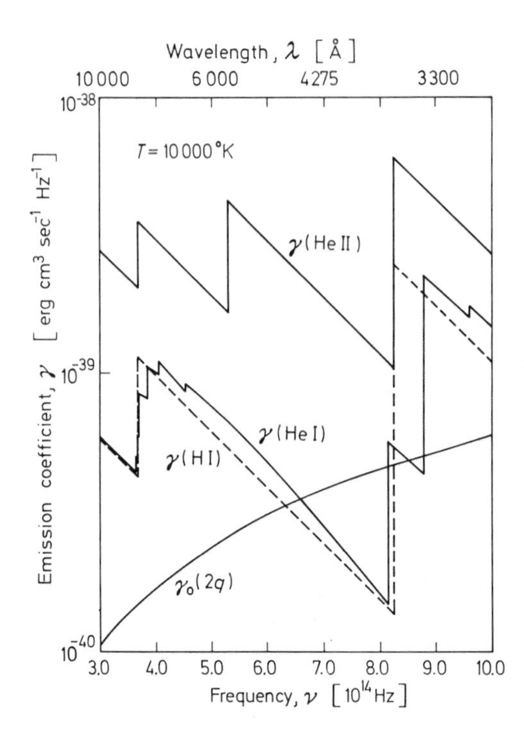

Fig. 9. Frequency variation of the continuous emission coefficients $\gamma(\text{HI})$ (dashed line), $\gamma(\text{He I})$, $\gamma(\text{He II})$, and $\gamma_0(2q)$ at $T = 10{,}000\,°\text{K}$. (After BROWN and MATHEWS, 1970, by permission of the American Astronomical Society and the University of Chicago Press)

The two photon emission process has been discussed by MAYER (1931), BREIT and TELLER (1940), and SPITZER and GREENSTEIN (1951). The emission coefficient is given by

$$\gamma(2q) = \frac{h\,A(y)\,y\,X\,\alpha_B(T)}{A_2} = g(\nu)\,X\,\alpha_B(T),$$

where $y = \nu/\nu_{12}$, the $h\nu_{12}$ is the excitation energy of the $2S_{1/2}$ level, $A_2 = 8.227\ \text{sec}^{-1}$ is the total probability per second of emitting a photon of energy $y h\nu_{12}$ as one member of the pair, X is the probability per recombination that two photon decay results, and the total recombination coefficient $\alpha_B(T)$ to excited levels of hydrogen is given by (HUMMER and SEATON, 1963)

$$\alpha_B(T) = 1.627 \times 10^{-13}\,t^{-1/2}\left[1 - 1.657 \log t + 0.584\,t^{1/3}\right] \quad \text{cm}^3\,\text{sec}^{-1},$$

where

$$t = 10^{-4}\,T.$$

The function $0.229\,A(y)$ is tabulated by SPITZER and GREENSTEIN (1951), and the function $g(\nu)$ is tabulated by BROWN and MATHEWS (1970). Values of X depend upon the population of atoms in the $2S_{1/2}$ and 2^2P states as well as the number density of $L\alpha$ photons (cf. COX and MATHEWS, 1969); and approximate values of X at optical wavelengths are tabulated by BROWN and MATHEWS (1970).

When the contributions of recombination radiation, bremsstrahlung, and two photon emission have been taken into account, the Balmer decrement, D_B, may be calculated using Eq. (2-111). It is defined as the logarithm of the intensity ratio on each side of the Balmer discontinuity at $\lambda = 3647$ Å, it is illustrated in Fig. 10, and is given by

$$D_B = \log\left[\frac{I(3647^-)}{I(3647^+)}\right], \tag{2-112}$$

where I denotes intensity and $-$ and $+$ denote, respectively, wavelengths lower and higher than 3647 Å. The intensity jump at the head of the Balmer series may be reduced by either raising the temperature or by lowering the electron density (which increases the contribution of the two-photon continuum emission). Consequently, a measurement of D_B leads to a measure of either the electron temperature, T_e, or the electron density, N_e. BOYARCHUK, GERSHBERG, and GODOVNIKOV (1968) have calculated D_B as a function of T_e and N_e by taking into account the recombination radiation, bremsstrahlung, and two photon emission of hydrogen, and their results are also illustrated in Fig. 10.

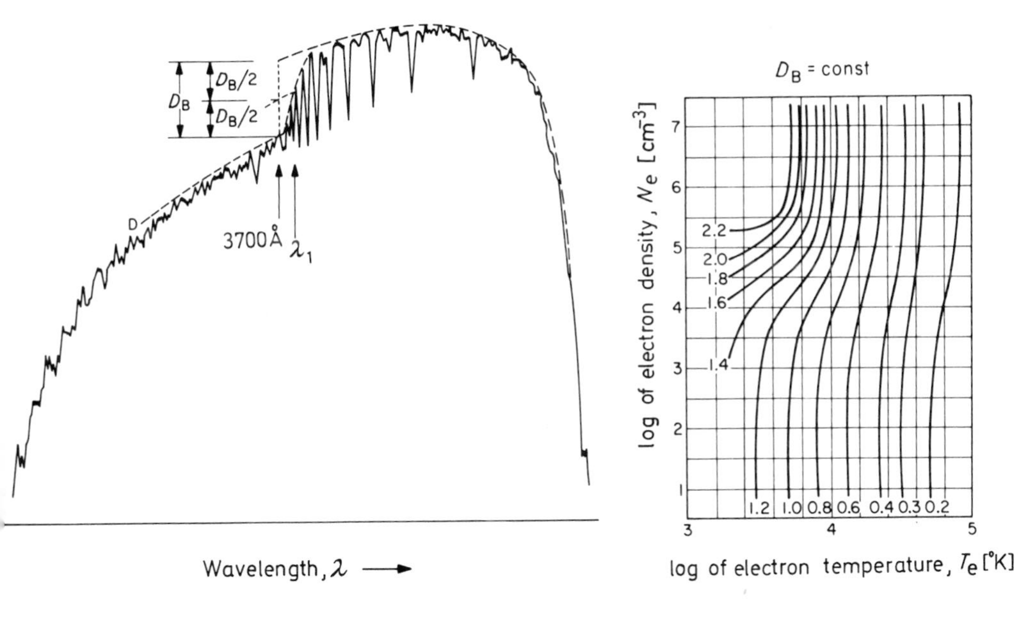

Fig. 10. The Balmer decrement, D_B, shown on the continuum spectrum of a star, and given as a function of electron density, N_e, and electron temperature, T_e. (After BOYARCHUK, GERSHBERG, and GODOVNIKOV, 1968)

BAHCALL and WOLF (1968) have shown that the presence or absence of absorption lines also give information on the density and temperature of the absorbing medium. Higher fine-structure states of a ground state multiplet are occupied provided that the density of the absorbing medium is higher than a critical value, N_{cr}, illustrated in Fig. 11 as a function of temperature for a variety of transitions and atoms or ions. The expected fine structure separations range from 0.4 to 18.9 Å.

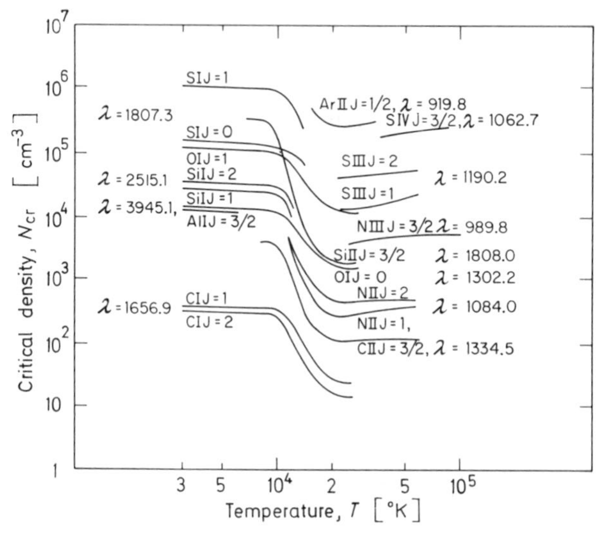

Fig. 11. The critical density, N_{cr}, below which the higher fine-structure states in absorption features are not heavily populated, given as a function of temperature (after BAHCALL and WOLF, 1969, by permission of the American Astronomical Society and the University of Chicago Press). Also given are the wavelengths, λ, in Angstroms for the ground resonant transitions whose fine structure separations, $\Delta\lambda$, are between 0.5 and 20 Å. Hydrogen is assumed to be ionized by collisions but deionized by radiative recombination. The critical density for a given state is defined by the condition that the rate of photon decay of the state is equal to the rate of collisional transitions to lower states

2.12. Atomic Recombination Lines and Ionized Hydrogen (H II) Regions

When a hydrogen-like atom undergoes a transition from an upper level m to a lower level n, it will radiate at the frequency, ν_{mn}, given by (RYDBERG, 1890; RITZ, 1908; PLANCK, 1910; BOHR, 1913)

$$\nu_{mn}=cR_A Z^2(n^{-2}-m^{-2})\approx 2cR_A Z^2 n^{-3}(m-n)\,, \tag{2-113}$$

where the velocity of light, $c=2.997925\times10^{10}$ cm sec^{-1}, Z is the "effective" charge of the nucleus (or ionic charge) given in units of the proton charge, and the atomic Rydberg constant, R_A, is given by

$$R_A=R_\infty\left(1+\frac{m_e}{M_A}\right)^{-1}\approx R_\infty\left(1-\frac{m_e}{M_A}\right), \tag{2-113a}$$

where the Rydberg constant for infinite mass is $R_\infty = 2\pi^2 m_e e^4/(ch^3) = 109{,}737.31$ cm^{-1}, the electron mass $m_e = 5.48597 \times 10^{-4}$ a.m.u., and the atomic mass is M_A. The atomic masses and the Rydberg constants for the more abundant atoms are given in Table 10.

Table 10. Atomic masses and Rydberg constants for the most abundant atoms

Atom	Atomic mass, M_A (a.m.u.)	Rydberg constant, R_A (cm^{-1})
Hydrogen, H^1	1.007825	109,677.6
Helium, He^4	4.002603	109,722.3
Carbon, C^{12}	12.000000	109,732.3
Nitrogen, N^{14}	14.003074	109,733.0
Oxygen, O^{16}	15.994915	109,733.5
Neon, Ne^{20}	19.99244	109,734.3

When an optically thin gas is observed with an antenna whose beamwidth is narrower than the angular extent of the gas cloud, the antenna temperature due to the line radiation, ΔT_L, is given by

$$\Delta T_L = \eta_A T_e \int_0^L \alpha_\nu \, dl , \tag{2-114}$$

where η_A is the antenna efficiency, T_e is the electron temperature of the ionized gas, L is the extent of the gas cloud along the line of sight, the integral is along the line of sight, and the quantum mechanical absorption coefficient, α_ν, is given by (Einstein, 1917, Eq. (2-65))

$$\alpha_\nu = \frac{c^2 N_n g_m}{8\pi \nu^2 g_n n_\nu^2} A_{mn} \left[1 - \frac{g_n}{g_m} \frac{N_m}{N_n} \right] \varphi_{mn}(\nu) , \tag{2-115}$$

where N_n is the population of level n, the A_{mn} is the Einstein coefficient for spontaneous transition from level m to level n, the statistical weight of level n is g_n, and the index of refraction of the gas is n_ν, which will be assumed to be unity in the following. When the gas is in local thermodynamic equilibrium (LTE), the population, N_n, of level n is given by (Saha, 1921, Eq. (3-129))

$$N_n = \frac{N_e N_i h^3}{(2\pi m_e k T_e)^{3/2}} \frac{g_n}{2} \exp(-\chi_n / k T_e) , \tag{2-116}$$

where N_e is the electron density, N_i is the density of the recombining ions, T_e is the electron temperature, and χ_n is the excitation energy of the nth level below the continuum. Under LTE conditions, $\exp(-h\nu_{mn}/kT) = g_n N_m/(g_m N_n)$ and at radio frequencies $h\nu_m \ll kT$. It then follows from Eqs. (2-71) and (2-113) to (2-116) that for hydrogen at the frequency ν_{mn} we have from Eq. (2-8) $\chi_n = -hcR_A Z^2/n^2$, and

$$\Delta T_{\mathrm{L}} = 4\pi \left(\frac{\ln 2}{\pi}\right)^{1/2} \frac{e^2 h^4 R_{\mathrm{A}} \eta_{\mathrm{A}} Z^2}{m_{\mathrm{e}} k (2\pi m_{\mathrm{e}} k T_{\mathrm{e}})^{3/2}} \frac{m-n}{\Delta \nu_{\mathrm{D}}} \frac{f_{nm}}{n} \exp\left(\frac{-\chi_n}{k T_{\mathrm{e}}}\right) \int_0^L N_{\mathrm{e}} N_{\mathrm{i}}\, d l$$

$$\approx 10^4 Z^2 \eta_{\mathrm{A}} \frac{m-n}{\Delta \nu_{\mathrm{D}}} \frac{f_{nm}}{n} T_{\mathrm{e}}^{-3/2} \exp\left(\frac{1.58 \times 10^5 Z^2}{n^2 T_{\mathrm{e}}}\right) \int_0^L N_{\mathrm{e}} N_{\mathrm{i}}\, dl \ {}^{\circ}\mathrm{K}, \tag{2-117}$$

where $g_n = 2n^2$, the full width to half maximum of the Doppler broadening is $\Delta \nu_{\mathrm{D}}$, and the numerical approximation is for $\Delta \nu_{\mathrm{D}}$ in kHz, and the line emission measure $E_{\mathrm{L}} = \int_0^L N_{\mathrm{e}} N_{\mathrm{i}}\, dl$ in pc cm^{-6}. The oscillator strength, f_{nm}, is related to the Einstein coefficient for spontaneous emission, A_{mn}, in Eq. (2-40), and the f_{nm} for the hydrogen atom are given by (KRAMERS, 1923; MENZEL and PEKERIS, 1935)

$$f_{nm} = \frac{-g_m}{g_n} f_{mn} = \frac{2^5}{3\sqrt{3}\pi} \frac{1}{m^2} \frac{1}{\left[\frac{1}{m^2} - \frac{1}{n^2}\right]^3} \left|\frac{1}{n^3} \frac{1}{m^3}\right| g_{\mathrm{bb}}, \tag{2-118}$$

where $g_n = 2n^2$ and g_{bb} is the Kramers-Gaunt factor for the bound-bound transition (KRAMERS, 1923; GAUNT, 1930; BAKER and MENZEL, 1938; BURGESS, 1958). Values of f_{mn} for low m and n are given by WIESE, SMITH, and GLENNON (1966). Approximation formulae for g_{bb} and numerical evaluations of f_{nm} at large n are given by GOLDWIRE (1968) and MENZEL (1969). Menzel obtains $f_{nm}/n = 0.194$, 0.0271, 0.00841, 0.00365, and 0.00191 for $m-n = 1, 2, 3, 4$ and 5 respectively. For machine calculation, MENZEL gives the formula

$$f_{n+c,n} = \frac{(2/c)^{2-2c} n}{3(c!)2} \left[\frac{(n+c)!}{n!\, n^c}\right]^2 \left[\frac{(1+c/n)^{2n+2}}{(1+c/2n)^{4n+2c+3}}\right] \left\{\ \right\}, \tag{2-119}$$

where

$$\{\ \} = {}_2F_1 \times {}_3F_2 \,.$$

The F's, in turn, are hypergeometric functions of four and six variables, respectively:

$${}_2F_1\left[-n, -n+1, c+1, -\frac{c^2}{4n(n+c)}\right] = 1 - \frac{n(n+1)}{1!(c+1)} \frac{c^2}{4n(n+c)} + \frac{n(n-1)^2(n-2)}{2!(c+1)(c+2)} \left[\frac{c^2}{4n(n+c)}\right]^2 - \cdots,$$

and

$${}_3F_2\left[-n, -n, \theta+1, c+1, \theta, -\frac{c^2}{4n(n+c)}\right] = 1 - \frac{n^2(\theta+1)}{1!(c+1)\theta} \frac{c^2}{4n(n+c)} + \frac{n^2(n-1)^2(\theta+2)}{2!(c+1)(c+2)\theta} \left[\frac{c^2}{4n(n+c)}\right]^2 - \frac{n^2(n-1)^2(n-2)^2(\theta+3)}{3!(c+1)(c+2)(c+3)\theta} \left[\frac{c^2}{4n(n+c)}\right]^3 + \cdots,$$

where $\theta = nc/(2n+c)$ and the transition is the $n+c \to n$ transition.

The important low n transitions for hydrogen were first observed by LYMAN (1906) for $n=1$, BALMER (1885) for $n=2$, PASCHEN (1908) for $n=3$, BRACKETT (1922) for $n=4$, PFUND (1924) for $n=5$, and HUMPHREYS (1953) for $n=6$. The wavelengths of these lines are given in Table 11 together with the PICKERING (1896, 1897) series for singly ionized helium.

Table 11. The wavelengths in Å of the $m \to n$ transitions of hydrogen for $n=1$ to 6, $m=2$ to 21, and $m=\infty$, and for the $n=4$ Pickering series for ionized helium (He II)[1]. Here the wavelengths are in Å where 1 Å $=10^{-8}$ cm

Series m	Lyman ($n=1$)	Balmer ($n=2$)	Paschen ($n=3$)	Brackett ($n=4$)	Pfund ($n=5$)	Humphreys ($n=6$)	Pickering (He^{+}, $n=4$)
2	1,215.67						
3	1,025.72	6,562.80					
4	972.537	4,861.32	18,751.0				
5	949.743	4,340.46	12,818.1	40.512.0			10,123.64
6	937.803	4,101.73	10,938.1	26,252.0	74,578		6,560.10
7	930.748	3,970.07	10,049.4	21,655.0	46,525	123,680	5,411.52
8	926.226	3,889.05	9,545.98	19,445.6	37,395	75,005	4,859.32
9	923.150	3,835.38	9,229.02	18,174.1	32,961	59,066	4,541.59
10	920.963	3,797.90	9,014.91	17,362.1	30,384	51,273	4,338.67
11	919.352	3,770.63	8,862.79	16,806.5	28,722	46,712	4,199.83
12	918.129	3,750.15	8,750.47	16,407.2	27,575	43,753	4,100.04
13	917.181	3,734.37	8,665.02	16,109.3	26,744	41,697	4,025.60
14	916.429	3,721.94	8,598.39	15,880.5	26,119	40,198	3,968.43
15	915.824	3,711.97	8,545.39	15,700.7	25,636	39,065	3,923.48
16	915.329	3,703.85	8,502.49	15,556.5	25,254	38,184	3,887.44
17	914.919	3,697.15	8,467.26	15,438.9	24,946	37,484	3,858.07
18	914.576	3,691.55	8,437.96	15,341.8	24,693	36,916	3,833.80
19	914.286	3,686.83	8,413.32	15,260.6	24,483	36,449	3,813.50
20	914.039	3,682.81	8,392.40	15,191.8	24,307	36,060	3,796.33
21	913.826	3,679.35					3,781.68
∞	911.5	3,646.0	8,203.6	14,584	22,788	32,814	3,644.67

[1] Data from WIESE, SMITH, and GLENNON (1966).

As first pointed out by WILD (1952) and KARDASHEV (1959), the $m-n$ transitions of hydrogen and helium may be observed in ionized hydrogen (H II) regions at radio frequencies when n is large. Confirming observations were first made by DRAVSKIKH and DRAVSKIKH (1964). The frequencies of these lines are specified by Eq. (2-113), and the transition is usually designated by the value of n together with a Greek suffix denoting the values of $m-n$. When $m-n=1, 2, 3, 4$, and 5, respectively, the appropriate symbols are α, β, γ, δ, and ε. Tables of hydrogen α, β, and γ line frequencies and helium α and β line frequencies have been tabulated by LILLEY and PALMER (1968) for $n=51$ to $n=850$. Observations of radio frequency hydrogen recombination lines result in the specification of the turbulent velocity, the kinematic distance, the electron temperature and the electron density of H II regions. The most extensive survey providing this information has been completed using the H 109α line at 5008.923 MHz (REIFENSTEIN *et al.*, 1970;

WILSON *et al.*, 1970). Observations of recombination lines may also be used to obtain the relative cosmic abundance of hydrogen and helium. The radiofrequency data (CHURCHWELL and MEZGER, 1970) indicate that the number ratio of singly ionized helium to ionized hydrogen is 0.08 ± 0.03. An example of the recombination lines from the H II region, Orion *A*, is shown in Fig. 12.

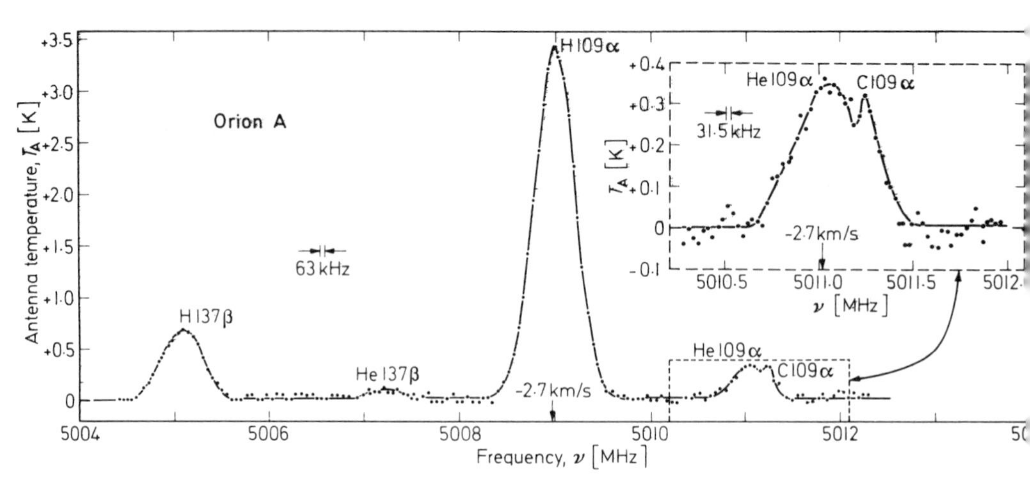

Fig. 12. Broadband spectrogram of the 109α region of the spectrum of the Orion Nebula. The frequency resolution is 63 kHz for the broadband spectrogram and 31.5 kHz for the narrow band spectrum centered on the He 109α line. (After CHURCHWELL and MEZGER, 1970, by permission of Gordon & Breach Science Publishers)

In order to illustrate how the parameters of an H II region are obtained from observations, we note that a recombination line is observed against the background continuum radiation of the gas cloud. It follows from Eq. (1-222) that for an optically thin gas whose angular extent is larger than that of the antenna beamwidth, the continuum antenna temperature, T_c, is given by

$$T_c = \frac{32\pi^2 e^6}{3\sqrt{2\pi m^3}\, c} \frac{\eta_A T_e}{n_\nu (2\pi\nu)^2} \left(\frac{m}{kT_e}\right)^{3/2} \int_0^L N_e \sum_j Z_j^2 N_j \ln \Lambda \, dl \tag{2-120}$$

$$\approx 3.01 \times 10^{-2} \eta_A T_e^{-1/2} \nu^{-2} \int_0^L N_e \sum_j Z_j^2 N_j \ln \Lambda \, dl \;{}^\circ\mathrm{K},$$

where η_A is the antenna efficiency, the summation, j, is over all ions, the index of refraction is n_ν, the extent of the gas cloud along the line of sight is L, and the logarithmic free-free Gaunt factor, Λ, is given in Table 2. The numerical approximation is for unity n_ν, a frequency ν in GHz, and the extent L in parsecs.

Table 12. Positions, angular extent, θ, flux density at 5000 MHz, S_{5000}, electron temperature, T_e, emission measure, E, diameter, $2R$, electron density, N_e, and mass, M_{HII}, for two hundred H II regions[1]

Source	G-Component	α(1950.0) h m s	δ(1950.0) ° ′ ″	θ_α ′	θ_δ ′	S_{5000} (f.u.)	T_e (°K)	E (cm^{-6} pc)	$2R$ (pc)	N_e (cm^{-3})	M_{HII} ($M_\odot$)
IC 1795	133.7 + 1.2	02 21 55	61 51 59	2.9	4.7	68.1	6,800 ± 800	5.4×10^2	3.3	330	450
Orion A	209.0 − 19.4	05 32 51	−05 25 39	3.8	4.3	382.2	7,000 ± 800	2.5×10^3	0.6	1.7×10^3	13
	208.9 − 19.3	05 33 03	−05 18 18	2.5	2.5	16.7	6,300 ± 1,000	2.8×10^5	0.7	613	3.1
	279.4 − 31.7	05 39 00	−69 06 42	4.0	5.0	33.7	9,400 ± 1,500	2.0×10^5	105.2	44	6.2×10^5
NGC 2024	206.5 − 16.4	05 39 12	−01 55 42	3.6	3.0	53.8	7,200 ± 890	5.5×10^2	0.6	790	6.7
	265.1 + 1.5	08 57 38	−43 33 36	2.3	2.5	23.8	7,400 ± 1,100	4.6×10^5	0.6	863	2.4
	267.8 − 0.9	08 57 40	−47 07 12	8.1	7.5	25.9	6,000 ± 2,600	4.3×10^4	6.0	85	220
	268.0 − 1.1	08 58 05	−47 20 12	1.3	2.1	172.3	7,900 ± 800	7.1×10^6	0.4	4,106	3.7
	282.0 − 1.2	10 04 53	−56 57 30	1.8	2.1	25.9	7,300 ± 1,200	7.6×10^5	5.5	371	730
	284.0 − 0.9	10 18 03	−57 49 36	6.3	9.2	7.7	3,500 ± 3,200	1.1×10^4	17.6	25	1,700
	284.1 − 0.4	10 22 44	−57 28 48	4.3	8.4	8.0	10,700 ± 8,900	2.8×10^4	7.2	62	2.8×10^2
	284.3 − 0.3	10 22 19	−57 32 00	4.7	6.2	178.8	6,800 ± 700	6.6×10^5	13.8	218	6.9×10^3
	285.3 + 0.0	10 29 36	−57 46 42	1.7	1.7	16.5	9,100 ± 2,000	6.8×10^5	0.1[2]	2,157	0.08
	287.2 − 0.7	10 39 48	−59 18 41	7.1	11.7	42.3	7,200 ± 2,100	5.5×10^4	10.5	72	1.0×10^3
	287.3 − 0.9	10 39 50	−59 29 11	6.3	4.3	15.1	5,900 ± 2,200	5.7×10^4	6.0	97	2.5×10^2
	287.4 − 0.6	10 41 36	−59 19 11	6.2	6.1	114.5	7,200 ± 800	3.3×10^5	7.1	216	9.2×10^2
	287.5 − 0.6	10 42 50	−59 22 59	5.7	5.3	86.0	6,100 ± 800	2.9×10^5	6.3	215	6.6×10^2
	287.6 − 0.9	10 42 09	−59 42 17	6.7	7.0	28.0	5,100 ± 2,000	5.8×10^4	7.9	86	5.1×10^2
	287.7 − 0.6	10 43 54	−59 29 29	4.2	2.2	10.4	6,700 ± 2,000	1.2×10^5	3.5	185	9.6×10^1
	287.8 − 0.8	10 43 43	−59 40 29	6.0	2.0	7.3	4,900 ± 2,000	5.9×10^4	4.0	121	9.3×10^1
	287.9 − 0.9	10 44 09	−59 50 17	7.9	5.4	23.9	4,400 ± 1,100	5.2×10^4	7.5	83	4.3×10^2
	287.9 − 0.8	10 44 53	−59 44 41	7.5	13.2	57.4	8,200 ± 2,500	6.6×10^4	11.5	76	1.4×10^3
	289.1 − 0.4	10 54 29	−59 49 48	3.2	3.1	7.8	13,700 ± 7,200	1.1×10^5	12.0	94	1.9×10^3
	289.8 − 1.1	10 56 51	−60 50 48	7.4	6.5	16.9	9,100 ± 4,600	4.1×10^4	25.8	40	8.2×10^3
	291.3 − 0.7	11 09 45	−61 02 36	0.9	0.9	97.4	7,700 ± 800	1.4×10^7	1.4	3,122	9.9×10^1
	291.6 − 0.5	11 12 53	−60 59 24	3.8	6.5	172.1	6,900 ± 800	7.5×10^5	17.4	208	1.3×10^4
	291.9 − 0.7	11 14 24	−61 13 18	0.0	1.8	3.0	7,000 ± 4,300	$>3.6\times10^5$	<3.9	>305	$<2.2\times10^2$

Table 12 (continued)

Source	G-Component	α(1950.0) h m s	δ(1950.0) ° ′ ″	θ_α ′	θ_δ ′	S_{5000} (f.u.)	T_e (°K)	E (cm^{-6} pc)	$2R$ (pc)	N_e (cm^{-3})	M_{HII} ($M_\odot$)
	295.1 − 1.6	11 37 47	−63 08 42	5.7	7.2	6.8	3,800 ± 1,800	1.5×10^4	6.6[2]	47	1.6×10^2
	295.2 − 0.6	11 41 05	−62 09 12	4.9	4.2	9.9	7,100 ± 3,900	5.1×10^4	24.2	46	7.9×10^3
	298.2 − 0.3	12 07 24	−62 33 18	1.8	2.1	30.6	8,100 ± 1,300	9.2×10^5	9.7	308	3.4×10^3
	298.8 − 0.3	12 12 38	−62 39 12	4.7	4.7	16.0	11,900 ± 8,200	9.2×10^4	23.1	63	9.4×10^3
	298.9 − 0.4	12 12 44	−62 44 48	3.8	2.8	32.7	6,000 ± 900	3.1×10^5	16.0[2]	140	6.9×10^3
	301.0 + 1.2	12 32 03	−61 22 54	0.0	0.0	2.8	6,100 ± 3,500	$>1.2\times10^6$	<1.1	$>1{,}022$	$<1.7\times10^1$
	301.1 + 1.0	12 33 11	−61 34 48	0.0	0.0	3.5	9,100 ± 5,400	$>1.7\times10^6$	<1.1	$>1{,}221$	$<2.0\times10^1$
	305.1 + 0.1	13 07 06	−62 22 36	5.0	5.7	14.2	5,200 ± 2,200	4.8×10^4	7.8[2]	79	4.4×10^2
	305.2 + 0.0	13 08 03	−62 29 12	3.6	3.9	28.0	5,200 ± 1,000	2.0×10^5	5.8[2]	184	4.2×10^2
	305.2 + 0.2	13 08 23	−62 17 30	4.9	6.8	52.0	5,700 ± 800	1.6×10^5	8.6[2]	135	1.0×10^3
	305.4 + 0.2	13 09 20	−62 18 48	3.0	2.5	39.4	5,200 ± 600	5.1×10^5	9.5	233	2.4×10^3
	305.6 + 0.0	13 11 06	−62 28 54	4.6	5.0	17.8	4,600 ± 1,100	7.3×10^4	8.8[2]	91	7.5×10^2
	308.6 + 0.6	13 36 37	−61 29 00	3.5	6.3	9.1	10,200 ± 8,500	5.0×10^4	9.4[2]	73	7.4×10^2
	308.7 + 0.6	13 37 18	−61 29 48	3.0	4.5	5.1	7,600 ± 4,400	4.2×10^4	6.1[2]	83	2.3×10^2
	311.5 + 0.4	14 00 01	−61 02 12	3.0	4.5	5.1	5,100 ± 3,000	3.6×10^4	10.4	59	8.0×10^2
	311.6 + 0.3	14 01 16	−61 05 36	0.0	0.0	2.8	3,300 ± 1,500	$>9.6\times10^5$	<1.4	>824	$<2.8\times10^1$
	311.9 + 0.1	14 03 52	−61 13 06	3.0	1.8	7.1	4,700 ± 1,400	1.2×10^5	9.7	113	1.2×10^3
	311.9 + 0.2	14 03 49	−61 05 18	3.0	3.5	7.8	4,300 ± 1,100	6.9×10^4	4.8[2]	119	1.6×10^2
	314.2 + 0.4	14 21 16	−60 09 12	2.7	2.8	5.2	3,000 ± 900	5.7×10^4	4.5[2]	113	1.2×10^2
	316.8 − 0.1	14 41 31	−59 36 54	2.3	1.3	28.7	5,900 ± 800	9.8×10^5	8.9	332	2.8×10^3
	317.0 + 0.3	14 41 45	−59 13 36	6.1	8.2	14.7	4,200 ± 1,600	2.7×10^4	34.0	28	1.3×10^4
	317.3 + 0.2	14 44 14	−59 08 18	6.2	8.6	6.7	2,700 ± 900	9.9×10^3	10.3[2]	31	4.1×10^2
	319.2 − 0.4	14 59 14	−58 51 42	5.1	4.6	8.2	11,200 ± 6,000	4.4×10^4	3.2[2]	117	4.6×10^1
	319.4 + 0.0	14 59 23	−58 24 30	5.0	2.3	8.2	7,400 ± 2,800	7.9×10^4	1.5[2]	233	8.5
	320.2 + 0.8	15 01 34	−57 19 24	1.3	1.3	7.7	8,400 ± 3,400	5.2×10^5	1.4	614	2.0×10^1
	320.3 − 0.3	15 06 17	−58 14 36	2.7	3.0	7.2	8,400 ± 3,100	1.0×10^5	6.1[2]	129	3.5×10^2
	320.3 − 0.2	15 06 14	−58 06 12	1.3	1.6	5.8	5,100 ± 1,600	2.7×10^5	0.4[2]	860	5.2×10^{-1}
	321.0 − 0.5	15 12 06	−58 00 30	3.2	2.3	7.8	4,900 ± 1,400	1.0×10^5	5.1[2]	141	2.3×10^2

321.1 − 0.5	15 12 45	−57 59 36	3.8	4.1	7.8	4,400 ± 1,500	4.6×10^4	6.7^2	83	3.0×10^2
322.2 + 0.6	15 14 49	−56 27 54	0.9	0.9	12.2	5,400 ± 1,100	1.5×10^6	1.4^2	1,029	3.5×10^1
324.2 + 0.1	15 29 02	−55 46 18	0.0	0.0	3.3	4,300 ± 1,800	$> 1.2 \times 10^6$	$< 1.5^2$	> 911	$< 3.5 \times 10^1$
326.5 + 0.9	15 38 33	−53 48 54	0.9	0.9	7.2	7,000 ± 2,300	9.7×10^5	0.8^2	1,094	6.9
326.7 + 0.6	15 40 56	−53 57 12	2.8	3.8	35.7	6,400 ± 800	3.5×10^5	4.5	281	3.0×10^2
327.3 − 0.5	15 49 12	−54 26 30	0.9	0.9	44.9	6,100 ± 700	5.8×10^6	1.3	2,069	6.0×10^1
327.6 − 0.4	15 50 02	−54 05 54	3.5	4.3	5.7	3,300 ± 1,200	3.1×10^4	8.5^2	61	4.5×10^2
328.0 − 0.1	15 50 54	−53 39 24	1.6	3.3	4.6	5,600 ± 2,400	8.7×10^4	3.2	165	6.5×10^1
328.3 + 0.4	15 50 18	−53 02 30	2.1	2.3	10.0	4,800 ± 1,200	2.0×10^5	8.8^2	150	1.2×10^3
330.9 − 0.4	16 06 27	−51 58 36	1.8	1.6	9.5	4,700 ± 1,000	3.1×10^5	3.0^2	322	1.1×10^2
331.0 − 0.2	16 06 20	−51 42 30	4.6	4.7	8.3	3,200 ± 900	3.2×10^4	13.1^2	49	1.3×10^3
332.2 − 0.5	16 08 23	−51 55 30	0.0	1.8	3.2	4,500 ± 1,600	$> 3.3 \times 10^5$	$< 2.0^2$	> 406	$< 4.0 \times 10^1$
331.3 − 0.2	16 07 36	−51 34 54	1.3	0.0	4.2	5,900 ± 3,400	$> 6.6 \times 10^5$	$< 2.1^2$	> 557	$< 6.5 \times 10^1$
331.3 − 0.3	16 08 33	−51 39 24	2.5	2.3	7.8	3,400 ± 1,000	1.2×10^5	4.9^2	155	2.1×10^2
331.4 + 0.0	16 07 18	−51 23 06	1.8	3.3	6.6	5,400 ± 1,600	1.1×10^5	6.0^2	135	3.6×10^2
331.5 − 0.1	16 08 22	−51 19 24	2.5	2.1	31.1	5,200 ± 600	5.8×10^5	10.9	232	3.5×10^3
332.1 − 0.5	16 12 52	−51 10 06	1.3	0.9	13.2	5,400 ± 1,100	1.1×10^6	1.9^2	771	6.3×10^1
332.5 − 0.1	16 13 17	−50 40 18	1.8	0.0	2.2	2,200 ± 1,000	$> 1.8 \times 10^5$	$< 1.7^2$	> 329	$< 1.9 \times 10^1$
332.7 − 0.6	16 15 58	−50 56 00	3.6	3.8	20.1	4,100 ± 600	1.3×10^5	5.7	153	3.4×10^2
332.8 − 0.6	16 16 25	−50 47 18	3.2	6.7	18.7	4,900 ± 1,300	8.3×10^4	8.5	99	7.3×10^2
333.0 − 0.4	16 16 51	−50 33 12	4.9	3.8	39.7	3,700 ± 400	1.9×10^5	7.7^2	156	8.7×10^2
333.1 − 0.4	16 17 16	−50 29 12	3.2	4.9	42.1	4,700 ± 600	2.5×10^5	22.7	106	1.5×10^4
333.2 − 0.1	16 15 56	−50 12 18	4.2	6.7	12.3	4,900 ± 1,300	4.1×10^4	15.2^2	52	2.2×10^3
333.3 − 0.4	16 17 45	−50 19 18	0.0	3.2	34.2	5,000 ± 600	2.1×10^6	2.1	994	1.1×10^2
333.6 − 0.1	16 17 53	−49 53 54	4.1	5.0	15.0	4,200 ± 1,200	6.7×10^4	7.9^2	92	5.5×10^2
333.6 − 0.2	16 18 26	−49 58 54	0.9	0.9	84.4	7,200 ± 800	1.1×10^7	1.4^2	2,835	9.8×10^1
333.7 − 0.5	16 19 58	−50 06 12	3.5	4.9	3.2	2,400 ± 1,200	1.5×10^4	6.9^2	46	1.8×10^2
335.8 − 0.2	16 27 27	−48 24 06	4.5	3.6	7.6	4,800 ± 1,700	4.4×10^4	7.2^2	78	3.6×10^2
336.4 − 0.2	16 29 52	−47 56 54	5.0	4.5	7.6	2,200 ± 500	2.5×10^4	10.8^2	48	7.2×10^2
336.4 − 0.3	16 30 33	−47 59 06	5.1	4.5	10.0	3,600 ± 1,100	3.8×10^4	14.1^2	52	1.8×10^3
336.5 + 0.0	16 29 33	−47 46 06	3.3	4.7	3.9	3,300 ± 1,300	2.1×10^4	8.4^2	50	3.6×10^2
336.5 − 0.2	16 30 37	−47 51 06	5.9	8.1	14.6	3,200 ± 1,000	2.5×10^4	19.5^2	36	3.2×10^3
336.5 − 1.5	16 36 22	−48 46 18	0.0	0.0	8.5	9,700 ± 3,200	4.1×10^6	0.4	3,030	3.3

Table 12 (continued)

Source	G-Component	α(1950.0) h m s	δ(1950.0) ° ′ ″	θ_α ′	θ_δ ′	S_{5000} (f.u.)	T_e (°K)	E (cm^{-6} pc)	$2R$ (pc)	N_e (cm^{-3})	M_{HII} ($M_\odot$)
	336.8 + 0.0	16 30 49	−47 30 24	10.7	6.1	57.5	6,000 ± 1,400	9.0×10^4	42.8	46	4.3×10^4
	336.9 − 0.1	16 32 09	−47 32 30	3.3	6.6	14.4	6,000 ± 1,700	6.7×10^4	11.4[2]	77	1.4×10^3
	337.1 − 0.2	16 33 01	−47 25 18	0.9	1.8	15.1	5,100 ± 800	9.1×10^5	3.1[2]	542	1.9×10^2
	337.1 − 0.1	16 33 27	−47 16 24	3.5	4.6	7.4	3,900 ± 1,900	4.1×10^4	7.5[2]	74	3.8×10^2
	337.6 + 0.0	16 34 30	−46 58 06	4.3	5.0	6.1	4,700 ± 1,900	2.7×10^4	8.9[2]	55	4.7×10^2
	337.9 − 0.5	16 37 27	−47 01 36	1.3	1.3	17.3	5,400 ± 1,000	1.0×10^6	1.9[2]	725	6.4×10^1
	338.0 − 0.1	16 34 14	−46 45 18	6.0	5.7	11.1	3,500 ± 1,400	2.8×10^4	11.0[2]	50	8.1×10^2
	338.1 + 0.0	16 36 14	−46 45 18	2.3	4.1	7.3	3,500 ± 1,100	6.7×10^4	4.7[2]	119	150
	338.1 − 0.2	16 36 58	−46 41 54	3.5	5.4	6.3	3,200 ± 1,800	2.8×10^4	7.6[2]	61	3.2×10^2
	338.4 − 0.2	16 38 00	−46 29 36	1.6	1.6	3.9	4,700 ± 1,900	1.5×10^5	0.3[2]	730	1.8×10^{-1}
	338.4 + 0.0	16 37 10	−46 18 18	5.0	8.8	46.6	4,900 ± 800	1.0×10^5	9.4[2]	104	1.0×10^3
	338.9 + 0.6	16 36 42	−45 34 24	3.2	3.5	9.9	5,000 ± 1,100	8.5×10^4	7.4[2]	107	530
	338.9 − 0.1	16 39 36	−46 00 48	1.8	1.8	3.4	5,700 ± 2,100	1.1×10^5	2.8[2]	196	50
	340.3 − 0.2	16 45 19	−45 04 06	1.8	2.3	4.9	5,400 ± 1,800	1.2×10^5	3.5[2]	184	93
	340.8 − 1.0	16 50 39	−45 12 12	1.8	2.7	12.1	5,000 ± 1,000	2.4×10^5	2.4[2]	320	50
	343.5 + 0.0	16 55 48	−42 30 12	3.5	3.6	12.6	7,800 ± 3,200	1.1×10^5	5.0[2]	150	220
	345.0 + 1.5	16 54 22	−40 19 36	6.9	11.3	14.8	5,700 ± 2,900	1.9×10^4	7.9[2]	49	290
	345.2 + 1.0	16 57 09	−40 29 18	1.3	1.6	10.4	4,800 ± 1,000	4.8×10^5	0.7[2]	806	4
	345.3 + 1.5	16 55 37	−40 09 24	3.8	3.3	12.6	5,800 ± 2,000	1.0×10^5	3.0[2]	184	61
	345.4 + 1.4	16 56 10	−40 07 00	1.3	1.3	13.1	4,500 ± 1,100	7.3×10^5	1.1[2]	833	12
	347.6 + 0.2	17 08 07	−39 04 42	8.9	5.4	23.4	3,800 ± 1,000	4.3×10^4	23.4[2]	43	6,600
	348.7 − 1.0	17 16 39	−38 54 36	1.6	2.1	33.4	5,300 ± 900	2.1×10^5	2.5[2]	240	140
	349.8 − 0.6	17 17 53	−37 44 04	5.1	5.2	9.2	4,900 ± 1,500	3.4×10^4	6.1[2]	61	530
	350.1 + 0.1	17 16 04	−37 07 39	2.4	3.4	6.5	8,200 ± 3,600	9.1×10^4	6.1[2]	100	870
NGC 6334	351.4 + 0.7	17 17 12	−35 45 58	7.0	10.2	171.4	6,400 ± 1,000	2.5×10^5	1.6	320	54
	351.6 − 1.2	17 25 52	−36 37 35	4.0	6.6	36.8	6,000 ± 1,000	1.4×10^5	3.9[2]	160	360
	351.6 + 0.2	17 19 54	−35 51 28	5.9	3.0	15.7	6,600 ± 1,900	9.4×10^4	7.6[2]	92	1,500

NGC 6357	353.1 + 0.3	17 23 19	−34 33 26	8.5	4.9	31.3	7,600 ± 1,900	8.4×10^4	1.9	170	47
	353.1 + 0.7	17 22 17	−34 18 19	11.7	13.4	259.4	6,000 ± 900	1.7×10^5	3.1	200	210
	353.2 + 0.9	17 21 30	−34 08 38	8.3	5.4	128.4	6,100 ± 830	3.0×10^5	1.8	330	77
	353.1 + 0.6	17 22 18	−34 20 06	5.5	5.7	111.3	6,000 ± 900	1.7×10^5	3.1	200	200
	353.4 − 0.4	17 27 08	−34 39 56	2.0	2.5	8.8	8,000 ± 4,100	2.0×10^5	2.1[2]	250	90
W 24	353.5 − 0.0	17 26 07	−34 23 24	2.0	2.0	3.3	3,300 ± 1,700	7.0×10^4	4.3[2]	100	310
	0.2 + 0.0	17 42 48	−28 45 53	16.9	5.0	180.0	7,600 ± 1,800	2.3×10^5	39.3	77	5.6×10^4
	0.5 + 0.0	17 43 56	−28 30 35	8.1	3.0	41.0	5,400 ± 800	1.7×10^5	21.1	89	1.0×10^4
	0.7 − 0.1	17 44 14	−28 23 17	5.0	2.1	59.8	7,900 ± 1,100	6.4×10^5	13.9	215	6.9×10^3
	0.5 − 0.0	17 43 49	−28 29 06	4.4	5.2	35.5	6,300 ± 1,100	1.6×10^5	13.9	90	9,100
	0.7 − 0.0	17 44 07	−28 21 36	2.9	3.6	47.8	6,900 ± 1,200	5.0×10^5	9.4	190	5,900
	1.1 − 0.1	17 45 25	−27 58 12	2.6	3.8	10.4	6,300 ± 1,800	1.1×10^5	9.1	91	2,600
AMWW 34	3.3 − 0.1	17 50 28	−26 10 23	8.2	8.0	11.0	4,200 ± 1,800	1.6×10^4	5.1[2]	45	230
AMWW 35	4.4 + 0.1	17 52 18	−25 06 07	3.8	4.2	6.2	5,200 ± 2,000	3.8×10^4	4.0[2]	80	200
W 28	5.9 − 0.4	17 57 34	−24 04 42	5.9	5.9	23.3	6,400 ± 1,300	7.1×10^4	5.1[2]	97	490
M 8	6.0 − 1.2	18 00 41	−24 23 20	8.5	7.6	85.1	7,300 ± 1,000	1.4×10^5	2.2	210	88
	6.1 − 0.1	17 56 52	−23 45 29	10.2	6.1	13.0	2,800 ± 800	1.7×10^4	11.1	39	640
W 28	6.5 + 0.1	17 57 05	−23 16 29	8.1	6.9	15.2	7,500 ± 5,500	3.0×10^4	11.8	50	1,000
	6.6 − 0.1	17 57 45	−23 18 59	3.9	5.8	11.4	5,900 ± 2,500	5.1×10^4	7.7	81	450
	6.6 − 0.3	17 58 39	−23 21 00	7.0	7.0	12.0	5,800 ± 1,700	2.5×10^4	7.3	48	710
	6.7 − 0.2	17 58 39	−23 18 59	6.9	7.7	20.4	10,000 ± 4,800	4.6×10^4	9.7	69	750
M 20	7.0 − 0.2	17 59 19	−23 02 09	5.4	5.8	13.3	7,300 ± 2,500	4.7×10^4	6.2	72	650
	8.1 + 0.2	18 00 00	−21 48 06	1.0	2.1	5.8	4,000 ± 1,200	2.5×10^5	1.6[2]	320	53
W 31	10.2 − 0.3	18 06 24	−20 19 53	3.3	3.6	51.8	5,900 ± 950	4.5×10^5	2.5[2]	350	210
	10.3 − 0.1	18 05 58	−20 06 05	3.2	1.9	13.6	7,000 ± 2,300	2.4×10^5	1.3[2]	360	30
	10.6 − 0.4	18 07 33	−19 56 32	5.7	4.2	10.2	5,400 ± 1,600	4.3×10^4	0.1[2]	450	<1
W 33	12.8 − 0.2	18 11 15	−17 57 02	4.9	3.6	44.9	7,800 ± 1,300	2.9×10^5	5.6[2]	190	1,200
	13.2 + 0.0	18 11 10	−17 29 28	4.0	1.0	4.7	4,700 ± 1,700	1.1×10^5	3.5[2]	150	240
	14.6 + 0.1	18 13 57	−16 14 38	11.0	9.0	24.3	4,500 ± 1,000	2.3×10^4	12.2[2]	36	2,500
M 17	15.0 − 0.7	18 17 37	−16 13 06	4.1	6.7	478.3	6,200 ± 600	1.8×10^6	5.2	594	970

Table 12 (continued)

Source	G-Component	α(1950.0) h m s	δ(1950.0) ° ′ ″	θ_α ′	θ_δ ′	S_{5000} (f.u.)	T_e (°K)	E (cm^{-6} pc)	2R (pc)	N_e (cm^{-3})	M_{HII} ($M_\odot$)
M 17	15.1 − 0.7	18 17 36	−16 12 17	4.9	7.3	534.6	6,400 ± 750	1.6×10^6	3.7	540	1,000
M 16	17.0 + 0.8	18 15 54	−13 47 32	14.1	18.2	107.8	6,100 ± 1,500	4.4×10^4	12.8	48	2.3×10^8
W 35	18.5 + 1.9	18 14 50	−11 57 20	27.6	29.1	102.6	3,000 ± 890	1.1×10^4	26.5	17	1.2×10^4
W 39	19.1 − 0.3	18 23 53	−12 29 02	12.4	4.2	17.0	4,000 ± 980	3.0×10^4	12.0[2]	41	2,700
	19.7 − 0.2	18 24 46	−11 53 42	5.8	5.9	12.4	5,300 ± 2,300	3.6×10^4	6.9[2]	60	730
	20.7 − 0.1	18 26 29	−10 54 47	5.6	7.2	14.5	4,400 ± 1,400	3.4×10^4	9.1[2]	50	1,400
W 41	22.8 − 0.3	18 31 01	−09 11 39	19.8	18.1	45.5	6,400 ± 4,300	1.3×10^4	34.4[2]	16	2.5×10^4
AMWW 47	23.4 − 0.2	18 31 58	−08 34 55	7.0	2.8	13.2	6,300 ± 1,700	7.1×10^4	9.6[2]	71	2,400
K 39	24.6 − 0.2	18 34 09	−07 30 09	7.1	6.2	8.2	4,500 ± 2,000	1.8×10^4	16.0	27	4,200
	24.8 + 0.1	18 33 30	−07 13 24	6.6	4.4	11.5	8,000 ± 2,700	4.5×10^4	14.2	46	5,100
W 42	25.4 − 0.2	18 35 32	−06 50 09	4.2	3.7	23.8	9,100 ± 2,800	1.8×10^5	5.3[2]	150	880
AMWW 48	25.8 + 0.2	18 34 55	−06 18 43	4.3	7.9	9.5	4,700 ± 2,100	2.7×10^4	15.3	35	4,600
	27.3 − 0.2	18 39 01	−05 08 29	6.3	13.7	7.7	5,000 ± 3,500	8.8×10^3	19.5[2]	18	4,900
	28.6 + 0.0	18 40 52	−03 52 38	8.3	14.3	14.4	6,700 ± 3,400	1.3×10^4	22.6[2]	20	8,700
W 40	28.8 + 3.5	18 28 51	−02 07 29	5.2	5.8	35.1	8,500 ± 2,400	1.4×10^5	0.2	760	<1
	29.9 − 0.0	18 43 29	−02 44 46	3.2	5.4	21.1	6,900 ± 1,500	1.3×10^5	8.8[2]	100	2,600
	30.2 − 0.2	18 44 26	−02 32 02	17.8	3.1	9.2	3,800 ± 1,700	1.5×10^4	17.6	24	4,900
W 43	30.8 − 0.0	18 45 01	−02 00 03	5.4	5.3	97.4	5,600 ± 860	3.4×10^5	10.9[2]	150	7,200
	31.1 + 0.0	18 45 18	−01 42 06	4.0	4.0	8.0	6,700 ± 3,200	5.4×10^4	9.9	61	2,200
W 44	34.3 + 0.1	18 50 48	01 11 07	1.5	2.5	15.0	7,500 ± 2,100	4.5×10^5	2.1[2]	380	140
W 48	35.2 − 1.7	18 59 15	01 09 04	1.0	1.8	15.3	6,500 ± 1,500	9.0×10^5	1.3	680	60
W 47	37.6 − 0.1	18 57 48	03 59 36	6.0	4.0	7.9	5,600 ± 2,200	3.4×10^4	5.6[2]	64	430
	37.9 − 0.4	18 59 19	04 09 26	19.3	4.4	24.4	6,500 ± 2,200	3.1×10^4	11.4[2]	43	2,400
W 49	43.2 − 0.0	19 07 54	09 01 01	1.5	2.0	49.8	7,700 ± 1,100	1.9×10^6	7.0	430	5,400
	48.6 + 0.0	19 18 07	13 49 18	4.6	4.8	10.6	5,900 ± 1,800	4.9×10^4	1.8[2]	140	29

W 51	49.0 − 0.3	19 20 08	14 02 49	12.2	13.0	111.3	7,400 ± 1,500	7.8×10^4	24.0	47	2.5×10^4
	49.1 − 0.4	19 20 32	14 03 30	4.0	5.6	15.0	4,300 ± 1,100	6.2×10^4	13.4	68	1,900
	49.2 − 0.3	19 20 42	14 13 57	2.0	2.0	12.0	4,700 ± 1,300	2.9×10^5	3.8	230	470
	49.4 − 0.3	19 20 53	14 18 36	3.5	3.5	37.2	7,000 ± 1,000	3.3×10^5	12.7	161	4,000
	49.5 − 0.4	19 21 23	14 24 29	4.7	3.0	117.4	7,600 ± 1,100	9.3×10^5	7.1	300	4,000
	51.2 − 0.1	19 23 10	16 08 14	20.9	22.4	37.0	4,200 ± 2,300	7.3×10^3	39.6	11	2.6×10^4
	61.5 + 0.1	19 44 42	25 05 04	1.2	1.0	6.1	3,400 ± 1,200	4.4×10^5	0.7^2	670	7
	63.2 + 0.4	19 47 12	26 43 08	2.7	2.5	4.8	3,700 ± 1,600	6.3×10^4	1.4^2	180	17
	70.3 + 1.6	19 59 51	33 24 27	3.4	2.2	14.5	7,200 ± 1,800	2.1×10^5	7.0	140	1,900
	75.8 + 0.4	20 19 47	37 19 53	1.7	6.2	14.1	6,200 ± 1,700	1.4×10^5	5.2	130	730
	76.4 − 0.6	20 25 33	37 12 55	1.8	2.2	13.5	6,800 ± 2,500	3.7×10^5	1.4	430	41
DR 9	78.0 + 0.0	20 27 44	38 52 12	3.9	2.8	5.6	6,100 ± 3,400	5.3×10^4	4.5	90	300
DR 6	78.0 + 0.6	20 25 22	39 15 56	3.4	3.4	9.8	7,200 ± 2,500	9.3×10^4	0.9^2	270	7
DR 4	78.1 + 1.8	20 20 44	40 02 20	6.9	14.0	25.6	9,700 ± 4,000	3.2×10^4	5.9	61	470
DR 13	78.2 − 0.4	20 29 52	38 48 00	8.6	8.8	24.3	8,100 ± 2,300	3.7×10^4	0.7^2	180	3
DR 5	78.5 + 1.2	20 24 08	40 00 20	8.4	26.2	40.2	5,800 ± 2,000	1.9×10^4	21.6	24	9,200
DR 15	79.2 + 0.3	20 30 22	40 03 00	5.0	5.0	7.5	7,000 ± 3,700	3.3×10^4	5.8	62	470
DR 7	79.3 + 1.3	20 26 22	40 41 39	3.7	3.6	15.6	7,300 ± 2,000	1.3×10^5	8.5	100	2,300
DR 10	80.0 + 1,5	20 27 53	41 24 34	21.1	35.5	73.4	6,300 ± 3,300	1.0×10^4	43.6	13	4.0×10^4
DR 19	80.1 − 0.1	20 34 48	40 31 26	23.8	8.1	30.0	8,900 ± 4,400	1.8×10^4	1.7^2	85	16
DR 18	80.4 + 0.5	20 33 15	41 05 00	4.4	32.2	29.0	9,100 ± 4,900	2.4×10^4	5.8	53	390
DR 20	80.9 + 0.4	20 35 10	41 26 02	3.5	4.6	7.6	5,600 ± 1,700	4.8×10^4	0.6^2	240	2
DR 22	80.9 − 0.2	20 37 42	41 07 35	7.4	6.4	21.7	6,000 ± 1,600	4.7×10^4	6.7	69	800
DR 17	81.3 + 1.2	20 33 23	42 15 17	9.1	15.5	41.8	6,500 ± 1,900	3.2×10^4	5.2	64	340
DR 23	81.5 + 0.0	20 39 01	41 42 49	16.4	21.8	64.5	7,000 ± 2,000	2.0×10^4	8.3	40	850
DR 21	81.7 + 0.5	20 37 13	42 08 51	0.3	0.4	19.0	7,200 ± 780	1.7×10^7	0.2	8,800	1.2×10^8
DR 16	82.4 + 2.5	20 30 53	43 55 39	5.7	23.8	17.2	6,300 ± 3,500	1.3×10^4	12.0	27	1,800
NGC 7027	84.8 − 3.5	21 04 37	41 57 51	<0.1	<0.1	6.4	15,000 ± 13,000	9.0×10^7	—	4.8×10^4	<1
NGC 7538	111.6 + 0.9	23 11 39	61 21 42	2.3	1.9	25.0	6,700 ± 1,300	6.1×10^2	3.0	370	380

[1] After REIFENSTEIN, WILSON, BURKE, MEZGER, and ALTENHOFF (1970) and WILSON, MEZGER, GARDNER, and MILNE (1970) (by permission of the European Southern Observatory).

[2] Radii marked with a superscript 2 are the smallest of two possible values.

When the radiofrequency approximation for optical depth given in Eq. (1-223) is used with Eq. (2-117), we obtain the relation

$$\frac{\Delta\nu_D \Delta T_L}{T_c} \approx 2.33\times 10^4\, \nu_{mn}^{2.1}\, T_e^{-1.15}\left(\frac{E_L}{E_c}\right) \text{ kHz}, \tag{2-121}$$

where the full width to half maximum of the line, $\Delta\nu_D$, is in kHz, the line frequency, ν_{mn}, is in GHz, the temperatures are in °K, and the ratio of the line emission measure, E_L, to the continuum emission measure, E_c, is given by

$$\frac{E_L}{E_c} \approx \frac{N_i/N_{H^+}}{1+(N_{He^+}+4N_{He^{++}})/N_{H^+}} \approx \frac{1}{1+N_{He^+}/N_{H^+}} \approx 0.93, \tag{2-122}$$

where N_i is the number density of the recombining ions, and N_{H^+}, N_{He^+}, and $N_{He^{++}}$ are, respectively, the number densities of ionized hydrogen, singly ionized helium, and doubly ionized helium. Eq. (2-121) indicates that the electron temperature, T_e, may be obtained by observing $\Delta\nu_D$, ΔT_L, and T_c. The emission measure $E_L=\int_0^L N_e N_i dl$ may then be obtained by using this value of T_e in Eq. (2-117). Eq. (2-121) applies when the ionized region is in local thermodynamic equilibrium (LTE). In non-LTE conditions, the formulae developed in Sect. 2.10 may be used to relate observed line parameters to the physical parameters of H II regions. Values of electron temperature and emission measure obtained using Eqs. (2-117) and (2-121) are given in Table 12 together with other parameters for two hundred H II regions.

2.13. Atomic Fine Structure

For an atom with several electrons, the fine structure is usually described by the quantum numbers based on Russell-Saunders (LS) coupling (RUSSELL and SAUNDERS, 1925). (The LS-coupling is valid for the elements with not very large Z, except for the noble gases (Ne, Ar, ...) and the highly excited states).

n = principal quantum number (cf. BOHR, 1913, and Eqs. (2-4), (2-8)).

L = total orbital quantum number in units of $h/2\pi$. The L is the vectoral sum of the orbital angular momentum, l, of all of the electrons. L is always integral and the $L=0, 1, 2, 3, 4$, and 5 is designated as S, P, D, F, G, and H, respectively. The l of the electron takes on $n-1$ values.

S = total spin angular momentum in units of $h/2\pi$. The S is the vectoral sum of the spin angular momentum, s, of all the electrons. An electron has $s=1/2$, and S takes on integral or half integral values for atoms having an even or odd number of electrons (cf. UHLENBECK and GOUDSMIT, 1925, 1926; THOMAS, 1926, 1927; PAULI, 1925, 1927).

J = the total angular momentum in units of $h/2\pi$. It is the vector sum $L+S$ for LS coupling. The number of possible J values is $2L+1$ or $2S+1$ according as $L<S$ or $L\geq S$.

Each pair of L and S designate a term, and each level of the term has the symbol $n^{2S+1}L_J$, where L is given by its alphabetical symbol. The possible transitions between all levels in two terms produce a neighboring set of lines called a multiplet, first observed by CATALÁN (1922). The transitions must follow the selection rules given in Table 13.

Table 13. Selection rules for atomic spectra (cf. PAULI, 1925; LAPORTE, 1924; SHORTLEY, 1940)

	Electric dipole (allowed)	Magnetic dipole (forbidden)	Electric quadrupole (forbidden)
(1)	$\Delta J=0, \pm 1$ $(0 \nleftrightarrow 0)$	$\Delta J=0, \pm 1$ $(0 \nleftrightarrow 0)$	$\Delta J=0, \pm 1, \pm 2$ $(0 \nleftrightarrow 0, \frac{1}{2} \nleftrightarrow \frac{1}{2}, 0 \nleftrightarrow 1)$
(2)	$\Delta M=0, \pm 1$	$\Delta M=0, \pm 1$	$\Delta M=0, \pm 1, \pm 2$
(3)	Parity change	No parity change	No parity change
(4)	One electron jump $\Delta l=\pm 1$ For $L-S$ coupling	No electron jump $\Delta l=0$ $\Delta n=0$	One or no electron jump $\Delta l=0, \pm 2$
(5)	$\Delta S=0$	$\Delta S=0$	$\Delta S=0$
(6)	$\Delta L=0, \pm 1$ $(0 \nleftrightarrow 0)$	$\Delta L=0$	$\Delta L=0, \pm 1, \pm 2$ $(0 \nleftrightarrow 0, 0 \nleftrightarrow 1)$

The symbol $\nleftrightarrow$ denotes "cannot combine with". For example, the $0 \nleftrightarrow 0$ following $\Delta J=0, \pm 1$ means that a level with $J=0$ cannot combine with another level with $J=0$. A parity change means that terms whose summed electronic l are even can only combine with those whose summed l are odd, and vice versa. Once the appropriate transitions have been determined, line frequencies ν_{mn}, and the appropriate A_{mn}, f_{nm}, or S_{mn} may be computed using the detailed formulae of quantum mechanics, or by measuring them in a laboratory. As an example, the fine structure energy, E_{fs}, arising between the interaction of the L and S of a hydrogen-like atom is given by (LANDÉ, 1923; THOMAS, 1926)

$$E_{\mathrm{fs}}=\alpha^2 \frac{h c R Z^4}{n^3}\left[\frac{J(J+1)-L(L+1)-S(S+1)}{L(L+1)(2L+1)}\right], \tag{2-123}$$

where the fine structure constant $\alpha=2\pi e^2/(h c) \approx 7.2973 \times 10^{-3}$, the Rydberg constant $R=R_{\mathrm{A}}=2\pi^2 \mu e^4/(c h^3)$ where R_{A} is given in Eq. (2-113a), the reduced mass $\mu=[m M_{\mathrm{A}}/(m+M_{\mathrm{A}})]$, and Z is the ionic charge. The frequency of any transition then follows from Eqs. (2-7) and (2-123). Eq. (2-123) expresses Landé's interval rule which states that the interval between the level of a given J and that of $J-1$ is proportional to J.

The energies and wavelengths are conveniently found from the Grotrian (1930) diagrams which graphically represent the relations of characteristic spectrum lines to the quantum energy levels of an atom or ion. The Grotrian diagrams for atoms and ions of astrophysical interest are given in Fig. 13. Multiplets are represented by straight lines: the line is solid for transitions between terms of the same multiplicity, dashed for transitions between terms of different multiplicity (intersystem multiplets) and dot-dashed for transitions whose upper terms are metastable (forbidden transitions between terms of the same parity). Wavelengths greater than 2000 Å are IA values in air, whereas those

Table 14. Corresponding lines in spectra of various elements—for use with Grotrian diagrams[1]

Key letter	Multiplet designation	λ	Mult. no.	λ	Mult. no.	λ	Mult. no.
		CI		NII		OIII	
A...	$2p^{2\,3}P - 2p^{2\,1}D$	(9,849.3 *P*) (9,823.1 *P*)	(1 F)	6,583.4 6,548.1	(1 F)	5,006.9 4,958.9	(1 F)
B...	$2p^{2\,1}D - 2p^{2\,1}S$	(8,727.6 *P*)	(3 F)	5,754.6	(3 F)	4,363.2	(2 F)
C...	$3s\ ^3P^\circ - 3p\ ^3P$	9,094.9 9,078.3 9,111.8 9,088.6 9,061.5 9,062.5	(3)	4,630.5 4,613.9 4,643.1 4,621.4 4,601.5 4,607.2	(5)	3,047.1 3,035.4 3,059.3 3,043.0 3,023.4 3,024.6	(4)
D...	$3s\ ^1P^\circ - 3p\ ^1P$	14,540.2		6,482.1	(8)	5,592.4	(5)
E...	$3s\ ^1P^\circ - 3p\ ^1D$	9,405.8	(9)	3,995.0	(12)	2,983.8	(6)
		CII		NIII		OIV	
A...	$2p\ ^2P^\circ - 3s\ ^2S$	858.6 858.1	(UV 4)	452.2 451.9	(UV 4)	279.9 279.6	(UV 4)
B...	$2p\ ^2P^\circ - 2p^{2\,2}D$	1,335.7 1,334.5	(UV 1)	991.6 989.8 991.5	(UV 1)	790.2 787.7 790.1	(UV 1)
C...	$3s\ ^2S - 3p\ ^2P^\circ$	6,578.0 6,582.8	(2)	4,097.3 4,103.4	(1)	3,063.5 3,071.7	(1)
D...	$3p\ ^2P^\circ - 3d\ ^2D$	7,236.2 7,231.1	(3)	4,640.6 4,634.2 4,641.9	(2)	3,411.8 3,403.6 3,413.7	(2)
		CIII		NIV		OV	
A...	$2s^{2\,1}S - 2p\ ^1P^\circ$	977.0	(UV 1)	765.1	(UV 1)	629.7	(UV 1)
B...	$2s^{2\,1}S - 3p\ ^1P^\circ$	386.2	(UV 2)	247.2	(UV 2)	172.2	(UV 2)
C...	$2p\ ^3P^\circ - 3s\ ^3S$	538.3 538.2 538.1	(UV 5)	322.7 322.6 322.5	(UV 4)	215.2 215.1 215.0	(UV 4)
D...	$2p\ ^1P^\circ - 2p^{2\,1}D$	2,296.9	(UV 8)	1,718.5	(UV 7)	1,371.3	(UV 7)
E...	$2p\ ^1P^\circ - 2p^{2\,1}S$	1,247.4	...	955.3	(UV 8)	774.5	(UV 8)
F...	$2p\ ^1P^\circ - 3s\ ^1S$	690.5	(UV 10)	387.4	(UV 9)	248.5	(UV 9)
G...	$2p\ ^1P^\circ - 3d\ ^1D$	574.3	(UV 11)	335.0	(UV 10)	220.4	(UV 10)
H...	$3s\ ^3S - 3p\ ^3P^\circ$	4,647.4 4,650.2 4,651.4	(1)	3,478.7 3,483.0 3,484.9	(1)	2,781.0 2,787.0 2,789.9	...
I ...	$3p\ ^1P^\circ - 3d\ ^1D$	5,696.0	(2)	4,057.8	(3)	3,144.7	(2)

Table 14 (continued)

Key letter	Multiplet designation	λ N I	Mult. no.	λ O II	Mult. no.	λ Ne IV	Mult. no.
A...	$2p^{3\,4}S^{\circ}-2p^{3\,2}D^{\circ}$	5,200.4 5,197.9	(1 F)	3,728.8 3,726.0	(1 F)	$(2{,}441 \pm P)$ $(2{,}438 \pm P)$	...
B...	$2p^{3\,2}D^{\circ}-2p^{3\,2}P^{\circ}$	(10,396.5 *P*) (10,406.2 *P*) (10,396.5 *P*) (10,406.2 *P*)	(3 F)	7,319.9 7,330.2 7,319.9 7,330.2	(2 F)	4,714.2 4,725.6 4,715.6 4,724.2	(1 F)
C...	$2p^{3\,4}S^{\circ}-2p^{3\,2}P^{\circ}$	(3,466.4 *P*) (3,466.4 *P*)	(2 F)	(2,470.4 *P*) (2,470.3 *P*)	...	$(1{,}609 \pm P)$ $(1{,}609 \pm P)$	...
D...	$3s\ ^4P-3p\ ^4D^{\circ}$	8,680.2 8,683.4 8,686.1 8,718.8 8,711.7 8,703.2	(1)	4,649.1 4,641.8 4,638.9 4,676.2 4,661.6 4,650.8	(1)	—	—
E...	$3s\ ^4P-3p\ ^4P^{\circ}$	8,216.3 8,210.6 8,200.3 8,242.3 8,223.1 8,184.8 8,188.0	(2)	4,349.4 4,336.9 4,325.8 4,367.0 4,345.6 4,319.6 4,317.1	(2)	(2,203.6 *P*) (2,192.4 *P*) (2,188.0 *P*) (2,220.3 *P*) (2,206.2 *P*) (2,176.1 *P*) (2,174.4 *P*)	...
F...	$3s\ ^4P-3p\ ^4S^{\circ}$	7,468.3 7,442.3 7,423.6	(3)	3,749.5 3,727.3 3,712.8	(3)	(1,875.2 *P*) (1,855.3 *P*) (1,842.4 *P*)	...

[1] After MOORE and MERRILL (1968).

less than 2000 Å are values in vacuum: In order to use one diagram for more than one element, corresponding lines in spectra of various elements are given in Table 14. Both the diagrams and the table are from data given by MOORE and MERRILL (1968).

The line strength for the electric dipole radiation in *LS* coupling is given by

$$S_{mn}=\mathcal{S}(\mathcal{M})\,\mathcal{S}(\mathcal{L})\,\sigma_{mn}^2\,, \tag{2-124}$$

where $\mathcal{S}(\mathcal{M})$ depends on the particular multiplet of the transition array, $\mathcal{S}(\mathcal{L})$ depends on the line of the multiplet, and

$$\sigma_{mn}^2=(4l^2-1)^{-1}\left\{\int_0^{\infty} P_m P_n r'\,dr'\right\}^2, \tag{2-125}$$

where P_m/r' and P_n/r' are the radial wave functions for the upper and lower states of the transition when the optical electron is in configurations with azimuthal quantum numbers l and $l-1$. Here the wave functions are normalized in atomic units (HARTREE, 1928) for which $h/2\pi = e = m = 1$. They are solutions to the time independent wave equation (SCHRÖDINGER, 1925, 1926)

$$H\psi = E\psi\,, \tag{2-126}$$

where the wave function has an amplitude, ψ, and energy state, E, and is operated on by the Hamiltonian operator, H, given by

$$H = \left[-\frac{h^2}{8\pi^2 m} \nabla^2 + V(r) \right], \tag{2-127}$$

for a particle of mass, m, in a scalar potential $V(r)$. Eqs. (2-126) and (2-127) follow directly from assuming that a particle of velocity v has a de Broglie wavelength (DE BROGLIE, 1923, 1925) given by $\lambda = h/(mv)$, that its quantized angular momentum is given by Eq. (2-4), and that the wave function satisfies a plane wave equation.

For a complete solution to the Russell-Saunders case, we must use the Hamiltonian

$$H = \sum_{i=1}^{p} \left[-\frac{\hbar^2}{2m} \nabla_i^2 - \frac{Ze^2}{r_i} \right] + \sum_{ij} \frac{e^2}{r_{ij}}, \tag{2-128}$$

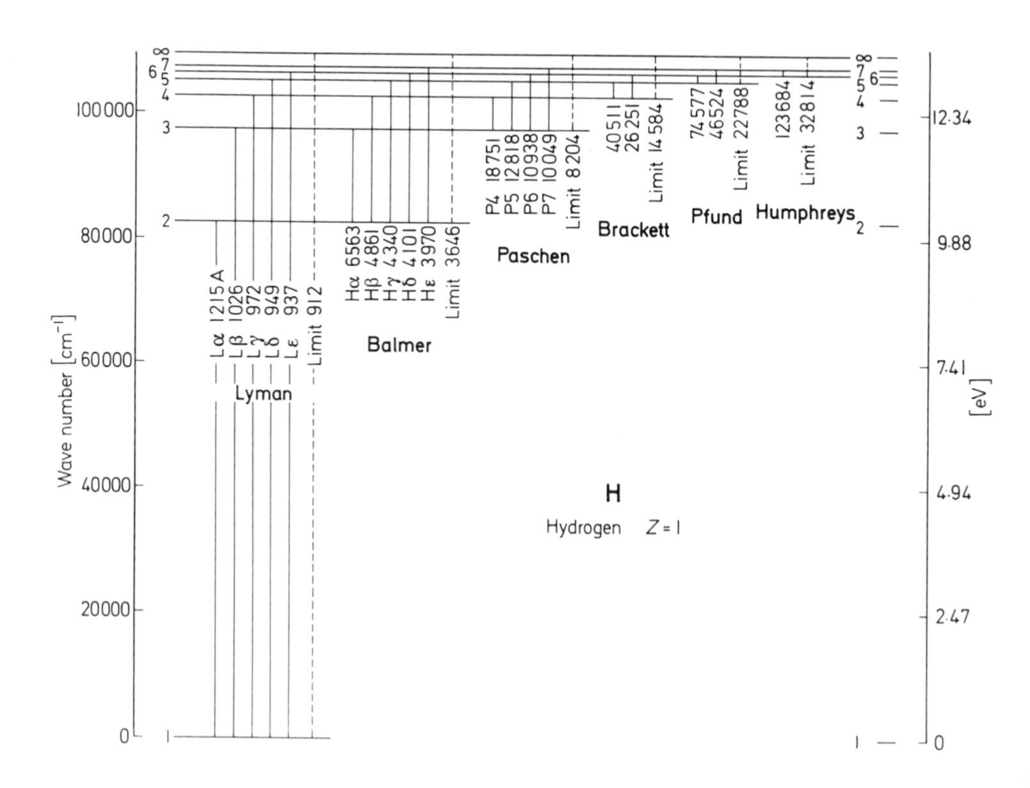

Fig. 13. Grotrian diagrams for H, He I, He II, C II, C III, C IV, O I, O II, O III, N II, Mg I, Mg II, Ca I, Ca II, Fe I, and Fe II. (After MOORE and MERRILL, 1968)

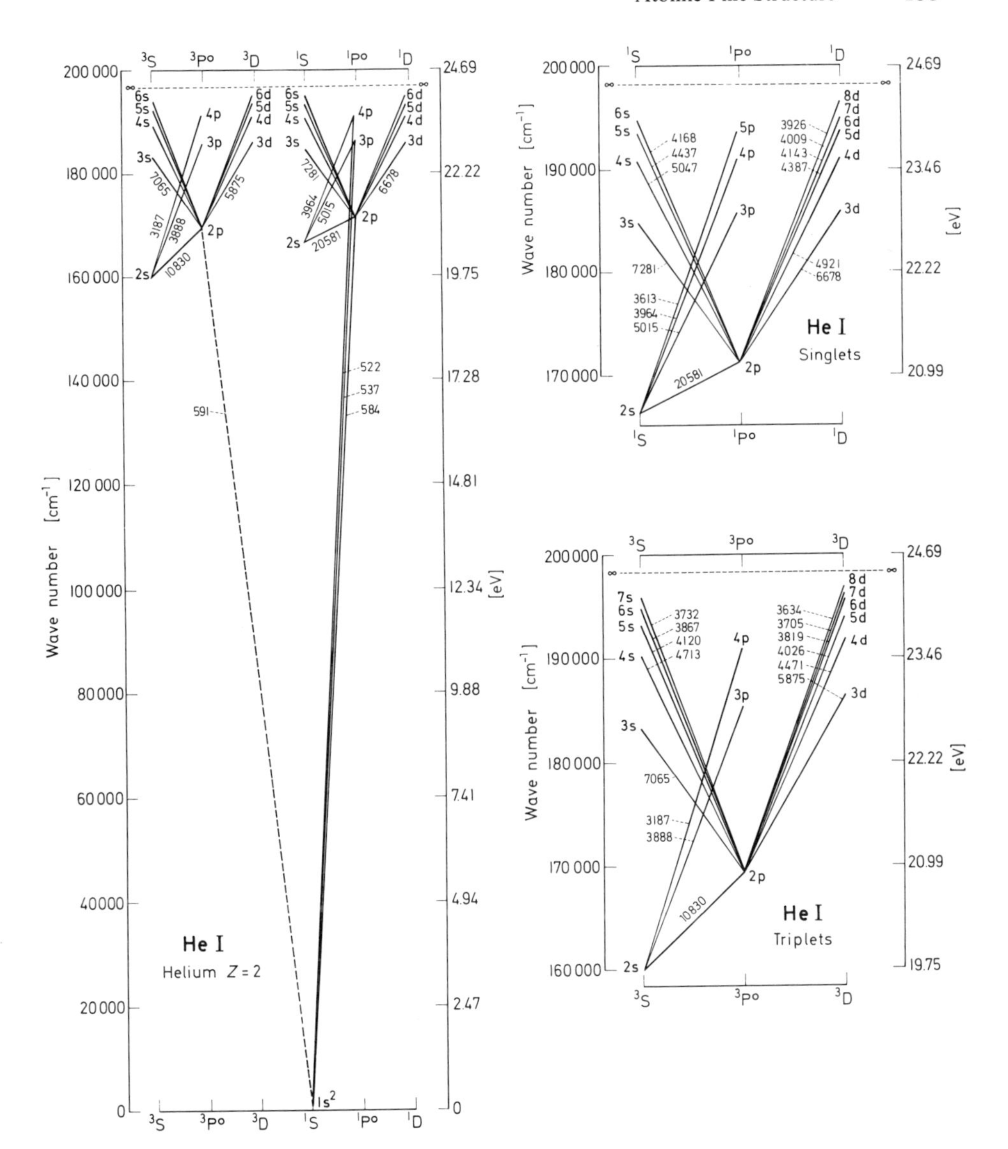

Fig. 13. Grotrian diagrams (continued)

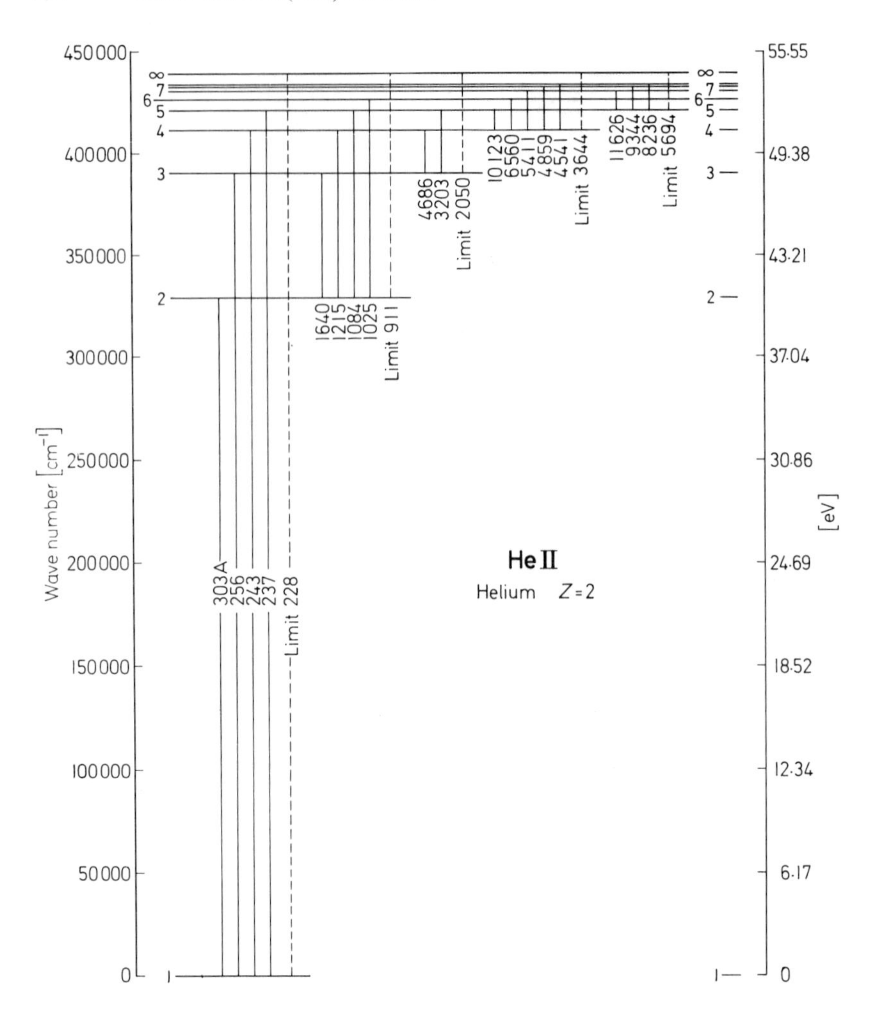

Fig. 13. Grotrian diagrams (continued)

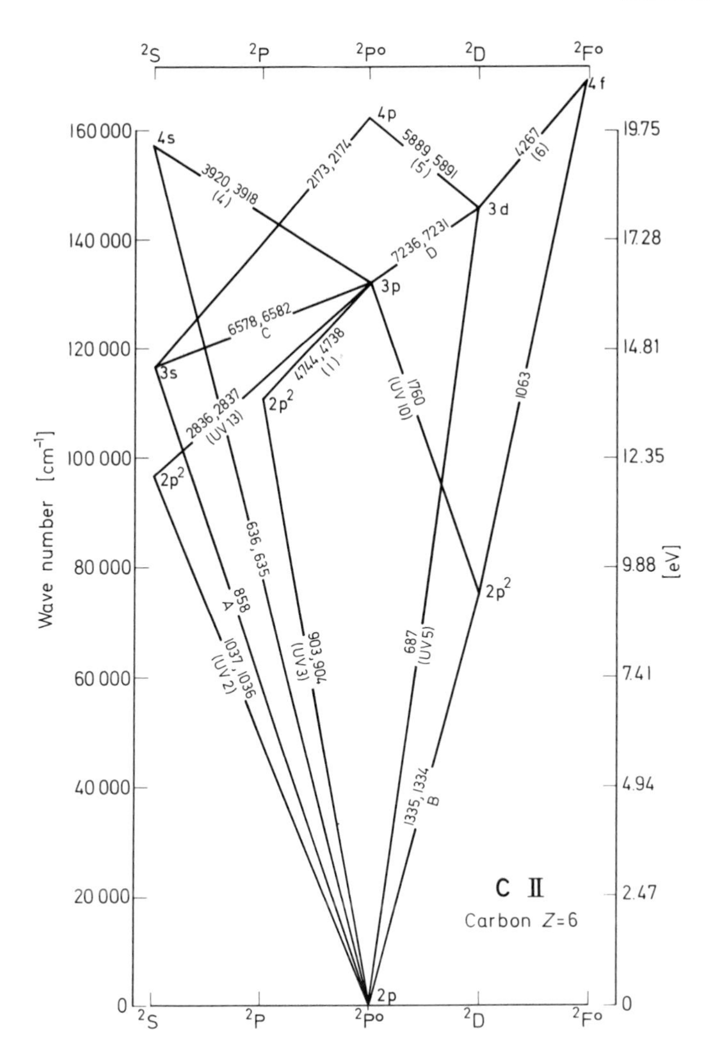

Fig. 13. Grotrian diagrams (continued)

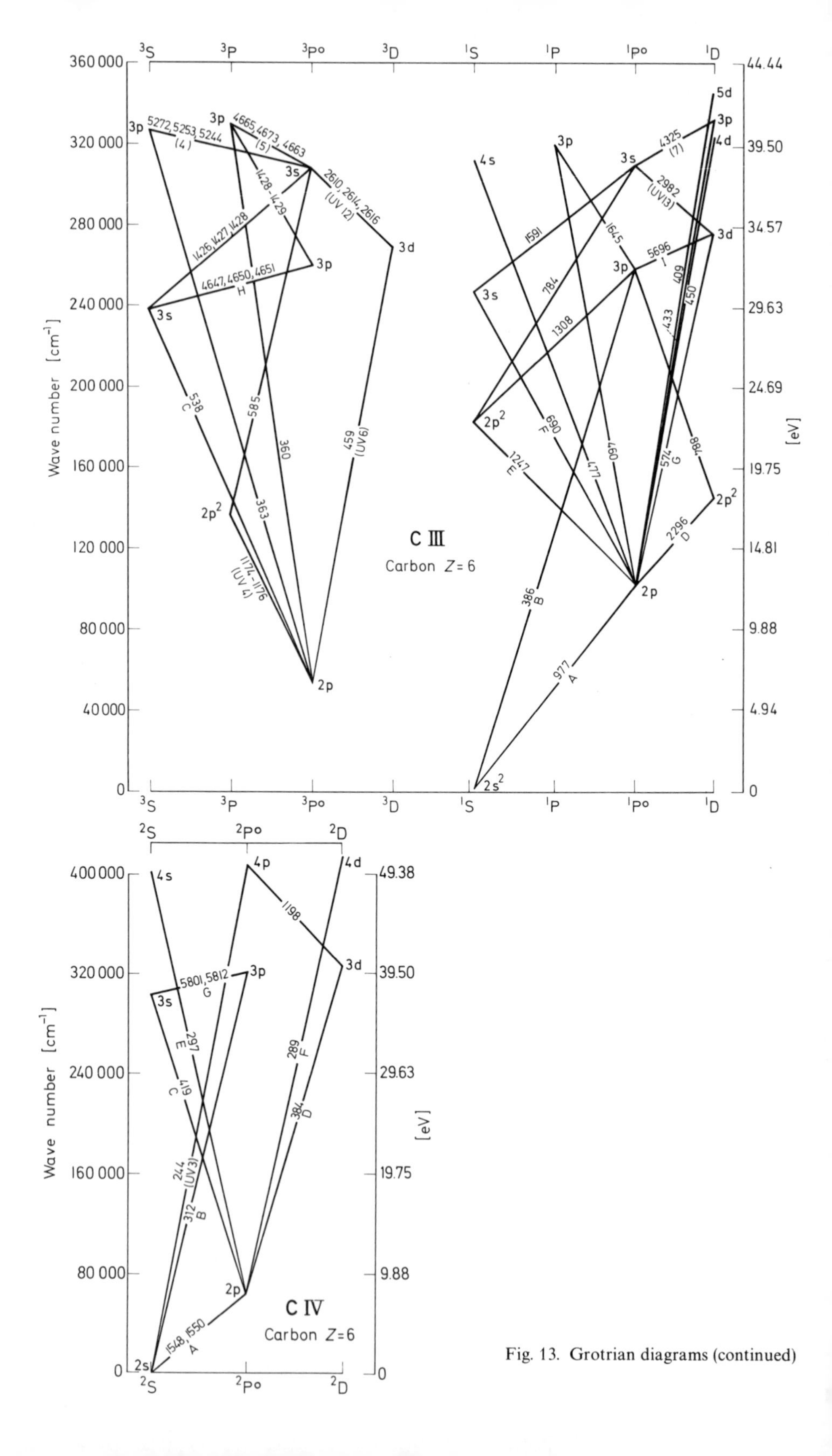

Fig. 13. Grotrian diagrams (continued)

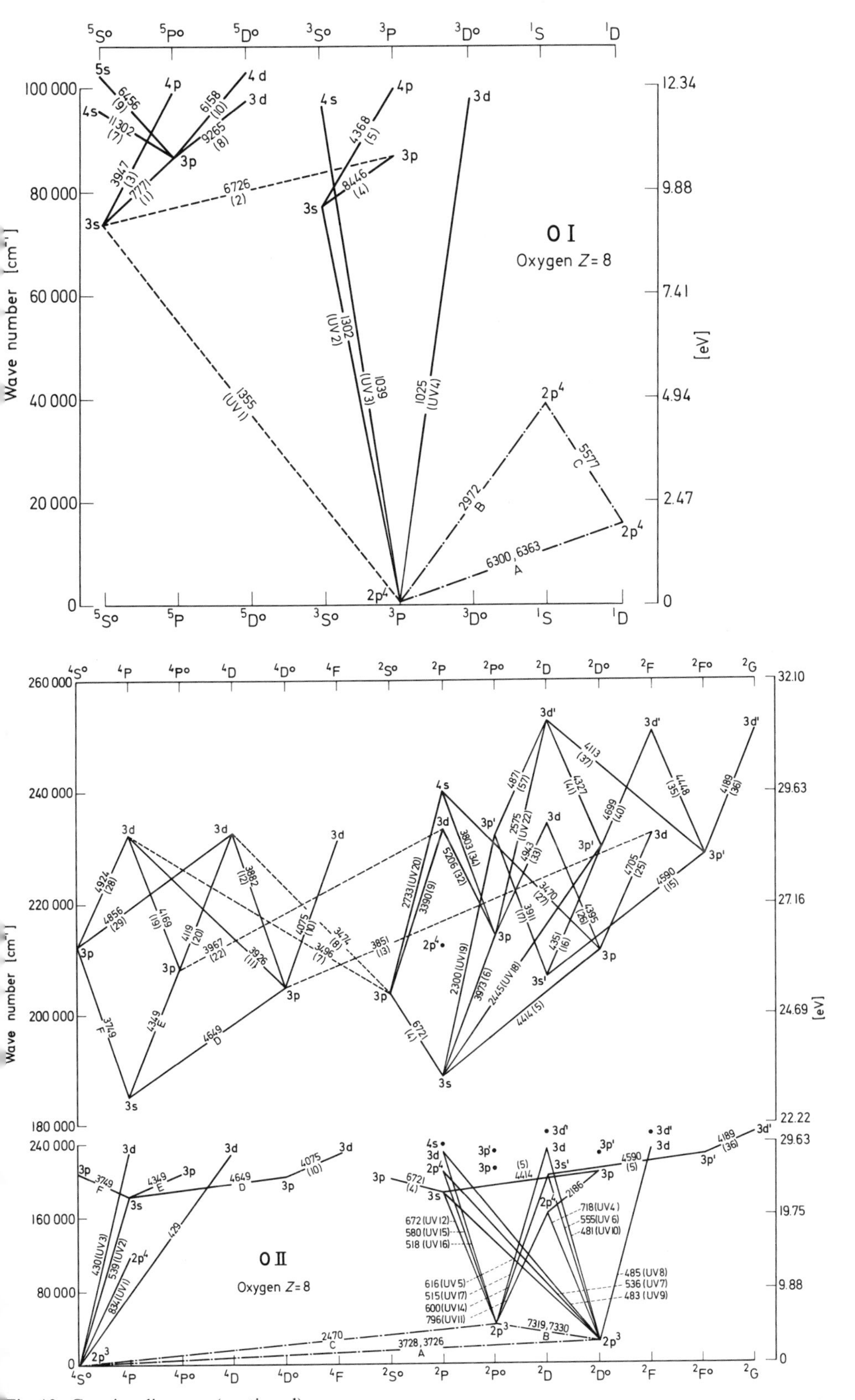

Fig. 13. Grotrian diagrams (continued)

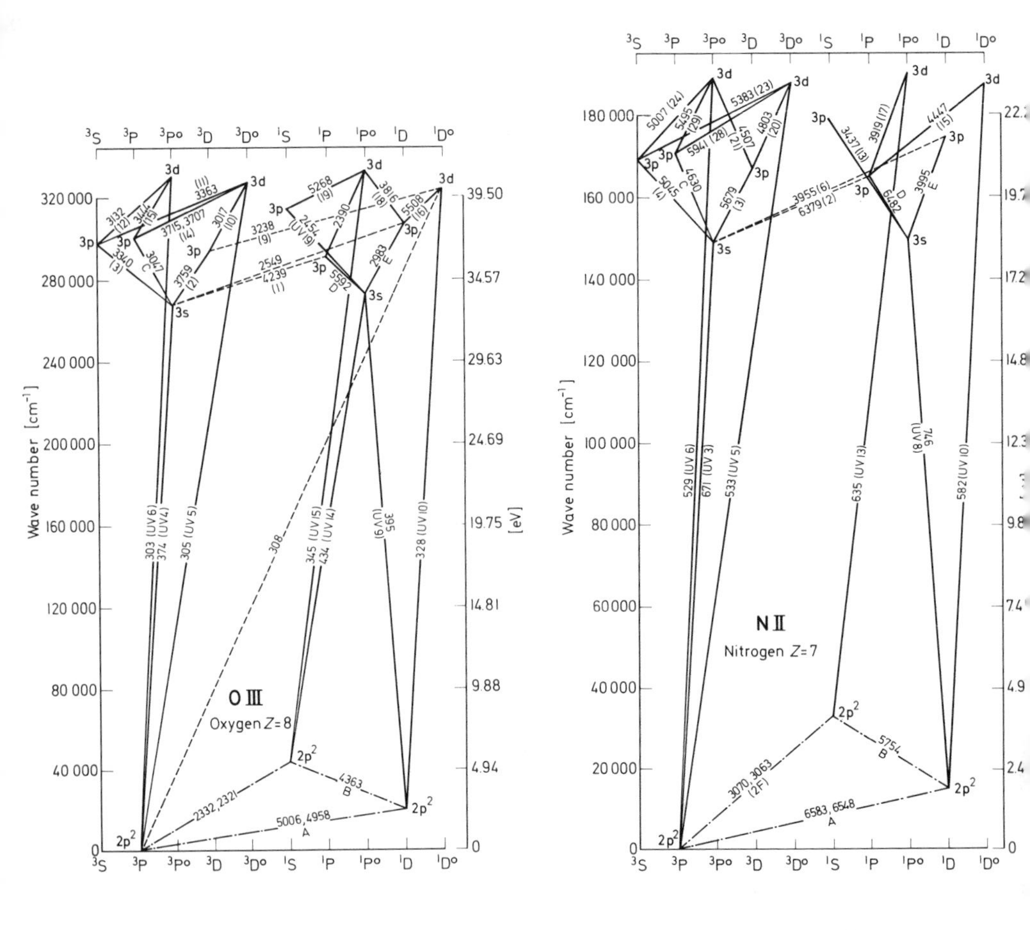

Fig. 13. Grotrian diagrams (continued)

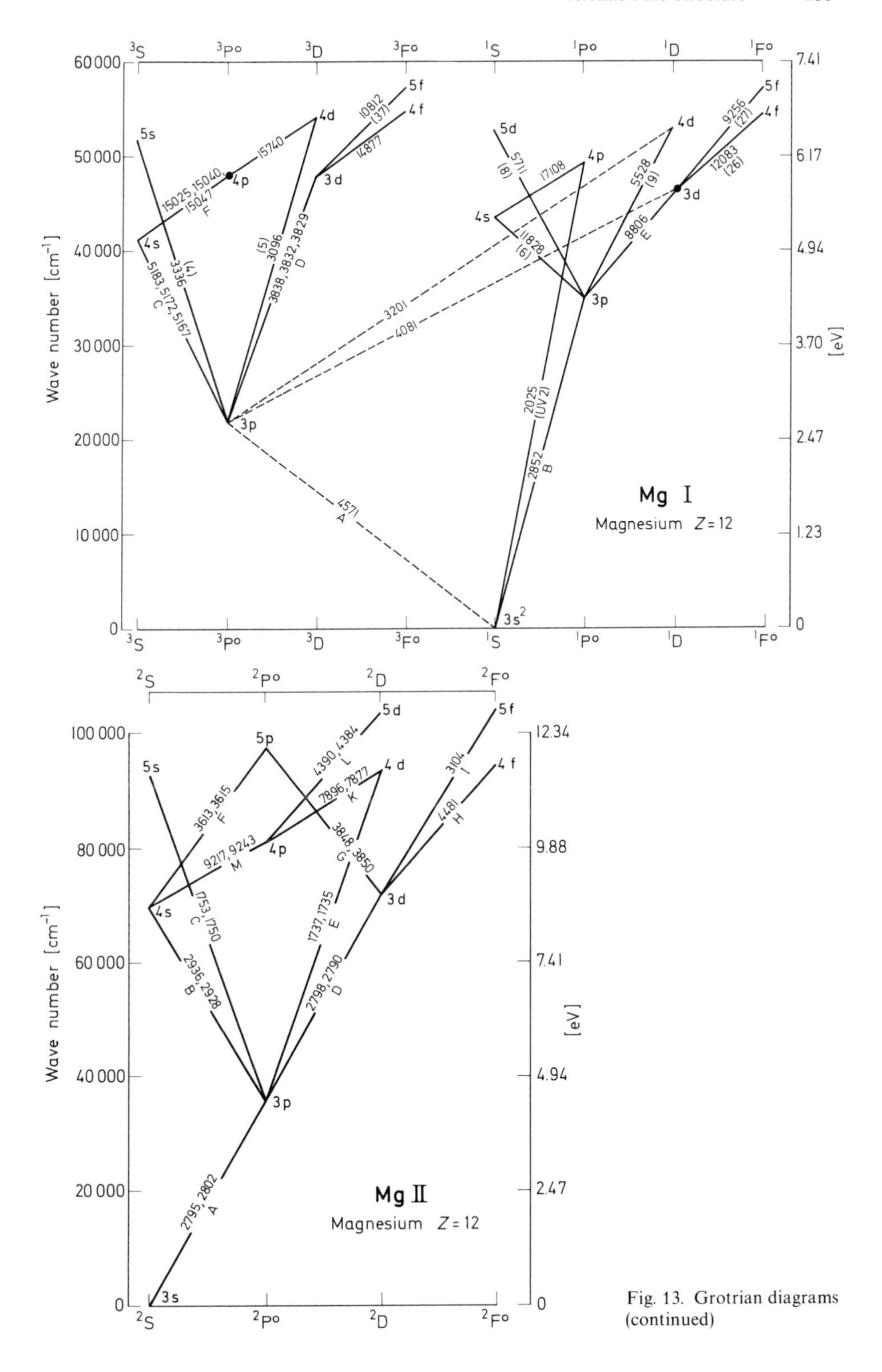

Fig. 13. Grotrian diagrams (continued)

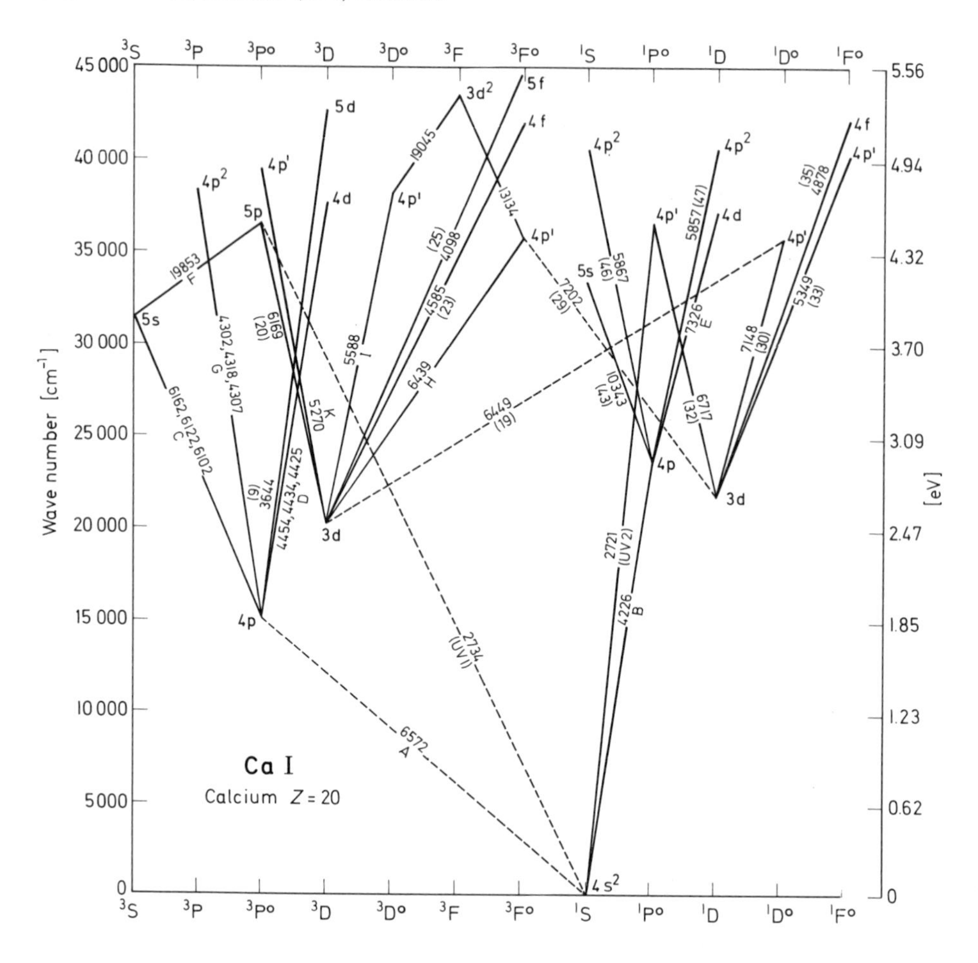

Fig. 13. Grotrian diagrams (continued)

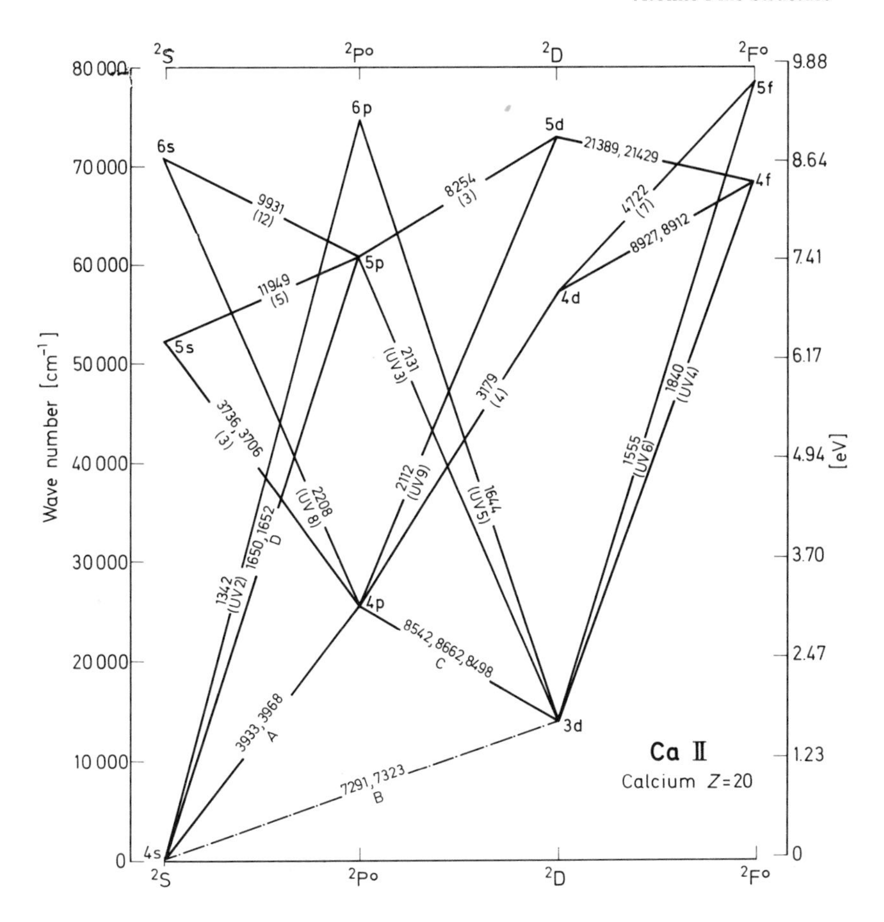

Fig. 13. Grotrian diagrams (continued)

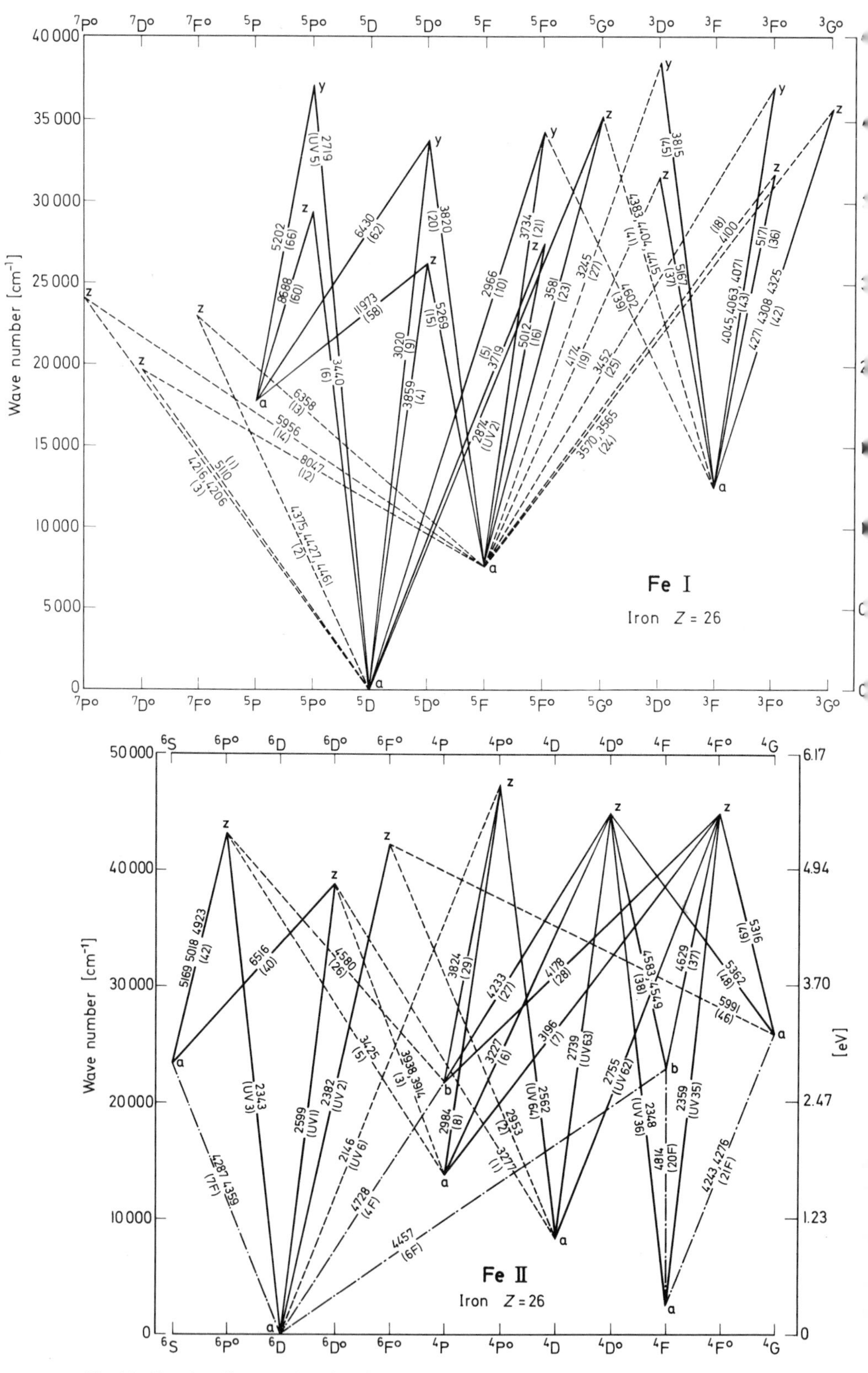

Fig. 13. Grotrian diagrams (continued)

where Z is the atomic number, p is the number of electrons, the term $\sum Ze^2/r_i$ accounts for the Coulomb interaction of the electrons with the nucleus, and the double sum accounts for the Coulomb interaction of the electrons themselves (hence including each pair only once, omitting $i=j$, and having $p(p-1)/2$ terms). A simpler Hamiltonian is obtained in the central field approximation (HARTREE, 1928) for which the Hamiltonian is given by

$$H=\sum_{i=1}^{p}\left[-\frac{\hbar^2}{2m}\nabla_i^2+V(r_i)\right], \tag{2-129}$$

where the potential $V(r)$ tends to $-Ze^2/r$ as $r\to 0$ and $-(Z-p+1)e^2/r$ for large r.

Solutions for the relative strengths of the lines within a multiplet were first computed by RUSSELL (1925) and DIRAC (1926). Useful tables for solutions are given by RUSSELL (1936). With reference to Eq. (2-124), BATES and DAMGAARD (1949) provide tables for evaluating σ_{mn}; whereas the relative multiplet strengths $\mathscr{S}(\mathscr{M})$ compiled by GOLDBERG (1935) and MENZEL and GOLDBERG (1936) and the relative line strengths $\mathscr{S}(\mathscr{L})$ compiled by WHITE and ELIASON (1933) and RUSSELL (1936) are given in the book by ALLEN (1963). Certain sum rules are useful for checking the relative strengths of lines. The Burger-Dorgelo-Ornstein sum rule (BURGER and DORGELO, 1924; ORNSTEIN and BURGER, 1924) states that the sum of the intensities of all lines in a LS multiplet which belong to the same initial or final state is proportional to the statistical weight, $2J+1$, of the initial or final state, respectively. The Reiche-Thomas-Kuhn f-sum rule (THOMAS, 1925; KUHN, 1925) states that the sum of the f values for all lines arising from a single atomic energy level shall equal the number of optical electrons (electrons which participate in giving the optical line and continuous spectrum). Additional useful sum rules have been derived by SHORTLEY (1935), MENZEL and GOLDBERG (1936), MENZEL (1947), and ROHRLICH (1959).

Tables of atomic energy levels are given by C. E. MOORE (1949, 1952, and 1958). Values of line frequencies are given by C. E. MOORE (1950–69) and P. W. MERRILL (1958). A bibliography of atomic forbidden line frequencies and strengths is given by GARSTANG (1962). A bibliography of f values and transition probabilities has been published by GLENNON and WIESE (1962). Some line strengths are given by VARSAVSKY (1961). Ultraviolet and forbidden line strengths are given by OSTERBROCK (1963), BURBIDGE and BURBIDGE (1967), and TARTER (1969). ALLEN (1963) has compiled the most complete catalog of allowed and forbidden atomic line frequencies together with their oscillator strengths and transition probabilities. Extensive bibliographies for atomic and molecular line frequencies and strengths are given in the reports of commission 14 of the International Astronomical Union (1955, 1958, 1961, 1964, 1967, 1970 ...).

The wavelengths of the strongest observed atomic lines from the most abundant elements [hydrogen (H), helium (He), oxygen (O), carbon (C), nitrogen (N), neon (Ne), silicon (Si), magnesium (Mg), sulpher (S), and argon (Ar)] are given in Table 15. The symbols I, II, III, ... denote atoms which are, respectively, neutral, singly ionized, doubly ionized, ... Absorption lines most likely to be found in the spectra of quasi-stellar objects are discussed by BAHCALL (1968) BAHCALL, GREENSTEIN, and SARGENT (1968) and SHIELDS, OKE, and SARGENT (1972). The wavelengths of these lines are given in Table 16.

Table 15. A table of the wavelengths, λ, of the strongest ultraviolet, optical and near-infrared lines from the most abundant elements in their most abundant stages of ionization. Forbidden transitions are denoted by brackets[1]

Ion	λ Å	Ion	λ Å	Ion	λ Å	Ion	λ Å	Ion	λ Å
N II	917	N IV	1,488	O III	3,047	[S II]	4,069, 4,076	Hα	6,563
Ar II	920, 932	C IV	1,549, 1,551	[N II]	3,063	N III	4,097	[N II]	6,583
S VI	933	[Ne V]	1,575	O III	3,133	He II	4,100	He I	6,678
C III	977	[Ne IV]	1,602	He II	3,203	Hδ	4,102	[S II]	6,717, 6,734
N III	990, 992	He II	1,640	O III	3,299, 3,312	He II	4,200	[Ar V]	7,006
He II	993	O III	1,661, 1,663	O III	3,341	Hγ	4,340	He I	7,065
Ar VI	992, 1,002	O III	1,667	[Ne III]	3,343	[O III]	4,363	[Ar III]	7,136
Ar VI	1,014, 1,023	N III	1,750	[Ne V]	3,346	He I	4,388, 4,471	[O II]	7,320, 7,325, 7,330
Ne VI	1,020	Si II	1,808, 1,817	O IV	3,412	He II	4,541	[Ar III]	7,751
O VI	1,035	[Ne III]	1,815	[Ne V]	3,426	Mg I	4,571	[Cl IV]	8,046
C II	1,037	Hγ, C III	1,909	O III	3,429, 3,444	N III	4,634, 4,641	He II	8,237
S IV	1,073	N II	2,141	H I	3,691, 3,697	C III	4,647	P_{13}	8,665
N II	1,084, 1,086	[O III]	2,321	H I	3,704, 3,712	C IV, [Fe III]	4,658	P_{12}	8,750
He II, N II	1,085	C II	2,326	H I, [S III]	3,722	He II	4,686	P_{11}	8,862
Si II	1,194	[Ne IV]	2,424, 2,426	[O II]	3,727, 3,729	[Ar IV], He I	4,711	P_{10}	9,014
S III	1,201	[O II]	2,470	H I	3,734, 3,750	[Ne IV]	4,724, 4,726	[S III]	9,069
Si III	1,207	He II, [Mg VII]	2,512	O III	3,760	[Ar IV]	4,740	P_9	9,229
Lyα, O V	1,216	[Mg VII]	2,632	H I	3,771, 3,798	Hβ	4,861	P_ε, [S III]	9,545
N V	1,240, 1,243	He II	2,734	He I	3,820	He I	4,922	C III	9,710
S II	1,261	[Mg V]	2,786	H I	3,835	[O III]	4,959, 5,007	Na I	9,961
Si II	1,265	Mg II	2,796, 2,799	[Ne III]	3,869	He II	5,412	P_δ	10,049
C II	1,336	Mg II	2,804	Hζ	3,889	[N II]	5,755	He II	10,120, 10,124
Si IV	1,394, 1,403	[Ar IV]	2,855	He I	3,965	He I	5,876	[S II]	10,320
O IV	1,402, 1,405	[Ar IV]	2,869	[Ne III]	3,968	[O I] [S III]	6,300	Si I	10,371, 10,603
O IV	1,406	[Mg V]	2,931	Hε, [Ne III]	3,970	[O I]	6,364	Si I	10,627, 10,689
O IV	1,410, 1,413	[Ne V]	2,973	He I, He II	4,026	[N II]	6,549	C I	10,691

[1] From Aller, Bowen, and Wilson (1963), O'Dell (1963), Osterbrock (1963), Burbidge and Burbidge (1967), and Tarter (1969).

Table 16. The wavelengths of the absorption lines most likely to be found in the spectra of quasi-stellar objects[1]

Ion[2]		Wavelength (Å)	Ion[2]		Wavelength (Å)
H I	(0.1)	949.74	C II		1,335.3
H I	(0.2)	972.54	Si IV	(2)	1,393.76
C III		977.03	Si IV	(1)	1,402.77
Si II	(0.15)	989.87	Si II	(0.15)	1,526.72
Si II[2]	(0.3)	992.68	Si II[2]	(0.3)	1,533.45
H I	(0.5)	1,025.72	C IV	(2)	1,548.20
O VI	(2)	1,031.95	C IV	(1)	1,550.77
O VI	(1)	1,037.63	Al II		1,670.81
N II		1,085.10	Al III	(2)	1,854.72
Fe III		1,122.53	Al III	(1)	1,862.78
Fe II		1,144.95	Fe II		2,344.2
Si II	(0.6)	1,193.28	Fe II		2,374.5
Si II[2]	(1.4)	1,194.50	Fe II		2,382.8
Si III		1,206.51	Fe II		2,586.7
H I	(3.3)	1,215.67	Fe II		2,600.2
N V	(2)	1,238.81	Mg II	(2)	2,795.50
N V	(1)	1,242.80	Mg II	(1)	2,802.70
Si II	(1)	1,260.42	He I	(0.7)	3,187.74
Si II[2]	(2)	1,264.76	He I	(2.5)	3,888.65

[1] After BAHCALL (1968) by permission of the American Astronomical Society and the University of Chicago Press.

[2] A superscript 2 is placed after the ion label for lines arising from excited line structure states. The expected relative strengths are given in parenthesis.

There are only a few atomic fine structure transitions which might be detectable in the radio frequency range. The most likely to be detected are the $n=2$ hydrogen $^2P_{3/2}-{}^2S_{1/2}$ transition, $\nu_{mn}=9.9126\times 10^9$ Hz and the $n=2$ hydrogen $^2S_{1/2}-{}^2P_{1/2}$ transition, $\nu_{mn}=1.0578\times 10^9$ Hz. As the levels are very unstable, these line widths will be very wide, $\Delta\nu_L \approx 10^8$ Hz.

2.14. Atomic Hyperfine Structure

When the spin of the nucleus is important, two additional quantum numbers are added to the description of the atom.

I = total spin angular momentum of the nucleus in units of $h/2\pi$. The I is positive and takes on integral or half integral values.

F = the total atomic angular momentum in units of $h/2\pi$. It is the vector sum of J and I, and takes on $2J+1$ or $2I+1$ values according as $J<I$ or $J\geq I$.

The nucleus has a magnetic moment, μ, given by

$$\mu=\mu_N g_N I, \tag{2-130}$$

where g_N is the Landé nuclear factor ($g_N \approx 1$) and $\mu_N = he/(4\pi m_p c)$ is the Bohr nuclear magneton for the proton of mass m_p. This magnetic moment interacts with the magnetic field produced by the electrons to give the hyperfine splitting. The energy, E_{hfs}, of the hyperfine structure of hydrogen like atoms is (FERMI, 1930; BETHE, 1933)

$$E_{hfs} = g_N \left(\frac{m}{m_p}\right) \frac{\alpha^2 hcRZ^3}{n^3} \left[\frac{F(F+1) - I(I+1) - J(J+1)}{J(J+1)(2J+1)}\right], \qquad (2\text{-}131)$$

where the fine structure constant $\alpha = 2\pi e^2/(hc) = 7.2973 \times 10^{-3}$, the Rydberg constant $R = R_A = 2\pi^2 \mu e^4/(ch^3)$ where R_A is given in Eq. (2-113a), the reduced mass $\mu = [mM_A/(m+M_A)]$, and Z is the ionic charge. The selection rules for the transitions are $\Delta J = 0, \pm 1$, $\Delta F = 0, \pm 1$, and $F = 0 \not\leftrightarrow F = 0$. The radiation is magnetic dipole with a dipole matrix element, μ, of about one Bohr magneton $\mu = eh/(4\pi mc) \approx 0.927 \times 10^{-20}$ erg gauss^{-1}. It follows from Eq. (2-44) that

$$A_{mn} \approx \frac{\pi^2 e^2 h}{c^5 m^2} \nu_{mn}^3 \approx 10^{-42} \nu_{mn}^3 \ \sec^{-1}. \qquad (2\text{-}132)$$

Detectable radio frequency radiation may arise from the $\Delta F = \pm 1$, $\Delta L = \Delta F = \Delta J = 0$ transitions of the abundant atoms or molecules. Radio frequency hyperfine transitions are given in Table 17.

Table 17. A table of hyperfine transitions at radio frequencies[1]

Atom or molecule		Spin	Transition	Frequency (Hz)	A_{mn} (sec^{-1})
H I	neutral hydrogen	$\frac{1}{2}$	$^2S_{1/2}$, $F = 0-1$	$1.420405751.786 \times 10^9 \pm 0.01$	2.85×10^{-15}
D	deuterium	1	$^2S_{1/2}$, $F = \frac{1}{2} - \frac{3}{2}$	$3.27384349 \times 10^8 \pm 5$	4.65×10^{-17}
He II	singly ionized helium	$\frac{1}{2}$	$^2S_{1/2}$, $F = 1-0$	$8.66566 \times 10^9 \pm 1.8 \times 10^5$	6.50×10^{-13}
N VII	ionized nitrogen	1	$^2S_{1/2}$, $F = \frac{1}{2} - \frac{3}{2}$	5.306×10^7	1.49×10^{-19}
N I	neutral nitrogen	1	$^4S_{3/2}$, $F = \frac{3}{2} - \frac{5}{2}$	2.612×10^7	1.78×10^{-20}
			$F = \frac{1}{2} - \frac{3}{2}$	1.567×10^7	3.84×10^{-21}
H_2^+	ionized molecular hydrogen	1	F_2, F $\frac{3}{2}, \frac{5}{2} - \frac{1}{2}, \frac{3}{2}$	$1.40430 \times 10^9 \pm 10^7$	2.75×10^{-15}
			F_2, F $\frac{3}{2}, \frac{3}{2} - \frac{1}{2}, \frac{3}{2}$	$1.41224 \times 10^9 \pm 10^7$	2.80×10^{-15}
Na I	neutral sodium	$\frac{3}{2}$	$^2S_{3/2}$, $F = 1-2$	1.77161×10^9	5.56×10^{-15}

[1] From TOWNES (1957), FIELD, SOMERVILLE, and DRESSLER (1966), and KERR (1968).

VAN DE HULST (1945) first predicted that emission from the 1420 MHz hyperfine transition of neutral hydrogen would be detected from interstellar gas, and this prediction was confirmed by EWEN and PURCELL (1951). Subsequent observations of the hydrogen line in emission led to a mapping of the spiral arms of the Galaxy (cf. KERR, 1968 and Chap. 5). Recent emission line surveys have been completed by WESTERHOUT (1969), WEAVER and WILLIAMS (1971), HEILES and HABING (1974), and HEILES and JENKINS (1974). Provided that the angular extent of the emitting gas cloud is larger than that of the antenna beam, and provided the gas is optically thin, measurements of the peak emission line antenna temperature, ΔT_L, leads to the column density, N_H, using the equation (MILNE, 1930; WILD, 1952 or Eqs. (2-69) and (2-75))

$$N_n = \frac{8\pi \nu_{mn}}{A_{mn} c^2} \frac{g_n}{g_m} \frac{k}{h} \frac{\Delta T_L \Delta \nu_L}{\eta_A} \tag{2-133}$$

or

$$N_H \approx \frac{4 N_1}{3} \approx 3.88 \times 10^{14} \Delta T_L \; \Delta \nu_L / \eta_A \text{ cm}^{-2},$$

where N_n is the line integrated density of atoms in the nth state, N_H is the total line integrated density of neutral hydrogen, ν_{mn} is the frequency of the $m-n$ transition, g_n is the statistical weight of the nth level, A_{mn} is the Einstein coefficient for spontaneous transition between levels m and n, the h and k are, respectively, Planck's and Boltzmann's constants, c is the velocity of light, $\Delta \nu_L$ is the half width of the observed emission line, and η_A is the antenna efficiency. The numerical approximation given in Eq. (2-133) is for the hyperfine ground state hydrogen transition for which $\nu_{mn} = 1420.40575$ MHz, $A_{mn} = 2.85 \times 10^{-15}$ sec^{-1}, $g_n = 2n^2$, and it is assumed that ΔT_L is given in degrees Kelvin and $\Delta \nu_L$ is in Hz. If the line width, $\Delta \nu_L$, is given in km sec^{-1}, Eq. (2-133) becomes

$$N_H \approx 1.823 \times 10^{18} \Delta T_L \Delta \nu_L / \eta_A \text{ cm}^{-2}. \tag{2-134}$$

The emission line measurements provide only a measure of the line integrated density of the neutral hydrogen. When absorption line measurements are combined with the emission line measurement, the temperature of the gas can also be determined. If T_s denotes the spin temperature of the gas, and τ denotes the optical depth determined from the absorption line study, then the spin temperature is related to the emission line temperature, ΔT_L, by the equation

$$T_s = \frac{\Delta T_L}{\eta_A [1 - \exp(-\tau)]} \approx \frac{\Delta T_L}{\tau \eta_A}. \tag{2-135}$$

For an optically thin gas seen in absorption against a source of effective brightness temperature, T_{0s}, the optical depth, τ, is given by

$$\tau = \frac{\Delta T_A}{\eta_A T_{0s}}, \tag{2-136}$$

where ΔT_A is the absorption line temperature (cf. Eq. (2-75)). Absorption line observations of neutral hydrogen in our Galaxy are given by CLARK (1965), HUGHES, THOMPSON, and COLVIN (1971), and RADHAKRISHNAN *et al.* (1972).

The excitation or spin temperature, T_s, is related to the kinetic temperature, T_k, of the gas and the radiation temperature, T_R, of the background radiation by the equation of statistical equilibrium (Eq. (2-82)). If R_{mn} denotes the probability coefficient for transitions induced by collisions, and A_{mn} denotes the Einstein coefficient for a spontaneous transition, then we have the relation

$$T_s = T_k \left[\frac{T_R + T_0}{T_k + T_0} \right], \tag{2-137}$$

where

$$T_0 = \frac{h\nu_{mn} R_{mn}}{k A_{mn}},$$

and ν_{mn} is the frequency of the transition. The kinetic temperature, T_k, may be measured from the Doppler broadening of the line. For the 21 cm hydrogen line we have $A_{mn} = 2.85 \times 10^{-15}\ \mathrm{sec}^{-1}$, and the probability coefficient for collision induced transitions is

$$R_{mn} = n_\mathrm{H} \sigma_\mathrm{F} \langle v \rangle = n_\mathrm{H} \sigma_\mathrm{F} \left(\frac{8kT_\mathrm{k}}{\pi m_\mathrm{H}} \right)^{1/2} \approx 7 \times 10^{-11} n_\mathrm{H} T_\mathrm{k}^{1/2}\ \mathrm{sec}^{-1}, \tag{2-138}$$

where n_H is the volume density of hydrogen atoms, the average effective cross section, σ_F, for mutual collisions between hydrogen atoms goes from $6.6 \times 10^{-15}\ \mathrm{cm}^2$ to $2.9 \times 10^{-15}\ \mathrm{cm}^2$ when T_k ranges from 2,000 to 10,000 °K, and $\langle v \rangle$ is the average thermal velocity. It follows from Eqs. (2-137) and (2-138) that the spin and kinetic temperatures will be equal when

$$n_\mathrm{H} > 6 \times 10^{-4}\, T_\mathrm{k}^{1/2}\ \mathrm{cm}^{-3}, \tag{2-139}$$

provided that hydrogen collision is the dominant collision process.

Attempts to detect intergalactic neutral hydrogen in emission (PENZIAS and WILSON, 1969) and in absorption (ALLEN, 1969) have been unsuccessful. These observations lead to the limits $n_\mathrm{H} < 3 \times 10^{-6}$ atoms cm^{-3} and $n_\mathrm{H}/T_\mathrm{s} < 2.7 \times 10^{-8}$ cm^{-3} °K^{-1} if the Hubble constant is assumed to be 75 km sec^{-1} Mpc^{-1}. These measurements also indicate that the maximum intergalactic neutral hydrogen density is $5 \times 10^{-30}\ \mathrm{g\ cm}^{-3}$. Nevertheless, neutral hydrogen in the bridges between galaxies has been detected (HINDMAN *et al.*, 1963; ROBERTS, 1968) and for this hydrogen $N_\mathrm{H} \approx 10^{20}\ \mathrm{cm}^{-2}$. For the intergalactic medium, electron-atom collisions and excitations induced by Lyman-α radiation are important contributions to Eq. (2-137). The detailed formulae for these contributions are given by FIELD (1958, 1959). Calculations of T_s as a function of electron density and kinetic temperature are given by KOEHLER (1966).

2.15. Line Radiation from Molecules

2.15.1. Energies and Frequencies of the Molecular Transitions

The interaction of electrons in molecules may be assumed to be similar to that of the atomic Russell-Saunders (LS) coupling discussed in Sect. 2.13. The molecular fine structure is usually described by the following quantum numbers.

J = the total angular momentum excluding nuclear spin in units of $h/2\pi$.

N = the total orbital angular momentum including rotation in units of $h/2\pi$.

K = the projection of N on the molecular axis.

O = the orbital angular momentum due to molecular rotation in units of $h/2\pi$.

Λ = the projection of the electronic orbital angular momentum on the molecular axis in units of $h/2\pi$. The molecular state is designated as $\Sigma, \Pi, \Delta, \varphi, \ldots$ according as $\Lambda = 0, 1, 2, 3, \ldots$ The Λ is always positive for $\Lambda = |M_\mathrm{L}|$ where

$M_L = L, L-1, \ldots, -L$ where L is the electronic angular momentum.

Σ = the projection of the electron spin angular momentum on the molecular axis in units of $h/2\pi$ (not to be confused with the symbol Σ for $\Lambda = 0$). The Σ takes on $2S+1$ values, where S is the electron spin.

Ω = the total electronic angular momentum about the molecular axis in units of $h/2\pi$. The $\Omega = |\Lambda + \Sigma|$, for the Hund coupling case (a).

Each pair of Λ and Σ designate a term which has the symbol ${}^{2S+1}\Lambda_{\Omega}$, where $2S+1$ is the numerical value of the multiplicity, the Λ is designated by its Greek symbol, and Ω is the numerical value of the total electronic angular momentum. A Σ state is called Σ^{+} or Σ^{-} according to whether its electronic eigenfunction remains unchanged or changes sign upon reflection in any plane passing through the internuclear axis. The electronic state is even (g) or odd (u) according to whether the electronic eigenfunction remains unchanged or changes sign for a reflection at the center of symmetry. A rotational level is positive $(+)$ or negative $(-)$ according to whether the total eigenfunction changes sign for reflection at the origin. When the nuclei are identical, the term is symmetric (s) or antisymmetric (a) according to whether the total eigenfunction remains unchanged or changes sign when the nuclei are exchanged. Transitions between levels must obey selection rules which depend on the quantum numbers, the symmetry properties, and the type of coupling. These rules are given in Table 18 for diatomic molecules.

The symbol $\nleftrightarrow$ denotes "cannot combine with". The coupling symbols (a) and (b) refer, respectively, to Hund's coupling cases (a) and (b). For case (a), the electronic motion is coupled strongly to the line joining the nuclei so that $\Omega = |\Lambda + \Sigma|$. For case (b) only the electronic orbital angular momentum, Λ, is strongly coupled to the molecular axis so that Λ and N couple to form K, and K and S couple to give J.

The total energy, E, of a molecule may be regarded as the superposition of the energy due to the electrons, E_e, the vibrational energy, E_v, and the rotational energy, E_r,

$$E = E_r + E_v + E_e \,. \tag{2-140}$$

Transitions between electronic states occur at optical and ultraviolet frequencies, the vibrational energies usually correspond to transitions in the infrared, and the rotational transitions are at radio frequencies. For end-over-end rotation of a rigid molecule, the rotational energy, E_r, solution to the Schrödinger equation (SCHRÖDINGER, 1925, 1926, Eq. (2-126)) is given by

$$E_r = h B J(J+1) \,, \tag{2-141}$$

where the rotational angular quantum number, J, is zero or an integer, the total angular momentum in units of $h/2\pi$ is $[J(J+1)]^{1/2}$, and the rotational constant, B, is given by

$$B = \frac{h}{8\pi^2 I} \,,$$

Table 18. Selection rules for diatomic molecular spectra[1]

Coupling	Electric dipole (allowed)	Magnetic dipole (forbidden)	Electric quadrupole (forbidden)
(1) General	$\Delta J=0, \pm 1$ $(0 \not\leftrightarrow 0)$	$\Delta J=0, \pm 1$ $(0 \not\leftrightarrow 0)$	$\Delta J=0, \pm 1, \pm 2$ $(0 \not\leftrightarrow 0,\ 0 \not\leftrightarrow 1,\ \frac{1}{2} \not\leftrightarrow \frac{1}{2})$
(2) General	$(+ \leftrightarrow -,\ + \not\leftrightarrow +,\ - \not\leftrightarrow -)$	$(+ \leftrightarrow +,\ - \leftrightarrow -,\ + \not\leftrightarrow -)$	$(+ \leftrightarrow +,\ - \leftrightarrow -,\ + \not\leftrightarrow -)$
(3) General	$(s \not\leftrightarrow a,\ s \leftrightarrow s,\ a \leftrightarrow a)$	$(s \leftrightarrow s,\ a \leftrightarrow a,\ s \not\leftrightarrow a)$	$(s \leftrightarrow s,\ a \leftrightarrow a,\ s \not\leftrightarrow a)$
(4) General	$(g \leftrightarrow u,\ g \not\leftrightarrow g,\ u \not\leftrightarrow u)$	$(g \leftrightarrow g,\ u \leftrightarrow u,\ g \not\leftrightarrow u)$	$(g \leftrightarrow g,\ u \leftrightarrow u,\ g \not\leftrightarrow u)$
(5) (a) and (b)	$\Delta S=0$	$\Delta S=0$	$\Delta S=0$
(6) (a)	$\Delta \Lambda=0, \pm 1$	$\Delta \Lambda=0$ if $\Delta \Sigma=\pm 1$; $\Delta \Lambda=\pm 1$ if $\Delta \Sigma=0$	$\Delta \Lambda=0, \pm 1, \pm 2$
(b)	$\Delta \Lambda=0, \pm 1$	$\Delta \Lambda=0, \pm 1$	$\Delta \Lambda=0, \pm 1, \pm 2$
(7) (a)	$\Delta \Sigma=0$	see (6) (a)	$\Delta \Sigma=0$
(8) (a)	$\Delta \Omega=0, \pm 1$	$\Delta \Omega=\pm 1$	$\Delta \Omega=0, \pm 1, \pm 2$
(9) (a)	$\Delta J \neq 0$ for $\Omega=0 \leftrightarrow \Omega=0$	—	$\Delta J \neq 1$ for $\Omega=0 \leftrightarrow \Omega=0$
(10) (b)	$\Delta K=0, \pm 1$	$\Delta K=0, \pm 1$	$\Delta K=0, \pm 1, \pm 2$
(11) (b)	$\Delta K \neq 0$ for $\Sigma \leftrightarrow \Sigma$ transitions	$\Delta K=0$ for $\Sigma \leftrightarrow \Sigma$ transitions	$\Delta K=0, \pm 1, \pm 2$ for $\Sigma \leftrightarrow \Sigma$ transitions
(12) (b)	$\Sigma^+ \not\leftrightarrow \Sigma^-$	$\Sigma^+ \not\leftrightarrow \Sigma^-$	$\Sigma^+ \not\leftrightarrow \Sigma^-$

[1] For (3) the nuclei are assumed to be identical, and the term is symmetric, s, or antisymmetric, a, according to whether the total eigenfunction remains unchanged or changes sign for an exchange of the nuclei. The g and u in (4) denote, respectively, even and odd electronic states for nuclei of equal charge. An electronic state is even or odd according to whether the electronic eigenfunction remains unchanged or changes sign for a reflection at the center of symmetry. For (5) through (12) the (a) and (b) denote, respectively, Hund's case a and case b. For (12) a Σ state is called Σ^+ or Σ^- according to whether its electronic eigenfunction remains unchanged or changes sign upon reflection in any plane passing through the internuclear axis.

where I is the moment of inertia of the molecule. For a diatomic molecule with atoms of respective masses m_1 and m_2, and separation r_{12},

$$I=\frac{r_{12}^2 m_1 m_2}{m_1+m_2}=r_{12}^2 \mu ,$$

where μ is the reduced mass. Similarly for a triatomic molecule

$$I=\frac{m_1 m_2 r_{12}^2+m_1 m_3 r_{13}^2+m_2 m_3 r_{23}^2}{m_1+m_2+m_3},$$

where r_{ij} represents the distance between the i and j atoms with respective masses m_i and m_j.

For Hund's coupling case (a) Eq. (2-141) becomes

$$E_r=hB\left[J(J+1)-\Omega^2\right]. \tag{2-142}$$

For the coupling case (b) and the $^2\Sigma$ states, we have (HUND, 1927)

$$E_r = hBK(K+1) + \tfrac{1}{2}h\gamma K \quad \text{for } J = K + \tfrac{1}{2}$$

and

$$E_r = hBK(K+1) - \tfrac{1}{2}h\gamma(K+1) \quad \text{for } J = K - \tfrac{1}{2}, \tag{2-143}$$

where the splitting constant γ is small compared to B.

When the molecule is not a rigid rotator and centrifugal stretching is taken into account, Eq. (2-141) becomes

$$E_r = hBJ(J+1) - hDJ^2(J+1)^2, \tag{2-144}$$

where for a vibrational radian frequency, ω, the change D in B due to centrifugal stretching is given by

$$D = \frac{4B^3}{\omega^2}.$$

Using the selection rule $\Delta J = \pm 1$ with Eq. (2-144), the allowed rotational transition frequencies, ν_r, are given by

$$\nu_r = 2B(J+1) - 4D(J+1)^3. \tag{2-145}$$

A symmetric top molecule has equal moments of inertia along the directions of two of the three principle axes of the molecule. The angular momentum has a fixed component along an internal molecular axis given by $Kh/2\pi$ where K takes on $2J+1$ values from $-J$ to J. The rotational energy, E_r, is given by

$$E_r = h[BJ(J+1) + (A-B)K^2], \tag{2-146}$$

where $A = h/(8\pi^2 I_a)$ and $B = h/(8\pi^2 I_b)$ when I_a and I_b denote the moments of inertia along the a and b axis. The selection rule is $\Delta K = 0$, however, and the frequency of the transition is the same as that resulting from Eq. (2-141). When the molecule is an asymmetric top molecule, the quantum number J is supplemented by the values of K_{-1} for the limiting prolate symmetric top, $I_a < I_b = I_c$, and K_{+1} for the limiting oblate symmetric top, $I_a = I_b < I_c$, where I_i denotes the moment of inertia along the axis i.

The solution to the time independent Schrödinger equation (2-126) for a harmonic oscillator of frequency, ν_0, and force constant k is

$$E_v = h\nu_0(v + \tfrac{1}{2}) = hc\omega_e(v + \tfrac{1}{2}), \tag{2-147}$$

where E_v is the vibrational energy for vibrational quantum number v, the $\omega_e = \nu_0/c$ is the term value, and

$$\nu_0 = \frac{1}{2\pi}\left(\frac{k}{m}\right)^{1/2},$$

where m is the mass of the oscillator. The quantum number, v, is zero or an integer and the selection rule is $\Delta v = \pm 1$, so that the vibration transition frequencies are equal to ν_0. BORN and OPPENHEIMER (1927) were the first to show that the massive nuclei could be assumed to be fixed in order to calculate the potential energy contribution of the electrons, after which the rotational-vibrational energy levels could be calculated. MORSE (1929) suggested a diatomic molecular potential energy, $U(r)$, given by

$$U(r) = D_e\{1 - \exp[-a(r - r_e)]\}^2, \tag{2-148}$$

where D_e is the dissociation energy of the molecule, r_e is the equilibrium distance between the nuclei, a is a constant, and r is the internuclear distance. The function $U(r)$ is shown in Fig. 14 where it is compared with the actual $U(r)$ for the hydrogen molecule together with its vibrational energy levels. When $U(r)$ and the potential energy due to rotation, $J(J+1)/r^2$, are substituted into the Schrödinger equation (2-126), the solution for the rotational-vibrational energy levels is given by

$$\frac{E_r+E_v}{h}=\omega_e(v+\tfrac{1}{2})-x_e\,\omega_e(v+\tfrac{1}{2})^2+B_e J(J+1)-D J^2(J+1)^2-\alpha_e(v+\tfrac{1}{2})J(J+1), \tag{2-149}$$

where

$$\omega_e=\frac{a}{2\pi}\left(\frac{2D_e}{\mu}\right)^{1/2},$$

$$x_e=\frac{h\,\omega_e}{4D_e},$$

$$B_e=\frac{h}{8\pi^2 I_e},$$

$$D=\frac{4B_e^3}{\omega_e^2}=\frac{h^3}{128\pi^6\mu^3\omega_e^2 r_e^6},$$

$$\alpha_e=6\left(\frac{x_e B_e^3}{\omega_e}\right)^{1/2}-\frac{6B_e^2}{\omega_e},$$

J and v are, respectively, the rotational and vibrational quantum numbers, and μ is the reduced mass of the molecule. Values of the rotation-vibration constants of Eq. (2-149) are given for most simple molecules of astrophysical interest in the appendices of the books by HERZBERG (1950) and TOWNES and SCHAWLOW (1955).

Although the rotational quantum number, J, obeys the simple selection rule $\Delta J=\pm 1$, there is no similar simple selection rule for the vibrational quantum number, v. Nevertheless, FRANCK (1925) and CONDON (1926, 1928) have developed a principle which relates the favored vibrational transitions to the potential energy curves of the two states under consideration. The favored transition is found to be the one in which there is no instantaneous change of nuclear momentum or position in the transition. This means that $\Delta v=0$ is favorable for states with similar potential energy curves, whereas for displaced curves a wide band of Δv are possible.

Theoretically there are two equivalent potential energy configurations separated by a barrier as shown in Fig. 14 for ammonia. This property reflects the inversion of the eigenfunction about its origin. Quantum mechanically the vibrating molecule can tunnel through the barrier, and the frequency of penetration, ν, is given by (DENNISON and UHLENBECK, 1932)

$$\nu=\frac{\nu_v}{\pi A^2}, \tag{2-150}$$

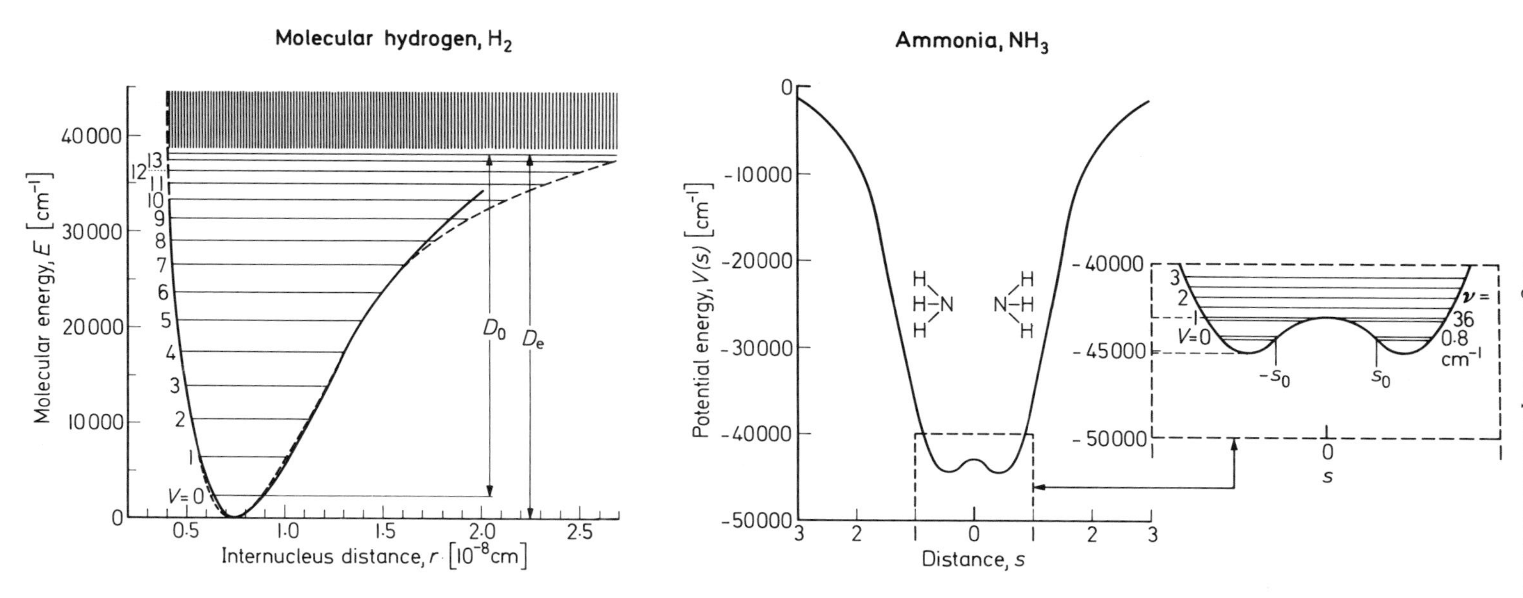

Fig. 14. Potential energy curves for the hydrogen molecule, H_2, and the ammonia molecule, NH_3 (after HERZBERG, 1950; TOWNES and SCHALOW, 1955, by permission, respectively, of the Van Nostrand Reinhold Co. and the McGraw-Hill Book Co.). The vibrational energy levels are denoted by horizontal lines for different values of the vibrational quantum number, V. For molecular hydrogen, the full curve is experimental data whereas the broken curve is the Morse curve. The variable, r, denotes the internuclear distance, and the continuous term spectrum above $V=14$ is indicated by vertical hatching. For the ammonia molecule, the variable, s, denotes the distance between the nitrogen and the plane of the hydrogens. The ammonia molecule resonates between the two potential minima, and the resonance splits the vibrational levels into the doublets shown in the figure

where ν_v is the vibration frequency for one of the potential minima, and the A represents the area under the potential hill. Here

$$A=\exp\left\{\frac{2\pi}{h}\int_0^{s_0}[2\mu(V-E_v)]^{1/2}\,ds\right\},$$

where μ is the reduced mass, V is the potential energy, ds is an element of the atomic separation, s, the s_0 is shown in Fig. 14, and E_v is the total vibrational energy. (The factor A^{-2} determines a probability of the tunnel effect.) For ammonia, the inversion barrier is sufficiently low that the vibrational levels have doublets whose frequency of separation lies in the radio frequency range (cf. Table 20). Detection of interstellar ammonia was first reported by CHEUNG *et al.* (1968).

Because positive and negative values of Λ represent rotation of the orbital electrons in one sense or the other, they have the same energy. Unless $\Lambda=0$ the energy levels are therefore doubly degenerate. For a rotating molecule with an unpaired electron, however, the electronic-rotational interaction removes this degeneracy and each energy level splits into two components. The most important transitions between the lambda doublet levels are those for the $\Lambda=1$ or Π state, and the energies, E, of these transitions are given by (VAN VLECK, 1929; TOWNES and SCHAWLOW, 1955)

$$E=q_\Lambda J(J+1) \tag{2-151}$$

for $S=0$ or $S=1$, $\Omega=1$

$$E=q_\Lambda N(N+1)$$

for Hund's coupling case (b) and $S=\frac{1}{2}$ or 1,

$$E=a(J+\tfrac{1}{2})$$

for Hund's coupling case (a) and $S=\frac{1}{2}$, $\Omega=\frac{1}{2}$,

$$E=b(J^2-\tfrac{1}{4})(J+\tfrac{3}{2})$$

for Hund's coupling case (a) and $S=\frac{1}{2}$, $\Omega=\frac{3}{2}$, and

$$E\approx\frac{q_\Lambda}{2}\left(J+\frac{1}{2}\right)\left[\left(2+\frac{A}{B}\right)\left(1+\frac{2-A/B}{X}\right)+\frac{4(J+\frac{3}{2})(J-\frac{1}{2})}{X}\right],$$

for intermediate coupling cases of the $^2\Pi$ states. Here we have

$$q_\Lambda=\frac{4B_v^2}{h\nu_e},\qquad a=\frac{4AB_v}{h\nu_e},\qquad b=\frac{8B_v^3}{Ah\nu_e},$$

and

$$X=\pm\left[\frac{A}{B}\left(\frac{A}{B}-4\right)+4\left(J+\frac{1}{2}\right)^2\right]^{1/2},$$

where B_v is the rotational constant, $h\nu_e$ is the separation between the ground and first excited electronic energy level, and A and B are the constants of interaction between the electronic and rotational states. The $^2\Pi_{3/2}$ and $^2\Pi_{1/2}$ states are given, respectively, by $X=+$ and $-$ or $-$ and $+$ according as A is positive or negative. Transitions between Λ doublets are electric dipole transitions with dipole moment, μ, and dipole matrix element, μ_{mn}, given by

$$|\mu_{mn}|^2 = \frac{\mu^2 \Lambda^2}{J(J+1)} \tag{2-152}$$

for Hund's coupling case (b). For case (a) Λ is replaced by Ω. Electron spin-nuclear spin interactions lead to further hyperfine splitting of each doublet level. For the $^2\Pi$ state, the energy, ΔE, of this splitting is given by (TOWNES, 1957)

$$\Delta E = \pm \frac{d(X+2-A/B_v)}{4XJ(J+1)} \left(J+\frac{1}{2}\right) I \cdot J\,, \tag{2-153}$$

where

$$I \cdot J = \frac{F(F+1)-I(I+1)-J(J+1)}{2}$$

and

$$d = 3\mu_0 \frac{\mu_I}{I} \left(\frac{\sin^2\theta}{r^3}\right).$$

Here A is the fine structure constant of energy $A\bar{s}\cdot\bar{\Lambda}$, B is the rotational constant, F is total molecular angular momentum including nuclear spin in units of $h/2\pi$, the I is the nuclear spin angular momentum in units of $h/2\pi$, the Bohr magneton is μ_0, the nuclear magnetic moment is μ_I, the distance from the nucleus to the electron is r, and θ is the angle between the molecular axis and the radius vector. Transitions between the hyperfine split, lambda doublet lines of interstellar OH were first detected by WEINREB, BARRETT, MEEKS, and HENRY (1963) for the $^2\Pi_{3/2}$ state. The frequencies of subsequently detected Λ-doublet OH lines are given in Table 20. TURNER (1970) gives the parameters of fifty OH regions associated with H II regions.

Extraterrestrial molecules were first observed as absorption bands in the optical spectra of planets (ADAMS and DUNHAM, 1932). Subsequent observations

Table 19. Gases identified in planetary and satellite atmospheres[1]

Planet or satellite	Gas
Mercury	No definite identifications
Venus	CO_2, CO, HCl, HF, H_2O
Earth	N_2, O_2, H_2O, A, CO_2, Ne, He, CH_4, Kr, N_2O, H_2, O, O_3, Xe
Mars	CO_2, CO, H_2O, O_2, O_3
Jupiter	H_2, CH_4, NH_3, C_2H_2, C_2H_6
Saturn	H_2, CH_4, NH_3 (?)
Uranus	H_2, CH_4
Neptune	H_2, CH_4
Pluto	No identifications
Jovian satellites	No definite identifications
Titan	CH_4, H_2
Triton	No identifications

[1] From *Planetary Atmospheres*, 1968, National Academy of Sciences. Publ. 1688, Washington, D.C., Dr. G. MUNCH, and S. T. RIDGWAY, in: Ap. J. *187*, L41 (1974).

Table 20. Wavelengths and classification of line radiation from interstellar molecules observed in absorption at optical wavelengths, together with the frequencies, antenna temperatures, T_A, and sources for the most intense line radiation from interstellar molecules observed at radio frequencies[2]

Molecules observed at optical wavelengths:

Wavelengths (Å)	Molecule	Classification
1,108	H_2 – molecular hydrogen	$B^1\Sigma_u - X^1\Sigma_u$, (0,0)
1,092		(1,0)
1,077		(2,0)
1,063		(3,0)
1,049		(4,0)
1,037		(5,0)
1,024		(6,0)
1,013		(7,0)
1,066.27	HD	[3]
1,054.29		
1,509.65	$C^{12}O^{16}$ – carbon monoxide	$A^1\Pi - X^1\Sigma^+$, (1,0)
1,477.46		(2,0)
1,447.26		(3,0)
1,418.97		(4,0)
1,367.56		(6,0)
1,322.10		(8,0)
1,301.37		(9,0)
1,281.83		(10,0)
1,478.8	$C^{13}O^{16}$	$A^1\Pi - X^1\Sigma^+$, (2,0)
1,449.3		(3,0)
1,421.4		(4,0)
1,370.8		(6,0)
3,876.84	CN – cyanogen radical	$B^2\Sigma^+ - X^2\Sigma^+$, (0,0), $P(3)$
3,876.30		(0,0), $P(2)$
3,875.77		(0,0), $P(1)$
3,874.61		(0,0), $R(0)$
3,874.00		(0,0), $R(1)$
4,232.54	$C^{12}H^+$	$A^1\Pi - X^1\Sigma$, (0,0), $R(0)$
3,957.70		(1,0), $R(0)$
3,745.31		(2,0), $R(0)$
3,579.02		(3,0), $R(0)$
3,447.08		(4,0), $R(0)$
4,232.08	$C^{13}H^+$	$A^1\Pi - X^1\Sigma$, (0,0), $R(0)$
4,300.30	CH	$A^2\Delta - X^2\Pi$, (0,0), $R_2(1)$
3,890.21		$B^2\Sigma^- - X^2\Pi$, (0,0), $P_{Q_{12}}(1)$
3,886.41		(0,0), $Q_2(1) + {}^QR_{12}(1)$
3,878.77		(0,0), $R_2(1)$
3,146.01		$C^2\Sigma^+ - X^2\Pi$, (0,0), $P_{Q_{12}}(1)$
3,143.20		(0,0), $Q_2(1) + {}^QR_{12}(1)$
3,137.53		(0,0), $R_2(1)$

Table 20 (continued)
Molecules observed at radio frequencies:

Frequency (MHz or GHz)	Molecule		T_A (°K)	Source
834.267 MHz	CH_3OH	– methyl alcohol	+ 0.58	Sgr A
1,065.075 MHz	$HCOCH_3$	– acetaldehyde	+ 0.25	Sgr A
1,538.135 MHz	$HCONH_2$	– formamide	+ 0.08	Sgr B2
1,538.693 MHz	$HCONH_2$	– formamide	+ 0.09	Sgr B2
1,539.295 MHz	$HCONH_2$	– formamide	+ 0.10	Sgr B2
1,539.570 MHz	$HCONH_2$	– formamide	+ 0.08	Sgr B2
1,539.851 MHz	$HCONH_2$	– formamide	+ 0.36	Sgr B2
1,541.018 MHz	$HCONH_2$	– formamide	+ 0.10	Sgr B2
1,612.231 MHz	OH	– hydroxyl	+ 200	NML Cyg
1,637.53 MHz	$O^{18}H$	– hydroxyl	– 0.22	Sgr A
1,638.805 MHz	HCOOH	– formic acid	+ 0.06	Sgr A
1,639.48 MHz	$O^{18}H$	– hydroxyl	– 0.18	Sgr A
1,665.401 MHz	OH	– hydroxyl	+ 100	W 49
1,667.358 MHz	OH	– hydroxyl	+ 34	W 49
1,720.533 MHz	OH	– hydroxyl	+ 28	W 28
3,139.38 MHz	H_2CS	– thioformaldehyde	– 0.3	Sgr B2
3,263.788 MHz	CH	– CH radical	+ 0.2	W 12
3,335.475 MHz	CH	– CH radical	+ 0.28	Cas A
3,349.185 MHz	CH	– CH radical	+ 0.2	Cas A
4,388.797 MHz	H_2CO^{18}	– formaldehyde	– 0.1	Sgr B2
4,593.089 MHz	$H_2C^{13}O$	– formaldehyde	– 0.1	Sgr B2
4,617.118 MHz	$HCONH_2$	– formamide	+ 0,07	Sgr B2
4,618.970 MHz	$HCONH_2$	– formamide	+ 0.13	Sgr B2
4,619.988 MHz	$HCONH_2$	– formamide	+ 0.03	Sgr B2
4,660.242 MHz	OH	– hydroxyl	+ 0.5	Sgr B2
4,750.656 MHz	OH	– hydroxyl	+ 0.3	Sgr B2
4,765.562 MHz	OH	– hydroxyl	+ 0.2	Sgr B2
4,829.660 MHz	H_2CO	– formaldehyde	– 9	Sgr B2
5,288.980 MHz	H_2CNH	– methanimine	+ 0.05	Sgr B2
5,289.786 MHz	H_2CNH	– methanimine	+ 0.15	Sgr B2
5,290.726 MHz	H_2CNH	– methanimine	+ 0.07	Sgr B2
5,291.646 MHz	H_2CNH	– methanimine	+ 0.07	Sgr B2
6,030.739 MHz	OH	– hydroxyl	+ 12	W 3
6,035.085 MHz	OH	– hydroxyl	+ 19	W 3
9,097.067 MHz	HC_3N	– cyanoacetylene	+ 0.32	Sgr B2
9,098.357 MHz	HC_3N	– cyanoacetylene	+ 0.87	Sgr B2
9,100.286 MHz	HC_3N	– cyanoacetylene		Sgr B2
13.441371 GHz	OH	– hydroxyl	+ 3.2	W 3
13.77886 GHz	$H_2C^{13}O$	– formaldehyde	—	—
14.48865 GHz	H_2CO	– formaldehyde	– 1	Sgr B2
18.195085 GHz	HC_3N	– cyanoacetylene	+ 0.4 (bracketed with next row)	Sgr B2
18.196275 GHz	HC_3N	– cyanoacetylene		Sgr B2
21.98052 GHz	HNCO	– isocyanic acid		Sgr B2
21.98146 GHz	HNCO	– isocyanic acid	0.2 (bracketed with adjacent rows)	Sgr B2
21.98206 GHz	HNCO	– isocyanic acid		Sgr B2
22.23508 GHz	H_2O	– water	+ 2,000	W 49
22.83417 GHz	NH_3	– ammonia	+ 0.11	Sgr B2

Table 20 (continued)
Molecules observed at radio frequencies:

Frequency (MHz or GHz)		Molecule		T_A (°K)		Source
23.09879	GHz	NH_3	– ammonia	+	0.29	Sgr B2
23.69448	GHz	NH_3	– ammonia	+	1.61	Sgr B2
23.72271	GHz	NH_3	– ammonia	+	1.32	Sgr B2
23.87011	GHz	NH_3	– ammonia	+	3.20	Sgr B2
24.13941	GHz	NH_3	– ammonia	+	0.61	Sgr B2
24.93347	GHz	CH_3OH	– methyl alcohol	+	1	Ori A
24.95908	GHz	CH_3OH	– methyl alcohol	+	1	Ori A
25.01814	GHz	CH_3OH	– methyl alcohol	+	1.0	Ori A
25.05604	GHz	NH_3	– ammonia	+	0.4	Sgr B2
25.12488	GHz	CH_3OH	– methyl alcohol	+	0.7	Ori A
25.29441	GHz	CH_3OH	– methyl alcohol	+	0.7	Ori A
28.974803	GHz	H_2CO	– formaldehyde	—		Sgr B2
36.16924	GHz	CH_3OH	– methyl alcohol	+	2	Sgr B2
43.96277	GHz	HNCO	– isocyanic acid	+	0.8	Sgr B2
46.247472	GHz	$C^{13}S$	– carbon monosulfide	+	0.148	Sgr B2
48.206948	GHz	CS^{34}	– carbon monosulfide	+	0.38	DR 21 (OH)
48.37260	GHz	CH_3OH	– methyl alcohol	+	3.0	Sgr B2
48.37709	GHz	CH_3OH	– methyl alcohol	+	3.0	Sgr B2
48.583264	GHz	CS^{33}	– carbon monosulfide			Sgr B2
48.585906	GHz	CS^{33}	– carbon monosulfide	+	0.12	Sgr B2
48.589068	GHz	CS^{33}	– carbon monosulfide			Sgr B2
48.991000	GHz	CS	– carbon monosulfide	+	3.53	Ori A
72.409095	GHz	H_2CO	– formaldehyde	+	0.1	Ori A
72.41325	GHz	DCN	– hydrogen cyanide	+	0.20	Ori A
72.41462	GHz	DCN	– hydrogen cyanide	+	0.25	Ori A
72.41668	GHz	DCN	– hydrogen cyanide	+	0.20	Ori A
72.783822	GHz	HC_3N	– cyanoacetylene	—		Sgr B2
72.837974	GHz	H_2CO	– formaldehyde	+	0.5	Ori A
72.97680	GHz	OCS	– carbonyl sulfide	+	0.4	Sgr B2
81.541	GHz	U 81.5	– unidentified	+	2.5	Sgr B2
81.883	GHz	HC_3N	– cyanoacetylene	—		Sgr B2
84.52121	GHz	CH_3OH	– methyl alcohol	+	2,8°	Sgr B2
85.13892	GHz	OCS	– carbonyl sulfide	+	0.7	Sgr B2
85.435	GHz	U 85.4	– unidentified	+	0.15	Sgr B2
85.44261	GHz	CH_3C_2H	– methylacetylene	+	0.15	Sgr B2
85.45729	GHz	CH_3C_2H	– methylacetylene	+	0.45	Sgr B2
86.05505	GHz	HCN^{15}	– hydrogen cyanide	+	0.15	Ori A
86.093938	GHz	SO	– sulfur monoxide	+	1	Ori A
86.24328	GHz	SiO	– silicon monoxide ($V=1$)	+	14	Ori A
86.33875	GHz	$HC^{13}N$	– hydrogen cyanide	+	0.5	Ori A
86.34005	GHz	$HC^{13}N$	– hydrogen cyanide	+	0.7	Ori A
86.34216	GHz	$HC^{13}N$	– hydrogen cyanide	+	0.3	Ori A
86.846859	GHz	SiO	– silicon monoxide	+	1	Ori A
87.3165	GHz	U 87.316	– unidentified	+	3.0	Ori A, M17
87.3283	GHz	U 87.328	– unidentified	+	1.5	Ori A, M17
87.4018	GHz	U 87.402	– unidentified	+	1.5	Ori A, M17

Table 20 (continued)
Molecules observed at radio frequencies:

Frequency (MHz or GHz)		Molecule		T_A (°K)		Source
87.4068	GHz	U 87.407	– unidentified	+	0.75	Ori A, M 17
87.92524	GHz	HNCO	– isocyanic acid	+	2.5	Sgr B2
88.23905	GHz	HNCO	– isocyanic acid	+	0.25	Sgr B2
88.6304157	GHz	HCN	– hydrogen cyanide	+	7	Ori A
88.6318473	GHz	HCN	– hydrogen cyanide	+	12	Ori A
88.6339360	GHz	HCN	– hydrogen cyanide	+	5	Ori A
89.189	GHz	"X-ogen"	– unidentified	+	10	Ori A
89.5058	GHz	CH_3OH	– methyl alcohol	+	0.3	Ori A
90.665	GHz	HNC	– hydrogen isocyanide	+	1.6	Ori A
90.981	GHz	HC_3N	– cyanoacetylene	—		Sgr B2
92.494084	GHz	$C^{13}S$	– carbon monosulfide	+	0.215	Ori A
96.412962	GHz	CS^{34}	– carbon monosulfide	+	0.62	Ori A
96.74142	GHz	CH_3OH	– methyl alcohol	—		—
97.171824	GHz	CS^{33}	– carbon monosulfide	}	0.2	Ori A
97.171843	GHz	CS^{33}	– carbon monosulfide			Ori A
97.30119	GHz	OCS	– carbonyl sulfide	+	0.85	Sgr B2
97.981007	GHz	CS	– carbon monosulfide	+	6.94	Ori A
99.29985	GHz	SO	– sulfur monoxide	+	2.4	Ori A
100.076392	GHz	HC_3N	– cyanoacetylene	+	1.0	Sgr B2
109.4628	GHz	OCS	– carbonyl sulfide	+	0.7	Sgr B2
109.782182	GHz	CO^{18}	– carbon monoxide	+	1	Ori A
109.90585	GHz	HNCO	– isocyanic acid	+	1.1	Sgr B2
110.201370	GHz	$C^{13}O$	– carbon monoxide	+	7	Ori A
110.33079	GHz	CH_3CN	– methyl cyanide	+	0.2	Sgr B2
110.34968	GHz	CH_3CN	– methyl cyanide	+	0.45	Sgr B2
110.36452	GHz	CH_3CN	– methyl cyanide	+	0.31	Sgr B2
110.37501	GHz	CH_3CN	– methyl cyanide	+	0.81	Sgr B2
110.38139	GHz	CH_3CN	– methyl cyanide	+	1.09	Sgr B2
110.38347	GHz	CH_3CN	– methyl cyanide	+	1.09	Sgr B2
110.709552	GHz	CH_3CN	– methyl cyanide (V_8)	+	0.4	Com. Kohoutek
110.712220	GHz	CH_3CN	– methyl cyanide (V_8)	+	0.6	Com. Kohoutek
112.358720	GHz	CO^{17}	– carbon monoxide	+	0.15	Ori A
112.358981	GHz	CO^{17}	– carbon monoxide	+	0.15	Ori A
112.360016	GHz	CO^{17}	– carbon monoxide	+	0.15	Ori A
113.1450	GHz	CN	– cyanogen radical	—		Ori A
113.1710	GHz	CN	– cyanogen radical	—		Ori A
113.1910	GHz	CN	– cyanogen radical	—		Ori A
113.4881	GHz	CN	– cyanogen radical	+	1	Ori A
113.4909	GHz	CN	– cyanogen radical	+	3	Ori A
113.4990	GHz	CN	– cyanogen radical	—		Ori A
113.5080	GHz	CN	– cyanogen radical	—		Ori A
115.271203	GHz	CO	– carbon monoxide	+	75[1]	Ori A
129.3613	GHz	SiO	– silicon monoxide ($V=1$)	+	0.9	Ori A
130.2684	GHz	SiO	– silicon monoxide	+	0.7	Sgr B2
138.1786	GHz	SO	– sulphur monoxide	+	2.06	NGC 2024
140.839529	GHz	H_2CO	– formaldehyde	+	3.3	Ori A
144.826810	GHz	DCN	– hydrogen cyanide	} +	0.65	Ori A
144.826841	GHz	DCN	– hydrogen cyanide			Ori A

Table 20 (continued)
Molecules observed at radio frequencies:

Frequency (MHz or GHz)		Molecule			T_A (°K)	Source
145.09375	GHz	CH_3OH	-methyl alcohol	+	1.25	Ori A
145.09747	GHz	CH_3OH	-methyl alcohol	+	1.45	Ori A
145.10323	GHz	CH_3OH	-methyl alcohol	+	1.35	Ori A
145.12441	GHz	CH_3OH	-methyl alcohol	+	1.45	Ori A
145.12637	GHz	CH_3OH	-methyl alcohol			Ori A
145.13188	GHz	CH_3OH	-methyl alcohol	+	1.25	Ori A
145.13346	GHz	CH_3OH	-methyl alcohol			Ori A
145.602971	GHz	H_2CO	-formaldehyde	+	1.9	Ori A
145.94679	GHz	OCS	-carbonyl sulfide	+	0.45	Sgr B2
146.96916	GHz	CS	-carbon monosulfide	+	3.0	Ori A
150.498359	GHz	H_2CO	-formaldehyde	+	2.7	Ori A
168.762762	GHz	H_2S	-hydrogen sulfide	+	2.3	Ori A
169.33534	GHz	CH_3OH	-methyl alcohol	+	0.7	Ori A
172.1081	GHz	HCN^{15}	-hydrogen cyanide	+	0.45	Ori A
172.6777	GHz	$HC^{13}N$	-hydrogen cyanide	+	0.91	Ori A
217.2384	GHz	DCN	-hydrogen cyanide		—	—
219.560369	GHz	CO^{18}	-carbon monoxide		—	—
220.398714	GHz	$C^{13}O$	-carbon monoxide	+	16[1]	Ori A
230.538001	GHz	CO	-carbon monoxide	+	76[1]	Ori A

[1] T_A corrected for all telescope and atmospheric losses.
[2] After SNYDER (1973) and private communication.
[3] Cf. SPITZER *et al.* (1973).

of spectral lines in the ultraviolet, visible, and infrared spectral regions have resulted in the identification of the atmospheric gases given in Table 19.

Interstellar molecules were first discovered in the absorption spectra of stars (DUNHAM and ADAMS, 1937). Earth-based optical observations resulted in the detection of interstellar CH^+, CH, CN, and NaH^- (cf. McNALLEY, 1968). Rocket techniques which permit observations in the extreme ultraviolet part of the spectrum have resulted in the detection of interstellar H_2 (CARRUTHERS, 1970) and CO (SMITH and STECHER, 1971). The first molecular line observed at radio frequencies with the OH line (WEINREB *et al.* 1963); and subsequently more than 1800 lines from 66 molecular species have been observed. The frequencies and wavelengths of the line radiation from observed interstellar molecules are given in Table 20. Laboratory measurements of the frequencies of the microwave spectral lines of many molecules are summarized by CORD, LOJKO, and PETERSON (1968).

2.15.2. Line Intensities and Molecular Abundances

It follows from Eqs. (2-72) and (2-74) that under conditions of local thermodynamic equilibrium (LTE), the line intensity observed for a line emitted by an optically thin gas is proportional to the line integral of the absorption coefficient per unit length, α_ν(LTE). For a rotational transition at frequency, ν_r, we have from Eq. (2-69)

$$\alpha_{\nu_r}(\mathrm{LTE}) = \frac{c^2}{8\pi\nu_r^2}\frac{N_J}{\Delta\nu_L}\exp\left(\frac{h\nu_r}{kT}\right)\left[1-\exp\left(\frac{-h\nu_r}{kT}\right)\right]A_J, \qquad (2\text{-}154)$$

where T is the excitation or spin temperature of the gas, $\Delta\nu_L$ is the full width to half maximum of the line, the Einstein coefficient for the spontaneous electric dipole transition is

$$A_J = \frac{64\pi^4\nu_r^3}{3hc^3}|\mu_J|^2,$$

where the electric dipole matrix element, μ_J, for symmetric molecules is given by

$$|\mu_J|^2 = \mu^2\frac{(J+1)}{(2J+1)} \quad \text{for the } J+1 \leftarrow J \text{ transition}$$

and

$$|\mu_J|^2 = \mu^2\frac{(J+1)}{(2J+3)} \quad \text{for the } J+1 \rightarrow J \text{ transition}.$$

Measured values of the electric dipole moment, μ, are given in Table 21 for the simple molecules of the most abundant elements.

The number of molecules, N_J, in the J state is given by the Boltzmann equation (3-125) and Eq. (2-141)

$$N_J = \frac{(2J+1)}{U}N\exp\left[\frac{-hBJ(J+1)}{kT}\right], \qquad (2\text{-}155)$$

where T is the excitation or spin temperature of the gas, the statistical weight of level J is $(2J+1)$, N is the total number of molecules, and the partition function, U, is given by

$$U = \sum_{J=0}^{\infty}(2J+1)\exp\left[\frac{-hBJ(J+1)}{kT}\right] \approx \frac{kT}{hB} \quad \text{for } hB \ll kT.$$

Eqs. (2-154) to (2-155) may be combined to give the relation

$$\alpha_{\nu_r}(\mathrm{LTE}) = \frac{8\pi^3|\mu_J|^2}{3hc}\left(\frac{h\nu_r}{kT}\right)^2\frac{BN(2J+1)}{\Delta\nu_L}\exp\left[\frac{-hBJ(J+1)}{kT}\right], \qquad (2\text{-}156)$$

for $hB \ll kT$ and $h\nu_r \ll kT$. For low values of J, the exponential factor in Eq. (2-156) can be approximated as unity. The line absorption coefficient given by Eq. (2-156) can be related to the observed antenna temperatures of the line and the antenna efficiency by Eqs. (2-74) to (2-77).

When vibrations are taken into account, Eq. (2-156) is modified by multiplication of N with the fraction, f_v, of the molecules in the vibrational state of energy $h\omega_e(v+\frac{1}{2})$. This fraction is given by

$$f_v = \exp\left(\frac{-vh\omega_e}{kT}\right)\left[1-\exp\left(\frac{-h\omega_e}{kT}\right)\right]. \qquad (2\text{-}157)$$

Table 21. Measured values of the electric dipole moments of simple molecules in the gas phase of the most abundant elements. The moments are in Debye units (1 Debye $=10^{-18}$ e.s.u.-cm)[1]

Formula	Compound name	Dipole moment
HNO_3	Nitric Acid	2.17
HN_3	Hydrogen Azide (Hydrazoic Acid)	(0.8)
HO	Hydroxyl Radical	1.66
$H_2N_2O_2$	Nitroamine (Nitramide)	(3.6)
H_2O	Water	1.85
H_2O_2	Hydrogen Peroxide	2.2
H_2S	Hydrogen Sulfide	0.97
H_3N	Ammonia	1.47
H_4N_2	Hydrazine	1.75
NO	Nitrogen Monoxide (Nitric Oxide)	0.153
NO_2	Nitrogen Dioxide	0.316
N_2O	Dinitrogen Oxide (Nitrous Oxide)	0.167
OS	Sulfur Monoxide	1.55
OS_2	Disulfur Monoxide	1.47
O_2S	Sulfur Dioxide	1.63
O_3	Ozone	0.53
CO	Carbon Monoxide	0.112
COS	Carbonyl Sulfide	0.712
CS	Carbon Monosulfide	1.98
CHN	Hydrogen Cyanide	2.98
CHNO	Hydrogen Isocyanate	(1.6)
CHNS	Hydrogen Isothiocyanate	(1.7)
CH_2N_2	Cyanogen Amide (Cyanamide)	4.27
CH_2N_2	Diazomethane	1.50
CH_2N_2	Diazirine	1.59
CH_2O	Methanal (Formaldehyde)	2.33
CH_2O_2	Methanoic Acid (Formic Acid)	1.41
CH_3NO	Hydroxyliminomethane (Formaldoxime)	0.44
CH_3NO	Formyl Amide (Formamide)	3.73
CH_3NOS	Methyl Sulfinylamine	1.70
CH_3NO_2	Nitromethane	3.46
CH_3NO_3	Methyl Nitrate	3.12
CH_3N_3	Methyl Azide	2.17
CH_4O	Methanol	1.70
CH_4S	Methanethiol (Methyl Mercaptan)	1.52
CH_5N	Methyl Amine	1.31
CH_6Si	Methyl Silane	0.735
C_2N_2S	Dicyano Sulfide	3.02
$C_2H_2N_2O$	1,2,5-Oxadiazole	3.38
$C_2H_2N_2O$	1,3,4-Oxadiazole	3.04
$C_2H_2N_2S$	1,2,5-Thiadiazole	1.56
$C_2H_2N_2S$	1,3,4-Thiadiazole	3.29
C_2H_2O	Methylene Carbonyl (Ketene)	1.42
C_2H_3N	Cyanomethane (Acetonitrile)	3.92
C_2H_3N	Isocyanomethane	3.85
C_2H_3NO	Methyl Isocyanate	(2.8)
C_2H_3NS	Methyl Thiocyanate	(4.0)
C_2H_4O	Oxirane (Ethylene Oxide)	1.89
C_2H_4O	Ethanal (Acetaldehyde)	2.69
$C_2H_4O_2$	Ethanoic Acid (Acetic Acid)	1.74
$C_2H_4O_2$	Methyl Methanoate (Methyl Formate)	1.77

Table 21 (continued)

Formula	Compound name	Dipole moment
C_2H_4S	Thiirane (Ethylene Sulfide)	1.85
C_2H_4Si	Silyl Acetylene	0.316
C_2H_5N	Iminoethane (Ethyleneimine)	1.90
C_2H_5N	Methyliminomethane ($CH_3N{=}CH_2$)	1.53
C_2H_5NO	Acetyl Amine (Acetamide)	3.76
C_2H_5NO	Methylaminomethanal (N-Methylformamide)	3.83
$C_2H_5NO_2$	Nitritoethane (Ethyl Nitrite)	2.40
$C_2H_5NO_2$	Nitroethane	3.65
C_2H_6O	Ethanol	1.69
C_2H_6O	Dimethyl Ether	1.30
C_2H_6OS	Dimethylsulfoxide	3.96
$C_2H_6O_2$	1,2-Ethanediol (Ethylene Glycol)	2.28
$C_2H_6O_2S$	Dimethyl Sulfoxylate (Dimethyl Sulfone)	4.49
C_2H_6S	Ethanethiol	1.58
C_2H_6S	Dimethyl Sulfide	1.50
C_2H_6Si	Silyl Ethylene	0.66
C_2H_7N	Aminoethane (Ethyl Amine)	1.22
C_2H_7N	Dimethyl Amine	1.03
C_2H_8Si	Dimethyl Silane	0.75
C_2H_8Si	Ethyl Silane	0.81
C_3HN	Cyanoacetylene	3.72
$C_3H_2N_2$	Dicyanomethane	3.73
C_3H_2O	Propynal	2.47
$C_3H_2O_3$	Vinylene Carbonate	4.55
C_3H_3N	Cyanoethylene	3.87
C_3H_3NO	Acetyl Cyanide	3.45
C_3H_3NS	Thiazole	1.62
C_3H_4	Cyclopropene	0.45
C_3H_4	Propyne	0.781
C_3H_4O	Ethylidene Carbonyl (Methyl Ketene)	1.79
C_3H_4O	Propenal, Trans Conformation (Acrolein)	3.12
$C_3H_4O_2$	2-Oxoöxetane (β-Propiolactone)	4.18
$C_3H_4O_2$	Vinyl Formate	1.49
C_3H_5N	Cyanoethane (Propionitrile)	4.02
C_3H_6	Propene	0.366
C_3H_6O	Methyl Oxirane (Propylene Oxide)	2.01
C_3H_6O	Propanone (Acetone)	2.88
C_3H_6S	Thietane (Trimethylene Sulfide)	1.85

[1] After NELSON, LIDE, and MARYOTT, 1967.

The effective excitation or spin temperature, T_s, is the appropriate temperature to use in Eq. (2-156). It is defined by the Boltzmann equation (3-126)

$$\frac{N_m}{N_n} = \frac{g_m}{g_n} \exp\left(\frac{-h\,\nu_{mn}}{k\,T_s}\right), \tag{2-158}$$

where N_m and g_m are, respectively, the number density and statistical weight of level m, and ν_{mn} is the frequency of the $m-n$ transition. Assuming a Maxwellian distribution of velocities, the gas is characterized by the kinetic temperature, T_k, which is related to the Doppler broadening of the line. The transition rate, R_{mn}, for transitions induced by collisions is given by

$$R_{mn} = \frac{1}{\tau_c} = N \langle v \rangle \sigma, \tag{2-159}$$

where τ_c is the collision lifetime, N is the density of the colliding particles, the average velocity, $\langle v \rangle$, is given by

$$\langle v \rangle = \left[\frac{8kT_k}{\pi \mu} \right]^{1/2},$$

where μ is the reduced mass of the molecule and the colliding particle, and the collision cross section, σ, is given by (PURCELL, 1952)

$$\sigma = \frac{16 e^2 |\mu_{mn}|^2}{3\hbar^2 \langle v \rangle^2} \ln \left[\frac{(0.706) 3\hbar^2 \pi \langle v \rangle^4}{32 e^2 |\mu_{mn}|^2 \omega^2} \right],$$

for $\langle v \rangle > 10^6 \text{ cm sec}^{-1}$. Here μ_{mn} is the dipole matrix element and ω is the transition frequency. Values of R_{mn} for strong collisions and lower values of $\langle v \rangle$ are given by ROGERS and BARRETT (1968) and GOSS and FIELD (1968).

For transitions induced by radiation of intensity, I_ν, the Einstein probability coefficients, A_{mn} and B_{mn}, are related by the equation

$$A_{mn} = \frac{1}{\tau_r} = I_\nu B_{mn} \left(\frac{h\nu}{kT_R} \right), \tag{2-160}$$

where the radiation lifetime is τ_r, and the radiation temperature, T_R, is related to the intensity, I_ν, by the Rayleigh-Jeans approximation

$$I_\nu = \frac{2kT_R \nu^2}{c^2} \quad \text{for } h\nu \ll kT_R. \tag{2-161}$$

The three temperatures T_s, T_k, and T_R are related by the equation

$$T_s = T_k \left[\frac{T_R + T_0}{T_k + T_0} \right], \tag{2-162}$$

where

$$T_0 = \frac{h\nu R_{mn}}{kA_{mn}} = \frac{h\nu \tau_R}{k \tau_c},$$

which follows from the equation of statistical equilibrium (Eq. (2-82)) and from Eqs. (2-159) to (2-161).

When the radiation from the interstellar OH molecule was found to be polarized, to have anomalous relative line intensities, and to imply very high brightness temperatures, it was postulated that the population of the levels had been inverted and that maser amplification occurred. The integrated flux density received from a masing source is given by

$$\int S_\nu \, d\nu = \frac{h \nu R_m}{D^2 \Omega_m}, \tag{2-163}$$

where S_ν is the flux density observed at frequency, ν, the distance to the source is D, the solid angle of the maser emission is Ω_m, and the time rate of microwave photons, R_m, is proportional to

$$R_m \propto \exp(\tau_\nu)$$

for an unsaturated maser, and

$$R_m \propto \tau_\nu$$

for a saturated maser. Detailed formulae for the proportionality constants are given by LITVAK (1969) for infrared pumping of the OH molecule. Here τ_ν is given by

$$\tau_\nu = \frac{h B_{mn} g_m \nu_{mn}}{\Delta \nu_L} \int \left(\frac{N_m}{g_m} - \frac{N_n}{g_n} \right) dl, \tag{2-164}$$

where B_{mn} is the coefficient for stimulated emission of the line, g_m is the degeneracy of the upper state, ν_{mn} is the line frequency, $\Delta \nu_L$ is the line width, and $\int (N_m/g_m - N_n/g_n) dl$ is the line integral of the population inversion. For an unsaturated maser, the observed Doppler-broadened lines will be narrowed and the line width, $\Delta \nu_L$, is given by

$$\Delta \nu_L = \frac{1}{\sqrt{\tau_\nu}} \left[\frac{1.67}{\lambda} \left(\frac{2 k T_K}{M} \right)^{1/2} \right], \tag{2-165}$$

where λ is the wavelength of the line, T_K is the kinetic temperature, M is the molecular mass, and the expression in square brackets is the thermal Doppler width of the line. Saturated masers do not exhibit line narrowing.

2.15.3. The Formation and Destruction of Molecules

The energetics of the various reactions which create or destroy molecules are determined by the dissociation energy and ionization potential of the molecule together with the energy of other reactants such as photons or charged particles. The dissociation energy of a stable electronic state of a diatomic molecule is that energy required to dissociate it into atoms from the lowest rotation-vibration level. The dissociation energy referred to the ground electronic state is termed D_0^0 when the dissociation products are normal atoms. The ionization potential of a molecule is defined as that energy necessary to remove an electron from the outermost filled molecular orbital of the ground state. WILKINSON (1963) has listed the ionization potentials and dissociation energies for 148 diatomic molecules of astrophysical interest. Some of these data are included in Table 22 together with other data taken from STIEF *et al.* (1972) and STIEF (1973).

Table 22. Dissociation energies and ionization potentials of abundant molecules and ions[2]

Molecule or ion	Dissociation energy, D_0^0 (eV)[1]	Ionization potential (eV)[1]
C_2	6.25 ± 0.2	12.0
C_2^+	(5.5 ± 0.5)	
CH	3.47	10.64
CH^+	4.09	
CH_4		13.0
CH_3C		10.36
CH_3OH		10.9
C_2H_2		11.4
C_6H_6		9.24
CN	7.5 ± 0.15	14.2 ± 0.3
CN^+	(5.56 ± 0.45)	
CO	11.09	14.013
CO^+	8.33	27.9 ± 0.5
CS	7.6	11.8 ± 0.2
CS^+	(6.2)	
H_2	4.47718 ± 0.00012	15.426
H_2^+	2.646	
HCN		13.6
H_2CO		10.9
H_2O		12.6
HCOOH		11.1
HD	4.51274 ± 0.00020	
NH	(3.76)	13.10
NH^+	(4.26)	
NH_3		10.2
NO	6.506	9.25
NO^+	10.87	30.6 ± 0.3
OCS		11.2
OH	4.395	13.36 ± 0.2
OH^+	(4.63 ± 0.2)	
O_2	5.115	12.075
O_2^+	6.65 ± 0.01	38.0 ± 0.5
SO	5.357	12.1 ± 0.3
SO^+	(3.4 ± 0.3)	

[1] Energies, E, in eV can be converted to wavelengths, λ, in Å by $\lambda = 12{,}396.3/E$.
[2] After WILKINSON (1963), STIEF *et al.* (1972), and STIEF (1973).

The basic reactions which create or destroy molecules are listed below in the order in which they are presented in the following text.

Reaction	*Process*
Photodissociation	A molecule is destroyed by a photon to form another molecule and/or component atoms.
Photoionization	A molecule is ionized by a photon to form an ion and an electron.
Gas exchange reaction	Atoms and molecules interact to form atoms or molecules.

Reaction	*Process*
Ion-Molecule reaction	Ions and molecules interact to form ions, molecules and/or atoms.
Associative detachment	Ions and atoms interact to form molecules and electrons.
Charge exchange reaction	Ions and atoms interact to form ions and/or molecules.
Surface recombination	Atoms combine on a grain surface to form a molecule which is then evaporated.
Radiative Association	Atoms combine to form a molecule and a photon.
Radiative attachment	Electrons and atoms interact to form an ion and a photon.
Photodetachment	A photon and ion interact to form an atom and an electron.
Dissociative recombination	An electron and an ion or molecule interact to form atoms and/or other molecules.
Radiative recombination	An electron and ion interact to form a molecule or atom and a photon.

The most important destruction mechanism for interstellar molecules is their photo-dissociation by the interstellar radiation field. The photodestruction rate, or dissociation probability, P, is given by

$$P = \frac{1}{h} \int_{912\,\text{Å}}^{\lambda_T} U_\lambda \sigma_\lambda \Phi \lambda \, d\lambda \quad \text{sec}^{-1}, \tag{2-166}$$

where h is Planck's constant, U_λ is the energy density of the radiation field at wavelength, λ, the absorption cross section is σ_λ, and Φ is the primary quantum yield for dissociation. The limits to the integration in this equation are $\lambda = 912$ Å where hydrogen is photoionized, and $\lambda_T = 12{,}396.3/E$ where E is the dissociation energy in eV. Threshold wavelengths, λ_T, lie in the ultraviolet range of wavelengths, and the radiation energy density in this range is given by (Habing, 1968)

$$U_\lambda = 4 \times 10^{-17}\ \text{erg cm}^{-3}\ \text{Å}^{-1} \quad \text{for } 912\ \text{Å} \le \lambda \le 2400\ \text{Å}. \tag{2-167}$$

When obscuring clouds are present, the radiation field is attenuated by the factor

$$[a_\lambda + (1 - a_\lambda) 10^{0.4 A_\lambda}]^{-1}, \tag{2-168}$$

where the grain albedo $a_\lambda \sim 0.5$, and the extinction, A_λ, is in magnitudes and is given by

$$A_\lambda = 2.3 \times 10^7 Q r_{\text{gr}}^2 N^{2/3} M^{1/3} \ \text{mag},$$

for the center of a cloud of M solar masses and average particle number density, N. Here the extinction efficiency $Q \approx 2.5$, the grain radius $r_{\text{gr}} \approx 0.12 \times 10^{-4}$ cm, and the dust-to-gas ratio $N_g/N \approx 10^{-12}$.

Unattenuated photodestruction rates for various ions and molecules have been compiled by KLEMPERER (1971), SOLOMON and KLEMPERER (1972), STIEF *et al.* (1972), and STIEF (1973). These rates are given in Table 23 together with a few other rates which come from the footnoted references.

Table 23. Photodestruction rates for abundant molecules and ions[5]

Reaction		Photodestruction rate (sec^{-1})
$C + h\nu \rightarrow C^+ + e$		1.4×10^{-10} [1]
$C_2^+ + h\nu \rightarrow C^+ + C$		8×10^{-12}
$CH + h\nu \rightarrow C + H$		1.1×10^{-10} [2]
$CH + h\nu \rightarrow CH^+ + e$		5.0×10^{-11}
$CH^+ + h\nu \rightarrow C^+ + H$		3.0×10^{-12}
$CH_4 + h\nu \rightarrow H_2 + CH_2$	dominant	0.75×10^{-9}
$CH_3C + h\nu \rightarrow CH_2C \rightarrow CH + H$		1.0×10^{-9}
$\rightarrow C_3 + 2H_2$		1.0×10^{-9}
$C_2H_2 + h\nu \rightarrow C_2H + H$	$\Phi = 0.2$	$0.16 \times 10^{-8}\, \Phi$
$\rightarrow C_2 + H_2$	$\Phi = 0.1$	$0.16 \times 10^{-8}\, \Phi$
$C_6H_6 + h\nu \rightarrow ?$		$0.66 \times 10^{-8}\, \Phi$
$CN + h\nu \rightarrow C + N$		4×10^{-11}
$CO + h\nu \rightarrow C + O$	$\Phi \sim 0.07$	$3.3 \times 10^{-10}\, \Phi$
$H^- + h\nu \rightarrow H + e$		2.4×10^{-7} [3]
$H_2 + h\nu \rightarrow H + H$		1.0×10^{-10} [4]
$HD + h\nu \rightarrow H + D$		1.0×10^{-10} [4]
$H_2CO + h\nu \rightarrow H_2 + CO$	$\Phi = 0.5$	1.0×10^{-9}
$\rightarrow 2H + CO$	$\Phi = 0.5$	1.0×10^{-9}
$H_2O + h\nu \rightarrow H + OH$		0.5×10^{-9}
$NH + h\nu \rightarrow N + H$		1.0×10^{-11}
$NH + h\nu \rightarrow NH^+ + e$		1.0×10^{-12}
$NH_3 + h\nu \rightarrow H + NH_2$		
$\rightarrow H_2 + NH$		0.75×10^{-9}
$\rightarrow 2H + NH$		
$NO + h\nu \rightarrow N + O$		3.3×10^{-10}
$OCS + h\nu \rightarrow CO + S$		3.3×10^{-9}
$\rightarrow CS + O$		
$OH + h\nu \rightarrow O + H$		4.0×10^{-12}
$OH + h\nu \rightarrow OH^+ + e$		1.0×10^{-12}

[1] After WERNER (1970), note that this and other photodetachment reactions require special calculations.
[2] After ELANDER and SMITH (1973).
[3] After DE JONG (1972), again a photodetachment reaction.
[4] After HOLLENBACH, WERNER, and SALPETER (1971). The HD rate is taken to be that of H_2 in this unattenuated case, but they differ in the attenuated case (cf. BLACK and DALGARNO, 1973).
[5] Unless otherwise noted, the rates are from KLEMPERER (1971) and STIEF (1973).

The astrophysically important gas exchange reactions are of the bimolecular type given by

$$A + BC \rightarrow AB + C,$$

where A and C are atoms and AB and BC are molecules. The number densities, N, of the reactants are related by the equation

$$-\frac{dN_A}{dt} = -\frac{dN_{BC}}{dt} = \frac{dN_{AB}}{dt} = \frac{dN_B}{dt} = \kappa N_A N_{BC},$$

where t is the time variable, and the rate constant, κ, is given by (POLANYI, 1962)

$$\kappa = P r_{AB}^2 \left(\frac{8\pi}{\mu} kT\right)^{1/2} \exp\left[-\frac{E_a}{kT}\right] \text{cm}^3 \text{sec}^{-1}$$

or (2-169)

$$\kappa = A \exp\left[-\frac{E_a}{kT}\right] \text{cm}^3 \text{sec}^{-1}.$$

Here the Arrhenius factor, A, increases slowly with temperature, and the activation energy, E_a, is the difference in internal energy between the activated and normal molecule. The steric factor, P, is an orientation parameter close to unity, the reagent molecules are assumed to approach each other to within a distance r_{AB} which is the mean of the gas kinetic collision parameters r_{AA} and r_{BB}, the reduced mass of the atom, A, and the molecule, BC, is μ, and T is the gas kinetic temperature. Most gas exchange reactions of astrophysical interest are exothermic, and therefore not very temperature dependent. The A factors measured at room temperature are summarized by KAUFMAN (1969), and for most atom molecule reactions of astrophysical interest we have

$$A \sim 4 \times 10^{-11} \text{cm}^3 \text{sec}^{-1}.$$

Specific rate constants are given in Table 24.

Table 24. Gas exchange reactions and rate constants for abundant molecules[1]

Reaction	Rate constant, A or κ ($\text{cm}^3 \text{sec}^{-1}$)
$C_2 + O \rightarrow CO + C$	3×10^{-11}
$C_2 + N \rightarrow CN + C$	3×10^{-11}
$CH + C \rightarrow C_2 + H$	4×10^{-11}
$CH + H \rightarrow C + H_2$	1×10^{-14}
$CH + O \rightarrow CO + H$	4×10^{-11}
$CH + N \rightarrow CN + H$	4×10^{-11}
$CN + O \rightarrow CO + N$	$10^{-11} \exp(-1{,}200/T)$
$CN + N \rightarrow C + N_2$	$\ll 1 \times 10^{-13}$
$OH + H \rightarrow O + H_2$	1×10^{-11} [2]
$N + OH \rightarrow NO + H$	7×10^{-11} at $T = 320\,^\circ$K
$O + OH \rightarrow O_2 + H$	5×10^{-11} at $T = 300\,^\circ$K
$N + NO \rightarrow N_2 + O$	2×10^{-11} at $T = 300\,^\circ$K

[1] After KLEMPERER (1971) and HERBST and KLEMPERER (Ap. J. **185,** 505 (1973)).
[2] After CARROLL and SALPETER (1966).

For ion-molecule reactions, associative detachment, and charge exchange reactions we have reactions of the form

$$A + B \rightarrow C + D,$$

where A is an ion, B is an atom or molecule, C is an atom or molecule, and D is an ion or a charged particle. A rate constant, κ, is defined by

$$-\frac{dN_A}{dt}=-\frac{dN_B}{dt}=\frac{dN_C}{dt}=\frac{dN_D}{dt}=\kappa N_A N_B,$$

where t is the time variable, and N_A, N_B, N_C, and N_D are, respectively, the number densities of A, B, C, and D. The interaction potential, V, between the ion, A, and the atom or molecule, B, is given by (RAPP and FRANCIS, 1962)

$$V=-\frac{\alpha e^4}{2r^4},$$

where α is the dipole polarizability of the atom or molecule, e is the charge of the electron, and r is the internuclear distance. According to GIOUMOUSIS and STEVENSON (1958), the critical impact parameter, r_{AB}, is given by

$$r_{AB}=\left[\frac{4e^2\alpha}{\mu v^2}\right]^{1/4},$$

so that the rate constant is given by

$$\kappa=\langle\sigma v\rangle=2\pi e f\left(\frac{\alpha}{\mu}\right)^{1/2}. \tag{2-170}$$

Here σ is the cross section for charge transfer, v is the relative velocity of the reactants, the angular brackets denote averaging over velocities, μ is the reduced mass of the reactants, and f is a statistical factor which takes into account the fact that not all collisions lead to charge transfer. For most ion-molecule, associative detachment, and charge exchange reactions of astrophysical interest we have

$$\kappa\sim 1\times 10^{-9}\ \mathrm{cm^3\ sec^{-1}}.$$

Specific rate constants are given in Table 25.

VAN DE HULST (1949) first suggested that molecules might form on the interstellar grains in a three body process where lattice vibrations in the grain absorbed the excess energy liberated when gas atoms combined to form molecules. This idea was developed as a mechanism for the formation of molecular hydrogen by GOULD and SALPETER (1963) and GOULD, GOLD, and SALPETER (1963). The efficiency, γ, at which atoms strike a grain surface and recombine to form a molecule is the product of two factors: the sticking coefficient, S, or probability that an atom hitting the grain surface from the interstellar gas becomes thermalized and sticks to the grain; and the recombination efficiency, γ', or probability that the first adsorbed atom will remain adsorbed and not evaporate before a second atom strikes the grain, becomes adsorbed, and recombines with the first atom. HOLLENBACH and SALPETER (1970, 1971) show that the sticking coefficient is given by

$$S=\frac{\Gamma^2+0.8\,\Gamma^3}{1+2.4\,\Gamma+\Gamma^2+0.8\,\Gamma^3} \tag{2-171}$$

if

$$k\,T_{\mathrm{gr}}<k\,T_{\mathrm{gas}}<D.$$

Table 25. Rate constants for ion-molecule, associative detachment, and charge exchange reactions[1]

Reaction	Rate constant, κ ($cm^3\ sec^{-1}$)
$C^+ + Ca \rightarrow C + Ca^+$	2.0×10^{-9}
$C^+ + CH \rightarrow H + C_2^+$	1.0×10^{-9}
$C^+ + K \rightarrow C + K^+$	2.0×10^{-9}
$C_2^+ + O \rightarrow CO + C^+$	1.0×10^{-9}
$C_2^+ + N \rightarrow CN + C^+$	1.0×10^{-9}
$CH^+ + C \rightarrow H + C_2^+$	1.0×10^{-9}
$CH^+ + H \rightarrow H_2 + C^+$	$7.5 \times 10^{-15} T^{5/4}$
$CH^+ + N \rightarrow CN + H^+$	1.0×10^{-9}
$CH^+ + O \rightarrow CO + H^+$	1.0×10^{-9}
$CH^+ + O \rightarrow H + CO^+$	1.0×10^{-9}
$CO^+ + H \rightarrow CO + H^+$	1.0×10^{-9}
$CN^+ + H \rightarrow CN + H^+$	1.0×10^{-9}
$D^+ + H_2 \rightarrow HD + H^+$	0.8×10^{-9}
$H^- + H^+ \rightarrow 2H$	$1.1 \times 10^{-6} T^{-0.4}$ or 2.3×10^{-7} at $T = 50\,°K$
$H^- + H \rightarrow H_2 + e$	1.3×10^{-9}
$H^+ + D \rightarrow H + D^+$	2.0×10^{-9}
$H^+ + O \rightarrow H + O^+$	0 for $T = 0\,°K$ [2] 0.01×10^{-9} for $T = 50\,°K$
$H_2^+ + H \rightarrow H_2 + H^+$	5.8×10^{-10}
$O^- + H_2 \rightarrow H_2O + e$	6.0×10^{-10}
$O^- + O \rightarrow O_2 + e$	1.4×10^{-10}
$O^- + N \rightarrow NO + e$	2.0×10^{-10}
$O^+ + H \rightarrow O + H^+$	0.76×10^{-9} at $T = 0\,°K$ [2]
$S^- + H_2 \rightarrow H_2S + e$	$< 1.0 \times 10^{-15}$

[1] After KLEMPERER (1971), DE JONG (1972), BLACK and DALGARNO (1973), and DALGARNO and MCCRAY (1973). Rate constants for other ion-molecule reactions, charge transfer reactions, and ion-electron recombinations are given by HERBST and KLEMPERER in Ap. J. **185**, 505 (1973).

[2] Cf. FIELD and STEIGMANN (1971).

Here $\Gamma = E_c/(k\,T_{gas})$ where E_c is the characteristic total energy transferred to the grain surface, T_{gas} is the gas temperature, T_{gr} is the grain temperature, and D is the binding energy for the adsorption ground state. For hydrogen atoms, $\Gamma \approx 1$ and $S \approx 0.3$ at $T_{gas} \approx 100\,°K$. For the heavier atoms C, N, and 0, we have $S \approx 1$ according to WATSON and SALPETER (1972). The recombination efficiency, γ', is unity if the time, t_s, for a new atom to strike the grain surface is much less than the time, t_{eV}, for an atom to evaporate from the surface. If N denotes the number density of atoms and V is their thermal velocity, then

$$t_s \approx [S\,N\,V\,\pi\,r_{gr}^2]^{-1}, \tag{2-172}$$

where r_{gr} is the grain radius, usually taken to be $r_{gr} \approx 0.17 \times 10^{-4}$ cm. The characteristic lattice vibration frequency of the grain is $\nu_0 \approx 10^{12}\ sec^{-1}$, and the evaporation time is

$$t_{eV} \approx \nu_0^{-1} \exp\left(\frac{D}{k\,T_{gr}}\right), \tag{2-173}$$

where D is the atom adsorption energy and T_{gr} is the grain temperature. Arguments about the efficiency of molecule formation have centered about different estimates of the grain temperature, T_{gr}, and the adsorption binding energy, D (cf. KNAAP *et al.* 1966; STECHER and WILLIAMS, 1968; WENTZEL, 1967; HOLLENBACH and SALPETER, 1970, 1971). Current arguments give $T_{gr} \lesssim 25\,°K$ and values of D such that $t_s \ll t_{eV}$ for most atoms, especially when dislocations and chemical impurity sites are included in calculating D.

The rate of formation of a molecule, AB, and the rate of depletion of the element, A, are governed by the equations

$$\frac{dN_{AB}}{dt} = \kappa N_A N_{gr}$$

and

$$\frac{dN_A}{dt} = \frac{-\alpha\kappa}{(1-\alpha)} N_A N_{gr},$$

where N_{AB}, N_A, and N_{gr} denote, respectively, the densities of the molecules, AB, atoms, A, and grains, gr, the time variable is t, the probability that an atom will become permanently locked to the grain is α, and $(1-\alpha)$ is the probability that a molecule will evaporate and return to the gas. The rate constant, κ, is given by

$$\begin{aligned}\kappa &= S(1-\alpha)\pi r_{gr}^2(1+\gamma Z)v \\ &\approx 6\times 10^{-9}(1-\alpha)(1+2.5\,Z)r_{gr}^2\, T^{1/2}\ \mathrm{cm}^3\,\mathrm{sec}^{-1},\end{aligned} \tag{2-174}$$

where S is the sticking coefficient, the grain radius $r_{gr} \approx 0.17\times 10^{-4}$ cm, γ is determined from the equation describing the balance of charge on the grain, Z is the charge of the element, A, and v is its thermal velocity given by

$$v = \left(\frac{8k\,T_{gas}}{\pi M_A}\right)^{1/2},$$

where the gas temperature is T_{gas} and the mass of element, A, is M_A. HOLLENBACH, WERNER, and SALPETER (1971) discuss the formation of molecular hydrogen, H_2, on grains. Because H_2 is a light saturated molecule with a low adsorption energy, it is easily evaporated thermally and $\alpha \approx 0$. In calculating molecular abundances, a constant gas to dust density ratio is used. Typical grains have a density $\rho_{gr} \approx 2\,\mathrm{gm\,cm}^{-3}$, a radius $r_{gr} \approx 0.17\times 10^{-4}$ cm, a mass $M_{gr} \approx 4\times 10^{-14}$ gm, and a number density, N_{gr}, given by

$$N_{gr} \approx 4\times 10^{-13} N_A,$$

where $N_A = N_H + 2N_2$ is the total number density of hydrogen atoms in atomic, N_H, and molecular, N_2, form. The formation of the heavier molecules of C, O, N and H is discussed by WATSON and SALPETER (1972). In this case $S=1$, but the mechanism by which the molecules return to the gas is unclear. For the saturated molecules such as CH_4, H_2O, and NH_3, the adsorption energy is low and they may be evaporated thermally.

At the low gas densities of the interstellar medium, some simple molecules may be formed by the radiative association process (SWINGS, 1942; KRAMERS and TER HAAR, 1946)

$$A + B \rightarrow AB + h\nu.$$

Here A and B are two ground state atoms which collide, and find themselves in the continuum of an excited molecular state. A molecule, AB, is formed if the excited complex relaxes and emits a photon of energy, $h\nu$, before the atoms separate. Radiative association routes for the diatomic molecules of the more abundant species are given in Table 26.

Table 26. Radiative association routes for abundant diatomic molecules[1]

Molecule	Radiative association route
CH	$C(^3P) + H(^2S) \rightarrow CH(B^2\Sigma^-) \rightarrow CH(X^2\Pi) + h\nu$
CH^+	$C^+(^2P) + H(^2S) \rightarrow CH^+(A^1\Pi) \rightarrow CH(X^1\Sigma^+) + h\nu$
CN	$C(^3P) + N(^4S) \rightarrow CN(A^2\Pi) \rightarrow CN(X^2\Sigma^+) + h\nu$
CO	$C(^3P) + O(^3P) \rightarrow CO(A^1\Pi) \rightarrow CO(X^1\Sigma^+) + h\nu$
CS	$C(^3P) + S(^3P) \rightarrow CS(A^1\Pi) \rightarrow CS(X^1\Sigma^+) + h\nu$

[1] After Lutz (1972).

Bates (1951) first gave the correct theory for calculating the radiative association rate constant, κ. If the number densities of atoms A and B, and the molecule AB are, respectively, N_A, N_B, and N_{AB}; then

$$\frac{dN_{AB}}{dt} = \kappa N_A N_B ,$$

where t is the time variable, and the rate constant for two atoms which meet with energy, E, is

$$\kappa = g \int \sigma(E) v(E) W(E) dE ,$$

where g is the probability that two atoms find themselves in the continuum of the required excited molecular state, $v(E)$ is the relative velocity of the colliding atoms, $\sigma(E)$ is the cross-section for radiative association, and $W(E)$ is the energy distribution of the atoms. The total cross-section for radiative association is (Lutz, 1972)

$$\sigma(E) = \frac{32 \times 2^{1/2} \pi^4 h^2}{3 \mu^{3/2} E^{1/2} c^3 G' v} \sum_{v''} \sum_{J''} |\langle \psi'' | D | \psi' \rangle|^2 \rho'(E) \nu(E, v'')^3 ,$$

where μ is the reduced mass of the two atoms, G' is the electronic degeneracy of the excited state, $|\psi'\rangle$, the density of initial continuum states is $\rho'(E)$, and the frequency $\nu(E, v'')$ is the frequency of the photon emitted for the transition from the continuum energy E to a vibrational level v'' of the lower state, $|\psi''\rangle$. For a Maxwellian velocity distribution, the integration over energy gives a rate constant, κ, of

$$\kappa = g \frac{64 \times 2^{1/2} \pi^{7/2} h^2}{3 \mu^{3/2} (kT)^{1/2} c^3 G'} S_e \sum_{v''} |\langle \psi''_{v''} | \psi'_E \rangle|^2 \rho'(E_0) \nu(E_0 v'')^3 , \qquad (2\text{-}175)$$

where E_0 is taken to be the mean energy at temperature, T, the molecular electronic dipole strength is S_e, and the $|\langle \psi''_{v''} | \psi'_E \rangle|^2$ are the Franck-Condon factors.

BATES and SPITZER (1951) first calculated the equilibrium abundance of CH and CH^+ by assuming that they were formed by the radiative association processes

$$C+H \rightarrow CH+h\nu$$

and

$$C^+ +H \rightarrow CH^+ +h\nu .$$

They concluded that the observed densities of CH and CH^+ demand hydrogen number densities of several hundred cm^{-3}. SOLOMON and KLEMPERER (1972) and SMITH, LISZT, and LUTZ (1973) have revaluated the rates for these reactions, the more recent values being

$$\kappa = 1.5 \times 10^{-17}\,\mathrm{cm^3\,sec^{-1}} \quad \text{for CH formation at } T=100\,°\mathrm{K},$$

and

$$\kappa = 5 \times 10^{-18}\,\mathrm{cm^3\,sec^{-1}} \quad \text{for CH}^+ \text{ formation at } T=100\,°\mathrm{K}.$$

JULIENNE, KRAUSS, and DONN (1971) showed that OH molecules could be formed by an indirect radiative association through resonance states in the molecular continuum. This process is called inverse predissociation. The reaction rate for indirect radiative association is given by (JULIENNE and KRAUSS, 1973)

$$\kappa = \hbar^2 \left(\frac{2\pi}{\mu k T}\right)^{3/2} \frac{1}{g_A g_B} \sum_n (2J_n+1) \frac{\Gamma_{nr}\Gamma_{np}}{\Gamma_{nr}+\Gamma_{np}} \exp\left(-\frac{E_n}{kT}\right), \tag{2-176}$$

where μ is the reduced mass of the reacting atoms, g_A and g_B are the degeneracies of the ground atomic levels, Γ_{nr} is the natural radiation width, Γ_{np} is the predissociation width, and E_n is the energy of the resonance level n above the lowest atomic fine structure asymptotic energy. Using this equation, JULIENNE, KRAUSS, and DONN (1971) obtain

$$\kappa \approx 2 \times 10^{-20}\,\mathrm{cm^3\,sec^{-1}} \quad \text{for OH formation at } T \geq 50\,°\mathrm{K}. \tag{2-177}$$

2.16. Line Radiation from Stellar Atmospheres—The Fraunhofer Spectrum and the Curve of Growth

As illustrated in Fig. 10, the spectrum of the Sun contains Fraunhofer (1817) lines of different wavelengths and intensities which are seen in absorption against the bright continuum. The wavelengths of the most notable of these lines are tabulated in Table 27, whereas the most intense solar emission lines are tabulated in Table 28. KIRCHHOFF and BUNSEN (1861) showed that the wavelengths of Fraunhofer lines correspond to certain transitions of elements observed on the earth; and the wavelengths of the lines for different atomic and ionic states of different elements are tabulated by ABT, MEINEL, MORGAN, and TABSCOTT (1969), MOORE, MINNAERT, and HOUTGAST (1966), MORGAN, KEENAN, and KELLMAN (1943), and STRIGANOV and SVENTITSKII (1968).

Observations of line intensities and widths may be compared with theoretical expectations in order to determine the excitation temperature, the turbulent velocity, the electron and gas pressures, the surface gravity, and the abundance of the elements in the stellar atmosphere. The two classical model atmospheres are the Schuster-Schwarzschild (SS) atmosphere (SCHUSTER, 1905; SCHWARZ-

Table 27. The most intense Fraunhofer lines from the Sun[1]

Wave-length (Å)	Equiv-alent width (m Å)	Element	Wave-length (Å)	Equiv-alent width (m Å)	Element	Wave-length (Å)	Equiv-alent width (m Å)	Element
2,795.4		Mg II	3,709.256	573	Fe I[2]	4,077.724	428	Sr II[2]
2,802.3		Mg II	3,719.947	1,664	Fe I	4,101.748	3,133	Hδ
2,851.6		Mg	3,734.874	3,027	Fe I	4,132.067	404	Fe I[2]
2,881.1		Si	3,737.141	1,071	Fe I	4,143.878	466	Fe I
3,067.262	663	Fe I[2]	3,745.574	1,202	Fe I[2]	4,167.277	200	Mg I
3,134.116	414	Ni I[2]	3,748.271	497	Fe I	4,202.040	326	Fe I
3,242.007	270	Ti II	3,749.495	1,907	Fe I	4,226.740	1,476	Ca I
3,247.569	246	Cu I	3,758.245	1,647	Fe I	4,235.949	385	Fe I[2]
3,336.689	416	Mg I	3,759.299	334	Ti II	4,250.130	342	Fe I[2]
3,414.779	816	Ni I	3,763.803	829	Fe I	4,250.797	400	Fe I[2]
3,433.579	492	Ni I[2]	3,767.204	820	Fe I	4,254.346	393	Cr I[2]
3,440.626	1,243	Fe I	3,787.891	512	Fe I	4,260.486	595	Fe I
3,441.019	634	Fe I	3,795.012	547	Fe I[2]	4,271.774	756	Fe I
3,443.884	655	Fe I	3,806.718	209	Fe I[2]	4,325.775	793	Fe I[2]
3,446.271	470	Ni I	3,815.851	1,272	Fe I	4,340.475	2,855	Hγ
3,458.467	656	Ni I	3,820.436	1,712	Fe I	4,383.557	1,008	Fe I
3,461.667	758	Ni I	3,825.891	1,519	Fe I	4,404.761	898	Fe I
3,475.457	622	Fe I	3,827.832	897	Fe I	4,415.135	417	Fe I[2]
3,476.712	465	Fe I[2]	3,829.365	874	Mg I	4,528.627	275	Fe I[2]
3,490.594	830	Fe I	3,832.310	1,685	Mg I	4,554.036	159	Ba II
3,492.975	826	Ni I	3,834.233	624	Fe I	4,703.003	326	Mg I
3,497.843	726	Fe I	3,838.302	1,920	Mg I	4,861.342	3,680	Hβ
3,510.327	489	Ni I	3,840.447	567	Fe I	4,891.502	312	Fe I
3,515.066	718	Ni I	3,841.058	517	Fe I[2]	4,920.514	471	Fe I[2]
3,521.270	381	Fe I	3,849.977	608	Fe I	4,957.613	696	Fe I[2]
3,524.536	1,271	Ni I	3,856.381	648	Fe I	5,167.327	935	Mg I[2]
3,554.937	404	Fe I	3,859.922	1,554	Fe I	5,172.698	1,259	Mg I
3,558.532	485	Fe I[2]	3,878.027	555	Fe I	5,183.619	1,584	Mg I
3,565.396	990	Fe I	3,886.294	920	Fe I	5,250.216	62	Fe I[3]
3,566.383	458	Ni I	3,899.719	436	Fe I	5,269.550	478	Fe I[2]
3,570.134	1,380	Fe I	3,902.956	530	Fe I[2]	5,328.051	375	Fe I
3,578.693	488	Cr I	3,905.532	816	Si I	5,528.418	293	Mg I
3,581.209	2,144	Fe I	3,920.269	341	Fe I	5,889.973	752	Na I (D_2)[2]
3,586.990	532	Fe I	3,922.923	414	Fe I[2]	5,895.940	564	Na I (D_1)
3,593.495	436	Cr I	3,927.933	187	Fe I	6,102.727	135	Ca I
3,608.869	1,046	Fe I	3,930.308	108	Fe I	6,122.226	222	Ca I
3,618.777	1,410	Fe I	3,933.682	20,253	Ca II[2]	6,162.180	222	Ca I
3,619.400	568	Ni I	3,944.016	488	Al I	6,302.499	83	Fe I[3]
3,631.475	1,364	Fe I[2]	3,961.535	621	Al I	6,562.808	4,020	Hα
3,647.851	970	Fe I[2]	3,968.492	15,467	Ca II[2]	8,498.062	1,470	Ca II
3,679.923	448	Fe I[2]	4,045.825	1,174	Fe I	8,542.144	3,670	Ca II
3,685.196	275	Ti II	4,063.605	787	Fe I[2]	8,662.170	2,600	Ca II
3,705.577	562	Fe I	4,071.749	723	Fe I	10,830		He I

[1] After MOORE, MINNAERT, and HOUTGAST (1966).
[2] Blended line.
[3] Magnetic sensitive line.

SCHILD, 1906), and the Milne-Eddington (ME) atmosphere (MILNE, 1921, 1930; EDDINGTON, 1917, 1926). In the SS approximation the continuum spectrum is assumed to be formed in the photosphere and the line spectrum is formed entirely in an overlying "reversing layer". In the ME approximation it is assumed that both the line and continuum spectrum are formed in the same layers in such a way that the ratio of the line and continuum absorption coefficients is a constant. Here we will outline a general approach which includes these classical solutions. A detailed discussion of this approach is given in the book by MIHALAS (1970). Model atmospheres which best fit the observed continuum and line data for the Sun are given by GINGERICH, NOYES, KALKOFEN, and CUNY (1971) and GINGERICH and DE JAGER (1968). The former is a non-LTE model whereas the latter is LTE.

The equation of transfer for the intensity of radiation, I_ν, at frequency, ν, is given by

$$\frac{\mu d I_\nu}{d \tau_\nu} = I_\nu - S_\nu . \tag{2-178}$$

For a plane parallel atmosphere, the equation of transfer has the emergent intensity solution

$$I_\nu(0, \mu) = \int_0^\infty S_\nu(t_\nu) \exp\left(\frac{-t_\nu}{\mu}\right) \frac{dt_\nu}{\mu}, \tag{2-179}$$

giving a total surface flux of

$$F_\nu = 2 \int_0^\infty S_\nu(t_\nu) E_2(t_\nu) dt_\nu , \tag{2-180}$$

where the exponential integral $E_n(x)$ is given by

$$E_n(x) = \int_1^\infty \frac{\exp(-xt)}{t^n} dt = x^{n-1} \int_x^\infty \frac{\exp(-t)}{t^n} dt . \tag{2-181}$$

Here $\mu = \cos\theta$ where θ is the angle between the surface normal and the direction of radiation, the optical depth, τ_ν, is given by

$$\tau_\nu = (\kappa + l_\nu + \sigma + \sigma_\nu) \rho dz , \tag{2-182}$$

where the absorption coefficients per unit mass for the continuum and line radiation are, respectively, κ and l_ν, the coefficients for the coherent scattering of the continuum and line radiation are, respectively, σ and σ_ν, the mass density is ρ, and z is the height measured normal to the plane of stratification of the atmosphere. If the continuum emission is thermal, then the source function, S_ν, is given by

$$S_\nu = \lambda_\nu B_\nu(T) + (1 - \lambda_\nu) J_\nu , \tag{2-183}$$

Table 28. The most intense emission lines from the Sun[1]

Ultraviolet emission lines from the corona, chromosphere, and the transition region between them[1]

Wavelength (Å)	Intensity at earth (erg cm^{-2} sec^{-1})	Element	Wavelength (Å)	Intensity at earth (erg cm^{-2} sec^{-1})	Element	Wavelength (Å)	Intensity at earth (erg cm^{-2} sec^{-1})	Element
284.2	0.017	Fe XV	625	0.011	Mg X	1,037.6	0.025	O VI
303.8	0.25	He II (Lyα)	629.7	0.045	O V	1,215.7	5.1	H I (Lyα)
335.0	0.012	Fe XVI	770.4	0.011	Ne VIII	1,548.2	0.11	C IV
361.7	0.005	Fe XVI	787.7	0.008	O IV	1,550.8	0.06	C IV
368.1	0.031	Mg IX	790.1	0.003	O IV	1,561.4	0.09	C I
465.2	0.005	Ne VII	832–835	0.013	O II, III	1,640.5	0.07	He II
499.3	0.006	Si XII	810–911	0.28	H I (Lyα)	1,657.0	0.16	C I
554	0.009	O IV	977.0	0.050	C III	1,808.0	0.15	Si II
584.3	0.053	He I	1,025.72	0.060	H I (Lyβ)	1,817.4	0.45	Si II
609.8	0.011	Mg X	1,031.9	0.020	O VI	1,892.0	0.10	Si III

Chromospheric emission lines observed during solar eclipse[2]

Wavelength (Å)	Integrated intensity at Sun (10^{11} erg sec^{-1} cm^{-1} ster^{-1})	Element	Wavelength (Å)	Integrated intensity at Sun (10^{11} erg sec^{-1} cm^{-1} ster^{-1})	Element	Wavelength (Å)	Integrated intensity at Sun (10^{11} erg sec^{-1} cm^{-1} ster^{-1})	Element
3,685.196	90	Ti II	3,835.39	228	H I (H 9)	4,861.342	1,632	H I (Hβ)
3,691.56	29	H I (H 18)	3,838.302	60	Mg I	5,015.67	6	He I
3,697.15	35	H I (H 17)	3,889.05	381	H I (H 8)	5,183.619	65	Mg I
3,703.86	43	H I (H 16)	3,933.66	818	Ca II	5,875.65	994	He I (D 3)
3,711.97	53	H I (H 15)	3,968.47	615	Ca II	6,562.808	4,738	H I (Hα)
3,721.94	73	H I (H 14)	3,970.076	306	H I (Hε)	7,065.18	138	He I
3,734.37	99	H I (H 13)	4,026.36	24	He I	7,771.954	91	O I
3,750.15	108	H I (H 12)	4,077.724	75	Sr II	7,774.177	75	O I
3,759.299	90	Ti II	4,101.748	459	H I (Hδ)	7,775.395	53	O I
3,761.320	82	Ti II	4,215.539	51	Sr II	8,498.02	512	Ca II
3,770.63	116	H I (H 11)	4,226.740	22	Ca I	8,542.09	1,362	Ca II
3,797.90	157	H I (H 10)	4,246.837	18	Sc II	8,545.38	23	H I (P 15)

Table 28 (continued)

Wavelength (Å)	Integrated intensity at Sun (10^{11} erg sec^{-1} cm^{-1} ster^{-1})	Element	Wavelength (Å)	Integrated intensity at Sun (10^{11} erg sec^{-1} cm^{-1} ster^{-1})	Element	Wavelength (Å)	Integrated intensity at Sun (10^{11} erg sec^{-1} cm^{-1} ster^{-1})	Element
3,819.61	5	He I	4,340.425	505	H I (Hγ)	8,598.39	26	H I (P14)
3,820.436	14	Fe I	4,471.69	121	He I	8,662.14	1,181	Ca II
3,829.365	20	Mg I	4,685.68	2	He I	8,665.02	34	H I (P13)
3,832.310	46	Mg I	4,713.14	9	He I	8,750.47	46	H I (P12)

Coronal emission lines[3]

Wavelength (Å)	Equivalent width (m Å)	Element	Wavelength (Å)	Equivalent width (m Å)	Element	Wavelength (Å)	Equivalent width (m Å)	Element
3,329	0.7	Ca XII	4,232.0	1.1	Ni XII	6,374.5	5	Fe X
3,388.0	10	Fe XIII	4,256.4	0.1	K XI	6,701.9	1.2	Ni XV
3,534.0	1	V X	4,351.0	0.1	Co XV	6,740	0.1	K XIV
3,600.9	1.3	Ni XVI	4,412.4	0.3	Ar XIV	7,059.6	0.8	Fe XV
3,642.8	0.4	Ni XIII	4,566.6	0.5	Cr IX	7,891.9	6	Fe XI
3,685	0.2	Mn XII	5,116.0	0.8	Ni XIII	8,024.2	0.3	Ni XV
3,800.7	0.5	Co XII	5,302.9	20	Fe XIV	10,776.8	50	Fe XIII
3,987.1	0.7	Fe XI	5,445.5	0.2	Ca XV	10,797.9	30	Fe XIII
3,998	0.1	Cr XI	5,536	0.3	Ar X			
4,086.5	0.4	Ca XIII	5,094.5	0.3	Ca XV			

[1] After HINTEREGGER (1965) by permission of the D. Reidel Publ. Co.
[2] After DUNN *et. al.* (1968) by permission of the American Astronomical Society and the University of Chicago Press.
[3] After ALLEN (1963) by permission of the Athlone Press—University of London.

where

$$\lambda_\nu = \frac{(1-\rho)+\varepsilon\eta_\nu}{1+\eta_\nu},$$

$$\eta_\nu = \frac{l_\nu}{\kappa+\sigma},$$

$$\rho = \frac{\sigma}{\kappa+\sigma},$$

the Planck function, $B_\nu(T)$, is given by (PLANCK, 1901)

$$B_\nu(T) = \frac{2h\nu^3}{c^2}\left[\exp\left(\frac{h\nu}{kT}\right)-1\right]^{-1}, \tag{2-184}$$

the temperature is T, the fraction of absorbed photons scattered to form line radiation is $(1-\varepsilon)$, the J_ν is the mean intensity, and κ is the mass absorption coefficient corrected for stimulated emission by multiplication with the factor $[1-\exp(-h\nu/kT)]$.

When the star is in radiative equilibrium the mean intensity is given by (SCHWARZSCHILD, 1906; MILNE, 1921)

$$J_\nu(\tau_\nu)=\tfrac{1}{2}\int_0^\infty S_\nu(t_\nu)E_1(t_\nu-\tau_\nu)\,dt_\nu=\tfrac{1}{2}\int_{-1}^{1} I_\nu(\tau_\nu,\mu)\,d\mu\,, \tag{2-185}$$

where the E_1 is given by Eq. (2-181) with $n=1$. Using the Eddington (1926) approximation

$$J_\nu(\tau_\nu)=\tfrac{3}{2}\int_{-1}^{1} I_\nu(\tau_\nu,\mu)\,\mu^2\,d\mu\,, \tag{2-186}$$

together with the Milne (1930) expansion of the Planck function

$$B_\nu(T)=a+b\tau=a+p_\nu\tau_\nu\,, \tag{2-187}$$

where τ and τ_ν are, respectively, the continuum and line optical depths, and

$$p_\nu = \frac{b}{1+\eta_\nu}; \tag{2-188}$$

the Milne-Eddington solution to the Schwarzschild-Milne equation (2-185) is obtained

$$J_\nu=a+p_\nu\tau_\nu+\left[\frac{p_\nu-\sqrt{3}a}{\sqrt{3}(1+\sqrt{\lambda_\nu})}\right]\exp(-\sqrt{3\lambda_\nu}\,\tau_\nu)\,. \tag{2-189}$$

When studying line radiation it is convenient to use the absorption depths A_ν and a_ν defined by

$$A_\nu=1-\frac{F_\nu^L}{F_\nu^C}$$

and (2-190)

$$a_\nu=1-\frac{I_\nu^L(0,\mu)}{I_\nu^C(0,\mu)},$$

where the I_ν and F_ν are given, respectively, by Eqs. (2-179) and (2-180), and the superscripts L and C denote, respectively, the line and continuum radiation. When the Milne-Eddington solution is used,

$$A_\nu = 1 - \left[\frac{p_\nu + \sqrt{3\lambda_\nu}\, a}{1 + \sqrt{\lambda_\nu}}\right]\left[\frac{1 + (1-\rho)^{1/2}}{b + a[3(1-\rho)]^{1/2}}\right], \tag{2-191}$$

and a_ν can be obtained from Eq. (2-190), using

$$I_\nu^L(0, \mu) = a + p_\nu \mu + \frac{p_\nu - \sqrt{3}\, a}{\sqrt{3}(1 + \sqrt{\lambda_\nu})} \frac{(1 - \lambda_\nu)}{(1 + \sqrt{3\lambda_\nu}\, \mu)} \tag{2-192}$$

and the $I_\nu^C(0, \mu)$ which is given by Eq. (2-192) with $\eta_\nu = 0$.

The equivalent width, W_ν, or W_λ, is given by

$$W_\nu = \int_0^\infty A_\nu \, d\nu = \frac{\nu_{mn}^2}{c} W_\lambda , \tag{2-193}$$

where ν_{mn} is the center frequency of the line. The equivalent width is a useful parameter because it is independent of the instrumental distortion of the profile, and because it provides a measure of the relative abundances of the elements in a stellar atmosphere. The equivalent width of a function is the area of the function divided by its central ordinate. Expressed differently, the equivalent width of a function is the width of the rectangle whose height is equal to that of the function. For a Gaussian function of standard deviation, σ, the equivalent width is 2.5066 σ and the full width to half maximum is 2.355 σ.

If we assume that line formation takes place in an isothermal layer in local thermodynamic equilibrium, then both the (ME) and (SS) approximations lead to the approximate relation (MENZEL, 1936)

$$\begin{aligned} \frac{W_\nu}{2A_0 \Delta\nu_D} &= \int_0^\infty \frac{\eta_0 H(a, b)}{1 + \eta_0 H(a, b)} \, db \\ &\approx \frac{\eta_0 \pi^{1/2}}{2} \quad \text{for } \eta_0 \ll 1 \\ &\approx (\ln \eta_0)^{1/2} \quad \text{for } 10 \le \eta_0 \le 1000 \\ &\approx \frac{(\pi a \eta_0)^{1/2}}{2} \quad \text{for } \eta_0 \ge 1000 . \end{aligned} \tag{2-194}$$

Here A_0 is the central depth of the line, $\Delta\nu_D$ is the Doppler broadened full width to half maximum, $H(a, b)$ is the Voigt function discussed in Sect. 2.22, the parameter a is given by

$$a = \frac{\Gamma}{4\pi \Delta\nu_D}, \tag{2-195}$$

where Γ is the damping constant for the line, and

$$\eta_0 = \frac{\sqrt{\pi}}{\kappa} \frac{e^2}{mc} \frac{f\lambda}{V} N_{r,s} , \tag{2-196}$$

where κ is the continuum opacity, the most probable velocity, V, is given by

$$V = \left[\frac{2kT_K}{M} + V_{tur}^2\right]^{1/2}, \tag{2-197}$$

the T_K is the gas kinetic temperature, the atomic mass is M, and V_{tur} is the most probable turbulent velocity. The oscillator strengths, f, are tabulated by CORLISS and BOZMAN (1962) and CORLISS and WARNER (1964), the wavelength is λ, and the number of atoms in the sth state of the rth ionization stage is given by the Saha equation (3-127) (SAHA, 1921)

$$N_{r,s} = \frac{N_r g}{U} \exp\left(\frac{-\chi}{kT_{exc}}\right), \tag{2-198}$$

where N_r is the total number of atoms in the rth stage of ionization, g is the statistical weight of the sth level, U is the partition function, and χ is the excita-

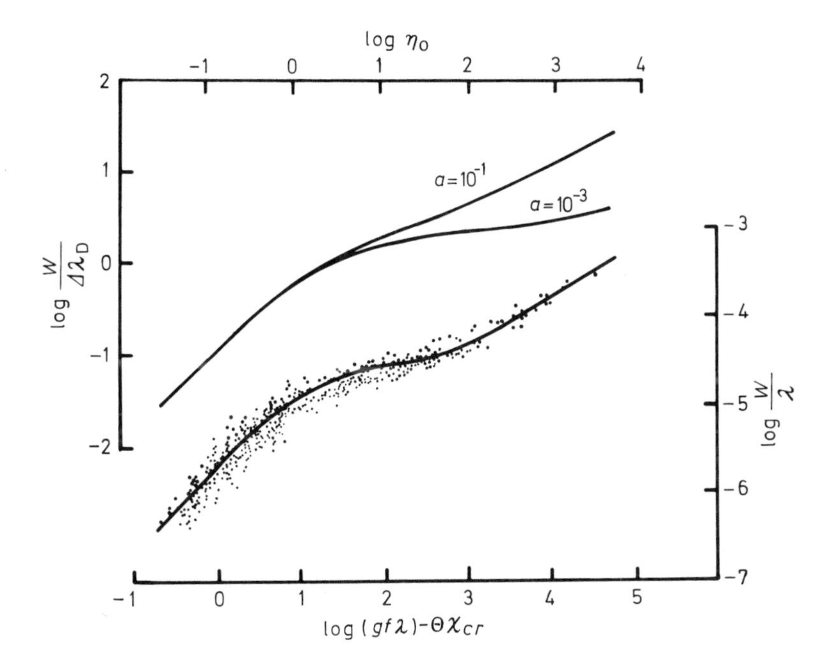

Fig. 15. Theoretical curves of growth [cf. Eq. (2-194)] and the solar curve of growth (after COWLEY and COWLEY, 1964, by permission of the American Astronomical Society and the University of Chicago Press). For the solar curve of growth, W is the equivalent width of the line, the designation χ_{cr} indicates that the abscissa is given in the chromium scale, and each point represents an individual line. An excitation temperature of 5143 °K was found to best approximate the data; and assuming that the excitation temperature equals the gas kinetic temperature, a turbulent velocity of 1.4 ± 0.2 km sec^{-1} is obtained. A comparison of the theoretical and empirical curves of growth also yields a value of 1.4 for the logarithm of the ratio of the observed, damping constant to the classical damping constant. The parameter $a = \Gamma/(4\pi \Delta\nu_D)$ where Γ is the damping constant and $\Delta\nu_D$ is the full width to half maximum of the Doppler broadened line profile. The variable, η_0, is related to the other parameters of the abscissa by Eq. (2-199)

tion potential of the sth level. We may then write Eq. (2-196) as

$$\log \eta_0 = \log(g f \lambda) - \frac{5040 \chi}{T_{\mathrm{exc}}} + \log C , \tag{2-199}$$

where $\log C = \log N_r + \log[\sqrt{\pi} e^2/(mc)] - \log V - \log U - \log \kappa$, and $\log[\sqrt{\pi} e^2/(mc)] = -1.826$. Theoretical curves of growth are plots of $\log(W_\lambda/\Delta\lambda_D)$ versus $\log \eta_0$, and an example is shown in Fig. 15. The curve of growth method was anticipated by STRUVE and ELVEY (1934). Curves of growth for the (ME) and (SS) approximations are given by VAN DER HELD (1931), WRUBEL (1950), WRUBEL (1954), and HUNGER (1956). Empirical curves of growth are plots of $\log(W_\lambda/\lambda)$ versus $\log(g f \lambda)$, and curves for the Sun are given by COWLEY and COWLEY (1964) and PAGEL (1965). As shown in Fig. 15 the ordinates of the empirical and theoretical curves are displaced by the amount

$$\log\left(\frac{V}{c}\right) = \log\left(\frac{W_\lambda}{\lambda}\right) - \log\left(\frac{W_\lambda}{\Delta\lambda_D}\right), \tag{2-200}$$

from which one can extract V. The excitation temperature, T_{exc}, is found by comparing an empirical curve of growth with various theoretical curves, or by comparing the empirical curves for lines of one excitation potential with similar curves for lines of another excitation potential. Provided that the excitation temperature is equal to the gas kinetic temperature, the turbulent velocity, V_{tur}, can be estimated from V and T_{exc} using Eq. (2-197). For the Sun, COWLEY and COWLEY (1964) obtain $T_{\mathrm{exc}} = 5143\,^\circ\mathrm{K}$, and $V_{\mathrm{tur}} = 1.4 \pm 0.2\ \mathrm{km\ sec^{-1}}$. The difference in abscissae between the theoretical and empirical curves leads to the $\log C$, which in turn leads to a measure of the relative abundances of the elements. The relative abundances of the elements in the atmosphere of the Sun and other normal stars are given by RUSSELL (1929), GOLDBERG *et al.* (1960), ALLER (1961), LAMBERT and WARNER (1968), UNSÖLD (1969), ROGERSON (1969), and MITLER (1970), and these abundances are given in Table 29. Abundances of peculiar A stars and the metal deficient G type subdwarfs are summarized, respectively, by SARGENT (1964) and ALLER and GREENSTEIN (1960). The value of the slope of the damping part of the curve of growth may lead to an estimate of the gas pressure, P_g, or electron pressure, P_e, according as the collision damping is dominated by collisions with neutral hydrogen or by the quadratic Stark effect. The gas pressure and electron pressure are related to the gas mass density, ρ, by the equation

$$\rho k T_K = (P_g - P_e) \mu m_H , \tag{2-201}$$

where m_H is the mass of hydrogen, μ is the mean molecular weight, and T_K is the gas kinetic temperature.

The mass absorption coefficient, κ_ν, is related to the mass density, ρ, by

$$\kappa_\nu = \frac{\alpha_\nu}{\rho} , \tag{2-202}$$

where the absorption coefficient per unit length, α_ν, is given by Eqs. (2-69) and (2-71), (1-237) and (1-221), and (1-219) and (1-221), respectively, for resonance absorption of line radiation, absorption by photoionization, and absorption by

Table 29. Relative abundances (densities) of the elements in the solar photosphere[1]

Atomic number	Element	Abundance $\log N(H)=12.00$	Abundance $N(Si)=10^6$
1	H	12.00	3.32×10^{10}
2	He	10.77	1.96×10^{9}
3	Li	0.96	0.30
4	Be	2.36	7.58
5	B	3.6	131.8
6	C	8.55	1.17×10^{7}
7	N	7.98	3.16×10^{6}
8	O	8.79	2.04×10^{7}
9	F	(4.87)	(2.45×10^{3})
10	Ne	7.87	2.46×10^{6}
11	Na	6.18	5.0×10^{4}
12	Mg	7.41	8.5×10^{5}
13	Al	6.40	8.3×10^{4}
14	Si	7.48	1.0×10^{6}
15	P	5.43	8.9×10^{3}
16	S	7.27	6.17×10^{5}
17	Cl	(5.23)	(5.7×10^{3})
18	Ar	(6.55)	(1.172×10^{5})
19	K	5.05	3.72×10^{3}
20	Ca	6.33	7.08×10^{4}
21	Sc	2.85	23.4
22	Ti	4.81	2.14×10^{3}
23	V	4.17	489.8
24	Cr	5.01	3.39×10^{3}
25	Mn	4.85	2.34×10^{3}
26	Fe	6.82	2.19×10^{5}
27	Co	4.70	1.66×10^{3}
28	Ni	5.77	1.95×10^{4}
29	Cu	4.45	933.3
30	Zn	3.52	109.6
31	Ga	2.72	17.4
32	Ge	2.49	10.2
33	As	(2.30)	(6.6)
34	Se	(3.31)	(67.2)
35	Br	(2.61)	(13.5)
36	Kr	(3.15)	(46.8)
37	Rb	2.48	10.0
38	Sr	3.02	34.7
39	Y	3.20	52.5
40	Zr	2.65	14.8
41	Nb	2.30	6.60
42	Mo	2.30	6.60
43	Tc	—	—
44	Ru	1.82	2.18
45	Rh	1.37	0.77
46	Pd	1.57	1.23
47	Ag	0.75	0.18

Table 29 (continued)

Atomic number	Element	Abundance $\log N(H)=12.00$	Abundance $N(S_i)=10^6$
48	Cd	1.54	1.14
49	In	1.45	0.93
50	Sn	1.54	1.14
51	Sb	1.94	2.88
52	Te	(2.29)	(6.42)
53	I	(1.52)	(1.09)
54	Xe	(2.21)	(5.38)
55	Cs	(1.07)	(0.387)
56	Ba	2.10	4.16
57	La	2.03	3.54
58	Ce	1.78	1.99
59	Pr	1.45	0.93
60	Nd	1.93	2.81
61	Pm	—	—
62	Sm	1.62	1.38
63	Eu	0.96	0.30
64	Gd	1.13	0.44
65	Tb	(0.22)	(0.055)
66	Dy	1.00	0.33
70	Yb	1.53	1.12
72	Hf	(0.80)	(0.21)
80	Hg	(1.08)	(0.40)
82	Pb	1.63	1.41

[1] The abundances of many of the lighter elements are from LAMBERT and WARNER (1968), ROGERSON (1969), and MITLER (1970), whereas those of the other light elements and the heavier elements are from GOLDBERG, MÜLLER, and ALLER (1960) and UNSÖLD (1969). Values in parenthesis are cosmic abundances given in Table 38.

free-free transitions of hydrogen like atoms. For hot stars, $T_K > 7000\,°K$, continuous absorption by hydrogen atoms predominates, whereas for cooler stars absorption by the negative hydrogen ion, H^-, predominates. The absorption coefficient for the latter transition is given by CHANDRASEKHAR and BREEN (1946), who give $\alpha(H^-) = 5.81 \times 10^{-26}\ cm^{-2}$ at 5600 °K and unit electron pressure. Absorption by He I and He II and Thomson and Rayleigh scattering (Eqs. (1-311) and (1-305)) are also important in some cases. More recent determinations of continuum absorption cross sections are given by MIHALAS (1970). When a frequency independent absorption coefficient is desired, the Rosseland mean opacity, κ_R, may be used (ROSSELAND, 1924)

$$\frac{1}{\kappa_R} = \frac{\pi \int\limits_0^\infty \frac{1}{\kappa_\nu} \frac{d B_\nu(T)}{d T} d\nu}{4 \sigma T^3}, \tag{2-203}$$

where the Stefan-Boltzmann constant $\sigma \approx 5.669 \times 10^{-5}$ erg cm^{-2} °K^{-4} sec^{-1}, and

$$\frac{dB_\nu(T)}{dT} = \frac{2h^2 \nu^4}{c^2 k T} \frac{\exp[h\nu/(kT)]}{\{\exp[h\nu/(kT)]-1\}^2}. \tag{2-204}$$

Rosseland opacities for population II compositions are given by COX and STEWART (1970).

The frequency independent absorption coefficient used near the surface of the star is the Planck mean opacity, κ_P, given by

$$\kappa_P = \frac{\pi}{\sigma T^4} \int_0^\infty \kappa_\nu B_\nu(T) d\nu, \tag{2-205}$$

where the Planck function, $B_\nu(T)$, is given by Eq. (2-184). If the continuum absorption coefficient is independent of frequency, then the temperature $T(\tau)$ will vary with the optical depth, τ, according to the gray body relation given by (MILNE, 1930)

$$T^4(\tau) = \tfrac{3}{4}(\tau + \tfrac{2}{3}) T_{\text{eff}}^4, \tag{2-206}$$

where T_{eff} is the effective temperature of a black body radiator which radiates a flux equal to that of the star.

2.17. Effects which Alter the Emitted Line Frequency

Normal Zeeman effect. For atoms with singlet lines (spin $S=0$), a magnetic field will split any term into $2J+1$ equally spaced levels (where J is the total angular momentum). The energies, E_M, of the levels are given by (ZEEMAN, 1896, 1897; LORENTZ, 1897)

$$E_M = E_n \pm \frac{ehHM}{4\pi mc} = E_n \pm ho M, \tag{2-207}$$

where M takes on integral values between 0 and J, E_n is the energy of the term without the magnetic field, H is the magnetic field strength, the factor $eh/(4\pi mc) = 9.2731 \times 10^{-21}$ erg gauss^{-1} is the magnetic moment of the Bohr magnetron, and $o = eH/(4\pi mc) = 1.400 \times 10^6 H$ Hz is the Larmor frequency of precession.

The selection rules for allowed transitions between the levels of different terms are $\Delta M = 0, \pm 1$ and a level with $M=0$ cannot combine with another level of $M=0$. Because there is equal splitting for all terms, the frequency of any singlet line, ν_{mn}, will be split into three components; the π component at frequency ν_{mn} and the two σ components at frequencies (LORENTZ, 1897)

$$\nu = \nu_{mn} \pm \frac{eH}{4\pi mc} = \nu_{mn} \pm 1.400 \times 10^6 H \text{ Hz}, \tag{2-208}$$

where the magnetic field strength H is given in gauss. The π component is plane polarized in the plane containing the line of sight and the vector of the magnetic field. The two σ components are elliptically polarized. For observation in the direction of the magnetic field, the central π component is absent and the two σ components are circularly polarized with opposed directions of rotation (the lower frequency having right hand rotation). Observation in a direction transverse to the field shows all three components, the undisplaced one has linear polarization parallel (π) to the field and the others show linear polarization perpendicular (σ), to the field.

In general, the observed intensity, $I(\nu)$, at frequency, ν, will depend on the intensity, $I_0(\nu)$, that would be produced if there were no magnetic field, the inclination, γ, of the magnetic vector to the line of sight, and the polarization reception characteristics of the antenna. The total intensity in a beam with right hand circular polarization is (SEARES, 1913)

$$I_R(\nu) = \tfrac{1}{8}(1-\cos\gamma)^2 I_0(\nu+\Delta\nu_0) + \tfrac{1}{4}\sin^2\gamma\, I_0(\nu) + \tfrac{1}{8}(1+\cos\gamma)^2 I_0(\nu-\Delta\nu_0), \tag{2-209}$$

where the Zeeman splitting frequency $\Delta\nu_0 = 1.4\times10^6 H$ Hz. The intensity in a beam with left hand polarization, $I_L(\nu)$, is obtained from $I_R(\nu)$ by changing the sign of the $\cos\gamma$ term.

When the magnetic fields are weak, the Zeeman splitting, $\Delta\nu_0 = 1.4\times10^6 H$, may be much smaller than the full width to half maximum, $\Delta\nu_L$, of the observed line. In that event, the line of sight component of the magnetic field, $H\cos\gamma$, may be detected by comparing the line profiles observed with right hand, $I_R(\nu)$, and left hand, $I_L(\nu)$, circularly polarized antennae. The two profiles will be displaced in frequency by $\Delta\nu = 1.4\times10^6 H\cos\gamma$ Hz. By adding and subtracting the two profiles the sum, $I(\nu)$, and difference, $D(\nu)$, profiles are obtained

$$\begin{aligned} I(\nu) &= I_L(\nu) + I_R(\nu) \\ D(\nu) &= I_L(\nu) - I_R(\nu) = 2.8\times10^6 I'(\nu) H\cos\gamma, \end{aligned} \tag{2-210}$$

where $I'(\nu)$ denotes the first derivative of $I(\nu)$ with respect to frequency. In terms of the peak antenna temperature, T_A, and half width, $\Delta\nu_L$, of $I(\nu)$, and the peak antenna temperature, ΔT_A, of $D(\nu)$, we have

$$H\cos\gamma \approx 10^{-6}\frac{\Delta\nu_L\,\Delta T_A}{4\,T_A} \text{ gauss.} \tag{2-211}$$

Effective Zeeman frequency displacements as small as $0.005\,\Delta\nu_L$ have been detected using this technique.

The Zeeman effect was first used at optical frequencies to measure the magnetic field in sunspots (HALE, 1908) and in peculiar A stars (BABCOCK, 1947); and then at 1420 MHz to measure the interstellar magnetic field (VERSCHUUR, 1968). Values of the magnetic field strength on the Sun range from one to several hundred gauss (BABCOCK and BABCOCK, 1955). The peculiar A stars have field strengths which range from the detection limit of one hundred gauss to 34,000 gauss for HD 215441 (BABCOCK, 1960; CAMERON, 1967). For the interstellar medium the Zeeman effect gives $H \approx 10^{-6}$ gauss, which is consistent with the Faraday rotation and dispersion measure results for pulsars. Magnetic field intensities for eighty nine magnetic stars are given together with other observational data in Table 30.

Table 30. Henry Draper Catalogue (HD) number, visual magnitude (v denotes variable), color index ($B-V$) where again v denotes variable, spectral type, extreme values of magnetic field, H, with a + sign denoting a field towards the observer, and periods of variation of l = light, s = spectrum, v = radial velocity, m = magnetic field, and b = spectroscopic binary, with eccentricity given in parenthesis (e) for eighty nine stars with strong magnetic fields[1]

HD number	Name	α(1950.0) h m s	δ(1950.0) ° ′	V (mag)	B − V (mag)	Spectral type	H extreme (gauss)		Period (days)
2453	BD+31°59	00 25 48	32 09	6.8	0.05 *v*	A1 *pv*	− 710	− 425	
4174	BD+39°167	00 41 54	40 24	7.5	—	M2 *epv*	−1,200	+1,100	
8441	BD+42°293	01 21 24	42 53	6.6	0.02	A2 *p*	− 750	+ 400	*b*: 106.3
9996	BD+44°341	01 35 30	45 09	6.3	0.13	A0 *p*	− 990	+ 135	
10221	43 Cas	01 38 36	67 47	5.6	−0.06	A0 *p*	−1,200	—	
10783	BD+7°275	01 43 06	08 18	6.6 *v*	−0.06	A2 *p*	+2,200	−1,200 *d*	$4.^d16$ *m*, *l*, *v*
11187	BD+54°393	01 48 12	54 40	7.2	−0.08	A0 *p*	+1,250	− 70 *d*	
15144A	ADS 1849A	02 23 36	−15 34	5.8 *v*	0.15	A4 *p*	−1,080	− 320	*b*: 2.9978
18296	21 Per	02 54 12	31 44	5.2 *v*	0.00 *v*	A0 *pv*	+1,350	−1,270 *d*	1.73, *l*, *m*?
19445	CC 209	03 05 30	26 09	8.0	0.45	A8: *p*	—	+ 415 *d*	
20210A	ADS 2433A	03 12 54	34 30	6.4 *v*	0.30 *v*	A6: *pm*	− 260	—	*b*: 5.5435 (0.04)
22374	9 Tau	03 34 00	23 03	6.7	0.13	A1 *p*	—	+ 140	
22649	B 826	03 37 48	63 03	5.3	1.6	S5	+ 450	+ 0	
24712	BD−12°752	03 52 54	−12 15	5.9	—	A8	+1,000	+ 570	
25267	36 τ^9 Eri	03 57 48	−24 09	4.7	−0.13	A0 *pv*	—	+ 400	*b*: 5.954
25354	BD+37°866	03 59 54	37 55	7.9 *v*	0.09 *v*	A0 *pv*	− 380	0	3.900 *l*
25823	41 Tau	04 03 30	27 28	5.2 *v*	−0.12	A0 *pv*	+ 700	− 530	11.9 *l*, *b*?
27962A	68 Tau A	04 22 36	17 49	4.3	0.05	A2 *m*	—	− 400	
30466	BD+29°742	04 46 06	29 29	7.2 *v*	0.05	A0 *p*	+2,320	—	
32633	BD+33°953	05 02 54	33 51	7.0 *v*	−0.06	B9 *p*	−5,870	+2,200	6.43 *l*, *m*?
33254	16h Ori	05 06 36	09 46	5.4	0.25	A2−*F*2m	− 420	+ 375	
33904	5 μ Lep	05 10 42	−16 16	3.3 *v*	−0.10 *v*	B9 *p*	+ 325	− 170	
42474	WY Gem	06 08 54	23 14	7.9 *v*	—	M3 *ep*	+ 540	—	
42616	BD+41°1392	06 10 12	41 43	6.9 *v*?	0.09	A2 *p*	− 840	0: *d*	
45677	MWC 142	06 26 00	−13 01	7.5	—	B2 *epv*	—	−1,600: *d*	
49976	BD−7°1592	06 48 18	−07 59	6.2 *v*	0.02	A0 *pv*	− 810	—	
50169	BD−1°1414	06 49 24	−01 35	8.9	—	A4 *pv*	+2,120	+ 670	

Table 30 (continued)

HD number	Name	α(1950.0) h m s	δ(1950.0) ° ′	V (mag)	B−V (mag)	Spectral type	H extreme (gauss)		Period (days)
53791	R Gem	07 04 18	22 47	6+*v*	—	Se	+ 400	+ 370*d*	370*l*
56495	BD−7°1851	07 14 30	−07 26	7.5	0.34	A3 *pmv*	+ 500	—	
60414,5	B 1985	07 31 30	−14 25	5.1	—	M2 *p*+*Be*	—	− 670	
65339 A	53 Cam A	07 57 30	60 28	6.0*v*	0.15*v*	A3 *pv*	−5,410	+3,760	8.026*m*, *l*, *s*
68351	15 Cnc	08 10 06	29 48	5.6*v*	−0.07*v*	A0 *p*	+strong	0	
71866	BD+40°2066	08 27 54	40 24	6.7*v*	0.07*v*	A0 *pv*	+2,460	−2,170*d*	6.798*m*, *l*, *s*
72968	3 Hya	08 33 00	−07 48	5.6	−0.01	A2 *pmv*	+ 740	+ 480*d*	
74521	49 b Cnc	08 42 00	10 16	5.6*v*	−0.09	A1 *p*	+1,450	− 200	
77350	69 *ν* Cnc	08 59 48	24 39	5.5*v*	−0.05	B9 *p*	+ 470	+ 105	
78316	76 *κ* Cnc	09 05 00	10 52	5.1	−0.10	B8 *p*	− 640	+ 460*d*	*b*: 6.3932 (0.14)
89822	B 2754	10 20 36	65 49	4.9	−0.06	A0 *p*	+ 340	− 290	*b*: 11.583 (0.38)
90569 A	45 Leo A	10 25 00	10 01	5.9*v*	−0.06	A1 *pv*	+ 400	—	2: *s*
98088 A	ADS 8115 A	11 14 24	−06 52	6.1 *v*	0.21	A9 *pv*	−1,188	+1,000*d*	5.905*v*, *s*, *m*
107168	8 Com	12 16 48	23 19	6.2	0.22	A6 *m*	—	− 150*d*	
108651	17 Com B	12 26 12	26 11	6.7	—	A4; *m*	—	moder.	
108662	17 Com A	12 26 24	26 11	5.4*v*	−0.05	A0 *pv*	−1,150	+ 360	
110066	BC+36°2295	12 36 54	36 14	6.3*v*	0.06	A3 *p*	+ 300	− 55	
110073	1 Cen	12 37 12	−39 43	4.8	−0.10*v*	B8 *p*	—	+ 580	*b*?
110380	29 γ^1 Vir (N)	12 39 06	−01 10	3.7	0.35	F0	—	− 390	
111133	BD+6°2660	12 44 30	06 13	6.4	−0.04	A2 *p*	−1,540	− 990	
112413	12 α^2 CVn	12 53 42	38 35	2.9*v*	−0.11	A0 *pv*	+1,600	−1,400*d*	5.4694*s*, *m*, *v*, *l*
115708	BD+27°2234	13 16 12	26 38	8.2*v*	0.25	A2 *p*	+ 740	0	
118022	78 o Vir	13 31 36	03 55	4.9	0.03	A2 *pv*	−1,680	− 140	
—	BD+46°1913	13 53 48	45 59	9.8	−0.09	Am(*p*)*v*	− 500 (est.)	—	
125248	B 3669	14 15 54	−18 29	5.8*v*	0.00*v*	A1 *pv*	+2,500	−2,080*d*	9.296*s*, *m*, *v*, *l*
126515	BD+1°2927	14 23 24	01 13	7.1	0.01	A2 *p*	+1,310	− 780	
129174	29 π^1 Boo	14 38 24	16 38	4.9	−0.1?	B9 *pv*	+ 190	− 75*d*	(2.2445?)*s*
130559 A	7 *μ* Lib A	14 46 36	−13 56	5.6		A1 *p*	−1,300	− 200	

133029A	ADS 9477A	14 58 54	47 28	6.2	−0.13 *v*	A0 *p*	3,270	1,150 *d*	
134793	BD+9°3006	15 09 06	08 43	8 *v*	0.14	A3 *pv*	− 530	+ 450	(2: ? *m*)
135297	BD+0°3322	15 11 48	00 33	8.0 *v*	−0.01	A0 *p*	−1,110	—	
137909	3 β Cr B	15 25 48	29 17	3.7	0.27 *v*	A9 *p*	+1,020	− 960	18.50 *m*+*b*: 3834 (0.4)
137949	33 Lib (ξ^2)	15 26 42	−17 16	7	0.38	F0 *p*	+1,120	—	
143807	14 ι Cr B	15 59 24	29 59	4.9 *v*	−0.05 *v*	A0 *p*	− 340	+ 75 *d*	
152107A	52 Her A	16 47 48	46 04	4.8	0.08	A3 *pv*?	+1,430	+ 840	
153286	BD+47°2407	16 54 42	47 27	6.9	—	A2−F *pm*	− 580	− 370	
153882A	ADS 10310A	16 59 18	15 01	6.2 *v*	0.04	A3 *pv*	+2,710	−1,330	6.008 *m*, *l*, *s*, *v*
165474	ADS 11056B	18 03 24	12 00	7.4	—	A6 *p*	—	+ 900	
171586A	ADS 11477A	18 33 06	04 54	6.6 *v*	0.08	A2 *p*	− 740	—	
173650	BD+21°3550	18 43 30	21 56	6.4 *v*	0.01 *v*	A0 *pv*	+ 700	− 540	10.1 *l*, *s*, *m*, *v*?
176232	10 Aql	18 56 30	13 50	5.9	0.25	A4 *p*	+ 440	− 315	
179761A	21 Aql A	19 11 12	02 12	5.1	−0.07	B8 *pv*	− 590	+ 170	
182989	RR Lyr	19 23 54	42 41	7.8 *v*	0.33	A3−F1 *v*	−1,580	+1,170	0.567 *l*, *v*, *s*
184552	51 h[1] Sgr	19 33 00	−24 50	5.7	—	A3−F5 *m*		− 230	*b*: 8.1158 (0.17)
184905	BD+43°3290	19 33 12	43 50	6.6 *v*	−0.03	A0 *pv*	−strong		~1−2? *s*
187474	B 5060	19 48 30	−40 01	5.4	−0.05 *v*	A0 *pv*	−2,000	+1,900	~2,500? *m*, b
188041	B 5074	19 50 42	−03 15	5.6	0.19 *v*	A6 *pv*	+1,470	− 230 *d*	226 *m*, *s*
190073	BD+5°4393	20 00 30	05 36	7.9	—	A0 *epv*	+strong	+ 130 *d*	
191742	BD+42°3616	20 08 06	42 24	7.8	0.22 *v*	A6 *p*	− 910	− 175	
192678	BD+53°2368	20 12 18	53 30	7.1	−0.03	A3 *p*	—	+2,000	
192913	BD+27°3668	20 14 24	27 37	6.7	−0.07	A0 *p*	− 670	+ 380 *d*	
196502	73 Dra	20 32 12	74 47	5.2 *v*	0.07	A2 *pv*	− 700	− 200	20.26 *s*, *l*, *m*
201601A	5 γ Equ A	21 07 54	09 56	4.8 *v*	0.27 *v*	A9 *pv*	+ 880	+ 180	
203006	θ^1 Mic	21 17 36	−41 01	4.9	0.02 *v*	A2 *pv*	− 650	—	
204075A	34 ξ Cap A	21 23 48	−22 38	3.9 *v*	—	G4 *p*	—	− 340	
207757	AG Peg	21 48 36	12 23	6.4–7.9 *v*	—	Bep+M	−1,800	+ 520	800:
208816	VV Cep	21 55 12	63 23	4.9–5.7 *v*	—	M2+B9 *p*	+ 850	− 360	*b*: 7,450 (0.25)
215038	BD+74°980	22 38 18	75 24	8.0 *v*	−0.03 *v*	A0 *p*	−3,000	—	2.036 *l*
215441	BD+54°2846	22 42 06	55 20	8.6 *v*	0.03 *v*	A0 *pv*	11,500	4,100	9.48 *l*
216533	BD+58°2497	22 50 36	58 33	7.9 *v*	0.08 *v*	A2 *pv*	− 700	+ 100 *d*	
221507	β Sci	23 30 18	−38 06	4.5	−0.11	B9 *p*	—	+ 660 *d*	
224801	BD+44°4538	23 58 12	44 58	6.3 *v*	−0.08 *v*	A0 *pv*	+2,300	—	3.74 *l*, *b*?

[1] After Ledoux and Renson (1966) by permission of Annual Reviews Inc.

Anomalous Zeeman effect. When the spin $S \neq 0$, and there is Russell-Saunders coupling, the magnetic field splits a term into $2J+1$ equally spaced levels. The energies, E_M, of the levels are given by (PRESTON, 1898; LANDÉ, 1920, 1923)

$$\begin{aligned} E_M &= E_n \pm \frac{ehHgM}{4\pi mc} = E_n \pm hogM \\ &= E_n \pm 9.27 \times 10^{-21}\, HgM \text{ erg}, \end{aligned} \quad (2\text{-}212)$$

where E_n is the energy of the term in the absence of the magnetic field, H is the magnetic field intensity in gauss, o is the Larmor frequency of precession, M is component of J along H, and the Landé g factor for LS coupling is given by (UHLENBECK and GOUDSMIT, 1925, 1926)

$$g = 1 + \frac{J(J+1)+S(S+1)-L(L+1)}{2J(J+1)}, \quad (2\text{-}213)$$

where J is the total angular momentum, S is the spin angular momentum, and L is the orbital angular momentum.

Because there is not equal line splitting between different terms, however, the number of line components will depend on the J, L and S of the upper and lower levels. The selection rule for allowed transitions between the levels of different terms is $\Delta M = 0$ (π components) and $\Delta M = \pm 1$ (σ components). Observations in a direction transverse to the field will show the π component linearly polarized parallel to the field and the σ components linearly polarized perpendicular to the field. For observations in a direction parallel to the field, only the two σ components are observed and they are circularly polarized in opposed directions of rotation (the lower frequency being right hand).

When the magnetic moment of the atomic nucleus is taken into account, the appropriate Landé g factor is given by (BACK and GOUDSMIT, 1928)

$$g(F) = g\,\frac{F(F+1)+J(J+1)-I(I+1)}{2F(F+1)}, \quad (2\text{-}214)$$

where g is the electronic g factor for the atom which is given above, I is the nuclear spin quantum number, J is the total angular momentum excluding nuclear spin, and F is the total angular momentum quantum number.

For molecules, the electronic Landé g factor is given by (TOWNES and SCHAWLOW, 1955)

$$g = \frac{(\Lambda + 2.002\,\Sigma)\,\Omega}{J(J+1)} = \frac{(\Omega + 1.002\,\Sigma)\,\Omega}{J(J+1)} \quad (2\text{-}215)$$

for Hund's coupling case (a), and

$$\begin{aligned} g = \frac{1}{2J(J+1)} \Big\{ &\Lambda^2 \frac{[N(N+1)-S(S+1)+J(J+1)]}{N(N+1)} \\ &+ 2.002\,[J(J+1)+S(S+1)-N(N+1)] \Big\} \end{aligned} \quad (2\text{-}216)$$

for Hund's coupling case (b). Here Λ is the projection of the electronic orbital angular momentum on the molecular axis, Σ is the projection of the electron spin angular momentum on the molecular axis, Ω is the projection of the total angular momentum excluding nuclear spin on the molecular axis, and N is the total angular momentum including rotation of the molecule.

Intensities of Zeeman components. The relative intensities, I, of the ordinary Zeeman components (weak fields) for observations at right angles to the magnetic field are:

For the $J \rightarrow J$ transition

$$\begin{aligned} &\pi \text{ components } (\Delta M=0) && I=4AM^2, \\ &\sigma \text{ components } (\Delta M=\pm 1) && I=A[J(J+1)-M(M\pm 1)]. \end{aligned} \tag{2-217}$$

For the $J \rightarrow J+1$ transition

$$\begin{aligned} &\pi \text{ components } (\Delta M=0) && I=4B[(J+1)^2-M^2], \\ &\sigma \text{ components } (\Delta M=\pm 1) && I=B[J\pm M+1][J\pm M+2]. \end{aligned} \tag{2-218}$$

Here A and B are constants for a given Zeeman pattern. The σ components will be observed at this same relative intensity when viewed parallel to the magnetic field. These formulae are given by CONDON and SHORTLEY (1963), and were first obtained empirically by ORNSTEIN and BURGER (1924) and theoretically by KRONIG (1925) and HÖNL (1925).

Zeeman effect for strong magnetic field—the Paschen-Back effect, the quadratic Zeeman effect, and the circular polarization of thermal radiation. When magnetic splitting becomes greater than the multiplet splitting, for magnetic fields stronger than a few thousand gauss, the anomalous Zeeman effect changes over to the normal Zeeman effect. This Paschen-Back effect (PASCHEN and BACK, 1912, 1913) is the result of magnetic uncoupling of L and S. To a first approximation the energies, E_M, of the split levels are given by

$$E_M = E_n + \frac{eh}{4\pi mc}(M_L + 2M_S)H, \tag{2-219}$$

where E_n is the energy of the term in the absence of the magnetic field, $eh/(4\pi mc) = 9.2731 \times 10^{-21}$ erg gauss^{-1}, the quantum numbers M_L and M_S take on integral values between $\pm L$ and $\pm S$, respectively, and the selection rules are $\Delta M_L = 0, \pm 1$ and $\Delta M_S = 0$.

If magnetic flux is conserved in gravitational collapse, it is expected that white dwarf and neutron stars will have magnetic fields of $\sim 10^6$ and 10^{12} gauss, respectively. For these very large field strengths, the quadratic Zeeman effect is expected to dominate over the linear effect (PRESTON, 1970). The quadratic Zeeman effect produces a displacement, $\Delta\lambda$, of spectral lines towards the short wavelengths. This displacement is given by (VAN VLECK, 1932; JENKINS and SEGRÈ, 1939)

$$\Delta\lambda = \frac{-e^2 a_0^2}{8mc^3 h}\lambda^2 n^4 (1+M^2) H^2 \approx -4.98 \times 10^{-23} \lambda^2 n^4 (1+M^2) H^2 \text{ Å} \tag{2-220}$$

for hydrogen. Here a_0 is the radius of the first Bohr orbit, n is the principal quantum number, M is the magnetic quantum number, H is the magnetic field intensity, and the numerical approximation is for $\Delta\lambda$ and λ in Angstroms ($1\,\text{Å} = 10^{-8}$ cm).

Unfortunately, white dwarfs and neutron stars show little spectral structure from which to observe the quadratic Zeeman effect. Any emission from a thermal source may, however, exhibit a diffuse spectral component which is circularly polarized. This may be seen by regarding the thermal radiation as a superposition of the emission from a collection of harmonic oscillators which emit circularly polarized emission at their Zeeman split frequencies, $\nu \pm eH/(4\pi m c)$. The net fractional polarization of the thermal radiation is given by (KEMP, 1970)

$$q(\nu) = \frac{I_+(\nu) - I_-(\nu)}{I_+(\nu) + I_-(\nu)} \approx \frac{-eH}{4\pi m \nu} \approx 4.2 \times 10^{16} \frac{H}{\nu}, \tag{2-221}$$

where $I(\nu)$ denotes the observed radiation intensity at frequency, ν, the + and − denote, respectively, right and left hand circular polarization, the H is the component of the magnetic field along the line of sight, and $q(\nu)$ is right handed for electrons. The percentage circular polarization will fall off faster than ν^{-1} for frequencies $\nu > \nu_p$, where the plasma frequency $\nu_p \approx 8.9 \times 10^3 N_e^{1/2}$ Hz. KEMP *et al.* (1970) and ANGEL and LANDSTREET (1971) have detected circular polarization in the optical emission of D.C. white dwarfs. They observe a $q(\nu)$ of a few percent indicating magnetic field strengths of 10^6 to 10^7 gauss. Circular polarization measurements are given together with other parameters of well known white dwarfs in Table 35.

Stark effect for hydrogen atoms. When a hydrogen atom is subjected to an electric field of strength, E, the alteration, ΔE, of the energy of a level is given by (STARK, 1913; SCHWARZSCHILD, 1916; EPSTEIN, 1916, 1926; BETHE and SALPETER, 1957)

$$\Delta E = AE + BE^2 + CE^3, \tag{2-222}$$

where

$$A = \frac{3h^2}{8\pi^2 me} n(n_2 - n_1),$$

$$B = \frac{-h^6 n^4}{2^{10}\pi^6 m^3 e^6} [17n^2 - 3(n_2 - n_1)^2 - 9m_l^2 + 19]$$

and

$$C = \frac{3h^{10}}{2^{15}\pi^{10} m^5 e^{11}} n^7 (n_2 - n_1) [23n^2 - (n_2 - n_1)^2 + 11m_l^2 - 71].$$

Here n is the principal quantum number, m_l is the component of the orbital quantum number, l, in the direction of the electric field, m_l takes on integral values between $-l$ and $+l$, and the parabolic quantum numbers n_2 and n_1 may assume integer values between 0 and $n-1$. The coefficients of A, B, and C are 1.27×10^{-20} erg cm volt^{-1}, 1.03×10^{-31} erg cm^2 volt^{-2} and 2.97×10^{-41} erg cm^3 volt^{-3}, respectively. The factor −71 given in brackets for the coefficient C is from CONDON and SHORTLEY (1963), whereas BETHE and SALPETER (1957) give this term as +39.

To the first approximation, each level of principal quantum number, n, is split into $2n-1$ equidistant levels with energies, E_s, given by

$$E_s = E_n + \frac{3h^2}{8\pi^2 me} n(n_2 - n_1)E$$
$$= E_n + 3.81 \times 10^{-18} n(n_2 - n_1)E \text{ erg}, \quad (2\text{-}223)$$

where E is given in e.s.u. (1.71×10^7 e.s.u. $= 5.142 \times 10^9$ volt cm^{-1}), E_n is the energy of the nth level in the absence of an electric field, and n_2 and n_1 may assume integer values between 0 and $n-1$. For a transition between an upper level m and a lower level n, the frequency is split into the frequencies $\nu_{m,n}$ given by

$$\nu_{m,n} = \nu_{mn} + \frac{3hE}{8\pi^2 m_e e}\{m(m_2 - m_1) - n(n_2 - n_1)\}$$
$$= \nu_{mn} + 1.91 \times 10^6 E\{m(m_2 - m_1) - n(n_2 - n_1)\} \text{ Hz}, \quad (2\text{-}224)$$

where the field strength E is in volt cm^{-1}, n_2 and n_1 may assume integer values between 0 and $n-1$, and m_2 and m_1 may assume integer values between 0 and $m-1$. The change in wavelength, $\Delta\lambda$, corresponding to $\nu_{m,n} - \nu_{mn}$, is

$$\Delta\lambda \approx 0.64 \times 10^{-4} \lambda^2 E\{m(m_2 - m_1) - n(n_2 - n_1)\} \text{ cm}, \quad (2\text{-}225)$$

where λ is in cm and E is in volt cm^{-1}.

If we define the new quantum numbers $|kn| = n - 1 - n_1 - n_2$ and $|km| = m - 1 - m_1 - m_2$, then for transverse observation the radiation is polarized parallel (π) to the electric field if $kn = km$, and perpendicular (σ) to the field if $km - kn = \pm 1$.

When considering astrophysical plasmas, the electric field is not uniform and constant; and we are concerned with the probability distribution of the electric field of any moving charged particles. For slowly moving ions, the net effect is to broaden and shift the line to higher frequencies (cf. Sect. 2.20).

Doppler shift. When line radiation of frequency, ν_L, is emitted from an object travelling at a velocity, V, with respect to an observer at rest, the observed frequency, ν_{obs}, is (DOPPLER, 1842; LORENTZ, 1904; EINSTEIN, 1905)

$$\nu_{obs} = \nu_L \left(1 - \frac{V}{c}\cos\theta\right)^{-1} \left(1 - \frac{V^2}{c^2}\right)^{1/2}$$
$$\approx \nu_L \left(1 - \frac{V_r}{c}\right) \quad \text{for } V \ll c \text{ and } \theta \ll \pi/2, \quad (2\text{-}226)$$

where θ is the angle between the velocity vector and the radiation wave vector. The radial velocity $V_r = -V\cos\theta$ is positive when the radiating object is moving away from the observer. Consequently, when receding objects are observed at optical frequencies, the lines are shifted towards the red (towards longer wavelengths). The redshift, z, is defined as

$$z = \frac{\lambda_{obs} - \lambda_L}{\lambda_L} = \frac{\nu_L - \nu_{obs}}{\nu_{obs}} \approx \frac{V_r}{c}, \quad (2\text{-}227)$$

where λ_{obs} and λ_{L} denote, respectively, the wavelengths of the observed and emitted line radiation, ν_{obs} and ν_{L} denote, respectively, the frequencies of the observed and emitted line radiation, and V_{r} is the radial velocity. For large velocities the special relativistic Doppler effect is expressed by the equations

$$1+z=\left[\frac{c+V_{\text{r}}}{c-V_{\text{r}}}\right]^{1/2} \tag{2-228}$$

and

$$\frac{V_{\text{r}}}{c}=\frac{(z+1)^2-1}{(z+1)^2+1}. \tag{2-229}$$

Measured velocities are often corrected for the orbital motion of the earth about the Sun, and the Sun's motion with respect to the local group of stars. The corrections to obtain this velocity with respect to the local standard of rest are discussed in Chap. 5. Bibliographies and catalogues of stellar radial velocities are given by ABT and BIGGS (1972) and BOYARCHUK and KOPYLOV (1964). Measured redshifts for optical galaxies are given by HUMASON, MAYALL, and SANDAGE (1956), MAYALL and DE VAUCOULEURS (1962), and in the book by DE VAUCOULEURS and DE VAUCOULEURS (1964). A comparison of radio and optical values of the redshifts of normal galaxies is given by FORD, RUBIN, and ROBERTS (1971). Observed redshifts for radio galaxies and quasi-stellar objects are given in Table 31.

Gravitational redshift. The energy, E, of a particle of mass, m, and velocity, v, at the surface of a body of mass, M, and radius, R, is given by

$$E=\frac{mv^2}{2}-\frac{GMm}{R}, \tag{2-230}$$

so that the velocity of escape, for which $E=0$, is given by

$$V_{\text{esc}}=\left(\frac{2GM}{R}\right)^{1/2}. \tag{2-231}$$

For a photon $V_{\text{esc}}=c$ and the limiting Schwarzschild radius, R_g, for its escape is given by

$$R_g=\frac{2GM}{c^2}. \tag{2-232}$$

When a photon of energy $h\nu=mc^2$ leaves the surface of the massive body it loses the energy

$$\Delta E=h\Delta\nu=\frac{GMm}{R}=\frac{GMh\nu}{Rc^2}, \tag{2-233}$$

so that the gravitational redshift is given by

$$z_g=\frac{\Delta\nu}{\nu}=\frac{GM}{Rc^2}=\frac{R_g}{2R}=1.47\times10^5\,\frac{M}{M_{\odot}R}, \tag{2-234}$$

Table 31. A table of positions, visual magnitudes, V, redshifts, z, flux density at 1400 MHz, S_{1400}, spectral indices, α, and largest angular size, θ, for 383 radio galaxies and quasi-stellar objects[1]

Object[2]	α(1950.0) h m s	δ(1950.0) ° ′ ″	V (mag.)	z	S_{1400} (f.u.)	α	θ (arc sec)
PHL 658	00 03 25.4	15 53 07	16.40	0.450	2.6	−0.6	14
3C 2	00 03 48.82	−00 21 06.5	19.35	1.037	3.83	−0.8	<6
3C 9	00 17 49.944	15 24 16.21	18.21	2.012	2.19	−1.05	9.6
PKS 0023−33[2]	00 23 04.0	−33 20 06.0	16.7	0.0497	1.3	−0.9	360
WK 10[2]	00 28 30.4	39 25 20	18.0	0.0728	0.1	—	—
4C 42.01	00 32 23.0	42 21 48	18.30	1.588[3]	—	—	<3
3C 15.0[2]	00 34 30.58	−01 25 32	15.34	0.0733	4.1	−0.65	52
3C 17.0[2]	00 35 47.13	−02 24 15	18.02	0.2201	6.0	−0.60	18
4C 03.01[2]	00 36 42.8	03 04 54	14.3	0.0145	1.5	−0.70	—
5C 03.100[2]	00 39 10.0	04 05 06	16.0	0.0718	(0.02)	—	—
5C 03.175[2]	00 44 48.7	39 48 02	16.0	0.1343	(0.01)	—	—
PKS 0045−25[2]	00 45 04.9	−25 33 33	7.0	0.00086	6.3	−0.90	420
3C 26.0[2]	00 51 35.67	−03 50 13.5	17.94	0.2106	2.0	−1.01	<15
3C 28.0[2]	00 53 08.70	26 08 21.4	17.54	0.1959	1.8	−1.20	29
3C 29.0[2]	00 55 01.0	−01 39 35	14.07	0.0450	4.6	−0.59	—
PHL 923	00 56 31.79	−00 09 18.4	17.33	0.717	2.6	−0.40	<7
PHL 938	00 58 12.0	01 56 00	17.16	1.930[3]			
PHL 957	01 00 36.0	13 00 00	16.60	2.720[3]			
3C 31.0[2]	01 04 42.6	32 08 26	12.14	0.0169	5.0	−0.79	150
PKS 0106+01	01 06 04.48	01 19 01.45	18.39	2.107	1.4	−0.70	<7
3C 33.0[2]	01 06 13.4	13 03 28	15.19	0.0600	13.2	−0.71	228
PKS 0108−14.2[2]	01 08 37.0	−14 13 54	15.8	0.0518	1.2	−0.9	42
3C 35.0[2]	01 09 05.04	49 12 48.9	15.0	0.0677	2.1	−0.95	558
PKS 0115+02	01 15 43.74	02 42 20.5	17.50	0.672	1.5	−0.94	9
OC 328[2]	01 16 47.27	31 55 06.4	14.5	0.0590	3.0	−0.40	<2
PKS 0119−04	01 19 55.9	−04 37 07	16.88	1.955[3]	1.3	−0.40	<7
PKS 0122−00	01 22 55.20	−00 21 30.7	16.70	1.070	1.5	−0.10	<6
4C 25.05	01 23 56.0	25 43 56.0	17.50	2.360[3]	0.7	−0.63	<6
3C 40.0[2]	01 23 27.8	−01 38 29.0	12.28	0.0177	4.4	−1.13	312
4C 18.06[2]	01 24 12.30	18 57 46.0	15.5	0.0437	1.4	−0.30	—
PHL 3375	01 28 24.0	07 28 00	18.02	0.390			
PHL 1027	01 30 30.0	03 22 00	17.04	0.363			
PHL 3424	01 31 12.0	05 32 00	18.25	1.847			
PKS 0131−36[2]	01 31 42	−36 44 36	14.2	0.030	5.6	—	—
3C 47	01 33 40.3	20 42 16	18.10	0.425	3.8	−0.99	60
3C 48	01 34 49.836	32 54 20.26	16.20	0.367	15.63	−0.8	<5
PHL 1078	01 35 29.0	−05 42 06	18.25	0.308	—	—	—
PHL 1093	01 37 22.7	01 16 18	17.07	0.258	1.4	−0.41	11
PHL 3632	01 39 54.0	06 10 00	18.15	1.479			
4C 33.03	01 41 18.0	33 57 00	17.50	1.455			
PHL 1127	01 41 30.0	05 14 00	18.29	1.990[3]			
PHL 1186	01 47 36.0	09 01 00	18.60	0.270			
PHL 1194	01 48 42.0	09 02 00	17.50	0.298[3]			
PHL 1222	01 51 12.0	04 48 00	17.63	1.910[3]			
PHL 1226	01 51 48.0	04 34 00	18.20	0.404			

Table 31 (continued)

Object[2]	α(1950.0) h m s	δ(1950.0) ° ′ ″	V (mag.)	z	S_{1400} (f.u.)	α	θ (arc sec)
PKS 0153+05[2]	01 53 46.0	05 23 00	13.2	0.0188	0.9	−0.60	113
PKS 0155−10	01 55 13.10	−10 58 16.6	17.09	0.616	1.9	−0.70	<6
3C 57	01 59 30.32	−11 46 58.4	16.40	0.680	3.2	−0.59	<3
DW 0202+31	02 02 09.69	31 57 11.3	18.00	1.466	1.0	—	<10
PKS 0202−17	02 02 34.47	−17 15 39.5	18.00	1.740	1.2	0.1	—
PKS 0203+05[2]	02 03 22	05 12 24	17.0	0.1300	0.5	−0.6	—
4C 35.03[2]	02 06 39.40	35 33 41.0	14.5	0.0374	2.2	−0.8	—
3C 61.1[2]	02 10 46.10	86 04 51.5	19.0	0.18	6.3	−1.00	180
PKS 0214+10	02 14 26.9	10 50 24	17.00	0.408	1.2	−0.6	58
3C 66.0[2]	02 20 02.4	42 46 27	12.9	0.0215	9.7	−0.74	498
PKS 0225−014	02 25 35.0	−01 29 12	18.00	0.685			
PHL 1305	02 26 21.0	−03 54 00	16.96	2.064	—	—	<7
PKS 0229+13	02 29 02.52	13 09 40.7	17.71	2.065[3]	1.2	0.20	<6
PKS 0231+022	02 31 14.6	02 16 18	18.00	0.322			
PHL 1377	02 32 36.6	−04 15 05	16.46	1.434[3]	1.5	−0.80	10.2
PKS 0237−23	02 37 52.62	−23 22 06.0	16.63	2.224[3]	7.2	−0.50	0.002
4C 08.11[2]	02 38 26.4	08 31 57	14.8	0.0216	1.2	—	—
3C 71.0[2]	02 40 07.09	−00 13 30.7	8.91	0.00344	4.9	−0.69	9
3C 75.0[2]	02 55 04.9	05 50 41.0	13.62	0.0241	5.8	−0.68	168
PKS 0256−005	02 56 54.4	−00 31 54.0	17.5	1.998			
4C 34.09[2]	02 58 35.6	35 00 13	14.0	0.0203	2.0	−0.44	—
3C 76.1[2]	03 00 27.4	16 14 31	14.86	0.0326	3.0	−0.87	45
3C 78.0[2]	03 05 49.11	03 55 13.8	12.84	0.0289	7.1	−0.53	80
3C 79.0[2]	03 07 10.6	16 54 42	19.0	0.2561	4.9	−1.02	60
4C 39.11[2]	03 09 13.1	39 05 06	14.0	0.0163	2.1	—	—
3C 83.1[2]	03 14 56.8	41 43 46	12.5	0.0181	8.4	−0.64	396
3C 84.0[2]	03 16 29.568	41 19 51.84	11.87	0.0176	13.5	−0.77	300
Fornax A[2]	03 20 42.0	−37 25 00	8.90	0.00577	108.0	−0.35	1,740
3C 88.0[2]	03 25 18.9	02 23 22	13.95	0.0302	4.4	−0.66	150
4C 39.12[2]	03 31 01.3	39 10 12	15.0	0.0209	0.5	—	—
NRAO 140	03 33 22.38	32 08 36.8	17.50	1.258	4.0	−0.05	<2
PKS 0336−35[2]	03 36 48.1	−35 33 03.0	10.9	0.0043	(2.0)	−0.8	330
CTA 26	03 36 58.97	−01 56 16.7		0.852	1.5	−0.10	<6
PKS 0349−14	03 49 09.5	−14 38 07	16.22	0.614	2.8	−1.09	117
PKS 0349−27[2]	03 49 32.6	−27 53 05	16.8	0.066	5.2	−0.80	252
4C 21.13[2]	03 49 45.1	21 17 32	16.0	0.1325	0.8	—	—
3C 94.0	03 50 05.41	−07 19 57.2	16.49	0.962	2.9	−0.98	6
3C 98.0[2]	03 56 10.4	10 17 34	14.45	0.0306	10.0	−0.87	192
PKS 0403−13	04 03 13.98	−13 16 17.95	17.17	0.571	4.0	−0.10	<6
PKS 0405−12	04 05 27.42	−12 19 30.3	17.07	0.574	3.3	−0.80	17.2
3C 109.0[2]	04 10 55.08	11 04 46	17.0	0.3056	4.1	−0.84	81
PKS 0424−13	04 24 48.0	−13 09 36	17.50	2.165[3]	1.5	−0.43	384
PKS 0427−53[2]	04 27 53.5	−53 56 06	13.2	0.0392	4.4	−0.7	—
3C 120.0[2]	04 30 31.605	05 15 00.19	15.0	0.0333	4.1	0.03	<3
PKS 0431−13.5[2]	04 31 49.30	−13 28 58.0	16.3	0.0360	(0.7)	−1.07	414
PKS 0449−17[2]	04 49 08.0	−17 35 12.0	13.7	0.0313	0.8	−0.7	—
PKS 0453−20[2]	04 53 14.16	−20 39 00.9	14.0	0.0354	3.7	−0.60	<15

Table 31 (continued)

Object[2]	α(1950.0) h m s	δ(1950.0) ° ′ ″	V (mag.)	z	S_{1400} (f.u.)	α	θ (arc sec)
PKS 0454+039	04 54 08.4	03 56 06	16.5	1.345			
3 C 135.0[2]	05 11 31.7	00 53 04.0	17.00	0.1270	2.0	−0.53	222
3 C 138.0	05 18 16.508	16 35 27.03	18.84	0.759	9.64	−0.37	0.4
PICT. A[2]	05 18 18.18	−45 49 39	15.7	0.0342	52.1	−0.80	420
PKS 0521−36[2]	05 21 13.21	−36 30 19	16.8	0.061	14.7	−0.70	14
3 C 147.0	05 38 43.464	49 49 43.22	17.80	0.545	22.24	−0.66	1
3 C 153.0[2]	06 05 44.384	48 04 49.23	18.5	0.2771	4.15	−0.71	6
PKS 0618−37[2]	06 18 18.0	−37 10 10	16.6	0.0326	2.4	−0.6	—
PKS 0620−52[2]	06 20 37.3	−52 40 01.0	15.5	0.0502	2.7	−0.7	—
PKS 0634−20[2]	06 34 23.1	−20 39 18	16.8	0.056	7.0	−0.90	270
3 C 171.0[2]	06 51 11.05	54 12 50.4	18.8	0.2387	3.89	−0.75	<9
4 C 23.18[2]	06 58 27.1	23 17 45	15.7	0.0905	1.2	—	—
3 C 175.0	07 10 15.60	11 51 20.5	16.6	0.768	2.61	−0.96	45
OI 318	07 11 05.62	35 39 52.6		1.620	1.8	−0.70	<2
PKS 0718−34[2]	07 18 57.2	−34 01 09	16.5	0.0297	1.7	1.7	144
3 C 178.0[2]	07 22 31.07	−09 34 03	13.0	0.0073	1.4	−0.97	—
3 C 181.0	07 25 20.35	14 43 46.3	18.92	1.382	2.37	−0.81	6
3 C 184.1 (Q)	07 29 22.0	81 52 36	17.50	1.022			
3 C 184.1 (G)[2]	07 34 28.2	80 33 32	17.0	0.1187	3.3	−0.71	168
PKS 0736+01	07 36 42.56	01 44 00.4	17.47	0.191	2.47	−0.40	<6
OI 363	07 38 00.19	31 19 02.2		0.63	2.2	0.10	<2
3 C 186.0	07 40 56.742	38 00 31.08	17.60	1.063	1.32	−1.19	<5
4 C 56.16[2]	07 45 45.2	56 02 08.1	16.0	0.0356	3.0	−0.90	—
4 C 37.21[2]	07 56 26.88	37 45 40.0	14.9	0.0433	1.5	−1.13	228
3 C 191.0	08 02 03.76	10 23 57.6	18.40	1.952[3]	1.7	−1.09	<3
3 C 192.0[2]	08 02 35.51	24 18 28.2	15.46	0.0596	4.8	−0.82	210
4 C 05.34	08 05 20.0	04 41 24	18.00	2.877			
3 C 196.0	08 09 59.386	48 22 07.71	17.79	0.871	14.23	−0.83	5.2
PKS 0812+02	08 12 47.20	02 04 13.4	18.50	0.402[3]	1.95	−0.90	39
3 C 197.1[2]	08 18 01.11	47 12 11.0	16.5	0.1302	0.3	−0.80	12
3 C 198.0[2]	08 19 50.77	06 06 35.0	16.78	0.0809	2.1	−0.93	168
4 C 37.24	08 27 55.09	37 52 16.9	18.11	0.914	2.45	−0.60	<6
3 C 204.0	08 33 18.025	65 24 04.44	18.21	1.112	1.3	−0.95	31.4
3 C 205.0	08 35 09.942	58 04 51.76	17.62	1.534[3]	2.43	−0.81	15.8
4 C 19.31	08 36 15.0	19 33 00	17.60	1.691	—	—	13.0
4 C 24.18[2]	08 37 02.3	24 14 36.0	17.0	0.0423	0.8	—	—
PKS 0837−12	08 37 28.0	−12 03 54	15.76	0.200	1.7	−1.06	150.0
3 C 207	08 38 01.84	13 23 06.1	18.15	0.684	2.59	−0.69	7
PKS 0843−33[2]	08 43 08.7	−33 36 21	12.5	0.00732	1.6	−0.6	—
4 C 54.17[2]	08 44 11.6	54 04 36.0	15.0	0.0449	(2.0)	—	—
LB 8707	08 46 01.0	14 32 00	18.10				
LB 8755	08 48 05.0	15 33 30	17.70	2.010			
3 C 208.0	08 50 22.98	14 04 18.9	17.42	1.110	2.26	−1.23	10.5
4 C 17.46	08 56 03.0	17 03 12	17.40	1.449	—	—	10.0
PKS 0859−14	08 59 55.0	−14 03 37	17.80	1.327	3.1	−0.40	<7
3 C 215.0	09 03 44.23	16 58 17.3	18.27	0.411	1.66	−0.86	25
3 C 218.0[2] Hyd A	09 15 41.27	−11 53 05.0	14.8	0.065	36.3	−0.96	16
3 C 219.0[2]	09 17 50.48	45 51 48.0	17.22	0.1745	7.2	−0.91	138

Table 31 (continued)

Object[2]	α(1950.0) h m s	δ(1950.0) ° ′ ″	V (mag.)	z	S_{1400} (f.u.)	α	θ (arc sec)
PKS 0922+14	09 22 22.3	14 57 26	17.96	0.896	0.7	−1.00	40
PKS 0922+005	09 22 35.7	00 32 06	18.07	1.720	—	—	<7
4C 39.25	09 23 55.33	39 15 23.5	17.86	0.698	1.1	−0.34	<4
PKS 0932+02	09 32 42.99	02 16 06	17.39	0.659[3]	0.7	−0.20	40
3C 223.0[2]	09 36 50.49	36 08 05	17.06	0.1367	3.1	−0.95	300
3C 223.1[2]	09 38 18.0	39 58 20.0	16.36	0.1075	1.9	−0.62	71
3C 227.0[2]	09 45 08.48	07 39 19.0	17.0	0.0855	7.3	−0.85	180
3C 231.0[2]	09 51 43.0	69 54 58.9	8.39	0.00107	8.1	−0.34	35
AO 0952+17	09 52 11.84	17 57 44.45	17.23	1.472	—	—	<6
3C 232.0	09 55 25.435	32 38 23.2	15.78	0.530	1.5	−0.57	<3
PKS 0957+00	09 57 43.8	00 19 50.0	17.57	0.907	1.0	−0.90	34
3C 234.0[2]	09 58 56.8	29 01 41.0	17.5	0.1846	5.3	−1.20	76
3C 236.0[2]	10 03 05.40	35 08 48.1	15.97	0.0988	4.3	—	360
PKS 1004+13	10 04 43.99	13 03 28.0	15.15	0.240	1.2	−0.60	28
TON 490	10 11 06.0	25 06 00	15.40	1.631			
4C 48.28	10 12 49.0	48 53 12	19.00	0.385	—	—	110
3C 245.0	10 40 06.00	12 19 15.1	17.29	1.029	3.17	−0.75	3.5
4C 14.37[2]	10 44 00.0	14 01 00.0	11.6	0.0092	0.4	—	—
PKS 1049−09	10 48 59.5	−09 02 12	16.79	0.344	2.2	−0.84	81
5C 02.10	10 49 41.0	48 55 54	18.00	0.478			
5C 02.56	10 55 18.0	49 55 36	20.00	2.380	0.6	—	—
PKS 1055+20	10 55 36.9	20 08 18	17.07	1.110	2.2	−0.60	21
3C 249.1	11 00 27.316	77 15 08.62	15.72	0.311	2.36	−0.71	19
QS 1108+285	11 08 26.5	28 57 56	20.00	2.192			
3C 254	11 11 53.11	40 53 41.1	17.98	0.734	3.14	−0.91	13.5
4C 29.41[2]	11 13 53.7	29 31 44	14.0	0.0481	2.0	—	—
PKS 1116+12	11 16 20.755	12 51 06.5	19.25	2.118[3]	2.42	−0.6	<6
3C 255.0[2]	11 16 52.61	−02 46 44	15.5	0.0238	1.6	−1.35	<5
PKS 1123−35[2]	11 23 28.99	−35 06 42.0	16.0	0.0336	2.1	−0.80	102
4C 25.35[2]	11 25 30.0	25 56 00	12.9	0.0088	1.0	—	—
PKS 1127−14	11 27 35.61	−14 32 54.0	16.90	1.187	6.2	0.10	<0.025
4C −03.43[2]	11 30 30.8	−03 44 32	15.5	0.0482	(1.0)	—	—
4C 21.32[2]	11 31 23.99	21 23 00.0	16.0	0.066	0.7	−1.36	—
3C 261	11 32 16.49	30 22 02.0	18.24	0.614	1.7	−0.56	10.8
PKS 1136−13	11 36 38.37	−13 34 09.3	17.80	0.554	4.5	−0.80	23
3C 263	11 37 09.296	66 04 27.04	16.32	0.652[3]	3.1	−0.81	47
4C 17.52[2]	11 37 40.4	17 59 48	13.3	0.0105	1.5	—	—
3C 264.0[2]	11 42 29.59	19 53 16.0	12.74	0.0206	5.1	−0.78	180
PKS 1148−00	11 48 10.17	−00 07 13.05	17.60	1.982	2.9	−0.20	0.0019
4C 49.22	11 50 48.00	49 47 48.0	16.10	0.334[3]	1.4	−0.70	7
4C 31.38	11 53 44.08	31 44 46.85	18.96	1.557	3.3	−0.40	<3
4C 29.45	11 56 58.18	29 33 24.0	15.60	0.729	2.0	−0.70	<7
3C 268.4	12 06 41.98	43 55 59.9	18.42	1.400	2.3	−0.73	9.4
MSH 12+04A[2]	12 15 01.20	03 54 57.0	17.0	0.0756	2.0	−0.80	270
MSH 12+04B, C[2]	12 15 01.20	03 54 57.0	17.0	0.0771	2.0	−0.80	270
3C 270.0[2]	12 16 50.07	06 06 09.0	10.40	0.00697	15.3	−0.60	282
PKS 1217+02	12 17 38.33	02 20 20.9	16.53	0.240	0.8	−0.90	150

Table 31 (continued)

Object[2]	α(1950.0) h m s	δ(1950.0) ° ′ ″	V (mag.)	z	S_{1400} (f.u.)	α	θ (arc sec)
3 C 270.1	12 18 03.90	33 59 48.2	18.61	1.519[3]	2.64	−1.09	7.5
4 C 21.35	12 22 23.48	21 40 00	17.50	0.435	1.5	−0.73	10
3 C 272.1[2]	12 22 31.48	13 09 46.0	9.36	0.00293	6.1	−0.63	90
TON 1530	12 23 12.0	22 51 00	17.00	2.051[3]			
4 C 25.40	12 23 12.0	25 14 30	16.00	0.268			
3 C 273.0	12 26 33.246	02 19 43.38	12.80	0.158	46.4	−0.24	20
3 C 274.0[2] Vir A	12 28 17.565	12 40 01.66	8.74	0.0041	197.0	−0.82	282
PKS 1229−02	12 29 25.88	−02 07 31.0	16.75	0.388[3]	1.7	−0.60	<8
TON 1542	12 29 48.0	20 25 00	15.30	0.064			
PKS 1233−24	12 32 59.40	−24 55 46.0	17.20	0.355	2.3	−0.90	—
3 C 275.1	12 41 27.56	16 39 18.0	19.00	0.557	2.94	−0.86	13.2
B 19	12 45 02.7	34 31 29	17.94	2.065			
PKS 1245−41[2]	12 46 03.86	−41 02 08.0	12.0	0.0086	3.2	−0.90	36
BSO 1	12 46 29.0	37 46 25	16.98	1.241[3]			
B 46	12 46 29.2	34 40 49	17.83	0.271			
BSO 2	12 48 17.7	33 47 04	18.64	0.186			
B 86	12 49 40.6	33 54 42	17.58	1.431			
PKS 1250−10[2]	12 50 15.0	−10 12 18.0	12.0	0.0138	0.9	−1.2	—
3 C 277.1	12 50 15.278	56 50 36.72	17.93	0.320	2.51	−0.57	<4
3 C 277.3[2] Coma A	12 51 46.10	27 53 45.0	15.94	0.0857	3.1	−0.54	34
3 C 278.0[2]	12 52 00.08	−12 17 07.0	11.46	0.014	6.6	−0.73	103
PKS 1252+11	12 52 07.78	11 57 20.8	16.64	0.870	1.2	−0.10	<4
B 114	12 52 57.7	35 55 26	17.92	0.221			
3 C 279	12 53 35.825	−05 31 07.65	17.75	0.536	11.2	−0.35	<1
B 142	12 54 55.0	37 03 29	17.84	0.280			
5 C 04.51[2]	12 54 58.8	27 46 02.0	14.2	0.0249	(0.08)	—	—
B 154	12 55 02.1	35 21 21	18.56	0.183			
B 185	12 55 40.2	37 15 19	18.12	1.530			
B 194	12 56 07.8	35 44 56	17.96	1.864[3]			
B 189	12 56 51.1	36 48 10	19.22	2.075			
5 C 04.81[2]	12 56 56.7	28 10 50	13.9	0.0224	0.37	—	—
5 C 04.85[2]	12 57 10.72	28 13 48.2	12.4	0.0239	0.18	—	—
B 201	12 57 26.6	34 39 34	16.79	1.375			
3 C 280.1	12 58 14.2	40 25 15	19.44	1.659	—	—	19.0
B 246	12 58 31.0	34 04 48	18.18	0.690			
B 196	12 58 42.0	35 38 48	18.28	0.323			
B 471	12 58 49.0	34 22 42	17.66	0.774			
B 228	12 59 21.4	36 46 22	17.83	1.194			
BSO 6	12 59 30.5	34 27 15	17.87	1.956			
B 264	12 59 30.9	32 21 58	16.89	0.095			
B 234	13 00 43.2	36 07 48	17.52	0.060			
B 272	13 01 35.0	37 30 48	17.25	0.036			
B 286	13 01 43.0	35 49 32	18.65	0.330			
PKS 1302−49[2]	13 02 33.1	−49 12 00	9.0	0.00184	5.8	−0.4	—
B 288	13 02 18.0	35 45 22	18.39	1.293			
B 340	13 04 47.1	34 40 39	16.97	0.184			
B 312	13 04 53.1	37 29 33	19.08	0.450			
3 C 281.9	13 05 22.5	06 58 16	17.02	0.599	1.2	−0.81	40.1

Table 31 (continued)

Object[2]	α(1950.0) h m s	δ(1950.0) ° ′ ″	V (mag.)	z	S_{1400} (f.u.)	α	θ (arc sec)
B 330	13 05 26.0	36 25 54	18.01	0.920			
B 337	13 05 30.2	35 17 50	17.62	0.300			
B 382	13 06 53.0	35 01 32	17.55	0.194			
B 360	13 08 17.6	38 14 17	17.56	2.090			
3 C 284.0[2]	13 08 41.39	27 44 03.1	18.0	0.2394	2.0	−0.75	162
BSO 8	13 09 15.0	34 03 08	17.43	1.750			
BSO 11	13 11 22.1	36 16 30	18.41	2.084[3]			
PKS 1317−00	13 17 04.5	−00 34 21	17.32	0.890[3]	1.8	−0.70	<6
TON 153	13 17 24.0	27 45 00	15.30				
TON 156	13 18 54.0	29 06 00	16.00				
3 C 285.0[2]	13 19 05.70	42 50 48.0	15.99	0.0797	2.3	−0.53	132
TON 157	13 21 00.0	29 27 00	16.00	0.960			
CENTA[2]	13 22 31.59	−42 45 24.0	6.98	0.00157	912.0	−0.60	—
4 C 31.42[2]	13 26 07.97	31 01 00.0	15.5	0.0241	0.7	−0.31	—
PKS 1327−21	13 27 23.2	−21 26 34	16.74	0.528	2.0	−0.70	16
3 C 287.0	13 28 15.925	25 24 37.7	17.67	1.055	7.31	−0.52	0.06
3 C 286.0	13 28 49.67	30 45 58.35	17.25	0.846	15.4	−0.38	0.05
3 C 287.1[2]	13 30 20.99	02 16 12.0	19.0	0.2156	2.7	−0.45	63
RS 12	13 31 11.0	28 08 24	18.61				
RS 13	13 31 30.0	27 45 48	17.94	1.287			
4 C 55.27	13 32 17.0	55 15 48	16.00	0.249	—	—	78
PKS 1333−33[2]	13 33 44	−33 43 00	11.9	0.0123	(3.0)	—	—
RS 23	13 33 55.0	28 40 18	18.74	1.908[3]			
PKS 1334−29[2]	13 34 09.5	−29 36 25	8.0	0.00112	2.1	−0.8	420
PKS 1335+023	13 35 07.3	02 22 06	18.00	0.610			
MSH 13−011	13 35 31.3	−06 11 57	17.68	0.625	3.2	−0.80	7.8
RS 32	13 36 03.0	26 29 06	18.91	0.341			
PKS 1340+05[2]	13 40 12.42	05 19 38.8	18.0	0.1333	1.8	−0.60	—
3 C 288.1	13 40 29.911	60 36 48.51	18.12	0.961	1.5	−1.08	6.4
PKS 1345+12[2]	13 45 06.17	12 32 20.1	17.0	0.1218	4.9	−0.50	<18
3 C 293.0[2]	13 50 03.23	31 41 32.6	14.32	0.0454	4.5	−0.75	130
PKS 1354+19	13 54 42.14	19 33 42.9	16.02	0.720	2.28	−0.70	13
PKS 1358−11[2]	13 58 58.37	−11 22 00.0	15.0	0.025	2.1	−0.70	300
PKS 1400−33[2]	14 00 58	−33 48 00	12.4	0.0143	(2.0)	—	—
OQ 208[2]	14 04 45.63	28 41 29.4	14.0	0.0768	(1.0)	—	—
4 C −01.32[2]	14 07 23.9	−01 01 12	14.8	0.0249	(0.4)	—	—
3 C 295.0[2]	14 09 33.50	52 26 13.0	20.11	0.4614	23.1	−0.75	4
3 C 296.0[2]	14 14 28.2	11 01 56	12.21	0.0237	4.4	−0.34	180
3 C 298.0	14 16 38.80	06 42 20.65	16.79	1.439[3]	5.96	−1.17	<5
PKS 1417−19[2]	14 17 01.78	−19 15 00.0	17.5	0.1192	1.7	−0.70	<5
4 C 20.33	14 22 37.0	20 13 49	17.86	0.871	1.8	−0.69	8
TON 202	14 25 18.0	26 46 00	15.68	0.366	3.2	−0.72	4
OQ 172	14 42 50.48	10 11 12.4		3.53	2.42	−0.09	—
3 C 305.0[2]	14 48 17.13	63 28 37.0	13.74	0.0416	3.2	—	—
3 C 306.0[2]	14 52 01.39	16 34 36.0	14.9	0.0458	0.7	−0.90	360
MSH 14−121	14 53 12.25	−10 56 51.5	17.37	0.938	3.90	−0.82	33
PKS 1454−06	14 54 02.68	−06 05 45.0	18.00	1.249	1.2	−0.80	(4)
3 C 309.1	14 58 56.476	71 52 11.42	16.78	0.905[3]	8.5	−0.48	<2

Table 31 (continued)

Object[2]	α(1950.0) h m s	δ(1950.0) ° ′ ″	V (mag.)	z	S_{1400} (f.u.)	α	θ (arc sec)
OR 103	15 02 00.18	10 41 21.3		0.572	1.2	0.40	<3
3 C 310.0[2]	15 02 48.59	26 12 29.0	15.24	0.0543	7.8	−1.14	252
3 C 311.0	15 02 58.0	60 12 36	18.00	1.022	1.7	−0.47	—
PKS 1508−05	15 08 14.95	−05 31 49.1	17	1.191	3.9	−0.70	—
PKS 1510−08	15 10 08.915	−08 54 47.2	16.52	0.361[3]	3.95	—	<0.05
3 C 315.0[2]	15 11 30.79	26 18 26.0	16.80	0.1086	3.9	−0.92	180
4 C 37.43	15 12 47.48	37 02 30	15.50	0.370	0.9	−0.81	56
4 C 00.56[2]	15 14 06.19	00 26 01.0	16.5	0.053	2.5	−0.60	—
3 C 317.0[2]	15 14 16.97	07 12 16.6	13.50	0.0351	5.2	−1.23	<20
4 C −06.41[2]	15 20 11.8	−06 19 42	18.0	0.1280	(0.5)	—	—
3 C 323.1	15 45 31.30	21 01 38.5	16.69	0.264	2.51	−0.77	68.1
3 C 327.0[2]	15 59 56.7	02 06 20	15.88	0.1041	9.8	−0.83	210
4 C 17.66[2]	16 02 54.0	17 51 00.0	15.1	0.0320	0.5	—	—
PKS 1610−60.8[2]	16 10 40.8	−60 47 46	12.8	0.0284	45.4	—	552
DA 406	16 11 47.93	34 20 20.9	17.50	1.401	2.9	0.04	<2
TON 256	16 12 08.7	26 13 00	15.41	0.131			
3 C 332.0[2]	16 15 47.27	32 29 45.0	16.0	0.1520	2.6	−0.84	54
3 C 334.0	16 18 06.81	17 43 37.7	16.41	0.555	2.2	−0.77	41
3 C 336.0	16 22 32.47	23 52 06.5	17.47	0.927	2.7	−0.79	21.7
3 C 338.0[2]	16 26 55.29	39 39 31.0	12.63	0.0303	3.6	−1.51	41
PKS 1634+26	16 34 21.4	26 54 18	17.75	0.561	1.3	−0.84	20
PKS 1637−77[2]	16 37 02.9	−77 09 57	16.0	0.0438	5.3	−0.7	156
3 C 345.0	16 41 17.625	39 54 10.99	15.96	0.594	7.0	−0.34	<0.025
3 C 348.0[2] Her A	16 48 40.80	05 04 36.0	16.90	0.1533	43.0	−1.05	84
4 C 02.44[2]	16 50 56.7	02 42 12.0	14.0	0.0250	(0.6)	—	—
PKS 1655−77[2]	16 55 11.1	−77 37 42	17.0	0.0663	2.0	−0.5	48
4 C 29.50	17 02 10.9	29 50 00	19.14	1.927	1.5	−0.72	(4)
3 C 351.0	17 04 04.48	60 48 50.3	15.28	0.371	3.3	−0.73	59
3 C 353.0[2]	17 17 55.59	−00 55 53.0	15.36	0.0307	49.0	−0.69	150
3 C 357.0[2]	17 26 27.41	31 48 23.9	15.5	0.1670	2.5	−0.78	74
PKS 1801+01	18 01 44.0	01 01 18	19.00	1.522	1.3	−0.60	—
3 C 371.0[2]	18 07 18.49	69 48 57.55	14.2	0.0508	2.7	0.10	29
3 C 380.0	18 28 13.485	48 42 40.15	16.81	0.691	14.4	−0.76	2
3 C 381.0[2]	18 32 24.29	47 24 48.0	17.24	0.1614	3.9	−0.63	60
3 C 382.0[2]	18 33 12.59	32 38 58.0	14.66	0.0586	5.8	−0.77	150
PKS 1834+19[2]	18 34 31.2	19 39 12	14.0	0.0166	1.7	−0.6	—
3 C 388.0[2]	18 42 35.49	45 30 22.4	15.32	0.0917	5.7	−0.87	20
3 C 390.3[2]	18 45 37.5	79 43 06	13.8	0.0569	12.3	−0.57	204
PKS 1928−34[2]	19 28 24.1	−34 01 42	17.0	0.0981	0.9	−0.7	<36
PKS 1934−63[2]	19 34 49	−63 49 12	16.0	0.182	13.0	0.2	<18
3 C 402.0[2]	19 40 22.5	50 29 29	14.0	0.0247	3.1	−0.71	342
3 C 403.0[2]	19 49 43.88	02 22 40.00	15.42	0.059	5.8	−0.77	96
3 C 403.1[2]	19 49 55.2	−01 25 07.2	16.0	0.0559	1.5	−0.73	246
PKS 1954−55[2]	19 54 19.7	−55 17 40	16.3	0.0598	5.5	−0.5	60
3 C 405.0[2] Cyg A	19 57 44.48	40 35 46.0	15.14	0.0570	1,255.0	−0.80	96
3 C 407.0	20 05 46.0	−04 27 18	18.00	1.864	1.0	−0.96	—
PKS 2006−56[2]	20 06 23.9	−56 38 58.0	16.0	0.0585	(3.0)	—	600

Table 31 (continued)

Object[2]	α(1950.0) h m s	δ(1950.0) ° ′ ″	V (mag.)	z	S_{1400} (f.u.)	α	θ (arc sec)
PKS 2040−26[2]	20 40 44.18	−26 43 50.0	13.5	0.0406	1.8	−0.50	180
PKS 2048−57[2]	20 48 06.0	−57 15 15.0	13.0	0.0110	2.1	−0.80	<30
PKS 2058−28[2]	20 58 40.60	−28 13 56.0	14.8	0.0394	1.7	−0.80	168
PKS 2058−13[2]	20 58 56.7	−13 30 38.0	15.2	0.0296	1.3	−0.70	—
PKS 2115−30	21 15 11.42	−30 31 55.3	16.47	0.980	2.4	−0.70	<11
3C 430.0[2]	21 17 01.88	60 35 34	15.0	0.0549	7.56	−0.78	57
3C 432.0	21 20 25.51	16 51 45.6	17.96	1.805	1.5	−1.08	13
3C 433.0[2]	21 21 30.57	24 51 17.9	16.24	0.1025	11.9	−0.90	52
PKS 2128−12	21 28 52.69	−12 20 19.7	15.99	0.501	1.8	0.20	<6
PKS 2130−53[2]	21 30 47.8	−53 49 57.0	14.0	0.0763	1.3	−0.80	162
PKS 2134+004	21 34 05.212	00 28 25.7	18.00	1.930	10.0	—	0.0015
PKS 2135−14	21 35 00.8	−14 46 27	15.53	0.200	3.0	−0.80	150
3C 436.0[2]	21 41 58.28	27 56 16.0	18.18	0.2154	3.3	−0.87	65
PKS 2144−17	21 44 17.7	−17 54 05	19.50	0.684	1.4	−1.20	—
PKS 2145+06	21 45 36.07	06 43 41.25	16.47	0.367	3.0	0.60	<3
PKS 2146−13	21 46 45.87	−13 18 11	20.00	1.800[3]	1.5	−0.90	(6)
PKS 2152−69[2]	21 53 01.6	−69 55 46	13.8	0.0266	25.9	−0.7	228
PKS 2153−204	21 53 48.0	−20 26 30	17.50	1.310	—	—	<7
3C 442.0[2]	22 12 20.40	13 35 40.0	13.66	0.0262	3.4	−1.05	270
PKS 2216−03	22 16 16.3	−03 50 41	16.38	0.901	0.9	0.20	<6
3C 445.0[2]	22 21 15.58	−02 21 54.0	17.0	0.0568	5.1	−0.86	276
3C 446.0	22 23 11.04	−05 12 17.2	18.39	1.404	5.85	−0.61	0.4
PHL 5200	22 25 50.58	−05 34 00	18.00	1.981[3]			
3C 449.0[2]	22 29 07.37	39 06 14.0	13.20	0.0181	3.7	−0.60	300
CTA 102	22 30 07.78	11 28 22.8	17.33	1.037	6.78	−0.60	<0.025
3C 452.0[2]	22 43 32.97	39 25 28.0	16.0	0.0820	10.7	−0.94	360
PKS 2247+11[2]	22 47 20.80	11 20 19.0	14.4	0.0268	2.5	−0.80	138
3C 454.0	22 49 07.65	18 32 43.7	18.40	1.757	2.1	−0.80	<3
3C 454.3	22 51 29.524	15 52 54.36	16.10	0.859	10.8	−0.16	<0.02
PKS 2251+11	22 51 40.6	11 20 39	15.82	0.323	1.6	−0.80	8
3C 455.0[2]	22 52 34.48	12 57 33.3	14.00	0.0331	3.0	−0.82	<7
PKS 2254+024	22 54 43.9	02 27 32	18.00	2.090	—	—	<6
PKS 2300−18[2]	23 00 22.80	−18 57 49.0	18.3	0.129	1.7	−0.60	—
PKS 2302−713	23 02 23.0	−71 19 12	17.5	0.384			
4C 07.61[2]	23 08 00.0	07 15 00.0	13.9	0.0448	1.6	—	—
3C 456.0[2]	23 09 56.60	09 03 10.2	18.54	0.2337	2.4	−0.77	<2
3C 459.0[2]	23 14 02.27	03 48 55.2	18.0	0.2205	4.3	−1.02	12
PKS 2318+07[2]	23 18 09.0	07 57 12	12.8	0.0119	0.7	−0.60	—
MSH 23−112[2]	23 22 42	−12 23 54	15.46	0.0825	1.9	−1.50	<15
CTD 141	23 25 43.0	29 20 00	17.3	1.015	1.1	−0.60	50
PKS 2326−477	23 26 32.7	−47 46 51	16	1.299	—	—	—
PKS 2329−384	23 29 17.9	−38 28 22	17.0	1.195			
OZ −252	23 31 18.0	−24 00 14	17.00	—			
3C 465.0[2]	23 35 56.57	26 44 20.0	13.29	0.0301	7.7	−0.96	540
PKS 2344+09	23 44 03.82	09 14 05.7	15.97	0.677	2.1	−0.50	<6
PKS 2345−16	23 45 27.63	−16 47 51.9	18.00	0.600	1.2	−0.20	—
PKS 2349−01[2]	23 49 22.50	−01 26 04.0	17.5	0.174	1.6	−0.70	—

Table 31 (continued)

Object[2]	α(1950.0) h m s	δ(1950.0) ° ′ ″	V (mag.)	z	S_{1400} (f.u.)	α	θ (arc sec)
PKS 2352−455	23 52 53.7	−45 30 08	17.5	1.868			
PKS 2353−68	23 53 31	−68 37 03	17	1.716	1.2	−0.6	—
PKS 2354−35[2]	23 54 26.2	−35 01 19	14.4	0.0487	1.0	−1.2	48
PKS 2354+14	23 54 44.7	14 29 26	18.18	1.810	1.4	−1.10	<6
PKS 2356−61[2]	23 56 24.1	−61 11 40	16.0	0.0959	19.2	−0.8	240
PKS 2357+00[2]	23 57 22.0	00 26 18	16.0	0.0844	0.5	—	—

[1] After Burbidge and Burbidge (1969), Burbidge (1970), Burbidge and Burbidge (1972), Burbidge and Strittmatter (1972), Cohen and Shaffer (1971), Ekers (1969), Fomalont and Moffet (1971), Miley (1971), Moffet (1974), Peterson and Bolton (1972), Wyndham (1966), and de Veny, Osborn, and Janes (1971).

[2] Radio galaxies are denoted by a superscript 2 after the object name. Positions are radio positions for compact sources where available and are usually accurate to ten times the last decimal quoted. Angular sizes are the largest angular size observed. Redshifts for an additional 51 radio galaxies in the 4 C catalogue are given by Sargent (1973).

[3] Redshifts marked with a superscript 3 have one or more absorption features in the spectrum.

where the solar mass $M_\odot \approx 2 \times 10^{33}$ grams. The more exact expression is given by (Einstein, 1911, 1916)

$$z_g = \frac{\nu_L - \nu_{obs}}{\nu_{obs}} = \left(1 - \frac{2GM}{Rc^2}\right)^{-1/2} - 1. \tag{2-235}$$

For the Sun, $z_g \approx 2 \times 10^{-6}$ which is a value which would be given by the Doppler effect due to a recessional velocity of about 0.64 km sec^{-1}, and as such is very difficult to measure. Nevertheless, such a measurement has been reported (Blamont and Roddier, 1961). White dwarf stars which are members of a binary system provide lines with large gravitational redshifts which can be compared with the unreddened lines of the less dense companion. Measurements of white dwarf spectra indicate a redshift equivalent to 19 km sec^{-1} for Sirius B (Adams, 1925) and $21 \pm 4 \text{ km sec}^{-1}$ for 40 Eridani B (Popper, 1954). More recent measurements of the gravitational redshift of these two objects give respective values of $89 \pm 16 \text{ km sec}^{-1}$ for Sirius B (Greenstein, Oke, and Shipman, 1971) and $23 \pm 5 \text{ km sec}^{-1}$ for 40 Eri B (Greenstein and Trimble, 1972). Trimble and Thorne (1969) have prepared a list of single-line spectroscopic binaries which may have a neutron star or collapsed star as the secondary companion.

Refraction effects. Although the frequency of a wave remains unchanged as it propagates through most media, its wavelength, λ, does not. When measuring a wavelength, λ_{air}, in air it must be compared with laboratory measurements in vacuum, λ_{vac}

$$\lambda_{air} = \frac{\lambda_{vac}}{n_{air}}, \tag{2-236}$$

where the index of refraction of air at optical frequencies (2000—7000 Å) is given by (EDLÉN, 1953)

$$n_{\text{air}} = 1 + 6432.8 \times 10^{-8} + \frac{2{,}949{,}810}{146 \times 10^8 - \tilde{\nu}^2} + \frac{25{,}540}{41 \times 10^8 - \tilde{\nu}^2} \tag{2-237}$$

and the wave number, $\tilde{\nu}$, is given by

$$\tilde{\nu} = \frac{\nu}{c} = \frac{1}{\lambda_{\text{vac}}}. \tag{2-238}$$

The measured optical redshift, z_{opt}, is usually calculated from the expression

$$z_{\text{opt}} = \frac{\lambda_{\text{air}} - \lambda_{\text{vac}}}{\lambda_{\text{vac}}}. \tag{2-239}$$

The true redshift, z, is given by the relation

$$1 + z = (1 + z_{\text{opt}})\, n_{\text{air}}, \tag{2-240}$$

where n_{air} is the index of refraction of air corresponding to λ_{air}, here λ_{air} is the observed wavelength, and the index of refraction, $n_{\text{air}} \approx 1.000297$.

2.18. Doppler Broadening (a Gaussian Profile)

When line broadening is caused by the motion of atoms or molecules in thermodynamic equilibrium, or by the random turbulent motion of the gas, Maxwell's distribution of velocities (MAXWELL, 1860, Eq. (3-114)) may be used together with the Doppler formula which relates frequency shift to radial velocity (DOPPLER, 1842, Eq. (2-226)). The resulting spectral line intensity distribution is given by (RAYLEIGH, 1889)

$$\varphi_{mn}(\nu) = \frac{1}{\sqrt{2\pi}\,\sigma} \exp\left(-\frac{(\nu - \nu_{mn})^2}{2\sigma^2}\right), \tag{2-241}$$

where ν is the frequency under consideration, ν_{mn} is the frequency of the $m-n$ transition, and

$$2\sigma^2 = \frac{\nu_{mn}^2}{c^2}\left(\frac{2kT_{\text{k}}}{M} + V^2\right), \tag{2-242}$$

where M is the atomic or molecular mass, T_{k} is the kinetic temperature of the moving atoms or molecules, and V is the most probable turbulent velocity. The turbulent velocities are assumed to have a Maxwellian distribution so that the r.m.s. turbulent velocity, $V_{\text{rms}} = (3/2)^{1/2}\, V$, and the mean turbulent velocity $\langle V \rangle = 2V/\pi^{1/2}$. The profile has a full width to half maximum given by

$$\Delta\nu_{\text{D}} = \frac{2\nu_{mn}}{c}\left[\ln 2\left(\frac{2kT_{\text{k}}}{M} + V^2\right)\right]^{1/2} = 2.3556\,\sigma \tag{2-243}$$

and the peak value of $2(\ln 2/\pi)^{1/2}(\Delta\nu_{\text{D}})^{-1}$. As shown in Sect. 2.15, the observed line may be narrower then $\Delta\nu_{\text{D}}$ if the radiator is a molecular maser.

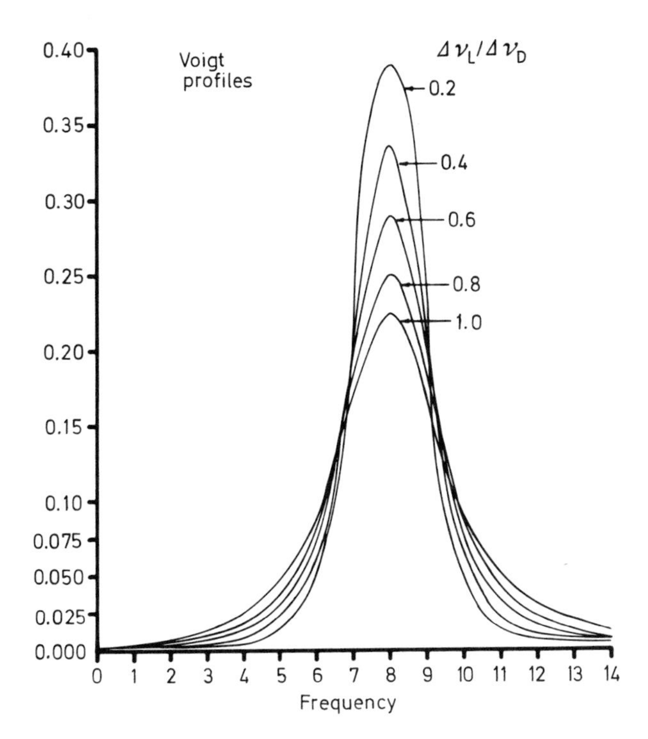

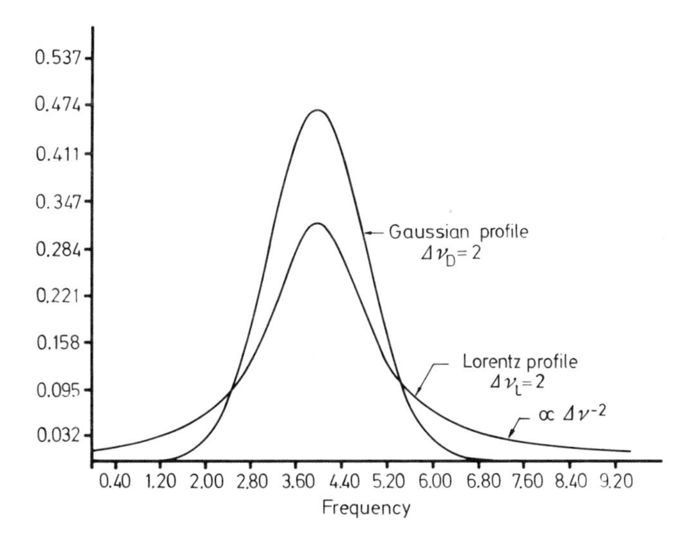

Fig. 16. The Doppler line profile with full width to half maximum, $\Delta\nu_D$, the Lorentz line profile with full width to half maximum, $\Delta\nu_L$, and their combined Voigt profile for various values of $\Delta\nu_L/\Delta\nu_D$

The effective Doppler temperature, T_D, is defined as

$$T_D = \frac{1}{8\ln 2}\frac{Mc^2}{k\nu_{mn}^2}(\Delta\nu_D)^2 \approx 1.17\times10^{36}\,M(\Delta\nu_D/\nu_{mn})^2\ ^\circ\mathrm{K}. \tag{2-244}$$

When a substantial fraction of the radiation of an atom or molecule is scattered by free electrons, the approximate half width of the observed line will be given by Eq. (2-243), where the mass, M, is the electron mass rather than the mass of the radiating atoms or molecules. A measure of the fraction of radiation which is scattered is the optical depth, τ, given by

$$\tau = \int_0^L \sigma_T N_e\, dl, \tag{2-245}$$

where the integral is a line of sight integral over the extent, L, of the radiating source, N_e is the density of free electrons, and the Thomson scattering cross section (Thomson, 1903) is given by

$$\sigma_T = \frac{8\pi}{3}\left(\frac{e^2}{mc^2}\right)^2 \approx 6.65\times10^{-25}\ \mathrm{cm}^2. \tag{2-246}$$

For optical depths greater than 0.5 almost all of the line radiation is in the wider scattered line as opposed to the narrower radiated line, whereas the fraction of the radiation energy in the wider line is about $\tau/(1-\tau)$ for optical depths less than 0.5. If we assume that the electrons have a Maxwellian velocity distribution, the Compton scattering probability $\sigma(\theta,\nu_1,\nu_2)$ that a photon is scattered by an angle θ, and that its frequency is shifted from ν_1 to ν_2 in this process is (Dirac, 1925)

$$\sigma(\theta,\nu_1,\nu_2) = \frac{3}{4}\sigma_T\frac{(1+\cos^2\theta)}{(1-\cos\theta)^{1/2}}\frac{1}{2\pi\Delta\nu_D}\exp\left[\frac{-(\nu_2-\nu_1)^2}{\Delta\nu_D^2(1-\cos\theta)}\right], \tag{2-247}$$

where $\Delta\nu_D$ is given by Eq. (2-243) when the mass, M, is the electron mass. A more complicated formula for the $\sigma(\theta,\nu_1,\nu_2)$ of Compton scattering by moving atoms is given by Henyey (1940) and Böhm (1960). The Doppler broadened line profile is illustrated in Fig. 16 together with the Lorentz and Voigts profiles.

2.19. Broadening due to Rotating or Expanding Sources

When a spherical source is rotating with an equatorial velocity, V_R, then the Doppler equation (2-226) gives a rotational line broadening of

$$\Delta\nu_R = \frac{\nu_{mn} V_R \sin i}{c}, \tag{2-248}$$

where ν_{mn} is the frequency of the $m-n$ transition and i is the inclination of the equator to the celestial plane. The quantity $V_R \sin i$ is called the projected linear equatorial velocity or the observed rotational velocity of the line. The determination of the axial rotation of stars from spectral line widths was first suggested by ABNEY (1877) and observed by SCHLESINGER (1909). Theoretical studies of rotational line broadening have been made by SHAPLEY and NICHOLSON (1919), SHAJN and STRUVE (1929), and STRUVE (1930), and are summarized by HUANG and STRUVE (1955, 1960). Early type stars, spectral types *B*, *A*, and *F*, have rotational velocities ranging from 200 to 350 km sec^{-1}, whereas the velocities drop rapidly after spectral type F5 (cf. SLETTEBAK, 1949, 1954, 1955; SLETTEBAK and HOWARD, 1955; ABT and HUNTER, 1962; KRAFT, 1965, 1970; SLETTEBAK, 1970, and Fig. 17).

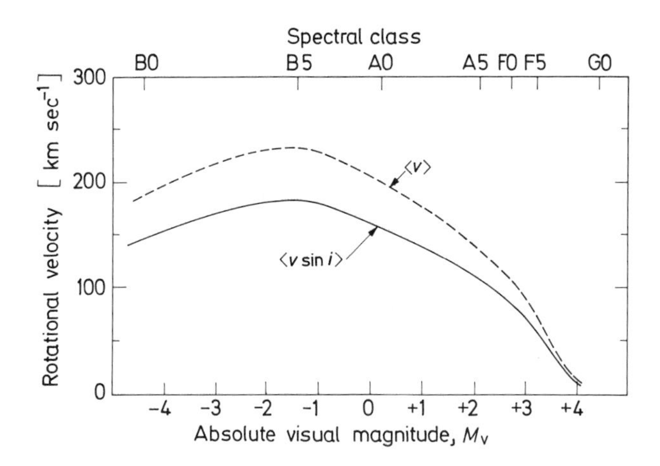

Fig. 17. The mean projected rotational velocity, $\langle V \sin i \rangle$, and the mean rotational velocity, $\langle V \rangle$, for stars as a function of spectral class and absolute visual magnitude (after ABT and HUNTER, 1962, by permission of the American Astronomical Society and the University of Chicago Press). The curve for $\langle V \rangle$ is derived from that for $\langle V \sin i \rangle$ on the assumption of a random orientation of rotational axis. The number of stars incorporated in the means for each spectral class was between fifteen and sixty-five

A radiating shell of gas which is expanding at a velocity, V_E, will give rise to a double line profile with peaks at

$$v = v_{mn}\left[1 \pm \frac{V_E}{c}\right], \tag{2-249}$$

or an asymmetrical line profile centered at

$$v = v_{mn}\left[1 + \frac{V_E}{c}\right], \tag{2-250}$$

according as the radiation from the back part of the shell is or is not allowed to pass through the part of the shell nearer the observer. Expansion velocities between 20 and 50 km sec^{-1} have been measured using this effect when observing planetary nebulae (WILSON, 1950, 1958).

2.20. Collision Broadening (Stark or Pressure Broadening)

2.20.1. Ion Broadening—The Quasi-Static Approximation

Ion broadening is calculated by assuming that the slow moving ions have a "quasi-static" electric field distribution, and then calculating the Stark splitting of the line considered. If each ion has an electric field distribution which is independent of the other ions and electrons in the plasma, the probability distribution of the ionic electric field is easily calculated. Each ion will have a Coulomb electric field strength given by $E = Ze/r^2$, where r is the distance from the ion of charge Ze. The probability, $P(r)$, that at least one ion will fall into a shell r to $r+dr$ from a radiating atom is given by

$$P(r)\,dr = \frac{3r^2}{r_0^3}\exp\left[-\left(\frac{r}{r_0}\right)^3\right]dr, \tag{2-251}$$

where the mean ion separation is

$$r_0 = [4\pi N_i/3]^{-1/3}, \tag{2-252}$$

and N_i is the density of ions. Thus, the probability, $P(E)$, of an emitting atom being subjected to the electric field of the nearest ion is (HOLTSMARK, 1919)

$$P(E) = W_H\left(\frac{E}{E_0}\right)d\left(\frac{E}{E_0}\right) = \frac{3}{2}\left(\frac{E}{E_0}\right)^{-5/2}\exp\left[-\left(\frac{E}{E_0}\right)^{-3/2}\right]d\left(\frac{E}{E_0}\right), \tag{2-253}$$

where $E = Ze/r^2$ and the mean electric field strength $E_0 = Ze/r_0^2 = 2.61\,eZN_i^{2/3}$. The $W_H(E/E_0)$ is called the Holtsmark ion field strength distribution for $E \gg E_0$ (HOLTSMARK, 1919). When the electric field, E, is written as the vector sum of the fields of many different ions rather than that of one ion, the Holtsmark distribution for the linear Stark effect is given by (HOLTSMARK, 1919; CHANDRASEKHAR, 1943)

$$\begin{aligned}
W_H(\beta) &= \frac{2}{\pi\beta}\int_0^\infty v\sin v\exp\left[-\left(\frac{v}{\beta}\right)^{3/2}\right]dv \\
&= \frac{4}{3\pi}\sum_{n=0}^\infty(-1)^n\Gamma\left(\frac{4n+6}{3}\right)\frac{\beta^{2n+2}}{(2n+1)!} \approx \frac{4\beta^2}{3\pi}(1-0.463\beta^2+0.1227\beta^4 \pm\cdots) \\
&\qquad \text{for } \beta \ll 1 \qquad\qquad (2\text{-}254) \\
&= \frac{2}{\pi}\sum_{n=1}^\infty(-1)^{n+1}\frac{\Gamma\left(\frac{3n+4}{2}\right)\sin\left(\frac{3\pi n}{4}\right)}{n!\,\beta^{(3n+2)/2}} \approx \frac{1.496}{\beta^{5/2}}\left(1+\frac{5.107}{\beta^{3/2}}+\frac{14.43}{\beta^3}+\cdots\right) \\
&\qquad \text{for } \beta \gg 1,
\end{aligned}$$

where $\beta = E/E_0$. A plot of $W_H(\beta)$ is shown in Fig. 18. For very large β, the $W_H(\beta) \to 1.5\,\beta^{-5/2}$ just as in Eq. (2-253).

The probability distribution of the ionic field will deviate from the Holtsmark distribution of Eq. (2-254) when ion correlations are important (ECKER and

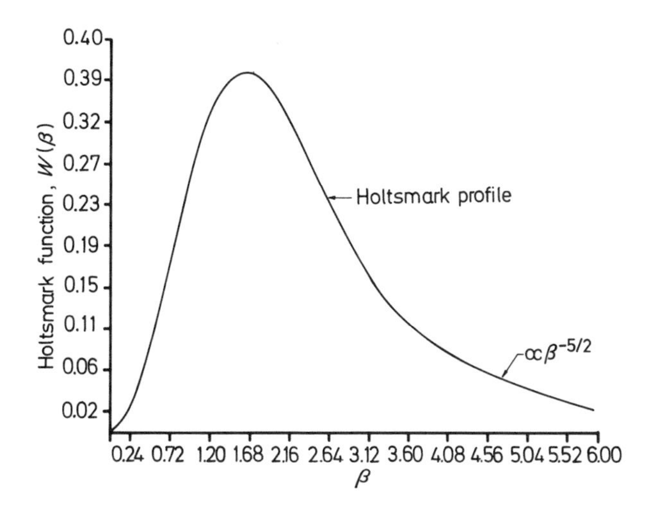

Fig. 18. Holtsmark distribution given by Eq. (2-254)

MÜLLER, 1958; MOZER and BARANGER, 1960; HOOPER, 1968), or when the radiating particle is an ion (LEWIS and MARGENAU, 1958). Ion correlations become important when r_0 is not much less than the Debye radius $r_D = 6.90(T/N_i)^{1/2}$ cm, where T is the kinetic temperature of the plasma, and N_i is the ion density. The general result is that $W_H(\beta)$ becomes more skewed towards lower β as r_0 approaches r_D. When the radiating particle is an ion, the wings of $W_H(\beta)$ are damped by the factor $\exp[-\frac{1}{3}(r_0/r_D)^2 \beta^{1/2}]$.

For hydrogen-like atoms the frequency displacement, $\Delta \nu$, caused by an electric field of strength, E, is a linear function of E (a relationship first observed by STARK, 1913). For other atoms it is proportional to E^2 (the quadratic Stark effect) for weak fields and to E for strong fields. Using Eq. (2-253), the spectral line intensity distribution for each Stark component is given by

$$\varphi(\Delta \nu) = \frac{1}{\Delta \nu_0} W_H\left(\frac{\Delta \nu}{\Delta \nu_0}\right) \quad \text{for the linear Stark effect} \tag{2-255}$$

and

$$\varphi(\Delta \nu) = \tfrac{1}{2} (\Delta \nu_0 \Delta \nu)^{-1/2} W_H[(\Delta \nu/\Delta \nu_0)^{1/2}] \quad \text{for the quadratic Stark effect,} \tag{2-256}$$

where $\Delta \nu_0$ is the Stark displacement for the mean electric field, E_0.

For hydrogen like atoms, the frequency, ν_{mn}, of a transition from an upper level m to a lower level n is split into the frequencies $\nu_{m,n}$ given by (SCHWARZSCHILD, 1916; EPSTEIN, 1916; BETHE and SALPETER, 1957)

$$\begin{aligned} \nu_{m,n} &= \nu_{mn} + \frac{3h}{8\pi^2 m_e e Z} E[m(m_2 - m_1) - n(n_2 - n_1)] \\ &\approx \nu_{mn} + 1.92 \times 10^6 \frac{E}{Z} [m(m_2 - m_1) - n(n_2 - n_1)] \text{ Hz}, \end{aligned} \tag{2-257}$$

where E is in units of volt cm^{-1}, and the frequency of the transition in the absence of the electric field is

$$\nu_{mn} = \frac{2\pi^2 m_e e^4 Z^2}{h^3}(n^{-2} - m^{-2}) \approx 3.288 \times 10^{15} Z^2 (n^{-2} - m^{-2}) \text{ Hz}, \quad (2\text{-}258)$$

m_2 and m_1 may take on integral values between 0 and $m-1$, and n_2 and n_1 may take on integral values between 0 and $n-1$. The frequency shift, $\Delta\nu_{mn}$, and the corresponding wavelength shift, $\Delta\lambda_{mn}$, are given by (STRUVE, 1929)

$$\Delta\nu_{mn} = 1.92 \times 10^6 \frac{E}{Z} X_{mn} \text{ Hz}, \quad (2\text{-}259)$$

and

$$\Delta\lambda_{mn} = E\, C_{mn} \text{ cm},$$

where

$$X_{mn} = [m(m_2 - m_1) - n(n_2 - n_1)]$$

and

$$C_{mn} = \frac{5.3 \times 10^{-15}}{Z^5}(n^{-2} - m^{-2})^{-2} X_{mn} \text{ cm}^2 \text{ volt}^{-1} = 6.4 \times 10^{-5} \frac{\lambda_{mn}^2}{Z} X_{mn} \text{ cm}^2 \text{ volt}^{-1},$$

and E is in units of volt cm^{-1}.

Tables of X_{mn} and C_{mn} are given for optical hydrogen lines by UNDERHILL and WADDELL (1959). The total line broadening will be a superposition of all of the Stark components given by Eq. (2-259) and appropriately weighted for the strength of the component. The mean values $\langle X \rangle$ of the weighted X_{mn} for the optical hydrogen lines are

Line	Lα	Lβ	Hα	Hβ	Hγ	Hδ	lim $m \gg n$
$\langle X \rangle$	2.0	4.0	2.24	5.96	11.8	15.9	$\frac{1}{2}m(m-1)$

Tables of the Stark broadening of the first four Lyman lines and the first four Balmer lines of hydrogen are given by VIDAL, COOPER, and SMITH (1973).

Using the mean electric field $E_0 = 2.6\, e Z N_i^{2/3}$, where N_i is the ion density, together with Eqs. (2-254), (2-255), and (2-259), we obtain the wing spectral line distribution function

$$\varphi(\Delta\nu) \approx N_i (\Delta\nu)^{-5/2} \langle X \rangle^{3/2}, \quad (2\text{-}260)$$

where $\langle X \rangle$ is the mean value of X_{mn}.

For a given series of lines (a fixed n), the frequency difference, $\Delta\nu_L$, between adjacent lines of frequencies ν_{m+1} and ν_m is, using Eq. (2-258),

$$\Delta\nu_L = \frac{4\pi^2 m_e e^4 Z^2}{h^3 m^3} \approx \frac{6.6 \times 10^{15}}{m^3} Z^2 \text{ Hz}. \quad (2\text{-}261)$$

The maximum Stark displacement, $\Delta\nu_{max}$, of a line with quantum number, m, is from Eq. (2-259)

$$\Delta\nu_{max} = \frac{3hm^2 E}{8\pi^2 m_e e Z} \approx 1.92 \times 10^6 \frac{E m^2}{Z} \text{ Hz}. \tag{2-262}$$

Assuming that E has its mean value $E_0 = 2.6 e Z N_i^{2/3}$, where N_i is the ion density; and equating $\Delta\nu_{max}$ to $\Delta\nu_L/2$, we obtain the limiting value of the quantum number, m, for which the lines of a series are observed to merge (INGLIS and TELLER, 1939)

$$N_i^{2/3} = \frac{16\pi^4 m_e^2 e^4 Z^2}{7.8 h^4 m^5} = 4.58 \times 10^{15} Z^2 m^{-5} \text{ cm}^{-2} \tag{2-263}$$

or

$$\log N_i \approx 23 - 7.5 \log m.$$

2.20.2. Electron Broadening—The Impact Approximation

In the impact approximation, a radiating atom is assumed to act as an unperturbed harmonic oscillator until it undergoes a collision with a perturbing particle. The effect of such a collision will be to change the phase and possibly the amplitude of the harmonic oscillation. When the time, t, is long enough to include collisions, the dipole moment of the harmonic oscillator is given by

$$d(t) = e x(t) = e A(t) \exp[i\omega_0 t + i\eta(t)], \tag{2-264}$$

where e is the charge of the electron, $x(t)$ denotes the linear displacement of the oscillator, $A(t)$ denotes the time dependent amplitude of the oscillation, ω_0 denotes the unperturbed frequency of the oscillator, and $\eta(t)$ denotes the phase shift induced by collision. The average total energy emitted per unit time by an oscillator in all directions is (LARMOR, 1897)

$$\langle I \rangle = \frac{2}{3c^3} \langle \ddot{d}(t)^2 \rangle \propto \langle |x(t)|^2 \rangle, \tag{2-265}$$

where $\langle\ \rangle$ denotes a time average, $\ddot{}$ denotes the second differential with respect to time, $|\ |$ denotes the absolute value, and $\propto$ means proportional to. We may express this average intensity in terms of the Fourier transform, $F(\omega)$, of $x(t)$ by using Rayleigh's theorem (RAYLEIGH, 1889)

$$\langle I(\omega) \rangle \propto \langle |x(t)|^2 \rangle = \langle |F(\omega)|^2 \rangle, \tag{2-266}$$

where the Fourier transform, $F(\omega)$, is given by

$$F(\omega) = \int_{-\infty}^{+\infty} x(t) \exp(-i\omega t) dt, \tag{2-267}$$

and $\langle\ \rangle$ denotes an average over time, t, or frequency, ω.

It is also convenient to express the average intensity in terms of the autocorrelation function, $\varphi(s)$, of $x(t)$ by using the Wiener-Khintchine theorem, (WIENER, 1930; KHINTCHINE, 1934)

$$I(\omega) \propto \langle |F(\omega)|^2 \rangle = \int_{-\infty}^{+\infty} \langle \varphi(s) \rangle \exp(-i\omega s) ds, \tag{2-268}$$

where the time averaged autocorrelation function, $\langle\varphi(s)\rangle$, is given by

$$\langle\varphi(s)\rangle = \left\langle \int_{-\infty}^{+\infty} x^*(t)\,x(t+s)\,dt \right\rangle, \tag{2-269}$$

where $x^*(t)$ denotes the complex conjugate of $x(t)$ and $\langle\ \rangle$ denotes a time average.

Let us suppose that one collision between an atom and a perturbing particle occurs at a time interval, T. Using Eqs. (2-264) and (2-267)

$$|F(\omega)|^2 = \frac{\sin^2[(\omega_0-\omega)T/2]}{[(\omega_0-\omega)/2]^2} = \left| \int_0^T \exp[i(\omega_0-\omega)t]\,dt \right|^2 . \tag{2-270}$$

If the mean time between collisions is τ, and if the collisions are independent and random, the probability that a collision occurs between time T and $T+dT$ is given by the Poisson distribution

$$P(T)\,dT = \frac{1}{\tau}\exp\left(-\frac{T}{\tau}\right)dT . \tag{2-271}$$

Eqs. (2-266), (2-270), and (2-271) then lead to the Lorentz spectral line intensity distribution (LORENTZ, 1906, 1909)

$$\varphi(\nu) = \frac{\Delta\nu_L/2\pi}{(\nu-\nu_0)^2+(\Delta\nu_L/2)^2} \propto \int_0^\infty |F(\omega)|^2\,P(T)\,dT, \tag{2-272}$$

where $\Delta\nu_L = (\pi\tau)^{-1} = \Gamma_{coll}/(2\pi)$ and we have normalized $\varphi(\nu)$ so that $\int_{-\infty}^{+\infty}\varphi(\nu)\,d\nu = 1$.

The mean collision time, τ, can be related to the mean density, N, of the perturbing particles by

$$\pi\Delta\nu_L = \pi b^2 v_0 N = (\tau)^{-1}, \tag{2-273}$$

where πb^2 is the effective collision cross section, b is the effective impact parameter, and the mean relative speed, v_0, between an atom of mass, m_1, and another of mass, m_2, is

$$v_0 = \left[\frac{8kT}{\pi}\left(\frac{1}{m_1}+\frac{1}{m_2}\right)\right]^{1/2}, \tag{2-274}$$

where T is the kinetic temperature of the gas.

WEISSKOPF (1933) related the critical impact parameter, b, of Eq. (2-273) to a measurable constant by noting that the frequency shift, $\Delta\nu$, of a line caused by a perturber at a distance, r, may be written

$$2\pi\Delta\nu = C_p r^{-p}, \tag{2-275}$$

where C_p is a measurable constant, and $p=2, 3, 4$, and 6, respectively, for the linear Stark effect, resonance broadening, the quadratic Stark effect, and the van der Waals interaction. The total phase shift, $\eta(b)$ induced by a perturbation in which the perturbing particle follows a straight line trajectory is

$$\eta(b) = C_p \int_{-\infty}^{+\infty} \frac{d\tau}{[r(\tau)]^p} = C_p \int_{-\infty}^{+\infty} \frac{d\tau}{(b^2 + v_0^2 \tau^2)^{p/2}} = \frac{\pi^{1/2} \Gamma[(p-1)/2] C_p}{\Gamma(p/2) v_0 b^{p-1}} = \frac{\psi_p C_p}{v_0 b^{p-1}}, \quad (2\text{-}276)$$

where the impact parameter, b, is the perpendicular distance between the radiating atom and straight line trajectory of the perturbing particle, and $\psi_p = \pi$, 2, $\pi/2$, and $3\pi/8$, respectively, for $p = 2, 3, 4$, and 6. The critical value of b to be used in Eq. (2-273) for the line width is the Weisskopf radius

$$b_{\mathrm{w}} = \left(\frac{C_p \psi_p}{\eta_0 v_0}\right)^{1/(p-1)}, \quad (2\text{-}277)$$

where η_0 is the minimum value of phase shift which contributes to line broadening. The constant η_0 was arbitrarily chosen to be unity by WEISSKOPF. This impact approximation holds as long as $b_{\mathrm{w}}/v_0 \ll \Delta \nu_{\mathrm{L}}$, and the frequency difference, $\Delta \nu$, satisfies the inequality (SPITZER, 1940)

$$\Delta \nu \ll \frac{v_0^{p/(p-1)}}{C_p^{1/(p-1)}}. \quad (2\text{-}278)$$

HOLSTEIN (1950) has shown that the static theory discussed in the previous section is valid when the inequality is reversed.

LINDHOLM (1941) and FOLEY (1946) showed that the impact approximation will generally lead to a line shift as well as a line broadening. They assumed that the amplitude, $A(t)$, of the harmonic oscillator is constant (an adiabatic theory) and noted that the number of collisions per unit time with oscillators whose impact parameters lie between b and $b + db$ is

$$N v_0 2\pi b\, db, \quad (2\text{-}279)$$

where N is the oscillator number density. Using Eqs. (2-264) and (2-279) in (2-269), the autocorrelation function, $\varphi(s)$, is obtained

$$\varphi(s) = \exp\{-N v_0(\sigma_{\mathrm{r}} - i\sigma_{\mathrm{i}})s\} \exp(-2\pi \nu_{mn} s), \quad (2\text{-}280)$$

where σ_{r} and σ_{i} are the real and imaginary parts of the integral

$$2\pi \int_0^{\infty} \{\exp[i\eta(b)] - 1\} b\, db. \quad (2\text{-}281)$$

Thus using Eq. (2-268), the spectral distribution of the line intensity is

$$\varphi_{mn}(\nu) = \frac{\Delta \nu_{\mathrm{L}}/2\pi}{[\nu - \nu_{mn} - \beta/(2\pi)]^2 + (\Delta \nu_{\mathrm{L}}/2)^2}, \quad (2\text{-}282)$$

where $\beta = N v_0 \sigma_{\mathrm{i}}$ and $\Delta \nu_{\mathrm{L}} = \Gamma/(2\pi) = N v_0 \sigma_{\mathrm{r}}/\pi$. Assuming that $\eta(t)$ is given by Eq. (2-276), we obtain for the linear Stark effect ($p = 2$)

$$\Gamma = 2\pi \Delta \nu_{\mathrm{L}} = \frac{2N\pi^3 C_2^2}{v_0} \left\{0.923 - \ln\left(\frac{\pi C_2}{v_0 b_{\max}}\right) + \frac{\pi^2 C_2^2}{24 v_0^2 b_{\max}^2} + \cdots\right\}, \quad (2\text{-}283)$$

where the maximum impact parameter, b_{max}, is often taken to be the Debye radius, and there is no line shift. For other values of p, we have:

p	Γ (sec^{-1})	β (radians)
3	$2\pi^2 C_3 N$	
4	$11.37\, C_4^{2/3} v_0^{1/3} N$	$9.85\, C_4^{2/3} v_0^{1/3} N$
6	$8.08\, C_6^{2/5} v_0^{3/5} N$	$2.94\, C_6^{2/5} v_0^{3/5} N$

(2-284)

The more exact theory of electron broadening takes into account the effects of inelastic collisions of the electrons (ANDERSON, 1949; KOLB and GRIEM, 1958; BARANGER, 1958; MARGENAU and LEWIS, 1959; BARANGER, 1962; GRIEM, 1964). This more exact theory is an impact theory in that it assumes the collisions are weak—that it takes many collision times for a light train to lose memory of phase. The interaction between perturbing particles and radiating atoms is, however, specified by a quantum mechanical formalism, and inelastic collisions are taken into account. The general conclusions of this non-adiabatic theory are that the $\Delta \nu_L$ given by Eq. (2-284) may be underestimated by a factor of two or three, and that $\Delta \nu_L / \beta$ is generally smaller than the value given. Detailed calculations have been made for hydrogen-like atoms and are given in the next section.

2.20.3. Wing Formulae for Collisional Broadening of Line Radiation from Hydrogen-like Atoms

At optical frequencies, ion broadening of hydrogen lines by the linear Stark effect is important. For a gas with ion density, N_i, the mean electric field strength is $E_0 \approx eZ(4\pi N_i/3)^{2/3}$, where eZ is the charge of the ion. Using this mean field strength, the wing spectral line distribution is, from Eqs. (2-255) and (2-259), the Holtsmark distribution

$$\varphi(\Delta\lambda) = E_0^{3/2} \langle C \rangle \Delta\lambda^{-5/2} \approx 4.4 \times 10^{-14} N_i \langle C \rangle \Delta\lambda^{-5/2}, \qquad (2\text{-}285)$$

where $\Delta\lambda$ is the wavelength displacement from the wavelength of the $m-n$ transition, and $\langle C \rangle$ is the average Stark displacement constant which is given in c.g.s. units in Table 32. They are from GRIEM, KOLB, and SHEN (1959) for Ly α, Ly β, H α, H β, H γ and H δ; GRIEM (1962) for He II (3203 Å and 4686 Å); and GRIEM (1960) and GRIEM (1967) for the higher order optical hydrogen transitions. More recent corrections to the Stark broadening of hydrogen and helium lines are given by KEPPLE and GRIEM (1968) and KEPPLE (1972).

The electron broadening corrections to the asymptotic wing formulae for the line intensities at optical frequencies have been calculated by GRIEM, KOLB,

Table 32. Parameters $\langle C_{mn}\rangle$ (in Å per cgs field strength units)$^{3/2}$ and $R(N_e, T)$ in Å$^{-1/2}$ for the asymptotic wing formulae for hydrogen and ionized helium lines[1]. The electron density, N_e, is in cm^{-3} and the electron temperature, T, is in °K

Lα ($\langle C_{mn}\rangle = 3.40\times10^{-6}$)

N_e \ T_e	5,000	10,000	20,000	40,000
10^{10}	2.11	1.93	1.45	1.09
10^{12}	2.01	1.54	1.17	0.89
10^{14}	1.45	1.14	0.89	0.69
10^{16}	0.88	0.74	0.61	0.49

Lβ ($\langle C_{mn}\rangle = 1.78\times10^{-5}$)

N_e \ T_e	5,000	10,000	20,000	40,000
10^{10}	4.30	3.29	2.47	1.86
10^{12}	3.31	2.56	1.96	1.50
10^{14}	2.29	1.83	1.45	1.14
10^{16}	1.26	1.11	0.94	0.77

Hα ($\langle C_{mn}\rangle = 1.30\times10^{-3}$)

N_e \ T_e	5,000	10,000	20,000	40,000
10^{10}	1.50	1.05	0.79	0.60
10^{12}	1.17	0.82	0.63	0.48
10^{14}	0.85	0.59	0.46	0.36
10^{16}	0.52	0.35	0.30	0.25

Hβ ($\langle C_{mn}\rangle = 3.57\times10^{-3}$)

N_e \ T_e	5,000	10,000	20,000	40,000
10^{10}	1.39	1.05	0.80	0.60
10^{12}	1.04	0.81	0.62	0.48
10^{14}	0.69	0.56	0.45	0.35
10^{16}	0.34	0.31	0.27	0.23

Hγ ($\langle C_{mn}\rangle = 6.00\times10^{-3}$)

N_e \ T_e	5,000	10,000	20,000	40,000
10^{10}	1.79	1.37	1.04	0.79
10^{12}	1.32	1.03	0.80	0.62
10^{14}	0.84	0.70	0.57	0.45
10^{16}	0.38	0.36	0.33	0.28

Hδ ($\langle C_{mn}\rangle = 9.81\times10^{-3}$)

N_e \ T_e	5,000	10,000	20,000	40,000
10^{10}	2.17	1.66	1.27	0.96
10^{12}	1.57	1.24	0.97	0.75
10^{14}	0.97	0.81	0.67	0.54
10^{16}	0.37	0.39	0.37	0.32

He II 4,686 Å ($\langle C_{mn}\rangle = 2.62\times10^{-4}$)

N_e \ T_e	5,000	10,000	20,000	40,000
10^{10}				
10^{12}	1.65	1.28	0.98	0.75
10^{14}	1.13	0.91	0.72	0.56
10^{16}			0.46	0.38

He II 3,203 Å ($\langle C_{mn}\rangle = 5.52\times10^{-4}$)

N_e \ T_e	5,000	10,000	20,000	40,000
10^{10}				
10^{12}	1.28	1.00	0.77	0.59
10^{14}	0.85	0.69	0.55	0.44
10^{16}			0.34	0.29

[1] After GRIEM (1964) by permission of the McGraw-Hill Book Co.

and SHEN (1959) and GRIEM (1962). The total broadened intensity, $I(\Delta\lambda)$, is given by (GRIEM, 1964)

$$I(\Delta\lambda) = \varphi(\Delta\lambda)\begin{cases}\{1 + [(\Delta\lambda_w)^{-1/2} + R(N,T)](\Delta\lambda)^{1/2}\}\\ \left\{1 + \left[(\Delta\lambda_w)^{-1/2} + R(N,T)\dfrac{\ln(\Delta\lambda_w/\Delta\lambda)}{\ln(\Delta\lambda_w/\Delta\lambda_p)}\right](\Delta\lambda)^{1/2}\right\},\\ \{1+1\}\end{cases} \quad (2\text{-}286)$$

for $\Delta\lambda < \Delta\lambda_p$, $\Delta\lambda_p < \Delta\lambda < \Delta\lambda_w$, and $\Delta\lambda_w < \Delta\lambda$, respectively. Here $\Delta\lambda_w$ and $\Delta\lambda_p$ are defined by

$$\Delta\lambda_w = \lambda^2 k T/(h m^2 c),$$

m being the principal quantum number of the upper level, and

$$\Delta\lambda_p = \lambda^2 [N_e e^2/(\pi m c^2)]^{1/2},$$

where T is the kinetic temperature of the plasma and N_e is the electron density. The $R(N, T)$ are given in Table 32 for some transitions of hydrogen and helium. Tables of the Stark broadening of the first four Lyman lines and the first four Balmer lines of hydrogen are also given by VIDAL, COOPER, and SMITH (1973).

At radio frequencies, most of the Stark broadening of hydrogen like atoms is due to inelastic collisions with electrons. In the wings of the line, the intensity distribution of the line is Lorentzian and is given by Eq. (2-272). The full width to half maximum of the line, $\Delta\nu_L$, for the hydrogen transition from an upper level with principal quantum number m to a lower level n is (GRIEM, 1960)

$$\Delta\nu_L = \frac{1}{18\pi}\left(\frac{8\pi m_e}{k T_e}\right)^{1/2} N_e \left(\frac{\hbar}{m_e}\right)^2 \int_{y_{min}}^{\infty} \frac{e^{-y}}{y}(m^5 + n^5)^2 \, dy, \tag{2-287}$$

where

$$y_{min} = \frac{4\pi}{3}\frac{N_e}{m_e}\left(\frac{e h m^2}{2\pi k T_e}\right)^2,$$

and the nuclear charge is e.

For the special case of radio frequency hydrogen α lines, $m = n + 1$ and $\nu_{mn} \approx 2cR_H n^{-3} = 6.576 \times 10^{15} n^{-3}$ Hz, where R_H is the Rydberg constant for hydrogen. Using the formulae of GRIEM (1967) for the Stark broadening of radio frequency Hα lines, the full width to half maximum of the line is

$$\begin{aligned} \Delta\nu_L &= \frac{5}{3(2\pi)^{5/2}}\left(\frac{h}{m_e}\right)^2 \left(\frac{m_e}{k T_e}\right)^{1/2} N_e n^4 \left[\frac{1}{2} + \ln\left(2.09 \frac{k T_e}{h \nu_{mn} n^2}\right)\right] \\ &\approx 2.4 \times 10^{-6} \frac{N_e n^4}{T_e^{1/2}}\left[\frac{1}{2} + \ln(6.64 \times 10^{-6} T_e n)\right] \text{Hz}, \end{aligned} \tag{2-288}$$

where N_e and T_e are, respectively, the electron density and temperature, and n is the principal quantum number of the lowest level of the α transition.

Using Eqs. (2-243) and (2-288) together with $\nu_{mn} \approx 6.576 \times 10^{15} n^{-3}$ Hz, we obtain

$$\begin{aligned} \frac{\Delta\nu_L}{\Delta\nu_D} &\approx \frac{3\sqrt{\pi}}{1.2012}\left(\frac{hc}{m}\right)\left(\frac{\hbar^2}{m_e e^2}\right)^2 \left(\frac{m_e}{k T_e}\right)^{1/2} \left(\frac{m_H}{k T_D}\right)^{1/2} N_e n^7 \left[\frac{1}{2} + \ln(6.64 \times 10^{-6} T_e n)\right] \\ &\approx 9.74 \times 10^{-16} \frac{N_e n^7}{(T_e T_D)^{1/2}} [-11.47 + \ln T_e + \ln(n)], \end{aligned} \tag{2-289}$$

where T_D is the effective Doppler temperature, $\Delta\nu_D$ is the full width to half maximum of the Doppler broadened line, the hydrogen mass, $m_H = 1.6733 \times 10^{-24}$ gm, N_e and T_e are, respectively, the electron density and temperature, and n is the principal quantum number of the lowest level of the α transition. The results obtained by taking accurate collision cross sections and degenerate energy levels into account are in close agreement with those of GRIEM (BROCKLEHURST and

LEEMAN, 1971; PEACH, 1972). Attempts to detect Stark broadening in H II regions at radio frequencies have failed (PEDLAR and DAVIES, 1971), but this may be explained if the electron density, N_e, is low (BROCKLEHURST and SEATON, 1971).

2.20.4. Van der Waals Broadening due to Collisions with Neutral Hydrogen Atoms

When line broadening is due to interactions between radiating atoms and neutral atoms, the adiabatic impact theory is appropriate. The spectral line intensity distribution is given by Eq. (2-282) with a full width to half maximum, $\Delta\nu_L$, and a line shift, β, which depend on the power, p, of the interaction potential (Eq. (2-275)). For atoms of neutral hydrogen, $p=6$ (van der Waals broadening). In this case (MARGENAU, 1939)

$$\Gamma = 2\pi\Delta\nu_L = 8.08\, C_6^{2/5}\, v_0^{3/5}\, N_H = 2.650\,\beta, \qquad (2\text{-}290)$$

where N_H is the density of neutral hydrogen, v_0 is given by Eq. (2-274), and the constant C_6 is given by

$$C_6 \approx 4.05\times 10^{-33}\,[R_m^2 - R_n^2],$$

where

$$R_m^2 \approx \frac{m^{*2}}{2Z^2}\,[5m^{*2}+1-3l(l+1)],$$

m^* is the effective principal quantum number of the level, l is the angular momentum quantum number, and

$$m^* = Z\left[\frac{\chi_H}{\chi_m-\chi_i}\right]^{1/2},$$

where $\chi_H = 13.6$ eV, and $\chi_m - \chi_i$ is the energy required to ionize the mth level. The numerical coefficient for C_6 is $\alpha e^2 a_0^2/\hbar$, where α is the polarizability of the atom and a_0 is the Bohr radius. For hydrogen $\alpha = 6.7\times 10^{-25}$ cm^2 and we obtain the coefficient 4.05×10^{-33}. For helium we have the polarizability $\alpha = 2.07\times 10^{-25}$ cm^2.

2.20.5. Resonance Broadening due to Interactions of Radiating and Ground State Atoms

Resonance broadening was first calculated by HOLTSMARK (1925) who assumed that the interaction between a radiating atom and a ground state atom could be represented by the coupling of two dipole fields. WEISSKOPF (1933) used the impact approximation with an interaction potential given by Eq. (2-275) with $p=3$. The full width to half maximum of the Lorentzian profile (Eq. (2-272)) is

$$\Delta\nu_L = \frac{Ne^2}{4m\nu_{mn}}\,f_a \approx 6\times 10^7\,\frac{Nf_a}{\nu_{mn}}\ \text{Hz}, \qquad (2\text{-}291)$$

where N is the density of atoms, ν_{mn} is the frequency of the transition, and $f_a = f_{nm}$ is the absorption oscillator strength. The most recent quantum mechanical calculation gives (ALI and GRIEM, 1965, 1966)

$$\Delta\nu_L \approx 1.9167\left(\frac{g_a}{g_e}\right)^{1/2}\frac{e^2 f_a}{m\nu_{mn}}N \approx 4.8542\times 10^7\,\frac{Nf_a}{\nu_{mn}}\left(\frac{g_a}{g_e}\right)^{1/2}\ \text{Hz}, \qquad (2\text{-}292)$$

where g_a and g_e are, respectively, the statistical weights of the absorbing and emitting states of the atom. For Lyman-α the appropriate constants are $f_a = 0.4162$ and $g_a/g_e = 1/3$.

2.21. Natural Broadening (a Lorentz Dispersion Profile)

The total power radiated by an electric dipole is (LARMOR, 1897)

$$P = \frac{2}{3}\frac{e^2}{c^3}[\ddot{x}(t)]^2, \qquad (2\text{-}293)$$

where $x(t)$ describes the linear displacement of a charge e during time t, and $\ddot{}$ denotes the second derivative with respect to time. The effective damping force created by this radiation is

$$F_{\text{rad}} = \frac{2}{3}\frac{e^2}{c^3}\dddot{x}(t), \qquad (2\text{-}294)$$

where $\dddot{}$ denotes the third derivative with respect to time. For a harmonic oscillator $x(t) = x_0\cos(2\pi\nu_0 t)$, where ν_0 is the resonant frequency of the oscillator. The equation of motion of the harmonic oscillator is, taking radiation damping into account,

$$m\ddot{x}(t) + m(2\pi\nu_0)^2 x(t) = \frac{2}{3}\frac{e^2}{c^3}\dddot{x}(t). \qquad (2\text{-}295)$$

Assuming that the damping force is small, this equation has the solution

$$x(t) = x_0 \exp[i2\pi\nu_0 t - \gamma_{\text{cl}} t/2], \qquad (2\text{-}296)$$

where the classical damping constant is

$$\gamma_{\text{cl}} = \frac{8\pi^2 e^2}{3mc^3}\nu_0^2 \approx 2.5\times 10^{-22}\nu_0^2 \sec^{-1}. \qquad (2\text{-}297)$$

Using Rayleigh's theorem (RAYLEIGH, 1889, Eq. (2-266)) with Eqs. (2-293) and (2-296), the average power radiated by the harmonic oscillator is

$$I(\omega) \propto \langle |x(t)|^2 \rangle = \langle |F(\omega)|^2 \rangle, \qquad (2\text{-}298)$$

where

$$F(\omega) = \int_{-\infty}^{+\infty} x(t)\exp(-i\omega t)\,dt = \frac{x_0}{i2\pi(\nu-\nu_0) + \gamma_{\text{cl}}/2}.$$

The optical line intensity distribution is therefore Lorentzian (LORENTZ, 1906, 1909) and is given by

$$\varphi(\nu) = \frac{\Delta\nu_{\text{L}}/2\pi}{(\nu-\nu_0)^2 + (\Delta\nu_{\text{L}}/2)^2}, \qquad (2\text{-}299)$$

where the full width to half-maximum $\Delta\nu_{\text{L}} = \gamma_{\text{cl}}/2\pi \approx 4\times 10^{-23}\nu_0^2$ Hz, and we have normalized $\varphi(\nu)$ so that $\int_{-\infty}^{+\infty}\varphi(\nu)\,d\nu = 1$. The halfwidth in wavelength units is $\Delta\lambda = c\gamma_{cl}/(2\pi\nu_0^2) \approx 1.18 \times 10^{-4}$ Å, which is independent of wavelength. The Lorentz profile given by Eq. (2-299) is illustrated in Fig. 16.

WEISSKOPF and WIGNER (1930) and WEISSKOPF (1933) have shown that the classical Lorentz solution of Eq. (2-299) is also a valid quantum mechanical solution if the transition frequency $\nu_{mn} = \nu_0$, and the half width $\Delta\nu_L = \Gamma_R/2\pi$ where the quantum mechanical damping constant, Γ_R, is given by

$$\Gamma_R = \Gamma_m + \Gamma_n = t_m^{-1} + t_n^{-1}, \tag{2-300}$$

where t_m is the mean lifetime of an atom in level m, and, under conditions of local thermodynamic equilibrium,

$$\Gamma_m = \sum_{n<m} A_{mn} + \sum_{n<m} B_{mn} I_\nu(mn) + \sum_{k>m} B_{mk} I_\nu(mk), \tag{2-301}$$

where

$$\frac{g_m}{g_n} A_{mn} = B_{nm} \frac{2h\nu^3}{c^2} \quad \text{and} \quad B_{nm} = \frac{g_m}{g_n} B_{mn}$$

relate the Einstein coefficients for spontaneous emission, A_{mn}, and stimulated emission, B_{mn}, and $I_\nu(mn)$ is the black body radiation intensity evaluated at the frequency ν_{mn}, of the $m-n$ transition. Here the first term is due to spontaneous emission to lower levels and the summation is taken over all levels to which a transition can occur, the second term is due to induced or stimulated emission (called negative absorption) to a lower level, and the third term is due to ordinary absorption processes and may include ionizations.

We may use the Wien and Rayleigh-Jeans approximations for the brightness of a black body (Eqs. (1-120) and (1-121)) to obtain the number of induced emissions per cm^3 per sec

$$\begin{aligned} B_{mn} I_\nu(mn) &\approx A_{mn} \exp(-h\nu/kT) \quad &\text{for } h\nu \gg kT \\ &\approx A_{mn} kT/h\nu &\text{for } h\nu \ll kT. \end{aligned} \tag{2-302}$$

Neglecting ordinary absorptions then,

$$\begin{aligned} \Gamma_m &\approx \sum_{n<m} A_{mn} \quad &\text{for } h\nu/kT \gg 1 \\ &\approx \frac{kT}{h\nu} \sum_{n<m} A_{mn} &\text{for } h\nu/kT \ll 1. \end{aligned} \tag{2-303}$$

Here $h\nu/kT \ll 1$ means $\nu \ll 10^{10}\,T$.

2.22. Combined Doppler, Lorentz, and Holtsmark Line Broadening (the Voigt Profile)

When the Doppler frequency shift, $\Delta\nu_D$, is taken into account in calculating a Lorentz dispersion profile, Eqs. (2-241) and (2-299) may be combined to give the spectral line intensity distribution function

$$\varphi(\nu) = \int_{-\infty}^{+\infty} \frac{\Delta\nu_L \exp(-\Delta\nu^2/\Delta\nu_D^2)\, d(\Delta\nu)}{2\pi\sqrt{\pi}\,\Delta\nu_D[(\nu-\nu_0-\Delta\nu)^2 + (\Delta\nu_L/2)^2]}, \tag{2-304}$$

where $\Delta\nu_D$ and $\Delta\nu_L$ are, respectively, the full width to half maximum of the Doppler and Lorentz line profiles. Following the development of VOIGT (1913),

$$\varphi(\nu) = \frac{2\sqrt{\ln 2}}{\sqrt{\pi}\,\Delta\nu_D} H(a,b), \tag{2-305}$$

where

$$a = \sqrt{\ln 2}\,\Delta\nu_L/(2\Delta\nu_D)$$
$$b = 2\sqrt{\ln 2}\,(\nu - \nu_0)/\Delta\nu_D$$

and

$$H(a,b) = \frac{a}{\pi} \int_{-\infty}^{+\infty} \frac{\exp(-y^2)\,dy}{(b-y)^2 + a^2}.$$

Tables of the Voigt function, $H(a,b)$ are given by HJERTING (1938), HARRIS (1948), and FINN and MUGGLESTONE (1965). Voigt profiles are illustrated in Fig. 16.

When there is a Holtsmark type of broadening as well as the Doppler and Lorentz types, the total spectral line intensity distribution is

$$I(\Delta\nu) = \int_{-\infty}^{+\infty} \varphi(\Delta\nu + b\,\Delta\nu_D) H(a,b)\,db, \tag{2-306}$$

where $\varphi(\Delta\nu)$ is given by Eqs. (2-255) and (2-256).

The Voigt function, $H(a,b)$, may be approximated by

$$H(a,b) \approx \exp(-b^2) + \frac{a}{\sqrt{\pi}\,b^2}, \tag{2-307}$$

illustrating the fact that in the wings the combined broadening practically coincides with a Lorentz broadening, whereas in the core it is Doppler broadening that dominates.

3. Gas Processes

"This notation may be perhaps further explained by conceiving the air near the earth to be such a heap of little bodies, lying one upon another as may be resembled to a fleece of wool. For this consists of many slender and flexible hairs, each of which may indeed, like a little spring, be easily bent or rolled up; but will also, like a spring, be still endeavouring to stretch itself out again."

R. BOYLE, (1660)

"O dark dark dark. They all go into the dark.
The vacant interstellar spaces, the vacant into the vacant."

T. S. ELIOT in East Coker III (1940)

3.1. Microstructure of a Gas

3.1.1. Boltzmann's Equation, the Fokker-Planck Equation, the B.B.G.K.Y. Hierarchy, Maxwell's Distribution Function, and the Vlasov Equation

The one particle probability distribution function, $f(\boldsymbol{r},\boldsymbol{p},t)$, is defined so that

$$f(\boldsymbol{r},\boldsymbol{p},t)\,dx\,dy\,dz\,dp_x\,dp_y\,dp_z = f(\boldsymbol{r},\boldsymbol{p},t)\,dV_r\,dV_p \tag{3-1}$$

is the probability that, at the time, t, a particle has momentum, $\boldsymbol{p}$, in the volume element dV_p at $\boldsymbol{p}$ and position, $\boldsymbol{r}$, in the volume element dV_r at $\boldsymbol{r}$. Similarly, the distribution function $f(\boldsymbol{r},\boldsymbol{v},t)$ is defined so that for an average particle density, N,

$$N\,f(\boldsymbol{r},\boldsymbol{v},t)\,dx\,dy\,dz\,dv_x\,dv_y\,dv_z = N\,f(\boldsymbol{r},\boldsymbol{v},t)\,dV_r\,dV_v \tag{3-2}$$

gives the probable number of particles in the six dimensional phase space $dV_r\,dV_v$ around position, $\boldsymbol{r}$, and velocity, $\boldsymbol{v}$. Boltzmann's equation for $f(\boldsymbol{r},\boldsymbol{p},t)$ may be written as (BOLTZMANN, 1872)

$$\frac{\partial f}{\partial t} + \frac{\boldsymbol{p}}{m}\cdot\nabla_r f - \nabla_r\varphi\cdot\nabla_p f = \left(\frac{df}{dt}\right)_{\text{coll}}, \tag{3-3}$$

where $\varphi(r)$ is the potential energy acting on every particle, $\boldsymbol{p}$ is the momentum, m is the particle mass, ∇_r is the gradient in position space, ∇_p is the gradient in momentum space, and $(df/dt)_{\text{coll}}$ is the rate of change in f due to collisions. Noting that $\dot{p} = -\nabla_r\varphi$, we may write Eq. (3-3) in Cartesian coordinates as

$$\frac{\partial f}{\partial t} + \dot{x}\frac{\partial f}{\partial x} + \dot{y}\frac{\partial f}{\partial y} + \dot{z}\frac{\partial f}{\partial z} + \dot{p}_x\frac{\partial f}{\partial p_x} + \dot{p}_y\frac{\partial f}{\partial p_y} + \dot{p}_z\frac{\partial f}{\partial z} = \left(\frac{df}{dt}\right)_{\text{coll}},$$

where $\dot{}$ denotes the first derivative with respect to time. The Boltzmann equation for $f(\boldsymbol{r},\boldsymbol{v},t)$ is

$$\frac{\partial f}{\partial t}+\boldsymbol{v}\cdot\nabla_r f+\frac{\boldsymbol{F}}{m}\cdot\nabla_v f=\left(\frac{df}{dt}\right)_{\text{coll}}, \tag{3-4}$$

where $\boldsymbol{v}$ is the velocity, $\boldsymbol{F}$ is the force acting on each particle, m is the particle mass, and ∇_r and ∇_v denote, respectively, gradients in position and velocity space. As an example of astrophysical forces, a particle of charge, q, and mass, m, experiences the force

$$\boldsymbol{F}=q\left(\boldsymbol{E}+\frac{1}{c}\boldsymbol{v}\times\boldsymbol{H}\right)-mg\,\boldsymbol{n}_r,$$

in the presence of an electric field of strength $\boldsymbol{E}$, a magnetic field of strength $\boldsymbol{H}$, and a gravitational field of acceleration g. Here $\boldsymbol{n}_r$ is a unit vector in the radial direction from the mass, m, to another mass, M, and the acceleration due to gravity is GM/r^2, where the gravitational constant $G=6.67\times10^{-8}$ dyn cm^2 g^{-2}, and r is the distance between the mass, M, and the particle of mass, m.

For a two particle system of types i and j, the collision term takes the Fokker-Planck form (Fokker, 1914; Rosenbluth, MacDonald, and Judd, 1957).

$$\frac{1}{\Gamma_i}\left(\frac{df_i}{dt}\right)_{\text{coll}}=-\frac{\partial}{\partial\boldsymbol{v}_i}\cdot\left(f_i\frac{\partial H_i}{\partial\boldsymbol{v}_i}\right)+\frac{1}{2}\frac{\partial^2}{\partial\boldsymbol{v}_i\partial\boldsymbol{v}_i}\left(f_i\frac{\partial^2 G_i}{\partial\boldsymbol{v}_i\partial\boldsymbol{v}_i}\right), \tag{3-5}$$

where

$$\Gamma_i=-\frac{4\pi e_i^4\ln\left(\sin\dfrac{\theta_{\min}}{2}\right)}{m_i^2},$$

e_i and m_i are, respectively, the charge and mass of the particle of type i, the minimum value of the scattering angle is $\theta_{\min}$,

$$H_i=\sum_j N_j\left(\frac{e_j}{e_i}\right)^2\left(\frac{m_i+m_j}{m_j}\right)\int d\boldsymbol{v}_j\frac{f_j}{g}$$

and

$$G_i=\sum_j N_j\left(\frac{e_j}{e_i}\right)^2\int d\boldsymbol{v}_j g\,f_j,$$

where the density of particle type i is N_i, and

$$g=|\boldsymbol{v}_i-\boldsymbol{v}_j|.$$

For a plasma at temperature, T, consisting of electrons of mass, m, and protons of mass, m_i,

$$g\approx\left(\frac{3kT}{m}\right)^{1/2}$$

and

$$\Gamma_i\approx\frac{4\pi e^4}{m_i^2}\ln\Lambda,$$

where

$$\Lambda=24\pi N R_{\mathrm{D}}^3=\frac{2}{\theta_{\min}},$$

the average particle density is N, the Debye radius, R_D, is given by (DEBYE and HÜCKEL, 1923)

$$R_D = \left(\frac{kT}{4\pi N e^2}\right)^{1/2} \approx \left(\frac{1}{Ng}\right)^{1/3}$$

and we have assumed that $N R_D^3 \gg 1$.

For an infinite single particle system the distribution function, f_s, for s particles is related to the f_{s+1} by the B.B.G.K.Y. hierarchy given by (BOGOLYUBOV, 1946; BORN and GREEN, 1949; GREEN, 1952; KIRKWOOD, 1947; YVON, 1935)

$$\frac{\partial f_s}{\partial t} + \left\{f_s; H_s\right\} = N \int \left\{\sum_{i=1}^{s} \varphi_{i,s+1}; f_{s+1}\right\} dX_{s+1}, \tag{3-6}$$

where the Poisson bracket of two quantities, A and B, which depend on $\boldsymbol{r}_i$ and $\boldsymbol{p}_i$, is defined by,

$$\{A; B\} = \sum_{i=1}^{s} \left[\frac{\partial A}{\partial \boldsymbol{r}_i} \cdot \frac{\partial B}{\partial \boldsymbol{p}_i} - \frac{\partial A}{\partial \boldsymbol{p}_i} \cdot \frac{\partial B}{\partial \boldsymbol{r}_i}\right],$$

where s is the total number of particles, and H_s is the Hamiltonian of s particles.

$$H_s = \sum_{i=1}^{s} \frac{p_i^2}{2m} + \sum_{i<j=1}^{s} \varphi_{ij} + \text{constant},$$

where $\varphi_{ij} = e^2/(|\boldsymbol{r}_i - \boldsymbol{r}_j|)$, for electrons of charge e, and the constant term represents the potential of the electrons moving in the field of ions. For an isolated system in thermal equilibrium, the first member of the hierarchy becomes the Maxwellian distribution function (MAXWELL, 1860)

$$f\,dp = 4\pi \left(\frac{1}{2\pi m k T}\right)^{3/2} \exp\left[-\frac{p^2}{2mkT}\right] p^2\,dp. \tag{3-7}$$

In the limit $g = 1/(N R_D^3) = 0$, the Boltzmann equation becomes collisionless, $(df/dt)_{coll} = 0$, and the B.B.G.K.Y. hierarchy becomes Vlasov's equation given by (VLASOV, 1938, 1945)

$$\frac{\partial f}{\partial t} + \boldsymbol{v}_1 \cdot \frac{\partial f}{\partial \boldsymbol{x}_1} - \left[\frac{N}{m} \int d\boldsymbol{x}_2\, d\boldsymbol{v}_2 \frac{\partial \varphi_{12}}{\partial \boldsymbol{x}_1} f(\boldsymbol{x}_2, \boldsymbol{v}_2)\right] \cdot \frac{\partial f(\boldsymbol{x}_1, \boldsymbol{v}_1)}{\partial \boldsymbol{v}_1} = 0, \tag{3-8}$$

or

$$\frac{\partial f}{\partial t} + \boldsymbol{v}_1 \cdot \frac{\partial f}{\partial \boldsymbol{x}_1} - \frac{e}{m} \boldsymbol{E} \cdot \frac{\partial f}{\partial \boldsymbol{v}_1} = 0,$$

where $\boldsymbol{E}$ is the electric field intensity. Linearized perturbation solutions to Vlasov's equation were given by LANDAU (1946), and such solutions lead to electron plasma oscillations discussed in Sect. 1.32, Eqs. (1-254) to (1-256); and by TONKS and LANGMUIR (1929) and BOHM and GROSS (1949). The general theory for calculating transport coefficients and velocity distribution functions from Boltzmann's equation is discussed by CHAPMAN (1916, 1917), ENSKOG (1917), and in the book by CHAPMAN and COWLING (1953). The main results of these calculations are given in the following subsections.

3.1.2. Collisions—The Mean Free Path and Mean Free Time between Collisions

For a gas of neutral atoms with number density, N, and effective collision radius, a_0, the mean free path, l, between collisions is (CLAUSIUS, 1858)

$$\begin{aligned} l &= (N\pi a_0^2)^{-1} \\ &\approx 10^{16} N^{-1}\ \text{cm} \quad \text{for hydrogen atoms,} \end{aligned} \tag{3-9}$$

where we have taken $a_0 \approx 0.5 \times 10^{-8}$ cm, the radius of the first Bohr orbit. The quantity $\sigma_c = \pi a_0^2$ is called the collision cross section.

When two charged particles collide, a measure of the collision radius is the perpendicular distance, b, between the slower particle and the original path of the faster particle. If $Z_1 e$ and $Z_2 e$ are the two charges, then this impact parameter is (RUTHERFORD, 1911)

$$b = \frac{Z_1 Z_2 e^2}{M_1 V_1^2 \tan(\theta/2)}, \tag{3-10}$$

where M_1 and V_1 are, respectively, the mass and velocity of the faster charge, and θ is its angle of deflection. Under conditions of thermal equilibrium at temperature, T, the root-mean-square velocity, V_{rms}, is given by

$$\begin{aligned} V_{\text{rms}} &= (3kT/M)^{1/2} \\ &\approx 6.7 \times 10^5\, T^{1/2}\ \text{cm sec}^{-1} \quad \text{for electrons} \\ &\approx 1.57 \times 10^4\, T^{1/2}\ \text{cm sec}^{-1} \quad \text{for hydrogen atoms.} \end{aligned} \tag{3-11}$$

For a gas composed mostly of neutral particles, Eqs. (3-9), (3-10), and (3-11) give the mean free path

$$l \approx \frac{M^2 V^4}{N_e Z^2 e^4} \approx 3.23 \times 10^6\, T^2 (Z^2 N_e)^{-1}\ \text{cm}, \tag{3-12}$$

where N_e is the free electron density, and an effective collision is assumed to be one for which the angle of deflection is 90°.

When a gas is composed mostly of charged particles, there is a maximum impact parameter given by

$$b_m = R_D = \left(\frac{kT}{4\pi N_e e^2}\right)^{1/2} \approx 6.9 \left(\frac{T}{N_e}\right)^{1/2}\ \text{cm}, \tag{3-13}$$

where R_D is the Debye radius, T is the temperature, and N_e is the electron density. In this case, the impact parameter must be integrated over its range of values to obtain (PINES and BOHM, 1952; SPITZER, 1962)

$$l \approx \frac{M^2 V_{\text{rms}}^4}{Z^2 N_e e^4 \ln \Lambda} \approx \frac{3.2 \times 10^6\, T^2}{Z^2 N_e \ln \Lambda}\ \text{cm}, \tag{3-14}$$

where

$$\Lambda = \frac{R_D}{b} = \frac{3}{2 Z_1 Z_2 e^3} \left(\frac{k^3 T^3}{\pi N_e}\right)^{1/2} \approx 1.3 \times 10^4 \frac{T^{3/2}}{N_e^{1/2}}. \tag{3-15}$$

It follows from Eq. (3-14) that the mean free path for collisions of electrons with electrons is the same as that for collisions of protons with protons provided that the electrons and protons have the same kinetic temperature. Using the Poisson distribution, the probability, $P_n(x/l)$, that a particle of a perfect gas will have n collisions after it has travelled the distance, x, is given by

$$P_n(x/l) = [(x/l)^n/n!]\exp(-x/l), \tag{3-16}$$

where l is the mean free path.

For a perfect gas in equilibrium at temperature, T, there is a Maxwellian distribution of momentum, p, and the mean free time, τ, between collisions is given by

$$\tau = 4\pi m l \int_0^\infty \frac{p}{(2\pi m k T)^{3/2}} \exp\left[\frac{-p^2}{2mkT}\right] dp = l\left(\frac{2m}{\pi k T}\right)^{1/2}, \tag{3-17}$$

or

$$\tau \approx l v_{\text{th}}^{-1},$$

where l is the mean free path, m is the particle mass, k is Boltzmann's constant, and the thermal velocity, $v_{\text{th}} \approx v_{\text{rms}} = (3kT/m)^{1/2}$. Using Eqs. (3-9), (3-11), (3-12), (3-14), and (3-17) we obtain

$$\begin{aligned} &\tau \approx 10^{12}\, T^{-1/2} N^{-1} \text{ sec} && \text{for hydrogen atoms} \\ &\tau \approx 4\, T^{3/2} N_e^{-1} \text{ sec} && \text{for electrons, } N_e \ll N \\ &\tau \approx 4\, T^{3/2} (N_e \ln \Lambda)^{-1} \text{ sec} && \text{for electrons, } N_e \gg N, \end{aligned} \tag{3-18}$$

where N is the density of neutral atoms, and N_e is the electron density. Using the Poisson distribution, we may specify the probability, $P_n(t/\tau)$, that a particle of a perfect gas will have n collisions in the time, t.

$$P_n(t/\tau) = [(t/\tau)^n/n!]\exp[-t/\tau]. \tag{3-19}$$

For conditions very near equilibrium, we may use the mean free time, τ, to give the approximate Boltzmann equation

$$\frac{\partial f}{\partial t} + \frac{\boldsymbol{p}}{m} \cdot \nabla_r f - \nabla_r \varphi \cdot \nabla_p f \approx \frac{1}{\tau}\left[f_0(r,p) - f\right], \tag{3-20}$$

or

$$f + \tau\left(\frac{\partial f}{\partial t}\right) = f_0 - \tau\left[\left(\frac{\boldsymbol{p}}{m}\right) \cdot \nabla_r f_0 + \boldsymbol{F} \cdot \nabla_p f_0\right],$$

where the equilibrium solution, $f_0(r,p)$, is given by Eq. (3-7).

When a stream of test particles of mass, M_+, charge, Z_+, and velocity, V_+, are injected into a plasma which is in thermal equilibrium, the characteristic time, τ_D, in which the particles are deflected through an angle of order $\pi/2$ is given by

$$\tau_D \approx \frac{M_+^2 V_+^3}{16\pi N e^4 Z_+^2 \ln \Lambda} \quad \text{for } V_+ \gg \left(\frac{kT}{m}\right)^{1/2}, \tag{3-21}$$

where Λ is given by Eq. (3-15), and N is the plasma number density.

3.1.3. Viscosity and the Reynolds Number

If a part of a gas which is in equilibrium is given a drift velocity, v_x, in the x direction, there will be a restoring force $\boldsymbol{F} = \mu \partial v_x / \partial z$ tending to reduce the rate of shear $\partial v_x / \partial z$. Here μ is called the coefficient of dynamic viscosity. The net rate of increase of momentum within a volume element may be obtained by considering the forces acting on all of the faces of the element. This net force is

$$\boldsymbol{F} = \mu \nabla^2 \boldsymbol{v}, \tag{3-22}$$

where $\boldsymbol{v}$ is the velocity, and μ is the coefficient of dynamic viscosity. In the absence of other forces, it follows directly from Eqs. (3-7), (3-20), and (3-22) that (cf. MAXWELL, 1866)

$$\mu \approx M l N v_{\mathrm{rms}} \approx M l N \left(\frac{3kT}{M}\right)^{1/2}, \tag{3-23}$$

where M is the atomic or molecular mass, l is the mean free path given by Eqs. (3-9), (3-12) or (3-14), N is the number density of atoms or molecules, v_{rms} is the thermal velocity, and T is the temperature. For hydrogen atoms, for example,

$$\mu \approx 5.7 \times 10^{-5}\, T^{1/2} \text{ gm cm}^{-1} \text{ sec}^{-1}, \tag{3-24}$$

and for a fully ionized gas

$$\mu \approx \frac{M_{\mathrm{i}}^{1/2} (3kT)^{5/2}}{Z^4 e^4 \ln \Lambda} \approx 2 \times 10^{-15} \frac{T^{5/2} A_{\mathrm{i}}^{1/2}}{Z^4 \ln \Lambda} \text{ gm cm}^{-1} \text{ sec}^{-1}, \tag{3-25}$$

where M_{i} is the ion mass, eZ is the ionic charge, A_{i} is the atomic weight of the positive ion, and the factor Λ is given by Eq. (3-15).

When a magnetic field of strength, B, is present, the effective viscosity in the direction parallel to the direction of B is given by Eq. (3-25). For the direction perpendicular to B, the viscosity is (SIMON, 1955)

$$\begin{aligned} \mu_{\perp} &= \frac{2}{5} \left(\frac{\pi}{M_{\mathrm{i}} k T_{\mathrm{i}}}\right)^{1/2} \frac{Z^4 e^4 N_{\mathrm{i}}^2 M_{\mathrm{i}}^2 c^2 \ln \Lambda}{Z^2 e^2 B^2} \\ &\approx 2.7 \times 10^{-26} \frac{A_{\mathrm{i}}^{3/2} Z^2 N_{\mathrm{i}}^2 \ln \Lambda}{T_{\mathrm{i}}^{1/2} B^2} \text{ gm cm}^{-1} \text{ sec}^{-1}, \end{aligned} \tag{3-26}$$

where M_{i} and N_{i} are, respectively, the ion mass and number density, A_{i} is the atomic weight of the ion, and T_{i} is the temperature of the ion.

The ability of a gas to damp out turbulent motion is measured by the Reynolds coefficient (STOKES, 1851; REYNOLDS, 1883). If a turbule of size, L, has velocity, v, the motion will not be damped out by viscous effects if the Reynolds number

$$\mathrm{Re} = L v \rho / \mu \tag{3-27}$$

is greater than unity. Here μ is the coefficient of dynamic viscosity and ρ is the mass density. Often the coefficient of kinematic viscosity, ν, is used, where

$$\nu = \mu / \rho, \quad \text{and } \mathrm{Re} = L v / \nu. \tag{3-28}$$

Using Eqs. (3-24), (3-25), and (3-26) in Eq. (3-27), we obtain

$$\mathrm{Re} \approx 2 \times 10^4 \rho L v T^{-1/2} \tag{3-29}$$

for an unionized gas,

$$\mathrm{Re} \approx 5 \times 10^{14} \frac{\rho Z^4 L v \ln \Lambda}{T^{5/2} A_i^{1/2}}, \tag{3-30}$$

for a fully ionized gas without a magnetic field, and

$$\mathrm{Re} \approx 3 \times 10^{25} \frac{\rho L v T^{1/2} B^2}{A_i^{3/2} Z^2 N_i^2 \ln \Lambda}, \tag{3-31}$$

for a fully ionized gas when the motion is perpendicular to the magnetic field.

3.1.4. Electrical Conductivity and Mobility

When an ionized gas is subject to an electric field of strength, E, the free electrons and ions will drift, causing a flow of current. When the number density of free electrons, N_e, is small compared with the total number density of neutral atoms or molecules, N, the net drift velocity, V, will be determined by collisions. In this case, the equation of motion of each free electron or ion is given by

$$Z e E = m V_{\mathrm{rms}} V/l = m V/\tau, \tag{3-32}$$

where eZE is the Lorentz force on the charge eZ, the particle mass is m, the r.m.s. particle velocity is V_{rms}, the mean free path is l, and τ is the mean time between collisions. If the number density of free electrons is N_e, then the net current density, J, is given by

$$J = \sigma E = N_e e V, \tag{3-33}$$

where the electrical conductivity, σ, is given by

$$\sigma = \frac{N_e e^2 l}{m V_{\mathrm{rms}}}. \tag{3-34}$$

Using Eq. (3-11) for V_{rms} and Eq. (3-12) for l in Eq. (3-34), we obtain

$$\sigma = \frac{(3kT)^{3/2}}{e^2 m^{1/2}} \approx 10^9 T^{3/2} \text{ e.s.u.} \tag{3-35}$$

for a partially ionized gas. For conversion, 1 mho-cm$^{-1} \approx 10^{-9}$ e.m.u. $\approx 10^{12}$ e.s.u. The corresponding expression for a fully ionized gas is (COWLING, 1945)

$$\sigma \approx 6.5 \times 10^6 T^{3/2} \text{ e.s.u.}, \tag{3-36}$$

where electron-electron encounters and electron shielding have been taken into account. As pointed out by COWLING (1945, 1946), a magnetic field of linear scale, L, has a mean lifetime, τ, of order

$$\tau \approx L^2 \sigma/c^2, \tag{3-37}$$

where σ is in e.s.u..

Detailed formulae for the resistivity, η, of an ionized gas are given by SPITZER and HARM (1953). They obtain

$$\eta = \frac{c^2}{\sigma} = \frac{\pi^{3/2} m^{1/2} Z e^2 c^2 \ln \Lambda}{2(2kT)^{3/2} \gamma} \approx 3.80 \times 10^3 \frac{Z \ln \Lambda}{\gamma T^{3/2}} \text{ ohm-cm}, \tag{3-38}$$

where Λ is given by Eq. (3-15) and γ is a Z dependent factor having the values $\gamma = 0.582$ for $Z = 1$, and $\gamma = 1$ for $Z = \infty$. For conversion, 1 ohm-cm $\approx 10^{-12}$ e.s.u. $\approx 10^9$ e.m.u.

From Eqs. (3-33) to (3-36), the net drift velocity, V, of the free electrons is

$$V \approx \frac{\sigma}{N_e e} E = \Omega E, \tag{3-39}$$

where N_e is the density of the free electrons, and the mobility, Ω, is given by

$$\Omega \approx 10^{19}\, T^{3/2} N_e^{-1} \text{ e.s.u.}, \tag{3-40}$$

for a partially ionized gas, and

$$\Omega \approx 10^{16}\, T^{3/2} N_e^{-1} \text{ e.s.u.},$$

for a fully ionized gas.

3.1.5. Diffusion and the Magnetic Reynolds Number

A spatial gradient, ∇N_i, in the ion density, N_i, will cause a current density, J, given by

$$J = e l Z V_{rms} \nabla N_i, \tag{3-41}$$

where the ions are assumed to move with the root mean square velocity, V_{rms}, for the mean free path, l. From the equation of continuity (Eq. (1-38)) for charge conservation, we obtain the diffusion equation

$$\frac{\partial N_i}{\partial t} = \frac{\nabla \cdot J}{eZ} = D \nabla^2 N_i, \tag{3-42}$$

where the diffusion constant

$$D \approx l V_{rms} = l \left(\frac{3kT}{m}\right)^{1/2} = \frac{kT}{e} \Omega, \tag{3-43}$$

where T is the temperature, and Ω is the mobility. The approximate mean free paths, l, are given in Eqs. (3-9), (3-12), and (3-14). A characteristic "deflection time", t_D, can be defined as

$$t_D = D/V_{rms}^2 = l/V_{rms}, \tag{3-44}$$

and a characteristic diffusion time for the length, L, is

$$t = L^2/D. \tag{3-45}$$

CHANDRASEKHAR (1943) has given formulae for the velocity dispersion of test particles which diffuse into a group of field particles. If the subscripts ||

and $\perp$ denote, respectively, directions parallel and perpendicular to the original motion of the test particles, then for a Maxwellian distribution of field particles

$$\langle \Delta v_{\parallel}^2 \rangle \approx \frac{v^3}{l}\left\{G\left[\left(\frac{mv^2}{2kT}\right)^{1/2}\right]\right\} \tag{3-46}$$

and

$$\langle \Delta v_{\perp}^2 \rangle \approx \frac{v^3}{l}\left\{\varphi\left[\left(\frac{mv^2}{2kT}\right)^{1/2}\right] - G\left[\left(\frac{mv^2}{2kT}\right)^{1/2}\right]\right\},$$

where $\varphi(x)$ is the error function, $G(x)=[\varphi(x)-x\varphi'(x)]/2x^2$, Δv denotes the dispersion in velocity, the factors in $\{\,\}$ are of order unity, the velocity of the field particles is v, and the mean free path, l, of the test particles is given by Eq. (3-14). It follows from Eq. (3-46) that a characteristic deflection time is

$$t_{\mathrm{D}} = \frac{v^2}{\langle \Delta v^2 \rangle} \approx \frac{l}{v}. \tag{3-47}$$

Maxwell's equations which relate the electric field intensity, $\boldsymbol{E}$, the magnetic induction, $\boldsymbol{B}$, and the current density, $\boldsymbol{J}$, are (cf. Eqs. (1-39) to (1-43))

$$\nabla \times \boldsymbol{E} + \frac{1}{c}\frac{\partial \boldsymbol{B}}{\partial t} = 0 \tag{3-48}$$

and

$$\nabla \times \boldsymbol{B} = \frac{4\pi}{c}\boldsymbol{J},$$

where displacement current has been neglected in this m.h.d. approximation. Using $\boldsymbol{J}=\sigma\boldsymbol{E}$ we obtain the diffusion equation

$$\frac{\partial \boldsymbol{B}}{\partial t} = D_{\mathrm{M}}\nabla^2 \boldsymbol{B}, \tag{3-49}$$

where the material velocity has been assumed to be negligible (cf. Eq. (3-258)), and the magnetic diffusion constant, D_{M}, is given by

$$D_{\mathrm{M}} = \frac{c^2}{4\pi\sigma} = \frac{\eta}{4\pi}. \tag{3-50}$$

The conductivity, σ, may be evaluated using Eq. (3-35) or (3-36). If L is the characteristic size of the spatial variation of $\boldsymbol{B}$, the decay time for the diffusion of the magnetic field is (Cowling, 1945, 1946)

$$\tau_{\mathrm{M}} = \frac{L^2}{D_{\mathrm{M}}} = \frac{4\pi\sigma L^2}{c^2}. \tag{3-51}$$

Magnetic lines of force will move with a fluid of velocity, v, and scale, L, if the magnetic Reynolds number

$$R_{\mathrm{M}} = v\,\tau_{\mathrm{M}}/L \tag{3-52}$$

is greater than unity. For $R_{\mathrm{M}}>1$, transport of the lines of force with a fluid or gas dominates over diffusion of the lines of force.

3.1.6. Heat Conductivity and the Prandtl Number

The flow of heat per unit area, Q, in the presence of a temperature gradient, ∇T, is given by

$$Q \approx -\kappa \nabla T, \tag{3-53}$$

where the coefficient of heat conductivity, κ, is given by

$$\kappa = \frac{5}{3} \frac{kTl}{V_{rms}} N \left(\frac{3}{2} \frac{k}{m}\right) = \frac{5}{3} \mu \left(\frac{3}{2} \frac{k}{m}\right) \tag{3-54}$$

for a gas at temperature, T, with a mean free path, l, root-mean-squared velocity $V_{rms} = (3kT/m)^{1/2}$, particle number density, N, particle mass, m, and viscosity, μ. The mean free path, l, or viscosity, μ, may be evaluated from Eqs. (3-9), (3-12), (3-14), (3-24), (3-25), or (3-26). The quantity $(3k/2m)$ is given by (C_v/M), where C_v and M are, respectively, the specific heat and the molecular weight of the gas. It follows from Eqs. (3-25), (3-26), and (3-54) that (cf. SPITZER, 1962)

$$\kappa \approx 2 \times 10^{-4} \frac{T^{5/2}}{Z^4 \ln \Lambda} \text{ erg sec}^{-1} \, {}^{\circ}\text{K}^{-1} \text{ cm}^{-1} \tag{3-55}$$

for a fully ionized gas, and

$$\kappa \approx 1.5 \times 10^{-17} \frac{A_i^{1/2} Z^2 N_i^2 \ln \Lambda}{T^{1/2} B^2} \text{ erg sec}^{-1} \, {}^{\circ}\text{K}^{-1} \text{ cm}^{-1} \tag{3-56}$$

for heat conduction in the direction perpendicular to the magnetic field.

For an incompressible fluid ($\rho = \text{constant}$), the change in temperature, T, with time, t, is given by the equation of thermal conduction:

$$\frac{\partial T}{\partial t} + \boldsymbol{v} \cdot \nabla T = \frac{\kappa}{\rho C_v} \nabla^2 T = \chi \nabla^2 T, \tag{3-57}$$

where κ is the coefficient of heat conductivity, ρ is the mass density, C_v is the specific heat at constant volume, and the constant χ is called the thermometric conductivity.

The relative magnitudes of viscosity and heat conductivity are measured by the Prandtl number, P_r, given by (PRANDTL, 1905)

$$P_r = \frac{C_p \mu}{\kappa} = \frac{C_p \nu \rho}{\kappa} = \frac{\nu}{\chi}, \tag{3-58}$$

where C_p is the specific heat at constant pressure, μ is the coefficient of dynamic viscosity, κ is the coefficient of heat conductivity, ν is the coefficient of kinematic viscosity, ρ is the mass density, and χ is the thermometric conductivity.

3.2. Thermodynamics of a Gas

3.2.1. First Law of Thermodynamics and the Perfect Gas Law

The first law of thermodynamics is the expression of the law of conservation of energy which takes into account the energy due to heat. If dQ is the amount of heat energy which is absorbed by a system from its surroundings, then this law states (MAYER, 1842; JOULE, 1847; HELMHOLTZ, 1847)

$$dQ = dU + dW, \tag{3-59}$$

where dU is the change in internal energy of the system when going from one equilibrium state to another, and dW is the amount of work done by the system on its surroundings. One of the simpler systems is the hydrostatic system in which the work consists of displacing the system boundaries against a uniform hydrostatic pressure, P. In this case, Eq. (3-59) becomes

$$dQ = dU + P\,dV, \tag{3-60}$$

where dV is the change in volume, V. For a perfect gas we have the additional useful relation

$$PV = nRT, \tag{3-61}$$

where n is the number of gram moles, the universal gas constant $R = 8.31434\,(35) \times 10^7$ erg mole^{-1} °K^{-1}, and T is the temperature. Eq. (3-61) follows from the observations that $P \propto T$ for constant V (GAY-LUSSAC, 1809), that $PV/T =$ constant (AVOGADRO, 1811), and that $P \propto V^{-1}$ for constant T (BOYLE, 1662; TOWNELY, 1662; POWER, 1663). The constant nR is related to other physical constants by equivalent forms of Eq. (3-61) given in Eq. (3-76).

3.2.2. Thermal (or Heat) Capacity, Molecular Heat, and Specific Heat

By definition the thermal or heat capacity, C, of a system is the amount of heat necessary to raise its temperature one degree under the conditions given. Letting the subscripts V and P denote, respectively, conditions of constant volume and pressure, it follows from Eq. (3-60) that for hydrostatic systems

$$C_v = \left(\frac{dQ}{dT}\right)_V = \left(\frac{\partial U}{\partial T}\right)_V \tag{3-62}$$

and

$$C_p = \left(\frac{dQ}{dT}\right)_P = \left(\frac{\partial U}{\partial T}\right)_P + P\left(\frac{\partial V}{\partial T}\right)_P, \tag{3-63}$$

where $\partial/\partial T$ denotes differentiation with respect to temperature. Noting that the internal energy, U, is only a function of pressure, P, temperature, T, and volume, V, we obtain from Eqs. (3-62) and (3-63) the relation

$$C_p - C_v = \left[\left(\frac{\partial U}{\partial V}\right)_T + P\right]\left(\frac{\partial V}{\partial T}\right)_P. \tag{3-64}$$

For an ideal gas, $(\partial U/\partial V)_T = 0$, and it follows from Eqs. (3-61) and (3-64) that

$$C_p - C_v = nR, \tag{3-65}$$

where n is the number of gram-moles and R is the universal gas constant. For a nonrelativistic, nondegenerate, monatomic ideal gas, $U = 3nRT/2$, and from

Eqs. (3-62) and (3-63), $C_p = 5nR/2$ and $C_v = 3nR/2$. Similarly, for an ideal diatomic gas, $U = 5nRT/2$, $C_p = 7nR/2$, and $C_v = 5nR/2$. The thermal capacity of one mole of a gas is called the molecular heat and is often given the symbol C, whereas the thermal capacity of one gram of a gas is called the specific heat and is usually given the symbol c.

3.2.3. Adiabatic Processes

An adiabatic process is one in which no heat is exchanged between a system and its surroundings. That is, the gain or loss of heat by conduction or radiation can be ignored in an adiabatic process. A measure of this process is the adiabatic index

$$\gamma = C_p/C_v. \tag{3-66}$$

The γ takes on the values 5/3, 7/5, and 4/3, respectively, for an ideal monatomic, diatomic and polyatomic gas. Using the adiabatic condition $dQ = 0$ in Eq. (3-60), it then follows from Eqs. (3-61), (3-62), and (3-63) that

$$PV^{\gamma} = \text{constant}, \quad TV^{\gamma-1} = \text{constant}, \quad \text{and } TP^{(1-\gamma)/\gamma} = \text{constant},$$

for an adiabatic process involving an ideal gas. These equations show, for example, that a gas is cooled or heated according as it undergoes an adiabatic expansion or contraction.

The requirement $dQ = 0$ imposes, through the first law of thermodynamics, one relation between the pressure, P, temperature, T, and the volume, V. Three adiabatic exponents are then defined by

$$\Gamma_1 = -\left(\frac{d\ln P}{d\ln V}\right)_{ad} = \left(\frac{d\ln P}{d\ln \rho}\right)_{ad},$$
$$\frac{\Gamma_2}{(\Gamma_2 - 1)} = \left(\frac{d\ln P}{d\ln T}\right)_{ad}, \tag{3-67}$$

and

$$\Gamma_3 - 1 = -\left(\frac{d\ln T}{d\ln V}\right)_{ad} = \left(\frac{d\ln T}{d\ln \rho}\right)_{ad},$$

where the specific volume $V = \rho^{-1}$, the mass density is ρ, and the subscript ad means the differential is along an adiabat for which $dS = dQ/T = 0$. Only two of the three gammas are independent and we have the general identity

$$\frac{\Gamma_1}{\Gamma_3 - 1} = \frac{\Gamma_2}{\Gamma_2 - 1}. \tag{3-68}$$

In the classical limit of a nonrelativistic, monatomic, ideal gas with no internal degrees of freedom, $\Gamma_1 = \Gamma_2 = \Gamma_3 = \gamma = C_p/C_v = 5/3$. In general Γ_1 is important for determining the conditions of dynamical instability of stars whereas Γ_2 and Γ_3 are important, respectively, in determining the conditions of convective and pulsational instability of stars.

3.2.4. Polytropic Processes

A polytropic process is defined as one in which the thermal capacity, C, remains constant during the entire process. The polytropic exponent, γ', is defined as

$$\gamma' = \frac{C_p - C}{C_v - C}, \tag{3-69}$$

where the thermal capacity $C = dQ/dT$ and C_p and C_v are defined by Eqs. (3-62) and (3-63). For an ideal gas, it follows from Eqs. (3-60), (3-61), (3-62), (3-63), and (3-68) that

$$C = \left(C_v - \frac{nR}{\gamma' - 1}\right) = \text{constant}$$

and

$$P\,V^{\gamma'} = \text{constant}, \quad T V^{\gamma'-1} = \text{constant}, \quad \text{and } T P^{(1-\gamma')/\gamma'} = \text{constant}.$$

If the index $n = 1/(\gamma' - 1)$, then the pressure, P, internal energy, E_{int}, gravitational energy, E_{grav}, total energy, E_{tot}, and central pressure, P_c, of a polytropic gas sphere of radius, R, and mass, M, are given by (Lane, 1870; Emden, 1907; Chandrasekhar, 1939)

$$P \propto \rho^{1+1/n},$$

$$E_{int} = \frac{GM^2}{R} \frac{n}{5-n},$$

$$E_{grav} = -\frac{GM^2}{R} \frac{3}{5-n}, \tag{3-70}$$

$$E_{tot} = E_{int} + E_{grav} = \frac{-GM^2}{R} \frac{3-n}{5-n}$$

and

$$P_c = K G M^{2/3} \rho_c^{4/3},$$

where ρ_c is the central mass density, and the constant $K = 0.488$, 0.396, and 0.364 for $n = 3/2$ ($\gamma = 5/3$), $n = 5/2$ ($\gamma = 7/5$), and $n = 3$ ($\gamma = 4/3$), respectively. For a non-relativistic monatomic gas $n = 3/2$, and for a relativistic gas $n = 4/3$. For $n > 3$ a star has no equilibrium state, and for $n = 3$, an equilibrium exists for any density as the energy is independent of density.

3.2.5. Second Law of Thermodynamics and the Entropy of a Gas

The entropy, S, of a system is defined so that a change in entropy is equal to the integral of dQ/T between the terminal states, where dQ is the amount of heat added to the system from the surroundings, and T is the temperature. The second law of thermodynamics states that, for an isolated system,

$$dS = \frac{dQ}{T} \geq 0, \tag{3-71}$$

where the equality sign holds for a reversible process, and the greater than sign holds for an irreversible process. As expressed by Clausius (1850, 1857, 1865), this law means that heat cannot pass by itself from a colder to a hotter body. For a reversible process which leaves a system in the same state as it was originally,

$$\int \frac{dQ}{T} = 0, \tag{3-72}$$

which was first stated by CARNOT (1824), and first used to define temperature by KELVIN (1848).

From Eqs. (3-60), (3-61), (3-62), (3-63), (3-65), and (3-71), the change in entropy, dS, for a process involving an ideal gas is

$$\begin{aligned} dS &= C_p \frac{dT}{T} - nR\frac{dP}{P} \\ &= C_v \frac{dT}{T} + nR\frac{dV}{V} \\ &= C_v \frac{dT}{T} - (C_p - C_v)\frac{d\rho}{\rho}, \end{aligned} \tag{3-73}$$

where C_p and C_v are the heat capacities at constant pressure, P, and volume, V, the mass density is ρ, the universal gas constant is R, and n is the number of gram-moles.

For a polytropic change of index $n = 1/(\gamma' - 1)$,

$$dS = C_v(\gamma' - \gamma)\frac{d\rho}{\rho} = C_v \frac{(\gamma' - \gamma)}{\gamma' - 1}\frac{dT}{T}, \tag{3-74}$$

where $\gamma = C_p/C_v$ is the adiabatic index.

In statistical mechanics, criteria are given for assigning to a given thermodynamical state the number, W, of corresponding dynamical states. The integer, W, is usually called the thermodynamic probability of the given thermodynamic state. Because the most stable state of a system is the state of highest probability consistent with the given total energy of the system, there is a functional relationship between the entropy, S, and the probability, W, given by (BOLTZMANN, 1868, 1877, 1896; PLANCK, 1901)

$$S = k \ln W, \tag{3-75}$$

where the conversion factor between ergs and degrees is given by Boltzmann's constant $k = 1.380622(59) \times 10^{-16}$ erg °K^{-1}. Because Boltzmann's constant is equal to the ratio R/N_A, the equation of state for a perfect gas, Eq. (3-61), may be written in the equivalent forms

$$P = NkT = \frac{nRT}{V} = \frac{nkN_A}{V}T = \frac{\rho RT}{\mu} = \frac{\rho kT}{\mu m_H}, \tag{3-76}$$

where N is the total particle number density, n is the number of gram moles, the universal gas constant $R \approx 8.31434(35) \times 10^7$ erg mole^{-1} °K^{-1}, Avogadro's number $N_A \approx 6.022169(40) \times 10^{23}$ mole^{-1}, the gas mass density is ρ, the mean molecular weight is μ, and $m_H \approx 1.67 \times 10^{-24}$ gm is the mass of the hydrogen atom. Avogadro's number is the number of molecules in a number of grams equal to the molecular weight. A related constant is Loschmidt's number $N_L \approx 2.68714 \times 10^{19}\ \mathrm{cm}^{-3}$ (LOSCHMIDT, 1865), which is the number of molecules in a cubic centimeter of an ideal gas at standard conditions ($V = 22{,}413.6\ \mathrm{cm}^3\ \mathrm{atm\ mole}^{-1}$, $T = 273.15\,^\circ\mathrm{K}$, and $P = 1\ \mathrm{atm} = 760\ \mathrm{mm}$ of $\mathrm{Hg} = 1.01325 \times 10^6\ \mathrm{dyn\ cm}^{-2}$). For a monatomic nonrelativistic gas, in the absence of ionization and excitation, the entropy (entropy per mole) for temperatures not near absolute zero is given by (SACKUR, 1911, 1913; TETRODE, 1912)

$$S = N_{\mathrm{tot}} k \ln\left[\frac{e^{5/2} g_0}{N}\left(\frac{2\pi M k T}{h^2}\right)^{3/2}\right], \tag{3-77}$$

where N_{tot} is the total number of gas atoms, $e = 2.7182812$ is the base of the natural logarithm, g_0 is the statistical weight of the ground state of the atom, N is the number density of the gas, and M is the mass of the gas atom. Eq. (3-77) may be written in the equivalent form

$$S = \frac{5}{2} k N_{\mathrm{tot}} + k N_{\mathrm{tot}} \ln\left[g_0 \left(\frac{4\pi M U}{3 h^2 N_{\mathrm{tot}}}\right)^{3/2} \frac{V}{N_{\mathrm{tot}}}\right],$$

where the volume is V, and the internal energy, U, of a monatomic gas is $3 k N_{\mathrm{tot}} T/2$. Because the joint probability is the product of the two separate probabilities, it follows from Eq. (3-75) that the entropy of a composite system is additive. The entropy of complex systems is often conveniently calculated using the Helmholtz free energy, and these calculations are discussed in the following section.

Another alternative form of Eq. (3-77) is (cf. Eq. (3-76))

$$S = -\frac{R}{\mu} \ln\left(\frac{\rho}{\mu M}\right) + \frac{3}{2}\left(\frac{R}{\mu}\right) \ln(kT) + \left(C + \frac{5}{2}\right)\left(\frac{R}{\mu}\right),$$

where S is the specific entropy (entropy per gram), R is the gas constant, μ is the mean molecular weight ($\mu = 1$ for hydrogen and $\mu = 4$ for ^{4}He), the gas mass density is ρ, and the constant $C = \ln[g_0 (2\pi M/h^2)^{3/2}]$ has the value $C = 20.204$ for atomic hydrogen and $C = 22.824$ for ^{4}He. For an ionized gas, $\mu = \mu_0/(1 + Z/A)$ where μ_0 is the molecular weight of the neutral gas. At full ionization $\mu = 0.5$ and 4/3 for hydrogen and ^{4}He, respectively.

The specific entropy (per gram) of a photon gas is given by

$$S = \frac{4 a T^3}{3 \rho}, \tag{3-78}$$

where the radiation density constant $a = 7.564 \times 10^{-15}\ \mathrm{erg\ cm}^{-3}\ {}^\circ\mathrm{K}^{-4}$, and the gas temperature and mass density are, respectively, T and ρ.

3.2.6. Combined First and Second Laws

The combined first and second law for a reversible process is

$$dU = T\,dS - dW, \tag{3-79}$$

where dU is the change in the internal energy of the system, dS is the change in entropy of the system, dW is the amount of work done by the system on its surroundings, and T is the temperature. For a hydrostatic system, this combined law becomes

$$dU = TdS - PdV, \tag{3-80}$$

where P is the pressure, and dV is the change in volume.

Useful functions which are combinations of thermodynamic variables include the enthalpy, H, given by

$$H = U + PV, \tag{3-81}$$

the Helmholtz function, F, (Helmholtz free energy) given by

$$F = U - TS, \tag{3-82}$$

and the Gibbs function, G, (Gibbs free energy) given by

$$G = U - TS + PV = F + PV = H - TS, \tag{3-83}$$

where U is the internal energy and S is the entropy.

A useful mnemonic diagram relating the thermodynamic variables is shown below

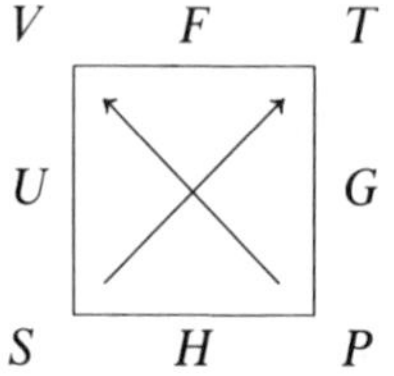

Each of the four thermodynamic potentials U, F, G, and H, is flanked by its natural independent variables. In writing the differential expression for each of the potentials in terms of its independent variables, the coefficient is designated by an arrow. An arrow pointing away from a natural variable implies a positive coefficient, whereas one pointing towards the variable implies a negative coefficient. Hence we have, for example,

$$dF = -PdV - SdT,$$
$$dG = -SdT + VdP,$$
$$dH = TdS + VdP,$$

and

$$dU = TdS - PdV.$$

The Maxwell relations

$$\left(\frac{\partial V}{\partial S}\right)_P = \left(\frac{\partial T}{\partial P}\right)_S,$$
$$\left(\frac{\partial S}{\partial P}\right)_T = -\left(\frac{\partial V}{\partial T}\right)_P,$$
$$\left(\frac{\partial P}{\partial T}\right)_V = \left(\frac{\partial S}{\partial V}\right)_T, \tag{3-84}$$

and

$$\left(\frac{\partial T}{\partial V}\right)_S = -\left(\frac{\partial P}{\partial S}\right)_V,$$

may also be read from the diagram by rotating about the four corners.

The Helmholtz free energy, F, is often called the free energy at constant volume, and the Gibbs free energy, G, is often called the free energy at constant pressure. For thermal equilibrium at constant temperature and volume, F is a minimum, whereas for thermal equilibrium at constant temperature and pressure, G is a minimum. When a system undergoes an isothermal transformation, the work performed by it can never exceed minus the variation, ΔF, of its Helmholtz free energy; and it is equal to $-\Delta F$ if the transformation is reversible. A consequence of this fact is that a thermodynamic system having the temperature of its environment is in a state of stable equilibrium when its Helmholtz free energy is a minimum. A thermodynamic system is equivalently described by the potentials U, H, F, or G depending on the choice of independent variables $U(V,S)$, $H(P,S)$, $F(V,T)$, or $G(P,T)$. The conjugate variables are described by the relations

$$\begin{aligned}
&T = \left(\frac{\partial U}{\partial S}\right)_V, \quad P = -\left(\frac{\partial U}{\partial V}\right)_S, \\
&T = \left(\frac{\partial H}{\partial S}\right)_P, \quad V = \left(\frac{\partial H}{\partial P}\right)_S, \\
&S = -\left(\frac{\partial F}{\partial T}\right)_V, \quad P = -\left(\frac{\partial F}{\partial V}\right)_T, \quad U = -T^2\left(\frac{\partial}{\partial T}\frac{F}{T}\right)_{V,N}, \\
\text{and} \quad &S = -\left(\frac{\partial G}{\partial T}\right)_P, \quad V = \left(\frac{\partial G}{\partial P}\right)_T,
\end{aligned} \tag{3-85}$$

where the subscripts V, T, P, or S denote, respectively, conditions of constant volume, temperature, pressure, or entropy.

According to statistical mechanics, a nongenerate, system consisting of N_{tot} particles has a free energy, F, given by

$$F = -kT \ln Q, \tag{3-86}$$

where the partition function, Q, of an ideal Boltzmann gas is given by (Gibbs, 1902)

$$Q = \sum_n \exp\left(\frac{-E_n}{kT}\right) = \left[Z\left(\frac{e}{N_{\text{tot}}}\right)\right]^{N_{\text{tot}}}, \tag{3-87}$$

where E_n is the energy of the nth state of the system, Z is the partition function for one particle, and $e = 2.7182812$ is the base of the natural logarithm. If to an energy level, E_n, there belong g_n permissable states, the level is said to be degenerate and g_n is called its statistical weight. In this case,

$$Q = \sum_n g_n \exp\left(\frac{-E_n}{kT}\right). \tag{3-88}$$

The probability, P, that a particle has energy, E_n, is given by

$$P = \frac{1}{Q} \exp\left(\frac{-E_n}{kT}\right). \tag{3-89}$$

The partition function for one molecule, Z, is given by

$$Z = Z_{\text{trans}} \cdot Z_{\text{rot}} \cdot Z_{\text{vib}} \cdot Z_{\text{el}}, \tag{3-90}$$

where the subscripts trans, rot, vib, and el denote, respectively, the translational, rotational, vibrational, and electronic partition functions. These partition functions are given by

$$Z_{\text{trans}} = \left(\frac{2\pi M k T}{h^2}\right)^{3/2} V, \tag{3-91}$$

where M is the particle mass and V is the volume occupied by the gas,

$$Z_{\text{rot}} = \frac{8\pi^2 I k T}{h^2 \sigma} = \frac{k T}{h \nu_{\text{rot}} \sigma}, \tag{3-92}$$

for a diatomic or linear polyatomic molecule. Here I is the moment of inertia given by $I = h/(8\pi^2 \nu_{\text{rot}})$ where ν_{rot} is the rotational frequency, and the symmetry factor, σ, is equal to one plus the number of transpositions of identical atoms ($\sigma = 2$ for a diatomic molecule of identical atoms and $\sigma = 1$ for one composed of different atoms),

$$Z_{\text{rot}} = \frac{8\pi^2}{\sigma}\left(\frac{2\pi I k T}{h^2}\right)^{3/2}, \tag{3-93}$$

for a nonlinear polyatomic molecule,

$$Z_{\text{vib}} = \left[1 - \exp\left(\frac{-h\nu_{\text{vib}}}{kT}\right)\right]^{-1}, \tag{3-94}$$

where ν_{vib} is the vibrational frequency, and

$$Z_{\text{el}} = \sum_n g_n \exp\left(\frac{-E_n}{kT}\right) = u \exp\left(\frac{-E_0}{kT}\right), \tag{3-95}$$

where g_n is the statistical weight of level n of energy, E_n, the E_0 is the zero point energy (the energy of the ground state), and the ionic partition function, u, is given by

$$u = \sum_k g_k \exp\left(\frac{-E_k}{kT}\right) = g_0 + g_1 \exp\left(\frac{-E_1}{kT}\right) + \cdots, \tag{3-96}$$

where $E_n = E_k - E_0$ is the excitation energy of the ion in the nth state, E_k is the energy of level k above the ground state energy, and g_k is the statistical weight of level k. An energy level with angular momentum, J, has $g_k = 2J+1$, and for hydrogen like atoms $g_k = 2k^2$. The statistical weight for the free electron is $g_e = 2$.

It follows from Eqs. (3-86) to (3-96) that the free energy of a nondegenerate, nonrelativistic, monatomic gas is given by

$$F = -N_{\text{tot}} k T \ln\left[\frac{e V g_0}{N_{\text{tot}}}\left(\frac{2\pi M k T}{h^2}\right)^{3/2}\right], \tag{3-97}$$

where g_0, the statistical weight of the ground level, is assumed to be equal to Z_{el}. Eq. (3-77) for the entropy of a monatomic gas follows from Eqs. (3-85) and (3-97). The free energy and entropy for more complex systems may be derived in a similar way using Eqs. (3-85) to (3-95). The internal energy, U, is given by

$$U = \tfrac{3}{2} N_{\text{tot}} k T \tag{3-98}$$

for a monatomic gas,

$$U_{rot} = N_{tot} k T \quad \text{and} \quad \tfrac{3}{2} N_{tot} k T,$$

for the rotational energy of diatomic or linear polyatomic molecules, and nonlinear polyatomic molecules, respectively, and

$$U_{vib} = N_{tot} \frac{h \nu_{vib}}{\left[\exp\left(\frac{h \nu_{vib}}{k T}\right) - 1\right]},$$

for the vibrational energy of N_{tot} identical oscillators (diatomic molecules).

For a system consisting of particles of one kind assumed to be fermions or bosons, we have

$$F = -k T \ln Z + \mu N_{tot} = -P V + \mu N_{tot}, \tag{3-99}$$

where the grand partition function, Z, is given by

$$\ln Z = \sum_k g(\varepsilon) \ln \{1 \pm \exp[(\mu - \varepsilon_k)/k T]\}^{\pm 1}, \tag{3-100}$$

where μ is the chemical potential for the kind of particle under consideration, ε_k is the energy of a single representative "particle" in the quantum state k, and $g(\varepsilon)$ is the statistical weight for the state k. The + sign is to be used for fermions, and the − sign for bosons. The total number of particles of the kind under consideration for the whole system is N_{tot}, the pressure is P, and V is the volume. If g_i denotes the statistical weight of the ith discrete state and $4\pi p^2 V dp/h^3$ is the statistical weight for the continuous translational states, then

$$g(\varepsilon) = g_i \frac{4\pi p^2 V dp}{h^3},$$

where p is the momentum and V is the volume.

3.2.7. Nernst Heat Theorem

The Nernst theorem (NERNST, 1906, 1926) states that the entropy of every thermodynamic system at a temperature of absolute zero can always be taken to be zero. In terms of the Boltzmann-Planck relation (3-75), $W=1$ for $S=0$ so that the thermodynamic state of a system at absolute zero corresponds to only one dynamical state. This ground state is that dynamical state of lowest energy which is compatible with the system. According to the Nernst theorem, in the limit as the temperature, T, approaches zero

$$S = \Delta S = C_p = C_v = 0 \quad \text{as} \quad T = 0. \tag{3-101}$$

Nernst's theorem can be related to the phenomenological principle (NERNST, 1926) that it is impossible to cool any system down to absolute zero.

3.2.8. Fluctuations in Thermodynamic Quantities

The thermal motion of the particles of a gas produce fluctuations in the temperature, T, pressure, P, volume, V, energy, U, entropy, S, and the total number of particles, N_{tot}, of the gas. In general for any physical quantity, X, these fluctuations are described by the mean square fluctuation $\langle(\Delta X)^2\rangle$ about the mean $\langle X\rangle$, where the brackets $\langle\ \rangle$ denote a time average.

The mean square fluctuations, $\langle(\Delta T)^2\rangle$ and $\langle(\Delta V)^2\rangle$, of temperature, T, and volume, V, respectively, are given by (LANDAU and LIFSHITZ, 1969)

$$\langle(\Delta T)^2\rangle = \frac{kT^2}{C_v},$$
$$\langle(\Delta V)^2\rangle = -kT\left(\frac{\partial V}{\partial P}\right)_T, \tag{3-102}$$

where C_v is the specific heat at constant volume and the differential is carried out at constant temperature. The mean square fluctuation, $\langle(\Delta N)^2\rangle$, in the total number of particles, N, is given by

$$\langle(\Delta N)^2\rangle = kT\left(\frac{\partial N}{\partial \mu}\right)_{T,V}, \tag{3-103}$$

where the chemical potential, μ, is given by

$$\mu = \left(\frac{\partial U}{\partial N}\right)_{S,V} = -T\left(\frac{\partial S}{\partial N}\right)_{U,V} = \left(\frac{\partial F}{\partial N}\right)_{V,T} = \left(\frac{\partial G}{\partial N}\right)_{P,T}. \tag{3-104}$$

For photons in thermodynamic equilibrium (black body radiation) we have $\mu=0$. For a nonrelativistic, monatomic gas we have

$$\mu = kT\ln\left[\frac{N_{tot}}{gV}\left(\frac{2\pi\hbar^2}{mkT}\right)^{3/2}\right] = kT\ln\left[\frac{P}{g(kT)^{5/2}}\left(\frac{2\pi\hbar^2}{m}\right)^{3/2}\right],$$

where N_{tot} is the total number of particles, m is the mass of one gas particle, and g is the statistical weight (degree of degeneracy) of the ground state. For example, for an electron gas $g=2$. For an ideal gas, Eq. (3-103) gives $\langle(\Delta N)^2\rangle = N$.

The mean square fluctuations, $\langle(\Delta S)^2\rangle$ and $\langle(\Delta P)^2\rangle$, of entropy, S, and pressure, P, are given by

$$\langle(\Delta S)^2\rangle = C_p$$

and

$$\langle(\Delta P)^2\rangle = -kT\left(\frac{\partial P}{\partial V}\right)_S. \tag{3-105}$$

It follows from Eq. (3-103) that for a Boltzmann gas the mean square fluctuation of the number of particles in the kth quantum state is equal to the number of particles in the state. For a Fermi gas, the average number $\langle N_k\rangle$ of particles in the kth quantum state is related to its mean square fluctuation $\langle(\Delta N_k)^2\rangle$ by the relation

$$\langle(\Delta N_k)^2\rangle = \langle N_k\rangle\,(1-\langle N_k\rangle), \tag{3-106}$$

whereas for an Einstein-Bose gas (a photon gas)

$$\langle(\Delta N_k)^2\rangle = \langle N_k\rangle\,(1+\langle N_k\rangle).$$

For the photon gas, the mean square fluctuation, $\langle(\Delta E)^2\rangle$, for the energy, E, in the frequency range ν to $\nu+d\nu$ is given by (EINSTEIN, 1925)

$$\langle(\Delta E)^2\rangle = h\nu E + \frac{c^3 E^2}{V 8\pi\nu^2 d\nu}, \tag{3-107}$$

where V is the volume.

3.3. Statistical Properties and Equations of State

3.3.1. The Nondegenerate, Perfect Gas

3.3.1.1. Maxwell Distribution Function for Energy and Velocity

The number density, $n(\varepsilon)$, of perfect gas particles with kinetic energy between ε and $\varepsilon+d\varepsilon$ is given by

$$n(\varepsilon) = \frac{g(\varepsilon)}{V}\exp\left(\frac{\mu}{kT}\right)\exp\left(-\frac{\varepsilon}{kT}\right)d\varepsilon, \tag{3-108}$$

where $g(\varepsilon)$ is the number of possible particle states of energy, ε, and is given by

$$g(\varepsilon)\,d\varepsilon = \frac{4\pi g V}{h^3}(2M^3)^{1/2}\varepsilon^{1/2}\,d\varepsilon = \frac{4\pi V g}{h^3}p^2\,dp, \tag{3-109}$$

where g is the statistical weight of the discrete energy states, M is the particle rest mass, the temperature is T, the volume is V, and p is the gas momentum. The chemical potential, or Fermi energy, μ, is given in Eq. (3-104). The total number density of particles, N, is given by

$$N = \int_0^\infty n(\varepsilon)\,d\varepsilon = \frac{g}{2\pi^2\hbar^3}\int_0^\infty \exp\left(\frac{\mu}{kT}\right)\exp\left(-\frac{\varepsilon}{kT}\right)p^2\,dp, \tag{3-110}$$

where the momentum, p, is related to the kinetic energy, ε, by the equation

$$\varepsilon = (p^2c^2 + M^2c^4)^{1/2} - Mc^2, \tag{3-111}$$

where M is the rest mass of the particle, and c is the velocity of light. For the nonrelativistic, nondegenerate, monatomic gas,

$$\exp\left[\frac{\mu}{kT}\right] = \frac{Nh^3}{g(2\pi MkT)^{3/2}} = \frac{P}{g(kT)^{5/2}}\left(\frac{2\pi\hbar^2}{M}\right)^{3/2}, \tag{3-112}$$

and Eq. (3-108) becomes

$$n(\varepsilon) = \frac{2N}{\pi^{1/2}(kT)^{3/2}}\varepsilon^{1/2}\exp\left(-\frac{\varepsilon}{kT}\right)d\varepsilon. \tag{3-113}$$

Eq. (3-113) was first obtained by MAXWELL (1860), who also gave the correct formula for the probability distribution, $f(v)dv$, of gas particles with speeds between v and $v+dv$. If the energy, ε, is the kinetic energy, $Mv^2/2$, it follows from Eq. (3-113) that the number of particles with speeds between v and $v+dv$, for a nonrelativistic, nondegenerate gas, is given by

$$N_{\mathrm{tot}} f(v) dv = N_{\mathrm{tot}} \left(\frac{2}{\pi}\right)^{1/2} \left(\frac{M}{kT}\right)^{3/2} v^2 \exp\left[-\frac{Mv^2}{2kT}\right] dv, \tag{3-114}$$

where M is the mass of the atom or molecule, and N_{tot} is the total number of atoms and molecules. Both Eqs. (3-113) and (3-114) follow directly from the Maxwell-Boltzmann solution to Boltzmann's equation (3-4).

From Eq. (3-114) we have the relation

$$N_{\mathrm{tot}} \int_0^\infty f(v) dv = N_{\mathrm{tot}}. \tag{3-115}$$

The most probable speed, v_{p}, the mean speed, $\langle v \rangle$, and the root-mean-square speed, v_{rms}, are easily calculated by the relations

$$\begin{aligned} v_{\mathrm{p}} &= (2kT/M)^{1/2}, \\ \langle v \rangle &= (8kT/\pi M)^{1/2}, \\ v_{\mathrm{rms}} &= (3kT/M)^{1/2} = \langle v^2 \rangle^{1/2}. \end{aligned} \tag{3-116}$$

The momentum distribution function for the nonrelativistic, nondegenerate gas may be obtained from Eq. (3-114) using $\varepsilon = p^2/(2m) = mv^2/2$. For the relativistic, nondegenerate gas the momentum $p = \varepsilon/c \gg mc$, and the number of particles with momenta between p and $p+dp$ is given by

$$N_{\mathrm{tot}} f(p) dp = \frac{N_{\mathrm{tot}} c^3}{8\pi (kT)^3} \exp\left(-\frac{pc}{kT}\right) \cdot 4\pi p^2 dp. \tag{3-117}$$

3.3.1.2. The Energy Density and Equation of State of a Perfect Gas

It follows directly from Eq. (3-108) that the energy density, U/V, of a nondegenerate perfect gas is given by

$$\frac{U}{V} = \int_0^\infty \varepsilon n(\varepsilon) d\varepsilon = \frac{g}{2\pi^2 \hbar^3} \int_0^\infty \exp\left[\left(\frac{\mu}{kT} - \frac{\varepsilon}{kT}\right)\right] \varepsilon p^2 dp, \tag{3-118}$$

where ε is the kinetic energy, the number density, $n(\varepsilon)$, is given by Eqs. (3-108) or (3-113), the Fermi constant, μ, is given by Eq. (3-112), the temperature is T, and the momentum, p, is related to ε by Eq. (3-111). For a nonrelativistic, nondegenerate gas, Eq. (3-118) becomes (MAXWELL, 1860)

$$U = \tfrac{3}{2} N_{\mathrm{tot}} kT, \tag{3-119}$$

where the total number of particles is N_{tot}. For the relativistic, nondegenerate gas,

$$U = 3 N_{\mathrm{tot}} kT. \tag{3-120}$$

The entropy, S, of a nonrelativistic, nondegenerate, monatomic gas is given in Eq. (3-77).

The entropy of a relativistic, nondegenerate gas is given by

$$S = k N_{\mathrm{tot}} \ln\left[\frac{T^4 g_0 8\pi k^4 e^4}{P c^3 h^3}\right] \tag{3-121}$$

where g_0 is the statistical weight of the ground state of the atom, $e=2.7182812$ is the base of the natural logarithm, and the pressure, P, of a relativistic, or non-relativistic, nondegenerate gas is given by

$$PV = N_{\text{tot}} kT .$$

For the relativistic, nondegenerate gas, $U=3PV$. Alternative forms of the equation of state are given in Eq. (3-76), the most useful being

$$P = \frac{R\rho T}{\mu} = 8.314 \times 10^7 \frac{\rho T}{\mu} \text{ dynes cm}^{-2}, \tag{3-122}$$

where here the mean molecular weight per free particle for an ionized gas is given by

$$\mu = \frac{\rho N_{\text{A}}}{N} = \left[\sum_i \frac{X_i(Z_i+1)}{A_i}\right]^{-1} \approx \frac{2}{1+3X+0.5Y}, \tag{3-123}$$

where X_i is the relative mass abundance (mass fraction) of the atom i whose atomic number and atomic weight are, respectively, Z_i and A_i. The weight fraction of hydrogen is X, and the weight fraction of helium is Y. For a hydrogen gas $\mu=0.5$, whereas for helium $\mu=4/3$ at full ionization. For population one stars $X \approx 0.60$ and $Y \approx 0.38$. The gas mass density is ρ and the temperature is T. Typical values of the mass density, temperature, and linear extent of various cosmic gases are given in Table 33.

Table 33. Values of mass density, ρ, temperature, T, and linear extent, R, of various cosmic gases

Region	ρ (g cm^{-3})	T (°K)	R (cm)
Ionosphere	10^{-20}–10^{-10}	200–1,500	6.4×10^8
Magnetosphere	10^{-21}	10^4	10^9–10^{11}
Sun (stars)	1.4 (mean)	10^4–10^7	10^{11}
Solar corona	10^{-19}–10^{-16}	10^6	10^{12}
Solar system	10^{-23}	10^5	10^{15}
Galactic nebulae	10^{-19}–10^{-16}	10^2–10^4	10^{18}–10^{21}
Galaxy	10^{-24}	10^2–10^4	10^{23}
Local cluster	10^{-27}	10^5?	3×10^{24}
Universe	10^{-29}	10^5–10^6?	3×10^{28}

3.3.1.3. Boltzmann Equation for the Population Density of Excited States

Under conditions of local thermodynamic equilibrium at a temperature, T, the number density, N_s, of atoms or molecules in the excited state, s, is (BOLTZMANN, 1872)

$$N_s = \frac{g_s}{g_0} N_0 \exp\left(\frac{-\chi_s}{kT}\right), \tag{3-124}$$

where χ_s is the excitation energy (energy above ground level) of the s state, N_0 is the number of molecules in the ground state, and g_s is the statistical weight of the s state. An energy level with total angular quantum number, J, has $g_s=2J+1$.

If the total number density of atoms or molecules is N_{tot},

$$N_s = g_s \frac{N_{\text{tot}}}{u} \exp\left(\frac{-\chi_s}{kT}\right), \tag{3-125}$$

where the partition function u is given by Eq. (3-96). For free particles, u becomes Z_{trans} given by Eq. (3-91).

The population of levels, m and n, with energies, E_m and E_n, are related by

$$\frac{N_n}{g_n} = \frac{N_m}{g_m} \exp\left(-\frac{E_n - E_m}{kT}\right). \tag{3-126}$$

3.3.1.4. The Saha-Boltzmann Ionization Equation

Under conditions of local thermodynamic equilibrium, the number density, N_r, of atoms in the rth stage of ionization is related to that of the $(r+1)$th stage, N_{r+1}, by (SAHA, 1920, 1921)

$$\frac{N_{r+1}}{N_r} N_e = \frac{u_{r+1}}{u_r} \frac{2(2\pi mkT)^{3/2}}{h^3} \exp\left(\frac{-\chi_r}{kT}\right), \tag{3-127}$$

where N_e is the free electron density, u_r is the partition function of the rth stage, and is given by Eq. (3-96), χ_r is the energy required to remove an electron from the ground state of the r-times ionized atom (its ionization potential), T is the temperature of thermal equilibrium, m is the electron mass, and h and k are, respectively, Planck's and Boltzmann's constants.

For computational purposes, it is useful to have Eq. (3-127) expressed in the logarithmic form

$$\log\left(\frac{N_{r+1} N_e}{N_r}\right) = \log\left(\frac{u_{r+1}}{u_r}\right) + 15.6826 + \frac{3}{2}\log T - \frac{5039.95}{T} I \tag{3-128}$$

or

$$\log\left(\frac{N_{r+1}}{N_r} P_e\right) = \log\left(\frac{2u_{r+1}}{u_r}\right) - 0.4772 + \frac{5}{2}\log T - \frac{5039.95}{T} I,$$

where T is in °K, the ionization potential, I, from the r to $r+1$st stage is in electron volts, and the electron pressure $P_e = N_e kT$. Numerical values for the ionization potentials are given in Table 34.

The population density, N_n, of the nth quantum level is given by (cf. Eq. (3-125) and Eq. (3-127))

$$N_n = N_e N_i \frac{h^3}{(2\pi mkT)^{3/2}} \frac{g_n}{2} \exp\left(\frac{\chi_r - \chi_n}{kT}\right), \tag{3-129}$$

where N_e and N_i are, respectively, the free electron and ion densities, g_n is the statistical weight of the nth level, and χ_n is the excitation energy of the nth level above ground level. For hydrogen like atoms, the excitation energy of the nth state is given by

$$\chi_n = I_H Z^2 \left(1 - \frac{1}{n^2}\right), \tag{3-130}$$

and the statistical weight is given by

$$g_n = 2n^2. \tag{3-131}$$

Here $I_H \approx 13.5\ \text{eV} = 21.36 \times 10^{-12}\ \text{erg} = \chi_r/Z^2 = 2\pi^2 e^4 m/h^2$ is the ionization potential of hydrogen.

For the first ionization of hydrogen and helium, Eq. (3-127) takes the numerical form

$$\frac{N_1 N_e}{N_0} \approx 2.4 \times 10^{15}\, T^{3/2} \exp(-1.58 \times 10^5\, T^{-1})\ \mathrm{cm}^{-3} \tag{3-132}$$

for hydrogen, and

$$\frac{N_1 N_e}{N_0} \approx 9.6 \times 10^{15}\, T^{3/2} \exp(-2.85 \times 10^6\, T^{-1})\ \mathrm{cm}^{-3},$$

for helium. It is sometimes convenient to rewrite Eq. (3-127) in the form relating particle concentrations $\alpha_r = N_r/N_{tot}$, where N_r and N_{tot} denote the total number of atoms in the rth state and the total number of atoms before ionization, respectively.

$$\frac{\alpha_{r+1}\alpha_e}{\alpha_r} = \frac{V}{N_{tot}}\,\frac{N_{r+1} N_e}{N_r}, \tag{3-133}$$

where V is gas volume and $N_{r+1} N_e/N_r$ is given by Eq. (3-127). For a singly ionized region $\alpha = \alpha_1 = \alpha_e$ and the degree of ionization is given by

$$\frac{\alpha^2}{1-\alpha} = 2\frac{u_1}{u_0}\,\frac{V}{N_{tot}}\left(\frac{2\pi m k T}{h^2}\right)^{3/2} \exp\left(\frac{-I}{kT}\right), \tag{3-134}$$

where $(2\pi m k T/h^2) = 2.4 \times 10^{15}\, T$, $u_1/u_0 = \frac{1}{2}$ for hydrogen and 2 for helium, I is the ionization potential, N_{tot} is the total number of atoms plus free protons for the case of hydrogen, and $\alpha = N_e/N_{tot}$ for hydrogen. The total number of particles is $N_{tot}\,(1+\alpha)$ so that the gas pressure, P, is given by

$$P = N_{tot}(1+\alpha)\frac{kT}{V}, \tag{3-135}$$

whereas the energy, U, is given by

$$U = \tfrac{3}{2} N_{tot}(1+\alpha) k T + N_{tot}\,\alpha(I + \chi_m), \tag{3-136}$$

where I is the ionization potential, and χ_m is the electron excitation energy of the ion.

3.3.1.5. Strömgren Radius for the Sphere of Ionization

In the immediate neighborhood of a source of ultraviolet radiation, the neutral interstellar hydrogen will become ionized to form an ionized hydrogen region (H II region) with a Strömgren radius, s_0, given by (Strömgren, 1939; Vandervoort, 1963)

$$s_0 = \left(\frac{3S}{4\pi\alpha N^2}\right)^{1/3}, \tag{3-137}$$

where α is the recombination coefficient of hydrogen to all states except the ground state, and is given by (Kaplan and Pikelner, 1970)

$$\alpha \approx 2.6 \times 10^{-13}\left(\frac{10^4}{T}\right)^{0.85}\ \mathrm{cm}^3\ \mathrm{sec}^{-1},$$

Table 34. Ionization potentials of the elements in electron volts (one eV = 1.6021×10^{-12} erg an

Z	Element	Stage of Ionization I	II	III	IV	V	VI	VII	VIII	IX	X
1	H	13.598									
2	He	24.587	54.416								
3	Li	5.392	75.638	122.451							
4	Be	9.322	18.211	153.893	217.713						
5	B	8.298	25.154	37.930	259.368	340.217					
6	C	11.260	24.383	47.887	64.492	392.077	489.981				
7	N	14.534	29.601	47.448	77.472	97.888	552.057	667.029			
8	O	13.618	35.116	54.934	77.412	113.896	138.116	739.315	871.387		
9	F	17.422	34.970	62.707	87.138	114.240	157.161	185.182	953.886	1,103.089	
10	Ne	21.564	40.962	63.45	97.11	126.21	157.93	207.27	239.09	1,195.797	1,362.16
11	Na	5.139	47.286	71.64	98.91	138.39	172.15	208.47	264.18	299.87	1,465.09
12	Mg	7.646	15.035	80.143	109.24	141.26	186.50	224.94	265.90	327.95	367.53
13	Al	5.986	18.828	28.447	119.99	153.71	190.47	241.43	284.59	330.21	398.57
14	Si	8.151	16.345	33.492	45.141	166.77	205.05	246.52	303.17	351.10	401.43
15	P	10.486	19.725	30.18	51.37	65.023	220.43	263.22	309.41	371.73	424.50
16	S	10.360	23.33	34.83	47.30	72.68	88.049	280.93	328.23	379.10	447.09
17	Cl	12.967	23.81	39.61	53.46	67.8	97.03	114.193	348.28	400.05	455.62
18	Ar	15.759	27.629	40.74	59.81	75.02	91.007	124.319	143.456	422.44	478.68
19	K	4.341	31.625	45.72	60.91	82.66	100.0	117.56	154.86	175.814	503.44
20	Ca	6.113	11.871	50.908	67.10	84.41	108.78	127.7	147.24	188.54	211.70
21	Sc	6.54	12.80	24.76	73.47	91.66	111.1	138.0	158.7	180.02	225.32
22	Ti	6.82	13.58	27.491	43.266	99.22	119.36	140.8	168.5	193.2	215.91
23	V	6.74	14.65	29.310	46.707	65.23	128.12	150.17	173.7	205.8	230.5
24	Cr	6.766	16.50	30.96	49.1	69.3	90.56	161.1	184.7	209.3	244.4
25	Mn	7.435	15.640	33.667	51.2	72.4	95	119.27	196.46	221.8	243.3
26	Fe	7.870	16.18	30.651	54.8	75.0	99	125	151.06	235.04	262.1
27	Co	7.86	17.06	33.50	51.3	79.5	102	129	157	186.13	276
28	Ni	7.635	18.168	35.17	54.9	75.5	108	133	162	193	224.5
29	Cu	7.726	20.292	36.83	55.2	79.9	103	139	166	199	232
30	Zn	9.394	17.964	39.722	59.4	82.6	108	134	174	203	238
31	Ga	5.999	20.51	30.71	64						
32	Ge	7.899	15.934	34.22	45.71	93.5					
33	As	9.81	18.633	28.351	50.13	62.63	127.6				
34	Se	9.752	21.19	30.820	42.944	68.3	81.70	155.4			
35	Br	11.814	21.8	36	47.3	59.7	88.6	103.0	192.8		
36	Kr	13.999	24.359	36.95	52.5	64.7	78.5	111.0	126	230.9	
37	Rb	4,177	27.28	40	52.6	71.0	84.4	99.2	136	150	277.1
38	Sr	5.695	11.030	43.6	57	71.6	90.8	106	122.3	162	177
39	Y	6.38	12.24	20.52	61.8	77.0	93.0	116	129	146.2	191
40	Zr	6.84	13.13	22.99	34.34	81.5					
41	Nb	6.88	14.32	25.04	38.3	50.55	102.6	125			
42	Mo	7.099	16.15	27.16	46.4	61.2	68	126.8	153		

[1] From C.E. Moore (1970).

$= 8.617 \times 10^{-5}$ eV °K^{-1})[1]

tage of Ionization

XI	XII	XIII	XIV	XV	XVI	XVII	XVIII	XIX	XX	XXI
,648.659										
,761.802	1,962.613									
442.07	2,085.983	2,304.080								
476.06	523.50	2,437.676	2,673.108							
479.57	560.41	611.85	2,816.943	3,069.762						
504.78	564.65	651.63	707.14	3,223.836	3,494.099					
529.26	591.97	656.69	749.74	809.39	3,658.425	3,946.193				
538.95	618.24	686.09	755.73	854.75	918	4,120.778	4,426.114			
564.13	629.09	714.02	787.13	861.77	968	1,034	4,610.955	4,933.931		
591.25	656.39	726.03	816.61	895.12	974	1,087	1,157	5,129.045	5,469.738	
249.832	685.89	755.47	829.79	926.00						
265.23	291.497	787.33	861.33	940.36						
255.04	308.25	336.267	895.58	974.02						
270.8	298.0	355	384.30	1,010.64						
286.0	314.4	343.6	404	435.3	1,136.2					
290.4	330.8	361.0	392.2	457	489.5	1,266.1				
305	336	379	411	444	512	546.8	1,403.0			
321.2	352	384	430	464	499	571	607.2	1,547		
266	368.8	401	435	484	520	557	633	671	1,698	
274	310.8	419.7	454	490	542	579	619	698	738	1,856
324.1										
206	374.0									

Table 34 (continued)

Z	Element	Stage of Ionization									
		I	II	III	IV	V	VI	VII	VIII	IX	X
43	Tc	7.28	15.26	29.54							
44	Ru	7.37	16.76	28.47							
45	Rh	7.46	18.08	31.06							
46	Pd	8.34	19.43	32.93							
47	Ag	7.576	21.49	34.83							
48	Cd	8.993	16.908	37.48							
49	In	5.786	18.869	28.03	54						
50	Sn	7.344	14.632	30.502	40.734	72.28					
51	Sb	8.641	16.53	25.3	44.2	56	108				
52	Te	9.009	18.6	27.96	37.41	58.75	70.7	137			
53	I	10.451	19.131	33							
54	Xe	12.130	21.21	32.1							
55	Cs	3.894	25.1								
56	Ba	5.212	10.004								
57	La	5.577	11.06	19.175							
58	Ce	5.47	10.85	20.20	36.72						
59	Pr	5.42	10.55	21.62	38.95	57.45					
60	Nd	5.49	10.72								
61	Pm	5.55	10.90								
62	Sm	5.63	11.07								
63	Eu	5.67	11.25								
64	Gd	6.14	12.1								
65	Tb	5.85	11.52								
66	Dy	5.93	11.67								
67	Ho	6.02	11.80								
68	Er	6.10	11.93								
69	Tm	6.18	12.05	23.71							
70	Yb	6.254	12.17	25.2							
71	Lu	5.426	13.9								
72	Hf	7.0	14.9	23.3	33.3						
73	Ta	7.89									
74	W	7.98									
75	Re	7.88									
76	Os	8.7									
77	Ir	9.1									
78	Pt	9.0	18.563								
79	Au	9.225	20.5								
80	Hg	10.437	18.756	34.2							
81	Tl	6.108	20.428	29.83							
82	Pb	7.416	15.032	31.937	42.32	68.8					
83	Bi	7.289	16.69	25.56	45.3	56.0	88.3				

the number density, N, of the H II region is $N=\rho/m_H$, where ρ is the gas mass density, m_H is the mass of the hydrogen atom, and S is the rate of emission of the ionizing photons from the exciting star

$$S \approx \frac{8\pi^2 R^2 kT}{hc^2} \nu_c^2 \exp\left(\frac{-h\nu_c}{kT}\right),$$

where R is the radius of the exciting star, T is the temperature of the star, and $\nu_c = 3.29 \times 10^{15}$ Hz is the frequency of the limit of the Lyman continuum. Eq. (3-137) indicates that s_0 is a function of N and the spectral class of the exciting star, and this dependence is given by HERSHBERG and PRONIK (1959). These results indicate that $s_0 = U N_e^{-2/3}$ where N_e is the electron density and U varies from 90 to 12 pc cm^{-2} as the spectral class goes from $O5$ to $B1$. STRÖMGREN (1939) first derived a detailed formula for s_0, which is

$$\log s_0 = -6.17 + \frac{1}{3}\log\left[(T_e/T)^{1/2}\left(\frac{2g_1}{g_0}\right)\right] - \frac{1}{3}\log a_u - \frac{1}{3}\theta I$$
$$+\frac{1}{2}\log T + \frac{2}{3}\log R - \frac{2}{3}\log N, \tag{3-138}$$

where s_0 is in parsecs (1 pc $= 3 \times 10^{18}$ cm), T_e is the electron temperature at s_0, the star's temperature is T, the absorption coefficient for the ionizing radiation per neutral gas atom is a_u, the factor $\theta = 5040/T$, the ionization potential is I, the stellar radius is R solar radii ($R_\odot = 6.96 \times 10^{10}$ cm), and N is the number density of neutral and ionized gas. Using the proper values for hydrogen, $I = 13.53$ volts and $a_u = 6.3 \times 10^{-18}$ cm^{-2}, Eq. (3-138) becomes

$$\log s_0 = -0.44 + \tfrac{1}{3}\log[(T_e/T)^{1/2}] - 4.51\,\theta + \tfrac{1}{2}\log T + \tfrac{2}{3}\log R - \tfrac{2}{3}\log N. \tag{3-139}$$

Calculations to determine the temperature of a H II region by taking into account energy gains and losses by photoionization, collisions, radiation transfer etc. have been carried out by SPITZER and SAVEDOFF (1950), OSTERBROCK (1965), HJELLMING (1966, 1968), and RUBIN (1968) for different exciting stars and hydrogen densities. The expansion of the H II region into the surrounding neutral hydrogen has been discussed by VANDERVOORT (1963), MATHEWS (1965), LASKER (1966, 1967), and MATHEWS and O'DELL (1969). The results indicate that the ionization front obtains the radius s_0 in about $(10^4/N)$ years and subsequently has a radius, r, given by Eq. (3-137) for s_0 when N is the density of the H II region. This means then $N = N_0 s_0^{3/2} r^{-3/2}$, where s_0 and N_0 are, respectively, the Strömgren radius and the density at time $t=0$. At a subsequent time, t, the radius, r, is given by

$$r = s_0\left(1 + \frac{7C_1 t}{4s_0}\right)^{4/7}, \tag{3-140}$$

and the density, N, is given by

$$N = N_0\left(1 + \frac{7C_1 t}{4s_0}\right)^{-6/7},$$

where the $C_1 = 2s/\sqrt{3}$ and s is the velocity of sound in the H II region. The kinetic energy, E_k, transferred to the surrounding neutral region is given by

$$E_k = \frac{4\pi}{3} N_0 m_H s^2 s_0^3 \left[\left(\frac{r}{s_0}\right)^{3/2} - 1\right] = \frac{2}{3}\left[1 - \left(\frac{r}{s_0}\right)^{-3/2}\right] E_T, \tag{3-141}$$

where E_T is the thermal energy in the H II region. Some of this energy goes into ionizing the neutral hydrogen. The resultant loss of energy, E_L, is given by

$$E_L = \frac{-4\pi}{7} N_0 m_H s^2 s_0^3 \ln\left[\left(\frac{7}{4} s s_0^{3/4} t + s_0^{7/4}\right) s_0^{-7/4}\right]. \tag{3-142}$$

The thickness, Δr, of the H I shell is given by

$$\Delta r = \frac{1}{6}\left[\left(\frac{7}{4} s s_0^{3/4} t + s_0^{7/4}\right)^{6/7} - s_0^{3/2}\right]\left(\frac{7}{4} s s_0^{3/4} t + s_0^{7/4}\right)^{-2/7} \approx \frac{r}{6} \quad \text{for } t \to \infty. \tag{3-143}$$

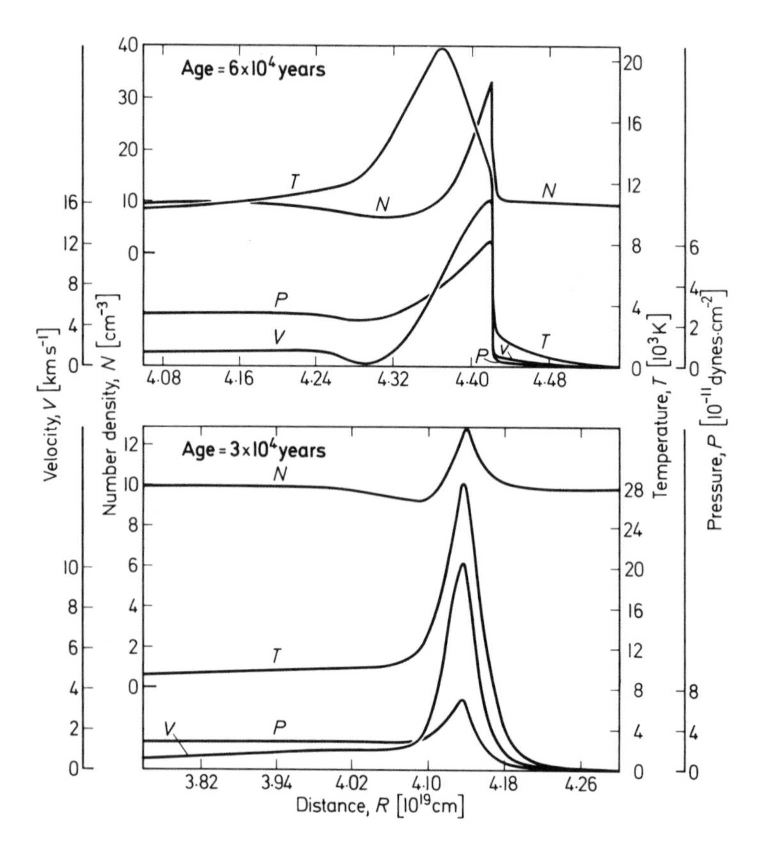

Fig. 19. Variation of velocity, V, number density, N, temperature, T, and pressure, P, in the vicinity of an ionization front as a function of distance, R, from an 0 star with temperature $T = 41{,}958\,°K$ after 3.07×10^4 and 6.16×10^4 years. The $30\,M_\odot$ star is assumed to be formed in an infinite medium of neutral hydrogen at rest at 100 °K, and the star's photon luminosity $L = 8.75 \times 10^{48}$ ultraviolet photons per second while on the main sequence (after MATHEWS, 1965, by permission of the American Astronomical Society and the University of Chicago Press)

The variation of velocity, number density, temperature, and pressure of an H II region as a function of distance from the exciting star are illustrated in Fig. 19 for an 0 star of $T = 42{,}000\,°\mathrm{K}$ and after 3×10^4 and 6×10^4 years.

3.3.2. The Degenerate Gas—Number Density, Energy Density, Entropy Density, and the Equation of State

3.3.2.1. Fermi-Dirac Statistics and Functions

Elementary particles whose spin is an odd multiple of $\hbar/2$ (electrons, protons, and nuclei with odd mass numbers) obey Fermi-Dirac statistics (FERMI, 1926; DIRAC, 1926). These particles obey Pauli's exclusion principle, which states that one system can never have two elements with exactly the same set of quantum numbers (cf. PAULI (1927)). Using this property, Fermi derived the following formula for the number density, $n(\varepsilon)$, of particles with kinetic energy between ε and $\varepsilon + d\varepsilon$ in a volume V

$$V n(\varepsilon) = \frac{g(\varepsilon)\, d\varepsilon}{\exp\left[\frac{-\mu}{kT} + \frac{\varepsilon}{kT}\right] + 1}, \tag{3-144}$$

where $g(\varepsilon)$ is the number of possible particle states of energy, ε, the temperature is T, and the Fermi constant or chemical potential, μ, is given by Eq. (3-104). For a non-relativistic particle of mass, m, Eq. (3-144) becomes

$$n(\varepsilon) = \frac{2\pi(2m)^{3/2}}{h^3} \frac{g\,\varepsilon^{1/2}\, d\varepsilon}{\exp\left[\frac{-\mu}{kT} + \frac{\varepsilon}{kT}\right] + 1}. \tag{3-145}$$

The total number of particles per unit volume, N, is given by

$$N = \int_0^\infty n(\varepsilon)\, d\varepsilon = \frac{g}{2\pi^2 \hbar^3} \int_0^\infty \frac{p^2\, dp}{\exp\left[\frac{-\mu}{kT} + \frac{\varepsilon}{kT}\right] + 1}, \tag{3-146}$$

where g is the statistical weight for discrete energy levels and ε is related to the momentum, p, by Eq. (3-111). For the nonrelativistic, degenerate gas ($\varepsilon = p^2/(2m)$, $p \ll mc$, $kT \ll mc^2$)

$$N = \frac{2\pi(2mkT)^{3/2} g}{h^3} \int_0^\infty \frac{x^{1/2}\, dx}{\exp\left[x + \frac{-\mu}{kT}\right] + 1}. \tag{3-147}$$

The total energy density, U/V, is given by

$$\frac{U}{V} = \int_0^\infty \varepsilon\, n(\varepsilon)\, d\varepsilon = \frac{g}{2\pi^2 \hbar^3} \int_0^\infty \frac{\varepsilon p^2\, dp}{\exp\left[\frac{-\mu}{kT} + \frac{\varepsilon}{kT}\right] + 1}. \tag{3-148}$$

For the nonrelativistic, degenerate gas,

$$\frac{U}{V} = \frac{4\pi}{h^3}(2m)^{3/2}(kT)^{5/2} F_{3/2}\left(\frac{\mu}{kT}\right) = \frac{3}{2} P_e, \tag{3-149}$$

where P_e is the electron pressure, and the Fermi-Dirac function

$$F_{3/2}\left(\frac{\mu}{kT}\right) = \int_0^\infty \frac{x^{3/2}\,dx}{\exp\left[x + \frac{-\mu}{kT}\right] + 1}.$$

Tables of $F_{3/2}(\mu/kT)$ are given by McDougal and Stoner (1938). Even at zero temperature, the energy is finite and is given by

$$\begin{aligned} U_0 &= \left(\frac{6\pi^2}{g}\right)^{2/3} \frac{\hbar^2}{2m} N_e^{2/3} = \mu, \\ &\approx 5.0 \times 10^{-27} N_e^{2/3} \text{ erg.} \end{aligned} \tag{3-150}$$

The chemical potential of a gas at absolute zero is the same as the energy of the electrons. Here N_e is the electron density, and we may write in general

$$\begin{aligned} U_0 = E_F &= \left\{\left[1.018\left(\frac{\rho}{\mu_e}\right)^{2/3} \times 10^{-4} + 1\right]^{1/2} - 1\right\} mc^2 \\ &= 0.509\left(\frac{\rho}{\mu_e}\right)^{2/3} \times 10^{-4}\, mc^2 \quad \text{for nonrelativistic electrons} \\ &= 1.009\left(\frac{\rho}{\mu_e}\right)^{1/3} \times 10^{-2}\, mc^2 \quad \text{for relativistic electrons.} \end{aligned}$$

Here ρ is the gas mass density, μ_e is the mean electron molecular weight, and the symbol E_F is used to indicate that this quantity is often called the Fermi energy of a completely degenerate gas. The maximum electron kinetic energy is E_F, whereas the average kinetic energy is $3E_F/5$. The electron density, N_e, may be written as

$$N_e = \frac{8\pi}{3h^3}(mc)^3 \left[\left(\frac{E_F}{mc^2} + 1\right)^2 - 1\right]^{3/2} \tag{3-151}$$

or

$$N_e = \frac{\rho N_A}{\mu_e} = \rho N_A \sum_i \frac{X_i Z_i}{A_i},$$

where ρ is the mass density of all ions and ionization electrons, N_A is Avogadro's number, and X_i, Z_i, and A_i are, respectively, the mass fraction, atomic number, and atomic weight of element i. The transition from Fermi-Dirac to Boltzmann statistics occurs at the degeneracy temperature given by

$$T_0 = \frac{E_F}{k} = \frac{1}{8}\left(\frac{3}{\pi}\right)^{2/3} \frac{h^2}{mk} N_e^{2/3} = 4.35 \times 10^{-11} N_e^{2/3}\ {}^\circ\text{K}. \tag{3-152}$$

The entropy, S, per unit volume, V, for the nonrelativistic, degenerate gas is given by

$$\frac{S}{V}=\frac{5P_e}{2T}-\frac{\mu}{kT}kN_e=kN_e\left[\frac{5}{2}\,\frac{\frac{2}{3}F_{3/2}\left(\frac{\mu}{kT}\right)}{F_{1/2}\left(\frac{\mu}{kT}\right)}-\frac{\mu}{kT}\right]. \tag{3-153}$$

For a relativistic degenerate gas we have for $\varepsilon=pc$, $p\gg mc$, and $kT\gg mc^2$, with

$$N_e=\frac{8\pi}{c^3h^3}(kT)^3F_2\left(\frac{\mu}{kT}\right),$$
$$\frac{U}{V}=\frac{8\pi}{c^3h^3}(kT)^4F_3\left(\frac{\mu}{kT}\right)=3P_e, \tag{3-154}$$

and

$$\frac{S}{V}=4\frac{P_e}{T}-kN_e\left(\frac{\mu}{kT}\right)=kN_e\left[4\times\frac{\frac{1}{3}F_3\left(\frac{\mu}{kT}\right)}{F_2\left(\frac{\mu}{kT}\right)}-\left(\frac{\mu}{kT}\right)\right].$$

where the Fermi-Dirac integral $F_n(\mu/kT)$ is given in Eq. (3-174).

3.3.2.2. Equation of State of a Degenerate Electron Gas—White Dwarf Stars

For densities higher than the critical density, ρ_{cd}, given by

$$\rho_{cd}=2.4\times10^{-8}\,T^{3/2}\,\mu_e\ \text{gm cm}^{-3}, \tag{3-155}$$

a gas is completely degenerate. Here T is the temperature, and the mean molecular weight per electron, μ_e, has the value 1 for hydrogen and 2 for the other elements. The gas pressure, P, for a nonrelativistic, completely degenerate gas, with $\rho_{cd}\leq\rho\leq\rho_{cr}$ and $\mu\gg kT$, is given by (FOWLER, 1926)

$$P=\frac{h^2}{20m}\left(\frac{3}{\pi}\right)^{2/3}N_e^{5/3}=3.12\times10^{12}\left(\frac{2}{g}\right)^{2/3}\left(\frac{2\rho}{\mu_e}\right)^{5/3}\ \text{dynes cm}^{-2}$$
$$=1.004\times10^{13}\left(\frac{\rho}{\mu_e}\right)^{5/3}\ \text{dynes cm}^{-2}=\frac{2}{3}\,\frac{U}{V}, \tag{3-156}$$

where N_e is the electron density, g is the statistical weight, ρ is the mass density, and the energy density is U/V. A gas is said to be relativistic when the density, ρ, is larger than the critical value, ρ_{cr} given by

$$\rho_{cr}=7.3\times10^6\,\mu_e\ \text{gm cm}^{-3}. \tag{3-157}$$

For the relativistic, completely degenerate gas, the pressure, P, is given by

$$P=\frac{c}{4}\left(\frac{3h^3}{8\pi}\right)^{1/3}N_e^{4/3}=4.56\times10^{14}\left(\frac{g}{2}\right)^{1/3}\left(\frac{2\rho}{\mu_e}\right)^{4/3}\ \text{dynes cm}^{-2}$$
$$\approx1.244\times10^{15}\left(\frac{\rho}{\mu_e}\right)^{4/3}\ \text{dynes cm}^{-2}=\frac{1}{3}\,\frac{U}{V}, \tag{3-158}$$

where $N_A\rho=\mu_eN_e$, and N_A is Avagadro's number. Relativistic effects were first considered by ANDERSON (1929) and STONER (1930).

For a partially degenerate gas in which the mass density, ρ, is near the degeneracy limit, ρ_{cd}, the pressure, P, is given by

$$P = \frac{g}{6\pi^2 \hbar^3} \int\limits_0^\infty \frac{p^3 \frac{\partial \varepsilon}{\partial p} dp}{\exp\left[\frac{-\mu}{kT} + \frac{\varepsilon}{kT}\right] + 1}, \tag{3-159}$$

where g is the statistical weight, the chemical potential, μ, is given by Eq. (3-104), and the momentum, p, is related to the kinetic energy, ε, by Eq. (3-111). Eq. (3-159) may be rewritten as Eq. (3-149)

$$P = \frac{8\pi k T}{3h^3} (2mkT)^{3/2} F_{3/2}\left(\frac{\mu}{kT}\right) = \frac{2}{3} \frac{U}{V}. \tag{3-160}$$

where the Fermi-Dirac function $F_{3/2}(\mu/kT)$ is given by Eq. (3-149). Similarly, the electron density, N_e, is given by

$$N_e = \frac{4\pi}{h^3} (2mkT)^{3/2} F_{1/2}\left(\frac{\mu}{kT}\right), \tag{3-161}$$

where the Fermi-Dirac integral $F_n(\mu/kT)$ is given in Eq. (3-174). Both Fermi-Dirac functions, $F_{3/2}(\mu)$ and $F_{1/2}(\mu)$, are tabulated by McDougal and Stoner (1938).

A star in which the degenerate gas pressure balances the force of gravitation is called a white dwarf because its surface temperature of $\approx 10^4$ °K implies a white colour. The stars are still considered cold, however, because temperature does not affect the equation of state. The properties of most of the well observed white dwarfs are given in Table 35.

At high densities the degenerate electron gas becomes relativistic, and when it is taken to be an ideal Fermi gas, dynamical instability will set in when the mass exceeds the critical mass, M_{cre}, given by (Landau, 1932; Chandrasekhar, 1935)

$$M_{cre} = \frac{3.1}{m'} \left(\frac{\hbar c}{G}\right)^{3/2} = \left(\frac{2}{\mu}\right)^2 1.4587\, M_\odot \approx 1.4587\, M_\odot, \tag{3-162}$$

where the mass per electron, m', is taken to be twice the proton mass, the gravitation constant $G \approx 6.67 \times 10^{-8}$ dyn cm^2 g^{-2}, the mean molecular weight per electron $\mu = A/Z$ is taken to be two, and the solar mass, $M_\odot \approx 1.989 \times 10^{33}$ gm. The details of this calculation are given in Sect. 3.4.3, where it is noted that an upper mass limit does not exist for nonspherical or differentially rotating stars. Detailed formulae for the radius, mass, and temperature of spherically symmetric, nonrotating white dwarfs are given by Chandrasekhar (1939).

In the standard treatment of the degenerate electron gas (given above), the mean molecular weight is assumed to be fixed and the electrons are assumed to be noninteracting and to form an ideal Fermi gas. For the range of densities considered here, $\rho_{cd} \le \rho \le \rho_{cr}$ or $500\ \text{g cm}^{-3} \le \rho \le 10^{11}\ \text{g cm}^{-3}$, the gas is in fact a plasma of electrons and nuclei. Salpeter (1961) has discussed the equation of state for a zero temperature plasma of electrons and nuclei of atomic weight, A, and charge, Z. The corrections include the classical Coulomb energy of an ion lattice with uniformly distributed electrons, the Thomas-Fermi deviations from a uniform charge distribution of electrons, the exchange energy between the electrons, and the spin-spin interactions between the electrons. The total energy per electron, E, is

$$E = E_0 + E_c + E_{TF} + E_{ex} + E_{cor}, \tag{3-163}$$

where the energy per electron for a Fermi gas of non-interacting electrons, E_0, is given by

$$E_0 = \frac{mc^2 g(x)}{8x^3},$$

where

$$g(x) \equiv 8x^3 [\sqrt{(1+x^2)} - 1] - x(2x^2-3)\sqrt{(1+x^2)} - 3\sinh^{-1} x,$$

and x is related to the gas mass density, ρ, by

$$\rho = 9.738 \times 10^5 \mu x^3 \text{ gm cm}^{-3},$$

where $\mu = A/Z$. The Coulomb correction, E_c, is given by

$$E_c = -\frac{9}{5}\frac{Z^{2/3}}{r_e} ry,$$

where one Rydberg $= ry = 2\pi^2 me^4/h^2 \approx 2.179 \times 10^{-11}$ erg ≈ 13.605 eV, and r_e is related to the mass density, ρ, by

$$\rho = 2.6787 \mu r_e^{-3} \text{ gm cm}^{-3}.$$

The Thomas-Fermi correction, E_{TF}, is given by

$$E_{TF} = -\frac{324}{175}\left(\frac{4}{9\pi}\right)^{2/3} \sqrt{(1+x^2)}\, Z^{4/3} ry.$$

The exchange energy correction, E_{ex}, is given by

$$E_{ex} = -\frac{3}{2}\left(\frac{9}{4\pi^2}\right)^{1/3} \frac{ry}{r_e} \quad \text{for } x \ll 1,$$

$$E_{ex} = -\left(\frac{3}{4\pi}\right) \alpha mc^2 x \varphi(x) \quad \text{for any } x,$$

where

$$\varphi(x) = \frac{1}{4x^4}\left[\frac{9}{4} + 3\left(\beta^2 - \frac{1}{\beta^2}\right)\ln\beta - 6(\ln\beta)^2 - \left(\beta^2 + \frac{1}{\beta^2}\right) - \frac{1}{8}\left(\beta^4 + \frac{1}{\beta^2}\right)\right],$$

and

$$\beta \equiv x + \sqrt{(1+x^2)}.$$

The correlation energy correction, E_{cor}, is given by

$$E_{cor} = (0.062 \ln r_e - 0.096) ry.$$

Similarly, the gas pressure can be written as the sum of five terms

$$P = P_0 + P_c + P_{TF} + P_{ex} + P_{cor}, \tag{3-164}$$

where the non-interacting Fermi electron pressure, P_0, is given by

$$P_0 = \frac{1}{24\pi^2} mc^2 \left(\frac{mc}{\hbar}\right)^3 f(x),$$

Table 35. Name, Eggen-Greenstein number, EG, position, parallax, π, proper motion, μ or $\mu_\alpha \cos\delta/\mu_\delta$, ity, g, mass, M, radius, R, and luminosity, L, for 153 white dwarfs. Magnetic dwarfs are marked by notes[1]. White dwarfs which are components of binaries are marked by [5]

Name	EG No.	Sp type	α(1950.0) h m s	δ(1950.0) ° ′	π (arc sec)	μ or $\mu_\alpha \cos\delta/\mu_\delta$ (″/year)
GD 2	203	DAwk	00 04 58	33 00.8		<0.1
LB 433	3	DB	00 17 24	13 36		0.036/0.008
L 1011 – 71 = G 1–7	4	DA	00 33 06	01 37		0.44/0.00
G 132–12	167	DA	00 36 23	31 15.2		0.28
GD 8	204	DAwk	00 37 15	31 16.1		<0.1
V Ma 2	5	DG	00 46 31	05 09.2	0.239 ± 0.004	1.26/ − 2.71
G 1–45	7	DAs	01 01 14	04 48.3		0.27/0.36
W 1516	9	DC	01 15 20	15 55.0	0.060 ± 0.003	0.639 ± 0.001
GD 13	205	DA	01 26 46	42 12.8		<0.1
R 548[3]	10	(DA)	01 33 42	−11 36		0.43/ − 0.07
L 870–2	11	DAs	01 35 26	−05 14.7	0.066 ± 0.013	0.57/ − 0.34
G 94–9	13	DA	01 43 56	21 39.8		−0.25/ − 0.12
GD 279	269	DAsp	01 48 56	46 45.2		0.07
GD 20	206	DA	01 55 04	06 57.5		<0.1
Oxf 25°6725	15	(DA)	02 05 56	25 00.1	0.029 ± 0.003	0.428 ± 0.002
G 74–7	168	DAs	02 08 13	39 41.5	0.064 ± 0.004	1.133 ± 0.002
h Per 1166	17	(DA)	02 14 00	56 53		0.16/ − 0.01
F 22	19	DA	02 27 39	05 02.6	0.045	0.095/ − 0.019
GD 31	207	DA	02 31 37	−05 24.9		0.20 − 0.26
F 24	20	DAwke	02 32 30	03 31		0.083/0.010
LB 3303	21	(DA)	03 10 04	−68 47.2	0.074 ± 0.006	0.038/ − 0.104
GD 45	208	DAs	03 16 34	34 31.6		0.20 − 0.26
L 587–77 A[5]	22	DAs	03 26 48	−27 34		0.71/0.36
G 37–44	23	DA	03 32 09	32 02.2	0.010 ± 0.004	0.375 ± 0.002
W 219	24	4670	03 41 36	18 18		0.47/ − 1.16
LB 1497	25	DAwk	03 49 06	24 47		0.038/ − 0.052
HZ 4	26	DA	03 52 06	09 37	0.025 ± 0.003	0.152/ − 0.008
LB 1240	28	DA	04 01 33	25 00.9	0.036 ± 0.003	0.227 ± 0.002
LB 227	29	DA	04 06 18	16 59	0.025 ± 0.003	0.116/ − 0.023
HZ 10	30	DA	04 07 18	17 54		0.070/ − 0.076
HZ 2	31	DA	04 10 06	11 45	0.025 ± 0.003	0.051/ − 0.086
40 Eri B[4] and [5]	33	DA	04 13 00	−07 44.1	0.202 ± 0.004	2.229/3.420
HL Tau 76[3]	265	DA	04 15 54	27 11		———
VR 7	36	(DA)	04 21 00	16 14		0.100/ − 0.020
VR 16	37	(DA)	04 25 42	16 52		0.009/ − 0.014
G 82–22–23	170	DFs	04 27 31	−03 10.2		0.29
HZ 7	39	DA	04 31 00	12 35	0.025 ± 0.003	0.089/0.001
L 879–14	41	4670	04 35 24	−08 53		0.23/ − 1.47
HZ 14	42	DAn	04 38 12	10 53	0.025 ± 0.003	0.085/0.010
GD 64	209	DA	04 53 50	41 51.5		0.20 − 0.26
G 191 B2B	247	DAwk	———	———		0.12
G 86–B1B[5,6]	43	DAe	05 18 24	33 19		0.17/ − 0.07
	288	DAwk	05 48 04	00 05.2		—
G 99–37*	248	4670p	05 48 46	−00 11.2		0.27
GD 71	210	DAwk	05 49 34	15 52.7		0.20 − 0.26

Footnotes 1 to 6 see pp. 262, 263.

visual, V, or photographic, m_{pg}, magnitudes, colors, $B-V$, $U-B$, temperatures, T_{eff}, surface grav-
a [2], variable dwarfs by [3], accurate redshift dwarfs by [4], and these parameters are given in the foot-

V or m_{pg} (mag)	$B-V$ (mag)	$U-B$ (mag)	T_{eff} (°K)	$\log g$	$\log(M/M_\odot)$	$\log(R/R_\odot)$	$\log(L/L_\odot)$
14.5	−0.29	−1.21					
15.22	−0.12	−0.97					
15.52	0.24	−0.68					
16.2	−1	—					
14.0	−0.22	−1.20					
12.37	0.56	0.04					
14.10	0.14	−0.50					
12.72	0.11	−0.80					
14.0	—	—					
14.10	0.20	−0.54					
12.84	0.34	−0.50	6,500 ± 500			−1.41 ± 0.15	−2.63 ± 0.40
15.05	0.25	−0.65					
(13.0)	(−1)	—					
14.5	(−1)	—					
10.53	−0.04	−0.85					
14.51	0.37	−0.43					
13.68	−0.12	−0.92					
12.65	−0.06	−0.83					
14.0	0.21	−0.68					
12.25	−0.23	−1.25	50,000	6.6			
11.40	—	—					
13.0	0.08	−0.68					
13.9	−0.1	—					
15.49	0.13	−0.51					
15.20	0.30	−0.52					
16.52	−0.20	−1.10					
14.47	0.15	−0.67	15,300 ± 400	8.1 ± 0.30	−0.09 ± 0.30	−1.92 ± 0.06	−2.16 ± 0.12
11.59	0.10	−0.58	13,800 ± 500	7.5 ± 0.30	−0.70 ± 0.30	−1.92 ± 0.05	−2.36 ± 0.10
15.35	0.09	−0.65	16,500 ± 500	8.3 ± 0.30	−0.28 ± 0.30	−2.12 ± 0.06	−2.43 ± 0.12
14.14	0.17	−0.58					
13.86	−0.05	−0.88	22,500 ± 500	8.0 ± 0.20	−0.20 ± 0.15	−1.92 ± 0.06	−1.48 ± 0.12
9.52	0.03	−0.68	17,000 ± 500	7.75 ± 0.15	−0.35 ± 0.15	−1.88 ± 0.02	−1.90 ± 0.10
15.2	0.20	−0.50					
14.29	−0.02	−0.84					
14.02	−0.09	−0.97					
14.70	0.72	−0.15					
14.18	−0.03	−0.89	23,000 ± 500	7.9 ± 0.20	−0.42 ± 0.20	−2.00 ± 0.06	−1.61 ± 0.12
14.10	0.14	−0.65					
13.83	−0.15	−1.04	34,000 ± 1,000	8.5 ± 0.40	−0.02 ± 0.40	−2.04 ± 0.06	−1.01 ± 0.12
13.77	0.23	−0.61					
11.8	−0.32	−1.20					
16.20	0.31	−0.64					
15.10	−0.30	−1.21					
14.58	0.46	−0.53	5,000(1)				
13.0	−0.25	−1.16					

Table 35 (continued)

Name	EG No.	Sp type	α(1950.0) h m s	δ(1950.0) ° ′	π (arc sec)	μ or $\mu_\alpha \cos\delta/\mu_\delta$ (″/year)
LP 658–2[6]	45	DKe	05 52 40	−04 09.2	0.155 ± 0.055	2.378 ± 0.004
G 99–47[2]	289	DC	05 53 47	05 22.0		1.07
GD 72	211	DA	06 06 51	28 15.1		—
L 1244–26	46	DA	06 12 24	17 45		−0.06/−0.40
GD 77	212	DA	06 37 26	47 47.3		0.20−0.26
α CMa B[4] and [5]	49	DA	06 42 57	−16 38.8	0.377 ± 0.004	−0.502/−1.215
He 3	50	DAn	06 44 15	37 34.9	0.068 ± 0.002	0.957 ± 0.002
G 87–29[5]	51	4670	07 06 52	37 45.4	0.043 ± 0.004	0.350 ± 0.003
G 89–10	171	DAs	07 15 22	12 35.5		0.30
GD 85	216	DBs	07 16 32	40 27.1		0.10−0.20
α CMi B[5]	53	—	07 36 42	05 21		0.708/−1.030
L 745–46A[5]	54	DF	07 38 02	−17 17.4	0.146 ± 0.007	1.136/−0.520
L 97–12	56	DC	07 52 48	−67 38	0.173 ± 0.030	1.45/−1.45
L 817–13	57	DA	07 52 48	−14 38		−0.13/−0.29
G 111–71[5]	58	DA	08 16 44	38 44.2		−0.20/−0.32
GD 91	217	DAs	08 26 42	45 30.6		—
G 51–16	249	DAs	08 27 32	32 52.3		0.64
L 532–81	62	DA	08 39 42	−32 47		−1.06/1.31
LDS 235 B[5]	63	DB	08 45 18	−18 48		−0.06/−0.93
GD 98	218	DA	08 54 16	40 27.9		—
G 47–18	182	4670	08 56 11	33 08.7	0.032	−0.40/0.00
GD 99	219	DA	08 58 41	36 19.0		—
G 195–19[2]	250	DC	09 12 27	53 38.6		1.56
G 116–16[5]	64	DAs	09 13 26	44 12.5		0.03/−0.28
LDS 275 AB[5]	66	DC	09 35 00	−37 07		−0.34/0.16
SA 29–130	67	DA	09 43 30	44 08.8	0.028 ± 0.003	0.297 ± 0.002
G 49–33	69	DA	09 55 00	24 47.4		−0.23/−0.30
L 825–14	70	DAn	10 31 24	−11 29		−0.32/−0.06
F 34	71	DO	10 36 42	43 22		0.034/−0.024
G 44–32[3]	72	DC	10 39 14	14 31.6		0.31
Ton 547	73	DA	10 46 48	28 20		—
GD 125	221	DAn	10 52 01	27 22.9		0.10−0.20
LB 253	75	DAn	11 04 48	60 13		−0.19/−0.17
L 970–30[5]	76	DA	11 05 30	−04 53	0.046	−0.43/0.00
Ton 573	77	DB	11 07 18	26 37		0.108/−0.167
R 627	79	DA, F	11 21 38	21 38.1	0.071 ± 0.003	1.029 ± 0.001
F 43	80	DA	11 26 30	38 28		−0.078/0.049
GD 140	184	DAwk	11 34 28	30 04.6		−0.15/0.00
F 46	81	DO–B	11 34 54	14 27		−0.054/0.044
L 145–141	82	4670	11 42 54	−64 34		2.66/−0.33
G 148–7[5]	83	DAn	11 43 21	32 06.2		−0.11/−0.26
HZ 21	86	DO	12 11 24	33 12		−0.096/0.042
HZ 28	90	DA	12 30 00	41 46		−0.124/0.015
HZ 29[3]	91	DBp	12 32 24	37 55		0.037/0.028
GD 148	186	DA	12 32 33	47 54.3		0.13/−0.19

V or m_{pg} (mag)	$B-V$ (mag)	$U-B$ (mag)	T_{eff} (°K)	$\log g$	$\log(M/M_{\odot})$	$\log(R/R_{\odot})$	$\log(L/L_{\odot})$
15.40	1.06	0.81					
14.12	0.62	−0.12					
15.0	0.00	−0.82					
13.40	−0.15	−0.98					
15.0	0.13	−0.64					
8.68	(−0.12)	(−1.03)	32,000 ± 1,000	8.65	+0.0086	0.0078 ± 0.0002	
11.21	−0.08	−0.92	25,000 ± 700	8.1 ± 0.15	−0.24 ± 0.22	−2.00 ± 0.07	−1.46 ± 0.14
13.71	0.30	−0.55					
16.3	0	—					
14.5	−0.11	−0.99					
10.7	—	—					
12.98	0.29	−0.61	5,000			−1.56(1)	−3.39(1)
14.5	0.4	—					
13.56	−0.02	−0.79					
16.58	0.36	−0.58					
16.0	0.20	−0.52					
15.7	0.35	−0.54					
12.0							
15.55	−0.06	−0.93	15,000 ± 1,000	8.0		−1.78 ± 0.05	−1.80 ± 0.20
15.0	−0.13	−0.94					
15.18	0.00	−0.89	10,000			−1.94(1)	−2.94(1)
15.0	0.19	−0.59					
13.79	0.30	−0.66					
15.32	0.24	−0.53					
15.0							
10.55	0.07	−0.54	14,500 ± 500	8.0 ± 0.30	+0.06 ± 0.40	−1.80 ± 0.10	−1.99 ± 0.30
15.09	0.25	−0.54					
12.97	−0.15	−1.02					
11.12	−0.30	−1.35					
16.55	0.29	−0.58					
15.40	0.16	−0.58					
14.0	−0.08	−0.98					
13.80	−0.02	−0.81					
12.92	0.09	−0.69	16,000 ± 500	7.9 ± 0.20	−0.27 ± 0.30	−1.91 ± 0.10	−2.05 ± 0.20
15.89	−0.07	−0.99	14,000 ± 1,000				
13.51	0.31	−0.52	7,000 ± 500			−1.86 ± 0.07	−3.38 ± 0.40
14.89	−0.13	−1.00					
12.50	−0.06	−1.13	23,000 ± 800	8.2 ± 0.50			
13.24	−0.30	−0.98					
11.44	0.19	−0.59					
13.60	0.06	−0.67					
14.22	−0.36	−1.25	70,000	8			
15.72	−0.04	−0.82					
14.18	−0.23	−1.01					
14.52	0.06	−0.68					

Table 35 (continued)

Name	EG No.	Sp type	α(1950.0) h m s	δ(1950.0) ° ′	π (arc sec)	μ or $\mu_\alpha \cos\delta/\mu_\delta$ (″/year)
G 61–17[5]	92	DAs	12 44 45	14 58.8		−0.38/0.22
HZ 34	93	DOp	12 53 00	37 49		−0.034/−0.051
GD 153	187	DAwk	12 54 35	22 18.2		0.00/−0.20
G 61–29[3,6]	—	DBe	13 03 16	18 17.0		0.38
HZ 43[5]	98	DAwk	13 14 00	29 22	0.021 ± 0.011	−0.149/−0.088
G 177–31	188	DAs	13 17 04	45 20.8		−0.50/0.18
W 485 A[5]	99	DA	13 27 40	−08 18.8	0.062 ± 0.003	1.181 ± 0.002
W 489	100	DK	13 34 13	03 57.0	0.117 ± 0.004	−3.738/−1.109
GW 70°5824	102	DA	13 37 48	70 33	0.027 ± 0.012	−0.40/−0.02
GD 163	189	DA	14 08 16	32 22.7		0.00/−0.25
F 93	107	DAwk	14 15 18	13 16		—
Ton 197	108	DAwk	14 21 30	31 50		—
GD 178	191	DAn	15 09 25	32 15.5		−0.13/0.08
GD 185	192	DAn	15 31 30	−02 17.2		−0.050/−0.126
GD 190	193	DBs	15 42 04	18 16.1		0.00/−0.075
1°3129 B[5]	113	DAwk	15 44 12	00 53		—
R 808	115	DA	15 59 33	36 57.2	0.034 ± 0.005	0.556 ± 0.002
C 2	116	DA	16 06 42	42 13		—
R 640	119	DFp	16 26 48	36 51	0.056 ± 0.007	−0.46/0.74
L 1491–27	120	DAs	16 37 24	33 32		−0.04/−0.43
G 169–34	197	DAss	16 55 01	21 31.7	0.042 ± 0.003	0.561 ± 0.001
G 185–B5B	198	DA	17 07 00	33 17		0.013/−0.130
L 845–70	122	DC	17 08 30	−14 45		0.27/−0.24
GD 363	240	DA	17 37 01	41 54.0		0.10−0.20
LP 9–231	199	DAss	17 57 00	82 44		−1.4/3.5
R 137	125	DA	18 24 45	04 02.0	0.014 ± 0.003	0.377 ± 0.002
L 993–18	126	DAs	18 26 30	−04 31		0.00/−0.29
GD 215	225	DAss	18 40 57	04 17.3		0.10−0.20
G 141–54	128	DAs	18 57 29	11 54.3		0.27
GW 70°8247[2]	129	4135	19 00 36	70 36	0.081 ± 0.010	0.101/0.508
G 142–B2B[5]	130	DAn	19 11 18	13 31		0.00/−0.10
LDS 678 B[5]	131	DAwk	19 17 54	−07 45.0	0.092 ± 0.004	0.194 ± 0.003
GD 219	201	DAn	19 19 23	14 34.9		0.000/−0.088
LDS 683 B[5]	132	DA	19 32 54	−13 36		−0.02/−0.14
L 1573–31	133	DB	19 40 24	37 24		0.02/−0.20
G 142–50	134	DA	19 43 16	16 20.4		0.00/−0.26
G 92–40	135	DAs	19 53 56	−01 10.2	0.082 ± 0.004	0.822 ± 0.002
W 1346	139	DA	20 32 18	24 53.8	0.066 ± 0.003	0.695 ± 0.002
GD 231	288	DA	20 32 58	18 49.1		0.18
L 711–10	141	DA	20 39 42	−20 16		0.32/−0.09
G 210–36	261	DA	20 47 10	37 16.9		0.27
GD 232	279	DA	20 58 58	18 09.1		0.08
L 24–52	142	(DA)	21 05 12	−82 01		0.08/−0.36
GD 394	244	DAwk	21 11 03	49 53.7		0.10−0.20
GW 73°8031	144	DA	21 26 43	73 25.9	0.045 ± 0.007	0.045/−0.310

V or m_{pg} (mag)	$B-V$ (mag)	$U-B$ (mag)	T_{eff} (°K)	$\log g$	$\log(M/M_{\odot})$	$\log(R/R_{\odot})$	$\log(L/L_{\odot})$
15.86	0.22	−0.53					
15.66	−0.28	−1.26					
13.42	−0.25	−1.18					
15.69	−0.10	−0.97					
12.86	−0.10	−1.14	50,000 ± 5,000	8.8 ± 1.0	−0.60 ± 1.0	−1.83 ± 0.25	−0.02 ± 0.50
14.13	0.03	−0.56					
11.27	0.08	−0.61	15,500 ± 500	8.5 ± 0.30	+0.37 ± 0.30	−1.90 ± 0.07	−2.09 ± 0.14
14.68	0.96	0.37					
12.79	−0.09	−0.84	21,000 ± 1,000	7.7 ± 0.30	−0.11 ± 0.50	−1.74 ± 0.25	−1.23 ± 0.40
14.06	−0.03	−0.81					
15.33	−0.23	−1.12					
15.30	−0.12	−1.00					
14.11	0.09	−0.65					
14.03	0.00	−0.81	19,000 ± 500	9.1 ± 0.50			
14.72	−0.10	−1.00	18,000 ± 2,000	8.5 ± 1.0			
15.27	−0.32	−1.19					
14.36	0.17	−0.56					
13.85	0.06	−0.54					
13.86	0.18	−0.68	6,000(1)			−1.53(1)	−3.00(1)
14.64	0.22	−0.58					
14.04	0.26	−0.52					
15.92	0.17	−0.55					
14.30	0.02	−0.68					
15.0	—	—					
14.51	0.41	−0.48					
9.65	0.05	−0.55					
14.54	0.24	−0.56					
15.0	0.25	−0.62					
15.52	0.20	−0.57					
13.19	0.05	−0.85	12,000			−2.08	−2.90
14.00	0.12	−0.60					
12.24	0.07	−0.84					
13.01	0.06	−0.66					
15.95	0.05	−0.76					
14.51	−0.09	−0.97	15,000 ± 1,000				
14.08	−0.06	−0.86					
13.71	0.30	−0.61					
11.54	−0.07	−0.87	20,000 ± 700	7.3 ± 0.30	−0.83 ± 0.15	−1.89 ± 0.07	−1.64 ± 0.14
15.3	−0.04	−0.80					
12.0	—	—					
13.0	+0.07	−0.68					
15.0	0.01	−0.75					
13.50	0.27	−0.59					
—	—	—					
12.88	0.01	−0.66	16,000 ± 500	7.7 ± 0.30	−0.40 ± 0.35	−1.88 ± 0.08	−2.00 ± 0.16

Table 35 (continued)

Name	EG No.	Sp type	α(1950.0) h m s	δ(1950.0) ° ′	π (arc sec)	μ or $\mu_\alpha \cos\delta/\mu_\delta$ (″/year)
LDS 749 B[5]	145	DB	21 29 36	00 00	0.021 ± 0.003	0.420/0.000
L 1002–62	146	DB	21 31 00	−04 46		0.090/−0.006
GW 82°3818	147	DA	21 36 42	82 49		0.31/0.56
L 1363–3	148	DC	21 40 22	20 46.7	0.074 ± 0.004	0.669 ± 0.002
L 930–80	149	DB	21 44 54	−07 58	0.011 ± 0.006	0.281 ± 0.002
G 93–48	150	DA	21 49 53	02 09.5	0.039 ± 0.002	0.268 ± 0.002
GD 398	281	DA	21 50 19	33 53.6		0.08
L 1003–16	151	DAs	21 51 30	−01 31		0.000/−0.275
LDS 785 A[5]	153	DB	22 24 36	−34 27		0.21/−0.02
GD 236	228	DA	22 26 38	06 07.1		0.20−0.26
G 28–13	154	DA, Fs	22 40 31	−01 43.4		−0.16/−0.30
GD 243	230	DBn	22 53 12	−06 16.9		<0.10
GD 244	231	DA	22 54 16	12 36.8		0.20−0.26
G 241–46	284	DA	23 07 59	63 41.8		0.45
F 108	157	DAs	23 13 36	−02 07		−0.054/−0.011
C 3	160	DAn	23 29 12	40 45		—
L 1512–34B[5]	162	DA	23 41 24	32 15	0.069 ± 0.015	−0.23/−0.06
L 362–81	165	(DAs)	23 59 36	−43 25		0.60/−0.67

[1] This table was prepared with the kind help of Professor J.L. GREENSTEIN of the California Institute of Technology. The EG numbers, spectral type, positions, proper motions, and colors come mainly from EGGEN and GREENSTEIN (1965a, b; 1967) and GREENSTEIN (1969, 1970). Positions of the G and GD objects are from GICLAS, BURNHAM, and THOMAS (1960 to 1970) and GICLAS, BURNHAM, and THOMAS (1965, 1967, 1970), respectively. Accurate positions and parallaxes are from GLIESE (1969), RIDDLE (1970), and ROUTLY (1972), and some values of parallax are from CHURNS and THACKERY (1971) and SHIPMAN (1972). Values of temperature, surface gravity, mass, radius, and luminosity are from GREENSTEIN, OKE, and SHIPMAN (1971) for α CMa B, and from SHIPMAN (1972) for the other objects.

[2] Circular polarization data indicate that:

Object	*Circular polarization* (%)	*Magnetic Field* (gauss)
G 99–37	0.63 ± 0.03	$\approx 10^7$
G 99–47	0.30 − 0.45	$\approx 10^6$
G 195–19	0.42 ± 0.04	$\approx 10^6$
GW 70° 8247	1 − 3	$\approx 10^7$

The circular polarization data are from KEMP *et al.* (1970), ANGEL and LANDSTREET (1971, 1972), KEMP *et al.* (1971), and LANDSTREET and ANGEL (1971). The circular polarization of G 195–19 is variable with a period of 1.34 days.

V or m_{pg} (mag)	$B-V$ (mag)	$U-B$ (mag)	T_{eff} (°K)	$\log g$	$\log(M/M_\odot)$	$\log(R/R_\odot)$	$\log(L/L_\odot)$
14.73	−0.07	−0.92	14,000 ± 1,000	7.5 ± 0.60	−0.60 ± 0.60	−1.83 ± 0.10	−2.12 ± 0.20
14.0	−0.05	−0.73					
13.02	−0.02	−0.72					
12.59	0.17	−0.72	6,000			−1.50(1)	−2.94(1)
14.83	−0.11	−0.94	16,500 ± 1,000	8.0 ± 0.50	−1.12	−2.34	−2.80
12.77	0.00	−0.78					
(15.0)	(−1)	—					
10.69	0.26	−0.51					
14.5	—	—					
15.5	—	—					
16.21	0.31	−0.60					
15.5	—	—					
16.0	—	—					
(14.4)	(−1)	—					
12.90	−0.28	−1.06					
13.82	0.03	−0.72					
12.90	0.15	−0.61	14,000 ± 300	7.9 ± 0.30	−0.49 ± 0.35	−2.03 ± 0.07	−2.52 ± 0.14
13.05	0.07	−0.87					

[3] Light Variable white dwarfs include:

Object	*Period* (sec)
R 548	212.834 ± 0.03
HL Tau 76	746.220 ± 0.180
G 44–32	1638
HZ 29	1051.118 ± 0.015
G 61–29	~20

Data on variable stars are given by Ostriker and Hesser (1968), Landolt (1968), Lasker and Hesser (1969, 1971), Warner and Nather (1972), and Warner (1972).

[4] Accurate gravitational redshifts, v_g, together with radial velocities, v_r, have been given by Greenstein, Oke, and Shipman (1971) for α Cma B (Sirius B) and by Greenstein and Trimble (1972) for 40 Eri B. They obtain $v_r = +81 \pm 16$ km sec^{-1}, $v_g = +89 \pm 16$ km sec^{-1} for Sirius B and $v_r = -18.7 \pm 5.3$ km sec^{-1}, $v_g = +23 \pm 5$ km sec^{-1} for 40 Eri B. Radial velocities for most of the sources given in this table are listed by Greenstein and Trimble (1967) and Trimble and Greenstein (1972), and are of a statistical use only.

[5] An additional nineteen binaries (EG 1, 12, 14, 18, 21b, 40, 44, 47, 52, 65, 68, 84, 89, 101, 112, 114, 123, 124, and 138) are listed in Eggen and Greenstein (1965a).

[6] Most emission lines originate in background nebulosity or in close companions; G 61–29 may be a close binary mass.

where $$f(x) = x(2x^2 - 3)(x^2+1)^{1/2} + 3\sinh^{-1} x,$$

the Coulomb and Thomas-Fermi terms are given by

$$P_C + P_{TF} = -mc^2 \left(\frac{mc}{\hbar}\right)^3 \left[\frac{\alpha Z^{2/3}}{10\pi^2}\left(\frac{4}{9\pi}\right)^{1/3} x^4 + \frac{162}{175}\frac{(\alpha Z^{2/3})^2}{9\pi^2}\left(\frac{4}{9\pi}\right)^{2/3} \frac{x^5}{\sqrt{(1+x^2)}}\right],$$

the exchange term is given by

$$P_{ex} = -\frac{\alpha}{4\pi^3} mc^2 \left(\frac{mc}{\hbar}\right)^3 \chi(x),$$

where

$$\chi(x) = \frac{1}{32}\left(\beta^4 + \frac{1}{\beta^4}\right) + \frac{1}{4}\left(\beta^2 + \frac{1}{\beta^2}\right) - \frac{9}{16} - \frac{3}{4}\left(\beta^2 - \frac{1}{\beta^2}\right)\ln\beta + \frac{3}{2}(\ln\beta)^2$$
$$-\frac{x}{3}\left[1 + \frac{x}{\sqrt{(1+x^2)}}\right]\left[\frac{1}{8}\left(\beta^3 - \frac{1}{\beta^5}\right) - \frac{1}{4}\left(\beta - \frac{1}{\beta^3}\right) - \frac{3}{2}\left(\beta + \frac{1}{\beta^3}\right)\ln\beta + \frac{3}{\beta}\ln\beta\right],$$

and the correlation term is

$$P_{cor} = -\frac{0.0311}{9\pi^2} mc^2 \left(\frac{mc}{\hbar}\right)^3 \alpha^2 x^3.$$

For extremely relativistic electrons, $x \gg 1$, these equations become

$$\frac{P}{P_0} = 1 + \frac{\alpha}{2\pi} - \frac{6}{5}(\alpha Z^{2/3})\left(\frac{4}{9\pi}\right)^{1/3} - \frac{216}{175}\left(\frac{4}{9\pi}\right)^{2/3}(\alpha Z^{2/3})^2$$
$$= 1.00116 - 4.56 \times 10^{-3} Z^{2/3} - 1.78 \times 10^{-5} Z^{4/3}.$$

Hamada and Salpeter (1961) have used the equation of state given immediately above to obtain a maximum radius of

$$R_{max} = 0.021\, R_\odot \tag{3-165}$$

with a mass of about $0.02\, M_\odot$. Here the solar radius, $R_\odot \approx 6.96 \times 10^{10}$ cm and the solar mass, $M_\odot \approx 1.989 \times 10^{33}$ gm. This value of R_{max} is near that first estimated by Kothari (1938). The value of the Chandrasekhar limiting mass is given by Eq. (3-162) with a μ depending on chemical composition. Hamada and Salpeter obtain a lowest possible value of M_{cre}, given by

$$M_{cre} = 1.015\, M_\odot. \tag{3-166}$$

Baym, Pethick, and Sutherland (1971) have included a lattice term in calculating the equilibrium species of nuclei present in the white dwarf density range of 10^4 g cm$^{-3} \lesssim \rho \lesssim 10^{11}$ g cm^{-3}. They obtain

$$M_{cre} = 1.00\, M_\odot$$

where

$$R_{max} = 2{,}140 \text{ km}.$$

3.3.2.3. Equation of State of a Degenerate Neutron Gas—Neutron Stars

The density of matter in a neutron star increases with depth from about 10^4 g cm^{-3} in the outermost layers to greater than 10^{15} g cm^{-3} in the inner core. To the first approximation, the matter is in its ground state and can be approximated by zero temperature equations of state, for the thermal energy of 10^{-4} to 10^{-2} MeV ($10^6\,°\mathrm{K} \lesssim T \lesssim 10^8\,°\mathrm{K}$) is much less than the typical excitation energies. In the outermost layers, 10^4 g cm$^{-3} \lesssim \rho \lesssim 4.3 \times 10^{11}$ g cm^{-3}, the constituents are electrons and nuclei and the equation of state may be calculated according to the procedures given in the previous Sect. 3.3.2.2 (cf. SALPETER, 1961 and BAYM, PETHICK, and SUTHERLAND, 1971). For densities in the range 4.3×10^{11} g cm$^{-3} \lesssim \rho \lesssim 3 \times 10^{14}$ g cm^{-3}, the neutron rich nuclei begin to "drip" neutrons and the matter consists of nuclei emersed in a neutron gas as well as an electron gas. OPPENHEIMER and VOLKOFF (1939) first considered the equation of state of an ideal Fermi gas of nonrelativistic neutrons. For a nonrelativistic, completely degenerate neutron gas of density, ρ_n, and a neutron mass, M_n, the pressure is given by (Eq. (3-156))

$$P = \frac{(3\pi^2)^{2/3}}{5}\,\frac{\hbar^2}{M_n^{8/3}}\,\rho_n^{5/3} \approx 10^{10}\,\rho_n^{5/3}\ \text{dynes cm}^{-2}, \qquad (3\text{-}167)$$

which corresponds to a kinetic energy per particle of

$$W(k,0) = \frac{3\hbar^2\,(2^{1/3}k)^2}{10\,M_n}, \qquad (3\text{-}168)$$

where the number density, N_n, of the neutron gas is

$$N_n = \frac{k^3}{1.5\,\pi^2}.$$

When more refined calculations for pure neutron matter are considered, the energy per particle is given by (SIEMANS and PANHARIPANDE, 1971)

$$W(k,0) \approx 19.74\,k^2 - k^3\,\frac{(40.4 - 1.088\,k^3)}{(1 + 2.545\,k)}\ \text{MeV} \qquad (3\text{-}169)$$

for $k \leq 1.5$ fm^{-1}, and for $W(k,0)$ in MeV and k in fm^{-1}. The first term in Eq. (3-169) corresponds to Eq. (3-168).

When actual neutron star matter is concerned, the energy and pressure must be a superposition of terms corresponding to the energy of neutrons, electrons, and nuclei as well as the lattice binding energy for the nuclei. In this case, the total energy per unit volume is given by (BAYM, BETHE, and PETHICK, 1971)

$$E_{tot} = N_N(W_N + W_L) + (1 - V_N N_N)\,E_N(N_n) + E_e(N_e), \qquad (3\text{-}170)$$

where N_N, N_n, and N_e are, respectively, the densities of nuclei, neutrons, and electrons, and $1 - V_N N_N$ is the fraction of the volume occupied by the neutron gas. The energy, W_N, of a nucleus is the sum of surface energy, W_s, and Coulomb energy terms, W_c, as well as a bulk energy term.

$$W_N = [(1-x)\,M_n c^2 + x\,M_p c^2 + W(k,x)]\,A + W_s + W_c,$$

where x is the fractional concentration of protons, M_p is the proton mass, and $W(k,x)$ is the energy per particle of bulk nuclear matter of density $N_N = k^3/(1.5\pi^2)$. BAYM, BETHE, and PETHICK give detailed formulae for $W(k,x)$, W_s, W_c and W_N. The energy density $E_N(N_n)/N_n$ of the neutron gas can, to first approximation, be given by $W(k,0)+M_n c^2$ where $W(k,0)$ is given by Eq. (3-169), and the energy density, $E_e(N_e)$, of the electrons is the value for a free electron gas given in the previous section. The lattice energy per nucleus is given by

$$W_L = -\frac{1.82\, Z^2 e^2}{a},$$

where the lattice constant $a = (2/N_N)^{1/3}$, and the Coulomb energy, W_c, is of order

$$W_c \approx \frac{3}{5} \frac{Z^2 e^2}{r_N},$$

where $4\pi N_N r_N^3 \approx 3$. BAYM, BETHE, and PETHICK (1971) give the formulae and values of neutron star matter for densities which range between 4.6×10^{11} and 2.4×10^{14} g cm^{-3}. BARKAT, BUCHLER, and INGBER (1972) use an extended nuclear Thomas-Fermi model to calculate the pressure as a function of density for neutron star matter in the range 10^8 g cm$^{-3} \lesssim \rho \lesssim 10^{14}$ g cm^{-3}. Additional work using the Fermi-Thomas model for nuclei is given by BUCHLER and BARKAT (1971) and BUCHLER and INGBER (1971). RAVENHALL, BENNET, and PETHICK (1972) employ a more reliable theory for the nuclear surface than did BAYM, BETHE, and PETHICK (1971), and find results similar to BARKAT, BUCHLER, and INGBER (1972).

OPPENHEIMER and SNYDER (1939) first pointed out that neutron stars with masses greater than some critical value, M_{crn}, will collapse indefinitely to form a closed trapped surface from which no electromagnetic radiation can escape. The critical value occurs at (OPPENHEIMER and VOLKOFF, 1939; TSURUTA and CAMERON, 1966)

$$M_{crn} \approx 0.69\, M_\odot \quad \text{to} \quad 2.0\, M_\odot,$$

which is at the minimum radius

$$R_{min} = 9.42 \text{ km}.$$

BAYM, PETHICK, and SUTHERLAND (1971) use the equation of state given by BAYM, BETHE, and PETHICK (1971) to obtain a maximum neutron star radius of

$$R_{max} = 164 \text{ km}, \tag{3-171}$$

a maximum stable mass of

$$M_{crn} \approx 1.41\, M_\odot$$

and a minimum stable mass

$$M_{min} = 0.0925\, M_\odot.$$

Neutron stars of mass $M \gtrsim M_{crn}$ are unstable against gravitational collapse, whereas those with $M \lesssim M_{min}$ are unstable against radial oscillations and possess no equilibrium configuration.

At densities higher than 2.4×10^{14} g cm^{-3}, the nuclei begin to touch each other, and the matter becomes a degenerate sea of neutrons in a sea of protons and electrons, each of which has about 4% of the neutron abundance. When the density surpasses 10^{15} g cm^{-3}, hyperons become important as well. Details of the equation of state in the density range 2—10×10^{14} g cm^{-3} are given by WANG, ROSE, and SCHLENKER (1970), and as shown in RUDERMAN'S (1969) review, the matter probably consists of superfluid neutrons together with a few per cent of superfluid protons and normal electrons. For a degenerate system of Fermi particles, superfluidity can exist when the particles are correlated in pairs with the same center of mass momentum. A new ground state of energy is established, and an energy gap, ε, exists in the energy spectrum at zero temperature. According to BARDEEN, COOPER, and SCHRIEFFER (1957),

$$\varepsilon = k T_c = \frac{4 N_p}{N_0} = 4 \hbar \omega \exp\left[-\frac{1}{N_0 V}\right], \tag{3-172}$$

where N_p is the number of electrons in pairs virtually excited above the Fermi surface, $N_0 \approx m\sqrt{2 m E_f}/(2\pi^2 \hbar^3)$ is the number of particle states of one spin per unit energy range at the Fermi surface, E_f is the Fermi energy, ω is the average lattice frequency and $-V$ is the interaction energy between pairs. At the transition temperature, T_c, thermal excitation has reduced the number of pairs to zero and the system is in a normal degenerate state. As emphasized by GINZBURG (1969) and RUDERMAN (1969), neutron star matter with $\rho \approx 10^{14} - 10^{15}$ g cm^{-3} may be superfluid. Details of the physics of the superfluidity of neutron star matter are given by YANG and CLARK (1971), CHAO, CLARK, and YANG (1972), and KROTSCHECK (1972). Assuming that changes in pulsar periods reflect changes in the rotation rate of a neutron star, the changes may be related to the physics of the superfluid core in a way which has been reviewed by PINES (1970).

The equation of state of matter at densities larger than 10^{15} g cm^{-3} has been discussed by HARRISON, THORNE, WAKANO, and WHEELER (1964), BUCHLER and INGBER (1971), PANHARIPANDE (1971), FRAUTSCHI, BAHCALL, STEIGMAN, and WHEELER (1971), and LEUNG and WANG (1971). One of the conclusions of these studies is that the neutron star must have a maximum mass which ranges between $0.1\,M_\odot$ and $1.5\,M_\odot$, and that this mass is dependent upon the details of the very high density core material (cf. CLARK, HEINTZMANN, HILLEBRANT, and GREWING, 1971).

3.3.2.4. The Neutrino Gas—Number Density, Energy Density, Entropy Density, and the Equation of State

The number density, N, of a neutrino gas with zero rest mass is given by (KUCHOWICZ, 1963 and private communication)

$$N = 8\pi g \left(\frac{kT}{hc}\right)^3 F_2\left(\frac{\mu}{kT}\right), \tag{3-173}$$

where $g \sim 1$ is the statistical weight of the neutrino, T is the gas temperature, and the Fermi-Dirac integral, $F_n(\mu/kT)$, is given by

$$F_n\left(\frac{\mu}{kT}\right) = \frac{1}{n!} \int_0^\infty \frac{x^n\, dx}{\exp\left(x - \frac{\mu}{kT}\right) + 1}, \tag{3-174}$$

where the neutrino chemical potential, μ, is given by Eq. (3-104).

The neutrino energy density, U/V, is given by

$$\frac{U}{V} = 3P = 24\pi g k T \left(\frac{kT}{hc}\right)^3 F_3\left(\frac{\mu}{kT}\right), \tag{3-175}$$

where P is the neutrino pressure, U is the energy, and V is the volume. The neutrino entropy density, S/V, is given by

$$\frac{S}{V} = 8\pi g k \left(\frac{kT}{hc}\right)^3 \left[4F_3\left(\frac{\mu}{kT}\right) - \left(\frac{\mu}{kT}\right) F_2\left(\frac{\mu}{kT}\right)\right], \tag{3-176}$$

where S is the entropy.

It follows from Eq. (3-175) that the pressure, P, of a neutrino gas with zero rest mass is given by

$$P = 8\pi g k T \left(\frac{kT}{hc}\right)^3 F_3\left(\frac{\mu}{kT}\right),$$

where $g \sim 1$ is the statistical weight of the neutrino, T is the gas temperature, μ is the chemical potential given by Eq. (3-104), and the Fermi-Dirac integral, $F_n(\mu/kT)$, is given in Eq. (3-174).

The asymptotic expressions for the equation of state are (Kuchowicz, 1963 and private communication):

For extreme degeneracy ($\mu \gg kT$):

$$P = \frac{\pi g k T}{3} \left(\frac{kT}{hc}\right)^3 \left\{\left[\frac{1}{2} y^4 + \frac{1}{3}\pi^6 y^2 + \frac{1}{27}\pi^{12} + D^{1/2}\right]^{1/3} + \left[\frac{1}{2} y^4 + \frac{1}{3}\pi^6 y^2 + \frac{1}{27}\pi^{12} - D^{1/2}\right]^{1/3} - \frac{\pi^4}{5}\right\} \approx \frac{\pi g}{3} \frac{(kT y^{1/3})^4}{(hc)^3}, \tag{3-177}$$

where

$$D = \frac{y^2}{4} + \frac{\pi^6}{27},$$

and

$$y = \frac{3N}{4\pi g} \left(\frac{hc}{kT}\right)^3.$$

Here N is the neutrino number density given in Eq. (3-173).

For weak degeneracy ($\mu \ll -0.4\,kT$):

$$P = \frac{16\pi g k T}{3}\left(\frac{kT}{hc}\right)^3\left[1+\frac{y}{24}-\left(1-\frac{y}{12}\right)^{1/2}\right]$$
$$\approx \frac{NkT}{3}\left[1+\frac{N}{128\pi g}\left(\frac{hc}{kT}\right)^3\right] \quad \text{for } N < T^3. \tag{3-178}$$

For zero chemical potential we obtain the neutrino analogue to the Stefan-Boltzmann law, the difference in statistics between photons and neutrinos being a $7g/16$ factor. This factor doubles for a neutrino-antineutrino gas.

3.3.3. The Photon Gas

3.3.3.1. Einstein-Bose Statistics

Elementary particles whose spin is an even multiple of $\hbar/2$ (photons and nuclei with even mass numbers) obey Einstein-Bose statistics (EINSTEIN, 1924; BOSE, 1924). These particles are indistinguishable from each other, and all values for the number of particles in a particular quantum state are equally likely. The number density, $n(\varepsilon)$, of particles with kinetic energy between ε and $\varepsilon+d\varepsilon$ is given by

$$Vn(\varepsilon) = \frac{g(\varepsilon)\,d\varepsilon}{\exp\left[\frac{-\mu}{kT}+\frac{\varepsilon}{kT}\right]-1}, \tag{3-179}$$

where $g(\varepsilon)$ is the number of possible particle states of energy, ε, and is given by Eq. (3-109), the volume is V, and the chemical potential, μ, is given by Eq. (3-104). For the special case of photons, $\mu=0$ and

$$Vn(\varepsilon) = \frac{g(\varepsilon)\,d\varepsilon}{\exp\left(\frac{\varepsilon}{kT}\right)-1}.$$

The total number of particles per unit volume, N, is given by

$$N = \int_0^\infty n(\varepsilon)\,d\varepsilon = \frac{g}{2\pi^2\hbar^3}\int_0^\infty \frac{p^2\,dp}{\exp\left[\frac{-\mu}{kT}+\frac{\varepsilon}{kT}\right]-1}, \tag{3-180}$$

where g is the statistical weight, and ε is related to the momentum, p, by Eq. (3-111). For the nonrelativistic gas ($\varepsilon=p^2/(2m)$, $p \ll mc$, $kT \ll mc^2$)

$$N = \frac{4\pi}{h^3}(2mkT)^{3/2}\int_0^\infty \frac{x^{1/2}\,dx}{\exp\left[\frac{-\mu}{kT}+x\right]-1}.$$

The total energy density, U/V, is given by

$$\frac{U}{V}=\frac{g}{2\pi^2\hbar^3}\int_0^\infty \frac{\varepsilon p^2\,dp}{\exp\left[\frac{-\mu}{kT}+\frac{\varepsilon}{kT}\right]-1},$$

which reduces to

$$\frac{U}{V}=\frac{4\pi}{h^3}(2m)^{3/2}(kT)^{5/2}\int_0^\infty \frac{x^{3/2}\,dx}{\exp\left[\frac{-\mu}{kT}+x\right]-1}, \tag{3-181}$$

for a nonrelativistic gas.

3.3.3.2. Equation of State, Energy Density, and Entropy of a Photon Gas

For densities lower than the critical density, ρ_{cp}, given by

$$\rho_{cp}=3.0\times10^{-23}\mu T^3\ \mathrm{gm\ cm^{-3}} \quad \text{or}\quad T=3.20\times10^7\left(\frac{\rho_{cp}}{\mu}\right)^{1/3}\ ^\circ K, \tag{3-182}$$

the pressure of a photon gas predominates over the pressure of the non-degenerate perfect gas. Here the constant μ the is mean molecular weight. As the photon energy is $h\nu$, it follows from Eq. (3-181) that the radiation pressure of the photon gas is

$$P=\frac{1}{3}\int_0^\infty h\nu N(\nu)\,d\nu=\frac{1}{3}\frac{U}{V}, \tag{3-183}$$

where U/V is the energy density and the number density of photons, $N(\nu)$, with frequencies between ν and $\nu+d\nu$ is

$$N(\nu)=\frac{8\pi\nu^2}{c^3}\frac{1}{\left[\exp\left(\frac{h\nu}{kT}\right)-1\right]}d\nu. \tag{3-184}$$

Substituting Eq. (3-184) into Eq. (3-183) we obtain

$$P=\frac{1}{3}\frac{U}{V}=\frac{1}{3}aT^4, \tag{3-185}$$

where the radiation constant, a, is given by

$$a=\frac{8\pi^5}{15}\frac{k^4}{h^3c^3}\approx 7.564\times10^{-15}\ \mathrm{erg\ cm^{-3}\ ^\circ K^{-4}}.$$

The equations of state of a gas at various conditions of density, ρ, and temperature, T, are given in Table 36. The radiation constant, a, is related to the Stefan-Boltzmann constant, σ, by the relation (STEFAN, 1879; BOLTZMANN, 1884)

$$\sigma=\frac{ac}{4}=5.67\times10^{-5}\ \mathrm{erg\ cm^{-2}\ ^\circ K^{-4}\ sec^{-1}}.$$

Further relations for a photon gas are given in Sect. 1.23 on thermal emission from a black body. For example, the entropy per unit volume is $S/V=4aT^3/3$, when the total energy per unit volume is $U/V=aT^4$ and the pressure is $P=aT^4/3$.

Table 36. Equations of state for different ranges of mass density, ρ, and temperature, T. The gas pressure is P, the mean molecular weight is μ, and μ_e is the mean molecular weight of the electron

State	Region (ρ in g cm^{-3})	Pressure, P (dynes cm^{-2})
Photon gas—radiation pressure	$\rho<3.0\times10^{-23}\mu T^3$	$P=2.521\times10^{-15}T^4$
Perfect nondegenerate gas	$3.0\times10^{-23}\mu T^3<\rho<2.4\times10^{-8}\mu_e T^{3/2}$	$P=8.314\times10^7\rho T/\mu$
Nonrelativistic, completely degenerate, electron gas	$2.4\times10^{-8}\mu_e T^{3/2}<\rho<7.3\times10^6\mu_e$	$P=1.004\times10^{13}\left(\frac{\rho}{\mu_e}\right)^{5/3}$
Relativistic, completely degenerate, electron gas	$7.3\times10^6\mu_e<\rho\lesssim10^{11}$	$P=1.244\times10^{15}\left(\frac{\rho}{\mu_e}\right)^{4/3}$
and	$2.4\times10^{-8}T^{3/2}\mu_e\lesssim\rho\lesssim10^{11}$	
Degenerate neutron gas	$10^{11}\lesssim\rho\lesssim10^{14}$	$P\approx10^{10}(\rho)^{5/3}$

3.4. Macrostructure of a Gas—The Virial Theorem

3.4.1. The Virial Theorem of Clausius

Consider the motion of a single molecule of mass, m. If $\boldsymbol{r}$ is its position vector measured from an arbitrary origin and $\boldsymbol{F}$ is the force acting on the molecule, then

$$m\frac{d^2\boldsymbol{r}}{dt^2}=\boldsymbol{F}.$$

Forming the scalar product with $\boldsymbol{r}$ and transforming the resulting equation we obtain

$$\frac{1}{2}m\frac{d^2r^2}{dt^2}=m\frac{d}{dt}\left(\boldsymbol{r}\cdot\frac{d\boldsymbol{r}}{dt}\right)=m\left(\frac{d\boldsymbol{r}}{dt}\right)^2+\boldsymbol{F}\cdot\boldsymbol{r}. \tag{3-186}$$

Letting $\sum$ denote an ensemble average over all molecules, and noting that a macroscopic system which is in equilibrium is stationary, Eq. (3-186) becomes the virial theorem (Clausius, 1870)

$$\sum m\left(\frac{dr}{dt}\right)^2+\sum\boldsymbol{F}\cdot\boldsymbol{r}=0, \tag{3-187}$$

where $\sum\boldsymbol{F}\cdot\boldsymbol{r}$ is called the virial. Eq. (3-187) is also known as Poincaré's theorem, as Poincaré derived it by assuming that the only force acting on a system of molecules is that of their mutual gravitational attraction. In this case, Eq. (3-187) becomes (POINCARÉ, 1811)

$$2T+\Omega=0, \tag{3-188}$$

where T is the total kinetic energy of the particles, and Ω is the total gravitational potential energy of the system.

As EDDINGTON (1916) noted, Eqs. (3-187) and (3-188) may be extended to include nonequilibrium situations by including the moment of inertia, I. In this nonstationary case,

$$\frac{1}{2}\frac{d^2 I}{dt^2}=\sum m\left(\frac{dr}{dt}\right)^2+\sum \boldsymbol{F}\cdot\boldsymbol{r}=2T+\Omega, \tag{3-189}$$

where $I=\sum m r^2$.

When the magnetic energy, $\mathscr{M}$, and the kinetic energy of mass motion, T_m, are taken into account, the equilibrium virial theorem becomes (CHANDRASEKHAR and FERMI, 1953)

$$2[T_m+T_k]+\Omega+\mathscr{M}=0, \tag{3-190}$$

where T_k is the kinetic energy of molecular motion, and Ω is the total gravitational potential energy of the system. For a perfect gas sphere of radius, R, mass, M, pressure, P, and temperature, T,

$$2T_k=\frac{3MkT}{\mu m_H}=4\pi PR^3=3(\gamma-1)U, \tag{3-191}$$

where μ is the mean molecular weight, m_H is the mass of hydrogen, γ is the adiabatic index, and U is the internal energy due to the molecular motion (the heat energy). For a gaseous mass rotating with angular velocity, ω, the kinetic energy of mass motion, T_m, is given by

$$T_m=\tfrac{1}{2}I\omega^2, \tag{3-192}$$

where I is the moment of inertia. For a sphere of mass, M, and radius, R,

$$T_m=\tfrac{1}{2}I\omega^2=\tfrac{1}{5}MR^2\omega^2. \tag{3-193}$$

The gravitational potential energy, Ω, of such a sphere is given by

$$\Omega=-\frac{3}{5}\frac{GM^2}{R}, \tag{3-194}$$

where the gravitational constant, $G=6.668\times10^{-8}$ dyn cm^2 g^{-2}, and the magnetic energy, $\mathscr{M}$, is given by

$$\mathscr{M}=\tfrac{1}{6}H^2R^3, \tag{3-195}$$

where H is the average magnetic field strength in the sphere.

3.4.2. Ritter's Relation

For an adiabatic process in a gas which is in hydrostatic equilibrium, it follows from the first law of thermodynamics that the change, dU, in the internal energy,

U, of the system is given by

$$dU = -P\,dV, \tag{3-196}$$

where P is the gas pressure, and dV is the change in volume, V. For an adiabatic process involving a perfect gas, the change, dT, in temperature, T, is given by

$$\frac{dT}{T} = -(\gamma-1)\frac{dV}{V}, \tag{3-197}$$

where γ is the adiabatic index. Using Eqs. (3-196) and (3-197) together with the ideal gas law, we obtain

$$dU = \frac{N_{\text{tot}} k\,dT}{(\gamma-1)},$$

and

$$U = \frac{N_{\text{tot}} k T}{(\gamma-1)} = \frac{2T_k}{3(\gamma-1)},$$

where k is Boltzmann's constant, N_{tot} is the total number of molecules, and T_k is the total internal energy of molecular motion. Using the virial theorem in Poincaré's form, it follows that

$$U = -\frac{\Omega}{3(\gamma-1)}, \tag{3-198}$$

which is Ritter's relation (RITTER, 1880). The total energy, E, is then given by

$$E = U + \Omega = \frac{3\gamma-4}{3(\gamma-1)}\Omega. \tag{3-199}$$

The energy, E, becomes negative for $\gamma<4/3$, and therefore a gas mass is unstable against adiabatic pulsations for $\gamma<4/3$. When the magnetic term, $\mathcal{M}$, given in Eq. (3-195) is included, the criterion for dynamical stability becomes $(3\gamma-4)(|\Omega|-\mathcal{M})>0$. Furthermore, CHANDRASEKHAR (1964) has shown that the Newtonian lower limit of 4/3 for the ratio of specific heats is increased by effects arising from general relativity. As long as γ is finite, dynamical instability occurs if the mass contracts and the radius falls below the critical value

$$R_c = \frac{K}{\gamma-4/3}\,\frac{2GM}{c^2} \approx 4.7\times10^{17}\ \text{cm}, \tag{3-200}$$

where the constant K ranges from 0.45 to 1.12 for polytropes whose index ranges from 0 to 4. The approximate value is for a polytrope of index 3.

3.4.3. Chandrasekhar Limiting Mass for Degenerate Matter

When only the kinetic energy of molecular motion and the gravitational potential energy are taken into account, the virial theorem for a spherical object of radius, R, becomes

$$3PV = 4\pi PR^3 = \frac{3}{5}\,\frac{GM^2}{R}, \tag{3-201}$$

where P is the gas pressure, V is the volume, M is the mass, and the gravitational constant $G = 6.668 \times 10^{-8}$ dyne cm^2 g^{-2}. For a relativistic, completely degenerate electron gas, P, is given by Eq. (3-158), which is

$$P = 1.244 \times 10^{15} \left(\frac{\rho}{\mu_e}\right)^{4/3} \text{dynes cm}^{-2}, \tag{3-202}$$

where ρ is the mass density and μ_e is the electron molecular weight. It follows from Eqs. (3-201) and (3-202) that, for a relativistic, completely degenerate, electron gas,

$$M = \left(\frac{2}{\mu_e}\right)^2 1.4587\, M_\odot = 1.4587\, M_\odot \quad \text{for hydrogen}, \tag{3-203}$$

where the solar mass, $M_\odot = 1.989 \times 10^{33}$ gm. The energy density of the relativistic gas is $3P$, however, and it follows from Eq. (3-201) that the total energy is zero. Thus, an object composed of a relativistic gas is not gravitationally bound, and the object is unstable. The M given by Eq. (3-203) is the maximum mass which may be supported by electron degeneracy. The existence of this mass limit was first predicted by LANDAU (1932) and the numerical calculations first worked out by CHANDRASEKHAR (1935). As shown in Sect. 3.3.2, corrections to the equation of state by HAMADA and SALPETER (1961) lead to a maximum mass between 1.01 and 1.40 $M_\odot$ depending on the chemical composition. MESTEL (1965) first noted that the upper mass limit applies to spherically symmetric, nonrotating stars and does not apply to nonspherical structures or to rotating structures. ROXBURGH (1965) obtains a critical mass of $6.13\, \mu_e^{-2}\, M_\odot$ for uniformly rotating stars; whereas OSTRIKER, BODENHEIMER, and LYNDEN-BELL (1966) showed that one can construct differentially rotating white dwarfs without any limit to mass. OSTRIKER and BODENHEIMER (1968) have presented a number of models of this type for which the mass ranges from 1.3 to 4.1 $M_\odot$. A magnetic field may also increase the value of the critical mass above which collapse takes place.

3.4.4. Conditions for Gravitational Contraction in the Presence of a Magnetic Field or an External Pressure—The Maximum Magnetic Field and the Maximum Mass for Dynamical Stability

Consider a perfect gas sphere of radius, R, and mass, M, with an internal magnetic field of average strength, H. If the gas is in equilibrium, it follows from the virial theorem (Eq. (3-190)) that

$$3(\gamma - 1)\, U + \Omega + \mathcal{M} = 0, \tag{3-204}$$

where the total energy

$$E = U + \Omega + \mathcal{M}, \tag{3-205}$$

and U, Ω, and $\mathcal{M}$ are, respectively, the heat energy of molecular motion, the total gravitational potential energy, and the magnetic energy. Eliminating U, we obtain

$$E = -\frac{3\gamma - 4}{3(\gamma - 1)} [|\Omega| - \mathcal{M}], \tag{3-206}$$

which must be greater than zero for dynamical stability. Thus, in the absence of a magnetic field, $\gamma > 4/3$ for dynamical stability. In the presence of a magnetic field. Eqs. (3-194) and (3-195) for Ω and $\mathscr{M}$ show that for masses larger than

$$M_c = \left[\frac{5}{18}\frac{H^2 R^4}{G}\right]^{1/2}$$

the magnetic field will not induce dynamical instability. In terms of the solar mass, $M_\odot$, and radius, $R_\odot$, the condition that $\mathscr{M} < |\Omega|$ gives

$$H < 2 \times 10^8 \frac{M}{M_\odot}\left(\frac{R_\odot}{R}\right)^2 \text{gauss}, \tag{3-207}$$

providing an upper limit to the magnetic field for dynamical stability.

When the surface of a perfect gas sphere of radius, R, mass, M, and temperature, T, is subjected to a uniform external pressure, P_e, the equilibrium virial equation (3-190) is

$$2T_k - 4\pi P_e R^3 + \Omega = 0 \tag{3-208}$$

or

$$P_e = \frac{3MkT}{4\pi R^3 \mu m_H} - \frac{3GM^2}{20\pi R^4}, \tag{3-209}$$

where μ is the molecular weight, and m_H is the mass of the hydrogen atom. As shown by McCrea (1957), P_e will reach a critical minimum value, P_{ec}, above which the cloud will begin to contract. Differentiating Eq. (3-209) with respect to radius, R, we obtain

$$P_{ec} = \frac{3}{16\pi G^3 M^2}\left(\frac{45}{12}\right)^3\left(\frac{kT}{\mu m_H}\right)^4. \tag{3-210}$$

All masses larger than the critical mass,

$$M_c = \frac{\sqrt{3}}{4\sqrt{\pi}\, G^{3/2} P_{ec}^{1/2}}\left(\frac{45}{12}\right)^{3/2}\left(\frac{kT}{\mu m_H}\right)^2, \tag{3-211}$$

will collapse. The corresponding critical radius below which collapse occurs is

$$R_c = \frac{12}{45}\frac{GM\mu m_H}{kT} \approx 6 \times 10^6 \frac{\mu}{T}\frac{M}{M_\odot} R_\odot \text{ cm}, \tag{3-212}$$

and the critical mass density above which collapse occurs is

$$\rho_c = \left(\frac{45}{12}\right)^3 \frac{3}{4\pi G^3 M^2}\left(\frac{kT}{m_H \mu}\right)^3 \approx 10^{-20}\left(\frac{T}{\mu}\right)^3\left(\frac{M_\odot}{M}\right)^2 \text{gm cm}^{-3}. \tag{3-213}$$

3.4.5. Gravitational Contraction, Hydrodynamic Time Scale, and the Kelvin-Helmholtz Contraction Time

Assuming free-fall and equating the kinetic energy of infall to the gravitational potential energy, the time for collapse, τ_H, is obtained. This hydrodynamic time scale is given by

$$\tau_{\mathrm{H}} \approx \left(\frac{1}{\rho}\left|\frac{d\rho}{dt}\right|^{-1}\right) \approx \frac{1}{(4\pi G\rho)^{1/2}} \approx \frac{R^{3/2}}{(GM)^{1/2}} \approx 2 \times 10^{3}\rho^{-1/2}\ \mathrm{sec} \quad (3\text{-}214)$$

for a sphere of density, ρ, radius, R, and mass, M. If there is an initial uniform density, ρ_0, and radius, R_0, then the radius, R, after a time, t, from the beginning of free-fall is given by the relation

$$\left(\frac{8\pi G}{3}\rho_0\right)^{1/2} t = \left(1-\frac{R}{R_0}\right)^{1/2}\left(\frac{R}{R_0}\right)^{1/2} + \sin^{-1}\left(1-\frac{R}{R_0}\right)^{1/2}. \quad (3\text{-}215)$$

If λ_J is the critical Jeans length, then the e-folding time for the gravitational condensation of a mass of diameter $D > \lambda_J$ is

$$\tau \approx \{4\pi G\rho[1-(\lambda_J/D)^2]\}^{-1/2}. \quad (3\text{-}216)$$

When the dissipation of energy by radiation is taken into account, the appropriate time scale is

$$\tau_{\mathrm{K-H}} \approx \frac{GM^2}{RL}, \quad (3\text{-}217)$$

where L is the luminosity of the sphere. This time is called the Kelvin-Helmholtz contraction time after HELMHOLTZ (1854) and KELVIN (1861, 1863) who first hypothesized that a star's lifetime might be determined by equating the energy of gravitational contraction to the radiated energy. Such calculations led to the conclusion that a nuclear energy source, as opposed to that of gravity, must keep a star shining.

3.4.6. Stable Equilibrium Ellipsoids of Rotating Liquid Masses

The equilibrium figures which arise when a homogeneous fluid rotates with a uniform angular velocity are derived from the virial equation. As discussed in detail by CHANDRASEKHAR (1969), and in the papers referenced therein, the virial equation is written in tensor form and includes additional terms corresponding to the distribution of pressure, the centrifugal potential, and the Coriolis acceleration. For a homogeneous liquid mass of mass density, ρ, and uniform rotational velocity, ω, the virial theorem becomes

$$\frac{d}{dt}\int_v \rho u_i x_j dx = 2T_{ij} + \Omega_{ij} + \omega^2 I_{ij} - \omega_i\omega_k I_{kj} + 2\varepsilon_{ilm}\omega_m \int_v \rho u_i x_j dx + \delta_{ij}\Pi, \quad (3\text{-}218)$$

where the integrals are volume integrals,

$$\frac{d}{dt}\int_v \rho u_i x_j dx = \frac{1}{2}\frac{d^2 I_{ij}}{dt^2}, \quad (3\text{-}219)$$

and becomes zero in the steady state, u_i and x_j denote the velocity and spatial components as measured from the center of mass, the moment of inertia tensor, I_{ij}, is given by

$$I_{ij} = \int_v \rho x_i x_j dx, \quad (3\text{-}220)$$

the kinetic energy tensor, T_{ij}, is given by

$$T_{ij}=\tfrac{1}{2}\int_{v}\rho u_i u_j dx\,, \tag{3-221}$$

the gravitational potential energy tensor, Ω_{ij}, is given by

$$\Omega_{ij}=-\frac{1}{2}\int_{v}\rho U_{ij}dx=\int_{v}\rho x_i\frac{\partial U_{ij}}{\partial x_j}dx\,, \tag{3-222}$$

where U_{ij} is the gravitational potential, the Krönecker delta function $\delta_{ij}=1$ if $i=j$ and is zero otherwise, the function $\varepsilon_{ilm}=+1$ if i, l, m are in cyclic order, -1 if i, l, m are in noncyclic order, and 0 if any two subscripts are repeated.

The term, Π, resulting from the distribution of pressure, P, is given by

$$\Pi=\int_{v}P\,dx\,. \tag{3-223}$$

If the x_3 axis is chosen to coincide with the direction of rotation, Eq. (3-218) becomes

$$\Omega_{ij}+\omega^2(I_{ij}-\delta_{i3}I_{3j})=-\delta_{ij}\Pi\,, \tag{3-224}$$

under static conditions.

For a homogeneous ellipsoid with semi-axes a_1, a_2, and a_3, the potential energy tensor, Ω_{ij}, is given by

$$\Omega_{ij}=-2\pi G\rho A_i I_{ij}\,, \tag{3-225}$$

where the moment of inertia tensor, I_{ij}, is given by

$$I_{ij}=\tfrac{1}{5}M a_i^2\delta_{ij}\,, \tag{3-226}$$

the mass, M, is given by

$$M=\tfrac{4}{3}\pi a_1 a_2 a_3\rho\,, \tag{3-227}$$

and

$$A_i=a_1 a_2 a_3\int_{0}^{\infty}\frac{du}{\Delta(a_i^2+u)}\,, \tag{3-228}$$

where

$$\Delta^2=(a_1^2+u)(a_2^2+u)(a_3^2+u)\,. \tag{3-229}$$

The potential, U, at an internal point, x_i, of a solid homogeneous ellipsoid is given by

$$U=\pi G\rho\left(I-\sum_{l=1}^{3}A_l x_l^2\right), \tag{3-230}$$

where

$$I=a_1 a_2 a_3\int_{0}^{\infty}\frac{du}{\Delta}=\sum_{i=1}^{3}a_i^2 A_i\,. \tag{3-231}$$

The potential, U, of a homogeneous ellipsoid at an external point, x_i, is given by

$$U=\pi G\rho a_1 a_2 a_3\int_{\lambda}^{\infty}\frac{du}{\Delta}\left(1-\sum_{i=1}^{3}\frac{x_i^2}{a_i^2+u}\right), \tag{3-232}$$

where the ellipsoidal coordinate of the point considered is λ, and is given by the positive root of the equation

$$\sum_{i=1}^{3} \frac{x_i^2}{a_i^2+\lambda} = 1 . \tag{3-233}$$

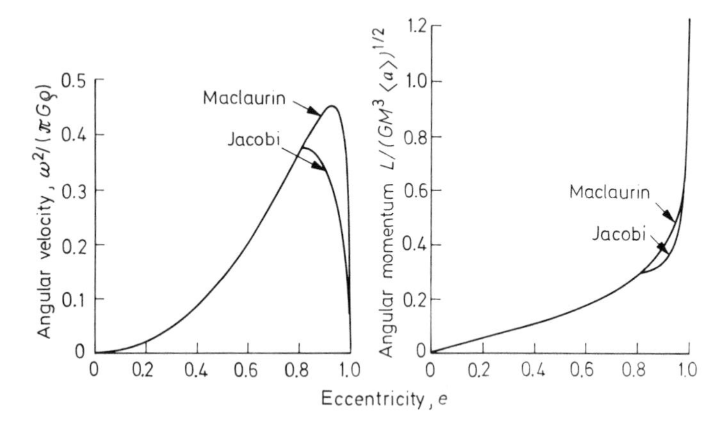

Fig. 20. The square of the angular velocity, ω, and the angular momentum, L, for the Maclaurin and Jacobi spheroids of different eccentricity, e. (After CHANDRASEKHAR, 1969, by permission of the Yale University Press). Here G is the Newtonian gravitational constant, M is the mass, ρ is the mass density, and $\langle a\rangle$ is the mean radius

It follows from Eqs. (3-224) and (3-225) that the angular rotation velocity, ω, for a Maclaurin spheroid for which $a_1=a_2$ is given by (MACLAURIN, 1742)

$$\omega = \left\{2\pi G\rho\left[\frac{(1-e^2)^{1/2}}{e^3}(3-2e^2)\sin^{-1}e - \frac{3}{e^2}(1-e^2)\right]\right\}^{1/2}, \tag{3-234}$$

where the eccentricity, e, is given by

$$e = \left(1-\frac{a_3^2}{a_1^2}\right)^{1/2} . \tag{3-235}$$

The angular momentum, L, of the Maclaurin spheroid is given by

$$L = (GM^3\langle a\rangle)^{1/2}\frac{\sqrt{3}}{5}\left(\frac{a_1}{\langle a\rangle}\right)^2 \frac{\omega}{\sqrt{\pi G\rho}}, \tag{3-236}$$

where

$$\langle a\rangle = (a_1^2 a_3)^{1/3}$$

is the radius of the sphere of the same mass, M, as the spheroid, and the mass M is given by Eq. (3-227). The angular velocity increases from zero at $e=0$ to a maximum, ω_{max}, given by

$$\omega_{max}^2 = 0.449331\,(\pi G\rho) \quad \text{at } e=0.92995 , \tag{3-237}$$

and then decreases again to zero as e approaches unity (cf. Fig. 20). The angular momentum, L, increases monotonically from zero to infinity as e varies from 0 to 1. Maclaurin spheroids become, however, secularly unstable via viscous processes for

$$0.81267 < e \leq 0.95289\,, \tag{3-238}$$

and they become dynamically unstable for

$$0.95289 \leq e \leq 1\,. \tag{3-239}$$

Jacobi (1834) first realized that equilibrium configurations are possible for ellipsoidal ($a_1 \neq a_2$) as well as for spheroidal ($a_1 = a_2$) configurations. It follows from Eqs. (3-224) and (3-225) that the angular rotation velocity, ω, for the Jacobi ellipsoid is given by (Jacobi, 1834)

$$\omega = \left[2\pi G \rho a_1 a_2 a_3 \int_0^\infty \frac{u\,du}{(a_1^2+u)(a_2^2+u)\Delta} \right]^{1/2}, \tag{3-240}$$

provided that the geometrical restriction

$$a_1^2 a_2^2 \int_0^\infty \frac{du}{(a_1^2+u)(a_2^2+u)\Delta} = a_3^2 \int_0^\infty \frac{du}{(a_3^2+u)\Delta} \tag{3-241}$$

is satisfied. This restriction allows a solution for a_3 which satisfies the inequality

$$\frac{1}{a_3^2} > \frac{1}{a_1^2} + \frac{1}{a_2^2}\,. \tag{3-242}$$

The angular momentum, L, of the Jacobi ellipsoid is given by

$$L = (G M^3 \langle a \rangle)^{1/2} \frac{\sqrt{3}}{10} \frac{a_1^2 + a_2^2}{\langle a \rangle^2} \frac{\omega}{\sqrt{\pi G \rho}}, \tag{3-243}$$

where

$$\langle a \rangle = (a_1 a_2 a_3)^{1/3}\,.$$

The Maclaurin spheroid is one solution of the Jacobi ellipsoidal equations for eccentricities, e, and angular velocities, ω, up to the bifurcation point given by

$$\omega = [0.37423\,\pi G \rho]^{1/2} \quad \text{and} \quad e = 0.81297\,. \tag{3-244}$$

Here $a_1 = a_2$, $a_3 = 0.582724\,a_1$ and $L = 0.303751\,(G M^3 \langle a \rangle)^{1/2}$. For $\omega \leq [0.37423\,\pi G \rho]^{1/2}$ and $e \geq 0.81297$, there are three equilibrium configurations possible, two Maclaurin spheroids and one Jacobi ellipsoid, whereas for $0.3742 \leq \omega^2/(\pi G \rho) \leq 0.4493$ only the Maclaurin figures are possible. Along the Jacobi sequence a_2/a_1, a_3/a_1, and ω decrease monotonically to zero as e increases, whereas the angular momentum increases monotonically to infinity (cf. Fig. 20). Poincaré (1855) showed that the Jacobian sequence bifurcates into a new sequence of pear-shaped configurations when

$$\frac{a_2}{a_1} = 0.432232, \quad \frac{a_3}{a_1} = 0.345069, \quad \text{and} \quad \omega = [0.284030\,\pi G \rho]^{1/2}. \tag{3-245}$$

At this point the Jacobi ellipsoid becomes unstable and in this respect is different from the Maclaurin sequence which remains stable on either side of the bifurcation point (CARTAN, 1924, LYTTLETON, 1953).

For an infinitesimal, homogeneous satellite of density, ρ, rotation about a rigid sherical planet of mass, M, in circular Keplerian orbit of radius, R, no equilibrium configurations are possible for rotational angular velocities, ω, which exceed the limit given by (ROCHE, 1847)

$$\frac{\omega^2}{\pi G \rho} = \frac{M}{\pi \rho R^3} \leq 0.090093 . \tag{3-246}$$

The lower limit to R set by Eq. (3-246) is called the Roche limit. When only the tidal forces between a satellite and its planet are taken into account, the following equilibrium condition is obtained (JEANS, 1917, 1919; CHANDRASEKHAR, 1969)

$$\frac{M}{\pi \rho R^3} \lesssim 0.12554 ,$$

and

$$e = 0.88303 ,$$

for $M = 0.12554 \pi \rho R^3$, where e is the eccentricity of the Jeans spheroid.

A large portion of the stars in our Galaxy ($\sim 50\%$) are binaries, and a substantial fraction of the binaries revolve at a distance which is about the same order as the sum of their radii. When the more massive star has depleted its core of hydrogen, it will expand, but its expansion is limited by Roche's criterion (Eq. (3-246)). Such a limit may be related to the tendency for close binary systems to favor more outbursts (cf. HACK, 1963; PACZŃSKI, 1971). If a star of mass, M_1, is in circular orbit about another star of mass, M_2, matter will flow outwards from one of the stars if its radius, R, becomes larger than the critical radius, R_{cr}, given by (PACZŃSKI, 1971)

$$\begin{aligned} R_{cr} &= \left[0.38 + 0.2 \log \left(\frac{M_1}{M_2} \right) \right] A \quad \text{for } 0.3 < \frac{M_1}{M_2} < 20 , \\ R_{cr} &= 0.46224 \, A \left(\frac{M_1}{M_1 + M_2} \right)^{1/3} \quad \text{for } 0 < \frac{M_1}{M_2} < 0.8 . \end{aligned} \tag{3-247}$$

Here A is the separation between the star centers and is related to the orbital period, P, by Kepler's law

$$\left(\frac{2\pi}{P} \right)^2 A^3 = G(M_1 + M_2) ,$$

where P is the orbital period and the gravitational constant $G = 6.668 \times 10^{-8}$ dyn cm^2 g^{-2}. Once a star has expanded beyond the critical radius, R_{cr}, the rate of mass outflow, dM/dt, is given by

$$\frac{dM}{dt} \approx \left(\frac{R - R_{cr}}{R} \right)^{n+1.5} ,$$

where R is the stellar radius and n is the polytropic index.

3.5. Gas Macrostructure—Hydrodynamics

3.5.1. The Continuity Equation for Mass Conservation

In many circumstances a gas may be regarded as a continuous fluid. The continuity equation for the conservation of gas mass for a fluid element having volume, V, is, in this case, expressed by

$$\frac{1}{\rho V}\frac{D}{Dt}(\rho V)=\frac{1}{\rho}\frac{D\rho}{Dt}+\frac{1}{V}\frac{DV}{Dt}=0. \tag{3-248}$$

where ρ is the mass density, and D/Dt denotes a differentiation following the fluid element. For a scalar, φ, and a vector, $\boldsymbol{F}$,

$$\frac{D\varphi}{Dt}=\frac{\partial\varphi}{\partial t}+\boldsymbol{v}\cdot\nabla\varphi \tag{3-249}$$

and

$$\frac{D\boldsymbol{F}}{Dt}=\frac{\partial\boldsymbol{F}}{\partial t}+(\boldsymbol{v}\cdot\nabla)\boldsymbol{F},$$

where $\boldsymbol{v}$ is the velocity of the moving fluid. Using Eq. (3-249) in Eq. (3-248), we obtain the continuity equation

$$\frac{\partial\rho}{\partial t}+\nabla\cdot(\rho\boldsymbol{v})=\frac{\partial\rho}{\partial t}+\boldsymbol{v}\cdot\nabla\rho+\rho\nabla\cdot\boldsymbol{v}=0. \tag{3-250}$$

For an incompressible fluid, Eq. (3-250) becomes $\nabla\cdot\boldsymbol{v}=0$.

3.5.2. Euler's Equation (the Navier-Stokes and Bernoulli's Equations)

The equation of motion of a fluid element is obtained from the law of conservation of momentum. Taking the time derivative of momentum we obtain Euler's (1755) force equation

$$\rho\frac{D\boldsymbol{v}}{Dt}=\rho\left[\frac{\partial\boldsymbol{v}}{\partial t}+(\boldsymbol{v}\cdot\nabla)\boldsymbol{v}\right]=\boldsymbol{F}-\nabla P, \tag{3-251}$$

where ρ is the mass density, $\boldsymbol{v}$ is the velocity, $\boldsymbol{F}$ is the external force (other than that due to gas pressure) acting on a unit volume, and P is the pressure. The contribution of gravity to the external force is

$$\boldsymbol{F}=-\rho\nabla\varphi, \tag{3-252}$$

where, from Poisson's equation, (Poisson, 1813), the gravitational potential, φ, is given by

$$\nabla^2\varphi=4\pi G\rho, \tag{3-253}$$

and the constant of gravity, $G=6.668\times10^{-8}$ dyn cm^2 g^{-2}. The viscous force of an incompressible fluid, one in which the mass density $\rho=$constant, is given by

$$\boldsymbol{F}=\mu\nabla^2\boldsymbol{v}, \tag{3-254}$$

where the coefficient of dynamic viscosity, μ, is given by Eqs. (3-23), (3-25), or (3-26). When Eq. (3-251) includes only the viscous force term, the resulting equation is called the Navier-Stokes equation (NAVIER, 1822; STOKES, 1845).

The force contribution due to the electric and magnetic fields of respective strengths $\boldsymbol{E}$ and $\boldsymbol{B}$ is

$$\boldsymbol{F} = \frac{1}{c} \boldsymbol{J} \times \boldsymbol{B}, \tag{3-255}$$

where the current $\boldsymbol{J}$ is given by

$$\boldsymbol{J} = \sigma \left[\boldsymbol{E} + \frac{\boldsymbol{v}}{c} \times \boldsymbol{B} \right], \tag{3-256}$$

and the electrical conductivity, σ, is given by Eqs. (3-35) or (3-36). For a highly conducting medium, the electromagnetic force term is

$$\boldsymbol{F} = \frac{-1}{4\pi} \boldsymbol{B} \times (\nabla \times \boldsymbol{B}), \tag{3-257}$$

where $\boldsymbol{B}$ and $\boldsymbol{v}$ are related by the diffusion equation

$$\frac{\partial \boldsymbol{B}}{\partial t} = \nabla \times (\boldsymbol{v} \times \boldsymbol{B}) + \frac{c^2}{4\pi\sigma} \nabla^2 B \approx \nabla \times (\boldsymbol{v} \times \boldsymbol{B}). \tag{3-258}$$

For the steady flow of a gas, $\partial \boldsymbol{v}/\partial t = 0$, and Euler's equation (3-251), in the absence of external forces, becomes

$$\rho \boldsymbol{v} \cdot \nabla \boldsymbol{v} = -\nabla P, \tag{3-259}$$

which is Bernoulli's equation for compressible flow (BERNOULLI, 1738). Here, ρ is the mass density, $\boldsymbol{v}$ is the velocity, and P is the pressure. Bernoulli's equation for compressible steady flow has the integral form

$$\frac{v^2}{2} + \int \frac{dP}{\rho} = \text{constant}. \tag{3-260}$$

The integral equation for incompressible steady flow is

$$\tfrac{1}{2} \rho v^2 + P = \text{constant}. \tag{3-261}$$

3.5.3. The Energy Equation

From the law of conservation of energy, the change per unit time in the total energy of the gas in any volume must equal the total flux of energy through the surface bounding that volume. When the effects of thermal conductivity and viscosity are unimportant, and in the absence of body forces like gravity, this relation may be expressed as

$$\frac{\partial}{\partial t} \int \left[\frac{1}{2} \rho v^2 + \rho u \right] dV = -\oint \rho \boldsymbol{v} \left(\frac{1}{2} v^2 + h \right) \cdot \boldsymbol{n} \, ds, \tag{3-262}$$

where t is the time variable, ρ is the mass density, v is the velocity, u is the internal energy per unit mass, $h = u + P/\rho$ is the enthalpy per unit mass, $\oint$ denotes a closed surface integral, $\boldsymbol{n}$ is a unit vector normal to the surface, ds is an element of area, and dV is an element of volume. In vector form, the energy equation (3-262) becomes

$$\frac{\partial}{\partial t}\left[\frac{1}{2}\rho v^2+\rho u\right]=-\nabla\cdot\left[\rho\boldsymbol{v}\left(\frac{1}{2}v^2+h\right)\right], \tag{3-263}$$

and the quantity $\rho\boldsymbol{v}[(v^2/2)+h]$ is called the energy flux density.

When thermal conductivity and viscous effects are important, the equation of energy balance can be written

$$\rho\frac{\partial}{\partial t}(c_v T)+\rho v_j\frac{\partial}{\partial x_j}(c_v T)=\frac{\partial}{\partial x_j}\left(\kappa\frac{\partial T}{\partial x_j}\right)-\rho P\frac{\partial v_j}{\partial x_j}+\varphi, \tag{3-264}$$

where c_v is the specific heat at constant volume, κ is the coefficient of heat conduction, and the rate at which energy is dissipated by viscosity is given by

$$\varphi=\frac{\mu}{2}\left[\left(\frac{\partial v_i}{\partial x_j}+\frac{\partial v_j}{\partial x_i}\right)^2-\frac{4}{3}\left(\frac{\partial v_j}{\partial x_j}\right)^2\right],$$

where μ is the coefficient of dynamic viscosity, and $\partial v_i/\partial x_i=0$ for an incompressible fluid. Under the Boussinesq approximation to be discussed later, Eq. (3-264) becomes Eq. (3-57):

$$\frac{\partial T}{\partial t}+\boldsymbol{v}\cdot\nabla T=\frac{\kappa}{\rho c_v}\nabla^2 T.$$

Eq. (3-263) is the energy balance equation which states that the rate of energy change per unit volume is equal to the amount of energy flowing out of this volume in unit time. If conduction, viscosity, and the kinetic energy term are ignored, and if all particles are at the same kinetic temperature, T, then Eq. (3-263) becomes, under conditions of constant pressure, (FIELD, 1965)

$$\rho\frac{d}{dt}\left(\frac{3}{2}NkT\right)-\frac{5}{2}\rho kT\frac{dN}{dt}=\rho(\Gamma-\Lambda), \tag{3-265}$$

where N is the number density of free particles, the left hand side of Eq. (3-265) is the rate of increase of thermal energy plus the work done by the gas, $\rho(\Gamma-\Lambda)$ is the energy input per gram per second, and Γ and Λ denote, respectively, the rate of energy gain or loss per unit volume. When $\Gamma=\Lambda$ the gas is in thermal equilibrium and has the equilibrium temperature, T_{E}. When T is different from T_{E}, then an effective cooling time, t_T, may be defined by the relation

$$\frac{d}{dt}\left(\frac{3}{2}NkT\right)=-\frac{3Nk(T-T_{\mathrm{E}})}{2t_T}. \tag{3-266}$$

Equilibrium temperatures of ionized hydrogen (H II) regions ($T_{\mathrm{E}}\approx 10^4\,^\circ$K), and neutral hydrogen (H I) regions ($T_{\mathrm{E}}\approx 10^2\,^\circ$K) are determined by equating the heat gained by photoionization of hydrogen, helium, or carbon to the cooling effects of the excitation of ions by electrons and the excitation of ions and hydrogen molecules by neutral atoms (cf. SPITZER, 1948, 1949, 1954; SPITZER and SAVEDOFF, 1950; SEATON, 1951, 1954, 1955; AXFORD, 1961, 1964; OSTERBROCK,

1965). The detailed formulae for determining T_E in these cases are given in SPITZER (1962). As an example, the energy balance equation for the exchange of energy between neutral hydrogen atoms, H, and electrons, e, is given by (SPITZER and SAVEDOFF, 1950)

$$\frac{d}{dt}\left(\frac{3}{2} N_H k T_H\right) = -\gamma N_e N_H \text{ erg cm}^{-3} \text{ sec}^{-1}, \tag{3-267}$$

where

$$\gamma = 8Q \frac{m_e}{M_H}\left(\frac{2kT_e}{\pi m_e}\right)^{1/2} k(T_H - T_e) \approx 1.16 \times 10^{-27} T_e^{1/2}(T_H - T_e) \text{ cm}^3 \text{ erg sec}^{-1},$$

and the elastic collision cross section $Q \approx 6.3 \times 10^{-15}$ cm^2 (SEATON, 1955). Eqs. (3-266) and (3-267) lead to a cooling time of

$$t_{TH} \approx 1.8 \times 10^{11} N_e^{-1} T_e^{-1/2} \text{ sec}, \tag{3-268}$$

or using $N_e \approx 2 \times 10^{-4} N_H$,

$$t_{TH} \approx 2.8 \times 10^7 N_H^{-1} T_e^{-1/2} \quad \text{years}. \tag{3-269}$$

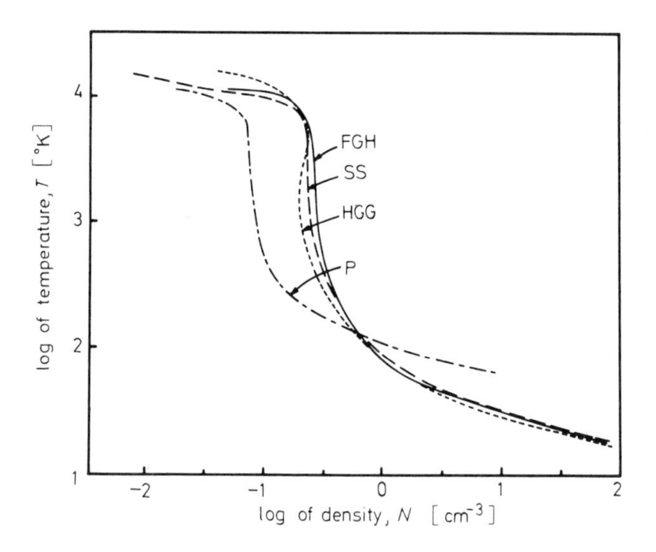

Fig. 21. The equilibrium temperature, T, of the interstellar medium as a function of the total number density, N, of nuclei (after PIKEL'NER, 1968, (P), FIELD, GOLDSMITH, and HABING, 1969 (FGH), SPITZER and SCOTT, 1969, (SS), and HJELLMING, GORDON, and GORDON, 1969 (HGG). The figure shows two thermally stable gas phases which coexist in pressure equilibrium; one for $T = 10^4$ °K and one for $T < 300$ °K. It is assumed that there is an equilibrium between heating by cosmic rays and cooling by inelastic thermal collisions. A cosmic ray energy density of $W = 6 \times 10^{-14}$ erg cm^{-3} in 2 MeV protons and an ionization rate of 4×10^{-16} sec^{-1} are assumed. The observed ionization rate is 1.5×10^{-15} sec^{-1}. For $T \approx 10^4$ °K, the electron density, $N_e \approx 0.016$ cm^{-3}, whereas the mean electron density for the interstellar medium is $N_e \approx 0.026$ cm^{-3}

For the interstellar medium, cosmic ray heating must be taken into account (PIKELŃER, 1968; SPITZER and SCOTT, 1969; FIELD, GOLDSMITH, and HABING 1969; HJELLMING, GORDON, and GORDON, 1969). The equilibrium temperatures obtained by equating cosmic ray heating to cooling by inelastic thermal collisions are shown in Fig. 21 as a function of nuclei number density. These results are normalized by assuming that the rate of primary ionization per neutral hydrogen atom is $4\times10^{-16}\,\mathrm{sec}^{-1}$.

If the cooling time defined in Eq. (3-266) is negative, then the gas is thermally unstable. In this case, T is greater than T_{E} and the kinetic energy grows until the cooling time changes. FIELD (1965) uses Eq. (3-265) to show that the gas will remain thermally stable if

$$T\frac{\partial}{\partial T}(\Gamma-\Lambda)-\rho\frac{\partial}{\partial\rho}(\Gamma-\Lambda)<0$$

or

$$\left[\frac{\partial}{\partial T}(\Gamma-\Lambda)\right]\left[1+\frac{\rho}{T_{\mathrm{E}}}\frac{dT_{\mathrm{E}}}{d\rho}\right]<0.$$

3.5.4. Atmospheres—Hydrostatic Equilibrium, the Barometric Equation, Scale Height, Escape Velocity, Stellar Winds, and the Solar Corona

In the absence of mass flow, Euler's equation (3-251) becomes

$$\boldsymbol{F}-\nabla P=0\,, \tag{3-270}$$

where $\boldsymbol{F}$ is the external force and P is the pressure. When the outward force of the gas pressure is just balanced by the inward force of gravity, Eq. (3-270) becomes

$$\frac{dP}{dr}=-\rho\frac{GM(r)}{r^2}, \tag{3-271}$$

which is the equation of hydrostatic equilibrium for a spherical mass of gas. Here r is the distance from the center of the sphere, P is the pressure, ρ is the mass density, the gravitational constant $G=6.67\times10^{-8}\,\mathrm{dyn\,cm^2\,g^{-2}}$, and the mass, $M(r)$, within a sphere of radius, r, is given by

$$M(r)=\int_0^r 4\pi r^2\rho\,dr\,. \tag{3-272}$$

For a perfect gas,

$$P=\frac{k\rho T}{\mu m_{\mathrm{H}}}, \tag{3-273}$$

where T is the temperature, μ is the mean molecular weight, Boltzmann's constant $k\approx1.380622(59)\times10^{-16}\,\mathrm{erg\,{}^{\circ}K^{-1}}$, and the mass of the hydrogen atom is $m_{\mathrm{H}} \approx 1.67 \times 10^{-24}$ gm. For a perfect gas atmosphere surrounding a spherical mass, M, of radius, R, Eqs. (3-271) and (3-273) give the barometric equation

$$P(r) = P(R) \exp\left[-\int_R^r \frac{dr}{H}\right], \tag{3-274}$$

$$= P(R) \exp\left[-\frac{(r-R)}{H}\right], \tag{3-275}$$

where $P(R)$ is the pressure at the surface of the mass, M, the scale height, H, is given by

$$H = \frac{kT}{\mu m_{\mathrm{H}} g}, \tag{3-276}$$

$$\approx \frac{kTR^2}{\mu m_{\mathrm{H}} GM},$$

and the g is the local acceleration of gravity given by

$$g = \frac{GM(r)}{r^2} \approx \frac{GM}{R^2}. \tag{3-277}$$

In order for a particle of mass, m, to escape from a larger mass, M, the particles must have kinetic energy, $mv^2/2$, greater than the gravitational potential energy mMG/R. This means the velocities, v, must be greater than the escape velocity, v_{esc}, given by

$$v_{\mathrm{esc}} = \left(\frac{2MG}{R}\right)^{1/2}. \tag{3-278}$$

The escape velocity for the Sun, for example, is 617.7 km sec^{-1}.

Observations of the optical continuum emission of the Sun during a solar eclipse show a corona of light extending beyond the visual extent of the uneclipsed Sun for several solar radii. Part of this light, the K corona, is thought to be due to Thomson scattering of solar photons by free electrons in the corona. Measurements of the intensity of the coronal light lead to measurements of the coronal electron density, $N_e(r)$, as a function of distance, r, from the Sun. Observations by BAUMBACH (1937) and ALLEN (1947), give a coronal model near the Sun $(r \approx R_\odot)$

$$N_e(r) \approx 1.55 \times 10^8 \left(\frac{r}{R_\odot}\right)^{-6} \left[1 + 1.93 \left(\frac{r}{R_\odot}\right)^{-10}\right] \mathrm{cm}^{-3}, \tag{3-279}$$

where the solar radius $R_\odot \approx 6.96 \times 10^{10}$ cm. More recent measurements using the scintillations of radio sources give (ERICKSON, 1964)

$$N_e(r) \approx 7.2 \times 10^5 \left(\frac{r}{R_\odot}\right)^{-2} \mathrm{cm}^{-3}, \tag{3-280}$$

for $r \gg R_\odot$.

Measurements of N_e are given as a function of r for the Sun in Fig. 22.

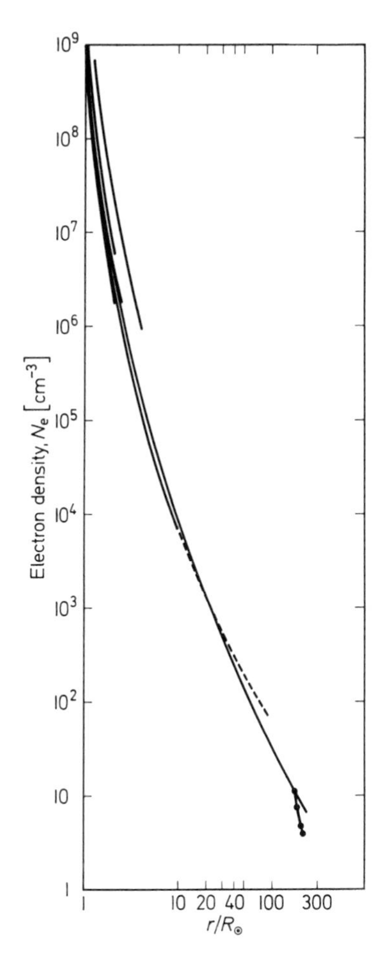

Fig. 22. Observed electron densities of the equatorial solar corona as a function of distance, r, from the Sun (after NEWKIRK, 1967, by permission of Annual Reviews Inc.). No attempt has been made to rectify data taken during different portions of the sunspot cycle. The long solid line corresponds to the theoretical model of WHANG, LIU, and CHANG (1966). The solar radius, $R_{\odot} \approx 6.96 \times 10^{10}$ cm

If the corona is isothermal and is in hydrostatic equilibrium, then the barometric equation (3-274) gives

$$N_e(r) = N_e(r_0) \exp\left[\frac{G M_{\odot} \mu m_H}{R_{\odot} k T}\left(\frac{1}{r} - \frac{1}{r_0}\right)\right], \tag{3-281}$$

whereas the equations of hydrostatic equilibrium and thermal conduction give

$$N_e(r) = N_e(r_0)\left(\frac{r}{r_0}\right)^{2/7} \exp\left[-\frac{7 r_0}{5H}\left\{1 - \left(\frac{r}{r_0}\right)^{-5/7}\right\}\right].$$

Here $N_e(r_0)$ is the electron density at some reference distance, r_0. When Eq. (3-281) is used together with measurements of $N_e(r)$ near the Sun, a coronal temperature of $T \approx 1.5 \times 10^6$ °K is obtained (DE JAGER, 1959). This temperature agrees with that obtained from radio observations of the Sun (BRACEWELL and PRESTON, 1956) and from the intensity and width of optical lines observed during solar eclipse (EDLÉN, 1942; BURGESS, 1964). The basic energy source for the corona is believed to be the dissipation of acoustic, gravity, or hydrodynamic waves from the photosphere (cf. Sects. 3.5.6 and 3.5.11).

CHAPMAN (1957, 1959) first showed that energy transport by thermal conduction in the outer solar atmosphere exceeds radiative energy loss, with the result that the 10^6 °K observed near the Sun must extend beyond the orbit of the Earth with little decline. The stationary heat flow equation is

$$\nabla \cdot (\kappa \nabla T) = 0\,, \tag{3-282}$$

where the coefficient of heat conductivity for a fully ionized gas is (Eq. (3-55))

$$\kappa \approx 6 \times 10^{-6}\, T^{5/2} \text{ erg cm}^{-1} \text{ sec}^{-1}\, {}^\circ\text{K}^{-1}\,. \tag{3-283}$$

It follows from Eqs. (3-282) and (3-283) that the temperature, $T(r)$, at a distance, r, from the Sun in a static solar atmosphere is given by

$$T(r) = \left(\frac{R_\odot}{r}\right)^{2/7} T(R_\odot)\,, \tag{3-284}$$

where $T(R_\odot) \approx 10^6$ °K and the solar radius, $R_\odot \approx 6.96 \times 10^{10}$ cm. The rate of thermal energy loss, Q, is given by (Eq. (3-53))

$$Q = -4\pi r^2 \kappa \nabla T = -\frac{8\pi}{7} R_\odot \kappa(T_\odot)\, T(R_\odot)\,. \tag{3-285}$$

From Eq. (3-284), the temperature at the Earth is $T(r \approx 215\, R_\odot) \approx 4 \times 10^5$ °K, and from Eq. (3-285) the thermal energy loss rate is $Q \approx 2 \times 10^{27}$ erg sec^{-1}. Measurements of solar wind protons and electrons near the Earth indicate a proton temperature of 4×10^4 °K and an electron temperature of 10^5 °K (LÜST, 1970).

BIRKELAND (1908) and CHAPMAN (1918, 1919) first suggested that particles could escape from the Sun and flow to the Earth causing geomagnetic storms and the aurorae. Then BIERMANN (1951, 1957) noted that comet tails never fail to point away from the Sun by an amount over and above that caused by the radiation pressure of the Sun, and that this must be due to a continuous radial flow of corpuscles with velocities of a few hundred km sec^{-1}. If all of the heat flux, Q, given by Eq. (3-285) went into an expanding spherical corona, then

$$Q \approx 4\pi r^2 M \frac{V}{2} N(r)(V^2 + V_{\text{esc}}^2)\,, \tag{3-286}$$

where $N(r)$ is the coronal density at distance, r, from the Sun, M is the particle mass, V is the constant solar wind velocity, and the escape velocity for the Sun is $V_{\text{esc}} \approx 617$ km sec^{-1}. Assuming $V \ll V_{\text{esc}}$, and using $Q = 2.5 \times 10^{27}$ erg sec^{-1}, we obtain a proton flux density of $N(r)\,V \approx 2.4 \times 10^8$ cm^{-2} sec^{-1} at the Earth orbit for which $r = 1.5 \times 10^{13}$ cm. The observed proton density near the Earth is 5 cm^{-3} implying $V \approx 500$ km sec^{-1}, near the 320 km sec^{-1} measured by satellites (LÜST, 1970). PARKER (1958) first suggested that the dynamical origin of the solar corpuscular radiation must be sought in the hydrodynamic equations describing an expanding solar atmosphere. For steady flow, Euler's equation (3-251) can be written

$$\rho \boldsymbol{v} \cdot \nabla \boldsymbol{v} = -\nabla P - \rho g = -\nabla P - \frac{\rho M_\odot G}{r^2}\,, \tag{3-287}$$

where ρ is the mass density, $\boldsymbol{v}$ is the velocity, P is the pressure, g is the acceleration due to gravity, the solar mass, $M_\odot \approx 1.989 \times 10^{33}$ g, the gravitational constant, $G \approx 6.668 \times 10^{-8}$ dyn cm^2 g^{-2}, and r is the radial distance from the center of the Sun. For a compressible gas flowing down a tube of varying cross sectional area, A,

$$A \rho v = \text{constant}, \tag{3-288}$$

and if the solar atmosphere expands with spherical symmetry, $A \propto r^2$. Hence using Eqs. (3-287) and (3-288) we obtain

$$\left[\frac{v^2}{s^2} - 1\right] \frac{dv}{v} = \left[2 - \frac{M_\odot G}{s^2 r}\right] \frac{dr}{r}, \tag{3-289}$$

where the speed of sound

$$s = \left(\frac{dP}{d\rho}\right)^{1/2} \approx \left(\frac{\gamma P}{\rho}\right)^{1/2} = \left(\frac{\gamma k T}{\mu m_{\mathrm{H}}}\right)^{1/2} \approx \left(8.5 \times 10^7 \frac{\gamma T}{\mu}\right)^{1/2} \approx 1.7 \times 10^7 \,\text{cm sec}^{-1}. \tag{3-290}$$

Here γ is the adiabatic constant, T is the temperature, μ is the mean molecular weight, and m_{H} is the mass of the hydrogen atom. Thus we see that the solar wind expands supersonically, $v > s$, for values of r greater than the critical value

$$r_c = \frac{M_\odot G}{2 s^2} \approx 3.5 R_\odot, \tag{3-291}$$

where the solar radius $R_\odot \approx 6.96 \times 10^{10}$ cm. The sound velocity may be compared with the thermal velocity of protons in the solar corona.

$$v_{\mathrm{th}} = \left(\frac{3kT}{m_H}\right)^{1/2} \approx 1.6 \times 10^7 \,\text{cm sec}^{-1}, \tag{3-292}$$

where m_H is the proton mass and the corona temperature is 10^6 °K. These velocities may also be compared with the solar escape velocity

$$v_{\mathrm{es}} \approx 6.18 \times 10^7 \,\text{cm sec}^{-1}. \tag{3-293}$$

Actual measurement of the wind velocity near the Earth indicate $v = 3 - 5 \times 10^7$ cm sec^{-1}.

As outlined above, PARKER (1958) showed that when the temperature of any corona declines less rapidly than r^{-1} the only steady equilibrium state is one of expansion to supersonic velocities at large radial distances from the surface of the parent body. A detailed summary of the velocity, density, and temperature of the stellar wind is given by PARKER (1965). Parker's model has been successfully applied to the modulation of cosmic rays, and reviews on this subject are given by FORBUSH (1966).

Satellite measurements (NESS, SCEARCE, and SEEK, 1964) indicate a magnetic field strength of $B \approx 5 \times 10^{-5}$ gauss for the solar wind near the Earth. If the flux lines are frozen into the expanding solar wind, we would expect $B \propto (R_\odot / r)^2$ to give a magnetic field of a few gauss at the solar surface. Such fields have been observed (BABCOCK and BABCOCK, 1955). A field of 5×10^{-5} gauss at the orbit of the Earth gives an Alfvén speed of $V_A \approx 30$ km sec^{-1} for a particle density of 5 protons cm^{-3}. The magnetic field applies a torque to the solar wind giving it an azimuthal velocity of about 1 km sec^{-1} at the orbit of the Earth, and completely decelerating the rotation of the Sun in about 10^{10} years (WEBER and DAVIS, 1967).

3.5.5. Convection-Schwarzschild Condition, Prandtl Mixing Length Theory, Rayleigh and Nusselt Numbers, Boussinesq Equations

When convective mass motion is absent, and when the temperature of a gas is not constant, the gas will remain stable and in mechanical equilibrium. When convective motion is present, currents appear which tend to mix the gas and equalize its temperature. When a gas is in mechanical equilibrium, Euler's equation (3-251) becomes

$$\nabla P = -g\rho\,, \tag{3-294}$$

where P is the pressure, g is the acceleration due to gravity, and ρ is the mass density. Following an idea first suggested by KELVIN (1862), let us suppose that the surface of constant pressure, P, is not changed by a small perturbation. For a perfect gas,

$$P = \frac{k\rho T}{\mu m_{\mathrm{H}}}, \tag{3-295}$$

where T is the temperature, k is Boltzmann's constant, μ is the mean molecular weight, and m_{H} is the mass of the hydrogen atom. For an adiabatic perturbation in a perfect gas, which is in mechanical equilibrium, we have the temperature gradient

$$\left(\frac{dT}{dr}\right)_{\mathrm{ad}} = \frac{T}{P}\left(1 - \frac{1}{\gamma}\right)\frac{dP}{dr} = -\frac{g}{C_p}, \tag{3-296}$$

where it has been assumed that the gas is a sphere of radius, r. Here, γ is the adiabatic index and C_p is the specific heat at constant pressure. As first pointed out by SCHWARZSCHILD (1906), convection will occur, in the absence of damping, if

$$\left(\frac{dT}{dr}\right)_{\mathrm{str}} > \left(\frac{dT}{dr}\right)_{\mathrm{ad}},$$

where $(dT/dr)_{\mathrm{str}}$ is the actual structural temperature gradient in the gas. The more general condition for convection in any kind of gas is

$$\left(\frac{dT}{dr}\right)_{\mathrm{str}} > -\frac{gT}{C_p V}\left(\frac{\partial V}{\partial T}\right)_p, \tag{3-297}$$

where V is the volume and $(\partial V/\partial T)_p$ denotes a differential at constant pressure.

SAMPSON (1894) and SCHWARZSCHILD (1906) first suggested that the transfer of heat by radiation was the predominant form of energy transfer in a star. For a star which is in radiative equilibrium, the equation for the temperature gradient is (EDDINGTON, 1917)

$$\left(\frac{dT}{dr}\right)_{\mathrm{str}} = \frac{-3\kappa\rho L(r)}{4acT^3 4\pi r^2}, \tag{3-298}$$

where the Rosseland mean opacity is κ, the radiation constant $a=7.56 \times 10^{-15}$ erg cm^{-3} °K^{-4}, and $L(r)$ is the luminosity at the radius, r. The net radiative energy flux, $F_{\rm rad}$, is given by

$$F_{\rm rad} = \frac{16\sigma T^3}{3\kappa} \frac{dT}{dr}, \tag{3-299}$$

where $\sigma \approx 5.669 \times 10^{-5}$ erg cm^{-2} °K^{-4} sec^{-1} is the Stefan-Boltzmann constant.

When convection occurs, the tendency is to reduce the structural temperature gradient until

$$\left(\frac{dT}{dr}\right)_{\rm str} = \left(\frac{dT}{dr}\right)_{\rm ad} = \frac{T}{P}\left(1 - \frac{1}{\gamma}\right)\frac{dP}{dr} = -\frac{g}{C_p}. \tag{3-300}$$

The equations which specify the convective motion of a gas are the continuity equation (3-250), the Navier-Stokes equation (3-254) to (3-251), and, for an incompressible fluid, the heat transfer equation (3-57)

$$\frac{\partial T}{\partial t} + \boldsymbol{v} \cdot \nabla T = \frac{\kappa}{\rho C_p} \nabla^2 T. \tag{3-301}$$

The solution to this equation involves a characteristic length, l, a velocity, v, and a temperature difference, ΔT, between the convective bubbles and their surroundings. When the convective bubble merges with its surroundings after travelling the distance, l, the length is called a "mixing-length". A typical value of l for a perfect gas in hydrostatic equilibrium is the scale height, H, given by

$$l \approx H = \frac{kT}{\mu m_{\rm H} g}, \tag{3-302}$$

where T is the temperature, μ is the mean molecular weight, $m_{\rm H}$ is the mass of the hydrogen atom, and g is the acceleration due to gravity. The temperature difference, ΔT, is given by

$$\Delta T = l\left(\left|\frac{dT}{dr}\right|_{\rm str} - \left|\frac{dT}{dr}\right|_{\rm ad}\right). \tag{3-303}$$

Assuming that the pressure remains constant, the equation of state for a perfect gas and Eq. (3-303) give

$$\Delta\rho = \frac{\rho\,\Delta T}{T} = l\left(\left|\frac{d\rho}{dr}\right|_{\rm str} - \left|\frac{d\rho}{dr}\right|_{\rm ad}\right), \tag{3-304}$$

where $\Delta\rho$ is the change in mass density, ρ, and $d\rho/dr$ denotes the radial gradient in ρ. For a bubble of volume, V, the buoyant force will be

$$F_{\rm b} = V g \Delta\rho = V\rho g \frac{\Delta T}{T} = \frac{V\rho g l}{T}\left(\left|\frac{dT}{dr}\right|_{\rm str} - \left|\frac{dT}{dr}\right|_{\rm ad}\right). \tag{3-305}$$

Assuming that all of the work done by this force goes into the kinetic energy of the bubble, the bubble velocity, v, is given by

$$\begin{aligned} v &= \frac{l}{2}\left(\frac{g}{T}\right)^{1/2}\left(\left|\frac{dT}{dr}\right|_{\rm str} - \left|\frac{dT}{dr}\right|_{\rm ad}\right)^{1/2} \\ &= \frac{l}{2}\left(\frac{g}{\rho}\right)^{1/2}\left(\left|\frac{d\rho}{dr}\right|_{\rm str} - \left|\frac{d\rho}{dr}\right|_{\rm ad}\right)^{1/2}. \end{aligned} \tag{3-306}$$

It then follows that the energy flux, F_{conv}, of the convective flow is

$$\begin{aligned} F_{\text{conv}} &= C_p \rho v \frac{l}{2} \left(\left| \frac{dT}{dr} \right|_{\text{str}} - \left| \frac{dT}{dr} \right|_{\text{ad}} \right) \\ &= \frac{C_p \rho l^2}{4} \left(\frac{g}{T} \right)^{1/2} \left(\left| \frac{dT}{dr} \right|_{\text{str}} - \left| \frac{dT}{dr} \right|_{\text{ad}} \right)^{3/2} . \end{aligned} \tag{3-307}$$

Eqs. (3-302) to (3-307) follow from the mixing-length theory first developed by PRANDTL (1952). The equations are put in a convenient form by using the relation

$$\frac{dT}{dr} = \frac{T}{H} \frac{d\ln T}{d\ln P} = \frac{T}{H} \nabla , \tag{3-308}$$

to obtain from Eqs. (3-299), (3-306), and (3-307) the relations (BÖHM-VITENSE, 1953, 1958; HENYEY, VARDYA, and BODENHEIMER, 1965)

$$\begin{aligned} F_{\text{rad}} &= \frac{16 \sigma T^4}{3 \kappa H f} \nabla , \\ v &= \left[\frac{g l^2}{4H} (\nabla - \nabla') \right]^{1/2} , \\ F_{\text{conv}} &= \frac{1}{2} C_p \rho v \frac{l}{H} T (\nabla - \nabla') , \end{aligned} \tag{3-309}$$

and

$$F = \frac{16 \sigma T^4}{3 \kappa H} \cdot \nabla_{\text{rad}} = \sigma T_e^4 = F_{\text{conv}} + F_{\text{rad}} ,$$

where ∇' is the logarithmic gradient for individual turbulent elements, ∇_{rad} is the gradient which would be required if the total flux, F, were carried away by radiation, f is a diffusion correction which is near unity, C_p is the specific heat per unit mass at constant pressure, and T_e is the effective temperature.

When a layer of gas is considered, convective instability occurs when the Rayleigh number, R, is greater than some critical value. For a layer of thickness, d, under an adverse temperature gradient, $\Delta T/d$, we have (RAYLEIGH, 1916)

$$R = \frac{g \alpha}{\chi \nu} \frac{\Delta T}{d} d^4 , \tag{3-310}$$

where g is the acceleration due to gravity, χ and ν are, respectively, the coefficients of thermometric conductivity and kinematic viscosity, and the coefficient of volume expansion, α, is given by the equation of state

$$\rho = \rho_0 [1 - \alpha (T - T_0)] , \tag{3-311}$$

where T_0 is the temperature for which the mass density $\rho = \rho_0$. For some gases and fluids, $\alpha \approx 10^{-3}$ to 10^{-4}, and we may treat ρ as a constant in all terms in the equation of motion except the one in external force. In this "Boussinesq" approximation Euler's equation and the equation of heat conduction become (BOUSSINESQ, 1903)

$$\frac{\partial}{\partial r} (\langle P \rangle + \rho \langle v^2 \rangle) = - g \rho \tag{3-312}$$

and

$$\frac{\partial}{\partial t}\langle T\rangle + \frac{\partial}{\partial r}(\langle v \Delta T\rangle) = \chi \frac{\partial^2 \langle T\rangle}{\partial r^2},$$

where ΔT is the difference in temperature from its mean $\langle T\rangle$ and $\langle\ \rangle$ is taken to denote a horizontal mean. These equations can be made to give the mixing length equations (3-302) to (3-307) by replacing the spatial derivative of fluctuating quantities by l^{-1} and dropping the pressure and time derivatives. The two characteristic numbers, the Peclet number, P_e, and the Reynolds number Re, are then given by

$$P_e = \frac{v l}{\chi} \approx (P_r R)^{1/2}, \qquad Re = \frac{v l}{\nu} = \left(\frac{R}{P_r}\right)^{1/2}, \tag{3-313}$$

where R and P_r denote, respectively, the Rayleigh and Prandtl numbers. These numbers measure the ratio of turbulent motion to the damping effects of thermal conductivity and viscosity. The Nusselt number is a convenient way of expressing the sum, Q, of the convective and conductive heat flux.

$$N = \frac{Q d}{\kappa \Delta T} \tag{3-314}$$

which becomes unity for conduction without convection. From the mixing length theory we have (SPIEGEL, 1971)

$$N = \left[\frac{\sqrt{1+4P_r} - 1}{2P_r}\right]^{2/3} \left(\frac{P_r R}{R_c}\right)^{1/3},$$

where the Prandtl number $P_r = \nu/\chi$ and is given by Eq. (3-58) and R_c is the critical Rayleigh number given below.

CHANDRASEKHAR (1961) has written an excellent text which includes a complete discussion of the convective instability of a layer heated from below. As first suggested by RAYLEIGH (1916), a layer of thickness, d, becomes unstable for

$$R > R_c\,, \tag{3-315}$$

where:

Surface	R_c	a	$2\pi/a$
Both free	657.511	2.2214	2.828
Both rigid	1707.762	3.117	2.016
One rigid and one free	1100.65	2.682	2.342

Here a disturbance of wavelength, λ, has wave number

$$a = \frac{2\pi d}{\lambda}.$$

For example, for two free boundaries

$$R = \frac{(\pi^2 + a^2)^3}{a^2}, \tag{3-316}$$

and the critical Rayleigh number for the onset of instability is set by the condition $\partial R/(\partial a^2)=0$ for which $a=2.214$ and $R=R_c=657.511$. The stability criteria for both incompressible and compressible fluids are also discussed by JEFFREYS (1926) and JEFFREYS (1930), respectively.

If rotation is introduced, convection is inhibited; and an inviscid, ideal fluid becomes stable with respect to the onset of convection for all adverse temperature gradients. This is a consequence of the Taylor-Proudman theorem (TAYLOR, 1921; PROUDMAN, 1916) which states that all steady slow motions in a rotating inviscid fluid are necessarily two dimensional for they cannot vary in the direction of rotation. For a viscous, rotating fluid, convection is possible and is characterized by the Taylor number

$$T=\frac{4\omega^2 d^4}{\nu^2}, \tag{3-317}$$

where ω is the angular velocity, ν is the coefficient of kinematic viscosity, and d is the layer thickness. For the case of two free boundaries we have (CHANDRASEKHAR, 1953; CHANDRASEKHAR and ELBERT, 1955)

$$R=\frac{1}{a^2}\left[(\pi^2+a^2)^3+\pi^2 T\right], \tag{3-318}$$

and for large T,

$$R_c=8.6956\,T^{2/3},$$

for which

$$a=1.3048\,T^{1/6}.$$

Magnetic fields also inhibit thermal convection; and the effect of a magnetic field of strength, H, is characterized by the parameter (THOMPSON, 1951; CHANDRASEKHAR, 1952, 1961)

$$Q=\frac{\mu^2 H^2 \sigma}{\rho\nu}d^2, \tag{3-319}$$

where μ is the magnetic permeability, ρ is the mass density, ν is the coefficient of kinematic viscosity, σ is the electrical conductivity, and d is the layer thickness. For the case of two free boundaries we have

$$R=\frac{\pi^2+a^2}{a^2}\left[(\pi^2+a^2)^2+\pi^2 Q\right], \tag{3-320}$$

and for large Q,

$$R_c=\pi^2 Q,$$

for which

$$a=[\pi^4 Q/2]^{1/6}.$$

3.5.6. Sound Waves—Velocity, Energy Density, and Solar Energy Flux

Assume that a gas is in a static, uniform equilibrium condition in which the velocity, v_0, is zero and the density, ρ_0, and pressure, P_0, are constant. Next assume a perturbation such that the density $\rho=\rho_0+\rho_1$ and the velocity $v=v_1$. Ignoring all external forces including that due to gravity, the continuity equation (3-250) and Euler's equation (3-251) become

$$\frac{\partial \rho_1}{\partial t} + \rho_0 \nabla \cdot v_1 = 0 \tag{3-321}$$

and

$$\frac{\partial v_1}{\partial t} = -\frac{1}{\rho_0} \nabla P_1 \, . \tag{3-322}$$

For an ideal gas we have

$$P_1 = \frac{k T}{\mu m_H} \rho_1 \, , \tag{3-323}$$

where T is the temperature, μ is the molecular weight, and m_H is the mass of the hydrogen atom. For an adiabatic or polytropic process involving an ideal gas

$$\nabla P_1 = \left(\frac{\gamma P_0}{\rho_0}\right) \nabla \rho_1 = \left(\frac{\partial P}{\partial \rho}\right) \nabla \rho_1 \, , \tag{3-324}$$

where γ is the adiabatic index or the polytropic exponent. Differentiating Eq. (3-321) with respect to time and using Eqs. (3-322) and (3-324), we obtain the wave equation

$$\frac{\partial^2 \rho_1}{\partial t^2} = \left(\frac{\partial P}{\partial \rho}\right) \nabla^2 \rho_1 \, , \tag{3-325}$$

which has the plane wave solution

$$\rho_1 \propto \exp\left[i\left(\frac{2\pi x}{\lambda} - \omega t\right)\right] , \tag{3-326}$$

where the frequency, ω, is related to the wavelength, λ, by the formula

$$\omega^2 = \left(\frac{2\pi}{\lambda}\right)^2 \left(\frac{\partial P}{\partial \rho}\right) . \tag{3-327}$$

Both the pressure, P_1, and the velocity, v_1, also satisfy the wave equation (3-325). The "sound" velocity, s, of the perturbation waves is given by

$$s = \left(\frac{\omega \lambda}{2\pi}\right) = \left(\frac{\partial P}{\partial \rho}\right)^{1/2} = \left(\frac{\gamma P_0}{\rho_0}\right)^{1/2} = \left(\frac{\gamma k T_0}{\mu m_H}\right)^{1/2} , \tag{3-328}$$

where the differentiation $\partial P/\partial \rho$ is carried out under conditions of constant entropy. Eq. (3-328) was first derived for the adiabatic case by LAPLACE (1816).

As the velocity, v_1, is in the direction of propagation, these sound waves are called longitudinal waves. The Mach number, M, of the wave is the ratio

$$M = v_1/s \, . \tag{3-329}$$

The energy density, U, of the plane wave of sound is

$$U = \rho_0 v_1^2/2 \, , \tag{3-330}$$

and the energy flux density, S, is given by

$$S = s \rho_0 v_1^2/2 \, . \tag{3-331}$$

The boundary conditions for the reflection and transmission of sound waves at a boundary between two gases are determined by holding the pressures and normal velocity components equal at the interface between the two gases.

As a sound wave propagates in a gas, its intensity falls off with distance, x, as $\exp[-2\gamma x]$, where the absorption coefficient, γ, is given by (LANDAU and LIFSHITZ, 1959)

$$\gamma = \frac{\omega^2}{2\rho c^3}\left[\left(\frac{4}{3}\mu+\xi\right)+\kappa\left(\frac{1}{C_v}-\frac{1}{C_p}\right)\right]. \tag{3-332}$$

Here ω is the frequency of the sound wave, ρ is the mass density, the coefficient of dynamic viscosity, μ, is given by Eqs. (3-23), (3-24), or (3-25), the second coefficient of viscosity, ξ, is usually of the same order of magnitude as μ, the coefficient of heat conductivity, κ, is given by Eqs. (3-55) or (3-56), and C_v and C_p are, respectively, the specific heats at constant volume and pressure. The cross section for scattering of sound waves are the same as those given in Sects. 1.36 to 1.40.

BIERMANN (1946, 1947), SCHWARZSCHILD (1948), and SHATZMAN (1949) first suggested that turbulent motion in the convective zone of a star might result in compressional waves whose energy could heat the star's corona. LIGHTHILL (1952, 1954) developed the theory for the generation of sound in a stratified atmosphere by fluid motions, and PROUDMAN (1952) used a Hiesenberg turbulence spectrum to give a sound wave emissivity of

$$\varepsilon_s = \frac{38\rho v^8}{l s^5}\ \text{erg cm}^{-3}\,\text{sec}^{-1}, \tag{3-333}$$

where the constant 38 is a result of the choice of turbulence spectrum, ρ is the gas mass density, v is the turbulent velocity, l is the turbulent length scale, and s is the speed of sound. OSTERBROCK (1961) used the Böhm-Vitense model for turbulent motion (cf. Sect. 3.5.5) to obtain a solar flux of

$$F_{s\odot} \approx 10^7 - 10^8\ \text{erg cm}^{-2}\,\text{sec}^{-1}. \tag{3-334}$$

WHITTAKER (1963), STEIN (1968), and STEIN and SCHWARTZ (1972) showed that the effects of gravity will damp out large scale turbulence, and will produce a gravity wave emissivity of order

$$\varepsilon_G \approx 10^2 \frac{\rho v^3}{l}\left(\frac{l}{H}\right)^5 \text{erg cm}^{-3}\,\text{sec}^{-1}, \tag{3-335}$$

where H is the scale height. Gravity waves cannot propagate in an isothermal atmosphere above the critical frequency, ω_G, given by

$$\omega_G = \frac{g}{s}(\gamma-1)^{1/2}, \tag{3-336}$$

whereas sound waves cannot propagate at frequencies below the critical frequency, ω_s, given by (LAMB, 1909)

$$\omega_s = \frac{\gamma g}{2s} = \frac{s}{2H}, \tag{3-337}$$

where H is the scale height. These considerations follow from the dispersion relation (HINES, 1960)

$$\omega^4 - \omega^2 s^2 (k_x^2 + K_z^2) + (\gamma - 1) g^2 k_x^2 + g i \gamma \omega^2 K_z = 0 \tag{3-338}$$

for an isothermal atmosphere whose density varies exponentially in the z direction with scale height $H = s^2/(\gamma g)$. The vertical wave number, K_z, has the complex form $K_z = k_z + i\gamma g/(2s^2)$ where k_z is real for propagating waves. For a horizontal wave number, $k_x = 0$, vertical waves with frequencies $\omega < \omega_s$ cannot propagate; whereas when $k_x \to \infty$, vertically propagating waves with frequencies $\omega > \omega_G$ cannot propagate. Vertically propagating compressional oscillations, modified by gravity, are possible for $\omega > \omega_s$; and vertically propagating gravity waves, modified by compression, are possible for $\omega < \omega_G$. Vertical velocity oscillations in the solar photosphere were first observed by LEIGHTON, NOYES, and SIMON (1962). They observed a velocity oscillation with a typical amplitude of 0.4 km sec^{-1} and a period of about 300 sec. For $\gamma = 1.2$ and $s = 7$ km sec^{-1}, values characteristic of the photosphere, Eq. (3-336) gives a period of 359 sec, whereas Eq. (3-337) gives a period of 267 sec. For a linear acoustic wave with a r.m.s. velocity oscillation, v_1, there will be a r.m.s. temperature oscillation, ΔT, given by

$$\Delta T = \frac{\gamma - 1}{s} v_1 T,$$

where T is the gas temperature. Values of $\Delta T = 225$ °K have been reported by NOYES and HALL (1972) at the temperature minimum in the solar atmosphere, whereas LANG (1974) has observed $\Delta T \approx 10^4$ °K in the radio emission coming from the chromosphere-corona transition region. For both cases the temperature oscillations had periods near 265 sec.

ALFVÉN (1947) first noted that magnetoacoustic or magnetohydrodynamic waves could also heat the solar corona, and these waves have been discussed by BARNES (1968, 1969) who shows that a total flux of only 10^3 erg cm^{-2} sec^{-1} propagates outwards into the solar corona in the fast-mode of magnetoacoustic waves (cf. Sect. 3.5.11).

3.5.7. Isentropic Flow—The Adiabatic Efflux of Gas

For the adiabatic nonconducting flow of a perfect gas, it follows directly from the first law of thermodynamics, Eq. (3-59), Eq. (3-61) for the perfect gas law, and Eq. (3-63) for specific heat, that

$$C_p dT + v dv = 0, \tag{3-339}$$

where C_p is the specific heat at constant pressure, dT is the change in temperature, T, and dv is the change in velocity, v. It also follows from Euler's equation (3-251) and from Eq. (3-339) that the change in entropy, dS, is

$$dS = C_p \frac{dT}{T} - \frac{R}{\mu} \frac{dP}{P} = 0, \tag{3-340}$$

where R is the universal gas constant. That is, the entropy is constant (isentropic) during the adiabatic flow of a perfect gas. If the subscript 0 denotes initial conditions, and no subscript denotes the condition at a later time, if follows from Eq. (3-340) that

$$\frac{P}{P_0}=\left(\frac{\rho}{\rho_0}\right)^{\gamma}=\left(\frac{T}{T_0}\right)^{\gamma/(\gamma-1)}, \tag{3-341}$$

where γ is the adiabatic index given by Eq. (3-66).

When a gas is initially static ($v_0=0$), it follows from Eqs. (3-341) and (3-339) that the following equations hold

$$\frac{v^2}{2}+C_p T=C_p T_0, \tag{3-342}$$

$$\frac{v^2}{2}+\frac{s^2}{\gamma-1}=\frac{s_0^2}{\gamma-1}, \tag{3-343}$$

$$\frac{s_0^2}{s^2}=\frac{T_0}{T}=1+\frac{\gamma-1}{2}M^2, \tag{3-344}$$

$$\frac{P_0}{P}=\left(1+\frac{\gamma-1}{2}M^2\right)^{\gamma/(\gamma-1)}, \tag{3-345}$$

and

$$\frac{\rho_0}{\rho}=\left(1+\frac{\gamma-1}{2}M^2\right)^{1/(\gamma-1)}, \tag{3-346}$$

where the speed of sound, s, is

$$s=\left(\frac{\partial P}{\partial \rho}\right)^{1/2}=\left(\frac{\gamma P}{\rho}\right)^{1/2},$$

and the Mach number, M, is

$$M=v/s.$$

A particularly useful point is that for which $M=1$ where

$$T=2T_0/(\gamma+1),$$

$$P=P_0\left(\frac{2}{\gamma+1}\right)^{\gamma/(\gamma-1)}, \tag{3-347}$$

and

$$\rho=\rho_0\left(\frac{2}{\gamma+1}\right)^{1/(\gamma-1)}.$$

As first pointed out by Reynolds (1876) equations (3-347) apply for the efflux of a gas into vacuum. The velocity of efflux in this case is

$$v=s_0\left(\frac{2}{\gamma+1}\right)^{1/2}. \tag{3-348}$$

For the adiabatic expansion of a radiation dominated gas, we have the relations (c.f. Eq. (3-78))

$$\frac{dT}{T} = \frac{1}{3}\frac{d\rho}{\rho}, \tag{3-349}$$

$$\rho \propto T^3 \propto R^{-3} \propto t^{-3}, \tag{3-350}$$

and

$$U = aT^4 \propto R^{-4} \propto t^{-4}, \tag{3-351}$$

where T is the temperature, ρ is the gas mass density, R is the radius of the gas cloud, U is the radiation energy density, the radiation constant $a = 7.564 \times 10^{-15}$ erg cm^{-3} °K^{-4}, and it is assumed that the gas cloud is in its early stages of expansion so that $R = vt$ where v is the constant velocity of expansion, and t is the time variable.

Similarly, the energy, E, of a relativistic particle in a gas cloud of radius, R, will go as

$$E \propto R^{-1} \propto t^{-1}, \tag{3-352}$$

and the conservation of magnetic flux means that

$$H \propto R^{-2} \propto t^{-2},$$

where H is the magnetic field intensity. The formulae of synchrotron radiation then lead to the conclusion that the synchrotron flux density, S, goes as (SHKLOVSKY, 1960; c.f. Eqs. (1-154) and (1-180))

$$S \propto R^{-2\gamma} \propto t^{-2\gamma}, \tag{3-353}$$

where γ is the power law index of the relativistic electron energy spectrum. Such a decrease in radio flux with time has been observed for Casseopeia A (cf. SHKLOVSKY, 1968). Additional formulae regarding the expansion of radio sources are given by KARDASHEV (1962) and VAN DER LAAN (1963).

In the later stages of expansion, a gas cloud will be slowed down by the interstellar medium. If the medium density is ρ, and the gas cloud has radius, R, mass, M, and initial velocity, V_0, then its velocity, V, is given by (OORT, 1946; SHKLOVSKY, 1968)

$$V = \frac{3MV_0}{4\pi\rho R^3}, \tag{3-354}$$

and

$$R = 4Vt = \left(\frac{3MV_0 t}{\pi\rho}\right)^{1/4}$$

provided that $\rho R^3 \gg M$.

For adiabatic compression or expansion the specific entropy (per gram), S, is constant, and we consequently have the relations

$$T = \left(\frac{3S\rho}{4a}\right)^{1/3} \tag{3-355}$$

and

$$P = \frac{3^{1/3} S^{4/3} \rho^{4/3}}{4^{4/3} a^{1/3}} \tag{3-356}$$

for a photon (radiation dominated) gas. Here T and P are, respectively, the temperature and pressure, and the radiation constant $a = 7.564 \times 10^{-15}$ erg cm^{-3} °K^{-4}.

3.5.8. Shock Waves

As first shown by RANKINE (1870), a surface of discontinuity in density, ρ, pressure, P, and temperature, T, may propagate as a "shock" wave in a gas. If u denotes the velocity of the shock front, and if the gas ahead of the shock is at rest, the gas behind the shock travels at the velocity, v_b, given by

$$v_b = u - v_2 = \frac{\rho_2 - \rho_1}{\rho_2} u , \tag{3-357}$$

where

$$u = v_1 = \left[\left(\frac{\rho_2}{\rho_1}\right) \frac{P_2 - P_1}{\rho_2 - \rho_1}\right]^{1/2} = \frac{\rho_1}{\rho_2} v_2 ,$$

$$\begin{aligned} v_1 &= M_1 s_1 = M_1 (\gamma P_1/\rho_1)^{1/2} , \\ v_2 &= M_2 s_2 = M_2 (\gamma P_2/\rho_2)^{1/2} , \end{aligned} \tag{3-358}$$

and

$$M_2^2 = \frac{1 + \frac{\gamma - 1}{2} M_1^2}{\gamma M_1^2 - \frac{\gamma - 1}{2}} .$$

Here M is the Mach number, s is the velocity of sound, γ is the adiabatic index, and subscripts 1 and 2 denote, respectively, the regions behind and in front of the shock. Eqs. (3-358) indicate that $M_1 > 1$ or < 1 according as $M_2 < 1$ or > 1, which means that either region 1 or 2 must be supersonic whereas the other region is subsonic.

The pressures, densities, and temperatures of the two regions behind (1) and in front (2) of the shock front are related by equations first derived by RANKINE (1870) and HUGONIOT (1889). These Rankine-Hugoniot relations, which follow directly from Euler's equation, the law of conservation of energy, and the perfect gas law, are

$$\frac{P_2}{P_1} = 1 + \frac{2\gamma}{\gamma + 1} [M_1^2 - 1] ,$$

$$\frac{\rho_2}{\rho_1} = \frac{v_1}{v_2} = \frac{(\gamma + 1) M_1^2}{(\gamma - 1) M_1^2 + 2} = \frac{1 + \frac{\gamma + 1}{\gamma - 1} \frac{P_2}{P_1}}{\frac{\gamma + 1}{\gamma - 1} + \frac{P_2}{P_1}} , \tag{3-359}$$

and

$$\frac{T_2}{T_1} = \frac{\{2\gamma M_1^2 - (\gamma - 1)\}\{(\gamma - 1) M_1^2 + 2\}}{(\gamma + 1)^2 M_1^2} = \frac{P_2}{P_1} \frac{\frac{\gamma + 1}{\gamma - 1} + \frac{P_2}{P_1}}{1 + \frac{\gamma + 1}{\gamma - 1} \frac{P_2}{P_1}} .$$

It follows from Eqs. (3-357) to (3-359) that:

$$v_b = \frac{s_1}{\gamma}\left(\frac{P_2}{P_1} - 1\right)\left[\frac{\dfrac{2\gamma}{\gamma+1}}{\dfrac{P_2}{P_1} + \dfrac{\gamma-1}{\gamma+1}}\right]^{1/2} \tag{3-360}$$

and

$$\frac{P_2 - P_1}{P_1} = \frac{2\gamma}{\gamma+1}(M_1^2 - 1) = \frac{\Delta P}{P_1}.$$

For a very weak shock, $\Delta P/P_1 \ll 1$, we have

$$\frac{\Delta\rho}{\rho_1} = \frac{1}{\gamma}\frac{\Delta P}{P_1} = \frac{v_b}{s_1}, \tag{3-361}$$

$$\frac{\Delta T}{T_1} = \frac{\gamma-1}{\gamma}\frac{\Delta P}{P_1},$$

and

$$v_1 \approx s_1.$$

For a very strong shock, $\Delta P/P_1 \gg 1$, we have

$$\frac{\rho_2}{\rho_1} = \frac{\gamma+1}{\gamma-1},$$

$$\frac{T_2}{T_1} = \frac{\gamma-1}{\gamma+1}\frac{P_2}{P_1},$$

$$v_1 = s_1\left[\frac{\gamma+1}{2\gamma}\frac{P_2}{P_1}\right]^{1/2}, \tag{3-362}$$

and

$$v_b = s_1\left[\frac{2}{\gamma(\gamma+1)}\frac{P_2}{P_1}\right]^{1/2}.$$

For a weak shock wave, the shock front thickness, Δx, is inversely proportional to the wave strength, and is given by

$$\Delta x \approx \frac{P_1 l}{P_2 - P_1} \approx \frac{V_1 l}{V_2 - V_1} \approx \frac{l}{M-1} \quad \text{for } M \ll 1, \tag{3-363}$$

where l is the molecular mean free path. For a strong shock for which $M \gg 1$, the shock front thickness is of the order of l (cf. MOTT-SMITH, 1951; TIDMAN, 1958). When the shock propagates through an ionized gas, $\Delta x \approx V_s \tau_i$, where the thermal velocity of the ions in the compression shock is of the order of the shock front velocity, V_s, and the ion collision time is τ_i (cf. Eq. (3-18)).

For a strong explosion of initial energy, E, in a perfect gas of mass density, ρ_1, the dimensionless variable

$$\xi = r\left(\frac{\rho_1}{E t^2}\right)^{1/5}$$

can serve as a similarity variable (cf. SEDOV, 1959). The shock front radius, R_s, and velocity, V_s, at a time, t, are given by

$$R_s = \xi_1 \left(\frac{E}{\rho_1}\right)^{1/5} t^{2/5} \quad \text{and} \quad V_s = \frac{2}{5} \xi_1^{5/2} \left(\frac{E}{\rho_1}\right)^{1/2} R_s^{-3/2},$$

where the parameter, ξ_1, is of order unity. For this strong shock case, the parameters for the shock front are given by

$$\rho_1 = \frac{\gamma+1}{\gamma-1}\rho_2, \quad V_1 = \frac{2}{\gamma+1}V_s,$$

$$P_1 = \frac{2}{\gamma+1}\rho_1 V_s^2 \approx \rho_2 \left(\frac{E}{\rho_1}\right)^{2/5} t^{-6/5}, \tag{3-364}$$

and

$$T_1 = \frac{2(\gamma-1)}{(\gamma+1)^2}\mu m_H V_s^2 \approx \frac{\gamma-1}{\gamma+1}\mu m_H \rho_2 \left(\frac{E}{\rho_1}\right)^{2/5} t^{-6/5}.$$

The implosion of a collapsing star often triggers a thermonuclear explosion which then leads to a relativistically expanding envelope (COLGATE and JOHNSON, 1960). There is a general increase in speed as the shock traverses the decreasing density of the stellar mantle. If the original density of the medium is

$$\rho = \rho_0 \left(\frac{x_0}{x}\right)^{7/4},$$

and the pressure driving the shock is

$$P_s = P_0 \left(\frac{t_0}{t}\right)^{5/3},$$

then the position of the shock front is (COLGATE and WHITE, 1966)

$$x_s = x_0 \left(\frac{t}{t_0}\right)^{4/3}, \tag{3-365}$$

and the velocity of the shock is

$$v_s = v_{s0} \left(\frac{x_s}{x_0}\right)^{1/4}. \tag{3-366}$$

Similarity solutions for the spherical and cylindrical shock waves produced by supernovae explosions are given in SEDOV's (1959) book.

The Rankine-Hugoniot relations expressed in Eq. (3-359) describe the discontinuity in fluid parameters across the shock front of width on the order of the molecular free path, l. When a radiation flux is present, the shock front still has a width of order l, but the radiation heats up the gas flowing into the temperature discontinuity. The shock becomes radiation dominated when the radiation flux, σT_2^4, obeys the inequality

$$\sigma T_2^4 \gg \frac{2}{\gamma^2 - 1}\rho_1 v_1^3 \tag{3-367}$$

or

$$\frac{\sigma}{R^4} \gg \left(\frac{\gamma+1}{2}\right)^4 \left(\frac{2}{\gamma^2-1}\right) \rho_1 v_1^{-5},$$

where $\sigma \approx 5.7 \times 10^{-5}$ erg cm^{-2} °K^{-4} sec^{-1} is the Stefan-Boltzmann constant, and the gas is assumed to be ideal with $P = \rho R T/\mu$ and an adiabatic index, γ. Bond, Watson, and Welch (1965) describe radiation dominated shocks by using the reduced variables

$$\tilde{v} = \frac{v}{c_1},$$

$$\tilde{T} = \frac{RT}{c_1^2} = (c_1 - v)\frac{v}{c_1^2}, \tag{3-368}$$

and

$$\tilde{J} = \frac{2(\gamma-1)}{\rho v c_1^2(\gamma+1)} J,$$

where J is the radiation flux, and the constant c_1 is given by

$$c_1 = \frac{\gamma+1}{\gamma} \frac{v_1+v_2}{2} \lesssim v_1 .$$

They then describe four classes of shock strength given by

$$\begin{aligned}
&\text{very weak shock:} && \frac{\gamma}{\gamma+1} > \tilde{v}_2 > \frac{1}{2} \\
&\text{weak shock:} && \frac{1}{2} > \tilde{v}_2 > \frac{1}{\gamma+1} \\
&\text{strong shock:} && \frac{1}{\gamma+1} > \tilde{v}_2 > \frac{3-\gamma}{2(\gamma+1)} \\
&\text{and very strong shock:} && \frac{3-\gamma}{2(\gamma+1)} > \tilde{v}_2 > \frac{\gamma-1}{\gamma+1} .
\end{aligned} \tag{3-369}$$

The result of the radiation is the establishment of a radiation precurser ahead of the shock density-velocity discontinuity, and a radiation tail behind it. If the subscripts + and − denote the conditions immediately in front and behind the discontinuity, and subscripts 1 and 2 denote distances behind and ahead for which the radiation flux has fallen to its unperturbed value, a radiation dominated shock has

$$\begin{aligned}
&\text{for very weak shocks:} && \tilde{T}_- = \tilde{T}_+ = \frac{\gamma}{\gamma+1} - \left(\frac{\gamma}{\gamma+1}\right)^2 = \tilde{T}_0 < \tilde{T}_2 \\
&\text{for weak shocks:} && \tilde{T}_- = \tilde{T}_+ = \tilde{T}_0 < \tilde{T}_2 < \tfrac{1}{4} \\
&\text{for strong shocks:} && \tilde{T}_- < \tilde{T}_+ \\
&\text{for very strong shocks:} && \tilde{T}_- \le \tilde{T}_2 < \tilde{T}_+ \le \tfrac{1}{4} .
\end{aligned} \tag{3-370}$$

It is only for the strong and very strong shocks that discontinuities in temperature and velocity occur.

3.5.9. Hydrodynamic Gravity Waves

The motion of an inviscid fluid executing small oscillations is irrotational. That is

$$\nabla \times \boldsymbol{v} = 0 , \tag{3-371}$$

where $\boldsymbol{v}$ is the velocity. Such flow is called potential flow. It follows from Eq. (3-371) that we may let

$$\boldsymbol{v} = \nabla \varphi , \tag{3-372}$$

where the velocity potential, φ, satisfies Poisson's equation

$$\nabla^2 \varphi = 0 . \tag{3-373}$$

If the fluid is incompressible ($\rho =$ constant), and the prevalent force is that of gravity whose direction is along the z axis, then it follows from Euler's equation (3-251) that

$$\left(\frac{\partial \varphi}{\partial z} + \frac{1}{g} \frac{\partial^2 \varphi}{\partial t^2} \right)_{z=0} = 0 , \tag{3-374}$$

where the wave equation is evaluated at the surface $z=0$, and g is the acceleration due to gravity. The solution to Eqs. (3-373) and (3-374) is

$$\varphi = A \cos(kz - \omega t) \exp[kz] , \tag{3-375}$$

where the gas occupies the region $z<0$, and the frequency, ω, is related to the wave vector, k, by the relation

$$\omega = (kg)^{1/2} . \tag{3-376}$$

The velocity of propagation, u, of this gravity wave is

$$u = \frac{\partial \omega}{\partial k} = \frac{1}{2} \left(\frac{g}{k} \right)^{1/2} = \frac{1}{2} \left(\frac{g \lambda}{2 \pi} \right)^{1/2} , \tag{3-377}$$

where the wave vector $k = 2\pi/\lambda$, and λ is the wavelength. It follows from Eqs. (3-375) and (3-377) that the velocity distribution in the moving fluid is given by

$$\begin{aligned} v_z &= -Ak \cos(kx - \omega t) \exp(kz) \\ v_x &= Ak \sin(kx - \omega t) \exp(kz) . \end{aligned} \tag{3-378}$$

For a viscous gas, the energy of the wave dies off as $\exp(-2\gamma t)$, where t is the time variable, and the damping coefficient, γ, is given by (LANDAU and LIFSHITZ, 1959)

$$\gamma = 2 \nu \omega^4 / (g^2) , \tag{3-379}$$

where ν is the coefficient of kinematic viscosity, ω is the wave frequency, and g is the acceleration of gravity.

The dispersion relation for gravity and sound waves in an isothermal, exponential atmosphere was given in Eq. (3-338); whereas the gravitational wave flux produced by a turbulent atmosphere was given in Eq. (3-335). The dispersion relation for the gravity waves in a viscous medium is (CHANDRASEKHAR, 1961)

$$Y^4 + 2Y^2 - 4Y + 1 + Q + Q^{1/3} S = 0 , \tag{3-380}$$

where

$$Y = \left[1 + \frac{\omega}{k^2 \nu}\right]^{1/2},$$

$$Q = \frac{g}{k^3 \nu^2},$$

and

$$S = \frac{T}{\rho (g \nu^4)^{1/3}}.$$

Here ω is the wave frequency, k is the wave number, ν is the coefficient of kinematic viscosity, T is the surface tension, g is the acceleration due to gravity, and ρ is the mass density. A different type of gravity wave which follows from Einstein's equations of general relativity is discussed in Sect. 5.9.

3.5.10. Jeans Condition for Gravitational Instability

Assume that a gas is in a static, uniform, equilibrium condition in which the velocity, v_0, is zero, and the density, ρ_0, and the pressure, P_0, are constant. Next assume a perturbation such that the density, $\rho = \rho_0 + \rho_1$ and the velocity $v = v_1$. Assuming that the only external force is that of the gravitational potential, φ, the continuity equation (3-250) and Euler's equation (3-251) become

$$\frac{\partial \rho_1}{\partial t} + \rho_0 \nabla \cdot v_1 = 0$$

and (3-381)

$$\frac{\partial v_1}{\partial t} = -\nabla \varphi_1 - \frac{\nabla P_1}{\rho_0}.$$

For an isothermal process in an ideal gas,

$$P_1 = \frac{kT}{\mu m_H} \rho_1, \tag{3-382}$$

where T is the temperature, μ is the mean molecular weight, and m_H is the mass of the hydrogen atom. From Poisson's equation

$$\nabla^2 \varphi_1 = 4\pi G \rho_1. \tag{3-383}$$

Eqs. (3-381) to (3-383) may be combined to form the wave equation

$$\frac{\partial^2 \rho_1}{\partial t^2} = 4\pi G \rho_0 \rho_1 + \frac{kT}{\mu m_H} \nabla^2 \rho_1, \tag{3-384}$$

which has the plane wave solution

$$\rho_1 = \exp\left[i\left(\frac{2\pi x}{\lambda} - \omega t\right)\right], \tag{3-385}$$

where the frequency, ω, is related to the wavelength, λ, by the formula

$$\omega^2 = \left(\frac{2\pi}{\lambda}\right)^2 \left(\frac{kT}{\mu m_H}\right) - 4\pi G \rho_0. \tag{3-386}$$

The velocity of propagation, V_J, of small density fluctuations is therefore given by (JEANS, 1902)

$$V_J = \frac{\lambda \omega}{2\pi} = s\left[1 - \frac{G \rho_0 \lambda^2}{\pi s^2}\right]^{1/2}, \tag{3-387}$$

where the velocity of sound, s, is given by

$$s = \left(\frac{P_0}{\rho_0}\right)^{1/2} = \left(\frac{k T}{\mu m_H}\right)^{1/2}.$$

A fluctuation with wavelength, λ, greater than the critical wavelength (JEANS, 1902)

$$\lambda_J = s\left(\frac{\pi}{G \rho_0}\right)^{1/2} = \left(\frac{\pi k T}{\mu m_H G \rho_0}\right)^{1/2} \approx 6 \times 10^7 \left(\frac{T}{\mu \rho_0}\right)^{1/2} \text{ cm} \tag{3-388}$$

will grow exponentially in time and the waves are unstable. The velocity, V_J, is called the Jeans velocity, and length, λ_J, is the Jeans length. For dimensions larger than λ_J, a mass will be gravitationally unstable and will contract continuously. The corresponding critical mass of a sphere whose diameter is λ_J is

$$M_J = \frac{\pi}{6} \rho_0 \lambda_J^3 \approx 10^{23} \frac{(T/\mu)^{3/2}}{\rho^{1/2}} \text{ gm}, \tag{3-389}$$

The extention of the Jeans criteria to adiabatic processes is discussed by CHANDRASEKHAR (1951) and PARKER (1952). The Jeans criteria for the gravitational instability of an infinite homogeneous medium is unaffected by the presence, separately, or simultaneously, of uniform rotation or a uniform magnetic field. The effects of rotation and a magnetic field are discussed by CHANDRASEKHAR and FERMI (1953), CHANDRASEKHAR (1953, 1954), and PACHOLCZYK and STÓDÓLKIEWICZ (1960).

When the direction of wave propagation is at right angles to the direction of rotation, the dispersion relation is

$$\omega^2 = k^2 s^2 + 4\Omega^2 - 4\pi G \rho,$$

where Ω is the rotation velocity, and the minimum unstable wavelength is

$$\lambda_J = \frac{\pi s}{(\pi G \rho - \Omega^2)^{1/2}}. \tag{3-390}$$

In this special case, gravitational instability cannot occur if $\Omega^2 > \pi G \rho$. For wave propagation in every other direction, gravitational instability will occur if Jean's condition $s^2 k^2 - 4\pi G \rho < 0$ is satisfied. For an infinite magnetic field strength, the Jeans disturbance can only propagate along the magnetic field and not perpendicular to it. The general dispersion relation when a magnetic field of strength, H, is present is given by

$$\omega^4 - \left(\frac{\pi H^2}{\lambda^2 \rho} + \frac{4\pi^2 s^2}{\lambda^2} - 4\pi G\rho\right)\omega^2 + \frac{\pi H^2}{\rho\lambda^2}\left(\frac{4\pi^2 s^2}{\lambda^2} - 4\pi G\rho\right)\cos^2\theta = 0,$$

where θ is the inclination of the direction of the magnetic field to the direction of wave propagation.

3.5.11. Alfvén Magneto-Hydrodynamic Waves

Assume that a gas is in a static, uniform, equilibrium condition in which the velocity, v_0, is zero and the density, ρ_0, and pressure, P_0, are constant. Next assume a perturbation such that the density $\rho = \rho_0 + \rho_1$ and the velocity $v = v_1$. Ignoring the external force due to gravity, but assuming a uniform magnetic field of strength, B_0, in a perfectly conducting medium, the continuity equation (3-250) and Euler's equation (3-251) become:

$$\frac{\partial \rho_1}{\partial t} + \rho_0 \nabla \cdot \boldsymbol{v}_1 = 0$$

and

$$\frac{\partial \boldsymbol{v}_1}{\partial t} = -\frac{\nabla P_1}{\rho_0} - \frac{\boldsymbol{B}_0}{4\pi\rho_0} \times (\nabla \times \boldsymbol{B}_1), \tag{3-391}$$

where B_1 is the perturbation in the magnetic field, and from Maxwell's equations

$$\frac{\partial B_1}{\partial t} = \nabla \times (\boldsymbol{v}_1 \times \boldsymbol{B}_0). \tag{3-392}$$

For an ideal gas,

$$\nabla P_1 = s^2 \nabla \rho_1,$$

where $s^2 = (\partial P / \partial \rho)$ is the square of the velocity of sound. Introducing the Alfvén velocity (ALFVÉN, 1942)

$$\boldsymbol{v}_A = \boldsymbol{B}_0 (4\pi\rho_0)^{-1/2}, \tag{3-393}$$

Eqs. (3-391) and (3-392) may be combined to give the equation

$$\frac{\partial^2 \boldsymbol{v}_1}{\partial t^2} - s^2 \nabla(\nabla \cdot \boldsymbol{v}_1) + \boldsymbol{v}_A \times \nabla \times [\nabla \times (\boldsymbol{v}_1 \times \boldsymbol{v}_A)] = 0. \tag{3-394}$$

As first pointed out by ALFVÉN (1942) and ASTROM (1950), Eq. (3-394) has the plane wave solution

$$v_1 = \exp[i(\boldsymbol{k} \cdot \boldsymbol{x} - \omega t)], \tag{3-395}$$

where ω is the wave frequency, and the wave vector

$$\boldsymbol{k} = \frac{2\pi}{\lambda}\boldsymbol{n},$$

where λ is the wavelength and $\boldsymbol{n}$ is a unit vector in the direction of propagation. When the wave vector $\boldsymbol{k}$ is perpendicular to $\boldsymbol{v}_A$, Eq. (3-394) has a plane wave solution with phase velocity

$$v_p = \frac{\lambda\omega}{2\pi} = (s^2 + v_A^2)^{1/2}, \tag{3-396}$$

which is a longitudinal wave. When $\boldsymbol{k}$ is parallel to $\boldsymbol{v}_A$, there are two plane wave solutions, a longitudinal wave with phase velocity, s, and a transverse wave with phase velocity, v_A.

For a medium of finite viscosity or conductivity, the Alfvén waves become damped and we have the dispersion relation (HERLOFSON, 1950; CHANDRASEKHAR, 1962)

$$(\omega - i\nu k^2)(\omega - i\eta k^2) = v_A^2 k^2$$

with the solutions

$$\begin{aligned} \omega &= \pm k\left[v_A^2 - \frac{1}{4}(\nu-\eta)^2 k^2\right]^{1/2} + \frac{1}{2} i(\nu+\eta)k^2 \\ &\approx \pm v_A k\left[1 - \frac{1}{8}\frac{(\nu-\eta)^2}{v_A^2}k^2\right] + \frac{1}{2} i(\nu+\eta)k^2 . \end{aligned} \tag{3-397}$$

Here ν is the coefficient of kinematic viscosity, the resistivity, $\eta = c^2/(4\pi\mu\sigma)$ where μ is the permeability and σ is the electrical conductivity, and the approximation is for $\nu \to 0$, and $\eta \to 0$. When viscosity is ignored, but conductivity is important, the condition for the existence of pronounced Alfvén waves is

$$\frac{8\pi\mu\sigma v_A^3}{\omega^2 c^2} = \frac{\mu v_A \lambda^2 \sigma}{c^2} \approx \frac{\sigma B_0 \lambda^2}{\rho_0^{1/2} c^2} \gg 1, \tag{3-398}$$

and the amplitude of an Alfvén wave is damped to $1/e$ of its initial value when it has travelled the distance (cf. ALFVÉN and FALTHAMMER, 1963)

$$z_0 = \frac{\sigma B_0 \lambda^2}{\pi^{3/2} \rho_0^{1/2} c^2}, \tag{3-399}$$

where $\lambda = 2\pi v_A/\omega$.

3.5.12. Turbulence

When studying the flow of fluids in a pipe, REYNOLDS (1883) found that turbulent motion is damped by viscosity when the Reynolds number

$$Re = \frac{Lv}{\nu} \tag{3-400}$$

falls below a critical value of approximately 2000. Here L is the pipe diameter, $\boldsymbol{v}$ is the average flow velocity, and ν is the coefficient of kinematic viscosity. Formulae for the viscosity and Reynolds number of a gas are given in Eqs. (3-23) to (3-31). As first pointed out by ROSSELAND (1929), the distance scale involved with astronomical objects is so large that the Reynolds number must often exceed the critical value for the onset of turbulence.

Turbulent motion is fluctuating and turbulent velocity must be treated as a random continuous function of position and time. The appropriate statistical theory assumes its simplest form when the medium is assumed to be homogeneous and isotropic. In this case the average properties of the motion are independent of position or the direction of the axis of reference. The turbulent velocity, v, is then described by the correlation function (TAYLOR, 1935)

$$R_{ij}(\boldsymbol{r}) = \langle v_i(\boldsymbol{x}) v_j(\boldsymbol{x}+\boldsymbol{r}) \rangle , \tag{3-401}$$

and its Fourier transform (TAYLOR, 1938)

$$\Phi_{ij}(\boldsymbol{k}) = \frac{1}{8\pi^3} \int_{-\infty}^{+\infty} R_{ij}(\boldsymbol{r}) \exp[-i\boldsymbol{k}\cdot\boldsymbol{r}] d\boldsymbol{r} .$$

Here the correlation is between two points, i, j, separated by the space vector, $\boldsymbol{r}$, the $\langle\ \rangle$ denotes a spatial average, and $\boldsymbol{k}$ is the wave number vector. When the turbulence is isotropic and homogeneous, the correlation function is of the form (VON KÁRMÁN, 1937; VON KÁRMÁN and HOWARTH, 1938)

$$R_{ij}(\boldsymbol{r}) = F(r) r_i r_j + G(r) \delta_{ij} , \tag{3-402}$$

where F and G are scalar functions of r, and δ_{ij} is the Dirac delta function. When the velocity correlation function is further devided into its longitudinal, $f(r)$, and lateral, $g(r)$, components, we have the relations

$$G = v^2 g$$

and

$$F = \frac{v^2}{r^2}(f - g) ,$$

where v^2 is the mean square turbulent velocity, and

$$g(r) = \frac{\langle v_n(\boldsymbol{x}) v_n(\boldsymbol{x}+\boldsymbol{r}) \rangle}{v^2}$$

$$f(r) = \frac{\langle v_p(\boldsymbol{x}) v_p(\boldsymbol{x}+\boldsymbol{r}) \rangle}{v^2} ,$$

where the subscripts p and n denote, respectively, the components of velocity parallel (longitudinal) and perpendicular (lateral) to $\boldsymbol{r}$.

Of particular physical interest is the energy spectrum function

$$E(k) = 2\pi k^2 \Phi_{ii}(k) . \tag{3-403}$$

The contribution of the total energy from that part of the wave number space between spheres of radii k and $k + dk$ is $E(k) dk$.

For the case in which turbulent velocity elements have a characteristic size $L = \alpha^{-1}$, we might take

$$v(x) = \alpha^2 x \exp\left[-\frac{\alpha^2 x^2}{2}\right] \tag{3-404}$$

to obtain

$$E(k) \propto k^4 \exp\left[-\frac{k^2}{\alpha^2}\right] .$$

This special case shows that $E(k)$ has a maximum at 1.4 α, a variance of $\sigma_k = 0.4\,\alpha$ and a wave number $k \approx L^{-1}$. That is, larger turbulent eddies have smaller wave numbers and vice versa.

The total turbulent kinetic energy density is given by

$$\frac{\rho}{2} R_{ii}(0) = \frac{\rho}{2} \langle v_i v_i \rangle = \frac{\rho}{2} \int\int\limits_{-\infty}^{\infty}\int \Phi_{ii}(\boldsymbol{k})\, d\boldsymbol{k} = \rho \int\limits_0^{\infty} E(k)\, dk\,. \qquad (3\text{-}405)$$

It follows from the Navier-Stokes equation that the time rate of change of kinetic energy is given by (cf. BATCHELOR, 1967)

$$\frac{\partial}{\partial t}\left[\frac{\rho R_{ii}(0)}{2}\right] = -2\rho\nu \int\limits_0^{\infty} k^2 E(k)\, dk = -\rho\varepsilon\,, \qquad (3\text{-}406)$$

where ν is the coefficient of kinematic viscosity, and ε is called the dissipation integral. It follows from Eq. (3-406) that kinetic energy is dissipated by viscosity into heat, and that most of this dissipation is done by the smaller eddies for which k is large.

When inertial forces are considered, it is seen that they do not change the total kinetic energy; but they do serve to transfer kinetic energy from larger turbulent elements to smaller ones. KOLMOGOROFF (1941) postulated that when the Reynolds number is large, the smaller scale components of turbulence are in statistical equilibrium. In this case, the size distribution of the smaller eddies depends only on the viscosity and the characteristics of the larger eddies. KOLMOGOROFF (1941) obtained the energy spectrum

$$E(k) \propto \varepsilon^{2/3} k^{-5/3}\,, \qquad (3\text{-}407)$$

where ε is the rate of transfer of energy from the bigger eddies to the smaller eddies and is given by Eq. (3-406). A simple way of interpreting the Kolmogoroff spectrum is to assume that a large eddy of scale, L_0, will exist for time

$$t_{L0} = L_0 / v_{L0}\,,$$

where v_{L0} is its velocity with respect to neighboring eddies. If there is a constant flow of energy to smaller eddies, we will have

$$v_L \propto L^{1/3}$$

and

$$t_L \propto L^{2/3}\,, \qquad (3\text{-}408)$$

where v_L and t_L denote, respectively, the velocity and lifetime of an eddy of size, L. When the velocities, v_L, are less than the Alfvén velocity, v_A, the magnetic field will damp them, whereas when v_L is larger than the velocity of sound, s, part of the kinetic energy may be dissipated as shock waves. In these cases we have (KAPLAN, 1954)

$$v_L \propto L \quad \text{for } v_L > s$$

$$v_L \propto L^{1/2} \quad \text{for } v_L < v_A\,. \qquad (3\text{-}409)$$

HEISENBERG (1948, 1949) calculated the distribution of energy over wave-numbers within the equilibrium range and found that (cf. CHANDRASEKHAR, 1949)

$$\begin{aligned} E(k) &= \frac{c\,\varepsilon^{2/3}k^{-5/3}}{[1+(8/3)(k/k_d)^4]^{4/3}} \\ &\propto \varepsilon^{2/3}k^{-5/3} \quad \text{for } k \ll k_d \\ &\approx k^{-7} \quad\quad\; \text{for } k \gg k_d\,. \end{aligned} \tag{3-410}$$

Here c is a constant, and the limiting wave number $k_d = (\varepsilon/v^3)^{1/4}$.

The suggestion that turbulence might play a role in astrophysics was put forth by ROSSELAND (1929) and CHANDRASEKHAR (1949). Turbulence theories were first applied to the motion of solar granules by RICHARDSON and SCHWARZSCHILD (1950); to the determination of the strength of the interstellar magnetic field by CHANDRASEKHAR and FERMI (1953); and to the formation of galaxies by WEIZSÄCKER (1951). The Kolmogoroff $k^{-5/3}$ energy spectrum is found to be valid for atmospheric turbulence (cf. LUMLEY and PANOFSKY, 1964). Spectra of the interplanetary magnetic field, velocity, and density all indicate that a Kolmogoroff spectrum is appropriate for the solar wind in the frequency range 10^{-5} Hz $\lesssim f \lesssim 10$ Hz (cf. JOKIPII, 1974). Here $f \approx v_w k/2\pi$, where k is the wave number and the solar wind velocity $v_w \approx 350$ km sec^{-1}. Furthermore, OZERNOI and CHIBISOV (1971) argue that a Kolmogoroff spectrum is valid for the scale sizes of galaxies.

3.5.13. Accretion

When a stationary seed object is imbedded in a gas which is in thermal equilibrium at temperature, T, the collision time, τ_c, between collisions with gas atoms is

$$\tau_c \approx (\pi r^2 N v_{th})^{-1} \approx \left[\pi r^2 N \left(\frac{kT}{m}\right)^{1/2}\right]^{-1},$$

where r is the radius of the seed object, v_{th} is the thermal velocity of the gas atoms, and N and m are, respectively, the number density and mass of the atoms. If every atom which hits a seed object stays with it, the rate of increase, A, of the mass, M, of the seed object is

$$A = \frac{dM}{dt} \approx \frac{m}{\tau_c} \approx \pi r^2 N (kTm)^{1/2}\,, \tag{3-411}$$

and the rate of increase of the radius is

$$\frac{dr}{dt} \approx \frac{N}{\rho}(kTm)^{1/2} \approx \frac{A}{\pi r^2 \rho},$$

where ρ is the gas mass density. This type of accretion in a gas cloud is thought to account for the formation of grains in interstellar matter (LINDBLAD, 1935; OORT and VAN DE HULST, 1946).

If a planet, star, or galaxy has a relative velocity, v, with respect to a gas cloud, the duration of their encounter is roughly

$$\tau \approx r/v\,, \tag{3-412}$$

where r is the smallest distance between the two objects. If the planet, star, or galaxy has mass, M, and the cloud has mass density, ρ, the gravitational force, F, exerted by the object on a unit volume of gas is

$$F = GM\rho/r^2\,, \tag{3-413}$$

and the momentum, p, transferred by this force is approximately

$$p = \tau F = \frac{GM\rho}{rv}\,. \tag{3-414}$$

The original momentum, ρv, becomes equal to p at the critical distance

$$R_A = \frac{GM}{v^2} = 1.3\times10^{14}\left(\frac{10\,\mathrm{km\,sec^{-1}}}{v}\right)^2\left(\frac{M}{M_\odot}\right)\,\mathrm{cm}. \tag{3-415}$$

As a seed object moves through a gas, it will capture or accrete a gas within the critical radius, R_A. Provided that R_A is larger than the mean free path of the gas, the captured gas will be maintained as a turbulent cloud surrounding the seed object. Turbulent friction will cause a loss of the gas kinetic energy, and most of the gas will eventually be united with the planet, star, or galaxy. As the total mass within this gaseous sphere is ρR_A^3, and as it is replenished in the time $\tau_A = R_A/v$, the rate of increase of mass, A, is (HOYLE and LYTTLETON, 1939; BONDI and HOYLE, 1944)

$$A = \frac{dM}{dt} \approx \pi\rho R_A^2 v = \pi R_A^2 \mu m_{\mathrm{H}} N v\,, \tag{3-416}$$

where μ is the molecular weight, m_{H} is the mass of the hydrogen atom, and N is the density of the gas cloud. The corresponding formula for the rate of increase of the radius is

$$\frac{dr}{dt} = \frac{v}{4}\frac{\rho}{\rho_s}\,, \tag{3-417}$$

where $\rho_s = 3M/(4\pi R^3)$ is the mass density of the planet, star, or galaxy.

If the average kinetic energy of the incoming gas is less than the gravitational energy of the planet, star, or galaxy, all of the gas will be pulled to the massive object by gravitational attraction. Eq. (3-416) is then written as

$$A = \pi\rho[R(R+R_A)]\,v\,, \tag{3-418}$$

where R is the radius of the planet, star, or galaxy. With $\rho = 10^{-24}\,\mathrm{g\,cm^{-3}}$, the value for the interstellar medium, $R = R_\odot$ and $v = 10\,\mathrm{km\,sec^{-1}}$, Eq. (3-418) gives $A \approx 10^{-18} M_\odot$ per year for $M = M_\odot$.

Accretion into a planet, star, or galaxy of mass, M, becomes supersonic when the object's radius, R, is less than the critical radius, R_s, given by

$$R_s = \frac{5-3\gamma}{4}\frac{GM}{s^2}\,, \tag{3-419}$$

where γ is the adiabatic index, and the velocity of sound of the gas at infinity, s, is given by

$$s=\left(\frac{\gamma P}{\rho}\right)^{1/2}\approx\left(\frac{kT}{m_{\mathrm{H}}}\right)^{1/2}, \tag{3-420}$$

where P and T are, respectively, the gas pressure and temperature, and m_{H} is the mass of the hydrogen atom. When $v<s$ and for accretion, then $R<R_s$. In this case the critical accretion radius, R_A, is given by (BONDI, 1952; ZELDOVICH and NOVIKOV, 1971)

$$R_A=\delta(\gamma)\frac{GM}{s^2}, \tag{3-421}$$

where

$$\delta(\gamma)=\frac{1}{2}\left[\frac{2}{(5-3\gamma)}\right]^{(5-3\gamma)/(3\gamma-3)},$$

and $\delta(\gamma)=0.5$ and 1 for $\gamma=5/3$ and 4/3, respectively. The accretion rate, A, is for the case of $v<s$,

$$A=4\pi R_s^2\,U_s\rho_s=\alpha(\gamma)\left(\frac{2GM}{c^2}\right)^2 c\rho\left(\frac{m_{\mathrm{H}}c^2}{kT}\right)^{3/2}\approx\pi R_A^2 s\rho, \tag{3-422}$$

where the velocity of sound, U_s, at the critical radius, R_s, is given by

$$U_s=s\left[\frac{2}{(5-3\gamma)}\right]^{1/2}, \tag{3-423}$$

and

$$\alpha(\gamma)=\frac{\pi}{4\gamma^{3/2}}\left[\frac{2}{(5-3\gamma)}\right]^{(5-3\gamma)/[2(\gamma-1)]},$$

which has the values of 1.5, 0.3 and 1.4 for $\gamma=1$, 5/3 and 4/3, respectively. Eqs. (3-421) and (3-422) are roughly the same as Eqs. (3-415) and (3-416) if v is replaced by s when v is less than s. For $\rho=10^{-24}$ g cm^{-3}, $M\approx M_\odot$ and $s=1$ km sec^{-1}, Eq. (3-422) gives $A\approx 5\times10^{-12}M_\odot$ per year.

The luminosity of a star due to accretion is given by (ZELDOVICH and NOVIKOV, 1971)

$$L\approx\varphi\frac{dM}{dt}\approx 2\times10^{31}\left(\frac{\varphi}{0.1\,c^2}\right)\left(\frac{M}{M_\odot}\right)^2\left(\frac{10^4}{T}\right)^{3/2} N\ \mathrm{erg\ sec}^{-1}, \tag{3-424}$$

where φ is the gravitational potential near the surface of the star, T and N are, respectively, the temperature and number density of the gas near R_A, and γ has been taken to be 4/3. Although the spectrum is difficult to estimate, one approach is to assume that the radiation emerges from the surface as a black body radiator of luminosity, L, and temperature, T. The Wien displacement law then gives the wavelength of maximum radiation once T is found from L and the star's radius, R. For a white dwarf star, $M\approx M_\odot$, $R\approx10^9$ cm, $\varphi\approx10^{-4}c^2$, and $L\approx10^{28}N$ erg sec^{-1}. In this case, $T\approx 500$–2000 °K for reasonable values of N, and the radiation is in the infrared region. For a neutron star, $M\approx M_\odot$ and $R\approx10^6$ cm so that the star radiates most of its energy in the ultraviolet wavelength range of

150—900 Å. Further formulae for the accretion of matter by neutron stars or black holes are given by MICHEL (1972).

A detailed discussion of the density distribution about an accreting object is given by DANBY and CAMM (1957). DANBY and BRAY (1967) show that when a gas is not cold, accretion may not operate as efficiently as indicated above. SPIEGEL (1970) gives formulae for the drag force, F_D, due to accretion shocks, and shows that the dissipation of gravitational energy by supersonically moving galaxies may be sufficient to heat a galactic cluster to around 10^8 °K. SPIEGEL gives

$$F_D = \pi R_A^2 \rho v \ln \left\{ \frac{\pi}{R k_J} \left[\frac{M}{(M^2-1)^{1/2}} \right] \right\}, \tag{3-425}$$

where

$$k_J = \left[\frac{4\pi G \rho}{s^2} \right]^{1/2},$$

and the mach number $M = v/s$ is greater than one. The rate of work by the supersonic object is $v F_D$.

3.5.14. Stellar Variability and Oscillation Theory

FABRICIUS (1594) first found that the light from the star, Mira, is variable; and GOODRICKE (1784) and BAILEY (1899), respectively, found that the light from δ Cephei and RR Lyrae is periodically variable. Chinese astronomers apparently first found stellar variability in 134 BC. Catalogues of thousands of variable stars have subsequently been prepared by KUKARKIN and PARENAGO (1969). In general, there are many different types of variables, with periods ranging from 0.3 to 1000 days. The RR Lyrae type variables have periods ranging from 0.3 to 0.9 days, whereas the classical Cepheid variables have periods ranging from 1 to 50 days.

VOGEL (1889) and BELOPOLSKI (1897) found that the radial velocities of classical Cepheids vary with the same period as the light variations; but the radial velocity curves tend to be mirror images of the light curves. If the radial velocity curves reflect the motion of the stellar surface, the phase lag between the light and velocity curves implies that the star is brightest when it is expanding through its equilibrium radius. HERTZSPRUNG (1905) showed that the pulsating stars occupy an "instability" strip in the Hertzsprung-Russell diagram, and that the Cepheid variable stars are "giant" stars in a particular stage of evolution (cf. Fig. 42). LEAVITT (1912) showed that the periods of classical Cepheids increase with their absolute luminosity and the details of this period-luminosity relation are discussed in Sect. 5.3.5. HERTZSPRUNG (1926) then showed that the light curves of Cepheids exhibit period correlated asymmetries and shapes. Furthermore, the total amplitude, $2K$, of the radial velocity curve has been shown to be linearly related to the amplitude, Apg, of the light curve (EGGEN, 1951)

$$2K \approx 35.3 \,\mathrm{Apg} \;\; \mathrm{km\,sec^{-1}} \quad \text{for Apg} < 1.2 \,\mathrm{mag}. \tag{3-426}$$

RITTER (1880, 1881) first advanced the hypothesis that adiabatic, radial oscillations of a star might account for their observed periodic variability. EDDINGTON

(1918, 1919, 1926) used the linearized forms of the continuity equation (3-250) and Euler's equation (3-251) together with the adiabatic relations

$$\frac{\Delta P}{P} = \Gamma_1 \frac{\Delta \rho}{\rho},$$

$$\frac{\Delta T}{T} = (\Gamma_3 - 1)\frac{\Delta \rho}{\rho},$$

$$\Gamma_1 = \beta + \frac{(4-3\beta)^2(\gamma-1)}{\beta + 12(\gamma-1)(1-\beta)}, \tag{3-427}$$

$$\Gamma_3 = 1 + \frac{\Gamma_1 - \beta}{4-3\beta},$$

and

$$\beta = \frac{P_G}{P} = \frac{N k \rho T}{N k \rho T + a T^4/3},$$

to obtain the wave equation for adiabatic radial oscillations.

$$-\frac{1}{\rho r^4}\frac{d}{dr}\left(\Gamma_1 P r^4 \frac{d\xi}{dr}\right) - \frac{1}{r\rho}\left\{\frac{d}{dr}[(3\Gamma_1 - 4)P]\right\}\xi = \sigma^2 \xi$$

or (3-428)

$$\frac{d^2\xi}{dx^2} + [4 - V(x)]\frac{1}{x}\frac{d\xi}{dx} + \frac{V(x)}{x^2}\left[\frac{4-3\gamma}{\gamma} + \frac{x^3 R^3 \sigma^2}{\gamma G M(x)}\right]\xi = 0.$$

Here a star has total pressure, P, gas pressure $P_G = N k \rho T$, where N is the total number of free electrons and ions per unit mass, radiation pressure $P_R = a T^4/3$, where $ac/4$ is the Stefan-Boltzmann constant, mass density, ρ, temperature, T, and adiabatic coefficient, γ. Linear radial oscillations result in a change, Δr, in radius, r, given by

$$\Delta r = r\xi \exp[i\sigma t], \tag{3-429}$$

where ξ and σ are, respectively, the amplitude and frequency of the oscillation. These oscillations are regarded as being superposed on a state of complete hydrostatic and thermal equilibrium. For an equilibrium radius, R, the variable $x = r/R$, and

$$V(x) = \frac{G\rho(x) M(x)}{r P(x)}, \tag{3-430}$$

where $M(x)$ is the mass interior to x, and is given by

$$M(x) = \int_0^x 4\pi r^2 \rho(r)\, dr.$$

The wave equation (3-428) is subject to the boundary conditions

$$\Delta r = 0 \quad \text{at } r = 0$$

and

$$\frac{\Delta P}{P} = -\left[4 + \frac{\sigma^2 R^3}{GM}\right]\xi \qquad \text{at } r = R, \tag{3-431}$$

or

$$\frac{d\xi}{dx} = \left[4 - 3\gamma + \frac{\sigma^2 R^3}{GM}\right]\frac{\xi}{\gamma} \quad \text{at } x = 1.$$

Eq. (3-428) admits eigenfunction solutions, ξ_i, for certain eigenvalues, σ_i^2, of the parameter σ^2, which increase as the integer i increases. SCHWARZSCHILD (1941) has computed amplitudes, ξ, for the first four modes, $i=0,1,2,3$, for $\Gamma_1 = 5/3$, and LEDOUX and WALRAVEN (1958) have summarized the properties of the first two modes of radial oscillation for different stellar models with $\Gamma_1 = 5/3$. In general, the fundamental mode, $i=0$, has the solution (LEDOUX and PEKERIS, 1941; LEDOUX, 1945)

$$(3\Gamma_1 - 4)\frac{4\pi G\rho}{3} \leq \sigma_0^2 \leq -(3\Gamma_1 - 4)\frac{\Omega}{I}, \tag{3-432}$$

where $\Omega \approx -GM^2/R$ is the gravitational potential energy and $I \approx MR^2$ is the moment of inertia. If a magnetic field of strength, H, is present, Ω must be replaced by $\Omega + \mathcal{M}$ (CHANDRASEKHAR and LIMBER, 1954) where the magnetic energy $\mathcal{M} = R^3 H^2/6$. For a homogeneous star we have (RITTER, 1881)

$$\sigma_0^2 = (3\Gamma_1 - 4)\frac{GM}{R^3}, \tag{3-433}$$

so that the period, Π_0, is given by

$$\Pi_0 = \frac{2\pi}{\sigma_0} \approx \frac{2R}{s} \propto \rho^{-1/2}, \tag{3-434}$$

where s is the velocity of sound, and the pulsation constant

$$Q_0 = \Pi_0 \left(\frac{\rho}{\rho_\odot}\right)^{1/2} \approx 0.116 \text{ days} \quad \text{for } \Gamma_1 = 5/3, \tag{3-435}$$

where the solar density $\rho_\odot \approx 1.41$ gm cm^{-3}. EPSTEIN (1950) showed that the fundamental mode solution to Eq. (3-428) is determined by a weighting function which peaks at

$$x = \frac{r}{R} \approx 0.75. \tag{3-436}$$

This means that the period of the fundamental mode is determined primarily by conditions in the envelope of the star, and is almost independent of conditions in the central regions where most of the mass is located. Period ratios, Π_i/Π_0, are given as a function of Q_0 by SCHWARZSCHILD (1941), BAKER and KIPPENHAHN (1965), and CHRISTY (1966). Observed data (CHRISTY, 1966) indicate that $\Pi_1/\Pi_0 \approx 0.77$ and $Q_0 = 0.033$ days. The homogeneous model gives $Q_0 \approx 0.12$ whereas the standard model for a polytrope of index, 3, has $Q_0 \approx 0.038$. CHRISTY (1966) gives

$$Q_0 \approx 0.022 \left(\frac{R}{R_\odot}\right)^{1/4} \left(\frac{M_\odot}{M}\right)^{1/4} \text{days}, \tag{3-437}$$

for a star of mass, M, and radius, R. Here the solar radius, $R_\odot \approx 6.96 \times 10^{10}$ cm and the solar mass $M_\odot \approx 2 \times 10^{33}$ gm.

Although the periods of variable stars are relatively insensitive to non-linear or non-adiabatic effects, adiabatic oscillations quickly decay and some non-linear or non-adiabatic effect is needed to explain their continual pulsation. Such effects are also needed to explain the observed phase shift between the oscillation of brightness and the stellar radius. EDDINGTON (1941) suggested that a possible cause of instability might be the non-adiabatic hydrogen ionization zone near the surface of the star; and ZHEVAKIN (1953, 1963) and COX and WHITNEY (1958) showed that stellar oscillations might be excited by the region of ionization of helium (He^+). Later linear, non-adiabatic calculations by BAKER and KIPPENHAHN (1962, 1965) and COX (1963) verified Zhevakin's conclusion. These calculations utilize the linear forms of the equation of motion

$$\frac{\partial^2 r}{\partial t^2} = -\frac{GM(r)}{r^2} - \frac{1}{\rho}\frac{\partial P}{\partial r}, \tag{3-438}$$

and the heat flow equation

$$\frac{\partial Q}{\partial t} = \frac{\partial E}{\partial t} - \frac{P}{\rho^2}\frac{\partial \rho}{\partial t} = \varepsilon - \frac{1}{4\pi r^2 \rho}\frac{\partial L}{\partial r}. \tag{3-439}$$

Here $M(r)$ is the mass interior to radius, r, $\partial Q/\partial t$ is the net rate of gain of heat per unit mass, E is the total internal energy per unit mass, ε is the rate per unit mass of thermonuclear energy generation

$$\varepsilon = \varepsilon_0 \rho^\lambda T^\nu$$

or (3-440)

$$\frac{\Delta\varepsilon}{\varepsilon} = \lambda\frac{\Delta\rho}{\rho} + \nu\frac{\Delta T}{T} \approx \frac{\Delta\rho}{\rho}[\lambda + \nu(\Gamma_3 - 1)],$$

and the "interior" luminosity, $L(r)$, is given by

$$L(r) = -4\pi r^2 \frac{4ac}{3}\frac{T^3}{\kappa\rho}\frac{\partial T}{\partial r}, \tag{3-441}$$

where the Rosseland mean opacity, κ, is tabulated by COX, STEWART, and EILERS (1965). Often the formula

$$\kappa = \kappa_0 \rho^n T^{-s}, \tag{3-442}$$

is employed to give

$$\frac{\Delta\kappa}{\kappa} = n\frac{\Delta\rho}{\rho} - s\frac{\Delta T}{T}$$

and (3-443)

$$\frac{\Delta L}{L} = 4\frac{\Delta r}{r} + (4+s)\frac{\Delta T}{T} - n\frac{\Delta\rho}{\rho} + \frac{1}{d\ln T/dx}\frac{\partial}{\partial x}\left(\frac{\Delta T}{T}\right).$$

Alternative forms of the heat flow equation (3-439) are

$$\frac{\partial P}{\partial t} = \frac{\Gamma_1 P}{\rho}\frac{\partial \rho}{\partial t} + \rho(\Gamma_3 - 1)\left[\varepsilon - \frac{1}{4\pi r^2 \rho}\frac{\partial L}{\partial r}\right],$$

and (3-444)

$$\frac{\partial \ln T}{\partial t} = (\Gamma_3 - 1)\frac{\partial \ln \rho}{\partial t} + \frac{1}{c_v T}\left[\varepsilon - \frac{1}{4\pi r^2 \rho}\frac{\partial L}{\partial r}\right].$$

When $\xi = \Delta r/r$ is used with the linear forms of Eqs. (3-438) and (3-439) or (3-444), we obtain the wave equation for non-adiabatic radial oscillations.

$$\ddot{\xi} - \frac{\dot{\xi}}{r\rho}\frac{d}{dr}[(3\Gamma_1 - 4)P] - \frac{1}{\rho r^4}\frac{\partial}{\partial r}\left[\Gamma_1 P r^4 \frac{\partial \dot{\xi}}{\partial r}\right]$$
$$= -\frac{1}{r\rho}\frac{\partial}{\partial r}\left[\rho(\Gamma_3 - 1)\Delta\left(\varepsilon - \frac{1}{4\pi r^2 \rho}\frac{\partial L}{\partial r}\right)\right], \tag{3-445}$$

where the dot over a symbol means the Stokes derivative $\partial/\partial t$. BAKER and KIPPENHAHN (1962, 1965), COX (1963), and ZHEVAKIN (1963) have given computer solutions to Eqs. (3-438), (3-439), (3-444), and (3-445), whereas LEDOUX (1963, 1965) has given integral solutions for the eigenvalues.

When general departures from an equilibrium spherical shape are considered, the radius, r, of the deformed surface of a fluid is given by

$$r = R + \varepsilon\, Y_l^m(\varphi, \theta), \tag{3-446}$$

where R is the equilibrium radius, Y_l^m is a spherical harmonic, and $\varepsilon = \varepsilon_0 \exp[-\sigma t]$ where ε_0 is a constant and the oscillatory period is $2\pi/\sigma$. Both radial and non-radial oscillations of an inviscid fluid have the Kelvin modes

$$\sigma^2 = -\frac{2l(l-1)}{(2l+1)}\frac{GM}{R^3} = -\frac{8}{3}\pi G\rho\frac{l(l-1)}{2l+1}. \tag{3-447}$$

When viscosity is introduced, the Kelvin modes are damped with a mean life-time of

$$\tau = \frac{R^2}{\nu(l-1)(2l+1)}, \tag{3-448}$$

where the kinematic viscosity, ν, is assumed to be vanishingly small. LAMB (1881) and CHANDRASEKHAR (1959) have considered the oscillations of a viscous globe. The general formulae for the radial and non-radial oscillations of gaseous masses have been given by CHANDRASEKHAR and LEBOVITZ (1962, 1963, and 1964), CHANDRASEKHAR (1964), TASSOUL and TASSOUL (1967), TASSOUL and OSTRIKER (1968), TASSOUL (1968), and LEBOVITZ and RUSSELL (1972). CHANDRASEKHAR and LEBOVITZ (1964), for example, give the fundamental frequencies of nonradial oscillations of polytropic gas spheres belonging to spherical harmonics of orders $l=1$ and $l=2$.

3.5.15. Instabilities in Fluids and Plasmas

A uniform plasma in thermodynamic equilibrium is stable; whereas the introduction of nonuniformities in plasma density, temperature, velocity, or magnetic field might cause instabilities. The instabilities are usually analyzed by assuming an initial equilibrium state, φ_0, subject to an infinitely small perturbation, φ_1, of the form

$$\varphi_1 \propto \exp i(\boldsymbol{k} \cdot \boldsymbol{x} - \omega t), \tag{3-449}$$

where $\boldsymbol{k}$ is the wave number vector and ω is the frequency. The equation of motion and the conservation equations are then linearized and assumed to have a solution $\varphi=\varphi_0+\varphi_1$. As a consequence, a dispersion relation between ω and $\boldsymbol{k}$ is found. The plasma is said to be linearly unstable if ω has a positive imaginary part for any real value of k. If all possible ω are real or have negative imaginary parts, then the plasma is said to be stable.

When an incompressible fluid has a nonuniform distribution of mass density, ρ, it may exhibit the RAYLEIGH (1900)-TAYLOR (1950) instability. For two vertically adjacent, hydrostatic, inviscid fluids, the lower fluid having density ρ_1 and the upper layer having density ρ_2, the dispersion relation is (RAYLEIGH, 1900)

$$\omega^2 = -gk\left(\frac{\rho_2-\rho_1}{\rho_2+\rho_1}\right), \tag{3-450}$$

where g is the gravitational acceleration. If $\rho_2<\rho_1$ the situation is stable, whereas it is unstable for all wave numbers if $\rho_2>\rho_1$. If there is a surface tension, T, however, the arrangement is stabilized for sufficiently short wavelengths. In this case, the fluid is unstable for all wave numbers, k, in the range $0<k<k_c$, where

$$k_c = \left[\frac{(\rho_2-\rho_1)}{T}g\right]^{1/2}. \tag{3-451}$$

When a horizontal magnetic field of strength, B, is introduced, the arrangement is also stabilized with an effective surface tension, T_{eff}, given by (KRUSKAL and SCHWARZSCHILD, 1954)

$$T_{\mathrm{eff}} = \frac{\mu B^2}{2\pi k}\cos^2\theta, \tag{3-452}$$

where μ is the magnetic permeability, and θ is the angle between the wave vector and the direction of B. If the fluid is rotating with angular velocity, Ω, the dispersion relation becomes (HIDE, 1956; CHANDRASEKHAR, 1961)

$$\omega^2\left[1-\frac{4\Omega^2}{\omega^2}\right]^{1/2} = \omega_0^2, \tag{3-453}$$

where ω_0^2 is the ω^2 given in Eq. (3-450). Rotation, therefore, puts a lower limit to $\omega=2\Omega$.

For a stratified medium of density $\rho=\rho_0\exp(-z/H)$, where H is the scale height, the Rayleigh-Taylor instability occurs for negative H, whereas stable gravity waves occur for positive H. When an inviscid fluid is confined between two rigid planes at $z=0$ and $z=d$, the dispersion relation is (RAYLEIGH, 1900; CHANDRASEKHAR, 1955)

$$\frac{g}{H\omega^2} = 1 + \frac{[d^2/(4H^2)]+m^2\pi^2}{k^2d^2}, \tag{3-454}$$

where m is an integer. Hence, for a gradient, ∇N, in density, N, we have instability for

$$g\nabla N < 0, \tag{3-455}$$

for which

$$\omega = \pm i(gH^{-1})^{1/2}, \tag{3-456}$$

where

$$H^{-1} = -\frac{d}{dz}[\ln N]. \tag{3-457}$$

Rotation stabilizes this arrangement for all wave numbers, k, less than a minimum wave number, $k_{\min}$, given by

$$k_{\min}^2 = -\frac{4\Omega^2 H}{g d^2}\left[\frac{d^2}{4H^2} + \pi^2\right], \tag{3-458}$$

where the angular rotational velocity is Ω, and the layer thickness is d.

When plasma particles move along curved magnetic field lines of radius, R, the flow creates a centrifugal acceleration

$$g \approx \frac{P}{\rho R}, \tag{3-459}$$

where the field lines are assumed to pass through an atmosphere of density, ρ, and pressure, P. For the ideal "flute mode" instability, we have

$$\omega^2 \approx g H^{-1} \approx \frac{P}{\rho R H}, \tag{3-460}$$

which is seen to be the same dispersion relation as that of the Rayleigh-Taylor instability. KULSRUD (1967) gives an approximate dispersion relation for the interchange instability

$$\omega^2 = -\left[\frac{\gamma P}{U}\frac{dU}{dP} + 1\right]\frac{P^2}{\rho U}\frac{dU}{dP}, \tag{3-461}$$

where it is assumed that the gas pressure, P, is much less than the magnetic pressure, $B^2/(8\pi)$, and the quantity

$$U = \int \frac{dl}{B},$$

where the integral is along a magnetic field line for a magnetic field of strength B. Instability occurs if

$$\frac{dP}{dU} > 0,$$

and stability results if

$$\frac{d}{dP}[P U^{\gamma}] < 0. \tag{3-462}$$

Here γ is the ratio of specific heats. The system is stable if the interchange of two equal flux tubes requires more compression of the plasma than expansion. HASEGAWA (1971) gives a general interchange dispersion relation

$$\frac{\omega_{ci}}{H k_\perp}\left[\frac{1}{\omega + k_\perp g/\omega_{ci}} - \frac{1}{\omega}\right] = 1 - \frac{M_i}{M_e}\left[\frac{k_\| \omega_{ci}}{k_\perp \omega}\right]^2, \tag{3-463}$$

where the gravitational acceleration, g, is taken to simulate the centrifugal force due to particle motion parallel to the curved field lines, $k_\perp$ and $k_\parallel$ denote, respectively, the two orthogonal wave vectors which are perpendicular to the field direction, the subscripts i and e denote, respectively, the ions and electrons, and the cyclotron frequency, ω_{cj}, for the jth particle type of charge eZ_j is

$$\omega_{cj} = \frac{Z_j e B}{M_j c} . \tag{3-464}$$

For the flute mode, $k_\parallel = 0$, and Eq. (3-463) becomes Eq. (3-460). For $k_\parallel \neq 0$, however, the instability is stabilized for a perturbation of any size if

$$R k_\parallel > 2\left(\frac{M_e}{M_i}\right)^{1/2} , \tag{3-465}$$

where R is the radius of curvature of the field line.

Even without the action of a gravitational field, a plasma with a density gradient becomes unstable for waves whose parallel phase velocity is between the thermal velocities of the electrons and ions. This drift wave instability has frequency, ω, given by the drift wave frequency (MOISEEV and SAGDEEV, 1963; HASEGAWA, 1971)

$$\omega = \frac{v_{Te}^2 k_\parallel}{H \omega_{ce}}$$

when (3-466)

$$v_{Te} > \frac{\omega}{k_\parallel} > v_{Ti} ,$$

where H is the scale height given by Eq. (3-457) and v_T is the thermal velocity.

When both a nonuniformity in density and shear flow are considered, instability between two fluids is found to be possible even when $\rho_2 < \rho_1$. If $v_1 - v_2 = v$ denotes the relative velocity between the two fluids, then this Kelvin (1910)-Helmholtz (1868) instability occurs for all wavenumbers

$$k > \frac{g(\rho_2 - \rho_1)(\rho_1 + \rho_2)}{\rho_1 \rho_2 (v_1 - v_2)^2 \cos^2 \varphi} , \tag{3-467}$$

where φ is the angle between the directions of $\boldsymbol{k}$ and $\boldsymbol{v}$. A surface tension, T, will suppress the instability if

$$(v_1 - v_2)^2 < \frac{2(\rho_1 + \rho_2)}{\rho_1 \rho_2} [T g(\rho_1 - \rho_2)]^{1/2} . \tag{3-468}$$

A magnetic field of strength B will also suppress the instability if the inequality in Eq. (3-468) is found to hold with T replaced by the T_{eff} given in Eq. (3-452).

An electrostatic Kelvin-Helmholtz instability exists in a plasma due to the shear flow caused by an E cross H drift. For an electron sheet of thickness $2a$ which is parallel to a uniform magnetic field of strength B, the dispersion relation becomes

$$\frac{4\omega^2}{\omega_0^2} = \left[1 - \frac{2 k v_0}{\omega_0}\right]^2 - \exp[-4ka] , \tag{3-469}$$

where

$$\omega_0 = 4\pi e c N/B = \omega_{pe}^2/\omega_{ce} ,$$

and the shear velocity, v_0, is given by

$$v_0 = E(a)/B,$$

where ω_{ce} and ω_{pe} are, respectively, given in Eqs. (3-464) and (3-471), and $E(a)$ is the electric field intensity at a. The wave becomes unstable for all wave numbers, k, satisfying the inequality

$$ka \lesssim 0.7, \tag{3-470}$$

which is sufficient to make the right side of Eq. (3-469) negative.

As was shown in Sect. 1.32, the basic frequency of oscillation, ω_{pj}, of a thermal equilibrium plasma of one species of particles, j, is given by (TONKS and LANGMUIR, 1929)

$$\omega_{pj} = \left[\frac{4\pi e^2 Z_j^2 N_j}{M_j}\right]^{1/2}, \tag{3-471}$$

where the particle charge is eZ_j, the mass is M_j, and the volume density is N_j. In the absence of collisions, these oscillations are damped with a damping constant given by (LANDAU, 1946)

$$\mathscr{I}(\omega) \approx -\omega_{pj}\sqrt{\frac{\pi}{8}}\left(\frac{\omega_{pj}}{k v_T}\right)^3 \exp\left(-\frac{\omega^2}{2k^2 v_T^2}\right), \tag{3-472}$$

where $\mathscr{I}$ denotes the imaginary part of the term in parenthesis, k is the wave number of the disturbance, and v_T is the root mean square thermal velocity. BOHM and GROSS (1949) pointed out that the physical mechanism of the damping described above is the trapping by the electric field of particles moving at approximately the phase velocity of the wave, with a consequent exchange of energy between particles and plasma oscillations.

It follows from the linearized Boltzmann equation that the general dispersion relation for a plasma is (VLASOV, 1945; LANDAU, 1946; BOHM and GROSS, 1949)

$$1 - \frac{\omega_{pj}^2}{k^2}\int \frac{\partial f_0/\partial v}{v-(\omega/k)}\, dv = 0, \tag{3-473}$$

where $f_0(v)$ is the equilibrium (unperturbed) distribution function of the velocity, v. If $f_0(v)$ has a single hump, it can be shown (JACKSON, 1960) that there can only be damped oscillations. In the case of a Maxwellian distribution, the dispersion relation is

$$\omega^2 \simeq \omega_{pj}^2\left[1 + \frac{3k^2 v_T^2}{\omega_{pj}^2}\right], \tag{3-474}$$

with a damping constant given by Eq. (3-472). However, if the velocity distribution is two humped, damped, steady state, or growing oscillations are possible. For example, for a plasma of j beams of velocity, $\boldsymbol{v}_j$, the dispersion relation for longitudinal oscillations becomes (BAILEY, 1948; PIERCE, 1948)

$$1 = \sum_j \frac{\omega_{pj}^2}{(\omega - \boldsymbol{k}\cdot\boldsymbol{v}_j)^2}, \tag{3-475}$$

where the plasma frequency, ω_{pj}, is given by Eq. (3-471). For one beam of velocity, v_s, in a stationary plasma, the dispersion relation becomes

$$1-\frac{\omega_p^2}{\omega^2}-\frac{\omega_{ps}^2}{(\omega-kv_s)^2}=0, \quad (3\text{-}476)$$

where ω_p and ω_{ps} denote, respectively, the plasma frequencies of the plasma and the stream. A "two-stream" instability can occur when the negative-energy wave of the stream $(\omega<kv_s)$ is coupled to the positive energy wave of the plasma. For weak coupling, four waves are possible with dispersion relations given by $\omega=kv_s\pm\omega_{ps}$ and $\omega=\pm\omega_p$; and the details of the stable and unstable oscillations for these cases are given by STURROCK (1958). One has an instability and growing oscillations for all wave numbers, k, such that

$$\frac{\omega}{v_s}\approx k\lesssim\frac{\omega_p}{v_s}. \quad (3\text{-}477)$$

The growth rate for this instability is given by

$$\mathscr{I}(\omega)=\frac{\omega_{ps}kv_s}{(\omega_p^2-k^2v_s^2)^{1/2}}, \quad (3\text{-}478)$$

if kv_s is not close to ω_p. For an electronic plasma consisting of two streams of electrons of the same density but equal and opposite velocities, v_s, growing oscillations occur for

$$k\leq 0.5\frac{\omega_p}{v_T}\approx\sqrt{2}\frac{\omega_p}{v_s}, \quad (3\text{-}479)$$

where v_T is the thermal velocity spread of the stream, and the wave frequency $\omega=kv_s$. For a relativistic stream, the appropriate mass to use in the equation for the plasma frequency is

$$M=\gamma^3M_0, \quad (3\text{-}480)$$

where M_0 is the rest mass and $\gamma=(1-v_s^2/c^2)^{-1/2}$. If the plasma is not cold, the two stream instability discussed above still occurs provided that the thermal velocity of the plasma, v_T, is much smaller than the stream velocity. In this case, however, the growth of the plasma wave is accompanied by a flattening of the velocity distribution and the stream loses its identity in the time (ZHELEZNYAKOV and ZAITSEV, 1970)

$$\tau=\frac{N_p}{N_s}\left(\frac{\Delta v_s}{v_s}\right)^2\frac{1}{\omega_p}, \quad (3\text{-}481)$$

where N_s, v_s, and Δv_s denote, respectively, the particle density, velocity, and velocity dispersion of the stream, and N_p and ω_p denote the density and plasma frequency of the plasma.

BUNEMAN (1958, 1959) and JACKSON (1960) have discussed the dispersion equation for an electron-ion plasma, which takes the general form

$$1-\frac{\omega_{pe}^2}{k^2}\int_{-\infty}^{+\infty}\frac{\partial f_{0e}/\partial v\,dv}{v-(\omega/k)}-\frac{\omega_{pi}^2}{k^2}\int_{-\infty}^{+\infty}\frac{\partial f_{0i}/\partial v\,dv}{v-(\omega/k)}=0, \quad (3\text{-}482)$$

where the subscripts e and i denote, respectively, the electrons and ions. For a Maxwellian distribution of electrons and ions with respective thermal velocities v_{Te} and v_{Ti}, growing oscillations occur for wave numbers, k, such that

$$k \lesssim \sqrt{2}\left[1+\left(\frac{M_{\mathrm{e}}}{M_{\mathrm{i}}}\right)^{1/2}\right]\frac{\omega_{\mathrm{pe}}}{v}, \tag{3-483}$$

where the velocity difference $v = v_{\mathrm{e}} - v_{\mathrm{i}}$, is the difference between the mean velocities of the electrons and ions and is assumed to be large. For ions streaming through cold electrons, or for electrons streaming through electrons, Eq. (3-483) becomes, respectively, Eqs. (3-477) or (3-479). For electrons moving thermally with respect to cold ions, however, the real part of the dispersion relation becomes

$$\omega = k s_{\mathrm{i}} \quad \text{for } \omega \ll \omega_{\mathrm{pi}}, \tag{3-484}$$

where the ion sound speed, s_{i}, is given by

$$s_{\mathrm{i}} = v_{\mathrm{Te}}\left(\frac{M_{\mathrm{e}}}{M_{\mathrm{i}}}\right)^{1/2}.$$

The wave represented by the dispersion relation given in Eq. (3-484) is called the ion acoustic wave, and unstable oscillations occur for (HASEGAWA, 1971)

$$v_{\mathrm{D}} > s_{\mathrm{i}} \quad \text{for } \omega \lesssim \omega_{\mathrm{pi}},$$

and

$$\frac{\partial f_{0\mathrm{e}}}{\partial v} > 0 \quad \text{for } \omega \sim \omega_{\mathrm{pe}}, \tag{3-485}$$

where v_{D} is the electron drift velocity, and $f_{0\mathrm{e}}$ is the unperturbed electron velocity distribution function.

When a magnetic field of strength, B, is present, velocity anisotropies lead to electromagnetic instabilities as well as the electrostatic instabilities discussed above. For a cold plasma in which j beams of velocity v_j occur, the dispersion relation is

$$k^2 = \frac{\omega^2}{c^2} + \frac{1}{c^2}\sum_j \frac{\omega_{\mathrm{p}j}^2\,\omega}{\omega + k v_j + \omega_{\mathrm{c}j}}, \tag{3-486}$$

where the cyclotron frequency, $\omega_{\mathrm{c}j}$, for the jth particle is given by Eq. (3-464), and the plasma frequency, $\omega_{\mathrm{p}j}$, is given in Eq. (3-471). For cold ions and drifting electrons, the dispersion relation becomes (HASEGAWA, 1971)

$$k^2 c^2 - \omega^2 + \omega_{\mathrm{pe}}^2 \frac{\omega - k v_{\mathrm{D}}}{\omega - k v_{\mathrm{D}} + \omega_{\mathrm{ce}}} + \omega_{\mathrm{ci}}^2 \frac{\omega}{\omega - \omega_{\mathrm{ci}}} = 0, \tag{3-487}$$

where v_{D} is the electron drift velocity, and the subscripts e and i denote, respectively, electrons and ions. Waves which satisfy the dispersion relation are called ion cyclotron waves, and they become unstable for $\omega \approx \omega_{\mathrm{ci}}$ or $\omega \approx 0$.

Anisotropies in pressure, P, may lead to instabilities as well as anisotropies in density or velocity. When a magnetic field is present and collision effects are negligible, the particle pressures along, $P_{\parallel}$, and perpendicular, $P_{\perp}$, to the magnetic field become decoupled and instabilities are possible. If $f_j(v)$ is the velocity distribution function of species, j,

and

$$P_{\|} = \sum M_j \int v_{\|}^2 f_j(v) d^3 v, \qquad P_{\perp} = \sum M_j \int v_{\perp}^2 f_j(v) d^3 v, \tag{3-488}$$

where the summation is over the species of particles. If we let $\beta = 8\pi P/B^2$ denote the ratio of the pressure of a plasma species to the pressure of the magnetic field, then the dispersion relation at low frequencies ($\omega \ll \omega_{ci}$) and large wavelengths ($k v_{Ti} \ll \omega_{ci}$) becomes (KUTSENKO and STEPANOV, 1960)

$$\frac{\omega^2}{k_{\|}^2 v_A^2} = 1 - \sum_{\text{species}} \frac{1}{2} (\beta_{\|} - \beta_{\perp}) \quad \text{for } \langle v_{\|} \rangle = 0, \tag{3-489}$$

and

$$k_{\|}^2 \left[1 + \sum_{\text{species}} \frac{\beta_{\perp} - \beta_{\|}}{2} \right] + k_{\perp}^2 \left[1 + \sum_{\text{species}} \beta_{\perp} \left(1 - \frac{\beta_{\perp}}{\beta_{\|}} \right) - i \frac{\beta_{\perp i}^2}{\beta_{\| i}} \frac{\omega}{k_{\|} \langle v_{\| i} \rangle} \left(\frac{\pi}{2} \right)^{1/2} \right] = 0 \tag{3-490}$$

for $\omega \ll k v_A$, where again the subscripts $\perp$ and $\|$ denote, respectively, directions perpendicular and parallel to the magnetic field, the subscript i denotes the ion component, and v_A is the Alfvén velocity. For $k \gg k_{\perp}$, an Alfvén wave propagates along (parallel) the lines of force, and the wave becomes unstable for

$$1 + \sum_j \frac{\beta_{\perp j} - \beta_{\| j}}{2} < 0$$

or

$$P_{\|} > P_{\perp} + \frac{B^2}{4\pi}. \tag{3-491}$$

This "fire hose" instability occurs for both the shear mode (no variation in the parallel component of the magnetic field (Eq. (3-489)) and for the compressional mode (variation in the parallel component ($k_{\|} > k_{\perp}$) of the magnetic field (Eq. (3-490)). For almost perpendicular propagation ($k_{\|} \ll k_{\perp}$), the compressional mode allows the "mirror" instability which occurs for

$$1 + \sum_j \beta_{\perp j} \left(1 - \frac{\beta_{\perp j}}{\beta_{\| j}} \right) < 0$$

or

$$\frac{P_{\perp}^2}{P_{\|}} > P_{\perp} + \frac{B^2}{8\pi}. \tag{3-492}$$

Of particular interest in the theory of solar flares is the "tearing" instability which occurs when two magnetic fields of opposite sign move against each other and reconnect. If a magnetic field of strength, B, is brought into a region of thickness, δ, and extension, L, along the field line, the outward diffusion of the field is balanced by the inward flow of the plasma. If v denotes the plasma velocity, then we have (SWEET, 1958; PARKER, 1963; FURTH, KILEEN, and ROSENBLUTH, 1963)

$$v = (4\pi\delta\sigma/c^2)^{-1}, \tag{3-493}$$

where σ is the electrical conductivity, and matter is squeezed out of the region at the Alfvén velocity

$$v_A = \frac{B}{(4\pi\rho)^{1/2}}. \tag{3-494}$$

Continuity of matter flow then requires that

$$vL = v_A\delta, \tag{3-495}$$

to give an instability lifetime of

$$\tau \approx \frac{L}{v} = \frac{LR_m^{1/2}}{v_A}, \tag{3-496}$$

where the magnetic Reynolds number, R_m, is given by

$$R_m = \left(\frac{L}{\delta}\right)^2 = \left(\frac{v_A}{v}\right)^2 = \frac{4\pi L\sigma v_A}{c^2}. \tag{3-497}$$

PETSCHEK (1964) matched the reconnection geometry to the external surroundings and found that the reconnection rate Mach number, M, is given by

$$M = \frac{v}{v_A} = \frac{\pi}{4}\,\frac{1}{\ln[2M^2R_m]}, \tag{3-498}$$

and in general

$$R_m^{-1/2} \lesssim \frac{v}{v_A} \lesssim 1. \tag{3-499}$$

In this case, the timescale, τ, of the flux annihilation is given by

$$\tau = \frac{L}{v} = \frac{4L}{\pi v_A}\ln\left[2R_m\left(\frac{L}{v_A\tau}\right)^2\right]. \tag{3-500}$$

As emphasized by GOLD and HOYLE (1960), the energy of the solar flare must come from the release of magnetic energy, $(\Delta B)^2/8\pi$, which would occur in the field annihilation-reconnection mechanism discussed above. STURROCK (1966) has proposed a flare mechanism which incorporates the Petscheck annihilation mode in the open field region of a magnetic loop.

One of the most extensively studied instabilities is the pinch instability in which an axial current flowing through a column of plasma with an axial magnetic field produces a transverse magnetic field which pinches the gas (TAYLER, 1957; KRUSKAL and TUCK, 1958; CHANDRASEKHAR, KAUFMAN, and WATSON, 1958). If B_z denotes the strength of the axial magnetic field within the plasma column, and B_θ denotes the transverse magnetic field, then instability occurs when

$$B_\theta^2 > 2B_z^2, \tag{3-501}$$

and the column forms a sausage pattern, when

$$B_\theta^2 \ln\left(\frac{L}{R}\right) > B_z^2, \tag{3-502}$$

the column of radius, R, is kinked or bent over a length, L, and when

$$B_\theta > \frac{2\pi R}{L} B_z, \tag{3-503}$$

the column is doubly bent to form a helical pattern.

4. Nuclear Astrophysics and High Energy Particles

"Certain physical investigations in the past year, make it probable to my mind that some portion of sub-atomic energy is actually being set free in the stars. F. W. Aston's experiments seem to leave no room for doubt that all the elements are constituted out of hydrogen atoms bound together with negative electrons. The nucleus of the helium atom, for example, consists of four hydrogen atoms bound with two electrons. But Aston has further shown that the mass of the helium atom is less than the sum of the masses of the four hydrogen atoms which enter into it. ... Now mass cannot be annihilated, and the deficit can only represent the energy set free in the transmutation. ... If only five per cent of a star's mass consists initially of hydrogen atoms, which are gradually being combined to form more complex elements, the total heat liberated will more than suffice for our demands, and we need look no further for the source of a star's energy. ... If, indeed, the sub-atomic energy in the star is being freely used to maintain their great furnaces, it seems to bring a little nearer to fulfillment our dream of controlling this latent power for the well being of the human race—or for its suicide."

A. S. EDDINGTON (1920)

"We therefore feel justified in advancing tentatively the hypothesis that cosmic rays are produced in the super-nova process. ... With all reserve we advance the view that a super-nova represents the transition of an ordinary star into a neutron star, consisting mainly of neutrons."

W. BAADE and F. ZWICKY (1934)

"When the conditions depart widely from being static, there is no necessary tendency towards equipartition, but the energy may instead become enormously concentrated into certain small parts of the system. Thus in the crack of a whip the tip of the lash is moving faster than the speed of sound, though the coachman's wrist never moves fast at all. Again, when a large sea-wave strikes the wall of a lighthouse, spray is thrown up to a great height, and this in spite of its later rise being much slowed by air resistance. ... It is suggested that cosmic rays may originate from some mechanism of this kind, and though there may be other possibilities, the most obvious source is from the stormy seas that must cover the surface of many of the stars."

C. DARWIN (1949)

4.1. Early Fundamental Particles, Symbols, and Definitions

4.1.1. The Electron, Proton, Neutron, and Photon and Their Antiparticles

At about the same time that THOMSON (1897) discovered that all atoms emit electrons, photons with energy in the range 1—500 keV, called X-rays, were observed by RÖNTGEN (1896). Photons with energy greater than 500 keV, called gamma (γ) rays, were subsequently observed by VILLARD (1900). EINSTEIN (1905) then suggested that a photon particle of energy, $h\nu$, and zero mass is an electromagnetic wave of frequency, ν, and vice versa. The nuclear theory of matter was

Table 37. Properties of the electron, proton, neutron, photon, and their antiparticles[1]

Particle	Symbol	Rest mass (grams)	(a.m.u.)	(MeV/c^2)	Charge (e.s.u.)	Spin	Magnetic moment[2]	Mean life (sec)
negatron or electron	β^- or e^-	$9.109558(54)E-28$	$5.485930(34)E-4$	0.5110041(16)	$-4.803250(21)E-10$	1/2	$1.0011596577(35)\mu_B$	∞
positron	β^+ or e^+	$9.109558(54)E-28$	$5.485930(34)E-4$	0.5110041(16)	$+4.803250(21)E-10$	1/2	$1.0011596577(35)\mu_B$	∞
proton	p	$1.672614(11)E-24$	1.00727661(8)	938.2592(52)	$+4.803250(21)E-10$	1/2	$2.792782(17)\mu_N$	∞
antiproton	$\bar{p}$	$1.672614(11)E-24$	1.00727661(8)	938.2592(52)	$-4.803250(21)E-10$	1/2	$2.792782(17)\mu_N$	∞
neutron	n	$1.674920(11)E-24$	1.00866520(10)	939.5527(52)	0.0	1/2	$-1.913148(66)\mu_N$	$9.35(14)E+2$[3]
antineutron	$\bar{n}$	$1.674920(11)E-24$	1.00866520(10)	939.5527(52)	0.0	1/2	$-1.913148(66)\mu_N$	
photon	γ	0.0	0.0	0.0	0.0	1	0.0	∞

[1] The digits in parentheses following each quoted value represent the standard deviation error in the final digit of the quoted value. Values are from TAYLOR *et al.* (1969) and SÖDING *et al.* (1972). The symbol $E-n$ means 10^{-n} where n is an integer. The conversion factors of mass are 1 gram $= 5.609538(24)E+26$ MeV and one atomic mass unit $=$ 1 a.m.u. $= 931.4812(52)$ MeV.

[2] The Bohr magnetron $\mu_B = 9.274096(65)E-21$ erg gauss^{-1} $= 0.5788381(18)E-14$ MeV gauss^{-1}, and the nuclear magnetron $\mu_N = 5.050951(50)E-24$ erg gauss^{-1} $= 3.152526(21)E-18$ MeV gauss^{-1}.

[3] The neutron mean life of $9.35 \pm 0.14 \times 10^2$ sec corresponds to a half life of 10.61 ± 0.16 min (cf. CHRISTENSEN *et al.*, 1967).

then introduced by RUTHERFORD (1911, 1914) who proposed that an atom, which has a radius of approximately 10^{-8} cm, actually consists of a swarm of electrons surrounding a positively charged nucleus whose radius is less than 10^{-12} cm. The subsequent discovery of the proton by RUTHERFORD and CHADWICK (1921) further confirmed the speculation that the nucleus contains positively charged particles. The neutron was then discovered (CHADWICK, 1932; CURIE and JOLIOT, 1932), and HEISENBERG (1932) proposed that the atomic nucleus contains the neutral neutrons as well as the protons. At about this time, studies of cosmic rays resulted in the discovery of the positron (ANDERSON, 1932), which differs from the electron only in that its charge is positive. Although similar antiparticles for the proton and neutron were expected on theoretical grounds, they were not observed until the advent of large accelerators (CHAMBERLAIN *et al.*, 1955). The properties of these early fundamental particles are given in Table 37.

4.1.2. Symbols, Nomenclature, and Units

A nucleus is defined by the numbers:

$$\begin{aligned} &\text{Atomic number} = Z = \text{number of protons} \\ &\text{Neutron number} = N = \text{number of neutrons} \\ &\text{Mass number} = A = N+Z = \text{number of nucleons} \\ &\text{Isotopic number} = N-Z = A-2Z. \end{aligned} \tag{4-1}$$

The nuclear mass, M_{nucl}, can be calculated from the atomic mass, $M_{A,Z}$, using the relation (FERMI, 1928; THOMAS, 1927)

$$M_{\text{nucl}} = M_{A,Z} - 5.48593 \times 10^{-4} Z + 1.67475 \times 10^{-8} Z^{7/3} \text{ a.m.u.,} \tag{4-2}$$

where Z is the atomic number of the nucleus. The second term on the right-hand side of Eq. (4-2) corrects for the electron mass, and the last term represents the Fermi-Thomas binding energy of $15.6\,Z^{7/3}$ eV. The mass number, A, is the integer nearest in value to the exact mass, M, expressed in atomic mass units.

Special names are given to nuclei having the same values of some of the numbers Z, N, and A.

$$\begin{aligned} &\text{Isotope} = \text{same } Z, \text{ different } N \\ &\text{Isotone} = \text{same } N, \text{ different } Z \\ &\text{Isobar} = \text{same } A, \text{ different } N, Z \\ &\text{Isomer} = \text{same } A, \text{ same } Z. \end{aligned} \tag{4-3}$$

Nuclei are given the symbols (Z,A) for unexcited nuclei, and $(Z,A)^*$ for excited nuclei.

A reaction in which a particle, a, interacts with a nucleus, X, to produce a nucleus, Y, and a new particle, b, is designated by

$$a + X \rightarrow Y + b + Q \quad \text{or} \quad X(a,b)\,Y, \tag{4-4}$$

where Q is the energy released in the reaction. An element, B, is given the symbol

$${}_ZB_N^A \quad \text{or} \quad {}_N^AB \quad \text{or} \quad {}^AB \quad \text{or} \quad B^A\,, \tag{4-5}$$

where A is the mass number, Z is the atomic number, and N is the neutron number. Elements appearing inside the parentheses of a reaction are given the symbols:

$$p \text{ for } H^1, \quad D \text{ for } H^2, \quad T \text{ for } H^3, \quad \tau \text{ for } He^3 \quad \text{and } \alpha \text{ for } He^4. \tag{4-6}$$

A fundamental particle such as the pion, π, which has positive, +, negative, −, or neutral, 0, charge is given the symbols

$$\pi^+, \pi^-, \quad \text{or } \pi^0 . \tag{4-7}$$

An antiparticle is denoted by a raised bar. For example, the antiproton is denoted by $\bar{p}$.

Typical units used in nuclear astrophysics are:

$$\begin{aligned}
&\text{One barn} && = 10^{-24}\,\text{cm}^2 \\
&\text{One a.m.u.} && = 931.4812(52)\ \text{MeV}/c^2 \\
& && = 1.660531(11) \times 10^{-24}\ \text{gram} \\
&\text{One MeV} && = 1.6021917(70) \times 10^{-6}\ \text{erg} \\
&\text{One Fermi} && = 10^{-13}\ \text{cm} \\
&\text{Boltzmann's constant,}\ k && = 8.61708(37) \times 10^{-11}\ \text{MeV}\ {}^\circ\text{K}^{-1} \\
& && = 1.380622(59) \times 10^{-16}\ \text{erg}\ {}^\circ\text{K}^{-1} \\
&\text{Planck's constant,}\ \hbar && = 1.0545919(80) \times 10^{-27}\ \text{erg sec} \\
& && = 0.6582183(22) \times 10^{-21}\ \text{MeV sec.}
\end{aligned} \tag{4-8}$$

The values of the physical constants are from TAYLOR, PARKER, and LANGENBERG (1969), and the number in parentheses corresponds to one standard deviation uncertainty in the last digits of the quoted value.

4.1.3. Binding Energy, Mass Defect, Mass Excess, Atomic Mass, Mass Fraction, Packing Fraction, Energy Release, Magic Numbers, and Mass Laws

The difference in energy between that due to the atomic mass and that of the atom's constituents is the binding energy, E_B, of the nucleus and the electrons.

$$\begin{aligned}
E_B(A,Z) &= c^2\,[Z M_H + (A-Z) M_N - M_{AZ}] \\
&= c^2\,[Z M_P + (A-Z) M_N - M_{\text{nucl}}] \\
&= c^2 \Delta M ,
\end{aligned} \tag{4-9}$$

where the velocity of light $c = 2.997924562(11) \times 10^{10}$ cm sec^{-1}, the atomic number is Z, the mass number is A, the mass defect is ΔM, the mass of the neutral hydrogen atom is M_H, the masses of the neutron and proton are, respectively, M_N and M_P, and the masses of the neutral atom and the nucleus are, respectively, M_{AZ} and M_{nucl}. When a semi-empirical mass law is used, the binding energy is given by Eq. (4-19). The separation energy, S_N, required to remove a neutron to infinity follows from EINSTEIN's (1905, 1906, 1907) energy-mass equivalence and is given by

$$S_N = [M_{A-1,Z} + M_N - M_{A,Z}]\,c^2 . \tag{4-10}$$

The atomic mass excess, ΔM_{AZ}, is given by

$$\Delta M_{AZ} = (M_{AZ} - A) \times 1 \text{ a.m.u.} = 931.481\,(M_{AZ} - A) \text{ MeV}, \tag{4-11}$$

where the atomic mass, M_{AZ}, is given in atomic mass units, and A is the mass number. The nuclear mass, M_{nucl}, is given by

$$M_{\text{nucl}} = A + \Delta M_{AZ} - 5.48593 \times 10^{-4} Z + 1.67475 \times 10^{-8} Z^{7/3} \text{ a.m.u.}, \tag{4-12}$$

where Z is the atomic number. Values of atomic mass excesses are given in Table 38. Nuclear masses, binding energies, and separation energies may be computed from the atomic mass excesses by using Eqs. (4-9) to (4-12).

Atomic masses may be calculated using these mass excesses and Eq. (4-11). Some frequently used atomic masses are:

$$\begin{aligned} A_{\text{N}} &= 1.008665 \text{ a.m.u.} \\ A_{\text{H}} &= 1.007825 \text{ a.m.u.} \\ A_{\text{D}} &= 2.014102 \text{ a.m.u.} \\ A_{\text{T}} &= 3.016050 \text{ a.m.u.} \\ A_{\text{He}^3} &= 3.016030 \text{ a.m.u.} \\ A_{\text{He}^4} &= 4.002603 \text{ a.m.u.}, \end{aligned} \tag{4-13}$$

where the symbol A_i is used to denote the atomic mass of element, i.

In describing the abundance of a given element, i, in a gas of mass density, ρ, the mass fraction, X_i, is often used

$$X_i = \frac{A_i N_i}{\rho N_{\text{A}}}, \tag{4-14}$$

where A_i is the mass in a.m.u., the number density of element i is N_i, and Avogadro's number, $N_{\text{A}} = 6.022169(40) \times 10^{23}\,\text{mole}^{-1}$.

The packing fraction, f, is given by

$$f = (M_{AZ} - A)/A, \tag{4-15}$$

where M_{AZ} is the atomic mass of the atom with mass number, A, and atomic number, Z.

The energy release, Q, in the reaction $a + X \to b + Y + d + Q$ follows from Einstein's (1905, 1906, 1907) relation, $E = Mc^2$, between energy, E, and mass, M, and is given by

$$\begin{aligned} Q &= E_{byd} - E_{ax} \\ &= c^2 (M_a + M_x - M_b - M_y - M_d) \\ &= 931.481\,(A_a + A_x - A_b - A_y - A_d) \text{ MeV}. \end{aligned} \tag{4-16}$$

Here E_{ax} and E_{byd} are the center-of-mass kinetic energies of the incident and outgoing particles, M_i is the mass of the particle i, the velocity of light is c, and A_i is the atomic mass of the particle, i, in atomic mass units. Provided that the number of nucleons is conserved in the reaction, we may also write

$$Q = \Delta M_b + \Delta M_y + \Delta M_d - \Delta M_a - \Delta M_x, \tag{4-17}$$

where ΔM_i denotes the mass excess of i in energy units. When a nuclear reaction includes the emission of a positron, it is customary to add the annihilation energy $2mc^2 = 1.022$ MeV $= 1.637 \times 10^{-6}$ erg to the value of Q given by Eqs. (4-16) or (4-17).

Nuclear binding energies, E_B, have a narrow range of values per nucleon, 7.4 MeV $\leq E_B/A \leq 8.8$ MeV for $A > 10$ (ASTON, 1927). Nevertheless, some nuclei are extremely stable when compared with others (ELSASSER, 1933), and especially stable nuclei are those with "magic number" values of Z or N. These numbers are (MAYER, 1948)

$$2, 8, 14, 20, 28, 50, 82, \quad \text{or} \quad 126.$$

A semi-empirical formula which gives the atomic mass, M_{AZ}, for a given value of A and Z was first derived by WEIZSÄCKER (1935). His mass law is

$$M_{AZ} = M_N A - (M_N - M_H) Z - E_B(A, Z)/c^2, \tag{4-18}$$

where the neutron mass, $M_N = 1.008665$ a.m.u., the mass of the hydrogen atom, $M_H = 1.007825$ a.m.u., the atomic mass unit is 1 a.m.u. $= 931.481$ MeV, and the nuclear binding energy, $E_B(A, Z)$, is given by

$$-E_B(A, Z) = -a_1 A + a_2 A^{2/3} + a_3 (Z^2 A^{-1/3}) + 0.25 a_4 (A - 2Z)^2 A^{-1}.$$

GREEN (1954) gives numerical values for the constants $a_1 = 16.9177$ MeV, $a_2 = 19.120$ MeV, $a_3 = 0.76278$ MeV, and $a_4 = 101.777$ MeV. An additional term of $\pm 132 A^{-1}$ MeV is often added to the binding energy expression, where the $+$ and $-$ of the $\pm$ sign correspond, respectively, to the cases where $N = A - Z$ and Z are both odd or both even. Modern attempts at deriving mass laws involve extrapolating from known nuclear masses to predict the masses and binding energies of yet unmeasured nuclei. Measured atomic masses are given by WAPSTRA and GOVE (1971), and these mass values are given in Table 38. Considerable effort has also gone into estimating theoretical values for atomic and nuclear masses (cf. MYERS and SWIATECKI, 1966; GARVEY *et al.*, 1969; TRURAN *et al.*, 1970; KODAMA, 1971). Using the semi-empirical mass formula and taking shell effects into account, the nuclear binding energy, $E_B(A, Z)$, to be used in Eq. (4-18) is (MYERS and SWIATECKI, 1966)

$$\begin{aligned} -E_B(A, Z) = & -c_1 A + c_2 A^{2/3} + c_3 (Z^2 A^{-1/3}) - c_4 Z^2 A^{-1} \\ & + [4E^3/(9F^2)] - [8E^3/(27F^2)], \end{aligned} \tag{4-19}$$

where

$$c_1 = 15.677 \left[1 - 1.79 \left(\frac{N - Z}{A}\right)^2\right] \text{MeV},$$

$$c_2 = 18.56 \left[1 - 1.79 \left(\frac{N - Z}{A}\right)^2\right] \text{MeV},$$

$$c_3 = 0.717 \text{ MeV},$$

$$c_4 = 1.21129 \text{ MeV},$$

$$E = \tfrac{2}{5} c_2 A^{2/3} (1 - x) \alpha_0^2,$$

$$F = \tfrac{4}{105} c_2 A^{2/3} (1 + 2x) \alpha_0^3,$$

$$x = c_3 Z^2/(2 c_2 A),$$

Table 38. The atomic number, Z, mass number, A, measured atomic mass excess, production class, solar system abundance, and neutron capture cross sections of the elements. The measured atomic mass excesses are normalized to $C^{12}=0.000$ with one a.m.u. $=931.504\ \mathrm{MeV}/c^2$, and are from WAPSTRA and GOVE (1971). The standard error in the mass excess value is less than ten times the rightmost digit. The production class is denoted by C for explosive carbon burning. E for nuclear statistical equilibrium, H for hydrogen burning, He for helium burning, N for nova explosions, O for explosive oxygen burning, P for the proton rich nuclides, R for rapid neutron capture, S for slow neutron capture, Si for explosive silicon burning, U for cosmological nucleosynthesis, and X for cosmic ray spallation. The production class and solar system abundances are from A. G. W. CAMERON (private communication—1973; Space Sci. Rev. **15**, 121, 1973). The abundance data are normalized to $\mathrm{Si}=10^6$, and come mainly from B. MASON's *Handbook of elemental abundances in meteorites* (Gordon and Breach, New York, 1971) and from G. L. WITHBROE's "The chemical composition of the photosphere and the corona" (in the *Menzel symposium on solar physics, atomic spectra, and gaseous nebulae*. Nat. Bur. of Stands.—Wash. Pub. No. 353, 127, 1971). The neutron capture cross sections are Maxwellian averaged cross sections measured at 30 keV ($T=3.48\times10^8$ °K), and are from ALLEN, GIBBONS, and MACKLIN (1971). Cross sections with a[1] following the quoted value are semiempirical estimates, whereas values in parenthesis are uncertain experimental values

Z	Element	A	Excess (MeV)	Class	Abundance	Cross section (millibarns)
0	N	1	8.07169			
1	H				3.18×10^{10}	
		1	7.28922		3.18×10^{10}	
		2	13.13627	P	5.2×10^{5}	
		3	14.95038			
		4	25.9			
		5	33.8			
2	He				2.21×10^{9}	
		3	14.93173	H, P	$\sim 3.7 \times 10^{5}$	
		4	2.42494	U, H	2.21×10^{9}	
		5	11.39			
		6	17.5973			
		7	26.111			
		8	31.65			
3	Li				49.5	
		4	25.13			
		5	11.68			
		6	14.0875	X	3.67	
		7	14.9086	P	45.8	
		8	20.9475			
		9	24.966			
		10	35.3			
		11	43.3			
4	Be				0.81	
		6	18.375			
		7	15.7703			
		8	4.9418			
		9	11.3484	X	0.81	
		10	12.6081			
		11	20.177			
		12	25.0			
		13	35.7			

Table 38 (continued)

Z	Element	A	Excess (MeV)	Class	Abundance	Cross section (millibarns)
5	B				350.0	
		7	27.94			
		8	22.9223			
		9	12.4157			
		10	12.0523	P	68.7	
		11	8.66795	P	281.3	
		12	13.3704			
		13	16.562			
		14	24.2			
		15	29.4			
6	C				1.18×10^{7}	0.2 ± 0.4
		9	28.912			
		10	15.7027			
		11	10.6502			
		12	0.00000	He	1.17×10^{7}	
		13	3.12527	N	1.31×10^{5}	
		14	3.01995			
		15	9.8735			
		16	13.693			
		17	17.6			
7	N				3.74×10^{6}	
		11	25.5			
		12	17.344			
		13	5.3457			
		14	2.86382	H	3.63×10^{6}	
		15	0.1018	N	1.33×10^{4}	
		16	5.6835			
		17	7.871			
		18	13.274			
		19	16.4			
8	O				2.15×10^{7}	
		13	23.106			
		14	8.00859			
		15	2.8611			
		16	− 4.73668	He	2.14×10^{7}	
		17	− 0.8074	N	8,040	
		18	− 0.78250	N, He	4.38×10^{4}	
		19	3.3323			
		20	3.800			
		21	10.7			
9	F				2,450	5.6 ± 0.4
		15	17.7			
		16	10.693			
		17	1.9518			
		18	0.8728			
		19	− 1.4861	P	2,450	
		20	− 0.0157			
		21	− 0.046			
		22	2.828			

Table 38 (continued)

Z	Element	A	Excess (MeV)	Class	Abundance	Cross section (millibarns)
10	Ne				3.44×10^6	
		17	16.48			
		18	5.319			
		19	1.7521			
		20	−7.0417	C	3.06×10^6	
		21	−5.7312	He, N	9,290	
		22	−8.0251	He, N	3.73×10^5	
		23	−5.1500			
		24	−5.948			
11	Na				6.0×10^4	
		19	12.98			
		20	6.84			
		21	−2.183			
		22	−5.1829			
		23	−9.5290	C	6.0×10^4	2.7 ± 0.4
		24	−8.4167			
		25	−9.356			
		26	−7.51			
		27	−6.6			
12	Mg				1.061×10^6	4.0 ± 1.0
		20	17.5			
		21	10.911			
		22	−0.384			
		23	−5.4724			
		24	−13.9313	C	8.35×10^5	
		25	−13.1915	C	1.07×10^5	
		26	−16.2134	C	1.19×10^5	
		27	−14.5847			
		28	−15.0170			
13	Al				8.5×10^4	
		22	18.0			
		23	6.77			
		24	−0.049			
		25	−8.9123			
		26	−12.2088			
		27	−17.1950	C	8.5×10^4	4.6 ± 0.8
		28	−16.8488			
		29	−18.213			
		30	−15.89			
14	Si				1.00×10^6	3.8 ± 1.0
		24	10.8			
		25	3.82			
		26	−7.147			
		27	−12.3854			
		28	−21.4911	O, Si	9.22×10^5	3.8 ± 1.0
		29	−21.8933	O	4.70×10^4	10.4[1]
		30	−24.4313	O	3.09×10^4	(1.9)
		31	−22.9479			
		32	−24.091			

Table 38 (continued)

Z	Element	A	Excess (MeV)	Class	Abundance	Cross section (millibarns)
15	P				9,600	
		27	0.2			
		28	− 7.154			
		29	−16.950			
		30	−20.2039			
		31	−24.4396	O	9,600	(7)
		32	−24.3042			
		33	−26.3370			
		34	−24.83			
16	S				5.0×10^5	3.0 ± 0.6
		29	− 3.2			
		30	−14.065			
		31	−18.998			
		32	−26.0143	O, Si	4.75×10^5	3.0 ± 0.6
		33	−26.5860	O, Si	3,800	
		34	−29.9292	O, Si	2.11×10^4	
		35	−28.8456			
		36	−30.6659	C, R	68	
		37	−26.907			
		38	−26.863			
17	Cl				5,700	11 ± 4
		31	− 7.2			
		32	−13.263			
		33	−21.0024			
		34	−24.4384			
		35	−29.0130	O, Si	4,310	13.5 ± 5
		36	−29.5218			
		37	−31.7615	O, Si	1,390	(3)
		38	−29.800			
		39	−29.802			
		40	−27.5			
18	Ar				1.172×10^5	
		33	− 9.4			
		34	−18.395			
		35	−23.0494			
		36	−30.2305	O, Si	9.87×10^4	
		37	−30.9474			
		38	−34.7144	O, Si	1.85×10^4	
		39	−33.240			
		40	−35.0392	C, R	~20?	(4.5)
		41	−33.0661			
		42	−34.42			
19	K				4,200	16 ± 2
		35	−11.2			
		36	−17.317			
		37	−24.7984			
		38	−28.792			
		39	−33.8053	O, Si	3,910	16 ± 2

Table 38 (continued)

Z	Element	A	Excess (MeV)	Class	Abundance	Cross section (millibarns)
19	K	40	−33.5341	O, R	5.76	
		41	−35.5583	O, Si	289	22 ±3
		42	−35.0214			
		43	−36.582			
		44	−35.801			
		45	−36.611			
		46	−35.426			
		47	−35.704			
20	Ca				7.21×10^4	10 ±1
		37	−13.23			
		38	−22.023			
		39	−27.283			
		40	−34.8457	O, Si	6.99×10^4	
		41	−35.1371			
		42	−38.5381	O, Si	461	
		43	−38.3990	C	105	
		44	−41.4636	O, Si	1,490	
		45	−40.8063			
		46	−43.138	C, R	2.38	
		47	−42.343			
		48	−44.222	C, R	133	
		49	−41.292			
		50	−39.578			
21	Sc				35	
		40	−20.521			
		41	−28.641			
		42	−32.1070			
		43	−36.1790			
		44	−37.814			
		45	−41.0631	C	35	44 ±6
		46	−41.7584			
		47	−44.3289			
		48	−44.495			
		49	−46.552			
		50	−44.545			
		51	−43.227			
22	Ti				2,775	20 ±
		42	−25.121			
		43	−29.320			
		44	−37.548			
		45	−39.0007			
		46	−44.1258	Si, E	220	34[1]
		47	−44.9292	C	202	92[1]
		48	−48.4856	Si, E	2,050	12[1]
		49	−48.5573	C	153	20[1]
		50	−51.4336	C, E	148	(2)
		51	−49.739			
		52	−49.470			

Table 38 (continued)

Z	Element	A	Excess (MeV)	Class	Abundance	Cross section (millibarns)
23	V				262	25 ±8
		45	−31.9			
		46	−37.0714			
		47	−42.0048			
		48	−44.4702			
		49	−47.9561			
		50	−49.2167	C	0.63	
		51	−52.1974	E	261	25 ±8
		52	−51.4369			
		53	−51.861			
		54	−49.93			
24	Cr				1.27×10^4	6.2±2
		47	−34.5			
		48	−42.816			
		49	−45.388			
		50	−50.2557	Si, E	547	31 ±4
		51	−51.4460			
		52	−55.4150	Si, E	1.06×10^4	3.8±1.0
		53	−55.2838	Si, E	1,210	40 ±5
		54	−56.9323	E	302	23[1]
		55	−55.121			
		56	−55.266			
25	Mn				9,300	
		49	−37.72			
		50	−42.6246			
		51	−48.240			
		52	−50.705			
		53	−54.6865			
		54	−55.557			
		55	−57.7100	Si, E	9,300	50 ±2
		56	−56.9087			
		57	−57.62			
		58	−56.06			
26	Fe				8.3×10^5	18 ±8
		52	−48.333			
		53	−50.942			
		54	−56.2517	Si, E	4.83×10^4	34 ±10
		55	−57.4784			
		56	−60.6094	Si, E	7.61×10^5	13.5±2.0
		57	−60.1838	E	1.82×10^4	30 ±5
		58	−62.1551	E	2,740	4.5[1]
		59	−60.6700			
		60	−61.435			
		61	−59.03			
27	Co				2,210	
		54	−48.002			
		55	−54.0124			
		56	−56.0412			

Table 38 (continued)

Z	Element	A	Excess (MeV)	Class	Abundance	Cross section (millibarns)
27	Co	57	−59.3470			
		58	−59.8472			
		59	−62.2357	E	2,210	35 ±10
		60	−61.6556			
		61	−62.920			
		62	−61.530			
		63	−61.863			
		64	−60.1			
28	Ni				4.80×10^4	12.4±2
		56	−53.908			
		57	−56.104			
		58	−60.2350	E	3.26×10^4	17 ±3
		59	−61.1626			
		60	−64.4792	E	1.26×10^4	7.5±2
		61	−64.2270	E	571	(30)
		62	−66.7519	C	1,760	6[1]
		63	−65.5215			
		64	−67.1093	C	518	(10)
		65	−65.133			
		66	−66.060			
		67	−63.20			
29	Cu				540	47±7
		58	−51.668			
		59	−56.363			
		60	−58.352			
		61	−61.9818			
		62	−62.805			
		63	−65.5874	E	373	49±14 (92)
		64	−65.4318			
		65	−67.2648	C	167	42±7 (18)
		66	−66.2598			
		67	−67.302			
		68	−65.42			
		69	−65.94			
30	Zn				1,244	41±10
		60	−54.193			
		61	−56.58			
		62	−61.115			
		63	−62.222			
		64	−66.0064	E	608	50[1]
		65	−65.9141			
		66	−68.8945	E	346	40[1]
		67	−67.8767	C	51.1	160[1]
		68	−70.0043	C	231	23±3
		69	−68.4162			
		70	−69.5597	C	7.71	16[1]
		71	−67.332			
		72	−68.131			

Table 38 (continued)

Z	Element	A	Excess (MeV)	Class	Abundance	Cross section (millibarns)
31	Ga				48	115 ± 20
		63	−56.7			
		64	−58.934			
		65	−62.655			
		66	−63.719			
		67	−66.876			
		68	−67.085			
		69	−69.3230	E	29.0	130 ± 30
		70	−68.9060			
		71	−70.1381	C	19.0	120 ± 30 (60)
		72	−68.5876			
		73	−69.74			
		74	−67.92			
32	Ge				115	74 ± 7
		65	−56.4			
		66	−61.617			
		67	−62.45			
		68	−66.698			
		69	−67.0975			
		70	−70.5595	E	23.6	84[1]
		71	−69.9030			
		72	−72.5807	E	31.5	65[1], 40
		73	−71.2932	C	8.92	270[1]
		74	−73.4224	E	42.0	34 ± 20
		75	−71.841			
		76	−73.2123	C	8.92	53 ± 10
		77	−71.16			
		78	−71.78			
		79	−69.39			
33	As				6.6	
		69	−63.13			
		70	−64.338			
		71	−67.894			
		72	−68.230			
		73	−70.954			
		74	−70.8587			
		75	−73.0297	S, R	6.6	490 ± 100
		76	−72.2862			
		77	−73.917			
		78	−72.76			
		79	−73.69			
		80	−71.76			
		81	−72.59			
34	Se				67.2	94 ± 8
		71	−62.9			
		72	−67.6			
		73	−68.214			
		74	−72.213	P	0.58	160[1]
		75	−72.1649			

Table 38 (continued)

Z	Element	A	Excess (MeV)	Class	Abundance	Cross section (millibarns)
34	Se	76	−75.2546	S	6.06	100[1]
		77	−74.6014	S, R	5.09	340[1]
		78	−77.0268	S, R	15.8	60[1]
		79	−75.933			
		80	−77.7570	S, R	33.5	20 ± 12
		81	−76.387			
		82	−77.587	R	6.18	36 ± 15
		83	−75.440			
		84	−75.92			
35	Br				13.5	600 ± 60
		73	−63.5			
		74	−65.2			
		75	−69.155			
		76	−70.2			
		77	−73.2369			
		78	−73.453			
		79	−76.0741	S, R	6.82	600 ± 150
		80	−75.8853			
		81	−77.974	S, R	6.68	460 ± 80
		82	−77.503			
		83	−79.018			
		84	−77.73			
		85	−78.67			
		86	−76.0			
		87	−74.2			
36	Kr				46.8	
		74	−62.1			
		75	−64.1			
		76	−69.2			
		77	−70.237			
		78	−74.147	P	0.166	250[1] 500
		79	−74.443			
		80	−77.896	S, P	1.06	140[1] 280
		81	−77.68			
		82	−80.591	S	5.41	80[1] 200
		83	−79.987	S, R	5.41	225[1] 670
		84	−82.4332	S, R	26.6	28[1] 60
		85	−81.4726			
		86	−83.2613	R	8.13	9[1] 20
		87	−80.700			
		88	−79.70			
		89	−76.560			
		90	−74.89			
		91	−71.5			
37	Rb				5.88	160 ± 20
		79	−70.92			
		80	−72.1			
		81	−75.42			
		82	−76.194			
		83	−78.949			

Table 38 (continued)

Z	Element	A	Excess (MeV)	Class	Abundance	Cross section (millibarns)
37	Rb	84	−79.753			
		85	−82.1596	S, R	4.16	215±20
		86	−82.7383			
		87	−84.5926	R	1.72	24±4
		88	−82.604			
		89	−81.710			
		90	−79.30			
		91	−78.00			
		92	−75.0			
		93	−73.1			
38	Sr				26.9	120±40
		81	−71.6			
		82	−75.6			
		83	−76.699			
		84	−80.6398	P	0.151	330[1]
		85	−81.096			
		86	−84.5094	S	2.65	74±7
		87	−84.8661	S	1.77	109±9
		88	−87.9076	S, R	22.2	6.9±2.5
		89	−86.196			
		90	−85.9279			
		91	−83.684			
		92	−82.92			
		93	−79.95			
		94	−78.74			
		95	−75.5			
39	Y				4.8	
		83	−72.2			
		84	−73.690			
		85	−77.836			
		86	−79.236			
		87	−82.984			
		88	−84.289			
		89	−87.6856	S, R	4.8	21±4
		90	−86.4739			
		91	−86.349			
		92	−84.834			
		93	−84.254			
		94	−82.26			
		95	−81.236			
		96	−78.6			
		97	−76.8			
40	Zr				28	25±10
		85	−72.9			
		86	−77.9			
		87	−79.484			
		88	−83.61			
		89	−84.851			

Table 38 (continued)

Z	Element	A	Excess (MeV)	Class	Abundance	Cross section (millibarns)
40	Zr	90	−88.7626	S, R	14.4	12 ± 2
		91	−87.8935	S, R	3.14	68 ± 8
		92	−88.4569	S, R	4.79	34 ± 6
		93	−87.1437			
		94	−87.2631	S, R	4.87	20 ± 2
		95	−85.666			
		96	−85.426	R	0.784	30 ± 12
		97	−82.933			
		98	−81.273			
		99	−78.4			
		100	−77.1			
		101	−72.9			
41	Nb				1.4	
		87	−74.3			
		88	−76.4			
		89	−80.98			
		90	−82.652			
		91	−86.632			
		92	−86.453			
		93	−87.2071	S, R	1.4	285 ± 30
		94	−86.3643			
		95	−86.7885			
		96	−85.609			
		97	−85.605			
		98	−83.51			
		99	−82.9			
		100	−80.2			
		101	−79.4			
		102	−76.2			
42	Mo				4.0	160 ± 20
		88	−71.8			
		89	−75.01			
		90	−80.165			
		91	−82.188			
		92	−86.8084	P	0.634	50[1]
		93	−86.809			
		94	−88.4099	P	0.362	80[1]
		95	−87.7133	S, R	0.629	430 ± 50
		96	−88.7959	S	0.661	90 ± 10
		97	−87.5402	S, R	0.378	350 ± 50
		98	−88.1109	S, R	0.951	150 ± 40, 110
		99	−85.956			
		100	−86.1851	R	0.385	100 ± 40
		101	−83.504			
		102	−83.6			
		103	−80.5			
		104	−80.2			

Table 38 (continued)

Z	Element	A	Excess (MeV)	Class	Abundance	Cross section (millibarns)
43	Tc					
		91	−76.6			
		92	−78.86			
		93	−83.623			
		94	−84.150			
		95	−86.012			
		96	−85.86			
		97	−87.195			
		98	−86.52			
		99	−87.328			800[1]
		100	−85.85			
		101	−86.325			
		102	−84.6			
		103	−84.90			
		104	−82.8			
		105	−82.53			
		106	−79.8			
44	Ru				1.9	(550)
		93	−77.4			
		94	−82.569			
		95	−83.450			
		96	−86.073	P	0.105	270 ± 60
		97	−86.04			
		98	−88.223	P	0.0355	300[1]
		99	−87.6202	S, R	0.242	1,240[1]
		100	−89.2219	S	0.240	290[1]
		101	−87.9557	S, R	0.324	1,120[1]
		102	−89.1002	S, R	0.601	330 ± 50
		103	−87.253			
		104	−88.094	R	0.353	120 ± 60
		105	−85.930			
		106	−86.323			
		107	−83.71			
		108	−83.7			
45	Rh				0.4	
		95	−78.4			
		96	−79.630			
		97	−82.55			
		98	−83.166			
		99	−85.568			
		100	−85.592			
		101	−87.402			
		102	−86.778			
		103	−88.016	S, R	0.4	900 ± 100
		104	−86.944			
		105	−87.847			
		106	−86.362			
		107	−86.86			
		108	−85.0			
		109	−85.1			
		110	−82.94			

Table 38 (continued)

Z	Element	A	Excess (MeV)	Class	Abundance	Cross section (millibarns)
46	Pd				1.3	440±40
		97	−77.7			
		98	−81.4			
		99	−82.163			
		100	−85.2			
		101	−85.412			
		102	−87.927	P	0.0125	320[1]
		103	−87.463			
		104	−89.411	S	0.143	270[1]
		105	−88.413	S, R	0.289	1,130[1]
		106	−89.902	S, R	0.355	230[1]
		107	−88.373			
		108	−89.526	S, R	0.347	200±60
		109	−87.606			
		110	−88.340	R	0.154	170±70
		111	−86.02			
		112	−86.28			
47	Ag				0.45	920±100
		99	−76.13			
		100	−77.9			
		101	−81.0			
		102	−82.367			
		103	−84.78			
		104	−85.311			
		105	−87.078			
		106	−86.928			
		107	−88.408	S, R	0.231	1,150±150
		108	−87.605			
		109	−88.7215	S, R	0.219	620±50
		110	−87.4555			
		111	−88.224			
		112	−86.58			
		113	−87.035			
		114	−85.0			
		115	−84.91			
		116	−82.4			
48	Cd				1.48	340±50
		101	−75.5			
		102	−79.5			
		103	−80.4			
		104	−84.0			
		105	−84.28			
		106	−87.1302	P	0.0180	210[1]
		107	−86.991			
		108	−89.2480	P	0.0130	210[1]
		109	−88.539			
		110	−90.3464	S	0.124	210[1]
		111	−89.2516	S, R	0.189	840[1]
		112	−90.5769	S, R	0.356	210[1]
		113	−89.0449	S, R	0.181	840[1]

Table 38 (continued)

Z	Element	A	Excess (MeV)	Class	Abundance	Cross section (millibarns)
48	Cd	114	−90.0142	S, R	0.427	200±40
		115	−88.090			
		116	−88.7150	R	0.112	220±40
		117	−86.408			
		118	−86.704			
		119	−84.21			
49	In				0.189	760±80
		102	−69.7			
		103	−73.8			
		104	−75.5			
		105	−79.2			
		106	−80.390			
		107	−83.50			
		108	−84.10			
		109	−86.520			
		110	−86.42			
		111	−88.426			
		112	−87.989			
		113	−89.342	P, S	0.008	220±70
		114	−88.584			
		115	−89.541	S, R	0.181	800±100
		116	−88.248			
		117	−88.929			
		118	−87.45			
		119	−87.714			
		120	−85.5			
		121	−85.82			
		122	−83.2			
		123	−83.42			
		124	−80.8			
50	Sn				3.6	95±15
		103	−65.9			
		104	−70.9			
		105	−72.6			
		106	−77.1			
		107	−78.2			
		108	−82.0			
		109	−82.7			
		110	−85.824			
		111	−85.918			
		112	−88.648	P	0.0346	180[1]
		113	−88.317			
		114	−90.565	P	0.0238	130[1]
		115	−90.027	P, S	0.0126	550[1]
		116	−91.5218	S	0.515	100±15
		117	−90.3926	S, R	0.274	420±30
		118	−91.6483	S, R	0.865	63±5
		119	−90.0616	S, R	0.309	260±40
		120	−91.0943	S, R	1.18	50±15
		121	−89.2027			
		122	−89.9356	R	0.170	23±5 (165)

Table 38 (continued)

Z	Element	A	Excess (MeV)	Class	Abundance	Cross section (millibarns)
50	Sn	123	−87.809			
		124	−88.229	R	0.214	23±4 (180)
		125	−85.890			
		126	−86.013			
		127	−83.5			
		128	−83.40			
51	Sb				0.316	490±50
		105	−63.2			
		106	−65.7			
		107	−70.1			
		108	−72.2			
		109	−76.0			
		110	−77.6			
		111	−81.0			
		112	−81.85			
		113	−84.419			
		114	−84.87			
		115	−86.997			
		116	−87.02			
		117	−88.64			
		118	−87.953			
		119	−89.483			
		120	−88.414			
		121	−89.5899	S, R	0.181	740±100
		122	−88.3256			
		123	−89.2191	S, R	0.135	440±50
		124	−87.6142			
		125	−88.262			
		126	−86.33			
		127	−86.708			
		128	−84.70			
		129	−84.591			
		130	−82.3			
		131	−82.1			
		132	−79.6			
		133	−79.0			
52	Te				6.42	97±9 (204)
		107	−60.0			
		108	−65.3			
		109	−67.4			
		110	−72.1			
		111	−73.6			
		112	−77.7			
		113	−78.5			
		114	−82.2			
		115	−82.46			
		116	−85.46			
		117	−85.15			
		118	−87.7			
		119	−87.189			

Table 38 (continued)

Z	Element	A	Excess (MeV)	Class	Abundance	Cross section (millibarns)
52	Te	120	−89.402	P	0.0057	400[1]
		121	−88.6			
		122	−90.3038	S	0.158	270 ± 30
		123	−89.1620	S	0.056	820 ± 30
		124	−90.5141	S	0.296	150 ± 20
		125	−89.0273	S, R	0.449	430 ± 30
		126	−90.0649	S, R	1.20	82 ± 8
		127	−88.289			
		128	−88.9889	R	2.04	32.5 ± 5
		129	−87.004			
		130	−87.3454	R	2.21	13.5 ± 2.0
		131	−85.191			
		132	−85.193			
		133	−82.90			
		134	−82.6			
		135	−77.8			
53	I				1.09	
		113	−71.3			
		114	−73.5			
		115	−77.0			
		116	−78.2			
		117	−80.84			
		118	−81.6			
		119	−84.0			
		120	−84.1			
		121	−86.2			
		122	−86.16			
		123	−87.96			
		124	−87.354			
		125	−88.8793			
		126	−87.914			
		127	−88.9814	S, R	1.09	760 ± 50
		128	−97.7351			
		129	−88.503			
		130	−86.888			
		131	−87.4432			
		132	−85.698			
		133	−85.86			
		134	−83.97			
		135	−83.776			
		136	−79.42			
		137	−76.8			
54	Xe				5.38	
		115	−69.5			
		116	−73.7			
		117	−74.8			
		118	−78.2			
		119	−79.0			
		120	−81.9			

Table 38 (continued)

Z	Element	A	Excess (MeV)	Class	Abundance	Cross section (millibarns)
54	Xe	121	−82.4			
		122	−85.1			
		123	−85.29			
		124	−87.45	P	0.00678	1,200[1]
		125	−87.14			
		126	−89.165	P	0.00619	800[1]
		127	−88.317			
		128	−89.8601	S	0.117	300[1]
		129	−88.694	S, R	1.48	760[1]
		130	−89.8801	S	0.229	100[1]
		131	−88.4140	S, R	1.15	250[1]
		132	−89.2784	S, R	1.40	36[1]
		133	−87.660			
		134	−88.123	R	0.547	13[1]
		135	−86.502			
		136	−86.423	R	0.451	5[1]
		137	−82.213			
		138	−80.1			
		139	−75.98			
		140	−73.2			
55	Cs				0.387	
		123	−81.1			
		124	−81.6			
		125	−84.07			
		126	−84.2			
		127	−86.227			
		128	−85.953			
		129	−87.6			
		130	−86.857			
		131	−88.059			
		132	−87.179			
		133	−88.087	S, R	0.387	700 ± 40
		134	−86.906			
		135	−87.659			
		136	−86.356			
		137	−86.561			
		138	−82.9			
		139	−80.78			
		140	−77.54			
		141	−74.9			
		142	−71.1			
56	Ba				4.8	61 ± 5
		125	−79.6			
		126	−82.4			
		127	−82.7			
		128	−85.2			
		129	−85.2			
		130	−87.297	P	0.00485	2,000[1]
		131	−86.719			
		132	−88.451	P	0.00466	650[1]

Table 38 (continued)

Z	Element	A	Excess (MeV)	Class	Abundance	Cross section (millibarns)
56	Ba	133	−87.572			
		134	−88.965	S	0.116	155[1]
		135	−87.868	S, R	0.316	315[1]
		136	−88.904	S	0.375	37[1]
		137	−87.734	S, R	0.543	76[1]
		138	−88.274	S, R	3.44	8 ± 2, 5
		139	−84.926			
		140	−83.241			
		141	−79.97			
		142	−77.77			
		143	−74.0			
		144	−71.8			
57	La				0.445	44 ± 4
		127	−77.7			
		128	−78.4			
		129	−81.2			
		130	−81.6			
		131	−83.76			
		132	−83.74			
		133	−85.7			
		134	−85.255			
		135	−86.83			
		136	−86.03			
		137	−87.2			
		138	−86.480	P	0.00041	
		139	−87.186	S, R	0.445	44 ± 4, 48
		140	−84.276			
		141	−82.969			
		142	−79.970			
		143	−78.21			
		144	−74.9			
		145	−72.9			
		146	−69.4			
58	Ce				1.18	35 ± 5
		131	−79.5			
		132	−82.3			
		133	−82.4			
		134	−84.7			
		135	−84.5			
		136	−86.462	P	0.00228	100[1]
		137	−86.0			
		138	−87.536	P	0.00295	30[1]
		139	−86.911			
		140	−88.042	S, R	1.04	3 ± 3, 12
		141	−85.399			
		142	−84.487	R	0.131	360 ± 60 (450)
		143	−81.593			
		144	−80.403			

Table 38 (continued)

Z	Element	A	Excess (MeV)	Class	Abundance	Cross section (millibarns)
58	Ce	145	−77.11			
		146	−75.74			
		147	−72.2			
		148	−70.7			
59	Pr				0.149	
		134	−78.6			
		135	−80.9			
		136	−81.3			
		137	−83.3			
		138	−83.099			
		139	−84.799			
		140	−84.654			
		141	−85.980	S, R	0.149	110 ± 20
		142	−83.752			
		143	−83.038			
		144	−80.719			
		145	−79.599			
		146	−76.82			
		147	−75.43			
		148	−72.5			
		149	−71.38			
		150	−68.7			
60	Nd				0.78	
		136	−78.8			
		137	−79.3			
		138	−81.8			
		139	−82.0			
		140	−84.18			
		141	−84.175			
		142	−85.916	S	0.211	70-
		143	−83.970	S, R	0.0949	425[1]
		144	−83.716	S, R	0.186	100[1]
		145	−81.404	S, R	0.0647	600[1]
		146	−80.898	S, R	0.134	150[1]
		147	−78.129			
		148	−77.381	R	0.0447	210 ± 80
		149	−74.377			
		150	−73.662	R	0.0438	240 ± 150
		151	−70.899			
		152	−70.126			
61	Pm					
		139	−77.5			
		140	−78.3			
		141	−80.45			
		142	−81.10			
		143	−82.901			
		144	−81.34			
		145	−81.234			
		146	−79.421			

Table 38 (continued)

Z	Element	A	Excess (MeV)	Class	Abundance	Cross section (millibarns)
61	Pm	147	−79.023			2,000[1]
		148	−76.852			
		149	−76.046			
		150	−73.53			
		151	−73.365			
		152	−71.35			
		153	−70.74			
		154	−68.4			
62	Sm				0.226	920 ± 50
		141	−76.1			
		142	−79.05			
		143	−79.422			
		144	−81.904	P	0.00698	120 ± 55
		145	−80.596			
		146	−80.947			
		147	−79.248	S, R	0.0349	1,150 ± 190
		148	−79.317	S	0.0254	260 ± 50
		149	−77.118	S, R	0.0313	1,620 ± 280
		150	−77.033	S	0.0168	370 ± 70
		151	−74.553			
		152	−74.749	R	0.0604	450 ± 50
		153	−72.544			
		154	−72.451	R	0.0513	380 ± 60
		155	−70.193			
		156	−69.359			
63	Eu				0.085	3,350 ± 150
		143	−74.42			
		144	−75.577			
		145	−77.876			
		146	−77.075			
		147	−77.486			
		148	−76.217			
		149	−76.4			
		150	−74.719			
		151	−74.629	S, R	0.0406	3,600 ± 500
		152	−72.863			
		153	−73.347	S, R	0.0444	2,700 ± 300
		154	−71.713			
		155	−71.818			
		156	−70.072			
		157	−69.461			
		158	−67.25			
		159	−65.92			
		160	−63.5			
64	Gd				0.297	940 ± 50
		144	−71.9			
		145	−72.9			
		146	−75.9			
		147	−75.158			

Table 38 (continued)

Z	Element	A	Excess (MeV)	Class	Abundance	Cross section (millibarns)
64	Gd	148	−76.207			
		149	−75.072			
		150	−75.728			
		151	−74.165			
		152	−74.691	P	0.000594	500[1]
		153	−73.106			
		154	−73.691	S	0.00639	520[1]
		155	−72.065	S, R	0.0437	2,280[1]
		156	−72.524	S, R	0.0608	470[1]
		157	−70.821	S, R	0.0466	2,070[1]
		158	−70.680	S, R	0.0739	540 ± 70
		159	−68.553			
		160	−67.934	R	0.0650	100 ± 30
		161	−65.494			
		162	−64.29			
65	Tb				0.055	
		146	−67.8			
		147	−70.6			
		148	−70.59			
		149	−71.375			
		150	−71.060			
		151	−71.557			
		152	−70.871			
		153	−71.3			
		154	−70.3			
		155	−71.220			
		156	−70.2			
		157	−70.757			
		158	−69.440			
		159	−69.503	S, R	0.055	2,200 ± 200
		160	−67.813			
		161	−67.445			
		162	−65.69			
		163	−64.67			
		164	−62.59			
66	Dy				0.36	730 ± 40
		148	−67.8			
		149	−67.5			
		150	−69.1			
		151	−68.552			
		152	−70.057			
		153	−69.090			
		154	−70.356			
		155	−69.121			
		156	−70.491	P	0.000189	870[1]
		157	−69.394			
		158	−70.384	P	0.000325	770[1]
		159	−69.138			
		160	−69.648	S	0.00826	650[1]

Table 38 (continued)

Z	Element	A	Excess (MeV)	Class	Abundance	Cross section (millibarns)
66	Dy	161	−68.027	S, R	0.0680	2,800 ± 300
		162	−68.151	S, R	0.0919	470 ± 50
		163	−66.351	S, R	0.0899	1,600 ± 300
		164	−65.934	S, R	0.101	180 ± 40
		165	−63.577			
		166	−62.563			
67	Ho				0.079	
		149	−61.6			
		150	−62.0			
		151	−63.5			
		152	−63.67			
		153	−64.832			
		154	−64.598			
		155	−65.8			
		156	−65.4			
		157	−66.9			
		158	−66.407			
		159	−67.4			
		160	−66.728			
		161	−67.21			
		162	−65.981			
		163	−66.342			
		164	−64.955			
		165	−64.873	S, R	0.079	1,250 ± 150 (2,000)
		166	−63.044			
		167	−62.298			
		168	−60.20			
		169	−58.75			
		170	−56.39			
68	Er				0.255	750 ± 50
		150	−57.9			
		151	−58.2			
		152	−60.4			
		153	−60.2			
		154	−62.4			
		155	−62.01			
		156	−63.7			
		157	−63.0			
		158	−64.9			
		159	−64.3			
		160	−65.9			
		161	−65.161			
		162	−66.299	P	0.000306	900[1]
		163	−65.134			
		164	−65.918	P, S	0.00351	750[1]
		165	−64.501			
		166	−64.904	S, R	0.0752	560[1]
		167	−63.268	S, R	0.0516	2,000[1]
		168	−62.968	S, R	0.0609	400[1]
		169	−60.899			

Table 38 (continued)

Z	Element	A	Excess (MeV)	Class	Abundance	Cross section (millibarns)
68	Er	170	−60.091	R	0.0335	250 ± 30
		171	−57.700			
		172	−56.480			
		173	−53.42			
69	Tm				0.034	
		151	−50.9			
		152	−51.9			
		153	−53.9			
		154	−54.5			
		155	−56.3			
		156	−56.7			
		157	−58.2			
		158	−58.3			
		159	−60.1			
		160	−60.1			
		161	−61.64			
		162	−61.60			
		163	−62.717			
		164	−61.956			
		165	−62.936			
		166	−61.869			
		167	−62.521			
		168	−61.27			
		169	−61.251	S, R	0.034	1,500 ± 200
		170	−59.773			
		171	−59.190			
		172	−57.369			
		173	−56.215			
		174	−53.87			
		175	−52.28			
		176	−49.34			
70	Yb				0.216	600 ± 50
		153	−47.3			
		154	−50.0			
		155	−50.4			
		156	−53.1			
		157	−53.2			
		158	−55.3			
		159	−55.1			
		160	−57.3			
		161	−57.1			
		162	−59.3			
		163	−59.0			
		164	−60.9			
		165	−60.184			
		166	−61.609			
		167	−60.566			
		168	−61.549	P	0.000292	700[1]
		169	−60.344			
		170	−60.741	S	0.00654	510[1]

Table 38 (continued)

Z	Element	A	Excess (MeV)	Class	Abundance	Cross section (millibarns)
70	Yb	171	−59.287	S, R	0.0309	1,320[1]
		172	−59.239	S, R	0.0471	380[1]
		173	−57.535	S, R	0.0348	990[1]
		174	−56.933	S, R	0.0688	275[1]
		175	−54.681			
		176	−53.485	R	0.0275	200 ± 50
		177	−50.975			
		178	−49.5			
71	Lu				0.036	1,400 ± 300 (3,700)
		155	−42.7			
		156	−43.9			
		157	−46.3			
		158	−47.1			
		159	−48.9			
		160	−49.5			
		161	−51.6			
		162	−52.2			
		163	−54.2			
		164	−54.5			
		165	−56.2			
		166	−56.1			
		167	−57.50			
		168	−57.19			
		169	−58.074			
		170	−57.301			
		171	−57.9			
		172	−56.7			
		173	−56.845			
		174	−55.562			
		175	−55.149	S, R	0.0351	1,460 ± 110
		176	−53.370	S	0.00108	2,250 ± 200
		177	−52.371			
		178	−50.17			
		179	−49.10			
		180	−46.47			
72	Hf				0.21	600 ± 50
		157	−39.0			
		158	−42.2			
		159	−42.7			
		160	−45.2			
		161	−45.5			
		162	−48.2			
		163	−48.6			
		164	−51.1			
		165	−51.1			
		166	−53.4			
		167	−53.2			
		168	−55.2			
		169	−54.70			
		170	−56.1			

Table 38 (continued)

Z	Element	A	Excess (MeV)	Class	Abundance	Cross section (millibarns)
72	Hf	171	−55.3			
		172	−56.3			
		173	−55.2			
		174	−55.760	P	0.00038	800[1]
		175	−54.542			
		176	−54.559	S	0.0109	640 ± 160
		177	−52.868	S, R	0.0389	110[1]
		178	−52.422	S, R	0.0570	370[1]
		179	−50.450	S, R	0.0289	960[1]
		180	−49.766	S, R	0.0740	290 ± 80
		181	−47.389			
		182	−45.90			
		183	−43.219			
73	Ta				0.021	800 ± 80
		163	−41.5			
		164	−42.5			
		165	−44.9			
		166	−45.6			
		167	−47.8			
		168	−48.2			
		169	−50.1			
		170	−50.1			
		171	−51.6			
		172	−51.3			
		173	−52.3			
		174	−51.8			
		175	−52.3			
		176	−51.46			
		177	−51.710			
		178	−50.51			
		179	−50.331			
		180	−48.840	P	0.00000258	
		181	−48.412	S, R	0.0210	800 ± 80
		182	−46.403			
		183	−45.259			
		184	−42.637			
		185	−41.38			
		186	−38.58			
74	W				0.16	290 ± 30
		165	−37.8			
		166	−40.6			
		167	−41.4			
		168	−44.0			
		169	−44.3			
		170	−46.6			
		171	−46.6			
		172	−48.5			
		173	−48.3			
		174	−49.9			

Table 38 (continued)

Z	Element	A	Excess (MeV)	Class	Abundance	Cross section (millibarns)
74	W	175	−49.3			
		176	−50.5			
		177	−49.7			
		178	−50.42			
		179	−49.2			
		180	−49.65	P	0.000216	270[1]
		181	−48.225			
		182	−48.208	S, R	0.0422	260 ± 30
		183	−46.327	S, R	0.0230	550 ± 50
		184	−45.667	S, R	0.0490	180 ± 20
		185	−43.345			
		186	−42.475	R	0.0454	220 ± 20
		187	−39.870			
		188	−38.634			
		189	−35.44			
75	Re				0.053	1,420 ± 100 (950)
		167	−33.7			
		168	−35.0			
		169	−37.5			
		170	−38.4			
		171	−40.7			
		172	−41.2			
		173	−43.2			
		174	−43.5			
		175	−45.0			
		176	−45.0			
		177	−46.1			
		178	−45.76			
		179	−46.5			
		180	−45.86			
		181	−46.4			
		182	−45.348			
		183	−45.771			
		184	−44.1			
		185	−43.774	S, R	0.0185	1,530 ± 200 (2,200)
		186	−41.881			
		187	−41.181	S, R	0.0341	1,570 ± 100 (780)
		188	−38.983			
		189	−37.942			
		190	−35.49			
		191	−34.5			
76	Os				0.75	300 ± 40
		169	−29.6			
		170	−32.6			
		171	−33.5			
		172	−36.1			
		173	−36.7			
		174	−39.1			
		175	−39.3			

Table 38 (continued)

Z	Element	A	Excess (MeV)	Class	Abundance	Cross section (millibarns)
76	Os	176	−41.4			
		177	−41.4			
		178	−43.0			
		179	−42.7			
		180	−44.0			
		181	−43.4			
		182	−44.2			
		183	−43.4			
		184	−44.158	P	0.000135	400[1]
		185	−42.759			
		186	−42.958	S	0.00968	330[1]
		187	−41.184	S	0.0088	900[1]
		188	−41.101	S, R	0.0998	275[1]
		189	−38.952	S, R	0.121	765[1]
		190	−38.674	S, R	0.198	230[1] (750)
		191	−36.362			
		192	−35.850	R	0.308	(200)
		193	−33.367			
		194	−32.397			
		195	−29.9			
77	Ir				0.717	1,120 ± 200
		171	−25.2			
		172	−26.6			
		173	−29.3			
		174	−30.4			
		175	−32.7			
		176	−33.6			
		177	−35.6			
		178	−36.0			
		179	−37.6			
		180	−37.9			
		181	−39.1			
		182	−39.0			
		183	−40.0			
		184	−39.44			
		185	−40.3			
		186	−39.127			
		187	−39.7			
		188	−38.268			
		189	−38.5			
		190	−36.62			
		191	−36.672	S, R	0.267	1,900 ± 300
		192	−34.799			
		193	−34.499	S, R	0.450	600 ± 80
		194	−32.494			
		195	−31.851			
		196	−29.46			
		197	−28.41			
		198	−25.51			

Table 38 (continued)

Z	Element	A	Excess (MeV)	Class	Abundance	Cross section (millibarns)
78	Pt				1.4	470 ± 60
		173	−20.8			
		174	−24.0			
		175	−25.0			
		176	−27.9			
		177	−28.7			
		178	−31.1			
		179	−31.6			
		180	−33.7			
		181	−33.8			
		182	−35.6			
		183	−35.5			
		184	−36.9			
		185	−36.5			
		186	−37.5			
		187	−36.8			
		188	−37.728			
		189	−36.5			
		190	−37.293	P	0.000178	770[1]
		191	−35.672			
		192	−36.256	S	0.0109	490[1]
		193	−34.438			
		194	−34.733	S, R	0.461	310[1]
		195	−32.786	S, R	0.473	780[1]
		196	−32.635	S, R	0.354	160 ± 40
		197	−30.414			
		198	−29.906	R	0.101	185 ± 20
		199	−27.406			
		200	−26.6			
		201	−23.50			
79	Au				0.202	
		177	−20.6			
		178	−21.9			
		179	−24.3			
		180	−25.4			
		181	−27.4			
		182	−28.1			
		183	−29.7			
		184	−30.1			
		185	−31.5			
		186	−31.5			
		187	−32.7			
		188	−32.4			
		189	−33.5			
		190	−32.9			
		191	−33.8			
		192	−32.742			
		193	−33.4			
		194	−32.224			
		195	−32.557			
		196	−31.153			

Table 38 (continued)

Z	Element	A	Excess (MeV)	Class	Abundance	Cross section (millibarns)
79	Au	197	−31.161	S, R	0.202	600 ± 50
		198	−29.602			
		199	−29.099			
		200	−27.31			
		201	−26.16			
		202	−23.85			
		203	−22.8			
		204	−20.19			
80	Hg				0.4	250 ± 60
		179	−16.1			
		180	−19.3			
		181	−20.1			
		182	−22.7			
		183	−23.4			
		184	−25.6			
		185	−25.9			
		186	−28.0			
		187	−27.9			
		188	−29.5			
		189	−29.4			
		190	−30.9			
		191	−30.5			
		192	−31.8			
		193	−31.1			
		194	−32.174			
		195	−31.2			
		196	−31.837	P	0.000584	360[1]
		197	−30.746			
		198	−30.975	S	0.0408	250[1], 125
		199	−29.552	S, R	0.0674	630[1]
		200	−29.509	S, R	0.0925	175[1]
		201	−27.662	S, R	0.0529	450[1]
		202	−27.346	S, R	0.119	50 ± 15
		203	−25.267			
		204	−24.686	R	0.0274	150 ± 50
		205	−22.282			
		206	−20.937			
81	Tl				0.192	70 ± 5
		187	−21.5			
		188	−21.9			
		189	−23.6			
		190	−23.9			
		191	−25.5			
		192	−25.5			
		193	−26.9			
		194	−26.7			
		195	−28.0			
		196	−27.4			
		197	−28.4			
		198	−27.51			

Table 38 (continued)

Z	Element	A	Excess (MeV)	Class	Abundance	Cross section (millibarns)
81	Tl	199	−28.1			
		200	−27.055			
		201	−27.25			
		202	−26.109			
		203	−25.758	S, R	0.0567	170 ± 30
		204	−24.342			
		205	−23.811	S, R	0.135	48 ± 10
		206	−22.244			
		207	−21.014			
		208	−16.749			
		209	−13.632			
		210	− 9.224			
82	Pb				4	4.6 ± 1.5
		189	−17.0			
		190	−19.2			
		191	−19.5			
		192	−21.5			
		193	−21.6			
		194	−23.5			
		195	−23.4			
		196	−25.0			
		197	−24.6			
		198	−26.0			
		199	−25.4			
		200	−26.4			
		201	−25.5			
		202	−26.059			
		203	−24.776			
		204	−25.105	S	0.0788	43 ± 5
		205	−23.768			
		206	−23.777	S, R	0.753	9.6 ± 3.0
		207	−22.446	S, R	0.824	8.7 ± 3.0
		208	−21.743	S, R	2.34	0.33 ± 0.07
		209	−17.609			
		210	−14.720			
		211	−10.463			
		212	− 7.544			
		213	− 3.1			
		214	− 0.147			
83	Bi				0.143	
		191	−12.4			
		192	−12.9			
		193	−15.0			
		194	−15.4			
		195	−17.3			
		196	−17.4			
		197	−19.1			
		198	−19.1			
		199	−20.5			
		200	−20.4			
		201	−21.5			

Table 38 (continued)

Z	Element	A	Excess (MeV)	Class	Abundance	Cross section (millibarns)
83	Bi	202	−20.9			
		203	−21.59			
		204	−20.7			
		205	−21.064			
		206	−20.125			
		207	−20.041			
		208	−18.875			
		209	−18.257	S, R	0.143	12.1 ± 4.0
		210	−14.783			
		211	−11.839			
		212	− 8.117			
		213	− 5.226			
		214	− 1.183			
		215	1.73			
		216	6.0			
84	Po					
		193	− 7.4			
		194	− 9.8			
		195	−10.3			
		196	−12.5			
		197	−12.8			
		198	−14.7			
		199	−14.9			
		200	−16.6			
		201	−16.4			
		202	−17.9			
		203	−17.4			
		204	−18.5			
		205	−17.7			
		206	−18.308			
		207	−17.132			
		208	−17.464			
		209	−16.364			
		210	−15.944			
		211	−12.429			
		212	−10.364			
		213	− 6.647			
		214	− 4.460			
		215	− 0.514			
		216	1.786			
		217	6.0			
		218	8.390			
85	At					
		196	− 3.3			
		197	− 5.5			
		198	− 6.1			
		199	− 8.1			
		200	− 8.4			
		201	−10.2			
		202	−10.4			

Table 38 (continued)

Z	Element	A	Excess (MeV)	Class	Abundance	Cross section (millibarns)
85	At	203	−11.9			
		204	−11.9			
		205	−13.0			
		206	−12.6			
		207	−13.29			
		208	−12.5			
		209	−12.882			
		210	−12.069			
		211	−11.637			
		212	− 8.624			
		213	− 6.578			
		214	− 3.409			
		215	− 1.254			
		216	2.260			
		217	4.398			
		218	8.117			
		219	10.54			
		220	14.2			
86	Rn					
		200	− 3.0			
		201	− 3.5			
		202	− 5.6			
		203	− 5.8			
		204	− 7.7			
		205	− 7.6			
		206	− 9.1			
		207	− 8.8			
		208	− 9.8			
		209	− 9.1			
		210	− 9.723			
		211	− 8.741			
		212	− 8.648			
		213	− 5.696			
		214	− 4.310			
		215	− 1.165			
		216	0.262			
		217	3.666			
		218	5.232			
		219	8.856			
		220	10.616			
		221	14.4			
		222	16.402			
87	Fr					
		203	1.6			
		204	1.1			
		205	− 0.8			
		206	− 1.1			
		207	− 2.6			
		208	− 2.7			
		209	− 3.8			

Table 38 (continued)

Z	Element	A	Excess (MeV)	Class	Abundance	Cross section (millibarns)
87	Fr	210	− 3.5			
		211	− 4.20			
		212	− 3.6			
		213	− 3.554			
		214	− 1.056			
		215	0.331			
		216	2.976			
		217	4.318			
		218	7.013			
		219	8.614			
		220	11.483			
		221	13.280			
		222	16.364			
		223	18.406			
		224	21.7			
88	Ra					
		206	4.3			
		207	3.9			
		208	2.0			
		209	2.0			
		210	0.5			
		211	0.7			
		212	− 0.3			
		213	0.2			
		214	− 0.026			
		215	2.547			
		216	3.257			
		217	5.890			
		218	6.662			
		219	9.392			
		220	10.279			
		221	12.974			
		222	14.336			
		223	17.257			
		224	18.828			
		225	22.011			
		226	23.694			
		227	27.201			
		228	28.962			
		229	32.7			
		230	34.6			
89	Ac					
		209	9.4			
		210	8.9			
		211	7.5			
		212	7.3			
		213	6.1			
		214	6.3			
		215	5.97			
		216	8.0			

Table 38 (continued)

Z	Element	A	Excess (MeV)	Class	Abundance	Cross section (millibarns)
89	Ac	217	8.7			
		218	10.746			
		219	11.582			
		220	13.748			
		221	14.529			
		222	16.569			
		223	17.821			
		224	20.231			
		225	21.639			
		226	24.327			
		227	25.871			
		228	28.907			
		229	30.7			
		230	33.8			
		231	35.93			
		232	39.2			
90	Th				0.058	
		213	12.2			
		214	10.7			
		215	10.8			
		216	10.2			
		217	12.0			
		218	12.3			
		219	14.4			
		220	14.7			
		221	16.943			
		222	17.215			
		223	19.271			
		224	20.008			
		225	22.319			
		226	23.212			
		227	25.827			
		228	26.770			
		229	29.604			
		230	30.886			
		231	33.829			
		232	35.467	R	0.058	500 ± 100
		233	38.752			
		234	40.645			
91	Pa					
		222	21.87			
		223	22.352			
		224	23.799			
		225	24.331			
		226	25.980			
		227	26.827			
		228	28.883			
		229	29.899			
		230	32.190			
		231	33.443	R		

Table 38 (continued)

Z	Element	A	Excess (MeV)	Class	Abundance	Cross section (millibarns)
91	Pa	232	35.953			
		233	37.508			
		234	40.382			
		235	42.33			
		236	45.56			
		237	47.71			
		238	51.30			
92	U				0.0262	
		227	28.9			
		228	29.236			
		229	31.216			
		230	31.628			
		231	33.80			
		232	34.608			
		233	36.937			330 ± 40
		234	38.168			610[1]
		235	40.934	R	0.0063	860 ± 80
		236	42.460			(680)
		237	45.407			
		238	47.335	R	0.0199	415 ± 50
		239	50.604			
		240	52.742			
93	Np					
		229	33.768			
		230	35.203			
		231	35.65			
		232	37.3			
		233	38.0			
		234	39.976			
		235	41.057			
		236	43.437			
		237	44.889			
		238	47.481			
		239	49.326			
		240	52.23			
		241	54.33			
		242	57.5			
94	Pu					
		232	38.36			
		233	40.057			
		234	40.363			
		235	42.19			
		236	42.900			
		237	45.113			
		238	46.186			
		239	48.602			580 ± 60
		240	50.140			600 ± 100
		241	52.972			
		242	54.742			

Table 38 (continued)

Z	Element	A	Excess (MeV)	Class	Abundance	Cross section (millibarns)
94	Pu	243	57.777			
		244	59.831			
		245	63.182			
		246	65.32			
95	Am					
		234	44.4			
		235	44.7			
		236	46.0			
		237	46.7			
		238	48.5			
		239	49.406			
		240	51.5			
		241	52.951			
		242	55.494			
		243	57.189			
		244	59.898			
		245	61.922			
		246	64.94			
		247	67.2			
		248	70.5			
96	Cm					
		236	48.0			
		237	49.2			
		238	49.419			
		239	51.1			
		240	51.721			
		241	53.723			
		242	54.827			
		243	57.196			
		244	58.469			
		245	61.020			
		246	62.641			
		247	65.556			
		248	67.417			
		249	70.776			
		250	73.1			
97	Bk					
		238	54.2			
		239	54.3			
		240	55.7			
		241	56.1			
		242	57.8			
		243	58.702			
		244	60.7			
		245	61.840			
		246	64.2			
		247	65.500			
		248	68.0			
		249	69.868			

Table 38 (continued)

Z	Element	A	Excess (MeV)	Class	Abundance	Cross section (millibarns)
97	Bk	250	72.970			
		251	75.3			
		252	78.6			
98	Cf					
		240	58.1			
		241	59.2			
		242	59.353			
		243	60.9			
		244	61.474			
		245	63.403			
		246	64.121			
		247	66.2			
		248	67.264			
		249	69.742			
		250	71.195			
		251	74.153			
		252	76.059			
		253	79.337			
		254	81.4			
99	Es					
		244	66.0			
		245	66.4			
		246	68.0			
		247	68.578			
		248	70.3			
		249	71.146			
		250	73.2			
		251	74.517			
		252	77.2			
		253	79.038			
		254	82.021			
		255	84.1			
		256	87.3			
100	Fm					
		246	70.151			
		247	71.6			
		248	71.900			
		249	73.5			
		250	74.094			
		251	76.0			
		252	76.842			
		253	79.373			
		254	80.934			
		255	83.821			
		256	85.518			
		257	88.628			

Table 38 (continued)

Z	Element	A	Excess (MeV)	Class	Abundance	Cross section (millibarns)
101	Md					
		249	77.3			
		250	78.7			
		251	79.1			
		252	80.6			
		253	81.3			
		254	83.4			
		255	84.9			
		256	87.5			
		257	89.1			
102	No					
		251	82.8			
		252	82.871			
		253	84.3			
		254	84.754			
		255	86.9			
		256	87.82			
		257	90.249			
103	Lr					
		256	91.8			
		257	92.7			

and $$\alpha_0^2 = 0.3645\, A^{-2/3}.$$

A recent discussion of some of the physics behind Eq. (4-19) is given by MYERS (1970), and individual mass values may be found from Garvey's law and from his mass table (cf. GARVEY *et al.* (1969)).

4.1.4. Alpha Decay and other Natural Nuclear Reactions

Following BECQUEREL'S (1896) discovery that uranium salts emit particles, RUTHERFORD (1899) showed that there were two such radioactive particles, the β and α particles which were, respectively, more or less penetrating. Subsequently, RUTHERFORD and SODDY (1902, 1903) showed that in the emission of an α particle the mass number, A, decreased by four and the atomic number, Z, decreased by two. It follows that two protons and two neutrons come together within a nucleus to form an α particle which is a He^4 nucleus with a charge $Z_\alpha = 2$ and a mass $M_\alpha = 4.002603$ a.m.u. RUTHERFORD and SODDY (1902, 1903) also showed that the mass number, A, stays the same when a nucleus emits a β particle, and that the charge, Z, changes by one. This meant that β particles were probably electrons and positrons, which was subsequently shown to be true. In fact, under different thermonuclear conditions a nucleus may emit an α particle, a positron, β^+, electrons, β^-, neutrons, n, protons, p, deuterons, d, tritons, t, and helium, He^3. The nucleus may also capture many of these particles. The theory of beta decay and electron capture is discussed in Sect. 4.3 on weak interactions, whereas the theory of alpha decay follows.

The α disintegration energies and half-lives of many elements are given in the chart of the nuclides available from the Educational Relations Department of the General Electric Company at Schenectady, New York. This chart also gives energies and lifetimes of other modes of decay, and is a useful tool in interpreting the flow patterns in the transmutation of the elements. The relative locations on the chart of nuclides of the products of various nuclear processes are illustrated in Fig. 23. Another useful source of nuclear data is the Table of Isotopes published by LEDERER *et al.* (1967). The nuclide chart shows that there are a total of thirty nuclides which α decay naturally and are found on the Earth. These nuclides are elements in three chains which begin with the elements $^{90}Th^{232}$, $^{92}U^{238}$, and $^{92}U^{235}$, terminate on elements with $Z \geq 82$, and have half lifetimes of 1.39×10^{10} years, 4.51×10^{9} years, and 7.13×10^{8} years, respectively.

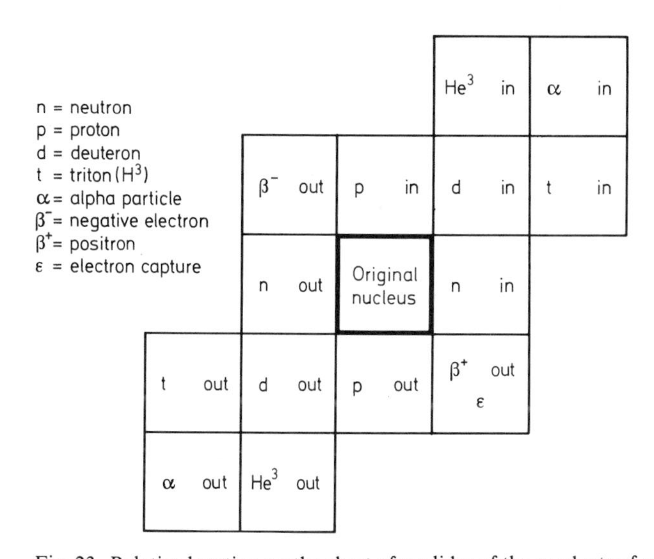

Fig. 23. Relative location on the chart of nuclides of the products of various nuclear processes

The α disintegration energy, E, which is defined as the sum of the kinetic energies of the α particle and the recoil nucleus, is given by

$$E = [M(Z,A) - M(Z-2,A-4) - M_\alpha]c^2 = E_\alpha[1 + M_\alpha/M(Z-2,A-4)], \quad (4\text{-}20)$$

where $M(Z,A)$ denotes the mass of the nucleus (Z,A), the kinetic energy and mass of the α particle are, respectively, E_α and M_α, and the mass equivalents of the electron binding energies have been ignored. Eq. (4-20) gives the maximum value of E, for the nucleus may sometimes α decay to an excited state and then radiate a gamma ray photon. Typical values of E range from 4 to 9 MeV.

Although the disintegration energy, E, is less than the energy of the Coulomb barrier of the nucleus, there will be a finite probability per second, λ_α, that the

α particle will escape the nucleus. Often the half-life, $\tau_{1/2}$, or mean lifetime, τ, are measured.

$$\lambda_\alpha = \frac{1}{\tau} = \frac{\ln 2}{\tau_{1/2}}, \tag{4-21}$$

where $\ln 2 \approx 0.69315$. If the number of nuclei which can undergo α decay is N_0 at time zero, then the number, N, at time, t, is given by

$$N(t) = N_0 \exp[-\lambda_\alpha t], \tag{4-22}$$

or

$$\frac{dN(t)}{dt} = -\lambda_\alpha N(t).$$

A theoretical formula for λ_α was first derived by GAMOW (1928) and GURNEY and CONDON (1928). They assumed that the potential energy, $V(r)$, of the α particle and the nucleus is given by $V(r) = -V_0$ for $r < R$ and $V(r) = 2(Z-2)e^2/r$ for $r > R$. Here r is the separation of the center of the nucleus and the α particle, and the nuclear radius, R, is defined as the greatest distance for which nuclear forces are significant. Experiments involving electron and neutron scattering indicate that

$$R \approx 1.2 \times 10^{-13} A^{1/3} \text{ cm}. \tag{4-23}$$

When $V(r)$ is used in the Schrödinger wave equation (Eq. (2-126)), and the continuity conditions are satisfied at $r = R$, it can be shown (PRESTON, 1962) that

$$\begin{aligned} \lambda_\alpha &= \frac{2V}{R} \frac{\tan\alpha_0}{f} \exp\left[\frac{-8(Z-2)e^2}{\hbar}\left(\frac{M}{2E}\right)^{1/2}(\alpha_0 - \sin\alpha_0\cos\alpha_0)\right] \\ &\approx \frac{V}{2R}\{\exp[2.97(Z-2)^{1/2}R^{1/2} - 3.95(Z-2)E^{-1/2}]\}, \end{aligned} \tag{4-24}$$

where, in the numerical approximation the disintegration energy, E, is in MeV and R is in Fermis (1 Fermi $= 10^{-13}$ cm), and the reduced mass, M, is given by

$$M = \frac{M_\alpha M_{Z-2}}{M_\alpha + M_{Z-2}},$$

where M_{Z-2} denotes the mass of the product nucleus, the final relative velocity of the α particle and the nucleus is given by

$$V = V_\alpha - V_{Z-2} \approx V_\alpha[1 + M_\alpha/M_{Z-2}], \tag{4-25}$$

where V_α and V_{Z-2} are the respective velocities of the α particle and the recoil nucleus,

$$\alpha_0 = \arccos\left[\frac{4(Z-2)e^2}{VR}\left(\frac{1}{2ME}\right)^{1/2}\right]^{-1/2} \approx \arccos\left[\frac{2(Z-2)e^2}{ER}\right]^{-1/2}$$

and

$$f = \operatorname{cosec}^2(KR) - \cot(KR)/KR,$$

where

$$K\cot(KR) = -\frac{MV}{\hbar}\tan\alpha_0,$$

and the energy of the state $E \approx 2(Z-2)e^2/R$. Experiments show that Eq. (4-24) holds quite well for nuclei with even Z and A with $R = 1.57 \pm 0.015 \times 10^{-13} A^{1/3}$ cm, $E + V_0 = 0.52 \pm 0.01$ MeV, and $KR = 2.986 \pm 0.005$.

4.2. Thermonuclear Reaction Rates

4.2.1. Definition and Reciprocity Theorem for Cross Sections

The cross section, σ, for any event is defined as the number of desired events per second divided by the number of particles incident per unit area per second. The cross section, σ_{12}, for the reaction, $1+2 \rightarrow 3+4$, is related to the cross section, σ_{34}, for the reverse reaction, $3+4 \rightarrow 1+2$, by the relation (BLATT and WEISSKOPF, 1952 as modified by FOWLER, CAUGHLAN, and ZIMMERMAN, 1967)

$$\frac{\sigma_{34}}{\sigma_{12}} = \frac{(1+\delta_{34}) g_1 g_2 A_1 A_2 E_{12}}{(1+\delta_{12}) g_3 g_4 A_3 A_4 E_{34}}, \tag{4-26}$$

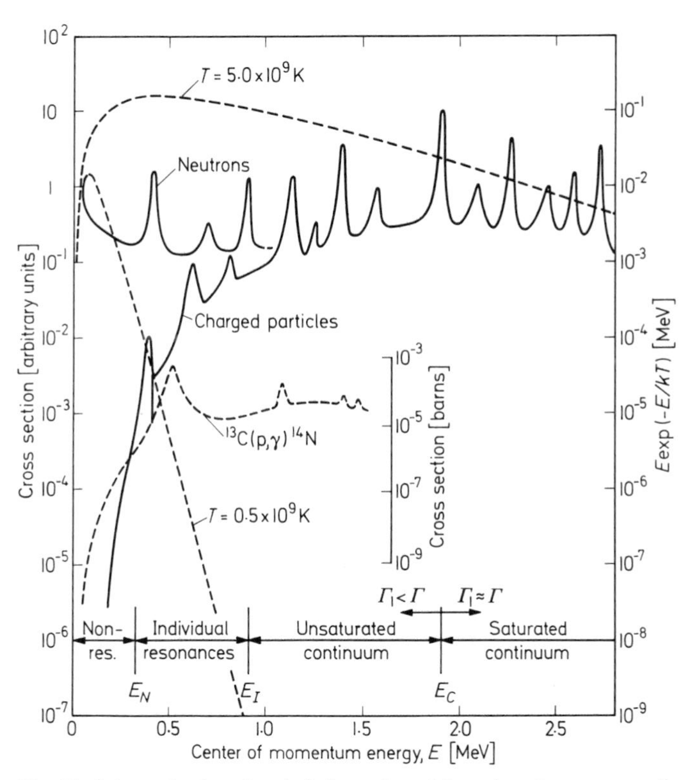

Fig. 24. Schematic plot of typical charged particle and neutron cross sections as a function of center-of-momentum energy, E, together with the Maxwell-Boltzmann distribution function (dashed lines). Also shown is the measured cross section in barns of the $^{13}C(p,\gamma)^{14}N$ reaction, where in this case E is the proton laboratory energy. The former curves are after WAGONER (1969, by permission of the American Astronomical Society and the University of Chicago Press), whereas the latter is after SEAGRAVE (1952)

where the Kronecker delta function, δ_{12}, is one if $1=2$ and zero if $1 \neq 2$, the statistical weight, g_i, of nucleus, i, is given by $g_i = 2I_i + 1$, where I_i is the spin of the nucleus, the mass number is A_i, and E_{12} and E_{34} are the kinetic energies, in the center of mass system, of the two sides of the nuclear reaction equation. At very high temperatures the nuclei may be in excited states which are in thermal equilibrium with their ground states, and in this case the g_i are replaced by the nuclear partition functions $G_i = \sum_j g_{ij} \exp(-E_j/kT)$ where the E_j is the excitation energy of the jth state. Schematic plots of typical charged particle and neutron cross sections are shown in Fig. 24 together with the Maxwell-Boltzmann distribution function and the experimental cross section for the $^{13}C(p,\gamma)^{14}N$ reaction.

4.2.2. Nonresonant Neutron Capture Cross Section

When a nucleus (Z,A) captures a neutron, n, it becomes the isotope $(Z,A+1)$ of the same element, and a photon, γ, can be radiated according to the reaction

$$(Z,A)+n \rightarrow (Z,A+1)+\gamma. \tag{4-27}$$

Under normal conditions in stellar interiors, the relative velocity, v, between a neutron and a nucleus is determined by the Maxwell-Boltzmann distribution. The effective cross section, $\langle\sigma\rangle$, for a Maxwell-Boltzmann distribution of particle velocities is given by

$$\langle\sigma\rangle = \frac{\langle\sigma v\rangle}{v_T} = \frac{\int_0^\infty \sigma v \varphi(v)\,dv}{v_T}, \tag{4-28}$$

where the Maxwell weighting function, $\varphi(v)$, is given by

$$\varphi(v) = \frac{4}{\pi^{1/2}} \left(\frac{v}{v_T}\right)^2 \exp\left[-\left(\frac{v}{v_T}\right)^2\right] \frac{dv}{v_T}, \tag{4-29}$$

the most probable velocity, v_T, is given by

$$v_T = \left(\frac{2kT}{M}\right)^{1/2} \approx 1.284 \times 10^4\, T^{1/2} \text{ cm sec}^{-1}, \tag{4-30}$$

the gas temperature is T, and the reduced neutron mass, $M = M_N M_A/(M_N + M_A) \approx M_N = 1.674920(11) \times 10^{-24}$ grams. Here M_N and M_A are, respectively, the mass of the neutron and the nucleus.

For heavy elements, nuclei more massive than the iron group nuclei, the effective neutron capture cross section, $\langle\sigma\rangle$, at the most probable energy $kT = 8.6167 \times 10^{-5}\, T$ eV, is given by

$$\langle\sigma\rangle = \sigma_T \quad \text{measured at } v_T = (1.648 \times 10^8\, T)^{1/2} \text{ cm sec}^{-1}. \tag{4-31}$$

Values of the cross section, σ_T, measured at 30 keV or $T = 3.48 \times 10^8$ °K or $v_T \approx 2.40 \times 10^8$ cm sec^{-1}, are given in Table 38. Values at other temperatures may be calculated using the relationship $\langle\sigma\rangle \propto v_T^{-1} \propto T^{-1/2}$. Other authors have given thermal neutron cross sections which are measured at $T = 293$ °K or $v_T = 2.198$ km sec^{-1} and $kT = 0.0252$ eV.

4.2.3. Nonresonant Charged Particle Cross Section

At low energies, nonresonant charged particle interactions are dominated by the Coulomb-barrier penetration factors first discussed by GAMOW (1928) and GURNEY and CONDON (1928, 1929). It is therefore convenient to factor out this energy dependence and express the cross section, $\sigma(E)$, by

$$\sigma(E) = \frac{S(E)}{E} \exp[-(E_G/E)^{1/2}], \tag{4-32}$$

where the kinetic energy, E, of the center-of-mass system is given by

$$E = \frac{M v^2}{2} = \frac{M_1 E_{L2}}{M_1 + M_2}, \tag{4-33}$$

the reduced mass, M, of interacting particles 1 and 2, is given by $M = M_1 M_2/(M_1 + M_2)$, the particles have masses, M_1 and M_2, and relative velocity, v, and E_{L2} is the laboratory energy of particle 2—i.e., the energy of particle 2 in the frame where particle 1 is at rest.

The Gamow energy, E_G, is given by

$$E_G = (2\pi\alpha Z_1 Z_2)^2 (M c^2/2) = [0.98948\, Z_1 Z_2 A^{1/2}]^2 \text{ MeV}. \tag{4-34}$$

Here the fine structure constant $\alpha = e^2/(\hbar c) = (137.03602)^{-1}$, the particle charges in units of the proton charge are Z_1 and Z_2, and the reduced nuclear mass $A = A_1 A_2/(A_1 + A_2)$ which differs little from the reduced atomic mass.

Far from a nuclear resonance, the cross section factor, $S(E)$, is a slowly varying function of the center of momentum energy, E, and it may be conveniently expressed in terms of the power series expansion

$$S(E) = S(0)\left[1 + \frac{S'(0)}{S(0)} E + \frac{1}{2}\frac{S''(0)}{S(0)} E^2\right], \tag{4-35}$$

where the prime denotes differentiation with respect to E. Experimental measurements of the constants in this expression are given in Sect. 4.4 for the more important thermonuclear reactions in stars. Additional values are given by FOWLER, CAUGHLAN, and ZIMMERMAN (1967, 1974). When the reaction proceeds through the wings of a resonance, $S(E)$ may be evaluated using measurements of the resonance at higher energies (cf. Eq. (4-49)).

4.2.4. Resonant Cross Sections for Neutrons and Charged Particles—Breit-Wigner Shapes

For a single resonance of energy, E_R, the cross section, $\sigma(E)$, of the nuclear reaction, $1 + 2 \rightarrow 3 + 4 + Q$, is given by (BREIT and WIGNER, 1936)

$$\sigma(E) = \pi\lambda\!\!\!^{-2} \frac{\omega \Gamma_{12}\Gamma_{34}}{(E - E_R)^2 + (\Gamma^2/4)} = \frac{0.657}{A E} \frac{\omega \Gamma_{12}\Gamma_{34}}{(E - E_R)^2 + (\Gamma^2/4)} \text{ barn}, \tag{4-36}$$

where the center of momentum energy, E, of 1 and 2 is in MeV in the numerical approximation, and the reduced de Broglie wavelength $\lambda\!\!\!^{-} = \hbar/(M v)$, where M is the reduced mass of 1 and 2 and v is their relative velocity. The statistical factor $\omega = (2J + 1)/[(2J_1 + 1)(2J_2 + 1)]$, where J is the angular momentum of the resonant state, and J_1 and J_2 are, respectively, the angular momenta of particles 1 and 2. The total width, Γ, of the resonant state is given by $\Gamma=\hbar/\tau=\Gamma_{12}+\Gamma_{34}+\cdots$, where τ is the effective lifetime of the state. The partial width Γ_{12} is the width for reemission of 1 with 2, and Γ_{34} is the width for emission of 3 and 4. Detailed formulae for evaluation of partial widths are given in Sect. 4.2.7.

4.2.5. Reaction Rate, Mean Lifetime, and Energy Generation

The reaction rate, r_{12}, between two nuclei, 1 and 2, with a relative velocity, v, is given by

$$r_{12}=\frac{N_1 N_2}{(1+\delta_{12})}\int_0^\infty v\sigma(v)\varphi(v)\,dv=\frac{N_1 N_2\langle\sigma v\rangle}{(1+\delta_{12})}\ \text{reactions cm}^{-3}\,\text{sec}^{-1}$$

$$=\frac{N_1 N_2}{(1+\delta_{12})}\int_0^\infty\left(\frac{2E}{M}\right)^{1/2}\sigma(E)\psi(E)\,dE\qquad\text{reactions cm}^{-3}\,\text{sec}^{-1}, \tag{4-37}$$

where the kinetic energy in the center of mass system is $E=Mv^2/2$, the reduced mass $M=M_1M_2/(M_1+M_2)$ where the nuclei have masses M_1 and M_2, the reaction cross section is $\sigma(v)$ or $\sigma(E)$, N_1 and N_2 are the number densities of 1 and 2, the relative velocity spectrum is $\varphi(v)dv=\psi(E)dE$, and the Kronecker delta function, δ_{12}, is one if $1=2$ and zero if $1\neq2$. For a gas of mass density, ρ, the number density, N_i, of the nuclide, i, is often expressed in terms of its mass fraction, X_i, by the relation

$$N_i=\rho N_A\frac{X_i}{A_i}\ \text{cm}^{-3}, \tag{4-38}$$

where Avogadro's number $N_A=6.022169\times10^{23}\,(\text{mole})^{-1}$, and A_i is the atomic mass of i in atomic mass units.

For a nondegenerate, nonrelativistic gas, the relative velocity spectrum is Maxwellian and is given by

$$\varphi(v)d^3v=\left(\frac{M}{2\pi kT}\right)^{3/2}\exp\left(-\frac{Mv^2}{2kT}\right)4\pi v^2\,dv \tag{4-39}$$

or

$$\psi(E)dE=\frac{2E}{\sqrt{\pi}\,kT}\exp\left(-\frac{E}{kT}\right)\frac{dE}{(kTE)^{1/2}},$$

where the gas temperature is T, the reduced mass is M, the relative velocity is v, and the kinetic energy in the center-of-mass system is E.

The mean lifetime, $\tau_2(1)$, of nucleus 1 for interaction with nucleus 2 is given by the relation

$$\lambda_2(1)=\frac{1}{\tau_2(1)}=N_2\langle\sigma v\rangle=\rho N_A\frac{X_2}{A_2}\langle\sigma v\rangle\ \text{sec}^{-1}, \tag{4-40}$$

where $\lambda_2(1)$ is the decay rate of 1 for interactions with 2. The energy generation, ε_{12}, for the forward reaction $1+2 \rightarrow 3+4+Q$ is given by

$$\varepsilon_{12} = r_{12} Q/\rho \text{ erg g}^{-1} \text{ sec}^{-1}, \tag{4-41}$$

where r_{12} is the reaction rate, Q is the energy release, and ρ is the gas mass density.

4.2.6. Nonresonant Reaction Rates

It follows from Eqs. (4-28), (4-30), (4-31), and (4-37) that the reaction rate, r_{1n}, for the nonresonant neutron capture reaction is given by

$$r_{1n} = N_1 N_n \sigma_T \left(\frac{2kT}{M}\right)^{1/2} = N_1 N_n \langle \sigma v \rangle, \tag{4-42}$$

where N_1 and N_n are, respectively, the number densities of the nuclei and the neutrons, σ_T is the neutron capture cross section at temperature, T (given in Table 38 for $kT = 30$ keV), and the reduced mass $M = M_1 M_N/(M_N + M_1) \approx M_N = 1.674920 \times 10^{-24}$ grams. The effective energy and energy spread of the neutron capture reaction is on the order of $kT \approx 0.08617\, T_9$ MeV, where T_9 is $T/10^9$ when T is in °K.

It follows from Eqs. (4-32), (4-33), (4-34), (4-35), and (4-37) that the reaction rate, r_{12}, for a nonresonant charged particle reaction in a nondegenerate, nonrelativistic gas is given by (BURBIDGE, BURBIDGE, FOWLER, and HOYLE (B^2FH), 1957; FOWLER, CAUGHLAN, and ZIMMERMAN, 1974)

$$r_{12} = \frac{N_1 N_2 \langle \sigma v \rangle}{(1+\delta_{12})} = \frac{N_1 N_2}{(1+\delta_{12})} \left(\frac{8}{\pi M}\right)^{1/2} \frac{1}{(kT)^{3/2}} \int S(E) \exp\left[-\left(\frac{E_G}{E}\right)^{1/2} - \frac{E}{kT}\right] dE$$

$$= \frac{N_1 N_2}{(1+\delta_{12})} \left(\frac{2}{M}\right)^{1/2} \frac{\Delta E_0}{(kT)^{3/2}} S_{\text{eff}} \exp\left[-\frac{3E_0}{kT} - \left(\frac{T}{T_0}\right)^2\right], \tag{4-43}$$

where N_1 and N_2 are the number densities of 1 and 2, the Kronecker delta function, δ_{12}, is one if $1=2$ and zero if $1 \neq 2$, the reduced mass $M = M_1 M_2/(M_1 + M_2)$, the temperature is T, the cutoff temperature, T_0, occurs when the nuclear cross section no longer varies according to the Gamow penetration factor given in Eq. (4-34) (FOWLER, CAUGHLAN, and ZIMMERMAN, 1974). This condition may occur when the effective energy, E_0, is larger than E_G (given in Eq. (4-34)), or when resonance or the continuum set in.

The integrand in Eq. (4-43) has its peak value at the effective thermal energy, E_0, (Gamow peak) given by (FOWLER and HOYLE, 1964)

$$E_0 = [\pi \alpha Z_1 Z_2 kT (Mc^2/2)^{1/2}]^{2/3} = 0.1220 (Z_1^2 Z_2^2 A)^{1/3} T_9^{2/3} \text{ MeV}, \tag{4-44}$$

where the fine structure constant $\alpha = e^2/\hbar c = (137.03602)^{-1}$, M is the reduced mass, Z_1 and Z_2 are, respectively, the charges of 1 and 2 in units of the proton charge, $A = A_1 A_2/(A_1 + A_2)$ is the reduced atomic mass, and $T_9 = T/10^9$, where

T is the temperature. The peak of the integrand in Eq. (4-43) has a full width at $1/e$ of the maximum value given by

$$\Delta E_0 = 4(E_0 k T/3)^{1/2} = 0.2368 (Z_1^2 Z_2^2 A)^{1/6} T_9^{5/6} \text{ MeV}, \tag{4-45}$$

which is the width of the effective range of thermal energy.[1] The effective value, S_{eff}, of the cross section factor, $S(E)$, is given by

$$S_{\text{eff}} = S(0)\left[1 + \frac{5kT}{36E_0} + \frac{S'(0)}{S(0)}\left(E_0 + \frac{35}{36}kT\right) + \frac{1}{2}\frac{S''(0)}{S(0)}\left(E_0^2 + \frac{89}{36}E_0 kT\right)\right] \text{MeV-barn}, \tag{4-46}$$

where the prime denotes differentiation with respect to the kinetic energy, E, in the center-of-mass system. Eqs. (4-38), (4-43), (4-44), and (4-46) may be combined to give the useful result

$$N_A\langle\sigma v\rangle = C_1 T_9^{-2/3} \exp[-C_2 T_9^{-1/3} - (T_9/T_0)^2]\{1 + C_3 T_9^{1/3} + C_4 T_9^{2/3} + C_5 T_9 + C_6 T_9^{4/3} + C_7 T_9^{5/3}\} \text{ cm}^3 \text{ sec}^{-1} (\text{mole})^{-1}, \tag{4-47}$$

where

$$C_1 = 7.8324 \times 10^9 (Z_1^2 Z_2^2 A)^{1/6} \frac{S(0)}{A^{1/2}},$$

$$C_2 = 4.2475 (Z_1^2 Z_2^2 A)^{1/3},$$

$$C_3 = 9.810 \times 10^{-2} (Z_1^2 Z_2^2 A)^{-1/3},$$

$$C_4 = 0.1220 \frac{S'(0)}{S(0)} (Z_1^2 Z_2^2 A)^{1/3},$$

$$C_5 = 8.377 \times 10^{-2} \frac{S'(0)}{S(0)},$$

$$C_6 = 7.442 \times 10^{-3} \frac{S''(0)}{S(0)} (Z_1^2 Z_2^2 A)^{2/3},$$

and

$$C_7 = 1.299 \times 10^{-2} \frac{S''(0)}{S(0)} (Z_1^2 Z_2^2 A)^{1/3}.$$

Here the temperature $T_9 = T/10^9$, where T is the temperature in °K, the effective cutoff temperature is T_0, the charges of 1 and 2 are Z_1 and Z_2 in units of the proton charge, the reduced atomic mass $A = A_1 A_2/(A_1 + A_2)$ is given in atomic mass units, and the quantities $S(0)$, $S'(0)$ and $S''(0)$ have respective units of MeV-barns, barns, and barns MeV^{-1}. Atomic masses may be determined from the constants given in Table 38 together with Eq. (4-11). Laboratory measurements of $S(0)$, $S'(0)$ and $S''(0)$ for many thermonuclear reactions are given by Fowler, Caughlan, and Zimmerman (1967, 1974) and Barnes (1971). Numerical values for the constants in Eq. (4-47) are given in Sect. 4.4 for the more important thermonuclear reactions in stars.

[1] Nuclei with the energies close to E_0 and spread over the energy range ΔE_0 contribute mainly to the total rate of the thermonuclear reaction.

4.2.7. Resonant Reaction Rates

It follows from Eqs. (4-36), (4-37), and (4-39) that the reaction rate, r_{12}, for a resonant reaction in a nonrelativistic, nondegenerate gas is given by

$$r_{12}=\frac{N_1 N_2}{(1+\delta_{12})}\langle\sigma v\rangle=\frac{N_1 N_2}{(1+\delta_{12})}\int_0^\infty \frac{0.657}{AE}\,\frac{\omega\Gamma_{12}\Gamma_{34}}{(E-E_{\mathrm{R}})^2+(\Gamma^2/4)} \times\left[\frac{2E}{M}\right]^{1/2}\left[\frac{2E}{\sqrt{\pi}\,kT}\exp\left(-\frac{E}{kT}\right)\frac{1}{(kTE)^{1/2}}\right]dE, \tag{4-48}$$

where N_1 and N_2 are the number densities of 1 and 2, the kinetic energy in the center-of-mass system is E, the reduced atomic mass $A=A_1A_2/(A_1+A_2)$, the reduced mass $M=M_1M_2/(M_1+M_2)$, the resonant energy is E_{R}, the gas temperature is T, the statistical factor $\omega=(2J+1)/[(2J_1+1)(2J_2+1)]$ where J is the angular momentum of the resonant state and J_1 and J_2 are, respectively, the angular momenta of 1 and 2, the width of the resonant state is Γ, the partial width Γ_{12} is the width for reemission of 1 with 2, and Γ_{34} is the width for emission of 3 and 4. The numerical constant of 0.657 is for a cross section in barns when the energy, E, is in MeV.

When the effective thermal energy, E_0, given in Eq. (4-44), is much smaller than the resonant energy, E_r, then the Breit-Wigner resonant cross section may be evaluated in its wings to give the cross section factor

$$S(E)=\frac{0.657}{A}\,\frac{\omega\Gamma_{12}(E)\Gamma_{34}}{(E-E_r)^2+(\Gamma^2/4)}\exp\left[0.98948\,Z_1Z_2A^{1/2}E^{-1/2}\right], \tag{4-49}$$

where $S(E)$ is in MeV barns, and E is in MeV. Eq. (4-49) may be evaluated using measurements of a resonance at high energies, and then used with the non-resonant formalism of Eq. (4-35), (4-43), and (4-46) to obtain reaction rates at low energies.

When the range of effective stellar energies[2] includes the resonance energy, then Eq. (4-48) must be used. Provided that the width of the resonance, Γ, is much less then the effective spread in energy of the interacting particles (less than kT when neutrons are involved and less than ΔE_0 for two charged particles). Eq. (4-48) may be integrated to give the useful result

$$N_A\langle\sigma v\rangle=C_8\,T_9^{-3/2}\exp(-C_9/T_9)\ \mathrm{cm^3\,sec^{-1}\,(g\ mole)^{-1}}, \tag{4-50}$$

where

$$C_8=1.53986\times10^{11}A^{-3/2}(\omega\gamma),$$
$$C_9=11.605\,E_{\mathrm{R}},$$

the temperature $T_9=T/10^9$, where the temperature T is in °K, the reduced atomic mass, A, is in atomic mass units, the factor $(\omega\gamma)=\omega\Gamma_{12}\Gamma_{34}/\Gamma$ is in MeV, and the resonant energy E_{R} is in MeV. Atomic masses may be determined from the constants given in Table 38 together with Eq. (4-11). Laboratory measurements of $(\omega\gamma)$ and E_{R} are given by Fowler, Caughlan, and Zimmerman (1967, 1974). Values of the constants C_8 and C_9 are given in Sect. 4.4 for many of the

[2] The effective energy range is of the order of kT for neutron interactions and is given by ΔE_0 from (4-45) centered at E_0 from (4-44) for charged particle interactions.

more important thermonuclear reactions in stars. When several resonances occur, the net reaction rate is determined by the superposition

$$N_A\langle\sigma v\rangle = \sum_n C_{8n} T_9^{-3/2} \exp(-C_{9n}/T_9), \tag{4-51}$$

where the constants C_{8n} and C_{9n} are the appropriate constants given by Eq. (4-50) for the nth resonance.

When the density of resonances lying within the Gamow peak becomes sufficiently large, we can replace the summation in Eq. (4-51) by the integration dE/D, where D is the average level distance within the Gamow peak. This occurs mainly in elements heavier than about Mg, and for energies in excess of a few MeV. Thus we have (HAYASHI, HOSHI, and SUGIMOTO, 1962)

$$r_{12} = \rho \frac{X_1}{A_1} N_2 N_A \langle\sigma v\rangle = \frac{(2\pi)^{3/2} \hbar^2 N_1 N_2}{(MkT)^{3/2}} \frac{\omega}{D} \int_0^\infty \frac{\langle\Gamma_1(E)\rangle \langle\Gamma_2(E)\rangle}{\langle\Gamma(E)\rangle} \exp\left(-\frac{E}{kT}\right) dE, \tag{4-52}$$

where $M = A$ (1 a.m.u.) is the reduced mass of the interacting nuclei, the reduced atomic mass number $A = A_1 A_2/(A_1 + A_2)$, the average spin factor is ω (the average of $(2J+1)/[(2I_1+1)(2I_2+1)]$) where J is the resonance value of angular momentum and I_1 and I_2 are the nuclear spins, Γ_1, Γ_2, and Γ denote, respectively, the absorption width, the emission width, and the total width, and the $\langle\,\rangle$ denotes an average value. Considering that the emission width is normally much larger than the absorption width in the relevant range of temperature, Eq. (4-52) becomes

$$\langle\sigma v\rangle = \frac{\sqrt{\pi}}{2} \left(\frac{2\pi\hbar^2}{MkT}\right)^{3/2} \frac{\Delta E_0}{\hbar} \frac{\omega \Gamma_1(E_0)}{D} \exp\left(-\frac{E_0}{kT}\right) \tag{4-53}$$

$$\approx \alpha T_9^{-2/3} \exp(-\beta T_9^{-1/3})\ \mathrm{cm^3\ sec^{-1}}, \tag{4-54}$$

where M is the reduced mass of the interacting nuclei, the effective thermal energy, E_0, is given in Eq. (4-44), the effective range of thermal energy, ΔE_0, is given in Eq. (4-45), and

$$\alpha = 1.78 \times 10^{-12} (Z_1^5 Z_2^5 A^{-11})^{1/6} \frac{\theta^2}{D R^{3/2}} \exp[1.05 (Z_1 Z_2 A R)^{1/2}],$$
$$\beta = 4.25 (Z_1^2 Z_2^2 A)^{1/3}, \tag{4-55}$$

and the average level distance, D, is measured in MeV. For $A \gtrsim 20$, we have $\omega \approx 1$ and $D \approx 0.1$ MeV. The reduced width, θ, is related to D by the equation

$$\theta^2 = \frac{2\lambda_0 M}{3\pi\hbar^2} R D, \tag{4-56}$$

where $\lambda_0 \approx 10^{-13}$ cm is the characteristic wavelength of nucleons inside the nucleus, and the interaction radius, R, is given by

$$R \approx 1.4 \times 10^{-13} [A_1^{1/3} + A_2^{1/3}]\ \mathrm{cm}. \tag{4-57}$$

Partial widths, Γ, may be evaluated using the expression

$$\Gamma = \frac{3\hbar v}{R} \theta_l^2 P_l, \tag{4-58}$$

where R is the interaction radius, $v=(2E/M)^{1/2}$ is the relative velocity of the particles, the reduced width, θ_l^2, for particles with relative angular momentum, l, usually lies in the range $0.01 \lesssim \theta_l^2 \lesssim 1.0$, and the penetration factor, P_l, is given by

$$P_l = [F_l^2(R) + G_l^2(R)]^{-1}, \tag{4-59}$$

where $F_l(R)$ and $G_l(R)$ are the regular and irregular solutions of the Schrödinger wave equation for a charged particle in a Coulomb field evaluated at the interaction radius, R. The Coulomb wave functions $F_l(R)$ and $G_l(R)$ are discussed and tabulated by FRÖBERG (1955) and HULL and BREIT (1959). Approximate values of the penetration factor are derived by using the WKB (WENTZEL, 1926; KRAMERS, 1926; BRILLOIN, 1926) solution to the Schrödinger equation (cf. BETHE, 1937, and VAN HORN and SALPETER, 1967). If the Coulomb barrier energy, E_c, is given by

$$E_c = \frac{Z_1 Z_2 e^2}{R} \approx 1.4 \frac{Z_1 Z_2}{R} \text{MeV}, \tag{4-60}$$

where R is in Fermis $(R \approx 1.4\, A^{1/3}\ \text{fm})$, then the WKB solution for the penetration factor is

$$\begin{aligned} P_l &\approx \left(\frac{E_c}{E}\right)^{1/2} \exp[-W_l] = \left(\frac{E_c}{E}\right)^{1/2} \exp\left[-\frac{2\pi Z_1 Z_2 e^2}{\hbar v} + 4\left(\frac{E_c}{\hbar^2/2MR^2}\right)^{1/2}\right. \\ &\quad \left. -2l(l+1)\left(\frac{\hbar^2/2MR^2}{E_c}\right)^{1/2}\right] \\ &\approx \left(\frac{E_c}{E}\right)^{1/2} \exp\left[-(E_G/E)^{1/2} + 1.05(ARZ_1Z_2)^{1/2} - 7.62\left(l+\frac{1}{2}\right)^2 (ARZ_1Z_2)^{-1/2}\right], \end{aligned} \tag{4-61}$$

where the center of momentum energy, E, is $\ll E_c$, the Gamow energy, E_G, is given in Eq. (4-34); M is the reduced mass of the interacting nuclei, all energies are in MeV and R is in Fermis. Substituting Eq. (4-61) into Eq. (4-58), we obtain

$$\begin{aligned} \Gamma &\approx 6.0\, \theta_l^2 \left(\frac{\hbar^2}{2MR^2} E_c\right)^{1/2} \exp(-W_l) \\ &\approx 33.0\, \theta_l^2 \left(\frac{Z_1 Z_2}{AR^3}\right)^{1/2} \exp(-W_l)\ \text{MeV}, \end{aligned} \tag{4-62}$$

where again R is in Fermis.

At high temperatures, where the energies are high enough to be on the order of resonance energies, the low lying nuclear states may be excited. For the special case of a thermal distribution of excited states of initial nuclei, and for reactions which proceed through resonances in the compound nucleus, we have the relation (BAHCALL and FOWLER, 1969)

$$N_A\langle\sigma v\rangle_e = \sum_n \frac{G_1^g}{G_1} N_A\langle\sigma v\rangle_g \left[1 + \sum_{e \neq 0} \frac{\Gamma_{12}^e}{\Gamma_{12}^g}\right], \tag{4-63}$$

where the summation is over n resonances, $N_A\langle\sigma v\rangle_g$ is the rate of the reaction when all target nuclei are assumed to be in their ground state, g, and is given by Eq. (4-50), the Γ_{12}^{e} is the partial width for the decay of the resonance by re-emission of the projectile 2 leaving nucleus 1 in the excited state, e, Γ_{12}^{g} is the decay width to the ground state of the target nucleus and is larger than Γ_{12}^{e} (for the same resonance) because of barrier-penetration factors, and the nuclear-partition functions, G_1, and G_1^g, are given by

$$G_1 = \sum_{e=0}^{n} G_1^e,$$

where

$$G_1^e = (2J_1^e + 1)\exp\left(-\frac{E_1^e}{kT}\right),$$

and

$$G_1^g = 2J_1^g + 1,$$

the summation is over n excited states $e = 0, 1, 2, \ldots, n$, the spin of the target nucleus, 1, is J_1^e, and the excitation energy for the excited state in question is E_1^e. Nuclear partition functions for nuclei with mass numbers $A \leq 40$ that have low lying excited states have been calculated as a function of temperature by BAHCALL and FOWLER (1970). The influence of excited states on thermonuclear reactions has been further discussed by MICHAUD and FOWLER (1970) and ARNOULD (1972). When the effective thermal energy, E_0, is larger than the Coulomb barrier energy, E_c, and the outgoing particle is a photon, γ, or if the penetration factor does not vary greatly over the range of interest, then the additional reaction rate term is given by (WAGONER, 1969)

$$N_A \langle \sigma v \rangle = C_{10} \exp(-C_{11}/T_9), \tag{4-64}$$

where the constant C_{10} may be measured experimentally, and $C_{11} = 11.605\, E_c$, where E_c is given by Eq. (4-60).

4.2.8. Inverse Reaction Rates and Photodisintegration

For the reaction $1 + 2 \rightarrow 3 + 4 + Q$, we have the reciprocity relation

$$N_A \langle 34 \rangle = C_{12} [\exp(-C_{13}/T_9)] N_A \langle 12 \rangle, \tag{4-65}$$

where $\langle 12 \rangle$ and $\langle 34 \rangle$ denote, respectively, the $\langle \sigma v \rangle$ for the forward and reverse reactions, and the constants C_{12} and C_{13} are given by (BLATT and WEISSKOPF, 1952; FOWLER, CAUGHLAN, and ZIMMERMAN, 1967)

$$C_{12} = \frac{(1+\delta_{34})}{(1+\delta_{12})} \frac{g_1 g_2}{g_3 g_4} \left[\frac{A_1 A_2}{A_3 A_4}\right]^{3/2}, \tag{4-66}$$

and $C_{13} = 11.605\, Q_6$, where the Kronecker delta function, δ_{12}, is one if $1 = 2$ and zero if $1 \neq 2$, the statistical weight, g_i, of the nucleus i is $g_i = (2I_i + 1)$, where I_i is the nuclear spin, A_i is the atomic mass of i in atomic mass units, and Q_6 is the energy release of the forward reaction in MeV. Similar reciprocity formulae for reactions involving five particles are given by FOWLER, CAUGHLAN, and ZIMMERMAN (1967) and BAHCALL and FOWLER (1969). At high temperatures, $T > 10^9$ °K, some nuclear states may be excited and Eq. (4-66) may be generalized by replacing $g_1 g_2/(g_3 g_4)$ by $G_1 G_2/(G_3 G_4)$, where G_i is the partition function for the ith nucleus (BAHCALL and FOWLER, 1969). Nuclear partition functions for nuclei with mass numbers $A \leq 40$, which have low lying excited states, have been calculated as a function of temperature by BAHCALL and FOWLER (1970).

For the special case of the radiative capture reaction $1+2 \rightarrow 3+\gamma+Q$, the lifetime, $\tau_\gamma(3)$, for the reverse photodisintegration is given by (FOWLER, CAUGHLAN, and ZIMMERMAN, 1967)

$$\lambda_\gamma(3) = \frac{1}{\tau_\gamma(3)} = C_{14}\, T_9^{3/2} [\exp(-C_{13}/T_9)]\, N_A \langle 12 \rangle \sec^{-1}, \tag{4-67}$$

where $\langle 12 \rangle$ is the $\langle \sigma v \rangle$ for the forward reaction, the constant C_{13} is given following Eq. (4-66), and the constant C_{14} is given by

$$C_{14} = 0.987 \times 10^{10} \frac{g_1 g_2}{(1+\delta_{12}) g_3} \left(\frac{A_1 A_2}{A_3} \right)^{3/2}. \tag{4-68}$$

Values of the constants C_{12}, C_{13}, and C_{14} for important thermonuclear processes in stars are given in Sect. 4.4.

4.2.9. Electron Shielding—Weak and Strong Screening

In a dense gas, each nucleus attracts neighboring electrons and repels neighboring nuclei, thus forming a screening cloud of electrons. The effect of this electron shield on reaction rates depends upon the density-temperature combination. Here we will discuss the weak and strong screening effects for which the reaction rate, r_{12}, determined from laboratory measurements, need only be multiplied by a screening factor, f_{12}, to obtain the correct reaction rate. At sufficiently high densities and low temperatures, the reaction rate is substantially altered, and this case is discussed in the next section on pycnonuclear reactions.

At low densities and high temperatures, the nuclei act as a classical ionized gas, and the effect of the background plasma may be calculated using the Debye-Hückel (1923) theory. At slightly higher densities and lower temperatures, the background nuclei are frozen into a Coulomb lattice, but the reacting nuclei may be considered free. The different regimes depend on the effective thermal energy, E_0, the thermal energy per particle, kT, and the electrostatic energy per nucleus, E_c. The effective thermal energy for nonresonant reactions is given by (cf. Eq. (4-44))

$$E_0 = \frac{1}{3} k T \left(\frac{27 \pi^2}{4} \frac{E^*}{kT} \right)^{1/3} \approx 0.122 (Z_1^2 Z_2^2 A)^{1/3}\, T_9^{2/3} \text{ MeV}, \tag{4-69}$$

where $T_9 = T/10^9$, the A is the reduced atomic weight of the reacting nuclei, $A = A_1 A_2/(A_1 + A_2)$, the number of protons of the two reacting nuclei are, respectively, Z_1 and Z_2, and their effective Coulomb energy is given by

$$E^* = \frac{Z_1 Z_2 e^2}{r^*} = 7.9462 \times 10^{-8} A (Z_1 Z_2)^2 \text{ erg}, \tag{4-70}$$

where the interaction radius, r^*, is given by

$$r^* = \frac{(A_1 + A_2)}{2 A_1 A_2 Z_1 Z_2} \frac{\hbar^2}{e^2 m_\mathrm{H}} = \frac{29.030}{A Z_1 Z_2} \text{ Fermi}, \tag{4-71}$$

where A is twice the reduced atomic weight, $A = 2A_1 A_2/(A_1 + A_2)$, and $m_H = 1.66044 \times 10^{-24} g$ is the atomic mass unit (a.m.u.) defined in Eq. (4-8). For a Wigner-Seitz (1934) sphere of radius, a_s, and negative charge, $-4\pi e a_s^3 N_e/3$, the electrostatic interaction energy per nucleus is given by

$$E_c = \frac{9}{10} \frac{Z^2 e^2}{a_s} = 1.8278 \lambda E^*, \tag{4-72}$$

where N_e is the electron density, and

$$\lambda = r^* \left(\frac{N_e}{2Z_1}\right)^{1/3} = \frac{A_1 + A_2}{2A_1 A_2 Z_1 Z_2} \left[\frac{1}{Z_1 \mu_e} \frac{\rho}{1.3574 \times 10^{11} \text{ g cm}^{-3}}\right]^{1/3}. \tag{4-73}$$

Here the gas mass density is ρ, and the mean electron molecular weight, μ_e, is given by

$$\mu_e^{-1} = \sum_i X_i \frac{Z_i}{A_i} \left[1 + \frac{Z_i m}{A_i m_H}\right]^{-1}, \tag{4-74}$$

where X_i, Z_i, and A_i are, respectively, the mass fraction, the number of protons, and the atomic weight of the nuclei of species, i, the electron mass is m, and $m_H \approx 1.6604 \times 10^{-24}$ g. The weak and strong screening conditions are given by

$$\begin{aligned} &E_c \ll kT \quad \text{and } E_0 \gg E_c \quad \text{(weak screening)} \\ &E_c \gg kT \quad \text{and } E_0 \gg E_c \quad \text{(strong screening).} \end{aligned} \tag{4-75}$$

The weak screening multiplicative correction factor for nonresonant reaction rates is given by (SALPETER, 1954; SCHATZMAN, 1948, 1958; SALPETER and VAN HORN, 1969)

$$f_{12} = \exp\left[\sqrt{3} \frac{Z_2}{Z_1} \left(\frac{\mu_e}{Z_1}\right)^{1/2} \xi \Gamma_{Z_1}^{3/2}\right] \tag{4-76}$$

$$\approx \exp[0.188 Z_1 Z_2 \xi \rho^{1/2} T_6^{-3/2}], \tag{4-77}$$

where the gas mass density is ρ, the $T_6 = T/10^6$, the factor, ξ, is given by

$$\xi^2 = \sum_i X_i \frac{Z_i^2}{A_i} \left[1 + \frac{Z_i}{A_i} \frac{m}{m_H}\right]^{-1} + \frac{f'}{f} \frac{1}{\mu_e} \tag{4-78}$$

$$\approx \sum_i \frac{X_i}{A_i} [Z_i^2 + Z_i], \tag{4-79}$$

where f'/f is the logarithmic derivative of the Fermi-Dirac distribution function, f, with respect to its argument, and is unity for a non-degenerate gas and zero for a fully degenerate gas of electrons. The dimensionless parameter, Γ_Z, which compares the Coulomb and thermal energies, is given by

$$\Gamma_Z = Z^{5/3} \frac{e^2}{kT} \left(\frac{4\pi N_e}{3}\right)^{1/3} \approx Z^{5/3} \frac{5.7562 \times 10^8 \text{ °K}}{T} \left(\frac{1}{\mu_e} \frac{\rho}{6.203 \times 10^{13} \text{ g cm}^{-3}}\right)^{1/3}, \tag{4-80}$$

for a nucleus of charge Z.

For strong screening, the multiplicative correction factor for nonresonant reaction rates is given by (SALPETER, 1954; SCHATZMAN, 1948, 1958; WOLF, 1965; SALPETER and VAN HORN, 1969)

$$f_{12} = \exp[U_{s0} + U_{s1}] \tag{4-81}$$

$$\approx \exp\left\{0.205\left[(Z_1+Z_2)^{5/3} - Z_1^{5/3} - Z_2^{5/3}\right]\left(\frac{\rho}{\mu_e}\right)^{1/3} T_6^{-1}\right\}, \tag{4-82}$$

where $T_6 = T/10^6$, the

$$U_{s0} = 0.9\left[\Gamma_{Z_1+Z_2} - \Gamma_{Z_1} - \Gamma_{Z_2}\right], \tag{4-83}$$

where $\Gamma_{Z_1+Z_2}$, Γ_{Z_1}, and Γ_{Z_2} are given by Eq. (4-80) with Z replaced, respectively, by (Z_1+Z_2), Z_1, and Z_2, and

$$\exp[U_{s1}] = \frac{(1+0.3\,\Gamma_{Z_1+Z_2} + 0.266\,\Gamma_{Z_1+Z_2}^{3/2})}{(1+0.3\,\Gamma_{Z_1} + 0.266\,\Gamma_{Z_1}^{3/2})(1+0.3\,\Gamma_{Z_2} + 0.266\,\Gamma_{Z_2}^{3/2})}. \tag{4-84}$$

VAN HORN and SALPETER (1969) have also discussed the screening corrections for resonant reactions, and give explicit correction formulae for the triple alpha reaction. They also give interpolation formulae connecting the weak and strong screening cases. Generalized screening functions for arbitrary charge conditions in a plasma screening nuclear reactions are given by DE WITT, GRABOSKE, and COOPER (1973) and GRABOSKE, DE WITT, GROOSMAN, and COOPER (1973).

4.2.10. Pycnonuclear Reactions

At sufficiently high densities and low temperatures, both the background nuclei and the reacting nuclei are bound in a Coulomb lattice and the reaction becomes pycnonuclear. A pycnonuclear reaction is one for which the reaction rates are relatively insensitive to temperature, but very sensitive to density. Pycnonuclear conditions hold when the effective thermal energy, E_0, given by Eq. (4-69) is much less than the electrostatic energy per nucleus, E_c, given by Eq. (4-72). That is

$$E_0 \ll E_c$$

and

$$\beta = 0.032234\,\lambda\left[A Z_1^2 Z_2^2\,\frac{7.6696\times 10^{10}\,{}^\circ\mathrm{K}}{T}\right]^{2/3} \gg 1 \tag{4-85}$$

for pycnonuclear reactions. Here the density parameter, λ, is given by

$$\lambda = \frac{1}{2 A Z_1 Z_2}\left(\frac{1}{Z_1 \mu_e}\,\frac{\rho}{1.3574\times 10^{11}\ \mathrm{g\ cm^{-3}}}\right)^{1/3}, \tag{4-86}$$

the reduced atomic weight $A = A_1 A_2/(A_1+A_2)$, the mass density is ρ, the μ_e is the mean electron molecular weight given in Eq. (4-74), and Z_1 and Z_2 are the respective number of protons in the two reacting nuclei.

Pycnonuclear reactions have been considered by WILDHACK (1940), ZELDOVICH (1958), CAMERON (1959), WOLF (1965), and SALPETER and VAN HORN (1969). The latter authors give a "zero temperature" reaction rate of

$$r = \begin{pmatrix} 3.90 \\ 4.76 \end{pmatrix} 10^{46} \frac{\rho}{2\mu_A} A^2 Z^4 S \lambda^{7/4} \exp\left[-\begin{pmatrix} 2.638 \\ 2.516 \end{pmatrix} \lambda^{-1/2} \right] \quad \text{reactions cm}^{-3} \text{ sec}^{-1}. \tag{4-87}$$

Here the upper and lower numbers in parentheses correspond to two different approximations to the potential function, the mass density, ρ, is in g cm^{-3}, and the cross section factor, S, is in units of MeV barns. The mean molecular weight, μ_A, is given by

$$\mu_A^{-1} = \sum_i \frac{X_i}{A_i} \left[1 + \frac{Z_i}{A_i} \frac{m}{m_H} \right]^{-1}, \tag{4-88}$$

where X_i, Z_i, and A_i are, respectively, the mass fraction, number of protons, and atomic weight of nuclei of species i, the electron mass is m, and $m_H = 1.66044 \times 10^{-24}$ g. The cross section factor, S, is defined in Eqs. (4-35) and (4-49), and the density factor, λ, is given in Eq. (4-86). It follows from Eqs. (4-86) and (4-87) that the reaction rate is a sensitive function of mass density, ρ, and goes as $r \propto \rho^{19/12} \exp[-C\rho^{-1/6}]$, where C is a constant. It follows from Eq. (4-41) that the energy generation, ε, is given by

$$\varepsilon = rQ/(\rho) \text{ erg g}^{-1} \text{ sec}^{-1}, \tag{4-89}$$

where r is given by Eq. (4-87) and Q is the energy release per reaction.

CAMERON (1959) has given approximate numerical expressions for the pycnonuclear reaction rates as functions of both temperature and density for the reactions

$$\begin{aligned} 3\,He^4 &\to C^{12} \\ C^{12} + He^4 &\to O^{16} + \gamma \\ N^{14} + He^4 &\to F^{18} + \gamma \\ &\quad\ \downarrow \\ &\quad\ O^{18} + \beta^+ + \nu_e \end{aligned} \tag{4-90}$$

and

$$O^{16} + He^4 \to Ne^{20} + \gamma,$$

as well as for the reactions of the heavy ions C^{12}, O^{16}, Ne^{24}, Mg^{32}, Si^{38}, S^{44}, A^{50} and Ca^{56} with themselves. He concludes that helium pycnonuclear reactions might ignite stellar explosions in the advanced stages of stellar evolution.

SALPETER and VAN HORN (1969) also give pycnonuclear corrections to the strong screening case when $E_0 \approx E_c$, and give a temperature corrected pycnonuclear reaction rate, $r(T)$.

$$\begin{aligned} \frac{r(T)}{r} - 1 = &\begin{pmatrix} 0.0430 \\ 0.0485 \end{pmatrix} \lambda^{-1/2} \left[1 + \begin{pmatrix} 1.2624 \\ 2.9314 \end{pmatrix} \exp(-8.7833\,\beta^{3/2}) \right]^{-1/2} \\ &\times \exp\left\{ -7.272\,\beta^{3/2} + \lambda^{-1/2} \begin{pmatrix} 1.2231 \\ 1.4331 \end{pmatrix} \exp(-8.7833\,\beta^{3/2}) \right. \\ &\left. \times \left[1 - \begin{pmatrix} 0.6310 \\ 1.4654 \end{pmatrix} \exp(-8.7833\,\beta^{3/2}) \right] \right\}, \end{aligned} \tag{4-91}$$

where r is given in Eq. (4-87) and λ and β are given, respectively, in Eqs. (4-86) and (4-85).

4.3. Weak Interaction Processes

4.3.1. Electron Neutrino, Mu Neutrino, Muons, Pions, and Weak Interaction Theory

PAULI (1930) first showed that the radioactive beta decay could only be explained if the proton, p, and the neutron, n, have the weak interactions

$$n \rightarrow p + \beta^- + \overline{\nu}_e,$$

and

$$p \rightarrow n + \beta^+ + \nu_e, \tag{4-92}$$

where the negative electron, β^-, is called the negatron, β^+ is called the positron, ν_e is the electron neutrino, and $\overline{\nu}_e$ is the antielectron neutrino. Properties of the negatron and positron are given in Table 37, whereas properties of the electron neutrino and the other particles introduced in this section are given in Table 39.

The strong interaction, which accounts for the interaction nuclear force which holds protons and neutrons together, was thought to be due to another fundamental particle called the pion or the π-meson (YUKAWA, 1935, 1937; OPPENHEIMER and SERBER, 1937). Just as a photon results from the quantization of the electromagnetic field, the pion results from the quantization of the field which explains the nuclear force. The strong interactions are given by

$$\begin{aligned} p &\rightarrow n + \pi^+ \\ n &\rightarrow p + \pi^- \\ p &\rightarrow p + \pi^0 \\ n &\rightarrow n + \pi^0, \end{aligned} \tag{4-93}$$

where π^+ and π^- are the charged pions, and the neutral pion, π^0, was introduced (YUKAWA, 1938) to explain the charge independence of the nuclear force.

Before Yukawa's theoretical arguments, extraterrestrial high energy particles, called cosmic rays, had been observed (HESS, 1911, 1912; KOLHÖRSTER, 1913). It was thought that the massive π meson of Yukawa's theory might be a component of the cosmic rays, and such a component was soon found (ANDERSON and NEDDERMEYER, 1936). Detailed examinations of the cross section and lifetime of the cosmic ray component (NEDDERMEYER and ANDERSON, 1937; STREET and STEVENSON, 1937; NISHINA, TAKEUCHI, and ICHIMIYA, 1937; ROSSI, 1939) indicated that this particle was different from the nuclear π meson. The cosmic ray meson was therefore called a muon or μ-meson, and a two meson theory was introduced (SAKATA and INOUE, 1946; MARSHAK and BETHE, 1947). According to this theory, the nuclear π meson and the cosmic ray μ meson were interrelated by the decay reactions

$$\begin{aligned} \pi^+ &\rightarrow \mu^+ + \nu_\mu \\ \pi^- &\rightarrow \mu^- + \overline{\nu}_\mu \\ \pi^0 &\rightarrow \gamma + \gamma, \end{aligned} \tag{4-94}$$

where ν_μ is the mu-neutrino, $\overline{\nu}_\mu$ is the anti-mu-neutrino, and γ denotes a gamma ray photon. The π-meson was then also found to be a cosmic ray component (LATTES, OCCHIALINI, and POWELL, 1947) as well as a product of particle ac-

Table 39. Properties of the electron neutrino, mu-neutrino, muons and pions[1]

Particle	Symbol	Spin	Rest mass (MeV)	Magnetic moment ($e\hbar/2m_\mu c$)	Mean life (sec)	Decay mode	Fraction mode	Q (MeV)
electron neutrino	$\nu_e, \bar{\nu}_e$	1/2	<0.00006	0	∞			
mu neutrino	$\nu_\mu, \bar{\nu}_\mu$	1/2	<1.2	0	∞			
muon, μ-meson	μ^+, μ^-	1/2	105.6594(4)	1.00116616(31)	2.1994(6) E–6	$\mu^\pm \to \nu_e/\bar{\nu}_e + \bar{\nu}_\mu/\nu_\mu + e^\pm$	$\sim 100\%$	105
						$\mu^\pm \to e^\pm + \gamma + \gamma$	<1.6 E–5	105
						$\mu^\pm \to 2e^\pm + e^\mp$	<6 E–9	104
						$\mu^\pm \to e^\pm + \gamma$	<2.2 E–8	105
pion, π-meson	π^+, π^-	0	139.576(11)	0	2.6024(24) E–8	$\pi^\pm \to \mu^\pm + \nu_\mu/\bar{\nu}_\mu$	$\sim 100\%$	34
						$\pi^\pm \to e^\pm + \nu_e/\bar{\nu}_e$	1.24(3) E–4	139
						$\pi^\pm \to \mu^\pm + \gamma + \nu_\mu/\bar{\nu}_\mu$	1.24(25) E–4	34
						$\pi^\pm \to e^\pm + \pi^0 + \nu_e/\bar{\nu}_e$	1.02(7) E–8	4
						$\pi^\pm \to e^\pm + \gamma + \nu_e/\bar{\nu}_e$	3.0(5) E–8	139
	π^0	0	134.972(12)	0	0.84(10) E–16	$\pi^0 \to \gamma + \gamma$	98.84(4) %	135
						$\pi^0 \to \gamma + e^+ + e^-$	1.16(4) %	134
						$\pi^0 \to 2e^+ + 2e^-$	3.47 E–5	133
						$\pi^0 \to \gamma + \gamma + \gamma$	<5 E–6	135

[1] The value in parentheses following each quoted value represents the standard error in the final digit of the quoted value. Parameter values are from Rosenfeld *et al.* (1968), and Söding *et al.* (1972). Both $\mu^\pm$ and $\pi^\pm$ have charges of $\pm$e where e is the charge of the electron. The decay mode nomenclature $\pi^\pm \to \mu^\pm + \nu_\mu/\bar{\nu}_\mu$ means $\pi^+ \to \mu^+ + \nu_\mu$ and $\pi^- \to \mu^- + \bar{\nu}_\mu$.

celerator bombardments (GARDNER and LATTES, 1948). Substituting Eq. (4-94) into Eq. (4-93) we obtain the reactions

$$n \rightarrow p + \mu^- + \overline{\nu}_\mu,$$
and
$$p \rightarrow n + \mu^+ + \nu_\mu, \tag{4-95}$$

which are similar to the electron neutrino reactions of Eq. (4-92). Apart from the interactions given in Eq. (4-95), the muon decays according to the reactions

$$\mu^- \rightarrow \beta^- + \overline{\nu}_e + \nu_\mu$$
and
$$\mu^+ \rightarrow \beta^+ + \nu_e + \overline{\nu}_\mu, \tag{4-96}$$

with a mean lifetime, $\tau \approx 2.199 \times 10^{-6}$ seconds.

The mu-neutrino, ν_μ, has the same physical properties as the electron neutrino, ν_e, except that it reacts with the muon instead of the electron. In fact, both neutrinos have been detected with interaction cross sections given by (REINES and COWAN, 1953)

$$\sigma \approx 10^{-44}\ \text{cm}^2 \quad \text{at 1 MeV}, \tag{4-97}$$

for the reaction $\overline{\nu}_e + p \rightarrow e^+ + n$, and (DANBY *et al.*, 1962)

$$\sigma \approx 10^{-38}\ \text{cm}^2 \quad \text{at 1 GeV}, \tag{4-98}$$

for the reactions $\overline{\nu}_\mu + p \rightarrow n + \mu^+$ and $\nu_\mu + n \rightarrow p + \mu^-$. These data are consistent with a universal coupling constant, g, applicable to both neutrino interactions. The interaction cross section, σ, is given by

$$\sigma \approx \frac{g^2 E_\nu^2}{\hbar^4 c^4} \approx 10^{-44} \left(\frac{E_\nu}{mc^2}\right)^2 \text{cm}^2, \tag{4-99}$$

where E_ν is the center of mass neutrino energy, and the Fermi constant, g, is given by

$$g = 1.4102 \pm 0.0012 \times 10^{-49}\ \text{erg cm}^3. \tag{4-100}$$

FERMI (1934) first realized that the β decay (Eq. (4-92)) was caused by a weak interaction which was different from the nuclear interaction; and postulated an interaction matrix element, v_{fi}, given by

$$v_{\text{fi}} = (\overline{\psi}_{\text{n}} O \psi_{\text{p}})(\overline{\psi}_{\text{e}} O_{\text{L}} \psi_{\nu_{\text{e}}}), \tag{4-101}$$

where the wave function ψ has subscripts n, p, e or ν_e denoting the neutron, n, the proton, p, the electron, β^- or e^- or e, and the neutrino, ν_e, and O and O_L denote, respectively, operators which operate on the nucleon (proton and neutron) and the lepton (electron and electron neutrino) wave functions. The transition probability, W, for a transition from an initial state, whose wave function is ψ_i, to a final state, whose wave function is ψ_{f}, is given by the golden rule of time dependent perturbation theory

$$\begin{aligned} W &= \frac{2\pi}{\hbar} v_{\text{fi}}^* v_{\text{fi}} \rho(E_{\text{f}}) \\ &= \frac{2\pi}{\hbar} \left|\int \psi_{\text{f}}^* H_{\text{int}} \psi_i \, d\tau\right|^2 \rho(E_{\text{f}}) \\ &= \frac{2\pi}{\hbar} |\langle \psi_{\text{f}} | H_{\text{int}} | \psi_i \rangle|^2 \rho(E_{\text{f}}), \end{aligned} \tag{4-102}$$

where v_{fi} is the interaction matrix element, H_{int} is the interaction Hamiltonian capable of matrix elements between the initial and final states, and $\rho(E_{\mathrm{f}})$ is the total number of final states in the energy range E_{f} to $E_{\mathrm{f}}+dE_{\mathrm{f}}$.

During Fermi's time, it was unknown what the exact form of the operator, O, was; but it was known that it must be some mixture of coupling constants which measure the strength of the interactions and the five possible sets of combinations of Dirac matrices (cf. PRESTON, 1962). It was also known that in nature the protons, neutrons, electrons, muons, and neutrinos interact according to the combinations given by (WHEELER, 1947; TIOMNO and WHEELER, 1949)

$$(\mathrm{p},\mathrm{n}) \leftrightarrow (\mu, \nu_\mu) \leftrightarrow (\mathrm{e}, \nu_\mathrm{e}) \leftrightarrow (\mathrm{p},\mathrm{n}). \tag{4-103}$$

FEYNMAN and GELL-MANN (1958) postulated the existence of a universal weak interaction in which any of the pairs given in Eq. (4-103) can interact with each other, as well as with themselves. Such self-weak interactions have a weak interaction Hamiltonian given by

$$H_{\mathrm{int}} = \sqrt{2}\, g_V J^* J, \tag{4-104}$$

where the weak interaction current, J, is given by

$$J = \bar{\psi}_\mathrm{n}(V-A)\psi_\mathrm{p} + \bar{\psi}_\mathrm{e}(V-A)\psi_{\nu_\mathrm{e}} + \bar{\psi}_\mu(V-A)\psi_{\nu_\mu}, \tag{4-105}$$

the wave function, ψ, has a subscript denoting the appropriate particle, and the vector J^* is the Hermitian conjugate of J. Here the operator $(V-A)$ reflects the experimental conclusion that only the vector, V, and axial vector, A, operators exist in the weak interaction; as well as Feynman and Gell-Mann's hypothesis that they appear in the mixture $(V-A)$. The vector coupling constant, C_V, is measured from the beta decay of ^{26}Al, and has the value (FREEMAN *et al.*, 1972)

$$g_v = C_V = 1.4102 \pm 0.0012 \times 10^{-49}\ \mathrm{erg\ cm}^3. \tag{4-106}$$

The ratio of the axial vector coupling constant, C_A, to C_V is measured from the neutron half life. CHRISTENSEN *et al.* (1967) measure a neutron half life of 10.61 ± 0.16 minutes which results in

$$\frac{C_A}{C_V} = \frac{g_A}{g_V} = 1.239 \pm 0.011. \tag{4-107}$$

This ratio follows from the neutron $(\mathrm{ft})_\mathrm{n}$ value given by

$$(ft)_\mathrm{n} = \frac{2\pi^3 (\ln 2)\hbar^7}{m^5 c^4 [C_V^2 |\mathcal{M}_V|^2 + C_A^2 |\mathcal{M}_A|^2]} = \frac{1.230627 \times 10^{-94}}{[C_V^2 + 3C_A^2]} \approx 1040\ \mathrm{sec}. \tag{4-108}$$

(For neutron $|\mathcal{M}_V|^2 = 1$ and $|\mathcal{M}_A|^2 = 3$.)

The current-current interaction results in a β decay interaction Hamiltonian, H_β, and an electron-neutrino self interaction Hamiltonian, H_W, given by

$$H_\beta = \frac{g_V}{\sqrt{2}} [\bar{\psi}_\nu \gamma_\alpha (1+\gamma_5)\psi_\mathrm{e}] [\bar{\psi}_\mathrm{n} \gamma_\alpha (C_V - C_A \gamma_5)\psi_\mathrm{p}] \quad \text{for } (\mathrm{p},\mathrm{n})(\mathrm{e},\nu), \tag{4-109}$$

and

$$H_W = \sqrt{2}\, g_V \left[\bar{\psi}_\mathrm{e} \gamma_\mu \frac{(1+\gamma_5)}{\sqrt{2}} \psi_\nu\right] \left[\bar{\psi}_\nu \gamma^\mu \frac{(1+\gamma_5)}{\sqrt{2}} \psi_\mathrm{e}\right] \quad \text{for } (\mathrm{e},\nu)(\mathrm{e},\nu). \tag{4-110}$$

Here the operator $(V-A)$ has been expressed in terms of combinations of Dirac matrices, γ, and the symbols have the meanings described in the texts of KONOPINSKI (1966) and PRESTON (1962). Eq. (4-110) is often rearranged to give

$$H_W = g_V [\overline{\psi}_e \gamma_\mu \psi_e] \left[\overline{\psi}_\nu \gamma^\mu \frac{(1+\gamma_5)}{\sqrt{2}} \psi_\nu \right] - g_V [\overline{\psi}_e \gamma_5 \gamma_\mu \psi_e] \left[\overline{\psi}_\nu \gamma^\mu \frac{(1+\gamma_5)}{\sqrt{2}} \psi_\nu \right]. \quad (4\text{-}111)$$

4.3.2. Beta Decay

As suggested by PAULI (1930), the particle called the electron neutrino, ν_e, must be present if reactions involving β decay satisfy the laws of conservation of energy and momentum. The nuclear reaction for negative beta, β^-, decay is

$$(Z-1,A) \rightarrow (Z,A) + \beta^- + \overline{\nu}_e, \quad (4\text{-}112)$$

where (Z,A) denotes a nucleus of charge, Z, and mass number, A, the β^- particle is an electron, and $\overline{\nu}_e$ denotes the antielectron neutrino. Negative beta decay will only occur if

$$M_{at}(Z-1,A) > M_{at}(Z,A), \quad (4\text{-}113)$$

or equivalently

$$M(Z-1,A) > M(Z,A) + m_e,$$

where $M_{at}(Z,A)$ denotes the mass of the atom, $M(Z,A)$ denotes the mass of the nucleus, and the electron mass $m_e = 5.485930 \times 10^{-4}$ a.m.u. $= 0.5110041$ MeV$/c^2$. Negative beta particles are usually ejected with energies ranging from a few keV to 15 MeV.

The nuclear reaction for positive beta, β^+, decay is

$$(Z+1,A) \rightarrow (Z,A) + \beta^+ + \nu_e, \quad (4\text{-}114)$$

where the positive beta particle, β^+, is the positron, and ν_e is the electron neutrino. Positive beta decay will only occur if

$$M_{at}(Z+1,A) > M_{at}(Z,A) + 2m_e, \quad (4\text{-}115)$$

or equivalently

$$M(Z+1,A) > M(Z,A) + m_e.$$

In general, it is found that for a given odd A only one nucleus is stable against beta decay, and for a given even A there are only two nuclei stable against beta decay.

It follows from Eqs. (4-102), (4-104), and (4-105) that the transition probability per unit time, W, for a single negative beta decay of a nucleus (Z,A) is given by

$$W = \frac{2\pi g_V^2}{\hbar} |\mathcal{M}|^2 \rho(E_f), \quad (4\text{-}116)$$

where the nuclear matrix element, $\mathcal{M}$, is given by

$$\mathcal{M} = \int (\overline{\psi}_{Z,A} (V-A) \psi_{Z+1,A}) (\overline{\psi}_e (V-A) \psi_{\nu_e}) \, d\tau \approx \int \overline{\psi}_{Z,A} (V-A) \psi_{Z+1,A} \, d\tau. \quad (4\text{-}117)$$

Here the integrals are over the volume of the nucleus, and the electron and neutrino wave functions have wave solutions with arguments so small that the wave functions contribute a constant factor near unity to the integral of Eq. (4-117). For the allowed transitions of nuclear β decay, the nuclear matrix element, $\mathcal{M}$, satisfies the equation

$$|\mathcal{M}|^2 = |\mathcal{M}_{\mathrm{F}}|^2 + \frac{g_A^2}{g_V^2}|\mathcal{M}_{\mathrm{GT}}|^2, \tag{4-118}$$

where $\mathcal{M}_{\mathrm{F}}$ and $\mathcal{M}_{\mathrm{GT}}$ denote, respectively, the matrix elements for the Fermi and the Gamow-Teller transitions. For neutron decay, for example, $|\mathcal{M}_{\mathrm{F}}| = |\mathcal{M}_V| = 1$ and $|\mathcal{M}_{\mathrm{GT}}| = |\mathcal{M}_A| = \sqrt{3}$ where we have included subscripts V and A for the vector and axial vector nuclear matrix elements. The Fermi transition corresponds to the vector interaction and has the selection rules $\Delta J = 0$, no change of parity; whereas the Gamow-Teller transition corresponds to the axial-vector interaction and has the selection rules $\Delta J = \pm 1, 0$, no $0 \to 0$ transition, and no change of parity. Here J is the total angular momentum of the nucleus.

As the product nucleus $(Z+1, A)$ may be assumed to be in a single state, only the electron and the neutrino contribute to the energy density of states, $\rho(E_{\mathrm{f}})$. If p_{e} and $p_{\nu_{\mathrm{e}}}$ denote, respectively, the momentum of the electron and the electron neutrino, and W_0 denotes the maximum energy release of the β decay, then it follows from Eq. (4-116) that

$$W = \frac{g_V^2|\mathcal{M}|^2}{2\pi^3\hbar^7 c^3}(W_0 - E)^2 p_{\mathrm{e}}^2\, dp_{\mathrm{e}}, \tag{4-119}$$

where the number of final states, $\rho(E)$, with energy between E and $E+dE$ is

$$\rho(E) = \frac{(W_0 - E)^2 p_{\mathrm{e}}^2\, dp_{\mathrm{e}}}{4\pi^4\hbar^6 c^3}.$$

Integration of Eq. (4-119) over all momenta up to the maximum corresponding to the maximum electron energy, W_0, available for beta decay gives the total transition probability per unit time, λ_β. This beta decay rate is given by

$$\lambda_\beta = \frac{\ln 2}{t} = \frac{f \ln 2}{f t} = \frac{1}{2\pi^3}\frac{m^5 c^4 g_V^2}{\hbar^7}|\mathcal{M}|^2 \int_1^{W_0} pE(W_0 - E)^2 F(Z,E)\, dE, \tag{4-120}$$

where the factor $m^5 c^4 g_V^2 |\mathcal{M}|^2/\hbar^7$ is of order unity. This factor assumes the value of one if the wave functions are normalized so that $2\pi|\mathcal{M}|^2$ is the probability of a transition per unit time per unit of energy and no state density factor is needed. The $\ln 2 = 0.693$, the half lifetime for beta decay is t, and the Fermi function, $F(Z,E)$ is given in detail by KONOPINSKI (1966) and PRESTON (1962) and has the approximate value

$$F(Z,E) \approx 2\pi\eta[1 - \exp(-2\pi\eta)]^{-1}, \tag{4-121}$$

where

$$\eta = \frac{Ze^2}{\hbar v},$$

and the velocity of the emitted electron is v. The quantity f is given by

$$f = f(Z, W_0) = \int_1^{W_0} F(Z,E)pE(W_0 - E)^2\, dE, \tag{4-122}$$

where E and W_0 are in terms of mc^2, and p is in terms of mc, and both $F(Z,E)$ and $f(Z,W_0)$ have been extensively tabulated by FEENBERG and TRIGG (1950), FIENGOLD (1951), DISMUKE *et al.* (1952), MAJOR and BIEDENHARN (1954), and AJZENBERG-SELOVE (1960). For an allowed transition (no parity change and $\Delta J = 0, \pm 1$), the product ft is well defined by the relation

$$\begin{aligned} ft &= \frac{2\pi^3 \hbar^7 \ln 2}{m^5 c^4} \left[g_V^2 |\mathcal{M}_V|^2 + g_A^2 |\mathcal{M}_A|^2\right]^{-1} \\ &\approx 1.230627 \times 10^{-94} \left[g_V^2 |\mathcal{M}_V|^2 + g_A^2 |\mathcal{M}_A|^2\right]^{-1} \text{sec}. \end{aligned} \tag{4-123}$$

Measured values of ft range from $10^{4.5}$ to $10^{6.0}$ sec, and therefore λ_β is often estimated using tabulated values of f together with an assumed value of $ft = 10^{5.5}$ sec.

As first suggested by HOYLE (1946), beta decay in stellar interiors might be inhibited due to the prepopulation of momentum states in the electron phase space into which the final electron would normally decay (called "exclusion principle inhibition"). For allowed decays (no parity change and $\Delta J = 0, \pm 1$) and for most first forbidden decays (parity change and $\Delta J = 0, \pm 1$), the stellar beta decay rate, λ_S, is related to the terrestrial value, λ_β, by the equation (PETERSON and BAHCALL, 1963)

$$\lambda_S = \lambda_\beta (1 - \delta), \tag{4-124}$$

where the exclusion principle inhibition factor, δ, is given by

$$\delta = \frac{\int S \, d\lambda}{\int d\lambda}, \tag{4-125}$$

where (cf. Eq. (4-120))

$$d\lambda = \frac{p}{2\pi^3} (E + mc^2)(W_0 - E)^2 F(Z,E) \, dE \propto (E + mc^2)^2 (W_0 - E)^2 \, dE,$$

the relativistic energy of the electron is E, the end point energy for decay is W_0, and the probability, S, that the electron states corresponding to an energy between E and $E + dE$ are occupied, is given by

$$S = \left[1 + \exp\left(\frac{E}{kT} - \frac{E_F}{kT}\right)\right]^{-1},$$

where E_F is the Fermi energy. We have the approximate relations

$$\begin{aligned} &\delta \approx 0 && \text{for a nondegenerate gas,} \\ \text{and} \quad &\delta \approx \left[1 - \exp\left(-\frac{E_F}{kT}\right)\right] \approx 1 && \text{for a degenerate gas and } W_0 \ll E_F. \end{aligned} \tag{4-126}$$

PETERSON and BAHCALL (1963) give a lengthy approximate relation for δ in the degenerate case of $W_0 \gg E_F$, where $0 \lesssim \delta \lesssim 1$.

As pointed out by HOYLE (1946) and CAMERON (1959), beta decay lifetimes are considerably shortened by decay from thermally populated excited states of the nuclear species in high temperature stellar interiors. The decay rate for these excited state reactions, λ_e, is given by (CAMERON, 1959)

$$\lambda_e = \frac{1}{P(T)} \sum_i \sum_j \lambda_{sij}(2J_i+1)\exp\left(-\frac{E_i}{kT}\right), \tag{4-127}$$

where λ_s is the stellar beta decay rate between parent states, i, and daughter states, j, the $E_i(J_i)$ is the excitation energy (spin) of the parent state, i, and the nuclear spin of state i is J_i. It is assumed that states i are populated according to a Boltzmann distribution, and the nuclear partition function, $P(T)$, is given by

$$P(T) = \sum_i (2J_i+1)\exp\left(-\frac{E_i}{kT}\right). \tag{4-128}$$

The partition function is often taken to be equal to the statistical weight of the ground state, $2I+1$, where I is the nuclear spin of the ground state. For example, for a two level nucleus with ground state spin J_1 and decay rate $\lambda_\beta(1)$, and an excited state of spin J_2 and decay rate $\lambda_\beta(2)$, the total beta decay rate, λ_β, is given by

$$\lambda_\beta = \frac{2J_1+1}{G}\lambda_\beta(1) + \frac{2J_2+1}{G}\lambda_\beta(2)\exp\left(-\frac{E^*}{kT}\right), \tag{4-129}$$

where

$$G = (2J_1+1)+(2J_2+1)\exp\left(-\frac{E^*}{kT}\right),$$

and E^* is the excitation energy of the excited state. The energy criterion for the occurence of this beta decay is

$$M(Z-1,A) > M(Z,A)+m_e+E^*/c^2, \tag{4-130}$$

where $M(Z,A)$ is the mass of the nucleus in the ground state. CAMERON (1959) has called the beta decay from thermally excited states "photobeta" decay because the excited state is caused by the exchange of energy between matter and radiation (photons) under conditions of statistical equilibrium. To calculate the excited state beta decay rates, level distributions and parameters must be known for the nuclei in question, and much of this data can be obtained from the nuclear energy levels given in the Nuclear Data Sheets of the Oak Ridge National Laboratory or from the Isotope Tables (LEDERER *et al.*, 1967). Otherwise, level parameters may be estimated using the nuclear level density formula of GILBERT and CAMERON (1965). Beta decay rates for both excited and unexcited nuclei are given by HANSEN (1968) and WAGONER (1969).

When excited state beta decay is unfavored due to the large excitation energies of all favorable excited states, another "photobeta" process may be important. In this process the photon can be considered to decay virtually into an electron-positron pair, with the positron being absorbed by the nucleus, giving rise to an antineutrino according to the reaction

$$\gamma+(Z,A) \rightarrow (Z+1,A)+e^- +\bar{\nu}_e, \tag{4-131}$$

subject to the energy criterion

$$\Delta = M(Z+1,A)+m_e-M(Z,A). \tag{4-132}$$

Here Δ is the atomic mass difference, $M(Z,A)$ is the nuclear mass of the nucleus (Z,A), and m_e is the electron mass. The decay rate, λ_p, of this photon induced beta decay is given by (SHAW, CLAYTON, and MICHEL, 1965)

$$\lambda_p = \frac{2\ln 2}{\pi}\frac{\alpha}{ft}\left(\frac{kT}{mc^2}\right)^{9/2}\exp\left[-\frac{\Delta}{kT}\right] \quad \text{for } kT<0.3mc^2 \quad \text{and } \Delta<2mc^2,$$

and (4-133)

$$\lambda_p = \frac{\ln 2}{ft}\frac{\alpha}{\pi}\left[5.78\left(\frac{kT}{mc^2}\right)^5 \ln\left(\frac{kT}{mc^2}\right)+3.10\left(\frac{kT}{mc^2}\right)^5+\cdots\right] \quad \text{for } kT\gg mc^2,\ \Delta,$$

where $\alpha=2\pi e^2/(hc)\approx(137.037)^{-1}$ is the fine structure constant, the ft value is that which characterizes the nuclear matrix element connecting the states (Z,A) and $(Z+1,A)$, the gas temperature is T, and the mass difference is Δ.

When the thermal energy of a star becomes greater than the electron rest mass energy, $T\gtrsim 10^9$ °K, the production of electron-positron pairs sets in. If there is a significant density of free positrons, even terrestrially stable nuclei might undergo beta transformation. Although this process may not be astrophysically important, we give here the probability per unit time, λ, that a nucleus (Z,A) will capture a positron with kinetic energy $W=(E-1)mc^2$ to form a nucleus $(Z+1,A)$. According to REEVES and STEWART (1965)

$$\lambda = \sum_i A_i\lambda_i, \qquad \text{(4-134)}^1$$

where the fractional population, A_j, of level j with spin J_j is

$$A_j = \frac{(2J_j+1)\exp[-E_j/kT]}{\sum_i (2J_i+1)\exp[-E_i/kT]},$$

the excitation energy of the ith level is E_i, the Σ is over all levels of the nucleus, and

$$\lambda_i = \frac{N^+\ln 2}{4\pi ft}\left(\frac{h}{mc}\right)^3 \int_{E_m}^{\infty}(W-W_m)^2 F(Z,W)N(E)\,dE \Bigg/ \int_1^{\infty} N(E)\,dE,$$

where W_m is the threshold energy of the reaction, $W=(E-1)mc^2$ is the kinetic energy of the positron, the positron density, N^+, is given by

$$N^+ = \{[(N_0^2/4)+4I^2/h^6]^{1/2}-N_0/2\}.$$

the residual electron density $N_0=\rho/(2M)$ where ρ is the gas mass density, M is the nuclear mass, and $I=4\pi(mc)^3Z^{-1}K_2(Z)$, where $Z=M_ec^2/(kT)$ and K_2 is the modified Bessel function of the second kind. The energy spectrum of the positrons is given by

$$N(E)\,dE = \frac{8\pi(mc/h)^3E(E^2-1)^{1/2}\,dE}{(2I/N^+h^3)\left[\exp(mc^2E/kT)+1\right]}. \qquad \text{(4-135)}$$

---

[1] The value of λ_i is overestimated by a factor of 2 (see footnote two on page 402).

4.3.3. Electron Capture

A nucleus (Z, A) which is unstable to positive beta, β^+, decay can instead capture an electron from the surrounding electrons (YUKAWA, 1935; MÖLLER, 1937). Moreover, every nucleus can capture an electron provided that the energy of electrons in stellar material is sufficiently high. The process of successive electron captures is called neutronization and it plays a fundamental role in the theory of stellar evolution. Under normal terrestrial conditions electron capture only occurs when

$$(Z+1, A) + e^- \rightarrow (Z, A) + \nu_e, \tag{4-136}$$

where the symbol e^- is used to denote the electron in the electron capture reaction. Electron capture will only occur if

$$M_{at}(Z+1, A) > M_{at}(Z, A) \tag{4-137}$$

or equivalently

$$M(Z+1, A) + m_e > M(Z, A),$$

where $M_{at}(Z, A)$ denotes the atomic mass of atom (Z, A), and $M(Z, A)$ denotes the mass of the nucleus (Z, A). It can be shown (FEENBERG and TRIGG, 1950) that the terrestrial decay rates for positron emission, λ_{β^+}, and the electron capture to the K shell, λ_K, are related by a function of the atomic number, Z, of the product nucleus and the maximum energy release, Q. For large Z and small Q electron capture becomes more probable than β^+ emission, whereas for small Z and large Q, positron emission is more probable. Precise values of the ratio $\lambda_K/\lambda_{\beta^+}$ are given in FEENBERG and TRIGG (1950) and PRESTON (1962). The f values for electron capture are also tabulated by MAJOR and BIEDENHARN (1954). As BAHCALL (1962) has pointed out, the atoms in stellar interiors are highly ionized and cannot easily capture electrons from bound orbits. Nuclei which decay terrestrially by electron capture from bound orbits can, however, decay in stars by the capture of free electrons from continuum orbits.

The stellar decay rate, λ_{ce}, for continuum electron capture on nuclei, which decay terrestrially by allowed and first-forbidden positron emission, is given by (BAHCALL, 1964)

$$\lambda_{ce} = \frac{K \ln 2}{f t}, \tag{4-138}$$

where ft is the terrestrially measured value, $\ln 2 = 0.693$, and the stellar phase-space function, K, is given by

$$K = \int_{p_0}^{\infty} p^2 q^2 F(Z, W) \left[1 - \exp\left(-\frac{\mu}{kT} + \frac{W}{kT}\right)\right]^{-1} dp, \tag{4-139}$$

where W is the total relativistic energy of the electron, μ is the chemical potential, $\mu \approx E_F$, where E_F is the Fermi energy, p is the electron momentum, $F(Z, W)$ is the Fermi function and $q = (W + W_0)/(mc^2)$ is the energy in mc^2 units (0.511 MeV) of the neutrino emitted in the electron capture process. If the difference between the initial and final nuclear masses is W_0, then the threshold value of momentum, p_0, is given by

$$p_0 = 0 \quad \text{if } W_0 \gtrsim -mc^2 \quad \text{(exoergic capture)}$$

and

$$p_0 = \frac{1}{mc^2}(W_0^2 - m^2c^4)^{1/2} \quad \text{if } W_0 \lesssim -mc^2 \quad \text{(endoergic capture).} \tag{4-140}$$

For nonrelativistic electron energies, the Fermi function, $F(Z,W)$, is given by

$$F(Z,W)=2\pi\eta[1-\exp(-2\pi\eta)]^{-1},$$

where

$$\eta=\frac{\alpha Z W}{p}=\frac{\alpha Z c}{v}\approx \alpha Z\left(\frac{mc^2}{3kT}\right)^{1/2},$$

$\alpha=2\pi e^2/(hc)\approx(137.037)^{-1}$, the velocity of the captured electron is v, and we assume that $\alpha^2 Z^2 \ll 1$. As FOWLER and HOYLE (1964) have pointed out, $F(Z,W)$ is a slowly varying function of p and can be factored out of the integral by assigning it a suitable average value $\langle F\rangle$, which they give. HANSEN (1968) has evaluated the resultant expression for K in the computer and gives values of λ_{ce} for several nuclei.

For both a degenerate and nondegenerate gas, and a beta transition threshold energy of W_0,

$$K\approx\pi^2\left(\frac{\hbar}{mc}\right)^3 N_e\langle F\rangle(W_0+1)^2 \quad \text{if } \langle E\rangle \ll . \tag{4-141}$$

Here N_e is the electron density, and $\langle F\rangle$ is the Fermi function evaluated at $W=1+\langle E\rangle$, where the average energy, $\langle E\rangle\approx kT$, for a nondegenerate gas, and $\langle E\rangle\approx E_F$, the Fermi energy, for a degenerate gas. For nonrelativistic and extremely relativistic electrons,

$$E_F=(3\pi^2)^{2/3}\frac{\hbar^2}{2m}N_e^{2/3} \quad \text{(nonrelativistic)}$$

and

$$E_F=(3\pi^2)^{1/3}\hbar c N_e^{1/3} \quad \text{(relativistic)}, \tag{4-142}$$

but note (4-141) is not applicable for a relativistic degenerate gas. For a nondegenerate gas for which Boltzmann statistics apply, BAHCALL (1962) gives

$$K\approx\pi^2 W_0^2 N_e \quad \text{if } \eta\ll 1,\quad \alpha^2Z^2\ll 1,\quad kT\ll mc^2 \quad\text{and}\quad W_0\ll kT. \tag{4-143}$$

For a degenerate gas, BAHCALL (1962) and TSURUTA and CAMERON (1965) give

$$K\approx\langle F\rangle\left[\tfrac{1}{5}P_F^5+\tfrac{1}{3}P_F^3(1+W_0^2)+\tfrac{1}{4}W_0\{2P_F E_F^3-P_F E_F-\ln(P_F+E_F)\}\right] \tag{4-144}$$

$$\text{if}\quad W_0>-mc^2 \text{ (exoergic) and } \eta_F\ll 1,$$

$$K\approx\langle F\rangle\{\tfrac{1}{5}(P_F^5-P_0^5)+\tfrac{1}{3}(P_F^3-P_0^3)(1+W_0^2)-\tfrac{1}{4}|W_0|[2(P_F E_F^3-P_0|W_0|^3) -(P_F E_F-P_0|W_0|)-\ln(P_F+E_F)+\ln(P_0+|W_0|)]\} \tag{4-145}$$

$$\text{if}\quad W_0<-mc^2 \text{ (endoergic) and } \eta_F\ll 1,$$

and for extreme degeneracy we obtain the simpler expressions

$$K\approx\frac{2\pi\alpha Z}{5}\left[\left(E_F^5+\frac{5}{2}W_0E_F^4+\frac{5}{3}W_0^2E_F^3\right)-\left(1+\frac{5}{2}W_0+\frac{5}{3}W_0^2\right)\right]$$

$$\text{if}\quad W_0>-mc^2 \text{ (exoergic) and } \eta_F\gg 1,$$

$$K\approx\frac{2\pi\alpha Z}{5}\left[(E_F-|W_0|)^5+\frac{5}{2}|W_0|(E_F-|W_0|)^4+\frac{5}{3}|W_0^2|^2(E_F-|W_0|)^3\right] \tag{4-146}$$

$$\text{if}\quad W_0<-mc^2 \text{ (endoergic) and } \eta_F\gg 1.$$

Here the energies E_F and W_0 are expressed in units of mc^2, and

and
$$\langle F\rangle = 2\pi\alpha Z \quad \text{for } Z>(2\pi\alpha)^{-1}\approx 23$$
$$\langle F\rangle = 1 \quad \text{if } Z<(2\pi\alpha)^{-1}\approx 23,$$

the fine structure constant is α, the Fermi momentum, $P_F = c^{-1}(E_F^2 - m^2c^4)^{1/2}$ where E_F is the Fermi energy, and $P_0 = c^{-1}(W_0^2 - m^2c^4)^{1/2}$, where W_0 is the threshold energy for the beta transition.

Finzi and Wolf (1967) have shown that contracting white dwarfs with magnesium rich cores or with calcium rich shells may lead to Type I supernovae. They show that the contractions are caused by the electron capture reactions of Mg^{24} and Ca^{40} and give formulae for the reaction lifetimes.

4.3.4. The URCA Processes

A nucleus (Z,A) may produce neutrinos, ν_e, through the beta decay reactions

and
$$(Z,A)\rightarrow(Z+1,A)+e^- + \bar{\nu}_e$$
$$(Z,A)\rightarrow(Z-1,A)+e^+ + \nu_e, \tag{4-147}$$

or by the electron capture reaction

$$(Z+1,A)+e^- \rightarrow (Z,A)+\nu_e. \tag{4-148}$$

In the ordinary URCA process, a nucleus alternately captures an electron, e^-, and undergoes negative beta decay, meanwhile emitting a neutrino, ν_e, and an antineutrino, $\bar{\nu}_e$, according to the reaction (Gamow and Schoenberg, 1941)

$$e^- + (Z,A)\rightarrow e^- + (Z,A)+\nu_e+\bar{\nu}_e. \tag{4-149}$$

Only nuclei of odd mass number can participate in the ordinary URCA process in a degenerate gas. This is because a nucleus with even mass number has a lower threshold for electron capture after capturing one electron. Hence only even-mass-number nuclei of even charge number can exist in a degenerate gas, and these will, in general, be stable against both electron capture and beta decay. As pointed out by Gamow and Schönberg (1941), neutrino pair emission by the URCA process may affect the rates of stellar evolution, and what is of interest in this case is the rate of energy loss by the neutrinos. By assuming that the nuclei are in statistical equilibrium, and that neutrinos and antineutrinos are produced continuously through negative beta decay and electron capture, neutrino luminosities can be calculated using the decay rates given in the two previous sections (cf. Bahcall, 1962, 1964; Bahcall and Peterson, 1963; Tsuruta and Cameron, 1965).

Tsuruta and Cameron (1965) consider conditions of nuclear statistical equilibrium in a degenerate gas and obtain an average neutrino energy production rate per nucleus in the excited level, i, given by

$$P_\nu = \lambda_i \omega_i = \frac{\ln 2\, f_i}{(ft)_i}\,\omega_i, \tag{4-150}$$

where for electron capture

$$\omega_i = \tfrac{5}{6}(E_{\mathrm{F}}+W_0)\left[\frac{(1-y^6)-\frac{12}{5}x(1-y^5)+\frac{3}{2}x^2(1-y^4)}{(1-y^5)-\frac{5}{2}x(1-y^4)+\frac{5}{3}x^2(1-y^3)}\right], \tag{4-151}$$

and

$$x = W_0/(E_{\mathrm{F}}+W_0)$$
$$y = (1+W_0)/(E_{\mathrm{F}}+W_0)$$

if

$$W_0 \gg -mc^2,$$

and

$$\omega_i = \tfrac{5}{6}(E_{\mathrm{F}}-|W_0|)\left[\frac{1+\frac{12}{5}x+\frac{3}{2}x^2}{1+\frac{5}{2}x+\frac{5}{3}x^2}\right], \tag{4-152}$$

where $x=|W_0|/(E_{\mathrm{F}}-|W_0|)$

if

$$W_0 \ll -mc^2.$$

For electron emission (beta decay)

$$\omega_i = \frac{D_2(W_0,W_0)}{D_1(W_0,W_0)}\left[1-\frac{D_2(E_{\mathrm{F}},W_0)}{D_2(W_0,W_0)}\right]\Big/\left[1-\frac{D_1(E_{\mathrm{F}},W_0)}{D_1(W_0,W_0)}\right], \tag{4-153}$$

where

$$D_1(x,y) = \tfrac{1}{5}(x^5-1)-\tfrac{1}{2}(x^4-1)y+\tfrac{1}{3}(x^3-1)y^2$$

and

$$D_2(x,y) = -\tfrac{1}{6}(x^6-1)+\tfrac{3}{5}(x^5-1)y-\tfrac{3}{4}(x^4-1)y^2+\tfrac{1}{3}(x^3-1)y^3.$$

Here W_0 is the threshold energy for the beta transition, E_{F} is the electron Fermi energy given in Eqs. (4-142), and W_0 and E_{F} are assumed to be in units of mc^2.

Under electron-degenerate conditions, the rate of the ordinary URCA process is only significant if the electron Fermi energy is near the electron capture threshold of the nucleus $(Z+1,A)$, and then only if the phase space is made available by thermal rounding of the Fermi surface, by a vibration of it, or by a macroscopic transport of nuclei via convection. Because of this restriction, the URCA process of a given pair of nuclei proceeds significantly only in a shell within a stellar core. A complete discussion of the thermal and vibrational energy losses due to URCA shells in stellar interiors is given by Tsuruta and Cameron (1970). The rate of neutrino energy loss at finite temperature due to the presence of a URCA shell is given by

$$L_\nu = 4\pi r_s^2 X\left[\left|\frac{dE_{\mathrm{F}}}{dr}\right|\right]^{-1}(F_1\xi_0^5+F_2T^5), \tag{4-154}$$

where r_s is the radius of the URCA shell, X is the abundance by mass fraction of the nuclear pair under consideration, E_{F} is the Fermi energy,

$$\left|\frac{dE_{\mathrm{F}}}{dr}\right| \approx 4\times10^{-8}\ \mathrm{MeV\ cm^{-1}},$$

T is the temperature, the relative amplitude of radial oscillation is $\xi_0=\Delta r/r_0$, where Δr is the displacement from the equilibrium radius, r_0, and F_1 and F_2 are tabulated by Tsuruta and Cameron (1970) for 132 pairs of nuclei. The F_1

and F_2 are characteristic of a given pair of nuclei and are determined by the ft value, threshold energy, and charge. TSURUTA and CAMERON show that in a stellar interior where URCA shells are present, the URCA neutrino energy losses dominate those of other neutrino energy losses in the temperature region up to about 2×10^9 °K. The effects of URCA shells on carbon ignition in degenerate stellar cores are discussed by BRUENN (1972) and PACZYNSKI (1972).

PINAEV (1964) first pointed out that when the thermal energy, kT, becomes greater than the electron rest mass energy, mc^2, electron-positron pairs are produced, and the positron capture reaction

$$(Z-1, A) + e^+ \rightarrow (Z, A) + \bar{\nu}_e \tag{4-155}$$

becomes important as well. Under equilibrium conditions, the neutrino energy loss when the reactions given in Eqs. (4-147), (4-148), and (4-155) are included, is given by (PINAEV, 1964)

$$P_\nu \approx \frac{4 \times 10^{18}}{A\,ft} \rho \left[\frac{Q + mc^2}{mc^2}\right]^2 \left(\frac{kT}{mc^2}\right)^4 \exp\left(-\frac{Q + mc^2}{kT}\right) \text{erg cm}^{-3}\,\text{sec}^{-1}, \tag{4-156}$$

for $kT \ll mc^2$, $kT \ll Q$, and

$$P_\nu \approx \frac{0.8 \times 10^{20}}{A\,ft} \rho \left(\frac{kT}{mc^2}\right)^6 \text{erg cm}^{-3}\,\text{sec}^{-1}, \qquad (4\text{-}157)^2$$

for $kT \gg mc^2$, $kT \gg Q$. Here ft is the experimental value, and Q is the threshold energy. BEAUDET, SALPETER, and SILVESTRO (1972) have given more detailed formulae which include the effects of capture of both positrons and electrons by both beta stable and beta unstable nuclei, as well as normal beta decay of electrons and positrons. They also give simple fitting formulae for all of these transitions as a function of mass density and temperature; and calculate the overall URCA energy loss rates for a number of common stable even-even nuclei under equilibrium conditions.

When there is an abundance of free protons, p, and neutrons, n, the appropriate URCA reactions are

$$\begin{aligned} p + e^- &\rightarrow n + \nu_e \\ n + e^+ &\rightarrow p + \bar{\nu}_e \end{aligned} \tag{4-158}$$

and

$$n \rightarrow p + e^- + \bar{\nu}_e .$$

HANSEN (1968) has extended Pinaev's (1964) results to take into account electron degeneracy, partial electron degeneracy, and electron-positron pair formation. For electron capture on protons, HANSEN obtains a capture rate

$$\lambda = 4.7 \times 10^{-4} \left(\frac{u^5}{5} + 8 \times 10^{-3} u^3 T_9^2 + 3.2 \times 10^{-3}\, T_9^5\right) \text{sec}^{-1} \text{ per proton,} \tag{4-159}$$

and a neutrino luminosity of

$$L_\nu = 3.1 \times 10^{-10} \left(\frac{u^6}{6} + 6 \times 10^{-2} u^4 T_9^2 + 3.2 \times 10^{-3}\, T_9^6\right) \text{erg sec}^{-1} \text{ per proton} \tag{4-160}$$

[2] In Pinaev's (1964) work and also in a number of works of that time, the overestimated cross sections of electron and positron captures were used. This error has been revealed by V. S. IMSHENNIK, D. K. NADYOZHIN and V. S. PINAEV (Astron. Zh., **44**, 768, 1977). Here we give the values for P_ν which are a factor of two larger than the correct values.

for $T_9 \gtrsim 10$ and $\mu \gtrsim kT$. Here $T_9 = T/10^9$, the chemical potential is μ, and

$$u = \frac{\mu}{mc^2} = \text{lesser of } 10^{-2}\left(\frac{\rho}{\mu_e}\right)^{1/3}, \quad 6\times 10^{-6}\left(\frac{\rho}{\mu_e}\right)T_9^{-2}. \tag{4-161}$$

For positron capture on neutrons,

$$\lambda = 10^{-6}\,T_9^5 \exp\left[-5.8\,\frac{(u+1)}{T_9}\right] \sec^{-1} \text{ per neutron,} \tag{4-162}$$

and

$$L_\nu = 7\times 10^{-13}\,T_9^6 \exp\left[-5.8\,\frac{(u+1)}{T_9}\right] \text{erg sec}^{-1} \text{ per neutron} \tag{4-163}$$

for $T_9 \gtrsim 5$. For the noninteracting gas model, a zero temperature neutron star would (BAHCALL and WOLF, 1965) have

$$N_n \approx 2\times 10^{38}\left(\frac{\rho}{\rho_{nucl}}\right) \text{cm}^{-3}, \tag{4-164}$$

and

$$N_e = N_p \approx 2\times 10^{36}\left(\frac{\rho}{\rho_{nucl}}\right)^2 \text{cm}^{-3}, \tag{4-165}$$

for $\rho \lesssim 2\rho_{nucl}$. Here N_n, N_e, and N_p denote, respectively, the number densities of neutrons, electrons, and protons, the gas mass density is ρ, and the nuclear matter density $\rho_{nucl} \approx 3.7\times 10^{14}$ g cm^{-3}. ARNETT (1967) applies Pinaev's analysis to the reactions given in Eq. (4-158) to obtain a total neutrino emissivity, Q_ν, given by

$$Q_\nu = 8.1\times 10^{16}\,X_n\left(\frac{T_9}{6}\right)^6 \text{erg g}^{-1}\sec^{-1}, \tag{4-166}$$ [3]

where $T_9 = T/10^9$, and X_n is related to the number density of protons, N_p, and neutrons, N_n, by

$$X_n = \frac{(N_p + N_n)}{6\times 10^{23}\,\rho}. \tag{4-167}$$

FASSIO-CANUTO (1969) has given the beta decay rate in a strong magnetic field, and CANUTO and CHOU (1971) have discussed the URCA energy loss rates in an intense magnetic field of strength, H. The results are given in terms of the parameter

$$\theta = \frac{H}{H_q} = \frac{e\hbar H}{m^2c^3} = \frac{H}{4.414\times 10^{13}\text{ gauss}}. \tag{4-168}$$

They give detailed formulae for the neutrino luminosity, L_H, and the beta decay and electron capture reaction rates under the conditions of statistical equilibrium and charge neutrality. Analytic approximations to these formulae show that when a neutron-proton gas is considered and $\theta \gg 1$, the neutrino luminosity L_H, from a nondegenerate gas may be up to 100 times the neutrino luminosity, L_0, when the field is absent. For a degenerate neutron-proton gas, however, $L_H \approx 10^{-2}L_0$ for $\theta \gtrsim 1$. When a degenerate gas of nuclei is considered, it is

[3] The numerical coefficient in (4-166) is overestimated (see footnote 2). The correct value is equal to 3.6×10^{16} (Imshennik et al., 1977).

found that the URCA energy loss rates are reduced up to a factor of 10^{-3} for $\theta \approx 1$ and $T \lesssim 10^9$ °K.

The interior of a neutron star has a large amount of free neutrons in equilibrium with a small amount of protons and electrons. Under these conditions, the neutrons can scatter and undergo the modified URCA reactions (CHIU and SALPETER, 1964)

$$\mathrm{n}+\mathrm{n} \rightarrow \mathrm{n}+\mathrm{p}+\mathrm{e}^- +\bar{\nu}_\mathrm{e} \tag{4-169}$$

and

$$\mathrm{n}+\mathrm{p}+\mathrm{e}^- \rightarrow \mathrm{n}+\mathrm{n}+\nu_\mathrm{e}. \tag{4-170}$$

If the neutron Fermi energy is greater than the muon rest energy, $m_\mu c^2$, muon neutrinos will also be formed according to the modified URCA reactions

$$\mathrm{n}+\mathrm{n} \rightarrow \mathrm{n}+\mathrm{p}+\mu^- +\bar{\nu}_\mu \tag{4-171}$$

and

$$\mathrm{n}+\mathrm{p}+\mu^- \rightarrow \mathrm{n}+\mathrm{n}+\nu_\mu. \tag{4-172}$$

If quasi-free pions are also present in neutron matter, there are the additional URCA type reactions

$$\begin{aligned} \pi^- +\mathrm{n} &\rightarrow \mathrm{n}+\mathrm{e}^- +\bar{\nu}_\mathrm{e} \\ \mathrm{n} +\mathrm{e}^- &\rightarrow \mathrm{n}+\pi^- +\nu_\mathrm{e} \end{aligned} \tag{4-173}$$

and

$$\begin{aligned} \pi^- +\mathrm{n} &\rightarrow \mathrm{n}+\mu^- +\bar{\nu}_\mu \\ \mathrm{n} +\mu^- &\rightarrow \mathrm{n}+\pi^- +\nu_\mu. \end{aligned} \tag{4-174}$$

BAHCALL and WOLF (1965) assume that the neutron gas is not superfluid and that the temperature and density are constant over the star. The independent particle model was then used to obtain the neutrino luminosities

$$L_\nu = 6\times 10^{38}\left(\frac{M}{M_\odot}\right)\left(\frac{\rho_{\mathrm{nucl}}}{\rho}\right)^{1/3} T_9^8 \text{ erg sec}^{-1}, \tag{4-175}$$

for the modified URCA reactions given in Eqs. (4-169) and (4-170),

$$\begin{aligned} L_\nu &= 6\times 10^{38}\left[1-2.25\left(\frac{\rho_{\mathrm{nucl}}}{\rho}\right)^{4/3}\right]^{1/2}\left(\frac{M}{M_\odot}\right)\left(\frac{\rho_{\mathrm{nucl}}}{\rho}\right)^{1/3} T_9^8 \text{ erg sec}^{-1} \quad \text{for } \rho > 1.8\,\rho_{\mathrm{nucl}} \\ &= 0 \quad \text{for } \rho < 1.8\,\rho_{\mathrm{nucl}}, \end{aligned} \tag{4-176}$$

for the modified URCA reactions given in Eqs. (4-171) and (4-172), and

$$L_\nu = 10^{46}\, T_9^6 \left(\frac{N_\pi}{N_b}\right)\left(\frac{M}{M_\odot}\right) \text{erg sec}^{-1}, \tag{4-177}$$

for the pion reactions given in Eqs. (4-173) and (4-174). Here M is the neutron star mass, the solar mass is $M_\odot$, the neutron star mass density is ρ, the density of nuclear matter $\rho_{\mathrm{nucl}} \approx 3.7\times 10^{14}$ g cm^{-3}, T_9 is the temperature of the stellar core in units of 10^9 °K, and N_π/N_b is the ratio of the number density of quasi-free pions to the number density of baryons. For comparison, for a neutron star of effective surface temperature, T_e, the photon luminosity, L_γ, is given by

$$L_\gamma = 7\times 10^{36}\, T_{e7}^4 R_{10}^2 \text{ erg sec}^{-1}, \tag{4-178}$$

where $T_{e7} = T_e/10^7$, and R_{10} is the radius of the star in units of 10 km. BAHCALL and WOLF (1965) also consider the cooling times for the loss of thermal energy from a neutron star which is in statistical equilibrium. FINZI (1965) pointed out that the exponential light curve of Type I supernovae might be due to the dissipation of the vibrational energy of a neutron star by the URCA process. When the neutron star is vibrating, equilibrium conditions are not satisfied, and the rate of dissipation of vibrational energy per unit mass, dw/dt, is given by (FINZI and WOLF, 1968)

$$\frac{dw}{dt} = K - Q_\nu, \tag{4-179}$$

where the neutrino emissivity due to the modified URCA process given in Eqs. (4-169) and (4-170) is

$$Q_\nu = 3 \times 10^5 \left(\frac{\rho_{\text{nucl}}}{\rho}\right) T_9^8 \text{ erg g}^{-1} \text{ sec}^{-1}, \tag{4-180}$$

and the constant K, ranges from 10^{-1} to 10^3 as the ratio of the chemical potential, μ, to the thermal energy, kT, ranges from one to sixteen.

4.3.5. Neutrino Pair Emission

As first suggested by PONTECORVO (1959), neutrino pair emission may rapidly accelerate the evolution of a star in its later stages. The regimes in which different neutrino pair emission processes dominate (other than the URCA process) are shown in Fig. 25 together with the predicted neutrino spectrum of the Sun.

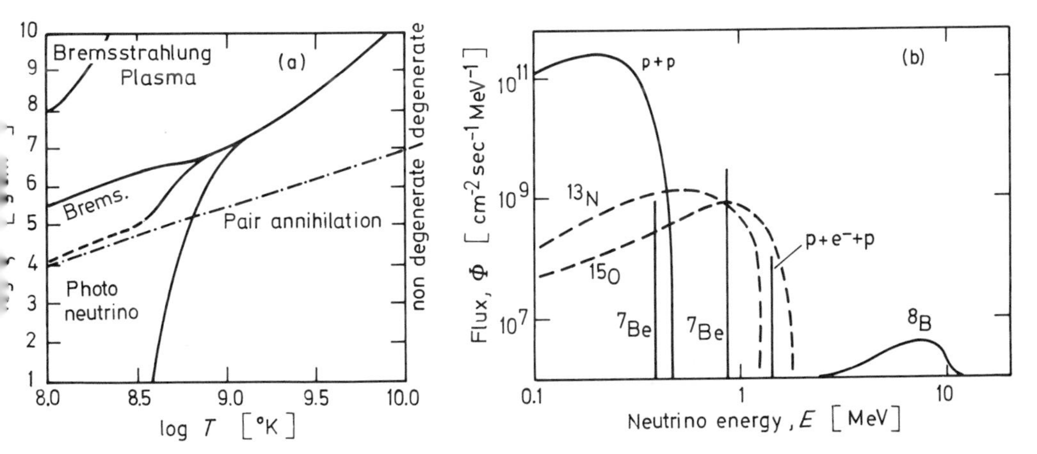

Fig. 25. Dominant regimes for various neutrino pair emission processes, (a), and the solar neutrino energy spectrum (b). The gas density and temperature are denoted, respectively, by ρ and T. The neutrino bremsstrahlung is calculated assuming the nuclear charge $Z = 26$, and lattice corrections are neglected. For the solar neutrino energy spectrum, solid lines denote the proton-proton chain and broken lines denote the CNO cycle. The neutrino fluxes are in units of number $\text{cm}^{-2} \text{ sec}^{-1} \text{ MeV}^{-1}$ for continuum sources, and number $\text{cm}^{-2} \text{ sec}^{-1}$ for the line sources. Fig. (a) is after FESTA and RUDERMAN (1969) whereas Fig. (b) is after BAHCALL and SEARS (1972), by permission of Annual Reviews, Inc.)

What is of interest to stellar evolution is the neutrino luminosity, and in what follows we give the luminosity of the major neutrino pair emission processes which have been outlined by FOWLER and HOYLE (1964).

4.3.5.1. *Neutrino Bremsstrahlung and Neutrino Synchrotron Radiation*

As first suggested by PONTECORVO (1959), neutrinos will be emitted by bremsstrahlung radiation according to the reaction

$$e^- + (Z, A) \rightarrow e^- + (Z, A) + \nu_e + \bar{\nu}_e. \tag{4-181}$$

Here electrons, e^-, collide with nuclei, (Z, A), to emit an electron, e^-, together with a neutrino, ν_e, and antineutrino, $\bar{\nu}_e$, pair. This process is analogous to ordinary photon bremsstrahlung with the neutrino pair replacing the usual photon emitted in inelastic electron scattering. GANDELMAN and PINAEV (1960) have considered the free-free neutrino bremsstrahlung of a nondegenerate gas, and obtain an effective cross section

$$\sigma = \sigma_0 Z^2 \left(\frac{E_e}{mc^2}\right)^3, \tag{4-182}$$

where E_e is the kinetic energy of the incident electron, and

$$\sigma_0 = \frac{8 r_0^2}{525 \pi^3} \left[\frac{g}{mc^2} \left(\frac{\hbar}{mc}\right)^3\right]^2 = 3.52 \times 10^{-52} \text{ cm}^2. \tag{4-183}$$

Here r_0 is the classical electron radius, and the weak interaction coupling constant $g \approx 1.41 \times 10^{-49}$ erg cm^3. If the electrons are assumed to have a Maxwellian distribution, and the nondegenerate gas is fully ionized, the electron neutrino luminosity density, P_ν, is given by

$$P_\nu = 2.75 \times 10^{-10} \frac{\rho^2}{\mu_e \nu} T^{4.5} \text{ erg cm}^{-3} \text{ sec}^{-1}, \tag{4-184}$$

where ρ is the gas mass density, the temperature, T, is in keV ($1 \text{ keV} = 1.1605 \times 10^7$ °K), and the mean electron molecular weight, μ_e, and the factor, ν, are given by

$$\mu_e^{-1} = \sum_i \frac{X_i Z_i}{A_i} \quad \text{and} \quad \nu^{-1} = \sum_i X_i \frac{Z_i^2}{A_i}, \tag{4-185}$$

where X_i, Z_i, and A_i are, respectively, the mass fraction, nuclear charge, and atomic weight of the nucleus of species, i. When the electrons have a Fermi-Dirac distribution and are nearly degenerate,

$$P_\nu \approx 0.82 \times 10^{-7} \frac{\rho}{\nu} T^6 \ln\left(0.89 \frac{E_0}{T}\right) \text{ erg cm}^{-3} \text{ sec}^{-1}, \tag{4-186}$$

where T is in keV, and

$$E_0 = 5.07 \times 10^{-5} mc^2 \left(\frac{\rho}{\mu_e}\right)^{2/3},$$

where $mc^2 = 511$ keV should be used if E_0 is to be expressed in keV.

The neutrino bremsstrahlung of a relativistic degenerate gas of electrons Coulomb scattering on nuclei has been considered by FESTA and RUDERMAN (1969). For a random gas of nuclei, the neutrino emissivity, Q_ν, is given by

$$\begin{aligned} Q_\nu &= 20 g^2 \frac{Z^2 e^4}{\pi^5} (kT)^6 \left\{ B_1 B_2 - (\beta^2 - 1) B_3 \left[\frac{2}{3} + \frac{1}{2} \right.\right. \\ &\quad \left.\left. \times \left(\beta - \beta \frac{(\beta^2+1)}{2} \ln \frac{\beta+1}{\beta-1} \right) \right] \right\} \text{erg nucleus}^{-1}\,\text{sec}^{-1} \\ &\approx 0.76 \frac{Z^2}{A} \left(\frac{T}{10^8\,{}^\circ\text{K}} \right)^6 \text{erg g}^{-1}\,\text{sec}^{-1} \quad \text{for } \rho \to \infty. \end{aligned} \tag{4-187}$$

Here $\beta = E_F/P_F$, where E_F and P_F denote, respectively, the Fermi energy and momentum, and the factors B_1, B_2, and B_3 are given by

$$B_1 = (1-\beta^2)\left[\frac{1}{2}\beta \ln\left(\frac{\beta+1}{\beta-1}\right) - 1\right] + \frac{1}{3},$$

$$B_2 = -4 + 2(1+\alpha^2)\ln\left(\frac{2+\alpha^2}{\alpha^2}\right),$$

and

$$B_3 = \ln\left(\frac{2+\alpha^2}{\alpha^2}\right) - \frac{2}{2+\alpha^2},$$

where $\alpha^2 = \beta/215$. When the nuclei are arranged in a zero temperature, rigid lattice, Q_ν is reduced at low Z going to zero at $Z=1$ and to $\gtrsim 0.4\, Q_\nu$ for $Z \gtrsim 25$. This suppression is less, however, when the temperature is finite. As it was illustrated in Fig. 25, neutrino pair emission by neutrino bremsstrahlung dominates over other pair emission processes at high density and moderate temperatures. The other processes are suppressed at high densities because of the absence of accessible unoccupied electron states, and because of the increase in plasma frequency. At low temperatures, the normally dominant photoneutrino emissivity decreases as T^9, whereas the neutrino bremsstrahlung emissivity decreases as T^6, making the bremsstrahlung dominant in the moderate density-temperature region shown in Fig. 25.

PINAEV (1964) has calculated the neutrino bremsstrahlung for recombination on the K shell of an atom. The total cross section, σ, for recombination on the K shell is given by

$$\sigma = \sigma_0 Z^5 \frac{c}{v} (E_e + I), \tag{4-188}$$

where

$$\sigma_0 = \frac{4\alpha^5}{15\pi^2} \frac{g^2 m^2}{\hbar^4} = 0.76 \times 10^{-56}\,\text{cm}^2,$$

the kinetic energy of the electron, E_e, and I, are in units of mc^2, the velocity of the initial electron is v, and the ionization potential of the K electron is

$I=\alpha^2 Z^2 mc^2/2$, where α is the fine structure constant. For a nondegenerate gas with a Maxwellian electron distribution, the free-bound neutrino luminosity density is

$$P_\nu = 1.54\times10^{-8} Z^6 A^{-2}\rho^2 T^2(1-f)\ \text{erg cm}^{-3}\ \text{sec}^{-1}, \qquad (4\text{-}189)$$

where Z is the nuclear charge, A is the atomic weight, ρ is the gas mass density, T is the temperature in keV, and

$$f=\left[1+320\,\frac{A\,T^{3/2}}{Z\rho}\right]^{-1}.$$

Equation (4-189) holds for $kT \gg I$.

For a degenerate electron gas, the free-bound neutrino luminosity density is

$$P_\nu = 1.45\times10^{-13} Z^4\left(\frac{Z\rho}{A}\right)^{10/3}\exp\left(-\frac{E_\mathrm{F}}{T}\right)\text{erg cm}^{-3}\ \text{sec}^{-1}. \qquad (4\text{-}190)$$

Equation (4-190) holds for $E_\mathrm{F} \gg kT \gg I$.

Neutrino bremsstrahlung of an electron accelerated in an intense magnetic field was first calculated by LANDSTREET (1967). For a magnetic field of strength, H, the neutrino synchrotron luminosity density is given by (LANDSTREET, 1967; CANUTO, CHIU, CHOU, and FASSIO-CANUTO, 1970)

$$P_\nu = 3\times10^{-44} H_8^6 T_7 \rho^4\ \text{erg cm}^{-3}\ \text{sec}^{-1} \qquad \text{if } H_8\rho^{2/3} \lesssim 8\times10^6\, T_7$$

and (4-191)

$$P_\nu = 4\times10^{-7}\, T_7^{19/3} \rho^{4/9} H_8^{2/3}\ \text{erg cm}^{-3}\ \text{sec}^{-1} \quad \text{if } H_8\rho^{2/3} \gtrsim 8\times10^6\, T_7,$$

for relativistic electrons in a degenerate gas of mass density, ρ, and temperature, T. Here $H_8=H/10^8$ and $T_7=T/10^7$. For a nondegenerate, nonrelativistic gas

$$P_\nu = 2\times10^{-46} H_8^6 N_\mathrm{e}\ \text{erg cm}^{-3}\ \text{sec}^{-1}, \qquad (4\text{-}192)$$

whereas for a degenerate, nonrelativistic gas

$$P_\nu = 1\times10^{-28}\, T_7 H_8^6 N_\mathrm{e}^{1/3}\ \text{erg cm}^{-3}\ \text{sec}^{-1}, \qquad (4\text{-}193)$$

where N_e is the electron number density. CANUTO, CHIU, CHOU, and FASSIO-CANUTO (1970) derive general expressions for the neutrino synchrotron radiation of a relativistic electron gas, and also give numerical approximations. These data agree with Landstreet's formulae at high densities, but disagree at lower densities.

4.3.5.2. Electron Pair Annihilation Neutrinos

The emission of neutrinos by the annihilation of electrons, e^-, and positrons, e^+, according to the reaction

$$e^- + e^+ \to \nu_e + \bar{\nu}_e, \qquad (4\text{-}194)$$

was considered by CHIU and MORRISON (1960), CHIU (1961), and CHIU and STABLER (1961). The neutrino luminosity density, P_ν, is given by

$$P_\nu = 4.90\times10^{18}\, T_9^3 \exp\left(-\frac{11.86}{T_9}\right)\text{erg cm}^{-3}\ \text{sec}^{-1}$$

(nondegenerate, nonrelativistic)

$$P_\nu = 4.22\times10^{15}\, T_9^9\ \text{erg cm}^{-3}\ \text{sec}^{-1}$$

(nondegenerate, relativistic)

$$P_\nu = 1.93 \times 10^{13}\left(\frac{\rho}{\mu_e}\right) T_9^{3/2} \exp\left[-\frac{(E_F + mc^2)}{kT}\right] \text{erg cm}^{-3}\,\text{sec}^{-1} \tag{4-195}$$

and (degenerate, nonrelativistic)

$$P_\nu = 1.44 \times 10^{11}\left(\frac{\rho}{\mu_e}\right)\left(\frac{E_F}{mc^2}\right)^2 T_9^4\left[1 + \frac{5kT}{E_F}\right]\exp\left(-\frac{E_F}{kT}\right) \text{erg cm}^{-3}\,\text{sec}^{-1}$$

(degenerate, relativistic),

where the gas mass density is ρ, the gas temperature, T, is in °K, the factor $T_9 = T/10^9$, the gas is degenerate if

$$\rho > 2.4 \times 10^{-8}\, T^{3/2} \mu_e \text{ g cm}^{-3} \quad \text{(degenerate)}, \tag{4-196}$$

and it is relativistic if

$$\rho > 7.3 \times 10^6 \mu_e \text{ g cm}^{-3} \quad \text{(relativistic)}, \tag{4-197}$$

the mean molecular weight per electron is

$$\mu_e^{-1} = \frac{N_e}{\rho N_A} = \sum \frac{X_Z Z}{A_Z}, \tag{4-198}$$

where N_e is the electron density, Avogadro's number $N_A \approx 6.022 \times 10^{23}$ (mole)$^{-1}$, the mass fraction is X_Z for an element whose atomic number is Z, the element mass number is A, and $E_F = \varepsilon_F + mc^2$ where the Fermi energy, ε_F, of a completely degenerate gas is given by

$$\begin{aligned}
\varepsilon_F &= mc^2\left\{\left[1.018 \times 10^{-4}\left(\frac{\rho}{\mu_e}\right)^{2/3} + 1\right]^{1/2} - 1\right\} \\
&= 0.509 \times 10^{-4}\left(\frac{\rho}{\mu_e}\right)^{2/3} m_e c^2 \quad \text{(nonrelativistic)} \\
&= 1.009 \times 10^{-2}\left(\frac{\rho}{\mu_e}\right)^{1/3} m_e c^2 \quad \text{(relativistic)}.
\end{aligned} \tag{4-199}$$

These equations can be used to give a pair annihilation neutrino emissivity $Q_\nu = P_\nu/\rho$, or

$$Q_\nu = \frac{4.8}{\rho}\left(\frac{T}{10^3}\right)^3 \exp\left(-\frac{2mc^2}{kT}\right) \text{erg g}^{-1}\,\text{sec}^{-1} \quad \text{for } kT \ll mc^2 \quad \text{(nondegenerate)} \tag{4-200}$$

$$Q_\nu = \frac{4.3 \times 10^{24}}{\rho}\left(\frac{T}{10^{10}}\right)^9 \text{erg g}^{-1}\,\text{sec}^{-1} \quad \text{for } kT \gg mc^2 \quad \text{(nondegenerate)}$$

$$Q_\nu = \frac{4.5}{\mu_e}\left(\frac{T}{10^4}\right)^{3/2} \exp\left[-\frac{(E_F + mc^2)}{kT}\right] \text{erg g}^{-1}\,\text{sec}^{-1} \quad \text{for } kT \ll mc^2,\ E_F \ll mc^2 \quad \text{(degenerate)}$$

and

$$Q_\nu = \frac{0.14}{\mu_e}\left(\frac{T}{10^6}\right)^4\left(\frac{E_F}{mc^2}\right)^2 \exp\left[-\frac{E_F}{kT}\right] \text{erg g}^{-1}\,\text{sec}^{-1} \quad \text{for } kT \gg mc^2 \quad \text{(degenerate)}.$$

A comparison of the neutrino energy loss rates of the pair annihilation process and the following photoneutrino and plasma neutrino processes is given by BEAUDET, PETROSIAN, and SALPETER (1967).

4.3.5.3. *Photoneutrino Process*

The emission of neutrinos by the collision of a photon, γ, and an electron, e^-, according to the photoneutrino reaction

$$\gamma + e^- \rightarrow e^- + \nu_e + \bar{\nu}_e \tag{4-201}$$

was first considered by CHIU and STABLER (1961) and RITUS (1962). PETROSIAN, BEAUDET, and SALPETER (1967) consider the neutrino energy loss rate due to photoneutrino processes in a hot plasma, including the contribution of positrons present in the black body radiation. They obtain the photoneutrino luminosity densities given by

$$P_\nu = 0.976 \times 10^8\, T_9^8 \left(\frac{\rho}{\mu_e}\right) \text{erg cm}^{-3}\,\text{sec}^{-1}$$

(nondegenerate, nonrelativistic)

$$P_\nu = 1.477 \times 10^{13}\, T_9^9 [\log T_9 - 0.536]\ \text{erg cm}^{-3}\,\text{sec}^{-1}$$

(nondegenerate, relativistic) (4-202)

$$P_\nu = 0.976 \times 10^8\, T_9^9 \left(\frac{\rho}{\mu_e}\right) \left[\frac{(\rho/\mu_e)}{3.504 \times 10^5}\right]^{-2/3} \text{erg cm}^{-3}\,\text{sec}^{-1}$$

(degenerate, nonrelativistic)

and

$$P_\nu = 1.514 \times 10^{13}\, T_9^9\ \text{erg cm}^{-3}\,\text{sec}^{-1}$$

(degenerate, relativistic),

where the conditions for degeneracy and the relativistic criterion are given in the previous section, $T_9 = T/10^9$, ρ is the gas mass density, and μ_e is the mean molecular weight per electron. For most applications, the photoneutrino luminosity density can be calculated from the expression (PETROSIAN, BEAUDET, and SALPETER, 1967)

$$P = 1.103 \times 10^{13}\, T_9^9 \exp\left(-\frac{5.93}{T_9}\right) + 0.976 \times 10^8\, T_9^8 (1 + 4.2\, T_9)^{-1} \left(\frac{\rho}{\mu_e}\right) \times \left[1 + \frac{6.446 \times 10^{-6}\, \rho}{\mu_e T_9 (1 + 4.2\, T_9)}\right]^{-1} \text{erg cm}^{-3}\,\text{sec}^{-1}. \tag{4-203}$$

A comparison of energy losses due to the pair annihilation, photoneutrino, and plasma neutrino processes is given by BEAUDET, PETROSIAN, and SALPETER (1967).

These equations can be used to give a pair annihilation neutrino emissivity $Q_\nu = P_\nu/\rho$, or

$$Q_\nu \approx \mu_e^{-1} \left(\frac{T}{10^8}\right)^8 \text{erg g}^{-1}\,\text{sec}^{-1} \quad \text{for } kT \ll mc^2$$

(nondegenerate, nonrelativistic)

$$Q_\nu \approx \frac{2.5 \times 10^{14}}{\mu_e} \left(\frac{T}{10^{10}}\right)^8 \left[\log\left(\frac{T}{10^{10}}\right) + 1.6\right] \text{erg g}^{-1}\,\text{sec}^{-1} \quad \text{for } kT \gg mc^2$$

(nondegenerate, relativistic)

$$Q_\nu \approx 1.5 \times 10^2 \left(\frac{T}{10^8}\right)^9 (\mu_e \rho^2)^{-1/3} \text{ erg g}^{-1} \text{ sec}^{-1} \quad \text{for } kT \ll mc^2 \quad \text{(degenerate, nonrelativistic)} \tag{4-204}$$

and

$$Q_\nu \approx \frac{6.3 \times 10^6}{\mu_e} \left[1 + 5\left(\frac{T}{10^9}\right)^2\right] \left(\frac{T}{10^9}\right)^7 \left(\frac{mc^2}{E_F}\right)^3 \text{ erg g}^{-1} \text{ sec}^{-1} \quad \text{for } kT \gg mc^2 \quad \text{(degenerate, relativistic)}.$$

4.3.5.4. Plasma Neutrino Process

When a photon propagates in an ionized gas, it creates virtual electron-hole pairs and behaves as if it had a rest mass, M, given by

$$M = \frac{\hbar \omega_p}{c^2}, \tag{4-205}$$

where the plasma frequency, ω_p, is given by

$$\begin{aligned} \omega_p^2 &= \frac{4\pi N_e e^2}{m} \quad \text{(nondegenerate)} \\ \omega_p^2 &= \frac{4\pi N_e e^2}{m} \left[1 + \left(\frac{\hbar}{mc}\right)^2 (3\pi^2 N_e)^{2/3}\right]^{-1/2} \quad \text{(degenerate)}. \end{aligned} \tag{4-206}$$

Such a particle, called a plasmon, may decay and emit neutrinos according to the reaction $\Gamma \to \nu_e + \bar{\nu}_e$. The plasmon, Γ, may propagate in both the longitudinal and transverse modes giving rise to respective neutrino luminosity densities, $P_{l\nu}$ and $P_{t\nu}$ given by (ADAMS, RUDERMAN, and WOO, 1963; INMAN and RUDERMAN, 1964)

$$P_{l\nu} = 1.224 \times 10^{13} T_9^9 x^9 (e^x - 1)^{-1} \text{ erg cm}^{-3} \text{ sec}^{-1}, \tag{4-207}$$

$$\begin{aligned} P_{t\nu} &= 3.214 \times 10^{14} T_9^9 x^9 F(x) \text{ erg cm}^{-3} \text{ sec}^{-1} \\ &\approx 7.4 \times 10^{21} \left(\frac{\hbar \omega_p}{mc^2}\right)^6 \left(\frac{mc^2}{kT}\right)^{-3} \text{ erg cm}^{-3} \text{ sec}^{-1} \quad \text{for } x \ll 1 \\ &\approx 3.85 \times 10^{21} \left(\frac{\hbar \omega_p}{mc^2}\right)^{7.5} \left(\frac{mc^2}{kT}\right)^{-1.5} \exp\left(-\frac{\hbar \omega_p}{kT}\right) \text{ erg cm}^{-3} \text{ sec}^{-1} \quad \text{for } x \gg 1, \end{aligned} \tag{4-208}$$

where

$$\begin{aligned} x = \frac{\hbar \omega_p}{kT} &= \left\{\frac{4\pi e^2 \rho \hbar^2}{m_p m \mu_e (kT)^2} \left[1 + \left(\frac{\hbar}{mc}\right)^2 \left(\frac{3\pi^3 \rho}{m_p \mu_e}\right)^{2/3}\right]^{1/2}\right\}^{1/2} \\ &\approx \frac{3.345 \times 10^{-4} (\rho/\mu_e)^{1/2}}{T_9 [1 + 1.0177 \times 10^{-4} (\rho/\mu_e)^{2/3}]^{1/4}}, \end{aligned}$$

and $F(x) = \sum_{n=1}^{\infty} [K_2(nx)/nx]$ where $K_2(nx)$ is the modified Bessel function of the second kind. Here we have incorporated the corrections given by ZAIDI (1965), who showed that the constants of ADAMS, RUDERMAN, and WOO (1963) for the longitudinal and transverse luminosity density were off by the respective factors of 3/8 and 1/4. BEAUDET, PETROSIAN, and SALPETER (1967) compare the neutrino luminosity due to pair annihilation, photoneutrinos, and the plasma neutrinos. They obtain the relations

$$\frac{P_{1\nu}}{P_{t\nu}} = 0.0158\left(\frac{\hbar\omega_p}{kT}\right)^2 \quad \text{for } \hbar\omega_p \ll kT$$
$$= 1.078 \quad \text{for } \hbar\omega_p \gg kT. \tag{4-209}$$

The emission of plasmon neutrinos in a strong magnetic field has been given by CANUTO, CHIUDERI, and CHOU (1970). The results are normalized in terms of the magnetic field parameter

$$H_q = \frac{m^2 c^3}{e\hbar} = 4.414 \times 10^{13} \text{ gauss}. \tag{4-210}$$

Although the transverse neutrino emission is not seriously altered unless the magnetic field strength, H, is $\gtrsim 10^{13}$ gauss; the longitudinal plasmons propagate in a new mode and give rise to a neutrino luminosity density, $P_{1\nu}$, given by

$$P_{1\nu} = 14.8\, T \rho_6^{-2}\left(\frac{H}{H_q}\right)^6 \text{erg cm}^{-3}\,\text{sec}^{-1}, \tag{4-211}$$

where $T \geq 10^9$ °K is the temperature, and $\rho_6 = \rho/10^6$ where ρ is the gas mass density. If $P_{1\nu H}$ and $P_{1\nu 0}$ indicate the luminosity densities in the presence and absence of a magnetic field, respectively, $P_{1\nu H} \gg 10^{10} P_{1\nu 0}$ for $\rho > 10^{11}$ g cm^{-3}; whereas $P_{1\nu 0} \gg 10^5 P_{1\nu H}$ for $\rho < 10^{11}$ g cm^{-3} if $H = 10^{11}$ gauss and $T \gtrsim 10^9$ °K.

4.3.5.5. *Photocoulomb and Photon-Photon Neutrinos*

MATINYAN and TSILOSANI (1962) and ROSENBERG (1963) have discussed the neutrino-pair production by photons, γ, in the Coulomb field of a nucleus (Z, A) according to the reaction

$$\gamma + (Z, A) \rightarrow (Z, A) + \nu_e + \bar{\nu}_e, \tag{4-212}$$

where ν_e and $\bar{\nu}_e$ denote, respectively, the electron neutrino and antielectron neutrino. ROSENBERG (1963) obtained the photocoulomb neutrino luminosity density, P_ν, given by

$$\dot{P_\nu} \approx \frac{4.6 \times 10^8 \rho}{\nu}\left(\frac{kT}{mc^2}\right)^{10} \text{erg cm}^{-3}\,\text{sec}^{-1} \tag{4-213}$$

where $\nu^{-1} = \sum_i X_i Z_i^2/A_i = 5\text{g}^{-1}$.

CHIU and MORRISON (1960), ROSENBERG (1963), and VAN HIEU and SHABALIN (1963) have discussed the conversion of a photon, γ, into a neutrino pair upon collision between two photons according to the reaction

$$\gamma + \gamma \rightarrow \gamma + \nu_e + \bar{\nu}_e. \tag{4-214}$$

VAN HIEU and SHABALIN obtained by means of detailed calculations a photon-photon neutrino luminosity density, P_ν, given by

$$P_\nu = 1.7 \times 10^{-28}\, T^{17} \text{erg cm}^{-3}\,\text{sec}^{-1}, \tag{4-215}$$

while according to a rough estimate by ROSENBERG (1963) we have

$$P_\nu \approx 1.6 \times 10^{-20} T^{13} \text{ erg cm}^{-3} \text{ sec}^{-1}, \tag{4-216}$$

where T is in keV (1 keV $\approx 1.16 \times 10^7$ °K).

4.3.5.6. The Muon and Pion Neutrino Processes

At temperatures exceeding 10^{11} °K (or $kT \approx 50$ MeV), the radiation field of a star can create muon, μ, or pion, π, pairs which can subsequently annihilate to form neutrino-antineutrino pairs according to the reactions

$$\begin{aligned} &\mu^- \rightarrow e^- + \bar{\nu}_e + \nu_\mu \\ &\mu^+ \rightarrow e^+ + \nu_e + \bar{\nu}_\mu, \end{aligned} \tag{4-217}$$

and

$$\begin{aligned} &\pi^- \rightarrow \mu^- + \bar{\nu}_\mu \\ &\pi^+ \rightarrow \mu^+ + \nu_\mu . \end{aligned} \tag{4-218}$$

According to ARNETT (1967), these processes may be important cooling mechanisms for highly evolved massive stars. The energy loss rate, Q_μ, for muon decay is given by

$$\begin{aligned} Q_\mu &= -\frac{2 N_{\text{pair}} E_{\text{av}}}{\rho \tau_\mu} \\ &\approx -9.9 \times 10^{38} \frac{\exp(-\beta)}{\rho \beta^{3/2}} \text{ erg g}^{-1} \text{ sec}^{-1}, \end{aligned} \tag{4-219}$$

where ρ is the mass density, the average decay energy, $E_{\text{av}} \approx 35$ MeV, the mean lifetime, $\tau_\mu \approx 2.2 \times 10^{-6}$ sec,

$$\beta \approx \frac{m_\mu c^2}{kT},$$

where $m_\mu c^2 \approx 105.659$ MeV is the muon neutrino rest mass energy. The number density of particle-antiparticle pairs, N_{pair}, is obtained by assuming equilibrium with the radiation field with $kT \ll mc^2$, and an equal concentration of particles and antiparticles

$$N_{\text{pair}} \approx \frac{1}{\sqrt{2}\pi^{3/2}} \left(\frac{mc}{\hbar}\right)^3 \frac{\exp(-\beta)}{\beta^{3/2}}. \tag{4-220}$$

The energy loss for pion decay, Q_π, is given by Eq. (4-219) with $E_{\text{av}} \approx 34$ MeV and $\tau_\pi \approx 2.60 \times 10^{-8}$ sec to obtain

$$Q_\pi \approx -9.6 \times 10^{40} \frac{\exp(-\beta)}{\rho \beta^{3/2}} \text{ erg g}^{-1} \text{ sec}^{-1}, \tag{4-221}[4]$$

where

$$\beta = \frac{m_\pi c^2}{kT},$$

and $m_\pi c^2 \approx 139.576$ MeV is the rest mass energy for pions. HANSEN (1968) has estimated the muon neutrino luminosity, L_μ, due to process $\mu^+ + \mu^- \rightarrow \nu_\mu + \bar{\nu}_\mu$, by

[4] This equation is obtained by the change of muon momentum, as compared with pion momentum in reactions (4-218) being neglected. With due regard for this effect, the numerical coefficient in (4-221) becomes 8.4×10^{40} (G. V. DOMOGATSKY, Preprint No. 96, Institute for Nuclear Research, Moscow, 1973).

assuming equal concentrations of μ^+ and μ^-. He obtains

$$L_\mu \approx 4 \times 10^{32} T_9^3 [1 + 4.75 \times 10^{-3} T_9 + 6.5 \times 10^{-6} T_9^2] \times \exp\left[-\frac{2.45 \times 10^3}{T_9}\right] \text{erg cm}^{-3} \text{sec}^{-1}, \qquad (4\text{-}222)$$

for $50 \leq T_9 \leq 500$, where $T_9 = T/10^9$. HANSEN (1968) also gives a lower limit to the muon neutrino luminosity due to the reaction $e^- + \mu^+ \rightarrow \nu_e + \bar{\nu}_\mu$.

4.3.6. Neutrino Opacities

Measured cross sections for the interaction of electron and mu neutrinos with matter are on the order of 10^{-44} cm^2 and were given in Eqs. (4-97) and (4-98). For the neutrino-electron and neutrino-muon scattering reactions given by

$$(\nu_e \text{ or } \bar{\nu}_e) + e^- \rightarrow (\nu'_e \text{ or } \bar{\nu}'_e) + e^{-\prime} \qquad (4\text{-}223)$$

and

$$(\nu_\mu \text{ or } \bar{\nu}_\mu) + \mu^- \rightarrow (\nu'_\mu \text{ or } \bar{\nu}'_\mu) + \mu^{-\prime}, \qquad (4\text{-}224)$$

BAHCALL (1964) finds cross sections which are proportional to

$$\sigma_{0e} = \frac{4}{\pi}\left(\frac{\hbar}{mc}\right)^{-4}\left(\frac{g^2}{m^2 c^4}\right) \approx 1.7 \times 10^{-44} \text{ cm}^2 \qquad (4\text{-}225)$$

and

$$\sigma_{0\mu} = \left(\frac{m_\mu}{m}\right)^2 \sigma_{0e} \approx 7.3 \times 10^{-40} \text{ cm}^2, \qquad (4\text{-}226)$$

where $g \approx 1.41 \times 10^{-49}$ erg cm^3 is the weak interaction coupling constant, and the subscripts e and μ denote, respectively, the scattering of electron neutrinos and mu neutrinos. For a neutrino of energy, E_ν, the cross sections for electron scattering in a nondegenerate gas are (BAHCALL, 1964)

$$\sigma = \frac{\sigma_0}{2}\left(\frac{E_\nu}{mc^2}\right) \quad \text{for } E_\nu \gg mc^2 \quad \text{and } kT \ll mc^2, \qquad (4\text{-}227)$$

$$\sigma = 1.6\left(\frac{kT}{mc^2}\right)\sigma_0\left(\frac{E_\nu}{mc^2}\right) \quad \text{for } E_\nu \gg mc^2 \quad \text{and } kT \gg mc^2, \qquad (4\text{-}228)$$

$$\sigma = \sigma_0\left(\frac{E_\nu}{mc^2}\right)^2 \quad \text{for } E_\nu \ll mc^2 \quad \text{and } kT \ll mc^2, \qquad (4\text{-}229)$$

and

$$\sigma = 17\left(\frac{kT}{mc^2}\right)^2 \sigma_0\left(\frac{E_\nu}{mc^2}\right)^2 \quad \text{for } E_\nu \ll mc^2 \quad \text{and } kT \gg mc^2, \qquad (4\text{-}230)$$

where the rest mass energy of the electron, $mc^2 \approx 8.2 \times 10^{-7}$ erg ≈ 0.51 MeV, and the thermal energy $kT \approx 0.86 \times 10^{-4}\, T$ eV. Eqs. (4-227) and (4-228) should be devided by three for antineutrino-electron scattering, whereas Eqs. (4-229) and (4-230) are valid for both neutrino-electron and antineutrino-electron scattering.

For a degenerate gas, the neutrino-electron scattering cross sections are given by (BAHCALL, 1964 as corrected by BAHCALL and WOLF, 1965)

$$\sigma=\sigma_0\left(\frac{E_\nu}{mc^2}\right)^2\left(\frac{E_\nu}{E_F}\right) \quad \text{for } E_\nu \ll E_F \tag{4-231}$$

and

$$\sigma=\sigma_0\left(\frac{E_\nu}{mc^2}\right)\left(\frac{E_F}{mc^2}\right) \quad \text{for } E_\nu \gg E_F\,. \tag{4-232}$$

Here E_F is the Fermi energy for the electron, and the equations should be multiplied by one third for antineutrino-electron scattering. The $E_\nu \approx 3kT$, and the Fermi energy is given by

$$\begin{aligned} E_F &= mc^2\left\{\left[\left(\frac{\rho}{10^6\,\mu_e}\right)^{2/3}+1\right]^{1/2}-1\right\} \\ &\approx mc^2\left(\frac{\rho}{10^6\,\mu_e}\right)^{1/3} = (3\pi^2)^{1/3}\hbar c N_e^{1/3} \quad \text{for } \rho \gg 10^6\,\mu_e\,, \end{aligned} \tag{4-233}$$

and the mass density, ρ, is related to the electron density, N_e, by the equation

$$\rho = N_A^{-1}\mu_e N_e = 10^6\,\mu_e x^3\,, \tag{4-234}$$

where

$$x^3 = 3\pi^2 N_e\left(\frac{\hbar}{mc}\right)^3,$$

$\mu_e \approx 2$ is the electron molecular weight, and $N_A \approx 6.02\times10^{23}\,\text{mole}^{-1}$ is Avogadro's number. For nonrelativistic conditions, $\rho \ll 10^6\,\mu_e$, we have

$$E_F \approx \frac{mc^2}{2}\left(\frac{\rho}{10^6\,\mu_e}\right)^{2/3} \approx (3\pi^2)^{2/3}\frac{\hbar^2}{2m}N_e^{2/3}\,. \tag{4-235}$$

HANSEN (1968) defines the chemical potential, μ, as the lesser of

$$\mu = 10^{-2}\,mc^2\left(\frac{\rho}{\mu_e}\right)^{1/3}$$

or

$$\mu = 6\times10^{-6}\,mc^2\,T_9^{-2}\left(\frac{\rho}{\mu_e}\right), \tag{4-236}$$

and obtains the electron-neutrino scattering cross sections

$$\sigma = \tfrac{2}{3}\sigma_0 E_\nu T_9 \quad \text{for } \mu \ll kT\,,$$

$$\sigma = \frac{\sigma_0}{2}E_\nu\left(\frac{\mu}{mc^2}\right) \quad \text{for } \mu \gg kT \quad \text{and } E_\nu \gg \mu\,, \tag{4-237}$$

and

$$\sigma = \frac{\sigma_0}{4}\frac{E_\nu}{\mu}mc^2\,T_9\left[1+\frac{3E_\nu^2}{T_9}\right]\left[1+\frac{T_9}{E_\nu}\right] \quad \text{for } \mu \gg kT \quad \text{and } E_\nu \ll \mu\,,$$

where E_ν is the neutrino energy in MeV, and $T_9 = T/10^9$. These results are quite close to those given above if we let $\mu = E_F + mc^2 \approx E_F$ and note that E_ν in $\text{MeV} \approx E_\nu/(mc^2)$ when $mc^2 = 0.51$ MeV. The exact expressions for the corresponding cross sections for $\mu \gg kT$, $E_\nu \gg kT$ are derived in (GERSHTEIN et al., 1975, 1976) and are given by

$$\sigma = \sigma_0 E_\nu^2\,\frac{2}{5}x\left\{1+\frac{2}{3}x+\frac{1}{7}x^2\right\} \quad \text{for } x = \frac{E_\nu}{\mu} < 1\,,$$

$$\sigma = 2\sigma_0 E_\nu \mu \left\{1 - \frac{2}{5}x - \frac{1}{5}x^2 + \frac{4}{105}x^3\right\} \quad \text{for } x = \frac{\mu}{E_\nu} < 1 ,$$

where E_ν and μ are in MeV. For $E_\nu < \mu$ the asymptotics practically coincides with that of Hansen, but for $E_\nu > \mu$ it is twice as large.

The opacity due to neutrino-nucleon scattering has been given by BAHCALL and FRAUTSCHI (1964), who give

$$\sigma \lesssim 10^{-2} \sigma_{0e} \left(\frac{E_\nu}{mc^2}\right)^2 , \tag{4-238}$$

for the scattering reactions

$$\begin{aligned} \nu_e + p &\to \nu_e' + p' , \\ \nu_e + n &\to \nu_e' + n' , \\ \bar{\nu}_e + p &\to \bar{\nu}_e' + p' , \\ \bar{\nu}_e + n &\to \bar{\nu}_e' + n' , \end{aligned} \tag{4-239}$$

where the n and p are, respectively, free neutrons and protons, E_ν is the neutrino energy, and σ_{0e} is given in Eqs. (4-227) to (4-230). BAHCALL and FRAUTSCHI (1964) also give the formulae for neutrino absorption by bound nucleons.

HANSEN (1968) has derived expressions for the total scattering rate, Q, by assuming that the incident neutrinos are distributed as a black body spectrum at the same temperature as the local electrons or nucleons. For neutrino-electron scattering, he obtains

$$Q \approx 9.1 \times 10^{-35} N_e N_\nu T_9^2 \ \text{sec}^{-1}\,\text{cm}^{-3} \quad \text{for } \mu \leq 4kT , \tag{4-240}$$

and

$$Q \approx 3.64 \times 10^{-34} \frac{N_e N_\nu}{(\mu/kT)} T_9^2 \ \text{sec}^{-1}\,\text{cm}^{-3} \quad \text{for } \mu \geq 4kT . \tag{4-241}$$

Here N_e is the electron number density, $T_9 = T/10^9$, the chemical potential, μ, is given by Eqs. (4-236), and the neutrino number density, N_ν, is given by

$$N_\nu = 7.65 \times 10^{27} T_9^3 \ \text{cm}^{-3} . \tag{4-242}$$

Similarly, for absorption by protons, p, or neutrons, n,

$$Q \approx 1.5 \times 10^{-6} N_p T_9^5 \ \text{sec}^{-1}\,\text{cm}^{-3} \quad \text{for } T_9 \gtrsim 10 , \tag{4-243}$$

and

$$Q \approx 1.5 \times 10^{-6} N_n T_9^5 \exp\left(-\frac{\mu}{kT}\right) \text{sec}^{-1}\,\text{cm}^{-3} \quad \text{for } T_9 \geq 5 , \tag{4-244}$$

where N_p and N_n are, respectively, the number densities of free protons and neutrons.

4.3.7. Solar Neutrinos

The hydrogen burning proton-proton chain is thought to provide most of the Sun's thermonuclear energy, and produces continuum neutrinos according to the reactions (BAHCALL, 1964, 1965)

$$\begin{aligned} H^1 + H^1 &\to H^2 + e^+ + \nu_e \quad (0.420\ \text{MeV}) , \\ B^8 &\to Be^{8*} + e^+ + \nu_e \quad (14.06\ \text{MeV}) , \\ N^{13} &\to C^{13} + e^+ + \nu_e \quad (1.20\ \text{MeV}) , \\ O^{15} &\to N^{15} + e^+ + \nu_e \quad (1.74\ \text{MeV}) . \end{aligned} \tag{4-245}$$

Here the numbers in parentheses are the maximum neutrino energies for the respective reaction. In addition to the continuum reactions of Eq. (4-245), neutrinos are also emitted by the following discrete energy reactions

$$\begin{aligned} H^1+H^1+e^- &\rightarrow H^2+\nu_e && (1.44\ \text{MeV}) \\ Be^7+e^- &\rightarrow Li^7+\nu_e && (0.861\ \text{MeV—}90\%) \\ & && (0.383\ \text{MeV—}10\%) \\ B^8+e^- &\rightarrow Be^8+\nu_e && (15.08\ \text{MeV}) \\ N^{13}+e^- &\rightarrow C^{13}+\nu_e && (2.22\ \text{MeV}) \\ O^{15}+e^- &\rightarrow N^{15}+\nu_e && (2.76\ \text{MeV})\,. \end{aligned} \tag{4-246}$$

Here the CNO bicycle neutrino producing reactions involving N^{13} and O^{15} are also included. What is of interest for neutrino detection experiments is the flux, φ, at the earth's surface which was illustrated in Fig. 25. Some of the neutrino fluxes are given in Table 40. In general the flux values are model dependent, and IBEN (1969) shows the dependence of neutrino flux on the assumed composition of the Sun. The values given below in Table 40 are for $Z/X=0.019$, $L_{\odot}=3.81\times10^{33}$ erg sec^{-1}, $M_{\odot}=1.989\times10^{33}$ gm, and a solar age of 4.7×10^9 years.

Table 40. Solar neutrino fluxes at the earth's surface for different neutrino sources[1]

Neutrino source	Flux ($cm^{-2}\ sec^{-1}$)
H^1+H^1	6.0×10^{10}
Be^7+e^-	4.5×10^9
N^{13}	3.3×10^8
O^{15}	2.7×10^8
$H^1+H^1+e^-$	1.5×10^8
B^8	5.4×10^5

[1] After BAHCALL and ULRICH (1971).

The present solar detection experiment is based on the reaction (DAVIS, HARMER, and HOFFMAN, 1968)

$$\nu_e + {}_{17}Cl^{37} \rightarrow e^- + {}_{18}Ar^{37} - 0.814\ \text{MeV}. \tag{4-247}$$

The experiment sets a limit to the summation of the products of the absorption cross section, σ_i, in cm^2 and the incident neutrino flux, φ_i, in $cm^{-2}\ sec^{-1}$ for the i neutrino sources. The present limit is

$$\sum_i \varphi_i\sigma_i \lesssim 1\ \text{SNU}\,, \tag{4-248}$$

where the solar neutrino unit is $1\,\mathrm{SNU}=10^{-36}$ neutrino absorptions per target atom per second. The absorption cross section for neutrinos to induce an allowed transition between states of a nucleus is given by (BAHCALL, 1964)

$$\sigma=\frac{2\pi^2(0.693)}{c(f\tau_{1/2})}\left(\frac{\hbar}{mc}\right)^3\left(\frac{W}{mc^2}\right)^2\frac{cp}{W}F(Z,W), \tag{4-249}$$

where p and W are the momentum and energy, respectively, of the electron which is produced, the quantity $f\tau_{1/2}$ is the measured ft value for the transition, and $F(Z, W)$ is the Fermi function which takes into account Coulomb and nuclear size corrections (KONOPINSKI, 1959; PRESTON, 1962). The cross section for Cl^{37} is

$$\sigma=1.91\times10^{-46}\frac{cpW}{(mc^2)^2}\left(\frac{F}{2\pi\alpha Z}\right)\mathrm{cm}^2, \tag{4-250}$$

where the fine structure constant $\alpha=1/137.037$, $W=W_\nu-0.814\,\mathrm{MeV}+0.511\,\mathrm{MeV}$, and the incident neutrino energy $W_\nu>0.814\,\mathrm{MeV}$. Values of this cross section are given in Table 41. When the neutrino fluxes given by BAHCALL and ULRICH (1971) are combined with these cross sections, we conclude that the incident neutrino level is 5.6 SNU, well above the detection limit of 1 SNU. If the thermonuclear energy for the Sun was supplied by the CNO cycle the incident neutrino level would be 35 SNU. Furthermore, the flux of about 1.5 SNU is expected from the proton-proton chain in spite of the stellar model used. It is generally concluded that a counting rate of <0.5 SNU is inconsistent with the view that thermonuclear reactions produce the observed solar luminosity. A possible explanation for this dichotomy was suggested by FOWLER (1972) who proposed that transient convective mixing took place in the Sun about 3×10^7 years ago—the time at which the surface luminosity we now see was generated. The neutrinos we now see were only generated about 8 minutes ago.

Table 41. The absorption cross sections of Cl^{37} for different neutrino sources[1]

Neutrino source	Cl^{37} cross section (cm^2)
Be^7	2.9×10^{-46}
N^{13}	2.1×10^{-46}
O^{15}	7.8×10^{-46}
$H^1+H^1+e^-$	1.7×10^{-45}
B^8	1.35×10^{-42}

[1] After BAHCALL (1964) and BAHCALL and SEARS (1972).

A summary of fifteen years of work on the solar neutrino problem has been given by BAHCALL and DAVIS (1976).

4.4. Nucleosynthesis of the Elements

4.4.1. Abundances of the Elements

The initial study of solar-system abundances (GOLDSCHMIDT, 1937) was followed by a detailed plot of the number distribution of the elements as a function of atomic weight (SUESS and UREY, 1956). This abundance curve was then used by BURBIDGE, BURBIDGE, FOWLER, and HOYLE (1957) in postulating the basic nucleosynthetic processes in stars. The abundance table of CAMERON (1973) is given as part of Table 38 together with the process by which most of each nuclide is made. The abundance curve is shown in Fig. 26. These data are based upon measurements of Type I carbonaceous chondrite meteorites (meteorites containing carbon compounds with a minimum of stony or metallic chrondrite material, and are thought to be a better representation than Suess and Urey's curve which was based on measurements of terrestrial, meteorite, and solar abundances.

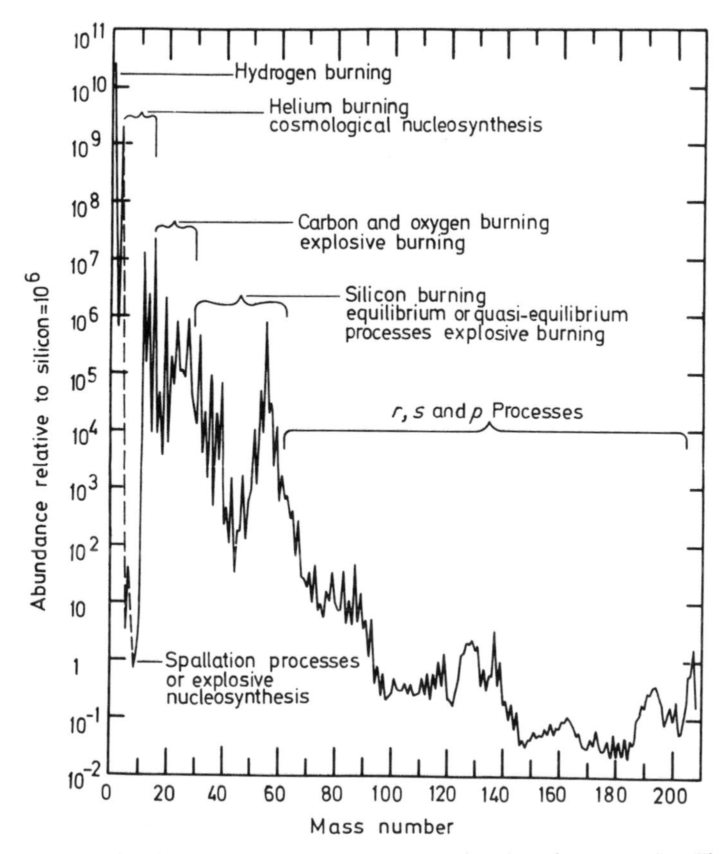

Fig. 26. Abundances of the nuclides plotted as a function of mass number. The figure also illustrates the different stellar nuclear processes which probably produce the characteristic features. The abundance data are given in Table 38

4.4.2. Nucleosynthetic Processes in Ordinary Stars—Energy Generation Stages and Reaction Rates

An examination of the abundance of the elements leads to a description of a series of stellar nucleosynthetic processes which are nearly the same as those postulated by BURBIDGE, BURBIDGE, FOWLER, and HOYLE (1957). These fundamental processes are shown on the nuclide abundance curve in Fig. 26 and are as follows:

a) Hydrogen burning (conversion of hydrogen to helium) temperature $T > 10^7$ °K and $\approx 10^{10}$ years duration.
b) Helium burning (conversion of helium to carbon, oxygen, etc.) temperature $T \gtrsim 10^8$ °K and $\approx 10^7$ years duration.
c) Carbon burning, temperature $T \gtrsim 6 \times 10^8$ °K, and oxygen burning, temperature $T \gtrsim 10^9$ °K (production of $16 \lesssim A \lesssim 28) \approx 10^5$ years duration, unless nucleosynthesis is explosive, then few seconds duration.
d) Silicon burning (production of $28 \lesssim A \lesssim 60$) temperature $T > 3$ or 4×10^9 °K and for the quasi-equilibrium and the e process ≈ 1 sec duration.
e) The s process (production of $A \geq 60$) temperature $T > 10^8$ °K and 10^3—10^7 years duration.
f) The r process (production of $A \geq 60$) temperature $T > 10^{10}$ °K and 10—100 seconds duration (uncertain).
g) The p process (production of the low abundance proton-rich heavy nuclei) temperature $T > 2$ to 3×10^9 °K and 10—100 seconds duration.

The hydrogen and helium burning stages govern most of the evolution paths observed on the Hertzsprung-Russell diagram, and the relation of thermonuclear reactions to stellar evolution is reviewed by HAYASHI, HOSHI, and SUGIMOTO (1962) and TAYLER (1966). The detailed formulae for evaluating these reaction rates in different mass zones are given by COX and GIULI (1969), SCHWARZSCHILD (1958), and SEARS and BROWNLEE (1965). Reaction rate constants for many of the important thermonuclear reactions in stars can be found in the papers by TRURAN, CAMERON, and GILBERT (1966), TRURAN (1972), FOWLER, CAUGHLAN, and ZIMMERMAN (1967, 1974), WAGONER (1969), and BARNES (1971). These authors adequately review the original papers from which their reviews of experimental data are extracted. In what follows, reaction rates are given for the nuclear processes suggested above. The component reactions for many of these processes will be specified together with the appropriate energy release and reaction rate constants. The reaction rate for the forward reaction can then be determined from the equation (cf. Eqs. (4-47), (4-51), (4-64), (4-65), and (4-67)).

$$\begin{aligned} N_A\langle 12\rangle = {} & C_1 T_9^{-2/3} \exp[-C_2 T_9^{-1/3} - (T_9/T_0)^2]\{1 + C_3 T_9^{1/3} + C_4 T_9^{2/3} \\ & + C_5 T_9 + C_6 T_9^{4/3} + C_7 T_9^{5/3}\} + \sum C_8 T_9^{-3/2} \exp(-C_9/T_9) \\ & + C_{10} \exp(-C_{11}/T_9)\ \text{cm}^3\ \text{sec}^{-1}\ (\text{g-mole})^{-1}, \end{aligned} \qquad (4\text{-}251)$$

and the reverse reaction rates from the equations

$$N_A\langle 34\rangle = C_{12}[\exp(-C_{13}/T_9)]\, N_A\langle 12\rangle\ \text{cm}^3\ \text{sec}^{-1}\ (\text{g-mole})^{-1} \qquad (4\text{-}252)$$

or

$$\lambda_\gamma(3) = C_{14} T_9^{3/2} [\exp(-C_{13}/T_9)]\, N_A\langle 12\rangle\ \text{sec}^{-1}. \qquad (4\text{-}253)$$

Here Avogadro's number $N_A = 6.022169 \times 10^{23}$ (g-mole)$^{-1}$, $\langle 12\rangle$ is the reaction rate $\langle \sigma v\rangle$ in the forward direction, $T_9 = T/10^9$, where T is the temperature in °K, the constants $C_1 - C_{14}$ are given in the tables, $\langle 34\rangle$ is the reaction rate $\langle \sigma v\rangle$

in the reverse direction, and $\lambda_\gamma(3)=[\tau_\gamma(3)]^{-1}$ is the inverse of the lifetime for photodisintegration. Stellar thermonuclear reaction rate constants are listed by FOWLER et al. (1975).

Hydrogen burning: the proton-proton chain. This process starts with the proton-proton reaction which has been discussed by WEIZSÄCKER (1937), BETHE and CRITCHFIELD (1938), BETHE (1939), SALPETER (1952), FOWLER (1959), REEVES (1963), and BAHCALL and MAY (1969). Subsequent reactions were suggested by SCHATZMAN (1951) and FOWLER (1958); and the roles of various parts of the chain have been discussed by PARKER, BAHCALL, and FOWLER (1964). The complete chain is

$$\begin{aligned}
&H^1+H^1 \rightarrow D^2+e^+ +\nu_e+1.442\ \text{MeV}-0.263\ \text{MeV}\\
&D^2+H^1 \rightarrow He^3+\gamma+5.493\ \text{MeV}\\
\text{or}\ &\begin{cases} He^3+He^3 \rightarrow He^4+2H^1+12.859\ \text{MeV}\\ He^3+He^4 \rightarrow Be^7+\gamma+1.586\ \text{MeV}\end{cases}\\
\text{or}\ &\begin{cases} Be^7+e^- \rightarrow Li^7+\nu_e+0.861\ \text{MeV}-0.80\ \text{MeV}\\ Li^7+H^1 \rightarrow He^4+He^4+17.347\ \text{MeV}\\ Be^7+H^1 \rightarrow B^8+\gamma+0.135\ \text{MeV}\end{cases}\\
&B^8 \rightarrow Be^8+e^+ +\nu_e+17.98\ \text{MeV}-7.2\ \text{MeV}\\
&Be^8 \rightarrow 2He^4+0.095\ \text{MeV}\,.
\end{aligned} \tag{4-254}$$

Here the energy release in MeV includes the annihilation energy of the positron when it is a reaction product, and the negative energy release denotes the average neutrino energy loss. The reaction rate constants for both the proton-proton chain and the C—N—O bi-cycle are given in Table 42.

Hydrogen burning: the C—N—O *bi-cycle.* For stars slightly more massive than the Sun, hydrogen will be fused into helium by the faster C—N cycle provided that carbon, nitrogen, or oxygen are present to act as a catalyst. These reactions are (WEIZSÄCKER, 1938; BETHE, 1939)

$$\begin{aligned}
C^{12}+H^1 &\rightarrow N^{13}+\gamma+1.944\ \text{MeV}\\
N^{13} &\rightarrow C^{13}+e^+ +\nu_e+2.221\ \text{MeV}-0.710\ \text{MeV}\\
C^{13}+H^1 &\rightarrow N^{14}+\gamma+7.550\ \text{MeV}\\
N^{14}+H^1 &\rightarrow O^{15}+\gamma+7.293\ \text{MeV}\\
O^{15} &\rightarrow N^{15}+e^+ +\nu_e+2.761\ \text{MeV}-1.000\ \text{MeV}\\
N^{15}+H^1 &\rightarrow C^{12}+He^4+4.965\ \text{MeV}\,.
\end{aligned} \tag{4-255}$$

Additional proton capture reactions, which may take place to form the complete C—N—O bi-cycle, are (FOWLER, 1954; SALPETER, 1955; MARION and FOWLER, 1957)

$$\begin{aligned}
N^{15}+H^1 &\rightarrow O^{16}+\gamma+12.126\ \text{MeV}\\
O^{16}+H^1 &\rightarrow F^{17}+\gamma+0.601\ \text{MeV}\\
F^{17} &\rightarrow O^{17}+e^+ +\nu_e+2.762\ \text{MeV}-0.94\ \text{MeV}\\
O^{17}+H^1 &\rightarrow N^{14}+He^4+1.193\ \text{MeV}\,.
\end{aligned} \tag{4-256}$$

Table 42. Reaction rates for hydrogen burning processes[1]

Proton–proton chain	C_1	C_2	T_0	C_3	C_4	C_5	C_6	C_7	C_8	C_9	C_{12}	C_{13}	C_{14}
$H^1(p,e^+\nu)D^2$	$4.214E{-15}$	3.380		0.123	1.09	0.938							
$D^2(p,\gamma)He^3$	$2.651E3$	3.720		0.112	1.99	1.56	0.162	0.324				63.752	$1.633E10$
$He^3(He^3,2p)He^4$	$5.964E10$	12.275		0.0339	−0.199	−0.0472	0.0316	0.0191				149.238	$3.395E{-10}\,T_9^{-3}$
$He^4(He^3,\gamma)Be^7$ [2]	$6.331E6$	12.826		0.0325	−0.350	−0.0797	0.0563	0.0325				18.419	$1.113E10$
$Li^7(p,\alpha)He^4$	$8.038E8$	8.471	30.068	0.0492	0.230	0.0793	−0.0265	−0.0232	$1.543E6$ $1.065E10$	4.479 30.443	4.690	201.308	
$Be^7(p,\gamma)B^8$	$4.345E5$	10.262		0.0406					$3.297E3$	7.306		1.564	$1.305E10$
C—N—O cycle													
$C^{12}(p,\gamma)N^{13}$	$2.043E7$	13.690	0.952	0.0304	1.190	0.254	2.06	1.120	$1.081E5$ $2.153E5$	4.925 18.179		22.557	$8.841E9$
$C^{13}(p,\gamma)N^{14}$	$8.006E7$	13.717	1.270	0.0304	0.958	0.204	1.39	0.753	$1.352E6$ $2.661E5$ $2.262E6$	5.978 11.987 13.463		87.615	$1.190E10$
$N^{14}(p,\gamma)O^{15}$	$4.207E7$	15.228	11.092	0.0274					$2.282E3$ $6.240E4$ $5.093E6$	3.010 11.476 28.147		84.688	$2.669E10$
$N^{15}(p,\gamma)O^{16}$	$4.185E8$	15.251	0.522	0.0273	2.97	0.569			$1.164E4$ $2.659E4$ $7.363E6$	3.676 4.665 10.983		140.720	$3.624E10$
$N^{15}(p,\alpha)C^{12}$	$8.157E11$	15.251	0.522	0.0273	6.74	1.29			$1.290E8$ $3.144E8$	3.676 7.974	0.706	57.614	
$O^{16}(p,\gamma)F^{17}$ [3]	$9.54\ E7$	16.693		0.00843								6.969	$3.034E9$
$O^{17}(p,\alpha)N^{14}$ [4]	$4.950E13$	16.712							$7.865E{-3}$ $8.302E6$ $4.891E7$	0.754 5.675 8.542	0.6757	13.842	

[1] After FOWLER, CAUGHLAN, and ZIMMERMAN (1974).

[2] $B_e^7(e^-,\nu)Li^7$ has $N_A\langle 12\rangle = 1.340\times 10^{-10}\times T_9^{-1/2}[1.0-0.537\,T_9^{1/3}+3.86\,T_9^{2/3}+1.2\,T_9+0.0027\,T_9^{-1}\exp(0.002515\,T_9^{-1})]$.

[3] $C_1\,T_9^{-2/3}=(C_1^*\,T_9^{-1})^{17/21}$ where C_1^* is listed as C_1 above.

[4] $C_1\,T_9^{-2/3}=C_1^*\,T_9^{-2/3}(1+19.986\,T_9)^{-5/6}$ where C_1^* is listed as C_1 above.

Here the energy release includes positron annihilation, and the negative energy release is the average neutrino energy loss. It is possible that the CNO cycle produces most of the N^{14} found in nature, and the details of nucleosynthesis by the CNO process are given by CAUGHLAN and FOWLER (1962) and CAUGHLAN (1965). During supernovae explosions, a rapid CNO cycle might take place in which the (n, p) reactions replace the beta decays in the cycle.

Helium burning: the triple alpha and alpha capture processes. The reactions assigned to the triple alpha process are (SALPETER, 1952, 1953, 1957; ÖPIK, 1951, 1954; HOYLE, 1954; FOWLER and GREENSTEIN, 1956; COOK, FOWLER, LAURITSEN, and LAURITSEN, 1957)

$$\begin{aligned} He^4 + He^4 &\rightarrow Be^8 - 0.0921 \text{ MeV} \\ Be^8 + He^4 &\rightarrow C^{12*} - 0.286 \text{ MeV} \\ C^{12*} &\rightarrow C^{12} + \gamma + 7.656 \text{ MeV} . \end{aligned} \tag{4-257}$$

The total reaction, $3\,He^4 \rightarrow C^{12} + \gamma + 7.274$ MeV, results in the mean lifetime, $\tau_{3\alpha}$, for the destruction of He^4 by the 3α process given by (FOWLER, CAUGHLAN, and ZIMMERMAN, 1967; CLAYTON, 1968; BARNES, 1971)

$$\begin{aligned} \lambda_{3\alpha} = \frac{1}{\tau_{3\alpha}} = \frac{3 r_{3\alpha}}{N_\alpha} &= 6.65 \times 10^{-10} (\rho X_\alpha)^2 T_9^{-3} \exp(-4.405/T_9) \\ &+ 1.66 \times 10^{-8} (\rho X_\alpha)^2 T_9^{-3} \exp(-27.443/T_9) \sec^{-1} , \end{aligned} \tag{4-258}$$

or for the forward triple alpha reaction:

$$\begin{aligned} N_A \langle \sigma v \rangle = 2.05 \times 10^{-8} T_9^{-3} \exp(-4.405/T_9) \\ + 5.315 \times 10^{-7} T_9^{-3} \exp(-27.433/T_9) \text{ cm}^3 \sec^{-1} \text{g-mole}^{-1} , \end{aligned}$$

where the first term holds for $0.03 \le T_9 \le 8$ and the second term is added for $4 \le T_9 \le 8$. Here $r_{3\alpha}$ is the reaction rate, N_α is the density of α particles, X_α is their mass fraction, and ρ is the gas mass density. It is now believed that helium burning results in the production of approximately equal amounts of ^{12}C and ^{16}O in stars in the wide range of masses from 0.5 to 50 $M_\odot$.

Once C^{12} is formed, O^{16} will be the product of the α capture process

$$C^{12} + He^4 \rightarrow O^{16} + \gamma + 7.161 \text{ MeV} . \tag{4-259}$$

This is another special reaction with the rate (FOWLER, CAUGHLAN, and ZIMMERMAN, 1974)

$$\begin{aligned} N_A \langle \sigma v \rangle = 1.90 \times 10^8 T_9^{-2} (1 + 0.046 T_9^{-2/3})^{-2} \exp[-32.12 T_9^{-1/3} - (T_9/3.270)^2] \\ + 3.338 \times 10^2 \exp(-26.316/T_9) \text{ cm}^3 \sec^{-1} \text{g-mole}^{-1} . \end{aligned} \tag{4-260}$$

This reaction rate is uncertain by a factor of five. Further α capture processes which then follow are

$$\begin{aligned} O^{16} + He^4 &\rightarrow Ne^{20} + \gamma + 4.730 \text{ MeV} \\ Ne^{20} + He^4 &\rightarrow Mg^{24} + \gamma + 9.317 \text{ MeV} \\ Mg^{24} + He^4 &\rightarrow Si^{28} + \gamma + 9.981 \text{ MeV} \\ Si^{28} + He^4 &\rightarrow S^{32} + \gamma + 6.948 \text{ MeV} \\ S^{32} + He^4 &\rightarrow Ar^{36} + \gamma + 6.645 \text{ MeV} . \end{aligned} \tag{4-261}$$

The reaction products of the C—N—O cycle might also produce neutrons by the α-capture reactions (CAMERON, 1954)

$$C^{13}+He^4 \rightarrow O^{16}+n+2.214\ \text{MeV}$$

or (CAMERON, 1960)

$$\begin{aligned} N^{14}+He^4 &\rightarrow F^{18}+\gamma+4.416\ \text{MeV} \\ F^{18} &\rightarrow O^{18}+e^+ +\nu_e \\ O^{18}+He^4 &\rightarrow Ne^{22}+\gamma+9.667\ \text{MeV} \\ O^{18}+He^4 &\rightarrow n+Ne^{21}-0.699\ \text{MeV} \\ Ne^{22}+He^4 &\rightarrow n+Mg^{25}-0.481\ \text{MeV}\,. \end{aligned} \tag{4-262}$$

Reaction rate constants for the α particle reactions are given in Table 43.

Carbon and oxygen burning. At the conditions of helium burning, the predominant nuclei are C^{12} and O^{16}. When temperatures greater than 8×10^8 °K are reached, carbon will begin to react with itself according to the reactions

$$\begin{aligned} C^{12}+C^{12} &\rightarrow Mg^{24}+\gamma+13.930\ \text{MeV} \\ &\rightarrow Na^{23}+p+2.238\ \text{MeV} \\ &\rightarrow Ne^{20}+He^4+4.616\ \text{MeV} \\ &\rightarrow Mg^{23}+n-2.605\ \text{MeV} \\ &\rightarrow O^{16}+2He^4-0.114\ \text{MeV}\,. \end{aligned} \tag{4-263}$$

At about 2×10^9 °K, oxygen will also react with itself according to the reactions

$$\begin{aligned} O^{16}+O^{16} &\rightarrow S^{32}+\gamma+16.539\ \text{MeV} \\ &\rightarrow P^{31}+p+7.676\ \text{MeV} \\ &\rightarrow S^{31}+n+1.459\ \text{MeV} \\ &\rightarrow Si^{28}+He^4+9.593\ \text{MeV} \\ &\rightarrow Mg^{24}+2He^4-0.393\ \text{MeV}\,. \end{aligned} \tag{4-264}$$

PATTERSON, WINKLER, and SPINKA (1968), PATTERSON, WINKLER, and ZAIDINS (1969), and SPINKA and WINKLER (1972) have measured the cross sections for several of these reactions. They find that the large interaction radius of the heavy ions requires an extra term in the cross section, so that the average cross section $\langle\sigma\rangle$ is given by

$$\langle\sigma\rangle = \frac{\langle S\rangle}{E}\exp\left[-(E_G/E)^{1/2}-gE\right], \tag{4-265}$$

where E is the kinetic energy in the center-of-mass system, the Gamow energy, E_G, is given by Eq. (4-34) and

$$g=0.122(A R_f^3/Z_1 Z_2)^{1/2}\ \text{MeV}^{-1}, \tag{4-266}$$

where A is the reduced mass of the two interacting particles in a.m.u., Z_1 and Z_2 are the particle charges in units of the proton charge, and the interaction radius, R_f, is given in Fermis. For the $C^{12}+C^{12}$ reaction $\langle S\rangle=2.9\times 10^{16}$ MeV-barns and $g=0.46\ \text{MeV}^{-1}$, whereas $\langle S\rangle=6\times 10^{27}$ MeV-barns and $g=0.84\ \text{MeV}^{-1}$

Table 43. Reaction rates for helium burning processes[1]

α particle reaction	C_1	C_2	T_0	C_3	C_8	C_9	C_{10}	C_{11}	C_{12}	C_{13}	C_{14}
$He^4(2\alpha,\gamma),C^{12}$	[2]									84.417	$2.003E20T_9^{3/2}$
$C^{12}(\alpha,\gamma)O^{16}$	[2]									83.106	$5.133E10$
$O^{16}(\alpha,\gamma)Ne^{20}$ [3]	$5.357E9$	39.756		0.0105	40.86 391.4 881.8	10.359 12.243 23.112				54.888	$5.652E10$
$Ne^{20}(\alpha,\gamma)Mg^{24}$	$1.235E18$	46.765	0.849				8,947.0	15.576		108.118	$6.011E10$
$Mg^{24}(\alpha,\gamma)Si^{28}$	$3.160E20$	53.301	0.670				1,848.0	15.162		115.833	$6.270E10$
$Si^{28}(\alpha,\gamma)S^{32}$	$1.675E21$	59.485	0.785				936.2	18.810		80.626	$6.465E10$
$S^{32}(\alpha,\gamma)Ar^{36}$	$8.132E20$	65.366	1.068				519.7	25.380		77.099	$6.620E10$
$C^{13}(\alpha,n)O^{16}$ [4]	$6.769E15$	32.329	0.941	0.0129	$2.119E6$ $1.921E7$ $4.749E6$	9.433 11.536 11.891			5.792	25.699	
$N^{14}(\alpha,\gamma)F^{18}$	$9.051E9$	36.031	0.863	0.0116	$2.364E{-10}$ 2.028	2.798 5.054	14.160	12.138		51.248	$5.419E10$
$O^{18}(\alpha,\gamma)Ne^{22}$ [5]	$1.504E15$	40.056		0.0104						112.189	$5.849E10$
$O^{18}(\alpha,n)Ne^{21}$ [6]	$2.661E18$	40.056							0.7841	8.113	

[1] After FOWLER, CAUGHLAN, and ZIMMERMAN (1974).
[2] See text for reaction rates.
[3] From TOEVS, FOWLER, BARNES, and LYONS (1971).
[4] $C_4=2.04$, $C_5=0.184$.
[5] $C_1 T_9^{-2/3}=C_1^* T_9^{-2/3}(1+0.098\,T_9)^{-5/6}$ and $\exp[-C_2 T_9^{-1/3}]=\exp[-C_2^* T_9^{-1/3}(1+0.098\,T_9)^{1/3}-(0.431\,T_9^{-1})^{3.89}]$, where C_1^* and C_2^* are the values of C_1 and C_2 listed above.
[6] $C_1 T_9^{-2/3}=C_1^* T_9^{-2/3}(1+0.098\,T_9)^{-5/6}$ and $\exp[-C_2 T_9^{-1/3}]=\exp[-C_2^* T_9^{-1/3}(1+0.098\,T_9)^{1/3}-(0.431\,T_9^{-1})^{3.98}]$, where C_1^* and C_2^* are the values of C_1 and C_2 listed above.

for the $O^{16}+O^{16}$ reactions above 7 MeV. These values result in a $C^{12}\rightarrow C^{12}$ reaction rate of (ARNETT and TRURAN, 1969)

$$\rho N_A\langle\sigma v\rangle = 3.24\times10^{26}\,T_9^{-2/3}\exp[-84.17\,T_9^{-1/3}(1+0.037\,T_9)^{1/3}]\ \mathrm{sec}^{-1}, \tag{4-267}$$

and an $O^{16}\rightarrow O^{16}$ reaction rate of (TRURAN and ARNETT, 1970)

$$\begin{aligned} N_A\langle\sigma v\rangle = {} & 6.6\times10^{37}\,T_9^{-2/3} \\ & \times\exp[-135.79\,T_9^{-1/3}(1+0.0530\,T_9)^{1/3}]\ \mathrm{cm^3\,sec^{-1}\,(g\text{-}mole)^{-1}}. \end{aligned} \tag{4-268}$$

Hydrostatic oxygen burning is treated by ARNETT (1972) and WOOSLEY, ARNETT and CLAYTON (1972), whereas explosive oxygen burning is considered by WOOSLEY, ARNETT, and CLAYTON (1973).

The α particles, protons, and neutrons which are products of the carbon and oxygen burning reactions given in Eqs. (4-263) and (4-264) will interact with the other products of the burning to form many of the other nuclides with $16\leq A\leq 28$. Experimental values for the reaction rate constants for many of these reactions are given by WAGONER (1969) and FOWLER, CAUGHLAN, and ZIMMERMAN (1967, 1974).

It is now thought that most of the carbon, oxygen, and silicon burning, which accounts for the observed solar system abundances for $20\leq A\leq 64$, occurs during fast explosions, and these explosive burning processes are discussed in Sect. 4.4.4.

Silicon burning. At the conclusion of carbon and oxygen burning, the most abundant nuclei will be S^{32} and Si^{28} with significant amounts of Mg^{24}. Because the binding energies for protons, neutrons and α particles in S^{32} are smaller than those in Si^{28}, the nuclide S^{32} will be the first to photodisintegrate according to the reactions

$$\begin{aligned} S^{32}+\gamma &\rightarrow P^{31}+p-8.864\ \mathrm{MeV} \\ P^{31}+\gamma &\rightarrow Si^{30}+p-7.287\ \mathrm{MeV} \\ Si^{30}+\gamma &\rightarrow Si^{29}+n \\ Si^{29}+\gamma &\rightarrow Si^{28}+n\,. \end{aligned} \tag{4-269}$$

The resulting reactions will leave little but Si^{28}. Silicon will then begin to photodisintegrate at temperatures greater than 3×10^9 °K according to the reactions

$$\begin{aligned} Si^{28}+\gamma &\rightarrow Al^{27}+p-11.583\ \mathrm{MeV} \\ Si^{28}+\gamma &\rightarrow Mg^{24}+\alpha-9.981\ \mathrm{MeV}\,. \end{aligned} \tag{4-270}$$

As the (γ,α) reaction has the lower threshold, it is the dominant reaction at low temperatures $T<2\times10^9$ °K; whereas the (γ, p) reaction has the shorter lifetime at higher temperatures. Further photodisintegrations lead to the build-up of lighter elements according to the reactions

$$\begin{aligned} Mg^{24}+\gamma &\rightarrow Na^{23}+p-11.694\ \mathrm{MeV} \\ Mg^{24}+\gamma &\rightarrow Ne^{20}+\alpha-9.317\ \mathrm{MeV} \\ Ne^{20}+\gamma &\rightarrow O^{16}+\alpha-4.730\ \mathrm{MeV} \\ O^{16}+\gamma &\rightarrow C^{12}+\alpha-7.161\ \mathrm{MeV}\,. \end{aligned} \tag{4-271}$$

Table 44. Reaction rate constants for some (γ,p) silicon burning reactions[1]

Reaction	C_1	C_2	T_0	C_{10}	C_{11}	C_{13}	C_{14}
$Na^{23}(p,\gamma)Mg^{24}$	2.263 *E* 13	20.766	0.315	4.938 *E* 4	3.564	135.708	7.489 *E* 10
$Al^{27}(p,\gamma)Si^{28}$	7.021 *E* 12	23.261	0.631	8.188 *E* 4	6.342	134.417	1.134 *E* 11
$Si^{30}(p,\gamma)P^{31}$	8.650 *E* 13	24.468	0.451	6.200 *E* 4	5.334	84.567	9.497 *E* 9
$P^{31}(p,\gamma)S^{32}$	6.942 *E* 13	25.629	0.442	1.692 *E* 4	5.514	102.867	3.805 *E* 10

[1] After FOWLER, CAUGHLAN, and ZIMMERMAN (1974).

Reaction rate constants for the (γ, α) reactions in Eqs. (4-269) to (4-271) are given in Table 43 whereas the (γ, p) reaction rate constants are given in Table 44. Resonance strengths for the various resonance energies of the $^{24}Mg(\alpha,\gamma)^{28}Si$ and $^{27}Al(p,\gamma)^{28}Si$ reactions are tabulated, respectively, by LYONS (1969) and LYONS, TOEVS, and SARGOOD (1969). Analytic fits to experimental data for the $^{20}Ne(\alpha,\gamma)^{24}Mg$ and $^{23}Na(p,\gamma)^{24}Mg$ reactions may lead to the rates (COUCH and SHANE, 1971; FOWLER, CAUGHLAN, and ZIMMERMAN, 1974).

For $Ne^{20}+\alpha\rightarrow Mg^{24}+\gamma$

$$N_A\langle\sigma v\rangle_{\alpha,\gamma}=\frac{1.235\times10^{18}}{T_9^{2/3}}\exp\left[\frac{-46.765}{T_9^{1/3}}-\left(\frac{T_9}{0.849}\right)^2\right] +8.947\times10^3\exp\left(\frac{-15.576}{T_9}\right)\mathrm{cm^3\,sec^{-1}\,(g\text{-}mole)^{-1}}. \tag{4-272}$$

For $Na^{23}+p\rightarrow Mg^{24}+\gamma$

$$N_A\langle\sigma v\rangle_{p,\gamma}=\frac{2.263\times10^{13}}{T_9^{2/3}}\exp\left[\frac{-20.766}{T_9^{1/3}}-\left(\frac{T_9}{0.315}\right)^2\right] +4.938\times10^4\exp\left(\frac{-3.564}{T_9}\right)\mathrm{cm^3\,sec^{-1}\,(g\text{-}mole)^{-1}}. \tag{4-273}$$

The corresponding photodisintegration rates for $1.0\leq T_9\leq5.0$ are tabulated by COUCH and SHANE (1971). For the $^{20}Ne(\alpha,\gamma)^{24}Mg$ reaction, we have $C_{13}=108.118$ and $C_{14}=6.011\times10^{10}$, whereas for the $^{23}Na(p,\gamma)^{24}Mg$ reaction, we have $C_{13}=135.708$ and $C_{14}=7.489\times10^{10}$. These constants can be used in Eq. (4-253) to determine photodisintegration rates.

The abundances of most of the nuclei in the range $28\lesssim A\lesssim60$ are thought to be determined by equilibrium or quasi-equilibrium processes in which the importance of many individual reaction rates is diminished, and these processes are discussed in the next Sect. 4.4.3. Abundances of the elements with $28\leq A\leq60$ have been discussed by HOYLE (1946), BURBIDGE, BURBIDGE, FOWLER, and HOYLE (1957), FOWLER and HOYLE (1964), CLIFFORD and TAYLER (1965), TRURAN, CAMERON, and GILBERT (1966), BODANSKY, CLAYTON, and FOWLER (1968). TRURAN (1968), and MICHAUD and FOWLER (1971). Most nuclear species between ^{28}Si and ^{59}Co, except the neutron-rich species (^{36}S, ^{40}Ar, ^{43}Ca, ^{46}Ca, ^{48}Ca, ^{51}Ti, ^{54}Cr, and ^{58}Fe), are generated by a quasi-equilibrium process in which the only important thermonuclear reaction rates are thought to be those of the "bottle-neck" nuclei ^{44}Ca, ^{45}Sc, and ^{45}Ti. The reaction rates for these nuclei may be estimated by the methods given by MICHAUD and FOWLER (1970). The abundances of the neutron-rich species could well be determined by the s or r processes discussed in the next section (SEEGER, FOWLER, and CLAYTON, 1965). They may, however, be formed in the explosive carbon burning stage (HOWARD, ARNETT, CLAYTON, and WOOSLEY, 1972).

The s, r and p processes. Because the binding energy per nucleon decreases with increasing A for nuclides beyond the iron peak ($A \gtrsim 60$), and because these elements have large Coulomb barriers, they are not likely to be formed by fusion or alpha and proton capture. It is thought that most of these heavier elements are formed by neutron capture reactions which start with the iron group nuclei (Cr, Mn, Fe, Co, and Ni). If the flux of neutrons is weak, most chains of neutron captures will include only a few captures before the beta decay of the product nucleus. Because the neutron capture lifetime is slower (s) than the beta decay lifetime, this type of neutron capture is called the s process. When there is a strong neutron flux, as it is believed to occur during a supernovae explosion, the neutron-rich elements will be formed by the rapid (r) neutron capture process. In this case, the neutron capture lifetime is much less than that of beta decay, and the neutron-rich elements only beta decay to beta stable elements after the capture processes are over. The details of the s and r processes are discussed by BURBIDGE, BURBIDGE, FOWLER, and HOYLE (1957), and the appropriate neutron capture rates and nuclide abundance equations are given in Sect. 4.4.5. Elements which are formed only by the r process, or the s process, or by both (r, s) are designated by the appropriate class symbol in Table 38, where neutron capture cross sections may also be found. Although the source of neutrons for the s process is uncertain, the most likely site of the s process synthesis is provided by the convective mixing of hydrogen and helium layers following a helium flash (SCHWARZSCHILD and HÄRM, 1962, 1967). In this case free neutrons are formed by the reactions (SANDERS, 1967; CAMERON and FOWLER, 1971)

$$C^{12}(p,\gamma)N^{13}(e^+\nu)C^{13}$$

and

$$C^{13}(\alpha,n)O^{16}\,. \tag{4-274}$$

The site of r process synthesis is thought to be related to the formation of a neutron star, and, perhaps, the formation of a supernova explosion (TRURAN, ARNETT, TSURUTA, and CAMERON, 1968). The proton rich heavy elements are much less abundant than the elements thought to be produced by the r and s processes, and are thought to be formed by a rapid proton capture process. A possible occasion for this process is the passage of a supernova shock wave through the hydrogen outer layer of a pre-supernova star. Possible reactions for this process include the (p,γ), (p,n), (γ,n) and $(n,2n)$ reactions and positron capture (BURBIDGE, BURBIDGE, FOWLER, and HOYLE, 1957; ITO, 1961; REEVES and STEWART, 1965; MACKLIN, 1970). The p process in explosive nucleosynthesis has been discussed by TRURAN and CAMERON (1972). Explosive r-process nucleosynthesis is considered by SCHRAMM (1973).

4.4.3. Equilibrium Processes

Under conditions of thermodynamic equilibrium, the composition of matter in a star may be calculated without determining individual reaction rates, and only the binding energies and partition functions of the various nuclear species need to be specified. Although this condition greatly simplifies calculations, it is only possible if the matter is in equilibrium with the radiation, and if every nucleus is transformable into any other nucleus. HOYLE (1946) showed that matter is in equilibrium with radiation for temperatures $T \approx 10^9$ °K, and that all known nuclei may be transformed into any other nucleus by nuclear reactions for $T \gtrsim 2 \times 10^9$ °K. Early investigations by TOLMAN (1922), UREY and BRADLEY (1931), POKROWSKI (1931), STERNE (1933), and CHANDRASEKHAR and HENRICH (1942) ruled out an equilibrium origin for most of the elements. Nevertheless, the abundance peak in the iron group, where the binding energy per nucleon is a maximum, can be accounted for by an equilibrium process (HOYLE, 1946; BURBIDGE, BURBIDGE, FOWLER, and HOYLE, 1957). More detailed calculations of the equilibrium concentrations in this region ($28 \leq A \leq 60$) have been carried out by FOWLER and HOYLE (1964), CLIFFORD and TAYLER (1965), and TRURAN, CAMERON, and GILBERT (1966) for strict equilibrium conditions; and by BODANSKY, CLAYTON, and FOWLER (1968), TRURAN (1968), MICHAUD and FOWLER (1972) and WOOSLEY, ARNETT, and CLAYTON (1973), for quasi-equilibrium conditions.

Under conditions of statistical equilibrium, the number density, N_i, of particles of the ith kind is given by

$$N_i = \frac{1}{V} \sum_r^n \mu_i [\pm \mu_i + \exp(\varepsilon_{ir}/kT)]^{-1}, \tag{4-275}$$

where V is the volume, μ_i is the chemical potential of the ith particle, the plus and minus signs refer to Fermi-Dirac and Bose-Einstein statistics, respectively, and the summation is over all energy states, ε_{ir}, which includes both internal energy levels and the kinetic energy. If an internal level has spin, J, then $2J+1$ states of the same energy must be included in the sum. When the nuclides are non-degenerate and non-relativistic, Maxwellian statistics can be employed to give

$$\begin{aligned} N_i &= \frac{\mu_i}{V}\left[\sum (2J_r+1)\exp\left(-\frac{\varepsilon_r}{kT}\right)\right]\left[\frac{4\pi V}{h^3}\int_0^\infty p^2 \exp\left(-\frac{p^2}{2M_i kT}\right) dp\right] \\ &= \mu_i \omega_i \left(\frac{2\pi M_i kT}{h^2}\right)^{3/2}, \end{aligned} \tag{4-276}$$

where p is the particle momentum, M_i is its mass, the partition function $\omega_i = \sum (2J_r+1)\exp(-\varepsilon_r/kT)$, and here ε_r refers to internal energy states only. For particles $p_i, p_j, \ldots$ which react according to

$$\alpha p_i + \beta p_j + \cdots \leftrightarrows \xi p_r + \eta p_s + \cdots, \tag{4-277}$$

the chemical potentials are related by the equation

$$\mu_i^\alpha \mu_j^\beta \cdots = \mu_r^\xi \mu_s^\eta \cdots \exp(-Q/kT), \tag{4-278}$$

where

$$Q = c^2 [\alpha M_i + \beta M_j + \cdots - \xi M_r - \eta M_s - \cdots] . \quad (4\text{-}279)$$

HOYLE (1946) and BURBIDGE, BURBIDGE, FOWLER, and HOYLE (1957) considered the condition of statistical equilibrium between the nuclei, (A, Z), and free protons, p, and neutrons, n. For a nucleus, there are Z protons and $(A-Z)$ neutrons, and the statistical weight of both protons and neutrons is two. It then follows from Eqs. (4-276) to (4-279) that for equilibrium between nuclides, protons, and neutrons, the number density, $N(A,Z)$, of the nucleus, (A,Z), is given by

$$N(A, Z) = \omega(A, Z) \left(\frac{A M_\mu k T}{2\pi\hbar^2} \right)^{3/2} \left(\frac{2\pi\hbar^2}{M_\mu k T} \right)^{3A/2} \frac{N_n^{(A-Z)} N_p^Z}{2^A} \exp\left[\frac{Q(A, Z)}{kT} \right], \quad (4\text{-}280)$$

where the partition function, $\omega(A,Z)$, of the nucleus, (A,Z), is given by

$$\omega(A, Z) = \sum_r (2I_r + 1) \exp\left(-\frac{E_r}{kT} \right), \quad (4\text{-}281)$$

where I_r and E_r are, respectively, the spin and energy of the rth excited level, the binding energy, $Q(A,Z)$, of the nucleus, (A,Z), is given by

$$Q(A, Z) = c^2 [(A-Z) M_N + Z M_p - M(A, Z)] , \quad (4\text{-}282)$$

where M_N, M_p and $M(A, Z)$ are, respectively, the masses of the free neutrons, free protons, and the nucleus, (A,Z), the factor

$$\left(\frac{2\pi\hbar^2}{M_\mu k T} \right)^{3/2} \approx 1.6827 \times 10^{-34} T_9^{-3/2} \text{ cm}^{-3} , \quad (4\text{-}283)$$

the atomic mass unit is M_μ, and N_n and N_p denote, respectively, the number densities of free neutrons and protons. Conservation of mass requires that the mass density, ρ, be given by

$$\rho = N_p M_p + N_n M_N + \sum_{A,Z} N(A, Z) M(A, Z) . \quad (4\text{-}284)$$

Conservation of charge further implies that

$$N_p + \sum_{A,Z} Z N(A, Z) = N_{e^-} - N_{e^+} , \quad (4\text{-}285)$$

where N_{e^-} and N_{e^+} denote, respectively, the number densities of electrons and positrons. CLIFFORD and TAYLER (1965) use the additional parameter

$$R = \frac{N_p + \sum_{A,Z} Z N(A,Z)}{N_n + \sum_{A,Z} (A-Z) N(A,Z)} = \frac{\langle Z \rangle}{\langle N \rangle} \quad (4\text{-}286)$$

to denote the ratio of the total number of protons to the total number of neutrons. They use the above equations to calculate nuclidic abundances as a function of ρ, T, and R for conditions with and without beta decay. TSURUTA and CAMERON (1965) consider nuclear statistical equilibrium at high densities and use T and the Fermi energy, E_F, as free parameters.

$$
\begin{aligned}
E_{\mathrm{F}} &= (3\pi^2)^{2/3} \frac{\hbar^2}{2m_{\mathrm{e}}} N_{\mathrm{e}}^{2/3} \quad \text{(nonrelativistic)} \\
E_{\mathrm{F}} &= (3\pi^2)^{1/3} \hbar c N_{\mathrm{e}}^{1/3} \quad \text{(relativistic)},
\end{aligned}
\tag{4-287}
$$

where N_e is the electron density. When positive and negative beta decay and electron capture rates are considered together with the steady state condition that the rate of neutrino emission is equal to the antineutrino emission rate, a specification of E_F gives the values of N_n and N_p to be used in Eq. (4-280). The appropriate beta decay and electron capture rates were given in Sect. 4.3.

TRURAN, CAMERON, and GILBERT (1966) noted that at temperatures greater than 3×10^9 °K the photodisintegration of silicon proceeds rapidly, releasing protons, neutrons and alpha particles. They calculated the synthesis of iron peak nuclei from ^{28}Si nuclei by solving thermonuclear reaction rates for all the neutron, proton and alpha particle reactions. BODANSKY, CLAYTON, and FOWLER (1968) then showed that the nuclei are in a quasi-equilibrium during silicon burning. It is a partial equilibrium in that ^{28}Si is not itself in equilibrium with the light particles, and the heavy nuclei are not in equilibrium among themselves; but nuclei heavier than ^{28}Si are in equilibrium with ^{28}Si under the exchange of photons, nucleons, and alpha particles. The light particles are ejected at a significant rate by the ^{28}Si$(\gamma, \alpha)^{24}$Mg and the ^{28}Si$(\gamma, p)(\gamma, p)(\gamma, n)(\gamma, n)^{24}$Mg reactions. The concentration of heavy nuclei builds up to such a value that the nuclei consume and liberate alpha particles at almost the same rate, and a quasi-equilibrium is established with the equilibrium reactions

$$
\begin{aligned}
&^{28}\mathrm{Si} + {}^{4}\mathrm{He} \leftrightarrows {}^{32}\mathrm{S} + \gamma \\
&^{32}\mathrm{S} + {}^{4}\mathrm{He} \leftrightarrows {}^{36}\mathrm{Ar} + \gamma \\
&\quad \vdots \\
&^{52}\mathrm{Fe} + {}^{4}\mathrm{He} \leftrightarrows {}^{56}\mathrm{Ni} + \gamma .
\end{aligned}
\tag{4-288}
$$

It follows from Eqs. (4-276) to (4-279) that the equilibria are characterized by Saha equations such as

$$
\frac{N(^{32}\mathrm{S})}{N(^{28}\mathrm{Si})} = N_\alpha \frac{\omega(^{32}\mathrm{S})}{\omega(^{28}\mathrm{Si})} \left[\frac{A(^{32}\mathrm{S})}{A(^{28}\mathrm{Si})A(^{4}\mathrm{He})}\right]^{3/2} \left(\frac{2\pi\hbar^2}{M_\mu k T}\right)^{3/2} \exp\left[\frac{B_\alpha(^{32}\mathrm{S})}{kT}\right] \tag{4-289}
$$

for

$$
^{28}\mathrm{Si} + {}^{4}\mathrm{He} \leftrightarrows {}^{32}\mathrm{S} + \gamma .
$$

Here N_α is the alpha particle number density, the partition functions, ω, are approximately unity for the alpha particle nuclei, the A's are atomic masses in units of the atomic mass unit, M_μ, and $B_\alpha(^{32}\mathrm{S})$ is the separation energy of an alpha particle from the ground state of ^{32}S. The equilibrium abundances, $N(A, Z)$, of other nuclei, (A, Z), depend on the number densities of ^{28}Si, as well as those of the alpha particles, neutrons, and protons. The abundance formula for this case is given in Sect. 4.4.5. BODANSKY, CLAYTON, and FOWLER (1968) have shown that quasi-equilibrium distributions account for the observed abundances of the $4n$ nuclei ($A = 4n$) for $A = 28$ through $A = 56$, as well as for the dominant nuclei in the lower part of the iron group ($A = 49$ through $A = 57$). TRURAN (1968) examined the effect of a variable initial composition on this silicon burning process, and MICHAUD and FOWLER (1972) examined the dependence of the synthesized abundances on the initial composition before silicon burning, as well as the effect of adding products from different burning zones. The latter authors show that with an initial neutron enhancement of 4×10^{-3} the natural abundances for nuclei with $28 \lesssim A \lesssim 59$ may be accounted for by quasi-equilibrium burning. In this case, a quasi-equilibrium between elements with $24 \leq A \leq 44$ and a separate equilibrium for elements with $46 \leq A \leq 60$ is assumed, and detailed nuclear reactions are given for the "bottleneck" at $A = 45$. This quasi-equilibrium silicon burning process must have taken place in a short time, $t \lesssim 1$ sec, and at high temperatures, $T \gtrsim 4.5 \times 10^9$ °K, suggesting the explosive burning processes discussed in the next section.

4.4.4. Explosive Burning Processes

As explained by BURBIDGE, BURBIDGE, FOWLER, and HOYLE (1957), the successive cycles of static nuclear burning and contraction, which successfully account for much of stellar evolution, must end when the available nuclear fuel is exhausted. They showed that the unopposed action of gravity in a helium exhausted stellar core leads to violent instabilities and to rapid thermonuclear reactions in the stellar envelope. SCHWARZSCHILD and HÄRM (1962, 1967) showed that even for a hydrogen exhausted core the thermonuclear reactions runaway in a helium flash. ARNETT (1968) then showed that when cooling by neutrino emission in a highly degenerate gas is considered, the $C^{12} + C^{12}$ reaction will ignite explosively at core densities of about 2×10^9 g cm^{-3}. The stellar material is instantaneously heated and then expands adiabatically so that the density, ρ, and temperature, T, are related by

$$\rho(t) \propto [T(t)]^3 , \tag{4-290}$$

for a $\Gamma_3 = 4/3$ adiabat, and a time variable, t. The appropriate expansion time is the hydrodynamic time scale, τ_{HD}, given by

$$\tau_{\mathrm{HD}} = (24 \pi G \rho)^{-1/2} \approx 446 \rho^{-1/2} \,\mathrm{sec} . \tag{4-291}$$

The initial temperature and density must be such that the mean lifetime, τ_{R}, for a nucleus undergoing an explosive reaction, R, must be close to τ_{HD}. For the interaction of nucleus 1 with a nucleus 2,

$$\tau_{\mathrm{R}} = \tau_2(1) = [N_2 \langle \sigma v \rangle]^{-1} = \left[\rho N_{\mathrm{A}} \frac{X_2}{A_2} \langle \sigma v \rangle \right]^{-1} , \tag{4-292}$$

where the mass density is ρ, the X_2, A_2 and N_2 are, respectively, the mass fraction, mass number, and number density of nucleus 2, and the $N_{\mathrm{A}} \langle \sigma v \rangle$ were given in Sect. 4.4.2 for various reactions (cf. Eq. (4-251)). ARNETT (1969) used a mean carbon nucleus lifetime, $\tau_{\mathrm{C}^{12}}$, given by

$$\log \tau_{\mathrm{C}^{12}} \approx 37.4 \, T_9^{-1/3} - 25.0 - \log_{10} \rho \approx \log_{10} \tau_{\mathrm{HD}} \tag{4-293}$$

for carbon burning to determine the initial conditions of explosive carbon burning. Known reaction rates were then used with Eqs. (4-290) and (4-293) to calculate expected abundances using the numerical scheme given in Sect. 4.4.5. Abundance ratios which closely approximate those of the solar system were found for

$$^{20}\mathrm{Ne},\ ^{23}\mathrm{Na},\ ^{24}\mathrm{Mg},\ ^{25}\mathrm{Mg},\ ^{26}\mathrm{Mg},\ ^{27}\mathrm{Al},\ ^{29}\mathrm{Si},\quad \text{and}\quad ^{30}\mathrm{Si}\,, \tag{4-294}$$

when it was assumed that a previous epoch of helium burning produced equal amounts of C^{12} and O^{16}, and that

$$\begin{aligned} T_p &= 2\times 10^9\ ^\circ\mathrm{K} \\ \rho_p &= 1\times 10^5\ \mathrm{g\,cm^{-3}} \end{aligned} \tag{4-295}$$

and

$$\eta = 0.002\,.$$

Here T_p and ρ_p denote, respectively, the peak values of temperature and mass density in the shell under consideration, and the neutron excess, η, is given by

$$\eta = \frac{N_n - N_p}{N_n + N_p}\,, \tag{4-296}$$

where N_n and N_p denote, respectively, the number densities of free neutrons and protons. HANSEN (1971) used a full energy generation equation to obtain similar results for explosive carbon burning. ARNETT, TRURAN, and WOOSLEY (1971), have integrated the composition changes in a ^{12}C-detonation to show that most of the solar system iron group abundances can be explained by explosive burning.

TRURAN and ARNETT (1970) have considered explosive oxygen burning for which the mean lifetime, $\tau_{O^{16}}$, of a ^{16}O nucleus is given by

$$\tau_{O^{16}} = [X_{16}\,\rho\, N_A\langle\sigma v\rangle/16]^{-1}\,, \tag{4-297}$$

where X_{16} is the fractional abundance by mass of ^{16}O,

$$N_A\langle\sigma v\rangle \approx \exp(86.804-\xi)\,T_9^{-2/3}\,, \tag{4-298}$$

where $T_9 = T/10^9$, and

$$\xi = 135.79(1+0.053\,T_9)^{1/3}\,T_9^{-1/3}\,. \tag{4-299}$$

By equating $\tau_{O^{16}}$ to τ_{HD} for initial conditions, assuming an adiabatic expansion, and calculating the appropriate relative abundances, they have shown that nuclei of intermediate mass ($28 \lesssim A \lesssim 42$) are created by explosive oxygen burning with abundance ratios close to those of the solar system. The appropriate conditions are

$$\begin{aligned} 3.5\times 10^9 \lesssim\ &T \lesssim 3.7\times 10^9\ ^\circ\mathrm{K} \\ 10^5 \lesssim\ &\rho \lesssim 10^6\ \mathrm{g\,cm^{-3}} \end{aligned} \tag{4-300}$$

and

$$\eta \approx 0.002\,.$$

Similarly, ARNETT (1969) has used the $\tau_{Si^{28}}$ of BODANSKY, CLAYTON, and FOWLER (1968) to calculate explosive silicon burning, and to account for abundance ratios in the range $44 \leq A \leq 62$. HOWARD, ARNETT, and CLAYTON (1971) have also considered explosive burning in helium zones and indicate that the nuclei ^{15}N, ^{18}O, ^{19}F, and ^{21}Ne are created primarily as ^{15}O, ^{18}F, ^{19}Ne, and ^{21}Ne in the explosive nuclear processes in helium zones. The general success of explosive carbon, oxygen and silicon burning in accounting for solar system abundance ratios for $20 \leq A \leq 64$ has been summarized by ARNETT and CLAYTON (1970). Explosive nucleosynthesis has been incorporated in models of galactic evolution by ARNETT (1971) and TRURAN and CAMERON (1971). WOOSLEY, ARNETT, and CLAYTON (1973) give a thorough treatment of explosive oxygen and silicon burning together with a treatment of the equilibrium process at high and low freeze-out densities. A general review of explosive nucleosynthesis in stars is given by ARNETT (1973).

4.4.5. Nuclide Abundance Equations

The equation governing the change in the number density, $N(A,Z)$, of the nucleus, (A,Z), is of the form

$$\frac{d}{dt}(N_i) = -\sum_j N_i N_j \langle \sigma v \rangle_{ij} + \sum_{kl} N_k N_l \langle \sigma v \rangle_{kl}, \tag{4-301}$$

where N_i is the number density of the ith species, $\langle \sigma v \rangle_{ij}$ is the product of cross section and the relative velocity for an interaction involving species i and j, the $N_m N_n$ is replaced by $N_n^2/2$ for identical particles, and the summations are over all reactions which either create or destroy the species, i. For numerical work it is convenient to deal with the parameter

$$Y_i = Y(A, Z) = \frac{N(A, Z)}{\rho N_{\mathrm{A}}} = \frac{N_i}{\rho N_{\mathrm{A}}}, \tag{4-302}$$

where ρ is the mass density of the gas under consideration, and Avogadro's number, $N_{\mathrm{A}} = 6.022169 \times 10^{23}$ (g-mole)$^{-1}$. The differential equation linking all reactions which create or destroy the ith nucleus is then given by

$$\frac{d}{dt}(Y_i) = -\sum_j f_{ij} + \sum_{kl} f_{kl}, \tag{4-303}$$

where the vector flow, f_{ij}, which contains nuclei i and j in the entrance channel, is given by

$$f_{ij} = \frac{N_i N_j \langle \sigma v \rangle_{ij}}{\rho N_{\mathrm{A}}} = Y_i Y_j \rho N_{\mathrm{A}} \langle \sigma v \rangle_{ij}, \tag{4-304}$$

and the $N_{\mathrm{A}} \langle \sigma v \rangle_{ij}$ are given in the previous sections. When all of the different types of reactions are taken into account, Eq. (4-303) becomes

$$\begin{aligned}
\frac{dY(A,Z)}{dt} = & -[\lambda_{\beta^-}(A,Z) + \lambda_{\beta^+}(A,Z) + \lambda_K(A,Z) + \lambda_\alpha(A,Z) \\
& + \lambda_\gamma(A,Z) + 2.48 \times 10^8 \sigma_T N_{\mathrm{n}} + \sum_j Y(A_j Z_j) \rho N_{\mathrm{A}} \langle \sigma v \rangle_j] \, Y(A,Z) \\
& + \lambda_{\beta^-}(A, Z-1) \, Y(A, Z-1) + \lambda_{\beta^+}(A, Z+1) \, Y(A, Z+1) \\
& + \lambda_K(A, Z+1) \, Y(A, Z+1) + \lambda_\alpha(A+4, Z+2) \, Y(A+4, Z+2) \\
& + 2.48 \times 10^8 \sigma_T N_{\mathrm{n}} \, Y(A-1, Z) + \sum_{ik} Y(A_i, Z_i) \, Y(A_k, Z_k) \rho N_{\mathrm{A}} \langle \sigma v \rangle_{ik},
\end{aligned} \tag{4-305}$$

where the symbol λ denotes the decay rate or the inverse mean lifetime, the subscripts β^-, β^+, K, α and γ denote, respectively, negative beta decay, positive beta decay, electron capture, alpha decay, and photodisintegration, σ_T is the cross section for neutron capture in cm^2 (given in Table 38), N_n is the number density of neutrons, the summation $\sum_j$ denotes all reactions between the nucleus (A, Z) and any other nucleus, the summation $\sum_{ik}$ denotes all reactions between two nuclei which have (A, Z) as a product, ρ is the gas mass density, and the reaction rate $N_A\langle\sigma v\rangle$ is given by the superposition of all rates given in the previous sections. When a reaction occurs between n nuclei of the same type, the reaction rates given in Sect. 4.3. must be multiplied by $n!$ when used in Eq. (4-305). For example, the reaction rates for the $C^{12}+C^{12}$ and $O^{16}+O^{16}$ reactions would be multiplied by two, whereas the reaction rate for the triple α process would be multiplied by six.

The numerical solution to the complex set of nuclear reaction networks has been given by ARNETT and TRURAN (1969). Assuming that the change in composition over some time interval $\Delta t = t^{n+1} - t^n$ is sufficiently small, the vector flows may be linearized to give

$$f_{ij}^{n+1} = Y_i^{n+1} Y_j^{n+1} [ij]^{n+1} \tag{4-306}$$

$$\approx (Y_i^n Y_j^n + \Delta_i Y_j^n + \Delta_j Y_i^n)[ij]^{n+1}, \tag{4-307}$$

where the beginning and end of the time interval are denoted by superscripts n and $n+1$,

$$\Delta_i = Y_i^{n+1} - Y_i^n,$$

and

$$[ij] = \rho N_A \langle\sigma v\rangle_{ij}. \tag{4-308}$$

Similarly, the approximate expression for the vector flow which is symmetric in time is given by

$$f_{ij}^{n+1/2} \approx \left(Y_i^n Y_j^n + \frac{\Delta_i}{2} Y_j^n + \frac{\Delta_j}{2} Y_i^n \right) [ij]^{n+1/2}. \tag{4-309}$$

When the time derivative of Y_i is replaced by

$$\frac{d}{dt}(Y_i) = \frac{\Delta_i}{\Delta t}, \tag{4-310}$$

we have the coupled set of linear equations (ARNETT and TRURAN, 1969)

$$Y_i^{n+1} a_{ii} + Y_j^{n+1} a_{ij} + Y_k^{n+1} a_{ik} + Y_l^{n+1} a_{il} = - Y_k^n a_{ik} - Y_i^n Y_j^n [ij], \tag{4-311}$$

where

$$a_{ii} = \frac{1}{\Delta t} + Y_j^n [ij]$$

$$a_{ij} = Y_i^n [ij]$$

$$a_{ik} = - Y_l^n [kl]$$

and

$$a_{il} = - Y_k^n [kl].$$

The solution of Eq. (4-311) by iteration is possible when $|a_{ii}|>|a_{ij}|+|a_{ik}|+|a_{il}|$.

In the buildup of nuclei by the s and the r processes, the important reactions are the (n, γ), (γ, n) and beta decay processes with respective inverse lifetimes of

$$\lambda_{\mathrm{n}}=\sigma_{\mathrm{n}} v_{\mathrm{n}} N_{\mathrm{n}} \approx 2.48 \times 10^{8} \sigma_{T} N_{\mathrm{n}} \sec^{-1}, \tag{4-312}$$

$$\lambda_{\gamma}=\sigma_{\gamma} c N_{\gamma}, \tag{4-313}$$

and

$$\lambda_{\beta}=\frac{f \ln 2}{f t} \approx 10^{-5.5} f \sec^{-1}, \tag{4-314}$$

where σ_{n} and σ_{γ} are, respectively, the cross sections for the (n,γ) and (γ,n) reactions, v_{n} and N_{n} are the velocity and number density of the neutrons responsible for the (n,γ) process, σ_T is given in units of cm^2 and is assumed to be measured at 30 keV as are the values in Table 38, N_γ is the density of photons, γ, and ft and f values are tabulated in the references given in Sect. (4.3.2). The general equation for the s process is

$$\begin{aligned}\frac{dN(A,Z)}{dt} &= \lambda_{\mathrm{n}}(A-1,Z)N(A-1,Z)-\lambda_{\mathrm{n}}(A,Z)N(A,Z)\\ &\quad +\lambda_{\beta}(A,Z-1)N(A,Z-1)-\lambda_{\beta}(A,Z)N(A,Z)\\ &\quad +\text{termination terms due to alpha decay at } A>209\,.\end{aligned} \tag{4-315}$$

Remembering that $\lambda_{\mathrm{n}} \ll \lambda_{\beta}$, we have for the s process

$$\frac{dN(A,Z)}{dt}=-\lambda_{\mathrm{n}}(A,Z)N(A,Z)+\lambda_{\mathrm{n}}(A-1,Z)N(A-1,Z)\,. \tag{4-316}$$

The general equation for the r process is

$$\begin{aligned}\frac{dN(A,Z)}{dt} &= \lambda_{\mathrm{n}}(A-1,Z)N(A-1,Z)-\lambda_{\mathrm{n}}(A,Z)N(A,Z)\\ &\quad +\lambda_{\beta}(A,Z-1)N(A,Z-1)-\lambda_{\beta}(A,Z)N(A,Z)\\ &\quad +\lambda_{\gamma}(A+1,Z)N(A+1,Z)-\lambda_{\gamma}(A,Z)N(A,Z)\\ &\quad +\text{termination terms due to fission for } A\geq 260\,.\end{aligned} \tag{4-317}$$

Assuming that equilibrium is reached between the rapid (n, γ) and (γ,n) processes, and that $\lambda_{\mathrm{n}} \gg \lambda_{\beta}$, we have for the r process

$$\frac{dN(A,Z)}{dt}=-\lambda_{\beta}(A,Z)N(A,Z)+\lambda_{\beta}(A,Z-1)N(A,Z-1)\,. \tag{4-318}$$

The general equation for the statistical balance of the r process is

$$\log\frac{N(A+1,Z)}{N(A,Z)}=\log N_{\mathrm{n}}-34.07-\frac{3}{2}\log T_9+\frac{5.04}{T_9}Q_{\mathrm{n}}, \tag{4-319}$$

where the neutron binding energy, Q_{n}, is given by

$$Q_{\mathrm{n}}(A,Z)=c^2[M_{\mathrm{N}}+M(A,Z)-M(A+1,Z)], \tag{4-320}$$

and is expressed in MeV, the $T_9=T/10^9$, and N_{n} is the number density of neutrons.

At sufficiently high temperatures, $T \gtrsim 3 \times 10^9$ °K, the reactions are so profuse that nearly all nuclei, (A, Z), are converted into other nuclei, (A', Z'), even when Z and Z' are large. When the rates of all nuclear reactions (excepting beta decays) are exactly equal to the rates of the inverse reactions, the nuclear abundances may be determined from statistical considerations similar to those which led to the Saha ionization equation of Sect. 3.3.1.4. In the condition of statistical equilibrium between the nuclei and free protons and neutrons, detailed reaction rates become unnecessary and the number density, $N(A, Z)$, of the nuclide, (A, Z), is given by (HOYLE, 1946; BURBIDGE, BURBIDGE, FOWLER, and HOYLE, 1957)

$$N(A, Z) = \omega(A, Z) \left(\frac{2\pi \hbar^2}{M_\mu k T} \right)^{3(A-1)/2} A^{3/2} \frac{N_{\mathrm{p}}^Z N_{\mathrm{n}}^{A-Z}}{2^A} \exp\left[\frac{E_{\mathrm{B}}(A, Z)}{kT} \right], \tag{4-321}$$

where the partition function or statistical weight factor, $\omega(A, Z)$ is given by

$$\omega(A, Z) = \sum_r (2I_r + 1) \exp\left(-\frac{E_r}{kT} \right), \tag{4-322}$$

where E_r is the energy of the excited state measured above the ground level and I_r is the spin, M_μ is the atomic mass unit, the factor $[2\pi^2\hbar^2/M_\mu kT]^{3/2} = 1.6827 \times 10^{-34} T_9^{-3/2}$ cm^{-3}, the N_{p} and N_{n} denote, respectively, the densities of free protons and neutrons, and the binding energy, $E_{\mathrm{B}}(A, Z)$, of the ground level of the nucleus, (A, Z), is given by

$$E_{\mathrm{B}}(A, Z) = c^2 [(A - Z) M_{\mathrm{N}} + Z M_{\mathrm{p}} - M(A, Z)], \tag{4-323}$$

where M_{N}, M_{p} and $M(A, Z)$ are the masses of the free neutron, free proton, and the nucleus (A, Z), respectively. Eq. (4-321) can be rewritten in the form

$$\log N(A, Z) = \log \omega(A, Z) + 33.77 + \frac{3}{2} \log(A T_9) + \frac{5.04}{T} E_{\mathrm{B}}(A, Z) + A(\log N_{\mathrm{n}} - 34.07 - \frac{3}{2} \log T_9) + Z \log\left(\frac{N_{\mathrm{p}}}{N_{\mathrm{n}}} \right), \tag{4-324}$$

where $T_9 = T/10^9$. Provided that nuclear equilibrium is achieved faster than any of the relevant decay rates, the ratio

$$\frac{\langle Z \rangle}{\langle N \rangle} = \frac{\sum Z N(A, Z) + N_{\mathrm{p}}}{\sum (A - Z) N(A, Z) + N_{\mathrm{n}}} \tag{4-325}$$

must be preserved so that the equilibrium $N(A, Z)$ is determined by the density, ρ, temperature, T, and $\langle Z \rangle / \langle N \rangle$. The equilibrium process has been used to determine abundances of the iron group nuclei $(46 \leq A \leq 60)$ by FOWLER and HOYLE (1964) and CLIFFORD and TAYLER (1965).

More recently, BODANSKY, CLAYTON, and FOWLER (1968) have shown that if equilibrium is reached between ^{28}Si and the iron group, but not between ^{28}Si and the alpha particles, a quasi-equilibrium condition prevails in which nuclei heavier than ^{28}Si are in equilibrium with ^{28}Si under exchange of protons, neutrons and alpha particles. In this case, the equilibrium number density, $N(A, Z)$, relative to that of ^{28}Si is given by

$$N(A, Z) = C(A, Z) N(^{28}\mathrm{Si}) N_\alpha^{\delta_\alpha} N_{\mathrm{p}}^{\delta_{\mathrm{p}}} N_{\mathrm{n}}^{\delta_{\mathrm{n}}}, \tag{4-326}$$

where

$$C(A,Z)=\frac{\omega(A,Z)}{\omega(^{28}\mathrm{Si})}2^{-(\delta_\mathrm{p}+\delta_\mathrm{n})}[U(A,Z)]^{3/2}\left(\frac{2\pi\hbar^2}{M_\mu kT}\right)^{3(\delta_\alpha+\delta_\mathrm{p}+\delta_\mathrm{n})/2}\exp\left[\frac{Q(A,Z)}{kT}\right],$$

and

$$U(A,Z)=\frac{A(A,Z)}{A(^{28}\mathrm{Si})}A(^4\mathrm{He})^{-\delta_\alpha}A_\mathrm{p}^{-\delta_\mathrm{p}}A_\mathrm{n}^{-\delta_\mathrm{n}},$$

where $A(A,Z)$ is the atomic mass of the nuclide, (A,Z), in a.m.u., A_i is the atomic weight of element, i,

$$Q(A,Z)=E_\mathrm{B}(A,Z)-E_\mathrm{B}(^{28}\mathrm{Si})-\delta_\alpha E_\mathrm{B}(^4\mathrm{He}), \tag{4-327}$$

where $E_\mathrm{B}(A,Z)$ is the binding energy of nuclide, (A,Z), given by Eqs. (4-9) or (4-19), and δ_α, δ_p and δ_n specify, respectively, the number of free alpha-particles, protons, or neutrons in (A,Z) in excess of the number in ^{28}Si. If the largest alpha-particle nucleus contains N^1 neutrons and Z^1 protons, then

$$\delta_\alpha=\tfrac{1}{4}(N^1+Z^1-28),\quad \delta_\mathrm{p}=Z-Z^1,\quad \text{and } \delta_\mathrm{n}=N-N^1. \tag{4-328}$$

The partition function, $\omega(A,Z)$, for the alpha-particles is unity, whereas it is two for the neutron and proton. Bodansky, Clayton, and Fowler (1968) have tabulated δ_α, δ_p and δ_n, $Q(A,Z)$, $\omega(A,Z)$ and $C(A,Z)$ for $24\leq A\leq 62$ and for $T_9=3.8$, 4.4 and 5.0.

4.4.6. Formation of the Rare Light Elements—Spallation Reactions and Cosmological Nucleosynthesis

The rare light nuclei, D^2, He^3, Li^6, Li^7, Be^7, Be^9, Be^{10}, B^{10} and B^{11}, are not generated in the normal course of stellar nucleosynthesis, and are, in fact, destroyed in hot stellar interiors. As it was illustrated in Fig. 26, this condition is reflected in the comparatively low abundances of these nuclei. The fact that the light elements do exist, however, means that they must arise in nuclear reactions of a non-thermonuclear character in relatively cool and moderately dense regions, or in nuclear reactions during the early stages of an expanding universe. Fowler, Burbidge, and Burbidge (1955) first suggested that light elements might be formed on stellar surfaces by the collision of electromagnetically accelerated protons with the more abundant M nuclei: C^{12}, N^{14}, O^{16} and Ne^{20}. Fowler, Greenstein, and Hoyle (1962) and Bernas, Gradsztajn, Reeves, and Shatzman (1967) advanced models in which the light elements were formed by the spallation reactions of solar protons during the early stages of the Sun's contraction. Bradt and Peters (1950) suggested that cosmic rays might interact with the M elements to form the light elements, and more recently Reeves, Fowler, and Hoyle (1970), Mitler (1970), and Meneguzzi, Audouze, and Reeves (1971) have shown that the stellar and solar system abundances of Li^6, Be^9, B^{10} and B^{11} can be accounted for if they are produced by the spallation reactions of cosmic ray particles with M nuclei over the lifetime of the Galaxy. Neither stellar nucleosynthesis nor spallation reactions produce sufficient D,

He^3, He^4 or Li^7 to account for the observed abundances (cf. HOYLE and TAYLER, 1964). WAGONER, FOWLER, and HOYLE (1967) have shown that the current solar system abundances of these elements could be produced in the low density, early stages of an expanding Friedmann universe. The anomalously high He^3/He^4 ratio found in some stars is accounted for by spallation reactions on their surfaces (NOVIKOV and SYUNYAEV, 1967; VAUCLAIR and REEVES, 1972).

As it is apparent from the previous discussion, the formation mechanism adopted to explain the light elements must be consistent with the observed abundances given in Table 45. For the planets we have the abundance ratios (MENEGUZZI, AUDOUZE, and REEVES, 1971)

$$\begin{aligned}
\frac{\mathrm{Li}}{\mathrm{H}} &= 1.2 \times 10^{-9} \\
\frac{\mathrm{Be}}{\mathrm{H}} &= 2 \times 10^{-11} \\
\frac{\mathrm{B}}{\mathrm{H}} & \quad \text{from } 2 \times 10^{-10} \quad \text{to } 8 \times 10^{-9} \\
\frac{{}^7\mathrm{Li}}{{}^6\mathrm{Li}} &= 12.5 \pm 0.3 \\
\frac{{}^{11}\mathrm{B}}{{}^{10}\mathrm{B}} &= 4 \pm 0.4 \\
\frac{\mathrm{D}}{\mathrm{H}} &= 1.6 \times 10^{-4} .
\end{aligned} \tag{4-329}$$

For the Sun we have the abundance ratios (GREENSTEIN, 1951; KINMAN, 1956; GREVESSE, 1968)

$$\begin{aligned}
\frac{\mathrm{He}^3}{\mathrm{He}^4} &\leq 0.02 \\
\frac{\mathrm{D}}{\mathrm{H}} &< 4 \times 10^{-5} \\
\frac{\mathrm{Li}}{\mathrm{H}} &= 10^{-11} \\
\frac{\mathrm{Be}}{\mathrm{H}} &= 10^{-11} \\
\frac{\mathrm{B}}{\mathrm{H}} &< 3 \times 10^{-10} .
\end{aligned} \tag{4-330}$$

For the special case of 3 CentauriA, we have (SARGENT and JUGAKU, 1961)

$$\frac{\mathrm{He}^3}{\mathrm{He}^4} = 5.3 , \tag{4-331}$$

and for the interstellar medium we have (WEINREB, 1962)

$$\frac{D}{H} < 8 \times 10^{-5} \tag{4-332}$$

or perhaps (CESARSKY, MOFFET, and PASACHOFF, 1973)

$$3 \times 10^{-5} < \frac{D}{H} < 5 \times 10^{-4},$$

and (AUDOUZE, LEQUEUX, and REEVES, 1973)

$$\frac{B}{H} < 2 \times 10^{-9}. \tag{4-333}$$

New observations of the deuterium and hydrogen Lyman lines along the lines of sight to bright stars indicate that $D/H = 1.8 \pm 0.4 \times 10^{-5}$ for the interstellar medium (YORK and ROGERSON, 1976).

For the F and G type stars, we have the abundance ratios (WALLERSTEIN and CONTI, 1969; MENEGUZZI, AUDOUZE, and REEVES, 1971)

$$\begin{aligned} \frac{\text{Li}}{\text{H}} &= 2 \times 10^{-10} \\ \frac{\text{Be}}{\text{H}} &= 3 \times 10^{-11} \\ \frac{^7\text{Li}}{^6\text{Li}} &= 2.5. \end{aligned} \tag{4-334}$$

Very young stars have high Li/H ratios (8×10^{-10} to 4×10^{-9}), whereas later type stars show smaller lithium abundances decreasing with spectral class for a given cluster and with age when stars of the same class are considered. As first suggested by GREENSTEIN and RICHARDSON (1951), and discussed by WEYMANN and SEARS (1965) and BODENHEIMER (1965), lithium depletion can be accounted for by the presence of convection zones in the later stages of stellar evolution.

Table 45. Solar system abundances of the rare light elements

Nuclei	Mass fraction
H	0.745
D	2.3×10^{-4}
He^3	5.0×10^{-5}
He^4	0.24
Li^6	4.4×10^{-10}
Li^7	6.5×10^{-9}
Be–B	1.6×10^{-9}

Although the abundances of elements formed by spallation reactions are dependent upon both reaction cross sections and the energy spectrum of the incident high energy particles, abundance ratios are relatively insensitive to the energy spectrum. For example, if only $p + {}^{16}O$ reactions are considered, measured cross sections give (YIOU, SEIDE, and BERNAS, 1969)

$$\begin{aligned}\frac{{}^{7}Li}{{}^{6}Li} &= 0.98 \pm 0.13 \\ \frac{{}^{9}Be}{{}^{7}Be} &= 0.34 \pm 0.08 \\ \frac{{}^{10}Be}{{}^{9}Be} &= 0.30 \pm 0.08\end{aligned} \tag{4-335}$$

and

$$\frac{{}^{11}B + {}^{11}C}{{}^{10}B + {}^{10}C} = 2.1 \pm 0.6\,,$$

for proton energies larger than 135 MeV. Furthermore, a proton flux capable of generating $Li/H = 10^{-9}$ will generate $D/H = 3 \times 10^{-7}$ and ${}^{3}He/{}^{4}He = 3 \times 10^{-6}$. Comparison with the observed abundances shows that spallation reactions might account for the Li^6, Be and B, but cannot account for the higher solar system abundances of Li^7 and D. For detailed calculations, the formation rate of an isotope, L, from the spallation of the M elements by a flux of protons with energy spectrum, $\varphi(E)$, is given by

$$\frac{dN_L}{dt} = \sum_M N_M \int \sigma(M, L, E)\, \varphi(E)\, dE\,, \tag{4-336}$$

where M is any of the M elements, C^{12}, N^{14}, O^{16}, and Ne^{20}, the number densities of the M and L elements are, respectively, N_M and N_L, the time variable is t, the proton energy is E, and the spallation reaction cross section is $\sigma(M, L, E)$. Observations of solar flares give a proton energy spectrum

$$\varphi(E) \propto E^{-\gamma}\,, \tag{4-337}$$

where γ ranges from 2 to 5, or more exactly,

$$\varphi(E) \propto \exp(-R/R_0)\,, \tag{4-338}$$

where the rigidity $R = pc/(Ze)$ is the ratio of momentum to charge, and R_0 is an experimentally determined parameter ranging from 40 to 400 MeV. If the proton rest mass energy is Mc^2, and its kinetic energy is E, then

$$R = \frac{A}{Ze}\left[2Mc^2 E + E^2\right]^{1/2}\,, \tag{4-339}$$

where A is the atomic number and Ze is the charge of the nucleus.

For cosmic ray protons, MENEGUZZI, AUDOUZE, and REEVES (1971) adopt

$$\varphi(E) = 12.5(E_0 + E)^{-2.6}\ \mathrm{cm^{-2}\ sec^{-1}\ GeV^{-1}}\,, \tag{4-340}$$

where $E_0 = 0.931$ GeV, and E is the proton kinetic energy in GeV. Production rates involve the integration of the spallation cross sections over the incident spectrum, and these rates are given in Table 46 for the cosmic ray spectrum.

Table 46. Production rates of rare light elements by the spallation reactions of abundant nuclei and cosmic ray protons (H) or alpha particles (He)[1]

Nucleus	Production rate in $g^{-1}\ sec^{-1}$ C^{12}	N^{14}	O^{16}	Ne^{20}
Li^6 (H)	21.4×10^{-6}	9.9×10^{-6}	39.3×10^{-6}	1.26×10^{-6}
Li^6 (He)	5.39	1.59	9.53	1.23
Li^7 (H)	54.3	15.4	57.0	1.80
Li^7 (He)	8.63	2.43	14.54	1.87
Be^9 (H)	11.3	1.58	7.36	0.24
Be^9 (He)	2.39	0.96	5.75	0.74
B^{10} (H)	47.6	2.90	43.8	1.40
B^{10} (He)	2.70	1.14	6.85	0.88
B^{11} (H)	170.7	23.9	85.0	2.70
B^{11} (He)	3.65	1.13	6.78	0.87

[1] After MITLER (1970). All values are $\times 10^{-6}$ as indicated for the first member of each column.

Detailed formulae for the energetics of cosmic ray spallation reactions are given by MENEGUZZI, AUDOUZE, and REEVES (1971), and SILBERBERG and TSAO (1973), whereas those for spallation reactions on stellar surfaces are given by RYTER, REEVES, GRADSZTAJN, and AUDOUZE (1970). Approximate relations may be derived using the formula

$$\frac{N_L}{N_M} = \langle \varphi \rangle \langle \sigma \rangle T, \tag{4-341}$$

where $\langle \varphi \rangle$ and $\langle \sigma \rangle$ denote, respectively, average fluxes and cross sections, and T denotes the duration of the spallation reaction. For energies greater than 30 MeV, for example, the galactic cosmic ray fluxes are (FRIER and WADDINGTON, 1968; COMSTOCK, FAN, and SIMPSON, 1969)

$$\begin{aligned} \varphi_H &= 3.6\ \text{cm}^{-2}\ \text{sec}^{-1} \\ \varphi_{He} &= 0.4\ \text{cm}^{-2}\ \text{sec}^{-1} \\ \varphi_{C,N,O} &= 25 \times 10^{-3}\ \text{cm}^{-2}\ \text{sec}^{-1} \\ \varphi_N &= 4 \times 10^{-3}\ \text{cm}^{-2}\ \text{sec}^{-1}, \end{aligned} \tag{4-342}$$

and the appropriate $T = 10^{10}$ years, the age of the Galaxy. MENEGUZZI, AUDOUZE, and REEVES (1971) have assumed a constant cosmic ray energy spectrum ($E^{-2.6}$) over 10^{10} years and obtain the abundance ratios

$$\frac{^6\text{Li}}{\text{H}} = 8 \times 10^{-11} \qquad \frac{^{10}\text{B}}{\text{H}} = 8.7 \times 10^{-11}$$

$$\frac{^7\text{Li}}{\text{H}} = 1.2 \times 10^{-10} \qquad \frac{^{11}\text{B}}{\text{H}} = 2 \times 10^{-10}$$

$$\frac{^7\text{Li}}{^6\text{Li}} = 1.5 \qquad \frac{^{11}\text{B}}{^{10}\text{B}} = 2.4$$

$$\frac{\mathrm{Li}}{\mathrm{H}} = 2 \times 10^{-10} \qquad \frac{\mathrm{B}}{\mathrm{H}} = 3.0 \times 10^{-10}$$
$$\frac{{}^{9}\mathrm{Be}}{\mathrm{H}} = 2 \times 10^{-11} \qquad \frac{\mathrm{B}}{\mathrm{Be}} = 15 \tag{4-343}$$
$$\frac{\mathrm{Li}}{\mathrm{Be}} = 10 \qquad \frac{\mathrm{B}}{\mathrm{Li}} = 1.4\,.$$

For detailed calculations, the spallation cross sections for protons on the M elements are given by AUDOUZE, EPHERRE, and REEVES (1967), GRADSZTAJN (1967), BERNAS, GRADSZTAJN, REEVES, and SCHATZMAN (1967), and REEVES (1971). The

Table 47. Proton spallation reaction Q values and cross sections in millibarns for the production of Li^6, Be, and B [1]

Reaction	Effective Q (MeV)	Reaction cross sections: Average, σ (mb)	Reaction cross sections: σ for $E>2$ GeV (mb)
$^{12}C\,(p,\ pn)\,^{11}C$	18.7	60 }	(50)
$^{12}C\,(p,\ 2p)\,^{11}B$	16.0	16 }	
$^{12}C\,(p,\ 2pn)\,^{10}B$	27.4	10	(15)
$^{12}C\,(p,\ 3pn)\,^{9}Be$	34	2	(6)
$^{12}C\,(p,\ p\alpha n)\,^{7}Be$	26.3	12	(9)
$^{12}C\,(p,\ 2p\alpha)\,^{7}Li$	24.6	7	(6)
$^{12}C\,(p,\ 2p\alpha n)\,^{6}Li$	31.9	10	(7.5)
$^{14}N\,(p,\ \alpha)\,^{11}C$	2.9	30 }	(45)
$^{14}N\,(p,\ 3pn)\,^{11}B$	28.4	10 }	
$^{14}N\,(p,\ p\alpha)\,^{10}B$	11.6	10	(12)
$^{14}N\,(p,\ 2p\alpha)\,^{9}Be$	18.2	2	(6)
$^{14}N\,(p,\ 2\alpha)\,^{7}Be$	10.5	10	(12)
$^{14}N\,(p,\ 3p\alpha n)\,^{7}Li$	37.1	10	(10)
$^{14}N\,(p,\ p2\alpha)\,^{6}Li$	16.1	10	(7)
$^{16}O\,(p,\ p\alpha n)\,^{11}C$	25.9	15 }	(25)
$^{16}O\,(p,\ \alpha 2p)\,^{11}B$	23.1	10 }	
$^{16}O\,(p,\ 2p\alpha n)\,^{10}B$	34.6	10	(15)
$^{16}O\,(p,\ 3p\alpha n)\,^{9}Be$	41.7	2	(3.5)
$^{16}O\,(p,\ p2\alpha n)\,^{7}Be$	33.4	4	(10)
$^{16}O\,(p,\ 2p\alpha)\,^{7}Li$	31.8	12	(14)
$^{16}O\,(p, 2p2\alpha n)\,^{6}Li$	39.0	10	(14)
$^{20}Ne\,(p,\ p2\alpha n)\,^{11}C$	30.6	12 }	(18)
$^{20}Ne\,(p,\ 2\alpha 2p)\,^{11}B$	27.8	12 }	
$^{20}Ne\,(p, 2p2\alpha n)\,^{10}B$	39.3	12	(14)
$^{20}Ne\,(p, 3p2\alpha n)\,^{9}Be$	45.9	2	(4)
$^{20}Ne\,(p,\ p3\alpha n)\,^{7}Be$	38.2	8	(10)
$^{20}Ne\,(p,\ 2p3\alpha)\,^{7}Li$	36.5	8	(10)
$^{20}Ne\,(p, 2p3\alpha n)\,^{6}Li$	43.8	8	(10)

[1] Effective Q values and average cross sections are from BERNAS, GRADSZTAJN, REEVES, and SHATZMAN (1967), whereas cross sections for $E>2$ GeV are from MENEGUZZI, AUDOUZE, and REEVES (1971). For conversion 1 MeV $=1.6022\times10^{-6}$ erg and 1 mb $=10^{-27}$ cm^2.

proton induced reaction cross sections for He^4, He^3, and D, are summarized by VAUCLAIR and REEVES (1972). Alpha induced spallation cross sections are given by CRANDALL, MILLBURN, PYLE, and BIRNBAUM (1956) and RADIN (1970). At high energies, the alpha spallation cross sections are two or three times larger than the equivalent proton induced cross sections. Because the helium to hydrogen ratio is about 0.1 for cosmic rays, the alpha induced contribution is about 20% of the proton induced cross section at high energies. At energies lower than about 20 MeV, the alpha induced reactions dominate. Partial cross sections for thermonuclear reactions for targets with $Z \leq 28$ are given by SILBER-

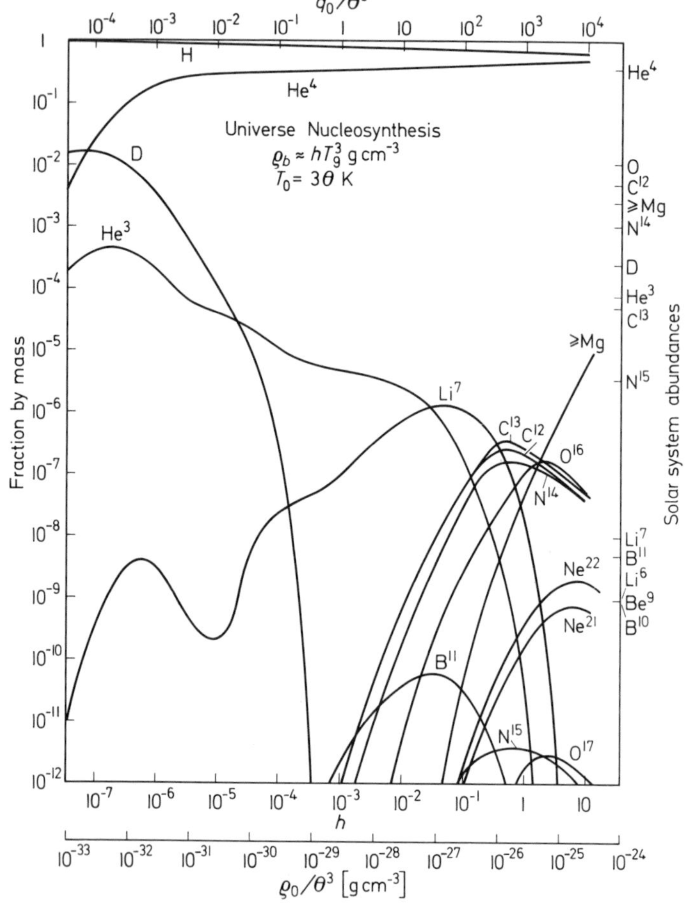

Fig. 27. Synthesis of the light elements in a universal expansion. The mass fractions of various light nuclei are shown as a function of the parameter h in the relation $\rho_B = h T_9^3$, where ρ_B is the baryon density and $T_9 = T/10^9$ where T is the universal black body temperature. The parameter h also determines q_0, the present value of the deceleration parameter of the expansion. Constraints imposed by the present gas density and temperature rule out the production of significant amounts of the elements heavier than lithium in such a universal process. The present photon temperature is T_0, the present value of the baryon density is ρ_0, and the symbol θ represents $T_0/3\,^\circ$K. Solar system abundances are given on the right-hand ordinate. From $A > 4$, Population II abundances are of order 10^{-2} of solar system values (after WAGONER, FOWLER, and HOYLE, 1967, by permission of the American Astronomical Society and the University of Chicago Press)

BERG and TSAO (1973). The proton spallation reactions which are mainly responsible for the production of Li^6, Be, and B are given in Table 47.

Here the "effective" energy threshold, Q, is the energy above which the cross section rises steeply and corresponds roughly to the threshold for the outgoing products having the maximum number of alpha particles. The general pattern of the cross sections is one in which the cross section reaches a maximum at some 10 to 20 MeV above the effective threshold, and then decreases to a constant or slowly decreasing function of energy for proton energies >150 MeV. The ratio of the maximum to the high energy cross section is usually between one and two. The average cross section, $\langle\sigma\rangle$, given above is the mean value for energies above the effective Q. Values of Q and $\langle\sigma\rangle$ are from BERNAS, GRADSZTAJN, REEVES, and SHATZMAN (1967). The proton induced reaction cross sections for proton energies greater than 2 GeV are given in parenthesis and are after MENEGUZZI, AUDOUZE, and REEVES (1971).

As discussed previously, appreciable amounts of D and He^3 are not formed in the cosmic ray spallations on the nuclei of the interstellar medium. These elements are not formed in appreciable amounts in stellar nucleosynthesis, and are destroyed by thermonuclear reactions at temperatures above 10^6 and 10^7 °K for D and He^3, respectively. GAMOW (1946), ALPHER, BETHE, and GAMOW (1948), and ALPHER and HERMAN (1950) first suggested that the observed abundances of the elements might be explained if they originated in the primeval fireball stage of the universe. WAGONER, FOWLER, and HOYLE (1967) and WAGONER (1973) have made detailed calculations of light element production in the early stages of a homogeneous and isotropic, expanding universe. Their results are illustrated in Fig. 27 which gives light element abundances as a function of the present value of the background radiation temperature, T_0, and the deceleration parameter, q_0. Measurements of the interstellar abundance of deuterium by YORK and ROGERSON (1976) give $D/H = 1.8 \times 10^{-5}$, which using Wagoner's work gives a present baryon density of 5×10^{-31} g cm^{-3}. Because this density is a factor of ten less than that required to close the expanding universe, an open, ever-expanding universe is inferred. Reasonable agreement with solar system abundances are obtained for D, He^3, He^4, and Li^7 if $T_0 = 3$ °K and $q_0 = 5 \times 10^{-3}$. This process cannot explain, however, the observed abundances of Li^6 and Be^9. Furthermore, the cosmological abundance of deuterium can be depleted in the convective envelopes of stars (BODENHEIMER, 1966), and convective helium burning shell flashes can enhance

Table 48. Abundances of deuterium, He^3, He^4, and Li^7 produced in the "big bang" origin of the universe as a function of the mass density of the universe[1]

Identification	Density ($T_0 = 2.7$ °K) in gm cm^{-3}						
	3×10^{-33}	10^{-32}	3×10^{-32}	10^{-31}	3×10^{-31}	10^{-30}	3×10^{-30}
H^1	0.95	0.89	0.81	0.76	0.75	0.74	0.73
He^4	0.032	0.098	0.19	0.24	0.25	0.26	0.27
$(D^2/H^1)/(D^2/H^1)_\odot$	53	34	13	2.7	0.40	0.054	1.2×10^{-3}
$(He^3/He^4)/(He^3/He^4)_\odot$	53	16	4.2	1.2	0.67	0.36	0.16
$(Li^7/H^1)/(Li^7/H^1)_\odot$	0.063	0.44	0.50	0.080	0.033	0.70	3.4

[1] After TRURAN and CAMERON (1971). Here the subscript $\odot$ denotes the solar system value.

the Li^7 abundance (SCHWARZSCHILD and HÄRM, 1967). The abundances of hydrogen, deuterium, He^3, He^4, and Li^7 produced in the WAGONER *et al.* (1967) model are given in Table 48.

4.4.7. Rapid Thermonuclear Reactions in Supernovae Explosions

BURBIDGE, BURBIDGE, FOWLER, and HOYLE (1957) argued that the gravitational collapse of a highly evolved star might lead to an imploding core whose heat might be sufficient to ignite the potentially explosive light nuclei such as C^{12} at temperatures of a few times 10^9 °K. The subsequent explosion of the stellar envelope was thought to coincide with nucleosynthesis by the equilibrium, *r*, and *p* processes. HOYLE and FOWLER (1960) then suggested two origins for supernovae explosions. The Type I explosions were thought to originate from the ignition of degenerate material in the core of stars of intermediate mass. More massive stars with nondegenerate cores are the site of Type II explosions which result from the implosion-explosion process. HAYAKAWA, HAYASHI, and NISHIDA (1960), TSUJI (1963), and TSUDA and TSUJI (1963) postulated that rapid thermonuclear reactions such as the rapid CNO process, *p* capture, *n* processes, and α capture could synthesize many of the elements in the range $20 \leq A \leq 60$ during supernovae explosions. FOWLER and HOYLE (1964) presented a detailed analysis of nucleosynthesis during the Type II supernova process. Detailed hydrodynamic models of the implosion-explosion process were then given by COLGATE and WHITE (1966) and ARNETT (1966, 1967). WAGONER (1969) showed that ^7Li, ^9Be, ^{10}B and ^{11}B could be produced in solar system abundances when a non-degenerate star explodes from very high temperatures ($T \approx 10^{10}$ °K). ARNETT (1969) has presented a supernova model in which stars of intermediate mass, $4 M_\odot \leq M \leq 9 M_\odot$, explosively ignite the carbon burning reaction in their degenerate interiors; and ARNETT, TRURAN, and WOOSLEY (1971) have shown that this model may successfully account for the solar system abundances of the iron group elements. The synthesis of elements by the *r* and *p* processes during supernovae explosions has been discussed by TRURAN, ARNETT, TSURUTA, and CAMERON (1968) and by TRURAN and CAMERON (1972), whereas a summary of the papers dealing with explosive burning processes was given in Sect. 4.4.4.

Before a supernova explosion, the mass density, ρ, and temperature, T, of a star are related by the equations of state which were discussed in detail in Chap. 3. For a completely degenerate relativistic gas we have a pressure, P, given by (HOYLE and FOWLER, 1960)

$$P \approx 1.243 \times 10^{15} \left(\frac{\rho}{\mu_e}\right)^{4/3} \left[1 + \frac{1}{x^2}\left(\frac{2\pi^2 k^2 T^2}{m^2 c^4} - 1\right)\right] \text{ dynes cm}^2, \quad (4\text{-}344)$$

where

$$\rho = 9.74 \times 10^5 \mu_e x^3 \text{ g cm}^{-3}. \quad (4\text{-}345)$$

Here the constant $1.243 \times 10^{15} = (3\pi^2)^{1/3} c\hbar/(4 M_\mu^{4/3})$, where M_μ is the atomic mass unit, the constant $9.74 \times 10^5 = M_\mu m^3 c^3/(3\pi^2 \hbar^3)$, and $x = \hbar(3\pi^2 N_e)^{1/3}/(mc)$ where N_e is the electron density. The pressure of a degenerate gas is relatively insensitive to temperature, and the temperature release by expansion or neutrino processes is insufficient to prohibit a temperature rise to explosive values during gravitational contraction. For example, the temperature and density of a degenerate gas are related by (HOYLE and FOWLER, 1960)

$$\left(\frac{T_9}{1.33}\right)^2 - 1 \approx \left(\frac{\rho}{9.74 \times 10^5 \mu_e}\right)^{2/3} \left[\left(\frac{M}{M_{cr}}\right)^{2/3} - 1\right], \tag{4-346}$$

where $T_9 = T/10^9$, and the Chandrasekhar critical mass is given by

$$M_{cr} = \frac{5.80}{\mu_e^2} M_\odot . \tag{4-347}$$

For a degenerate relativistic gas, $\rho > 7.3 \times 10^6 \mu_e \text{ g cm}^{-3}$, explosive temperatures for carbon burning, $T_9 \approx 1$, are realizable for M on the order of M_{cr}. ARNETT (1969) has followed the evolution of a carbon-oxygen stellar core for stars of intermediate mass, $4M_\odot \leq M \leq 9M_\odot$, and shows that explosive ignition of carbon burning does indeed occur for degenerate core densities on the order of 10^9 g cm^{-3}. The explosion generates a strong shock wave, called a detonation wave, which progresses outward through the unburned stellar envelope momentarily increasing its temperature as well. Prior to ignition of this fuel, the pressure and specific volume before (P_1, V_1) and after (P_2, V_2) the shock front are related by

$$\frac{V_2}{V_1} = \left(\frac{2}{M_1^2} + \gamma - 1\right) / (\gamma + 1) , \tag{4-348}$$

and

$$\frac{P_2}{P_1} = (2\gamma M_1^2 - \gamma + 1)/(\gamma + 1) , \tag{4-349}$$

where

$$M_1^2 = \left(\frac{V_1}{S_1}\right)^2 = 1 + [1 + (1 + 2\beta)^{1/2}]/\beta ,$$

$$\beta = \gamma P_1 V_1 / [Q(\gamma^2 - 1)] ,$$

Q is the energy release per gram for the reaction, $M_1 = (V_1/S_1)$ is the ratio of the shock front speed to the velocity of sound before the front, and γ is the adiabatic index. Typical velocities are $V \approx 20{,}000 \text{ km sec}^{-1}$, and the raise in temperature, T_2, can be calculated from P_2, V_2, and the equation of state. Explosive temperatures are found for T_2, and even higher temperatures are found to result from the ignition of stellar fuel after the passage of the detonation wave. These temperatures are sufficiently high to eventually set up complete nuclear statistical equilibrium following the passage of the wave. Initial nuclear abundances are then determined from the equilibrium equations (Sect. 4.4.5). During the subsequent expansion, the temperature and density are related by the adiabatic condition, and abundances depend only on the neutron-proton ratio and the rate of expansion. The same conditions also follow for the nondegenerate objects exploding from temperatures sufficiently high to initially establish nuclear equilibrium (cf. WAGONER, 1969).

The initial condition of a nondegenerate, nonrelativistic stellar core is, of course, different from the degenerate case, with a pressure, P, given by (FOWLER and HOYLE, 1964)

$$P = \left[\frac{3R^4(1-\beta)}{a\mu^4\beta^4}\right]^{1/3} \rho^{4/3}, \tag{4-350}$$

and a mass density, ρ, given by

$$\rho = \frac{a\mu\beta}{3R(1-\beta)} T^3. \tag{4-351}$$

Here $R = 8.314 \times 10^{16}$ erg mole^{-1} $(10^9\ ^\circ\mathrm{K})^{-1}$ is the gas constant, the radiation constant $a = 7.565 \times 10^{21}$ erg cm^{-3} $(10^9\ ^\circ\mathrm{K})^{-4}$, the mean molecular weight is μ, and β, the ratio of gas pressure to total pressure, satisfies the relation

$$1-\beta = 0.0030\left(\frac{2M}{3M_\odot}\right)^2 \mu^4\beta^4. \tag{4-352}$$

For a massive star, $M \approx 30\,M_\odot$, we have $\beta \approx 0.40$, and with $\mu \approx 2.1$ and $\beta \approx 0.4$ we have

$$\rho \approx 4.3 \times 10^4\, T_9^3 \text{ g cm}^{-3} \tag{4-353}$$

and in general

$$\rho \approx 1.016 \times 10^7 \left(\frac{M_\odot}{M_c}\right)^2 \left(\frac{T}{\mu\beta}\right)^3 \text{g cm}^{-3} \tag{4-354}$$

for a core of mass M_c. The implosion-explosion phenomenon which follows from a contracting core is described by FOWLER and HOYLE (1964); whereas the details of the hydrodynamic processes following implosion are given by COLGATE and WHITE (1966), ARNETT (1966, 1967, 1968), and ARNETT and CAMERON (1967). As pointed out by FOWLER and HOYLE (1964), the raise in temperature following the outward moving shock wave is sufficient to cause explosive oxygen and hydrogen burning, and other rapid processes on the explosion time scale of ≈ 100 sec.

4.5. High Energy Particles

4.5.1. Creation of High Energy Particles

4.5.1.1. Creation of Electron-Positron Pairs by Gamma Ray Absorption in the Presence of a Nucleus

A photon of energy, $h\nu$, greater than the threshold energy $2mc^2 = 1.022$ MeV, may form an electron-positron pair when passing through the Coulomb field of a nucleus of charge eZ. When all energies under consideration are large compared with mc^2, the cross section, $\sigma(E_0)$, for the pair creation of a positron and electron of respective energies, E_0 and E_1, is given by (BETHE and HEITLER, 1934)

$$\sigma(E_0)\,dE_0 = 4\alpha Z^2 r_0^2 \frac{[E_0^2 + E_1^2 + \frac{2}{3}E_0E_1]}{(h\nu)^3}\left[\ln\left(\frac{2E_0E_1}{h\nu mc^2}\right) - \frac{1}{2}\right] dE_0$$

for relativistic energies, and no screening

$$\frac{2E_0 E_1}{h\nu} \ll \frac{mc^2}{\alpha Z^{1/3}},$$

$$\sigma(E_0)\,dE_0 = \frac{4\alpha Z^2 r_0^2}{(h\nu)^3}\left[\left(E_0^2+E_1^2+\frac{2}{3}E_0E_1\right)\ln\left(\frac{191}{Z^{1/3}}\right)-\frac{E_1E_0}{9}\right]dE_0 \qquad (4\text{-}355)$$

for relativistic energies and complete screening

$$\frac{2E_0 E_1}{h\nu} \gg \frac{mc^2}{\alpha Z^{1/3}},$$

where the fine structure constant $\alpha \approx 1/137.037$, and the classical electron radius $r_0 = e^2/(mc^2) \approx 2.818\times 10^{-13}$ cm. The total cross section, $\sigma(h\nu)$, for the creation of electron-positron pairs is obtained by integrating Eq. (4-355) from $E_0 = mc^2$ to $E_0 = h\nu - mc^2$.

$$\sigma(h\nu) = 4\alpha Z^2 r_0^2\left[\frac{7}{9}\ln\left(\frac{2h\nu}{mc^2}\right)-\frac{109}{54}\right] \quad \text{for } \frac{2E_0E_1}{h\nu} \ll \frac{mc^2}{\alpha Z^{1/3}}$$

$$\sigma(h\nu) = 4\alpha Z^2 r_0^2\left[\frac{7}{9}\ln\left(\frac{191}{Z^{1/3}}\right)-\frac{1}{54}\right] \quad \text{for } \frac{2E_0E_1}{h\nu} \gg \frac{mc^2}{\alpha Z^{1/3}}, \qquad (4\text{-}356)$$

where $4\alpha r_0^2 \approx 2.318\times 10^{-27}$ cm². When electron velocities, v, are so small that $Ze^2/(hv) \gg 1$, a Coulomb correction factor, $C(Z) \approx (\alpha Z)^2$, must be subtracted from the terms in the square brackets in Eq. (4-356). The cross sections are also suppressed at very high energies in a very dense medium (Landau and Pomeranchuk, 1953; Migdal, 1956), and in a crystalline medium (Überall, 1956, 1957). For a completely ionized gas, Eq. (4-355) may be used.

4.5.1.2. Creation of Electron-Positron Pairs by Charged Particles

The electric field of a fast charged particle has an associated "virtual" photon which in turn may create an electron-positron pair. When a heavy charged particle of mass, M_0, charge, eZ_0, and kinetic energy, E_0, collides with a heavy particle of mass, M, and charge, eZ, the total cross section, σ, for the creation of electron-positron pairs of any energy is given by (Heitler and Nordheim, 1934)

$$\sigma \approx \frac{(\alpha Z Z_0 r_0 mc^2)^2}{M_0 c^2 E_0}\left[\frac{ZM_0 - Z_0 M}{M}\right]^2 \quad \text{for } 2mc^2 < E_0 \ll M_0c^2, \qquad (4\text{-}357)$$

where the fine structure constant $\alpha = 1/137.037$, the classical electron radius $r_0 = e^2/(mc^2) \approx 2.818\times 10^{-13}$ cm, and mc^2 is the rest mass energy of the electron.

When the energy, E_0, of the incident particle is greater than M_0c^2, and the other particle is at rest, the total cross section is given by (Bhabha, 1935)

$$\sigma \approx \frac{28}{27\pi}(\alpha Z Z_0 r_0)^2 \ln^3\left(\frac{E_0}{M_0c^2}\right) \quad \text{for } E_0 \gg M_0c^2. \qquad (4\text{-}358)$$

If the particle at rest is an atom, and $E_0 \gg M_0 c^2/(\alpha Z^{1/3})$ for complete screening, the total cross section is (NISHINA, TOMONAGA, and KOBAYASI, 1935; HEITLER, 1954)

$$\sigma = \frac{28}{27\pi}(\alpha Z Z_0 r_0)^2 \ln\left(\frac{1}{\alpha Z^{1/3}}\right)\left[3\ln\left(\frac{E_0}{M_0 c^2}\right)\ln\left(\frac{E_0 Z^{1/3}}{191 M_0 c^2}\right) + \ln^2\left(\frac{191}{Z^{1/3}}\right)\right] \tag{4-359}$$

for $E_0 > M_0 c^2/\alpha Z^{1/3}$.

Eqs. (4-358) and (4-359) are thought to be accurate to within a factor of two. More complicated expressions accurate to twenty percent are given by MUROTA, UEDA, and TANAKA (1956) and HAYAKAWA (1969). When the incident particle is an electron, Eqs. (4-358) and (4-359) are valid with $Z_0 = 1$ and $M_0 = m$, the electron mass.

4.5.1.3. Creation of Electron-Positron Pairs by Two Photon Collision

An electron-positron pair may be produced in the collision of a photon of energy, E_1, with a photon of energy, E_2, provided that $E_1 E_2 > (mc^2)^2$, where $mc^2 \approx 0.511$ MeV is the rest mass energy of the electron. The pair creation cross section, $\sigma(E_1, E_2)$, is given by (DIRAC, 1930; HEITLER, 1954)

$$\sigma(E_1, E_2) = \frac{\pi r_0^2}{2}(1-\beta^2)\left[2\beta(\beta^2-2)+(3-\beta^4)\ln\left(\frac{1+\beta}{1-\beta}\right)\right], \tag{4-360}$$

where

$$\beta = \left[1 - \frac{(mc^2)^2}{E_1 E_2}\right]^{1/2},$$

the velocity of the outgoing electron in the center-of-mass system is βc, and the classical electron radius $r_0 = e^2/(mc^2) \approx 2.818 \times 10^{-13}$ cm. Applications of Eq. (4-360) to gamma rays are given by NIKISHOV (1962), GOLDREICH and MORRISON (1964), GOULD and SCHRÉDER (1966), and JELLEY (1966).

4.5.1.4. Creation of μ-Meson Pairs by Gamma Rays in the Presence of a Nucleus

A photon, γ, of energy, $h\nu$, greater than the threshold energy $2m_\mu c^2 \approx 211$ MeV may form a μ^- meson pair, μ^+ and μ^-, when passing through the Coulomb field of a nucleus. The cross section for pair creation will be given by Eq. (4-356) for the electron-positron pair creation, with m_e replaced by m_μ, and a slight modification due to the large momentum transfer to the nucleus during meson pair production. The cross section is reduced below that given by Eq. (4-356) by the ratio $(m/m_\mu)^2 \approx (1/207)^2$. Detailed calculations of meson pair production cross sections are given by RAWITSCHER (1956). Electron-positron pairs will be formed by the decay of the muon pairs, whose decay modes and lifetimes were given in Table 39.

4.5.1.5. Creation of Recoil (Knock-on) Electrons by Charged Particle Collision

The cross section, $\sigma(E_0, W_r)$, for the production of a recoil electron of kinetic energy, W_r, by the collision of a charged particle of total energy, E_0, with another charged particle of charge, eZ, is given by (BHABHA, 1936; HAYAKAWA, 1969)

$$\sigma(E_0, W_r)\,dW_r = 2\pi Z^2 r_0^2 \frac{mc^2}{\beta_0^2}\frac{dW_r}{W_r^2} \times \begin{cases} \left(1-\beta_0^2\dfrac{W_r}{W_m}\right) & \text{for spin } 0\,, \\ \left[1-\beta_0^2\dfrac{W_r}{W_m}+\dfrac{1}{2}\left(\dfrac{W_r}{E_0}\right)^2\right] & \text{for spin } \tfrac{1}{2}\,, \\ \left[\left(1-\beta_0^2\dfrac{W_r}{W_m}\right)\left(1+\dfrac{1}{3}\dfrac{mW_r}{M^2}\right)+\dfrac{1}{3}\left(\dfrac{W_r}{E_0}\right)^2 \right. & \\ \qquad \left. \times\left(1+\dfrac{1}{2}\dfrac{mW_r}{M^2}\right)\right] & \text{for spin } 1\,, \end{cases}$$

$$\approx 2\pi r_0^2 \frac{mc^2 Z^2}{\beta_0^2 W_r^2}\,dW_r\,, \tag{4-361}$$

where the spin is that of the incident particle, the classical electron radius $r_0 = e^2/(mc^2) \approx 2.82\times 10^{-13}$ cm, the velocity of the incident particle is $c\beta_0$, the electron mass is m, the mass of the incident particle is M, and the maximum energy that can be transferred in a direct collision to a free electron, W_m, is given by

$$W_m = \frac{2mp^2}{m^2+M^2+2mE_0} \approx p \quad \text{for } m \ll M \quad \text{and } E_0 \gg \frac{M^2}{m}$$
$$\approx 2m\beta_0^2\gamma^2 = 2m\left(\frac{p}{M}\right)^2 \quad \text{for } m \ll M \quad \text{and } p \ll \frac{M^2}{m}\,, \tag{4-362}$$

where the momentum and total energy of the incident particle are, respectively, p and E_0.

For the special case of relativistic protons of total energy, E_p, and velocity, $c\beta_p$, Eq. (4-361) becomes

$$\sigma(E_p, W_r)\,dW_r = 2\pi r_0^2 \frac{mc^2}{\beta_p^2 W_r^2}\left[1-\frac{m^2c^2W_r}{2mE_p^2}+\frac{1}{2}\left(\frac{W_r}{E_p}\right)^2\right]dW_r\,. \tag{4-363}$$

4.5.1.6. *Creation of Photons by Positron Annihilation*

A positron, e^+, may collide with an electron, e^-, to produce two gamma ray photons according to the reaction $e^- + e^+ \to \gamma + \gamma$. One photon will have a high energy and, if the electron is at rest, the other photon will have an energy on the order of $mc^2 = 0.511$ MeV. If the energy of the positron is given by γmc^2, where here γ is taken to denote an energy factor, the cross section, σ, for two photon annihilation with a free electron at rest is given by (DIRAC, 1930)

$$\sigma = \frac{\pi r_0^2}{\gamma+1}\left[\frac{\gamma^2+4\gamma+1}{\gamma^2-1}\ln\left(\gamma+\sqrt{\gamma^2-1}\right)-\frac{\gamma+3}{\sqrt{\gamma^2-1}}\right]$$
$$\approx \frac{\pi r_0^2}{\gamma}\left[\ln(2\gamma)-1\right] \quad \text{for } \gamma \gg 1 \tag{4-364}$$
$$\approx \frac{\pi r_0^2}{\beta} \qquad \text{for } \beta \ll 1\,,$$

where the classical electron radius $r_0 = e^2/(mc^2) \approx 2.818 \times 10^{-13}$ cm, and $\beta = v/c$, where v is the velocity of the positron.

The positron may also be annihilated by emitting only one photon when colliding with an electron which is bound to an atom. For the collision of a positron with an electron in the K shell of an atom, the one photon annihilation cross section is given by (FERMI and UHLENBECK, 1933; BETHE and WILLS, 1935)

$$\begin{aligned} \sigma &= \frac{4\pi Z^5 \alpha^4 r_0^2}{\beta\gamma(\gamma+1)^2}\left[\gamma^2 + \frac{2\gamma}{3} + \frac{4}{3} - \frac{\gamma+2}{\beta\gamma}\ln[(1+\beta)\gamma]\right] \\ &\approx \frac{4\pi Z^5 \alpha^4 r_0^2}{\gamma} \qquad \text{for } \gamma \gg 1 \\ &\approx \frac{4\pi}{3} Z^5 \alpha^4 r_0^2 \beta \qquad \text{for } \beta \ll 1, \end{aligned} \tag{4-365}$$

where the positron has an energy γmc^2 and velocity $v = \beta c$, the atom has charge, eZ, the fine structure constant $\alpha = e^2/(\hbar c) \approx 1/137.037$, and the classical electron radius, $r_0 = e^2/(mc^2) \approx 2.818 \times 10^{-13}$ cm.

4.5.1.7. *Creation of π-Mesons, μ-Mesons, Positrons, Electrons, Photons, and Neutrinos by Nuclear Interaction*

The important reactions for the production of π-mesons are

$$\begin{aligned} &\mathrm{p+p \rightarrow p+p+n_1(\pi^+ + \pi^-) + n_2\pi^0} \\ &\mathrm{p+p \rightarrow p+n+\pi^+ + n_3(\pi^+ + \pi^-) + n_4\pi^0} \\ &\mathrm{p+p \rightarrow n+n+2\pi^+ + n_5(\pi^+ + \pi^-) + n_6\pi^0} \\ &\mathrm{p+p \rightarrow D+\pi^+ + n_7(\pi^+ + \pi^-) + n_8\pi^0}, \end{aligned} \tag{4-366}$$

where p denotes a proton, n a neutron, D a deuteron, and $\mathrm{n_1 - n_8}$ are positive integers. Once π-mesons are produced, μ-mesons, electrons, positrons, photons, and neutrinos may also be produced through the decay reactions

$$\begin{aligned} &\pi^\pm \rightarrow \mu^\pm + \nu_\mu/\bar{\nu}_\mu \\ &\pi^0 \rightarrow \gamma + \gamma \\ &\mu^\pm \rightarrow \mathrm{e}^\pm + \nu_e/\bar{\nu}_e + \bar{\nu}_\mu/\nu_\mu. \end{aligned} \tag{4-367}$$

In the proton energy region from 290 MeV to 1 GeV, only the reactions $\mathrm{p + p \rightarrow p + n} + \pi^+$ or $\mathrm{p + p} + \pi_0$ or $\mathrm{D} + \pi^+$ are important. Other pion reactions are given in Sect. 4.5.1.12 on solar gamma rays.

The minimum kinetic energy, W_{min}, for a nuclear proton to produce a total of Y π-mesons is given by

$$W_{min} = \frac{Y^2 m_\pi^2 c^4}{2 m_p c^2} + 2 Y m_\pi c^2 \approx Y(280 + 10\,Y)\ \mathrm{MeV}, \tag{4-368}$$

where $m_\pi c^2$ and $m_p c^2$ are, respectively, the rest mass energies of the π meson and the proton.

Experimental measurements (POLLACK and FAZIO, 1963; RAMATY and LINGENFELTER, 1966), indicate that the cross section, $\sigma(E_p)$, for the production of π-mesons by protons of energy E_p is

$$\sigma(E_p) \approx 2.7 \times 10^{-26}\,\mathrm{cm}^2 \quad \text{for } E_p > 2\,\mathrm{GeV}, \tag{4-369}$$

with multiplicity

$$m_{\pm} = 2E_p^{0.25}$$
$$m_0 = 0.5\,m_{\pm}.$$

4.5.1.8. Emission of a High Energy Photon by the Inverse Compton Effect

When a free electron with kinetic energy $E = \gamma mc^2$ collides with a photon of energy, $h\nu$, the Compton scattering cross section, σ_c, is given by (KLEIN and NISHINA, 1929)

$$\sigma_c = \frac{3}{4}\sigma_T\left\{\frac{1+q}{q^3}\left[\frac{2q(1+q)}{1+2q} - \ln(1+2q)\right] + \frac{1}{2q}\ln(1+2q) - \frac{1+3q}{(1+2q)^2}\right\}$$
$$\approx \sigma_T\left(1 - \frac{2\gamma h\nu}{mc^2}\right) \quad \text{for } \gamma h\nu \ll mc^2 \quad \text{or } q \ll 1 \tag{4-370}$$
$$\approx \frac{3}{8}\sigma_T\frac{mc^2}{\gamma h\nu}\left[\frac{1}{2} + \ln\left(\frac{2\gamma h\nu}{mc^2}\right)\right] \quad \text{for } \gamma h\nu \gg mc^2 \quad \text{or } q \gg 1,$$

where $q = \gamma h\nu/(mc^2)$, the classical electron radius $r_0 = e^2/(mc^2) \approx 2.818 \times 10^{-13}$ cm, and the Thomson (1906) scattering cross section $\sigma_T = 8\pi e^4/(3m^2c^4) \approx 6.65 \times 10^{-25}\,\mathrm{cm}^2$. Tables and graphs of the Compton cross section are given by DAVISSON and EVANS (1952) and NELMS (1953).

The frequency, ν_0, of the emitted photon is given by

$$\nu_0 \approx \gamma^2 \nu \qquad \text{for } \gamma h\nu \ll mc^2$$

and

$$\nu_0 \approx \gamma mc^2/h \quad \text{for } \gamma h\nu \gg mc^2. \tag{4-371}$$

This process has been discussed in detail by FEENBERG and PRIMAKOFF (1948), DONAHUE (1951), and FELTEN and MORRISON (1963, 1966). When relativistic electrons produce synchrotron radiation and then undergo Compton scattering from it, the resulting Compton-synchrotron radiation may have photon energies nearly as high as that of the electrons, and a spectrum the same as that of the synchrotron radiation (GOULD, 1965). This process is discussed in detail in Sect. 4.5.2.3.

4.5.1.9. High Energy Photon Emission by the Bremsstrahlung of a Relativistic Electron or Muon

In the collision of a relativistic electron of energy, E, with a nucleus of charge, eZ, the differential cross section, $\sigma_B(E, E_\gamma)\,dE_\gamma$, for the bremsstrahlung of a photon of energy, E_γ, is given by (BETHE and HEITLER, 1934)

$$\sigma_B(E, E_\gamma)\,dE_\gamma \approx \frac{4\alpha r_0^2 Z^2}{E_\gamma}\ln\left(\frac{2E}{mc^2}\right)dE_\gamma \quad \text{for } mc^2 \ll E \ll \frac{mc^2}{\alpha Z^{1/3}} \tag{4-372}$$

for the relativistic case and no screening. Here the fine structure constant $\alpha = e^2/(\hbar c) \approx 1/137.037$, the classical electron radius $r_0 = e^2/(mc^2) \approx 2.818 \times 10^{-13}$ cm, and the rest mass energy of the electron is $mc^2 \approx 0.511$ MeV. In the extreme relativistic but complete screening case we have

$$\sigma_B(E, E_\gamma) dE_\gamma \approx \frac{M}{XE_\gamma} dE_\gamma \quad \text{for } E \gg \frac{mc^2}{\alpha Z^{1/3}}, \tag{4-373}$$

where M is the mass of the target nucleus in grams, and the radiation length, X, in gram cm^{-2} is given by

$$\frac{1}{X} = 4\alpha r_0^2 Z^2 N_A A^{-1} \ln(183 Z^{-1/3}), \tag{4-374}$$

where Avogadro's number $N_A \approx 6.022 \times 10^{23}$ (gram mole)$^{-1}$, and the mass number is A. Values of the radiation length in gm cm^{-2} are (NISHIMURA, 1967)

$$\begin{aligned} X_H &= 62.8, & X_{He} &= 93.1, & X_{Li} &= 83.3, \\ X_C &= 43.3, & X_N &= 38.6, & X_O &= 34.6, \\ X_{Al} &= 24.3, & X_{Si} &= 22.2, & X_{Fe} &= 13.9, \end{aligned} \tag{4-375}$$

where the subscript denotes the element under consideration. The cross sections for electron-atom bremsstrahlung are given by Eqs. (4-372) and (4-373) if Z^2 is replaced by $Z(Z+1)$.

When accuracies greater than a few percent are required, the bremsstrahlung cross section may be calculated from the expressions (ROSSI, 1952)

$$\sigma_B(E, E_\gamma) dE_\gamma = \frac{4\alpha r_0^2 Z^2}{E_\gamma} F(E, E_\gamma) dE_\gamma, \tag{4-376}$$

where the function $F(E, E_\gamma)$ is a slowly varying function of the relativistic electron energy, E, and the photon energy E_γ. Extreme values of $F(E, E_\gamma)$ are

$$F(E, E_\gamma) = \left\{1 + \left(1 - \frac{E_\gamma}{E}\right)^2 - \frac{2}{3}\left(1 - \frac{E_\gamma}{E}\right)\right\}\left\{\ln\left[\frac{2E}{mc^2}\frac{(E - E_\gamma)}{E_\gamma}\right] - \frac{1}{2}\right\} \tag{4-377}$$

for a bare nucleus ($\xi \gg 1$), and

$$F(E, E_\gamma) = \left[1 + \left(1 - \frac{E_\gamma}{E}\right)^2 - \frac{2}{3}\left(1 - \frac{E_\gamma}{E}\right)\right] \ln\left(\frac{183}{Z^{1/3}}\right) + \frac{1}{9}\left(1 - \frac{E_\gamma}{E}\right) \tag{4-378}$$

for complete screening ($\xi \approx 0$). Here the screening parameter, ξ, is given by

$$\xi = \frac{mc^2}{\alpha E} \frac{E_\gamma}{E - E_\gamma} Z^{-1/3},$$

where the fine structure constant $\alpha = e^2/(\hbar c) \approx 1/137.037$, and the rest mass energy of the electron is $mc^2 \approx 0.511$ MeV.

For muons of mass, m_μ, the bremsstrahlung cross section for collisions with a bare nucleus is given by

$$\begin{aligned} \sigma_B(E, E_\gamma) dE_\gamma = {} & \frac{4\alpha r_0^2 Z^2}{E_\gamma}\left(\frac{m}{m_\mu}\right)^2 \left\{1 + \left(1 - \frac{E_\gamma}{E}\right)^2 - \frac{2}{3}\left(1 - \frac{E_\gamma}{E}\right)\right\} \\ & \times \left\{\ln\left[\frac{2E}{m_\mu c^2}\frac{\hbar}{m_\mu Rc}\left(\frac{E - E_\gamma}{E_\gamma}\right)\right] - \frac{1}{2}\right\} dE_\gamma, \end{aligned} \tag{4-379}$$

where m is the electron mass, and the nuclear radius $R \approx 1.2 \times 10^{-13} A^{1/3}$ cm, where A is the mass number.

4.5.1.10. *Photon Emission by the Synchrotron Radiation of a Relativistic Electron (Magnetobremsstrahlung)*

When a relativistic electron of energy $E = \gamma mc^2$ moves through a region with a uniform magnetic field of strength H, it will emit high energy photons by synchrotron radiation, which is discussed in detail in Sect. 1.25. Here we note that although the emitted photon spectrum is continuous, in most situations it is sufficient to assume that the emitted photon has a frequency

$$\nu = \frac{3\gamma^2}{4\pi} \frac{eH}{mc} \approx 4.6 \times 10^{-6} H E^2 \text{ Hz}, \quad (4\text{-}380)$$

where E is in electron volts and H is in gauss.

4.5.1.11. *Photon Emission from Nuclear Reactions*

A nucleus (Z, A) can emit a photon by the following reactions

$$\begin{aligned} (Z, A) + \text{n} &\rightarrow (Z, A+1) + \gamma \\ (Z, A) + \text{p} &\rightarrow (Z+1, A) + \gamma \\ (Z, A) + \alpha &\rightarrow (Z+2, A+4) + \gamma \end{aligned} \quad (4\text{-}381)$$

or

$$(Z, A)^* \rightarrow (Z, A) + \gamma .$$

Details of the formation of gamma rays by collisions of nuclei with protons and neutrons are discussed in Sect. 4.5.1.12 on solar gamma rays.

4.5.1.12. *Solar X and Gamma Ray Radiation*

Following the first balloon-borne observations of hard X-rays from solar flares (PETERSON and WINKLER, 1959), the continuum radiation from solar X-ray flares has been divided into two categories according to the energy of the emission. At energies below ten keV, the soft X-rays, the emission is thought to be due to the thermal bremsstrahlung of a Maxwellian distribution of electrons with temperature $T \approx 10^6$ to 10^8 °K. For energies between ten and a few hundred keV, the hard X-rays, the time profiles are often coincident with those observed at microwave frequencies (8,000 to 17,000 MHz); and both the hard X-rays and the microwaves are thought to be due to the nonthermal radiation of accelerated electrons. The microwave radiation is partially circularly polarized, and is therefore thought to be due to the gyrosynchrotron radiation of mildly relativistic electrons. The hard X-rays are thought to be due to the collisional bremsstrahlung of the accelerated electrons in a coronal plasma, and models differ in the way the electrons are injected into the plasma. In the thin target model, that of DEJAGER and KUNDU (1963), the electrons are continuously accelerated and injected into the dense chromosphere losing all of their energy immediately. In the thick target model, that of TAKAKURA and KAI (1966), the electrons are impulsively injected into a magnetic bottle high in the chromosphere between bipolar magnetic spots; and the observed burst decay is thought to be due to the slow collisional energy loss of the trapped non-thermal electrons.

The theory of thermal bremsstrahlung, which accounts for the soft X-ray radiation, was discussed in detail in Sect. 1.30. The photon energy, $h\nu$, is given by

$$h\nu \approx kT \approx 8.6\times10^{-8}\,T\,\mathrm{keV}\,, \tag{4-382}$$

where ν is the frequency of the radiation and T is the temperature of the plasma in °K. The free-free flux, η_{FF}, per unit volume of the solar corona, at the earth's distance, is given by (cf. Eq. (1-219))

$$\eta_{FF} = 7.15\times10^{-50}\,N_e \sum_z N_z Z^2 \left[\exp\left(-\frac{143.89}{\lambda T}\right)\right] \times \frac{\langle g(Z,T,c/\lambda)\rangle}{T^{1/2}\lambda^2}\,d\lambda\ \mathrm{erg\ cm^{-2}\ sec^{-1}\ \AA^{-1}}\,. \tag{4-383}$$

Here N_e is the free electron density, N_z is the number density of ions of charge Z, the photon wavelength $\lambda = c/(2\pi\nu)$ is assumed to be given in Å (1 Å $=10^{-8}$ cm), the electron temperature, T, is in units of 10^6 °K, and the $\langle g(Z,T,c/\lambda)\rangle \approx 1$ is the temperature averaged free-free Gaunt factor. In different units, the free-free X-ray flux at the earth is given by

$$F_{FF}(h\nu) = \frac{4.78\times10^{-39}}{T^{1/2}}\,\frac{\exp[-h\nu/(kT)]}{h\nu}\left[\int_v N_e N_i\,dv\right]\langle g(T,\nu)\rangle \quad \mathrm{photons\ cm^{-2}\ sec^{-1}\ keV^{-1}}\,, \tag{4-384}$$

where $h\nu$ is the photon energy, and the emission measure, $\int N_e N_i\,dv$ is the volume integral of the product of the free electron and ion densities. The free-free Gaunt factor, as well as the free-bound radiation formula, is given in Sect. 1.31, and numerical evaluations of the thermal emissivity are given in the references listed in this section. For wavelengths in the range 1 Å to 30 Å and at temperatures in the range 0.8×10^6 to 10^8 °K, the free-free and free-bound emissivity have been evaluated by CULHANE (1969). The temperature, T, is evaluated from the spectrum of the soft X-ray data, and the emission measure is evaluated by the observed flux at a given wavelength. Typical values are $T \approx 10^7$ °K and $N_e N_i V \approx 7\times10^{47}\ \mathrm{cm^{-3}}$.

The microwave burst radiation is thought to be the gyrosynchrotron radiation of mildly relativistic electrons, kinetic energy $E \approx mc^2$, spiralling about a magnetic field of strength, H. The radiation flux is a maximum at about twice the gyrofrequency, ν_H, given by

$$\nu_H = \frac{eH}{2\pi mc\gamma}\sin\psi \approx 2.8\times10^6\,H\sin\psi\ \mathrm{Hz}\,, \tag{4-385}$$

where $\gamma = E/mc^2 = [1-(v/c)^2]^{-1/2} \approx 1$ for an electron of velocity, v, and kinetic energy, $E \approx mc^2$, and ψ is the pitch angle between the direction of the magnetic field and the direction of the electron motion. Detailed formulae for gyrosynchrotron radiation were given in Sect. 1.25, and their application to solar flares is given by TAKAKURA (1967), RAMATY (1969), and RAMATY and PETROSIAN (1972).

To a first approximation, the total gyrosynchrotron emission, P, from one electron is given by

$$P = \frac{8\pi^2}{3} \frac{e^2}{c} v_H^2 \langle \sin^2 \psi \rangle [\gamma^2 + 2\gamma] \tag{4-386}$$

for an electron with kinetic energy $E = \gamma m c^2$. The change in electron kinetic energy due to both gyrosynchrotron radiation and collisions with thermal electrons is given by (TAKAKURA, 1967)

$$\frac{d\gamma}{dt} = \frac{1}{mc^2} \frac{dE}{dt} = -3.8 \times 10^{-9} H^2 \langle \sin^2 \psi \rangle \left[\frac{\gamma^2}{2} + \gamma \right] + 1.5 \times 10^{-16} N_e \gamma^{-3/2}, \tag{4-387}$$

where H is the magnetic field intensity in gauss and N_e is the electron number density in cm^{-3}.

The hard X-ray emission is thought to be due to the bremsstrahlung emission of nonrelativistic electrons, and BROWN (1971) has evaluated the electron-proton bremsstrahlung cross section which is appropriate for the thick target case. For an X-ray spectrum with spectral index, γ, we have an X-ray energy flux spectrum given by

$$I(h\nu) = A(h\nu)^{-\gamma} \quad \text{photons cm}^{-2}\,\text{sec}^{-1}\,\text{keV}^{-1}, \tag{4-388}$$

where A is a time varying constant. The injected electron spectrum is also power law with a spectral index $\alpha = \gamma + 1$. In the X-ray energy range of 20—100 keV, the instantaneous nonthermal electron spectrum, $N(E)$, present in the X-ray emitting region and the instantaneous the nonthermal electron flux spectrum, $F(E) \propto E^{-(\gamma+1)}$, are given by (BROWN, 1971)

$$N(E) = 3.61 \times 10^{41} \gamma(\gamma - 1)^3 \beta(\gamma - 1/2, 3/2) \frac{E_c^{\gamma-1} E^{-(\gamma-1/2)}}{N_e} I(E_c) \quad \text{electrons keV}^{-1}, \tag{4-389}$$

and

$$F(E) = 2.68 \times 10^{33} \gamma^2 (\gamma - 1)^3 \beta(\gamma - 1/2, 3/2) E_c^{\gamma-1} E^{-(\gamma+1)} I(E_c) \quad \text{electrons keV}^{-1}\,\text{sec}^{-1}, \tag{4-390}$$

where the Beta function $\beta(p, q)$ is given by

$$\beta(p, q) = \int_0^1 u^{p-1} (1-u)^{q-1} du, \tag{4-391}$$

$N_e \approx N_p$ is the mean value of the electron or proton density in cm^{-3} in the emitting volume, E_c is the lower cutoff energy in the electron spectrum in keV, E is the kinetic energy of the electron in keV, and $I(E_c)$ is the X-ray photon flux at energies $E \gtrsim E_c$ in units of photons $cm^{-2}\,sec^{-1}$.

In the thick target, impulsive injection case, it is the electron spectrum, $N(E)$, in the X-ray emitting region which is important. Although $N(E) \propto E^{-(\gamma-1/2)}$, the spectrum of the injected electrons goes as $E^{-(\gamma+1)}$ in this thick target case. The power input in the form of kinetic energy of electrons with energy $E \geq E_c$ is given by

$$P_{\text{thick}}(E_c, \gamma) = 4.29 \times 10^{24} A \gamma^2 (\gamma - 1) \beta(\gamma - 1/2, 3/2) E_c^{-(\gamma-1)} \text{erg sec}^{-1}, \tag{4-392}$$

where A is the constant in Eq. (4-388), LIN and HUDSON (1971) discuss the relationships between nonthermal X-rays and electrons detected at the earth. Details of the directivity and polarization of thick target bremsstrahlung are given by BROWN (1972), and the emission from partially ionized thick targets is given by BROWN (1973). HUDSON (1972) has related thick target processes and white light flares.

When the electron energy loss is due to collisions in a plasma of electron density, N_e, the electron energy loss is given by

$$\frac{d\gamma}{dt} \approx -\frac{\gamma}{\tau_c}, \tag{4-393}$$

where $\gamma = E/(mc^2)$, and τ_c is the deflection time due to Coulomb collisions with thermal particles and is given by (TRUBNIKOV, 1965)

$$\tau_c \approx 6.3 \times 10^{-20} v^3 N_e \sec^{-1}, \tag{4-394}$$

where v is the electron velocity. TAKAKURA (1969) gives

$$\frac{d\gamma}{dt} \approx -10^{-12} N_e^{-1} \sec, \tag{4-395}$$

for $0.2 < \gamma < 0.4$ or $100 \text{ keV} < E < 2 \text{ MeV}$, and

$$\frac{d\gamma}{dt} \approx -\frac{1.2 \times 10^{-12}}{2^{3/2}} N_e \gamma^{-1/2} \sec^{-1}$$

or (4-396)

$$\frac{dE}{dt} \approx -4.9 \times 10^{-9} N_e E^{-1/2} \text{ keV} \sec^{-1}$$

for $\gamma < 0.2$ or $E < 100$ keV. The decay of microwave and hard X-ray bursts in this case is discussed by BROWN (1972) and TAKAKURA (1969).

For the thin target case, the electrons are continuously accelerated and the X-ray spectrum is nearly unchanged from the acceleration or injection spectrum. For a power law X-ray spectrum at the earth, given by Eq. (4-388), the instantaneous nonthermal electron spectrum is given by

$$N(E) = 1.05 \times 10^{42} \frac{(\gamma - 1)}{N_e V} A E^{-(\gamma - 1/2)} \text{ electrons cm}^{-3} \text{ keV}^{-1}, \tag{4-397}$$

where N_e is the free electron density and V is volume of the plasma. The total collisional energy transfer, P, is obtained by integrating equation (4-396) from E_c to infinity over the spectrum given in Eq. (4-397)

$$P_{\text{thin}} = 8.19 \times 10^{24} A E_c^{-(\gamma - 1)} \text{ erg sec}^{-1}. \tag{4-398}$$

In this case, it is the instantaneous flux spectrum, $F(E)$, which is important in determining the X-ray emission, which is independent of the conditions in the "emitting" region. Here A is the constant in Eq. (4-388).

Detection at the earth of high energy protons coming from solar flares led to the suggestion that secondary particles and gamma rays might be produced by the nuclear interaction of protons in the solar atmosphere. Detailed calculations of expected particle and gamma ray yields were given by LINGENFELTER and RAMATY (1967) and LINGENFELTER (1969), and were summarized by CHUPP (1971). CHUPP (1973) has subsequently reported the detection of the 0.511 MeV, 2.20 MeV, 4.43 MeV and 6.14 MeV gamma ray lines. The most important nuclei are the most abundant, H^1, He^4, C^{12}, N^{14}, and O^{16}, whose relative solar abundances are $H:He:C:N:O=1:10^{-1}:5.3\times10^{-4}:10^{-4}:9.2\times10^{-4}$. The most important secondary particles produced by the interaction of protons with these nuclei are: neutrons which subsequently interact with hydrogen to produce the 2.20 MeV line, excited nuclei of C^{12}, O^{16} and N^{14}, which subsequently decay to emit the 4.43, 6.14 and 1.63 or 2.14 MeV lines, respectively, and pions and C^{11}, N^{13} and O^{15}, which all subsequently emit positrons which annihilate to form the 0.511 MeV line. Many other gamma ray sources which yield lines which might be observed, are summarized in Table 49.

Table 49. Energies and sources of gamma-ray lines which might be observed[1]

Energy (MeV)	Source	Cross section (135°) (mb/sr)	Source half-life
0.060	^{241}Am, *r*-process		458 y
0.18	^{251}Cf, *r*-process		800 y
0.34	^{249}Cf, *r*-process		360 y
0.39	^{249}Cf, *r*-process		360 y
0.472	Nucleosynthesis (^{56}Ni) and universal background		6.1 d
0.478	Cosmic-ray ^{7}Li		—
0.511	Pair annihilation		48 y, 77 d[2]
0.61	^{214}Bi (^{226}Ra) *r*-process		1,602 y
0.748	Nucleosynthesis (^{56}Ni) and universal background		6.1 d
0.812	Nucleosynthesis (^{56}Ni) and universal background		6.1 d
0.847	Nucleosynthesis (^{56}Co), universal background, and cosmic rays		77 d
0.857 0.870 0.875	Nuclear de-excitation of ^{56}Fe, 0.857 MeV is a cosmic ray	12 ± 5 44 ± 14 30 ± 15	
0.983	Nucleosynthesis (^{48}V)		16 d
1.03	Nucleosynthesis (^{56}Co), universal background, and cosmic rays		77 d
1.156	Nucleosynthesis (^{44}Sc)		48 y
1.173	Nucleosynthesis (^{60}Co)		3×10^5 y
1.24	Nucleosynthesis (^{56}Co), universal background, and cosmic rays		77 d
1.24 1.28 1.34	Nuclear de-excitation of ^{56}Fe, 1.24 MeV is a cosmic ray	131 ± 53 25 ± 14 61 ± 18	

Table 49 (continued)

Energy (MeV)	Source	Cross section (135°) (mb/sr)	Source half-life
1.312	Nucleosynthesis (^{48}V)		16 d
1.33	Nucleosynthesis (^{60}Co)		3×10^5 y
1.37	Nuclear de-excitation of ^{28}Si		
1.393	Nuclear de-excitation of ^{24}Mg, 1.39 MeV is a cosmic ray	18 ± 9	
1.56	Nucleosynthesis (^{56}Ni) and universal background		61 d
1.63	Nuclear de-excitation of ^{20}Ne		
1.675	Nuclear de-excitation of ^{24}Mg	16 ± 9	
1.76	Nucleosynthesis (^{56}Co), universal background, and cosmic rays		77 d
1.78	Nuclear de-excitation of ^{28}Si		
1.98 1.99 2.00 2.03	Nuclear de-excitation of ^{12}C	1.8 ± 0.9 0.5 ± 0.13 2.3 ± 2.0 < 1.0	
2.02	Nucleosynthesis (^{56}Co), universal background, and cosmic rays		77 d
2.23	Nuclear de-excitation of deuteron		
2.19 2.23 2.35	Nuclear de-excitation of ^{16}O	< 1.0 1.0 ± 0.4 0.26 ± 0.10	
2.24	Nuclear de-excitation of ^{32}S		
2.31	Nuclear de-excitation of ^{14}N		
2.60 2.66 2.73	Nuclear de-excitation of ^{16}O	0.9 ± 0.3 0.6 ± 0.3 2.0 ± 0.6	
2.80	Nuclear de-excitation of ^{12}C	0.8 ± 0.4	
3.26	Nucleosynthesis (^{56}Co), universal background, and cosmic rays		77 d
3.32	Nuclear de-excitation of ^{24}Mg	< 1.0	
3.625 3.72 3.78	Nuclear de-excitation of ^{16}O	0.8 ± 0.5 0.22 ± 0.10 0.8 ± 0.5	
3.82	Nuclear de-excitation of ^{12}C	2.3 ± 0.4	
3.95	Nuclear de-excitation of ^{14}N		
4.27 4.34 4.37 4.43	Nuclear de-excitation of ^{16}O	4.0 ± 0.4 8.7 ± 0.7 9.0 ± 0.7 1.26 ± 0.46	

Table 49 (continued)

Energy (MeV)	Source	Cross section (135°) (mb/sr)	Source half-life
4.34 4.36 4.37 4.38 4.47	Nuclear de-excitation of ^{12}C and ^{13}C, Cosmic rays	49 ±4 12.7 ±0.6 8.2 ±0.5 3.5 ±0.3 0.89 ±0.23	
5.09 5.19 5.25 5.26	Nuclear de-excitation of ^{16}O	4.1 ±0.5 5.8 ±0.5 6.9 ±0.6 0.96 ±0.39	
5.25	Nuclear de-excitation of ^{12}C	1.9 ±0.3	
6.01 6.09 6.21 6.22 6.29	Nuclear de-excitation of ^{16}O, Cosmic rays	6.8 ±0.5 9.9 ±0.8 4.7 +0.5 10.5 ±1.2 4.4 ±1.6	
6.52 6.60 6.65 6.75	Nuclear de-excitation of ^{12}C	2.6 ±0.2 0.9 ±0.1 1.8 ±0.3 0.24 ±0.09	
6.80 6.95 6.98 7.08 7.10	Nuclear de-excitation of ^{16}O from the 7.12 MeV transition. 7.12 MeV is a cosmic ray	2.2 ±0.2 3.7 ±0.3 2.5 ±0.3 4.6 ±0.6 0.98 ±0.35	
8.87	Nuclear de-excitation of ^{16}O		
12.7	Nuclear de-excitation of ^{12}C		

[1] Prepared from the references: CLAYTON and CRADDOCK (1965) for *r*-process sources, CLAYTON, COLGATE, and FISHMAN (1969) and CLAYTON (1971) for nucleosynthesis sources, RAMATY, STECKER, and MISRA (1970) for cosmic ray sources of the 0.511 MeV line, CLAYTON and SILK (1969) and FISHMAN and CLAYTON (1972) for universal background and cosmic ray sources, and FOWLER, REEVES, and SILK (1970) for cosmic ray sources. The possibility of detecting lines from excited nuclei in the Sun is discussed by LINGENFELTER and RAMATY (1967), LINGENFELTER (1969) and CHUPP (1971). CHUPP *et al.* (1973) have reported the detection of the 0.511 MeV pair annihilation, 2.20 MeV deuteron, 4.43 MeV carbon, and the 6.14 oxygen lines from a solar flare. The gamma ray energies and cross sections for the nuclear de-excitation lines are mainly from ZOBEL, MAIENSCHEIN, TODD, and CHAPMAN (1967).

[2] Lifetimes of 48 y and 77 d are for positrons from nucleosynthesis (^{44}Sc and ^{56}Co).

For detailed calculations, the proton energy spectrum is taken to be

$$N(P)=P_0^{-1}\exp\left(-\frac{P}{P_0}\right), \tag{4-399}$$

where for a nucleus of mass number, A, and charge, Ze, the rigidity, $P=A[2M_p c^2 E_p+E_p^2]^{1/2}/(Ze)$ for protons of energy, E_p, and rest mass, M_p. The rigidity is the ratio of momentum to charge, P_0 is a characteristic rigidity, $P_0 \approx 80$ to 200 MeV, and $N(P)$ may be normalized at the Sun to one particle of rigidity greater than zero. The most important neutron producing reactions are given in Table 50. The total neutron production cross sections, σ_n, are (LINGENFELTER and RAMATY, 1967) $\sigma_n \approx 20$, 200, and 1000 mb, respectively, for the pp, pα, and pCNONe reactions when the proton energy, E_p, is greater than 1 BeV. For reference, one millibarn $= 10^{-27}$ cm^2. The differential energy spectrum, $f(E_n, E_p)$, for incident protons of energy, E_p, and secondary neutrons of energy, E_n, is given by

$$f(E_n, E_p) = \frac{1}{E_p(1-K)} \quad \text{for } 0 < E_n < E_p(1-K) \quad \text{and the (p, p) reactions}$$

$$f(E_n, E_p) = \frac{1}{E_p - Q} \quad \text{for } 0 < E_n < E_p - Q \quad \text{and the (p, } \alpha\text{) reactions}$$

$$f(E_n, E_p) = E_n \theta^{-2} \exp(-E_n/\theta) \quad \text{for the pCNONe reactions}. \tag{4-400}$$

Here $K \approx 0.25$ is the inelasticity defined as the fraction of the incident proton energy going into total pion energy, Q is the kinetic energy converted into rest mass in the (p, α) reaction, and $\theta \approx 1.5$ MeV is the temperature of the excited nucleus. The neutrons so produced interact with hydrogen according to the reaction $n + p \rightarrow D + \gamma$ to produce the 2.23 MeV gamma ray line. The relative probability of neutron capture by other nuclei in the solar atmosphere is less than 10^{-3} of that of hydrogen.

Table 50. Neutron producing reactions for the abundant nuclei

Reaction	Threshold energy (MeV/nucleon)
H^1 (p, nπ^+) H^1	292.3
He^4 (p, pn) He^3	25.9
He^4 (p, 2pn) H^2	32.8
He^4 (p, 2p2n) H^1	35.6
C^{12} (p, n ...)	19.8
N^{14} (p, n ...)	6.3
O^{16} (p, pn ...)	16.5
Ne^{20} (p, pn ...)	17.7

The proton excited states of the C^{12}, N^{14}, O^{16} and Ne^{20} nuclei result in the emission of the lines given by

Excited nucleus	Principal Lines (MeV)
Ne^{20}	1.632
N^{14}	2.312
N^{14}	3.945
C^{12}	4.433
O^{16}	6.14
O^{16}	7.12

(4-401)

A more detailed list including less predominant lines was given in Table 49.

The principal pion producing reactions are

$$
\begin{aligned}
p+p \rightarrow\ & D+\pi^+ \\
& p+p+a(\pi^+ + \pi^-)+b\pi^0 \\
& p+n+\pi^+ + a(\pi^+ + \pi^-)+b\pi^0 \\
& 2n+2\pi^+ + a(\pi^+ + \pi^-)+b\pi^0 \\
p+He^4 \rightarrow\ & p+He^4+a(\pi^+ + \pi^-)+b\pi^0 \\
& p+He^3+n+a(\pi^+ + \pi^-)+b\pi^0 \\
& 2p+H^2+n+a(\pi^+ + \pi^-)+b\pi^0 \\
& 4p+n+\pi^- + a(\pi^+ + \pi^-)+b\pi^0 \\
& 3p+2n+a(\pi^+ + \pi^-)+b\pi^0 \\
& 2p+3n+\pi^+ + a(\pi^+ + \pi^-)+b\pi^0 \\
& p+4n+2\pi^+ + a(\pi^+ + \pi^-)+b\pi^0 \,.
\end{aligned}
\qquad (4\text{-}402)
$$

The measured cross sections for these reactions have been summarized by LINGENFELTER and RAMATY (1967), and approximate values for the production of π pions are $\sigma_{\pi^+} \approx 20$ and 100 mb for the pp and pα reactions, and $\sigma_{\pi^0} \approx 9 E_p$ and $90 E_p$ mb for a proton energy, E_p, in GeV and greater than one GeV. The differential energy spectrum, $f(E_\pi, E_p)$, for incident protons of energy, E_p, and secondary pions of energy, E_π, is given by

$$f(E_\pi, E_p) = \delta(E_\pi - \langle E_\pi \rangle), \qquad (4\text{-}403)$$

where δ is the Dirac delta function and

$$\langle E_\pi \rangle = 0.985\, E_p^{3/4} \quad \text{in MeV for the (pp) reaction}\,,$$

and

$$\langle E_\pi \rangle = 0.754\, E_p^{3/4} \quad \text{in MeV for the (p}\alpha\text{) reaction}\,.$$

The principal positron emitters are given in Table 51 for the most abundant nuclei.

Table 51. The principal positron emitters for the abundant nuclei[1]

β^+ emitter and decay mode	Maximum β^+ energy (MeV)	Half life (min)	Production mode	Threshold energy (MeV)	Production cross section (mb)
$C^{11} \rightarrow B^{11} + \beta^+ + \nu$	0.97	20.5	C^{12}(p, pn)C^{11}	20.2	50
			N^{14}(p, 2p2n)C^{11}	13.1	30
			N^{14}(p, α)C^{11}	2.9	—
			O^{16}(p, 3p3n)C^{11}	28.6	10
$N^{13} \rightarrow C^{13} + \beta^+ + \nu$	1.19	9.96	N^{14}(p, pn)N^{13}	11.3	10
			O^{16}(p, 2p2n)N^{13}	5.54	8
$O^{14} \rightarrow N^{14} + \beta^+ + \nu$	1.86	1.18	N^{14}(p, n)O^{14}	6.4	—
$O^{15} \rightarrow N^{15} + \beta^+ + \nu$	1.73	2.07	O^{16}(p, pn)O^{15}	16.54	50

[1] Here the data are from LINGENFELTER and RAMATY (1967) and the cross sections are approximate values for energies greater than 100 MeV. Both the π^+ and β^+ particles lead to the 0.51 MeV annihilation line.

LINGENFELTER and RAMATY (1967) assumed that protons are accelerated in an isotropic manner and interact with the solar atmosphere isotropically. They calculated the total number of secondary components, $Q(E_s)$, of energy, E_s, produced during the acceleration of protons, and also when the protons are released in the solar material and stop. The acceleration and slowing down phases may be taken, respectively, to be equivalent to the thin and thick target models. Assuming that the spectral energy density, $\langle N(E)\rangle$, of accelerated particles is constant over the acceleration time, t_1, the total number of secondaries produced during the acceleration phase is (LINGENFELTER and RAMATY, 1967; CHUPP, 1971)

$$Q(E_s)=N\,t_1\int_0^\infty \sigma(E)\,f(E_s,E)\,V(E)\langle N(E)\rangle\,dE\ \mathrm{MeV}^{-1}, \tag{4-404}$$

where $Q(E_s)$ is in particles MeV^{-1} when the number density of target nuclei, N, is in cm^{-3}, the acceleration time, t_1, is in seconds, the cross section, $\sigma(E)$, at energy, E, for secondary production is in cm^2, $f(E_s, E)$ is the normalized differential energy spectrum, $V(E)\langle N(E)\rangle$ is the flux of accelerated nuclei in $\mathrm{cm}^{-2}\,\mathrm{sec}^{-1}\,\mathrm{MeV}^{-1}$, and the proton kinetic energy and the proton velocity are, respectively, E and $V(E)$. The $f(E_s, E)$ is the probability that a neutron or gamma ray which results from interaction with a proton of energy, E, will have an energy, dE_s, around E_s. The proton energy spectrum, $\langle N(E)\rangle$, is given by Eq. (4-399) and is normalized so that it gives the total number of accelerated particles at energy, E.

In the slowing down phase, the total number of particles per MeV is given by

$$Q(E_s)=\frac{N}{\rho}\int_0^\infty \sigma(E)\,f(E_s,E)\,\frac{dR(E)}{dE}\,dE\int_E^\infty N(E_1,t_1)\times\exp\left(-\frac{[R(E_1)-R(E)]}{L}\right)dE_1\ \mathrm{MeV}^{-1}, \tag{4-405}$$

where N/ρ is the number of target atoms per gram of solar material, ρ is the mass density in $\mathrm{g\,cm}^{-3}$ seen in slowing down, $dR(E)/dE$ is the slope of the stopping range, $R(E)$, versus energy, E, curve, E_1 is the energy the proton must have at t_1 in order to slow down to an energy, E, at time, t, and L is the mean attenuation length of protons over energy, E_1, to energy, E. Both $R(E)$ and L are in units of $\mathrm{g\,cm}^{-2}$.

If during acceleration one half of all the neutrons produced escape the Sun, and the neutron production rate is an impulse function in time, the time dependent neutron flux at the earth, $\varphi(t)$, is given by

$$\varphi(t)=\frac{Q(E_n)}{2\pi R^2}\,\frac{dE_n}{dt}\,P_e(E_n)\,P_s(E_n)\ \mathrm{cm}^{-2}\,\mathrm{sec}^{-1}, \tag{4-406}$$

where $Q(E_n)$ is given by Eqs. (4-404) or (4-405), R is the Sun-earth distance $\approx 1\,A.\,U.$, dE_n/dt is the rate of change of neutron energy at $1\,A.\,U.$ for a delta function production at the Sun, $P_e(E_n)=0.5$, the neutron survival probability at $1\,A.\,U$ is $P_s(E_n)$ and $Q(E_n)$ is evaluated at $E_n(t)=mc^2\{[1-(R/ct)^2]^{-1/2}-1\}$ where $R=vt$ and v is the velocity of the neutron.

4.5.2. Energy Loss Mechanisms for High Energy Particles

4.5.2.1. Charged Particle Energy Loss by Ionization

When a charged particle passes through matter, it loses energy by exciting and ionizing atoms. The loss in the energy, E, of a particle of charge, eZ, is given by (BETHE, 1930, 1932; BLOCH, 1933)

$$-\frac{dE}{dx}=\frac{2\pi Z^2 e^4 N}{mv^2}\left[\ln\left(\frac{2mv^2\gamma^2 W_{\mathrm{m}}}{I^2}\right)-2\beta^2+f\right] \tag{4-407}$$

$$\approx 2.54\times 10^{-19} Z^2 N\sqrt{2/(\gamma-1)}\,[\ln(\gamma-1)+11.8]\ \mathrm{eV\ cm^{-1}}$$

for atomic hydrogen and $\gamma \ll 1$

$$\approx 2.54\times 10^{-19} Z^2 N\left[3\ln\gamma+\ln\left(\frac{M}{m}\right)+19.5\right]\mathrm{eV\ cm^{-1}}$$

for atomic hydrogen and $\gamma \gg M/m$,

where dx is an element of unit length in the direction of particle motion, m is the electron mass, M is the mass of the incident particle whose velocity is v, the $\beta=v/c$, the $\gamma=(1-\beta^2)^{-1/2}=E/(Mc^2)$, the number density of electrons in the material is N, the ionization potential is I, and the maximum energy transfer, W_{m}, is given by Eq. (4-362). The density effect term, f, was first suggested by FERMI (1939, 1940), and is tabulated together with other constants in Eq. (4-407) by STERNHEIMER (1956) and HAYAKAWA (1969). Ionization potentials were given in Table 34.

For a completely ionized gas and an ultrarelativistic incident particle, we have (GINZBURG, 1969)

$$f=\ln(1-\beta^2)+\ln\left(\frac{I^2}{\hbar^2\omega_{\mathrm{p}}^2}\right)+1, \tag{4-408}$$

where the plasma frequency, ω_{p}, is given by

$$\omega_{\mathrm{p}}=(4\pi e^2 N_{\mathrm{e}}/m)^{1/2},$$

and N_{e} is the free electron density. For the case of an ionized gas and an ultra-relativistic incident particle, Eq. (4-407) becomes

$$-\frac{dE}{dx}\approx 2.54\times 10^{-19} Z^2 N_{\mathrm{e}}\left[\ln\left(\frac{W_{\mathrm{m}}}{mc^2}\right)-\ln N_{\mathrm{e}}+74.1\right]\mathrm{eV\ cm^{-1}}, \tag{4-409}$$

where $W_{\mathrm{m}}=E$ if $\gamma \gg (M/m)$ and $W_{\mathrm{m}}=2E^2/(mc^2)$ for $1 \ll \gamma \ll (M/m)$.

For the special case of an electron incident upon neutral atoms,

$$-\frac{dE}{dx}=\frac{2\pi e^4 N}{mv^2}\left[\ln\left(\frac{mv^2\gamma^2 W_{\mathrm{m}}}{I^2}\right)+\frac{9}{8}-\beta^2+f\right] \tag{4-410}$$

$$\approx 2.54\times 10^{-19} N[3\ln\gamma+20.2]\ \mathrm{eV\ cm^{-1}} \quad \text{for hydrogen and } \gamma \ll 1.$$

When ultrarelativistic electrons are incident upon a fully ionized gas,

$$-\frac{dE}{dx}\approx 2.54\times 10^{-19} N_{\mathrm{e}}[\ln\gamma-\ln N_{\mathrm{e}}+73.4]\ \mathrm{eV\ cm^{-1}}. \tag{4-411}$$

Electron energy losses by ionization and other processes are illustrated in Fig. 28 for the Galaxy and the intergalactic medium.

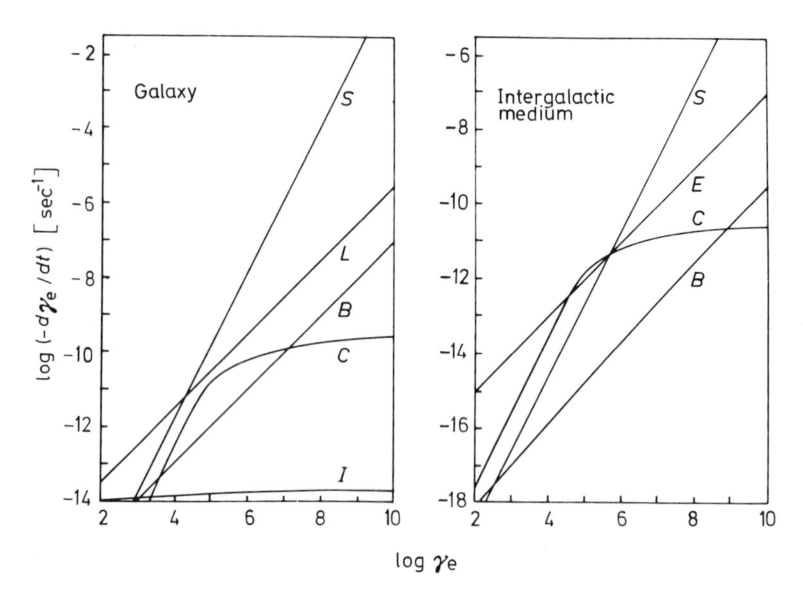

Fig. 28. Electron energy loss rates in the Galaxy and the intergalactic medium by synchrotron emission (S), cosmic expansion (E), leakage out of the halo (L), bremsstrahlung emission (B), Compton scattering (C), and ionization (I), (after BURBIDGE, 1966, by permission of Academic Press, Inc.). Here $\gamma = E/mc^2$ where E is the total energy of the electron. All energy loss rates are given in Sect. 4.5 except for the leakage loss rate $d\gamma/dt = \gamma lc/R^2$, and the expansion loss rate $d\gamma/dt = \gamma H_0$. Here the radius of the galactic halo is $R \approx 5 \times 10^{22}$ cm, the mean free path for Brownian motion of interstellar gas clouds is $l \approx 1$ kiloparsec $= 3 \times 10^{21}$ cm, and the Hubble constant gives an age of $H_0^{-1} \approx 10^{17}$ sec

4.5.2.2. *Electron Energy Loss by Bremsstrahlung*

When a relativistic electron of energy, E, collides with nuclei of charge, eZ, it emits photons of energy, E_γ, and loses energy according to the equation (BETHE and HEITLER, 1934)

$$-\frac{dE}{dx} = N \int_0^{E - mc^2} E_\gamma \sigma_B(E, E_\gamma) dE_\gamma, \tag{4-412}$$

where N is the number density of the nuclei, and the differential cross section $\sigma_B(E, E_\gamma)$ is given by Eqs. (4-372), (4-373), or (4-376). Using Eqs. (4-372) and (4-373) in Eq. (4-412), we obtain

$$-\frac{dE}{dx} \approx 4N\alpha r_0^2 E Z^2 \ln\left(\frac{2E}{mc^2}\right) \quad \text{for } mc^2 \ll E \ll \frac{mc^2}{\alpha Z^{1/3}}, \tag{4-413}$$

$$-\frac{dE}{dx} \approx \frac{MNE}{X} \quad \text{for } E \gg \frac{mc^2}{\alpha Z^{1/3}}. \tag{4-414}$$

Here N is the number density of target nuclei, the fine structure constant $\alpha = e^2/(\hbar c) \approx 1/137.037$, the classical electron radius $r_0 = e^2/(mc^2) \approx 2.818 \times 10^{-13}$ cm, the rest mass energy of the electron is $mc^2 \approx 0.511$ MeV, the mass of the target nucleus is M, and the radiation length X is given by Eqs. (4-374) or (4-375). For electron-atom bremsstrahlung, the energy losses are given by Eqs. (4-413) and (4-414) with Z^2 replaced by $Z(Z+1)$. For the special case of hydrogen atoms, $Z=1$, $M=1.67 \times 10^{-24}$ grams, $X=62.8$ gram cm^{-2}, and

$$-\frac{dE}{dx} \approx 2.6 \times 10^{-26} N E \text{ eV cm}^{-1} \qquad \text{for } E \gg \frac{mc^2}{\alpha} \qquad (4\text{-}415)$$

$$\approx 2.3 \times 10^{-27} N E \ln\left(\frac{2E}{mc^2}\right) \text{eV cm}^{-1} \quad \text{for } mc^2 \ll E \ll \frac{mc^2}{\alpha Z^{1/3}}. \qquad (4\text{-}416)$$

When the gas is completely ionized, the no screening effect equations are appropriate. Bremsstrahlung, synchrotron radiation, and Compton scattering of high energy electrons traversing a dilute gas are reviewed by BLUMENTHAL and GOULD (1970).

4.5.2.3. *Electron Energy Loss by Compton Scattering—The Inverse Compton Effect*

When a relativistic electron of energy, $E = \gamma mc^2$, collides with photons of energy, $h\nu$, it produces a photon of energy, E_γ, and loses energy according to the equation

$$-\frac{dE}{dx} = \int \sigma_c(E_\gamma, h\nu) N(h\nu) E_\gamma dE_\gamma d(h\nu), \qquad (4\text{-}417)$$

where the KLIEN-NISHINA (1929) scattering cross section, $\sigma_c(E_\gamma, h\nu)$, is given in Eqs. (4-370), and $N(h\nu)$ is the total number density of incident photons of energy, $h\nu$. From Eqs. (4-370) and (4-371) we obtain

$$E_\gamma \approx \gamma^2 h\nu \quad \text{and } \sigma_c \approx \sigma_T \quad \text{for } \gamma h\nu \ll mc^2$$

and

$$E_\gamma \approx \gamma mc^2 \quad \text{and } \sigma_c \approx \frac{3}{8}\sigma_T\left(\frac{mc^2}{\gamma h\nu}\right)\left[\ln\left(\frac{2\gamma h\nu}{mc^2}\right) + \frac{1}{2}\right] \quad \text{for } \gamma h\nu \gg mc^2, \qquad (4\text{-}418)$$

where the Thomson scattering cross section, $\sigma_T = 8\pi e^4/(3m^2c^4) \approx 6.65 \times 10^{-25}$ cm^2. Substituting Eqs. (4-418) into Eq. (4-417) we obtain

$$-\frac{dE}{dx} \approx \sigma_T U \gamma^2 \qquad \text{for } \gamma h\nu \ll mc^2$$

and

$$-\frac{dE}{dx} \approx \frac{3}{8}\sigma_T U \left(\frac{mc^2}{h\nu}\right)^2 \ln\left(\frac{2\gamma h\nu}{mc^2}\right) \quad \text{for } \gamma h\nu \gg mc^2, \qquad (4\text{-}419)$$

where the photon energy density, U, is given by

$$U = \int h\nu N(h\nu) d(h\nu) \approx h\nu N(h\nu) \qquad (4\text{-}420)$$

and the photon energy is $h\nu$. Various contributions to U have been discussed by FEENBERG and PRIMAKOFF (1948), DONAHUE (1951), and FELTEN and MORRISON

(1963, 1966) for the Galaxy and for intergalactic objects. The energy density of electromagnetic radiation in various spectral regions is illustrated in Fig. 29. Bremsstrahlung, synchrotron radiation, and Compton scattering of high energy electrons traversing a dilute gas are reviewed by BLUMENTHAL and GOULD (1970).

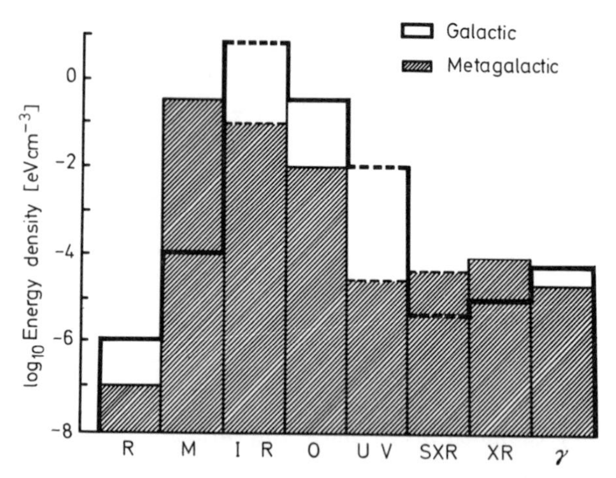

Fig. 29. Estimates of the flux of electromagnetic radiation in different spectral regions, plotted in histogram form and in energy density units of eV cm^{-3} = 1.6×10^{-12} erg cm^{-3}. The spectral regions considered are the radio ($R \approx 10^8$–10^{10} Hz), microwave ($M \approx 10^{10}$–10^{12} Hz), infrared ($I-R \approx 10^{12}$–10^{14} Hz), optical ($0 \approx 10^{13}$–10^{15} Hz), ultraviolet ($UV \approx 10^{15}$–10^{17} Hz), soft X-ray ($SXR \approx 10^{17}$ Hz), X-ray ($XR \approx 10^{17}$–10^{19} Hz) and γ-ray ($\gamma \approx 10^{18}$–10^{20} Hz) frequencies. Fluxes are shown in the thatched regions for the galactic contribution in the solar neighbourhood, and under the heavy line for an average point in intergalactic space. This histogram compares, in effect, the surface brightness of the Milky Way (after subtracting the surface brightness of the galactic poles) with that at the galactic poles. Estimates for those spectral regions where considerable uncertainty still exists are shown as dashed lines. Only in the microwave region, where the 2.7°K black body photon radiation contributes ≈ 0.3 eV cm^{-3}, and in the X-ray region, where there is $\approx 10^{-4}$ eV cm^{-3} in hard photons, does the metagalactic contribution greatly exceed the galactic contribution (after SILK, 1970, by permission of D. Reidel Publishing Co.)

4.5.2.4. Electron Energy Loss by Synchrotron (Magneto-Bremsstrahlung) Radiation

When a relativistic electron of energy $E = \gamma m c^2$ moves at a velocity v in a magnetic field of strength H, it loses energy according to the equation (SCHOTT, 1912) (cf. Eq. (1-155)).

$$-\frac{dE}{dx} = \frac{1}{c}\frac{dE}{dt} = \frac{2e^4}{3m^2c^4}\gamma^2 H^2 \approx 6.6 \times 10^{-14} E^2 H^2 \text{ erg cm}^{-1}, \qquad (4\text{-}421)$$

where $\beta = v/c$, E is in ergs, and H is in gauss. For conversion, 1 eV = 1.6021 $\times 10^{-12}$ erg.

4.5.2.5. Photon Energy Loss by the Photoelectric Effect, Compton Scattering, and Pair Formation

A beam of photons of original intensity, I_0, will have an intensity, I, given by

$$I = I_0 \exp[-\tau] \qquad (4\text{-}422)$$

after passing through a given distance, L, of a medium. The optical depth, τ, is given by

$$\tau = \int_0^L N\sigma\, dL \approx N\sigma L\,, \tag{4-423}$$

where N is the number density of scattering or absorbing particles and σ is the cross section for the scattering or absorbing process.

When the quantum energy, $h\nu$, of a proton is large compared with the ionization energy, I, of the K shell electrons of an atom, K shell electrons will be ejected when the photon and atom collide. The cross section, σ_K, for this photoelectric effect, in the nonrelativistic case, $h\nu \ll mc^2$, and the ejection of two K shell electrons, is given by (STOBBE, 1930; HEITLER, 1954)

$$\begin{aligned}\sigma_K &= \frac{128\pi\sigma_T}{\alpha^3 Z^2}\left(\frac{I}{h\nu}\right)^4 \frac{\exp(-4\eta\cot^{-1}\eta)}{[1-\exp(-2\pi\eta)]} && \text{for } \nu \gtrsim \nu_K \text{ and } h\nu \ll mc^2 \\ &= \frac{64\sigma_T}{\alpha^3 Z^2}\left(\frac{I}{h\nu}\right)^{7/2} = 4\sqrt{2}\,\alpha^4 Z^5 \left(\frac{mc^2}{h\nu}\right)^{7/2} && \text{for } \nu \gg \nu_K \text{ and } h\nu \ll mc^2,\end{aligned} \tag{4-424}$$

where the Thomson scattering cross section, σ_T, is given by (THOMSON, 1906)

$$\sigma_T = \frac{8\pi}{3}\left(\frac{e^2}{mc^2}\right)^2 = 6.652453(61) \times 10^{-25}\ \text{cm}^2,$$

the constant, η, is given by

$$\eta = \left(\frac{I}{h\nu - I}\right)^{1/2} = \frac{Ze^2}{\hbar v},$$

where v is the velocity of the ejected electron, Z is the atomic number of the atom, and the ionization energy, I, is related to the frequency, ν_K, of the absorption edge and the frequency, ν, by

$$I = h\nu_K = h\nu - mv^2/2 = mc^2\alpha^2 Z^2/2\,. \tag{4-425}$$

Cross sections for other shells may be evaluated by changing Z and I in Eq. (4-424). Ionization potentials are given in Table 34 of Sect. 3.3. The total cross section, σ, due to the photoelectric effect of all shells is $\sigma \approx 5\sigma_K/4$ for frequencies $\nu > \nu_K$, the frequency of the K shell absorption edge. Empirical values of photoelectric cross sections are given by DAVISSON and EVANS (1952).

For the special case of the hydrogen-like atom, the cross section, σ_n, for photoionization of the nth energy level is given by (KRAMERS, 1923) (cf. Eq. (1-230))

$$\sigma_n = \frac{32\pi^2 e^6 R_\infty Z^4}{3^{3/2} h^3 \nu^3 n^5} \approx 2.8 \times 10^{29} \frac{Z^4}{\nu^3 n^5}\ \text{cm}^2, \tag{4-426}$$

where the frequency, ν, is larger than $2\pi^2 me^4 Z^2/(h^2 n^2) \approx 3.3 \times 10^{15} Z^2 n^{-2}$ Hz, and the Rydberg constant for infinite mass, $R_\infty = 2\pi^2 me^4/(ch^3) \approx 1.097 \times 10^5$ cm^{-1}.

In the relativistic case, the photoelectric cross section, σ_K, for the two K shell electrons is given by (SAUTER, 1931; HEITLER, 1954)

$$\sigma_{\mathrm{K}} = \frac{3\sigma_{\mathrm{T}} Z^5 \alpha^4}{2} \left(\frac{mc^2}{h\nu}\right)^5 (\gamma^2-1)^{3/2} \left\{\frac{4}{3} + \frac{\gamma(\gamma-2)}{\gamma+1}\left[1 - \frac{1}{2\gamma\sqrt{\gamma^2-1}} \ln\left(\frac{\gamma+\sqrt{\gamma^2-1}}{\gamma-\sqrt{\gamma^2-1}}\right)\right]\right\}$$
$$\approx \frac{3\sigma_{\mathrm{T}} Z^5 \alpha^4}{2} \left(\frac{mc^2}{h\nu}\right) \quad \text{for } h\nu \gg mc^2, \tag{4-427}$$

where

$$\gamma = \left[1 - \left(\frac{v}{c}\right)^2\right]^{-1/2} = \frac{h\nu + mc^2}{mc^2},$$

the velocity of the ejected electron is v, the fine structure constant $\alpha = e^2/(\hbar c) \approx 1/(137.037)$, the rest mass energy of the electron is $mc^2 \approx 0.511$ MeV, the atomic number of the atom is Z, and $h\nu$ is the quantum energy of the photon of frequency, ν.

The total Compton scattering cross section, σ_{c}, for photon scattering from a free electron of energy, $E = \gamma mc^2$, is given by Eq. (4-370) with $q = \gamma h\nu/(mc^2)$. The asymptotic forms of this equation are (Klein and Nishina, 1929)

$$\sigma_{\mathrm{c}} \approx \sigma_{\mathrm{T}} \quad \text{for } \gamma h\nu \ll mc^2$$
$$\sigma_{\mathrm{c}} \approx \frac{3}{8}\sigma_{\mathrm{T}} \frac{mc^2}{\gamma h\nu}\left[\frac{1}{2} + \ln\left(\frac{2\gamma h\nu}{mc^2}\right)\right] \quad \text{for } \gamma h\nu \gg mc^2, \tag{4-428}$$

where a photon of frequency, ν, has quantum energy, $h\nu = 4.106 \times 10^{-21}\,\nu$ MeV, the rest mass energy of the electron is $mc^2 \approx 0.511$ MeV, and the Thomson (1906) scattering cross section $\sigma_{\mathrm{T}} = 8\pi e^4/(3m^2c^4) \approx 6.65 \times 10^{-25}\ \mathrm{cm}^2$. The photon is, of course, not completely absorbed but scattered with the new frequency, ν_0, given by (Compton, 1923)

$$\nu_0 = \nu\left[1 + \left(\frac{h\nu}{mc^2}\right)(1-\cos\varphi)\right]^{-1} \quad \text{for } \gamma \ll 1 \text{ and } \gamma h\nu \ll mc^2$$
$$= 4\gamma^2\nu/3 \quad \text{for } \gamma \gg 1 \text{ and } \gamma h\nu \ll mc^2, \tag{4-429}$$
$$= \gamma mc^2/h \quad \text{for } \gamma \gg 1 \text{ and } \gamma h\nu \gg mc^2,$$

where ν is the frequency of the incident photon and φ is its deflection angle.

A photon may be absorbed in creating electron-positron pairs when it collides with another photon or when it passes through the Coulomb field of a nucleus. The total cross sections for these two processes are given by Eqs. (4-360) and (4-356). Discussions of various photon absorption processes in astrophysical objects have been made by Feenberg and Primakoff (1948), Donahue (1951), Felten and Morrison (1963, 1966), and Goldreich and Morrison (1964). Absorption of X-ray photons in the energy range 0.1 to 8 keV is dominated by photoionization and optical depths as large as unity may arise in the interstellar medium. Calculations of photoelectric absorption of X-rays in the interstellar medium are given by Brown and Gould (1970). For X-ray energies above 8 keV, Compton scattering predominates with a maximum cross section of $5 \times 10^{-25}\ \mathrm{cm}^2$ at 100 keV, but optical depths for galactic and metagalactic space are less than unity. Pair production interactions with starlight photons give sufficient optical depth to absorb gamma ray photons in the energy range 100 GeV to 5000 GeV

(NIKISHOV, 1962), whereas all metagalactic gamma rays with energies larger than 10^5 GeV will be absorbed by pair production interaction with the universal 3 °K black body photon gas (GOULD and SCHRÉDER, 1966; JELLEY, 1966).

4.5.3. The Origin of High Energy Particles

4.5.3.1. Energy Spectrum of Cosmic Ray Electrons, Protons, and Positrons

When observing high energy particles, the differential energy flux, $J(E)$, is often measured. If $dJ(E)$ denotes the number of particles in the energy range, dE, which pass a unit area per unit time per unit solid angle, then

$$J(E) = \frac{dJ(E)}{dE} \quad \text{particles (cm}^2\text{ sec ster MeV or GeV)}^{-1}\,. \tag{4-430}$$

Detectors are often sensitive to particles whose energies lie above a certain threshold energy, and therefore the integral energy spectrum

$$J(>E) = \int_E^\infty J(E)\,dE \quad \text{particles (cm}^2\text{ sec ster)}^{-1}\,, \tag{4-431}$$

is also often measured. The differential energy spectrum of the cosmic ray electrons (first observed by EARL, 1961) is shown in Fig. 30. This spectrum is well represented by the following approximations:

$$\begin{aligned}
J(E) &= 1.32\times10^{-2}E^{-1.75}\,(\text{cm}^2\text{ sec ster MeV})^{-1} && \text{for } 3\text{ MeV}\le E\le 20\text{ MeV}\\
J(E) &= 1.00\times10^{-5}\,(\text{cm}^2\text{ sec ster MeV})^{-1} && \text{for } 30\text{ MeV}\le E\le 300\text{ MeV}\\
J(E) &= 3.5\times10^{-3}E^{-1.6}\,(\text{cm}^2\text{ sec ster GeV})^{-1} && \text{for } 0.5\text{ GeV}\le E\le 3\text{ GeV}\\
J(E) &= 1.16\times10^{-2}E^{-2.6}\,(\text{cm}^2\text{ sec ster GeV})^{-1} && \text{for } 3\text{ GeV}\le E\le 300\text{ GeV}\,.
\end{aligned} \tag{4-432}$$

The primary proton spectrum is parallel to the high energy ($3\le E\le 300$ GeV) electron spectrum with a displacement to the right by a factor of about twenty in energy. The differential energy spectrum of protons and nuclei (first observed by FREIER *et al.*, 1948 and BRADT and PETERS, 1948) is given by (RYAN, ORMES, and BALASUBRAHMANYAN, 1972)

$$\begin{aligned}
J(E) &= 2.0\pm0.2\,E^{-2.75\pm0.03}\ \text{protons (cm}^2\text{ sec ster GeV)}^{-1}\\
J(E) &= 8.6\pm1.4\times10^{-2}\,E^{-2.77\pm0.05}\ \text{He (cm}^2\text{ sec ster GeV/nucleon)}^{-1}\\
&\qquad \text{for}\quad 10\text{ GeV}\le E\le 10^3\text{ GeV}
\end{aligned} \tag{4-433}$$

for protons and helium (He) nuclei. Here the energy, E, is in GeV. The integral energy spectrum is given by (HAYAKAWA, 1969)

$$\begin{aligned}
J(>E) &= 1.6\times10^{-5}\left(\frac{E}{10^3}\right)^{-1.6}(\text{cm}^2\text{ sec ster})^{-1} && \text{for } 10\text{ GeV}\le E\le 10^4\text{ GeV}\\
J(>E) &= 2.0\times10^{-14}\left(\frac{E}{10^8}\right)^{-2.2}(\text{cm}^2\text{ sec ster})^{-1} && \text{for } 10^6\text{ GeV}\le E\le 10^9\text{ GeV}\,,
\end{aligned} \tag{4-434}$$

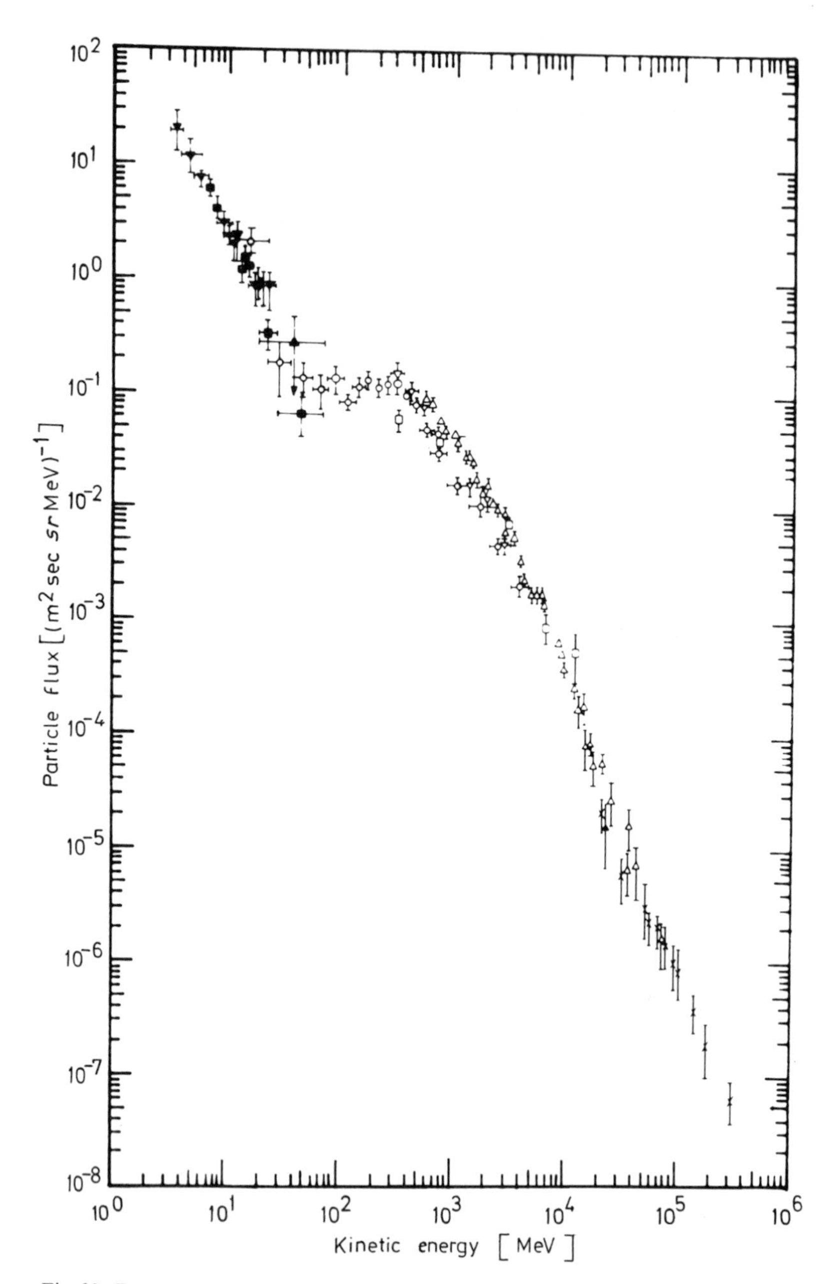

Fig. 30. Energy spectrum of the primary cosmic ray electrons (after FANSELOW *et al.*, 1969, by permission of the American Astronomical Society and the University of Chicago Press). Each symbol denotes a different set of observations, and these are referenced by FANSELOW *et al.*

where E is in GeV. The spectrum and composition of the primary cosmic radiation and the heavy cosmic ray nuclei are reviewed by WEBBER (1967) and SHAPIRO and SILBERBERG (1970), respectively.

In the low energy range (3 to 20 MeV), the cosmic ray electrons lose most of their energy by ionization of interstellar hydrogen, and are thought to originate from the knock-on collisions of higher energy protons (ABRAHAM, BRUNSTEIN, and CLINE, 1966; SIMNETT and MCDONALD, 1969). Observations of the electron-positron positive fraction, R, indicate that higher energy electrons are directly accelerated in sources and do not originate as secondaries of the protons. The positive fraction, R, is given by

$$R = \frac{J^+(E)}{J^+(E)+J^-(E)}, \qquad (4\text{-}435)$$

where the superscripts + and − denote, respectively, positrons and electrons. Observed values of R are given in Table 52. The high energy electrons must be produced within our Galaxy because otherwise the X-ray background would be much larger than that observed (FELTEN and MORRISON, 1966). The question of the origin of the high energy particles reduces, then, to a suitable mechanism for accelerating particles within our Galaxy.

Table 52. Fraction of positrons in the primary cosmic ray electron component[1]

Energy interval (BeV)	$R = \frac{J^+(E)}{J^+(E)+J^-(E)}$
0.053– 0.088	0.31 ±0.19
0.088– 0.173	0.45 ±0.08
0.173– 0.44	0.29 ±0.09
0.44 – 0.86	0.10 ±0.07
0.86 – 1.70	0.083±0.024
1.70 – 4.2	0.046±0.018
4.2 – 8.4	0.013±0.05
8.4 –14.3	0.15 ±0.18

[1] After FANSELOW *et al.* (1969).

4.5.3.2. *Acceleration Mechanisms for High Energy Particles*

When a charged particle encounters a moving region of higher magnetic field, the particle's energy will increase or decrease according as the collision is head-on or overtaking. Statistical arguments (FERMI, 1949, 1954) lead to the conclusion that a head-on collision of a particle with a randomly moving magnetic field is more likely than an overtaking one, and that on the average the particle will be accelerated. The interesting part of Fermi's acceleration mechanism is that it leads to the power-law spectrum like that observed for cosmic ray electrons and synchrotron radio sources. It follows from the law of conservation of momentum that if $\boldsymbol{W}$ is the velocity of a region of high magnetic field (a magnetic mirror), then for one collision of a relativistic particle the change in particle energy, ΔE, is given by

$$\Delta E = -2E \frac{\boldsymbol{W} \cdot \boldsymbol{V}_{11}}{c^2}, \tag{4-436}$$

where E is the particle energy and V_{11} is the component of particle velocity parallel to $\boldsymbol{W}$. As the probability of a head on collision is proportional to $V+W$, and that for an overtaking one is proportional to $V-W$, the average gain in energy per collision, $\langle \Delta E \rangle$, is given by

$$\langle \Delta E \rangle \approx \frac{V+W}{2V} \Delta E - \frac{V-W}{2V} \Delta E \approx 2 \frac{W^2}{c^2} E. \tag{4-437}$$

If t_F is the mean time between collisions, then the average rate of energy gain is given by

$$\frac{dE}{dt} \approx \frac{2W^2}{t_F c^2} E. \tag{4-438}$$

Once the particle energy gain exceeds that caused by ionization and other losses, it will be accelerated until it leaves the accelerating region or undergoes a nuclear collision. If T denotes the average acceleration time, then the probability, $P(t)dt$, of finding a particle with an age between t and $t+dt$ is given by

$$P(t)dt = \frac{1}{T} \exp\left(-\frac{t}{T}\right) dt. \tag{4-439}$$

It follows from Eq. (4-438) that

$$t = \frac{t_F c^2}{2W^2} \ln\left(\frac{E}{mc^2}\right), \tag{4-440}$$

where mc^2 is taken to be the initial particle energy. It then follows from Eqs. (4-439) and (4-440) that the number of particles, $N(E)dE$, with an energy between E and $E+dE$ is given by

$$N(E)dE \propto E^{-\gamma} dE, \tag{4-441}$$

where the spectral index, γ, is given by

$$\gamma = 1 + \frac{t_F c^2}{2TW^2}.$$

Variants of the Fermi acceleration mechanism employing turbulence, hydrodynamic waves, and plasma waves are discussed by MORRISON, OLBERT, and ROSSI (1954), PARKER (1955, 1958), DAVIS (1956), BURBIDGE (1957), WENTZEL (1963, 1964), and TSYTOVICH (1964, 1966). WENTZEL (1963), for example, shows that a shock of high magnetic field strength, H, moving across a closed, curved magnetic line of force will accelerate a particle of energy, E_0, to

$$E_f = E_0 \left[1 + \frac{2}{3} \frac{\cos^3\theta}{\sin^2\theta}\right]\left[1 - \frac{1}{3}\cos^2\theta\right], \tag{4-442}$$

where
$$\sin^2\theta = \frac{H_0}{H},$$
the undisturbed magnetic field strength along the line of force is H_0, and it is assumed that the particle is moving in front of the shock with a speed greater than that of the reflecting magnetic field.

BAADE and ZWICKY (1934) first noted that only a small fraction of the rest mass energy of a collapsing star is needed to account for the energy of the cosmic rays as well as for the radiant energy and kinetic energy of expansion of a supernova. They therefore advanced the hypothesis that supernovae emit cosmic rays. The fact that supernovae are sources of high energy particles was confirmed when their radio radiation was detected, and it was shown that this radiation was the synchrotron emission of relativistic electrons (SHKLOVSKI, 1953). PACINI (1967) first showed that a supernova could obtain its radiation from its remnant neutron star. When a normal star of radius $R_0 \approx 10^{11}$ cm collapses to form a neutron star of $R \approx 10^6$ cm, flux will be conserved and the stars surface magnetic field will increase from $B_0 \approx 100$ gauss to
$$B \approx B_0 \left(\frac{R_0}{R}\right)^2 \approx 10^{12} \text{ gauss}. \tag{4-443}$$
The luminosity, L, of a rotating magnetic dipole is given by
$$L = \frac{dT}{dt} = \frac{d}{dt}\left[\frac{I\omega^2}{2}\right] = \frac{2m_\perp^2\omega^4}{3c^3}, \tag{4-444}$$
where the rotational kinetic energy is T, and is given by
$$T = \frac{M\omega^2 R^2}{2}, \tag{4-445}$$
the moment of inertia is I, the component of the magnetic dipole moment perpendicular to the rotation axis is $m_\perp \approx BR^3$, where B is the surface magnetic field strength at radius, R, the angular velocity, ω, of the star can be found from the conservation of angular momentum, J, where
$$J = M\omega R^2. \tag{4-446}$$
For example, a normal star with $\omega_0 \approx 10^{-6}$ rad sec^{-1} will have the angular velocity $\omega \approx 10^4$ rad sec^{-1} after collapse to a neutron star. Eq. (4-444) then gives $L = 10^{40}$ erg sec^{-1} if $B = 10^{12}$ gauss and $R = 10$ km. This is about the observed luminosity of supernovae remnants. The source of this energy is the rotational kinetic energy of the star, $T \approx 10^{53}$ ergs. Following the discovery of pulsars (HEWISH *et al.*, 1968), it was suggested by GOLD (1968) that pulsars were rotating neutron stars, and that their periods, P, would be observed to slow down at a rate dP/dt. If the neutron star has a dipole magnetic field, then a characteristic pulsar age, τ_0, is given by (OSTRIKER and GUNN, 1969)
$$\tau_0 = \left[\frac{1}{P}\frac{dP}{dt}\right]^{-1} \approx \frac{3c^3 M}{2B^2 R^4 \omega^2}. \tag{4-447}$$
A list of pulsar parameters was given in Table 5 of Chapter 1.

Observations of pulsar periods and their rates of slowing down are also compatible with the observed luminosity of supernovae if Eqs. (4-444) and (4-447) are used. OSTRIKER and GUNN (1969) further proposed that cosmic ray particles were accelerated by the low frequency waves of the rotating magnetic dipole. This mechanism makes use of the fact that, in sufficiently strong wave fields, a charged particle is accelerated in the propagation direction to nearly the velocity of light in a small fraction of a wavelength, and thereafter rides the wave at essentially constant phase, slowly taking energy from the wave. The maximum energy, E_{max}, available for an electron is given by

$$E_{max} \approx mc^2 \left[2\omega_B r_0 c^{-1} \ln \frac{r_c}{r_0} \right]^{2/3}, \tag{4-448}$$

where r_c is the radius at which the wave energy is exhausted, r_0 is the inner wave radius, $r_0 \approx c/\omega$, and ω_B is the cyclotron frequency, $eB/(mc)$, in the magnetic field at maximum amplitude at r_0. For example, if $B = 10^6$ gauss at $r_0 = 10^8$ cm, then $E_{max} = 10^{13}$ eV if $r_c = 10^{12}$ cm. For a particle of atomic mass, A, and atomic number, Z,

$$E_{max} \approx 3.6 \times 10^{13} A \left(\frac{Z}{A}\right)^{2/3} \left(\frac{\omega}{10^2}\right)^{4/3} \left(\frac{B}{10^{12}}\right)^{2/3} \left(\frac{R}{10^6}\right)^2 \text{ eV}, \tag{4-449}$$

where R is the radius of the star. GOLDREICH and JULIAN (1969) first showed that cosmic ray particles could initially come from the surface of a neutron star, and that substantial acceleration might take place on the surface as well. They showed that the potential difference, V, between the pole and equator of the neutron star is given by (cf. Eq. (1-98))

$$V \approx \frac{\omega R^2 B}{c} \approx 10^{16} \omega \left(\frac{B}{10^{12}}\right) \left(\frac{R}{10^6}\right)^2 \text{ volts}, \tag{4-450}$$

an equation first applied to a rotating star by ALFVÉN (1938). Thus a potential difference as large as 10^{14} volts is possible. This acceleration is more than sufficient to allow the particles to escape the surface gravity of the star and to flow out into space along the magnetic field lines, along which the particle is accelerated as well.

The rate of injection of particles, r, is given by

$$r = \frac{B\omega^2 R^3}{Zec} \approx 7 \times 10^{32} \left(\frac{\omega}{10^2}\right)^2 \left(\frac{B}{10^{12}}\right) \left(\frac{R}{10^6}\right)^3 Z^{-1} \text{ particles sec}^{-1}. \tag{4-451}$$

It follows from Eqs. (4-444), (4-447), and (4-450) that the number of particles, $N(E)dE$, with an energy between E and $E + dE$, is given by

$$N(E)dE = r \left[\left(\frac{dE}{d\omega}\right) \left(\frac{d\omega}{dt}\right) \right]^{-1} dE \propto E^{-2.5} dE, \tag{4-452}$$

a spectrum remarkably close to the observed spectral index of -2.6 for the high energy cosmic rays. The number density of cosmic rays depends upon the birthrate of pulsars, and therefore the birthrate of supernovae which is thought to be about one every thirty years. Within the uncertainty of this number, the observed number of high energy cosmic ray particles may be accounted for. In the manner stated above, this theory can only account for high energy cosmic rays with energies in the range 10^{18}—10^{23} eV. Nevertheless, the wave energy is probably not exhausted in accelerating particles near the neutron star and may accelerate moderately relativistic particles (10^9 to 10^{16} eV) in the nebulous debris of the supernova explosion (KULSRUD, OSTRIKER, and GUNN, 1972).

Following Baade and Zwicky's (1934) suggestion that a supernova represents the transition of an ordinary star into a neutron star, BURBIDGE, BURBIDGE, FOWLER, and HOYLE (1957) and HOYLE and FOWLER (1960) suggested a mechanism for the associated supernova explosion. At the end of its evolution, an ordinary star has used up its nuclear fuel and the density of its central region increases much more rapidly than its temperature. As a result, the virial theorem (Eq. (3-188)) is violated and a catastrophic implosion of the core must take place. The dynamical energy of the subsequent inward motion of the outer shell is converted into heat lifting the temperatures of the lighter elements to an explosive state. COLGATE and WHITE (1966) then showed that much more energy is released by the emission and deposition of neutrinos than the available thermonuclear energy. The present description of a supernova explosion involves the initial composition and mass of the pre-supernova star, the equations of state of an ideal gas, a degenerate gas, and a photon gas (Sect. 3.3 and Table 36), the production rate of neutrinos (Sect. 4.3) the cross sections for the absorption and scattering of neutrinos (Sect. 4.4.3.6) and the hydrodynamic equations for adiabatic flow and for shock waves (Sects. 3.5.6 and 3.5.7). A list of supernova remnants is given in Table 53.

Once the mass, M, of a star is greater than the limiting mass for stable electron degeneracy, $M \gtrsim 1.4587\, M_{\odot}$, an implosion will occur because neutrino emission by electron capture will carry away heat faster than quasi-static contraction can supply it. At nuclear densities, the core will halt its collapse because of the degenerate neutron gas pressure, and the kinetic energy of the infalling matter is transformed into thermal energy by an ingoing shock wave. The heat generated behind the shock wave is emitted by neutrinos whose energy is deposited in the outer stellar envelope. The deposition of neutrino energy then gives rise to an outgoing shock followed by the adiabatic expansion of the stellar envelope (COLGATE and JOHNSON, 1960; COLGATE and WHITE, 1966; ARNETT, 1966, 1967, 1968; ARNETT and CAMERON, 1967). The neutrino-deposited energy can be as large as the binding energy, GM^2/R, of the star, or about 10^{50} ergs for $M \approx M_{\odot}$ and $R \approx 10^8$ cm. The radial differential in neutrino-deposited energy gives rise to a shock wave whose initial velocity, V_0, is given by $V_0 \approx [10^{50}/M_{\odot}]^{1/2} \approx 0.05\, c$. For a strong plane shock propagating through an ideal gas with a specific heat ratio of $\gamma = 5/3$, and a density, $\rho(X)$, given as a function of distance, X, from the star center by

$$\rho(X) = \rho_0 \left(\frac{X_0}{X}\right)^{7/4}, \tag{4-453}$$

where ρ_0 is the density at distance X_0, the velocity of the shock, V, is given by

$$V = V_0 \left(\frac{X_s}{X_0}\right)^{1/4}, \tag{4-454}$$

Table 53. Name, G number, position, angular size, flux density, spectral index, surface brightness and distance data for ninety seven supernova remnants[1]

Name	Galactic source number	$\alpha(1950)$ h m s	$\delta(1950)$ ° ′	Angular size $\varphi_1 \times \varphi_2$ (min arc)	Flux density at 1 GHz S_0 (f.u.)	Spectral index at 1 GHz α	Surface brightness at 1 GHz[2] Σ ($W\,m^{-2}\,Hz^{-1}\,sr^{-1}$)	Distance from Sun d (kpc)
CTA 1	G 119.5 + 10.0	00 02 35	+72 20	125′ diam.	38	−0.2	2.89 [−22]	
Tycho's Nova A.D. 1572, 3C 10	G 120.1 + 1.4	00 22 28	+63 51.9	6.0 × 7.0	58	−0.74	1.64 [−19]	5.0
HB 3	G 132.4 + 2.2	02 14	+63 15	80′ diam.	36	−0.7	6.70 [−22]	1.7
CTB 13	G 156.4 − 1.2	04 24	+47 00	300 × 120	225	−0.5	7.43 [−22]	
HB 9	G 160.5 + 2.8	04 57 30	+46 30	120 × 140	150	−0.35	1.06 [−21]	
VRO 42.05.01	G 166.2 + 4.3	05 23 06	+42 50	45′ diam.	6.6	(−0.5)	3.88 [−22]	
OA 184	G 166.3 + 2.4	05 15	+41 40	75′ diam.	10.0	−0.5	2.12 [−22]	
S 147	G 180.0 − 1.7	05 36	+28 00	180′diam.	120	−0.3?	4.41 [−22]	
Crab Neb. A.D. 1054, Tau A	G 184.6 − 5.8	05 31 31	+21 59.0	3.0 × 4.2	1,000	−0.25	9.45 [−18]	2.02
IC 443	G 189.1 + 2.9	06 14	+22 30	40′ diam.	160	−0.45	1.19 [−20]	1.5 to 2.0
PKS 0607 + 17	G 193.3 − 1.5	06 06	+16 40	80′ diam.	27	−0.5	5.03 [−22]	
Monoceros Neb.	G 205.5 + 0.2	06 35	+06 30	210′ diam.	150	−0.5	4.05 [−22]	
Puppis A	G 260.4 − 3.4	08 20 30	+42 50	55′ diam.	145	−0.5	5.70 [−21]	
PKS 0902 − 38	G 261.9 + 5.5	09 02 22	−38 29	35′ diam.	10.0	−0.38	9.70 [−22]	
Vela X, Y, Z	G 263.4 − 3.0	08 32	−45 00	220 × 180	1,800	−0.30	5.40 [−21]	0.5
MSH 10 − 53	G 284.2 − 1.8	10 15 40	−58 40.5	23 × 12	25	−0.46	1.05 [−20]	
	G 289.1 − 0.4	10 54 28	−59 49.6	3.0 × 3.2	20	−0.65	2.48 [−19]	
MSH 11 − 61 B[3]	G 289.9 − 0.8	10 59 02	−60 33.8	4.1 × 5.2	15	−0.6	8.37 [−20]	
MSH 11 − 61 A[3]	G 290.1 − 0.8	11 00 45	−60 38.0	13 × 12	80	−0.6	6.10 [−20]	
MSH 11 − 62	G 291.0 − 0.1	11 09 49	−60 22.0	4.2 × 5.4	30	−0.65	1.57 [−19]	
MSH 11 − 54	G 292.0 + 1.8	11 22 21	−58 59.4	2.8 × 2.7	15	−0.36	2.37 [−19]	
Kes 16	G 295.1 − 0.6	11 40 57	−62 10.4	13 × 14	29	−0.6	1.87 [−20]	
PKS 1209 − 51/52	G 296.3 + 10.0	12 06	−52 10	86 × 75	49	−0.52	9.05 [−22]	
Kes 17	G 304.6 + 0.1	13 02 39	−62 26.7	5.2 × 6.1	15	−0.50	5.62 [−20]	

	G 307.1 + 1.2	13 22 59	−61 07.4	1.3 × 0.1	13	−0.72	1.19 [−17]	
	G 307.6 − 0.3	13 29 08	−62 32	3.7′ diam.	24	−0.45	2.09 [−19]	
13S6A B[3]	G 309.6 + 1.7	13 42 42	−60 13.9	6.3′ diam.	47	−0.60	1.41 [−19]	
13S6A A[3]	G 309.7 + 1.8	13 43 33	−60 08.3	7.1 × 5.4	83	−0.60	2.58 [−19]	
Kes 20B[3]	G 310.6 − 0.3	13 54 41	−61 56.6	7.5′ diam.	9	−0.6	1.91 [−20]	
Kes 20A[3]	G 310.8 − 0.4	13 56 32	−62 00.8	8 × 11	20	−0.35	2.71 [−20]	
MSH 14−63, RCW 86	G 315.4 − 2.3	14 39 08	−62 15	40′ diam.	33	−0.5	2.46 [−21]	2.5
MSH 14−57	G 316.3 + 0.0	14 37 43	−59 47	17′ diam.	42	−0.5	1.73 [−20]	
MSH 15−52A, Kes 23[3]	G 320.4 − 1.0	15 09 39	−58 49	8′ diam.	40	−0.6	7.44 [−20]	
MSH 15+52B[3]	G 320.3 − 1.4	15 11 20	−59 07	14.5 × 12.5	25	(−0.5)	1.65 [−20]	
Kes 24	G 322.3 − 1.2	15 23 03	−57 55.6	2.6′ diam.	4.3	−0.7	7.56 [−20]	
MSH 15−56	G 326.2 − 1.7	15 48 26	−56 02.8	11 × 8.6	145	−0.24	1.84 [−19]	
Kes 27	G 327.4 + 0.4	15 45 24	−53 39.9	10′ diam.	26	−0.78	3.10 [−20]	
PKS 1459−41, SN 1006 A.D.	G 327.6 + 14.5	15 00 00	−41 45	30 × 22	25	−0.63	4.52 [−21]	1.3
MSH 15−57, Kes 29	G 328.4 + 0.2	15 51 47	−53 08.4	3.7′ diam.	18	−0.7	1.56 [−19]	
Lupus Loop	G 330.0 + 15.0	15 09	−40	270′ diam.	340	−0.3	5.56 [−22]	
	G 332.0 + 0.1	16 10 32	−50 51	—	—	—	—	
MSH 16−51, Kes 32	G 332.5 + 0.1	16 11 48	−50 33.2	22′ diam.	25	−0.45	6.15 [−21]	
PKS 1613−50, RCW 103	G 332.4 − 0.4	16 13 44	−50 56.0	7.0 × 7.9	22	−0.34	4.73 [−20]	3.9
	G 336.7 + 0.5	16 28 30	−47 16.0	–	—	—	—	
Part of CTB 33, MHR 55, Kes 39	G 337.0 − 0.1	16 32 10	−47 30	9.0 × 4.0	31	−0.8	1.02 [−19]	
Kes 40	G 337.3 + 1.0	16 29 06	−46 29.5	11′ diam.	15	−0.2	1.48 [−20]	
Kes 41	G 337.8 − 0.1	16 35 22	−46 51.9	5.0 × 4.0	15	−0.2	8.93 [−20]	
MSH 16−48, MHR 58, Kes 45	G 342.1 + 0.1	16 50 11	−43 30.3	30′ diam.	54	−0.5	7.15 [−21]	
CTB 37A[3]	G 348.5 + 0.1	17 11 12	−38 26.7	8.1 × 5.3	84	−0.5	2.33 [−19]	
CTB 37B[3]	G 348.7 + 0.3	17 10 47	−38 07.7	3.8 × 6.0	47	−0.5	2.44 [−19]	
	G 349.7 + 0.2	17 14 37	−37 23.3	1.9 × 2.6	23	−0.6	5.54 [−19]	
NGC 6383	G 355.2 + 0.1	17 30 09	−32 52.7	7.5 × 9.5	30	−0.6	5.01 [−20]	
MSH 17−39	G 357.7 − 0.1	17 37 05	−30 56.9	4.4 × 3.4	38	−0.6	3.02 [−19]	
	G 359.4 − 0.1	17 41 24	−29 24.0	5.7 × 6.7	29	−0.4	9.05 [−20]	
Kepler's Nova A.D. 1604, 3C 358	G 4.5 + 6.8	17 27 43	−21 26.0	2.2′ diam.	20.0	−0.58	4.92 [−19]	6.7 to 10.0

Table 53 (continued)

Name	Galactic source number	$\alpha(1950)$ h m s	$\delta(1950)$ ° ′	Angular size $\varphi_1 \times \varphi_2$ (min arc)	Flux density at 1 GHz S_0 (f.u.)	Spectral index at 1 GHz α	Surface brightness at 1 GHz[2] Σ ($W m^{-2} Hz^{-1} sr^{-1}$)	Distance from Sun d (kpc)
A 4	G 5.3 − 1.1	17 58 44	−24 54	12 × 18	38	−0.3	2.06 [−20]	
A 1, W 28	G 6.5 − 0.1	17 57 30	−23 25	30′ diam.	300	−0.42	3.97 [−20]	
	G 11.2 − 0.4	18 08 32	−19 26.9	2.7 × 3.7	13	−0.2	1.55 [−19]	
MSH 18 − 18, Kes 67	G 18.9 + 0.3	18 21 21	−12 22.2	7.0 × 14	35	−0.57	4.25 [−20]	
MSH 18 − 113, Kes 69	G 21.8 − 0.5	18 30 16	−10 13.0	16′ diam.	64	−0.70	2.97 [−20]	
W 41, Kes 70	G 23.1 + 0.0	18 30 38	−08 48	45 × 60	350	−0.22	1.54 [−20]	
	G 27.3 + 0.0	18 38 30	−05 01	32′ diam.	41	−0.4	4.77 [−21]	
Kes 75, 4C −03.70	G 29.7 − 0.2	18 43 49	−03 02.1	2.2 × 1.2	7.3	−0.5	3.29 [−19]	
PKS 1846 − 00, 3C 391	G 31.9 + 0.0	18 46 47	−00 58.7	3.5′ diam.	22	−0.50	2.13 [−19]	
3C 396.1	G 32.0 − 4.9	19 04 32	−03 06.5	60′ diam.	19	−0.45	6.28 [−22]	
	G 33.0 + 0.1	18 48 50	−00 05	31 × 10	25	−0.4?	9.61 [−21]	
4C +00.70	G 33.7 + 0.0	18 50 04	+00 35.9	7.3 × 5.6	12	−0.5	3.49 [−20]	
W 44, 3C 392	G 34.6 − 0.5	18 53 45	+01 13.0	28 × 35	230	−0.40	2.79 [−20]	
	G 35.6 − 0.0	18 55 18	+02 06	10′ diam.	25	−0.3	2.98 [−20]	
W 47, CTB 64	G 37.6 − 0.1	18 57 42	+04 04	30′ diam.	48	−0.6	6.35 [−21]	
NRAO 593, Part of 3C 396, CTB 65	G 39.2 − 0.3	19 01 31	+05 22.5	4.6′ diam.	16	−0.3	9.00 [−20]	
W 50, CTB 69	G 39.7 − 2.0	19 08 24	+05 04	50′ diam.	48	−0.7	2.29 [−21]	
3C 397	G 41.1 − 0.3	19 05 05	+07 03.6	10′ × 8′	17	−0.3	2.53 [−20]	
CTB 72	G 41.9 − 4.1	19 20	+06 00	165 × 140	150	−0.5?	7.74 [−21]	
W 49 B	G 43.3 − 0.2	19 08 39	+09 00.5	5.4 × 3.0	33	−0.33	2.43 [−19]	14.0
Kuzmin 47, Part of CTB 70	G 45.5 + 0.1	19 11 59	+11 04.3	3.9 × 2.9	23	−0.4	2.42 [−19]	
	G 46.8 − 0.3	19 15 48	+12 06	12′ × 5′	18	−0.5	3.57 [−20]	

CTB 63	G 47.6 + 6.1	18 54	+15 45	60′ diam.	12	(−0.5)	3.97 [−22]	
Part of W 51	G 49.0 − 0.3	19 20	+14 00	24′ diam.	180	−0.25	3.72 [−20]	
3C 400.2	G 53.7 − 2.2	19 36 20	+17 13	20′ diam.	8.5	−0.6	2.53 [−21]	
Cygnus Loop	G 74.0 − 8.6	20 49	+30 30	200 × 160	180	−0.45	6.70 [−22]	0.77
	G 74.8 + 0.6	20 15 49	+36 36.1	4.7 × 3.9	(10)	(−0.5)	6.50 [−20]	
	G 74.9 + 1.2	20 14 04	+37 03.8	9.3 × 4.7	(18)	(−0.5)	4.89 [−20]	
W 66, CTB 91, DR 4	G 78.1 + 1.8	20 20 44	+40 02.3	30 × 20	230	−0.7	4.56 [−20]	
DR 3	G 78.3 + 2.5	20 18 16	+40 36.5	13.3 × 9.5	18	−0.2	1.71 [−20]	
DR 12	G 78.5 − 0.1	20 29 31	+39 13.1	9.5 × 8.0	8	−0.2	1.25 [−20]	
	G 78.6 + 1.0	20 25 25	+39 56.1	12 × 15	54	−0.5	3.57 [−20]	
DR 1	G 78.9 + 3.7	20 14 37	+41 45.1	37 × 38	480	−0.2	1.52 [−20]	
DR 11	G 79.8 + 1.2	20 28 32	+41 04.3	13 × 29	39	−0.2	1.23 [−20]	
W 63	G 82.2 + 5.4	20 17 15	+45 24.6	80 × 55	130	−0.5	3.52 [−21]	
HB 21	G 89.1 + 4.7	20 45	+50 30	120′ diam.	175	−0.15	1.44 [−21]	
CTB 104	G 93.6 − 0.3	21 27 11	+50 30	54′ diam.	45	−0.69	1.83 [−21]	
Cas A	G 111.7 − 2.1	23 21 11	+58 32.8	4.0 × 3.8	3,000	−0.72	2.34 [−17]	3.4
CTB 1	G 117.3 + 0.1	23 59 48	+62 11	130′ diam.	55	−0.5	3.87 [−21]	
N 49 { 3 SNR's	L.M.C.	05 26 00	−66 08.0	1.12′ diam.	3.5	−1.01	3.33 [−19]	55
N 63A { in	L.M.C.	05 35 39	−66 03.5	0.45′ diam.	1.9	−0.50	1.12 [−18]	55
N 132D { L.M.C.	L.M.C.	05 25 32	−69 41.0	0.37′ diam.	4.6	−0.50	4.00 [−18]	55

[1] After Milne (1970) by permission of the Commonwealth Scientific and Industrial Research Organization. The flux densities are in flux units, where 1 f.u. $= 10^{-23}$ erg sec^{-1} cm^{-2} Hz^{-1}.

[2] For each value, the number in square brackets gives the power of 10 of the multiplier.

[3] Possibly components of the one source.

where the distance of the shock, X_s, after time, t, is given by

$$X_s = X_0 \left(\frac{t}{t_0}\right)^{4/3} . \tag{4-455}$$

Eq. (4-454) indicates that shock velocities $V \approx c$ may be achieved as a shock propagates outwards through the density gradient of the stellar mantle. Such a process might account for the acceleration of many moderately relativistic cosmic rays. Nevertheless, high energy electrons have synchrotron lifetimes much shorter than the lifetime of a supernova. If the supernova has been radiating by the synchrotron emission mechanism over its lifetime, then the electrons must be continuously accelerated and cannot be the survivors of an original explosion.

4.5.3.3. *The Origin of High Energy Photons*

Following the prediction that cosmic X-rays and γ-rays might be observed (MORRISON, 1958), several groups have observed a diffuse, isotropic, extragalactic X and γ ray background together with discrete, galactic X-ray sources. The differential energy spectrum of the diffuse background, shown in Fig. 31, is given by (FRIEDMAN, 1973)

$$\begin{aligned}
&J(E) \propto E^{-1} \text{ to } E^{-2} && (\text{cm}^2 \text{ sec ster keV})^{-1} && \text{for } 100 \text{ eV} \le E \le 1 \text{ keV} \\
&J(E) \propto E^{-0.4} && (\text{cm}^2 \text{ sec ster keV})^{-1} && \text{for } 1 \text{ keV} \le E \le 10 \text{ keV} \\
&J(E) \propto E^{-0.75} && (\text{cm}^2 \text{ sec ster keV})^{-1} && \text{for } 20 \text{ keV} \le E \le 40 \text{ keV} \\
&J(E) \propto E^{-1.3} && (\text{cm}^2 \text{ sec ster keV})^{-1} && \text{for } 40 \text{ keV} \le E \le 1 \text{ MeV} \\
&J(E) \propto E^{-2} && (\text{cm}^2 \text{ sec ster keV})^{-1} && \text{for } 1 \text{ MeV} \le E \le 10 \text{ MeV},
\end{aligned} \tag{4-456}$$

or in integral form (SILK, 1973)

$$\begin{aligned}
&dJ(E) = 90 && \text{keV}(\text{cm}^2 \text{ sec ster keV})^{-1} && \text{for } E = 0.25 \text{ keV} \\
&dJ(E) = 12E^{-0.5} && \text{keV}(\text{cm}^2 \text{ sec ster keV})^{-1} && \text{for } 1 \text{ keV} \le E \le 20 \text{ keV} \\
&dJ(E) = 40E^{-1.2} && \text{keV}(\text{cm}^2 \text{ sec ster keV})^{-1} && \text{for } 20 \text{ keV} \le E \le 1000 \text{ keV}.
\end{aligned}$$

The diffuse X-rays between 10 and 100 keV are isotropic to better than five per cent over angular sizes of ten degrees (SCHWARTZ, 1970), and are probably extragalactic. The soft X-ray flux below 1 keV rises in intensity towards the galactic pole, which may be due to absorption effects.

HENRY *et al.* (1968) have proposed that the steep spectrum below 1 keV may be due to the thermal bremsstrahlung of a hot, ionized intergalactic gas with a density of 10^{-5} to 10^{-6} g cm^{-3} and a temperature of 3 to 8×10^5 °K. The specific intensity, I, received from the galactic plane when the optical depth is greater than unity is given by

$$I = \frac{S}{4\pi N_H \sigma}, \tag{4-457}$$

where N_H is the neutral hydrogen column density along the line of sight, the absorption cross section, σ, for cosmic abundances is given by (BROWN and GOULD, 1970)

$$\sigma \approx 0.7 \times 10^{-22} \left(\frac{h\nu}{1 \text{ keV}}\right)^{-3} \text{cm}^2 \quad \text{up to } 532 \text{ eV},$$

where the photon energy is $h\nu$, and the source function, S, for the X-ray free-free emissivity depends on the cosmological model, the temperature, T_0, of the intergalactic gas, and the density, N_0, of intergalactic matter. If H_0 is the Hubble parameter, then the specific intensity, I_0, in the absence of absorption by our

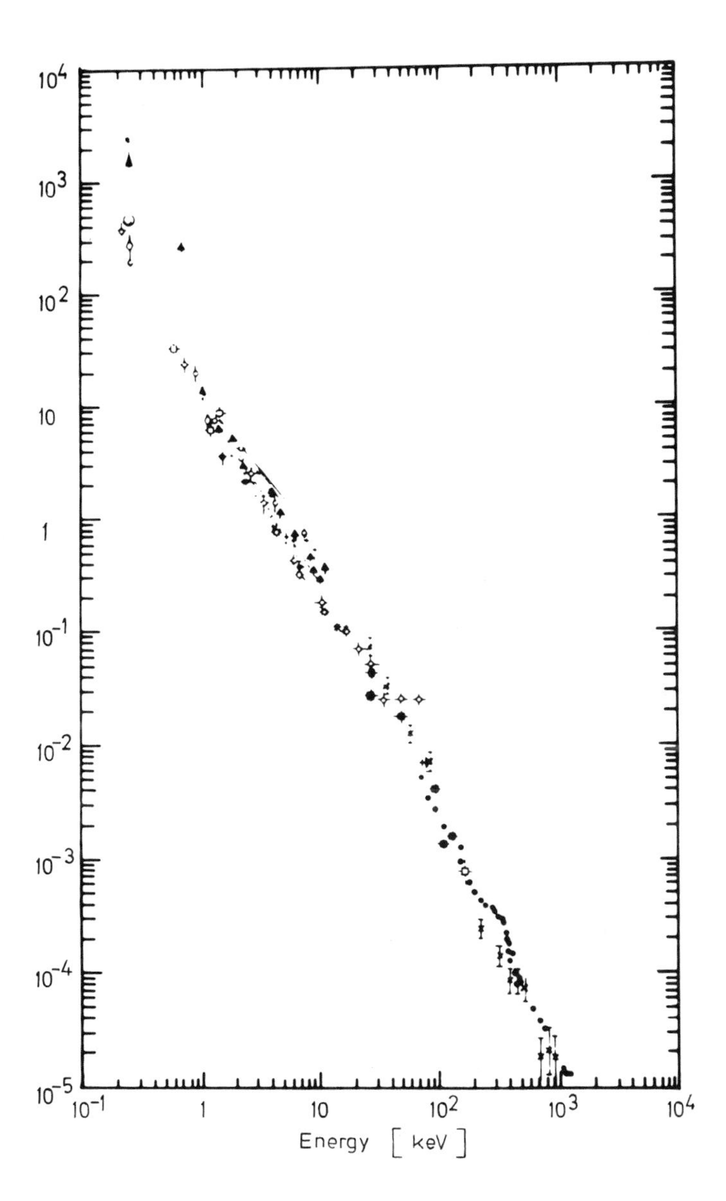

Fig. 31. The differential energy spectrum of the diffuse X- and γ-ray background from 0.27 keV to 1000 keV (after ODA, 1970, by permission of the International Astronomical Union). The data in the energy range 1–100 keV are represented by a power law spectrum $E^{-1.7 \pm 0.2}$, where E is the energy. Each symbol denotes a different set of observations, and these are referenced by ODA

Galaxy is (FIELD and HENRY, 1964; WEYMANN, 1966, 1967)

$$I_0 \approx 5.44 \times 10^{-39} N_0^2 T_0^{-1/2} c H_0^{-1} \text{ erg cm}^{-3} \text{ sec}^{-1} \text{ ster}^{-1} \text{ Hz}^{-1},$$

where

$$N_0 = \frac{3 H_0^2 q_0}{4\pi G m_H} \approx 2.2 \times 10^{-5} q_0 \text{ cm}^{-3}, \tag{4-458}$$

and q_0 is the deceleration parameter. For the steady-state case $q_0 = 0.5$ in the formula for N_0.

The numerical coefficient of 5.44×10^{-39} erg cm^{-3} sec^{-1} ster^{-1} Hz^{-1} takes on the equivalent value of 8.2×10^{-13} keV (cm^3 sec ster keV)$^{-1}$. In Eq. (4-458), the I_0 is the intensity of the thermal bremsstrahlung of a slab of gas of thickness $c H_0^{-1}$; and $N_0^2 \approx C \Sigma \langle N_e \rangle \langle N_i \rangle Z_i^2$ where the clumpiness factor $C = \langle N^2 \rangle / \langle N \rangle^2$, the $\langle N_e \rangle$ and $\langle N_i \rangle$ denote, respectively, the average values of the electron density and ion density along the line of sight, and the ith ion has charge $e Z_i$. Detailed formulae for the thermal bremsstrahlung of a hot, ionized, intergalactic gas depend on the cosmological model (cf. FIELD and HENRY, 1964; WEYMANN, 1966, 1967; BENJAMINI, LONDRILLO, and SETTI, 1967; and BERGERON, 1969). For example, a comparison of the observed X-ray spectrum in the energy range 1—20 keV with theoretical models gives the constraint (SILK, 1973)

$$\langle g \rangle C \left(\frac{\rho}{\rho_{\text{crit}}}\right)^2 \left(\frac{H_0}{50 \text{ km sec}^{-1} \text{ Mpc}^{-1}}\right)^3 f \lesssim 0.38 \left(\frac{E}{kT}\right)^{-1/2} \exp\left(\frac{E}{kT}\right), \tag{4-459}$$

where $\langle g \rangle \approx 1$ is the average Gaunt factor along the line of sight, the clumpiness factor $C = \langle \rho^2 \rangle / \langle \rho \rangle^2$, the intergalactic gas mass density is ρ, the critical mass density for a closed universe is $\rho_{\text{crit}} = 3 H_0^2 / (8\pi G)$, the Hubble constant is H_0, the energy of the X-ray under consideration is E, the present value for the temperature of the intergalactic gas is T, and

$$f = \int_0^z \frac{(1+z)^{5/2 - 3/2\gamma}}{[1 + (\rho z / \rho_{\text{crit}})]^{1/2}} [1 - (1+z)^{4-3\gamma}] \exp\left(\frac{E}{kT}\right) dz,$$

where γ is the adiabatic index, the redshift is z, and $f \approx z$ for $z \ll 1$. As mentioned previously, FIELD and HENRY (1964) have shown that the soft X-ray emission (100 eV to 1 keV) may be due to the thermal bremsstrahlung of intergalactic gas with $\rho \approx 10^{-6}$ g cm^{-3} and $T \approx 10^6$ °K. Evidence against this interpretation comes from the lack of absorption when X-rays are viewed through the small Magellanic cloud (MCCAMMON *et al.*, 1971). COWSIK and KOBETICH (1972) have revived the hot intergalactic gas hypothesis by showing that the background X-ray spectrum in the energy range 10 keV to 1 MeV fits the bremsstrahlung spectrum of an intergalactic gas with $\rho = 3 \times 10^{-6}$ g cm^{-3} and $T = 3 \times 10^8$ °K if $H_0 = 50$ km sec^{-1} Mpc^{-1}.

Models which generate a diffuse, isotropic X-ray background by the Compton scattering of the photons of the universal 2.7 °K microwave background on cosmic ray electrons have been proposed by FELTEN and MORRISON (1966) and BRECHER and MORRISON (1969). The popularity of this model was partly due to the rough agreement between the spectral slope of the high energy X-rays

(50 to 300 keV) with that of the electrons observed in our Galaxy. In the collision of a fast electron of total energy $E=\gamma m c^2$ with a low energy photon of energy $\varepsilon=h\nu$, the energy, ε_c, of the scattered photon is given by (cf. Eq. (4-371))

$$\varepsilon_c = 4\gamma^2 \varepsilon/3 \quad \text{for } \gamma h\nu \ll mc^2, \tag{4-460}$$

and the Compton power scattered by one electron is (cf. Eq. (4-419))

$$P_c = \tfrac{4}{3}\gamma^2 \sigma_T c\, U = 2.66\times 10^{-14}\gamma^2 U \text{ eV sec}^{-1} \quad \text{for } \gamma h\nu \ll mc^2, \tag{4-461}$$

where $\sigma_T \approx 6.65\times 10^{-25}$ cm^2 is the Thomson scattering cross section, and the energy density of the isotropic radiation field, U, is given in eV cm^{-3} in the numerical approximation. The lifetime, τ_c, of an electron against loss by inverse Compton radiation is

$$\tau_c \approx \frac{\gamma m c^2}{P_c} \approx 2\times 10^{19}\gamma^{-1} U^{-1} \text{ sec}, \tag{4-462}$$

where again U is taken to be in units of eV cm^{-3}.

Values of the energy density, U, of electromagnetic radiation in various spectral regions were illustrated in Fig. 29. The distribution of relativistic electrons, $N(\gamma)$, is normally taken to be power law of spectral index, m,

$$N(\gamma)d\gamma = N_0 \gamma^{-m} d\gamma. \tag{4-463}$$

The low energy photons are assumed to have a black body energy distribution with average energy $\langle\varepsilon\rangle = 2.7kT$, where T is the black body temperature. Assuming that both the electron and photon distributions are homogeneous and isotropic over the region of interaction, then the specific intensity, I_c, of the Compton scattered radiation is given by (Felten and Morrison, 1966)

$$\begin{aligned} I_c(h\nu) &\approx 1.06\times 10^{-15}(3.66\times 10^{-6})^{3-m} N_0 D\, U\, T^{(m-3)/2} \nu^{(1-m)/2} \text{ erg sec}^{-1} \text{ cm}^{-2} \\ &\qquad\qquad \text{ster}^{-1} \text{ Hz}^{-1} \\ &\approx 10^3 (56.9)^{3-m} N_0 D\, U\, T^{(m-3)/2} (h\nu)^{(1-m)/2} \text{ eV (cm}^2 \text{ sec ster eV)}^{-1}, \end{aligned} \tag{4-464}$$

where D is the distance to the source, the parameters are in c.g.s. units in the first numerical approximation, and in the second numerical approximation N_0 is in cm^{-3}, D is in light years (1 l.y. $= 9.460\times 10^{17}$ cm), U is in eV cm^{-3}, T is in °K and $h\nu$ is in eV. The inverse Compton model requires an ultrarelativistic electron density of 10^{-4} eV cm^{-3} throughout intergalactic space to provide the observed background by scattering with the 2.7 °K black body photon radiation of $U \approx 0.3$ eV cm^{-3}.

If the relativistic electrons also give rise to synchrotron radiation at frequency, ν_s, then the energy ε_c, of the inverse Compton photon is given by (Felten and Morrison, 1966)

$$\varepsilon_c \approx 0.9\times 10^2 \frac{T\nu_s}{H}, \tag{4-465}$$

where ε_c is in eV, the magnetic field intensity, H, in the synchrotron source is in microgauss, the temperature, T, of the low energy photons is in °K, and the frequency, ν_s, is in megacycles. The ratio of the synchrotron radiation intensity,

I_s, and the inverse Compton radiation intensity, I_c, is given by (FELTEN and MORRISON, 1966)

$$\frac{I_s}{I_c} = \frac{H^2}{8\pi U}\left(\frac{2\times 10^4 T}{H}\right)^{(3-m)/2} \tag{4-466}$$

in cgs units. Here the thermal photons have temperature, T, and energy density, U, and the electron spectral index is m. The synchrotron radiation would have a frequency spectral index of $(1-m)/2$. Estimates of the energy density of the electromagnetic radiation in different spectral regions were shown in Fig. 29 for the Galaxy and the metagalaxy.

If radio galaxies radiate for 10^9 years and produce about 10^{60} ergs of relativistic electrons, then they would provide a relativistic electron energy density of about 10^{-5} eV cm^{-3}.

A detailed comparison of the observed X-ray spectrum with the theory of inverse Compton scattering requires a knowledge of the angular distribution of the 2.7 °K black body photons, the Planck number density distribution of photon energies,

$$N(\varepsilon) = \frac{8\pi}{h^3 c^3}\frac{\varepsilon^2}{\left[\exp\left(\frac{\varepsilon}{kT}\right)-1\right]} \tag{4-467}$$

for photon energy, ε, and $T = 2.7$ °K, and the energy spectrum of the Compton scattering process (cf. COWSIK and KOBETICH, 1972). The exact expression for the differential Compton scattering cross section is (KLEIN and NISHINA, 1929)

$$\frac{d\sigma}{d\Omega} = \frac{r_0^2(1+\cos^2\theta)}{2[1+2q\sin^2(\theta/2)]^2}\left\{1 + \frac{4q^2\sin^4(\theta/2)}{[1+\cos^2\theta][1+2q\sin^2(\theta/2)]}\right\} \text{cm}^2\,\text{ster}^{-1}, \tag{4-468}$$

where $r_0 = e^2/(mc^2) \approx 2.818\times 10^{-13}$ cm is the classical electron radius, θ is the deflection angle for the photon, and $q = \varepsilon/(mc^2)$ is the incident photon energy in units of mc^2. The scattering cross section for the emission of a X-ray photon of energy, ε_x, in the collision of an electron of energy, E, with a photon of energy, ε, integrated over the angular distribution of incoming and outgoing photons, is (BLUMENTHAL and GOULD, 1970)

$$\sigma = \frac{8DB}{E^2\varepsilon}(1+x-2x^2+2x\ln x) \quad \text{for } x \lesssim 1, \tag{4-469}$$

where

$$B = (mc^2)^2, \quad D = \pi r_0^2/4, \quad \text{and } x = B\varepsilon_x/(4E^2\varepsilon).$$

It is also possible that a substantial fraction of the isotropic X-ray background is due to the superposition of discrete X-ray sources. For a discrete source of X-ray luminosity, L, and space density, N, the observed X-ray background flux will be (SILK, 1973)

$$F = \frac{cNL}{4\pi H_0}[g(z)-g(o)]\ \text{keV}\,(\text{cm}^2\,\text{sec ster})^{-1}, \tag{4-470}$$

where it is assumed that the discrete source spectrum is flat over the energy range 2 keV to $10(1+z)$ keV, the F is for the energy range 2 to 10 keV, and

$$g(z) = \frac{1}{(1-\Omega)^{1/2}} \ln\left[\frac{(1+\Omega z)^{1/2}-(1-\Omega)^{1/2}}{(1+\Omega z)^{1/2}+(1+\Omega)^{1/2}}\right] \quad \text{if } \Omega<1$$

$$= -2(1+z)^{-1/2} \quad \text{if } \Omega=1$$

and

$$= \frac{1}{(\Omega-1)^{1/2}} \tan^{-1}\left[\frac{1+\Omega z}{\Omega-1}\right]^{1/2} \quad \text{if } \Omega>1$$

where $\Omega=\rho/\rho_{\text{crit}}$ is the ratio of the intergalactic gas mass density to the critical value $\rho_c=3H_0^2/(8\pi G)$. The space densities of small galaxies, normal galaxies, radio galaxies, Seyfert galaxies, clusters of galaxies, and quasars are, respectively, $10N_0$, N_0, $10^{-3}N_0$, $0.02N_0$, 10^{-5} to $5\times10^{-6}N_0$, and $3\times10^{-8}N_0$, where $N_0=0.03\ \text{Mpc}^{-3}$.

Discrete X-ray sources have been catalogued by GIACCONI *et al.* (1972), and many of these sources coincide in position with well known astronomical objects whose positions and other parameters are given in the various tables of this book. The discrete X-ray and gamma ray sources may be devided into the following representative categories.

Radio galaxies and quasars which are X-ray sources (SILK, 1970, 1973)

Source	Radio luminosity (0.3 mm to 30 m)	X-ray luminosity (2 to 10 keV)
NGC 1275 (Perseus A)		4×10^{45} erg sec^{-1}
NGC 4151 (Seyfert)	1×10^{41} erg sec^{-1}	2×10^{41}
NGC 4486 (M 87)	1×10^{41}	3×10^{43}
NGC 5158 (Centaurus A)	2×10^{41}	8×10^{41}
3C 273 (Quasar)	2×10^{46}	7×10^{45}

Clusters of galaxies (SILK, 1973)

Cluster	X-ray luminosity (2 to 10 keV)
Virgo	1.5×10^{43} erg sec^{-1}
Centaurus	$4\ \times10^{43}$
Coma	$5\ \times10^{44}$
Perseus	$1\ \times10^{45}$
Abell 2256	$8\ \times10^{44}$

Supernova remnant X-ray sources (FRIEDMAN, 1973)

Supernova remnant	Age (Years)	X-ray luminosity (erg sec^{-1})	Energy range (keV)
Crab Nebula	900	10^{37}	1—200
Cassiopeia A	300	5×10^{36}	1—10
Tycho 1572	400	5×10^{36}	1—10
Puppis A	10^4—10^5	10^{36}	0.2—3
Vela X, Y, Z	10^4—10^5	10^{36}	0.2—3
Cygnus Loop	10^4—10^5	2×10^{36}	0.2—1
IC 443		2×10^{34}	2—10
MSH 15-52A		5×10^{35}	2—10

Normal and small galaxies which are X-ray sources (SILK, 1973)

Galaxy	X-ray luminosity (2—10 keV)
LMC	4×10^{38} erg sec^{-1}
SMC	1×10^{38}
M 31	3×10^{39}
Our Galaxy	5×10^{39}

Variable and/or star like X-ray sources (FORMAN *et al.*, 1972; GIACCONI *et al.*, 1972; SCHREIER *et al.*, 1972; and TANANBAUM *et al.*, 1972)

Source	α(1950.0) h m s	δ(1950.0) ° ′ ″	Eclipse period (days)	Pulsation period (sec)
2U (0115-73)	01 15 02	−73 41 24	4.0	—
P 0531 + 21	05 31 31	21 58 55	—	0.033129
2U (0900-40)	09 00 19	−40 22 48	8.95	—
Centaurus X-3	11 19 00	−60 19 12	2.08712(4)	4.843496(7)
2U (1543-47)	15 43 50	−47 33 36	—	Nova like
Sco X-1	16 17 04.3	−15 31 13	—	—
Hercules X-1	16 56 42	35 25	1.70017(4)	1.237772(1)
2U (1700-37)	17 00 19	−37 48 00	3.4	—
Cygnus X-1	19 56 22.0	35 03 36	5.6	$\gtrsim 0.1$
Cygnus X-3	20 30 37.6	40 47 13	—	17,280.0
Cygnus X-2	21 42 36.91	38 05 27.9	—	—

The X-ray luminosity of the pulsar NP 0532 is 10^{36} erg sec^{-1} in the energy range 1 to 10^6 keV, whereas that for most of the eclipsing variables is 10^{36} to 10^{37} erg sec^{-1} in the energy range 1 to 10 keV.

Most recently, bursts of hard X-ray or gamma ray radiation in the energy range 0.1 to 1.2 MeV have been observed. These bursts have the following characteristics (KLEBESADEL, STRONG, and OLSON, 1973; CLINE, DESAI, KLEBESADEL, and STRONG, 1973; and WHEATON *et al.*, 1973)

Characteristics of γ ray bursts

Energy range	0.2 to 1.5 MeV
Duration	0.1 to 30 sec
Time integrated flux density	$\Phi = \int E \frac{dN}{dE} dE = 10^{-5}$ to 2×10^{-4} erg cm^{-2}
Energy spectrum	$\frac{dN}{dE} = I_0 \exp\left(\frac{-E}{E_0}\right)$ photons cm^{-2} keV^{-1} burst^{-1}

where $100\,\text{keV} \le E_0 \le 300\,\text{keV}$.

Discrete X-ray sources associated with radio galaxies, quasars, and, perhaps, clusters of galaxies and normal galaxies, might be explained by the inverse Compton mechanism. The appropriate equations relating photon energies, electron energies, electron distribution, synchrotron radiation, and X-ray radiation are given in Eqs. (4-460) to (4-466). Detailed inverse Compton formulae are given in Eqs. (4-467) to (4-469). The X-ray emission from clusters of galaxies and supernova remnants have a possible explanation in models which incorporate the bremsstrahlung emission of a hot gas. The free-free emission (bremsstrahlung) of an optically thin plasma of temperature, T, electron density, N_e, and radius, R, is given by (cf. Eq. (1-219))

$$I(\nu) = 5.4 \times 10^{-39} \langle g \rangle R N_e^2 T^{-1/2} \exp\left(\frac{-h\nu}{kT}\right) \text{erg sec}^{-1}\,\text{cm}^{-2}\,\text{ster}^{-1}\,\text{Hz}^{-1}, \tag{4-471}$$

$$I(E) = 800 \langle g \rangle \left(\frac{R}{10^{28}\,\text{cm}}\right)\left(\frac{N_e}{10^{-5}\,\text{cm}^{-3}}\right)^2 \left(\frac{T}{10^6}\right)^{-1/2} \exp\left(\frac{-h\nu}{kT}\right) \text{keV (cm}^2\ \text{sec ster keV})^{-1}$$

where $E = h\nu$ is the X-ray photon energy, and $\langle g \rangle \approx 1$ is the average Gaunt factor. As an example, cluster X-ray luminosities of 10^{43}—10^{45} erg sec^{-1} in the energy range $2\,\text{keV} \le E \le 10\,\text{keV}$ may be accounted for by a hot gas with $T = 10^7$ to 10^8 °K, $N_e = 10^{-4}$ to 10^{-3} cm^{-3}, and $R \approx 1$ Mpc.

Several of the variable X-ray sources have been identified with binary pairs in which accretion may take place by mass transfer from a visible primary (a blue supergiant) to an invisible secondary (a neutron star or a black hole). The primary star is large enough, and the pair is close enough, that the Roche critical equipotential surface is filled (cf. Eqs. (3-246) and (3-247)). For example, the physical parameters of the binary source for which Cen X-3 is the X-ray companion are given by (SCHREIER *et al.*, 1972)

$$415\,\text{km sec}^{-1} < v < 588\,\text{km sec}^{-1}$$

$$1.19 \times 10^{12}\,\text{cm} < \frac{A}{2} < 1.69 \times 10^{12}\,\text{cm} \tag{4-472}$$

$$30.7 \times 10^{34}\,\text{g} < \frac{M^3}{(M+m)^2} < 8.81 \times 10^{34}\,\text{g}$$

and

$$0.84 \times 10^{12}\,\text{cm} < R < \frac{A}{2}\left(1 + \frac{m}{M}\right),$$

where v is the orbital velocity, $A/2$ is the orbital radius in the center of mass system, R is the radius of the central occulting object, and m and M are the masses of the two objects. Similar data may be derived from the light curve for Hercules X-1. If i denotes the inclination of the orbit, T the orbital period, $\Delta\tau$ the half-amplitude of the period variations, and τ the average pulsation period, then the derived parameters are (TANANBAUM *et al.*, 1972)

	Cen X–3	Her X–1
$v \sin i = \Delta\tau c/T$	415.1 ± 0.4 km sec^{-1}	169.2 ± 0.4 km sec^{-1}
$A \sin i/2 = Tv \sin i/2\pi$	$1.191 \pm 0.001 \times 10^{12}$ cm	$3.95 \pm 0.01 \times 10^{11}$ cm
$\frac{M^3 \sin^3 i}{(m+M)^2} = \frac{4\pi^2 (A \sin i)^3}{8 T^2 G}$	$3.074 \pm 0.008 \times 10^{34}$ g	$1.69 \pm 0.01 \times 10^{33}$ g

The criterion for mass flow between the binary pair is (cf. Eq. (3-247))

$$\frac{R}{A} > 0.38 + 0.2 \log\left(\frac{M}{m}\right). \tag{4-473}$$

If a neutron star of mass, m, is travelling at a speed, v, through a gas of mass density, ρ, then material occupying the radius (cf. Eq. (3-415))

$$R_A = \frac{Gm}{v^2} = 1.3 \times 10^{14} \left(\frac{10 \text{ km sec}^{-1}}{v}\right)^2 \left(\frac{M}{M_\odot}\right) \text{cm} \tag{4-474}$$

falls inward upon the neutron star, and the mass accretion rate for spherically symmetric accretion is (cf. Eq. (3-416))

$$\frac{dm}{dt} = \pi \xi R_A^2 v \rho. \tag{4-475}$$

Here the factor, ξ, which is less then unity, is intended to correct for radiation pressure. The X-ray luminosity, L_x, scattered by electrons in an optically thin gas, produces an inverse square repulsive force which effectively reduces the gravity of the X-ray source (EDDINGTON, 1926).

$$\xi \approx \left[1 - 10^{-4.6} \left(\frac{L_x}{L_\odot}\right) \left(\frac{M_\odot}{m}\right)\right]^2, \tag{4-476}$$

where $L_\odot$ and $M_\odot$ denote, respectively, the luminosity and mass of the Sun. This effect places an upper limit to the X-ray luminosity provided by accretion

$$L_x < 10^{38} \text{ erg sec}^{-1}, \tag{4-477}$$

which is called the Eddington limit. The luminosity, L_x, provided by accretion is given by (cf. Eq. (3-424))

$$L = \left(\frac{dm}{dt}\right) \left(\frac{Gm}{R}\right). \tag{4-478}$$

Because the gas which is drawn to the compact companion has a high angular momentum, the accretion will not be spherically symmetric and gas will circulate in an accretion disc about the compact source. In this case, the appropriate velocity to use in equation (4-475) is the velocity

$$v \approx \left(\frac{2Gm}{R}\right)^{1/2} \approx 1.15 \times 10^{10} \left(\frac{m}{M_\odot}\right)^{1/2} \left(\frac{R}{10^6}\right)^{-1/2} \text{ cm sec}^{-1}. \quad (4\text{-}479)$$

Viscosity removes the angular momentum and heats the gas to provide the X-ray bremsstrahlung. The dynamics of gas flow are governed by both stellar rotation and the stellar magnetic field. For example, for a neutron star the accreting matter is channeled along the magnetic field lines towards the magnetic poles. In this case, the X-ray emission comes not from the disc, but from hot spots near the magnetic poles. Details for these kind of effects are discussed by Prendergast and Burbidge (1968), Pringle and Rees (1972), Davidson and Ostriker (1973), and Lamb, Pethick, and Pines (1973).

5. Astrometry and Cosmology

"There is not, in strictness speaking, one fixed star in the heavens, ... there can hardly remain a doubt of the general motion of all the starry systems, and consequently of the solar one amongst the rest."

Sir F. W. HERSCHEL, 1753

"A luminous star, of the same density as the Earth, and whose diameter should be two hundred and fifty times larger than that of the Sun, would not, in consequence of its attraction, allow any of its rays to arrive at us. It is therefore possible that the largest luminous bodies in the universe may, through this cause, be invisible."

P. S. LAPLACE, 1798

"The laws of physical phenomena must be the same for a fixed observer as for an observer who has a uniform motion of translation relative to him.... There must arise an entirely new kind of dynamics which will be characterized above all by the rule that no velocity can exceed the velocity of light."

H. POINCARÉ, 1904

5.1. Position

5.1.1. Figure of the Earth

For reference purposes, the earth's surface at sea level may be represented by revolving an ellipse of eccentricity, e, and major axis, a_e, about the polar axis. The flattening factor, f, or ellipticity, is related to the eccentricity, e, by the equation

$$e^2 = 2f - f^2 . \tag{5-1}$$

The flattening factor is alternatively called the ellipticity or the oblateness factor. The major axis, a_e, is the equatorial radius, which is related to the polar radius, a_p, by the equation

$$a_p = a_e(1-f) = a_e(1-e^2)^{1/2} , \tag{5-2}$$

and the mean radius, $\langle a \rangle$, is given by

$$\langle a \rangle = (a_e^2 a_p)^{1/3} . \tag{5-3}$$

Until 1964 the internationally accepted parameters for the reference ellipsoid were those of the Hayford ellipsoid (HAYFORD, 1910). For this reference, $a_e = 6.378388 \times 10^8$ cm and $f = 0.0033670 = 1/297.0$. The new system of astronomical constants (CLEMENCE, 1965) includes $a_e = 6.378160 \times 10^8$ cm and a derived value of $f = 0.0033529 = 1/298.25$.

For a homogeneous fluid ellipsoid of mass, M, and angular velocity ω, we have the relation (NEWTON, 1687)

$$\frac{a_e - a_p}{\langle a \rangle} = \frac{5}{4} \frac{\omega^2 \langle a \rangle^3}{GM}, \tag{5-4}$$

where the Newtonian gravitational constant

$$G = 6.668 \times 10^{-8} \text{ dyn cm}^2 \text{ gm}^{-2}.$$

For the earth we have the values

$$\begin{aligned} M &= 5.977 \times 10^{27} \text{ g} \\ \omega &= 7.292115 \times 10^{-5} \text{ rad sec}^{-1} \\ a_e &= 6.378160 \times 10^8 \text{ cm} \\ \langle a \rangle &= 6.371030 \times 10^8 \text{ cm}. \end{aligned} \tag{5-5}$$

The earth's surface may be represented as a geoid of constant potential, U, given by

$$U = V - \tfrac{1}{2}\omega^2 r^2 \cos^2 \varphi, \tag{5-6}$$

where r and φ are, respectively, the radius and geocentric latitude of a point on the surface, and the gravitational potential, V, satisfies Laplace's equation in polar coordinates (LAPLACE, 1782, 1816, 1817).

$$\nabla^2 V = \frac{1}{r^2} \frac{\partial}{\partial r}\left(r^2 \frac{\partial V}{\partial r}\right) + \frac{1}{r^2 \sin\theta} \frac{\partial}{\partial \theta}\left(\sin\theta \frac{\partial V}{\partial \theta}\right) + \frac{1}{r^2 \sin^2\theta} \frac{\partial^2 V}{\partial \lambda^2} = 0, \tag{5-7}$$

where $\theta = \pi/2 - \varphi$, and the term involving the longitude, λ, may be discounted by assuming rotational symmetry about the z axis. Laplace's equation (5-7) may then be solved to give (LEGENDRE, 1789, 1817)

$$V = -\frac{GM}{r} \sum_{n=0}^{\infty} J_n \left(\frac{a_e}{r}\right)^n P_n(\theta), \tag{5-8}$$

where the zonal harmonics, J_n, are dimensionless constants, and the $P_n(\theta)$ are Legendre polynomials. Assuming that the origin of the coordinate system is the center of mass, $J_1 = 0$, and taking $J_0 = 1$, Eq. (5-8) takes the approximate form

$$V \approx -\frac{GM}{r} + \frac{GM a_e^2 J_2}{2r^3}(3\sin^2\varphi - 1). \tag{5-9}$$

Under the assumption that U is everywhere a constant, it follows from Eqs. (5-6) and (5-9) that

$$f = \frac{a_e - a_p}{a_e} = \frac{3}{2} J_2 + \frac{1}{2} m, \tag{5-10}$$

where the ratio of centrifugal acceleration at the equator to the gravitational acceleration at the equator is given by

$$m = \frac{\omega^2 a_e^3}{GM}. \tag{5-11}$$

For the earth, we have $$m = 3.4678 \times 10^{-3}, \tag{5-12}$$
and satellite measurements give (CLEMENCE, 1965)
$$J_2 = 1.08270 \times 10^{-3}, \tag{5-13}$$
which results in
$$f = 3.35289 \times 10^{-3} = \frac{1}{298.25}. \tag{5-14}$$

The J_2 can be derived from observations of the secular motion of the right ascension of the ascending node, Ω, and the argument of the perigree, ω, of artificial satellites. To first order, we have (KRAUSE, 1956; LECAR, SORENSON, and ECKELS, 1959)

$$\begin{aligned} \frac{\partial \Omega}{\partial t} &= -3\pi \left(\frac{a_e}{a_o}\right)^2 \frac{\cos i_o J_2}{(1-e_o^2)^2} && \text{radians per period} \\ &\approx -9.97 \left(\frac{a_e}{a_o}\right)^{3/2} (1-e_o^2)^{-2} \cos i_o && \text{degrees per day} \end{aligned}$$

and (5-15)

$$\begin{aligned} \frac{\partial \omega}{\partial t} &= 6\pi \left(\frac{a_e}{a_o}\right)^2 \frac{(1-\frac{5}{4}\sin^2 i_o)}{(1-e_o^2)^2} J_2 && \text{radians per period} \\ &\approx 4.98 \left(\frac{a_e}{a_o}\right)^{3/2} (1-e_o^2)^{-2} (5\cos^2 i_o - 1) && \text{degrees per day,} \end{aligned}$$

where t is the time variable, $a_e \approx 6.378165 \times 10^8$ cm is the earth's radius, the semi-major axis, a_o, of the satellite orbit is related to the orbital period, P, and the anomalistic mean motion, n, by the equations

$$P = \frac{2\pi a_o^{3/2}}{(GM)^{1/2}}$$

and

$$n^2 a_o^3 = GM \left\{1 + \frac{3}{4} \frac{J_2(1-e_o^2)^{1/2}}{a_o^2(1-e_o^2)^2} (1 - 3\cos^2 i_o)\right\},$$

where for the earth the product of the gravitational constant, G, and the mass, M, is

$$GM = 3.986032 \times 10^{20} \text{ cm}^3 \text{ sec}^{-2},$$

or $\sqrt{GM} = 17.043570$ if the mean motion is in revolutions per day and the semi-major axis is in earth radii. The eccentricity, e_o, and inclination, i_o, of the satellite orbit are defined together with a, Ω, and ω in Sect. 5.4.2. For an artificial satellite these variables are defined with respect to the center of the earth, the equatorial plane of the earth, and the vertical at the time of launch.

Satellite measurements indicate that the harmonic coefficients, J_n, of Eq. (5-8) take on the values

$$J_2 = 1082.63 \times 10^{-6}, \qquad J_3 = -2.54 \times 10^{-6},$$

$$J_4 = -1.61 \times 10^{-6}, \qquad J_5 = -\underset{\pm 25}{0.210} \times 10^{-6},$$

$$J_6 = \underset{\pm 30}{0.646 \times 10^{-6}}, \qquad J_7 = \underset{\pm 39}{-0.333 \times 10^{-6}},$$

$$J_8 = \underset{\pm 50}{-0.270 \times 10^{-6}}, \qquad J_9 = \underset{\pm 60}{-0.053 \times 10^{-6}},$$

$$J_{10} = \underset{\pm 50}{-0.054 \times 10^{-6}}, \qquad J_{11} = \underset{\pm 35}{0.302 \times 10^{-6}},$$

$$J_{12} = \underset{\pm 47}{-0.357 \times 10^{-6}}, \qquad J_{13} = \underset{\pm 84}{-0.114 \times 10^{-6}},$$

$$J_{14} = \underset{\pm 63}{-0.179 \times 10^{-6}}.$$

Detailed formulae for measuring the higher coefficients are given by (LECAR, SORENSON, and ECKELS, 1959; O'KEEFE, ECKELS, and SQUIRES, 1959; KING-HELE, 1961; KOZAI, 1959, 1961, 1966), and in the book by KAULA (1966).

The surface gravitational acceleration, g, is given by

$$g = -\nabla U. \tag{5-16}$$

It follows from Eqs. (5-6), (5-9), and (5-16) that

$$g = g_e\left[1 + \left(\frac{5m}{2} - f\right)\sin^2\varphi\right], \tag{5-17}$$

where the equatorial gravity is given by

$$g_e = \frac{-GM}{a_e^2}\left(1 + \frac{3}{2}J_2 - m\right). \tag{5-18}$$

Eq. (5-17) is known as Clairaut's theorem after Clairaut's (1743) derivation of the result by assuming that the rotating body is a homogeneous fluid ellipsoid. By retaining higher terms, the following result is obtained

$$g = g_e\left[1 + \left(\frac{5}{2}m - f - \frac{17}{14}mf\right)\sin^2\varphi + \left(\frac{f^2}{8} - \frac{5}{8}mf\right)\sin^2 2\varphi\right]. \tag{5-19}$$

Using the values of f, m, J_2, and a_e given in the previous equations for the earth, we obtain,

$$g_e = 978.03090 \text{ cm sec}^{-2}, \tag{5-20}$$

and

$$g = (978.03090 + 5.18552\sin^2\varphi - 0.00570\sin^2 2\varphi) \text{ cm sec}^{-2}. \tag{5-21}$$

The surface of the earth geoid is specified by the equation

$$r = a_e(1 - f\sin^2\varphi). \tag{5-22}$$

Another measure of position on the earth is the astronomical latitude, φ, which is defined as the smallest angle between the direction of a plumb line at the point on the earth and the equatorial plane. It follows from Eq. (5-17) that the astronomical latitude at an elevation, h, is greater than that at sea level by the amount, $\Delta\varphi$, given by

$$\Delta\varphi = \frac{-h}{a_e g_e}\frac{\partial g}{\partial\varphi} \approx -\frac{h}{a_e}\left(\frac{5m}{2} - f\right)\sin 2\varphi$$
$$\approx -1.72\times 10^{-4} h \sin 2\varphi \text{ seconds of arc,} \tag{5-23}$$

where the numerical approximation is for h in meters, assuming that one radian = 206,264.806 seconds of arc.

When the astronomical latitude has been reduced to the sea level value, and corrected for local gravity anomalies, it is called the geodetic latitude, φ. The geocentric radial vector, ρ, and latitude, φ', are related to the geodetic latitude, φ, by the equations

$$\tan\varphi' = (1-f)^2 \tan\varphi \tag{5-24}$$

or

$$\varphi' - \varphi = -11'32.7430'' \sin 2\varphi + 1.1633'' \sin 4\varphi - 0.0026'' \sin 6\varphi$$

and

$$\rho = a_e(0.998327073 + 0.001676438 \cos 2\varphi$$
$$- 0.000003519 \cos 4\varphi + 0.000000008 \cos 6\varphi).$$

Series expressions relating h and φ to ρ and φ' are given by MORRISON and PINES (1961).

5.1.2. Coordinates on the Celestial Sphere

Celestial coordinates are defined by referring to the following points and great circles on the celestial sphere.

North and south celestial poles: The respective points of intersection of the celestial sphere with the northward and southward prolongations of the earth's axis of rotation.

Zenith and nadir: The respective points of intersection of the celestial sphere with the upward and downward prolongations of the observer's plumb line.

Celestial equator: The great circle defined by the intersection of the celestial sphere and the projection of the plane of the earth's equator.

Ecliptic: The apparent great-circle path of the Sun on the celestial sphere during the course of a year. The plane of the ecliptic is inclined at an angle of about 23°27′ to the plane of the celestial equator. More exact values of this inclination are tabulated as the obliquity of the ecliptic in the ephemeris.

Vernal equinox: That point of intersection of the ecliptic and the celestial equator which occurs when the Sun is going from south to north. The vernal equinox has the symbol γ and is sometimes called the first point of Aries. The hour angle and transit time of the vernal equinox at Greenwich observatory are tabulated in the ephemeris.

Horizon: The great circle defined by the intersection of the celestial sphere and that plane which is perpendicular to the observer's plumb line at the position of the observer.

Meridian: The great circle which is perpendicular to the horizon and passes through the zenith and the north celestial pole.

North point: The northward point of intersection of the meridian and the horizon.

Vertical circle: Any great circle which passes through the zenith and nadir and is therefore perpendicular to the horizon.

Hour circle: Any great circle which passes through the north and south celestial poles and is therefore perpendicular to the celestial equator.

Circle of celestial latitude: Any great circle which passes through the poles of the ecliptic, and is therefore perpendicular to the ecliptic.

Galactic equator: The great circle defined by the intersection of the celestial sphere and the projection of the plane of our Galaxy. The plane of the galactic equator is inclined at an angle of 62°36′ to the plane of the celestial equator.

North galactic pole: The point of intersection of the celestial sphere with the northward prolongation of the rotation axis of the Galaxy. The equatorial coordinates of this point are $\alpha = 12^h 49^m$ and $\delta = 27°24'$ for the equinox 1950.0 (Blaauw, Gum, Pawsey, and Westerhout, 1960).

Galactic center: The center of our Galaxy has the equatorial coordinates $\alpha = 17^h 42^m 24^s$ and $\delta = -28°55'$ for the equinox 1950.0 (Blaauw *et al.*, 1960).

Circle of galactic latitude: Any great circle which passes through the galactic poles and is therefore perpendicular to the galactic equator.

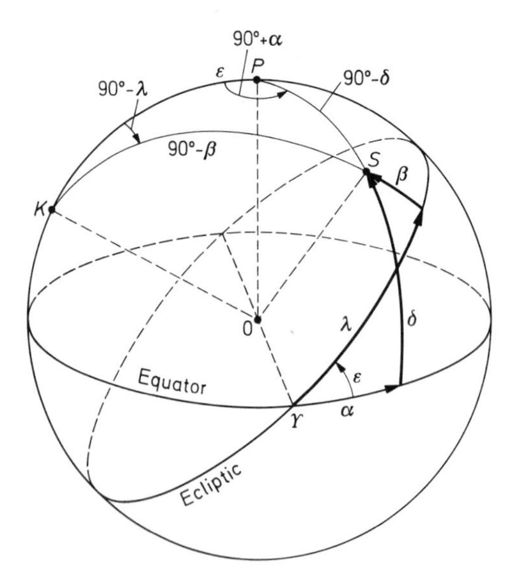

Fig 32. Graphic representation of the equatorial and heliocentric coordinates on the celestial sphere. Here α and δ denote, respectively, the right ascension and declination, λ and β denote, respectively, celestial longitude and latitude, γ denotes the vernal equinox, ε denotes the obliquity of the ecliptic, and K, P, S, and O denote, respectively, the ecliptic pole, the north celestial pole, the position of the stellar object, and the observer. The ecliptic is the apparent path of the Sun across the celestial sphere during the course of the year, and the celestial equator is the intersection of the celestial sphere and the projection of the equator of the earth.

The celestial coordinate systems are:

The horizontal or horizon coordinate system: The altitude, a, of the celestial object is the angle in degrees from the horizon to the object as measured along the vertical circle which passes through the object. The altitude is positive or negative, respectively, when measured above or below the horizon. The zenith distance, z, is $90° - a$. The azimuth, A, is the angle in degrees measured from the south point through west along the horizon to the foot of the vertical circle which passes through the object.

The equatorial coordinate system: The declination, δ, of the celestial object is the angle in degrees from the celestial equator to the object as measured along the hour circle which passes through the object. The declination is positive or negative, respectively, when measured north or south of the celestial equator. The right ascension, α, is the angle in degrees (or hours) measured from the vernal equinox along the celestial equator toward the east to the foot of the hour circle which passes through the object. The hour angle, h, is the angular distance in degrees (or hours) measured along the celestial equator from the foot of the meridian toward the west to the foot of the hour circle which passes through the object.

The ecliptic coordinate system: The celestial latitude, β, of the celestial object is the angle in degrees from the ecliptic to the object as measured along the circle of celestial latitude which passes through the object. The celestial latitude is positive or negative, respectively, when measured north or south of the ecliptic. The celestial longitude, λ, is the angle in degrees measured eastward along the ecliptic from the vernal equinox to the foot of the circle of celestial latitude which passes through the object.

The galactic coordinate system: The galactic latitude, b^{II}, of the celestial object is the angle in degrees from the galactic equator to the object as measured along the circle of galactic latitude which passes through the object. The galactic latitude is positive or negative, respectively, when measured north or south of the galactic equator. The galactic longitude, l^{II}, is the angle in degrees measured eastward along the galactic equator from the galactic center to the foot of the circle of galactic latitude which passes through the object. The old system of galactic coordinates (OHLSSON, 1932) had its north galactic pole at $\alpha = 12^{\mathrm{h}}40^{\mathrm{m}}$, $\delta = 28°$ for the equinox 1900.0, and the old zero of longitude was taken at the intersection of the galactic plane and the celestial equator at the equinox 1900.0. The old galactic coordinates of the galactic center are $l^{\mathrm{I}} = 327.69°$ and $b^{\mathrm{I}} = -1.40°$, and those of the new pole are $l^{\mathrm{I}} = 347.7°$ and $b^{\mathrm{I}} = 88.51°$. The inclination between the two systems is 1.4866°.

A graphic representation of the equatorial and heliocentric coordinates on the celestial sphere is given in Fig. 32.

5.1.3. Precession, Nutation, Aberration, and Refraction

The earth's flattening, combined with the obliquity of the ecliptic, results in a slow turning of the celestial equator on the ecliptic due to the differential gravitational effect of the moon and the Sun. This effect causes the equinox to move westward along the ecliptic at the rate of about 50″ per year. The gravitational couples between the center of mass of the earth and the other planets cause an additional motion of the equinox eastward of about 0.12″ per year as well as a diminution of the obliquity of the ecliptic of about 0.47″ per year. The net result of both the luni-solar and planetary effects is a general precession in right ascension, m, declination, n, and longitude, χ, which was first observed by HIPPARCHUS (125 B.C.). The theoretical values are (NEWCOMB, 1892, 1906)

$$
\begin{aligned}
m &= 3.07234 + 0.00186\,T \quad \text{seconds of time per year} \\
m &= 46.0851 + 0.0279 T \quad \text{seconds of arc per year} \\
n &= 20.0468 - 0.0085 T \quad \text{seconds of arc per year} \\
\chi &= 50.2564 + 0.0222 T \quad \text{seconds of arc per year,}
\end{aligned}
\tag{5-25}
$$

where T is the number of tropical centuries since 1900.0. The obliquity of the ecliptic for the reference epoch 1900.0 is 23°27′08.26″. The increase in right ascension, $\Delta\alpha$, and declination, $\Delta\delta$, are given by

$$\Delta\alpha = [m + n \sin\alpha \tan\delta]\,N$$

and

$$\Delta\delta = [n \cos\alpha]\,N, \tag{5-26}$$

where α and δ are, respectively, the right ascension and declination for the reference epoch, and N is the number of years between the desired date and the reference date. From Eq. (5-25), for example, the values of m and n for the reference epoch 1950.0 are $m = 3.07327$ seconds of time per year and $n = 20.0426$ seconds of arc per year.

BRADLEY (1748) first observed periodic nutation terms in the precession which are caused by the periodic motions of the moon in orbit around the earth and the earth in orbit around the Sun. The principal term depends on the longitude of the node of the moon's orbit and had a period of 18.6 years and an amplitude $N = 9.210''$ in 1900.0. The constant N is known as the constant of nutation. The nutation may be resolved into a correction, $\Delta\psi$, to the Sun's longitude, and a correction, $\Delta\varepsilon$, to the mean obliquity of the ecliptic by using the theory given by WOOLARD (1953). The nutation in longitude is given in the ephemeris, and the nutation in the obliquity, $\Delta\varepsilon$, is $-B$, where the Besselian day number, B, is tabulated in the ephemeris. The first-order nutation corrections to the right ascension and declination are given by

$$\Delta\alpha = (\cos\varepsilon + \sin\varepsilon \sin\alpha \tan\delta)\,\Delta\psi - \cos\alpha \tan\delta\,\Delta\varepsilon$$

and

$$\Delta\delta = \sin\varepsilon \cos\alpha\,\Delta\psi + \sin\alpha\,\Delta\varepsilon, \tag{5-27}$$

where α and δ are, respectively, the right ascension and declination for the reference epoch, and ε is the obliquity of the ecliptic which is tabulated in the ephemeris.

BRADLEY (1728) first observed aberration, which is a tilting of the apparent direction of a celestial object toward the direction of motion of the observer. Its magnitude, $\Delta\theta$, depends on the ratio of the velocity of the observer, v, to the velocity of light, c, and the angle, θ, between the direction of observation and the direction of motion. The displacement, $\Delta\theta$, in the sense of apparent minus mean place, is given by

$$\tan\Delta\theta = \frac{v \sin\theta}{c + v\cos\theta} \tag{5-28}$$

or

$$\Delta\theta \approx \frac{v}{c}\sin\theta - \frac{1}{2}\left(\frac{v}{c}\right)^2 \sin 2\theta + \left(\frac{v}{c}\right)^3 (\sin\theta\cos^2\theta - 0.33\sin^3\theta) + \cdots.$$

The coefficient in the second term is about 0.001 arc seconds for the earth's motion. For small $\Delta\theta$ but high velocities, v, the displacement is given by

(LORENTZ, 1904; EINSTEIN, 1905)

$$\Delta\theta \approx \left[\frac{1+\frac{v}{c}\cos\theta}{\sqrt{1-\left(\frac{v}{c}\right)^2}} - 1\right]\frac{\sin\theta}{\cos\theta}. \tag{5-29}$$

For the orbital motion of the earth about the Sun, the first-order annual aberration corrections to the right ascension, α, and declination, δ, are given by

$$\Delta\alpha = \frac{1}{\cos\delta}\left[-k\sin\lambda\sin\alpha - k\cos\lambda\cos\varepsilon\cos\alpha\right]$$

and

$$\Delta\delta = -k\sin\lambda\cos\alpha\sin\delta + k\cos\lambda\left[\cos\varepsilon\sin\alpha\sin\delta - \sin\varepsilon\cos\delta\right], \tag{5-30}$$

where λ is the true longitude of the Sun, ε is the obliquity of the ecliptic, and the constant of aberration is given by

$$k = \frac{2\pi a}{Pc}(1-e^2)^{-1/2} = 20.496 \text{ seconds of arc}. \tag{5-31}$$

Here $a = 1.49598 \times 10^{13}$ cm is the astronomical unit (the mean distance to the Sun), $e = 0.01675$ is the mean eccentricity of the earth's orbit, $P = 3.1558 \times 10^7$ seconds is the length of the sidereal year, $c = 2.997925 \times 10^{10}$ cm sec^{-1} is the velocity of light, and it has been assumed that one radian $= 2.062648 \times 10^5$ seconds of arc. The corrections given in Eq. (5-30) are in the sense that $\Delta\alpha$ and $\Delta\delta$ are the apparent coordinate minus the mean coordinate, and the correction can be as large as 20.47 seconds of arc.

The diurnal aberration caused by the rotation of the earth about its axis has an aberration constant, k_d, given by

$$\begin{aligned} k_d &= \frac{\omega a_e}{c}\rho\cos\varphi' \\ &= 0.3198\ \rho\cos\varphi' \text{ seconds of arc} \\ &= 0.02132\,\rho\cos\varphi' \text{ seconds of time}, \end{aligned} \tag{5-32}$$

where the angular velocity of the earth is $\omega = 7.2921 \times 10^{-5}$ radians per second, the radius of the earth is $a_e = 6.37816 \times 10^8$ cm, c is the velocity of light, ρ is the magnitude of the geocentric radius vector of the observer given in units of a_e, and φ' is the geocentric latitude of the observer. The diurnal aberration corrections to right ascension and declination, in the sense of apparent minus mean position, are given by

$$\Delta\alpha = k_d\cos h\sec\delta$$

and

$$\Delta\delta = k_d\sin h\sin\delta, \tag{5-33}$$

where the mean hour angle, h, is the local sidereal time minus the mean right ascension, α, and δ is the mean declination of the celestial object.

Aberration corrections for the motion of the earth, with respect to both the earth-moon center of mass and the earth-Jupiter center of mass, can be as large as 0.008 seconds of arc, and the formulae for these corrections are given by WOOLARD and CLEMENCE (1966).

The total corrections to the right ascension and declination for precession, nutation, and annual aberration may be made using the independent day numbers in the ephemeris. The mean position, α_0 and δ_0, for the nearest beginning of a year is related to the apparent position, α and δ, by the formulae

$$15.0(\alpha-\alpha_0)=f+g\sin(G+\alpha_0)\tan\delta_0+h\sin(H+\alpha_0)\sec\delta_0+J\tan^2\delta_0 \quad (5\text{-}34)$$

and

$$\delta-\delta_0=g\cos(G+\alpha_0)+h\cos(H+\alpha_0)\sin\delta_0+i\cos\delta_0+J'\tan\delta_0, \quad (5\text{-}35)$$

where the independent day numbers f, g, G, H, h, i, J, and J' are given in seconds of arc in the ephemeris. Additional corrections using Eqs. (5-26), (5-27), and (5-30) are necessary to correct the position from the true equinox of the day number reference to some standard equinox. Approximate day numbers for such a reduction to the standard equinox of 1950.0 are also given in the ephemeris.

Observed positions also differ from the true position due to the refraction of radiation in the atmosphere. At optical wavelengths, the index of refraction, n, is given by (EDLÉN, 1953, cf. Eq. (2-237))

$$(n-1)\times10^7=643.28+\frac{294981}{146-\sigma^2}+\frac{2554.0}{41-\sigma^2}, \quad (5\text{-}36)$$

where σ is the wave number in vacuum expressed in reciprocal microns (one micron $=10^{-4}$ cm). The wavelength in air, λ_{air}, is related to that in vacuum, λ_{vac}, by the relation

$$\lambda_{vac}-\lambda_{air}=\lambda_{air}(n-1). \quad (5\text{-}37)$$

Often the approximate value of $n=1.000297$ is used. At optical frequencies, the changes in right ascension, $\Delta\alpha$, and declination, $\Delta\delta$, caused by refraction are given by

$$\Delta\alpha=(n-1)\cos\varphi\sin h\,\mathrm{cosec}\,z\sec\delta=\pm R\sec\delta\sin q$$

and

$$\Delta\delta=(n-1)\left[\frac{\sin\varphi}{\cos\delta\cos z}-\tan\delta\right]=\pm R\cos q, \quad (5\text{-}38)$$

where the constant of refraction $R=n-1=0.0002927=60.4$ arc seconds, the parallactic angle, q, is given by $\sin q=\cos\varphi\sin h/\sin z$, φ is the observer's latitude, z is the zenith distance, h is the hour angle, and δ is the declination. Here $\Delta\alpha$ and $\Delta\delta$ are the true minus the observed values, and the $\pm$ signs are used as follows: $\Delta\alpha$ is $-$ or $+$, respectively, for $0^h<h<12^h$ or $12^h<h<24^h$, and $\Delta\delta$ is $-$ or $+$, respectively, for $\varphi>0$ or $\varphi<0$. At radio frequencies, positions are also displaced by refraction in the ionosphere. Detailed formulae for the apparent displacement of source position due to ionospheric refraction are given by KOMESAROFF (1960).

Further corrections to position are needed to account for the proper motion and parallax effects which are discussed in the following sections.

5.1.4. Spherical Trigonometry

For a spherical triangle with angles A, B, and C, and sides a, b, and c respectively opposite these angles, the following formulae apply.

Sine formula

$$\frac{\sin A}{\sin a} = \frac{\sin B}{\sin b} = \frac{\sin C}{\sin c}. \tag{5-39}$$

Cosine formulae

$$\begin{aligned} \cos a &= \cos b \cos c + \sin b \sin c \cos A \\ \cos b &= \cos a \cos c + \sin a \sin c \cos B \\ \cos c &= \cos a \cos b + \sin a \sin b \cos C. \end{aligned} \tag{5-40}$$

Extended cosine formulae

$$\begin{aligned} \sin a \cos B &= \cos b \sin c - \sin b \cos c \cos A \\ \sin a \cos C &= \cos c \sin b - \sin c \cos b \cos A \\ \sin b \cos A &= \cos a \sin c - \sin a \cos c \cos B \\ \sin b \cos C &= \cos c \sin a - \sin c \cos a \cos B \\ \sin c \cos A &= \cos a \sin b - \sin a \cos b \cos C \\ \sin c \cos B &= \cos b \sin a - \sin b \cos a \cos C. \end{aligned} \tag{5-41}$$

Haversine formulae

$$\begin{aligned} \cos\frac{A}{2} &= \left[\frac{\sin s \sin(s-a)}{\sin b \sin c}\right]^{1/2} \\ \sin\frac{A}{2} &= \left[\frac{\sin(s-b)\sin(s-c)}{\sin b \sin c}\right]^{1/2} \\ \tan\frac{A}{2} &= \left[\frac{\sin(s-b)\sin(s-c)}{\sin s \sin(s-a)}\right]^{1/2}, \end{aligned} \tag{5-42}$$

where $s = (a+b+c)/2.$

These formulae may be extended by replacing the sides and angles of the spherical triangle with the corresponding supplements of the angles and sides to obtain:

$$\begin{aligned} \cos A &= \sin B \sin C \cos a - \cos B \cos C \\ \cos B &= \sin A \sin C \cos b - \cos A \cos C \\ \cos C &= \sin A \sin B \cos c - \cos A \cos B, \end{aligned} \tag{5-43}$$

and

$$\begin{aligned} \sin A \cos b &= \cos B \sin C + \sin B \cos C \cos a \\ \sin A \cos c &= \cos C \sin B + \sin C \cos B \cos a \\ \sin B \cos a &= \cos A \sin C + \sin A \cos C \cos b \\ \sin B \cos c &= \cos C \sin A + \sin C \cos A \cos b \\ \sin C \cos a &= \cos A \sin B + \sin A \cos B \cos c \\ \sin C \cos b &= \cos B \sin A + \sin B \cos A \cos c. \end{aligned} \tag{5-44}$$

5.1.5. Coordinate Transformations

The formulae connecting the azimuth, A, and altitude, a, to the hour angle, h, and declination, δ, are

$$\begin{aligned}\cos a \sin A &= -\cos\delta \sin h\\ \cos a \cos A &= \sin\delta \cos\varphi - \cos\delta \cos h \sin\varphi\\ \sin a &= \sin\delta \sin\varphi + \cos\delta \cos h \cos\varphi\\ \cos\delta \cos h &= \sin a \cos\varphi - \cos a \cos A \sin\varphi\\ \sin\delta &= \sin a \sin\varphi + \cos a \cos A \cos\varphi,\end{aligned} \tag{5-45}$$

where φ is the observer's latitude, and the hour angle, h, is related to the right ascension, α, by

$$h = \text{local sidereal time} - \alpha. \tag{5-46}$$

The formulae connecting the longitude, λ, and latitude, β, in the ecliptic to the right ascension, α, and declination, δ, are

$$\begin{aligned}\cos\delta \cos\alpha &= \cos\beta \cos\lambda\\ \cos\delta \sin\alpha &= \cos\beta \sin\lambda \cos\varepsilon - \sin\beta \sin\varepsilon\\ \sin\delta &= \cos\beta \sin\lambda \sin\varepsilon + \sin\beta \cos\varepsilon\\ \cos\beta \sin\lambda &= \cos\delta \sin\alpha \cos\varepsilon + \sin\delta \sin\varepsilon\\ \sin\beta &= \sin\delta \cos\varepsilon - \cos\delta \sin\alpha \sin\varepsilon,\end{aligned} \tag{5-47}$$

where ε is the obliquity of the ecliptic which is tabulated in the ephemeris.

The formulae connecting the new galactic longitude, l^{II}, and new galactic latitude, b^{II}, to the right ascension, α, and declination, δ, are

$$\begin{aligned}\cos b^{\text{II}} \cos(l^{\text{II}} - 33^\circ) &= \cos\delta \cos(\alpha - 282.25^\circ),\\ \cos b^{\text{II}} \sin(l^{\text{II}} - 33^\circ) &= \cos\delta \sin(\alpha - 282.25^\circ)\cos 62.6^\circ + \sin\delta \sin 62.6^\circ,\\ \sin b^{\text{II}} &= \sin\delta \cos 62.6^\circ - \cos\delta \sin(\alpha - 282.25^\circ)\sin 62.6^\circ,\\ \cos\delta \sin(\alpha - 282.25^\circ) &= \cos b^{\text{II}} \sin(l^{\text{II}} - 33^\circ)\cos 62.6^\circ - \sin b^{\text{II}} \sin 62.6^\circ,\\ \sin\delta &= \cos b^{\text{II}} \sin(l^{\text{II}} - 33^\circ)\sin 62.6^\circ + \sin b^{\text{II}} \cos 62.6^\circ,\end{aligned} \tag{5-48}$$

where it is assumed that α and δ are for the equinox 1950.0. A graphical conversion chart between equatorial (1950.0) and new galactic coordinates is shown in Fig. 33. Tables for the conversion of equatorial coordinates to new galactic coordinates, the conversion of new galactic coordinates to equatorial coordinates, and the conversion of old galactic coordinates to new galactic coordinates have been published by the Lund Observatory (1961).

The formulae for connecting the old galactic longitude, l^{I}, and the old galactic latitude, b^{I}, to the new galactic longitude, l^{II}, and the new galactic latitude, b^{II}, are

$$\begin{aligned}\cos b^{\text{II}} \cos(l^{\text{II}} - 109.9497^\circ) &= \cos b^{\text{I}} \cos(l^{\text{I}} - 77.6500^\circ),\\ \cos b^{\text{II}} \sin(l^{\text{II}} - 109.9497^\circ) &= \cos b^{\text{I}} \sin(l^{\text{I}} - 77.6500^\circ)\cos 1.4866^\circ\\ &\quad + \sin b^{\text{I}} \sin 1.4866^\circ,\\ \sin b^{\text{II}} &= -\cos b^{\text{I}} \sin(l^{\text{I}} - 77.6500^\circ)\sin 1.4866^\circ\\ &\quad + \sin b^{\text{I}} \cos 1.4866^\circ.\end{aligned} \tag{5-49}$$

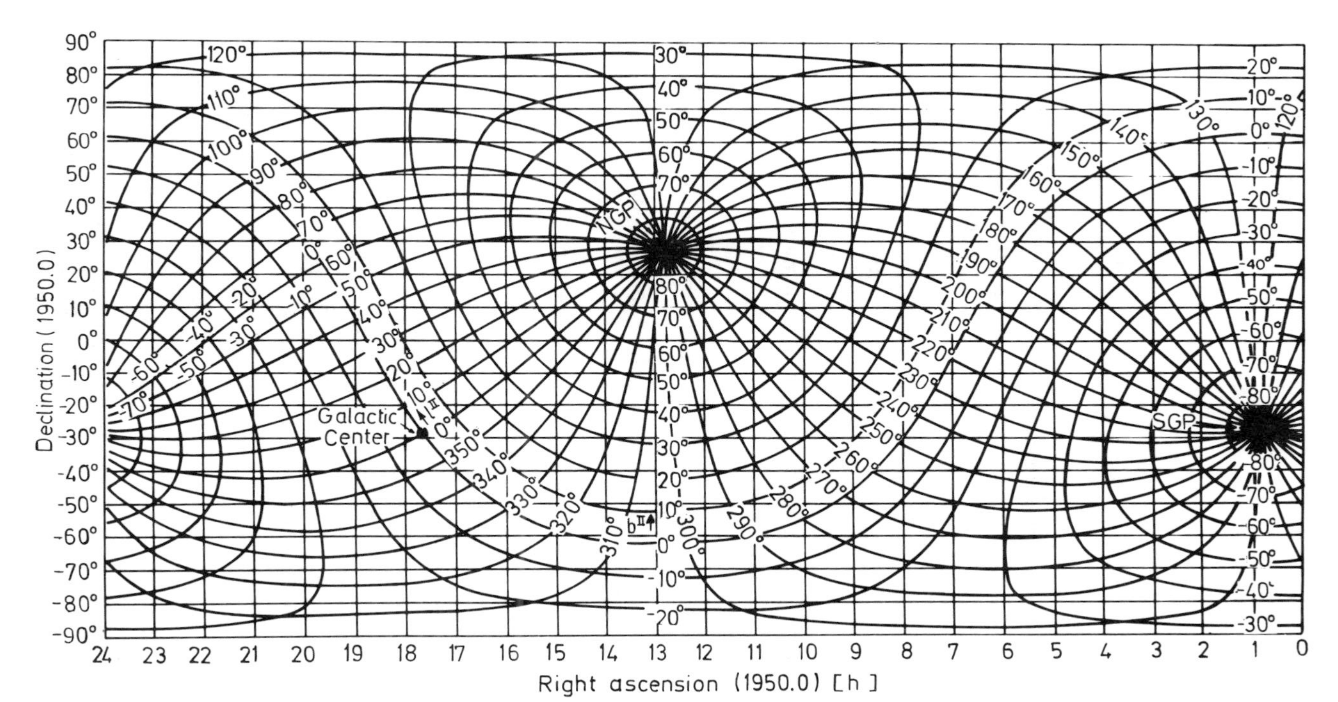

Fig. 33. Chart for the conversion of equatorial (1950.0) coordinates to new galactic coordinates ($l^{\mathrm{II}}, b^{\mathrm{II}}$) or vice versa

These formulae follow from an assumed inclination of the new galactic system to the old galactic system of 1.4866°, and the ascending node coordinates of $b^{\mathrm{I}} = b^{\mathrm{II}} = 0$, $l^{\mathrm{I}} = 77.6500°$, and $l^{\mathrm{II}} = 109.9497°$.

The natural reference coordinate system for external galaxies is the super-galactic coordinate system whose north pole is at $l^{\mathrm{I}} = 15°$, $b^{\mathrm{I}} = 5°$ and whose origin is at $l^{\mathrm{I}} = 105°$, $b^{\mathrm{I}} = 0°$.

5.2. Time

5.2.1. Ephemeris Time

The time which is the independent variable in Newcomb's (1898) gravitational theory of the Sun and planets is called ephemeris time, E.T. In this theory a fictitious mean Sun is assumed to have a uniform sidereal motion for all time equal to the mean sidereal motion that the true Sun had at the epoch 1900 Jan 0.5 days E.T. Newcomb's expression for the longitude, L, of the mean apparent motion of the Sun around the earth, measured from the mean equinox of date, is

$$L = 279°41'48.04'' + 129{,}602{,}768.13'' T_{\mathrm{E}} + 1.089'' T_{\mathrm{E}}^2, \tag{5-50}$$

where T_{E} is the number of Julian centuries of ephemeris time elapsed since the epoch of 12^{h} E.T. on 1900 January 0. The right ascension, α_{E}, of the fictitious mean Sun is given by

$$\alpha_{\mathrm{E}} = 18^{\mathrm{h}}38^{\mathrm{m}}45.836^{\mathrm{s}} + 8{,}640{,}184.542^{\mathrm{s}}\, T_{\mathrm{E}} + 0.0929^{\mathrm{s}}\, T_{\mathrm{E}}^2. \tag{5-51}$$

By definition, one Julian century has 36,525 days and each day has 86,400 ephemeris seconds. It then follows from Eq. (5-51) that:

$$\begin{aligned} \text{one tropical year} &= 31{,}556{,}925.9747 \text{ ephemeris seconds} \\ &= 365.24219879 \text{ days} \end{aligned} \tag{5-52}$$

at the epoch 12^{h} E.T. on 1900 January 0. The number of days in the tropical year centered at time T_{E} is given by

$$\text{one tropical year} = 365.24219879 - 0.00000614\, T_{\mathrm{E}} \text{ days}. \tag{5-53}$$

To facilitate dating, the Julian day number is defined as the number of ephemeris days that have elapsed, at the preceeding 12^{h} E.T., since 12^{h} E.T. on January 1, 4713 B.C. The fundamental epoch 1900 January $0^{\mathrm{d}}12^{\mathrm{h}}$ E.T. has the Julian day number 2,415,020.0, whereas for $0^{\mathrm{d}}12^{\mathrm{h}}$ E.T. on January 1950.0 the Julian day number is 2,433,282.

5.2.2. Universal Time

Universal time, U.T., is defined as 12 hours plus the Greenwich hour angle of Newcomb's fictitious mean Sun. That is, universal time is 12^{h} plus the Greenwich hour angle of the point on the celestial equator whose right ascension, α_{U}, measured from the mean equinox of date, is (Newcomb, 1895)

$$\alpha_{\mathrm{U}} = 18^{\mathrm{h}}38^{\mathrm{m}}45.836^{\mathrm{s}} + 8{,}640{,}184.542^{\mathrm{s}}\, T_{\mathrm{U}} + 0.0929^{\mathrm{s}}\, T_{\mathrm{U}}^2, \tag{5-54}$$

where T_U is the number of Julian centuries of 36,525 days of universal time elapsed since the epoch of Greenwich mean noon, 12^h U.T., on 1900 January 0.[1] Because this time is based on the position of the mean Sun, it is also called mean solar time or Greenwich mean time (G.M.T.). For an observatory with geodetic longitude, λ, the local mean solar time, L.M.T., is given by

$$\text{L.M.T.} = \text{U.T.} - \lambda, \tag{5-55}$$

which is the local hour angle of the point whose right ascension is α_U.

5.2.3. Sidereal Time

The apparent sidereal time is defined as the hour angle of the true vernal equinox (first point of Aries). When the position of the true equinox is corrected for nutation of the earth's axis, the resultant sidereal time is called the mean sidereal time. The difference between the apparent and mean sidereal times is called the equation of the equinoxes. The Greenwich mean sidereal time, G.M.S.T., is given in the ephemeris for 0 hours U.T. of each day. It follows from Eq. (5-54) that

$$\text{G.M.S.T.} \quad \text{at } 0^h \text{ U.T.} = 6^h 38^m 45.836^s + 8{,}640{,}184.542^s\, T_U + 0.0929^s\, T_U^2. \tag{5-56}$$

The local mean sidereal time, L.M.S.T., at an observatory whose geodetic longitude is λ, is given by

$$\text{L.M.S.T.} = \text{G.M.S.T.} - \lambda. \qquad (5\text{-}57)^2$$

A celestial object crosses an observer's meridian when the local sidereal time is equal to the right ascension of the object. The mean sidereal time is affected only by precession. The motion of the earth around the sun causes mean sidereal time to gain $3^m 56.55536^s$ per day on mean solar time. The two times are related by the expressions

$$\begin{aligned} \frac{\text{mean sidereal day}}{\text{mean solar day}} &= 0.997269566414 - 0.586\, T_U \times 10^{-10} \\ \frac{\text{mean solar day}}{\text{mean sidereal day}} &= 1.002737909265 + 0.589\, T_U \times 10^{-10}, \end{aligned} \tag{5-58}$$

where T_U is the number of Julian centuries of 36,525 days of universal time elapsed since the epoch of Greenwich mean noon, 12^h U.T., on 1900 January 0. The equivalent measures of the lengths of days are

$$\begin{aligned} &\text{mean sidereal day} = 23^h 56^m 04.09054^s \text{ of mean solar time} \\ &\text{mean solar day} \quad = 24^h 03^m 56.55536^s \text{ of mean sidereal time.} \end{aligned} \tag{5-59}$$

5.2.4. Measurement of Time—Atomic Time

Practical measurements of time are best made by observing the transit time of the moon and comparing this with the gravitational theory of the moon given by Brown (1919) and Jones (1939). This theory indicates that a difference, ΔT, between E.T. and U.T. is given by

$$\Delta T = \text{E.T.} - \text{U.T.} \approx 24.349^s + 72.318^s\, T_U + 29.950^s\, T_U^2, \tag{5-60}$$

[1] 1900 January 0 corresponds to 1899 December 31.

[2] If the longitude to the west from Greenwich is assumed to be positive.

where T_U is the number of Julian centuries of 36,525 days of universal time elapsed since 1900 January 0 Greenwich mean noon. Values of ΔT_0 are given in the ephemeris, where

$$\Delta T_0 = \text{E.T.} - \text{U.T.2}, \tag{5-61}$$

and U.T.2 is the universal time (U.T.) corrected for the observed motion of the geographic poles and for the extrapolated annual variation in the rate of rotation of the earth. Uncorrected universal time is designated as U.T.0, whereas U.T.1 designates universal time corrected for the movement of the poles. Time may be measured to a precision of one part in 10^{12} by observing the frequency, ν, of the hyperfine transition of caesium, for which (ESSEN, 1969)

$$\nu_c = 9{,}192{,}631{,}770 \text{ Hz}. \tag{5-62}$$

The epoch of atomic time, A.T., is taken to be $0^h0^m0^s$ U.T.2 on January 1, 1958, at which time A.T. = U.T.2 and E.T. = A.T. + 32.15^s. Other hyperfine transitions which may be used are those of hydrogen and rubidium which have respective frequencies of 1,420,405,751 Hz and 6,834,682,614 Hz.

Individual observatory clocks are regulated by checking them against satellite transmissions of atomic time. These radio frequency time pulses are called universal coordinated time (U.T.C.), and are thought to be accurate to a few microseconds. Observatory clocks are further regulated by correction for the difference U.T.2 − U.T.C. which is issued weekly by the U.S. Naval Observatory.

5.2.5. The Age of Astronomical Objects

If the radioactive decay constant is λ, then the number of radioactive nuclei, N_t, at time, t, is given by

$$N_t = N_0 \exp(-\lambda t) = N_0 \exp(-0.693\, t/\tau_{1/2}), \tag{5-63}$$

where N_0 is the original number of nuclei at time $t=0$, and $\tau_{1/2}$ is the half-lifetime of the radioactive nucleus. Decay constants and half-lifetimes of some long-lived radioactive nuclei are given in Table 54. HOLMES and LAWSON (1926) pointed out that the observed lead to uranium abundance ratio established a lower limit to the age of the earth of 1.3×10^9 years; and RUTHERFORD (1929) and ASTON (1929) pointed out that if uranium were formed with equal isotopic abundances of U^{235} and U^{238}, then the present abundance ratio indicates a decay age of 3×10^9 years.

As first pointed out by NIER (1939), a measurement of the relative abundance of the long-lived parent and the stable daughter nuclei of a radioactive decay chain can be used to determine the age of the sample material. For example, if the original abundance of lead isotopes is the same for two meteorites, then the time, t, of the formation of the isotopes is given by the equation

$$\left(\frac{Pb^{206}}{Pb^{204}}\right)_a - \left(\frac{Pb^{206}}{Pb^{204}}\right)_b = \left[\left(\frac{U^{238}}{Pb^{204}}\right)_a - \left(\frac{U^{238}}{Pb^{204}}\right)_b\right] [\exp(\lambda_{238}\, t) - 1], \tag{5-64}$$

which relates the present abundance of the isotopic species of Pb^{206} and U^{238} for the two meteorites at time, t. Here samples from the two meteorites are

Table 54. Decay constants and half-lifetimes of long-lived radioactive decay chains. Mean lifetimes, $T=\tau_{1/2}/0.693=\lambda^{-1}$

Parent	Daughter	Decay constant, λ $(yr)^{-1}$	Half-lifetime, $\tau_{1/2}$ (yr)
Rb^{87}	Sr^{87}	1.391×10^{-11}	4.982×10^{10}
Re^{187}	Os^{187}	1.613×10^{-11}	4.297×10^{10}
Th^{232}	Pb^{208}	4.990×10^{-11}	1.389×10^{10}
U^{238}	Pb^{206}	1.537×10^{-10}	4.508×10^{9}
U^{235}	Pb^{207}	9.718×10^{-10}	0.713×10^{9}
I^{129}	I^{127}	4.082×10^{-8}	1.697×10^{7}

respectively denoted by subscripts a and b, and the abundances are measured relative to that of Pb^{204} at time $t=0$. If the meteorite became a closed system at time $t=0$, then it follows from Eq. (5-64) that the present abundances of lead and uranium are related by the equation

$$\left(\frac{Pb^{206}}{Pb^{204}}\right)_t=\left(\frac{U^{238}}{Pb^{204}}\right)_t\left[\exp(\lambda_{238}t)-1\right]+\left(\frac{Pb^{206}}{Pb^{204}}\right)_0, \tag{5-65}$$

where the subscripts t and 0 denote the present and initial abundances, respectively. If all of the meteorites have the same initial Pb^{206}/Pb^{204} abundances, and if all meteorites have the same age, t, then a plot of $(Pb^{206}/Pb^{204})_t$ against $(U^{238}/Pb^{204})_t$ should lie in a straight line of slope $[\exp(\lambda_{238}t)-1]$. Such an isochrone plot defines a straight line with an age of $4.55 \pm 0.07 \times 10^9$ years (PATTERSON, 1956). Examination of the isotopic abundances of other elements leads to similar meteorite age estimates (ANDERS, 1963). The terrestrial abundance ratios of lead and uranium indicate that primordial lead on the earth had the same isotopic composition as that of the meteorites $4.54 \pm 0.02 \times 10^9$ years ago (OSTIC, RUSSELL, and REYNOLDS, 1963).

The age of the Galaxy, T_G, is given by (SCHRAMM and WASSERBURG, 1970) $T_G=T+\Delta+t$, where T is the duration of nucleosynthesis in the Galaxy prior to the formation of the solar system, $\Delta=1$ to 2×10^8 years is the time between the last nucleosynthesis and solidification, and $t=4.6\pm0.1\times10^9$ years is the age of solid bodies of the solar system. The age, T_G, is estimated by assuming that the Galaxy initially contained no heavy elements, that the heavy elements are produced in supernovae explosions according to the r process, and that the frequency of supernovae explosions has decayed exponentially with time (BURBIDGE, BURBIDGE, FOWLER, and HOYLE, 1957; FOWLER and HOYLE, 1960; CLAYTON, 1964; SEEGER, FOWLER, and CLAYTON, 1965; CLAYTON, 1969; DICKE, 1969; SEEGER and SCHRAMM, 1970; SCHRAMM and WASSERBURG, 1970; FOWLER, 1972). Values of T_G are model dependent and current estimates (FOWLER, 1972; SCHRAMM, 1972) indicate 2×10^9 years $\leq T \leq 10\times10^9$ years and 7×10^9 years $\leq T_G \leq 15\times10^9$ years.

The age of a star is defined as the time interval between the beginning of the main sequence and the position of the star on the Hertzsprung-Russell plot of luminosity versus temperature shown in Fig. 34 (HERTZSPRUNG, 1905; RUSSELL, 1912, 1914). The beginning of the main sequence is defined as that stage when

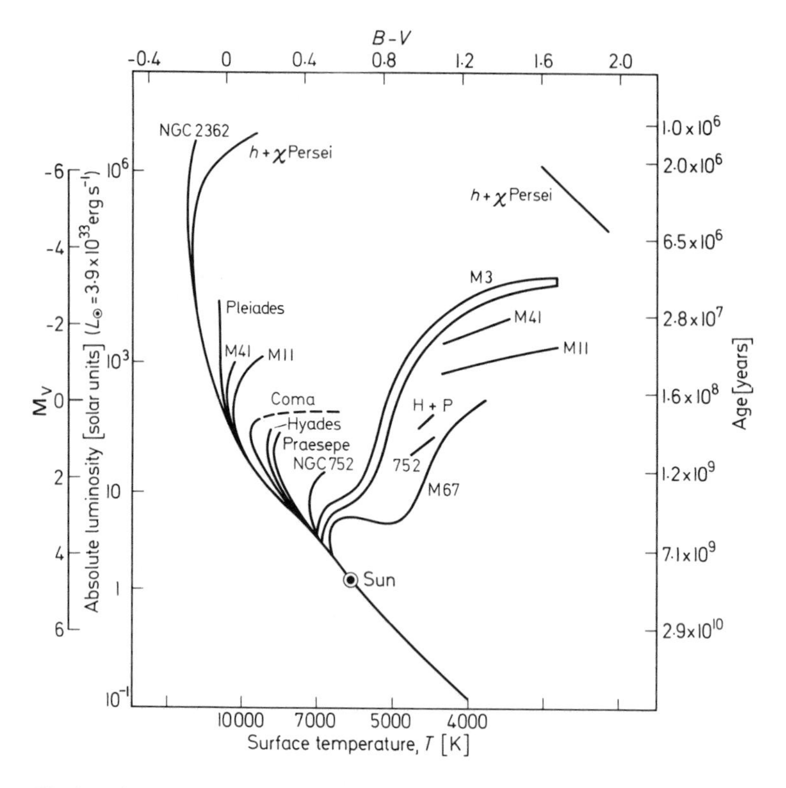

Fig. 34. The composite Hertzsprung-Russell diagram for ten galactic clusters and one globular cluster. The age of each cluster whose main sequence termination point is at a given absolute visual magnitude, M_V, is shown on the right. (After SANDAGE, 1957, by permission of the American Astronomical Society and the University of Chicago Press)

the nuclear energy production just begins to balance the luminosity. According to the Vogt-Russell theorem (VOGT, 1926; RUSSELL, DUGAN, and STEWART, 1926), a star of a given mass, age, and initial composition occupies a unique position on the H—R diagram. The equations which determine this position are:

The mass equation:

$$\frac{dM(r)}{dr} = 4\pi r^2 \rho \quad \text{or} \quad \frac{dr}{dM(r)} = \frac{1}{4\pi r^2 \rho}, \tag{5-66}$$

where $M(r)$ is the mass within radius, r, and ρ is the mass density given by

$$\rho = (2X + \tfrac{3}{4}Y + \tfrac{1}{2}Z)^{-1} N m_H \tag{5-67}$$

for a fully ionized gas of total number density, N, respective mass fractions X, Y, and Z for hydrogen, helium and the heavier elements, and where $m_H = 1.673 \times 10^{-24}$ grams is the mass of the hydrogen atom. Representative values of X, Y, and Z are $X=0.61$, $Y=0.37$, and $Z=0.02$ for population I stars, and $X=0.90$, $Y=0.10$, and $Z=0.001$ for population II stars. The presently accepted values for the initial abundances of the Sun are $X=0.762$, $Y=0.223$, and $Z=0.015$.

The equation of hydrostatic equilibrium:

$$\frac{dP}{dr} = -\rho \frac{GM(r)}{r^2} \quad \text{or} \quad \frac{dP}{dM(r)} = -\frac{GM(r)}{4\pi r^4}, \tag{5-68}$$

where P is the gas pressure and G is the gravitation constant.

The equation for energy balance:

$$\frac{dL(r)}{dr} = 4\pi r^2 \rho \varepsilon \quad \text{or} \quad \frac{dL(r)}{dM(r)} = \varepsilon, \tag{5-69}$$

where $L(r)$ is the energy flux from within r, and ε is the rate of energy generation per unit mass per unit time. Values of ε are discussed in Chapter 4.

The equation of radiation energy transfer:

$$\frac{dT}{dr} = \frac{-3\kappa\rho L(r)}{4acT^3 4\pi r^2} \quad \text{or} \quad \frac{dT}{dM(r)} = \frac{-3\kappa L(r)}{64\pi^2 ac T^3 r^4}, \tag{5-70}$$

where T is the temperature, the radiation density constant $a=7.564\times10^{-15}$ erg cm^{-3} °K^{-4}, and κ is the opacity to radiation in area per unit mass (cf. Cox and Stewart, 1970).

The equation of convective energy transport:

$$\frac{dT}{dr} = \left(\frac{\Gamma-1}{\Gamma}\right)\frac{T}{P}\frac{dP}{dr} \quad \text{or} \quad \frac{dT}{dM(r)} = \left(\frac{\Gamma-1}{\Gamma}\right)\frac{T}{P}\frac{dP}{dM(r)}, \tag{5-71}$$

where Γ is the adiabatic exponent (c.f. Eq. (3-300)).

Given an initial composition and a mass. Eqs. (5-66) to (5-71) determine an age for an observed luminosity and temperature. For the Sun, for example, Sears (1964) uses $X=0.71$, $Y=0.27$ and $Z=0.02$ together with $M_\odot=1.989\times10^{33}$ g, $L_\odot=3.90\times10^{33}$ erg sec^{-1}, and $T_\odot=5784$ °K, to give a solar age of 4.5×10^9 years. This age is provisional, however, until an exact theory of convective transport is worked out.

A star will leave the main sequence when it has exhausted hydrogen to 12% of its original mass (Schönberg and Chandrasekhar, 1942), and the evolutionary track is then described by Hoyle's (1959) theory. The order of magnitude age, T, of a star when it leaves the main sequence is given by (Sandage and Schwarzschild, 1952; Sandage, 1957)

$$T = 1.1\times10^{10}\, M/L \text{ years}, \tag{5-72}$$

where the mass, M, and absolute luminosity, L, are given in solar units. Using the mass-luminosity law of Eq. (5-180), Eq. (5-72) becomes

$$T \approx 10^{10}\left(\frac{M_\odot}{M}\right)^2 \approx 10^{10}\left(\frac{L_\odot}{L}\right)^{2/3} \quad \text{for } L<L_\odot,$$

where $M_\odot$ and $L_\odot$ denote, respectively, the solar mass and luminosity. Approximate ages for different stars are indicated on the Hertzsprung-Russell diagram in Fig. 34. Eq. (5-72) is easily derived by assuming that the position of a star on the main sequence is governed by hydrogen burning, that the fraction of mass converted into energy in hydrogen burning is $\Delta M/(4m_H)=0.007$ for a mass difference ΔM in the reaction $4H^1 \rightarrow He^4 + 2e^+ + 2\nu_e$, and that ten percent of the mass of a star is available for hydrogen burning. We then have $T=0.10\,\Delta M c^2 M/(L m_H)$. If a star shines by its store of thermal energy, T_k, then its lifetime is given by the Kelvin-Helmholtz time (KELVIN (1863), HELMHOLTZ (1854))

$$T_{K-H} = \frac{T_k}{L} \approx \frac{3GM^2}{10RL} \approx 2\times 10^7 \left(\frac{M}{M_\odot}\right)^2 \left(\frac{L_\odot}{L}\right)\left(\frac{R_\odot}{R}\right) \text{years}, \qquad (5\text{-}73)$$

where from the virial theorem (hydrostatic equilibrium), the kinetic energy of gas motion is $T_k = -\Omega/2$, and the gravitational potential energy $\Omega \approx -3GM^2/(5R)$. If the star is not in hydrostatic equilibrium, then the relevant lifetime is the free-fall time, T_{f-f}, given by

$$T_{f-f} = \left(\frac{R^3}{GM}\right)^{1/2} = 2\times 10^3 \rho^{-1/2}\ \text{sec}, \qquad (5\text{-}74)$$

where ρ is the gas mass density.

Under the hypothesis that all stars which are in a cluster of stars were formed at the same time, the age of a cluster can be found by matching a family of evolutionary tracks of various masses, initial compositions, and ages to the observed H—R diagram of the cluster. The ages of the globular clusters M3, M13, M15, and M92 lie between 9.4 and 13.4×10^9 years (SANDAGE, 1970; IBEN and ROOD, 1970; DEMARQUE, MENGEL, and AIZENMAN, 1971). SCHWARZSCHILD (1970) reports that globular clusters have ages of $1.0\pm 0.4\times 10^{10}$ years. The astronomical parameters of well-known galactic (open) and globular (closed) clusters are given in Tables 55 and 56.

The dynamics of clusters of stellar objects leads to reference times which may be related to the age of clusters. For a uniform cluster of stars of constant particle density, n, situated in a potential well with a flat bottom and steep sides, the relaxation time, T_R, is defined as (SPITZER and HÄRM, 1958)

$$T_R = \frac{(3/2)^{3/2} V^3}{2\pi G^2 M^2 n \ln N},$$

where G is the gravitation constant, and there are a total of N stars of mass, M, and r.m.s. velocity, V. If the virial theorem is used to evaluate V, then

$$T_R \approx 8\times 10^5 \frac{N^{1/2} R^{3/2}}{M^{1/2} \log_{10} N} \text{years}, \qquad (5\text{-}75)$$

where the cluster radius, R, is in parsecs, and the mass, M, is expressed in units of solar mass. Under these simplified assumptions, one percent of the stars will escape the cluster in a time T_R, and the cluster will collapse to a point in $40\,T_R$ (SPITZER, 1971). For most clusters the density is not uniform, and a radial density gradient results in larger relaxation times. Another useful time is the collision

Table 55. Positions, apparent angular diameter, θ, distance modulus, $m-M$, distance, D, reddening, E_{B-V}, turn up color, $(B-V)_T$, of the absorption corrected color index of the earliest or bluest stars on the cluster main sequence, the apparent magnitude, V_B, of the brightest or fifth brightest star, and the cluster type for ninety seven galactic (population I) clusters[1]

Cluster NGC	(Name)	α(1950.0) h m	δ(1950.0) ° ′	θ (arc min)	m − M (mag)	D (pc)	$E_{(B-V)}$ (mag)	$(B-V)_T$ (mag)	V_B (mag)	Type
103		00 22.5	61 04	4	12.4	2,980	0.46	−0.13	12.4	B 6 I 2*p*
129		00 27.1	59 57	13	11.1	1,660	0.58	−0.16	8.56	B 5 IV 2*pU*
225		00 40.5	61 31	14	9.0	630	0.29	−0.11	9.24	B 8 III 1*p*
457		01 15.9	58 04	12	12.3	2,880	0.47	−0.25	6.96	B 2 I 3*r*
581	(M 103)	01 29.9	60 27	6.5	12.1	2,600	0.37	−0.21	7.26	B 3 II 3*m*
654		01 40.6	61 38	5.5	10.9	1,550	0.90	−0.30	7.24	O9.5 I 2*p*
663		01 42.6	60 59	14	11.8	2,300	0.85	−0.28	8.29	B 0 IV 2*m*
744		01 55.2	55 14	14	10.8	1,490	0.41	−0.08	10.13	B 9 I 2*p*
752		01 54.8	37 26	45	7.9	380	0.03	+0.35	9.6	F 0 III 1*m*
869										
884	(h + χ Per)	02 15.5	56 55	30	11.8	2,360	0.56	−0.28	9.5	B 0.5 IV 3*r*
IC 1805		02 29.0	61 13	20	11.6	2,100	0.82	−0.32	8.12	O 5 IV 3*mN*
957		02 30.0	57 18	9	11.7	2,250	0.80	−0.26	8.05	B 1 II 3*mU*
1027		02 38.8	61 20	21	10.5	1,280	0.40	−0.02	9.33	A 0 IV 3*m*
1039	(M 34)	02 38.8	42 34	30	8.2	440	0.09	−0.11	8.6	B 8 I 3*m*
IC 1848		02 47.3	60 14	22	11.7	2,200	0.61	−0.29	8.24	O9 IV 3*mN*
1245		03 11.3	47 03	7	11.8	2,300	0.28	+0.16	7.95	A 7 III 2*r*
1342		03 28.4	37 10	15	8.7	550	0.28	+0.05	8.46	A 3 II 2*m*
Pleiades	(M 45)	03 44.1	23 57	120	5.5	126	0.04	−0.12	4.2	B 7 II 3*rN*
1444		03 45.6	52 30	3.5	10.0	1,000	0.70	−0.32	10.37	O II 3*p*
1502		04 03.1	62 11	8	9.7	880	0.77	−0.29	7.92	B 0 II 3*p*
1528		04 11.6	51 07	22	9.5	800	0.29	−0.02	8.75	A 0 II 2*m*
Hyades	(Taurus Cl.)	04 17	15 30	400	3.0	40	0.00	+0.10	4	A 2 II 3*m*
1545		04 17.1	50 08	15	9.5	800	0.36	−0.08	8.08	B 9 I 2*p*
1647		04 43.1	18 59	35	8.7	550	0.39	−0.09	8.59	B 9 III 2*m*
1662		04 45.7	10 51	14	8.1	420	0.34	−0.04	8.30	B 9.5 II 2*p*

Table 55 (continued)

Cluster NGC	(Name)	α(1950.0) h m	δ(1950.0) ° ′	θ (arc min)	m − M (mag)	D (pc)	$E_{(B-V)}$ (mag)	$(B-V)_T$ (mag)	V_B (mag)	Type
1664		04 47.4	43 37	13	10.2	1,100	0.20	—	7.50	— II 2*m*
1778		05 04.7	36 59	10	10.7	1,380	0.34	−0.10	9.05	B 8 II 2*p*
1893		05 19.4	33 21	15	13.0	4,000	0.59	−0.34	9.04	O 5–O 6 I 3*m*
1907		05 24.7	35 17	5	10.7	1,380	0.38	0.00	9.30	A 0 I 2*m*
1912	(M 38)	05 25.3	35 48	18	10.6	1,320	0.27	−0.09	9.13	B 8 II 2*r*
1960	(M 36)	05 32.8	34 06	16	10.5	1,260	0.24	−0.24	8.7	— I 3*m*
2099	(M 37)	05 49.1	32 32	24	10.5	1,280	0.31	−0.10	9.93	B 8 I 1*r*
2129		05 58.1	23 18	7	11.6	2,100	0.67	−0.27	7.36	B 1 II 3*p*
2168	(M 35)	06 05.8	24 21	29	9.7	870	0.23	−0.17	8.57	B 5 III 3*r*
2169		06 05.6	13 58	6	10.2	1,100	0.18	−0.27	8.13	B 1 II 3*p*
2215		06 18.4	−07 16	8.5	10.0	1,000	0.10	+0.10	11.2	A 3 II 2*p*
2244		06 29.7	04 54	27	11.1	1,660	0.55	−0.32	8.9	O 5 IV 3*m N*
2251		06 32.0	08 24	7	11.0	1,580	0.20	−0.16	9.11	B 5 II 2*p*
2264	(S Mon)	06 38.3	09 56	30	9.4	750	0.10	−0.30	8.9	O 9.5 II 3*p N*
2287	(M 41)	06 44.9	−20 41	32	9.1	670	0.00	−0.20	7.06	B 3 I 3*r*
2301		06 49.2	00 32	15	9.5	790	0.04	−0.14	8.00	B 6 I 3*m*
2323	(M 50)	07 00.6	−08 16	16	9.8	910	0.26	−0.19	7.90	B 3 I 2*m*
2324		07 01.6	01 08	8	12.4	3,000	0.11	+0.15	8.25	A 5 I 2*m*
2353		07 12.2	−10 13	20	10.1	1,050	0.12	−0.25	9.17	B 1 I 3*m U*
2362	(τ Can Maj)	07 16.7	−24 51	7	10.9	1,500	0.11	−0.28	9.4	B 0.5 I 3*p*
2422		07 34.3	−14 23	30	8.4	480	0.08	−0.18	7.82	B 4 II 3*m*
2439		07 39.0	−31 32	9	11.0	1,610	0.25	−0.11	10.6	B 7 II 3*m U*
2447	(M 93)	07 42.5	−23 45	18	10.2	1,100	0.06	−0.06	9.7	B 9 I 3*r*
2516		07 57.4	−60 44	50	7.8	365	0.10	−0.07	10.1	B 9 I 3*r*
2632	(Praesepe, M 44)	08 37.2	20 10	90	6.0	158	0.00	+0.15	7.5	A 5 I 2*r*

Cluster NGC	(Name)	$\alpha(1950.0)$ h m	$\delta(1950.0)$ ° ′	θ (arc min)	$m-M$ (mag)	D (pc)	$E_{(B-V)}$ (mag)	$(B-V)_T$ (mag)	V_B (mag)	Type
2682	(M 67)	08 47.8	12 00	18	9.6	830	0.06	+0.40	10.8	F 2 II 2r
3330	(Harv 4)	10 36.6	−53 53	8	10.7	1,390	0.18	−0.10	10.2	B 8 II 3m
4103		12 04.0	−60 58	9	10.4	1,200	0.34	−0.24	9.4	B 2 I 3m
Coma		12 22.5	26 23	300	4.5	80	0.00	+0.05	5	A 2 II 3p
4755	(κ Cru)	12 50.7	−60 04	12	9.6	830	0.31	−0.20	7	B 3 I 3r
6405	(M 6)	17 36.8	−32 11	26	9.0	630	0.16	−0.18	8.3	B 4 II 3m
IC 4665		17 43.8	05 44	50	7.6	330	0.17	−0.15	7.62	B 5 II 2p
6475	(M 7)	17 50.6	−34 48	50	7.0	250	0.08	−0.10	7.4	B 8 I 3m
6494	(M 23)	17 54.0	−19 01	27	9.1	660	0.38	−0.06	8.25	B 9 I 2r
6530		18 01.7	−24 20	14	10.7	1,400	0.32	−0.32	9.3	O 5 II 2mN
6531	(M 21)	18 01.6	−22 30	12	10.5	1,250	0.27	−0.30	7.24	O 9.5 I 3p
6611	(M 16)	18 16.0	−13 48	8	12.0	2,500	0.85	−0.32	8.09	O 5 II 3mN
6633		18 25.1	06 32	25	7.5	320	0.17	+0.15	8.0	A 5 I 2pE
IC 4725	(M 25)	18 28.8	−19 17	35	8.9	600	0.50	−0.20	9.3	B 3 IV 3r
6664		18 34.0	−08 16	20	—	—	0.60	−0.10	11.3	B 8 IV 2m
6694	(M 26)	18 42.6	−09 27	9	10.9	1,500	0.58	−0.16	9.19	B 5 II 2m
6705	(M 11)	18 48.4	−06 20	12.5	11.2	1,740	0.40	−0.10	12	B 8 II 2r
6709		18 49.1	10 17	12	9.8	910	0.30	−0.16	8.60	B 5 II 2p
6755		19 05.3	04 08	15	11.3	1,820	0.93	−0.10	10.23	B 8 IV 2m
6802		19 28.4	20 10	5	10.2	1,100	0.81	−0.14	9.23	B 6 III 1mE
6823		19 41.0	23 11	7	11.1	1,650	0.80	−0.30	8.75	O 9 IV 3p
6830		19 48.9	22 57	10	11.2	1,740	0.53	−0.29	9.06	B 0 IV 2m
6834		19 50.2	29 17	7	12.4	3,030	0.72	−0.23	8.50	B 2 I 2m
6866		20 02.1	43 51	10	10.4	1,200	0.14	+0.06	9.82	A 3 I 2p
6871		20 04.0	35 38	25	11.2	1,740	0.46	−0.24	6.81	B 0 IV 3p
6882/5		20 09.6	26 24	22	8.9	600	0.08	+0.36	9.8	F 2 I–IV 2–3p
6910		20 21.3	40 37	13	11.1	1,650	1.05	−0.28	7.37	B 0 IV 3p
6913	(M 29)	20 22.1	38 22	7	10.3	1,150	1.02	−0.28	8.57	B 0 III 3p
6940		20 32.5	28 08	26	9.5	800	0.26	+0.15	8.58	A 7 III 1m
7031		21 05.7	50 38	7	10.3	1,150	0.93	—	11.30	— II 3pU

Table 55 (continued)

Cluster NGC	(Name)	α(1950.0) h m	δ(1950.0) ° ′	θ (arc min)	$m-M$ (mag)	D (pc)	$E_{(B-V)}$ (mag)	$(B-V)_T$ (mag)	V_B (mag)	Type
7062		21 21.4	46 10	6	11.2	1,760	0.74	−0.18	8.05	B 4 II 2p
7063		21 22.4	36 17	8	9.0	630	0.08	−0.10	8.90	B 8 II 2p
7067		21 22.4	47 48	2.5	12.3	2,900	0.84	−0.25	9.46	B 1 II 2p
7086		21 28.8	51 22	7.5	10.7	1,400	0.72	−0.10	10.72	B 8 I 2m
7092	(M 39)	21 30.4	48 13	32	7.0	250	0.00	−0.06	6.5	B 9 II 2p
7128		21 42.3	53 29	3.2	12.0	2,500	1.10	−0.24	7.84	B 2 II 2p
7142		21 44.7	65 34	11	10.0	1,000	0.18	−0.40	10.14	F 2 II 1m
7160		21 52.3	62 22	10	9.6	840	0.30	−0.30	6.70	O9.5 I 3p
7209		22 03.2	46 15	20	9.8	910	0.15	0.00	8.55	A 0 II 2p
7235		22 10.8	57 02	4.5	12.4	3,030	0.92	−0.28	8.79	B 0 II 3p
7261		22 18.6	57 50	7	9.4	700	0.58	−0.10	9.61	B 8 II 2p
7380		22 45.0	57 50	9	11.6	2,100	0.58	−0.30	7.57	O9 III 2p
7510		23 09.4	60 18	3	12.6	3,300	1.08	−0.30	8.75	O9 II 2mU
7654	(M 52)	23 22.0	61 19	13	11.6	2,100	0.68	−0.20	7.76	B 3 II 2r
7686		23 27.8	48 51	13	—	—	—	—	7.74	— I 3p
7789		23 54.5	56 27	19	11.4	1,870	0.28	0.30	12.5	F 0 III 1r
7790		23 55.9	60 56	4.5	12.8	3,600	0.52	−0.18	11.7	B 4 II 2p

[1] Positions and apparent angular diameters are from HOGG (1959), and the latter data are compiled from the lists of TRUMPLER (1930) and COLLINDER (1931). The distance moduli, the distances, the reddening, and the colors are from JOHNSON *et al.* (1963). The magnitude of the brightest star is given to two decimal places and is from HOAG *et al.* (1961), and otherwise V_B is the magnitude of the fifth brightest star given by SHAPLEY (1933) and HOGG (1959). The spectral type of the cluster is the spectral type corresponding to $(B-V)_T$ and is from JOHNSON *et al.* (1963), whereas the Trumpler type (TRUMPLER, 1930) is designated by I–IV for decreasing concentration and by p, m, and r for clusters with less than 50, 50 to 100, and greater than 100 stars, respectively.

time, T_c, which is the mean time between the collision of any one star with another. According to SPITZER (1971)

$$T_c = \frac{10^{22} R^{7/2}}{N^{3/2} (M^{1/2} r^2) \left(1 + 8.8 \times 10^7 \dfrac{R}{r N}\right)} \text{ years}, \tag{5-76}$$

where the cluster radius, R, is in parsecs, and the mass, M, and radius, r, of the member stars are given in solar units. Well ordered clusters must be older than about $1.0\, T_R$ in order to have been formed, but must be younger than about $1000\, T_R$ or $1000\, T_c$ in order to be still seen. The younger galactic clusters have $T_R \approx 10^6$ years whereas most globular clusters have $T_R \approx 10^9$—10^{10} years.

The radio sources describe an evolutionary sequence in which young, small, luminous and bright objects evolve into older, larger, less luminous, and less bright objects with a steeper spectrum (HEESCHEN, 1966; KELLERMANN, 1966; SIMON and DRAKE, 1967). A radio source with an initially flat spectrum will have a steeper spectrum at higher frequencies because of synchrotron radiation losses. An electron of energy, E, in a magnetic field of strength, H, radiates most of its synchrotron radiation at the frequency, ν, given by (SCHOTT, 1912) (c.f. Eq. (1-154))

$$\nu = 1.61 \times 10^7 H E^2 \text{ MHz}, \tag{5-77}$$

whereas the electron loses its energy due to synchrotron radiation loss at the rate (c.f. Eq. (1-163))

$$\frac{dE}{dt} = -6.08 \times 10^{-9} H^2 E^2 \text{ erg sec}^{-1}, \tag{5-78}$$

where E is in GeV and H is in gauss in both Eqs. (5-77) and (5-78). It is clear from Eqs. (5-77) and (5-78) that the older electrons radiate at the lower frequencies. KARDASHEV (1962) showed that the radiation spectrum of electrons with an energy distribution, $N(E)dE \propto E^{-\gamma} dE$, will have spectral indices, α, given by

$$\begin{aligned} \alpha &= \frac{-(1-\gamma)}{2} && \text{for } \nu < \nu_c \\ \alpha &= \frac{-(2\gamma+1)}{3} && \text{for } \nu > \nu_c \text{ and instantaneous injection} \\ \alpha &= \frac{-\gamma}{2} && \text{for } \nu > \nu_c \text{ and continuous injection,} \end{aligned} \tag{5-79}$$

where the critical frequency, ν_c, is given by

$$\nu_c = 340 H^{-3} T^{-2} \text{ MHz}, \tag{5-80}$$

T is the age of the radio source in years, and H is the magnetic field intensity in gauss. Observations of radio sources at low frequencies indicate that their magnetic fields are on the order of $10^{-4 \pm 1}$ gauss (KELLERMANN and PAULINY-TOTH, 1969). Therefore observations of the high frequency cutoff, ν_c, beyond which the radio source flux decreases will provide an age according to the equation

$$T = [10^{16} \nu_c^{-1}]^{1/2} \text{ years}, \tag{5-81}$$

Table 56. Name, concentration class, position, absorption, $A = 0.24 \csc b^{\text{II}}$, apparent angular diameter, θ, mean apparent B magnitude of the twenty five brightest stars, apparent distance modulus, $(m-M)$, assuming $M = -0.8$, distance, D, absorption corrected color, $(B-V)_0$, spectral type, radial velocity, V_r, and the number of variable stars for 119 globular (population II) clusters[1]

NGC	Cluster (Name)	Concen. class	α(1950.0) h m	δ(1950.0) °	A (mag)	θ (arc min)	Mag 25 Br. Star	$(m-M)$ (mag)	D (kpc)	$(B-V)_0$	Type	V_r (km sec^{-1})	No. vars.
104	(47 Tuc)	III	00 21.9	−72 21	0.34	(44.0)	13.54	13.75	5.0	—	G 3	− 24	11
288		X	00 50.2	−26 52	0.24	(12.4)	14.98	15.78	14.8	—	—	− 47	1
362	(Δ 62)	III	01 00.6	−71 07	0.33	(17.7)	14.23	15.03	9.7	—	F 8	+221	14
1261		II	03 10.9	−55 25	0.31	(4.0)	—	—	(29.0)	—	F 8	+ 46	—
Pal 1		XII	03 25.7	79 28	0.70	(1.3)	19.6	20.4	87.1	—	—	—	0
Pal 2		IX	04 43.1	31 23	1.53	(1.7)	—	—	—	—	—	—	—
1841		—	04 52.5	−84 05	0.48	(2.4)	—	—	—	—	—	—	—
1851	(Δ 508)	II	05 12.4	−40 05	0.42	(11.5)	—	—	(17.0)	—	F 7	+309	2
1904	(M 79)	V	05 22.2	−24 34	0.50	(7.8)	15.52	16.32	16.5	—	F 6	+196	5
2298		VI	06 47.2	−35 57	0.87	(4.2)	—	—	(30.0)	—	F 7	+ 64	6
2419		II	07 34.8	39 00	0.55	(6.2)	18.32	20.0	83.2	0.57	F 6	+ 14	36
2808		I	09 10.9	−64 39	1.26	(18.8)	15.09	15.89	9.1	—	F 8	+101	4
Pal 3		XII	10 03.0	00 18	0.36	(2.2)	—	20.3	100.0	—	—	—	1
3201	(Δ 445)	X	10 15.5	−46 09	1.53	(29.3)	—	—	(4.0)	—	—	+493	77
Pal 4		XII	11 26.6	29 15	0.25	(2.5)	20.59	20.3	100.0	—	—	—	2
4147		VI	12 07.6	18 49	0.25	3.4	16.94	16.55	18.7	0.50	A 6	+191	16
4372		XII	12 23.0	−72 24	1.38	(19.8)	—	—	(6.0)	—	—	+ 66	0
4590	(M 68)	X	12 36.8	−26 29	0.40	6.4	14.98	15.78	14.0	0.57	A 7	−116	38
4833		VIII	12 56.0	−70 36	1.72	(12.7)	—	—	(5.0)	—	—	+204	10
5024	(M 53)	V	13 10.5	18 26	0.24	8.3	15.28	16.7	20.0	0.52	F 4	−112	43
5053		XI	13 13.9	17 57	0.24	(8.9)	15.86	16.25	16.4	0.57	(F 5)	—	10
5139	(ω Cen)	VIII	13 23.8	−47 03	0.93	(65.4)	—	14.28	5.2	—	F 7	+230	165
5272	(M 3)	VI	13 39.9	28 38	0.25	9.3	14.35	15.32	10.6	0.58	F 7	−153	189
5286	(Δ 388)	V	13 43.0	−51 07	1.26	(13.6)	—	—	(12.0)	—	F 8	+ 45	0
5466		XII	14 03.2	28 46	0.25	(9.2)	15.99	16.79	21.3	0.65	(F 5)	—	22
5634		IV	14 27.0	−05 45	0.32	(3.7)	16.65	17.45	26.8	0.56	F 5	− 63	7
5694		VII	14 36.7	−26 19	0.48	2.3	17.17	17.97	31.6	0.57	F 0	−187	0

I 4499		XI	14 52.7	−82 02	0.67	(6.2)	—	—	—	—	—	—	—
5824		I	15 00.9	−32 53	0.64	3.3	—	—	(46.0)	0.55	F 5	− 58	27
Pal 5		XII	15 13.5	00 05	0.34	(10.3)	17.27	18.07	35.2	—	—	—	5
5897		XI	15 14.5	−20 50	0.48	(8.7)	15.49	16.2	14.5	0.58	(F 5)	—	4
5904	(M 5)	V	15 16.0	02 16	0.33	10.7	14.07	14.80	8.1	0.62	F 6	+ 49	97
5927		VIII	15 24.5	−50 29	2.75	(12.0)	—	—	(3.0)	—	G 2	− 96	—
5946		IX	15 31.8	−50 30	3.44	(2.6)	—	—	—	—	—	—	—
5986	(Δ 552)	VII	15 42.8	−37 37	0.99	8.6	—	—	(14.0)	0.57	G 1	+ 2	5
A–v dB	(Pal 14)	—	16 08.8	15 02	0.36	(8.4)	—	20.0	83.2	—	—	—	0
6093	(M 80)	II	16 14.1	−22 52	0.74	8.6	15.07	15.87	12.6	0.64	F 9	+ 18	8
6101		X	16 20.0	−72 06	0.87	(14.6)	—	—	—	—	—	—	—
6121	(M 4)	IX	16 20.6	−26 24	0.87	22.6	13.21	14.01	4.3	0.76	(G 0)	+ 65	43
6139		II	16 24.3	−38 44	1.97	(2.6)	—	—	(13.0)	—	F 8	+ 20	—
6144		XI	16 24.2	−25 56	0.93	11.1	16.04	16.84	16.9	0.66	(G 0)	—	1
6171		X	16 29.7	−12 57	0.61	12.8	15.75	16.55	17.1	0.88	G 3	−147	24
6205	(M 13)	V	16 39.9	36 33	0.37	12.9	13.85	14.3	6.3	0.56	F 6	−241	10
6218	(M 12)	IX	16 44.6	−01 52	0.55	21.5	14.07	14.87	7.4	0.70	F 8	− 16	1
6229		IV	16 45.6	47 37	0.37	3.6	16.50	17.30	24.7	0.62	F 8	−150	22
6235		X	16 50.4	−22 06	1.07	(1.9)	16.56	17.36	19.4	—	—	—	2
6254	(M 10)	VII	16 54.5	−04 02	0.61	16.2	14.17	14.42	6.2	0.71	G 1	+ 71	3
6266	(M 62)	IV	16 58.1	−30 03	1.97	8.8	16.16	16.96	11.7	0.52	G 2	− 75	83
6273	(M 19)	VIII	16 59.5	−26 11	1.53	9.3	14.98	15.78	7.1	0.58	F 3	+102	4
6284		IX	17 01.5	−24 41	1.38	5.7	16.37	17.17	16.3	0.62	G 2	+ 22	6
6287		VII	17 02.1	−22 38	1.26	5.8	16.39	17.19	17.0	0.88	(G 5)	—	3
6293		IV	17 07.1	−26 30	1.72	6.2	15.63	16.43	9.8	0.50	F 0	− 73	5
6304		VI	17 11.4	−29 24	2.75	8.0	—	—	(6.0)	0.60	G 4	− 98	11
6316		III	17 13.4	−28 05	2.75	7.2	—	—	—	0.55	(G 5)	—	—
6325		IV	17 15.0	−23 42	1.72	4.9	—	—	—	1.18	(G 5)	—	—
6333	(M 9)	VIII	17 16.2	−18 28	1.38	7.9	15.75	16.55	12.8	0.59	F 2	+224	13
6341	(M 92)	IV	17 15.6	+43 12	0.42	12.3	13.96	14.80	7.9	0.50	F 1	−118	16
6342		IV	17 18.2	−19 32	1.53	4.7	—	—	—	0.89	(G 5)	—	—
6352		XI	17 21.6	−48 26	1.97	(8.9)	—	—	—	—	—	—	—
6355		—	17 20.9	−26 19	2.75	6.1	—	—	—	0.80	(G 5)	—	—

Table 56 (continued)

NGC	Cluster (Name)	Concen. class	α(1950.0) h	α(1950.0) m	δ(1950.0) °	δ(1950.0) ′	A (mag)	θ (arc min)	Mag 25 Br. Star	$(m-M)$ (mag)	D (kpc)	$(B-V)_0$	Type	V_r (km sec^{-1})	No. vars.
6356		II	17	20.7	−17	46	1.38	6.3	16.72	17.4	19.1	0.77	G 4	+ 31	5
HP 1		IX	17	24.9	−29	57	4.59	(1.3)	—	—	—	—	—	—	—
6362	(Δ 225)	X	17	26.6	−67	01	0.82	(8.5)	—	—	(7.0)	—	—	− 18	31
6366		XI	17	25.1	−05	02	0.87	(5.8)	16.06	16.86	17.4	—	—	—	2
6380		—	17	31.9	−39	02	3.44	—	—	—	—	—	—	—	—
6388		III	17	32.6	−44	43	1.97	(6.8)	—	—	(13.0)	—	G 3	+ 81	—
6397	(Δ 366)	IX	17	36.8	−53	39	1.15	(19.0)	12.71	13.51	2.9	—	F 5	+ 11	3
6401		—	17	35.6	−23	53	3.44	(1.0)	—	—	—	—	(G 5)	—	—
6402	(M 14)	VIII	17	35.0	−03	15	0.99	10.8	15.68	16.48	14.5	0.94	G 1	−129	72
Pal 6		XI	17	40.6	−26	12	13.75	(1.8)	20.7	21.5	—	—	—	—	—
6426		IX	17	42.4	03	12	0.87	(2.2)	—	—	(21.0)	0.70	(G 0)	—	12
6440		V	17	45.9	−20	21	4.59	5.8	—	—	(4.0)	0.77	G 5	−133	—
6441		III	17	46.8	−37	02	2.75	(3.0)	—	—	(9.0)	—	G 2	− 70	—
6453		IV	17	48.0	−34	37	3.44	(3.6)	—	—	—	0.24	(G 5)	—	—
6496		XII	17	55.5	−44	13	1.38	(12.7)	—	—	—	—	—	—	—
6517		IV	17	59.1	−08	57	2.30	7.7	—	—	—	1.19	(G 0)	—	—
6522		VI	18	00.4	−30	02	3.44	4.1	—	—	(12.0)	0.22	F 8	—	9
6528		V	18	01.6	−30	04	2.75	3.0	—	—	—	0.66	(G 5)	—	0
6535		XI	18	01.3	−00	18	1.38	(1.3)	16.27	17.07	15.6	—	—	—	1
6539		X	18	02.1	−07	35	2.30	13.8	—	—	—	1.99	(G 0)	—	1
6541	(Δ 473)	III	18	04.4	−43	44	1.26	(23.2)	13.45	14.25	4.0	—	F 6	−148	1
6544		—	18	04.3	−25	01	6.88	(1.0)	—	—	(5.0)	—	G 2	− 12	—
6553		XI	18	06.3	−25	56	4.59	8.2	—	—	—	0.32	(G 5)	—	6
6558		—	18	07.0	−31	47	2.30	—	—	—	—	—	—	—	9
I 1276		XII	18	08.0	−07	14	2.30	(6.0)	18.5	19.3	25.1	—	—	—	5
6569		VIII	18	10.4	−31	50	1.97	6.6	—	—	—	0.69	(G 5)	—	5
6584	(Δ 376)	VIII	18	14.6	−52	14	0.82	(9.7)	—	—	(13.0)	—	F 7	+160	0
6624		VI	18	20.5	−30	23	1.72	4.2	—	—	(13.0)	0.63	G 5	+ 69	—
6626	(M 28)	IV	18	21.5	−24	54	2.30	9.1	14.90	15.7	4.8	0.45	G 1	+ 1	16
6637	(M 69)	V	18	28.1	−32	23	1.26	6.8	—	—	(7.0)	0.62	G 5	+ 95	5

6638		VI	18 27.9	−25 32	1.72	4.3	16.54	17.34	15.2	0.59	G 4	− 14	—
6642		—	18 28.4	−23 30	2.30	(0.8)	—	—	—	—	—	—	—
6652		VI	18 32.5	−33 02	1.15	4.2	—	—	(16.0)	0.56	G 4	−124	—
6656	(M 22)	VII	18 33.3	−23 58	1.72	26.2	13.03	13.7	3.0	0.50	F 7	−144	24
Pal 8		X	18 38.5	−19 52	1.97	(1.6)	19.6	20.4	47.9	—	—	—	—
6681	(M 70)	V	18 40.0	−32 21	1.07	5.1	—	—	(20.0)	0.40	G 3	+198	2
6712		IX	18 50.3	−08 47	2.75	12.3	15.59	15.85	5.7	0.41	G 5	−131	12
6715	(M 54)	III	18 52.0	−30 32	0.93	4.8	—	—	(17.0)	0.87	F 8	+122	80
6717		VIII	18 52.1	−22 47	1.26	(2.6)	16.3	17.1	14.7	—	—	—	—
6723	(Δ 573)	VII	18 56.2	−36 42	0.78	11.7	14.32	15.12	7.4	0.49	G 4	− 3	19
6749		—	19 02.6	01 48	6.88	—	—	—	—	—	—	—	—
6752	(Δ 295)	VI	19 06.4	−60 04	0.55	(41.9)	13.36	14.16	5.3	—	F 6	− 39	1
6760		IX	19 08.6	00 57	3.44	8.9	17.15	17.95	8.4	0.78	(G 0)	—	4
6779	(M 56)	X	19 14.6	30 05	1.53	10.1	15.54	16.34	10.5	0.43	F 6	−145	12
Pal 10		XII	19 16.0	18 28	6.88	(3.1)	20.7	21.5	8.3	—	—	—	—
Anon		—	19 25.6	−30 27	0.70	—	—	—	—	—	—	—	—
6809	(M 55)	XI	19 36.9	−31 03	0.59	21.1	13.68	14.48	6.0	0.48	(F 5)	+170	6
Pal 11		XI	19 42.6	−08 09	0.87	(2.8)	17.4	18.2	28.8	—	—	—	0
6838	(M 71)	—	19 51.5	18 39	2.75	10.2	15.00	14.15	2.6	0.32	G 6	− 80	4
6864	(M 75)	I	20 03.2	−22 04	0.55	4.9	17.47	18.27	35.1	0.64	G 2	−198	11
6934		VIII	20 31.7	07 14	0.74	3.3	16.06	16.86	18.3	0.51	G 0	−360	51
6981	(M 72)	IX	20 50.7	−12 44	0.44	6.4	16.15	16.95	21.1	0.54	G 3	−255	39
7006		I	20 59.1	16 00	0.74	3.0	17.51	18.6	39.8	0.50	F 2	−348	49
7078	(M 15)	IV	21 27.6	11 57	0.53	9.4	14.44	15.50	10.5	0.54	F 2	−107	103
7089	(M 2)	II	21 30.9	−01 03	0.41	6.8	14.77	15.75	12.3	0.55	F 4	− 5	17
7099	(M 30)	V	21 37.5	−23 25	0.33	6.8	14.79	15.59	13.3	0.48	A 7	−174	4
Pal 12		XII	21 43.7	−21 28	0.32	(2.1)	17.4	18.2	38.0	—	—	—	3
Pal 13		XII	23 04.2	12 28	0.36	(4.0)	19.6	20.4	100.0	—	—	—	4
7492		XII	23 05.7	−15 54	0.27	(4.3)	17.20	18.00	34.7	—	—	—	1

[1] Concentration class, positions, absorption, mean B magnitudes, distance moduli, distances, color indices, spectral types, and radial velocities are from ARP (1965). The concentration class is after the system of SHAPLEY and SAWYER (1927) in which numbers I to XII denote equal classes of decreasing concentration of the central stars. The angular diameters not enclosed in parenthesis are after KRON and MAYALL (1960) and denote the diameter within which ninety percent of the light is contained. Angular diameters in parenthesis are from HOGG (1963), and these values are compiled from the lists of SHAPLEY and SAYER (1935) and MOWBRAY (1946). The distances in parenthesis and the number of variable stars are from HOGG (1963), and the radial velocities are from MAYALL (1946).

where ν_c is in MHz, and it is assumed that $H = 0.3 \times 10^{-4}$ gauss. If $\nu_c = 100$ MHz, $T = 10^7$ years, whereas $T = 10^6$ years for $\nu_c = 10{,}000$ MHz. A detailed model of radio sources which relates age to electron parameters is given by VAN DER LAAN (1963). SCHMIDT (1966) notes that strong radio sources are associated with bright elliptical galaxies whose ages are on the order of 10^{10} years. Statistical data on the space density of radio sources and elliptical galaxies then indicates that quasars and radio galaxies are, respectively, about 10^6 and 10^9 years old.

5.2.6. The Cosmic Time Scale

Following the first measurement of the line spectrum of a galaxy (HUGGINS, 1864) all galaxies were found to have line spectra which were shifted to the red (SLIPHER, 1914, 1917, 1925) indicating a general recession of the galaxies away from the earth. HUBBLE (1929) and HUBBLE and HUMASON (1931, 1934) then showed that the velocity of recession, V, was linearly related to the distance, D, when the distances to the galaxies were determined from the luminosities of the brightest stars in the galaxies or the brightest galaxies in clusters of galaxies. The Hubble diagram, then, indicates a cosmic age, T, given by

$$T = \frac{D}{V} = \frac{1}{H_0} = 0.98 \times 10^{+10} h^{-1} \text{ years}, \tag{5-82}$$

where the Hubble constant $H_0 = 100h$ km sec^{-1} Mpc^{-1}, and h is a constant. The detailed value of h depends on the distance scale discussed in Sect. 5.3, and current estimates give $h \approx 0.5$ (see footnote 4). As pointed out by BOK (1946) and SANDAGE (1971) radioactive decay ages, stellar evolution ages, and the Hubble expansion age all give an age of 3×10^9 to 10^{10} years. For reference, the observed Hubble diagram for 474 optical galaxies (HUMASON, MAYALL, and SANDAGE, 1956) and for radio galaxies and quasi-stellar objects (SANDAGE, 1971) is shown in Fig. 35.

The linear form of the Hubble law, except at large redshifts, is a necessary consequence of a homogeneous, isotropic, expanding cosmological model. If k denotes the curvature index of the Robertson-Walker metric, q_0 denotes the deceleration parameter, and H_0 denotes Hubble's constant, then the Friedmann time, T_0, (the age of the universe) is given by

$$T_0 = \frac{1}{H_0}\left[\frac{1}{(1-2q_0)} - \frac{q_0}{(1-2q_0)^{3/2}} \cosh^{-1}\left(\frac{1}{q_0} - 1\right)\right] \tag{5-83}$$

for $k = -1$, $\rho_0 < \rho_c$ and $0 < q_0 < 0.5$,

$$T_0 = \tfrac{2}{3} H_0^{-1} \tag{5-84}$$

for $k = 0$, $\rho_0 = \rho_c$, and $q_0 = 0.5$, and

$$T_0 = \frac{q_0}{H_0(2q_0 - 1)^{3/2}}\left[\cos^{-1}\left(\frac{1}{q_0} - 1\right) - \frac{1}{q_0(2q_0-1)^{1/2}}\right] \tag{5-85}$$

for $k = 1$, $\rho_0 > \rho_c$ and $q_0 > 0.5$. Here ρ_0 denotes the present value of the matter and radiation energy density of the universe, and the critical density, ρ_c, is given by

$$\rho_c = \frac{3H_0^2}{8\pi G} = 1.9 \times 10^{-29} h^2 \text{ g cm}^{-3}. \tag{5-86}$$

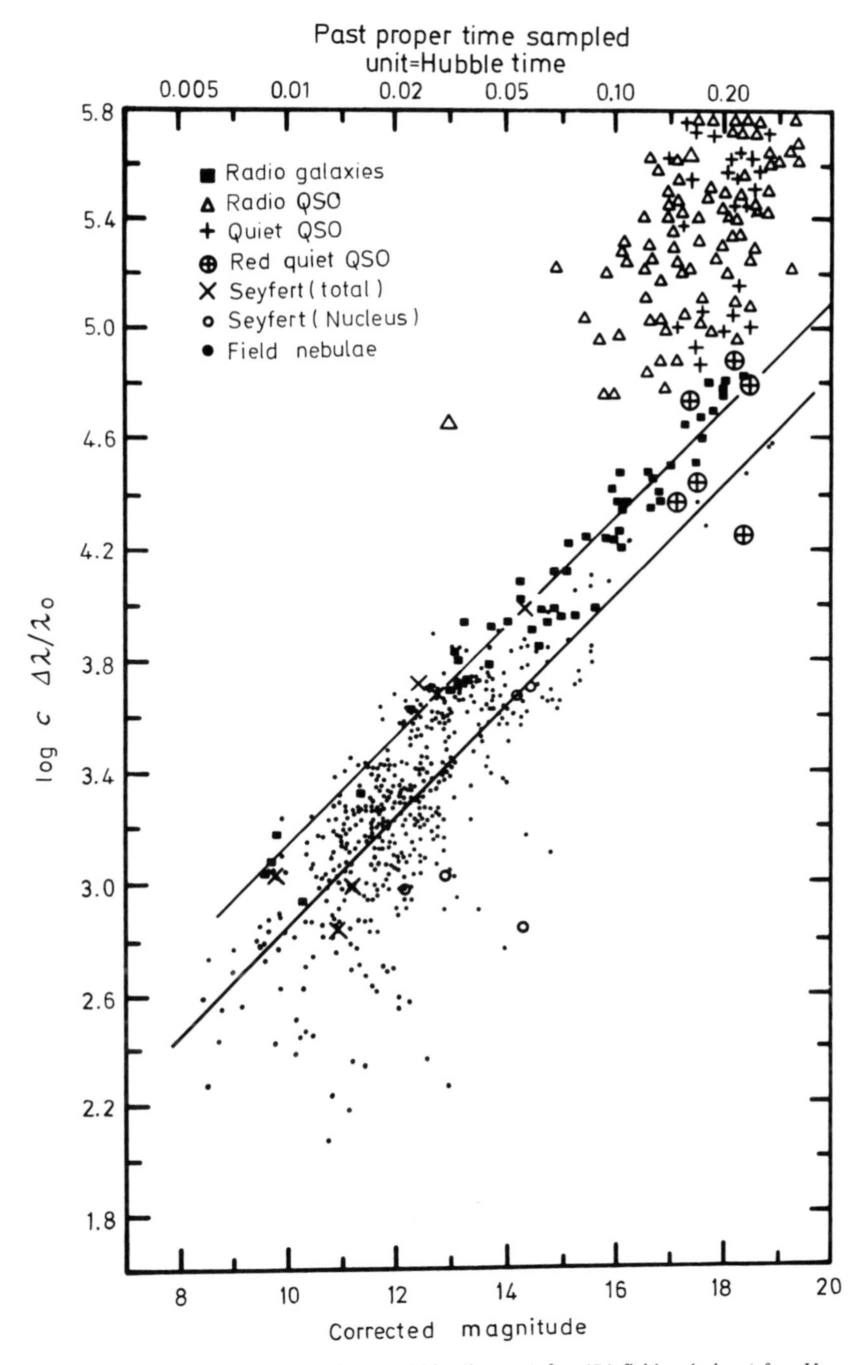

Fig. 35. The redshift-magnitude relation (Hubble diagram) for 474 field nebulae (after HUMASON, MAYALL, and SANDAGE, 1956, by permission of the American Astronomical Society and the University of Chicago Press), and for radio galaxies, quasi-stellar objects, QSO's, and for Seyfert galaxies (after SANDAGE, 1971, by permission of Pontificia Academia Scientarium-Vatican City). The ordinate is the product $c\Delta\lambda/\lambda_0$, where $z=\Delta\lambda/\lambda_0$ is the redshift, and the abscissa is the apparent visual magnitude corrected for aperature effect, K dimming, and absorption within our own Galaxy. The top solid line is for Sandage's data on the brightest galaxy of a cluster, whereas the bottom solid line is the best linear fit to the field nebulae data

The cosmological parameters k, q_0, and H_0 are described in detail in Sect. 5.7.1. The universe is closed and oscillatory, Euclidean and expanding, or open and expanding according as $k=1$, 0, or -1. For $k=0$ the $q_0=0.5$ and the Friedmann time $T_0=2H_0^{-1}/3=1.3\times 10^{10}$ years for $H_0=50$ km sec^{-1} Mpc^{-1}.

5.3. Distance

5.3.1. Distance of the Moon, Sun, and Planets

The Greeks first estimated the parallax of the moon, $\pi_☾$, by observing the angular breadth, φ, of the earth's shadow during a lunar eclipse.

$$\pi_☾ \approx \pi_☾ + \pi_\odot = (\theta + \varphi)/2, \tag{5-87}$$

where the solar parallax, $\pi_\odot$, was taken to be negligible compared to the lunar parallax, and θ is the angular extent of the Sun. HIPPARCHUS (125 B.C.) measured $\theta/2 = 16'36.9''$, $\varphi \approx 2.5\,\theta$ and $\pi_☾ = 3489''$. The presently accepted value is (CLEMENCE, 1965)

$$\pi_☾ = \frac{a_e}{a_☾} = 3422.451 \text{ seconds of arc}, \tag{5-88}$$

where the earth radius, $a_e = 6.378160 \times 10^8$ cm, and the mean distance to the moon is

$$a_☾ = 3.8440 \times 10^{10} \text{ cm}. \tag{5-89}$$

As first observed by KEPLER (1619) and explained by NEWTON (1687), the orbital period, P, of a planet about the Sun is related to the semi-major axis, a, of its elliptical orbit by the equation (Kepler's third law)

$$P^2 = \frac{4\pi^2 a^3}{G(m+M_\odot)}, \tag{5-90}$$

where the Newtonian gravitational constant $G \approx 6.668 \times 10^{-8}$ dyn cm^2 g^{-2}, m is the mass of the planet, and the solar mass $M_\odot \approx 2 \times 10^{33}$ gm. It follows from Eq. (5-90) that the semi-major axis of the earth's orbit, the astronomical unit, is given by

$$a = \left(\frac{m_e + M_\odot}{m_p + M_\odot}\right)^{1/3} \left(\frac{P}{P_p}\right)^{2/3} a_p \approx \left(\frac{P}{P_p}\right)^{2/3} a_p, \tag{5-91}$$

where m_e and P are, respectively, the mass and orbital period of the earth, m_p and P_p are the same quantities for another planet, and a_p is the semi-major axis of the orbit of the other planet. Observations of Mars (CASSINI, 1672) and Venus (ENCKE, 1835) led to estimates of the solar parallax $\pi_\odot = a_e/a$ of $\pi_\odot = 9.5''$ and $8.6''$, respectively, by using Eq. (5-91). The presently accepted value of

$$\pi_\odot = 8.79405 \text{ seconds of arc} \tag{5-92}$$

is inferred from accurate radar measurements of the astronomical unit, a. The most accurate value is (ASH, SHAPIRO, and SMITH, 1967; MUHLEMAN, 1969)

$$a = 1.49597892(1) \times 10^{13} \text{ cm}. \tag{5-93}$$

Radar measurements of the orbital motions of the planets result in the parameters given in Table 57. These data roughly follow the Titius (1764)-Bode (1772) law for planetary distances. If a_n denotes the mean distance of planet, n, then the Titius-Bode law states that

$$a_n = 0.1\,[4 + 3 \times 2^n], \tag{5-94}$$

where $n = -\infty, 0, 1, \ldots, 7$, and the distances are given in astronomical units. BLAGG (1913) suggests that the law which applies to the solar system is given by

$$a_n = A(1.7275)^n\,[B + f(\alpha + n\beta)] \tag{5-95}$$

where A and B are positive numerical constants, α and β are angular constants, and f is a periodic function of 2π radians. A historical and theoretical discussion of the Titius-Bode law is given by NIETO (1972).

5.3.2. Distance to the Nearby Stars—Annual, Secular, and Dynamic Parallax, and the Solar Motion

BESSEL (1839) first measured the annual parallax of 61 Cygni showing it to be 0.3136″; but he was closely followed by Henderson's (1839) and Struve's (1840) measurements of the parallax of α Centauri and Vega, respectively. The annual parallax, π, is related to the distance, D, of the star and the astronomical unit, a, by the equation

$$\tan\pi \approx \pi = \frac{a}{D}. \tag{5-96}$$

The distance of the star in parsecs is given by

$$D = \pi^{-1} \text{ parsecs}, \tag{5-97}$$

when π is in seconds of arc. The parsec unit of distance is given as

$$1 \text{ parsec} = 3.0856 \times 10^{18} \text{ cm} = 3.2615 \text{ light years}, \tag{5-98}$$

where a light year is the distance that light travels in one year. The distances and parallax of the stars nearer than five parsecs are given in Table 58, whereas those for the 1049 stars nearer than twenty parsecs are given in GLIESE'S (1969) catalog.

HERSCHEL (1753) first observed that the Sun is moving with respect to the local group of stars, and fixed its apex within a few degrees of the presently accepted value. The standard solar motion relative to the majority of stars in the general catalogs of radial velocity and proper motion is (DELHAYE, 1965)

$$V_\odot = 19.5 \text{ km sec}^{-1}, \tag{5-99}$$

towards

$$\begin{aligned} \alpha_\odot &= 18.0^{\mathrm{h}} & \delta_\odot &= 30^\circ \\ l_\odot^{\mathrm{II}} &= 56^\circ & b_\odot^{\mathrm{II}} &= 23^\circ, \end{aligned} \tag{5-100}$$

at epoch 1900.0. The basic solar motion relative to stars in the solar neighborhood is

$$V_\odot = 15.4 \text{ km sec}^{-1}, \tag{5-101}$$

Table 57. Distance, radius, inverse mass, sidereal period, rotation period, eccentricity, inclination, density, albedo, and theoretical temperatures for the planets of our solar system[1]

Object	Symbol	Mean distance a in (a.u.) (1900.0)	Radius (km)	Inverse mass	Sidereal period (tropical years)	Rotation period (days)	Eccentricity e(1900.0)	Inclination i(1900.0) ° ′ ″	Density (gm cm^{-3})	Albedo	Temperature[3] (°K)
Sun	☉	—	695,980	1.0		25.36	—	—	1.409	—	5,778
Moon	☽	[2]	1,738	27,069,696	[2]		0.05490	05 08 ±09′	3.34	0.068	~441
Mercury	☿	0.3870984	2,439	6,025,000	0.241	59.7	0.20561421	07 00 10.37	5.42	0.058	441
Venus	♀	0.7233299	6,050	408,520	0.615	−243.16	0.00682069	03 23 37.07	5.25	0.76	230
Earth	⊕	1.0000038	6,378.16	332,945.4	1.000	1.00	0.01675104	00 00 00.00	5.517	0.39	246
Mars	♂	1.5237	3,394	3,098,000	1.881	1.03	0.09331290	01 51 01.20	3.96	0.15	218
Jupiter	♃	5.2037	70,850	1,047.4	11.865	0.4135	0.0484	01 18	1.33	0.51	102
Saturn	♄	9.5803	60,000	3,498.5	29.650	0.43	0.0557	02 29	0.68	0.50	76
Uranus	⛢	19.1410	24,500	22,900	83.744	0.89	0.0472	00 46	1.60	0.66	49
Neptune	♆	30.1982	25,100	19,400	165.51	0.53	0.0086	01 46	1.65	0.62	40
Pluto	♇	39.4387	3,200	566	247.687	6.39	0.250	17 10′	—	0.16	42

[1] The inverse mass of the Earth + Moon is 328.900.1. The ratio of the Earth's mass to that of the Moon is $\mu^{-1} = 81.3035$ (Anderson, Efron, and Wong, 1970). The other inverse mass values are from Ash, Shapiro, and Smith (1971) and the references therein. The radii of Venus and Mars are radar values (Ash, Shapiro, and Smith, 1967; Melbourne, Muhleman, and O'Handley, 1968) whereas those for Jupiter, Saturn, Uranus, Neptune, and Pluto are optical (Bixby and van Flandern, 1969; Dollfus, 1970). Sidereal periods and orbital elements are from the Explanatory Supplement to the Ephemeris (1961). Rotation periods of Venus and Mercury are from radar data (Dyce, Pettengill, and Shapiro, 1967), Jupiter's rotation period is from observations of its decametric radiation (Duncan, 1971).

[2] The Moon's mean distance is 384,400 km. The Moon has a sidereal period of 27.32166 days.

[3] Temperatures of the planets are for a spherical black body.

towards

$$\alpha_{\odot} = 17.8^{\mathrm{h}} \qquad \delta_{\odot} = 25^{\circ}$$
$$l_{\odot}^{\mathrm{II}} = 51^{\circ} \qquad b_{\odot}^{\mathrm{II}} = 23^{\circ}, \tag{5-102}$$

at epoch 1900.0. When measurements are corrected for solar motion, they are said to be referred to the local standard of rest. Early discussions of the solar motion with respect to the stars are given by Boss (1910), Eddington (1910), and Smart and Green (1936).

The secular parallax, π_s, is defined in terms of the distance the Sun moves towards its apex in the course of a year.

$$\pi_s = \frac{V_{\odot}}{4.74}\,\pi \approx 4.1\,\pi \text{ seconds of arc,} \tag{5-103}$$

where π is the annual parallax in seconds of arc, and one astronomical unit per year is equivalent to $4.74\,\mathrm{km\,sec^{-1}}$. Here the numerical approximation is for $V_{\odot} = 19.4\,\mathrm{km\,sec^{-1}}$, but $V_{\odot}$ is variable up to $100\,\mathrm{km\,sec^{-1}}$ depending on the reference.

Herschel (1803) was also the first to observe binary stars rotating about each other. If the stars are moving in an elliptical orbit, then it follows from Kepler's third law (Eq. (5-90)) that the dynamical parallax, π_D, of binaries is given by

$$\pi_D = \frac{\alpha}{(M_1 + M_2)^{1/3}\,P^{2/3}} \text{ seconds of arc,} \tag{5-104}$$

where α is the angular size of the semi-major axis of the orbit in seconds of arc, M_1 and M_2 are the stellar masses in solar mass units, and P is the orbital period in years. Because π_D is weakly dependent on mass, $M_1 + M_2$ is often taken to be two.

If radial velocities can be observed, the orbital elements can be measured. To first order, the radial velocity, V_r, is given by (Doppler, 1842)

$$V_r = \frac{c\,\Delta\lambda}{\lambda}, \tag{5-105}$$

where $\Delta\lambda$ is the shift of the wavelength, λ, from its rest value, and V_r is taken to be positive for receding objects (redshift) and negative for approaching ones. Pickering (1889) and Vogel (1890) nearly simultaneously reported the observation of spectroscopic binaries in which the separation of lines was observed to change with time. The observed orbital radial velocity, V_r, is related to the orbital elements by

$$V_r = \frac{2\pi a_1 \sin i}{P\sqrt{1-e^2}}(e\cos\omega + u) = K(e\cos\omega + u), \tag{5-106}$$

where here $\pi = 3.14159$, the orbital period, P, is the period of the variations of V_r, a_1 is the semi-major axis of the orbit of component 1, i is its inclination, $a\sin i$ is the projection of the semi-major axis on the line of sight, e is the eccentricity, ω is the angle formed by the line of nodes and the line of apsides, and u is the

Table 58. Positions, proper motion, position angle, radial velocity, parallax, distance, apparent mag-

No.	Gliese No.	Name	α (1950.0) (h m)	δ (1950.0) (° ′)	Proper motion (″)	Position angle (°)	Radial velocity (km sec⁻¹)	Parallax (″)	Distance (light yrs.)
1		Sun							
2	559, 551	α Centauri[2]	14 36.2	−60 38	3.68	281	− 22	0.760	4.3
3	699	Barnard's star	17 55.4	+04 33	10.31	356	−108	0.552	5.9
4	406	Wolf 359	10 54.1	+07 19	4.71	235	+ 13	0.431	7.6
5	411	BD +36°2147	11 00.6	+36 18	4.78	187	− 84	0.402	8.1
6	244	Sirius	06 42.9	−16 39	1.33	204	− 8	0.377	8.6
7	65	Luyten 726–8	01 36.4	−18 13	3.36	80	+ 30	0.365	8.9
8	729	Ross 154	18 46.7	−23 53	0.72	103	− 4	0.345	9.4
9	905	Ross 248	23 39.4	+43 55	1.58	176	− 81	0.317	10.3
10	144	ε Eridani	03 30.6	−09 38	0.98	271	+ 16	0.305	10.7
11	866	Luyten 789–6	22 35.7	−15 36	3.26	46	− 60	0.302	10.8
12	447	Ross 128	11 45.1	+01 06	1.37	153	− 13	0.301	10.8
13	820	61 Cygni	21 04.7	+38 30	5.22	52	− 64	0.292	11.2
14	845	ε Indi	21 59.6	−57 00	4.69	123	− 40	0.291	11.2
15	280	Procyon	07 36.7	+05 21	1.25	214	− 3	0.287	11.4
16	725	Σ 2398	18 42.2	+59 33	2.28	324	+ 5	0.284	11.5
17	15	BD +43°44	00 15.5	+43 44	2.89	82	+ 17	0.282	11.6
18	887	CD −36°15693	23 02.6	−36 09	6.90	79	+ 10	0.279	11.7
19	71	τ Ceti	01 41.7	−16 12	1.92	297	− 16	0.273	11.9
20	273	BD +5°1668	07 24.7	+05 23	3.73	171	+ 26	0.266	12.2
21	825	CD −39°14192	21 14.3	−39 04	3.46	251	+ 21	0.260	12.5
22	191	Kapteyn's star	05 09.7	−45 00	8.89	131	+245	0.256	12.7
23	860	Krüger 60	22 26.3	+57 27	0.86	246	− 26	0.254	12.8
24	234	Ross 614	06 26.8	−02 46	0.99	134	+ 24	0.249	13.1
25	628	BD −12°4523	16 27.5	−12 32	1.18	182	− 13	0.249	13.1
26	35	van Maanen's star	00 46.5	+05 09	2.95	155	+ 54	0.234	13.9
27	473	Wolf 424	12 30.9	+09 18	1.75	277	− 5	0.229	14.2
28		G158–27	00 04.2	−07 48	2.06	204		0.226	14.4
29	1	CD −37°15492	00 02.5	−37 36	6.08	113	+ 23	0.225	14.5
30	380	BD +50°1725	10 08.3	+49 42	1.45	249	− 26	0.217	15.0
31	674	CD −46°11540	17 24.9	−46 51	1.13	147		0.216	15.1
32	832	CD −49°13515	21 30.2	−49 13	1.81	185	+ 8	0.214	15.2
33	682	CD −44°11909	17 33.5	−44 17	1.16	217		0.213	15.3
34	83.1	Luyten 1159–16	01 57.4	+12 51	2.08	149		0.212	15.4
35	526	BD +15°2620	13 43.2	+15 10	2.30	129	+ 15	0.208	15.7
36	687	BD +68°946	17 36.7	+68 23	1.33	194	− 22	0.207	15.7
37	440	L145–141	11 43.0	−64 33	2.68	97		0.206	15.8
38	876	BD −15°6290	22 50.6	−14 31	1.16	125	+ 9	0.206	15.8
39	166	40 Eridani	04 13.0	−07 44	4.08	213	− 43	0.205	15.9
40	388	BD +20°2465	10 16.9	+20 07	0.49	264	+ 11	0.202	16.1
41	768	Altair	19 48.3	+08 44	0.66	54	− 26	0.196	16.6
42	702	70 Ophiuchi	18 02.9	+02 31	1.13	167	− 7	0.195	16.7
43	445	AC +79°3888	11 44.6	+78 58	0.89	57	−119	0.194	16.8
44	873	BD +43°4305	22 44.7	+44 05	0.83	237	− 2	0.193	16.9
45	169.1	Stein 2051	04 26.8	+58 53	2.37	146		0.192	17.0

[1] After VAN DE KAMP (1971) by permission of Annual Reviews Inc. The symbol b denotes an unseen

[2] The position of α Centauri C ("Proxima") is $14^h26.3^m$, $-62°28'$; $2°11'$ from the center of mass of

nitude, spectrum, absolute magnitude, and luminosity of the stars nearer than five parsecs[1]

No.	Visual apparent magnitude and spectrum A	B	C	Visual absolute magnitude A	B	C	Visual luminosity A	B	C	No.
1	−26.8 G2			4.8			1.0			1
2	0.1 G2	1.5 K6	11 M5e	4.5	5.9	15.4	0.3	0.36	0.00006	2
3	9.5 M5	b		13.2	b		0.00044	b		3
4	13.5 M8e			16.7			0.00002			4
5	7.5 M2	b		10.5	b		0.0052	b		5
6	− 1.5 A1	8.3 DA		1.4	11.2		23.0	0.0028		6
7	12.5 M6e	13.0 M6e		15.3	15.8		0.00006	0.00004		7
8	10.6 M5e			13.3			0.0004			8
9	12.2 M6e			14.7			0.00011			9
10	3.7 K2			6.1			0.30			10
11	12.2 M6			14.6			0.00012			11
12	11.1 M5			13.5			0.00033			12
13	5.2 K5	6.0 K7	b	7.5	8.3	b	0.083	0.040	b	13
14	4.7 K5			7.0			0.13			14
15	0.3 F5	10.8		2.6	13.1		7.6	0.0005		15
16	8.9 M4	9.7 M5		11.2	12.0		0.0028	0.0013		16
17	8.1 M1	11.0 M6		10.4	13.3		0.0058	0.00040		17
18	7.4 M2			9.6			0.012			18
19	3.5 G8			5.7			0.44			19
20	9.8 M4	b		11.9	b		0.0014	b		20
21	6.7 M1			8.8			0.025			21
22	8.8 M0			10.8			0.0040			22
23	9.7 M4	11.2 M6		11.7	13.2		0.0017	0.00044		23
24	11.3 M5e	14.8		13.3	16.8		0.0004	0.00002		24
25	10.0 M5			12.0			0.0013			25
26	12.4 DG			14.2			0.00017			26
27	12.6 M6e	12.6 M6e		14.4	14.4		0.00014	0.00014		27
28	13.8 m			15.5			0.00005			28
29	8.6 M3			10.4			0.00058			29
30	6.6 K7			8.3			0.040			30
31	9.4 M4			11.1			0.0030			31
32	8.7 M3			10.4			0.0058			32
33	11.2 M5			12.8			0.00063			33
34	12.3 M8			13.9			0.00023			34
35	8.5 M2			10.1			0.0076			35
36	9.1 M3.5	b		10.7	b		0.0044	b		36
37	11.4			12.6			0.0008			37
38	10.2 M5			11.8			0.0016			38
39	4.4 K0	9.5 DA	11.2 M4e	6.0	11.2	12.8	0.33	0.0027	0.00063	39
40	9.4 M4.5	b		10.9	b		0.0036	b		40
41	0.8 A7			2.3			10.0			41
42	4.2 K1	6.0 K6		5.7	7.5		0.44	0.083		42
43	11.0 M4			12.4			0.0009			43
44	10.1 M5e	b		11.5	b		0.0021	b		44
45	11.1 M5	12.4 DC		12.5	13.8		0.0008	0.0003		45

companion.

α Centauri A and B. The proper motion of C is 3.84″ in position angle 282°. One light year $= 9.4605 \times 10^{17}$ cm.

angle between the radius vector and the line of nodes. The constant, K, is obtained directly from the maximum, V_{max}, and minimum, V_{min}, radial velocities

$$K = \frac{V_{max} - V_{min}}{2}. \tag{5-107}$$

The constants e and ω can be determined by comparing the shape of a plot of V_r versus time with curves of $(e\cos\omega + u)$ for a variety of e and ω (KING, 1920). The $a\sin i$ is then found from K, P, and e. The mass of component 1 is given by

$$M_1 \sin^3 i = \frac{(a\sin i)^3}{P^2\left(1 + \dfrac{a_1 \sin i}{a_2 \sin i}\right)}, \tag{5-108}$$

where $a\sin i = a_1 \sin i + a_2 \sin i$ can be determined from the radial velocity plots of the two components. We also have the relations $a^3 = GP^2(M_1 + M_2)/(4\pi^2)$ and $a_1 M_1 = a_2 M_2$.

5.3.3. Distance to Moving Clusters and Statistical Parallax

HALLEY (1718) first noted that Sirius, Aldebaran, Betelgeux, and Arcturus were occupying positions different from those given by Ptolemy in the Syntaxis. If V_T is the component of the stellar space velocity perpendicular to the line of sight, then the observed proper motion, μ, is given by

$$\mu = \frac{V_T}{4.74\, D} \text{ seconds of arc per year,} \tag{5-109}$$

where V_T is in km sec^{-1}, and the distance to the star, D, is in parsecs. Measurements of positions and proper motions accurate to 0.05″ and 0.2″ per century, respectively, are given by FRICKE and KOPFF (1963) for 1535 fundamental stars and 1987 supplementary stars. This fundamental catalogue is called the FK4 catalogue and is used as the reference catalogue for the Smithsonian Astrophysical Observatory (1966) catalogue of the positions and proper motions of 258,997 stars for epoch and equinox 1950.0. These data are available on computer magnetic tape.

When a group of stars appears to converge to a common point, the parallax, π, is given by

$$\pi = \frac{4.74\, \mu}{V_r \tan\theta}, \tag{5-110}$$

where V_r is the component of velocity along the line of sight in km sec^{-1}, and is obtainable from measurements of line spectra (Eq. (5-105)), and θ is the angular distance between a given star and the convergence point. Stellar radial velocities are tabulated by WILSON (1953) and ABT and BIGGS (1972). Radial velocities for the stars nearer than five parsecs were given in Table 58.

KAPTEYN (1905, 1912) first noted that the stars drift in two streams with the respective apex coordinates of $\alpha = 91°$, $\delta = -15°$ and $\alpha = 288°$, $\delta = -64°$ at epoch 1900.0 (cf. EDDINGTON, 1910). For a single drift of stars, it is convenient to follow Kapteyn's practice and divide the proper motion into components, ν and τ, along and at right angles to a great circle through the star and the position of the solar apex. If μ_α and μ_δ denote, respectively, the components of proper motion in right ascension and declination, then

$$\nu = \mu_\alpha \cos\delta \sin\Psi - \mu_\delta \cos\Psi$$

and (5-111)

$$\tau = \mu_\delta \sin\Psi + \mu_\alpha \cos\delta \sin\Psi,$$

where the declination of the star is δ, and Ψ is the angle between the great-circle arcs joining the star to the solar apex and to the north celestial pole. The angular distance, λ, between the star and the apex of solar motion is given by

$$\cos\lambda = \sin\delta \sin\delta_\odot + \cos\delta \cos\delta_\odot \cos(\alpha - \alpha_\odot), \tag{5-112}$$

where α, δ are the equatorial coordinates of the star at epoch 1900.0, and $\alpha_\odot$, $\delta_\odot$ are the coordinates of the solar apex given in Eq. (5-100). The angle, Ψ, is given by the equation

$$\sin\lambda \cos\Psi = \sin\delta_\odot \cos\delta - \sin\delta \cos\delta_\odot \cos(\alpha - \alpha_\odot). \tag{5-113}$$

The statistical parallax, π, is then given by

$$\pi = \frac{4.74}{V_\odot} \frac{\langle \nu \sin\lambda \rangle}{\langle \sin^2\lambda \rangle} \text{ seconds of arc}, \tag{5-114}$$

where $V_\odot$ is given by Eq. (5-99), ν is in units of seconds of arc per year, and $\langle\ \rangle$ denotes an average over the stars in the group. If the group of stars has velocities which are randomly oriented, then the statistical parallax may be determined from the equation

$$\pi = \frac{4.74 \langle |\tau| \rangle}{\langle |V_R + V_\odot \cos\lambda| \rangle} \text{ seconds of arc}, \tag{5-115}$$

where V_R is the radial velocity in km $\sec^{-1}$, τ is in units of seconds of arc per year, and $|\ |$ denotes the absolute value. These statistical parallax techniques can yield accurate distances out to about 500 parsecs, whereas the annual parallax technique discussed in the previous section yields accurate distances out to about 30 parsecs.

5.3.4. Galactic Rotation and Kinematic Distance

Slipher (1914) first observed rotation in external galaxies, and Shapley (1918) first showed that the system of globular clusters is centered at our galactic center rather than the Sun. Lindblad (1927) then suggested that the high apparent velocities of clusters could be accounted for if both the Sun and the globular clusters were rotating about the galactic center. Oort (1927, 1928) provided some observational evidence for differential galactic rotation, and put forth the following formulae for circular motion about the galactic center. The radial velocity, V_r, of a galactic object at the distance, R, from the galactic center is given by

$$V_r = R_0[\omega(R) - \omega(R_0)] \sin l^{\mathrm{II}} \cos b^{\mathrm{II}}$$

or (5-116)

$$V_r \approx R_0 \left(\frac{d\omega}{dR}\right)_{R=R_0} (R - R_0) \sin l^{\mathrm{II}} \cos b^{\mathrm{II}} \quad \text{for } R - R_0 \ll R_0,$$

where R_0 is the distance of the local standard of rest from the galactic center, $\omega(R)$ is the circular angular velocity of the Galaxy at R, and l^{II} and b^{II} denote, respectively, the galactic longitude and latitude. Conversion formulae from equatorial to galactic coordinates were given in Sect. 5.1.5. The proper motion, μ, of the celestial object in galactic longitude is

$$\mu = \frac{1}{4.74}\left[-\frac{1}{2}R_0\left(\frac{d\omega}{dR}\right)_{R=R_0}\cos 2l^{\mathrm{II}} - \frac{1}{2}R_0\left(\frac{d\omega}{dR}\right)_{R=R_0} - \omega(R_0)\right], \quad (5\text{-}117)$$

where μ is in seconds of arc per year. Oort's constants of differential galactic rotation, A and B, are given by (SCHMIDT, 1965)

$$A = -\frac{1}{2}R_0\left(\frac{d\omega}{dR}\right)_{R=R_0} = 15\ \mathrm{km\ sec^{-1}\ kpc^{-1}}$$

and

$$B = -\frac{1}{2}R_0\left(\frac{d\omega}{dR}\right)_{R=R_0} - \omega(R_0) = -10\ \mathrm{km\ sec^{-1}\ kpc^{-1}}, \quad (5\text{-}118)$$

where the numerical values are the internationally accepted values. Using Eqs. (5-118) in (5-116) and (5-117) we obtain

$$V_r = -2A(R-R_0)\sin l^{\mathrm{II}}\cos b^{\mathrm{II}} \quad \text{for } R-R_0 \ll R_0, \quad (5\text{-}119)$$

and

$$\mu = \frac{1}{4.74}\left[B + A\cos 2l^{\mathrm{II}}\right]. \quad (5\text{-}120)$$

Along a direction for which $|l^{\mathrm{II}}| < 90°$ and in the galactic plane, the maximum radial velocity, $V_{r\,max}$, will occur at that point which is closest to the galactic center. At this point, $R = R_0|\sin l^{\mathrm{II}}|$, and Eq. (5-116) becomes

$$V_{r\,max} = 2AR_0\sin l^{\mathrm{II}}(1-|\sin l^{\mathrm{II}}|) = R_0\left[\frac{\theta_c(R)}{R} - \frac{\theta_c(R_0)}{R_0}\right]\sin l^{\mathrm{II}}, \quad (5\text{-}121)$$

where $R\omega(R) = \theta_c(R)$ is the circular velocity at the distance, R, from the galactic center. The internationally accepted values for R_0 and $\theta_c(R_0)$ are

$$R_0 = 10\ \mathrm{kpc}$$

and

$$\theta_c(R_0) = R_0(A-B) = 250\ \mathrm{km\ sec^{-1}}. \quad (5\text{-}122)$$

It follows from Eq. (5-116) that a measurement of the radial velocity, V_r, may lead to a measurement of the distance, D, of a galactic object from the local standard of rest. For an object near the Sun, for example, Eq. (5-116) becomes

$$V_r = AD\sin 2l^{\mathrm{II}}\cos^2 b^{\mathrm{II}}, \quad (5\text{-}123)$$

and for objects in the galactic plane ($b^{\mathrm{II}} \approx 0$),

$$R^2 = R_0^2 + D^2 - 2R_0 D\cos l^{\mathrm{II}}, \quad (5\text{-}124)$$

where R may be determined from V_r and Eq. (5-119). For $R < R_0$, however, a single value of radial velocity, V_r, corresponds to two values of R, and there is an ambiguity in determining the distance to a galactic object. This ambiguity is resolved by using Eq. (5-116) to obtain $\omega(R)$ from V_r, and then comparing $\omega(R)$ to models which give $\omega(R)$ as a function of R. By using observations of the neutral hydrogen line (KWEE, MULLER, and WESTERHOUT, 1954; KERR, 1962), and theoretical models of the galactic mass distribution (SCHMIDT, 1956, 1965), a plot of circular velocity $\theta_c(R) = R\,\omega(R)$ versus distance, R, from the galactic center may be formed (cf. Figs. 36 and 37). Distances obtained from radial velocities and such a rotation curve are called kinematic distances. The outer part of the rotation curve for our Galaxy is not observationally determined, and it is based on an assumed mass model given in the caption to Fig. 36. ROBERTS and ROTS (1973) have shown that the rotation curves for some nearby spiral galaxies fall off less rapidly than the curve shown in Fig. 36 for distances larger than 15 kpc. Their data indicate that a significant amount of matter may exist at large distances from the center of spiral galaxies, and that these galaxies are larger than the extent found from photometric measurements.

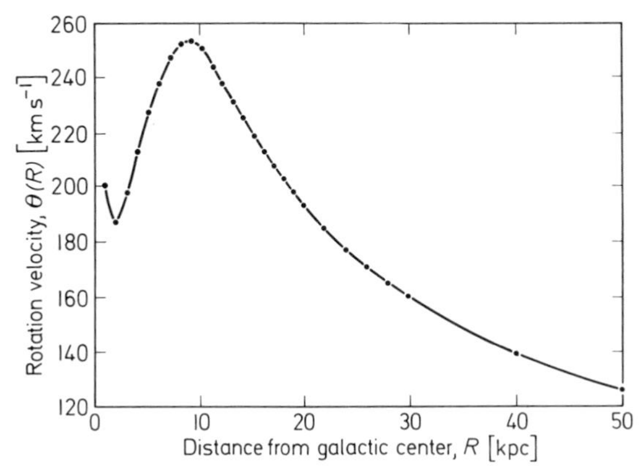

Fig. 36. Galactic rotation curve for a mass model in which the mass density, ρ, is given by $\rho = 3930\,R^{-1} - 0.02489\,R$ for $R \le R_0$, and $\rho = 1449.2\,R^{-4}$ for $R > R_0$. Here the mass density is in units of solar masses per cubic parsec, and R is the distance from the galactic center in parsecs. Near the Sun, $R = R_0 = 10$ kiloparsecs and $\rho = 0.145$ solar masses per cubic parsec. (After SCHMIDT, 1965, by permission of the University of Chicago Press)

LINDBLAD (1927) first suggested that the observed persistence of galactic spiral arms against differential rotation would be explained if the material arms coexist with a density wave pattern propagating around the Galaxy with a pattern speed, $\Omega_p(R)$, given by

$$\Omega_c(R) - \frac{\kappa(R)}{m} < \Omega_p(R) < \Omega_c(R) + \frac{\kappa(R)}{m}, \tag{5-125}$$

where $\Omega_c(R)$ and $\Omega_p(R)$ are the respective angular velocities of the material and the density wave about the galactic center at a distance, R, therefrom, $\kappa(R)$ is the epicyclic frequency, and m is an integer denoting the number of spiral arms. The epicyclic frequency is defined by the relation

$$[\kappa(R)]^2 = [2\Omega_c(R)]^2 \left\{1 + \frac{R}{2\Omega_c(R)} \frac{d\Omega_c(R)}{dR}\right\}, \tag{5-126}$$

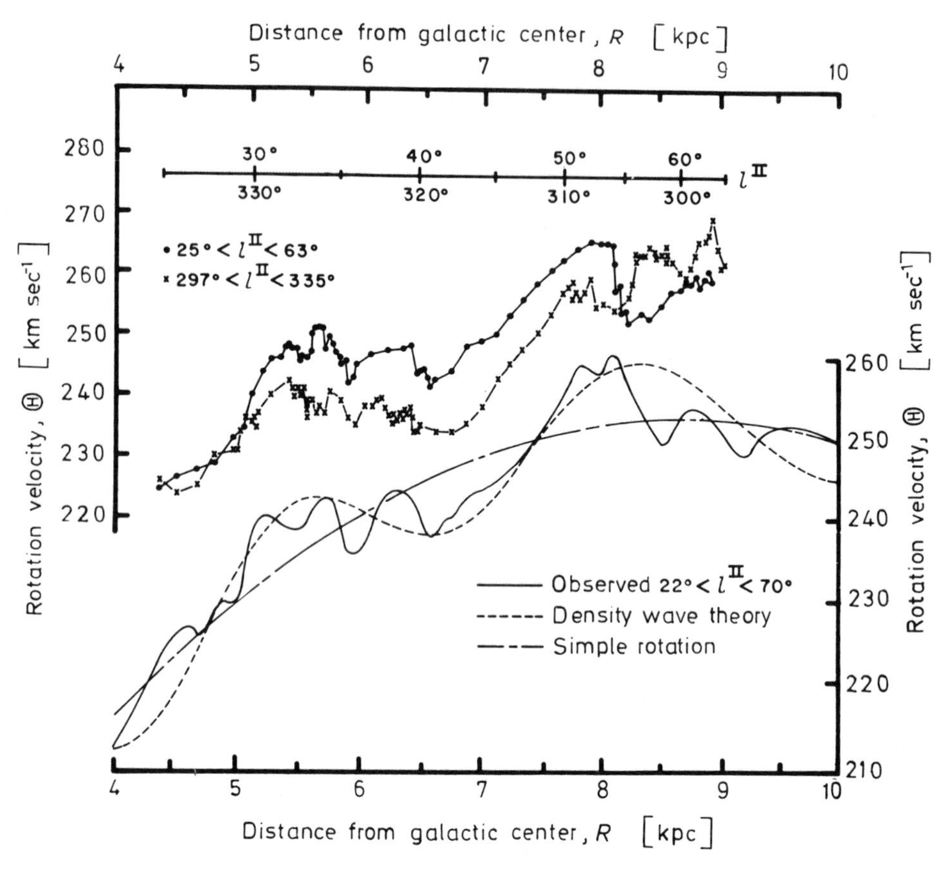

Fig. 37. Apparent galactic rotation curves. The top curves are for sections of the Milky Way north and south of the galactic center, and they are derived from tangential-point observations of H I profiles assuming circular rotation and no streaming (after KERR, 1969, by permission of Annual Reviews Inc.). The bottom solid curve is derived from similar observations in the longitude range $22° < l^{\mathrm{II}} < 70°$ (after SHANE and BIEGER-SMITH, 1966, by permission of the Astronomical Institute of the Netherlands and the North Holland Publ. Co.). The long dashed curve is the basic rotation curve free of streaming motions, whereas the short dashed curve is the basic rotation curve perturbed by density-wave theory streaming (after BURTON, 1971, 1974, by permission of Springer-Verlag)

and is related to the spacing, λ_*, between the spiral arms by the equation

$$\lambda_* \approx \frac{4\pi^2 G\sigma(R)}{[\kappa(R)]^2}, \tag{5-127}$$

where G is the gravitational constant, and $\sigma(R)$ is the projected mass surface density. In the presence of a density wave, the outer and inner sides of a concentration of hydrogen will have, respectively, higher and lower speeds than would have been observed in its absence. The rotation curve will then exhibit oscillations which are in fact observed (cf. Fig. 37), and kinematic distances must be corrected for the density wave effect. Such corrections may be made using the density wave theory of LIN, YUAN, and SHU (1969). They derive a dispersion relation between the arm spacing, λ, and the angular frequency, v, at which stars see the wave pattern

$$v = \frac{m[\Omega_c(R) - \Omega_p(R)]}{\kappa(R)}. \tag{5-128}$$

They show that λ varies smoothly from $0.55\,\lambda_*$ to $0.05\,\lambda_*$ as $|v|$ varies from zero to one, and indicate that $\Omega_p(R_0) = 13.5$ km sec^{-1} kpc^{-1}. BURTON (1971, 1974) has used the density wave theory of LIN *et al.* (1969) together with the SCHMIDT (1965) like rotation curve of

$$\begin{aligned} \theta_c(R) &= 250.0 + 4.05(10-R) - 1.62(10-R)^2 \text{ km sec}^{-1} \quad \text{for } 4\text{ kpc} \le R \le 10\text{ kpc} \\ \theta_c(R) &= 885.44\,R^{-1/2} - 30{,}000\,R^{-3} \text{ km sec}^{-1} \quad \text{for } 10\text{ kpc} \le R \le 14\text{ kpc} \end{aligned} \tag{5-129}$$

to obtain curves relating radial velocity to the distance from the Sun, R, and galactic longitude, l^{II} (cf. Fig. 38). Here the radial velocity is defined with respect to the local standard of rest, so that the observed radial velocity is corrected for the components of radial motion due to the solar motion, $V_{r\odot}$, and the rotation of the earth about the Sun. If the source right ascension, α, and declination, δ, are given at the same epoch as that of those of the solar apex, $\alpha_\odot$ and $\delta_\odot$, (Eq. (5-99)) then

$$\begin{aligned} V_{r\odot} = 19.5(&\cos\alpha_\odot\cos\delta_\odot\cos\alpha\cos\delta + \sin\alpha_\odot\cos\delta_\odot\sin\alpha\cos\delta \\ &+ \sin\delta_\odot\sin\delta) \text{ km sec}^{-1}. \end{aligned} \tag{5-130}$$

Radial velocities are defined to be positive for motion away from an observer, and $V_{r\odot}$ must be added to the observed radial velocity to obtain the radial velocity corrected to the local standard of rest. The radial velocity correction, V_{re}, for the earth's orbital motion about the Sun is given by (SCHLESINGER, 1899)

$$V_{re} = -V\cos\beta\sin(\lambda_s - \lambda) + Ve\sin(\Gamma - \lambda)\cos\beta, \tag{5-131}$$

where λ and β are the ecliptic coordinates of the celestial object, λ_s is the longitude of the Sun at the epoch for which λ and β are determined, Γ is the longitude of the Sun at perigee given by

$$\Gamma = 281^\circ 13' 15.00'' + 6{,}189.03''\,T + 1.63''\,T^2 + 0.012''\,T^3, \tag{5-132}$$

where T is the number of tropical centuries since 1900.0 (to the relevant epoch), $e = 0.0167$ is the eccentricity of the earth's orbit, and V is given by

$$V = \frac{2\pi a}{P(1-e^2)^{1/2}} \approx 29.974 \times 10^5 \text{ cm sec}^{-1}, \tag{5-133}$$

where the semi-major axis of the earth's orbit is $a = 1.49597892 \times 10^{13}$ cm, and the number of seconds in a sidereal year is $P = 31{,}470{,}758$ seconds. The radial velocity correction, V_{rer}, for the rotation of the earth about its axis is given by

$$V_{rer} = V_e \sin h \cos\delta \cos\varphi', \tag{5-134}$$

where the equatorial rotational velocity of the earth is $V_e = 0.465$ km sec^{-1}, the hour angle, h, of the celestial object is taken to be positive for objects east of the meridian and negative for those west of the meridian, δ is the declination of the celestial object, and φ' is the geocentric latitude of the observatory. Additional corrections to radial velocity which are caused by refraction effects were discussed in Sect. 5.1.3.

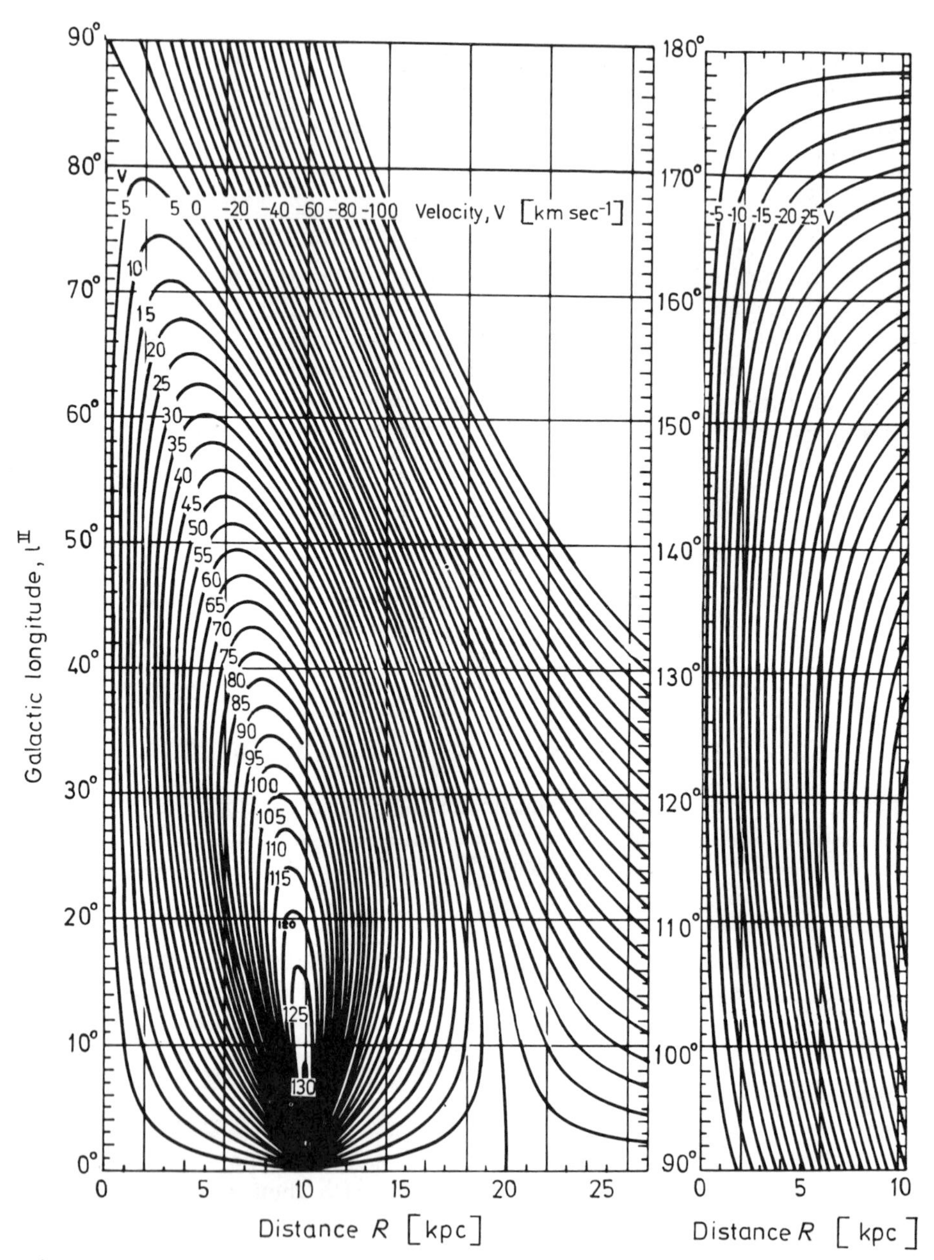

Fig. 38. Curves relating radial velocity, V, to the distance from the Sun, R, and to the galactic longitude, l^{II}. Here the radial velocity is defined with respect to the local standard of rest (after BURTON, 1974, by permission of Springer-Verlag)

5.3.5. Distance to the Variable Stars

As pointed out by CLERKE (1905), the stars whose intrinsic luminosity varies may be devided into two classes—those with short periods (RR Lyrae variables) of a few hours to about one day, and those with long periods (classical Cepheid or δ Cepheid variables) of two to sixty days. BAADE (1944) subsequently showed that RR Lyrae variables belong to a general class of stars called population II, whereas classical Cepheid variables belong to population I. Population II stars are found in globular clusters, whereas population I stars are found in galactic clusters and the spiral arms of galaxies. Population II stars have lower metal content and greater age than population I stars. The RR Lyrae variables exhibit a narrow range in absolute visual magnitude, M_v, given by (ARP, 1962; SANDAGE, 1962; CHRISTY, 1966)

$$0.2 \lesssim M_v \lesssim 1.0 \quad \text{for RR Lyrae variables.} \tag{5-135}$$

Once a RR Lyrae variable is recognized by its short period, its distance, D, may be determined from its observed apparent visual magnitude, m_v, by the relation

$$m_v - M_v = 5 \log D - 5, \tag{5-136}$$

where the distance, D, is in parsecs (1 parsec $\approx 3.086 \times 10^{18}$ cm). However, RR Lyrae stars are not bright enough to be detected at distances beyond about 3×10^5 parsecs.

Classical Cepheid variables are about 350 times as bright as RR Lyrae variables, but exhibit a wide range in absolute luminosity. LEAVITT (1912), however, showed that the apparent luminosity of twenty-five Cepheids in the Magellanic clouds increased roughly linearly with period. The distance to the Magellanic clouds was not known, but RUSSELL (1913), HERTZSPRUNG (1913) and SHAPLEY (1918, 1919) calibrated the period-luminosity relation by using statistical parallax techniques to estimate the distance of other nearby variable stars. Following the discovery of Cepheids in extragalactic nebulae by DUNCAN (1922) and HUBBLE (1922), the period-luminosity relation was extended to determine the extragalactic distance scale (cf. HUBBLE, 1936). JEANS (1929) provided a theoretical framework for the empirical relation by arguing that the period, P, of an oscillatory star must vary as $\rho^{-1/2}$, where ρ is the mass density of the star (c.f. Eq. (3-434)). If M_b denotes the absolute bolometric magnitude and T_e the stars effective temperature, then it follows from $P \propto \rho^{-1/2}$ that

$$\log P + 0.23\, M_b + 3 \log T_e = \text{constant}, \tag{5-137}$$

and if all variables have the same mass and temperature, T_e,

$$\log P + 0.30\, M_v = \text{constant}, \tag{5-138}$$

where M_v is the absolute visual magnitude. Unhappily, early efforts to determine the constant in Eqs. (5-137) and (5-138) were confused by not realizing that there were two types of variables, and that yet another type of variable (W Virginus variables) belongs to population II, and has long periods, but is about four times less bright than classical Cepheids of the same period.

BAADE (1952) first revised the period-luminosity relation by using only classical Cepheid variables, and this has been put on a firmer foundation by Kraft's (1961) summary of Arp's and Sandage's measurements of distances to clusters which contain classical Cepheids. The latest form of the period-luminosity relation is (SANDAGE and TAMMANN, 1971)

$$M_B^0 = -3.534 \log P + 3.647(\langle B\rangle^0 - \langle V\rangle^0) - 2.469, \tag{5-139}$$

where

$$\langle B\rangle^0 - \langle V\rangle^0 = 0.323 \log P + 0.290,$$

and that given by KRAFT (1961) is

$$\begin{aligned} M_V^0 &= -1.67 - 2.54 \log P \\ M_B^0 &= -1.33 - 2.25 \log P. \end{aligned} \tag{5-140}$$

Here M_V^0 and M_B^0 denote, respectively, the absolute magnitudes in the visual ($\lambda \approx 5480$ Å) and the blue ($\lambda \approx 4400$ Å) corrected for reddening (superscript 0), and $\langle B\rangle - \langle V\rangle$ is the average apparent magnitude difference, $m_B - m_V$. The observations (SANDAGE and TAMMANN, 1968) indicate a dispersion of ± 0.60 mag in M_B and ± 0.50 mag in M_V, which is thought to be due to the narrow range of effective temperatures for which a star will become unstable and oscillate (cf. SANDAGE, 1958).

The intrinsic luminosity, L, of individual Cepheid stars can be determined from the Stefan-Boltzmann formula (STEFAN, 1879; BOLTZMANN, 1884)

$$L = 4\pi R^2 \sigma T_e^4, \tag{5-141}$$

where $\sigma = 5.669 \times 10^{-5}$ erg sec^{-1} cm^{-2} °K^{-4}, the stellar radius is R, and T_e is the effective temperature. The effective temperature can be determined by measuring the stellar spectrum and comparing it with theoretical black body spectra (cf. Chapt. 1). As first pointed out by BAADE (1926), it follows from Eq. (5-141) that the change, Δl, in apparent luminosity, l, is given by

$$\frac{\Delta l}{l} = \frac{2\Delta R}{R} = p \frac{\Delta V_r \Delta t}{R}, \tag{5-142}$$

where ΔR is the change in radius, R, the $\Delta V_r \Delta t$ denotes the time integral of the change in observed radial velocity, and the factor p lies between 3/2 and 4/3. WESSELINK (1949) used observations of changes in radial velocity and brightness of δ Cephei to determine ΔR and R, and observations of the color index to give T_e. Once the intrinsic luminosity is known, the distance to the star follows from the observed luminosity and the well known fact that luminosity falls off as the square of the distance.

5.3.6. The Cosmic Distance Scale

The classical Cepheids are bright enough to be used to determine distances out to about 4×10^6 parsecs, but not bright enough to give the distance of the nearest cluster of galaxies. HUBBLE (1936) assumed that the brightest stars in galaxies all have the same absolute magnitude, M_s, given by

$$M_s = -6.1 \pm 0.4. \tag{5-143}$$

Some of Hubble's stars were H II regions, however, and recent estimates give

$$M_s \approx -9.3, \tag{5-144}$$

which allows distance estimates to about 3×10^7 parsecs. SANDAGE (1968) gives the absolute magnitude of the brightest globular cluster in a galaxy, M_C, as

$$M_C = -9.8 \pm 0.3, \tag{5-145}$$

and gives the absolute magnitude, M_G, of the brightest elliptical galaxy in the Virgo cluster as

$$M_G = -21.68. \tag{5-146}$$

If all clusters have brightest elliptical galaxies of the same absolute magnitude, then this would lead to distance estimates out to about 10^{10} parsecs. SCOTT (1957) has pointed out, however, that as we look further into space we pick out increasingly rich clusters of galaxies for study, and if there is no upper limit to galactic luminosity the brightest galaxies will have increasingly larger absolute magnitudes.

HUBBLE (1929) first pointed out that the luminosity distance, D_L, is related to galaxy redshift, z, by

$$D_L = \frac{zc}{H_0}, \tag{5-147}$$

where H_0 is the Hubble constant. If the luminosity distance is defined as that distance for which the apparent intensity of light falls off as the square of the distance, then to first order we have (HECKMANN, 1942)

$$D_L = H_0^{-1} c[z + 0.5(1 - q_0)z^2], \tag{5-148}$$

where q_0 is the deceleration parameter. The exact relation is given by (TERRELL, private communication)

$$D_L = \frac{cz}{H_0}\left\{1 + \frac{z(1 - q_0)}{\left[(1 + 2q_0 z)^{1/2} + 1 + q_0 z\right]}\right\}. \tag{5-149}$$

Cosmological distances may be measured by luminosity, D_L, (WHITTACKER, 1931; WALKER, 1933), apparent angular size, D_A (WHITTACKER, 1931; ETHERINGTON, 1933) or by parallax, D_p (MCCREA, 1935; TEMPLE, 1938). The angular diameter distance, D_A, is related to D_L by

$$D_A = (1 + z)^{-2} D_L. \tag{5-150}$$

The parallax distance, D_p, is given by

$$D_p = \frac{c[zq_0 + (q_0 - 1)(-1 + \sqrt{2q_0 z + 1})]}{H_0[q_0^4(1 + z)^2 - (2q_0 - 1)\{zq_0 + (q_0 - 1)(-1 + \sqrt{2q_0 z + 1})\}^2]^{1/2}}. \tag{5-151}$$

An additional distance, D_M, given by observations of proper motion is related to D_L by

$$D_M = (1 + z)^{-1} D_L. \tag{5-152}$$

For a discussion of these distances see WEINBERG (1972).

5.4. Mass

5.4.1. Inertial and Gravitational Mass

The equation of motion of a particle of inertial mass, M_i, is given by (NEWTON, 1687)

$$F = M_i a, \tag{5-153}$$

whereas the law of gravitation is given by

$$F = -M_g g, \tag{5-154}$$

where M_g is the gravitational mass. Here F is the force producing an acceleration a, or the force on a mass caused by the gravitational acceleration, g. For a distance, R, from a spherical mass, we have (NEWTON, 1687)

$$g = \frac{m_g G}{R^2}, \tag{5-155}$$

where m_g is the gravitational mass, and $G = 6.674 \times 10^{-8}$ dyn cm^2 g^{-2} is the Newtonian constant of gravitation. In an important experiment EÖTVOS (1890) showed that the difference between M_i and M_g for wood and platinum was less than 10^{-9}. ROLL, KROTOV, and DICKE (1964) and BRAGINSKY and PANOV (1971) have improved this limit to respective values of 10^{-11} and 10^{-12}. It has also been shown that molecules (ESTERMAN *et al.*, 1938) and neutrons (DABBS *et al.*, 1965) fall with the same acceleration as atoms, and that the gravitational force on electrons in copper is the same as that on free electrons (WITTEBORN and FAIRBANK, 1967). Further evidence for the equality of inertial and gravitational mass has been given by BEALL (1970).

All of the available experimental evidence, then, allows us to follow Newton's example and write the equation of motion of a system of particles interacting gravitationally as

$$M_n \frac{d^2 X_n}{dt^2} = G \Sigma_m \frac{M_n M_m (X_m - X_n)}{|X_m - X_n|^3}, \tag{5-156}$$

where M_n is the mass of the nth particle, and X_n is its Cartesian position vector in an inertial reference frame at time t.

5.4.2. Mass of the Sun and the Satellites of the Solar System

It follows from Eq. (5-156) that the equation of motion of a particle in orbit about another particle is given by

$$r^2 \frac{d\theta}{dt} = h, \tag{5-157}$$

where r is the radius vector of the orbiting planet as measured from the focus of the orbit, the angle, θ, is the angular distance along the orbit from a reference axis of a spherical coordinate system, and h is a constant of integration. The constant h is equal to twice the rate at which area is swept over by the radius vector. Eq. (5-157) expresses Kepler's second law (KEPLER, 1619) which states that the line joining the center of mass of the Sun and a planet covers an area that increases at a constant rate as the planet moves in its orbit. As the area of an ellipse is πab, where a and b are, respectively, the semi-major and semi-minor axis, we have an areal constant, h, of

$$h = \frac{2\pi ab}{P} = \frac{2\pi a^2(1-e^2)^{1/2}}{P}, \tag{5-158}$$

where P is the orbital period, and $b^2 = a^2(1-e^2)$ where e is the eccentricity of the orbit.

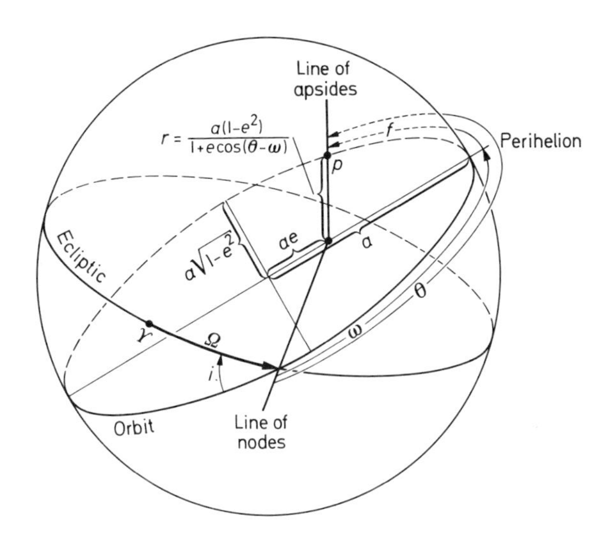

Fig. 39. Graphic representation of the elements of a planetary orbit. Here a is the semimajor axis, e is the eccentricity, i is the inclination, Ω is the longitude of the ascending node, ω is the argument of the perihelion, and γ denotes the vernal equinox

The polar equation of an orbit is given by

$$r = \frac{a(1-e^2)}{1+e\cos(\theta-\omega)}, \tag{5-159}$$

where the argument of the perihelion, ω, is the value of θ for the perihelion, which is the closest point on the orbit to the occupied focus (cf. Fig. 39). The orbit will be elliptical, parabolic, or hyperbolic according as $e<1$, $e=1$, or $e>1$.

Newton's (1687) law of gravitation gives

$$F = -G\frac{mM}{r^2}, \tag{5-160}$$

for the force of attraction between a mass, m, and another mass, M, where r is the separation of the masses, and laboratory measurements give (CAVENDISH, 1798; HEYL, 1930; ROSE *et al.*, 1969)

$$G = 6.674 \pm 0.012 \times 10^{-8} \text{ dyn cm}^2 \text{ g}^{-2}. \tag{5-161}$$

where the 0.012 represents three standard deviations. It follows from Eqs. (5-156) and (5-160) that the velocity, V, of mass, m, in a nonrotating set of axis which moves with M, is given by

$$V^2 = G(m+M)\left[\frac{2}{r} - \frac{1}{a}\right], \tag{5-162}$$

where a is the semi-major axis of the elliptical orbit of m. The orbit will be elliptical, parabolic, or hyperbolic according as $V < V_{esc}$, $V = V_{esc}$, and $V > V_{esc}$, where the escape velocity, V_{esc}, is given by

$$V_{esc} = \left[\frac{2G(m+M)}{r}\right]^{1/2}. \tag{5-163}$$

It follows from Eqs. (5-157), (5-158), and (5-160) that for an elliptical orbit

$$\frac{P^2}{a^3} = \frac{4\pi^2}{G(m+M)}, \tag{5-164}$$

which is Kepler's third law (KEPLER, 1619). When the orbit is elliptical, its plane is specified by the angle of inclination, i, between the orbit plane and some reference plane. The line of intersection of the planes is called the line of nodes, and the ascending node is the point where the orbiting object passes from the southern to northern hemisphere. If the reference plane is the plane of the ecliptic, the orbit is further specified by the longitude of the ascending node, Ω. This angle is the arc of the ecliptic between the vernal equinox and the ascending node. The closest point on the orbit to the occupied focus, the perihelion, is specified by the angular distance, ω, from the ascending node as measured along the orbit. The angle, ω, is called the argument of the perihelion, and is sometimes given as $\tilde{\omega} = \omega + \Omega$, which is called the longitude of the perihelion. A particle is then located on the ellipse by specifying the true anomaly, f, which is its angular distance along the orbit from the perihelion. Orbital elements of the planets were given in Table 57, whereas those of the natural satellites of the planets are given in Table 59.

For the motion of a satellite of mass, M_s, about the earth, Eq. (5-164) becomes

$$G M_{\oplus}\left(1 + \frac{M_s}{M_{\oplus}}\right) = n^2 a_s^3 \approx G M_{\oplus}, \tag{5-165}$$

where the subscripts $\oplus$ and s denote, respectively, the earth and the satellite, and $n = 2\pi/P$ is the sidereal motion in radians per second of the satellite. When observations of artificial satellites are used with Eq. (5-165), one obtains (CLEMENCE, 1965)

$$G M_{\oplus} = 3.98603 \times 10^{20} \text{ cm}^3 \text{ sec}^{-2}, \tag{5-166}$$

Table 59. Mean distance, sidereal period, orbit inclination and eccentricity, diameter, mass and density of the satellites of the planets of our solar system[1]

		Mean distance (Equatorial radius of planet = 1)	Mean distance (km)	Sidereal period (days)	Orbit inclination to: planet's equator[2] proper plane[3]	Orbit eccentricity	Diameter (km)	Mass (planet = 1)	Mass (g)	Density (g cm^{-3})
Moon		60.268	384,405	27.321661	28°35′ to 18°19′ [2]	0.05490	3,476	0.01229	7.344×10^{25}	3.34
Mars I	Phobos	2.82	9,380	0.318910	1°8′	0.0210	—	—	—	—
Mars II	Deimos	7.06	23,500	1.262441	1°46′	0.0028	—	—	—	—
Jupiter V		2.531	180,500	0.498179	0°24′	0.003	—	—	—	—
Jupiter I	Io	5.911	421,600	1.769138	0°2′ [3]	0.0000	3,240	0.0000381 ± 30	7.23×10^{25}	4.06
Jupiter II	Europa	9.405	670,800	3.551181	0°28′	0.0003	2,830	0.0000248 ± 5	4.70×10^{25}	3.95
Jupiter III	Ganymede	15.003	1,070,000	7.154553	0°11′	0.0015	4,900	0.0000817 ± 10	1.55×10^{26}	2.52
Jupiter IV	Callisto	26.388	1,882,000	16.689018	0°15′	0.0075	4,570	0.0000509 ± 40	9.66×10^{25}	1.94
Jupiter VI		160.8	11,470,000	250.57	27°38′	0.15798	—	—	—	—
Jupiter X		166	11,850,000	263.55	29°01′	0.13029	—	—	—	—
Jupiter VII		165	11,800,000	259.65	24°46′	0.20719	—	—	—	—
Jupiter XII		297	21,200,000	631.1	147°	0.16870	—	—	—	—
Jupiter XI		316	22,600,000	692.5	164° [2]	0.20678	—	—	—	—
Jupiter VIII		329	23,500,000	738.9	145°	0.378	—	—	—	—
Jupiter IX		332	23,700,000	758	153°	0.275	—	—	—	—
Jupiter XIII		156	11,110,000		20°42′	0.147	—	—	—	—
Saturn I	Mimas	3.07	185,400	0.942422	1°31′	0.0201	—	6.68×10^{-8}	—	—
Saturn II	Enceladus	3.95	237,900	1.370218	0°01′	0.00444	<940	1.51×10^{-7}	—	—
Saturn III	Tethys	4.88	294,500	1.887802	1°06′	0.0000	—	1.14×10^{-6}	—	—
Saturn IV	Dione	6.25	377,200	2.736916	0°01′	0.00221	—	1.85×10^{-6}	—	—
Saturn V	Rhea	8.73	526,700	4.517503	0°21′ [2]	0.00098	1,350	4×10^{-6}	2.27×10^{24}	1.76
Saturn VI	Titan	20.2	1,221,000	15.945452	0°20′	0.02890	4,950	2.48×10^{-4}	1.41×10^{26}	2.21
Saturn VII	Hyperion	24.53	1,479,300	21.276665	0°26′	0.1042	—	2×10^{-7}	—	—
Saturn VIII	Iapetus	59.01	3,558,400	79.33082	14°43′	0.02813	—	2.7×10^{-6}	—	—
Saturn IX	Phoebe	215	12,945,500	550.45	150°	0.16326	—	—	—	—
Saturn X	Janus	1.0	159,000	0.7490	0°	0	300	—	—	—
Uranus V	Miranda	5.10	123,000	1.414	0°	0	—	—	—	—
Uranus I	Ariel	7.938	191,700	2.520383	0°	0.0028	—	—	—	—
Uranus II	Umbriel	11.06	267,000	4.144183	0° [2]	0.0035	—	—	—	—
Uranus III	Titania	18.14	438,000	8.705876	0°	0.0024	—	—	—	—
Uranus IV	Oberon	24.26	585,960	13.463262	0°	0.0007	—	—	—	—
Neptune I	Triton	15.5	353,400	5.876833	159°57′ [2]	0.000	—	0.00128 ± 23	1.3×10^{26}	—
Neptune II	Nereid	245	5,560,000	359.881	27°48′ [2]	0.749	—	—	—	—

[1] After BLANCO and MCCUSKEY (1961) by permission of Addison-Wesley Publ. Co.

to give

$$M_{\oplus} = 5.972 \pm 0.013 \times 10^{27}\ \text{gm}. \tag{5-167}$$

When the orbit of the moon about the earth and the orbit of the earth about the Sun are considered, Eq. (5-164) can be used to obtain

$$\frac{M_{\odot}}{M_{\oplus}} \approx \left(\frac{P_{\text{☾}}}{P_{\oplus}}\right)^2 \left(\frac{a_{\oplus}}{a_{\text{☾}}}\right)^3 \left(1 + \frac{M_{\text{☾}}}{M_{\oplus}}\right). \tag{5-168}$$

Measurements of the parallax of a planet's position due to the motion of the earth about the center of mass of the earth-moon system during the course of a month give

$$\frac{M_{\text{☾}}}{M_{\oplus}} \approx \frac{1}{81.30} \approx 0.01230. \tag{5-169}$$

Measurements of the orbital periods and axis are then used with Eq. (5-168) to obtain

$$\frac{M_{\odot}}{M_{\oplus}} \approx 332{,}958, \tag{5-170}$$

or

$$M_{\odot} \approx 1.989 \pm 0.004 \times 10^{33}\ \text{g} \approx \frac{4\pi^2 a_{\oplus}^3}{G P_{\oplus}^2}. \tag{5-171}$$

Because the constant, G, is only known to three significant figures, the ephemeris of the planets are calculated using the Gaussian constant of gravitation, k, given by

$$k^2 = \frac{4\pi^2 a_{\oplus}^3}{P_{\oplus}^2 (M_{\oplus} + M_{\odot})}. \tag{5-172}$$

Gauss (1857) took $a_{\oplus} = 1.0$, $P_{\oplus} = 365.2563835$ days, and $M_{\oplus}/M_{\odot} = 1/354{,}710$ to obtain

$$k = 0.01720209895\ (\text{a.u.})^{3/2}\ (\text{ephemeris day})^{-1}\ (\text{solar mass})^{-1/2}. \tag{5-173}$$

This value of k is presently used as a defining constant for the ephemeris, and it is customary to give the planet distances in astronomical units (a.u.) and the planet mass, M_p, in inverse mass units, $M_{\odot}/M_p$. The distances and masses of the planets of our solar system were given in Table 57.

The angular momentum of a particle orbiting the Sun is progressivly destroyed by reradiating solar energy. This Poynting-Robertson effect will eventually result in the particle falling into the Sun. If a particle has initial orbital radius, a_0, and an orbital radius, a, after time, t, then its mass, M, is given by (Poynting, 1904; Robertson, 1937)

$$M = \frac{4 S A a_{\oplus}^2 t}{c^2 (a_0^2 - a^2)}, \tag{5-174}$$

where the energy flux of the Sun per unit area at the orbit of the earth is $S = 1.3553 \times 10^6\ \text{erg cm}^{-2}\ \text{sec}^{-1}$, and A is the cross sectional area of the particle.

5.4.3. Mass and Density of the Stars

The sum of the masses of the two stars of a visual binary star system can be calculated from Kepler's third law (Eq. (5-164)). If π is the dynamical parallax in seconds of arc and P is the orbital period in years, then the sum of the masses of the two stars, M_1+M_2, in solar mass units, is given by

$$M_1+M_2=\frac{\alpha^3}{\pi^3 P^2}, \tag{5-175}$$

where α is the angular size of the semi-major axis of the orbit in seconds of arc. When the orbits of the two stars around the center of gravity of their system is observed, the individual masses may be determined from Eq. (5-175) and the relation

$$a_1 M_1 = a_2 M_2, \tag{5-176}$$

where a_1 and a_2 are the values of the semi-major axes of the two ellipses.

The mass of components of a spectroscopic binary system can be estimated from the formula

$$M_1 \sin^3 i = \frac{(a \sin i)^3}{P^2\left(1+\dfrac{a_1 \sin i}{a_2 \sin i}\right)}, \tag{5-177}$$

where i is the inclination of the orbital plane, P is the orbital period in years, and the projection of the semi-major axis on the line of sight, $a \sin i$, is in astronomical units and is given by

$$a \sin i = a_1 \sin i + a_2 \sin i. \tag{5-178}$$

As described in Sect. 5.3.2, the projected components of the semi-major axes of the two ellipses, $a_1 \sin i$ and $a_2 \sin i$, can be measured from observations of the periodic variations of the radial velocity.

As first stated by HALM (1911) and observed by HERTZSPRUNG (1919), there is an empirical relationship between the absolute luminosity, L, and the mass, M, of the main sequence stars. When the absolute luminosity is expressed in terms of the bolometric magnitude, M_{bol}, the mass-luminosity relation takes the form (HARRIS, STRAND, and WORLEY, 1963)

$$\begin{aligned} M_{\text{bol}} &= 4.6-10.0 \log\left(\frac{M}{M_\odot}\right) \quad \text{for} \quad 0 \le M_{\text{bol}} \le 7.5 \\ M_{\text{bol}} &= 5.2-6.9 \log\left(\frac{M}{M_\odot}\right) \quad \text{for} \quad 7.5 \le M_{\text{bol}} \le 11, \end{aligned} \tag{5-179}$$

where the solar mass, $M_\odot \approx 1.989 \times 10^{33}$ grams. Eq. (5-179) leads to the approximate relations

$$\begin{aligned} \log\left(\frac{L}{L_\odot}\right) &\approx 4 \log\left(\frac{M}{M_\odot}\right) \quad \text{for } L>L_\odot \\ \log\left(\frac{L}{L_\odot}\right) &\approx 2.8 \log\left(\frac{M}{M_\odot}\right) \quad \text{for } L<L_\odot, \end{aligned} \tag{5-180}$$

where the solar absolute luminosity, $L_{\odot} = 3.9 \times 10^{33}$ erg sec^{-1}. EDDINGTON (1924) used purely theoretical considerations to show that the absolute luminosity was proportional to $M^{3.5}$ for a perfect gas sphere which is in radiative equilibrium.

MICHELSON (1890) first showed how the diameter of a star may be measured from the visibility of the fringes from a two element interferometer. MICHELSON and PEASE (1921) and PEASE (1931) used this technique to show that the radii of M type stars are a few hundred solar radii. HANBURY BROWN, and TWISS (1958) showed that an intensity interferometer would not be limited by atmospheric scintillation effects, and such an interferometer has been used to measure the angular size of several A and B type stars (cf. HANBURY BROWN *et al.*, 1967).

If the light curve of an eclipsing binary star system is normalized to unity between eclipses, then the ratio of the radii of the two stars is given by

$$\left(\frac{R_1}{R_2}\right)^2 = \frac{l_2}{1-l_1}, \tag{5-181}$$

where l is the loss of light observed when one component passes in front of the other. The light curve also gives the inclination of the orbit, i, and the ratios R_1/a and R_2/a of the individual radii to the semi-major axis, a, of the orbit (cf. IRWIN, 1962, 1963). Radial velocity measurements can give $a \sin i$, and individual radii may then be obtained. Effective temperatures, luminosities, radii, masses and densities obtained for some representative stars are given in Table 60.

Table 60. Effective temperature, T_e; luminosity, L; radius, R; mass, M; and density, ρ; for stars of different spectral type[1]

Star	MKK Sp. type	T_e (°K)	$L/L_{\odot}$	$R/R_{\odot}$	$M/M_{\odot}$	ρ (gm cm^{-3})	Class
α Sco A (Antares)	M0 Ib	3,300	34,000	530	19	1.7×10^{-7}	Supergiant
α Boo (Arcturus)	K2 III	3,970	130	26	4.2	3.2×10^{-4}	Giant
η Ori	B1 V	23,000	13,000	7.2	13.7	0.052	Main sequence
α Ca Ma A (Sirius)	A1 V	9,700	61	2.4	3.3	0.81	Main sequence
Sun	G2 V	5,740	1	1	1	1.410	Main sequence
Barnard's star	M5 V	3,000	0.015	0.50	0.38	4.3	Main sequence
α C Ma B (Sirius B)	A5 V	8,200	0.0026	0.026	0.96	7.7×10^4	White dwarf

[1] After COX and GIULI (1968) by permission of Gordon and Breach Science Publ. Here the T_e are those of the MKK spectral types and not necessarily those of the actual stars listed, and the solar values are $L_{\odot} \approx 3.9 \times 10^{33}$ erg sec^{-1}, $R_{\odot} \approx 6.96 \times 10^{10}$ cm, and $M_{\odot} \approx 1.989 \times 10^{33}$ g.

5.4.4. Mass and Density of Galaxies

If the entire mass, M, under whose influence the Sun describes its circular orbit, were concentrated at the galactic center, then M would be given by

$$M=\frac{R_0\,\theta_c^2(R_0)}{G}=2.9\times10^{44}\ \text{grams}=1.5\times10^{11}\,M_\odot, \tag{5-182}$$

where G is the gravitational constant, $R_0=10$ kpc is the distance of the Sun from the galactic center, and $\theta_c(R_0)=250\ \text{km sec}^{-1}$ is the circular velocity of the galaxy at R_0. A variety of observations (OORT, 1932; HILL, 1960; OORT, 1960) indicate that the force per unit mass, K_z, exerted by the Galaxy in the direction, z, perpendicular to the galactic plane is given by

$$\frac{\partial K_z}{\partial z}=-0.3\times10^{-10}\ \text{cm sec}^{-2}\ \text{pc}^{-1}, \tag{5-183}$$

where $1\ \text{pc}=3.0856\times10^{18}$ cm. It follows from Poisson's equation that the mass density, $\rho(z)$, at distance, z, is given by

$$\rho(z)=-\frac{1}{4\pi G}\left[\frac{\partial K_R}{\partial R}+\frac{K_R}{R}+\frac{\partial K_z}{\partial z}\right]=-\frac{1}{4\pi G}\left[2(A-B)(A+B)+\frac{\partial K_z}{\partial z}\right], \tag{5-184}$$

where K_R is the force per unit mass in the R direction, R is the distance from the galactic center, $G=6.68\times10^{-8}\ \text{dyn cm}^2\ \text{gm}^{-2}$ is the Newtonian gravitational constant, and Oort's constants, A and B, are $A=4.7\times10^{-16}\ \text{sec}^{-1}=15\ \text{km sec}^{-1}\ \text{kpc}^{-1}$ and $B=-3.1\times10^{-16}\ \text{sec}^{-1}=-10\ \text{km sec}^{-1}\ \text{kpc}^{-1}$. Near the galactic plane equation (5-184) gives

$$\rho(0)=10^{-23}\ \text{g cm}^{-3}=0.15\,M_\odot\ \text{pc}^{-3}.$$

For comparison, in the solar neighborhood we have a mass density of neutral hydrogen (HI) of (WESTERHOUT, 1957; SCHMIDT, 1957)

$$\rho(\text{H I})=0.7\ \text{atoms cm}^{-3}=1.2\times10^{-24}\ \text{g cm}^{-3}=0.018\,M_\odot\ \text{pc}^{-3}$$

and a local star density of (cf. Sect. 5.5.3)

$$\rho(\text{star})=0.120\,M_\odot(m/M_\odot)\ \text{pc}^{-3},$$

where m is the average stellar mass.

As first pointed out by OEPIK (1922), the mass of external galaxies may be determined by measuring radial velocity as a function of position in the galaxy. For solid body rotation, for example, the rotation velocity, $\theta_c(r)$, will increase with increasing distance, r, from the galaxy center until the edge of the galaxy is reached. At this point $\theta_c(r)$ has the turnover value θ_T and r has the value r_T. The total mass may then be determined by using an equation similar to Eq. (5-182). For a galaxy in which the matter is concentrated in a thin disc, one form for $\theta_c(r)$ is given by (BRANDT, 1960)

$$\theta_c(r)=\frac{3^{3/2n}\,\theta_T\,r}{r_T\left[1+2\left(\frac{r}{r_T}\right)^n\right]^{3/2n}}, \tag{5-185}$$

where n is an arbitrary integer which may be adjusted to fit the observed rotation curve. The total mass in the galaxy is then given by (BRANDT and BELTON, 1962)

$$M=\left(\frac{3}{2}\right)^{3/n}\frac{r_T\,\theta_T^2}{G}, \tag{5-186}$$

where G is the gravitational constant. Rotation curves are obtained by observing emission lines of ionized gas at optical frequencies or the 21 cm line of neutral hydrogen at radio frequencies (BURBIDGE and BURBIDGE, 1974; ROGSTAD, ROUGOOR, and WHITEOAK, 1967; WRIGHT, 1971; ROBERTS and ROTS, 1973). The masses of individual galaxies are given in Table 61 together with other integral properties of the galaxies.

The average mass, $\langle M \rangle$, and the mass-luminosity ratio, $\langle M/L \rangle$, of double galaxies may be determined using the regression formula (PAGE, 1960, 1962)

$$(\Delta V_i)^2 - \frac{\sigma_\Delta^2}{W_i} = \frac{2.96}{h}\left\langle \frac{M}{L} \right\rangle \left[\left(10^{-4}\frac{V_i}{S_i}\right) + 0.19 \times 10^{-8}\, V_i^2\right] \sum_{j=1}^{N_i} 10^{0.4(20.26 - m_j)}, \quad (5\text{-}187)$$

where $(\Delta V_i)^2$ is the square of the differential radial velocity, σ_Δ^2 is the mean square error in ΔV for weight $W=1$, W_i is the proper weight given by PAGE (1961), V_i is the average radial velocity of the ith pair, h is the ratio of the Hubble constant to $100 \text{ km sec}^{-1} \text{ Mpc}^{-1}$, S_i is the separation of the ith pair in minutes of arc, m_j is the apparent photographic magnitude corrected for galactic absorption, N_i is the number of galaxies involved in the ith pair, and $\langle M/L \rangle$ is in solar units. Page obtains the results

$$\begin{aligned} h\langle M \rangle &= 59.3, \quad \left\langle \frac{M}{hL} \right\rangle = 94 \quad \text{for 13 pairs of } E \text{ and } So \\ h\langle M \rangle &= 2.13, \quad \left\langle \frac{M}{hL} \right\rangle = 0.33 \quad \text{for 14 pairs of spirals and } Ir, \end{aligned} \quad (5\text{-}188)$$

where the average mass, $\langle M \rangle$, is in 10^{10} solar masses, and $\langle M/L \rangle$ is in solar units. More recent data for irregular and spiral galaxies indicate that (ROBERTS, 1969)

$$\begin{aligned} 10^{10} &\lesssim \frac{M}{M_\odot} \lesssim 5 \times 10^{11} \\ 10^{9} &\lesssim \frac{L}{L_\odot} \lesssim 10^{11} \end{aligned} \quad (5\text{-}189)$$

and

$$\log\left(\frac{L}{L_\odot}\right) \approx \frac{1}{7.5} \log\left(\frac{M}{M_\odot}\right),$$

where the solar mass, $M_\odot \approx 1.989 \times 10^{33}$ grams, and the solar luminosity, $L_\odot \approx 3.90 \times 10^{33} \text{ erg sec}^{-1}$. For elliptical galaxies, GENKIN and GENKINA (1970) obtain

$$10^{10} \lesssim \frac{M}{M_\odot} \lesssim 2 \times 10^{12}$$

and

$$\log\left(\frac{L}{L_\odot}\right) \approx \frac{1}{30} \log\left(\frac{M}{M_\odot}\right).$$

When the tidal forces between a dwarf galaxy and our Galaxy are considered, the orbit has a tidal limit, r_T, given by (VON HOERNER, 1957)

$$r_T = R_G \left(\frac{m}{3 M_G}\right)^{1/3}, \quad (5\text{-}190)$$

Table 61. Name, position, type, distance, D, inclination, radius, R, color, absolute magnitude, hydrogen mass, and total mass for 106 galaxies[1]

NGC	α(1950.0) h m s	δ(1950.0) ° ′ ″	Type	D (Mpc)	Incl. (°)	R (kpc)	Color C_0	Absolute magnitude (M_0) pg	Hydrogen mass $(10^9 M_\odot)$	Total mass $(10^{10} M_\odot)$
45	00 11 31	−23 27 24	Sd	2.4	50	4.2	—	−16.45 #	—	2.1
55	00 12 24	−39 28 00	SBM	2.4	84	19.2 #	—	−20.38 #	8.7	2.9
157	00 32 14	−08 40 18	Sbc	15.1	59	12.7	0.55	−20.24	—	5.0[2]
221	00 40 00	40 36	E	0.69	—	—	—	−15.7	—	0.21
224	00 40 00	41 00	Sb	0.69	77	19.8	0.62	−21.81	8.5	31.0
247	00 44 39	−21 01 48	Sd	2.4	75	9.8	0.43 #	−18.48	1.6	1.8
253	00 45 06	−25 34	Sc	2.4	78	9.9 #	—	−20.31 #	3.7	12.0[2]
300	00 52 31	−37 57 24	Sd	2.4	42	8.6 #	—	−17.70 #	3.4	3.2
428	01 10 22.6	00 42 56	Sm	7.9	38	6.3	0.21	−18.12	1.2	2.1
598	01 31 06	30 24	Scd	0.72	57	8.7	0.29	−18.70	1.7	3.9[2]
613	01 32 00	−29 40 12	SBbc	15.0	47	17.7 #	0.52 #	−20.65 #	—	13.0[2]
628	01 34 00.5	15 31 38	Sc	7.8	35	13.6	0.29	−20.13	10.6	3.9
681	01 45 42	−10 40	Sab	17.3	81	8.8 #	0.59 #	−20.14 #	—	1.8[2]
772	01 56 35.2	18 45 57	Sb	16.6	55	25.4	0.48	−20.70	8.5	23.0
925	02 24 18	35 22	Sd	6.8	53	13.8	0.21	−19.45	4.4	4.9
972	02 31 18	29 06	IO	16.6	66	11.3 #	0.55 #	−20.07 #	—	1.3[2]
1055	02 39 12	00 16	Sb	11.0	77	19.0	0.60	−20.48	8.2	6.6
1084	02 43 30	−07 47	Sc	14.2	65	9.5 #	0.28 #	−20.41 #	—	1.5[2]
1097	02 44 11	−30 29 12	Sb	12.1	50	20.1 #	0.52 #	−20.87 #	11.6	64.0
1140	02 52 12	−10 14	Ir	15.0	62	5.0 #	0.15 #	−18.98 #	3.7	—
1156	02 56 46.5	25 02 21	Ir	6.3	52	5.4	0.23	−17.85	1.1	0.63
1365	03 31 42	−36 18 18	SBb	15.1	66	28.6 #	0.32 #	−21.94 #	5.7	17.0
1569	04 26 00	−64 45	Ir	2.5	65	2.5	0.23	−16.93	0.21	—
1637	04 38 58	02 57 06	Sc	5.6	35	6.3	0.35	−18.04	0.72	1.3
1744	04 57 55	−26 05 42	SBd	10.0	49	11.8 #	0.17 #	−18.89 #	5.7	9.6
1792	05 03 30	−38 02 42	Sbc	10.4	64	8.8 #	—	−19.89 #	—	1.8[2]
1808	05 05 59	−37 34 42	SOa	7.4	67	8.7 #	—	−19.71 #	—	2.4 #[2]

Table 61 (continued)

NGC	α(1950.0) h m s	δ(1950.0) ° ′ ″	Type	D (Mpc)	Incl. (°)	R (kpc)	Color C_0	Absolute magnitude (M_0) pg	Hydrogen mass ($10^9 M_\odot$)	Total mass ($10^{10} M_\odot$)
2366	07 23 36	69 08	Ir	3.3	64	4.8	0.25	−16.79	0.68	0.16
2403	07 32 00	65 43	Scd	3.3	54	13.9	0.24	−19.24	4.9	7.3
2683	08 49 36	33 38	Sb	5.8	85	10.2	0.61	−20.02	0.96	2.8
2835	09 15 37	−22 08 30	SBc	7.6	43	10.1 #	—	−19.24 #	1.4	5.8
2841	09 18 36	51 12	Sb	6.0	67	9.9	0.55	−19.82	1.1	4.2
2903	09 29 20.2	21 43 14	Sbc	7.0	70	14.2	0.29	−20.64	4.8	6.1[2]
3031	09 51 30	69 18	Sab	3.3	55	16.8	0.69	−20.31	2.3	19.0[2]
3034	09 51 54	69 56	IO	3.3	82	6.4	0.71	−19.67	1.7	1.0[2]
3079	09 58 36	55 57	SBm	12.0	83	19.2	0.37	−20.67	5.7	5.7
3109	10 00 47	−25 54 54	Ir	2.2	90	5.9 #	—	−18.16 #	2.2	0.60
3198	10 16 42	45 49	SBc	9.6	73	16.6	0.21	−20.07	12.5	6.0
3227	10 20 46.6	20 07 07	Sap	16.5	51	14.6 #	0.64 #	−20.25 #	—	3.6[2]
3319	10 36 24	41 56	SBcd	9.2	58	11.8	0.18	−18.68	2.0	1.5
3344	10 40 46.9	25 11 06	Sbc	7.9	25	10.7	0.22	−19.43	2.0	3.4
3359	10 43 24	63 30	SBc	10.0	51	13.1	0.23	−19.60	4.8	10.1
3379	10 45 11.1	12 50 49	E	7.5	—	—	—	−19.1	—	19
3432	10 49 42	36 54	SBm	9.6	89	11.4	0.19	−19.63	6.4	3.2
3504	11 00 30	28 15	Sab	16.5	60	9.4 #	0.47 #	−20.35 #	—	1.0[2]
3521	11 03 15.4	00 14 12	Sbc	6.9	66	13.6	0.66	−20.07	4.5	8.6[2]
3556	11 08 42	55 57	SBcd	8.2	84	13.2	0.32	−20.33	6.6	4.3
3623	11 16 18.8	13 21 53	SBa	7.0	76	12.1	0.55	−20.66	—	10.5[2]
3627	11 17 38.4	13 15 47	Sb	7.6	57	15.3	0.49	−20.39	2.0	33.0
3628	11 17 39.5	13 52 06	Sb	7.6	89	19.9	0.52	−20.78	7.3	4.3
3631	11 18 18	53 28	Sc	14.5	32	15.2	0.35	−20.24	3.7	3.4
3646	11 19 05.4	20 26 42	Sbcp	20.9	60	15.2	0.33	−20.33	—	18.6[2]
3718	11 29 54	53 21	SBap	14.5	57	17.1	0.41	−20.22	3.0	8.4
3726	11 30 42	47 19	Sc	11.5	46	13.7	0.24	−19.86	6.3	—
3938	11 50 12	44 24	Sc	8.9	28	8.8	0.27	−19.27	2.2	1.3

3992	11 55 00	53 39	SBb	14.5	51	20.2	0.57	−20.72	3.7	—
4214	12 13 06	36 36	Ir	3.8	45	5.9	0.19	−18.16	1.3	0.38
4236	12 14 18	69 45	SBm	3.3	75	12.5	0.11	−18.68	—	3.0
4244	12 15 00	38 05	Scd	3.8	86	9.9	0.17	−18.71	5.0	7.1
4258	12 16 30	47 35	Sbc	4.0	64	14.0	0.37	−19.93	—	7.5[2]
4406	12 23 39.8	13 13 21	E	12.0	—	—	—	−20.6	—	137.0
4449	12 25 48	44 22	Ir	3.8	51	5.6	0.13	−18.42	3.2	2.4
4472	12 27 14.3	08 16 39	E	12.0	—	—	—	−21.5	—	116.0
4486	12 28 17.6	12 40 04	E	12.0	—	—	—	−21.1	—	208.0
4490	12 28 18	41 55	SBdp	8.0	47	10.4	0.19	−19.82	6.0	2.7
4559	12 33 30	28 14	Scd	9.6	67	16.2	0.18	−20.34	12.5	5.2
4605	12 37 48	61 53	SBcp	3.3	67	3.3 #	0.13 #	−17.59 #	—	0.10[2]
4621	12 39 30.9	11 55 14	E	12.0	—	—	—	−20.1	—	72.0
4631	12 39 48	32 49	SBd	4.0	85	11.1	0.31	−19.58	4.0	6.8
4656	12 41 36	32 26	SBm	4.0	85	8.4	0.13	−18.55	3.2	1.5
4697	12 46 00	−05 31 42	E	12.0	—	—	—	−20.4	—	49.0
4736	12 48 36	41 23	Sab	4.0	35	8.7	0.52	−19.46	—	31.0
5005	13 08 30	37 19	Sbc	8.7	66	10.2	0.47	−20.07	—	5.9[2]
5033	13 11 12	36 51	Sc	9.5	59	17.0	0.29	−19.80	6.8	6.5
5055	13 13 30	42 17	Sbc	4.6	59	10.7	0.47	−19.71	—	4.3[2]
5194	13 27 48	47 27	Sc	4.6	35	9.5	0.43	−19.76	1.1	6.4[2]
5204	13 28 18	58 40	Sm	4.6	53	5.4	0.16	−17.17	0.64	0.40
5236	13 34 10	−29 37 00	Sc	4.0	46	8.3 #	0.43 #	−20.70 #	7.3	28.0
5248	13 35 02.8	09 08 31	Sbc	10.5	55	12.1	0.29	−20.23	—	4.2[2]
5383	13 55 00	42 05	SBbp	24.0	40	19.9 #	0.38 #	−20.48 #	—	4.6[2]
5457	14 01 24	54 35	Sc	4.6	27	18.7	0.21	−20.43	9.3	16.0
5474	14 03 12	53 54	Scd	4.6	20	4.8	0.23	−17.40	0.64	5.1
5585	14 18 00	56 57	Sd	4.6	50	5.8	0.25	−17.52	0.82	1.7
5668	14 30 53.8	04 40 12	Sd	15.0	20	11.6 #	0.42 #	−19.25 #	4.3	5.6
5713	14 37 37.8	−00 04 32	Sbc	13.2	22	7.9 #	0.34 #	−19.32 #	3.2	—
5907	15 14 36	56 31	Sc	4.6	87	10.5	0.43	−18.62	1.8	15.0
6015	15 50 42	62 28	Sc	9.8	67	9.0	0.27	−19.07	1.8	1.4
6181	16 30 09.6	19 55 54	Sc	24.0	60	14.3 #	0.25 #	−20.24 #	—	6.0[2]

Table 61 (continued)

NGC	α(1950.0) h m s	δ(1950.0) ° ′ ″	Type	D (Mpc)	Incl. (°)	R (kpc)	Color C_0	Absolute magnitude (M_0) pg	Hydrogen mass ($10^9 M_\odot$)	Total mass ($10^{10} M_\odot$)
6217	16 34 48	78 18	SBbc	16.5	40	9.4 #	0.33 #	−19.83 #	5.8	2.4
6503	17 49 54	70 10	Scd	4.6	74	7.5	0.38	−18.76	2.3	1.0
6643	18 21 12	74 33	Sc	16.5	61	12.2	0.30	−20.31	2.4	—
6822	19 42 07	−14 53 30	Ir	0.50	43	1.5	—	−15.19	0.15	0.24
6946	20 33 54	59 58	Scd	4.2	22	8.8	0.40	−19.92	3.9	23.0
7331	22 34 48	34 10	Sbc	7.9	69	15.5	0.44	−20.69	5.7	6.1[2]
7626	23 18 10.4	07 56 36	E	35.6	—	—	—	−20.5	—	137.0
7640	23 19 42	40 35	SBc	4.4	89	8.6	0.00	−18.71	3.7	5.7
7741	23 41 22.9	25 47 55	SBcd	10.1	45	10.6	0.27	−18.96	7.6	4.2
7793	23 55 16	−32 52 30	Sd	2.4	47	4.1 #	0.49 #	−17.20 #	0.36	—
IC 10	00 17 36	59 02	Ir	1.25	—	—	—	—	0.31	0.12
IC 342	03 41 54	67 57	Sd	3.3	35	—	—	—	9.1	5.2
IC 1613	01 02 14.0	01 51 09	Ir	0.66	32	2.2	0.36	−14.27	0.051	0.025
IC 2574	10 25 00	68 43	Sm	3.3	69	7.7	0.17	−17.54	2.1	1.7
Ho II	08 14 06	70 52	Ir	3.3	40	5.3	0.21	−17.00	0.91	0.96
LMC	05 24 00	−69 48	SBm	0.052	27	—	0.30 #	−18.37 #	0.54	1.0[2]
Sex A	10 08 48	−04 28	Ir	1.0	36	1.4	0.12	−13.91	0.070	0.14
SMC	00 51 00	−73 06	Ir	0.06	60	—	0.24 #	−16.70 #	0.48	0.15

[1] Accurate positions are from GLANFIELD and CAMERON (1967) and GALLOUET and HEIDMANN (1971), whereas the other positions are from DEVAUCOULEURS and DEVAUCOULEURS (1964). The other data are from ROBERTS (1969) and GENKIN and GENKINA (1970). Linear radii are from HOLMBERG (1958) or by conversion of the deVaucouleurs data. The latter are denoted by a #. Masses based on optically derived data are marked by a superscript 2.

[2] A Hubble constant of 100 km $\sec^{-1}$ Mpc^{-1} is used for distances derived from radial velocities. The inclination, i, is the angle between the plane of the sky and the principal plane of the galaxy. Colors are corrected for redshifts.

where m is the mass of the dwarf galaxy, M_G is the mass of our Galaxy, and R_G is the distance between the galaxies. HODGE (1966) gives the mass and orbital parameters for several dwarf galaxies.

ZWICKY (1933) and SMITH (1936) first estimated the mass of galaxies from the virial theorem for a cluster of galaxies. If T is the kinetic energy of the cluster, then according to the virial theorem

$$2T + \Omega = 0, \tag{5-191}$$

where Ω is the gravitational potential energy, and it is assumed that the cluster is dynamically stable (not expanding or contracting). For a spherically symmetric cluster, the virial theorem mass, M_{VT}, which follows from Eq. (5-191), is given by

$$M_{\mathrm{VT}} = \frac{3 R_{\mathrm{e}} \sigma_{\mathrm{v}}^2}{G}, \tag{5-192}$$

where R_{e} is the effective radius of the cluster, G is the Newtonian constant of gravitation, and σ_{v} is the velocity dispersion observed along the line of sight. The effective radius, R_{e}, is given by (SCHWARZSCHILD, 1954)

$$R_{\mathrm{e}} = \frac{2\left(\int_0^R S\, d\delta\right)^2}{\int_0^R S^2\, d\delta}, \tag{5-193}$$

where $S(\delta)\, d\delta$ is the number of galaxies appearing in projection in a linear strip of width $d\delta$, and R is the cluster radius.

If m_i and v_{ri} denote, respectively, the mass and radial velocity of the ith component, then (LIMBER and MATHEWS, 1960)

$$T = \frac{3}{2} \sum m_i v_{\mathrm{r}i}^2 = \frac{M \sigma_{\mathrm{v}}^2}{2}$$

and

$$\Omega = -\frac{2}{\pi} \sum_{\mathrm{pairs}} \frac{G m_i m_j}{r'_{ij}} = -\frac{M^2 G}{R}, \tag{5-194}$$

where $M = \sum m_i$ is the total "visible" mass of the cluster, r'_{ij} is the projected separation of components i and j, and the $\sum_{\mathrm{pairs}}$ denotes a summation over all pairs of objects. Substitution of Eqs. (5-194) into (5-192) results in the relation

$$\frac{M_{\mathrm{VT}}}{M} = \frac{3\pi}{2} \left(\sum m_i v_{\mathrm{r}i}^2\right) \left[\sum_{\mathrm{pairs}} \frac{G m_i m_j}{r'_{ij}}\right]^{-1} = \frac{3R}{M^2 G} \sum_{\mathrm{r}} m_i v_{\mathrm{r}i}^2, \tag{5-195}$$

where it has been assumed that $R_{\mathrm{e}} = 3R$.

As first noted by ZWICKY (1933), the virial mass of many clusters of galaxies is about $10^{12} M_{\odot}$, whereas the total mass in the optically visible galaxies is about $10^{10.5} M_{\odot}$, where the solar mass $M_{\odot} \approx 2 \times 10^{33}$ grams. The missing mass may be due to neutral or ionized hydrogen, or, as suggested by AMBARTSUMIAN (1958), the clusters may be dynamically unstable. The luminosity, L, and M_{VT}/L are given in Table 62 together with other parameters for fifteen clusters of galaxies.

Table 62. Name, Abell number, position, angular extent, θ, distance, D, radius, R, mean radial velocity, $\langle V_0 \rangle$, dispersion in radial velocity, σ_v, optical luminosity, L, ratio of virial theorem mass, M_{VT}, to luminosity, the brightest galaxy of the cluster, the radio source, and its flux density S_{1400}, at 1,400 MHz for fifteen clusters of galaxies[1]

Cluster	Abell No.	α(1950.0) h m	δ(1950.0) ° ′	θ (degrees)	D (Mpc)	R (Mpc)	$\langle V_0 \rangle$ (km sec^{-1})	σ_v (km sec^{-1})	L ($10^{11} L_\odot$)	M_{VT}/L ($M_\odot/L_\odot$)	Br. Gal (NGC)	Radio source	S_{1400} (f.u.)
Pisces		01 20	33 00	10	66	0.47	4,373	339	4.2	258	499	—	—
NGC 541	194	01 23	−01 38	0.3	53	0.35	5,321	406	3.9	300	541	3C 40	4.4
Perseus	426	03 16	41 20	4	97	0.44	5,460	1,420	10.0	461	1270	3C 84	13.5
Fornax		03 30	−36 00	6	10	0.75	1,464	260	—	—	1316	Fornax	108.0
Cancer		08 18	21 14	3	80	0.54	4,800	501	2.4	1,156	2563	—	—
Ursa Major		11 45	55 59	0.7	11	1.31	1,222	407	7.1	619	—	—	—
Virgo		12 28	12 40	12	19	1.07	1,141	666	12.0	668	4486	3C 274	197.0
Canes Venatici		12 48	21 43	19	8	1.23	437	135	1.5	300	4736	—	—
Coma	1656	12 57	28 14	4	113	2.63	6,645	977	49.0	1,020	4489	5C 04.80	0.37
Centaurus		13 22	−42 45	2	250	—	—	—	—	—	5128	Centaurus	912.0
Cr B		15 20	27 54	0.5	190	2.81	21,600	1,202	53.0	1,580	—	—	—
Hercules	2151	16 03	17 55	0.1	175	1.25	10,737	631	18.0	556	—	4C 17.66	0.5
Abell 2199	2199	16 27	39 38	0.2	—	1.07	9,028	864	20.0	775	6166	3C 338	3.6
Pegasus II	2634	23 08	07 20	2	—	0.89	—	537	8.5	598	7720	4C 07.61	1.6
Pegasus I		23 18	07 55	1	65	0.30	3,606	246	2.0	182	7619	PKS 2318+07	0.8

[1] Angular extents are from Zwicky (1959) and Page (1965), distances are from deVaucouleurs (1961) and Allen (1963), radii and velocity dispersions are from Rood, Rothman, and Turnrose (1970). The velocity data for NGC 541, Perseus, Virgo, Canes Venatici, Hercules, and Abell 2199 are taken from, respectively, Zwicky and Humason (1964), Chincarini and Rood (1971), Tammann (1972), van der Bergh (1961), Corwin (1971), and Minkowski (1961). Other values of $\langle V_0 \rangle$ are from Zwicky (1959) and Page (1965). Values of luminosity, L, and M_{VT}/L are from Karachetsev (1966).

5.4.5. Mass-Radius-Density Relations

As first pointed out by SCHWARZSCHILD (1916), no visible, static, self supporting sphere can have a radius smaller than the Schwarzschild radius

$$R_g = \frac{2GM}{c^2}, \tag{5-196}$$

or a mass density greater than

$$\rho_g = \frac{3c^2}{8\pi G R_g^2}. \tag{5-197}$$

Here c is the velocity of light, G is the gravitation constant, and M is the mass of the object. If the mass densities, ρ, of all objects were equal to the Schwarzschild limiting density, then they would have a density-radius relation given by

$$\log\rho = 27.2 - 2\log R, \tag{5-198}$$

where R is the radius of the object. According to DE VAUCOULEURS (1970), stellar objects are observed to obey the relation

$$\log\rho \approx 29.7 - 2.7\log R, \tag{5-199}$$

whereas systems of stars such as galaxies and clusters of galaxies obey the relation

$$\log\rho \approx 15.2 - 1.7\log R. \tag{5-200}$$

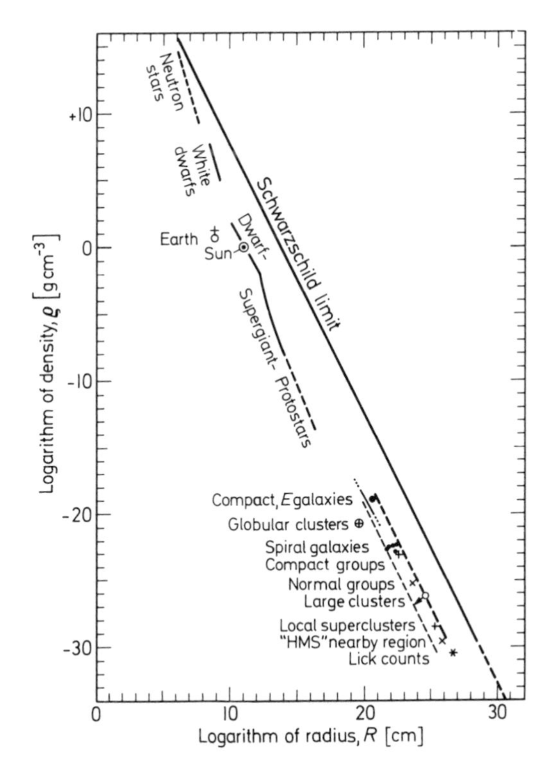

Fig. 40. Universal density-radius relation for observed astronomical objects. The dashed lines denote the range of density obtained from the virial theorem for stellar and galaxy clusters. The solid line is the Schwarzschild limit for which the density $\rho = 3c^2/(8\pi G R^2)$. (After DE VAUCOULEURS, 1970, by permission of the American Association for the Advancement of Science)

The mass, M, radius, R, and density, ρ, of various astronomical objects are given in Table 63 and illustrated in Fig. 40.

Also given in Table 63 is the filling factor, φ, which is the ratio of the observed density to the Schwarzschild limiting density. If all of the objects have

Table 63. Mass, M, radius, R, density, ρ, and filling factor, φ, for a variety of astronomical objects[1]

Class of objects	Examples	$\log M$ (g)	$\log R$ (cm)	$\log \rho$ (g cm^{-3})	$\log \varphi$
Neutron stars	Pulsars	33.16	5.93	14.75	−0.6?
		32.54	7.44	9.60	−2.5
White dwarfs	L 930–80	33.45	8.3	7.93	−2.7
	α CMaB	33.30	8.77	6.37	−3.2
	vM 2	32.90	9.05	4.13	−5.0
Main sequence stars	dM 8	32.2	9.95	1.76	−5.6
	Sun	33.30	10.84	0.15	−5.5
	A0	33.85	11.25	− 0.55	−4.7
	O5	34.9	12.1:	− 2.0	−5.0:
Supergiant stars	F0	34.4	12.65	− 4.2	−6.1
	K0	34.4	13.15	− 5.7	−6.6
	M 2	34.7	13.75	− 7.2	−6.9
Protostars	IR	35.3?	16.2?	−13.9?	−8.7?
Compact dwarf elliptical galaxies	M 32, core	41.0	19.5?	−18.1	−6.3
	M 32, effective	42.5	20.65	−20.0	−5.9
	N 4486–B	43.4	20.5	−18.75	−5.0
Spiral galaxies	LMC	43.2	21.75	−22.65	−6.3
	M 33	43.5	21.8	−22.5	−6.1
	M 31	44.6	22.3	−22.9	−5.5
Giant elliptical galaxies	N 3379	44.3	22.0	−22.35	−5.6
	N 4486	45.5	22.4	−22.3	−4.7
Compact groups of galaxies	Stephan	45.5	22.6	−23.1	−4.7
Small groups of spirals	Sculptor	46.2	24.1	−26.7	−5.7
Dense groups of ellipticals	Virgo E, core Fornax I	46.5	23.7	−25.2	−5.0
Small clouds of galaxies	Virgo S Ursa Major	47.0	24.3	−26.5	−5.1
Small clusters of galaxies	Virgo E	47.2	24.3	−26.3	−4.9
Large clusters of ellipticals	Coma	48.3	24.6	−26.1	−4.9
Superclusters	Local	48.7:	25.5	−28.4	−4.7
HMS sample to m ≃ 12.5			26.0	−29.6	−4.6
Lick Observatory counts to m ≃ 19.0			26.8	−30.5	−4.1

[1] From DE VAUCOULEURS (1970) by permission of the American Association for the Advancement of Science.

virial mass densities, ρ, and velocity dispersions, σ_v, in the range $100 \text{ km sec}^{-1} \leq \sigma_v \leq 1{,}000 \text{ km sec}^{-1}$, then

$$\log \rho = 6.5 + 2 \log \sigma_v - 2 \log R$$
$$\approx 21.5 - 2 \log R \tag{5-201}$$

and

$$\log \varphi \approx -5.7.$$

For elliptical galaxies, observational data indicate that the potential energies are proportional to the three halves power of their masses (Fish, 1964). For an elliptical galaxy of mass, M, and radius, R, Fish's observational law is

$$\Omega = 9.6 \times 10^{-8} M^{3/2} \text{ ergs}$$
$$\approx G M^{5/3} \rho^{1/3} \quad \text{for } 3 \times 10^9 M_\odot \leq M \leq 3 \times 10^{12} M_\odot \tag{5-202}$$

or

$$M R^{-2} = \text{constant} \approx 2 \text{ g cm}^{-2},$$

where Ω is the gravitational potential energy, the mass, M, is in grams, the solar mass, $M_\odot \approx 2 \times 10^{33}$ grams, and the mass density $\rho = 3M/(4\pi R^3)$.

5.4.6. Mass Density of the Universe

Hubble (1934, 1936) showed that the large scale distribution of galaxies across the sky was uniform, and that they were uniformly distributed in depth as well. He counted the number $N(<m_v)$ of nebulae per unit solid angle brighter than the visual magnitude, m_v, and obtained the relation

$$N(<m_v) = 10^{-9.09} \times 10^{0.6 m_v} \text{ galaxies per square degree,} \tag{5-203}$$

which was found to hold down to the 21st magnitude. Holmberg's (1958) re-calibration of the number magnitude relation gives

$$N(<m_v) = 1.43 \times 10^{-5} \times 10^{0.6 m_v} \text{ galaxies per steradian,} \tag{5-204}$$

where $1 \text{ steradian} = 3{,}280$ square degrees. Hubble assumed a mean galaxy absolute magnitude of $M_v = -15.1$ to obtain the volume density of

$$N = 7 \times 10^{-18} \text{ galaxies pc}^{-3}. \tag{5-205}$$

Noting that galaxy masses, M, lie in the range $10^9 M_\odot \lesssim M \lesssim 10^{11} M_\odot$, where the solar mass $M_\odot \approx 2 \times 10^{33}$ grams, Hubble concluded that the mass density of visible galaxies, ρ_G, is given by

$$10^{-30} \text{ g cm}^{-3} \leq \rho_G \leq 10^{-28} \text{ g cm}^{-3}. \tag{5-206}$$

Oort (1958) introduced the idea that ρ_G could be calculated by determining the total number of galaxies in a certain magnitude range, and assuming a constant mass to luminosity ratio, M/L, for the galaxies. That is,

$$\rho_G = L_T \left(\frac{M}{L} \right), \tag{5-207}$$

where the total luminosity per unit volume, L_T, due to galaxies is given by (KIANG, 1961)

$$L_T = \int \frac{dN(<M_v)}{dM_v} L(M_v) dM_v \approx 3.0 \times 10^8 h L_\odot \text{ Mpc}^{-3}, \tag{5-208}$$

where $h = H_0/100$ km sec^{-1} Mpc^{-1}, Hubble's constant is H_0, and the solar luminosity, $L_\odot \approx 4 \times 10^{33}$ erg sec^{-1}. Assuming that $M/L \lesssim 20 h M_\odot/L_\odot$, Eqs. (5-207) and (5-208) give

$$\rho_G \lesssim 4 \times 10^{-31} h^2 \text{ g cm}^{-3}, \tag{5-209}$$

which may be compared with the critical mass density, ρ_c, needed to close the universe

$$\rho_c = \frac{3 H_0^2}{8 \pi G} = 1.9 \times 10^{-29} h^2 \text{ g cm}^{-3}. \tag{5-210}$$

In Eq. (5-208), the mean number of galaxies per unit volume brighter than absolute magnitude, M_v, was taken to be (ABELL, 1962; PEEBLES, 1971)

$$N(<M_v) = 0.015 h^3 10^{0.75(M_v - M_v^*)} \text{ nebulae Mpc}^{-3} \quad \text{for } M_v < M_v^*$$

and

$$N(<M_v) = 0.015 h^3 10^{0.25(M_v - M_v^*)} \text{ nebulae Mpc}^{-3} \quad \text{for } M_v > M_v^*, \tag{5-211}$$

where

$$M_v^* = -19.5 + 5 \log h \text{ magnitudes},$$

and the luminosity, $L(M_v)$, of a galaxy with absolute magnitude, M_v, is given by

$$L(M_v) = 10^{0.4(4.79 - M_v)} L_\odot, \tag{5-212}$$

where $L_\odot \approx 4 \times 10^{33}$ erg sec^{-1}.

When absorption of coefficient, α, is taken into account, and both ellipticals, E, and spirals, S, are included, the mass density due to optically visible galaxies, ρ_G, is found to be (SHAPIRO, 1971)

$$\rho_G = \left[3.3 \times 10^7 (\alpha + 0.41) h^2 \left(\frac{M}{L}\right)_E + 4.7 \times 10^8 (\alpha + 0.43) h^2 \left(\frac{M}{L}\right)_S \right] M_\odot \text{ Mpc}^{-3}. \tag{5-213}$$

With $h = 0.5$, $\alpha = 0.24$, $(M/L)_E = 30$, and $(M/L)_S = 7.5$, Eq. (5-213) gives

$$\rho_G = 0.75 \times 10^9 M_\odot \text{ Mpc}^{-3} = 0.5 \times 10^{-31} \text{ g cm}^{-3}, \tag{5-214}$$

which again is considerably below ρ_c.

5.5. Luminosity

5.5.1. Magnitude, Color Index, Flux Density, and Brightness

According to Ptolemy, the Greek astronomers classified the brightness of stars into six magnitudes according to which the nth magnitude was 2.5^{m-n} brighter than the mth magnitude. POGSON (1856) suggested that observations be calibrated by using the constant $2.512 = 10^{2/5}$, whose common logarithm is exactly 0.4. In this system, the apparent magnitudes, m_1 and m_2, of two objects are related by the equations

$$m_1 - m_2 = -2.50 \log\left(\frac{f_1}{f_2}\right)$$

or

$$\frac{f_1}{f_2} = 2.512^{-(m_1 - m_2)}, \tag{5-215}$$

where f denotes the observed energy flux and has the units of ergs $\text{cm}^{-2}\,\text{sec}^{-1}$. The luminosity, L, of a source at distance, D, is given by

$$L = f D^2, \tag{5-216}$$

and the absolute magnitude, M, is given by the equation

$$m - M = 5 \log D - 5 = -5(1 + \log \pi), \tag{5-217}$$

where m is the apparent magnitude, $m - M$ is called the distance modulus, D is the distance of the star in parsecs, and π is the annual parallax in seconds of arc. The absolute magnitude is the apparent magnitude the object would have if it were at a distance of ten parsecs.

In actual practice, the energy received from an object is measured over a finite band of wavelengths, and apparent magnitudes, m, are designated by a subscript which denotes the center wavelength of the filter used. The subscripts R, V, pg, B, and U denote, respectively, the red, yellow or visual, photographic, blue, and ultraviolet bands, and have center wavelengths given by

$$\begin{array}{ll} m_{\text{R}} & 6800\,\text{Å} \\ m_{\text{v}} = V & 5550\,\text{Å} \\ m_{\text{pg}} & 4300\,\text{Å} \\ m_{\text{B}} = B & 4350\,\text{Å} \\ m_{\text{u}} = U & 3500\,\text{Å}. \end{array} \tag{5-218}$$

The blue and photographic apparent magnitudes are related by the equation

$$m_{\text{B}} = m_{\text{pg}} + 0.11. \tag{5-219}$$

The red filter is used to study galaxies (SANDAGE, 1961), and

$$m_{\text{V}} = m_{\text{R}} + 0.3 \tag{5-220}$$

for the nearby galaxies.

JOHNSON and MORGAN (1953) introduced the three color U, B, V system which uses filters about 1000 Å in width and with the center wavelengths given in Eq. (5-218). The fundamental standards of the U, B, V system are given in Table 64. Additional standard stars for the U, B, V system are given by JOHNSON (1963). For the Sun we have $M_{\text{U}} = 5.51$, $M_{\text{B}} = 5.41$, $M_{\text{V}} = 4.79$, $m_{\text{U}} = -26.06$, $m_{\text{B}} = -26.16$, $m_{\text{V}} = -26.78$, and $m_{\text{B}} - m_{\text{V}} = 0.62$.

The color index, C, is defined by the equation

$$C = m_{\text{pg}} - m_{\text{v}} = B - V - 0.11, \tag{5-221}$$

Table 64. The ten primary standard stars of the U, B, V system

HD No.	Name	V	$B-V$	$U-B$	Sp. type
12929	α Ari	2.00	1.151	1.12	K 2 III
18331	HR 875	5.17	0.084	0.05	A 1 V
69267	β Cnc	3.52	1.480	1.78	K 4 III
74280	η Hya	4.30	−0.195	−0.74	B 3 V
135742	β Lib	2.61	−0.108	−0.37	B 8 V
140573	α Ser	2.65	1.168	1.24	K 2 III
143107	ε CrB	4.15	1.230	1.28	K 3 III
147394	τ Her	3.89	−0.152	−0.56	B 5 IV
214680	10 Lac	4.88	−0.203	−1.04	O 9 V
219134	HR 8832	5.57	1.010	0.89	K 3 V

where the apparent magnitudes are measured above the atmosphere. By convention, the colors $B-V$ and $U-B$ are zero above the atmosphere for the mean of six dwarf stars of spectral type A 0: α Lyr, γ UMa, α CrB, 109 Vir, γ Oph, and HR 3314. The colors of the main sequence stars of other spectral types are given in Table 66. For galaxies, HOLMBERG (1958) found that

$$C = 0.87 \quad \text{(for galaxies)},$$

whereas

$$C = 0.00 \text{ to } 0.30 \quad \text{(for most quasars)}. \tag{5-222}$$

The total energy flux, f, is the integral over all wavelengths of the flux density, $S(\lambda)$, given by

$$S(\lambda) = \frac{f(\lambda)}{\Delta\lambda} \text{ erg cm}^{-2} \text{ sec}^{-1} \text{ Å}^{-1},$$

where $f(\lambda)$ is the energy flux measured with a filter centered at wavelength, λ, and whose width is $\Delta\lambda$. The zero points of the magnitude systems are fixed by the equations (MATTHEWS and SANDAGE, 1963)

$$\log S(\lambda) = -0.4\, m_{\mathrm{v}} - 8.42 \quad \text{for } \lambda = 5500 \text{ Å}$$

and

$$\log S(\lambda) = -0.4\, m_{\mathrm{pg}} - 8.3 \quad \text{for } \lambda = 4300 \text{ Å}. \tag{5-223}$$

Radio astronomers measure the flux density, $S(\nu)$, at frequency, ν, where

$$S(\nu) = \frac{f(\nu)}{\Delta\nu} \text{ erg cm}^{-2} \text{ sec}^{-1} \text{ Hz}^{-1}. \tag{5-224}$$

Here $f(\nu)$ is the energy flux at frequency, ν, and $\Delta\nu$ is the bandwidth of the filter used. The conventional unit for radio flux density is the flux unit, f.u., given by

$$1 \text{ f.u.} = 10^{-26} \text{ watt m}^{-2} \text{ Hz}^{-1} = 10^{-23} \text{ erg cm}^{-2} \text{ sec}^{-1} \text{ Hz}^{-1}. \tag{5-225}$$

Flux densities at 1400 MHz were given in Table 31 for radio galaxies and quasars. Solar radio astronomers often use the solar flux unit, s.f.u., given by

$$1 \text{ s.f.u.} = 10^4 \text{ f.u.} \tag{5-226}$$

Flux densities for the Sun are given in Table 65 for radio frequencies. The observed flux density, $S_o(\nu)$, depends on the source flux density, $S_s(\nu)$, the antenna efficiency, η_A, the antenna solid angle, Ω_A, and the solid angle extent of the source, Ω_s.

$$S_o(\nu) = \eta_A S_s(\nu) \qquad \text{for } \Omega_s \ll \Omega_A$$
$$S_o(\nu) = \eta_A \frac{\Omega_A}{\Omega_s} S_s(\nu) \quad \text{for } \Omega_s \gtrsim \Omega_A . \tag{5-227}$$

For a nonthermal radiator of spectral index, α, the total luminosity, L, is given by

$$L = 4\pi D^2 S_s(\nu_0) \nu_0^\alpha \int_{\nu_1}^{\nu_2} \nu^{-\alpha} d\nu , \tag{5-228}$$

where the source distance is D, the source flux density at frequency, ν_0, is $S_s(\nu_0)$, and the source flux density is assumed to be negligible for frequencies, ν, such that $\nu \lesssim \nu_1$ and $\nu \gtrsim \nu_2$.

Table 65. The flux density of the Sun, S, and the maximum brightness temperature, T_{max}, at radio and microwave frequencies[1]

Frequency (MHz)	Wavelength (cm)	S_{min} (s.f.u.)	S_{max} (s.f.u.)	T_{max} (°K)
50	600	0.48	0.6	1.15×10^6
100	300	2.1	2.4	1.15×10^6
200	150	6.8	9.5	1.14×10^6
300	100	11.2	16	8.5×10^5
600	50	21	31	4.1×10^5
1,000	30	29	47	2.25×10^5
1,500	20	40	64	1.36×10^5
3,000	10	69	105	5.6×10^4
5,000	6	110	150	2.85×10^4
10,000	3	248	270	1.29×10^4
25,000	1.2	1,120	1,120	8.56×10^3
35,000	0.86	2,080	2,080	8.1×10^3

[1] Here the flux densities are in solar flux units where 1 s.f.u. $= 10^{-19}$ erg cm^{-2} sec^{-1} Hz^{-1}.

When a radiator is thermal, its brightness, $B_\nu(T)$, at frequency, ν, is given by (Planck, 1901) (cf. Eqs. (1-119) to (1-121))

$$B_\nu(T) = \frac{S_s(\nu)}{\Omega_s} = \frac{2h\nu^3}{c^2} \frac{1}{\left[\exp\left(\frac{h\nu}{kT}\right) - 1\right]} \text{ erg cm}^{-2} \text{ sec}^{-1} \text{ Hz}^{-1} \text{ ster}^{-1}, \tag{5-229}$$

where Ω_s is the solid angle extent of the source, and T is the brightness temperature of the source. The observed brightness, $B_0(\nu)$, is related to the source brightness, $B_s(\nu)$, by the relations

$$\begin{aligned} B_0(\nu) &= \eta_A \frac{\Omega_s}{\Omega_A} B_s(\nu) = \frac{S_0(\nu)}{\Omega_A} \quad \text{for } \Omega_s \ll \Omega_A \\ B_0(\nu) &= \eta_A B_s(\nu) \quad = \frac{S_0(\nu)}{\Omega_A} \quad \text{for } \Omega_s \gtrsim \Omega_A, \end{aligned} \tag{5-230}$$

where Ω_s and Ω_A denote, respectively, the solid angle extent of the source and the antenna, and η_A is the antenna efficiency. The temperature calculated from Eqs. (5-230) and (5-229) is called the equivalent brightness temperature, or the color temperature. It is the temperature of a black body radiator which has the same relative intensity distribution, or color, as the source under observation. If the source is a black-body radiator, then the total brightness, B, is given by integrating equation (5-229) over all frequencies to obtain (Stefan, 1879; Boltzmann, 1884)

$$B = 4\pi n^2 \sigma T^4 \text{ erg cm}^{-2} \text{ sec}^{-1}, \tag{5-231}$$

where n is the index of refraction, an isotropic source radiates over 4π steradians, and

$$\sigma = 2\pi^5 k^4/(15 c^2 h^3) \approx 5.67 \times 10^{-5} \text{ erg sec}^{-1} \text{ cm}^{-2} \text{ °K}^{-4}.$$

The luminance, $\mathscr{L}$, and luminosity, L, of a source at distance, D, are then given by

$$\mathscr{L} = \frac{B}{\Omega_s} \approx \left(\frac{D}{R}\right)^2 B \text{ erg cm}^{-2} \text{ sec}^{-1} \text{ ster}^{-1} \tag{5-232}$$

and

$$L = 4\pi R^2 \sigma T_e^4 \text{ erg sec}^{-1} = R^2 B \text{ erg sec}^{-1}, \tag{5-233}$$

where R is the source radius, and Ω_s is its solid angle extent.

5.5.2. The Solar Constant, Bolometric Corrections, and Stellar Temperatures and Radii

The total energy flux received from the Sun outside the atmosphere of the earth, the solar constant, is given by (Thekaekhra and Drummond, 1971)

$$f_\odot = 1.3533(18) \times 10^6 \text{ erg cm}^{-2} \text{ sec}^{-1}, \tag{5-234}$$

to give a luminosity of

$$L_\odot = 3.826(8) \times 10^{33} \text{ erg sec}^{-1}. \tag{5-235}$$

The adopted apparent visual magnitude, m_v, and absolute visual magnitude, M_v, are (Harris, 1961)

$$\begin{aligned} m_{v\odot} &= -26.77 \\ M_{v\odot} &= \quad 4.79. \end{aligned} \tag{5-236}$$

It follows from Eqs. (5-215), (5-216), (5-217), and (5-236) that the luminosity, L, of an object with absolute visual magnitude, M_v, is given by

$$L = 10^{0.4(4.79 - M_v)} L_\odot. \tag{5-237}$$

The bolometric system of magnitudes is defined by measurements of the total energy flux as measured above the atmosphere, and is designated by the subscript b. The bolometric correction, BC, is defined by

$$\mathrm{BC} = M_b - M_v = m_b - m_v, \tag{5-238}$$

where the subscript v denotes visual magnitude, and M and m denote, respectively, absolute and apparent magnitudes. Bolometric corrections are discussed by KUIPER (1938), HARRIS (1963), and JOHNSON (1965), and are given in Table 66 for stars of various spectral types. The absolute luminosity, L, and the apparent energy flux, f, of a source are related to the absolute bolometric magnitude, M_b, and the apparent bolometric magnitude, m_b, by

$$L = 3.02 \times 10^{35} \times 10^{-2M_b/5} \text{ erg sec}^{-1},$$

and

$$f = 2.52 \times 10^{-5} \times 10^{-2m_b/5} \text{ erg cm}^{-2} \text{ sec}^{-1}. \tag{5-239}$$

For the Sun we have $M_b = 4.72$ and $m_b = -26.85$, with the distance modulus $M_b - m_b = -31.57$.

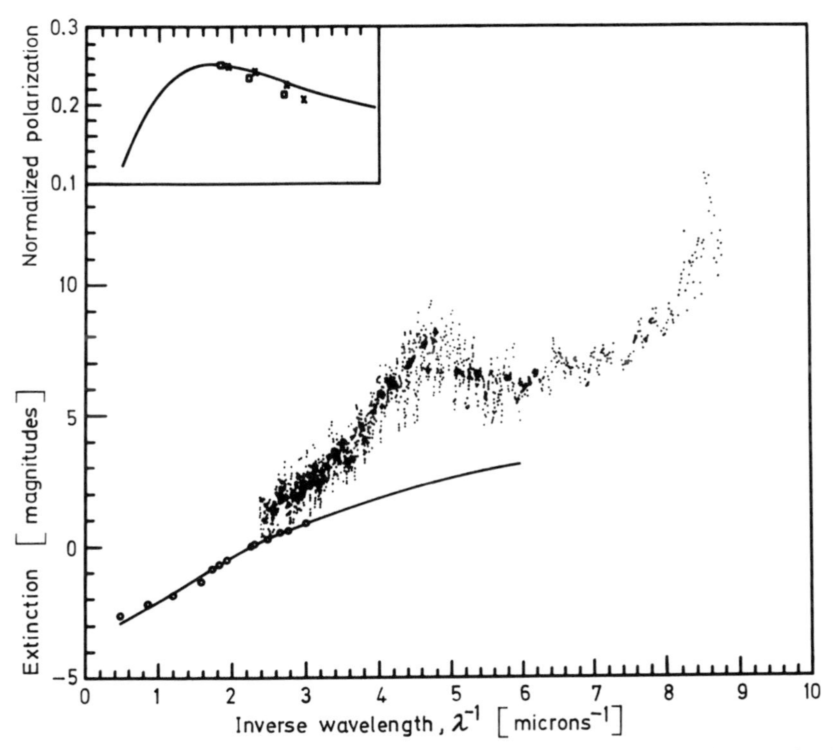

Fig. 41. The observed mean value for interstellar extinction. The open circles are after BORGMAN and BOGGESS (1964), and the other points are from STECHER (1969) (both by permission of the American Astronomical Society and the University of Chicago Press). These data are normalized so that $B - V = 1$ magnitude and $V = 0$. Also shown are polarization observations (after GREENBERG, 1970, by permission of the International Astronomical Union). The solid lines denote theoretical polarization and extinction for a size distribution $n(a) = 49 \exp[-5(a/0.2)^3] + \exp[-5(a/0.6)^3]$ of particles with radius, a, and index of refraction $m = 1.33 - 0.55\,i$

Table 66. Bolometric corrections, BC, effective temperatures, T_e, and colors $B-V$, $U-B$ for stars of various spectral types[1]

MK	BC	T_e	$B-V$	$U-B$
O 5	—	38,000	−0.32	−1.15
O 7	—	37,000	−0.32	−1.14
O 9	−3.34	31,900	−0.31	−1.12
B 0	−3.17	30,000	−0.30	−1.08
B 1	−2.50	24,200	−0.26	−0.93
B 2	−2.23	22,100	−0.24	−0.86
B 3	−1.77	18,800	−0.20	−0.71
B 5	−1.39	16,400	−0.16	−0.56
B 6	−1.21	15,400	−0.14	−0.49
B 7	−1.04	14,500	−0.12	−0.42
B 8	−0.85	13,400	−0.09	−0.30
B 9	−0.66	12,400	−0.06	−0.19
A 0	−0.40	10,800	0.00	0.00
A 1	−0.32	10,200	0.03	
A 2	−0.25	9,730	0.06	
A 3	−0.20	9,260	0.09	
A 5	−0.15	8,620	0.15	
A 7	−0.12	8,190	0.20	
F 0	−0.08	7,240	0.33	
F 2	−0.06	6,930	0.38	
F 5	−0.04	6,540	0.45	
F 6	−0.04	6,450	0.47	
F 7	−0.04	6,320	0.50	
F 8	−0.05	6,200	0.53	
G 0	−0.06	5,920	0.60	
G 2	−0.07	5,780	0.64	
G 5	−0.10	5,610	0.68	
G 8	−0.15	5,490	0.72	
K 0	−0.19	5,240	0.81	
K 2	−0.25	4,780	0.92	
K 3	−0.35	4,590	0.98	
K 5	−0.65	4,410	1.15	
K 7	−0.90	4,160	1.30	
M 0	−1.20	3,920	1.41	
M 1	−1.48	3,680	1.48	
M 2	−1.76	3,500	1.52	
M 3	−2.03	3,360	1.55	
M 4	−2.31	3,230	1.56	
M 5	−2.62	3,120	1.61	
M 8	−4.2	2,660	2.00	

[1] Data from HARRIS (1963) and JOHNSON (1965).

To obtain the actual magnitude of an object, the observed magnitudes must also be corrected for interstellar extinction (TRUMPLER, 1930; STEBBINS, 1933), which is illustrated in Fig. 41 as a function of inverse wavelength, λ^{-1}. The photoelectric systems commonly used to study interstellar reddening are (STEBBINS and WHITFORD, 1943, 1948; WHITFORD, 1958)

Band	U	B	V	G	R	I
λ(Å)	3500	4350	5550	5750	6800	8250
$\lambda^{-1}(\mu^{-1})$	2.857	2.299	1.802	1.739	1.470	1.212

In this wavelength region the extinction is proportional to λ^{-1}. The total absorption in magnitudes, A_λ, at wavelength, λ, is given by

$$A_\lambda = 1.086\, N \sigma_e D \text{ magnitudes}, \tag{5-240}$$

where N is the volume density of absorbing or scattering particles, the extinction cross section is given in Eq. (1-349), and D is the distance to the source. In the visual, V, band, we have

$$\sigma_e \approx 10^{-9}\,\text{cm}^2 \quad \text{at } \lambda \approx 4880\,\text{Å}, \tag{5-241}$$

where the grain radius $a = 0.2 \times 10^{-4}$ cm $\approx \lambda$, and we assume that $\sigma_e = \pi a^2$. The volume density of grains, N_{gr}, is given by

$$N_{gr} \approx 4 \times 10^{-13} N_A \,\text{cm}^{-3},$$

where $N_A = N_H + 2N_2$ is the total number density of hydrogen atoms in atomic, N_H, and molecular, N_2, form.

The bolometric absolute magnitude, M_b, of a black body radiator is related to its radius, R, by the equation

$$M_b = 42.31 - 5\log\left(\frac{R}{R_\odot}\right) - 10\log T_e \text{ magnitudes}, \tag{5-242}$$

where the effective temperature, T_e, is the temperature given in Eq. (5-231), and it has been assumed that

$$\begin{aligned} M_{b\odot} &= 4.69 \text{ magnitudes} \\ T_{e\odot} &= 5784\,^\circ\text{K}, \end{aligned} \tag{5-243}$$

the radius, R, is in cm, and

$$R_\odot = 6.9598 \times 10^{10}\,\text{cm}.$$

By using observations of spectral lines together with the theory of stellar atmospheres, the effective temperature can be related to spectral type (KEENAN and MORGAN, 1951; MIHALAS, 1964), and these temperatures are also given in Table 66.

5.5.3. Stellar Luminosity and Spectral Type

SECCHI (1868) and RUTHERFORD (1893) first noted that stars of different colors have very different spectra. SECCHI divided the stars into four groups with the colors white, yellow, orange, and red, and with the respective spectral features of strong absorption lines of hydrogen, strong calcium lines, strong metallic lines, and broad absorption lines which are more luminous on the violet side. When larger numbers of stars were observed, letters were introduced to denote a wider variety of color and spectral types (PICKERING and FLEMING, 1897; PICKERING, 1889). The development of the Henry Draper Catalogue of 225,330 stars (MAURY and PICKERING, 1897; CANNON and PICKERING, 1918—1924), together with additions and improvements by MORGAN, KEENAN, and KELLMAN (1943), has led to the main classes designated by

$$\text{Q, P, W, O, B, A, F, G, K, M, S, R, and N.} \tag{5-244}$$

Each class is divided into ten groups designated by a digit 0,1,2,...,9 placed after the capital letter for the class. When the spectra do not seem to form a continuous sequence, lower case letters are used in place of numbers. Peculiarities

Table 67. The "Henry Draper" (HD) classification types O–M[1]

Type	Main characteristics	Sub-types	HD Criteria	Typical stars
O	Hottest stars, continuum strong in UV	Oa	O II λ 4650 dominates	BD +35°4013
		Ob	He II λ 4686 dominates } emission lines	BD +35°4001
		Oc	Lines narrower } emission lines	BD +36°3987
		Od	Absorption lines dominate; only He II, O II in emission	ξ Pup, λ Cep
		Oe	Si IV λ 4089 at maximum	29 CMa
		Oe 5	O II λ 4649, He II λ 4686 strong	τ CMa
B	Neutral helium dominates	B 0	C III λ 4650 at maximum	ε Ori
		B 1	He I λ 4472 > O II λ 4649	β CMa, β Cen
		B 2	He I lines are maximum	δ Ori, α Lup
		B 3	He II lines are disappearing	π^4 Ori, α Pav
		B 5	Si λ 4128 > He λ 4121	19 Tau, φ Vel
		B 8	λ 4472 = Mg λ 4481	β Per, δ Gru
		B 9	He I λ 4026 just visible	λ Aql, λ Cen
A	Hydrogen lines decreasing from maximum at A 0	A 0	Balmer lines at maximum	α CMi
		A 2	Ca II K = 0.4 Hδ	S CMa, ι Cen
		A 3	K = 0.8 Hδ	α PsA, τ^3 Eri
		A 5	K > Hδ	β Tri, α Pic
F	Metallic lines becoming noticeable	F 0	K = H + Hδ	δ Gem, α Car
		F 2	G band becoming noticeable	π Sgr
		F 5	G band becoming continuous	α CMi, ρ Pup
		F 8	Balmer lines slightly stronger than in sun	β Vir, α For
G	Solar-type spectra	G 0	Ca λ 4227 = Hδ	α Aur, β Hya
		G 5	Fe λ 4325 > Hγ on small-scale plates	κ Gem, α Ret
K	Metallic lines dominate	K 0	H and K at maximum strength	α Bov, α Phe
		K 2	Continuum becoming weak in blue	β Cnc, υ Lib
		K 5	G band no longer continuous	α Tau
M	TiO bands	Ma	TiO bands noticeable	α Ori, α Hya
		Mb	Bands conspicuous	ρ Per, γ Cru
		Mc	Spectrum fluted by the strong bands	W Cyg, RX Aqr
		Md	Mira variables, Hγ, Hδ in emission	χ Cyg, o Cet

[1] After KEENAN (1963) by permission of the University of Chicago Press.

in the spectra are designated by the prefix c, g, or d according as the lines are narrow, and the stars are giants or dwarfs. The suffix letters, n, s, e, ev, v, k and pec denote, respectively, wide and diffuse (nebulous) lines, sharp lines, emission lines, variable emission lines, variable spectrum, H and K lines of Ca^+, and peculiar lines which tend to remove it from the class given. The criteria for the classification of the "Henry Draper" (HD) classification and the Yerkes (MKK) classification are given in Tables 67 and 68.

It can be seen from Table 67 that when an element is found both in an ionized and neutral state in the spectral series, the ionized atom and then the neutral atom are found in going down the sequence from O to M. Furthermore, the intensity of each line of successive stages of ionization of one element is a maximum at lower stages in the sequence. For example, Si IV, Si III, Si II, and Si I have maximum line intensities at the respective classes O 9, B 1, A 0 and G 5. As first pointed out by SAHA (1921), the energy needed to excite different stages of ionization decreases with increasing spectral type, and the spectral sequence must therefore be associated with decreasing temperatures in the layers of the stellar atmosphere where the lines are formed. As illustrated in Fig. 34, and Table 66, the Henry Draper classification is a one dimensional system of decreasing effective temperature, T_e. If all stars have the same radii this sequence must also correspond to decreasing absolute luminosity. However, it soon became apparent that stars of a given spectral type could have widely differing intrinsic luminosities. The fainter stars were called dwarfs, the brighter stars giants, and the brightest supergiants. This led to a description by luminosity class as well as spectral type with the designations I for supergiants, II for bright giants, III for giants, IV for subgiants, V for the main sequence and dwarfs, and VI for subdwarfs and white dwarfs.

HERTZSPRUNG (1905) and RUSSELL (1914) showed that, in fact, the absolute luminosity did decrease with spectral type, but that two main groups of stars are formed, one intrinsically brighter than the other. ADAMS and KOHLSCHÜTTER (1914) and ADAMS *et al.* (1935) observed that the relative strengths of closely spaced pairs of spectral lines are linearly related to the absolute magnitude of the star. Such data soon led to the discovery of many different types of objects, and to the luminosity classification of MORGAN, KEENAN, and KELLMAN (1943). This wide variety of stellar objects is shown in the Hertzsprung-Russell diagram of Fig. 42.

The position of an object on the Hertzsprung-Russell diagram is a function of luminosity, radius, initial composition, mass, and age. The detailed evolutionary tracks may be calculated using Eqs. (5-66) to (5-71) together with the thermonuclear reaction rate constants given in Chap. 4. The details are reviewed by BURBIDGE and BURBIDGE (1958), HAYASHI, HOSHI, and SUGIMOTO (1962), CHRISTY (1962, 1966), IBEN (1967), and TAYLER (1968).

SALPETER (1955) has suggested that stellar evolution might account for the observed luminosity function, $\varphi(M_v)$, given by

$$dN = \varphi(M_v)\, dM_v, \tag{5-245}$$

where dN is the total number of stars of all spectral classes per cubic parsec with absolute visual magnitude lying between M_v and $M_v + dM_v$. The observed

Table 68. The Morgan, Keenan, Kellman (MKK) or Yerkes classification[1]

	Criteria for spectral types	Criteria for increasing luminosity	Type stars
	Ratios of Si III/Si IV, Si II/Si III, Si II/He I		
B0	4552 Si III/4089 Si IV (<1)	4089 Si IV/4009 He I	B0 Ia ε Ori, V δ Sco
B1	.... " .... (>1)	3995 N II/4009, 4552/4387 He I	B1 Iab ζ Per
B2–B3	4128–31 Si II/4121 He I. In B3 the K line appears.	.... "	B2 III γ Ori, V ζ Cas; B3 V η UMa
B5	.... 4144 He I	Balmer lines sharper	B5 III δ Per
			B7 V α Leo; B8 Iab β Ori
B9	4481 Mg II $\gg$ 4471 He I	.... "	B9 III γ Lyr
	Increasing intensity of metal lines	.... "	
A0	He I very weak or absent; Fe II very weak	(Fe II a little stronger)	A0 III α Dra, IV γ Gem, V α Lyr
A1	4030–34 Mn I appears, 4385 (blend)/4481 Mg II		A1 V α CMa
A2–A5	.... /4128–31, 4300 (blend)/4385	4416 (blend)/4481, 4416/4300	A2 Ia α Cyg; A3 III β Tri
			A4 III α Oph; A5 V δ Cas
			A7 III γ Boo
F0	.... "	.... "	F0 Ib α Lep; V γ Vir
F2	G band shades off toward the red (CH)	.... " .... and 4172/4226 Ca I	F2 IV β Cas
F5	G band is intensified, 4045 Fe I/Hδ, 4226 Ca I/Hγ	4077 Sr II/4226, 4045, 4063, 4250 Fe I	F5 Ib γ Cyg, V β Vir

G0	.... "	.... "	G0 Ib α Aqr, II α Sag, IV η Boo G2 V the sun, 16 Cyg A
G5	4030–34/4300 violet side of G, 4325 Fe I/Hγ	4077/4062 Fe I, 4085, 4144, 4250 Fe I CN bands stronger 4215–4144	G5 IV μ Her; G8 II ζ Cyg G8 III δ Boo, IV β Aql, V ζ Boo A
K0	........"........, 4290 (blend)/4300, 4096/Hδ	4077/4063, 4077/4071 CN band 4216	K0 III ε Cyg, IV η Cep; K1 IV γ Cep K2 Ib ε Peg, III α Ari, V ε Eri
K5	4226 Ca I/4325 Fe I, ... "	..... " ....., 4215 Sr II/4250 Fe I	K3 Ib η Per, II γ Agl, III δ And K5 II ζ Cyg, III α Tau, V 61 Cyg A
M0–M5	Increasing intensity of TiO bands Band heads: 4762, 4954, 5168, 5445, 5763, 5816, 5857 (farther out 6651, 7054, 7589)	Increasing intensity of H lines For giants and supergiants: decreasing intensity of 4226 Ca I. 4077 Sr II/4045 Fe I, 4215 Sr II/4250 Fe I	M0 III β And M2 Ia μ Cep, Ib α Ori M5 II α Her

Blend = mixture of many lines of the same element or different elements.

[1] After DUFAY (1964, by permission of George Newnes, Ltd.).

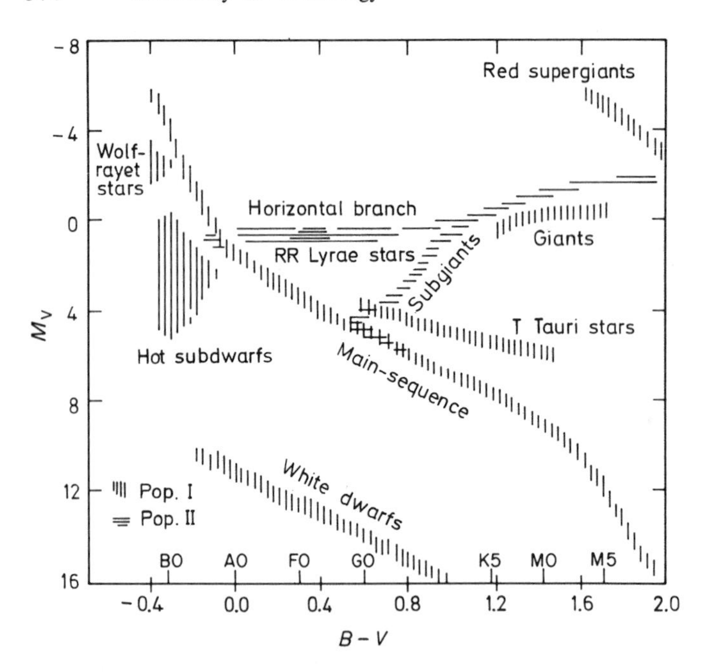

Fig. 42. Terminology of sequences and regions in the Hertzsprung-Russell diagram. The ordinate is absolute visual magnitude, M_v, whereas the abscissa is $B-V$, the color index. The spectral types above the $B-V$ axis are for main-sequence stars. (After HAYASHI, HOSHI, and SUGIMOTO, 1962, by permission of the Research Institute for Fundamental Physics, the Physical Society of Japan, and the Nissha Publishing Co.)

function is given by VAN RHIJN (1925, 1936), LUYTEN (1941), and MCCUSKEY (1956) for our Galaxy, and by SANDAGE (1957) for several galactic clusters. The observed $\varphi(M_v)$ for main sequence stars is given in Table 69, together with an initial luminosity function, $\psi(M)$, which represents the relative frequency with which stars populate the main sequence at the time of their arrival there. The initial luminosity function was calculated by SALPETER (1955) who assumed a constant mass for stars on the main sequence evolving toward the SCHÖNBERG-CHANDRASEKHAR (1942) limit. The main sequence lifetime, τ, of a star of luminosity, L, and mass, M, is given by (SANDAGE, 1957; cf. Eq. (5-72))

$$\tau = 1.10 \times 10^{10} \frac{L_\odot}{L} \frac{M}{M_\odot} \text{ years}, \tag{5-246}$$

where $M_\odot$ and $L_\odot$ denote, respectively, the mass and luminosity of the Sun. Assuming a constant rate of star formation, dN/dt, for a time, T, we have

$$\begin{aligned} \psi(M_v) &= \frac{dN(M_v)}{dt} T = \varphi(M_v) \frac{M_L}{M} \frac{L}{L_L} \quad \text{for } \tau \leq T \\ &= \varphi(M_v) \qquad\qquad \text{for } \tau > T, \end{aligned} \tag{5-247}$$

where M_L and L_L are the values of mass and luminosity which give $\tau = T$ in Eq. (5-246). Integration of the $\psi(M_v)$ in Table 69 gives 0.120 stars pc^{-3} for the number density of stars created in the lifetime of the Galaxy in the solar neighborhood. If $T = 6 \times 10^9$ years, this gives an annual rate of 2×10^{-11} stars pc^{-3} year^{-1}. SCHMIDT (1963) however, has investigated the rate of formation of stars of different mass as a function of time, and concludes that in the past relatively more bright stars were formed.

Table 69. The observed main sequence luminosity function, $\varphi(M_v)$, and the "zero age" luminosity function, $\psi(M_v)$, as a function of absolute visual magnitude, M_v [1]

M_v	$\log\varphi + 10$	$\log\psi + 10$	M_v	$\log\varphi + 10$	$\log\psi + 10$
−6	1.29	4.71	+ 8	7.66	7.66
−5	2.43	5.59	+ 9	7.72	7.72
−4	3.18	6.08	+10	7.81	7.81
−3	3.82	6.41	+11	7.95	7.95
−2	4.42	6.68	+12	8.11	8.11
−1	5.04	6.92	+13	8.22	8.22
0	5.60	7.10	+14	8.21	8.21
+1	6.17	7.26	+15	8.12	8.12
+2	6.60	7.25	+16	7.98	7.98
+3	7.00	7.23	+17	7.76	7.76
+4	7.30	7.30	+18	7.40	7.40
+5	7.45	7.45	+19	6.58	6.58
+6	7.56	7.56	+20	5.28	5.28
+7	7.63	7.63			

[1] After BURBIDGE and BURBIDGE (1958) by permission of Springer-Verlag.

5.5.4. The Luminosity of the Night Sky

If space is Euclidean and galaxies are uniformly distributed in that space, then the apparent luminosity, l, of the radiation from any spherical shell at a distance, r, and of thickness, dr, is

$$l = 4\pi r^2 \frac{dr}{r^2} \rho(L) L \, dL, \qquad (5\text{-}248)$$

where $\rho(L)$ denotes the number of sources of luminosity, L, per unit volume. When the contributions of all spherical shells are added, l becomes infinite, contradicting the observed fact that the sky is dark at night. This paradox, often called Olbers' paradox, was first noted by HALLEY (1720), DE CHESEAUX (1744), and OLBERS (1826). DICKSON (1968) and JAKI (1969) give historical reviews of this paradox.

For a homogeneous, isotropic, expanding universe, an observer only detects radiation from distances out to cH_0^{-1}, where H_0 is Hubble's constant, and Olbers' paradox is avoided. For galaxies of absolute luminosity, L, and space density, $\rho(L)$, for example, the observed brightness, B_G, will be given by

$$B_G = \frac{c L \rho(L)}{4\pi H_0} \text{ erg cm}^{-2} \text{ sec}^{-1} \text{ ster}^{-1}. \qquad (5\text{-}249)$$

The luminosity function for galaxies is given by KIANG (1961) and is illustrated in Fig. 43. SCHMIDT (1971) gives the data

M_B (mag)	L (erg sec^{-1})	$\rho(L)$ Mpc^{-3}
<-21	10^{44}	$10^{-3.5}$
<-20	10^{44}	$10^{-2.5}$
<-19	10^{44}	$10^{-1.5}$
<-18	10^{43}	10^{-1}

for optical galaxies of absolute magnitude, M_B. ROACH and SMITH (1968) and PEEBLES (1971)[3] give:

	B at $\lambda=5300$ Å
Zodiacal light	146 S_{10}
Integrated starlight	105 S_{10}
Airglow	48 S_{10}
Galaxies	0.73 S_{10}

Here S_{10} is the number of equivalent 10th magnitude stars per square degree. $1\,S_{10}=18.89$ mag (arc minute)$^{-2}=27.78$ mag (arc second)$^{-2}$, whereas $300\,S_{10}=12.70$ mag (arc minute)$^{-2}=21.59$ mag (arc second)$^{-2}$. To convert the surface brightness into c.g.s. units, multiply by $1\,S_{10}=6.86\times10^{-6}$ erg cm^{-2} sec^{-1} steradian^{-1}. The mean intensity of the night sky at the galactic poles as observed at Palomar at $\lambda=6500$ Å is (SHECTMAN, 1973)

$$4.5\times10^{-18}\ \text{erg sec}^{-1}\,\text{cm}^{-2}\,\text{ster}^{-1}\,\text{Hz}^{-1}.$$

PARTRIDGE and PEEBLES (1967) give detailed formulae for the integrated background of the night sky as a function of cosmological model, and suggest that measurements might test models for the formation and evolution of galaxies.

RYLE (1958) noted that Olbers' problem may be expressed by the fact that the integrated emission from a given class of objects must not exceed the observed background emission. As the integrated emission in a static Euclidean universe is proportional to $L\rho(L)$, and the number of sources at a particular flux density is proportional to $L^{3/2}\rho(L)$, we have the condition

$$L\rho(L)<KI$$

and

$$L>KI^{-2}, \tag{5-250}$$

where the constant, K, is given by

$$K=L^{3/2}\rho(L),$$

and I is the integrated emission. These formulae may be used together with observations to derive limits to the luminosities and space densities of radio

[3] New measurements of the optical background are made by DUBE et al. (1977).

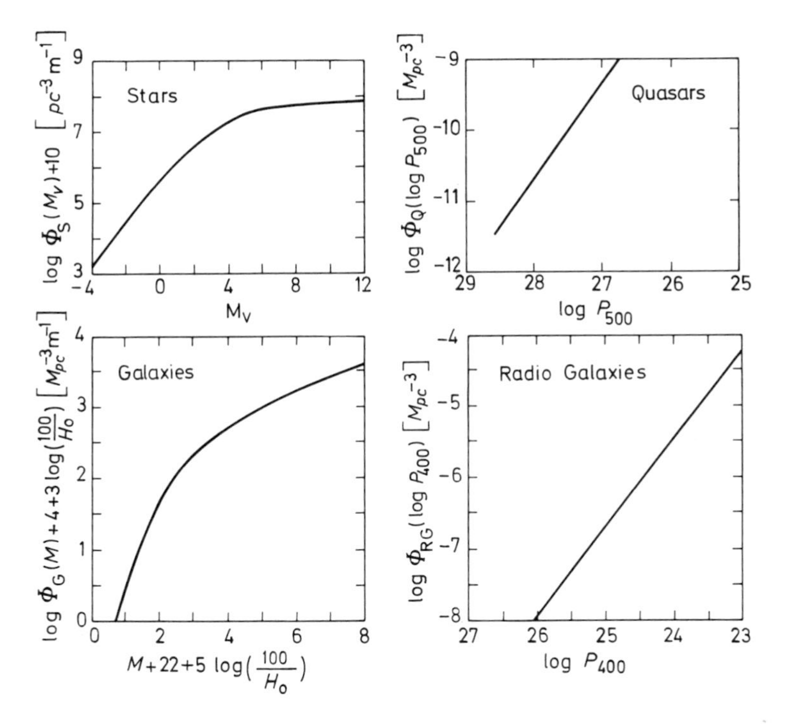

Fig. 43. Respective luminosity functions $\phi_S(M_v)$, $\phi_G(M)$, $\phi_Q(\log P_{500})$ and $\phi_{RG}(\log P_{400})$ for stars of absolute magnitude, M_v, galaxies of absolute magnitude, M, quasars emitting monochromatic power, P_{500}, watt Hz^{-1} at 500 MHz and radio galaxies emitting monochromatic power, P_{400}, watt Hz^{-1} steradian^{-1} at 400 MHz. The data are after BURBIDGE and BURBIDGE (1958), KIANG (1961), SCHMIDT (1968), and MERKELIJN (1971), respectively. Integration of the curve for stars gives 0.120 stars per cubic parsec, whereas the number of galaxies in the first 8 magnitudes of the luminosity range is found to be $0.46\,(H_0/100)^3$ per cubic megaparsec, where H_0 is the Hubble constant. The quasar data show that their space density is about 10^{-9} per cubic megaparsec for the luminosity range indicated. The radio galaxy data in the range $\log P_{400}=23$ to 26 can be represented by the equation $\log[N(\log P_{400})] = -1.06\log P_{400}+19.96$ where N is the number of radio galaxies per cubic megaparsec per unit interval of 0.4 in $\log P_{400}$

objects. SCHMIDT (1968) and MERKELIJN (1971) give the radio luminosity functions for quasi-stellar objects and radio galaxies, respectively; and these functions are shown in Fig. 43. SCHMIDT (1971) gives the total radio luminosity function for radio galaxies and quasi-stellar objects as

L (erg sec^{-1})	$\rho(L)$ Mpc^{-3}
10^{45}	$10^{-10.5}$
10^{44}	10^{-9}
10^{43}	10^{-8}
10^{42}	10^{-7}

whereas MERKLIJN gives

$$\log[N(P_{400})] = -1.06 \log P_{400} + 19.96,$$

where N is the number of radio galaxies per cubic megaparsec per unit interval of 0.4 in $\log P_{400}$, and P_{400} is the monochromatic power at an emission frequency of 400 MHz in watt Hz^{-1} $\mathrm{steradian}^{-1}$.

CHARLIER (1908, 1922) first noted that the light of the night sky would be finite if a hierarchy of clusters of objects in a Euclidean universe satisfies the condition

$$\frac{R_{i+1}}{R_i} > [N_{i+1}]^{1/2}, \tag{5-251}$$

where R_i is the radius of the system of order i, and N_{i+1} is the number of systems of order i in the system $i+1$. SHAPLEY (1933), ZWICKY (1957), ABELL (1958, 1962, 1965), and ZWICKY *et al.* (1961) have observed that galaxies do form in clusters, and that these clusters may form superclusters. The hierarchial clustering of galaxies has been revived as a solution to Olbers' paradox by DE VAUCOULEURS (1970). A modification of this hypothesis has been suggested by ALFVÉN (1971) and KLEIN (1971) who suggest that the galaxies about us are in a finite, expanding metagalactic system.

5.6. The Theory of General Relativity and Its Observational Verification

5.6.1. Formulae of General Relativity

According to the general theory of relativity (EINSTEIN, 1916, 1917, 1919) curvature in the geometry of spacetime manifests itself as gravitation. Gravitation works on the separation of nearby world particle lines. In turn, particles and other sources of mass energy cause curvature in the geometry of spacetime. Thus, gravity manifests itself in the geometry of spacetime, which is described by the line element, ds, given by

$$ds^2 = g_{ik} dx^i dx^k, \tag{5-252}$$

where dx^i is a differential of the coordinate x^i, a space coordinate for $i=1,2,3$, and the time coordinate, ct, for $i=0$, the velocity of light is c, summation is implied when an index appears twice, and the metric tensor, g_{ik}, is itself a function of the spacetime coordinates.

The amount of mass energy in a unit volume is determined by the stress-energy, or energy-momentum, tensor, T_{ik}. For a gas or perfect fluid

$$T^{ik} = (\varepsilon + P) u^i u^k - P g^{ik}, \tag{5-253}$$

where $\varepsilon = \rho c^2$ is the energy density of the matter as measured in its rest frame, ρ is the mass density, P is the pressure, and u^i is the four-velocity of the gas. The stress-energy tensor for the electromagnetic field is given by

$$T^{ik} = -\frac{g_{lm}}{4\pi} F^{il} F^{km} + \frac{g^{ik}}{16\pi} F_{lm} F^{lm}, \tag{5-254}$$

where F_{lm} is the electromagnetic field tensor. In this case, T^{ik} gives a tension $(E^2+B^2)/8\pi$ along the field lines, and a pressure $(E^2+B^2)/8\pi$ perpendicular to the field lines, where $(E^2+B^2)/8\pi$ is the energy density of the electromagnetic field.

The differential form for the conservation of energy-momentum is given by

$$\nabla \cdot T = 0,$$

where $\nabla\cdot$ denotes the covariant divergence, in components,

$$\frac{1}{\sqrt{-g}} \frac{\partial \sqrt{-g}\, T_i^k}{\partial x^k} - \frac{T^{km}}{2} \frac{\partial g_{km}}{\partial x^i} = 0, \tag{5-255}$$

where the fundamental determinant $g = |g_{ik}|$.

For a perfect fluid in flat space, this equation in Cartesian coordinates is

$$\frac{\partial T^{ik}}{\partial x^k} = \frac{\partial P}{\partial x_i} + \frac{\partial}{\partial x^k}[(\rho+P)u^i u^k] = 0, \tag{5-256}$$

and by substituting $u^i = v^i u^0$ it becomes Euler's equation

$$\frac{\partial v}{\partial t} + (v\cdot\nabla)v = -\frac{(1-v^2)}{(\rho+P)}\left(\nabla P + v\frac{\partial P}{\partial t}\right). \tag{5-257}$$

The Riemann curvature tensor, R^i_{klm}, is the agent by which curves in space-time produce the relative acceleration of geodesics.

$$R^i_{klm} = \frac{\partial \Gamma^i_{km}}{\partial x^l} - \frac{\partial \Gamma^i_{kl}}{\partial x^m} + \Gamma^i_{nl}\Gamma^n_{km} - \Gamma^i_{nm}\Gamma^n_{kl}. \tag{5-258}$$

Here the Christoffel symbols Γ^l_{mn} are defined by the relations

$$\Gamma^i_{kl} = \frac{g^{im}}{2}\left(\frac{\partial g_{mk}}{\partial x^l} + \frac{\partial g_{ml}}{\partial x^k} - \frac{\partial g_{kl}}{\partial x^m}\right) \tag{5-259}$$

where

$$\Gamma^i_{ik} = \frac{1}{2}\frac{\partial}{\partial x^k}(\ln -g) = \frac{1}{g^{1/2}}\frac{\partial}{\partial x^k}(-g^{1/2}),$$

and g is the determinant $|g_{ik}|$. Using the Christoffel symbols, the equation expressing the conservation of energy-momentum becomes

$$T^{ik}_{;m} = \frac{\partial T^{ik}}{\partial x^m} + \Gamma^i_{lm}T^{lk} + \Gamma^k_{lm}T^{il},$$

or

$$T^{ik}_{;i} = \frac{\partial T^{ik}}{\partial x^i} + \Gamma^i_{il}T^{lk} + \Gamma^k_{il}T^{il}$$

$$= \frac{1}{-g^{1/2}}\frac{\partial}{\partial x^i}(-g^{1/2}T^{ik}) + \Gamma^k_{il}T^{il},$$

where the ; denotes the covariant divergence. For a contravariant vector, V^i, and a covariant vector, V_i, the covariant derivatives are

$$V^i_{;k} = \frac{\partial V^i}{\partial x^k} + \Gamma^i_{lk}V^l,$$

and

$$V_{i;k} = \frac{\partial V_i}{\partial x^k} - \Gamma^l_{ik} V_l,$$

where for any vector

$$V^i_{;i} = \frac{1}{-g^{1/2}} \frac{\partial}{\partial x^i} (-g^{1/2} V^i).$$

The production of curvature by mass-energy is specified by Einstein's field equations (EINSTEIN, 1916, 1919)

$$R_{ik} - \frac{1}{2} R g_{ik} - \Lambda g_{ik} = \frac{-\kappa}{c^2} T_{ik}, \tag{5-260}$$

where the cosmological constant, Λ, is usually taken to be zero, and Einstein's gravitational constant, $\kappa = 8\pi G/c^2 \approx 1.86 \times 10^{-27}$ dyn $\text{g}^{-2}\text{sec}^2$. Sometimes $\kappa_0 = \kappa/c^2 \approx 2.07 \times 10^{-48}$ cm^{-1} gm^{-1} sec^2 is called the Einstein gravitational constant as well. The Newtonian constant of gravitation is G, and the Ricci tensor, R_{ik}, is given by

$$R_{km} = R^i_{klm} g^l_i = R^i_{kim}, \tag{5-261}$$

where R^i_{klm} is the Riemann curvature tensor, and the scalar curvature of spacetime, R, is given by the contraction of the Ricci tensor

$$R = R_{ik} g^{ik} = R^i_i. \tag{5-262}$$

Einstein's field equations follow from the assumptions that the ratio of gravitational and inertial mass is a universal constant, that the laws of nature are expressed in the simplest possible set of equations that are covariant for all systems of spacetime coordinates, and that the laws of special relativity hold locally in a coordinate system with a vanishing gravitational field.

In a gravitational field, a particle moves along a geodesic line which is specified by the differential equation

$$\frac{d^2 x^i}{ds^2} + \Gamma^i_{kl} \frac{dx^k}{ds} \frac{dx^l}{ds} = 0. \tag{5-263}$$

Light rays are represented by null geodesics for which $ds = 0$. An observer's proper time, τ, is governed by the metric along his world line

$$d\tau = \sqrt{ds^2} = \sqrt{g_{ik} dx^i dx^k}, \tag{5-264}$$

where the world line is described by any time parameter, t,

$$x^i = x^i(t). \tag{5-265}$$

In any infinitesimal neighborhood of any point in spacetime, the proper time intervals must satisfy the laws of special relativity. That is, locally the line element, ds, becomes the Minkowski metric given by (MINKOWSKI, 1908; HARGREAVES, 1908)

$$ds^2 = c^2 dt^2 - dx^2 - dy^2 - dz^2 \tag{5-266}$$

in rectangular coordinates,

$$ds^2 = c^2 dt^2 - dr^2 - r^2 d\theta^2 - r^2 \sin^2\theta \, d\varphi^2 \tag{5-267}$$

in spherical coordinates, and

$$ds^2 = c^2 dt^2 - dr^2 - r^2 d\varphi^2 - dz^2 \tag{5-268}$$

in cylindrical coordinates, where in each case the coordinates are measured in an inertial frame of reference. In Minkowski spacetime, ds is invariant under the Lorentz transformation, geometry is Euclidean, space is flat, and the interval of proper time, $d\tau$, is given by

$$d\tau = \left[1 - \frac{v^2}{c^2}\right]^{1/2} dt, \tag{5-269}$$

which is the time interval read by a clock moving at the velocity, $v = [(dx^2 + dy^2 + dz^2)/dt^2]^{1/2}$.

5.6.2. The Schwarzschild Metric and Classical Tests of General Relativity

For a spherically symmetric gravitational field outside a massive non-rotating body in vacuum (where $R_{ik} = 0$), the line element, ds, becomes the Schwarzschild metric, ds, given by (SCHWARZSCHILD, 1916)

$$ds^2 = \left[1 - \frac{2GM}{c^2 r}\right] c^2 dt^2 - \frac{dr^2}{\left[1 - \frac{2GM}{c^2 r}\right]} - r^2 d\theta^2 - r^2 \sin^2\theta \, d\varphi^2. \tag{5-270}$$

Here, r, θ, and φ are spherical coordinates whose origin is at the center of the massive object, and M is the mass which determines the Newtonian gravitational field GM/r, where G is the Newtonian gravitational constant. The equivalent "Newtonian" force felt by a freely falling test particle of mass, m, with low velocity, $v \ll c$, has a magnitude

$$F = \frac{-GMm}{r^2 \left[1 - \frac{2GM}{c^2 r}\right]^{1/2}}. \tag{5-271}$$

However, this interpretation is only valid at radii much greater than the Schwarzschild or gravitational radius, r_g, given by

$$r_g = \frac{2GM}{c^2}. \tag{5-272}$$

The Schwarzschild line element may be expressed in the isotropic form

$$ds^2 = \frac{\left(1 - \frac{GM}{2c^2 \rho}\right)^2}{\left(1 + \frac{GM}{2c^2 \rho}\right)^2} c^2 dt^2 - \left(1 + \frac{GM}{2c^2 \rho}\right)^4 [d\rho^2 + \rho^2 (d\theta^2 + \sin^2\theta \, d\varphi^2)], \tag{5-273}$$

using the transformation $r = \rho[1 + GM/(2c^2 \rho)]^2$.

A more general post-Newtonian metric is now used, however, to allow comparisons of observations with other gravitation theories such as the scalar-tensor theory of BRANS and DICKE (1961) and the C-field theory of HOYLE and NARLIKAR (1966). The post-Newtonian metric of a spherically symmetric gravitational field outside of a massive non-rotating body in a vacuum is given by (EDDINGTON, 1923; ROBERTSON, 1962)

$$\begin{aligned} ds^2 &= \left[1 - 2\alpha\left(\frac{GM}{c^2\rho}\right) + 2\beta\left(\frac{GM}{c^2\rho}\right)^2 + \cdots\right] c^2 dt^2 \\ &\quad - \left[1 + 2\gamma\left(\frac{GM}{c^2\rho}\right) + \cdots\right] [d\rho^2 + \rho^2(d\theta^2 + \sin^2\theta\, d\varphi^2)], \end{aligned} \tag{5-274}$$

where an isotropic form has been used. Using the transformation $r = \rho[1 + \gamma GM/(c^2\rho)]$ we obtain the equivalent "standard" form

$$\begin{aligned} ds^2 &= \left[1 - 2\alpha\left(\frac{GM}{c^2 r}\right) + 2(\beta - \alpha\gamma)\left(\frac{GM}{c^2 r}\right)^2 + \cdots\right] c^2 dt^2 \\ &\quad - \left[1 + 2\gamma\left(\frac{GM}{c^2 r}\right)\right] dr^2 - r^2(d\theta^2 + \sin^2\theta\, d\varphi^2). \end{aligned}$$

A further generalization of Eq. (5-274) is given by THORNE and WILL (1971) who include non-spherical, non-Newtonian terms. The Einstein field equations give

$$\alpha = \beta = \gamma = 1, \tag{5-275}$$

whereas the Brans-Dicke theory gives

$$\alpha = \beta = 1, \qquad \gamma = \frac{\omega + 1}{\omega + 2}, \tag{5-276}$$

where ω is an unknown dimensionless parameter of the theory. The parameter α must be 1.0 because planetary orbits agree well with the Newtonian theory.

A light ray passing a minimum distance, R_0, from the center of an object of mass, M, will be deflected from a straight line path by the angle (EINSTEIN, 1911)

$$\varphi = 2(\alpha + \gamma)\frac{GM}{R_0 c^2} \text{ radians}, \tag{5-277}$$

where G is the Newtonian gravitational constant. For the Schwarzschild metric, $\varphi = 4GM/(R_0 c^2)$, which is twice that predicted from the Newtonian theory of gravitation. At the limb of the Sun, $\varphi = 1.749$ arc seconds if $\alpha = \gamma = 1$. EDDINGTON (1919) and DYSON, EDDINGTON, and DAVIDSON (1920) extrapolated optical observations to obtain $\varphi = 1.98 \pm 0.16$ arc seconds. Interferometry measurements of radio sources by SEIELSTAD, SRAMEK, and WEILER (1970), MUHLEMAN, EKERS, and FOMALONT (1970), SRAMEK (1971), and HILL (1971) lead to respective values of $\varphi = 1.77 \pm 0.20$, $1.82^{+0.24}_{-0.17}$, 1.57 ± 0.08 and 1.87 ± 0.3 arc seconds.

The perihelion of an orbiting planet will advance by the amount

$$\Delta\theta = 2\pi[2\alpha(\alpha + \gamma) - \beta]\frac{GM}{c^2 a(1 - e^2)} \text{ radians per revolution}, \tag{5-278}$$

where M is the mass of the central object, and a and e are, respectively, the semi-major axis and the eccentricity of the orbital ellipse. Theoretical, $\Delta\theta_t$, and observed, $\Delta\theta_0$ advances in seconds of arc per century are (WHITROW and MORDUCH, 1965) $\Delta\theta_t=43.03$, $\Delta\theta_0=43.11\pm0.45$ for Mercury, $\Delta\theta_t=8.63$ $\Delta\theta_0=8.4\pm4.8$ for Venus, and $\Delta\theta_t=3.84$, $\Delta\theta_0=5.0\pm1.2$ for Earth. The values for Icarus are (SHAPIRO *et al.*, 1971) $\Delta\theta_t=10.3$ and $\Delta\theta_0=9.8\pm0.8$. Here the theoretical values are for $\alpha=\beta=\gamma=1$. Radar measurements of the perihelion advance for Mercury by SHAPIRO *et al.* (1972) give $(2+2\gamma-\beta)/3=1.005\pm0.007$, where α has been assumed to be unity, 0.007 represents the statistical standard error, and the second zonal harmonic of the Sun has been assumed to be negligible (no oblateness). The value of the precession of the perihelion of Mercury is only 7 seconds of arc per century if the equations of the relativistic motion of a particle in a Newtonian square law force are used (GOLDSTEIN, 1950). The one percent agreement between Einstein's prediction and the observed anomalous precession would not, however, confirm his theory if the Sun is oblate (DICKE, 1964). Such an oblateness has been inferred (DICKE and GOLDENBERG, 1967), but the question of solar oblateness is still a matter of controversy. For example, CHAPMAN and INGERSOLL (1973) show that faculae, which are brightest at the solar equator, could account for the observed oblateness.

The received wavelength at infinity, λ, of a spectral line emitted at a distance, R, from the center of a spherical body of mass, M, is given by the equations

$$\lambda=\lambda_0+\Delta\lambda=\lambda_0\left[1-\frac{2GM}{c^2R}\right]^{-1/2}, \tag{5-279}$$

and

$$z=\frac{\Delta\lambda}{\lambda_0}\approx\frac{GM}{c^2R}\approx1.47\times10^5\left(\frac{M}{M_\odot}\right)R^{-1}\quad\text{if } GM\ll Rc^2, \tag{5-280}$$

where z is the gravitational redshift, the radius, R, is in cm in the numerical approximation, the solar mass $M_\odot=2\times10^{33}$ gm, and λ_0 is the wavelength emitted by the atom when the gravitational field is negligible. The gravitational redshift is not a definitive test of general relativity, however, for the Newtonian theory of gravity predicts the same redshift as that given in Eq. (5-280), assuming that the photon energy is $h\nu$, and that the principle of equivalence and the law of conservation of energy hold. Although measurement of the gravitational redshift does provide empirical evidence for the equivalence of gravitational and inertial mass, no new information with regard to other gravitational theories may be obtained from the gravitational redshift. This is because it depends only on α which must be 1.0 in order for planetary orbits to agree with even the Newtonian theory. For light originating from the solar limb, $\Delta\lambda/\lambda_0=2.17\times10^{-6}$. Measurements of the gravitational redshift for atoms on the Sun and the more dense white dwarfs are discussed by ADAMS (1925), ST. JOHN (1928), POPPER (1959), BLAMONT and RODDIER (1961), BRAULT (1963), FORBES (1970), GREENSTEIN, OKE, and SHIPMAN (1971), GREENSTEIN and TRIMBLE (1967, 1972), and TRIMBLE and GREENSTEIN (1972). The discovery of the narrow gamma ray spectral lines of nuclei (MÖSSBAUER, 1958) has led to the detection of the terrestrial gravitational redshift. Using the Mössbauer effect, POUND and REBKA (1959, 1960) have measured a value of $\Delta\lambda/\lambda_0=2.57\pm0.26\times10^{-15}$ as opposed to the predicted value of 2.46×10^{-15}. The agreement between theory and experiment has since been improved to about one percent (POUND and SNIDER, 1964).

The two way time delay of a radar signal from a planet is increased when the ray path passes near the Sun. The additional coordinate time delay, $\Delta\tau$, is given by (SHAPIRO, 1964, 1968)

$$\Delta\tau = \frac{2GM}{c^3}(1+\gamma)\ln\left[\frac{r_e+r_p+R}{r_e+r_p-R}\right], \tag{5-281}$$

where r_e and r_p are the respective radial distances of the earth and the planet from the sun, and R is the earth-planet distance. From elongation to superior conjunction, $\Delta\tau$ increases from 15 to 240 microseconds for Mercury. The actual delay measured by SHAPIRO *et al.* (1971) gave $\gamma = 1.03 \pm 0.04$, where the theoretical general relativity value is 1.00.

A satellite in orbit about a planet of mass, M, will exhibit a gravitational precession, Ω, given by (DE SITTER, 1916, 1920; SCHIFF, 1960)

$$\Omega = \frac{\alpha+2\gamma}{2r^3}\,\frac{GM}{c^2}(\boldsymbol{r}\times\boldsymbol{v}), \tag{5-282}$$

where $\boldsymbol{r}$ and $\boldsymbol{v}$ are vectors denoting the distance and velocity of the satellite. If a gyroscope is in circular orbit about the earth at an orbital radius, r, then the precession averaged over one revolution is given by

$$\Omega = \frac{(\alpha+2\gamma)}{2r^{5/2}}\left(\frac{GM_\oplus}{c^2}\right)^{3/2} \approx 8.4\,\frac{\alpha+2\gamma}{3}\left(\frac{R_\oplus}{r}\right)^{5/2} \text{ seconds of arc per year,} \tag{5-283}$$

where $M_\oplus$ and $R_\oplus$ denote, respectively, the mass and radius of the earth. If $r \approx R_\oplus$, the geodetic precession, Ω, given in Eqs. (5-282) and (5-283) is several orders of magnitude larger than the Thomas precession caused by the interaction of the spin orbital angular momenta of the earth and the gyroscope.

As pointed out by DIRAC (1937, 1938), the age of the universe, H_0^{-1}, in terms of the atomic unit of time, $e^2/(mc^3)$, is of order 10^{39}, as is the ratio, $e^2/(Gm_p m)$, of the electrical force between the electron and proton to the gravitational force between the same particles. This, according to Dirac, suggested that both ratios are functions of the age of the universe, and that the gravitational constant, G, might vary with time. For example, Dirac chose

$$\left(\frac{\dot{G}}{G}\right)_0 = -3H_0,$$

where $\dot{}$ denotes the first derivative with respect to time, the subscript 0 denotes the present value, and H_0 is the present value of the Hubble constant. The cosmological model of BRÅNS and DICKE (1961), DICKE (1962), gives (WEINBERG, 1972)

$$\begin{aligned}\left(\frac{\dot{G}}{G}\right)_0 &= \frac{-3q_0H_0}{\omega+2} && \text{for } q_0 \ll 1\\ &= \frac{-H_0}{\omega+2} && \text{for } q_0 = 0.5\\ &= \frac{-1.71H_0}{\omega+2} && \text{for } q_0 = 1\end{aligned}$$

and

$$= \frac{-3.34\, H_0\, q_0^{1/2}}{\omega+2} \quad \text{for } q_0 \gg 1,$$

where ω is a dimensionless constant of the cosmological model (cf. Eq. (5-276)), q_0 denotes the present value of the deceleration parameter, and

$$H_0 = [0.98 \times 10^{10} h^{-1}]^{-1} \text{ years}^{-1},$$

if $H_0 = 100 h$ km sec^{-1} Mpc^{-1}. Radar measurements of the orbits of the inner planets give (SHAPIRO *et al.*, 1971)

$$\left(\frac{\dot{G}}{G}\right)_0 = +0 \pm 4 \times 10^{-10} \text{ years}^{-1}.$$

VAN FLANDERN (1975) reports that observations of the orbits of the Sun and moon suggest that $(\dot{G}/G)_0 = -8 \pm 5 \times 10^{-11}$ year^{-1}.

5.7. Cosmological Models and Their Observational Verification

5.7.1. The Homogeneous and Isotropic Universes

HUBBLE (1926), HUBBLE and HUMASON (1931), and SHAPLEY and AMES (1932) observed that the total number of galaxies to various limits of total magnitude vary directly with the volumes of space represented by the limits, and concluded that the density of observable matter is constant. Their additional observation that the distribution of nebulae is spatially isotropic across the sky led to the conclusion that the universe is filled with a spatially homogeneous and isotropic distribution of matter. WEYL (1923, 1930) first postulated that the galaxies are on a bundle of geodesics in spacetime which converges toward the past, and showed that the line element is given by

$$ds^2 = c^2 dt^2 + g_{ik} dx^i dx^k, \tag{5-284}$$

where the cosmic time, t, measures the proper time of an observer following the geodesic, and $i, k = 1, 2, 3$. Following the first derivation of the line element of a nonstationary universe (LANCZOS, 1922, 1923), it was shown that the general form for the line element of a spatially homogeneous and isotropic universe is given by (FRIEDMANN, 1922, 1924)

$$ds^2 = c^2 dt^2 - R^2(t)\left[\frac{dr^2}{1-kr^2} + r^2(d\theta^2 + \sin^2\theta\, d\varphi^2)\right], \tag{5-285}$$

where the curvature index $k = 0, \pm 1$, and $R(t)$ is called the radius of curvature or the scale factor of the universe. The r coordinate has zero value for some arbitrary fundamental observer, the surface $r = \text{constant}$ has the geometry of the surface of a sphere, and θ, φ are polar coordinates. Although this metric was first used by FRIEDMANN (1922, 1924) for $k = \pm 1$, it is often called the Robertson-Walker metric after its rigorous deduction by ROBERTSON (1935, 1936) and WALKER (1936) (cf. also ROBERTSON, 1928, 1933).

When the Robertson-Walker metric is used with Einstein's field equations, the following equations are obtained (LEMAÎTRE, 1927; RAYCHAUDHURI, 1955, 1957)

$$\frac{3\dot{R}^2}{R^2} + \frac{3kc^2}{R^2} = \kappa\rho c^2 + \Lambda c^2, \tag{5-286}$$

$$\frac{2\ddot{R}}{R} + \frac{\dot{R}^2}{R^2} + \frac{kc^2}{R^2} = -\kappa P + \Lambda c^2 \tag{5-287}$$

or

$$\frac{\ddot{R}}{R} = \frac{\Lambda c^2}{3} - \frac{4\pi G}{3}\left[\rho + \frac{3P}{c^2}\right],$$

where $R = R(t)$ is the radius of curvature or the scale factor of the universe, the $\dot{}$ denotes differentiation with respect to the cosmic time, t, Einstein's gravitational constant $\kappa = 8\pi G/c^2 = 1.86 \times 10^{-27}$ dyn g^{-2} sec^2, the mean density of matter and energy in the universe is $\rho(t)$ (the energy density is ρc^2), the isotropic hydrodynamic pressure of matter and radiation is $P(t)$, the cosmological constant is Λ, and k/R^2 is the Riemannian space curvature. The index k takes values of $+1$, -1, or 0 according to whether the space is closed, open, or Euclidean. Following ROBERTSON (1955) and HOYLE and SANDAGE (1956), it is customary to define different cosmological models in terms of the present value of the Hubble expansion parameter, H_0, and the present value of the deceleration parameter, q_0.

$$H_0 = \frac{\dot{R}_0}{R_0}, \tag{5-288}$$

and

$$q_0 = -\frac{\ddot{R}_0}{R_0 H_0^2} = -\frac{\ddot{R}_0 R_0}{\dot{R}_0^2} = \frac{-1}{H_0^2}\left[\frac{\Lambda c^2}{3} - 4\pi G\left(\frac{\rho_0}{3} + \frac{P_0}{c^2}\right)\right], \tag{5-289}$$

where the subscript zero is used to denote the present epoch, and R_0 is the present value of $R(t)$. The $R(t)$ curves for various expanding universes with $\Lambda = 0$ and various values of q_0 are shown in Fig. 44 together with the $R(t)$ for the steady state and the Lemaître-Eddington universes. The formulae which relate H_0 and q_0 to the age of the universe, T_0, were given in Eqs. (5-83) to (5-85); whereas the various formulae for distance as a function of H_0 and q_0 were given in Eqs. (5-148) to (5-152).

Observational evidence (PEACH, 1970) indicates that the cosmological constant, Λ, is either zero or very small for $\Lambda \lesssim 2 \times 10^{-55}$ cm^{-2}. Isotropic, homogeneous universes with zero cosmological constant are often called Friedmann universes after the first solution for this case by FRIEDMANN (1922, 1924). For those models for which the cosmological constant is zero, the present value of the density of matter and radiation, ρ_0, the present value of the cosmic pressure, P_0, and the present value of the spatial curvature, kc^2/R_0^2, are related by the equations

$$\rho_0 + \frac{3P_0}{c^2} = \frac{3H_0^2 q_0}{4\pi G}, \tag{5-290}$$

$$\frac{kc^2}{R_0^2} = \frac{4\pi G}{3q_0}\left[\rho_0(2q_0 - 1) - \frac{3P_0}{c^2}\right]. \tag{5-291}$$

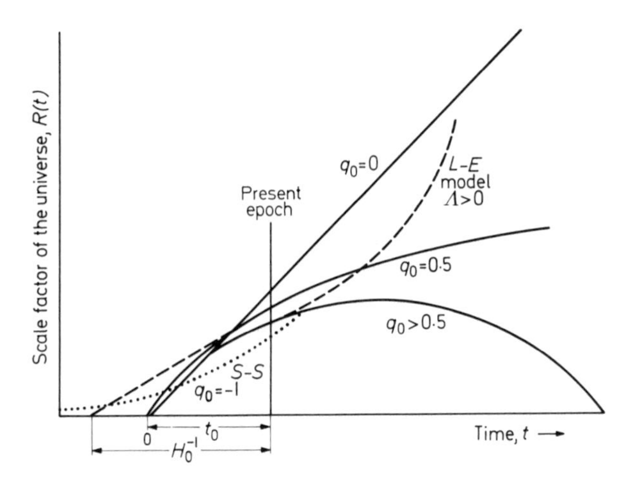

Fig. 44. The scale factor, $R(t)$, of the universe for three exploding models with cosmological constant $\Lambda=0$, for a generalized Lemaître-Eddington model with $\Lambda>0$, and for the steady-state model. Here the deceleration parameter is q_0, the Hubble constant is H_0, and the "age of the universe" is t_0

If the present value of P_0 is small compared with ρ_0, these equations reduce to

$$\rho_0=\frac{3q_0H_0^2}{4\pi G}=3.8\times10^{-33}\,q_0H_0^2\ \mathrm{g\,cm^{-3}}, \tag{5-292}$$

where in the numerical approximation H_0 is in km sec^{-1} Mpc^{-1} and 1 Mpc $=3.18\times10^{24}$ cm, and

$$\frac{kc^2}{R_0^2}=H_0^2(2q_0-1). \tag{5-293}$$

The critical density, ρ_c, is given by

$$\rho_c=\frac{3H_0^2}{8\pi G}=1.9\times10^{-29}\,h^2\ \mathrm{g\,cm^{-3}}, \tag{5-294}$$

where $H_0=100\,h$ km sec^{-1} Mpc^{-1} and h is a constant. For model universes with negligible cosmic pressure, $P_0=0$, and zero cosmological constant, $\Lambda=0$, we have

If $k=1$ then $q_0>\frac{1}{2}$ and $\rho_0>\rho_c$ for elliptical closed space and an oscillating universe,

If $k=0$ then $q_0=\frac{1}{2}$ and $\rho_0=\rho_c$ for a flat, Euclidean space and an ever-expanding Einstein-deSitter universe (Einstein and deSitter, 1932),

If $k=-1$ then $\frac{1}{2}>q_0>0$ and $\rho_0<\rho_c$ for a hyperbolic open space and an ever-expanding Milne universe (Milne, 1935).

LEMAÎTRE (1927, 1931) proposed a homogeneous, isotropic universe with positive curvature, $k=1$, but with a nonzero cosmological constant, Λ, given by

$$\Lambda = (1+\varepsilon)\Lambda_E, \tag{5-295}$$

where $0<\varepsilon \ll 1$ and the cosmological constant, Λ_E, appropriate for the static Einstein universe is given by (EINSTEIN, 1917)

$$\Lambda_E = \frac{\kappa\rho}{2} = \frac{4\pi G\rho}{c^2} = \frac{1}{R^2}, \tag{5-296}$$

where for the Einstein universe R was a constant. LEMAÎTRE, like EINSTEIN, considered the pressure $P=0$, so that

$$\rho R^3 = \frac{\alpha c^2}{4\pi G\sqrt{\lambda_E}} = \text{constant}, \tag{5-297}$$

where the constant α is the amount by which ρR^3 differs from the Einstein value. In the Lemaître model $R(t)$ begins to expand at $t=0$ like $t^{2/3}$, but eventually slows down and spends some time at the value $R=\alpha^{1/3}/\sqrt{\lambda_E}$, at which R is a minimum. This property of the model has been used to explain the possible concentration of quasar redshifts around $z=2$ (KARDASHEV, 1967; PETROSIAN, SALPETER, and SZEKERES, 1967). In addition to showing that the Einstein static universe is unstable, EDDINGTON (1930) presented a limiting case of the Lemaître models. In this Lemaître-Eddington universe, $R(t)$ expands asymptotically from the Einstein value of $1/\sqrt{\Lambda_E}$ at $t=0$ to the value

$$R(t) = \exp[Ht] \tag{5-298}$$

at $t=\infty$. This asymptotic value of the radius of curvature is that of the deSitter static universe (DESITTER, 1917) for which $\rho=P=0$, $k=1$, and $\Lambda=\Lambda_e$.

For tests of the Lemaître type models it is convenient to define the dimensionless density parameter, σ, which has the present value of

$$\sigma_0 = \frac{4\pi G\rho_0}{3H_0^2}, \tag{5-299}$$

where ρ_0 is the present density of matter and radiation in the universe, and H_0 is the present value of the Hubble parameter. For a universe with negligible present cosmic pressure, $P_0 \approx 0$, the cosmological constant is given by

$$\Lambda = 3H_0^2(\sigma_0 - q_0), \tag{5-300}$$

and the spatial curvature is given by

$$\frac{kc^2}{R_0^2} = H_0^2(3\sigma_0 - q_0 - 1), \tag{5-301}$$

where q_0 is the present value of the deceleration parameter, and the pressure term is neglected. PEACH (1970) obtains the limits $2\times10^{-55}\,\text{cm}^{-2} \geq \Lambda \geq -2\times10^{-55}\,\text{cm}^{-2}$.

For the special case where the universe presents the same large scale view to all fundamental observers at all times (the "perfect cosmological principle"), the line element takes the steady state form (BONDI and GOLD, 1948; HOYLE, 1948)

$$ds^2 = c^2 dt^2 - \exp(2Ht)\left[dr^2 + r^2(d\theta^2 + \sin^2\theta \, d\varphi^2)\right], \quad (5\text{-}302)$$

where the constant H is that appearing in the deSitter (1917) static universe

$$H = \frac{\dot{R}}{R} = \text{constant}. \quad (5\text{-}303)$$

In this model, $k=0$, $q_0=-1$, space is Euclidean, and the density of matter is constant.

5.7.2. The Redshift-Magnitude Relation

As discussed in Sect. 5.2.6 and 5.3.6, observations of the redshift, z, of optical galaxies support the linear relation

$$V = cz = \frac{c\Delta\lambda}{\lambda_0} = H_0 D_{\mathrm{L}} = c\left[\frac{R(t_0)}{R(t_1)} - 1\right], \quad (5\text{-}304)$$

where V is the velocity of recession, $\Delta\lambda$ is the amount by which the observed wavelength of a spectral line exceeds its laboratory value, λ_0, the constant H_0 is the present value of the Hubble parameter, D_{L} is the luminosity distance, $R(t)$ is the radius of curvature at time, t, the time of observation is t_0, and t_1 is the time at which the light was emitted from the galaxy. The linear dependence of velocity on distance was first predicted by WEYL (1923) using the static deSitter model; discussed by LANCZOS (1923) for the nonstationary $k=+1$ model; given as a function of luminosity distance by TOLMAN (1930); and shown by MILNE (1935) to be an immediate consequence of the assumed homogeneity and isotropy of the universe. The observed redshift-magnitude relation was shown in Fig. 35 of Sect. 5.2.6.

Because the luminosity distance depends on the deceleration parameter, q_0, and the redshift, z, it follows that the apparent magnitude is also a function of q_0 and z. For a spatially homogeneous and isotropic universe with nonzero cosmological constant, the apparent bolometric magnitude, m_{bol}, is given by (SOLHEIM, 1966)

$$m_{\mathrm{bol}} = 5\log R_0 s_k(\omega)(1+z) + M_{\mathrm{bol}} + 25, \quad (5\text{-}305)$$

where

$$\begin{aligned} s_k(\omega) &= \sin\omega && \text{for } k = 1 \\ &= \omega && \text{for } k = 0 \\ &= \sinh\omega && \text{for } k = -1, \end{aligned}$$

$$\omega = \frac{c}{H_0 R_0}\int_1^{1+z} (2\sigma_0 v^3 + (1+q_0-3\sigma_0)v^2 + \sigma_0 - q_0)^{-1/2}\, dv,$$

and

$$\sigma_0 = \frac{4\pi G \rho_0}{3H_0^2}.$$

Here M_{bol} is the absolute bolometric magnitude, ρ_0 is the mean density of matter, and H_0 is the Hubble parameter. When the cosmological constant $\Lambda=0$, this relation becomes (MATTIG, 1958, 1959)

$$m_{bol} = 5\log \frac{c}{H_0 q_0^2} \{q_0 z + (q_0 - 1) [(1 + 2 q_0 z)^{1/2} - 1]\} + M_{bol} + 25 \quad (5\text{-}306)$$

for $q_0 > 0$. An expansion of this relation in powers of z gives the equation (HECKMANN, 1942)

$$m_{bol} = 5\log\left(\frac{cz}{H_0}\right) + 1.086(1 - q_0)z + \cdots + M_{bol} + 25 \text{ magnitudes,} \quad (5\text{-}307)$$

where H_0 is in km sec^{-1} Mpc^{-1} and cz is in km sec^{-1}. For the steady state model, we have

$$m_{bol} = 5\log\left[\frac{cz}{H_0}(1+z)\right] + M_{bol} + 25. \quad (5\text{-}308)$$

The bolometric distance modulus, $m_{bol} - M_{bol}$, is related to the observed distance modulus, $m - M$, by the equation

$$m_{bol} - M_{bol} = m - M - K - A, \quad (5\text{-}309)$$

where the K correction accounts for the redshift of the energy curve of the observed galaxy, and A accounts for interstellar absorption. The K correction is given by

$$K = 2.5\log(1+z) + 2.5\log \frac{\int_0^\infty I(\lambda)s(\lambda)d\lambda}{\int_0^\infty I\left(\frac{\lambda}{1+z}\right)s(\lambda)d\lambda} \text{ magnitudes,} \quad (5\text{-}310)$$

where the first term arises from the narrowing of the photometer pass-band in the rest frame of the galaxy by the factor $(1+z)$, and the second term is due to the fact that the radiation received by the observer at wavelength, λ, is emitted by the galaxy at wavelength $\lambda/(1+z)$. Here $I(\lambda)$ is the incident energy flux per unit wavelength observed at wavelength, λ, and corrected for absorption, and $s(\lambda)$ is the photometer response function. HUBBLE (1936) first calculated the K correction by assuming the $I(\lambda)$ is that of a black body radiator, and the first observational correction was given by HUMASON, MAYALL, and SANDAGE (1956) using the z dependent intensity spectrum observed by STEBBINS and WHITFORD (1948). Intensity distributions, $I(\lambda)$, have been observed for giant elliptical galaxies by OKE and SANDAGE (1968), SCHILD and OKE (1971); WHITFORD (1971), and OKE (1971). The K corrections K_B, K_V and K_R for the blue, B, visual, V, and red, R, wavelength regions are tabulated in these papers as a function of redshift, z. For visual, V, and blue, B, magnitudes, the absorption term is given by (SANDAGE, 1968)

$$A_B = 0.18(\operatorname{cosec} b^{II} - 1) + 0.25 = A_V + 0.25 \approx 0.25 \operatorname{cosec} b^{II} \text{ magnitudes,} \quad (5\text{-}311)$$

where b^{II} is the galactic latitude.

Values of H_0 depend on the distance scale (cf. Sect. 5.3). Measurements by different observers using different distance estimates give

$$\begin{aligned}
&H_0 \approx 530\ \mathrm{km\ sec^{-1}\ Mpc^{-1}} && (\text{Hubble, 1929})\\
&H_0 \approx 100\ \mathrm{km\ sec^{-1}\ Mpc^{-1}} && (\text{Baade and Swope, 1955})\\
&H_0 = 98 \pm 15\ \mathrm{km\ sec^{-1}\ Mpc^{-1}} && (\text{Sandage, 1962})\\
&H_0 \lesssim 75\ \mathrm{km\ sec^{-1}\ Mpc^{-1}} && (\text{Sandage, 1968})\\
&H_0 \approx 50\ \mathrm{km\ sec^{-1}\ Mpc^{-1}} && (\text{Sandage, 1971}).
\end{aligned} \tag{5-312}$$

Values of q_0 depend on the deviation from linearity of the observed redshift-magnitude diagram. Comparisons of the observed diagrams with families of curves specified by putting various values of q_0 in Eq. (5-306) give

$$\begin{aligned}
&q_0 = 2.6 \ \pm 0.8 && (\text{Humason, Mayall, and Sandage, 1956})\\
&q_0 = 1.0 \ \pm 0.5 && (\text{Baum, 1961})\\
&q_0 = 1.5 \ \pm 0.4 && (\text{Peach, 1970})\\
&q_0 = 0.65 {+0.5 \atop -0.3} && (\text{Sandage, 1971})\\
&q_0 = 0.03 \pm 0.4 && (\text{Peach, 1972}).
\end{aligned} \tag{5-313}$$

The observed redshift-magnitude relation was shown in Fig. 35 of Sect. 5.2.6.[4]

5.7.3. The Angular Diameter-Redshift Relation

A spherical source of linear diameter, l, and redshift, z, will have the apparent angular diameter, θ, given by

$$\theta = \frac{l(1+z)^2}{D_L} = \frac{l H_0}{c} \frac{q_0^2 (1+z)^2}{\{q_0 z + (q_0 - 1)\,[(1+2 q_0 z)^{1/2} - 1]\}}, \tag{5-314}$$

for a homogeneous, isotropic universe with zero cosmological constant, luminosity distance, D_L, Hubble constant, H_0, and deceleration parameter, q_0. The smallest angle, θ_{min}, at a given z is given for $q_0 = 0$.

$$\theta_{min} = \frac{l H_0}{c} \frac{(1+z)^2}{z + z^2/2}. \tag{5-315}$$

For the steady state universe

$$\theta = \frac{l H_0}{c}\left(1 + \frac{1}{z}\right). \tag{5-316}$$

If extragalactic objects have some standard linear size, then observations of the angular diameter and redshift could help to determine the correct cosmological model. For the steady state universe, for example, θ will decrease with increasing z to a minimum constant value, whereas θ will decrease to a minimum and then increase with increasing z for the Einstein-deSitter universe ($q_0 = 0.5$). In the past this test has not been used because of the serious errors in measuring the diameter of optical galaxies. Baum (1972) has introduced a technique whereby optical and atmospheric effects are cancelled when observing clusters of galaxies.

---

[4] For a recent discussion of the values of H_0 and q_0 see Proc. IAU Symp. No. 63 (Krakov, 1973) and Proc. IAU Symp. No. 79 (Tallinn, 1977).

His results are shown in Fig. 45, for four clusters of galaxies, and they indicate that $q_0 \approx 0.3$. MILEY (1971) has prepared an apparent diameter-redshift diagram using the largest angular separation of the radio components of quasi-stellar sources and radio galaxies. The diagram shows a decrease in angular diameter with increasing redshift and indicates a clear continuity between the angular size-redshift properties of radio galaxies and quasars. He finds $q_0 < 0.5$ for linear sizes between 200 and 500 kiloparsecs.

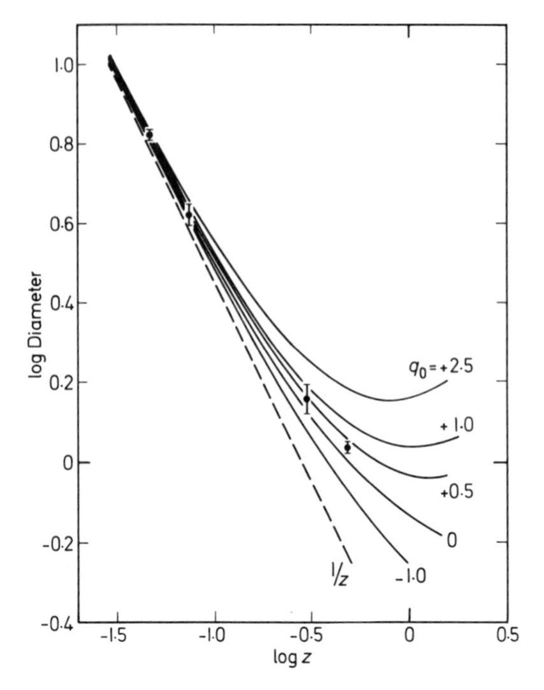

Fig. 45. Comparison of diameter-redshift data with world models. Each cluster is represented here by a single point. Error bars represent standard errors based on the scatter of the points for the individual clusters. The curve for $q_0 = -1.0$ represents the steady-state model, the range between $q_0 = 0$ and $q_0 = +0.5$ represents the open-universe case of the exploding model, while values of q_0 greater than $+0.5$ represent the closed-universe case of the exploding model. (After BAUM, 1972, by permission of the International Astronomical Union)

5.7.4. Number Counts of Optical Galaxies and Radio Sources

HUBBLE (1936) first showed that counts of optical galaxies to successively larger apparent magnitudes can be used to help show that the universe is homogeneous and isotropic on the large scale. If galaxies are uniformly distributed and have the same absolute luminosity, L, then the number $N(>f)$ of galaxies with apparent energy flux greater than f, is given by

$$N(>f) = \frac{4\pi}{3} D_L^3 n = \frac{4\pi}{3} n \left(\frac{L}{4\pi f}\right)^{3/2} \propto f^{-3/2}, \qquad (5\text{-}317)$$

where n is the number density of galaxies, and the observed energy flux, f, is given by

$$f = \frac{L}{4\pi D_L^2} \text{ erg cm}^{-2} \text{ sec}^{-1}, \tag{5-318}$$

where D_L is the luminosity distance. SHAPLEY and AMES (1932) and HUBBLE (1934, 1936) found rough agreement with Eq. (5-317) for counts of optical galaxies down to the limiting magnitude $m_{pg} = 20.7$ or $m_V = 19.8$.

In addition to showing that the distribution of galaxies is uniform on a large scale, number counts might lead to tests of cosmological models. For homogeneous, isotropic expanding universes with a cosmological constant of zero, the number of galaxies, $N(m)$, brighter than apparent magnitude, m, is given by (MATTIG, 1959; SANDAGE, 1962)

$$\begin{aligned} N(m) &= \frac{2\pi n}{H_0^3}(1-2q_0)^{-3/2}\left[p(1+p^2)^{1/2} - \operatorname{arc\,sinh} p\right] \\ &\approx \frac{2\pi n}{H_0^3}(1-2q_0)^{-3/2}\left[\frac{2p^3}{3} - \frac{p^5}{5} + \frac{12}{112}p^7 + \cdots\right] \end{aligned} \tag{5-319}$$

for $k = -1$, and

$$\begin{aligned} N(m) &= \frac{2\pi n}{H_0^3}(2q_0 - 1)^{-3/2}\left[\arcsin p - p(1-p^2)^{1/2}\right] \\ &\approx \frac{2\pi n}{H_0^3}(2q_0-1)^{-3/2}\left[\frac{2p^3}{3} + \frac{p^5}{5} + \frac{12}{112}p^7 + \cdots\right] \end{aligned} \tag{5-320}$$

for $k = +1$. Here k is the curvature constant, n is the number density of galaxies, H_0 is the Hubble parameter, q_0 is the deceleration parameter, we have assumed that all galaxies have the same absolute luminosity, L, and the parameter p is given by

$$p = \frac{A[k(2q_0 - 1)]^{1/2}}{q_0(1+A) - (q_0 - 1)(1+2A)^{1/2}},$$

where

$$A = 10^{0.2(m_R - K_R - C)},$$

m_R is the apparent magnitude in the red, K_R is the K correction for redshifting the energy distribution curve through the receiving filter and is given by the last part of Eq. (5-310) (cf. OKE, 1971) and

$$C = M + 25 + 5\log\left(\frac{c}{H_0}\right)$$

for galaxies with absolute magnitude, M. For the Euclidean case where $k=0$ we have

$$N(m) = \frac{4\pi n A^3}{3H_0^3}\left\{\frac{1}{2}\left[1 + A + (1+2A)^{1/2}\right]\right\}^{-3}, \tag{5-321}$$

and for the steady state model we have (BONDI and GOLD, 1948)

$$N(m) \propto \left[\ln(1+z) - \frac{z(2+3z)}{2(1+z)^2}\right], \tag{5-322}$$

where z is the redshift and is related to the apparent magnitude, m, by

$$m = 5\log z + 5\log(1+z) + M + 25 + 5\log\left(\frac{c}{H_0}\right). \tag{5-323}$$

Unhappily, the difference in the observed $N(m)$ for small ranges of q_0 between 0 and 1 is less than the probable observation errors when observing as low as $m_R = 22$.[5]

Radio source counts, however, offer sampling to much greater redshifts than are available to optical observers, and statistical analysis of radio sources reveal no anisotropy over angular sizes larger than a few minutes of arc (cf. SHAKESHAFT *et al.*, 1955; LESLIE, 1961; HOLDEN, 1966; HUGHES and LONGAIR, 1967). For comparisons with different model universes, the flux density, $S(\nu_0)$, measured at the frequency, ν_0, is related to the fractional energy spectrum, $I(\nu)$, emitted at frequency $\nu = (1+z)\nu_0$, where z is the redshift. For a radio source with spectral index, α, we have $I(\nu) \propto \nu^{-\alpha}$, and

$$S(\nu_0) = \frac{L I(\nu_0)}{4\pi D_L^2}(1+z)^{1-\alpha} \text{ erg cm}^{-2}\text{ sec}^{-1}\text{ Hz}^{-1}\text{ ster}^{-1}, \tag{5-324}$$

where L is the absolute luminosity of the source, D_L is its luminosity distance, and $I(\nu)$ is normalized so that $\int_0^\infty I(\nu)d\nu = 1$. PAULINY-TOTH, KELLERMANN, and DAVIS (1972) have shown that the mean spectral index $\alpha = 0.82$ down to a flux density of 0.25 flux units (1 f.u. $= 10^{-26}$ watt m^{-2} Hz^{-1} ster^{-1} $= 10^{-23}$ erg sec^{-1} cm^{-2} Hz^{-1} ster^{-1}) at 178 MHz.

What is actually observed in radio source counts is the number $N(>S)$ of radio sources per steradian with flux densities greater than S. Because the volume of a sphere of radius, r, is $4\pi r^3/3$, the number of sources within such a sphere is proportional to r^3 provided that the sources are uniformly distributed. If all the sources have the same intrinsic luminosity, then the faintest source visible from the center of the sphere will have a flux, S, proportional to r^{-2}. It follows that the number $N_0(>S)$ of sources per steradian with flux densities greater than S in an isotropic, static, Euclidean universe in which no evolution takes place is

$$N_0(>S) = \frac{n_0 P_0^{3/2}}{3} S^{-3/2} \propto S^{-3/2}, \tag{5-325}$$

where n_0 is the mean number density of sources, and P_0 is their mean brightness at the frequency under consideration. Sometimes a luminosity function, $\rho(P)$, is used, so that

$$N_0(>S) = \frac{S^{-3/2}}{3}\int P^{3/2}\rho(P)dP. \tag{5-326}$$

Number count data are then plotted logarithmically and the observed slope is compared to the -1.5 slope expected from a static Euclidean universe. Early integral counts of radio sources (RYLE and CLARKE, 1961) show a logarithmic slope of about -1.8 for strong radio sources. Observations to fainter flux levels

[5] Recent counts of galaxies using the 6-meter telescope observations demonstrate appreciable evolutionary effects (KOPYLOV and KARACHENTSEV, 1978).

(RYLE, 1968) show that the slope of the $\log N(>S) - \log S$ plot gradually flattens to a slope of -0.8 at about 0.01 flux units. The steep observed slope was taken to indicate an excess of weak radio sources, and to require an evolving universe in which radio sources were more luminous or more numerous in the past (LONGAIR, 1966). HOYLE (1968) argued that the data may be interpreted as a local deficiency of strong sources. Because the data obtained in integral counts are not generally independent, the best information is obtained using differential plots (CRAWFORD, JAUNCEY, and MURDOCH, 1970). When this is done it is found that when only the very strong sources are excluded the slope of the number-flux relation is close to the Euclidean value of -1.5 between 5 and 5000 sources per steradian (KELLERMANN, DAVIS, and PAULINY-TOTH, 1970). Furthermore, the differential count at 1400 MHz indicates that the exponent of the number-flux density relation is only greater than 1.5 at very low source densities (BRIDLE *et al.*, 1972), arguing against the ability of radio source number counts to test cosmological models.

Nevertheless, we set forth here the formulae which relate radio source number counts to nonevolutionary cosmological models. The Robertson-Walker metric may be written in the form (SCHEUER, 1974; LONGAIR, 1971)

$$ds^2 = dt^2 - \frac{R^2(t)}{c^2}\left[dr^2 + \left(\frac{\sin A r}{A}\right)^2 (d\theta^2 + \sin^2\theta\, d\varphi^2)\right], \tag{5-327}$$

where r is the radial co-moving distance coordinate, A^2 is the curvature of the world model, and the constant A is related to the Hubble constant, H_0, and the deceleration parameter, q_0, by the relations

$$A^2 = \frac{1}{c^2}\left[H_0^2\left(\frac{\rho_0}{\rho_c} - 1\right) + \frac{\Lambda c^2}{3}\right], \tag{5-328}$$

and

$$q_0 = \frac{\rho_0}{2\rho_c} - \frac{\Lambda c^2}{3 H_0^2}, \tag{5-329}$$

where c is the velocity of light, ρ_0 is the present value of the density of matter and radiation, ρ_c is the critical density given by Eq. (5-294), and Λ is the cosmological constant. The luminosity distance, D_L, is given by

$$D_L = \left(\frac{\sin A r}{A}\right)(1+z), \tag{5-330}$$

and the luminosity distance is given in terms of H_0 and q_0 in Eq. (5-149). The number, $N(>S)$, of radio sources per steradian with flux densities greater than S is given by

$$\frac{N(>S)}{N_0(>S)} = \frac{3}{4}\left[\frac{2Ar - \sin 2Ar}{(\sin A r)^3 (1+z)^{3(1+\alpha)/2}}\right], \tag{5-331}$$

where $N_0(>S)$ is the static Euclidean universe value of $N(S)$ given by Eqs. (5-325) or (5-326), and α is the mean spectral index of the sources counted. For model universes in which the cosmological constant $\Lambda = 0$, Eq. (5-331) may be evaluated using the relations

$$A = (2q_0 - 1)^{1/2} H_0/c$$

and

$$\sin A r = A D_L/(1+z),$$

where the luminosity distance, D_L, is given by Eq. (5-149).

For the steady state cosmology we have

$$\frac{N(>S)}{N_0(>S)} = \frac{3}{z^3(1+z)^{3(1+\alpha)/2}} \int_0^z \frac{z^2 dz}{1+z}, \tag{5-332}$$

which is a decreasing function of z for all $\alpha > -5/3$.

5.8. Supermassive Objects, Gravitational Collapse, and Black Holes

In order to account for the tremendous luminosity, L, of radio galaxies and quasars, $L \approx 10^{42}$ erg sec^{-1}, HOYLE and FOWLER (1963) suggested that supermassive objects might emit large amounts of energy. When there is sufficient entropy to support the star against rotational forces, a nearly spherical object is formed, and we have the relations (EDDINGTON, 1926)

$$P_r = \tfrac{1}{3} a T^4$$

$$P_g = \rho R T/\mu$$

$$1-\beta = 0.00298 \left(\frac{M}{M_\odot}\right)^2 (\mu\beta)^4 \tag{5-333}$$

or

$$\beta = \frac{P_g}{P} \approx \frac{4.28}{\mu} \left(\frac{M_\odot}{M}\right)^{1/2} \ll 1,$$

where M is the stellar mass, the solar mass, $M_\odot \approx 2 \times 10^{33}$ grams, the total pressure $P = P_r + P_g$, the radiation pressure is P_r, the gas pressure is P_g, the radiation constant $a = 7.564 \times 10^{-15}$ erg cm^{-3} °K^{-4}, the gas temperature is T, the gas mass density is ρ, the gas constant $R = 8.317 \times 10^7$ erg °K^{-1} mole^{-1}, and the mean molecular weight is μ. The radiation pressure is dominant, and the luminosity for a polytrope of index, $n=3$, is given by (HOYLE and FOWLER, 1963)

$$L = \frac{4\pi c G M(1-\beta)}{\kappa} = \frac{6.9 \times 10^4}{(1+X_H)} \frac{M}{M_\odot} L_\odot$$
$$\approx \frac{2.51 \times 10^{38}}{(1+X_H)} \frac{M}{M_\odot} \text{erg sec}^{-1}, \tag{5-334}$$

where it has been assumed that the temperature is high enough so that the opacity, κ, is due to electron scattering near the surface, $\kappa = 0.198(1+X_H)$ cm^2 g^{-1}, the mass fraction of hydrogen is X_H, the solar luminosity, $L_\odot \approx 4 \times 10^{33}$ erg sec^{-1}, and the solar mass, $M_\odot \approx 2 \times 10^{33}$ g. The radius, R, pressure, P, density, ρ, effective surface temperature, T_e, and binding energy, E_B, are given by (HOYLE and FOWLER, 1963; FOWLER, 1966; WAGONER, 1969)[6]

$$R = \frac{5.8 \times 10^9}{(T_9)_c} \left(\frac{M}{M_\odot}\right)^{1/2} \text{cm}$$

[6] For a review of properties of supermassive objects and their relation to the problem of energy sources in active galaxies and quasars, see OZERNOY (1976).

$$P \approx \frac{1}{3} a T^4 = 3.85 \times 10^{14} \left(\frac{M}{M_\odot}\right)^{2/3} \rho^{4/3} \text{ erg cm}^{-3}$$

$$\rho = 1.3 \times 10^5 \left(\frac{M_\odot}{M}\right)^{1/2} T_9^3 \text{ g cm}^{-3} \tag{5-335}$$

$$T_e = \frac{3.19 \times 10^5}{(1+X_H)^{1/4}} (T_9)_c^{1/2} \text{ °K}$$

$$E_B = -(M - M_0)c^2 = M c^2 \left[\frac{3\beta x}{8} + 1.265 x^2 (f-1)\right],$$

where the gravitational parameter

$$x = \frac{2GM}{Rc^2} = 5.04 \times 10^{-5} \left(\frac{M}{M_\odot}\right)^{1/2} (T_9)_c, \tag{5-336}$$

M_0 is the rest mass, the $(T_9)_c$ is the central temperature, T_c, devided by 10^9, and the constant f is the ratio of the rotational energy to the "general relativistic" energy.

$$f = \frac{0.0989}{K^2} \left(\frac{cJ}{GM^2}\right)^2, \tag{5-337}$$

where the constant K depends on the distribution of the angular velocity, ω (for $\omega =$ constant, KR is the radius of gyration and $K^2 = 0.0755$ for small ω), and the angular momentum is J (for constant ω, $J = K^2 M \omega R^2$). The rotational period, P_R, at the periphery of the supermassive star is given by

$$P_R \approx \frac{\pi K R^2 c}{GM(2f)^{1/2}} = 2.2 \times 10^{-23} R^2 (M_\odot/M) \text{ years} = 7.5 \times 10^{14} T_c^{-2} \text{ years}. \tag{5-338}$$

For $f \geq 1$ the binding energy increases until the object begins to lose kinetic energy while it is still dynamically stable, and the lifetime, τ, is given by

$$\tau = \frac{E_B}{L} \approx 2.3 \times 10^8 (1 + X_H) \frac{E_B}{Mc^2} \text{ years}, \tag{5-339}$$

as long as nuclear reactions have not begun. If the central temperature rises sufficiently for hydrogen burning to commence, the rest mass can be reduced enough to supply $0.007\,Mc^2$ of energy, which gives an age $\tau \approx 10^6$ years.

When there is insufficient entropy to support the object against rotational forces, a highly flattened rotating disk is formed. The object can attain relativistic speeds and the equations of general relativity then become important. In this case the binding energy is given by (BARDEEN and WAGONER, 1969)

$$E_B = M_0 c^2 \left[\frac{z_c}{1+z_c} - \frac{2\omega J}{M_0 c^2}\right], \tag{5-340}$$

where z_c is the redshift of photons emitted from the center of the object, M_0 is the rest mass, c is the velocity of light, ω is the angular velocity, and J is the angular momentum. Further formulae for thin, uniformly rotating, relativistic disks are given by BARDEEN and WAGONER (1971) and SCHARLEMANN and WAGONER (1972).

As first proposed by LANDAU (1932), and rigorously calculated by CHANDRASEKHAR (1935), there is no equilibrium state at the endpoint of thermonuclear evolution for a star containing more than about twice the number of baryons in the Sun. BAADE and ZWICKY (1934) hypothesized that the unstable, contracting star ejects mass to form a supernovae, and the details of the implosion and subsequent explosion have been calculated by BURBIDGE, BURBIDGE, FOWLER, and HOYLE (1957), HOYLE and FOWLER (1960), COLGATE and WHITE (1966), MAY and WHITE (1966), ARNETT (1966, 1967, and 1968), and ARNETT and CAMERON (1967). If the residual mass, M, after the explosion is greater than the Chandrasekhar limit, M_c, given by

$$M_c = \left(\frac{2}{\mu_e}\right)^2 1.4587\, M_\odot\,, \tag{5-341}$$

it will collapse to form a neutron star. If the mass is greater than the upper limit for a neutron star (2—$3\,M_\odot$), then the collapse must continue to form a closed, trapped surface from which no electromagnetic signal can escape—a black hole. Here μ_e is the mean electron molecular weight, $\mu_e = 2$ for hydrogen, and the solar mass, $M_\odot \approx 2\times 10^{33}$ grams. The radius, R, of a self-supporting body must also be larger than the critical radius, R_c, given by (OPPENHEIMER and VOLKOFF, 1939; CHANDRASEKHAR, 1964)

$$R_c = \frac{6.74}{\beta} R_g \approx 3.4\times 10^5 \left(\frac{M}{M_\odot}\right)^{3/2} \text{cm}\,, \tag{5-342}$$

for an equilibrium state of a polytrope of index three with no rotation. Here $R_g = 2GM/c^2$ is the Schwarzschild radius, and β has been evaluated using $\mu = 0.73$ for a mixture of 50 percent hydrogen, 47 percent helium, and 3 percent heavy elements by mass. For a massive polytrope of index $n = 3$, this radius corresponds to a critical temperature, T_c, above which instability for contraction sets in

$$T_c \approx 1.7\times 10^{13} \left(\frac{M_\odot}{M}\right) {}^\circ\text{K}\,. \tag{5-343}$$

When rotation is taken into account, the minimum critical radius and the maximum temperature for rotational stability of a polytrope of index 3 are (FOWLER, 1966)

$$R_c \approx 1.5\times 10^6 \frac{f}{K^2} \left(\frac{M}{M_\odot}\right) \frac{GM}{\omega^2 R^3} \text{cm} \tag{5-344}$$

and

$$T_c \approx 3.9\times 10^{12} \frac{K^2}{f} \left(\frac{M_\odot}{M}\right)^{1/2} \frac{R^3 \omega^2}{GM} {}^\circ\text{K}\,, \tag{5-345}$$

where ω is the rotational velocity, and the rotation parameters f and K are defined in Eq. (5-337).

OPPENHEIMER and SNYDER (1939) first analyzed the optical appearance of a collapsing star, and showed that the object collapses to a singularity in the time

$$\tau = \left(\frac{3\pi}{32 G \rho_0}\right)^{1/2} \approx \frac{GM}{c^3} \approx 10^{-5} \left(\frac{M}{M_\odot}\right) \text{sec}\,, \tag{5-346}$$

where ρ_0 is the initial stellar density before collapse, and the radius has been taken to be the Schwarzschild radius, $R_g = 2GM/c^2$. A distant observer will observe the luminosity to decay exponentially with the time constant, τ, whereas the photons will arrive with exponentially increasing redshifts of the same time constant. The object will be observed to collapse in infinite time to an infinite redshift and a radius equal to the Schwarzschild radius. When both rotation and a polar magnetic field are included in the Newtonian hydrodynamics of a collapsing star, it may be (LEBLANC and WILSON, 1970) that the gravitational energy is converted into rotational and magnetic energy which eventually leads to the formation of a double jet of material similar to that observed in many strong radio sources.

For the region immediately outside a collapsing, nonrotating spherical star, the appropriate metric is the Schwarzschild metric given in Eq. (5-270). As first recognized by LEMAÎTRE (1933), however, the singular behavior of the Schwarzschild metric at the Schwarzschild radius, R_g, is not physical, but is due to the choice of coordinates. EDDINGTON (1924) first constructed a metric which is nonsingular at R_g, but simpler expressions have been obtained using the KRUSKAL (1960)-SZEKERES (1960) metric. They independently made the coordinate transformation

$$u = \left(\frac{r}{R_g} - 1\right)^{1/2} e^{r/(2R_g)} \cosh\left(\frac{ct}{2R_g}\right) \qquad \text{for } r > R_g$$
$$v = \left(\frac{r}{R_g} - 1\right)^{1/2} e^{r/(2R_g)} \sinh\left(\frac{ct}{2R_g}\right)$$

and

$$u = \left(1 - \frac{r}{R_g}\right)^{1/2} e^{r/(2R_g)} \sinh\left(\frac{ct}{2R_g}\right) \qquad \text{for } r < R_g, \tag{5-347}$$
$$v = \left(1 - \frac{r}{R_g}\right)^{1/2} e^{r/(2R_g)} \cosh\left(\frac{ct}{2R_g}\right)$$

to obtain from the Schwarzschild metric the metric

$$ds^2 = \frac{4R_g^3}{r} e^{-r/R_g}(du^2 - dv^2) + r^2(d\theta^2 + \sin^2\theta \, d\varphi^2). \tag{5-348}$$

Here we have

$$R_g = \frac{2GM}{c^2}$$

and

$$u^2 - v^2 = \left(\frac{r}{R_g} - 1\right) e^{r/R_g} \tag{5-349}$$

so that the physical singularity at $r = 0$ is located at $u^2 - v^2 = -1$.

More generally, a black hole may be rotating or charged, or both. When only rotation is considered, the metric becomes the Kerr metric given by KERR (1963). NEWMAN *et al.* (1965) included charge, and BOYER and LINDQUIST (1967) introduced generalized coordinates to obtain the metric

$$ds^2 = -\frac{\Delta}{\Sigma}[dt - a\sin^2\theta \, d\varphi]^2 + \frac{\sin^2\theta}{\Sigma}[(r^2 + a^2) d\varphi - a\, dt]^2 + \frac{\Sigma}{\Delta} dr^2 + \Sigma \, d\theta^2, \tag{5-350}$$

where

$$\Delta = r^2 - 2Mr + a^2 + Q^2$$

$$\Sigma = r^2 + a^2 \cos^2\theta$$

and

$$a = J/M .$$

The corresponding electromagnetic field tensor is given by

$$\begin{aligned} \overline{F} &= \frac{Q}{\Sigma^2}(r^2 - a^2\cos^2\theta)\overline{dr} \wedge [\overline{dt} - a\sin^2\theta\,\overline{d\varphi}] \\ &+ \frac{2Q}{\Sigma^2}\arccos\theta\,\sin\theta\,\overline{d\theta} \wedge [(r^2+a^2)\overline{d\varphi} - a\,\overline{dt}], \end{aligned} \tag{5-351}$$

where the component F_{ij} is the coefficient of $\overline{dx^i} \wedge \overline{dx^j} = -\overline{dx^j} \wedge \overline{dx^i}$. Here the mass is M, the charge is Q, the angular momentum is J, and geometrized units have been employed. In geometrized units, the Newtonian gravitational constant, G, the velocity of light, c, and Boltzmann's constant, k, all have the value unity. To convert to physical units, employ the relations (cf. MISNER, THORNE, and WHEELER, 1973)

Length:	$l_{\text{geom}} = l_{\text{phys}}$
Time:	$t_{\text{geom}} = c\,t_{\text{phys}} = 2.997925 \times 10^{10}\, t_{\text{phys}}$
Velocity:	$v_{\text{geom}} = \frac{v_{\text{phys}}}{c} = 0.3336 \times 10^{-10}\, v_{\text{phys}}$
Mass:	$M_{\text{geom}} = \frac{G}{c^2} M_{\text{phys}} = 0.742 \times 10^{-28}\, M_{\text{phys}}$
Angular momentum:	$J_{\text{geom}} = \frac{G}{c^3} J_{\text{phys}} = 0.247 \times 10^{-38}\, \omega_{\text{phys}} R^2_{\text{phys}} M_{\text{phys}}$
Energy:	$E_{\text{geom}} = \frac{G}{c^4} E_{\text{phys}} = 0.826 \times 10^{-49}\, E_{\text{phys}}$
Temperature:	$T_{\text{geom}} = \frac{Gk}{c^4} T_{\text{phys}} = 1.140 \times 10^{-65}\, T_{\text{phys}}$ (5-352)
Magnetic field:	$H_{\text{geom}} = \frac{c^2}{G^{1/2}} H_{\text{phys}} = 3.48 \times 10^{24}\, H_{\text{phys}}$
Entropy:	$S_{\text{geom}} = \frac{S_{\text{phys}}}{k} = 7.2435 \times 10^{15}\, S_{\text{phys}}$
Luminosity:	$L_{\text{geom}} = \frac{G}{c^5} L_{\text{phys}} = 2.754 \times 10^{-60}\, L_{\text{phys}}$
Angular momentum per unit mass:	$a_{\text{geom}} = \frac{a_{\text{phys}}}{c} = 0.3336 \times 10^{-10}\, a_{\text{phys}}$
Charge:	$Q_{\text{geom}} = \frac{\sqrt{G}}{c^2} Q_{\text{phys}} = 2.874 \times 10^{-25}\, Q_{\text{phys}}$,

where the subscripts geom and phys denote, respectively, quantities given in geometrical units expressed in cm, and physical or c.g.s. units.

The static limit, r_0, defines a surface at and in which an observer cannot remain at rest and must orbit the black hole in the same direction in which the hole rotates. It is given by

$$r = r_0 = M + (M^2 - Q^2 - a^2 \cos^2\theta)^{1/2} . \tag{5-353}$$

The horizon, r_+, defines a surface at and in which an observer cannot communicate with the exterior. It is given by

$$r = r_+ = M + (M^2 - Q^2 - a^2)^{1/2} . \tag{5-354}$$

Here r_0 and r_+ are given in geometrized units. Particles and photons can fall inward through the horizon, but nothing can emerge outward through it. The ergosphere is the region of spacetime between the horizon and the static limit.

The Kerr metric in Boyer-Lindquist coordinates and geometrized units is given by

$$\begin{aligned} ds^2 = & -\left(1 - \frac{2Mr}{\Sigma}\right) dt^2 - \left(4Mar\frac{\sin^2\theta}{\Sigma}\right) dt\, d\varphi \\ & + \frac{\Sigma}{\Delta} dr^2 + \Sigma\, d\theta^2 + \left(r^2 + a^2 + 2Ma^2 r\frac{\sin^2\theta}{\Sigma}\right)\sin^2\theta\, d\varphi^2 . \end{aligned} \tag{5-355}$$

Here Δ, Σ, and a are given in Eq. (5-350) with $Q=0$, and Eq. (5-355) reduces to the Schwarzschild metric when $a=0$. Detailed properties of circular orbits around a Kerr black hole are given by BARDEEN, PRESS, and TEUKOLSKY (1972). The coordinate radius of the innermost stable orbit is given by

$$r = r_{ms} = \text{outermost root} \quad \text{of } r^2 - 6Mr \pm 8aM^{1/2} r^{1/2} - 3a^2 = 0 . \tag{5-356}$$

For $a=0$, we have $r_{ms}=6M$, and for $a=M$ we have $r_{ms}=M$ or $9M$. The innermost stable orbit is the maximally bound orbit. The binding energy of the maximally bound orbit varies from

$$\begin{aligned} & \left(1 - \frac{2\sqrt{2}}{3}\right) mc^2 \quad \text{for Schwarzschild } (a=0) \\ \text{to} \quad & \\ & \left(1 - \frac{\sqrt{3}}{3}\right) mc^2 \quad \text{for extreme Kerr } (a=M), \end{aligned} \tag{5-357}$$

where m is the rest mass of the orbiting particle.

The behavior of black holes is governed by the following set of theorems and laws (cf. for example MISNER, THORNE, and WHEELER, 1973):

a) *Birkhoff (1923) theorem:* Any spherically symmetric geometry of a given region of spacetime which is a solution to the Einstein field equations in vacuum is necessarily a piece of the Schwarzschild geometry. That is, a spherically symmetric gravitational field in empty space must be static (zero angular momentum) with a Schwarzschild metric.

b) *Isreal (1967, 1968) theorem:* Any static black hole with event horizon of spherical topology has external fields determined uniquely by its mass, M, and charge, Q, and moreover those external fields are the Schwarzschild solution if $Q=0$ and the Reissner-Nordstrom solution if $Q \neq 0$. HAWKING had already shown that a stationary black hole must have a horizon with spherical topology, and that it must either be static or axially symmetric, or both. The REISSNER (1916)-NORDSTROM (1918) metric is given by

$$ds^2 = -\left(1 - \frac{2M}{r} + \frac{Q^2}{r^2}\right) dt^2 + \left(1 - \frac{2M}{r} + \frac{Q^2}{r^2}\right)^{-1} dr^2 + r^2 (d\theta^2 + \sin^2\theta \, d\varphi^2), \tag{5-358}$$

and the electric field, E, is given by

$$E = \frac{Q}{r^2}, \tag{5-359}$$

where the charge is Q. Here geometrized units have been employed.

c) *Penrose (1969) theorem:* By injecting matter into a Kerr-Newman black hole in the proper way one can extract energy from the hole. That is, when certain particles traverse the field of a black hole they may gain energy at the expense of the rotational energy of the black hole.

d) *Carter (1971) theorem:* All uncharged, stationary, axially symmetric black holes with event horizons of spherical topology fall into disjoint families not deformable into each other. The black holes in each family have external gravitational fields determined uniquely by two parameters—the mass, M, and the angular momentum, J. The only such family presently known is the family of Kerr metrics.

e) *Black holes obey the first law of thermodynamics*—total energy is conserved. Total momentum, angular momentum, and charge are also conserved.

f) *Hawking's (1972) second law of black hole dynamics:* For any process involving black holes, the sum of the surface areas, A, of the event horizons of all black holes involved can never decrease. The surface area, A, of the event horizon of a Kerr-Newman black hole is given by

$$A = 8\pi [M^2 + M(M^2 - a^2 - Q^2)^{1/2} - Q^2/2], \tag{5-360}$$

which becomes

$$A = 8\pi [M^2 + (M^4 - J^2)^{1/2}] \tag{5-361}$$

for a Kerr black hole. Here, again, geometrized units have been employed for the mass, M, angular momentum, J, and for $a = J/M$.

5.9. Gravitational Radiation

It follows directly from Einstein's field equations that any nonspherical, dynamically changing system must produce gravitational waves. These waves produce radiation-reaction forces in their source, and these forces extract energy from the source at the same rate as the rate at which the gravitational waves carry off energy. The waves carry energy at the speed of light, c, so that the energy flux is the product of c and the energy density. As with electromagnetic waves, the amplitude falls off as the inverse of the distance, and the flux falls off as the inverse square of the distance travelled.

For slow-motion, weak gravitational field systems, the flux, F, emitted in a given direction (unit vector n_j) is given by (LANDAU and LIFSHITZ, 1962; ZELDOVICH and NOVIKOV, 1971; PRESS and THORNE, 1972)

$$F = \frac{1}{8\pi D^2}\frac{G}{c^5}\sum_{j,k}\langle(\dddot{I}_{jk}^{TT})^2\rangle_{ret}, \tag{5-362}$$

where the third derivative with respect to time is denoted by $\dddot{}$, and the

$$I_{jk}^{TT} = \sum_{l,m}(\delta_{jl} - n_j n_l) I_{lm} (\delta_{mk} - n_m n_k), \tag{5-363}$$

where δ_{ij} is the Kronecker delta function, n_j is a unit vector in the j direction, and the reduced quadrupole moment tensor, I_{jk}, is given by

$$I_{jk} = \int \rho x_j x_k d^3x - \tfrac{1}{3}\delta_{ik}\int \rho D^2 d^3x, \tag{5-364}$$

where ρ is the mass density.

Slow motion, weak gravitational field systems are those for which the radius, R, and internal velocity, v, satisfy the inequalities

$$R \gg \frac{GM}{c^2}$$

and (5-365)

$$v \ll c,$$

where M is the system mass, $2GM/c^2$ is the Schwarzschild radius and c is the velocity of light. The total luminosity, L, radiated in quadrupole gravitational waves is given by

$$L = \frac{1}{5}\frac{G}{c^5}\sum_{j,k}\langle(\dddot{I}_{jk})^2\rangle \approx L_0\left(\frac{2GM_{\mathrm{eff}}}{c^2 R}\right)^2\left(\frac{R}{\lambda}\right)^6, \tag{5-366}$$

where M_{eff} is the effective mass for which amplitude changes in $I_{jk} = M_{\mathrm{eff}} R^2$, the wavelength $\lambda = cR/v$, and the natural unit of luminosity, L_0, is given by

$$L_0 = \frac{c^5}{G} = 3.63\times10^{59}\,\mathrm{erg\,sec^{-1}} = 2.03\times10^5\,M_\odot c^2\,\mathrm{sec^{-1}}, \tag{5-367}$$

where $M_\odot \approx 2\times10^{33}$ grams is the solar mass. For a binary star system of orbital period, P, semi-major axis, a, reduced mass $\mu = M_1 M_2/(M_1+M_2)$, where M_1 and M_2 are the respective masses of the two components, and total mass $M = M_1 + M_2$, the gravitational wave luminosity is given by (PETERS and MATHEWS, 1963)

$$\begin{aligned} L &= \frac{32}{5}\frac{G^5}{c^{10}}\frac{\mu^2 M^3}{a^5} L_0 f(e) \\ &\approx 3.0\times10^{33}\,\mathrm{erg\,sec^{-1}}\left(\frac{\mu}{M_\odot}\right)^2\left(\frac{M}{M_\odot}\right)^{4/3}\left(\frac{P}{1\,\mathrm{hour}}\right)^{-10/3} f(e), \end{aligned} \tag{5-368}$$

where

$$f(e) = \frac{1 + \frac{73}{24}e^2 + \frac{37}{96}e^4}{(1-e^2)^{7/2}},$$

and e is the orbital eccentricity. The radiation is emitted at a fundamental frequency equal to twice the orbital frequency, and at increasing harmonics according as e increases. If gravitational radiation is the dominant force in changing the orbital period, and if the orbit is nearly circular,

$$\frac{1}{P}\frac{dP}{dt} = -\frac{96}{5}\frac{G^3}{c^5}\frac{\mu M^2}{a^4} = \left(\frac{1}{2.8\times 10^7\,yr}\right)\left(\frac{M}{M_\odot}\right)^{2/3}\left(\frac{\mu}{M_\odot}\right)\left(\frac{1\,hr}{P}\right)^{8/3}. \quad (5\text{-}369)$$

For a slightly deformed, homogeneous neutron star with moment of inertia $I=2MR^2/5$, mass, M, radius, R, rotation period, P, and ellipticity, ε, the gravitational wave luminosity is (Ostriker and Gunn, 1969)

$$\begin{aligned} L &= \frac{32}{5}\frac{G}{c^5} I^2 \varepsilon^2 \left(\frac{2\pi}{P}\right)^6 \\ &\approx 10^{38}\,\mathrm{erg\,sec^{-1}} \left(\frac{I}{4\times 10^{44}\,\mathrm{g\,cm^2}}\right)^2 \left(\frac{P}{0.033\,\mathrm{sec}}\right)^{-6} \left(\frac{\varepsilon}{10^{-3}}\right)^2, \end{aligned} \quad (5\text{-}370)$$

and at a time, t, after its birth,

$$L \approx 10^{45}\,\mathrm{erg\,sec^{-1}} \left(\frac{4\times 10^{44}\,\mathrm{g\,cm^2}}{I}\right)^{1/2} \left(\frac{10^{-3}}{\varepsilon}\right) \left(\frac{10^6\,\mathrm{sec}}{t+10^4\,\mathrm{sec}}\right)^{3/2}. \quad (5\text{-}371)$$

Additional sources of gravitational radiation which might exist are reviewed by Press and Thorne (1972).

The gravitational wave force, F, which acts on a mass, M, is given by

$$F = -M\nabla\varphi = -\sum_j \sum_k M R_{joko} x_k\,, \quad (5\text{-}372)$$

where φ is the equivalent Newtonian gravitational potential of the gravitational wave, and the components of the Riemann curvature tensor, R_{joko}, are given by

$$R_{joko} = \frac{\partial^2 \varphi}{\partial x_j \partial x_k} = -\frac{G}{c^4 D}\left[\frac{d^4 I_{jk}^{TT}}{dt^4}\right]_{ret}, \quad (5\text{-}373)$$

where D is the distance to the source, and I_{jk}^{TT} is given by Eq. (5-363). When a gravitational wave acts on a pair of free, or almost free, masses separated by a distance, l, it will produce a change in distance, Δl, given by

$$\frac{\Delta l}{l} = h(t) = [h_+^2(t) + h_x^2(t)]^{1/2}\,, \quad (5\text{-}374)$$

where $h(t)$ is the dimensionless field strength, and h_+ and h_x are the magnitudes of the perturbations in the metric tensor, g_{ik}. For a resonant type of gravitational wave detector, the displacement is between the two ends of an elastic solid, and

$$\frac{\Delta l}{l} \approx \pi \nu_0 h(\nu_0)\,, \quad (5\text{-}375)$$

where ν_0 is the resonant frequency, and $h(\nu_0)$ is the Fourier transform of $h(t)$ evaluated at ν_0.

The gravitational wave flux, F, is given by

$$F = \frac{c^3}{16\pi G} \langle [\dot{h}(t)]^2 \rangle \,\mathrm{erg\,cm^{-2}\,sec^{-1}}, \tag{5-376}$$

where $\langle\ \rangle$ denotes an average over a few wavelengths, and $\dot{}$ denotes differentiation with respect to time. Integrating Eq. (5-376) over time and using Parseval's theorem, we obtain a total energy, E, per burst of

$$E = 4\pi \frac{c^3 D^2}{16\pi G} \int_{-\infty}^{+\infty} (2\pi\nu)^2 h^2(\nu)\, d\nu, \tag{5-377}$$

where D is the distance of the source, we assume that the source radiates isotropically, and $h(\nu)$ is the Fourier transform of $h(t)$. If the gravitational wave has a flat frequency spectrum out to the resonant frequency, ν_0, a resonant detector gives the limit

$$\begin{aligned} E &\gtrsim \frac{c^3}{4G}\left(\frac{2\Delta l}{l}\right)^2 \nu_0 D^2\,\mathrm{erg} \\ &\gtrsim 5.6\times 10^{-17}\left(\frac{2\Delta l}{l}\right)^2 \nu_0 D^2 (M_\odot c^2)\,\mathrm{erg}, \end{aligned} \tag{5-378}$$

where the solar mass $M_\odot \approx 2\times 10^{33}$ gm.

WEBER (1960) pioneered detection schemes for gravitational wave radiation, and since 1969 has reported the detection of bursts of radiation (cf. Weber, 1970). For a typical burst, $\Delta l \approx 5\times 10^{-15}$ cm for aluminum cylinders with resonant frequency $\nu_0 = 1661$ Hz and length $l = 153$ cm. Assuming an isotropic radiator at distance, D, Eq. (5-378) indicates that WEBER is detecting gravitational wave energy equivalent to the destruction of a mass, M, given by

$$M \approx 0.5\, M_\odot \left(\frac{D}{10\,\mathrm{kpc}}\right)^2. \tag{5-379}$$

Here D is the distance to the source in kiloparsecs, and 10 kpc is the distance to the galactic center. Weber detects about three events per day, so that a source at the galactic center would be emitting energy equivalent to a mass loss of about $500\, M_\odot$ per year. If the efficiency of Weber's detector is taken into account, and if the radiator is nonisotropic, then as much as 10^3 to $10^5\, M_\odot c^2$ of gravitational wave energy may be being emitted each year. Using a detector similar to that of Weber's, however, LEVINE and GARWIN (1973) have failed to detect gravity wave events suggesting that Weber's events were not produced by gravity waves or that such waves do not exist in similar numbers and intensity in 1973. PRESS and THORNE (1972) have reviewed the prospects for future detection of sources of gravitational waves from known astronomical phenomena such as binary stars, neutron stars, and supernovae.

5.10. The Background Radiation, Helium Synthesis, the Intergalactic Gas, and the Formation of Galaxies

The presently observed values for the mean densities of matter and radiation are given by (cf. Fig. 29)

$$
\begin{aligned}
\rho_G &\approx 4\times10^{-31}h^2\ \mathrm{g\ cm^{-3}} \approx 10^{-31}\ \mathrm{g\ cm^{-3}} \approx 10^{-10}\ \mathrm{erg\ cm^{-3}}\\
\rho_B &\approx 5\times10^{-34}\ \mathrm{g\ cm^{-3}} \approx 4.8\times10^{-13}\ \mathrm{erg\ cm^{-3}}\\
\rho_S &\approx 10^{-34}\ \mathrm{g\ cm^{-3}} \approx 10^{-13}\ \mathrm{erg\ cm^{-3}}\\
\rho_{GS} &\approx 10^{-35}\ \mathrm{g\ cm^{-3}} \approx 10^{-14}\ \mathrm{erg\ cm^{-3}}\\
\rho_X &\approx 10^{-37}\ \mathrm{g\ cm^{-3}} \approx 10^{-16}\ \mathrm{erg\ cm^{-3}}\\
\rho_R &\approx 10^{-40}\ \mathrm{g\ cm^{-3}} \approx 10^{-19}\ \mathrm{erg\ cm^{-3}},
\end{aligned}
\tag{5-380}
$$

where ρ_G denotes the density of matter in galaxies, and the ρ with subscripts B, S, GS, X, and R denote, respectively, the radiation energy density of the background radiation, the optically visible starlight, the optical background due to galaxies, the X-ray background, and the radio wave background. At the present time, then, we have

$$\rho=\Omega\rho_c=\Omega\frac{3H_0^2}{c^2\kappa}=1.9\times10^{-29}\,\Omega h^2\ \mathrm{g\ cm^{-3}}, \tag{5-381}$$

where

$$\Omega_{\mathrm{rad}}\approx10^{-3}\,\Omega_{\mathrm{gal}}\approx10^{-5}.$$

Here $H_0=100\,h\ \mathrm{km\ sec^{-1}\ Mpc^{-1}}$ is the Hubble constant, and Einstein's gravitational constant, $\kappa=8\pi G/c^2\simeq1.86\times10^{-27}\ \mathrm{dyn\ gm^{-2}\ sec^2}$. The critical density, ρ_c, needed to close the universe is given by

$$\rho_c=1.9\times10^{-29}h^2\ \mathrm{g\ cm^{-3}},$$

and choosing $H_0=50\ \mathrm{km\ sec^{-1}\ Mpc^{-1}}$ or $h=0.5$, this critical density corresponds to:

Mass density	$\rho_c=4.7\times10^{-30}\ \mathrm{g\ cm^{-3}}$,
Particle density	$N_p=2.8\times10^{-6}\ \mathrm{cm^{-3}}$
Hydrogen density (He/H = 0.1)	$N_{\mathrm{HI}}=2.0\times10^{-6}\ \mathrm{cm^{-3}}$
Helium density	$N_{\mathrm{He}}=2.0\times10^{-7}\ \mathrm{cm^{-3}}$
Electron density—complete ionization	$N_e=2.4\times10^{-6}\ \mathrm{cm^{-3}}$
Weighted ion density	$\sum N_i Z_i^2=2.8\times10^{-6}\ \mathrm{cm^{-3}}$
Bremsstrahlung parameter	$\sum N_e N_i Z_i^2=6.8\times10^{-12}\ \mathrm{cm^{-6}}$.

There is clearly much more matter observed than radiation, and we are presently in a matter dominated era for which the pressure term in Einstein's field equations can be omitted. The present mean density of matter is then given by

$$
\begin{aligned}
\rho_{m0}&=\frac{3q_0H_0^2}{4\pi G}=\frac{3q_0kc^2}{4\pi G R_0^2(2q_0-1)}\\
&\approx3.8\times10^{-29}q_0h^2\ \mathrm{g\ cm^{-3}},
\end{aligned}
\tag{5-382}
$$

where the curvature constant $k=\pm 1,0$, the present value of the scale factor, $R(t)$, is R_0, and the present value of the deceleration parameter is q_0. The matter density, $\rho_m(t)$, radiation energy density, $\rho_r(t)$, and radiation temperature, $T_r(t)$, in a universe dominated by matter go as (EINSTEIN, 1917; TOLMAN, 1934)

$$\begin{aligned} \rho_m(t) &\propto [R(t)]^{-3} \\ \rho_r(t) &\propto [R(t)]^{-4} \\ T_r(t) &\propto [R(t)]^{-1}, \end{aligned} \tag{5-383}$$

where the scale factor, $R(t)$, for the $k=0$ case is given by

$$R(t) \propto t^{2/3} \propto (1+z)^{-1}. \tag{5-384}$$

Here t is the time since the origin of the universe, and the redshift z corresponds to epoch, t.

It follows from Eqs. (5-383) and (5-384) that in the past the universe was a dense, hot, fireball, and GAMOW (1946) pointed out that the relic of this primeval fireball might be observed as part of the background radiation. PENZIAS and WILSON (1965) found that the background radiation at a wavelength of 3 cm corresponded to a brightness temperature of 2.7 °K, which was much more intense than that expected from extrapolations of measurements of other types of radiation at lower frequencies. DICKE *et al.* (1965) were quick to point out that the new type of radiation might be the remnant of Gamow's fireball. Because radiation and matter were in thermal equilibrium when the fireball was formed, the radiation should have a black body spectrum which would be preserved for all time. Such a spectrum was found for the microwave background radiation, and is illustrated in Fig. 46. Detailed references to measurements of the background radiation may be found in LONGAIR (1971), PEEBLES (1971), or WEINBERG (1972).

Following the pioneering observations of CONKLIN and BRACEWELL (1967) and WILKINSON and PARTRIDGE (1967), the microwave background radiation has been found to be isotropic to 0.1 percent on all angular scales greater than one arc minute (cf. LONGAIR, 1971).[7] Furthermore, the isotropic X-ray background discovered by GIACCONI *et al.* (1962) is isotropic to 3 percent for all angular sizes greater than 15° (SCHWARTZ, 1970). These observations provide added support for the Robertson-Walker or Lemaître metrics. In fact EHLERS, GEREN, and SACHS (1968) have shown that if the isotropy of the background is exact, then the universe is exactly a Robertson-Walker universe. The discovery of the background radiation has led not so much to tests of cosmological models, however, but rather to an understanding of how the universe may have evolved physically. GAMOW (1946), for example, pointed out that in the distant past the universe was dominated by radiation, and (cf. TOLMAN, 1934)

$$\begin{aligned} T_r(t) &= 2.7(1+z) \propto [R(t)]^{-1} \\ \rho_r(t) &= a\, T_r^4(t) \propto [R(t)]^{-4} \\ \rho_m(t) &\propto [R(t)]^{-3}, \end{aligned} \tag{5-385}$$

[7] Recent measurements of the anisotropy of the cosmic blackbody radiation at 33 GHz (0.9 cm) show an anisotropy which is interpreted in terms of a motion of the earth relative to the background with a velocity of 390 ± 60 km sec^{-1}. (SMOOT et al., 1977). Excluding this component the cosmic blackbody radiation is isotropic to one part in three thousand.

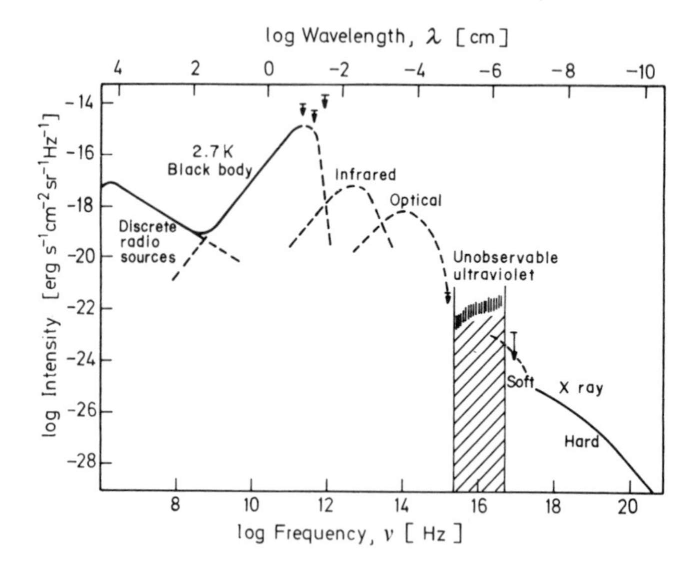

Fig. 46. The spectrum of the isotropic background radiation in the frequency range 10^6 to 10^{21} Hz (after LONGAIR and SUNYAEV, 1969, by permission of Gordon and Breach Science Publ.). Full lines indicate those wavelength regions in which an isotropic background has been measured, whereas broken lines are theoretical estimates

where we have provided the present value of $T_r = 2.7\,°\text{K}$, and the radiation constant

$$a = \frac{8\pi^5 k^4}{15c^3h^3} = 7.5641 \times 10^{-15}\,\text{erg cm}^{-3}\,°\text{K}^{-4}.$$

The scale factor, $R(t)$, is given by

$$R(t) \propto t^{1/2} \propto (1+z)^{-1}, \tag{5-386}$$

and P_r, ρ_r, and T_r denote, respectively, the radiation pressure, density, and temperature. When we note that the present values at time t_0, are

$$\rho_r(t) = \rho_{r0} = 4.4 \times 10^{-34}\,\text{g cm}^{-3},$$

and

$$T_r(t) = T_{r0} = 2.7\,°\text{K}, \tag{5-387}$$

it follows from Eqs. (5-382) and (5-383) that the matter and radiation energy densities are equal at the critical temperature, T_{rc}, given by

$$T_{rc} = \frac{\rho_{m0}}{\rho_{r0}} T_{r0} \approx 2.3 \times 10^5 q_0 h^2\,°\text{K}, \tag{5-388}$$

which corresponds to the redshift, z_c, given by

$$z_c = \frac{T_{rc}}{T_{r0}} - 1 \approx 10^5 q_0 h^2 , \tag{5-389}$$

and a critical time, t_c, given by

$$t_c = \left(\frac{T_{r0}}{T_{rc}}\right)^2 t_0 \approx 10^{-10.5} q_0^{-2} h^{-4} t_0 . \tag{5-390}$$

It further follows from Eqs. (5-286) and (5-385) that

$$t_c \approx \left[\frac{c^2}{32\pi G a T_c^4}\right]^{1/2} \approx 5 \times 10^2 q_0^{-2} h^{-4} \text{ years} . \tag{5-391}$$

After time, t_c, the expansion obeys Eqs. (5-383) and (5-384), whereas before time, t_c, the expansion obeys Eqs. (5-385) and (5-386).

GAMOW (1946) proposed that the present abundance of the elements could be explained if the elements were built up in the primeval fireball, and details were worked out by ALPHER, BETHE, and GAMOW (1948), ALPHER and HERMAN (1948, 1950), and ALPHER, FOLLIN, and HERMAN (1953). As described in Chap. 4, stellar nucleosynthesis adequately explains the abundance of the heavier elements, whereas hydrogen, helium, and perhaps some of the lighter elements may have been produced in the early stages of the universe. HOYLE and TAYLER (1964), for example, point out that the observed luminosity of the Galaxy implies that only 2 or 3 percent of the hydrogen in stars could have been converted into helium by nucleosynthesis over the lifetime of the Galaxy. The measured helium abundance in stellar atmospheres and interiors, planetary nebulae, novae, the interstellar medium, and extragalactic systems all indicate, however, a high uniform helium abundance of about 25 percent by mass (DANZIGER, 1970). Following GAMOW's (1948) argument for the primeval production of helium, we have from Eq. (5-391) an expansion time of

$$t = \left(\frac{c^2}{32\pi G a T^4}\right)^{1/2} \approx 10^{20} T^{-2} \text{ sec} . \tag{5-392}$$

At high temperatures, the universe is in thermal equilibrium and contains photons, neutrinos, antineutrinos, electrons, and positrons. As the universe cools, muons will first annihilate at about 10^{12} °K, then electrons and positrons will annihilate forming neutrons and protons. The neutrons and protons will be in equilibrium with the other particles up until

$$t \gtrsim \tau \approx 10^{50} T^{-5} \text{ sec}, \tag{5-393}$$

where τ is the weak interaction time scale. From Eq. (5-392), we have $T \approx 10^{10}$ °K at this time, and the neutron-proton ratio is given roughly by the equilibrium value

$$\frac{N_n}{N_p} = \exp\left(-\frac{\Delta m c^2}{kT}\right) = 0.22 , \tag{5-394}$$

where Δm is the mass difference between neutrons and protons, and N_n and N_p denote, respectively, the number density of neutrons and protons. The neutrons and protons will interact to form deuterons, which at a later time will result in the synthesis of helium. Eventually most of the neutrons are consumed, each pair of neutrons giving one helium nucleus and the abundance ratio is expected to be

$$\frac{N(\mathrm{He}^4)}{N(\mathrm{He}^4)+N(H)}=0.36\,, \tag{5-395}$$

where $N(\mathrm{He}^4)$ and $N(\mathrm{H})$ denote, respectively, the number densities of helium and hydrogen. A mass fraction of helium of $Y \approx 28\%$ is obtained if the mass fraction of all other heavier elements is $Z=0.02$. Detailed calculations involving the many possible primeval nuclear reactions give similar values for Y (cf. HOYLE and TAYLER, 1964; PEEBLES, 1966; WAGONER, FOWLER, and HOYLE, 1967; and Fig. 27).

Another key point in the past history of the universe is the critical time, t_r, for which the temperature has lowered enough for hydrogen to recombine. The relevant temperature is 4000 °K and from Eq. (5-392), (cf. PEEBLES, 1968)

$$t_r \approx 10^{13}\ \mathrm{sec} \approx 10^5\ \mathrm{years}\,, \tag{5-396}$$

at which time $z \approx 1500$. After the time t_r, the universe becomes transparent, and the motions of matter are unaffected by the black body radiation. Galaxies may have been formed before t_r during which time matter and radiation must have interacted strongly via Compton scattering of photons by electrons. It is, however, difficult to imagine a galaxy formation process which is so efficient that all of the matter has condensed into galaxies, and it has been hypothesized that an intergalactic hydrogen gas exists with density

$$\rho_{\mathrm{IGG}}=\Omega\rho_c \approx 10^{-5}\,\Omega h^2 \quad \text{particles cm}^{-3}\,, \tag{5-397}$$

where

$$q_0=\frac{1}{2}\Omega\left(1+\frac{3P_0}{\rho_0 c^2}\right)\approx\frac{\Omega}{2}$$

and

$$\rho_c=1.9\times 10^{-29}\,h^2\ \mathrm{g\ cm^{-3}}\,,$$

where $H_0=100\,h\ \mathrm{km\ sec^{-1}\ Mpc^{-1}}$. Values of the critical mass density, ρ_c, needed to close the universe are given following Eq. (5-381) for hydrogen, helium, and electrons. Methods of detecting neutral intergalactic hydrogen are discussed in Chap. 2, and we have the following limits

$$\Omega_{\mathrm{HI}} \leq 0.4\,h^{-1} \quad \text{or} \quad N_{\mathrm{HI}} \leq 5\times 10^{-6}\,h\ \mathrm{cm^{-3}} \tag{5-398}$$

from 21 cm emission studies (PENZIAS and WILSON, 1969),

$$\Omega_{\mathrm{HI}} \leq 2.7\times 10^{-3}\,T_s h^{-1} \quad \text{or} \quad N_{\mathrm{HI}} \leq 3\times 10^{-8}\,T_s h\ \mathrm{cm^{-3}} \tag{5-399}$$

from 21 cm absorption studies (PENZIAS and SCOTT, 1968), where $3<T_s<30$ according to KOEHLER (1966), (cf. FIELD (1959, 1962) and PEEBLES (1971) for a discussion of T_s)

$$\Omega_{\mathrm{HI}} \leq 4\times 10^{-7}\,h^{-1} \quad \text{or} \quad N_{\mathrm{HI}} \leq 4.7\times 10^{-12}\,h\ \mathrm{cm^{-3}} \quad \text{for } z \geq 2 \tag{5-400}$$

from the absence of absorption of the continuum radiation of quasars below the redshifted Lyman α line at $\lambda = 1216$ Å (SHKLOVSKII, 1964; SCHEUER, 1965; GUNN and PETERSON, 1965), and

$$\Omega_{H2} \leq 7 \times 10^{-4} h^{-1} \quad \text{or} \quad N_{H2} \leq 4 \times 10^{-9} h \,\text{cm}^{-3} \quad \text{for } z \geq 2 \qquad (5\text{-}401)$$

for molecular hydrogen (FIELD, SOLOMON, and WAMPLER, 1966). All of these tests indicate that if there is intergalactic hydrogen gas, it is likely to be nearly one hundred percent ionized. As discussed in Chap. 4, the soft X-ray background may be due to the bremsstrahlung of such a gas (cf. FIELD and HENRY, 1964; WEYMANN, 1966, 1967; BOWYER, FIELD, and MACK, 1968; HENRY *et al.*, 1968; BERGERON, 1969). If $\Omega_{IGG} \approx 1$, and the soft X-ray background originates at redshift, z, then we have the temperature limit

$$T_{IGG}(z) \leq 10^6 (1+z)\,^\circ\text{K}, \qquad (5\text{-}402)$$

where $T_0 = 10^6\,^\circ$K is taken to be the temperature at $z = 0$. Detailed formulae for the bremsstrahlung of a high temperature intergalactic gas may be found in the references given immediately above as well as in PEEBLES (1971), WEINBERG (1972), FIELD (1972), and SCHEUER (1974).

In a Robertson-Walker universe filled with an absorbing medium of cross section, $\sigma_a(\nu)$, radiation emitted at frequency, ν_1, and received at frequency, ν_0, will have undergone absorption at frequency, ν_a, provided that

$$\nu_0 < \nu_a < \nu_1$$

and (5-403)

$$I_a = \frac{1}{\nu_a} \int \sigma_a(\nu)\, d\nu \approx \frac{\sigma(\nu_a)}{\nu_a}.$$

The optical depth, $\tau(\nu_0)$, at the received frequency, ν_0, is given by

$$\begin{aligned} \tau(\nu_0) &= \frac{\nu_0 N(t_a)}{\nu_a H_0} I_a \left[1 - 2q_0 + 2q_0 \frac{\nu_a}{\nu_0}\right]^{-1/2} \left[1 - \exp\left(-\frac{h\nu_a}{kT(t_a)}\right)\right] \\ &\approx \frac{N(t_0)\sigma(\nu_a)}{H_0 \nu_a} \frac{(1+z)^2}{(1+2q_0 z)^{1/2}} \left[1 - \exp\left(-\frac{h\nu_a}{k(1+z)T(t_0)}\right)\right] \qquad (5\text{-}404) \\ &\quad \text{for } \nu_a(1+z)^{-1} \leq \nu_0 \leq \nu_a \quad \text{where } z = (\nu_1/\nu_0) - 1, \\ &\approx 0 \quad \text{for } \nu_0 > \nu_a \quad \text{and } \nu_0 < \nu_a(1+z)^{-1}. \end{aligned}$$

Here the number density, $N(t)$, of absorbing atoms at time, t, is related to the number density, $N(t_0)$, at the present time by the relation

$$N(t) = N(t_0) \left[\frac{\nu}{\nu_0}\right]^3 = N(t_0)[1+z]^3, \qquad (5\text{-}405)$$

where z is the redshift corresponding to frequency, ν. Similarly, the temperature, $T(t)$, is given by

$$T(t) = T(t_0) \left[\frac{\nu}{\nu_0}\right] = T(t_0)[1+z]. \qquad (5\text{-}406)$$

The present value of hydrogen density, $N(t_0)$, is related to the critical density, ρ_c, and the parameter, Ω, by the relation

$$N(t_0) = \frac{\Omega \rho_c}{m_H} = \frac{3 \Omega H_0^2}{8 \pi G m_H} \approx 2 \times 10^{-6} \Omega \left(\frac{H_0}{50 \text{ km sec}^{-1} \text{ Mpc}^{-1}} \right)^2, \quad (5\text{-}407)$$

where $\Omega = 2q_0$. Measurements of the expected absorption trough in the frequency range $\nu_a(1+z)^{-1} \leq \nu_0 \leq \nu_a$ lead to estimates of $N(t_0)$ given in Eqs. (5-399) and (5-400).

If the intergalactic medium is ionized, then the optical depth, $\tau(z)$, due to Thomson scattering of the radiation from a source at redshift, z, is given by (BAHCALL and SALPETER, 1965; BAHCALL and MAY, 1968; REES, 1969)

$$\tau(z) = \frac{\sigma_T c N_{e0}}{3 H_0 q_0^2} \{(3 q_0 + q_0 z - 1)(1 + 2 q_0 z)^{1/2} - (3 q_0 - 1)\}, \quad (5\text{-}408)$$

where $q_0 = \Omega/2 \neq 0$, and

$$\tau(z) = \frac{\sigma_T c N_{e0}}{2 H_0} (1+z)^2,$$

when $q_0 = 0$. Here the Thomson scattering cross-section $\sigma_T \approx 6.652 \times 10^{-25}$ cm^2, the N_{e0} is the electron density given by Eq. (5-407), and the coefficient for $\tau(z)$ is given by

$$\frac{H_0 \sigma_T c}{4 \pi G m_H q_0} = \frac{\sigma_T c N_{e0}}{3 H_0 q_0^2} \approx \frac{0.035}{q_0} \left(\frac{H_0}{75 \text{ km sec}^{-1} \text{ Mpc}^{-1}} \right). \quad (5\text{-}409)$$

An ionized intergalactic medium would delay signals as well as scatter them, and the delay, Δt, due to intergalactic dispersion of a signal at frequency ν_0, is given by (HADDOCK and SCIAMA, 1965; WEINBERG, 1972)

$$\begin{aligned} \Delta t &= \frac{[\nu_p(t_0)]^2}{2 q_0 \nu_0^2 H_0} \{[1 + 2 q_0 z]^{1/2} - 1\} \\ &\approx 10^{20} \frac{1}{\nu_0^2} \{(1+z)^{1/2} - 1\} \text{ sec} \quad \text{for } \Omega = 1, \end{aligned} \quad (5\text{-}410)$$

where z is the redshift of the source and $\Omega = 2q_0$. Here the present value of the plasma frequency, $\nu_p(t_0)$, is given by

$$\nu_p(t_0) = \left[\frac{e^2 N_e(t_0)}{\pi m_e} \right]^{1/2} \approx 8.97 \times 10^3 [N_e(t_0)]^{1/2} \text{ Hz}, \quad (5\text{-}411)$$

where $N_e(t_0)$ is the present value of the electron density in cm^{-3}, and is given by Eq. (5-407).

Free-free absorption by the intergalactic gas might be important at the low radio frequencies, and if the electron temperature, T_e, were constant back to redshift, z, then the optical depth, $\tau(z)$, is given by (SCHEUER, 1974; LONGAIR, 1971)

$$\begin{aligned} \tau(z) = \frac{2 \tau_0}{\Omega^3} \Bigg[& (1 + \Omega z)^{1/2} \left\{ \frac{1}{5} (1 + \Omega z)^2 + \frac{2}{3} (\Omega - 1)(1 + \Omega z) + (\Omega - 1)^2 \right\} \\ & - \Omega^2 + \frac{4}{3} \Omega - \frac{8}{15} \Bigg], \end{aligned} \quad (5\text{-}412)$$

where

$$\tau_0 = 2 \times 10^{11} \left(\frac{\Omega}{\nu}\right)^2 \left(\frac{T_e}{10^4}\right)^{-3/2},$$

and $\Omega \simeq 2q_0$. For large redshifts, $\tau(z)$ varies as $(1+z)^{5/2}$. SCHEUER (1974) gives similar formulae for the optical depth and temperature due to the bremsstrahlung of steady state, adiabatically expanding, and isothermally expanding universes.

In his famous correspondence with BENTLEY, NEWTON (1692) first suggested that space in an infinite universe would convene by its gravity to form the observed masses. As discussed in detail in Chap. 3, JEANS (1902) applied linear perturbation theory to an ideal fluid to show that fluctuations with wavelengths, λ, greater than

$$\lambda_J = \left(\frac{\pi s^2}{G\rho}\right)^{1/2} = \left(\frac{5}{3}\frac{\pi k T}{G\rho m_H}\right)^{1/2}, \tag{5-413}$$

are unstable to gravitational collapse. Here the velocity of sound, s, is $[5kT/(3m_H)]^{1/2}$ for a monatomic hydrogen gas of temperature, T, hydrogen mass, m_H, and gas density, $\rho = N m_H$. For masses, M, greater than the Jeans mass

$$M_J = \frac{\pi}{6}\rho\lambda_J^3 \approx \left(\frac{\pi}{3}\right)^{5/2}\left(\frac{5kT}{G}\right)^{3/2} N^{-1/2} m_H^{-2}, \tag{5-414}$$

the self-gravitational attraction of matter in a perturbation cannot be supported by the internal pressure of the gas, and gravitational collapse is the outcome. Fluctuations with scales $\lambda < \lambda_J$ will be stabilized by pressure gradients and will oscillate like sound waves, whereas fluctuations with $\lambda > \lambda_J$ will grow exponentially with time, t, at the maximum rate

$$\frac{\Delta\rho}{\rho} \propto \exp\left[\frac{st}{\lambda_J}\right], \tag{5-415}$$

where $\Delta\rho$ is the perturbation in density ρ.

The JEANS (1902) theory is applicable to the formation of galaxies in a static universe. LEMAÎTRE (1931) first studied gravitational instability in an expanding universe for which the cosmological constant, $\Lambda \neq 0$, whereas LIFSHITZ (1946) presented the classical treatment of linear perturbations in homogeneous, isotropic, expanding universes with $\Lambda = 0$. For the Einstein-deSitter universe with $k=0$, LIFSHITZ obtained the result

$$\frac{\Delta\rho}{\rho} \propto t^{2/3} \propto R(t) \propto (1+z)^{-1}, \tag{5-416}$$

which has been derived using the Newtonian approximation by BONNER (1957). This result is applicable for times, t, greater than the recombination time, $t_r \simeq 10^5 q_0^{-1/2}$ years, for which the universe is the EINSTEIN (1917) matter dominated, "dust" universe. In this case, Eqs. (5-383) and (5-384) are applicable and we have

$$\lambda_J \approx 0.046\, T^{1/2} h^{-1}\,\mathrm{Mpc} \propto [R(t)]^{-1/2} \propto t^{-1/3} \quad \text{for } t > t_r$$

and

$$M_J \approx 10^5\, T^{3/2} h^{-1} M_\odot \propto [R(t)]^{-3/2} \propto t^{-2} \quad \text{for } t > t_r, \tag{5-417}$$

where $1\,\text{Mpc}=3\times10^{24}$ cm, the solar mass $M_\odot\approx2\times10^{33}$ gram, the Hubble constant $H_0=100h\,\text{km sec}^{-1}\,\text{Mpc}^{-1}$, and we have assumed that $\rho=\rho_c=2\times10^{-29}h^2\,\text{g cm}^{-3}$. Here M_J is the minimum mass able to condense freely after the recombination time, t_r. As LIFSHITZ (1946) pointed out, the problem is not so much in providing masses as large as those of most astronomical objects, but rather that the present value of $\Delta\rho/\rho\approx1$ together with Eq. (5-416) require large values of $\Delta\rho/\rho$ in the past. If z_0 denotes the epoch at which $\Delta\rho/\rho=1$, then the expected temperature fluctuations at the epoch of recombination are (SUNYAEV and ZELDOVICH, 1970)

$$\frac{\Delta T}{T}\approx10^{-5}\left[\frac{M\Omega^{1/2}}{10^{15}M_\odot}\right]^{5/6}(1+z_0) \tag{5-418}$$

for masses $M<10^{15}\Omega^{1/2}M_\odot$, with angular scales of

$$\theta\approx10\left(\frac{M\Omega^2}{10^{15}M_\odot}\right)^{1/3}\text{ arc minutes}. \tag{5-419}$$

For times, t, before the time, $t_c\approx500\,q_0^{-2}$ years, at which matter and radiation decouple, the velocity of sound, s, is given by

$$s=\frac{c}{\sqrt{3}}\left[1+\frac{3}{4}\frac{\rho_m}{\rho_r}\right]^{-1/2}\approx\frac{c}{\sqrt{3}}, \tag{5-420}$$

where ρ_m is the density of matter, and $\rho_r=aT^4$ is the radiation energy density. For this radiation dominated era, perturbations, $\Delta\rho$, in density, ρ, grow with time, t, as

$$\frac{\Delta\rho}{\rho}\propto t\propto R^2(t)\propto(1+z)^{-2}. \tag{5-421}$$

Again we are left with the inconsistency that perturbations grow only linearly with time, but that $\Delta\rho/\rho=1$ at present, and $\Delta\rho/\rho\ll1$ in the past because of the uniformity of the background radiation. Nevertheless, for completeness we give the Jeans mass (WEINBERG, 1972)

$$\begin{aligned}M_J&=\frac{2\pi^{5/2}k^2\sigma^2}{9a^{1/2}m_H^2G^{3/2}(1+\sigma kT/m_H)^3}\quad\text{for } t<t_r\\&\approx9.06\,\sigma^2\left(1+\frac{\sigma kT}{m_H}\right)^{-3}M_\odot,\end{aligned} \tag{5-422}$$

where

$$\sigma=\frac{4aT^3}{3Nk},$$

the radiation constant is a, the hydrogen mass is m_H, the matter density $\rho_m=Nm_H$, and T is the radiation temperature. Before $t=t_c\simeq500\,q_0^{-2}$ years, the Jeans mass, M_J, is just less than the mass, M_H, within the particle horizon.

$$\begin{aligned}M_J\approx M_H&\approx10^{11}t^{3/2}M_\odot\quad\text{for } t<t_c\\&\approx10^{31}T^{-3}M_\odot\quad\text{for } t<t_c,\end{aligned} \tag{5-423}$$

where t is the time scale of the expansion in years, and T is the radiation temperature in °K.

Between times t_c and t_r, the Jeans mass is constant with a value given by

$$M_J \approx 10^{15} q_0^{-2} M_\odot \quad \text{for } t_c < t < t_r, \tag{5-424}$$

and the fluctuations undergo acoustic oscillations. During this stage, dissipative processes damp small scale fluctuations, and oscillations with mass, M, less than the critical mass (SILK, 1968; WEINBERG, 1972)

$$M_c \approx \frac{32\pi^4}{3}\left(\frac{4m_H}{45\sigma_T}\right)^{3/2}(15.5\pi a T^4 G)^{-3/4}(N_r m_H)^{-1/2} \approx 10^{12} q_0^{-5/4} M_\odot, \tag{5-425}$$

are severely attenuated by the time hydrogen begins to recombine. Here $\sigma_T = 6.652 \times 10^{-25}\ \text{cm}^2$ is the Thomson scattering cross section, and the subscript, r, denotes the time of hydrogen recombination. For instance, if the present mass density is $10^{-30}\ \text{g cm}^{-3}$, fluctuations surviving at the time of recombination will have a minimum mass of about $10^{12} M_\odot$, the mass of a large galaxy.

LIFSHITZ (1946) pointed out that as long as the universe is radiation dominated and viscous damping is ignored, any primeval turbulent velocity will be preserved to be released when matter and radiation decouple. WEIZSÄCKER (1951) and GAMOW (1952) were the first to argue that galaxy formation might take place in a turbulent medium. Such arguments are needed to overcome the slow growth rate of the adiabatic perturbations discussed in the previous sections. It is generally argued that during the radiation dominated era, large scale motions existed with subsonic velocity, and the Reynolds number, Re, is given by (OZERNOY and CHIBISOV, 1971)

$$Re = v r \nu^{-1} \approx 5 \times 10^5 \Omega^{+2}\left(\frac{v_c}{c}\right)^2\left(\frac{z}{z_c}\right), \tag{5-426}$$

where v is the turbulent velocity, r is the physical scale, z is the redshift, Ω is the ratio of the total metagalactic density to the critical density of the Friedmann models, the subscript, c, denotes the time at which the density of matter, ρ_m, and radiation, ρ_r, were equal, $z_c \approx 2 \times 10^4 \Omega$, and the coefficient of kinematic viscosity, ν, is given by (LEDOUX and WALRAVEN, 1958)

$$\nu = \frac{4}{15}\frac{m_p}{\sigma_T \rho_m}\frac{c\rho_r}{\rho_r + \rho_m}, \tag{5-427}$$

where σ_T/m_p is the opacity. Eddy motions with scales smaller than $(\nu t)^{1/2}$ are dissipated by viscosity, so that survived eddies at time, t, satisfy the mass requirement (SATO, MATSUDA, and TAKEDA, 1970)

$$\begin{aligned} M &\gtrsim \frac{4\pi}{3}\rho_m(\nu t)^{3/2} \\ &\gtrsim 10^{9.8}\Omega^{-11/4} M_\odot \quad \text{at } t = t_r \quad \text{and for } \Omega > 10^{-1.1} \\ &\gtrsim 10^{12.2}\Omega^{-1/2} M_\odot \quad \text{at } t = t_r \quad \text{and for } \Omega < 10^{-1.1}. \end{aligned} \tag{5-428}$$

WEIZSÄCKER (1951) and OZERNOI and CHERNIN (1968) postulate that a Kolmogorov spectrum of eddy scale sizes, λ, is built up during the radiation dominated era, and that perturbations, $\Delta\rho$, in density, ρ, and turbulent velocities, v, satisfy

$$\frac{\Delta\rho}{\rho} \propto \lambda^{-2}, \quad v \propto \lambda^{1/3}. \tag{5-429}$$

Further discussions of turbulent theories of galaxy formation are given by OORT (1969), HARRISON (1970), and PEEBLES (1971). Alternative theories for galaxy formation involving matter and anti-matter are discussed by HARRISON (1967), ALFVÉN (1971), KLEIN (1971), and OMNÈS (1971).[8]

[8] For recent reviews of theories for galaxy formation, see JONES (1976) (general ideas); DOROSHKEVICH et al. (1976) (adiabatic perturbations); OZERNOY (1976) (whirl perturbations); GOTT (1977) (entropy perturbations).

References

General References

Alfvén, H., Fälthammar, C.: Cosmical electrodynamics. London: Oxford at the Clarendon Press 1963

Allen, C. W.: Astrophysical quantities. University of London, London: Athlone Press 1963.

Aller, L. H.: Astrophysics—The atmospheres of the sun and stars. New York: Ronald Press 1963.

Bekefi, G.: Radiation processes in plasmas. New York: Wiley 1966.

Blanco, V. M., McCuskey, S. M.: Basic physics of the solar system. Reading, Mass.: Addison-Wesley 1961.

Bond, J. W., Watson, K. M., Welch, J. A.: Atomic theory of gas dynamics. Reading, Mass.: Addison-Wesley 1965.

Born, M., Wolf, E.: Principles of optics. New York: Pergamon Press 1964.

Bracewell, R. N.: The Fourier transform and its applications. New York: McGraw-Hill 1965.

Brandt, J. C., Hodge, P. W.: Solar system astrophysics. New York: McGraw-Hill 1964.

Chandrasekhar, S.: Hydrodynamic and hydromagnetic stability. London: Oxford at the Clarendon Press 1961.

Chandrasekhar, S.: Ellipsoidal figures of equilibrium. New Haven, Conn.: Yale Univ. Press 1969.

Chiu, H. Y.: Stellar physics. Waltham, Mass.: Blaisdell 1968.

Clayton, D. D.: Principles of stellar evolution and nucleosynthesis. New York: McGraw-Hill 1968.

Condon, E. U., Shortley, G. H.: The theory of atomic spectra. Cambridge University Press 1963.

Cox, J. P., Giuli, R. T.: Principles of stellar structure. New York: Gordon and Breach 1968.

Dufay, J.: Introduction to astrophysics—The stars. London: G. Newnes 1964.

Evans, J. V., Hagfors, T.: Radar astronomy. New York: McGraw-Hill 1968.

Explanatory supplement to the astronomical ephemeris and nautical almanac. London: H. M. S. O. 1961.

Ginzburg, V. L.: Propagation of electromagnetic waves in plasma. New York: Gordon and Breach 1961.

Ginzburg, V. L., Syrovatskii, S. I.: The origin of cosmic rays. New York: Pergamon Press 1964.

Griem, H. R.: Plasma spectroscopy. New York: McGraw-Hill 1964.

Herzberg, G.: Atomic spectra and atomic structure. New York: Dover 1964.

Hietler, W.: The quantum theory of radiation. London: Oxford at the Clarendon Press 1960.

Jackson, J. D.: Classical electrodynamics. New York: Wiley 1962.

Kaplan, S. A., Pikelner, S. B.: The interstellar medium. Cambridge, Mass.: Harvard University Press 1970.

Kraus, J. D.: Radio astronomy. New York: McGraw-Hill 1966.

Landau, L. D., Lifshitz, E. M.: Fluid mechanics. New York: Pergamon Press 1959.

Landau, L. D., Lifshitz, E. M.: Statistical physics. Reading, Mass: Addison-Wesley 1969.

Landau, L. D., Lifshitz, E. M.: The classical theory of fields. New York: Pergamon Press 1971.

McVittie, G. C.: General relativity and cosmology. Urbana, Ill.: University Illinois Press 1965.

Mihalas, D.: Galactic astronomy. San Francisco: W. H. Freeman 1968.

Mihalas, D.: Stellar atmospheres. San Francisco: W. H. Freeman 1970.

Misner, C. W., Thorne, K. S., Wheeler, J. A.: Gravitation. San Francisco: W. H. Freeman 1973.

North, J. D.: The measure of the universe: A history of modern cosmology. London: Oxford University Press 1965.

Pacholczyk, A. G.: Radio astrophysics. San Francisco: W. H. Freeman 1970.

Peebles, P. J. E.: Physical cosmology. Princeton, N. J.: Princeton University Press 1971.

Preston, M. A.: Physics of the nucleus. Reading, Mass.: Addison-Wesley 1965.

ROBERTSON, H. P., NOONAN, T. W.: Relativity and cosmology. Philadelphia: W. B. Saunders 1968.
SMART, W. M.: Spherical astronomy. Cambridge University Press 1956.
SOMMERFELD, A.: Thermodynamics and statistical mechanics. New York: Academic Press 1964.
SPITZER, L.: Diffuse matter in space. New York: Wiley 1968.
STIX, T. H.: The theory of plasma waves. New York: McGraw-Hill 1962.
TOWNES, C. H., SCHAWLOW, A. L.: Microwave spectroscopy. New York: McGraw-Hill 1955.
TRUMPLER, R. J., WEAVER, H. F.: Statistical astronomy. Berkeley, Calif.: University Calif. Press 1953.
VAN DE KAMP, P.: Principles of astrometry. San Francisco, Calif.: W. H. Freeman 1967.
WEINBERG, S.: Gravitation and cosmology: Principles and applications of the general theory of relativity. New York: Wiley 1972.
WOOLARD, E. W., CLEMENCE, G. M.: Spherical astronomy. New York: Academic Press 1966.
ZELDOVICH, Y. B., NOVIKOV, I. D.: Relativistic astrophysics. Chicago, Ill.: University of Chicago Press 1971.

For Historical References Consult:

CAMPBELL, W. W.: Stellar motions. New Haven, Conn.: Yale University Press 1913.
CLERKE, A. M.: A popular history of astronomy during the nineteenth century. London: Black 1885.
SHAPLEY, H. (ed.): Source book in astronomy 1900–1950. Cambridge, Mass.: Harvard University Press 1960.
TODHUNTER, I.: A history of the mathematical theories of attraction and the figure of the earth. New York: Dover 1962.
WHITTAKER, E. T.: A history of the theories of aether and electricity. London: Thomas Nelson 1953.

Chapter 1

AARONS, J., WHITNEY, H. E., ALLEN, R. S.: Global morphology of ionospheric scintillations. Proc. I.E.E.E. **59** (2), 159 (1971).
AIRY, G. B.: On the diffraction of an object-glass with circular aperture. Trans. Camb. Phil. Soc. **5**, 283 (1835).
ALFVÉN, H., HERLOFSON, N.: Cosmic radiation and radio stars. Phys. Rev. **78**, 616 (1950).
ALTENHOFF, W., MEZGER, P. G., STRASSL, H., WENDKER, H., WESTERHOUT, G.: Radio astronomical measurements at 2.7 KMHz. Veroff Sternwarte, Bonn **59**, 48 (1960).
AMPÈRE, A. M.: Sur l'action mutuelle d'un aiment et d'un conduct. voltaique. Ann. Chim. Phys. **37** (1827).
APPLETON, E. V.: Wireless studies of the ionosphere. J. Inst. Elec. Engrs. (London) **71**, 642 (1932).
BEAUJARDIÈRE, O. DE LA: L'évolution du spectre de rayonnement synchrotron. Ann. Astrophys. **29**, 345 (1966).
BEKEFI, G.: Radiation processes in plasmas. New York: Wiley 1966.
BERGE, G. L., GREISEN, E. W.: High-resolution interferometry of Venus at 3.12-cm wavelength. Ap. J. **156**, 1125 (1969).
BETHE, H. A.: The influence of screening on the creation and stopping of electrons. Proc. Camb. Phil. Soc. **30**, 524 (1930).
BETHE, H. A., HEITLER, W.: On the stopping of fast particles and on the creation of positive electrons. Proc. Roy. Soc. London A **146**, 83 (1934).
BIOT, J. B., SAVART, F.: Sur le magnétisme de la pile de Volta. Ann. Chimie **15**, 222 (1820).
BLUMENTHAL, G. R., GOULD, R. J.: Bremsstrahlung, synchrotron radiation, and Compton scattering of high-energy electrons traversing dilute gases: Rev. Mod. Phys. **42** (2), 237 (1970).
BOHM, D., GROSS, E. P.: Theory of plasma oscillations: A. Origin of medium-like behavior; B. Excitation and damping of oscillations. Phys. Rev. **75**, 1851, 1854 (1949).
BOHR, N.: On the decrease of velocity of swiftly moving electrified particles in passing through matter. Phil. Mag. **30**, 581 (1915).
BOLTZMANN, L. VON: Über eine von Hrn. Bartoli entdeckte Beziehung der Wärmestrahlung zum zweiten Hauptsatze (On a relation between thermal radiation and the second law of thermodynamics, discovered by Bartoli). Ann. Phys. **22**, 31 (1884).
BOND, G. P.: Light of the moon and of Jupiter. Mem. Amer. Ac. **8**, 221 (1863).
BOOKER, H. G.: A theory of scattering by nonisotropic irregularities with application to radar reflections from the aurora. J. Atmos. Terr. Phys. **8**, 204 (1956).

BOOKER, H. G., GORDON, W. E.: A theory of radio scattering in the troposphere. Proc. I.R.E. **38**, 401 (1950).

BOOKER, H. G., RATCLIFFE, J. A., SHINN, D. H.: Diffraction from an irregular screen with applications to ionospheric problems. Phil. Trans. Roy. Soc. London A **242**, 75 (1950).

BORN, M., WOLF, E.: Principles of optics. New York: Pergamon Press 1970.

BREWSTER, D.: On the laws which regulate the polarization of light by reflection from transparent bodies. Phil. Trans. **15**, 125 (1815).

BRIGGS, B. H., PHILLIPS, G. J., SHINN, D. H.: The analysis of observations on spaced receivers of the fading of radio signals. Proc. Roy. Soc. London B **63**, 106 (1950).

BROWN, R. L., MATHEWS, W. G.: Theoretical continuous spectra from gaseous nebulae. Ap. J. **160**, 939 (1970).

BRUSSARD, P. J., HULST, H. C. VAN DE: Approximation formulas for non-relativistic bremsstrahlung and average Gaunt factors for a Maxwellian electron gas. Rev. Mod. Phys. **34** (3), 507 (1962).

BUNEMAN, O.: Scattering of radiation by the fluctuations in a nonequilibrium plasma. J. Geophys. Res. **67** (5), 2050 (1962).

BURBIDGE, G. R.: Estimates of the total energy in particles and magnetic field in the non-thermal radio sources. Ap. J. **129**, 849 (1959).

BURGESS, A.: The hydrogen recombination spectrum. M.N.R.A.S. **118**, 477 (1958).

CANUTO, V., CHIU, H. Y., FASSIO-CANUTO, L.: Electron bremsstrahlung in intense magnetic fields. Phys. Rev. **185**, 1607 (1969).

CANUTO, V., CHIU, H. Y.: Nonrelativistic electron bremsstrahlung in a strongly magnetized plasma. Phys. Rev. A **2**, 518 (1970).

CANUTO, V., CHIU, H. Y.: Intense magnetic fields in astrophysics. Space Sci. Rev. **12**, 3 (1971).

CANUTO, V., LODENQUAI, J., RUDERMAN, M.: Thomson scattering in a strong magnetic field. Phys. Rev. D **3**, 2303 (1971).

CERENKOV, P. A.: Visible radiation produced by electrons moving in a medium with velocities exceeding that of light. Phys. Rev. **52**, 378 (1937). (Cf. J. V. JELLEY: Cerenkov Radiation and its Applications. New York: Pergamon Press 1958.)

CHANDRASEKHAR, S.: A statistical basis for the theory of stellar scintillation. M.N.R.A.S. **112**, 475 (1952).

CHANDRASEKHAR, S., BREEN, F. H.: On the continuous absorption coefficient of the negative hydrogen ion. Ap. J. **104**, 430 (1946).

CHIU, H. Y., CANUTO, V., FASSIO-CANUTO, L.: Nature of radio and optical emissions from pulsars. Nature **221**, 529 (1969).

CHIU, H. Y., FASSIO-CANUTO, L.: Quantized synchrotron radiation in intense magnetic fields. Phys. Rev. **185**, 1614 (1969).

CLARK, B. G., KUZ'MIN, A. D.: The measurement of the polarization and brightness distribution of Venus at 10.6-cm wavelength. Ap. J. **142**, 23 (1965).

CLARKE, R. W., BROTEN, N. W., LEGG, T. H., LOCKE, J. L., YEN, J. L.: Long baseline interferometer observations at 408 and 448 MHz II—The interpretation of the observations. M.N.R.A.S. **146**, 381 (1969).

COHEN, M. H.: Radiation in a plasma. I. Cerenkov effect. Phys. Rev. **123** (3), 711 (1961).

COHEN, M. H., GUNDERMANN, E. J., HARDEBECK, H. E. SHARP, L. E.: Interplanetary scintillations, II. Observations. Ap. J. **147**, 449 (1967).

COMPTON, A. H.: A quantum theory of the scattering of X-rays by light elements. Phys. Rev. **21**, 207, 483 (1923).

COMPTON, A. H.: The spectrum of scattered X-rays. Phys. Rev. **22** (5), 409 (1923).

COOPER, J.: Plasma spectroscopy. Rpt. Prog. Phys. **29**, 2 (1966).

COULOMB, C. A.: Sur l'électricité et le magnétisme. Mem. l'Acad. (1785).

CULHANE, J. L.: Thermal continuum radiation from coronal plasmas at soft X-ray wavelengths. M.N.R.A.S. **144**, 375 (1969).

DAVY, SIR H.: Further researches on the magnetic phenomena produced by electricity. Phil. Trans. **111**, 425 (1821).

DEBYE, P. VON: Der Lichtdruck auf Kugeln von beliebigem Material (The light pressure on spheres of arbitrary material). Ann. Phys. **30**, 57 (1909).

DEBYE, P. VON, HÜCKEL, E.: Zur Theorie der Elektrolyte: I. Gefrierpunktserniedrigung und verwandte Erscheinungen; II. Das Grenzgesetz für die elektrische Leitfähigkeit (On the theory of electrolytes: I. Lowering of the freezing point and related phenomena; II. The limiting laws for the electrical conductivity). Phys. Z. **24**, 185 (1923).

DENNISON, P. A., HEWISH, A.: The solar wind outside the plane of the ecliptic. Nature **213**, 343 (1967).

DICKEL, J. R., DEGIOANNI, J. J., GOODMAN, G. C.: The microwave spectrum of Jupiter. Radio Science **5** (2), 517 (1970).

DITCHBURN, R. W., ÖPIK, U.: Photoionization processes. In: Atomic and molecular processes (ed. D. R. BATES). New York: Academic Press 1962.

DOMBROVSKI, V. A.: On the nature of the radiation from the Crab nebula. Dokl. Acad. Nauk. S.S.S.R. **94**, 1021 (1954).

DONAHUE, T. M.: The significance of the absence of primary electrons for theories of the origin of the cosmic radiation. Phys. Rev. **84** (5), 972 (1951).

DOPPLER, C.: Über das Farbige Licht der Doppelsterne u.s.w. (On the colored light of double stars, etc.). Abhandlungen d.k. Böhmischen Gesell. d. Wiss. **2**, 467 (1843).

DOUGHERTY, J. P., FARLEY, D. T.: A theory of incoherent scattering of radio waves by a plasma. Proc. Roy. Soc. London A **259**, 79 (1960).

EIDMAN, V. YA.: Investigation of the radiation of an electron moving in a magnetoactive medium. Sov. Phys. J.E.T.P. **7**, 91 (1958). (Corrections in Sov. Phys. J.E.T.P. **9**, 947 (1959)).

EINSTEIN, A. VON: Über einen die Erzeugung und Verwandlung des Lichtes betreffenden heuristischen Gesichtspunkt (On a heuristic point of view concerning the generation and transformation of light). Ann. Physik **17**, 132 (1905).

ELWERT, G. VON: Die weiche Röntgenstrahlung der ungestörten Sonnenkorona (The soft X-ray radiation of the undisturbed solar corona). Z. Naturforschung, **9**, 637 (1954).

EPSTEIN, R. I., FELDMAN, P. A.: Synchrotron radiation from electrons in helical orbits. Ap. J. **150**, L 109 (1967).

ERBER, T.: High-energy electromagnetic conversion processes in intense magnetic fields. Rev. Mod. Phys. **38**, 626 (1966).

EVANS, J. V.: Radar studies of planetary surfaces. Ann. Rev. Astron. Astrophys. **7**, 201 (1969).

FARADAY, M.: On static electrical inductive action. Phil. Mag. **22**, 200 (1843).

FARADAY, M.: Experimental researches in electricity. London: R. Taylor 1844. Repr. New York: Dover 1952.

FARLEY, D. T.: A theory of incoherent scattering of radio waves by a plasma: 4. The effect of unequal ion and electron temperatures. J. Geophys. Res. **17**, 4091 (1966).

FEENBERG, E.: The scattering of slow electrons by neutral atoms. Phys. Rev. **40**, 40 (1932).

FEENBERG, E., PRIMAKOFF, H.: Interaction of cosmic-ray primaries with sunlight and starlight. Phys. Rev. **73** (5), 449 (1948).

FEJER, J. A.: The diffraction of waves in passing through an irregular refracting medium. Proc. Roy. Soc. London A **220**, 455 (1953).

FEJER, J. A.: Scattering of radio waves by an ionized gas in thermal equilibrium. Can. J. Phys. **38**, 1114 (1960).

FEJER, J. A.: Radio-wave scattering by an ionized gas in thermal equilibrium. J. Geophys. Res. **65**, 2635 (1960).

FELTEN, J. E., MORRISON, P.: Recoil photons from scattering of starlight by relativistic electrons. Phys. Rev. Lett. **10** (10), 453 (1963).

FELTEN, J. E., MORRISON, P.: Omnidirectional inverse Compton and synchrotron radiation from cosmic distributions of fast electrons and thermal photons. Ap. J. **146**, 686 (1966).

FERMAT, P.: Oeuvres de Fermat. Paris **2**, 354 (1891).

FERMI, E.: The ionization loss of energy in gases and in condensed materials. Phys. Rev. **57**, 485 (1940).

FRESNEL, A. J.: Explication de la réfraction dans le système des ondes. Mem. de l'Acad. **11**, 893 (1822).

GAUNT, J. A.: Continuous absorption. Phil. Trans. Roy. Soc. London A **229**, 163 (1930).

GINZBURG, V. L.: The origin of cosmic rays and radio astronomy. Usp. Fiz. Nauk. **51**, 343 (1953).

GINZBURG, V. L.: Propagation of electromagnetic waves in a plasma. New York: Gordon and Breach 1961.

GINZBURG, V. L.: Elementary processes for cosmic ray astrophysics. New York: Gordon and Breach 1969.

GINZBURG, V. L., SYROVATSKII, S. I.: The origin of cosmic rays. New York: Macmillan 1964.

GINZBURG, V. L., SYROVATSKII, S. I. Cosmic magnetobremsstrahlung (Synchrotron radiation). Ann. Rev. Astron. Astrophys. **3**, 297 (1965).

GINZBURG, V. L.: Elementary processes for cosmic ray astrophysics. New York: Gordon and Breach 1969.

GINZBURG, V. L., SYROVATSKII, S. I.: The origin of cosmic rays. New York: Macmillan 1964.

GINZBURG, V. L., SYROVATSKII, S. I.: Cosmic magnetobremsstrahlung (Synchrotron radiation). Ann. Rev. Astron. Astrophys. **3**, 297 (1965).

GINZBURG, V. L., ZHELEZNIAKOV, V. V.: On the possible mechanisms of sporadic solar radio emission (Radiation in an isotropic plasma). Soviet Astron. **2**, 653 (1958).

GINZBURG, V. L., ZHELEZNIAKOV, V. V.: On the mechanisms of sporadic solar radio emission. In: Paris symposium on radio astronomy (ed. R. N. BRACEWELL). Stanford, Calif.: Stanford Univ. Press 1959.

GOLDREICH, P., MORRISON, P.: On the absorption of gamma rays in intergalactic space. Sov. Phys. J.E.T.P. **18**, 239 (1964).

GORDON, Doklady Akad. Nauk. **94**, 413 (1954).

GOULD, R. J., SCHRÉDER, G.: Opacity of the universe to high-energy photons. Phys. Rev. Lett. **16** (6), 252 (1966).

GREEN, P. E.: Radar measurements of target scattering properties. In: Radar astronomy (ed. J. V. EVANS, T. HAGFORS). New York: McGraw-Hill 1968.

HAGFORS, T.: Some properties of radio waves reflected from the moon and their relation to the lunar surface. J. Geophys. Res. **66** (3), 777 (1961).

HAGFORS, T.: Density fluctuations in a plasma in a magnetic field, with applications to the ionosphere. J. Geophys. Res. **66** (6), 1699 (1961).

HAGFORS, T.: Backscattering from an undulating surface with applications to radar returns from the moon. J. Geophys. Res. **69** (18), 3779 (1964).

HAGFORS, T.: Remote probing of the moon by infrared and microwave emissions and by radar. Radio Science **5**, 189 (1970).

HARRIS, D. E., ZEISSIG, G. A., LOVELACE, R. V.: The minimum observable diameter of radio sources. Astron. Astrophys. **8**, 98 (1970).

HARTREE, D. R.: The propagation of electromagnetic waves in a refracting medium in a magnetic field. Proc. Camb. Phil. Soc. **27**, 143 (1931).

HEAVISIDE, O.: On the electromagnetic effects due to the motion of electrification through a dielectric. Phil. Mag. **27**, 324 (1889).

HEAVISIDE, O.: The waste of energy from a moving electron. Nature **67**, 6 (1902).

HEAVISIDE, O.: The radiation from an electron moving in an elliptic, or any other orbit. Nature **69**, 342 (1904).

HEILES, C. E., DRAKE, F. D.: The polarization and intensity of thermal radiation from a planetary surface. Icarus **2**, 291 (1963).

HEITLER, W.: The quantum theory of radiation. Oxford: Oxford University Press 1954.

HERTZ, H.: Über die Beziehungen zwischen Licht und Elektrizität (On the relations between light and electricity). Gesammelte Werke **1**, 340 (1889).

HERTZ, H.: Die Kräfte elektrischer Schwingungen, behandelt nach der Maxwell'schen Theorie (The force of electrical oscillations treated with the Maxwell theory). Ann. Phys. **36**, 1 (1889).

HEWISH, A.: The diffraction of radio waves in passing through a phase-changing ionosphere. Proc. Roy. Soc. London A **209**, 81 (1951).

HJELLMING, R. M., CHURCHWELL, E.: An analysis of radio recombination lines emitted by the Orion nebula. Astrophys. Lett. **4**, 165 (1969).

HOYLE, F., BURBIDGE, G. R., SARGENT, W. L. W.: On the nature of the quasi-stellar sources. Nature **209**, 751 (1966).

HULST, H. C. VAN DE: On the attenuation of plane waves by obstacles of arbitrary size and form. Physica **15**, 740 (1949).

HULST, H. C. VAN DE: Light scattering by small particles. New York: Wiley 1957.

JACKSON, J. D.: Classical electrodynamics. New York: Wiley 1962.

JAUCH, J. M., ROHRLICH, F.: The theory of protons and electrons. Reading, Mass.: Addison-Wesley 1955.

JEANS, SIR J. H.: On the partition of energy between matter and aether. Phil. Mag. **10**, 91 (1905).

JEANS, SIR J. H.: Temperature-radiation and the partition of energy in continuous media. Phil. Mag. **17**, 229 (1909).

JOKIPII, J. R., HOLLWEG, J. V.: Interplanetary scintillations and the structure of solar-wind fluctuations. Ap. J. **160**, 745 (1970).

KARDASHEV, N. S.: Nonstationariness of spectra of young sources of nonthermal radio emission. Soviet Astron. **6**, 317 (1962).

KARZAS, W. J., LATTER, R.: Electron radiative transitions in a Coulomb field. Ap. J. Suppl. **6**, 167 (1961).

KELLERMANN, K. I.: The radio source 1934–63. Austr. J. Phys. **19**, 195 (1966).

KELLERMANN, K. I.: Thermal radio emission from the major planets. Radio Science **5** (2), 487 (1970).

KELLERMANN, K. I., PAULINY-TOTH, I. I. K.: The spectra of opaque radio sources. Ap. J. **155**, L 71 (1969).

KELLERMANN, K. I., PAULINY-TOTH, I. I. K., WILLIAMS, P. J. S.: The spectra of radio sources in the revised 3C catalogue. Ap. J. **157**, 1 (1969).

KELVIN, LORD (W. THOMSON): A mathematical theory of magnetism. Phil. Mag. **37**, 241 (1850).

KELVIN, LORD (W. THOMSON): Transient electric currents. Proc. Glascow Phil. Soc. **3**, 281 (1835).

KIRCHHOFF, H.: On the simultaneous emission and absorption of rays of the same definite refrangibility. Phil. Mag. **19**, 193 (1860).

KIRCHHOFF, H.: On a new proposition in the theory of heat. Phil. Mag. **21**, 240 (1860). See also: Gesammelte Abhandlungen. Leipzig: J. A. Barth 1882.

KLEIN, O., NISHINA, Y.: Über die Streuung von Strahlung durch freie Elektronen nach der neuen relativistischen Quantendynamik von Dirac (On the scattering of radiation by free electrons according to the new relativistic quantum dynamics by Dirac). Z. Physik **52**, 853 (1929).

KRAMERS, H. A.: On the theory of X-ray absorption and of the continuous X-ray spectrum. Phil. Mag. **46**, 836 (1923).

KRAMERS, H. A.: The law of dispersion and Bohr's theory of spectra. Nature **113**, 673 (1924).

KRONIG, R. DE L., KRAMERS, H. A.: Zur Theorie der Absorption und Dispersion in den Röntgenspektren (Theory of absorption and dispersion in X-ray spectra). Z. Physik **48**, 174 (1928).

LADENBURG, R. VON: Die Quantentheoretische Deutung der Zahl der Dispersionselektronen (The quantum theory interpretation of the number of scattering electrons). Z. Physik **4**, 451 (1921).

LAMBERT, J. H.: Photometria, sive de mensura et gradibus luminis, colorum et umbrae. Augsburg 1760.

LANDAU, L. D.: On the vibrations of the electronic plasma. J. Phys. (U.S.S.R.) **10**, 25 (1946).

LANDAU, L. D., LIFSHITZ, E. M.: The classical theory of fields. Reading, Mass.: Addison-Wesley 1962.

LANG, K. R.: Interstellar scintillation of pulsar radiation. Ap. J. **164**, 249 (1971).

LANG, K. R., RICKETT, B. J.: Size and motion of the interstellar scintillation pattern from observations of CP 1133. Nature **225**, 528 (1970).

LARMOR, SIR J.: On the theory of the magnetic influence on spectra; and on the radiation from moving ions. Phil. Mag. **44**, 503 (1897).

LEGG, M. P. C., WESTFOLD, K. C.: Elliptic polarization of synchrotron radiation. Ap. J. **154**, 499 (1968).

LE ROUX, E.: Étude théorique de rayonnement synchrotron des radiosources. Ann. Astrophys. **24**, 71 (1961).

LIEMOHN, H. B.: Radiation from electrons in a magnetoplasma. J. Res. Nat. Bur. Stands. USNC-URSI Radio Science **69D**, 741 (1965).

LIÉNARD, A. M.: Théorie de Lorentz, Théorie de Larmor et celle de Lorentz. Éclairage électr. **14**, **16** (1898).

LORENTZ, H. A.: Über die Beziehung zwischen der Fortpflanzungsgeschwindigkeit des Lichtes und der Körperdichte (Concerning the relation between the velocity of propagation of light and the density and composition of media). Ann. Phys. **9**, 641 (1880).

LORENTZ, H. A.: La théorie électromagnétique de Maxwell et son application aux corps mouvants. Archives Neerl, **25**, 363 (1892).

LORENTZ, H. A.: Electromagnetic phenomena in a system moving with any velocity smaller than that of light. Proc. Amst. Acad. Sci. **6**, 809 (1904). Reprod. in H. A. LORENTZ, A. EINSTEIN, H. MINKOWSKI, H. WEYL: The principle of relativity. New York: Dover 1952.

LORENTZ, H. A.: The theory of electrons. 1909. Republ. New York: Dover 1952.

LORENZ, L.: Über die Refractionsconstante (On the constant of refraction). Wiedem. Ann. **11**, 70 (1881).

LOVELACE, R. V. E., SALPETER, E. E., SHARP, L. E., HARRIS, D. E.: Analysis of observations of interplanetary scintillations. Ap. J. **159**, 1047 (1971).

MARGENAU, H.: Conduction and dispersion of ionized gases at high frequencies. Phys. Rev. **69**, 508 (1946).

MAXWELL, J. C.: Illustrations of the dynamical theory of gases: Part I. On the motions and collisions of perfectly elastic spheres. Phil. Mag. **19**, 19 (1860). Part. II. On the process of diffusion of two or more kinds of moving particles among one another. Phil. Mag. **20**, 21 (1860).

MAXWELL, J. C.: On physical lines of force: Part. I. The theory of molecular vortices applied to magnetic phenomena; Part II. The theory of molecular vortices applied to electric currents. Phil. Mag. **21**, 161, 281 (1861).

MAXWELL, J. C.: The theory of anomalous dispersion. Phil. Mag. **48**, 151 (1899).

MAXWELL, J. C.: A treatise on electricity and magnetism. 1873. Republ. New York: Dover 1954.

MAYER, C. H.: Thermal radio emission of the planets and moon. In: Surfaces and interiors of planets and satellites (ed. A. DOLLFUSS). New York: Academic Press 1970.

MAYER, C. H., MCCULLOUGH, T. P.: Microwave radiation of Uranus and Neptune. Icarus **14**, 187 (1971).

MCCRAY, R.: Synchrotron radiation losses in self-absorbed radio sources. Ap. J. **156**, 329 (1969).

MELROSE, D. B.: On the degree of circular polarization of synchrotron radiation. Astrophys. and Space Sci. **12**, 172 (1971).

MENZEL, D. H., PEKERIS, C. L.: Absorption coefficients and hydrogen line intensities. M.N.R.A.S. **96**, 77 (1935). Reproduced in : Selected papers on physical processes in ionized plasmas (ed. D. H. MENZEL). New York: Dover 1962.

MERCIER, R. P.: Diffraction by a screen causing large random phase fluctuations. Proc. Camb. Phil. Soc. **58**, 382 (1962).

MERCIER, R. P.: The radio-frequency emission coefficient of a hot plasma. Proc. Phys. Soc. **83**, 819 (1964).

MEZGER, P. G., HENDERSON, A. P.: Galactic H II regions I. Observations of their continuum radiation at the frequency 5 GHz. Ap. J. **147**, 471 (1967).

MIE, G. VON: Beiträge zur Optik trüber Medien, speziell Kolloidaler Metallösungen (Contributions to the optics of opaque media, especially colloide metal solutions). Ann. Physik **25**, 377 (1908).

MILNE, E. A.: Radiative equilibrium in the outer layers of a star: the temperature distribution and the law of darkening. M.N.R.A.S. **81**, 361 (1921).

MILNE, E. A.: Thermodynamics of the stars. Handbuch der Astrophysik **3**, 80 (1930). Reproduced in: Selected papers on the transfer of radiation (ed. D. H. MENZEL). New York: Dover 1966.

MORRISON, D.: Thermophysics of the planet Mercury. Space Sci. Rev. **11**, 271 (1970).

MOSENGEIL, K.: Theorie der stationären Strahlung in einem gleichförmig bewegten Hohlraum (Theory of stationary radiation in a uniformly moving cavity). Ann. Phys. **22**, 867 (1907).

MOTT, N. F., MASSEY, H. S. W.: The theory of atomic collisions. Oxford: Oxford at the Clarendon Press 1965.

MUHLEMAN, D. O.: Microwave opacity of the Venus atmosphere. Astron. J. **74**, 57 (1969).

O'DELL, S. L., SARTORI, L.: Limitation on synchrotron models with small pitch angles. Ap. J. **161**, L 63 (1970).

OHM, G. S.: Versuch einer Theorie der durch galvanische Kräfte hervorgebrachten elektroskopischen Erscheinungen (An attempt to a theory of the galvanic forces generated through electroscopic phenomena). Ann. Phys. **6**, 459; **7**, 45 (1826).

OORT, J. H., WALRAVEN, T.: Polarization and composition of the Crab nebula. B.A.N. **12**, 285 (1956).

OSTER, L.: Effects of collisions on the cyclotron radiation from relativistic particles. Phys. Rev. **119**, 1444 (1960).

OSTER, L.: Emission, absorption, and conductivity of a fully-ionized gas at radio frequencies. Rev. Mod. Phys. **33** (4), 525 (1961).

OSTER, L.: The free-free emission and absorption coefficients in the radio frequency range at very low temperatures. Astron. Astrophys. **9**, 318 (1970).

PLANCK, M.: Über das Gesetz der Energieverteilung im Normalspectrum (On the theory of thermal radiation). Ann. Physik **4**, 553 (1901).

PLANCK, M.: The theory of heat radiation. 1913. Reprod. New York: Dover 1959.

POINCARÉ, H.: Électricité—sur la dynamique de l'électron. Comptes Rendus Acad. Sci. Paris **140**, 1504 (1905).

POISSON, S. D.: Remarques sur une équation qui se présente dans la théorie des attractions des spéroides. Bull. de la Soc. Philomathique **3**, 388 (1813).

POLLACK, J. B., MORRISON, D.: Venus: Determination of atmospheric parameters from the microwave spectrum. Icarus **12**, 376 (1970).

POYNTING, J. H.: On the transfer of energy in the electromagnetic field. Phil. Trans. **175**, 343 (1884).

RABI, I. I. VON: Das freie Elektron im homogenen Magnetfeld nach der Diracschen Theorie (The free electron in a homogeneous magnetic field according to the Dirac theory). Z. Physik **49**, 507 (1928).

RAMATY, R.: Gyrosynchrotron emission and absorption in a magnetoactive plasma. Ap. J. **158**, 753 (1969).
RATCLIFFE, J. A.: The magneto-ionic theory and its applications to the ionosphere. Cambridge: Cambridge Univ. Press 1959.
RAYLEIGH, LORD: On the light from the sky, its polarization and colour. Phil. Mag. **41**, 107, 274 (1871).
RAYLEIGH, LORD: Investigations in optics, with special reference to the spectroscope. Phil. Mag. **8**, 403 (1879).
RAYLEIGH, LORD: On the transmission of light through an atmosphere containing small particles in suspension, and on the origin of the blue of the sky. Phil. Mag. **47**, 375 (1899).
RAYLEIGH, LORD: Remarks upon the law of complete radiation. Phil. Mag. **49**, 539 (1900).
RAYLEIGH, LORD: The dynamical theory of gases and of radiation. Nature, **72**, 54 (1905).
RAZIN, V. A.: The spectrum of nonthermal cosmic radio emission. Radiophysica **3**, 584, 921 (1960).
REES, M. J.: Appearance of relativistically expanding radio sources. Nature **211**, 468 (1966).
RICKETT, B. J.: Interstellar scintillation and pulsar intensity variations. M.N.R.A.S. **150**, 67 (1970).
RUSSELL, H. N.: On the albedo of the planets and their satellites. Ap. J. **43**, 173 (1916).
RUTHERFORD, E.: The scattering of α and β particles by matter and the structure of the atom. Phil. Mag. **21**, 669 (1911).
SALPETER, E. E.: Electron density fluctuations in a plasma. Phys. Rev. **120**, 1528 (1960).
SALPETER, E. E.: Interplanetary scintillations I. Theory. Ap. J. **147**, 433 (1967).
SAUTER, F. VON: Zur unrelativistischen Theorie des kontinuierlichen Röntgenspektrums (On the non-relativistic theory of the continuous X-ray spectrum). Ann. Physik **18**, 486 (1933).
SCHEUER, P. A. G.: The absorption coefficient of a plasma at radio frequencies. M.N.R.A.S. **120**, 231 (1960).
SCHEUER, P.A.G.: Synchrotron radiation formulae. Ap. J. **151**, L 139 (1968).
SCHOTT, G. A.: Electromagnetic radiation: Cambridge: Cambridge University Press 1912.
SCHWINGER, J.: On the classical radiation of accelerated electrons. Phys. Rev. **75**, 1912 (1949).
SHAPIRO, I. I.: Theory of the radar determination of planetary rotations. Astron. J. **72**, 1309 (1967).
SHKLOVSKI, I. S.: On the nature of the radiation from the Crab nebula. Dokl. Akad. Nauk. S.S.S.R. **90**, 983 (1953).
SIMON, M.: Asymptotic form for synchrotron spectra below Razin cutoff. Ap. J. **156**, 341 (1969).
SITENKO, A. G., STEPANOV, K. N.: On the oscillations of an electron plasma in a magnetic field. Sov. Phys. J.E.T.P. **4**, 512 (1957).
SLISH, V. I.: Angular size of radio stars. Nature **199**, 682 (1963).
SMERD, S. F., WESTFOLD, K. C.: The characteristics of radio-frequency radiation in an ionized gas, with applications to the transfer of radiation in the solar atmosphere. Phil. Mag. **40**, 831 (1949).
SMERD, S. F., WILD, J. P., SHERIDAN, K. V.: On the relative position and origin of harmonics in the spectra of solar radio bursts of spectral types II and III. Austr. J. Phys. **15**, 180 (1962).
SNELL, W.: 1621, unpubl.
SOMMERFELD, A.: Über die Beugung und Bremsung der Elektronen (On the deflection and deceleration of electrons). Ann. Phys. **11**, 257 (1931).
STEFAN, A. J.: Beziehung zwischen Wärmestrahlung und Temperatur (Relation between thermal radiation and temperature). Wien. Ber. **79**, 397 (1879).
STOKES, G. G.: On the composition and resolution of streams of polarized light from different sources. Trans. Camb. Phil. Soc. **9**, 399 (1852).
TAKAKURA, T.: Synchrotron radiation from intermediate energy electrons and solar radio outbursts at microwave frequencies. Publ. Astron. Soc. Japan **12**, 325, 352 (1960).
TERZIAN, Y., PARRISH, A.: Observations of the Orion nebula at low radio frequencies. Astrophys. Lett. **5**, 261 (1970).
THOMSON, J. J.: Conduction of electricity through gases. Cambridge: Cambridge University Press 1903. Republ. New York: Dover 1969.
TONKS, L., LANGMUIR, I.: Oscillations in ionized gases. Phys. Rev. **33**, 195 (1929).
TROITSKII, V. S.: On the possibility of determining the nature of the surface material of Mars from its radio emission. Radio Science **5** (2), 481 (1970).
TRUBNIKOV, B. A.: Plasma radiation in a magnetic field. Sov. Phys. "Doklady" **3**, 136 (1958).
TRUBNIKOV, B. A.: Particle interactions in a fully ionized plasma. Rev. Plasma Phys. **1**, 105 (1965).
TRUMPLER, R. J.: Absorption of light in the galactic system. P.A.S.P. **42**, 214 (1930).
TSYTOVICH, V. N.: The problem of radiation by fast electrons in a magnetic field in the presence of a medium. Vestn. Mosk. Univ. **11**, 27 (1951).

TWISS, R. Q.: On the nature of discrete radio sources. Phil. Mag. **45**, 249 (1954).

VLADIMIRSKI, V. V.: Influence of the terrestial magnetic field on large Auger showers. Zh. Exp. Teor. Fiz. **18**, 392 (1948).

VLASOV, A. A.: On the kinetic theory of an assembly of particles with collective interaction. J. Phys. (U.S.S.R.) **9**, 25 (1945).

WESTFOLD, K. C.: The polarization of synchrotron radiation. Ap. J. **130**, 241 (1959).

WHEELER, J. A., LAMB, W. E.: Influence of atomic electrons on radiation and pair production. Phys. Rev. **55**, 858 (1939).

WHITFORD, A. E.: The law of interstellar reddening. Astron. J. **63**, 201 (1958).

WIECHERT, J. E.: Elektrodynam. Elementargesetze (Fundamental electrodynamic laws). Arch. Néerl **5**, 1 (1901).

WIEN, W.: Eine neue Beziehung der Strahlung schwarzer Körper zum zweiten Hauptsatz der Wärmetheorie (One new relation between the radiation of blackbodies and the second law of thermodynamics). Sitz. Acad. Wiss. Berlin **1**, 55 (1893).

WIEN, W.: On the division of energy in the emission-spectrum of a black body. Phil. Mag. **43**, 214 (1894).

WILD, J. P., HILL, E. R.: Approximation of the general formulae for gyro and synchrotron radiation in a vacuum and isotropic plasma. Austr. J. Phys. **24**, 43 (1971).

WILD, J. P., SMERD, S. F., WEISS, A. A.: Solar bursts. Ann. Rev. Astron. Astrophys. **1**, 291 (1963).

WILLIAMS, P. J. S.: Absorption in radio sources of high brightness temperature. Nature **200**, 56 (1963).

ZHELEZNYAKOV, V. V.: Radio emission of the sun and planets. New York: Pergamon Press 1970.

Chapter 2

AANNESTAD, P. A.: Molecule formation: I. In normal H I clouds; II. In interstellar shock waves. Ap. J. Suppl. No. 217, **25**, 205 (1973).

ABNEY, W. DE W.: Effect of a star's rotation on its spectrum. M.N.R.A.S. **37**, 278 (1877).

ABT, H. A., BIGGS, E. S.: Bibliography of stellar radial velocities. New York: Latham Press and Kitt Peak Nat. Obs. 1972.

ABT, H. A., HUNTER, J. H.: Stellar rotation in galactic clusters. Ap. J. **136**, 381 (1962).

ABT, H. A., MEINEL, A. B., MORGAN, W. W., TABSCOTT, J. W.: An atlas of low-dispersion grating stellar spectra. Kitt Peak Nat. Obs., Steward Obs., Yerkes Obs., 1969.

ADAMS, W. S.: The relativity displacement of the spectral lines in the companion of Sirius. Proc. Nat. Acad. Sci. (Wash.) **11**, 382 (1925).

ADAMS, W. S., DUNHAM, T.: Absorption bands in the infra-red spectrum of Venus. P.A.S.P. **44**, 243 (1932).

ALI, A. W., GRIEM, H. R.: Theory of resonance broadening of spectral lines by atom-atom impacts. Phys. Rev. **140 A**, 1044 (1965).

ALI, A. W., GRIEM, H. R.: Theory of resonance broadening of spectral lines by atom-atom impacts. Phys. Rev. **144**, 366 (1966).

ALLEN, C. W.: Astrophysical quantities. Univ. of London, London: Athalone Press 1963.

ALLEN, R. J.: Intergalactic hydrogen along the path to Virgo A. Astron. Astrophys. **3**, 382 (1969).

ALLER, L. H.: Gaseous nebulae. London: Chapman and Hall 1956.

ALLER, L. H.: The Abundance of the elements. New York: Interscience Publ. 1961.

ALLER, L. H., BOWEN, I. S., WILSON, O. C.: The spectrum of NGC 7027. Ap. J. **138**, 1013 (1963).

ALLER, L. H., CZYZAK, S. S.: The chemical composition of planetary nebulae. In: Planetary nebulae (ed. D. E. OSTERBROCK, C. R. O'DELL). I.A.U. Symp. No. 34. Dordrecht, Holland: D. R. Reidel. Berlin-Heidelberg-New York: Springer 1968.

ALLER, L. H., FAULKER, D. J.: Spectrophotometry of 14 southern planetary nebulae. In: The Galaxy and the Magellanic clouds—I.A.U. Symp. No. 20 Aust. Acad. Sci., Canberra, Austr. 1964.

ALLER, L. H., GREENSTEIN, J. L.: The abundances of the elements in G-type subdwarfs. Ap. J. Suppl. No. 46, **5**, 139 (1960).

ALLER, L. H., LILLER, W.: Planetary nebulae. In: Nebulae and interstellar matter—Stars and stellar systems VII (ed. B. M. MIDDLEHURST, L. H. ALLER). Chicago, Ill.: Univ. of Chicago Press 1960.

ANDERSON, P. W.: Pressure broadening in the microwave and infra-red regions. Phys. Rev. **76**, 647 (1949).

Andrews, M. H., Hjellming, R. M.: Intensities of radio recombination lines (II). Astrophys. L. **4**, 159 (1969).

Angel, J. R. P., Landstreet, J. D.: Detection of circular polarization in a second white dwarf. Ap. J. **164**, L 15 (1971).

Babcock, H. W.: Zeeman effect in stellar spectra. Ap. J. **105**, 105 (1947).

Babcock, H. W.: The 34-kilogauss magnetic field of HD 215441. Ap. J. **132**, 521 (1960).

Babcock, H. W., Babcock, H. D.: The sun's magnetic field, 1952–1954. Ap. J. **121**, 349 (1955).

Back, E. von, Goudsmit, S.: Kernmoment und Zeeman-Effekt von Wismut (Nuclear moment and Zeeman effect in bismuth). Z. Physik **47**, 174 (1928).

Bahcall, J. N.: A systematic method for identifying absorption lines as applied to PKS 0237–23. Ap. J. **153**, 679 (1968).

Bahcall, J. N., Greenstein, J. L., Sargent, W. L. W.: The absorption-line spectrum of the quasi-stellar radio source PKS 0237–23. Ap. J. **153**, 689 (1968).

Bahcall, J. N., Wolf, R. A.: Fine-structure transitions. Ap. J. **152**, 701 (1968).

Baker, J. G., Menzel, D. H.: Physical processes in gaseous nebulae III. The Balmer decrement. Ap. J. **88**, 52 (1938). Reprod. in: Selected papers on physical processes in ionized plasmas (ed. D. H. Menzel). New York: Dover 1962.

Balmer, J. J.: Notiz über die Spectrallinien des Wasserstoffs (A note on the spectral lines of hydrogen). Ann. Phys. Chem. **25**, 80 (1885).

Baranger, M.: Problem of overlapping lines in the theory of pressure broadening. Phys. Rev. **111**, 494 (1958).

Baranger, M.: General impact theory of pressure broadening. Phys. Rev. **112**, 855 (1958).

Baranger, M.: Spectral line broadening in plasmas. In: Atomic and molecular processes (ed. D. R. Bates). New York: Academic Press 1962.

Bates, D. R.: Rate of formation of molecules by radiative association. M. N. R. A. S. **111**, 303 (1951).

Bates, D. R., Dalgarno, A.: Electronic recombination. In: Atomic and molecular processes (ed. D. R. Bates). New York: Academic Press 1962.

Bates, D. R., Damgaard, A.: The calculation of the absolute strengths of spectral lines. Phil. Trans. Roy. Soc. London A **242**, 101 (1949).

Bates, D. R., Spitzer, L.: The density of molecules in interstellar space. Ap. J. **113**, 441 (1951).

Berman, L.: The effect of reddening on the Balmer decrement in planetary nebulae. M. N. R. A. S. **96**, 890 (1936).

Bethe, H. A.: Quantenmechanik der Ein- und Zwei-Elektronen-Probleme (Quantum mechanics of one- and two-electron problems). In: Handbuch der Physik (ed. H. Geiger, K. Scheel), Vol. XXIV, p. 273. Berlin: Springer 1933.

Bethe, H. A., Salpeter, E. E.: Quantum mechanics of one and two electron atoms. Berlin-Göttingen-Heidelberg: Springer 1957.

Black, J. H., Dalgarno, A.: The cosmic abundance of deuterium. Ap. J. **184**, L 101 (1973).

Blamont, J. E., Roddier, F.: Precise observation of the profile of the Fraunhofer strontium resonance line. Evidence for the gravitational redshift of the Sun. Phys. Rev. Lett. **7**, 437 (1961).

Böhm, K. H.: Basic theory of line formation. In: Stars and stellar systems (ed. J. Greenstein) vol. 6. Chicago, Ill.: Univ. of Chicago Press 1960.

Böhm, K. H.: The atmospheres and spectra of central stars. In: Planetary nebulae—I. A. U. Symp. No. 34. Dordrecht, Holland: D. Reidel. Berlin-Heidelberg-New York: Springer 1968.

Böhm, D., Aller, L. H.: The electron velocity distribution, in gaseous nebulae and stellar envelopes, Ap. J. **105**, 131 (1947). Reprod. In: Selected papers on physical processes in ionized plasmas (ed. D. H. Menzel). New York: Dover 1962.

Bohr, N.: On the constitution of atoms and molecules. Phil. Mag. **26**, 1, 10, 476, 857 (1913).

Born, M., Oppenheimer, R.: Zur Quantentheorie der Molekeln (On the quantum theory of molecules). Ann. Phys. **84**, 457 (1927).

Bowen, I. S.: The origin of the nebular lines and the structure of the planetary nebulae. Ap. J. **67**, 1 (1928).

Boyarchuk, A. A., Kopylov, I. M.: A general catalogue of rotational velocities of 2558 stars. Bull. Crimean. Ap. Obs. **31**, 44 (1964).

Boyarchuk, A. A., Gershberg, R. E., Godovnikov, N. V.: Formulae graphs, and tables for a quantitative analysis of the hydrogen radiation of emission objects. Bull. Crimean. Ap. Obs. **38**, 208 (1968).

BOYARCHUK, A. A., GERSHBERG, R. E., GODOVNIKOV, N. V., PRONIK, V. I.: Formulae and graphs for a quantitative analysis of the forbidden line radiation of emission objects. Bull. Crimean Ap. Obs. **39**, 147 (1969).

BRACKETT, F. S.: Visible and infra-red radiation of hydrogen. Ap. J. **56**, 154 (1922).

BRIET, G., TELLER, E.: Metastability of hydrogen and helium levels. Ap. J. **91**, 215 (1940).

BROCKLEHURST, M.: Level populations of hydrogen in gaseous nebulae. M. N. R. A. S. **148**, 417 (1970).

BROCKLEHURST, M., LEEMAN, S.: The pressure broadening of radio recombination lines. Astrophys. Lett. **9**, 35 (1971).

BROCKLEHURST, M., SEATON, M. J.: The profiles of radio recombination lines. Astrophys. Lett. **9**, 139 (1971).

DE BROGLIE, L.: Waves and quanta. Nature **112**, 540 (1923).

DE BROGLIE, L.: Recherches sur la théorie des quanta. Ann. Physique **3**, 22 (1925).

BROWN, R. L., MATHEWS, W. G.: Theoretical continuous spectra for gaseous nebulae. Ap. J. **160**, 939 (1970).

BURBIDGE, G. R.: Nuclei of galaxies. Ann. Rev. Astron. Astrophys. **8**, 369 (1970).

BURBIDGE, G. R., BURBIDGE, E. M.: Quasi-stellar objects. San Francisco, Calif.: W. H. Freeman, 1967.

BURBIDGE, G. R., BURBIDGE, E. M.: Red-shifts of quasi-stellar objects and related extragalactic systems. Nature **222**, 735 (1969).

BURBIDGE, E. M., BURBIDGE, G. R.: Optical observations of southern radio sources. Ap. J. **172**, 37 (1972).

BURBIDGE, E. M., STRITTMATTER, P. A.: Redshifts of twenty radio galaxies. Ap. J. **172**, L 37 (1972)

BURGER, H. C. VON, DORGELO, H. B.: Beziehung zwischen inneren Quantenzahlen und Intensitäten von Mehrfachlinien (Relations between inner quantum numbers and intensities of multiple lines). Z. Physik **23**, 258 (1924).

BURGESS, A.: The hydrogen recombination spectrum. M. N. R. A. S. **118**, 477 (1958).

BURGESS, A.: Dielectronic recombination and the temperature of the solar corona. Ap. J. **139**, 776 (1964).

BURGESS, A.: A general formula for the estimation of dielectronic recombination coefficients in low density plasmas. Ap. J. **141**, 1588 (1965).

BURGESS, A., SEATON, M. J.: The ionization equilibrium for iron in the solar corona. M. N. R. A. S. **127**, 355 (1964).

CAMERON, R. C.: The magnetic and related stars. Baltimore, Maryland: Mono Book 1967.

CAPRIOTTI, E. R. The hydrogen radiation spectrum in gaseous nebulae. Ap. J. **139**, 225 (1964).

CAPRIOTTI, E. R., DAUB, C. T.: Hβ and [O III] fluxes from planetary nebulae. Ap. J. **132**, 677 (1960).

CARROLL, T. D., SALPETER, E. E.: On the abundance of interstellar OH. Ap. J. **143**, 609 (1966).

CARRUTHERS, G. R.: Rocket observation of interstellar molecular hydrogen. Ap. J. **161**, L 81 (1970).

CATALÁN, M. A.: Series and other regularities in the spectrum of manganese. Phil. Trans. Roy. Soc. London **A 223**, 127 (1922).

CHANDRASEKHAR, S.: Stochastic problems in physics and astronomy. Rev. Mod. Phys. **15**, 1 (1943). Reprod. in: Selected papers on noise and stochastic processes (ed. N. WAX). New York: Dover 1954.

CHANDRASEKHAR, S.: Radiative transfer. New York: Dover 1960.

CHANDRASEKHAR, S., BREEN, F. H.: On the continuous absorption coefficient of the negative hydrogen ion. III. Ap. J. **104**, 430 (1946).

CHEUNG, A. C., RANK, D. M., TOWNES, C. H., THORTON, D. D., WELCH, W. J.: Detection of NH^3 molecules in the interstellar medium by their microwave emission. Phys. Rev. Lett. **21**, 1701 (1968).

CHURCHWELL, E., MEZGER, P. G.: On the determination of helium abundance from radio recombination lines. Astrophys. Lett. **5**, 227 (1970).

CLARK, B. G.: An interferometer investigation of the 21-centimeter hydrogen-line absorption. Ap. J. **142**, 1398 (1965).

COHEN, M. H., SHAFFER, D. B.: Positions of radio sources from long-baseline interferometry. Astron. J. **76**, 91 (1971).

COMPTON, A. H.: A quantum theory of the scattering of X-rays by light elements. Phys. Rev. **21**, 483 (1923).

CONDON, E. U.: A theory of intensity distribution in band systems. Phys. Rev. **28**, 1182 (1926).

CONDON, E. U.: Nuclear motions associated with electron transitions in diatomic molecules. Phys. Rev. **32**, 858 (1928).

CONDON, E. U., SHORTLEY, G. H.: The theory of atomic spectra. Cambridge: Cambridge University Press 1963.

CORD, M. S., LOJKO, M. S., PETERSEN, J. D.: Microwave spectral tables—Spectral lines listing. Nat. Bur. Stands. (Wash.) Mon. 79, 1968.

CORLISS, C. H., BOZMAN, W. R.: Experimental transition probabilities for spectral lines of seventy elements. Nat. Bur. Stands. (Wash.) Mon. 53, 1962.

CORLISS, C. H., WARNER, B.: Absolute oscillator strengths for Fe I. Ap. J. Suppl. **8**, 395 (1964).

COWLEY, C. R., COWLEY, A. P.: A new solar curve of growth. Ap. J. **140**, 713 (1964).

COX, D. P., MATHEWS, W. G.: Effects of self-absorption and internal dust on hydrogen-line intensities in gaseous nebulae. Ap. J. **155**, 859 (1969).

COX, A. N., STEWART, J. N.: Rosseland opacity tables for population II compositions. Ap. J. Suppl. No. 174, **19**, 261 (1970).

CURTIS, H. D.: I. Descriptions of 762 nebulae and clusters, III. The planetary nebulae. Lick Obs. Bull. **13**, 11, 57 (1918).

CZYZAK, S. J., KRUEBER, T. K., MARTINS, P. DE A. P., SARAPH, H. E., SEATON, M. J., SHEMMING, J.: Collision strengths for excitation of forbidden lines. In: Planetary nebulae (ed. D. E. OSTERBROCK, C. R. O'DELL). I.A.U. Symp. No. 34. Dordrecht, Holland: D. Reidel. Berlin-Heidelberg-New York: Springer 1968.

DALGARNO, A., MCCRAY, R. A.: The formation of molecules from negative ions. Ap. J. **181**, 85 (1973).

DAVIES, P. C. W., SEATON, M. J.: Radiation damping in the optical continuum. J. Phys. B. (Atom. Molec. Phys.) **2**, 757 (1969).

DENNISON, D. M., UHLENBECK, G. E.: The two-minima problem and the ammonia molecule. Phys. Rev. **41**, 313 (1932).

DIRAC, P. A. M.: The effect of Compton scattering by free electrons in a stellar atmosphere. M.N.R.A.S. **85**, 825 (1925).

DIRAC, P. A. M.: The elimination of the nodes in quantum mechanics, Sec. 10 The relative intensities of the lines of a multiplet. Proc. Roy. Soc. London **A 111**, 302 (1926).

DOPPLER, C.: Über d. farbige Licht d. Doppelsterne usw. (On the colored light of double stars, etc.). Abhandlungen d. k. Böhmischen Gesell. d. Wiss. **2**, 467 (1842).

DRAVSKIKH, Z. V., DRAVSKIKH, A. F.: An attempt of observation of an excited hydrogen radio line Astron. Tsirk **282**, 2 (1964).

DUNHAM, T., ADAMS, W. S.: Interstellar neutral potassium and neutral calcium. P.A.S.P. **49**, 26 (1937).

DUNN, R. B., EVANS, J. W., JEFFERIES, J. T., ORRALL, F. Q., WHITE, O. R., ZIRKER, J. B.: The chromospheric spectrum at the 1962 Eclipse. Ap. J. Suppl. No. 139, **15**, 275 (1968).

DUPREE, A. K.: Radiofrequency recombination lines from heavy elements: Carbon. Ap. J. **158**, 491 (1969).

ECKER, G., MÜLLER, K. G.: Plasmapolarisation und Trägerwechselwirkung (Plasma polarization and carrier interaction). Z. Physik **153**, 317 (1958).

EDDINGTON, A. S.: On the radiative equilibrium of the stars. M.N.R.A.S. **77**, 16 (1917).

EDDINGTON, A.S.: The internal constitution of the stars. Cambridge: Cambridge University Press 1926, Reprod. New York: Dover.

EDDINGTON, A. S.: Interstellar matter. Obs. **60**, 99 (1937).

EDLÉN, B.: The dispersion of standard air. J. Opt. Soc. Amer. **43**, 339 (1953).

EINSTEIN, A.: Über einen die Erzeugung und Verwandlung des Lichtes betreffenden heuristischen Gesichtspunkt (On a heuristic point of view concerning the generation and transformation of light). Ann. Phys. **17**, 132 (1905).

EINSTEIN, A.: Über den Einfluß der Schwerkraft auf die Ausbreitung des Lichtes (On the influence of gravity on the propagation of light). Ann. Phys. **35**, 898 (1911). Trans. In: The principle of relativity, H. A. LORENTZ, A. EINSTEIN, H. MINKOWSKI, H. WEYL. New York: Dover 1952.

EINSTEIN, A.: Die Grundlage der allgemeinen Relativitätstheorie (The foundations of the general theory of relativity). Ann. Physik **49**, 769 (1916). Trans. in: The principle of relativity, H. A. LORENTZ, A. EINSTEIN, H. MINKOWSKI, H. WEYL. New York: Dover 1952.

EINSTEIN, A.: Zur Quantentheorie der Strahlung (On the quantum theory of radiation). Phys. Z. **18**, 121 (1917).

EKERS, J. A.: The Parkes catalogue of radio sources—declination zone $+20°$ to $-90°$. Austr. J. Phys. Suppl. No. 7 (1969).

ELANDER, N., SMITH, W. H.: Predissociation in the $C^2\Sigma^+$ state of CH and its astrophysical importance. Ap. J. **184**, 663 (1973).

EPSTEIN, P. S.: Zur Theorie des Starkeffektes (The theory of the Stark effect). Ann. Physik **50**, 489 (1916).
EPSTEIN, P. S.: The Stark effect from the point of view of Schroedinger's quantum theory. Phys. Rev. **28**, 695 (1926).
EWEN, H. I., PURCELL, E. M.: Radiation from galactic hydrogen at 1,420 MHz. Nature **168**, 356 (1951).
FERMI, E.: Über die magnetischen Momente der Atomkerne (On the magnetic moments of atomic nuclei). Z. Physik **60**, 320 (1930).
FIELD, G. B.: Excitation of the hydrogen 21-cm line. Proc. I. R. E. **46**, 240 (1958).
FIELD, G. B.: The spin temperature of intergalactic neutral hydrogen. Ap. J. **129**, 536 (1959).
FIELD, G. B., STEIGMAN, G.: Charge transfer and ionization equilibrium in the interstellar medium. Ap. J. **166**, 59 (1971).
FIELD, G. B., SOMERVILLE, W. B., DRESSLER, K.: Hydrogen molecules in astronomy. Ann. Rev. Astron. Astrophys. **4**, 207 (1966).
FINN, G. D., MUGGLESTONE, D.: Tables of the line broadening function H (a, v). M.N.R.A.S. **129**, 221 (1965).
FOLEY, H. M.: The pressure broadening of spectral lines. Phys. Rev. **69**, 616 (1946).
FOMALONT, E. B., MOFFET, A. T.: Positions of 352 small diameter sources. Astron. J. **76**, 5 (1971).
FORD, W. K., RUBIN, V. C., ROBERTS, M. S.: A comparison of 21-cm radial velocities and optical radial velocities of galaxies. Astron. J. **76**, 22 (1971).
FRANCK, J.: Elementary processes of photochemical reactions. Trans. Far. Soc. **21**, 536 (1925).
FRAUNHOFER, J.: Auszug davon in. Gilb. Ann. Physik **56**, 264 (1817).
GARSTANG, R. H.: Forbidden transitions. In: Atomic and molecular processes (ed. D. R. BATES). New York: Academic Press 1962.
GARSTANG, R. H.: Transition probabilities for forbidden lines. In: Planetary nebulae (ed. D. E. OSTERBROCK, C. R. O'DELL). I.A.U. Symp. No. 34. Dordrecht, Holland: D. Reidel. Berlin-Heidelberg-New York: Springer 1968.
GAUNT, J. A.: Continuous absorption. Phil. Trans. Roy. Soc. London **A 229**, 163 (1930).
GINGERICH, O., NOYES, R. W., KALKOFEN, W., CUNY, Y.: The Harvard-Smithsonian reference atmosphere. Solar Phys. **18**, 347 (1971).
GINGERICH, O., DE JAGER, C.: The Bilderberg model of the photosphere and low chromosphere. Solar Phys. **3**, 5 (1968).
GIOUMOUSIS, G., STEVENSON, D. P.: Reactions of gaseous molecule ions with gaseous molecules V. Theory. J. Chem. Phys. **29**, 294 (1958).
GLENNON, B. M., WIESE, W. L.: Bibliography on atomic transition probabilities. Nat. Bur. Stands. (Wash.). **Mon. 50**, 1962.
GOLDBERG, L.: Relative multiplet strengths in LS coupling. Ap. J. **82**, 1 (1935).
GOLDBERG, L.: Stimulated emission of radio-frequency lines of hydrogen. Ap. J. **144**, 1225 (1966).
GOLDBERG, L., DUPREE, A. K.: Population of atomic levels by dielectronic recombination. Nature **215**, 41 (1967).
GOLDBERG, L., MÜLLER, E. A., ALLER, L. H.: The abundances of the elements in the solar atmosphere. Ap. J. Suppl. No. 45, **5**, 1 (1960).
GOLDWIRE, H. C.: Oscillator strengths for electric dipole transitions of hydrogen. Ap. J. Suppl. No. 152, **17**, 445 (1968).
GOSS, W. M., FIELD, G. B.: Collisional excitation of low-energy permitted transitions by charged particles. Ap. J. **151**, 177 (1968).
GOULD, R. J., GOLD, T., SALPETER, E. E.: The interstellar abundance of the hydrogen molecule II. Galactic abundance and distribution. Ap. J. **138**, 408 (1963).
GOULD, R. J., SALPETER, E. E.: The interstellar abundance of the hydrogen molecule I. Basic processes. Ap. J. **138**, 393 (1963).
GREEN, L. C., RUSH, P. P., CHANDLER, C. D.: Oscillator strengths and matrix elements for the electric dipole moment of hydrogen. Ap. J. Suppl. No. 26, **3**, 37 (1957).
GREENSTEIN, J. L., OKE, J. B., SHIPMAN, H. L.: Effective temperature, radius, and gravitational redshift of Sirius B. Ap. J. **169**, 563 (1971).
GREENSTEIN, J. L., TRIMBLE, V. L.: The gravitational redshift of 40 Eridani B. Ap. J. **175**, L 1 (1972).
GRIEM, H. R.: Stark broadening of higher hydrogen and hydrogen-like lines by electrons and ions. Ap. J. **132**, 883 (1960).
GRIEM, H. R.: Wing formulae for Stark-broadened hydrogen and hydrogenic lines. Ap. J. **136**, 422 (1962).
GRIEM, H. R.: Plasma spectroscopy. New York: McGraw-Hill 1964.

GRIEM, H. R.: Corrections to the asymptotic Holtsmark formula for hydrogen lines broadened by electrons and ions in a plasma. Ap. J. **147**, 1092 (1967).

GRIEM, H. R.: Stark broadening by electron and ion impacts of n α hydrogen lines of large principal quantum number. Ap. J. **148**, 547 (1967).

GRIEM, H. R., KOLB, A. C., SHEN, K. Y.: Stark broadening of hydrogen lines in a plasma. Phys. Rev. **116**, 4 (1959).

GROTRIAN, W.: Gesetzmäßigkeiten in den Serienspektren (Regularities in the series spectra). In: Handbuch der Astrophysik. Berlin: Springer 1930.

HABING, H. J.: The interstellar radiation density between 912 Å and 2400 Å. B. A. N. **19**, 421 (1968).

HALE, G. E.: On the probable existence of a magnetic field in Sun-spots. Ap. J. **28**, 315 (1908).

HARRIS, D. L.: On the line-absorption coefficient due to Doppler effect and damping. Ap. J. **108**, 112 (1948).

HARTREE, D. R.: The wave mechanics of an atom with a non-Coulomb central field. Part. 1. Theory and methods. Proc. Camb. Phil. Soc. **24**, 89 (1928).

HEBB, M. H., MENZEL, D. H.: Physical processes in gaseous nebulae X. collisional excitation of nebulium. Ap. J. **92**, 408 (1940). Reprod. in: Selected papers on physical processes in ionized plasmas (ed. D. H. MENZEL). New York: Dover 1962.

HEILES, C., HABING, H. J.: To be publ. Astron. and Astrophys. (1974).

HEILES, C., JENKINS, E.: To be publ. Astron. and Astrophys. (1974).

HELD, E. F. M. VAN DER: Intensity and natural widths of spectral lines. Z. Phys. **70**, 508 (1931).

HENYEY, L. G.: The Doppler effect in resonance lines. Proc. Nat. Acad. Sci. **26**, 50 (1940).

HERZBERG, G.: Molecular spectra and molecular structure I. Spectra of diatomic molecules. New York: Van Nostrand 1950.

HIGGS, L. A.: A survey of microwave radiation from planetary nebulae. M. N. R. A. S. **153**, 315 (1971).

HIGGS, L. A.: A catalogue of radio observations of planetary nebulae and related optical data. Nat. Res. Council (Canada), NRC 12129, Astrophys. Branch, Ottawa, 1971.

HINDMAN, J. V., KERR, F. J., MCGEE, R. X.: A low resolution hydrogen-line survey of the Magellanic system. Austr. J. Phys. **16**, 570 (1963).

HINTEREGGER, H. E.: Absolute intensity measurements in the extreme ultraviolet spectrum of solar radiation. Space Sci. Rev. **4**, 461 (1965).

HJELLMING, R. M., ANDREWS, M. H., SEJNOWSKI, T. J.: Intensities of radio recombination lines. Astrophys. Lett., **3**, 111 (1969).

HJELLMING, R. M., GORDON, M. A.: Radio recombination lines and non-LTE theory: A reanalysis. Ap. J. **164**, 47 (1971).

HJERTING, F.: Tables facilitating the calculation of line absorption coefficients. Ap. J. **88**, 508 (1938).

HOLLENBACH, D. J., SALPETER, E. E.: Surface adsorption of light gas atoms. J. Chem. Phys. **53**, 79 (1970).

HOLLENBACH, D. J., SALPETER, E. E.: Surface recombination of hydrogen molecules. Ap. J. **163**, 155 (1971).

HOLLENBACH, D. J., WERNER, M. W., SALPETER, E. E.: Molecular hydrogen in H I regions. Ap. J. **163**, 165 (1971).

HOLSTEIN, T.: Pressure broadening of spectral lines. Phys. Rev. **79**, 744 (1950).

HOLTSMARK, J. VON: Über die Verbreiterung von Spektrallinien (On the broadening of spectral lines). Ann. Physik **58**, 577 (1919).

HOLTSMARK, J. VON: Über die Absorption in Na-Dampf (On the absorption in sodium vapor). Z. Physik **34**, 722 (1925).

HÖNL, H. VON: Die Intensitäten der Zeeman-Komponenten (The intensity of the Zeeman components). Z. Physik **31**, 340 (1925).

HOOPER, C. F. Low-frequency component electric microfield distributions in plasmas. Phys. Rev. **165**, 215 (1968).

HUANG, S. S., STRUVE, O.: Study of Doppler velocities in stellar atmospheres. The spectrum of Alpha Cygni, Ap. J. **121**, 84 (1955).

HUANG, S. S., STRUVE, O.: Stellar rotation and atmospheric turbulence. In: Stars and stellar systems, vol. 6 (ed. J. GREENSTEIN). Chicago, Ill.: Univ. of Chicago Press 1960.

HUGHES, M. P., THOMPSON, A. R., COLVIN, R. S.: An absorption line study of the galactic neutral hydrogen at 21 centimeters wavelength. Ap. J. Suppl. No. 200, **23**, 323 (1971).

HULST, H. C. VAN DE: Radiogloven mit hat wereldrium: I. Ontvangst der radiogloven; II. Herkomst der radiogloven (Radio waves from space: I. Reception of radiowaves; II. Origin of radiowaves). Ned. Tijdsch. Natuurk **11**, 201 (1945).

HULST, H. C. VAN DE: The solid particles in interstellar space. Rech. Astr. Obs. Utrecht XI, part 2 (1949).

HUMASON, M. L., MAYALL, N. U., SANDAGE, A. R.: Redshifts and magnitudes of extragalactic nebulae. Astron. J. **61**, 97 (1956).

HUMMER, D. G., SEATON, M. J.: The ionization structure of planetary nebulae: I. Pure hydrogen nebulae. M.N.R.A.S. **125**, 437 (1963).

HUMPHREYS, C. J.: The sixth series in the spectrum of atomic hydrogen. J. Res. Nat. Bur. Stands., **50**, 1 (1953).

HUND, F. VON: Zur Deutung der Molekelspektren II (On the interpretation of molecular spectra II). Z. Physik **42**, 93 (1927).

HUNGER, K.: The theory of curves of growth. Z. Ap. **39**, 38 (1956).

INGLIS, D. R., TELLER, E.: Ionic depression of series limits in one-electron spectra. Ap. J. **90**, 439 (1939).

JEFFERTS, K. B., PENZIAS, A. A., WILSON, R. W.: Deuterium in the Orion nebula. Ap. J. **179**, L 57 (1973).

JEFFRIES, J. T.: Spectral line formation. Waltham, Mass.: Blaisdell 1968.

JENKINS, F. A., SEGRE, É.: The quadratic Zeeman effect. Phys. Rev. **55**, 52 (1939).

DE JONG, T.: The density of H_2 molecules in dark interstellar clouds. Astron. Astrophys. **20**, 263 (1972).

JULIENNE, P. S., KRAUSS, M.: Molecule formation by inverse predissociation. In: Molecules in the galactic environment (ed. M. A. GORDON, L. E. SNYDER). New York: Wiley 1973.

JULIENNE, P. S., KRAUSS, M., DONN, B.: Formation of OH through inverse predissociation. Ap. J. **170**, 65 (1971).

KALER, J. B.: Chemical composition and the parameters of planetary nebulae. Ap. J. **160**, 887 (1970).

KARDASHEV, N. S.: On the possibility of detection of allowed lines of atomic hydrogen in the radio-frequency spectrum. Soviet Astron. **3**, 813 (1959).

KAUFMAN, F.: Elementary gas processes. Ann. Rev. Phys. Chem. **20**, 46 (1969).

KEMP, J. C.: Circular polarization of thermal radiation in a magnetic field. Ap. J. **162**, 169 (1970).

KEMP, J. C., SWEDLUND, J. B., LANDSTREET, J. D., ANGEL, J. R. P.: Discovery of circularly polarized light from a white dwarf. Ap. J. **161**, L 77 (1970).

KEPPLE, P. C.: Improved Stark-profile calculations for the He II lines at 256, 304, 1085, 1216, 1640, 3203 and 4686 Å. Phys. Rev. A **6**, 1 (1972).

KEPPLE, P. C., GRIEM, H. R.: Improved Stark profile calculations for the hydrogen lines $H\alpha$, $H\beta$, $H\gamma$ and $H\delta$. Phys. Rev. A **173**, 317 (1968).

KERR, F. J.: Radio line emission and absorption by the interstellar gas. In: Stars and stellar systems, vol. 7 (ed. G. P. KUIPER), Chicago, Ill.: Univ. of Chicago Press 1968.

KHINTCHINE, A.: Korrelationstheorie der stationären stochastischen Prozesse (Correlation theory of stationary stochastic process). Math. Annalen **109**, 604 (1934).

KIRCHHOFF, H.: On the simultaneous emission and absorption of rays of the same definite refrangibility. Phil. Mag. **19**, 193 (1860).

KIRCHHOFF, G., BUNSEN, R.: Chemical analysis by spectrum-observations. Phil. Mag. **22**, 329, 498 (1861).

KLEMPERER, W.: Interstellar molecule formation, radiative association and exchange reactions. In: Highlights of astronomy, vol. 2 (ed. C. DE JAGER). Dordrecht, Holland: D. Reidel (1971).

KNAPP, H. F. P., VAN DEN MEIJDENBERG, C. J. N., BEENAKKER, J. J. M., VAN DE HULST, H. C.: Formation of molecular hydrogen in interstellar space. B.A.N. **18**, 256 (1966).

KOEHLER, J. A.: Intergalactic atomic neutral hydrogen II. Ap. J. **146**, 504 (1966).

KOLB, A. C., GRIEM, H.: Theory of line broadening in multiplet spectra. Phys. Rev. **111**, 514 (1958).

KRAFT, R. P.: Studies in stellar rotation I. Comparison of rotational velocities in the Hyades and Coma clusters. Ap. J. **142**, 681 (1965).

KRAFT, R. P.: Stellar rotation. In: Spectroscopic astrophysics—An assessment of the contributions of Otto Struve. Berkeley: University of California Press 1970.

KRAMERS, H. A.: On the theory of X-ray absorption and of the continuous X-ray spectrum. Phil. Mag. **46**, 836 (1923).

KRAMERS, H. A.: The law of dispersion and Bohr's theory of spectra. Nature **113**, 783 (1924).

KRAMERS, H. A., TER HAAR, D.: Condensation in interstellar space. B.A.N. **10**, 137 (1946).

KRONIG, R. VON DE L.: Über die Intensität der Mehrfachlinien und ihrer Zeeman-Komponenten (On the intensity of multiple lines and their Zeeman components). Z. Physik **31**, 885 (1925).

KUHN, W. VON: Über die Gesamtstärke der von einem Zustande ausgehenden Absorptionslinien (On the total intensity of the absorption lines originating at one level). Z. Physik **33**, 408 (1925).

LADENBURG, R. VON: Die quantentheoretische Deutung der Zahl der Dispersionselektronen (The quantum theory interpretation of the number of scattering electrons). Z. Physik **4**, 451 (1921).

LAMBERT, D. L., WARNER, B.: The abundances of the elements in the solar photosphere II., III. M.N.R.A.S. **138**, 181, 213 (1968).

LANDAU, L. D., LIFSHITZ, E. M.: Classical theory of fields. New York: Pergamon Press 1962.

LANDÉ, A. VON: Über den anomalen Zeeman-Effekt (On the anomalous Zeeman effect). Z. Physik **5**, 231 (1920).

LANDÉ, A. VON: Termstruktur und Zeeman-Effekt der Multipletts (Term structure and Zeeman effect in multiplets). Z. Physik **19**, 112 (1923).

LAPORTE, O.: Die Struktur des Eisenspektrums (The structure of the iron spectrum). Z. Physik **23**, 135 (1924).

LARMOR, SIR J.: On the theory of magnetic influence on spectra and on the radiation from moving ions. Phil. Mag. **44**, 503 (1897).

LEDOUX, P., RENSON, P.: Magnetic stars. Ann. Rev. Astron. Astrophys. **4**, 326 (1966).

LEWIS, M., MARGENAU, H.: Statistical broadening of spectral lines emitted by ions in a plasma. Phys. Rev. **109**, 842 (1958).

LILLER, W.: The photoelectric photometry of planetary nebulae. Ap. J. **122**, 240 (1955).

LILLER, W., ALLER, L. H.: Photoelectric spectrophotometry of planetary nebulae. Ap. J. **120**, 48 (1955).

LILLEY, A. E., PALMER, P.: Tables of radio-frequency recombination lines. Ap. J. Suppl. No. 144, **16**, 143 (1968).

LINDHOLM, E.: Zur Theorie der Verbreiterung von Spektrallinien (On the theory of broadening of spectral lines). Arkiv. Mat. Astron. Physik **28** B, No. 3 (1941).

LITVAK, M. M.: Infrared pumping of interstellar OH. Ap. J. **156**, 471 (1969).

LORENTZ, H. A.: Influence du champ magnétique sur l'emission lumineuse (On the influence of magnetic forces on light emission). Rev. d'electricité **14**, 435 (1898) and also Ann. Physik **63**, 278 (1897).

LORENTZ, H. A.: Electromagnetic phenomena in a system moving with any velocity smaller than that of light. Proc. Am. Acad. Sci. **6**, 809 (1904). Reprod. in: The principle of relativity (H. A. LORENTZ, A. EINSTEIN, H. MINKOWSKI, H. WEYL). New York: Dover 1952.

LORENTZ, H. A.: The absorption and emission lines of gaseous bodies. Proc. Am. Acad. Sci. **8**, 591 (1906).

LORENTZ, H. A.: The theory of electrons. 1909. Republ. New York: Dover 1952.

LUTZ, B. L.: Radiative association and formation of diatomic molecules. Ann. N. Y. Acad. Sci. **194**, 29 (1972).

LYMAN, T.: The spectrum of hydrogen in the region of extremely short wave-length. Ap. J. **23**, 181 (1906).

MARGENAU, H.: Van der Waals forces. Rev. Mod. Phys. **11**, 1 (1939).

MARGENAU, H., LEWIS, M.: Structure of spectral lines from plasmas. Rev. Mod. Phys. **31**, 579 (1959).

MAXWELL, J. C.: Illustrations of the dynamical theory of gases—Part I. On the motions and collisions of perfectly elastic spheres. Phil. Mag. **19**, 19 (1860).

MAYALL, N. U., DE VAUCOULEURS, A.: Redshifts of 92 galaxies. Astron. J. **67**, 363 (1962).

MAYER, M. G.: Über Elementarakte mit zwei Quantensprüngen (On elementary processes with two quantum emissions). Ann. Phys. **9**, 273 (1931).

MCNALLEY, D.: Interstellar molecules. In: Advances in astronomy and astrophysics, vol. 6 (ed. Z. KOPAL). New York: Academic Press 1968.

MEGGERS, W. F., CORLISS, C. H., SCRIBNER, B. F.: Tables of spectral line intensities. Nat. Bur. Stands. Mon. 32 (Wash.), 1961.

MENZEL, D. H.: The theoretical interpretation of equivalent breadths of absorption lines. Ap. J. **84**, 462 (1936).

MENZEL, D. H.: Extended sum rules for transition arrays. Ap. J. **105**, 126 (1947).

MENZEL, D. H.: Oscillator strengths, f, for high-level transitions in hydrogen. Ap. J. Suppl. No. 161, **18**, 221 (1969).

MENZEL, D. H., GOLDBERG, L.: Multiplet strengths for transitions involving equivalent electrons. Ap. J. **84**, 1 (1936).

MENZEL, D. H., PEKERIS, C. L.: Absorption coefficients and hydrogen line intensities. M.N.R.A.S. **96**, 77 (1935). Reprod. In: Selected papers on physical processes in ionized plasmas (ed. D. H. MENZEL). New York: Dover 1962.

MERRILL, P. W.: Lines of chemical elements in astronomical spectra. Washington, D. C.: Carnegie Inst. of Wash. Publ. 610, 1958.

MILEY, G. K.: The radio structure of quasars—a statistical investigation. M.N.R.A.S. **152**, 477 (1971).

MILNE, D. ALLER, L. H.: to be publ.

MILNE, E. A.: Radiative equilibrium in the outer layers of a star: The temperature distribution and the law of darkening. M.N.R.A.S. **81**, 361 (1921).

MILNE, E. A.: Thermodynamics of the stars. In: Handbuch der Astrophysik. **3**, 80 (1930). Reprod. In: Selected papers on the transfer of radiation (ed. D. H. MENZEL). New York: Dover 1966.

MIHALAS, D.: Stellar atmospheres. San Francisco, Calif.: W. H. Freeman 1970.

MITLER, H. E.: The solar light element abundances and primeval helium. Smithsonian Astrophys. Obs. Rpt. 323 (1970).

MOFFET, A.: Strong nonthermal radio emission from galaxies. In: Galaxies and the universe—Stars and stellar systems, vol. IX (ed. A. R. SANDAGE). Chicago, Ill.: Univ. of Chicago Press 1974.

MOORE, C. E.: Atomic energy levels—Vol. I, II, III. Nat. Bur. Stands. (Wash.) 1949, 1952, 1959.

MOORE, C. E.: An ultraviolet multiplet table. Nat. Bur. Stands. (Wash.) 1950–69.

MOORE, C. E., MINNAERT, M. G. J., HOUTGAST, J.: The solar spectrum 2935 Å to 8770 Å. Nat. Bur. Stands. (Wash.) Mon. 61, 1966.

MOORE, C. E., MERRILL, P. W.: Partial Grotrian diagrams of astrophysical interest. Nat. Bur. Stands. (Wash.) NSRDS-NBS. 23, 1968.

MORGAN, W. W., KEENAN, P. C., KELLMAN, E.: An atlas of stellar spectra. Chicago, Ill.: University of Chicago Press 1943.

MORSE, P. M.: Diatomic molecules according to the wave mechanics: II. Vibrational levels. Phys. Rev. **34**, 57 (1929).

MOTTEMANN, J.: Thesis, Dept. of Astron., U.C.L.A. 1972.

MOZER, B., BARANGER, M.: Electric field distributions in an ionized gas. Phys. Rev. **118**, 626 (1960).

NELSON, R. D., LIDE, D. R., MARYOTT, A. A.: Selected values of electric dipole moments for molecules in the gas phase, NSRDS-NBS 10. Nat. Bur. Stands. (Wash.), 1967.

O'DELL, C. R.: A distance scale for planetary nebulae based on emission line fluxes. Ap. J. **135**, 371 (1962).

O'DELL, C. R.: Photoelectric spectrophotometry of planetary nebulae. Ap. J. **138**, 1018 (1963).

ORNSTEIN, L. S., BURGER, H. C.: Strahlungsgesetz und Intensität von Mehrfachlinien (The radiation laws and intensities of multiple lines). Z. Physik **24**, 41 (1924).

ORNSTEIN, L. S., BURGER, H. C.: Nachschrift zu der Arbeit Intensität der Komponenten im Zeeman-Effekt (Postscript to the work on intensities of the components in the Zeeman effect). Z. Physik **29**, 241 (1924).

OSTERBROCK, D. E.: Expected ultraviolet emission spectrum of a gaseous nebulae. Planet Space Sci. **11**, 621 (1963).

PAGEL, B. E. J., Revised abundance analysis of the halo red-giant HD 122563. Roy. Obs. Bul.. No. 104 (1965).

PALMER, P., ZUCKERMAN, B., PENFIELD, H., LILLEY, A. E., MEZGER, P. G.: Detection of a new microwave spectral line. Nature **215**, 40 (1967).

PASCHEN, F. VON: Zur Kenntnis ultraroter Linienspektra I (On the knowledge of infrared line spectra I). Ann. Physik **27**, 537 (1908).

PASCHEN, F. VON, BACK, E.: Normale und anomale Zeeman-Effekte (Normal and anomalous Zeeman effect). Ann. Physik **39**, 929 (1912), Ann. Physik **40**, 960 (1913).

PAULI, W.: Über den Zusammenhang des Abschlusses der Elektronengruppen im Atom mit der Komplexstruktur der Spektren (On the connection of the termination of electron groups in atoms with the complex structure of spectra). Z. Physik **31**, 765 (1925).

PAULI, W.: Zur Quantenmechanik des magnetischen Elektrons (On the quantum mechanics of the magnetic electron). Z. Physik **43**, 601 (1927).

PEACH, G.: The broadening of radio recombination lines by electron collisions. Astrophys. Lett. **10**, 129 (1972).

PEDLAR, A., DAVIES, R. D.: Stark broadening in high quantum number recombination lines of hydrogen. Nature-Phys. Sci. **231**, 49 (1971).

PENGALLY, R. M.: Recombination spectra I. Calculations for hydrogenic ions in the limit of low densities. M.N.R.A.S. **127**, 145 (1964).

PENZIAS, A. A., WILSON, R. W.: Intergalactic H I emission at 21 centimeters. Ap. J. **156**, 799 (1969).

PEREK, L.: Photometry of southern planetary nebulae. Czech. Astr. Bull. No. 22, 103 (1971).

PEREK, L., KOHOUTEK, L.: Catalogue of planetary nebulae. Prague: Czech. Acad. Sci. 1967.

PETERSON, B. A., BOLTON, J. G.: Redshifts of southern radio sources. Ap. J. **173**, L 19 (1972).

PFUND, A. H.: The emission of nitrogen and hydrogen in the infrared. J. Opt. Soc. Am. **9**, 193 (1924).

PICKERING, E. C.: Stars having peculiar spectra, new variable stars in Crux and Cygnus. Ap. J. **4**, 369 (1896).

PICKERING, E. C.: The spectrum of ξ Puppis. Ap. J. **5**, 92 (1897).

PLANCK, M.: Über das Gesetz der Energieverteilung im Normalspectrum (On the law of energy distribution in a normal spectrum). Ann. Physik **4**, 553 (1901).

PLANCK, M.: Zur Theorie der Wärmestrahlung (On the theory of thermal radiation). Ann. Physik **31**, 758 (1910).

PLANCK, M.: The theory of heat radiation. 1913. Reprod. New York: Dover 1969.

Planetary atmospheres. Pub. 1688 of Nat. Acad. Sci. (Wash.) 1968.

POLANYI, J. C.: Chemical processes. In: Atomic and molecular processes (ed. D. R. BATES). New York: Academic Press 1962.

POPPER, D. M.: Red shift in the spectrum of 40 Eridani B. Ap. J. **120**, 316 (1954).

PRESTON, G. W.: The quadratic Zeeman effect and large magnetic fields in white dwarfs. Ap. J. **160**, L 143 (1970).

PRESTON, T.: Radiation phenomena in the magnetic field. Phil. Mag. **45**, 325 (1898).

PURCELL, E. M.: The lifetime of the $2^2S_{1/2}$ state of hydrogen in an ionized atmosphere. Ap. J. **116**, 457 (1952).

RADHAKRISHNAN, V., BROOKS, J. W., GOSS, W. M., MURRAY, J. D., SCHWARZ, U. J.: The Parkes survey of 21-centimeter absorption in discrete-source spectra. Ap. J. Suppl. No. 203, **24**, 1 (1972).

RAPP, D., FRANCIS, W. E.: Charge exchange between gaseous ions and atoms. J. Chem. Phys. **37**, 2631 (1962).

RAYLEIGH, LORD: On the limit to interference when light is radiated from moving molecules. Phil. Mag. **27**, 298 (1889).

RAYLEIGH, LORD: On the visibility of faint interference bands. Phil. Mag. **27**, 484 (1889).

RAYLEIGH, LORD: On the widening of spectrum lines. Phil. Mag. **29**, 274 (1915).

REIFENSTEIN, E. C., WILSON, T. L., BURKE, B. F., MEZGER, P. G., ALTENHOFF, W. J.: A survey of H 109 α recombination line emission in galactic H II regions of the northern sky. Astron. Astrophys. **4**, 357 (1970).

RITZ, W.: On a new law of series spectra. Ap. J. **28**, 237 (1908).

ROBERTS, M. S.: Neutral hydrogen observations of the binary galaxy system NGC 4631/4656. Ap. J. **151**, 117 (1968).

ROGERS, A. E. E., BARRETT, A. H.: Excitation temperature of the 18-cm line of OH in H I regions. Ap. J. **151**, 163 (1968).

ROGERSON, J.: On the abundance of iron in the solar photosphere. Ap. J. **158**, 797 (1969).

ROHRLICH, F.: Sum rules for multiplet strengths. Ap. J. **124**, 449 (1959).

ROSSELAND, S.: Note on the absorption of radiation within a star, M.N.R.A.S. **84**, 525 (1924). Reprod. in: Selected papers on the transfer of radiation (ed. D. H. MENZEL). New York: Dover 1966.

RUSSELL, H. N.: The intensities of lines in multiplets. Nature **115**, 835 (1925).

RUSSELL, H. N.: On the composition of the Sun's atmosphere. Ap. J. **70**, 11 (1929).

RUSSELL, H. N.: Tables for intensities of lines in multiplets. Ap. J. **83**, 129 (1936).

RUSSELL, H. N., SAUNDERS, F. A.: New regularities in the spectra of the alkaline earths. Ap. J. **61**, 38 (1925).

RYDBERG, J. R.: On the structure of the line-spectra of the chemical elements. Phil. Mag. **29**, 331 (1890).

RYDBERG, J. R.: The new elements of Clevèite gas. Ap. J. **4**, 91 (1896).

SAHA, M. N.: On the physical theory of stellar spectra. Proc. Roy. Soc. London A **99**, 135 (1921).

SARGENT, W. L. W.: The atmospheres of the magnetic and metallic-line stars. Ann. Rev. Astron. Astrophys. **2**, 297 (1964).

SARGENT, W. L. W.: Redshifts for 51 galaxies identified with radio sources in the 4C catalog. Ap. J., **182**, L 13 (1973).

SCHLESINGER, F.: The algol-variable δ Libre. Pub. Allegheny Obs. **1**, 125 (1909).

SCHRÖDINGER, E. VON: Quantisierung als Eigenwertproblem (Quantisation as an eigenvalue problem). Ann. Physik **79**, 361, **80**, 437, **81**, 109 (1925, 1926).

SCHUSTER, A.: Radiation through a foggy atmosphere, Ap. J. **21**, 1 (1905) Reprod. in: Selected papers on the transfer of radiation (ed. D. H. MENZEL). New York: 1966.

SCHWARZSCHILD, K. VON: Über das Gleichgewicht der Sonnenatmosphäre (On the equilibrium of the solar atmosphere). Nach. Ges. Gott. **195**, 41 (1906). Engl. trans. in: Selected papers on the transfer of radiation (ed. D. H. MENZEL). New York: Dover 1966.

SCHWARZSCHILD, K. VON: Über Diffusion und Absorption in der Sonnenatmosphäre (On scattering and absorption in the solar atmosphere). Sitz. Akad. Wiss. 1183 (1914), Engl. trans. in: Selected paper on the transfer of radiation (ed. D. H. MENZEL). New York: Dover 1966.

SCHWARZSCHILD, K. VON: Zur Quantenhypothese (On the quantum hypothesis). Sitz. Akad. Wiss. 548 (1916).

SEARES, F. H.: The displacement-curve of the sun's general magnetic field. Ap. J. **38**, 99 (1913).

SEATON, M. J.: Thermal inelastic collision processes. Rev. Mod. Phys. **30**, 979 (1958).

SEATON, M. J.: Planetary nebulae. Rep. Prog. Phys. **23**, 314 (1960).

SEATON, M. J.: The theory of excitation and ionization by electron impact. In: Atomic and molecular processes (ed. D. R. Bates). New York: Academic Press 1962.

SEATON, M. J.: Recombination spectra. M. N. R. A. S. **127**, 177 (1964).

SEJNOWSKI, T. J., HJELLMING, R. M.: The general solution of the b_n problem for gaseous nebulae. Ap. J. **156**, 915 (1969).

SHAJN, G., STRUVE, O.: On the rotation of the stars. M. N. R. A. S. **89**, 222 (1929).

SHAPLEY, H., NICHOLSON, S. B.: On the spectral lines of a pulsating star. Comm. Nat. Acad. Sci. Mt. Wilson Obs. **2**, 65 (1919).

SHIELDS, G. A., OKE, J. B., SARGENT, W. L. W.: The optical spectrum of the Seyfert galaxy 3C 120. Ap. J. **176**, 75 (1972).

SHORTLEY, G. H.: Line strengths in intermediate coupling. Phys. Rev. **47**, 295, 419 (1935).

SHORTLEY, G. H.: The computation of quadrupole and magnetic-dipole transition probabilities. Phys. Rev. **57**, 225 (1940).

SLETTEBAK, A.: On the axial rotation of the brighter O and B stars. Ap. J. **110**, 498 (1949).

SLETTEBAK, A.: The spectra and rotational velocities of the bright stars of Draper types B8–A2. Ap. J. **119**, 146 (1954).

SLETTEBAK, A.: The spectra and rotational velocities of the bright stars of Draper types A3–G0. Ap. J. **121**, 653 (1955).

SLETTEBAK, A., ed.: Proceedings I.A.U. colloquium on stellar rotation. New York: Gordon and Breach 1970.

SLETTEBAK, A., HOWARD, R. F.: Axial rotation in the brighter stars of Draper types B2–B5. Ap. J. **121**, 102 (1955).

SMERD, S. F., WESTFOLD, K. C.: The characteristics of radio-frequency radiation in an ionized gas, with applications to the transfer of radiation in the solar atmosphere. Phil. Mag. **40**, 831 (1949).

SMITH, A. M., STECHER, T. P.: Carbon monoxide in the interstellar spectrum of Zeta Ophiuchi. Ap. J. **164**, L 43 (1971).

SMITH, W. H., LISZT, H. S., LUTZ, B. L.: A reevaluation of the diatomic processes leading to CH and CH^+ formation in the interstellar medium. Ap. J. **183**, 69 (1973).

SNYDER, L. T., in: M.T.P. International review of science, sec. 1, **3**, ch. 6, 1973.

SOLOMON, P., KLEMPERER, W.: The formation of diatomic molecules in interstellar clouds. Ap. J. **178**, 389 (1972).

SOMMERFELD, A. VON: Zur Quantentheorie der Spektrallinien (On the quantum theory of spectral lines). Ann. Phys. **50**, 385, **51**, 1 (1916).

SPITZER, L.: Impact broadening of spectral lines. Phys. Rev. **58**, 348 (1940).

SPITZER, L., GREENSTEIN, J. L.: Continuous emission from planetary nebulae. Ap. J. **114**, 407 (1951).

SPITZER, L., DRAKE, J. F., JENKINS, E. B., MORTON, D. C., ROGERSON, J. B., YORK, D. G.: Spectrophotometric results from the Copernicus satellite IV molecular hydrogen in interstellar space. Ap. J. **181**, L 116 (1973).

STARK, J. VON: Beobachtungen über den Effekt des elektrischen Feldes auf Spektrallinien (Observations of the effect of the electric field on spectral lines). Sitz. Akad. Wiss. **40**, 932 (1913).

STECHER, P., WILLIAMS, D. A.: Interstellar molecule formation. Ap. J. **146**, 88 (1968).

STIEF, L. J., DONN, B., GLICKER, S., GENTIEU, E. P., MENTALL, J. E.: Photochemistry and lifetimes of interstellar molecules. Ap. J., **171**, 21 (1972).

STIEF, L. J.: Photochemistry of interstellar molecules. In: Molecules in the galactic environment (ed. M. A. GORDON, L. E. SNYDER). New York: Wiley 1973.

STRIGANOV, A. R., SVENTITSKII, N. S.: Tables of spectral lines of neutral and ionized atoms. New York: IFI/Plenum 1968.

STRUVE, O.: The Stark effect in stellar spectra. Ap. J. **69**, 173 (1929).

STRUVE, O.: On the axial rotation of stars. Ap. J. **72**, 1 (1930).

STRUVE, O., ELVEY, C. T.: The intensities of stellar absorption lines. Ap. J. **79**, 409 (1934).

SWINGS, P.: Considerations regarding cometary and interstellar molecules. Ap. J., **95**, 270 (1942).

TARTER, C. B.: Ultraviolet and forbidden-line intensities. Ap. J. Suppl. No. 154, **18**, 1 (1969).

TERZIAN, Y.: Observations of planetary nebulae at radio wavelengths. In: Planetary nebulae (ed. D. E. OSTERBROCK, C. R. O'DELL), I.A.U. Symp. No. 34. Dordrecht, Holland: D. Reidel. Berlin-Heidelberg-New York: Springer 1968.

THOMAS, L. H.: The motion of the spinning electron. Nature **117**, 514 (1926).

THOMAS, L. H.: The kinematics of an electron with an axis. Phil. Mag. **1**, 1 (1927).

THOMAS, W.: Über die Zahl der Dispersionselektronen die einem stationären Zustande zugeordnet sind (On the number of scattering electrons which are associated with a stationary state). Naturwiss. **13**, 627 (1925).

THOMSON, J. J.: Conduction of electricity through gases. Cambridge: Cambridge University Press, 1903. Republ. New York: Dover 1969.

TOWNES, C. H.: Microwave and radiofrequency resonance lines of interest to radio astronomy. In: I.A.U. Symp. No. 4 (ed H. C. VAN DE HULST). Cambridge: Cambridge University Press 1957.

TOWNES, C. H., SCHAWLOW, A. L.: Microwave spectroscopy. New York: McGraw-Hill 1955.

TRIMBLE, V. L., THORNE, K. S.: Spectroscopic binaries and collapsed stars. Ap. J. **156**, 1013 (1969).

TURNER, B. E.: Fifty new OH sources associated with H II regions. Astrophys. Lett **6**, 99 (1970).

UHLENBECK, G. E., GOUDSMIT, S.: Zuschriften und vorläufige Mitteilungen (Replacement of the hypothesis of nonmechanical constraint by a requirement referring to the inner properties of each individual electron). Naturwiss. **13**, 953 (1925).

UHLENBECK, G. E., GOUDSMIT, S.: Spinning electrons and the structure of spectra. Nature **117**, 264 (1926).

UNDERHILL, A. B., WADDELL, J. H.: Stark broadening functions for the hydrogen lines. Nat. Bur. Stands. (Wash.), Circ. 603, 1959.

UNSÖLD, A.: Stellar abundances and the origin of the elements. Science **163**, 1015 (1969).

VAN VLECK, J. H.: On σ-type doubling and electron spin in the spectra of diatomic molecules. Phys. Rev. **33**, 467 (1929).

VAN VLECK, J. H.: Theory of electric and magnetic susceptibilities. London: Oxford University Press, 1932.

VARSAVSKY, M.: Some atomic parameters for ultraviolet lines. Ap. J. Suppl. **6**, 75 (1961).

DE VAUCOULEURS, G., DE VAUCOULEURS, A.: Reference catalogue of bright galaxies. Austin, Texas: Univ. of Texas Press 1964.

DE VENY, J. B., OSBORN, W. H., JANES, K.: A catalogue of quasars. P.A.S.P. **83**, 611 (1971).

VERSCHUUR, G. L. Positive determination of an interstellar magnetic field by measurement of the Zeeman splitting of the 21-cm hydrogen line. Phys. Rev. Lett. **21**, 775 (1968).

VIDAL, C. R., COOPER, J., SMITH, E. W.: Hydrogen Stark—broadening tables. Ap. J. Suppl. No. 214, **25**, 37 (1973).

VOIGT, W.: Über die Intensitätsverteilung innerhalb einer Spektrallinie (Distribution of intensity within a spectrum line. Phys. Zeits. **14**, 377 (1913).

WAMPLER, E. J., ROBINSON, L. B., BALDWIN, J. A., BURBIDGE, E. M.: Redshift of OQ 172. Nature **243**, 336 (1973).

WATSON, W. D., SALPETER, E. E.: Molecule formation on interstellar grains. Ap. J. **174**, 321 (1972).

WATSON, W. D., SALPETER, E. E.: On the abundances of interstellar molecules. Ap. J. **175**, 659 (1972).

WEAVER, H. F., WILLIAMS, D. R. W.: The Berkeley low latitude survey of neutral hydrogen part I. Profiles. Astron. Astrophys. Suppl. **8**, 1 (1973).

WEINREB S., BARRETT, A. H., MEEKS, M. L., HENRY, J. C.: Radio observations of OH in the interstellar medium. Nature **200**, 829 (1963).

WEISSKOPF, V. VON: Die Breite der Spektrallinien in Gasen (The width of spectral lines in gases). Phys. Zeits. **34**, 1 (1933).

WEISSKOPF, V. VON: The intensity and structure of spectral lines. Observatory **56**, 291 (1933).

WEISSKOPF, V. VON, WIGNER, E.: Berechnung der natürlichen Linienbreite auf Grund der Diracschen Lichttheorie (Calculation of the natural line width on the basis of Dirac's theory of light). Z. Phys. **63**, 54 (1930).

WENTZEL, D. G.: An upper limit on the abundance of H_2 formed by chemical exchange. Ap. J. **150**, 453 (1967).

WERNER, M. W.: Ionization equilibrium of carbon in interstellar clouds. Astrophys. Lett. **6**, 81 (1970).

WESTERHOUT, G.: Maryland-Greenbank 21-cm line survey. College Park, Maryland: Univ. Maryland 1969.

WHITE, H. E., ELIASON, A. Y.: Relative intensity tables for spectrum lines. Phys. Rev. **44**, 753 (1933).

WHITFORD, A. E.: An extension of the interstellar absorption-curve. Ap. J. **107**, 102 (1948).

WHITFORD, A. E.: The law of interstellar reddening. Astron. J. **63**, 201 (1958).

WIENER, N.: Generalized harmonic analysis. Acta. Math. Stockholm **55**, 117 (1930).

WIESE, W. L., SMITH, M. W., GLENNON, B. M.: Atomic transition probabilities, vol. 1. Hydrogen through helium. Nat. Bur. Stands. (Wash.), NSRDS-NBS 4 1966.

WILD, J. P.: Radio-frequency line spectrum of atomic hydrogen and its applications in astronomy. Ap. J. **115**, 206 (1952).

WILKINSON, P. G.: Diatomic molecules of astrophysical interest: Ionization potentials and dissociation energies. Ap. J. **138**, 778 (1963).

WILSON, O. C.: A survey of internal motions in the planetary nebulae. Ap. J. **111**, 279 (1950).

WILSON, O. C.: Kinematic structures of gaseous envelopes—internal kinematics of the planetary nebulae. Rev. of Mod. Phys. **30**, 1025 (1958).

WILSON, T. L., MEZGER, P. G., GARDNER, F. F., MILNE, D. K.: A survey of H 109 α recombination line emission in galactic H II regions of the southern sky. Astron. Astrophys. **6**, 364 (1970).

WILSON, R. W., PENZIAS, A. A., JEFFERTS, K. P., SOLOMON, P. M.: Interstellar deuterium: The hyperfine structure of DCN. Ap. J. **179**, L 107 (1973).

WOOLLEY, R. V. D. R.: The solar corona. Suppl. Austr. J. of Sci. **10**, 2 (1947).

WRUBEL, M. H.: Exact curves of growth for the formation of absorption lines according to the Milne-Eddington model: II. Center of the disk. Ap. J. **111**, 157 (1950).

WRUBEL, M. H.: Exact curves of growth: III. The Schuster-Schwarzschild model. Ap. J. **119**, 51 (1954).

WYNDHAM, J. D.: Optical identification of radio sources in the 3 C revised catalogue. Ap. J. **144**, 459 (1966).

ZANASTRA, H.: An application of the quantum theory to the luminosity of diffuse nebulae. Ap. J. **65**, 50 (1927).

ZEEMAN, P.: On the influence of magnetism on the nature of the light emitted by a substance. Phil. Mag. **43**, 226 (1896).

ZEEMAN, P.: Doublets and triplets in the spectrum produced by external magnetic forces I., II. Phil. Mag. **44**, 55, 255 (1897).

Chapter 3

ALFVÉN, H.: On the existence of electromagnetic-hydrodynamic waves. Arkiv. f. Mat., Astron., Physik. **29**, 1 (1942).

ALFVÉN, H.: The existence of electromagnetic-hydrodynamic waves. Nature **150**, 405 (1942).

ALFVÉN, H.: Granulation, magneto-hydrodynamic waves, and the heating of the solar corona. M.N.R.A.S. **107**, 211 (1947).

ALFVÉN, H., FALTHAMMER, C. G.: Cosmical electrodynamics. Oxford: Oxford at the Clarendon Press 1963.

ALLEN, C. W.: Interpretation of electron densities from corona brightness. M.N.R.A.S. **107**, 426 (1947).

ANDERSON, W. VON: Gewöhnliche Materie und strahlende Energie als verschiedene "Phasen" eines und desselben Grundstoffes (Common matter and radiated energy as different "phases" of the same chemical element). Z. Physik **54**, 433 (1929).

ANGEL, J. R. P., LANDSTREET, J. D.: Detection of circular polarization in a second white dwarf. Ap. J. **164**, L 15 (1971).

ANGEL, J. R. P., LANDSTREET, J. D.: Discovery of circular polarization in the red degenerate star G 99-47. Ap. J. **178**, L 21 (1972).

ASTRÖM, E.: On waves in an ionized gas. Arkiv. Physik **2**, 443 (1950).

AVOGADRO, A.: Essay on a manner of determining the relative masses of the elementary molecules of bodies and the proportions in which they enter into these compounds. J. de. Physique **73**, 58 (1811). Engl. trans. in: Foundations of the molecular theory. Alemic Club. Repr. No. 4. Edinburgh: Bishop 1950.

AXFORD, W. I.: Ionization fronts in interstellar gas: The structure of ionization fronts. Phil. Trans. Roy. Soc. London A **253**, 301 (1961).

AXFORD, W. I.: The initial development of H II regions. Ap. J. **139**, 761 (1964).

BABCOCK, H. W., BABCOCK, H. D.: The Sun's magnetic field, 1952–1954. Ap. J. **121**, 349 (1955).

BAILEY, S. I.: The periods of the variable stars in the cluster Messier 5. Ap. J. **10**, 255 (1899).

BAILEY, V. A.: Plane waves in an ionized gas with static electric and magnetic fields present. Austr. J. Sci. Res. A **1**, 351 (1948).

BAKER, N., KIPPENHAHN, R.: The pulsation of models of δ Cephei stars. Z. Ap. **54**, 114 (1962).

BAKER, N., KIPPENHAHN, R.: The pulsations of models of Delta Cephei stars II. Ap. J. **142**, 868 (1965).

BARDEEN, J., COOPER, L. N., SCHRIEFFER, J. R.: Theory of superconductivity. Phys. Rev. **108**, 1175 (1957).

BARKAT, Z., BUCHLER, J. R., INGBER, L.: Equation of state of neutron-star matter at subnuclear densities. Ap. J. **176**, 723 (1972).

BARNES, A.: Collisionless heating of the solar-wind plasma: I. Theory of the heating of collisionless plasma by hydromagnetic waves. Ap. J. **154**, 751 (1968).

BARNES, A.: Collisionless heating of the solar-wind plasma: II. Application of the theory of plasma heating by hydromagnetic waves. Ap. J. **155**, 311 (1969).

BATCHELOR, G. K., The theory of homogeneous turbulence. Cambridge: Cambridge University Press 1967.

BAUMBACH, S. VON: Strahlung, Ergiebigkeit und Elektronendichte der Sonnenkorona (Radiation, abundance, and electron density of the solar corona). Astron. Nach. **263**, 120 (1937).

BAYM, G., BETHE, H. A., PETHICK, C. J.: Neutron star matter. Nuclear Phys. A **175**, 225 (1971).

BAYM, G., PETHICK, C. J., SUTHERLAND, P.: The ground state of matter at high densities: Equation of state and stellar models. Ap. J. **170**, 299 (1971).

BÉLOPOLSKY, A.: Researches on the spectrum of the variable star η Aquilae. Ap. J. **6**, 393 (1897).

BERNOULLI, D.: Hydrodynamia Argentorati (Hydrodynamics) 1738.

BIERMANN, L.: Zur Deutung der chromosphärischen Turbulenz und des Exzesses der UV-Strahlung der Sonne (An explanation of chromospheric turbulence and the UV excess of solar radiation). Naturwiss. **33**, 118 (1946).

BIERMANN, L.: Über die Ursache der chromosphärischen Turbulenz und des UV-Exzesses der Sonnenstrahlung (About the cause of the chromospheric turbulence and the UV excess of the solar radiation). Z. Ap. **25**, 161 (1947).

BIERMANN, L.: Kometenschweife und solare Korpuskular-Strahlung (The tails of comets and the solar corpuscular radiation). Z. Ap. **29**, 274 (1951).

BIERMANN, L.: Solar corpuscular radiation and the interplanetary gas. Observatory **77**, 109 (1957).

BIRKELAND, K.: The Norwegian aurora polar expedition 1902–1903: I. On the cause of magnetic storms and the origin of terrestial magnetism. Christiana: H. Aschehong 1908.

BOGOLYUBOV, N. N.: Problems of a dynamical theory in statistical physics. State Tech. Press, Moscow 1946.

BOHM, D., GROSS, E. P.: Theory of plasma oscillations: A. Origin of medium-like behavior; B. Excitation and damping of oscillations. Phys. Rev. **75**, 1851, 1854 (1949).

BÖHM-VITENSE, E.: Die Wasserstoffkonvektionszone der Sonne (The hydrogen convection zone of the Sun). Z. Ap. **32**, 135 (1953).

BÖHM-VITENSE, E.: Über die Wasserstoffkonvektionszone in Sternen verschiedener Effektivtemperaturen und Leuchtkräfte (About the hydrogen convection zone in stars of various effective temperatures and luminosities). Z. Ap. **46**, 108 (1958).

BOLTZMANN, L.: Studien über das Gleichgewicht der lebendigen Kraft zwischen bewegten materiellen Punkten (Studies of the equilibrium and the life force between material points). Wien. Ber. **58**, 517 (1868).

BOLTZMANN, L.: Weitere Studien über das Wärmegleichgewicht unter Gasmolekülen (Further studies on the thermal equilibrium of gas molecules). Sitz. Acad. Wiss. **66**, 275 (1872). Engl. trans. Brush (1966).

BOLTZMANN, L.: Über die Beziehung eines allgemeinen mechanischen Satzes zum zweiten Hauptsatz der Wärmetheorie (On the relation of a thermal mechanical theorem to the second law of thermodynamics). Sitz. Acad. Wiss. **75**, 67 (1877). Engl. trans. Brush (1966).

BOLTZMANN, L.: Vorlesungen über Gastheorie (Lectures in gas theory). 1896. Eng. trans. by Brush. Berkeley: Univ. of Calif. Press 1967.

BOND, J. W., WATSON, K. M., WELCH, J. A.: Atomic theory of gas dynamics. Reading, Mass.: Addison-Wesley 1965.

BONDI, H.: On spherically symmetrical accretion. M. N. R. A. S. **112**, 195 (1952).

BONDI, H., HOYLE, F.: On the mechanism of accretion by stars. M.N.R.A.S. **104**, 273 (1944).

BORN, M., GREEN, H. S.: A general kinetic theory of liquids. Cambridge: Cambridge University Press 1949.

BOSE, S. N.: Plancks Gesetz und Lichtquantenhypothese (Plancks law and the light quantum hypothesis). Z. Physik **26**, 178 (1924).

BOUSSINESQ, J.: Théorie analytique de la chaleur (Analytic theory of heat). **2**, 172. Paris: Gauthier-Villars 1903.

BOYLE, R.: New experiments physico-mechanical, touching the spring of air and its effects, made for the most part in a new pneumatical engine. Oxford, 1660—reprod. Brush (1965).

BOYLE, R.: A defense of the doctrine touching the spring and weight of the air, ... Oxford 1662. Repr. in Boyle's Works (ed. T. BIRCH). London 1772.

BRACEWELL, R. N., PRESTON, G. W.: Radio reflection and refraction phenomena in the high solar corona. Ap. J. **123**, 14 (1956).

BRUSH, S. G.: Kinetic theory, vol. I. New York: Pergamon Press 1965.

BRUSH, S. G.: Kinetic theory, vol. II. New York: Pergamon Press 1966.

BUCHLER, J. R., BARKAT, Z.: Properties of low density neutron star matter. Phys. Rev. Lett. **27**, 48 (1971).

BUCHLER, J. R., INGBER, L.: Properties of the neutron gas and applications to neutron stars. Nucl. Phys. A **170**, 1 (1971).

BUNEMAN, O.: Instability, turbulence, and conductivity in current-carrying plasma. Phys. Rev. Lett. **1**, 8 (1958).

BUNEMAN, O.: Dissipation of currents in ionized media. Phys. Rev. **115**, 503 (1959).

BURGESS, A.: Dielectronic recombination and the temperature of the solar corona. Ap. J. **139**, L 776 (1964).

CARATHÉODORY, C.: Untersuchungen über die Grundlagen der Thermodynamik (Investigation of the foundations of thermodynamics). Math. Ann. **67**, 355 (1909).

CARNOT, PAR S.: Reflexions sur la puissance motrice du feu et sur les machines (Reflections on the motivating force of fire and the machines). Bachelier, Paris 1824. Eng. trans. in: Reflections on the motive power of fire, ... (ed. E. MENDOZA). New York: Dover 1960.

CARTAN, H.: Sur la stabilite ordinare des ellipsoides de Jacobi (On the ordinary stability of Jacobian ellipsoids). Proc. Int. Math. Cong. Toronto **2**, 2, 1924 (cf. Lyttleton 1953).

CHANDRASEKHAR, S.: Stellar configurations with degenerate cores. M. N. R. A. S. **95**, 226 (1935).

CHANDRASEKHAR, S.: The highly collapsed configurations of a stellar mass. (2nd paper). M. N. R. A. S. **95**, 207 (1935).

CHANDRASEKHAR, S.: An introduction to the study of stellar structure. Chicago, Ill.: University of Chicago Press 1939, and New York: Dover.

CHANDRASEKHAR, S.: Dynamical friction: I. General considerations: The coefficient of dynamical friction. Ap. J. **97**, 255 (1943).

CHANDRASEKHAR, S.: On Heisenberg's elementary theory of turbulence. Proc. Roy. Soc. London A **200**, 20 (1949).

CHANDRASEKHAR, S.: Turbulence—A physical theory of astrophysical interest. Ap. J. **110**, 329 (1949).

CHANDRASEKHAR, S.: The fluctuations in density in isotropic turbulence. Proc. Roy. Soc. London A **210**, 18 (1951).

CHANDRASEKHAR, S.: The gravitational instability of an infinite homogeneous turbulent medium. Proc. Roy. Soc. London A **210**, 26 (1951).

CHANDRASEKHAR, S.: On the inhibition of convection by a magnetic field. Phil. Mag. **43**, 501 (1952).

CHANDRASEKHAR, S.: The instability of a layer of fluid heated from below and subject to coriolis force. Proc. Roy. Soc. London A **217**, 306 (1953).

CHANDRASEKHAR, S.: The virial theorem in hydromagnetics. M. N. R. A. S. **113**, 667 (1953).

CHANDRASEKHAR, S.: The gravitational instability of an infinite homogeneous medium when coriolis force is acting and when a magnetic field is present. Ap. J. **119**, 7 (1954).

CHANDRASEKHAR, S.: Hydrodynamic turbulence: II. An elementary theory. Proc. Roy. Soc. London A **233**, 330 (1955).

CHANDRASEKHAR, S.: The character of the equilibrium of an incompressible heavy viscous fluid of variable density. Proc. Camb. Phil. Soc. **51**, 162 (1955).

CHANDRASEKHAR, S.: The oscillations of a viscous liquid globe. Proc. London Math. Soc. **9**, 141 (1959).

CHANDRASEKHAR, S.: Hydrodynamics and hydromagnetic stability. Oxford: Oxford at the Clarendon Press 1961.

CHANDRASEKHAR, S.: A general variational principle governing the radial and non-radial oscillations of gaseous masses. Ap. J. **139**, 664 (1964).

CHANDRASEKHAR, S.: The dynamical instability of gaseous masses approaching the Schwarzschild limit in general relativity. Ap. J. **140**, 417 (1964).

CHANDRASEKHAR, S.: Ellipsoidal figures of equilibrium. New Haven, Conn.: Yale University Press 1969.

CHANDRASEKHAR, S., ELBERT, D. D.: The instability of a layer of fluid heated from below and subject to coriolis force II. Proc. Roy. Soc. London A **231**, 198 (1955).

CHANDRASEKHAR, S., FERMI, E.: Problems of gravitational stability in the presence of a magnetic field. Ap. J. **118**, 116 (1953).

CHANDRASEKHAR, S., KAUFMAN, A. N., WATSON, K. M.: The stability of the pinch. Proc. Roy. Soc. London A **245**, 435 (1958).

CHANDRASEKHAR, S., LIMBER, D. N.: On the pulsation of a star in which there is a prevalent magnetic field. Ap. J. **119**, 10 (1954).

CHANDRASEKHAR, S., LEBOVITZ, N. R.: On the oscillations and the stability of rotating gaseous masses. Ap. J. **135**, 248 (1962).

CHANDRASEKHAR, S., LEBOVITZ, N. R.: On the oscillations and the stability of rotating gaseous masses: II. The homogeneous compressible model. Ap. J. **136**, 1069 (1962).

CHANDRASEKHAR, S., LEBOVITZ, N. R.: Non-radial oscillations and the convective instability of gaseous masses. Ap. J. **138**, 185 (1963).

CHANDRASEKHAR, S., LEBOVITZ, N. R.: Non-radial oscillations of gaseous masses. Ap. J. **140**, 1517 (1964).

CHAO, N. C., CLARK, J. W., YANG, C. H.: Proton superfluidity in neutron star matter. Nucl. Phys. **179**, 320 (1972).

CHAPMAN, S.: On the law of distribution of molecular velocities and the theory of viscosity and thermal conductivity in a non-uniform simple monatomic gas. Phil. Trans. Roy. Soc. London A **216**, 279 (1916).

CHAPMAN, S.: On the kinetic theory of a gas: Part. II. A composite monatomic gas, diffusion, viscosity, and thermal conduction. Phil. Trans. Roy. Soc. London A **217**, 115 (1917).

CHAPMAN, S.: The energy of magnetic storms. M.N.R.A.S. **79**, 70 (1918).

CHAPMAN, S.: An outline of a theory of magnetic storms. Proc. Roy. Soc. London A **95**, 61 (1919).

CHAPMAN, S.: Noise in the solar corona and the terrestial ionosphere. Smithsonian Contr. Astrophys. **2**, 1 (1957).

CHAPMAN, S.: Interplanetary space and the earth's outermost atmosphere. Proc. Roy. Soc. London A **253**, 462 (1959).

CHAPMAN, S., COWLING, T. G.: The mathematical theory of non-uniform gases. Cambridge: Cambridge University Press 1953.

CHRISTY, R. F.: Pulsation theory. Ann. Rev. Astron. Astrophys. **4**, 353 (1966).

CHRISTY, R. F.: A study of pulsation in RR Lyrae models. Ap. J. **144**, 108 (1966).

CHURNS, J., THACKERY, A. D.: Trigonometric parallaxes of LB 3303 and LB 3459. In: White dwarfs—I.A.U. Symp. No. 42 (ed. W. J. LUYTEN). Dordrecht, Holland: D. Reidel 1971.

CLARK, J. W., HEINTZMANN, H., HILLEBRANDT, W., GREWING, M.: Nuclear forces, compressibility of neutron matter, and the maximum mass of neutron stars. Astrophys. Lett. **10**, 21 (1971).

CLAUSIUS, R. VON: Über die bewegende Kraft der Wärme und die Gesetze, die sich daraus für die Wärmelehre selbst ableiten lassen (On the moving force of heat and the laws of thermodynamics that can be deduced from it). Ann. Phys. **79**, 368 (1850). Engl. trans. in Phil. Mag. **2**, 1 (1851) and in: Reflections on the motive power of fire, ... (ed. E. MENDOZA). New York: Dover 1960.

CLAUSIUS, R. VON: Über die Art der Bewegung, welche wir Wärme nennen (The nature of the motion which we call heat). Ann. Phys. **100**, 353 (1857). Eng. trans. in Phil. Mag. **14**, 108 (1857) and Brush (1965).

CLAUSIUS, R. VON: Über die mittlere Länge der Wege, welche bei Molekularbewegung gasförmigen Körper von den einzelnen Molekülen zurückgelegt werden, nebst einigen anderen Bemerkungen über die mechanischen Wärmetheorie (On the mean length of the paths described by the separate molecules of gaseous bodies). Ann. Phys. **105**, 239 (1858). Eng. trans. in Phil. Mag. **17**, 81 (1859) and Brush (1965).

CLAUSIUS, R. VON: Über verschiedene für die Anwendung bequeme Formen der Hauptgleichungen der mechanischen Wärmetheorie (On different, convenient to use, forms of the main equations of mechanical heat theory). Ann. Phys. u. Chem. **125**, 353 (1865).

CLAUSIUS, R. VON: Über einen auf die Wärme anwendbaren mechanischen Satz (On a mechanical theorem applicable to heat). Sitz. Nidd. Ges. 114 (1870). Engl. trans. in Phil. Mag. **40**, 122 (1870) and Brush (1965).

COLGATE, S. A., JOHNSON, M. H.: Hydrodynamic origin of cosmic rays. Phys. Rev. Lett. **5**, 235 (1960).

COLGATE, S. A., WHITE, R. H.: The hydrodynamic behavior of supernovae explosions. Ap. J. **143**, 626 (1966).

COWLING, T. G.: The electrical conductivity of an ionized gas in a magnetic field, with applications to the solar atmosphere and the ionosphere. Proc. Roy. Soc. London A **183**, 453 (1945).

COWLING, T. G.: On the Sun's general magnetic field. M.N.R.A.S. **105**, 166 (1945).

COWLING, T. G.: The growth and decay of the sunspot magnetic field. M.N.R.A.S. **106**, 218 (1946).

COX, A. N., STEWART, J. N., EILERS, D. D.: Effects of bound-bound absorption on stellar opacities. Ap. J. Suppl. **11**, 1 (1965).

COX, J. P. On second helium ionization as a cause of pulsational instability in stars. Ap. J. **138**, 487 (1963).

COX, J. P., WHITNEY, C.: Stellar pulsation: IV. A semitheoretical period-luminosity relation for classical Cepheids. Ap. J. **127**, 561 (1958).

DANBY, J. M. A., BRAY, T. A.: Density of interstellar matter near a star. Astron. J. **72**, 219 (1967).

DANBY, J. M. A., CAMM, G. L.: Statistical dynamics and accretion. M.N.R.A.S. **117**, 50 (1957).

DEBYE, P. VON, HÜCKEL, E.: Zur Theorie der Elektrolyte: I. Gefrierpunktserniedrigung und verwandte Erscheinungen; II. Das Grenzgesetz für die elektrische Leitfähigkeit (On the theory of electrolytes: I. Lowering of the freezing point and related phenomena; II. The limiting laws for the electrical conductivity). Phys. Z. **24**, 185, 305 (1923).

DIRAC, P. A. M.: On the theory of quantum mechanics. Proc. Roy. Soc. London A **112**, 661 (1926).

EDDINGTON, A. S.: The kinetic energy of a star cluster. M.N.R.A.S. **76**, 525 (1916).

EDDINGTON, A. S.: On the radiative equilibrium of the stars. M.N.R.A.S. **77**, 16 (1917).

EDDINGTON, A. S.: On the pulsations of a gaseous star and the problem of the Cepheid variables. M.N.R.A.S. **79**, 2 (1918).

EDDINGTON, A. S.: The pulsations of a gaseous star and the problem of the Cepheid variables. M.N.R.A.S. **79**, 177 (1919).

EDDINGTON, A. S.: Internal constitution of the stars. Cambridge: Cambridge University Press 1926.

EDDINGTON, A. S.: On the cause of Cepheid pulsation. M.N.R.A.S. **101**, 182 (1941).

EDLÉN, B.: Die Deutung der Emissionslinien im Spektrum der Sonnenkorona (The interpretation of the emission line spectrum of the solar corona). Z. Ap. **22**, 30 (1942).

EGGEN, O. J.: Photoelectric studies: V. Magnitudes and colors of classical Cepheid variable stars. Ap. J. **113**, 367 (1951).

EGGEN, O. J., GREENSTEIN, J. L.: Spectra, colors, luminosities, and motions of the white dwarfs. Ap. J. **141**, 83 (1965).

EGGEN, O. J., GREENSTEIN, J. L.: Observations of proper-motion stars II. Ap. J. **142**, 925 (1965).

EGGEN, O. J., GREENSTEIN, J. L.: Observations of proper-motion stars III. Ap. J. **150**, 927 (1967).

EINSTEIN, A. VON: Zum gegenwärtigen Stand des Strahlungsproblems (Additional new opinions on radiation problems). Phys. Z. **10**, 185 (1909).

EINSTEIN, A. VON: Quantentheorie des einatomigen idealen Gases (The quantum theory of the monatomic perfect gas). Preuss. Acad. Wiss. Berl. Berlin Sitz. **22**, 261 (1924).

EINSTEIN, A. VON: Quantentheorie des einatomigen idealen Gases (The quantum theory of the monatomic perfect gas). Preuss. Acad. Wiss. Berl. Berlin Sitz. **1**, 3, 18 (1925).

EMDEN, R.: Gaskugeln (Gas spheres). Leipzig: Teuber 1907.

ENSKOG, D.: Kinetische Theorie der Vorgänge in Mässing verdünnten Gasen (Kinetic theory of processes in massive, dilute gases). Uppsala: Almqvist and Wiksell 1917.

EPSTEIN, I.: Pulsation properties of giant-star models. Ap. J. **112**, 6 (1950).

ERICKSON, W. C.: The radio-wave scattering properties of the solar corona. Ap. J. **139**, 1290 (1964).

EULER, L.: Principes generaux du movement des fluides (General principles of the movement of fluids). Hist. de l'acad. de Berlin 1755.

FABRICIUS, D.: Vierteljahrsschrift (Fourth year book). **4**, 290 (1594).

FERMI, E. VON: Zur Quantelung des idealen einatomigen Gases (On quantisation of the ideal monatomic gas). Z. Physik **36**, 902 (1926).

FIELD, G. B.: Thermal instability. Ap. J. **142**, 531 (1965).

FIELD, G. B., GOLDSMITH, D. W., HABING, H. J.: Cosmic-ray heating in the interstellar gas. Ap. J. **155**, L 149 (1969).

FOKKER, A. D.: Die mittlere Energie rotierender elektrischer Dipole im Strahlungsfeld (Mean energy of a rotating electric molecule in a radiation field). Ann. Physik **43**, 810 (1914).

FORBUSH, S. E.: Time-variations of cosmic rays. In: Handbuch der Physik, vol. XLIX/1: Geophysics III. Berlin-Heidelberg-New York: Springer 1966.

FOWLER, R. H.: On dense matter. M.N.R.A.S. **87**, 114 (1926).

FRAUTSCHI, S., BAHCALL, J. N., STEIGMAN, G., WHEELER, J. C.: Ultradense matter. Comm. Ap. and Space Phys. **3**, 121 (1971).

FURTH, H. P., KILLEEN, J., ROSENBLUTH, M. N.: Finite resistivity instabilities of a sheet pinch. Phys. Fluids **6**, 459 (1963).

GAY-LUSSAC, M.: Memoir on the combination of gaseous substances with each other. Mem. de la société d'arcueil **2**, 207 (1809). Eng. trans. in: Foundations of the molecular theory. Alemic Club Repr. No. 4. Edinburgh: Bishop 1950.

GIBBS, J. W.: Elementary principles in statistical mechanics. New York 1902.

GICLAS, H. L., BURNHAM, R., THOMAS, N. G.: Lowell proper motions I to XV. Lowell Obs. Bull. **89**, **102**, **112**, **120**, **122**, **124**, **129**, **136**, **140**, **144**, **150**, **151**, **152** (1960 to 1970).

GICLAS, H. L., BURNHAM, R., THOMAS, N. G.: A list of white dwarf suspects I, II, III. Lowell Obs. Bull. **125**, **141**, **153** (1965, 1967, 1970).

GINZBURG, V. L.: Superfluidity and superconductivity in astrophysics. Comments Astrophys. and Space Phys. **1**, 81 (1969).

GLIESE, W.: Catalogue of nearby stars. Veroff-Astron. Rechen-Institut Heidelberg: Braun, Karlsruhe 1969.

GOLD, T., HOYLE, F.: On the origin of solar flares. M.N.R.A.S. **120**, 89 (1960).

GOODRICKE, J.: A series of observations on, and a discovery of, the period of the variation of the light of the bright star in the head of Medufa, called Algol. Phil. Trans. **73**, 474 (1783).

GREEN, H. S.: The molecular theory of fluids. New York: Interscience 1952.

GREENSTEIN, J. L.: The Lowell suspect white dwarfs. Ap. J. **158**, 281 (1969).

GREENSTEIN, J. L.: Some new white dwarfs with peculiar spectra VI. Ap. J. **162**, L 55 (1970).

GREENSTEIN, J. L., OKE, J. B., SHIPMAN, H. L.: Effective temperature, radius, and gravitational redshift of Sirius B. Ap. J. **169**, 563 (1971).

GREENSTEIN, J. L., TRIMBLE, V. L.: The Einstein redshift in white dwarfs. Ap. J. **149**, 283 (1967).

GREENSTEIN, J. L., TRIMBLE, V. L.: The gravitational redshift of 40 Eridani B. Ap. J. **175**, L 1 (1972).

HACK, M.: The evolution of close binary systems. In: Star evolution—Proc. Int. Sch. Phys. Enrico Fermi—Course 38 (ed. L. GRATTON). New York: Academic Press 1963.

HAMADA, T., SALPETER, E. E.: Models for zero-temperature stars. Ap. J. **134**, 683 (1961).

HARRISON, B. K., THORNE, K. S., WAKANO, M., WHEELER, J. A.: Gravitation theory and gravitational collapse. Chicago, Ill. University Chicago Press 1964.

HASEGAWA, A.: Plasma instabilities in the magnetosphere. Rev. of Geophys. and Space Sci. **9**, 703 (1971).

HEISENBERG, W.: Zur statischen Theorie der Turbulenz (The statistical theory of turbulence). Z. Phys. **124**, 628 (1948).

HEISENBERG, W.: On the theory of statistical and isotropic turbulence. Proc. Roy. Soc. **195**, 402 (1949).

HELMHOLTZ, H. VON: Über die Erhaltung der Kraft (The conservation of force). Berlin: G. Riemer 1847. Engl. trans. in Brush, 1965.

HELMHOLTZ, H. VON: Popular lectures (1854).

HELMHOLTZ, H.: Über diskontinuierliche Flüssigkeitsbewegungen (On discontinuities in moving fluids). Wiess Abhandlungen 146, J. A. Barth, 1882. Engl. trans. in Phil. Mag. **36**, 337 (1868).

HENYEY, L., VARDYA, M. S., BODENHEIMER, P.: Studies in stellar evolution: III. The calculation of model envelopes. Ap. J. **142**, 841 (1965).

HERLOFSON, N.: Magneto-hydrodynamic waves in a compressible fluid conductor. Nature **165**, 1020 (1950).

HERSHBERG, R. E., PRONIK, V. I.: The theory of the Strömgren zone. Astron. Zh. **36**, 902 (1959).

HERTZSPRUNG, E.: Zur Strahlung der Sterne (Giants and dwarfs). Z. Wiss. Photog. 3 (1905). Eng. trans. in: Source book in astronomy (ed. H. SHAPLEY). Cambridge, Mass.: Harvard University Press 1960.

HERTZSPRUNG, E.: On the relation between period and form of the light curve of variable stars of the δ Cephei type. B.A.N. **3**, 115 (1926).

HIDE, R.: The character of the equilibrium of a heavy, viscous, incompressible fluid of variable density: I. General theory; II. Two special cases. Quart. J. Math. Appl. Math. **9**, 22, 35 (1956).

HINES, C. O.: Internal atmospheric gravity waves at ionospheric heights. Can. J. Phys. **38**, 1441 (1960).

HJELLMING, R. M.: Physical processes in H II regions. Ap. J. **143**, 420 (1966).

HJELLMING, R. M.: The effects of star formation and evolution on the evaluation of H II regions, and theoretical determinations of temperatures in H II regions. In: Interstellar ionized hydrogen (ed. Y. TERZIAN). New York: W. A. Benjamin 1968.

HJELLMING, R. M., GORDON, C. P., GORDON, K. J.: Properties of interstellar clouds and the intercloud medium. Astron. Astrophys. **2**, 202 (1969).

HOYLE, F., LYTTLETON, R. A.: The evolution of the stars. Proc. Camb. Phil. Soc. **35**, 592 (1939).

HUGONIOT, PAR H.: Sur la propagation du mouvement dans les corps et specialement dans les gaz parfarts (On the propagation of the movement of bodies, and especially of the perfect gas). J. de l'Ecole Polytechnique **57**, 1 (1887), **59**, 1 (1889).

JACKSON, J. D.: Longitudinal plasma oscillation. J. Nucl. Energy C, **1**, 171 (1960).

JACOBI, C. G. J.: Über die Figur des Gleichgewichts (On the figure of objects of the same specific gravity). Ann. Phys. u. Chem. **33**, 229 (1834).

DE JAEGER, C.: Structure and dynamics of the solar atmosphere. In: Handbuch der Physik, vol. LII: Astrophysics IV: The solar system (ed. S. FLÜGGE). Berlin-Heidelberg-New York: Springer 1959.

JEANS, SIR J. H.: On the stability of a spherical nebula. Phil. Trans. Roy. Soc. London **199**, 1 (1902).

JEANS, SIR J. H.: The motion of tidally-distorted masses, with special reference to theories of cosmogony. Mem. R.A.S. London **62**, 1 (1917).

JEANS, SIR J. H.: Problems of cosmogony and stellar dynamics. Cambridge: Cambridge University Press 1919.

JEANS, SIR J. H.: Astronomy and cosmogony. Cambridge: Cambridge University Press 1929.

JEFFREYS, H.: The stability of a layer of fluid heated below. Phil. Mag. **2**, 833 (1926).

JEFFREYS, H.: The instability of a compressible fluid heated below. Proc. Camb. Phil. Soc. **26**, 170 (1930).

JOKIPII, J. R.: Turbulence and scintillations in the interplanetary plasma. Ann. Rev. Astron. Astrophys. **11** (1974).

JOULE, J. P.: On matter, living force, and heat. (1847), lecture repr. in Joule's scientific papers and Brush (1965).

KAPLAN, S. A.: A system of spectral equations of magneto-gas-dynamic isotropic turbulence. Dokl. Acad. Nauk. SSSR **94**, 33 (1954).

KAPLAN, S. A., PIKELNER, S. B.: The interstellar medium. Cambridge, Mass.: Harvard University Press 1970.

KARDASHEV, N. S.: Nonstationariness of spectra of young sources of nonthermal radio emission. Sov. Astron. **6**, 317 (1962).

KÁRMÁN, T. VON: On the statistical theory of turbulence. Proc. Nat. Acad. Sci. **23**, 98 (1937).

KÁRMÁN, T. VON, HOWARTH, L.: On the statistical theory of isotropic turbulence. Proc. Roy. Soc. London A **164**, 192 (1938).

KELVIN, LORD (W. THOMSON): On an absolute thermometric scale founded on Carnot's theory of the motive power of heat. Proc. Camb. Phil. Soc. **1**, 66 (1848).

KELVIN, LORD (W. THOMSON): Physical considerations regarding the possible age of the Sun's heat Phil. Mag. **23**, 158 (1862) Brit. Assoc. Rpt. 27 (1861), Math. and Phys. Papers **5**, 141 (1861).

KELVIN, LORD (W. THOMSON): On the convective equilibrium of temperature in the atmosphere. Math. and Phys. Papers **3**, 255 (1862).

KELVIN, LORD (W. THOMSON): Sur le refroidissement seculaire du soleil. De la température actuelle due soleil. De l'origine et de la somme totale de la chaleur solaire (On the age of the Sun's heat, the actual temperature of the Sun, and the origin of the sum total of the Sun's heat). Les Mondes, **3**, 473 (1863).

KELVIN, LORD (W. THOMSON): Hydrokinetic solutions and observations, on the motion of free solids through a liquid. Phil. Mag. **42**, 362 (1871).

KEMP, J. C., SWEDLUND, J. B., LANDSTREET, J. D., ANGEL, J. R. P.: Discovery of circularly polarized light from a white dwarf. Ap. J. **161**, L 77 (1970).

KEMP, J. C., SWEDLUND, J. B., WOLSTENCROFT, R. D.: Confirmation of the magnetic white dwarf G 195-19. Ap. J. **164**, L 17 (1971).

KIRKWOOD, J. G.: The statistical mechanical theory of transport processes: II. Transport in gases. J. Chem. Phys. **15**, 72, 155 (1947).

KOLMOGOROFF, A. N.: The local structure of turbulence in incompressible viscous fluids for very large Reynolds numbers. Compt. Rend. Acad. Sci. (SSSR) **30**, 301 (1941).

KOLMOGOROFF, A. N.: Dissipation of energy in the locally isotropic turbulence. Compt. Rend. Acad. Sci. (SSSR) **32**, 16 (1941).

KOTHARI, D. S.: The theory of pressure-ionization and its applications. Proc. Roy. Soc. London A **165**, 486 (1938).

KROTSCHECK, E.: Superfluidity in neutron matter. Z. Phys. **251**, 135 (1972).

KRUSKAL, M., SCHWARZSCHILD, M.: Some instabilities of a completely ionized plasma. Proc. Roy. Soc. London A **223**, 348 (1954).

KRUSKAL, M., TUCK, J. L.: The instability of a pinched fluid with a longitudinal magnetic field. Proc. Roy. Soc. London A **245**, 222 (1958).

KUCHOWICZ, B.: Neutrino gas statistics. Bull. Acad. Pol. Sci. Ser. Sci. Mat. Astr. et Phys. **11**, 317 (1963).

KUCHOWICZ, B.: Neutrinos in superdense matter: I. A tentative statistical approach. Inst. Nucl. Res. Warsaw Rpt. 384 (1963).

KUKARIN, B. V., PARENAGO, P. P.: General cataloque of variable stars. 3rd ed. Moscow 1969.

KULSRUD, R. M.: Plasma instabilites. In: Plasma astrophysics—Proc. Int. Sch. Phys. Enrico Fermi—Course 39 (ed. P. A. STURROCK). New York: Academic Press 1967.

KUTSENKO, A. B., STEPANOV, K. N.: Instability of plasma with anisotropic distributions of ion and electron velocities. Sov. Phys. JETP **11**, 1323 (1960).

LAAN, H. VAN DER: Radio galaxies: I. The interpretation of radio source data. M. N. R. A. S. **126**, 519 (1963).

LAMB, H.: On the oscillations of a viscous spheroid. Proc. London Math. Soc. **13**, 51 (1881).

LAMB, H.: On the theory of waves propagated vertically in the atmosphere. Proc. London Math. Soc. **7**, 122 (1909).

LAMB, H.: Hydrodynamics. Cambridge: Cambridge University Press 1916. Republ. New York: Dover.

LANDAU, L. D.: On the theory of stars. Phys. Z. Sowjetunion **1**, 285 (1932).

LANDAU, L. D.: On the vibrations of the electronic plasma. J. Phys. (U. S. S. R.) **10**, 25 (1946).

LANDAU, L. D., LIFSHITZ, E. M.: Fluid mechanics. New York: Pergamon Press 1959.

LANDAU, L. D., LIFSHITZ, E. M.: Statistical physics. Reading, Mass.: Addison-Wesley 1969.

LANDSTREET, J. D., ANGEL, J. R. P.: Discovery of circular polarization in the white dwarf G 99-37. Ap. J. **165**, L 67 (1971).

LANDOLT, A. U.: A new short-period blue variable. Ap. J. **153**, 151 (1968).

LANE, J. H.: On the theoretical temperature of the Sun, under the hypothesis of a gaseous mass maintaining its volume by its internal heat and depending on the laws of gases as known to terrestial experiment. Amer. J. Sci. **50**, 57 (1870).

LANG, K. R.: The small scale, quasi-periodic, disk component of solar radio radiation. To be published Ap. J. Sept. 15 (1974).

LAPLACE, P. S. MARQUIS DE: Sur la vitesse du son dans l'air et dan l'eau (On the velocity of sound in the air and the water). Ann. Chem. Phys. **3**, 238 (1816).

LASKER, B. M.: An investigation of the dynamics of old H II regions. Ap. J. **143**, 700 (1966).

LASKER, B. M.: Ionization fronts for H II regions with magnetic fields. Ap. J. **146**, 471 (1966).

LASKER, B. M.: The energization of the interstellar medium by ionization limited H II regions. Ap. J. **149**, 23 (1967).

LASKER, B. M., HESSER, J. E.: High frequency stellar oscillations: II. G 44-32 A new short period blue variable star. Ap. J. **158**, L 171 (1969).

LASKER, B. M., HESSER, J. E.: High frequency stellar oscillations: VI. R 548 A periodically variable white dwarf. Ap. J. **163**, L 89 (1971).

LEAVITT, H. S.: Periods of 25 variable stars in the small Magellanic cloud. Harvard Circular No. **173** (1912). Repr. in: Source book in astronomy (ed. H. SHAPLEY). Cambridge, Mass.: Harvard University Press 1960.

LEBOVITZ, N. R., RUSSELL, G. W.: The pulsations of polytropic masses in rapid, uniform rotation. Ap. J. **171**, 103 (1972).

LEDOUX, P.: On the radial pulsation of gaseous stars. Ap. J. **102**, 143 (1945).

LEDOUX, P.: Stellar stability and stellar evolution. In: Star evolution—Proc. Int. Sch. Phys. "Enrico Fermi" Course 28. New York: Academic Press 1963.

LEDOUX, P.: Stellar stability. In: Stellar structure—Stars and stellar systems VIII (ed. L. H. ALLER and D. B. MCLAUGHLIN). Chicago, Ill.: University of Chicago Press 1965.

LEDOUX, P., PEKERIS, C. L.: Radial pulsations of stars. Ap. J. **94**, 124 (1941).

LEDOUX, P., WALRAVEN, T.: Variable stars. In: Handbuch der Physik, vol. LI: Astrophysics II—Stellar structure (ed. S. FLÜGGE). Berlin-Heidelberg-New York: Springer 1958.

LEIGHTON, R. B., NOYES, R. W., SIMON, G. W.: Velocity fields in the solar atmosphere: I. Preliminary report. Ap. J. **135**, 474 (1962).

LEUNG, Y. C., WANG, C. G.: Properties of hadron matter: II. Dense baryon matter and neutron stars. Ap. J. **170**, 499 (1971).

LIGHTHILL, M. J.: On sound generated aerodynamically: I. General theory. Proc. Roy. Soc. London **A 211**, 564 (1952).

LIGHTHILL, M. J.: On sound generated aerodynamically: II. Turbulence as a source of sound. Proc. Roy. Soc. London **A 222**, 1 (1954).

LINDBLAD, B.: A condensation theory for meteoric matter and its cosmological significance. Nature **135**, 133 (1935).

LOSCHMIDT, J.: Zur Größe der Luftmoleküle (The size of atmospheric molecules). Wien. Ber. **52**, 395 (1865).

LUMLEY, J., PANOFSKY, H.: The structure of atmospheric turbulence. New York: Interscience Publ. 1964.

LÜST, R.: The solar wind. In: Interstellar gas dynamics (ed. H. J. HABING). Dordrecht, Holland: D. Reidel 1970.

LYTTLETON, R. A.: The stability of rotating liquid masses. Cambridge: Cambridge University Press 1955.

MACLAURIN, C.: A treatise on fluxions. (1742. Cf.: History of the mathematical theories of attraction, and the figure of the Earth by I. Todhunter. Macmillan 1873. Repr. New York: Dover 1962.

MATHEWS, W. G.: The time evolution of an H II region. Ap. J. **142**, 1120 (1965).

MATHEWS, W. G., O'DELL, C. R.: Evolution of diffuse nebulae. Ann. Rev. Astron. Astrophys. **1**, 67 (1969).

MAXWELL, J. C.: Illustrations of the dynamical theory of gases: Part I. On the motions and collisions of perfectly elastic spheres; Part II. On the process of diffusion of two or more kinds of moving particles among one another. Phil. Mag. **19**, 19, **20**, 21 (1860), repr. Brush (1965).

MAXWELL, J. C.: Viscosity or internal friction of air and other gases. Phil. Trans. Roy. Soc. London **156**, 249 (1866).

MAYER, R.: The forces of inorganic nature. Ann. Chem. and Pharm. **42**, 233 (1842). Engl. trans. in Phil. Mag. **24**, 371 (1862) reprod. Brush (1965).

MCCREA, W. H.: The formation of population I stars: Part I. Gravitational contraction. M.N.R.A.S. **117**, 562 (1957).

MCDOUGALL, J., STONER, E. C.: The computation of Fermi-Dirac functions. Phil. Trans. Roy. Soc. London **237**, 67 (1938).

MESTEL, L.: The theory of white dwarfs. In: Stellar structure—Stars and stellar systems VIII (ed. L. H. ALLER and D. MCLAUGHLIN). Chicago, Ill.: University of Chicago Press 1965.

MICHEL, F. C.: Accretion of matter by condensed objects. Astrophys. Space Sci. **15**, 153 (1972).

MOISEEV, S. S., SAGDEEV, R. Z.: On the Bohm diffusion coefficient. Sov. Phys. JETP **17**, 515 (1963).

MOORE, C. E.: Ionization potentials and ionization limits derived from the analysis of optical spectra. Nat. Bur. Stands. (Wash.) rpt. NSRDS-NBS 34 (1970).

MOTT-SMITH, H. M.: The solution of the Boltzmann equation for a shock wave. Phys. Rev. **82**, 885 (1951).

NAVIER, C. L. M. H.: Mem. de l'acad. Sci. **6** (1822).

NERNST, W.: Über die Beziehungen zwischen Wärmeentwicklung und maximaler Arbeit bei kondensierten Systemen (About the heat and maximum work of a condensed system). Sitz. Berl. **1**, 933 (1906).

NERNST, W.: The new heat theorem—Its foundations in theory and experiment. 1926. Repr. New York: Dover 1969.

NESS, N. F., SCEARCE, C. S., SEEK, J. B.: Initial results of the Imp 1 magnetic field experiment. J. Geophys. Res. **69**, 3531 (1964).

NEWKIRK, G.: Structure of the solar corona. Ann. Rev. Astron. Astrophys. **5**, 213 (1967).

NOYES, R. W., HALL, D. N. B.: Thermal oscillations in the high solar photosphere. Ap. J. **176**, L 89 (1972).

ONSAGER, L.: The distribution of energy in turbulence. Phys. Rev. **68**, 286 (1945).

OORT, J. H.: Some phenomena connected with interstellar matter. M.N.R.A.S. **106**, 159 (1946).

OORT, J. H., VAN DE HULST, H. C.: Gas and smoke in interstellar space. B.A.N. **10**, 187 (1946).

OPPENHEIMER, J. R., SNYDER, H.: On continued gravitational contraction. Phys. Rev. **56**, 455 (1939).

OPPENHEIMER, J. R., VOLKOFF, G. M.: On massive neutron cores. Phys. Rev. **55**, 374 (1939).

OSTERBROCK, D. E.: The heating of the solar chromosphere, plages, and corona by magnetohydrodynamic waves. Ap. J. **134**, 347 (1961).

OSTERBROCK, D. E.: Temperature in H II regions and planetary nebulae. Ap. J. **142**, 1423 (1965).

OSTRIKER, J. P., BODENHEIMER, P., LYNDEN-BELL, D.: Equilibrium models of differentially rotating zero-temperature stars. Phys. Rev. Lett. **17**, 816 (1966).

OSTRIKER, J. P., BODENHEIMER, P.: Rapidly rotating stars: II. Massive white dwarfs. Ap. J. **151**, 1089 (1968).

OSTRIKER, J. P., HESSER, J. E.: Ultrashort-period stellar oscillations: II. The period and light curve of HZ 29. Ap. J. **153**, L 151 (1968).

OZERNOY, L. M., CHIBISOV, G. V.: Galactic parameters as a consequence of cosmological turbulence. Astrophys. Lett. **7**, 201 (1971).

PACHOLCZYK, A. G., STODOLKIEWICZ, J. S.: On the gravitational instability of some magnetohydrodynamical systems of astrophysical interest. Acta. Astron. (Polska Akad. Nauk) **10**, 1 (1960).

PACZYNSKI, B.: Evolutionary processes in close binary systems. Ann. Rev. Astron. Astrophys. **9**, 183 (1971).

PANHARIPANDE, V. R.: Hyperonic matter. Nucl. Phys. **A 178**, 123 (1971).

PARKER, E. N.: Gravitational instability of a turbulent medium. Nature **170**, 1030 (1952).

PARKER, E. N.: Dynamics of the interplanetary gas and magnetic fields. Ap. J. **128**, 664 (1958).

PARKER, E. N.: The solar-flare phenomenon and the theory of reconnection and annihilation of magnetic fields. Ap. J. Suppl. Ser. 77, **8**, 177 (1963).

PARKER, E. N.: Dynamical theory of the solar wind. Space Sci. Rev. **4**, 666 (1965).

PAULI, VON W.: Über Gasentartung und Paramagnetismus (Gas degeneration and paramagnetism). Z. Physik **41**, 81 (1927).

PETSCHEK, H. E.: Annihilation of magnetic fields. Proc. AAS-NASA conference on physics of solar flares (ed. W. N. HESS). NASA SP-50, Wash. D.C. 1964.

PIKELNER, S.: Ionization and heating of the interstellar gas by subcosmic rays, and the formation of clouds. Sov. Astron. **11**, 737 (1968).

PIERCE, J. R.: Possible fluctuations in electron streams due to ions. J. Appl. Phys. **19**, 231 (1948).

PINES, D.: Inside neutron stars. In: Proc. 12th int. conf. on low temperature physics. Academic Press Japan 1970.

PINES, D., BOHM, D.: A collective description of electron interactions: II. Collective vs. individual particle aspects of the interactions. Phys. Rev. **85**, 338 (1952).

PLANCK, M.: Über das Gesetz der Energieverteilung im Normalspektrum (On the theory of thermal radiation). Ann. Phys. **4**, 553 (1901).

POINCARÉ, H.: Lecons sur les hypothèses cosmogoniques (Lessons on cosmological hypothesis). Paris: Librarie Scientifique, A. Hermann 1811.

POINCARÉ, H.: Lecons l'équilibre d'une masse fluide animeé d'un mouvement de rotation (Lesson on the equilibrium of rotating fluid masses). Acta. Math. **7**, 259 (1855).

POISSON, S. D.: Remarques sur une equation qui se présente das la theorie des attractions des spheroides (Remarks on an equation which presents itself in the theory of spheroidal attractions) Bull. de la Soc. Philomatique **3**, 388 (1813).

POWER, H.: Experimental philosophy in three books containing new experiments, microscopical, mercurial, magnetical. London (1663), cf. I. B. Cohen. Newton, Hooke, and Boyle's law. Nature **204**, 618 (1964).

PRANDTL, L.: Verhandlungen des Dulten Internationalen Mathematiker-Kongresses (Transactions of the international mathematical congress). 484 (1905), see also The physics of solids and fluids by P. P. EWALD and L. PRANDTL. London: Blakie 1930.

PRANDTL, L.: Essentials of fluid dynamics. London: Blakie 1952.

PROUDMAN, J.: On the motion of solids in a liquid possesing vorticity. Proc. Roy. Soc. London **A 92**, 408 (1916).

PROUDMAN, J.: The generation of noise by isotropic turbulence. Proc. Roy. Soc. London **A 214**, 119 (1952).

RANKINE, W. J. M.: On the thermodynamic theory of waves of finite longitudinal disturbance. Phil. Trans. Roy. Soc. London **160**, 277 (1870).

RAVENHALL, D. G., BENNETT, C. D., PETHICK, C. J.: Nuclear surface energy and neutron star matter. Phys. Rev. Lett. **78**, 978 (1972).

RAYLEIGH, LORD: Investigation of the character of the equilibrium of an incompressible heavy fluid of variable density. Proc. London Math. Soc. **14**, 170 (1883).

RAYLEIGH, LORD: On convective currents in a horizontal layer of fluid when the higher temperature is on the under side. Phil. Mag. **32**, 529 (1916).

REYNOLDS, O.: On the force caused by the communication of heat between a surface and a gas and on a new photometer. Phil. Mag. **23**, 1 (1876).

REYNOLDS, O.: An experimental investigation of the circumstances which determine whether the motion of water shall be direct or sinuous, and of the law of resistance in parallel channels. Phil. Trans. Roy. Soc. London **174**, 935 (1883).

RICHARDSON, R. S., SCHWARZSCHILD, M.: On the turbulent velocities of solar granules. Ap. J. **111**, 351 (1950).

RIDDLE, R. K.: First catalogue of trigonometric parallaxes of faint stars. Publ. U. S. Naval Obs. **120**, part 3 (1970).

RITTER, A. VON: Untersuchungen über die Höhe der Atmosphäre und die Konstitution gasförmiger Weltkörper (Investigations on the height of the atmosphere and the constitution of gaseous celestial bodies). Ann. Phys. u. Chem. **8**, 157 (1880).

RITTER, A. VON: Untersuchungen über die Höhe der Atmosphäre und die Konstitution gasförmiger Weltkörper (Investigation of the height of the atmosphere and the constitution of gaseous celestial bodies). Ann. Phys. u. Chem. **13**, 360 (1881).

ROCHE, M.: Mémoire sur la figure d'une masse fluide (Soumise à l'attraction d'un point eloigne), Memoir on the figure of a fluid mass (subject to the attraction of a distant point). Acad. des Sci. de Montpellier **1**, 243, 333 (1847).

ROSENBLUTH, M., MACDONALD, W. M., JUDD, D. L.: Fokker-Planck equation for an inverse square force. Phys. Rev. **107**, 1 (1957).

ROSSELAND, S.: Viscosity in the stars. M. N. R. A. S. **89**, 49 (1929).

ROUTLY, P. M.: Second catalogue of trigonometric parallaxes of faint stars. Publ. U. S. Naval Obs. **20**, part 6 (1972).

ROXBURGH, I. W.: On models of nonspherical stars: II. Rotating white dwarfs. Z. Ap. **62**, 134 (1965).

RUBIN, R. H.: The structure and properties of H II regions. Ap. J. **153**, 761 (1968).

RUDERMAN, M.: Superdense matter in stars. J. Phys. Suppl. 11, **30**, 152 (1969).

RUTHERFORD, E.: The scattering of α and β particles by matter and the structure of the atom. Phil. Mag. **21**, 669 (1911).

SACKUR, O.: Die Anwendung der Kinetischen Theorie der Gase auf chemische Probleme (The application of the kinetic theory of gases to chemical problems). Ann. Phys. **36**, 958 (1911).

SACKUR, O.: Die universelle Bedeutung des sog. elementaren Wirkungsquantums (Universal significance of the elementary working-quantum). Ann. Phys. **40**, 67 (1913).

SAHA, M. N.: Ionization in the solar chromosphere. Phil. Mag. **40**, 472 (1920).

SAHA, M. N.: On the physical theory of stellar spectra. Proc. Roy. Soc. London **A 99**, 135 (1921).

SALPETER, E. E.: Energy and pressure of a zero temperature plasma. Ap. J. **134**, 669 (1961).

SAMPSON, R. A.: On the rotation and mechanical state of the Sun. Mem. R.A.S. **51**, 123 (1894).

SCHWARZSCHILD, K.: Über das Gleichgewicht der Sonnenatmosphäre (On the equilibrium of the Sun's atmosphere). Gott. Nach. **1**, 41 (1906).

SCHWARZSCHILD, M.: Overtone pulsations for the standard model. Ap. J. **94**, 245 (1941).
SCHWARZSCHILD, M.: On noise arising from the solar granulation. Ap. J. **107**, 1 (1948).
SEATON, M. J.: The chemical composition of the interstellar gas. M.N.R.A.S. **111**, 368 (1951).
SEATON, M. J.: Electron temperatures and electron densities in planetary nebulae. M.N.R.A.S. **114**, 154 (1954).
SEATON, M. J.: The kinetic temperature of the interstellar gas in regions of neutral hydrogen. Ann. Astrophys. **18**, 188 (1955).
SEDOV, L. I.: Similarity and dimensional methods in mechanics. New York: Academic Press 1959 and Course in continuum mechanics, vol. 1–4. Groningen: Wolter-Noordhoff 1971.
SHATZMAN, E.: The heating of the solar corona and chromosphere. Ann. Astrophys. **12**, 203 (1949).
SHIPMAN, H. L.: Masses and radii of white dwarfs. Ap. J. **177**, 723 (1972).
SHKLOVSKY, I. S.: Secular variations in the flux and intensity of radio emission from discrete sources. Sov. Astron. **6**, 317 (1960).
SHKLOVSKY, I.S.: Supernovae. New York: Wiley Interscience 1968.
SIEMANS, P. J., PANDHARIPANDE, V. R.: Neutron matter computations in Brueckner and variational theory. Nucl. Phys. **A 173**, 561 (1971).
SIMON, A.: Diffusion of like particles across a magnetic field. Phys. Rev. **100**, 1557 (1955).
SPIEGEL, E. A.: The gas dynamics of accretion. In: Interstellar gas dynamics (ed. H. J. HABING). Dordrecht, Holland: D. Reidel 1970.
SPIEGEL, E. A.: Convection in stars: Part. I. Basic Boussinesq convection. Ann. Rev. Astron. Astrophys. **9**, 223 (1971).
SPITZER, L.: The temperature of interstellar matter I. Ap. J. **107**, 6 (1948).
SPITZER, L.: The temperature of interstellar matter II. Ap. J. **109**, 337 (1949).
SPITZER, L.: Behavior of matter in space. Ap. J. **120**, 1 (1954).
SPITZER, L.: Physics of fully ionized gases. New York: Wiley 1962.
SPITZER, L., HÄRM, R.: Transport phenomena in a completely ionized gas. Phys. Rev. **89**, 977 (1953).
SPITZER, L., SAVEDOFF, M. P.: The temperature of interstellar matter III. Ap. J. **111**, 593 (1950).
SPITZER, L., SCOTT, E. H.: Heating of H I regions by energetic particles: II. Interaction between secondaries and thermal electrons. Ap. J. **158**, 161 (1969).
STEFAN, A. J.: Beziehung zwischen Wärmestrahlung und Temperatur (Relation between thermal radiation and temperature). Wien. Ber. **79**, 397 (1879).
STEIN, R. F.: Waves in the solar atmosphere: I. The acoustic energy flux. Ap. J. **154**, 297 (1968).
STEIN, R. F., SCHWARTZ, R. A.: Waves in the solar atmosphere: II. Large amplitude acoustic pulse propagation. Ap. J. **177**, 807 (1972).
STOKES, G. G.: On the theories of the internal friction of fluids in motion, and of the equilibrium and motion of elastic solids. Trans. Camb. Phil. Soc. **8**, 287 (1845).
STONER, C.: The equilibrium of dense stars. Phil. Mag. **9**, 944 (1930).
STRÖMGREN, B.: The physical state of interstellar hydrogen. Ap. J. **89**, 526 (1939).
STURROCK, P. A.: Kinematics of growing waves. Phys. Rev. **112**, 1488 (1958).
STURROCK, P. A.: Model of the high-energy phase of solar flares. Nature **211**, 695 (1966).
SWEET, P. A.: The neutral point theory of solar flares. In: Proc. I.A.U. Symp. on electromagnetic phenomenon in cosmical physics (ed. B. LEHNERT). Cambridge: Cambridge Univ. Press 1958.
TASSOUL, J. L.: Adiabatic pulsations and convective instability of gaseous masses III. M.N.R.A.S. **138**, 123 (1968).
TASSOUL, J. L., Ostriker, J. P.: On the oscillations and stability of rotating stellar modes I. Ap. J. **154**, 613 (1968).
TASSOUL, M., TASSOUL, J. L.: Adiabatic pulsations and convective instability of gaseous masses I, II. Ap. J. **150**, 213, 1031 (1967).
TAYLER, R. J.: Hydrodynamic instabilities of an ideally conducting fluid. Proc. Phys. Soc. London **B 70**, 31 (1957).
TAYLOR, G. I.: Experiments with rotating fluids. Proc. Roy. Soc. London **A 100**, 114 (1921).
TAYLOR, G. I.: Statistical theory of turbulence. Proc. Roy. Soc. London **A 151**, 421 (1935).
TAYLOR, G. I.: The spectrum of turbulence. Proc. Roy. Soc. London **A 164**, 1476 (1938).
TAYLOR, G. I.: The instability of liquid surfaces when accelerated in a direction perpendicular to their planes I. Proc. Roy. Soc. London **A 201**, 192 (1950).
TETRODE, H. VON: Die chemische Konstante der Gase und das elementare Wirkungsquantum (Chemical gas constants and the elementary action quantum). Ann. Phys. **38**, 434 (1912).
THOMPSON, W. B.: Thermal convection in a magnetic field. Phil. Mag. **42**, 1417 (1951).

TIDMAN, D. A.: Structure of a shock wave in fully ionized hydrogen. Phys. Rev. **111**, 1439 (1958).
TONKS, L., LANGMUIR, I.: Oscillations in ionized gases. Phys. Rev. **33**, 195 (1929).
TOWNELY, R.: (1662), cf. I. B. COHEN. Newton, Hooke, and Boyle's law. Nature **204**, 618 (1964).
TRIMBLE, V. L., GREENSTEIN, J. L.: The Einstein redshift in white dwarfs III. Ap. J. **177**, 441 (1972).
TSURUTA, S., CAMERON, A. G. W.: Some effects of nuclear forces on neutron-star models. Can. J. Phys. **44**, 1895 (1966).
VANDERVOORT, P. O.: The formation of H II regions. Ap. J. **137**, 381, **138**, 426 (1963).
VANDERVOORT, P. O.: The stability of ionization fronts and the evolution of H II regions. Ap. J. **138**, 599 (1963).
VLASOV, A. A.: The oscillation properties of an electron gas. Zhur. Eksp. Theor. Fiz. **8**, 291 (1938).
VLASOV, A. A.: On the kinetic theory of an assembly of particles with collective interaction. J. Phys. (U.S.S.R.) **9**, 25 (1945).
VOGEL, H. C.: Spectrographische Beobachtungen an Algol (Spectroscopy of Algol). Astron. Nach. **123**, 289 (1889).
WANG, C. G., ROSE, W. K., SCHLENKER, S. L.: Models for neutron-core stars based on realistic nuclear-matter calculations. Ap. J. **160**, L 17 (1970).
WARNER, B.: Observations of rapid blue variables X. G 61-29. M.N.R.A.S. **159**, 315 (1972).
WARNER, B., NATHER, R. E.: Observations of rapid blue variables II—HL Tau 76. M.N.R.A.S. **156**, 1 (1972).
WEBER, E. J., DAVIS, L.: The angular momentum of the solar wind. Ap. J. **148**, 217 (1967).
WEIZSÄCKER, C. F. VON: The evolution of galaxies and stars. Ap. J. **114**, 165 (1951).
WHANG, Y. C., LIU, C. K., CHANG, C. C.: A viscous model of the solar wind. Ap. J. **145**, 255 (1966).
WHITTAKER, W. A.: Heating of the solar corona by gravity waves. Ap. J. **137**, 914 (1963).
YANG, C. H., CLARK, J. W.: Superfluid condensation energy of neutron matter. Nucl. Phys. **A 174**, 49 (1971).
YVON, J.: La theorie des fluides et l'equation d'état: Actualites scientifiques et industrielles (The theory of fluids and the equation of state: Scientific and industrial actualities). Paris: Hermann and Cie 1935.
ZELDOVICH, Y. B., NOVIKOV, I. D.: Relativistic astrophysics, vol. 1: Stars and relativity. Chicago, Ill.: University of Chicago Press 1971.
ZHELEZNYAKOV, V. V., ZAITSEV, V. V.: Contribution to the theory of type III radio bursts I. Sov. Astron. A. J. **14**, 47 (1970).
ZHEVAKIN, S. A.: Theory of Cepheids. Astron. J. Sov. Union **30**, 161 (1953).
ZHEVAKIN, S. A.: Pulsation theory of variable stars. Ann. Rev. Astron. Astrophys. **1**, 367 (1963).

Chapter 4

ABRAHAM, P. B., BRUNSTEIN, K. A., CLINE, T. L.: Production of low-energy cosmic-ray electrons. Phys. Rev. **150**, 1088 (1966).
ADAMS, J. B., RUDERMAN, M. A., WOO, C. H.: Neutrino pair emission by a stellar plasma. Phys. Rev. **129**, 1383 (1963).
AJZENBERG-SELOVE, F.: Nuclear spectroscopy. New York: Academic Press 1960.
ALFVÉN, H.: On the sidereal time variation of the cosmic radiation. Phys. Rev. **54**, 97 (1938).
ALLEN, B. J., GIBBONS, J. H., MACKLIN, R. L.: Nucleo-synthesis and neutron-capture cross sections. In: Advances in nuclear physics, vol. 4 (ed. M. BARANGER and E. VOGT). New York: Plenum Press 1971.
ALPHER, R. A., BETHE, H. A., GAMOW, G.: The origin of chemical elements. Phys. Rev. **73**, 803 (1948).
ALPHER, R. A., HERMAN, R, C,: Theory of the origin and relative abundance distribution of the elements. Rev. Mod. Phys. **22**, 153 (1950).
ANDERSON, C. D.: Energies of cosmic-ray particles. Phys. Rev. **41**, 405 (1932).
ANDERSON, C. D., NEDDERMEYER, S. H.: Cloud chamber observations of cosmic rays at 4300 meters elevation and near sea-level. Phys. Rev. **50**, 263 (1936).
ARNETT, W. D.: Gravitational collapse and weak interactions. Can. J. Phys. **44**, 2553 (1966).
ARNETT, W. D.: Mass dependence in gravitational collapse of stellar cores. Can. J. Phys. **45**, 1621 (1967).
ARNETT, W. D.: On supernova hydrodynamics. Ap. J. **153**, 341 (1968).
ARNETT, W. D.: A possible model of supernovae: Detonation of C^{12}. Astrophys. and Space Sci. **5**, 180 (1969).
ARNETT, W. D.: Explosive nucleosynthesis in stars. Ap. J. **137**, 1369 (1969).

ARNETT, W. D.: Galactic evolution and nucleosynthesis. Ap. J. **166**, 153 (1971).

ARNETT, W. D.: Hydrostatic oxygen burning in stars: I. Oxygen stars. Ap. J. **173**, 393 (1972).

ARNETT, W. D.: Explosive nucleosynthesis in stars. Ann. Rev. Astron. Astrophys. **11**, 73 (1973).

ARNETT, W. D., CAMERON, A. G. W.: Supernova hydrodynamics and nucleosynthesis. Can. J. Phys. **45**, 2953 (1967).

ARNETT, W. D., CLAYTON, D. D.: Explosive nucleosynthesis in stars. Nature **227**, 780 (1970).

ARNETT, W. D., TRURAN, J. W.: Carbon-burning nucleosynthesis at constant temperature. Ap. J. **157**, 339 (1969).

ARNETT, W. D., TRURAN, J. W., WOOSLEY, S. E.: Nucleosynthesis in supernova models II. The ^{12}C detonation model. Ap. J. **165**, 87 (1971).

ARNOULD, M.: Influence of the excited states of target nuclei in the vicinity of the iron peak on stellar reaction rates. Astron. Astrophys. **19**, 92 (1972).

ASTON, F. W.: A new mass-spectrograph and the whole number rule. Proc. Roy. Soc. London **A 115**, 487 (1927).

AUDOUZE, J., EPHERRE, M., REEVES, H.: Survey of experimental cross sections for proton-induced spallation reactions in He^4, C^{12}, N^{14}, and O^{16}. In: High energy nuclear reactions in astrophysics (ed. B. SHEN). New York: W. A. Benjamin 1967.

AUDOUZE, J., LEQUEUX, J., REEVES, H.: On the cosmic boron abundance. Astron. and Astrophys. **28**, 85 (1973).

BAADE, W., ZWICKY, F.: On super-novae. Proc. Nat. Acad. Sci. (Wash.) **20**, 254 (1934).

BAADE, W., ZWICKY, F.: Cosmic rays from super-novae. Proc. Nat. Acad. Sci. (Wash.) **20**, 259 (1934). Reprod. in: Selected papers on cosmic ray origin theories (ed. S. ROSEN). New York: Dover 1969.

BAADE, W., ZWICKY, F.: Remarks on super-novae and cosmic rays. Phys. Rev. **46**, 76 (1934). Reprod. in: Selected papers on cosmic ray origin theories (ed. S. ROSEN). New York: Dover 1969.

BAHCALL, J. N.: Beta decay in stellar interiors. Phys. Rev. **126**, 1143 (1962).

BAHCALL, J. N.: The exclusion principle and photobeta reactions in nucleogenesis. Ap. J. **136**, 445 (1962).

BAHCALL, J. N.: Electron capture in stellar interiors. Ap. J. **139**, 318 (1964).

BAHCALL, J. N.: Solar neutrino cross sections and nuclear beta decay. Phys. Rev. **135**, B 137 (1964).

BAHCALL, J. N.: Neutrino opacity: I. Neutrino-lepton scattering. Phys. Rev. **136**, B 1164 (1964).

BAHCALL, J. N.: Observational neutrino astronomy. Science **147**, 115 (1965).

BAHCALL, J. N., DAVIS, R., Jr.: Science **191**, 264 (1976).

BAHCALL, J. N., FRAUTSCHI, S. C.: Neutrino opacity: II. Neutrino-nucleon interactions. Phys. Rev. **136**, B 1547 (1964).

BAHCALL, J. N., MAY, R. M.: The rate of the proton-proton reaction and some related reactions. Ap. J. **155**, 501 (1969).

BAHCALL, J. N., SEARS, R. L.: Solar neutrinos. Ann. Rev. Astron. Astrophys. **10**, 25 (1972).

BAHCALL, J. N., ULRICH, R. K.: Solar neutrinos: III. Composition and magnetic field effects and related inferences. Ap. J. **170**, 593 (1971).

BAHCALL, J. N., WOLF, R. A.: Neutron stars. Phys. Rev. Lett. **14**, 343 (1965).

BAHCALL, J. N., WOLF, R. A.: Neutron stars: I. Properties at absolute zero temperature. Phys. Rev. **140**, B 1445 (1965).

BAHCALL, J. N., WOLF, R. A.: Neutron stars: II. Neutrino cooling and observability. Phys. Rev. **140**, B 1452 (1965).

BAHCALL, N. A., FOWLER, W. A.: The effect of excited nuclear states on stellar reaction rates. Ap. J. **157**, 645 (1969).

BAHCALL, N. A., FOWLER, W. A.: Nuclear partition functions for stellar reaction rates. Ap. J. **161**, 119 (1970).

BARDIN, R. K., BARNES, C. A., FOWLER, W. A., SEEGER, P. A.: ft value of O^{14} and the universality of the Fermi interaction. Phys. Rev. **127**, 583 (1962).

BARNES, C. A.: Nucleosynthesis by charged-particle reactions. In: Advances in nuclear physics, vol. 4 (ed. M. BARANGER, E. VOGT). New York: Plenum Press 1971.

BEAUDET, G., PETROSIAN, V., SALPETER, E. E.: Energy losses due to neutrino processes. Ap. J. **150**, 979 (1967).

BEAUDET, G., SALPETER, E. E., SILVESTRO, M. L.: Rates for URCA neutrino processes. Ap. J. **174**, 79 (1972).

BECQUEREL, H.: Sur les radiations émises par phosphorescence (On the radiation emitted by phosphorescence). Compt. Rend. **122**, 420 (1896).

BENJAMINI, R., LONDRILLO, P., SETTI, G.: The cosmic black-body radiation and the inverse Compton effect in the radio galaxies: The X-ray background. Nuovo Cimento **B 52**, 495 (1967).

BERGERON, J.: Étude des possibilités d'existence d'un plasma intergalactique dense. Astron. Astrophys. **3**, 42 (1969).

BERNAS, R., GRADSZTAJN, E., REEVES, H., SHATZMAN, E.: On the nucleosynthesis of lithium, beryllium, and boron. Ann. Phys. (N. Y.) **44**, 426 (1967).

BETHE, H. A.: Zur Theorie des Durchgangs schneller Korpuskularstrahlen durch Materie (On the theory of the penetration of matter by fast (nuclear) particle beams). Ann. Phys. **5**, 325 (1930).

BETHE, H. A.: Bremsformel für Elektronen relativistischer Geschwindigkeit (A braking formula for relativistic electrons). Z. Phys. **76**, 293 (1932).

BETHE, H. A.: Nuclear physics B. Nuclear dynamics, theoretical. Rev. Mod. Phys. **9**, 69 (1937).

BETHE, H. A.: Energy production in stars. Phys. Rev. **55**, 103, 434 (1939).

BETHE, H. A., CRITCHFIELD, C. L.: The formation of deutrons by proton combination. Phys. Rev. **54**, 248 (1938).

BETHE, H. A., HEITLER, W.: On the stopping of fast particles and on the creation of positive electrons. Proc. Roy. Soc. London **A 146**, 83 (1934).

BETHE, H. A., WILLS, H. H.: On the annihilation radiation of positrons. Proc. Roy. Soc. London **A 150**, 129 (1935).

BHABHA, H. J.: The creation of electron pairs by fast charged particles. Proc. Roy. Soc. London **A 152**, 559 (1935).

BHABHA, H. J.: On the calculation of pair creation by fast charged particles and the effect of screening. Proc. Camb. Phil. Soc. **31**, 394 (1935).

BHABHA, H. J.: The scattering of positrons by electrons with exchange on Dirac's theory of the positron. Proc. Roy. Soc. London **A 154**, 195 (1936).

BLATT, J. M., WEISSKOPF, V. F.: Theoretical nuclear physics. New York: Wiley 1952.

BLIN-STOYLE, R. J., FREEMAN, J. M.: Coupling constants and electromagnetic radiative corrections in beta-decay and the mass of the intermediate vector boson. Nucl. Phys. **A 150**, 369 (1970).

BLOCH, F. VON: Bremsvermögen von Atomen mit mehreren Elektronen (Braking capabilities of multi-electron atoms). Z. Phys. **81**, 363 (1933).

BLUMENTHAL, G. R., GOULD, R. J.: Bremsstrahlung, synchrotron radiation, and Compton scattering of high energy electrons traversing dilute gases. Rev. Mod. Phys. **42** (2), 237 (1970).

BODANSKY, D., CLAYTON, D. D., FOWLER, W. A.: Nucleosynthesis during silicon burning. Phys. Rev. Lett. **20**, 161 (1968).

BODANSKY, D., CLAYTON, D. D., FOWLER, W. A.: Nuclear quasi-equilibrium during silicon burning. Ap. J. Suppl. No. 148, **16**, 299 (1968).

BODENHEIMER, P.: Studies in stellar evolution: II. Lithium depeletion during the pre-main sequence contraction. Ap. J. **142**, 451 (1965).

BODENHEIMER, P.: Depletion of deuterium and beryllium during pre-main sequence evolution. Ap. J. **144**, 103 (1966).

BRADT, H. L., PETERS, B.: Investigation of the primary cosmic radiation with nuclear photographic emulsion. Phys. Rev. **74**, 1828 (1948).

BRADT, H. L., PETERS, B.: The heavy nuclei of the primary cosmic radiation. Phys. Rev. **77**, 54 (1950).

BRECHER, K., MORRISON, P.: Leakage electrons from normal galaxies: The diffuse cosmic X-ray source. Phys. Rev. Lett. **23**, 802 (1969).

BREIT, G., WIGNER, E.: Capture of slow neutrons. Phys. Rev. **49**, 519 (1936).

BRILLOIN, L.: Les principes de la nouvelle mécanique ondulatoire (Principles of the undulatory mechanics). J. Phys. et le Radium **7**, 321 (1926). Remarques sur la mécanique ondulatoire (Notes on undulatory mechanics). J. Phys. Radium **7**, 353 (1926).

BROWN, J. C.: The deduction of energy spectra of non-thermal electrons in flares from the observed dynamic spectra of hard X-ray bursts. Solar Phys. **18**, 489 (1971).

BROWN, J. C.: The decay characteristics of models of solar hard X-ray bursts. Solar Phys. **25**, 158 (1972).

BROWN, J. C.: The directivity and polarization of thick target X-ray bremsstrahlung from solar flares. Solar Phys. **26**, 441 (1972).

BROWN, J. C.: Thick target X-ray bremsstrahlung from partially ionized targets in solar flares. Solar Phys. **28**, 151 (1973).

BROWN, R. L., GOULD, R. J.: Interstellar absorption of cosmic X-rays. Phys. Rev. D 1, **1**, 2252 (1970).

BRUENN, S. W.: The effect of URCA shells on the density of carbon ignition in degenerate stellar cores. Ap. J. **177**, 459 (1972).

BURBIDGE, G. R.: Acceleration of cosmic-ray particles among extragalactic nebulae. Phys. Rev. **107**, 269 (1957). Reprod. in: Selected papers on cosmic ray origin theories (ed. S. ROSEN). New York: Dover 1969.

BURBIDGE, G. R.: X-ray and γ-ray sources. In: High energy astrophysics—Int. school of physics Enrico Fermi course 35 (ed. L. GRATTON). New York: Academic Press 1966.

BURBIDGE, E. M., BURBIDGE, G. R., FOWLER, W. A., HOYLE, F.: Synthesis of the elements in stars. Rev. Mod. Phys. **29**, 547 (1957).

CAMERON, A. G. W.: Origin of anomalous abundances of the elements in giant stars. Phys. Rev. **93**, 932 (1954).

CAMERON, A. G. W.: Origin of anomalous abundances of the elements in giant stars. Ap. J. **121**, 144 (1955).

CAMERON, A. G. W.: Photobeta reactions in stellar interiors. Ap. J. **130**, 452 (1959).

CAMERON, A. G. W.: Pycnonuclear reactions and nova explosions. Ap. J. **130**, 916 (1959).

CAMERON, A. G. W.: New neutron sources of possible astrophysical interest. Astron. J. **65**, 485 (1960).

CAMERON, A. G. W.: Abundances of the elements in the solar system. Space Sci. Rev. **15**, 121 (1973).

CAMERON, A. G. W., FOWLER, W. A.: Lithium and the *s*-Process in red-giant stars. Ap. J. **164**, 111 (1971).

CANUTO, V., CHOU, C. K.: Neutrino luminosity by the ordinary URCA process in an intense magnetic field. Astrophys. and Space Sci. **10**, 246 (1971).

CANUTO, V., CHIU, H. Y., CHOU, C. K.: Neutrino bremsstrahlung in an intense magnetic field. Phys. Rev. **D 2**, 281 (1970).

CANUTO, V., CHIUDERI, C., CHOU, C. K.: Plasmon neutrino emission in a strong magnetic field: I. Transverse plasmons; II. Longitudinal plasmons. Astrophys. and Space Sci. **7**, 407, **9**, 453 (1970).

CAUGHLAN, G. R.: Approach to equilibrium in the CNO bi-cycle. Ap. J. **141**, 688 (1965).

CAUGHLAN, G. R., FOWLER, W. A.: The mean lifetimes of carbon, nitrogen, and oxygen nuclei in the CNO bi-cycle. Ap. J. **136**, 453 (1962).

CESARSKY, D. A., MOFFET, A. T., PASACHOFF, J. M.: 327 MHz observations of the galactic center: Possible detection of a deuterium absorption line. Ap. J. **180**, L 1 (1973).

CHADWICK, J.: The existence of a neutron. Proc. Roy. Soc. London **A 136**, 692 (1932).

CHAMBERLAIN, O., SEGRE, E., WIEGAND, C., YPSILANTIS, T.: Observation of antiprotons. Phys. Rev. **100**, 947 (1955).

CHANDRASEKHAR, S., HENRICH, L. R.: An attempt to interpret the relative abundances of the elements and their isotopes. Ap. J. **95**, 288 (1942).

CHIU, H. Y.: Neutrino emission processes, stellar evolution, and supernovae, part I, part II. Ann. Phys. (N. Y.) **15**, 1, **16**, 321 (1961).

CHIU, H. Y.: Annihilation process of neutrino production in stars. Phys. Rev. **123**, 1040 (1961).

CHIU, H. Y.: Stellar physics. Waltham, Mass.: Blaisdell 1968.

CHIU, H. Y., MORRISON, P.: Neutrino emission from black-body radiation at high stellar temperatures. Phys. Rev. Lett. **5**, 573 (1960).

CHIU, H. Y., SALPETER, E. E.: Surface X-ray emission from neutron stars. Phys. Rev. Lett. **12**, 413 (1964).

CHIU, H. Y., STABLER, R. C.: Emission of photoneutrinos and pair annihilation neutrinos from stars. Phys. Rev. **122**, 1317 (1961).

CHUPP, E. L.: Gamma ray and neutron emissions from the Sun. Space Sci. Rev. **12**, 486 (1971).

CHUPP, E. L., FORREST, D. J., HIGBIE, P. R., SURI, A. N., TSAI, C., DUNPHY, P. P.: Solar gamma ray lines observed during the solar activity of August 2 to August 11, 1972. Nature (GB) **241**, 335 (1973).

CHRISTENSEN, C. J., NIELSEN, A., BAHNSEN, H., BROWN, W. K., RUSTAD, B. M.: The half-life of the free neutron. Phys. Lett. **26 B**, 11 (1967).

CLAYTON, D. D.: Cosmoradiogenic chronologies of nucleosynthesis. Ap. J. **139**, 637 (1964).

CLAYTON, D. D.: Principles of stellar evolution and nucleosynthesis. New York: McGraw-Hill 1968.

CLAYTON, D. D.: Isotopic composition of cosmic importance. Nature **224**, 56 (1969).

CLAYTON, D. D.: New prospect for gamma-ray line astronomy. Nature **234**, 291 (1971).

CLAYTON, D. D., COLGATE, S. A., FISHMAN, G. J.: Gamma-ray lines from young supernova remnants. Ap. J. **155**, 75 (1969).

CLAYTON, D. D., CRADDOCK, W. L.: Radioactivity in supernova remnants. Ap. J. **142**, 189 (1965).

CLAYTON, D. D., SILK, J.: Measuring the rate of nucleosynthesis with a gamma-ray detector. Ap. J. **158**, L 43 (1969).
CLIFFORD, F. E., TAYLER, R. J.: The equilibrium distribution of nuclides in matter at high temperatures. Mem. R. A. S. **69**, 21 (1965).
CLINE, T. L., DESAI, U. D., KLEBESADEL, R. W., STRONG, I. B.: Energy spectra of cosmic gamma-ray bursts. Ap. J. **185**, L 1 (1973).
COLGATE, S. A., JOHNSON, M. H.: Hydrodynamic origin of cosmic rays. Phys. Rev. Lett. **5**, 235 (1960).
COLGATE, S. A., WHITE, R. H.: The hydrodynamic behavior of supernovae explosions. Ap. J. **143**, 626 (1966).
COMPTON, A. H.: A quantum theory of the scattering of X-rays by light elements. Phys. Rev. **21**, 207 (1923).
COOK, C. W., FOWLER, W. A., LAURITSEN, C. C., LAURITSEN, T.: B^{12}, C^{12}, and the red giants. Phys. Rev. **107**, 508 (1957).
COMSTOCK, G. M., FAN, C. Y., SIMPSON, J. A.: Energy spectra and abundances of the cosmic-ray nuclei helium to iron from the OGOI satellite experiment. Ap. J. **155**, 609 (1969).
COUCH, R. G., SHANE, K. C.: The photodisintegration rate of ^{24}Mg. Ap. J. **169**, 413 (1971).
COWSIK, R., KOBETICH, E. J.: Comment on inverse Compton models for the isotropic X-ray background and possible thermal emission from a hot intergalactic gas. Ap. J. **177**, 585 (1972).
COX, J. P., GIULI, R. T.: Principles of stellar structure, vol. I, vol. II. New York: Gordon and Breach 1968.
CRANDALL, W. E., MILLBURN, G. P., PYLE, R. W., BIRNBAUM, W.: $C^{12}(x, xn)C^{11}$ and $A^{27}(x, x2pn)Na^{24}$ cross sections at high energies. Phys. Rev. **101**, 329 (1956).
CULHANE, J. L.: Thermal continuum radiation from coronal plasmas at soft X-ray wavelengths. M. N. R. A. S. **144**, 375 (1969).
CURIE, I., JOLIOT, F.: Émission de protons de grande vitesse par les substances hydrogénées sous l'influence des rayons γ trés pénétrants (Emission of very fast protons by hydrogen substances under the influence of very penetrating γ-rays). Comp. Rend. **194**, 273 (1932).
DANBY, G., GAILLARD, J.-M., GOULIANOS, K., LEDERMAN, L. M., MISTRY, N., SCHWARTZ, M., STEINBERGER, J.: Observation of high-energy neutrino reactions and the existence of two kinds of neutrinos. Phys. Rev. Lett. **9**, 36 (1962).
DARWIN, SIR C.: Source of the cosmic rays. Nature **164**, 1112 (1949). Reprod. in: Selected papers on cosmic ray origin theories (ed. S. ROSEN). New York: Dover 1966.
DAVIDSON, K., OSTRIKER, J. P.: Neutron star accretion in a stellar wind: Model for a pulsed X-ray source. Ap. J. **179**, 585 (1973).
DAVIS, L.: Modified Fermi mechanism for the acceleration of cosmic rays. Phys. Rev. **101**, 351 (1956).
DAVIS, R., HARMER, D. S., HOFFMAN, K. C.: Search for neutrinos from the Sun. Phys. Rev. Lett. **20**, 1205 (1968).
DAVISSON, C. M., EVANS, R. D.: Gamma-ray absorption coefficients. Rev. Mod. Phys. **24**, 79 (1952).
DEBYE, P. VON, HÜCKEL, E.: Zur Theorie der Elektrolyte: I. Gefrierpunktserniedrigung und verwandte Erscheinungen; II. Das Grenzgesetz für die elektrische Leitfähigkeit (On the theory of electrolytes: I. Lowering of the freezing point and related phenomena; II. The limiting laws for electrical conductivity). Phys. Z. **24**, 185 (1923).
DIRAC, P. A. M.: On the annihilation of electrons and protons. Proc. Camb. Phil. Soc. **26**, 361 (1930).
DISMUKE, N., ROSE, M. E., PERRY, C. L., BELL, P. R.: Tables for the analysis of beta spectra. Nat. Bur. Stands. (Wash.) App. Math. Ser. **13** (1952).
DONAHUE, T. M.: The significance of the absence of primary electrons for theories of the origin of the cosmic radiation. Phys. Rev. **84**, 972 (1951).
EARL, J. A.: Cloud-chamber observations of primary cosmic-ray electrons. Phys. Rev. Lett. **6**, 125 (1961).
EBERHARDT, P., GEISS, J., GRAF, H., GROGLER, N., KRAHENBUHL, U., SCHWALLER, H., SCHWARZMÜLLER, J., STETTLER, A.: Trapped solar wind noble gases, Kr^{81}/Kr exposure ages, and K/Ar ages in Apollo 11 lunar material. Science **167**, 558 (1970).
EDDINGTON, A. S.: The internal constitution of the stars. Nature **106**, 14 (1920).
EDDINGTON, A. S.: The internal constitution of the stars. Cambridge: Cambridge University Press 1926. Republished New York: Dover Publ. 1959.
EINSTEIN, A. VON: Über einen die Erzeugung und Verwandlung des Lichtes betreffenden heuristischen Gesichtspunkt (On a heuristic point of view concerning the generation and propagation of light). Ann. Phys. **17**, 132 (1905).

EINSTEIN, A.: Ist die Trägheit eines Körpers von seinem Energieinhalt abhängig? (Does the inertia of a body depend on its energy?). Ann. Phys. **18**, 639 (1905).

EINSTEIN, A.: Das Prinzip von der Erhaltung der Schwerpunktsbewegung und die Trägheit der Energie (Conservation of the motion of the mass center). Ann. Phys. **20**, 627 (1906).

EINSTEIN, A.: Über die vom Relativitätsprinzip geforderte Trägheit der Energie (Variation of inertia with energy on the principle of relativity). Ann. Phys. **23**, 371 (1907).

ELSASSER, W. M.: Sur le principe de Pauli dans les noyaux (On Pauli's principle within the nucleus). J. Phys. Radium **4**, 549 (1933).

FANSELOW, J. L., HARTMAN, R. C., HILDEBRAND, R. H., MEYER, P.: Charge composition and energy spectrum of primary cosmic-ray electrons. Ap. J. **158**, 771 (1969).

FASSIO-CANUTO, L.: Neutron beta decay in a strong magnetic field. Phys. Rev. **187**, 2141 (1969).

FEENBERG, E., PRIMAKOFF, H.: Interaction of cosmic-ray primaries with sunlight and starlight. Phys. Rev. **73**, 449 (1948).

FEENBERG, E., TRIGG, G.: The interpretation of comparative half-lives in the Fermi theory of beta-decay. Rev. Mod. Phys. **22**, 399 (1950).

FEINGOLD, A. M.: Table of ft values in beta-decay. Rev. Mod. Phys. **23**, 10 (1951).

FELTEN, J. E., MORRISON, P.: Recoil photons from scattering of starlight by relativistic electrons. Phys. Rev. Lett. **10**, 453 (1963).

FELTEN, J. E., MORRISON, P.: Omnidirectional inverse Compton and synchrotron radiation from cosmic distributions of fast electrons and thermal photons. Ap. J. **146**, 686 (1966).

FERMI, E.: Eine statistische Methode zur Bestimmung einiger Eigenschaften des Atoms und ihre Anwendung auf die Theorie des periodischen Systems der Elemente (A statistical method for determining the eigenstate of atoms and its application to the theory of the periodical system of the elements). Z. Physik **48**, 73 (1928).

FERMI, E.: Versuch einer Theorie der β-Strahlen I (An attempt at a theory of β-rays I). Z. Physik **88**, 161 (1934).

FERMI, E.: The absorption of mesotrons in air and in condensed materials. Phys. Rev. **56**, 1242 (1939).

FERMI, E.: The ionization loss of energy in gases and in condensed materials. Phys. Rev. **57**, 485 (1940).

FERMI, E.: On the origin of cosmic radiation. Phys. Rev. **75**, 1169 (1949). Reprod. in: Selected papers on cosmic ray origin theories (ed. S. ROSEN). New York: Dover 1969.

FERMI, E.: Galactic magnetic fields and the origin of cosmic radiation. Ap. J. **119**, 1 (1954). Reprod. in: Selected papers on cosmic ray origin theories (ed. S. ROSEN). New York: Dover 1969.

FERMI, E., UHLENBECK, G. E.: On the recombination of electrons and positrons. Phys. Rev. **44**, 510 (1933).

FESTA, G. G., RUDERMAN, M. A.: Neutrino-pair bremsstrahlung from a degenerate electron gas. Phys. Rev. **180**, 1227 (1969).

FEYNMAN, R. P., GELL-MANN, M.: Theory of the Fermi interaction. Phys. Rev. **109**, 193 (1958).

FIELD, G. B., HENRY, R. C.: Free-free emission by intergalactic hydrogen. Ap. J. **140**, 1002 (1964).

FINZI, A.: Vibrational energy of neutron stars and the exponential light curves of type I supernovae. Phys. Rev. Lett. **15**, 599 (1965).

FINZI, A.: Cooling of a neutron star by the URCA process. Phys. Rev. **137**, B 472 (1965).

FINZI, A., WOLF, R. A.: Type I supernovae. Ap. J. **150**, 115 (1967).

FINZI, A., WOLF, R. A.: Hot, vibrating neutron stars. Ap. J. **153**, 835 (1968).

FISHMAN, G. J., CLAYTON, D. D.: Nuclear gamma rays from ^{7}Li in the galactic cosmic radiation. Ap. J. **178**, 337 (1972).

FORMAN, W., JONES, C. A., LILLER, W.: Optical studies of UHURU sources: III. Optical variations of the X-ray eclipsing system HZ Herculis. Ap. J. **177**, L 103 (1972).

FOWLER, W. A.: Experimental and theoretical results on nuclear reactions in stars. Mem. Soc. Roy. Sci. Liege **14**, 88 (1954).

FOWLER, W. A.: Completion of the proton-proton reaction chain and the possibility of energetic neutrino emission by hot stars. Ap. J. **127**, 551 (1958).

FOWLER, W. A.: Experimental and theoretical results on nuclear reactions in stars II. Mem. Soc. Roy. Sci. Liege Ser 5, **3**, 207 (1959).

FOWLER, W. A.: What cooks with solar neutrinos. Nature **238**, 24 (1972).

FOWLER, W. A., BURBIDGE, G. R., BURBIDGE, E. M.: Nuclear reactions and element synthesis in the surfaces of stars. Ap. J. Suppl. No. 17, **2**, 167 (1955).

FOWLER, W. A., CAUGHLAN, G. R., ZIMMERMAN, B. A.: Thermonuclear reaction rates. Ann. Rev. Astron. Astrophys. **5**, 525 (1967).

FOWLER, W. A., CAUGHLAN, C. R., ZIMMERMAN, B. A.: Ann. Rev. Astron. Astrophys. **13**, 69 (1975).
FOWLER, W. A., CAUGHLAN, G. R., ZIMMERMAN, B. A.: Thermonuclear reaction rates. Priv. Comm. (1974).
FOWLER, W. A., GREENSTEIN, J. L.: Element building reactions in stars. Proc. Nat. Acad. Sci. (Wash.) **42**, 173 (1956).
FOWLER, W. A., GREENSTEIN, J. L., HOYLE, F.: Nucleosynthesis during the early history of the solar system. J. Geophys. Res. R.A.S. **6**, 148 (1962).
FOWLER, W. A., HOYLE, F.: Nuclear cosmochronology. Ann. Phys. **10**, 280 (1960).
FOWLER, W. A., HOYLE, F.: Neutrino processes and pair formation in massive stars and supernovae. Ap. J. Suppl. **9**, 201 (1964).
FOWLER, W. A., REEVES, H., SILK, J.: Spallation limits on interstellar fluxes of low-energy cosmic rays and nuclear gamma rays. Ap. J. **162**, 49 (1970).
FREEMAN, J. M., CLARK, G. J., RYDER, J. S., BURCHAM, W. E., SGUIER, G. T. A., DRAPER, J. E.: Present values of the weak interaction coupling constants. U.K. Atom. Energy Res. Group, 1972.
FREIER, P., LOFGREN, E. J., NEY, E. P., OPPENHEIMER, F., BRADT, H. L., PETERS, B.: Evidence for heavy nuclei in the primary cosmic radition. Phys. Rev. **74**, 213, 1818 (1948).
FREIER, P. S., WADDINGTON, C. J.: Singly and doubly charged particles in the primary cosmic radiation. J. Geophys. Res. **73**, 4261 (1968).
FRIEDMAN, H.: Cosmic X-ray sources: A progress report. Science **181**, 395 (1973).
FRÖBERG, C. E.: Numerical treatment of Coulomb wave functions. Rev. Mod. Phys. **27**, 399 (1955).
GAMOW, G.: Zur Quantentheorie des Atomkernes (On the quantum theory of the atomic nucleus). Z. Physik **51**, 204 (1928).
GAMOW, G.: Expanding universe and the origin of the elements. Phys. Rev. **70**, 572 (1946).
GAMOW, G., SCHOENBERG, M.: Neutrino theory of stellar collapse. Phys. Rev. **59**, 539 (1941).
GANDELMAN, G. M., PINAEV, V. S.: Emission of neutrino pairs by electrons and the role played by it in stars. Sov. Phys. JETP **10**, 764 (1960).
GARDNER, E., LATTES, C. M. G.: Production of mesons by the 184 inch Berkeley cyclotron. Science **107**, 270 (1948).
GARVEY, G. T., GERACE, W. J., JAFFE, R. L., TALMI, I., KELSON, I.: Set of nuclear-mass relations and a resultant mass table. Rev. Mod. Phys. **41**, S 1 (1969).
GERSHTEIN, S. S., IMSHENNIK, V. S., NADYOZHIN, D. K., FOLOMESHKIN, V. N., KHLOPOV, M. Yu., CHECHETKIN, V. M., ERAMZHYAN, R. A.: Zh. ETP **69**, 1473 (1975); Phys. Lett. **62B**, 100 (1976).
GIACCONI, R., MURRAY, S., GURSKY, H., KELLOGG, E., SCHREIER, E., TANANBAUM, H.: The UHURU catalog of X-ray sources. Ap. J. **178**, 281 (1972).
GILBERT, A., CAMERON, A. G. W.: A composite nuclear-level density formula with shell corrections. Can. J. Phys. **43**, 1446 (1965).
GINZBURG, V. L.: Elementary processes for cosmic ray astrophysics. New York: Gordon and Breach 1969.
GOLD, T.: Rotating neutron stars as the origin of the pulsating radio sources. Nature **218**, 731 (1968).
GOLDREICH, P., JULIAN, W. H.: Pulsar electrodynamics. Ap. J. **157**, 869 (1969).
GOLDREICH, P., MORRISON, P.: On the absorption of gamma rays in intergalactic space. Soviet Phys. JETP **18**, 239 (1964).
GOLDSCHMIDT, V. M.: Geochemische Verteilungsgesetze der Elemente: IX. Die Mengenverhältnisse der Elemente und der Atom-Arten (Geochemical distribution laws of the elements: IX. The abundances of elements and of types of atoms). Skrift. Norske Videnskaps-Akad. Oslo I. Mat. Nat. No. 4 (1937).
GOULD, R. J.: High-energy photons from the Compton-synchrotron process in the Crab nebula. Phys. Rev. Lett. **15**, 577 (1965).
GOULD, R. J., SCHRÉDER, G.: Opacity of the universe to high-energy photons. Phys. Rev. Lett. **16**, 252 (1966).
GRABOSKE, H. C., DE WITT, H. E., GROSSMAN, A. S., COOPER, M. S.: Screening factors for nuclear reactions: II. Intermediate screening and astrophysical applications. Ap. J. **181**, 457 (1973).
GRADSZTAJN, E.: Production of Li, Be, and B isotopes in C, N, and O. In: High energy nuclear reactions in astrophysics (ed. B. SHEN). New York: W. A. Benjamin 1967.
GREEN, A. E. S.: Coulomb radius constant from nuclear masses. Phys. Rev. **95**, 1006 (1954).
GREENSTEIN, J. L.: A search for He^3 in the Sun. Ap. J. **113**, 531 (1951).
GREENSTEIN, J. L., RICHARDSON, R. S.: Lithium and the internal circulation of the Sun. Ap. J. **113**, 536 (1951).
GREVESSE, N.: Solar abundances of lithium, beryllium, and boron. Solar Phys. **5**, 159 (1968).

GURNEY, R. W., CONDON, E. U.: Wave mechanics and radioactive disintegration. Nature **122**, 493 (1928).

GURNEY, R. W., CONDON, E. U.: Quantum mechanics and radioactive disintegration. Phys. Rev. **33**, 127 (1929).

HANSEN, C. J.: Some weak interaction processes in highly evolved stars. Astrophys. and Space Sci. **1**, 499 (1968).

HANSEN, C. J.: Explosive carbon-burning nucleosynthesis. Astrophys. and Space Sci. **14**, 389 (1971).

HAYAKAWA, S.: Cosmic ray physics. New York: Wiley 1969.

HAYAKAWA, S., HAYASHI, C., NISHIDA, M.: Rapid thermonuclear reactions in supernova explosions. Suppl. Prog. Theor. Phys. (Japan) **16**, 169 (1960).

HAYASHI, C., HOSHI, R., SUGIMOTO, D.: Evolution of stars. Prog. Theor. Phys. (Japan) Suppl. **22**, 1 (1962).

HEISENBERG, W.: Über den Bau der Atomkerne I (On the structure of atomic nuclei I). Z. Physik **77**, 1 (1932).

HEITLER, W.: The quantum theory of radiation. Oxford: Oxford University Press 1954.

HEITLER, W., NORDHEIM, L.: Sur la production des paires par des chocs de particules lourdes (On the production of pairs by the collision of heavy particles). J. Phys. Radium **5**, 449 (1934).

HENRY, R. C., FRITZ, G., MEEKINS, J. F., FRIEDMAN, H., BYRAM, E. T.: Possible detection of a dense intergalactic plasma. Ap. J. Lett. **153**, L 11 (1968).

HESS, V. F.: Über die Absorption der γ-Strahlen in der Atmosphäre (On the absorption of γ-rays in the atmosphere). Phys. Z. **12**, 998 (1911).

HESS, V. F.: Über Beobachtungen der durchdringenden Strahlung bei sieben Freiballonfahrten (On observations of the penetrating radiation during seven balloon flights). Phys. Z. **13**, 1084 (1912).

HEWISH, A., BELL, S. J., PILKINGTON, J. D. H., SCOTT, P. F., COLLINS, R. A.: Observation of a rapidly pulsating radio source. Nature **217**, 709 (1968).

VAN HIEU, N., SHABALIN, E. P.: Role of the $\gamma+\gamma\to\gamma+\nu+\bar{\nu}$ process in neutrino emission by stars. Sov. Phys. JETP **17**, 681 (1963).

VAN HORN, H. M., SALPETER, E. E.: WKB approximation in three dimensions. Phys. Rev. **157**, 751 (1967).

HOWARD, W. M., ARNETT, W. D., CLAYTON, D. D.: Explosive nucleosynthesis in helium zones. Ap. J. **165**, 495 (1971).

HOWARD, W. M., ARNETT, W. D., CLAYTON, D. D., WOOSLEY, S. E.: Nucleosynthesis of rare nuclei from seed nuclei in explosive carbon burning. Ap. J. **175**, 201 (1972).

HOYLE, F.: The synthesis of elements from hydrogen. M.N.R.A.S. **106**, 343 (1946).

HOYLE, F.: On nuclear reactions occuring in very hot stars: I. The synthesis of elements from carbon to nickel. Ap. J. Suppl. **1**, 121 (1954).

HOYLE, F., FOWLER, W. A.: Nucleosynthesis in supernovae. Ap. J. **132**, 565 (1960).

HOYLE, F., TAYLER, R. J.: The mystery of the cosmic helium abundance. Nature **203**, 1108 (1964).

HUDSON, H. S.: Thick-target processes and white light flares. Solar Phys. **24**, 414 (1972).

HULL, M. H., BRIET, G.: Coulomb wave functions. In: Handbuch der Physik, vol. XLI: Nuclear reactions; II: Theory (ed. S. FLÜGGE), p. 408. Berlin-Göttingen-Heidelberg: Springer 1959.

IBEN, I.: The Cl^{37} solar neutrino experiment and the solar helium abundance. Ann. Phys. (N. Y.) **54**, 164 (1969).

INMAN, C. L., RUDERMAN, M. A.: Plasma neutrino emission from a hot, dense electron gas. Ap. J. **140**, 1025 (1964).

ITO, K.: Stellar synthesis of the proton-rich heavy elements. Prog. Theor. Phys. (Japan) **26**, 990 (1961).

DE JAGER, C., KUNDU, M. R.: A note on bursts of radio emission and high energy (> 20 KeV) X-rays from solar flares. Space Res. **3**, 836 (1963).

JELLEY, J. V.: High-energy γ-ray absorption in space by a 3.5° K microwave field. Phys. Rev. Lett. **16**, 479 (1966).

KÄLLÉN, G.: Radioactive corrections to β-decay and nucleon form factors. Nucl. Phys. **B 1**, 225 (1967).

KINMAN, T. D.: An attempt to detect deuterium in the solar atmosphere. M.N.R.A.S. **116**, 77 (1956).

KLEBESADEL, R. W., STRONG, I. B., OLSON, R. A.: Observations of gamma-ray bursts of cosmic origin. Ap. J. **182**, L 85 (1973).

KLEIN, O., NISHINA, Y.: Über die Streuung von Strahlung durch freie Elektronen nach der neuen relativistischen Quantendynamik von Dirac (On the scattering of radiation by free electrons according to the new relativistic quantum dynamics by Dirac). Z. Physik **52**, 853 (1929).

KODAMA, T.: β-stability line and liquid-drop mass formulas. Prog. Theor. Phys. (Japan) **45**, 1112 (1971).

KOLHÖRSTER, W.: Messungen der durchdringenden Strahlung im Freiballon in größeren Höhen (Measurements of the penetrating radiation from a balloon at greater altitudes). Phys. Z. **14**, 1153 (1913).

KONOPINSKI, E. J.: The experimental clarification of the laws of β-radioactivity. Ann. Rev. Nucl. Sci. **9**, 99 (1959).

KONOPINSKI, E. J.: The theory of beta radioactivity. Oxford: Oxford at the Clarendon Press 1966.

KRAMERS, H. A.: On the theory of X-ray absorption and of the continuous X-ray spectrum. Phil. Mag. **46**, 836 (1923).

KRAMERS, H. A.: Wellenmechanik und halbzahlige Quantisierung (Wave mechanics and semi-numerical quantization). Z. Phys. **39**, 828 (1926).

KUCHOWICZ, B.: Nuclear astrophysics: A bibliographical survey. New York: Gordon and Breach 1967.

KULSRUD, R. M., OSTRIKER, J. P., GUNN, J. E.: Acceleration of cosmic rays in supernova remnants. Phys. Rev. Lett. **28**, 636 (1972).

LAMB, F. K., PETHICK, C. J., PINES, D.: A model for compact X-ray sources: Accretion by rotating magnetic stars. Ap. J. **184**, 271 (1973).

LANDAU, L., POMERANCHUK, I.: Limits of applicability of the theory of bremsstrahlung electron and pair production for high energies. Dokl. Akad. Nauk. USSR **92**, 535 (1953).

LANDSTREET, J. D.: Synchrotron radiation of neutrinos and its astrophysical significance. Phys. Rev. **153**, 1372 (1967).

LATTES, C. M. G., OCCHIALINI, G. P. S., POWELL, C. F.: Observations of the tracks of slow mesons in photographic emulsions. Nature **160**, 453 (1947).

LEDERER, C. M., HOLLANDER, J. M., PERLMAN, J.: Table of isotopes. New York: Wiley 1967.

LIN, R. P., HUDSON, H. S.: 10–100 KeV electron acceleration and emission from solar flares. Solar Phys. **17**, 412 (1971).

LINGENFELTER, R. E.: Solar flare optical, neutron, and gamma-ray emission. Solar Phys. **8**, 341 (1969).

LINGENFELTER, R. E., RAMATY, R.: High energy nuclear reactions in solar flares. In: High energy nuclear reactions in astrophysics (ed. B. SHEN). New York: W. A. Benjamin 1967.

LYONS, P. B.: Total yield measurements in ^{24}Mg$(\alpha, \gamma)^{28}$Si. Nucl. Phys. **A 130**, 25 (1969).

LYONS, P. B., TOEVS, J. W., SARGOOD, J. W.: Total yield measurements in ^{27}Al$(p, \gamma)^{28}$Si. Phys. Rev. **C 2**, 22041 (1969).

MACKLIN, R. L.: Were the lightest stable isotopes produced by photodissociation? Ap. J. **162**, 353 (1970).

MAJOR, J. K., BIEDENHARN, L. C.: Sargent diagram and comparative half-lives for electron capture transitions. Rev. Mod. Phys. **26**, 321 (1954).

MARION, J. B., FOWLER, W. A.: Nuclear reactions with the neon isotopes in stars. Ap. J. **125**, 221 (1957).

MARSHAK, R. E., BETHE, H. A.: On the two-meson hypothesis. Phys. Rev. **72**, 506 (1947).

MATINYAN, S. G., TSILOSANI, N. N.: Transformation of photons into neutrino pairs and its significance in stars. Sov. Phys. JETP **14**, 1195 (1962).

MAYER, M. G.: On closed shells in nuclei. Phys. Rev. **74**, 235 (1948).

MCCAMMON, D., BUNNER, A. N., COLEMAN, P. L., KRAUSHAAR, W. L.: A search for absorption of the soft X-ray diffuse flux by the small Magellanic cloud. Ap. J. **168**, L 33 (1971).

MENEGUZZI, M., AUDOUZE, J., REEVES, H.: The production of the elements Li, Be, B by galactic cosmic rays in space and its relation with stellar observations. Astron. Astrophys. **15**, 337 (1971).

MICHAUD, G., FOWLER, W. A.: Thermonuclear reaction rates at high temperatures. Phys. Rev. **C 2**, 22041 (1970).

MICHAUD, G., FOWLER, W. A.: Nucleosynthesis in silicon burning. Ap. J. **173**, 157 (1972).

MIGDAL, A. B.: Bremsstrahlung and pair production in condensed media at high energies. Phys. Rev. **103**, 1811 (1956).

MILNE, D. K.: Nonthermal galactic radio sources. Austr. J. Phys. **23**, 425 (1970).

MITLER, H. E.: Cosmic-ray production of deuterium, He^3, lithium, beryllium, and boron in the Galaxy. Smithsonian Ap. Obs. Spec. Rpt. 330 (1970).

MÖLLER, C.: On the capture of orbital electrons by nuclei. Phys. Rev. **51**, 84 (1937).

MORRISON, P.: On gamma-ray astronomy. Nuovo Cimento **7**, 858 (1958).

MORRISON, P., OLBERT, S., ROSSI, B.: The origin of cosmic rays. Phys. Rev. **94**, 440 (1954).

MUROTA, T., UEDA, A.: On the foundation and the applicability of Williams-Weizsäcker method. Prog. Theor. Phys. **16**, 497 (1956).

MUROTA, T., UEDA, A., TANAKA, H.: The creation of an electron pair by a fast charged particle. Prog. Theor. Phys. **16**, 482 (1956).

MYERS, W. D.: Droplet model nuclear density distributions and single-particle potential wells. Nucl. Phys. **A 145**, 387 (1970).

MYERS, W. D., SWIATECKI, W. J.: Nuclear masses and deformations. Nucl. Phys. **81**, 1 (1966).

NEDDERMEYER, S. H., ANDERSON, C. D.: Note on the nature of cosmic-ray particles. Phys. Rev. **51**, 884 (1937).

NELMS, A. T.: Graphs of the Compton energy-angle relationship and the Klein-Nishina formula from 10 KeV to 500 MeV. Cir. Nat. Bur. Stands. No. 542 (1953).

NISHIMURA, J.: Theory of cascade showers. In: Handbuch der Physik, vol. XLVI/2, p. 1. Berlin-Heidelberg-New York: Springer 1967.

NISHINA, Y., TAKEUCHI, M., ICHIMIYA, T.: On the nature of cosmic ray particles. Phys. Rev. **52**, 1198 (1937).

NISHINA, Y., TOMONAGA, S., KOBAYASI, M.: On the creation of positive and negative electrons by heavy charged particles. Sci. Pap. Inst. Phys. Chem. Res. Japan **27**, 137 (1935).

NIKISHOV, A. I.: Absorption of high-energy photons in the universe. Soviet Phys. J.E.T.P. **14**, 393 (1962).

NOVIKOV, I. D., SYUNYAEV, R. A.: An explanation of the anolmalous helium abundance in the star 3 Cen A. Soviet Astr. A. J. **11**, 2, 252 (1967).

ODA, M.: Observational results on diffuse cosmic X-rays. In: Non-solar X- and gamma ray astronomy—I.A.U. Symp. No. 37 (ed. L. GRATTON). Dordrecht, Holland: D. Reidel 1970.

ÖPIK, E. J.: Stellar models with variable composition: II. Sequences of models with energy generation proportional to the fifteenth power of temperature. Proc. Roy. Irish Acad. **A 54**, 49 (1951).

ÖPIK, E. J.: The chemical composition of white dwarfs. Mem. Soc. Roy. Sci. Liege **14**, 131 (1954).

OPPENHEIMER, J. R., SERBER, R.: Note on the nature of cosmic-ray particles. Phys. Rev. **51**, 1113 (1937).

OSTRIKER, J. P., GUNN, J. E.: On the nature of pulsars: I. Theory. Ap. J. **157**, 1395 (1969).

PACINI, F.: Energy emission from a neutron star. Nature **216**, 567 (1967).

PACZYŃSKI, B.: Carbon ignition in degenerate stellar cores. Astrophys. Lett. **11**, 53 (1972).

PARKER, E. N.: Hydromagnetic waves and the acceleration of cosmic rays. Phys. Rev. **99**, 241 (1955). Reprod. in: Selected papers on cosmic ray origin theories (ed. S. ROSEN). New York: Dover 1969.

PARKER, E. N.: Origin and dynamics of cosmic rays. Phys. Rev. **109**, 1328 (1958). Reprod. in: Selected papers on cosmic ray origin theories (ed. S. ROSEN). New York: Dover 1969.

PARKER, P. D., BAHCALL, J. N., FOWLER, W. A.: Termination of the proton-proton chain in stellar interiors. Ap. J. **139**, 602 (1964).

PATTERSON, J. R., WINKLER, H., SPINKA, H. M.: Experimental investigation of the stellar nuclear reaction $^{16}O+^{16}O$ at low energies. Bull. Am. Phys. Soc. **13**, 1495 (1968).

PATTERSON, J. R., WINKLER, H., ZAIDINS, C. S.: Experimental investigation of the stellar nuclear reaction $^{12}C+^{12}C$ at low energies. Ap. J. **157**, 367 (1969).

PAULI, W.: Les théories quantities du magnétisive l'électron magnetique (The theory of magnetic quantities: The magnetic electron). Rpt. Septiene Couseil. Phys. Solvay, Bruxelles, 1930.

PETERSON, L. E., WINCKLER, J. R.: Gamma ray burst from a solar flare. J. Geophys. Res. **64**, 697 (1959).

PETERSON, V. L., BAHCALL, J. N.: Exclusion principle inhibition of beta decay in stellar interiors. Ap. J. **138**, 437 (1963).

PETROSIAN, V., BEAUDET, G., SALPETER, E. E.: Photoneutrino energy loss rates. Phys. Rev. **154**, 1445 (1967).

PINAEV, V. S.: Some neutrino pair production processes in stars. Sov. Phys. JETP **18**, 377 (1964).

POKROWSKI, G. I.: Versuch der Anwendung einiger thermodynamischer Gesetzmäßigkeiten zur Beschreibung von Erscheinungen in Atomkernen (Thermodynamical principles of nuclear phenomena). Phys. Z. **32**, 374 (1931).

POLLACK, J. B., FAZIO, G. G.: Production of π mesons and gamma radiation in the Galaxy by cosmic rays. Phys. Rev. **131**, 2684 (1963).

PONTECORVO, B.: The universal Fermi interaction and astrophysics. Sov. Phys. JETP **9**, 1148 (1959).

PRENDERGAST, K. H., BURBIDGE, G. R.: On the nature of some galactic X-ray sources. Ap. J. **151**, L 83 (1968).

PRESTON, M. A.: Physics of the nucleus. Reading, Mass.: Addison-Wesley 1962.

PRINGLE, J. E., REES, M. J.: Accretion disc models for compact X-ray sources. Astron. and Astrophys. **21**, 1 (1972).

RADIN, J.: Cross section for $C^{12}(\alpha, \alpha n)C^{11}$ at 920 MeV. Phys. Rev. **C 2**, 793 (1970).

RAMATY, R.: Gyrosynchrotron emission and absorption in a magnetoactive plasma. Ap. J. **158**, 753 (1969).

RAMATY, R., LINGENFELTER, R. E.: Galactic cosmic-ray electrons. J. Geophys. Res. **71**, 3687 (1966).

RAMATY, R., PETROSIAN, V.: Free-free absorption of gyrosynchrotron radiation in solar microwave bursts. Ap. J. **178**, 241 (1972).

RAMATY, R., STECKER, F. W., MISRA, D.: Low-energy cosmic ray positrons and 0.51-MeV gamma rays from the Galaxy. J. Geophys. Res. **75**, 1141 (1970).

RAWITSCHER, G. H.: Effect of the finite size of the nucleus on μ-pair production by gamma rays. Phys. Rev. **101**, 423 (1956).

REEVES, H.: Stellar energy sources. In: Stellar structure-stars and stellar systems VIII (ed. L. H. ALLER and D. B. MCLAUGHLIN). Chicago, Ill.: University of Chicago Press 1963.

REEVES, H.: Nuclear reactions in stellar surfaces and their relations with stellar evolution. New York: Gordon and Breach 1971.

REEVES, H., FOWLER, W. A., HOYLE, F.: Galactic cosmic ray origin of Li, Be and B in stars. Nature **226**, 727 (1970).

REEVES, H., STEWART, P.: Positron-capture processes as a possible source of the p elements. Ap. J. **141**, 1432 (1965).

REINES, F., COWAN, C. L.: Detection of the free neutrino. Phys. Rev. **92**, 830 (1953).

RITUS, V. I.: Photoproduction of neutrinos on electrons and neutrino radiation from stars. Sov. Phys. JETP **14**, 915 (1962).

RÖNTGEN, W. C.: On a new kind of rays. Nature **103**, 274 (1896).

ROSENBERG, L.: Electromagnetic interactions of neutrinos. Phys. Rev. **129**, 2786 (1963).

ROSENFELD, A. H., BARASH-SCHMIDT, N., BARBARO-GALTIERI, A., PRICE, L. R., SÖDING, P., WOHL, C. G., ROOS, M., WILLIS, W. J.: Data on particles and resonant states. Rev. Mod. Phys. **40**, 77 (1968).

ROSSI, B.: The disintegration of mesotrons. Rev. Mod. Phys. **11**, 296 (1939).

ROSSI, B.: High energy particles. New York: Prentice-Hall 1952.

RUTHERFORD, E.: Uranium radiation and the electrical conduction produced by it. Phil. Mag. **47**, 109 (1899).

RUTHERFORD, E.: The scattering of α and β particles by matter and the structure of the atom. Phil. Mag. **21**, 669 (1911).

RUTHERFORD, E.: The structure of the atom. Phil. Mag. **27**, 488 (1914).

RUTHERFORD, E., CHADWICK, J.: The disintegration of elements by α-particles. Nature **107**, 41 (1921).

RUTHERFORD, E., SODDY, F.: The cause and nature of radioactivity, part I, part II. Phil. Mag. **4**, 370, 569 (1902).

RUTHERFORD, E., SODDY, F.: The radioactivity of uranium. A comparative study of the radioactivity of radium and thorium. Condensation of the radioactive emanations. The products of radioactive change and their specific material nature. Phil. Mag. **5**, 441, 445, 561, 576 (1903).

RYAN, M. J., ORMES, J. F., BALASUBRAHMANYAM, V. K.: Cosmic-ray proton and helium spectra above 50 GeV. Phys. Rev. Lett. **28**, 985 (1972).

RYTER, C., REEVES, H., GRADSZTAJN, E., AUDOUZE, J.: The energetics of L nuclei formation in stellar atmospheres and its relevance to X-ray astronomy. Astron. Astrophys. **8**, 389 (1970).

SAKATA, S., INOUE, T.: On the correlations between mesons and Yukawa particles. Prog. Theor. Phys. **1**, 143 (1946).

SALPETER, E. E.: Nuclear reactions in the stars: I. Proton-proton chain. Phys. Rev. **88**, 547 (1952).

SALPETER, E. E.: Nuclear reactions in stars without hydrogen. Ap. J. **115**, 326 (1952).

SALPETER, E. E.: Energy production in stars. Ann. Rev. Nucl. Sci. **2**, 41 (1953).

SALPETER, E. E.: Electron screening and thermonuclear reactions. Austr. J. Phys. **7**, 373 (1954).

SALPETER, E. E.: Nuclear reactions in stars: II. Protons on light nuclei. Phys. Rev. **97**, 1237 (1955).

SALPETER, E. E.: Nuclear reactions in stars. Buildup from helium. Phys. Rev. **107**, 516 (1957).

SALPETER, E. E., VAN HORN, H. M.: Nuclear reaction rates at high densities. Ap. J. **155**, 183 (1969).

SANDERS, R. H.: S-process nucleosynthesis in thermal relaxation cycles. Ap. J. **150**, 971 (1967).

SARGENT, W. L. W., JUGAKU, J.: The existence of He^3 in 3 Centauri A. Ap. J. **134**, 777 (1961).

SAUTER, F.: Über den atomaren Photoeffekt bei großer Härte der anregenden Strahlung (Atomic photoelectric effect excited by very hard rays). Ann. Phys. **9**, 217 (1931).

SCHATZMAN, E.: Les reactions thermonucléaires aux grandes densités, gaz dégénérés et non dégénérés (Thermonuclear reactions at large densities, degenerate and nondegenerate gases). J. Phys. Radium **9**, 46 (1948).

SCHATZMAN, E.: L'isotope ^{3}He das les étoiles. Application a la théorie des novae et des naines blanches (The helium three isotope in stars. Application of the theory of novae and of white dwarfs). Compt. Rend. **232**, 1740 (1951).

SCHATZMAN, E.: White dwarfs. Amsterdam: North Holland 1958.

SCHOTT, G. A.: Electromagnetic radiation and the mechanical reactions arising from it. Cambridge: Cambridge Univ. Press 1912.

SCHRAMM, D. N.: Explosive *r*-process nucleosynthesis. Ap. J. **185**, 293 (1973).

SCHREIER, E., LEVINSON, R., GURSKY, H., KELLOGG, E., TANANBAUM, H., GIACCONI, R.: Evidence for the binary nature of Centaurus X-3 from UHURU X-ray observations. Ap. J. **172**, L 79 (1972).

SCHWARTZ, D. A.: The isotropy of the diffuse cosmic X-rays determined by OSO-III. Ap. J. **162**, 439 (1970).

SCHWARZSCHILD, M.: Structure and evolution of stars. Princeton, N. J.: Princeton Univ. Press 1958. Reprod. New York: Dover 1965.

SCHWARZSCHILD, M., HÄRM, R.: Red giants of population II. Ap. J. **136**, 158 (1962).

SCHWARZSCHILD, M., HÄRM, R.: Hydrogen mixing by helium-shell flashes. Ap. J. **150**, 961 (1967).

SEAGRAVE, J. D.: Radiative capture of protons by C^{13}. Phys. Rev. **85**, 197 (1952).

SEARS, R. L., BROWNLEE, R. R.: Stellar evolution and age determinations. In: Stellar structure—Stars and stellar systems VIII (ed. L. H. ALLER and D. B. MCLAUGHLIN). Chicago, Ill.: Univ. of Chicago Press 1965.

SEEGER, P. A., FOWLER, W. A., CLAYTON, D. D.: Nucleosynthesis of heavy elements by neutron capture. Ap. J. Suppl. No. 9, **11**, 121 (1965).

SEEGER, P. A., SCHRAMM, D. N.: *r*-Process production ratios of chronologic importance. Ap. J. **160**, L 157 (1970).

SHAPIRO, M. M., SILBERBERG, R.: Heavy cosmic ray nuclei. Ann. Rev. Nucl. Sci. **20**, 323 (1970).

SHAW, P. B., CLAYTON, D. D., MICHEL, F. C.: Photon-induced beta decay in stellar interiors. Phys. Rev. **140**, B 1433 (1965).

SHKLOVSKY, I. S.: On the origin of cosmic rays. Dokl. Akad. Nauk (SSSR) **91**, 475 (1953).

SILBERBERG, R., TSAO, C. H.: Partial cross-sections in high energy nuclear reactions and astrophysical applications: I. Targets with $Z \leq 28$. Ap. J. Suppl. No. 220, **25**, 315 (1973).

SILK, J.: Diffuse cosmic X and gamma radiation: The isotropic component. Space Sci. Rev. **11**, 671 (1970).

SILK, J.: Diffuse X and gamma radiation. Ann. Rev. Astron. Astrophys. **11**, 269 (1973).

SIMNETT, G. H., MCDONALD, F. B.: Observations of cosmic-ray electrons between 2.7 and 21.5 MeV. Ap. J. **157**, 1435 (1969).

SÖDING, P., BARTELS, J., BARBARO-GALTIERI, A., ENSTROM, J. E., LASINSKI, T. A., RITTENBERG, A., ROSENFELD, A. H., TRIPPE, T. G., BARASH-SCHMIDT, N., BRICMAN, C., CHALORIPKA, V.: Review of particle properties. Phys. Lett. **39 B**, 1 (1972).

SPINKA, H., WINKLER, H.: Experimental investigation of the $^{16}O + ^{16}O$ total reaction cross section at astrophysical energies. Ap. J. **174**, 455 (1972).

STERNE, T. E.: The equilibrium theory of the abundance of the elements: A statistical investigation of assemblies in equilibrium in which transmutations occur. M.N.R.A.S. **93**, 736 (1933).

STERNHEIMER, R. M.: Density effect for the ionization loss in various materials. Phys. Rev. **103**, 511 (1956).

STOBBE, M. VON: Zur Quantenmechanik photoelektrischer Prozesse (On the quantum mechanics of photoelectric processes). Ann. Phys. **7**, 661 (1930).

STREET, J., STEVENSON, E. C.: New evidence for the existence of a particle of mass intermediate between the proton and electron. Phys. Rev. **52**, 1003 (1937).

SUESS, H. E., UREY, H. C.: Abundances of the elements. Rev. Mod. Phys. **28**, 53 (1956).

TAKAKURA, T.: Implications of solar radio bursts for the study of the solar corona. Space Sci. Rev. **5**, 80 (1966).

TAKAKURA, T.: Theory of solar bursts. Solar Phys. **1**, 304 (1967).

TAKAKURA, T.: Interpretation of time characteristics of solar X-ray bursts referring to associated microwave bursts. Solar Phys. **6**, 133 (1969).

TAKAKURA, T., KAI, K.: Energy distribution of electrons producing microwave impulsive bursts and X-ray bursts from the Sun. Pub. Astron. Soc. Japan **18**, 57 (1966).

TANANBAUM, H., GURSKY, H., KELLOGG, E. M., LEVINSON, R., SCHREIER, E., GIACCONI, R.: Discovery of a periodic pulsating binary X-ray source from UHURU. Ap. J. **174**, L 143 (1972).

TAYLER, R. J.: The origin of the elements. Prog. Theor. Phys. **29**, 490 (1966).

TAYLOR, B. N., PARKER, W. H., LANGENBERG, D. N.: Determination of e/h, using macroscopic quantum phase coherence in superconductors: Implications for quantum electrodynamics and the fundamental physical constants. Rev. Mod. Phys. **41**, 375 (1969).

THOMAS, L. H.: The calculation of atomic fields. Proc. Camb. Phil. Soc. **23**, 542 (1927).

THOMSON, J. J.: Conductivity of a gas through which cathode rays are passing. Phil. Mag. **44**, 298 (1897).

THOMSON, J. J.: Conduction of electricity through gases. Cambridge: Cambridge University Press 1906. Republ. New York: Dover 1969.

TIOMNO, J., WHEELER, J. A.: Charge-exchange reaction of the μ-meson with the nucleus. Rev. Mod. Phys. **21**, 153 (1949).

TOEVS, J. W., FOWLER, W. A., BARNES, C. A., LYONS, P. B.: Stellar rates for the $^{28}Si(\alpha, \gamma)^{32}S$ and $^{16}O(\alpha, \gamma)^{20}Ne$ reactions. Ap. J. **169**, 421 (1971).

TOLMAN, R. C.: Thermodynamic treatment of the possible formation of helium from hydrogen. J. Am. Chem. Soc. **44**, 1902 (1922).

TRUBNIKOV, B. A.: Particle interactions in a fully ionized plasma. Rev. of Plasma Phys. **1**, 105 (1965).

TRURAN, J. W.: The influence of a variable initial composition on stellar silicon burning. Astrophys. and Space Sci. **2**, 384 (1968).

TRURAN, J. W.: Charged particle thermonuclear reactions in nucleosynthesis. Astrophys. and Space Sci. **18**, 306 (1972).

TRURAN, J. W., ARNETT, W. D.: Nucleosynthesis in explosive oxygen burning. Ap. J. **160**, 181 (1970).

TRURAN, J. W., ARNETT, W. D., TSURUTA, S., CAMERON, A. G. W.: Rapid neutron capture in supernova explosions. Astrophys. and Space Sci. **1**, 129 (1968).

TRURAN, J. W., CAMERON, A. G. W.: Evolutionary models of nucleosynthesis in the Galaxy. Astrophys. and Space Sci. **14**, 179 (1971).

TRURAN, J. W., CAMERON, A. G. W.: The *p* process in explosive nucleosynthesis. Ap. J. **171**, 89 (1972).

TRURAN, J. W., CAMERON, A. G. W., GILBERT, A.: The approach to nuclear statistical equilibrium. Can. J. Phys. **44**, 576 (1966).

TRURAN, J. W., CAMERON, A. G. W., HILF, E.: In: Proc. inst. conf. prop. nucl. far from regular beta-stability Geneva report CERN 70-30 (1970).

TSUDA, H., TSUJI, H.: Synthesis of Fe-group elements by the rapid nuclear process. Prog. Theor. Phys. **30**, 34 (1963).

TSUJI, H.: Synthesis of 4 N and their neighboring nuclei by the rapid nuclear process. Prog. Theor. Phys. **29**, 699 (1963).

TSURUTA, S., CAMERON, A. G. W.: Composition of matter in nuclear statistical equilibrium at high densities. Can. J. Phys. **43**, 2056 (1965).

TSURUTA, S., CAMERON, A. G. W.: URCA shells in dense stellar interiors. Astrophys. and Space Sci. **7**, 374 (1970).

TSYTOVICH, V. N.: Acceleration by radiation and the generation of fast particles under cosmic conditions. Sov. Astron. A. J. **7**, 471 (1964). Reprod. in: Selected papers on cosmic ray origin theories (ed. S. ROSEN). New York: Dover 1969.

TSYTOVICH, V. N.: Statistical acceleration of particles in a turbulent plasma. Usp. Fiz. Nauk. **89**, 89 (1966).

ÜBERALL, H.: High-energy interferance effect of bremsstrahlung and pair production in crystals. Phys. Rev. **103**, 1055 (1956).

ÜBERALL, H.: Polarization of bremsstrahlung from monocrystalline targets. Phys. Rev. **107**, 223 (1957).

UREY, H. C., BRADLEY, C. A.: On the relative abundances of the isotopes. Phys. Rev. **38**, 718 (1931).

VAUCLAIR, S., REEVES, H.: Spallation processes in stellar surfaces: Anomalous helium ratios. Astron. Astrophys. **18**, 215 (1972).

VILLARD, M. P.: Sur le rayonnement du radium (On the radiation of radium). Compt. Rend. **130**, 1178 (1900).

WAGONER, R. V.: Synthesis of the elements within objects exploding from very high temperatures. Ap. J. Suppl. No. 162, **18**, 247 (1969).

WAGONER, R. V.: Big bang nucleosynthesis revisited. Ap. J. **179**, 343 (1973).

WAGONER, R. V., FOWLER, W. A., HOYLE, F.: On the synthesis of elements at very high temperatures. Ap. J. **148**, 3 (1967).

WALLERSTEIN, G., CONTI, P. S.: Lithium and beryllium in stars. Ann. Rev. Astr. Astrophys. **7**, 99 (1969).

WAPSTRA, A. H., GOVE, N. B.: The 1971 atomic mass evaluation. Nuclear Data Tables **9**, 265 (1971).

WEBBER, W. R.: The spectrum and charge composition of the primary cosmic radiation. In: Handbuch der Physik, vol. XLVI/2: Cosmic Rays II, p. 181 (ed. K. SITTE). Berlin-Heidelberg-New York: Springer 1967.

WEINREB, S.: A new upper limit to the galactic deuterium-to-hydrogen ratio. Nature **195**, 367 (1962).

WEIZSÄCKER, C. F. VON: Zur Theorie der Kernmassen (On the theory of nuclear masses). Z. Physik **96**, 431 (1935).

WEIZSÄCKER, C. F. VON: Über Elementumwandlungen im Innern der Sterne I (On transformation of the elements in stellar interiors I). Phys. Z. **38**, 176 (1937).

WEIZSÄCKER, C. F. VON: Über Elementumwandlungen im Innern der Sterne II (On transformation of the elements in stellar interiors II). Phys. Z. **39**, 633 (1938).

WENTZEL, D. G.: Fermi acceleration of charged particles. Ap. J. **137**, 135 (1963).

WENTZEL, D. G.: Motion across magnetic discontinuities and Fermi acceleration of charged particles. Ap. J. **140**, 1013 (1964).

WENTZEL, G.: Eine Verallgemeinerung der Quantenbedingungen für die Zwecke der Wellenmechanik (An overall description of the quantum requirements for the use of the wave mechanics). Z. Physik **38**, 518 (1926).

WEYMANN, R.: The energy spectrum of radiation in the expanding universe. Ap. J. **145**, 560 (1966).

WEYMANN, R.: Possible thermal histories of intergalactic gas. Ap. J. **147**, 887 (1967).

WEYMANN, R., SEARS, R. L.: The depth of the convective envelope on the lower main sequence and the depletion of lithium. Ap. J. **142**, 174 (1965).

WHEATON, W. A., ULMER, M. P., BAITY, W. A., DATLOWE, D. W., ELCAN, M. J., PETERSON, L. E., KLEBESADEL, R. W., STRONG, I. B., CLINE, T. L., DESAI, V. D.: The direction and spectral variability of a cosmic gamma-ray burst. Ap. J. **185**, L 57 (1973).

WHEELER, J. A.: Mechanism of capture of slow mesons. Phys. Rev. **71**, 320 (1947).

WIGNER, E., SEITZ, F.: On the constitution of metallic sodium II. Phys. Rev. **46**, 509 (1934).

WILDHACK, W. A.: The proton-deuteron transformation as a source of energy in dense stars. Phys. Rev. **57**, 81 (1940).

DE WITT, H. E., GRABOSKE, H. C., COOPER, M. S.: Screening factors for nuclear reactions: I. General theory. Ap. J. **181**, 439 (1973).

WOLF, R. A.: Rates of nuclear reactions in solid-like stars. Phys. Rev. **137**, B 1634 (1965).

WOOSLEY, S. E., ARNETT, W. D., CLAYTON, D. D.: Hydrostatic oxygen burning in stars: II. Oxygen burning at balanced power. Ap. J. **175**, 731 (1972).

WOOSLEY, S. E., ARNETT, W. D., CLAYTON, D. D.: The explosive burning of oxygen and silicon. Ap. J. Suppl. No. 231, **26**, 231 (1973).

YIOU, F., SEIDE, C., BERNAS, R.: Formation cross sections of lithium, beryllium, and boron isotopes produced by spallation of oxygen by high-energy protons. J. Geophys. Res. **74**, 2447 (1969).

YORK, D. G., ROGERSON, J. B.: Astrophys. J. **203**, 378 (1976).

YUKAWA, H.: On the interaction of elementary particles I. Proc. Phys. Math. Soc. Japan **17**, 48 (1935).

YUKAWA, H.: On a possible interpretation of the penetrating component of the cosmic ray. Proc. Phys. Math. Soc. Japan **19**, 712 (1937).

YUKAWA, H.: On the interaction of elementary particles IV. Proc. Phys. Math. Soc. Japan **20**, 720 (1938).

ZAIDI, M. H.: Emission of neutrino-pairs from a stellar plasma. Nuovo Cimento **40 A**, 502 (1965).

ZELDOVICH, I. B.: Nuclear reactions in super-dense cold hydrogen. Sov. Phys. JETP **6**, 760 (1968).

ZOBEL, W., MAIENSCHEIN, F. C., TODD, J. H., CHAPMAN, G. T.: ONRL 4183 (1967).

Chapter 5

ABELL, G. O.: The distribution of rich clusters of galaxies. Ap. J. Suppl. No. 31, **3**, 211 (1958).

ABELL, G.O.: Membership of clusters of galaxies. In: Problems in extra-galactic research (ed. G. C. MCVITTIE). New York: Macmillan 1962.

ABELL, G. O.: Clustering of galaxies. Ann. Rev. Astron. and Astrophys. **3**, 1 (1965).

ABT, H. A., BIGGS, E. S.: Bibliography of stellar radial velocities. New York: Latham Press and Kitt Peak National Observatory 1972.

ADAMS, W. S.: The relativity displacement of the spectral lines in the companion of Sirius. Proc. Nat. Acad. Sci. **11**, 382 (1925).

ADAMS, W. S., KOHLSCHÜTTER, A.: Some spectral criteria for the determination of absolute stellar magnitudes. Ap. J. **40**, 385 (1914).

ADAMS, W. S., JOY, A. H., HUMASON, M. L., BRAYTON, A. M.: The spectroscopic absolute magnitudes and parallaxes of 4179 stars. Ap. J. **81**, 188 (1935).

ALFVÉN, H.: Plasma physics applied to cosmology. Phys. Today **24**, 28 (1971).

ALLEN, C. W.: Astrophysical quantities. University London: Athlone Press 1963.

ALPHER, R. A., BETHE, H. A., GAMOW, G.: The origin of the chemical elements. Phys. Rev. **73**, 803 (1948).

ALPHER, R. A., HERMAN, R. C.: Evolution of the universe. Nature **162**, 774 (1948).

ALPHER, R. A., HERMAN, R. C.: Theory of the origin and relative abundance distribution of the elements. Rev. Mod. Phys. **22**, 153 (1950).

ALPHER, R. A., FOLLIN, J. W., HERMAN, R. C.: Physical conditions in the initial stages of the expanding universe. Phys. Rev. **92**, 1347 (1953).

AMBARTSUMIAN, V. A.: On the evolution of galaxies. In: La structure et l'evolution de l'univers—Institut International de Physique Solvay (ed. R. STOOPS). Bruxelles: Coodenberg 1958.

American ephemeris and nautical almanac. U.S. Gov't Print. Off. Wash., 1971.

ANDERS, E.: Meteorite ages. In: The moon, meteorites, and comets—The solar system IV (ed. B. M. MIDDLEHURST and G. P. KUIPER). Chicago, Ill.: Univ. of Chicago Press 1963.

ANDERSON, J. D., EFRON, J. D., WONG, S. K.: Martian mass and earth-moon mass ratio from coherent s-band tracking of mariners 6 and 7. Science **167**, 277 (1970).

ARNETT, W. D.: Gravitational collapse and weak interactions. Can. J. Phys. **44**, 2553 (1966).

ARNETT, W. D.: Mass dependence in gravitational collapse of stellar cores. Can. J. Phys. **45**, 1621 (1967).

ARNETT, W. D.: On supernova hydrodynamics. Ap. J. **153**, 341 (1968).

ARNETT, W. D., CAMERON, A. G. W.: Supernova hydrodynamics and nucleosynthesis. Can. J. Phys. **45**, 2953 (1967).

ARP, H. C.: The Hertzsprung-Russell diagram. In: Handbuch der Physik, vol. LI: Astrophysics II—Stellar structure (ed. S. FLÜGGE) p. 75. Berlin-Göttingen-Heidelberg: Springer 1958.

ARP, H.: The globular cluster M 5. Ap. J. **135**, 311 (1962).

ARP, H.: The effect of reddening on the derived ages of globular clusters and the absolute magnitudes of RR Lyrae Cepheids. Ap. J. **135**, 971 (1962).

ARP, H.: Globular clusters in the Galaxy. In: Galactic structure-stars and stellar systems V (ed. A. BLAAUW and M. SCHMIDT). Chicago, Ill.: University of Chicago Press 1965.

ASH, M. E., SHAPIRO, I. I., SMITH, W. B.: Astronomical constants and planetary ephemeris deduced from radar and optical observations. Astron. J. **72**, 338 (1967).

ASH, M. E., SHAPIRO, I. I., SMITH, W. B.: The system of planetary masses. Science **174**, 551 (1971).

ASTON, F. W.: The mass-spectrum of uranium lead and the atomic weight of protactinium. Nature **123**, 313 (1929).

BAADE, W.: Über eine Möglichkeit die Pulsationstheorie der δ Cephei-Veränderlichen zu prüfen (A possible check of the pulsation theory of the variable δ-Cephei). Astron. Nach. **228**, 359 (1926).

BAADE, W.: The resolution of Messier 32, NGC 205, and the central region of the Andromeda nebula. Ap. J. **100**, 137 (1944).

BAADE, W.: Commission des nebuleuses (Commission of the nebulae). Trans. Int. Astron. Union **8**, 397 (1952).

BAADE, W., SWOPE, H. H.: The Palomar survey of variables in M 31. Astron. J. **60**, 151 (1955).

BAADE, W., ZWICKY, F.: On super-novae. Proc. Nat. Acad. Sci. (Wash.) **20**, 254 (1934).

BAHCALL, J. N., MAY, R. M.: Electron scattering and tests of cosmological models. Ap. J. **152**, 37 (1968).

BAHCALL, J. N., SALPETER, E. E.: On the interaction of radiation from distant sources with the intervening medium. Ap. J. **142**, 1677 (1965).

BARDEEN, J. M., PRESS, W. H., TEUKOLSKY, S. A.: Rotating black holes: Locally nonrotating frames, energy extraction, and scaler synchrotion radiation. Ap. J. **178**, 347 (1973).

BARDEEN, J. M., WAGONER, R. V.: Uniformly rotating disks in general relativity. Ap. J. **158**, L 65 (1969).

BARDEEN, J. M., WAGONER, R. V.: Relativistic disks: I. Uniform rotation. Ap. J. **167**, 359 (1971).

BAUM, W. A.: Meeting of the Royal Astronomical Society Friday May 12, 1961. Observatory **81**, 114 (1961).

BAUM, W. A.: The diameter-red shift relation. In: External galaxies and quasi-stellar objects—I.A.U. Symp. No. 44 (ed. D. S. EVANS). Dordrecht, Holland: D. Reidel 1972.

BEALL, E. F.: Measuring the gravitational interaction of elementary particles. Phys. Rev. **D 1**, 961 (1970).

BERGERON, J. E.: Étude des Possibilités d'existence d'un Plasma Intergalactique Dense (A study of the possibilities of the existence of a dense intergalactic plasma). Astron. Astrophys. **3**, 42 (1969).

VAN DEN BERGH, S.: A preliminary luminosity classification for galaxies of type Sb. Ap. J. **131**, 558 (1961).

BESSEL, F. W.: Bestimmung der Entfernung des 61 Sterns des Schwans (A new calculation of the parallax of 61 Cygni). Astron. Nach. **16**, No. 365 (1839).

BIRKHOFF, G. D.: Relativity and modern physics. Cambridge, Mass.: Havard University Press 1923.

BIXBY, J. E., VAN FLANDERN, T. C.: The diameter of Neptune. Astron. J. **74**, 1220 (1969).

BLAAUW, A., GUM, C. S., PAWSEY, J. L., WESTERHOUT, G.: The new I.A.U. system of galactic coordinates. M.N.R.A.S. **121**, 123 (1960).

BLAGG, A.: M.N.R.A.S. **73**, 414 (1913).

BLAMONT, J. E., RODDIER, F.: Precise observation of the profile of the Fraunhofer strontium resonance line, evidence for the gravitational red shift on the Sun. Phys. Rev. Lett. **7**, 437 (1961).

BLANCO, V.M., MCCUSKEY, S.W.: Basic physics of the solar system. Reading, Mass.: Addison-Wesley 1961.

BODE, J. E.: Anleitung zur Kenntnis des gestirnten Himmels (Guide to a knowledge of the heavenly bodies in the sky). 1772.

BOK, B. J.: The time-scale of the universe. M.N.R.A.S. **106**, 61 (1946).

BOLTZMANN, L. VON: Über eine von Hrn. Bartoli entdeckte Beziehung der Wärmestrahlung zum zweiten Hauptsatze (On a relation between thermal radiation and the second law of thermodynamics, discovered by Bartoli). Ann. Phys. **31**, 291 (1884).

BONDI, H., GOLD, T.: The steady-state theory of the expanding universe. M.N.R.A.S. **108**, 252 (1948).

BONNER, W. B.: Jeans' formula for gravitational instability. M.N.R.A.S. **117**, 104 (1957).

BORGMAN, J., BOGGESS, A.: Interstellar extinction in the middle ultraviolet. Ap. J. **140**, 1636 (1964).

BOSS, L.: Precession and solar motion. Astron. J. **26**, 95 (1910).

BOWYER, C. S., FIELD, G. B., MACK, J. E.: Detection of an anisotropic soft X-ray background flux. Nature **217**, 32 (1968).

BOYER, R. H., LINDQUIST, R. W.: Maximal analytic extension of the Kerr metric. J. Math. Phys. **8**, 265 (1967).

BRADLEY, J.: A new apparent motion in the fixed stars discovered, its cause assigned, the velocity and equable motion of light deduced. Phil. Trans. **6**, 149 (1728).

BRADLEY, J.: An apparent motion in some of the fixed stars. Phil. Trans. **10**, 32 (1748).

BRAGINSKY, V. B., PANOV, V. I.: Verification of equivalency of inertial and gravitational mass. Zh. E.T.F. **61**, 875 (1971).

BRANDT, J. C.: On the distribution of mass in galaxies: I. The large scale structure of ordinary spirals with application to M 31; II. A discussion of the mass of the Galaxy. Ap. J. **131**, 293, 553 (1960).

BRANDT, J. C., BELTON, M. J. S.: On the distribution of mass in galaxies: III. Surface densities. Ap. J. **136**, 352 (1962).

BRANS, C., DICKE, R. H.: Mach's principle and a relativistic theory of gravitation. Phys. Rev. **124**, 925 (1961).

BRAULT, J.: Gravitational redshift of solar lines. Bull. Am. Phys. Soc. **8**, 28 (1963).

BRIDLE, A. H., DAVIS, M. M., FORMALONT, E. B., LEQUEUX, J.: Counts of intense extragalactic radio sources at 1,400 MHz. Nature-Phys. Sci. **235**, 135 (1972).

BROWN, E. W.: Tables of the motion of the moon. New Haven, 1919 (cf. An introductory treatise on lunar theory by E. W. BROWN, New York: Dover 1960).

BURBIDGE, E. M., BURBIDGE, G. R., FOWLER, W. A., HOYLE, F.: Synthesis of the elements in stars. Rev. Mod. Phys. **29**, 547 (1957).

BURBIDGE, G. R., BURBIDGE, E. M.: Stellar evolution. In: Handbuch der Physik, Vol. LI: Astrophysics II: Stellar Structure, p. 134 (ed. S. FLÜGGE). Berlin-Göttingen-Heidelberg: Springer 1958.

BURBIDGE, G.R., BURBIDGE, E.M.: In: Galaxies and the universe—stars and stellar systems IX (ed. A. R. SANDAGE). Chicago, Ill.: Univ. of Chicago Press 1974.

BURTON, W. B.: Galactic structure derived from neutral hydrogen observations using kinetic models based on the density wave theory. Astron. Astrophys. **10**, 76 (1971).

BURTON, W. B.: The large-scale distribution of neutral hydrogen. In: Galactic and extra-galactic radio astronomy (ed. G. VERSCHUUR and K.I. KELLERMANN). Berlin-Heidelberg-New York: Springer 1974.

CANNON, A. J., PICKERING, E. C.: The Henry Draper catalogue. Ann. Harvard Coll. Obs. **91–99** (1918–1924).

CARTER, B.: Axisymmetric black hole has only two degrees of freedom. Phys. Rev. Lett. **26**, 331 (1971).

CASSINI, M.: "Recherche de la parallaxe du Soliel par le moyen de celle de Mars observé à mesme temps à Paris and en Caiene" in Ouvrage D'Astronomie par. M. CASSINI (1672).

CAVENDISH, H.: Experiments to determine the density of the earth. Phil. Trans. Roy. Soc. **88**, 469 (1798).

CHANDRASEKHAR, S.: The highly collapsed configurations of a stellar mass. M.N.R.A.S. **95**, 207 (1935).

CHANDRASEKHAR, S.: Dynamical instability of gaseous masses approaching the Schwarzschild limit in general relativity. Phys. Rev. Lett. **12**, 114 (1964).

CHANDRASEKHAR, S.: The dynamical instability of gaseous masses approaching the Schwarzschild limit in general relativity. Ap. J. **140**, 417 (1964).

CHAPMAN, G. A., INGERSOLL, A. P.: Photospheric faculae and the solar oblateness: A reply to "Faculae and the solar oblateness" by R. H. DICKE. Ap. J. **183**, 1005 (1973).

CHARLIER, C. V. L.: Wie eine unendliche Welt aufgebaut sein kann (How an infinite universe can be constructed). Ark. Mat. Astron. Phys. **4**, No. 24 (1908).

CHARLIER, C. V. L.: How an infinite world may be built up. Arkiv. Mat. Astron. Phys. **16**, No. 22 (1922).

DE CHÉSEAUX, J. P. DE L.: Traité de la cométe qui a paru en Decembre 1743 et ene Janvier, Fevrier et Mars 1744 (Treatise of the comet that appeared in December 1743 and in January, February and March 1944). Lussanne et Geneve—Michel Bouguet et Compagnie, 1744—relevant appendix reprod. in Dickson (1968), Jaki (1969).

CHINCARINI, G., ROOD, H. J.: Dynamics of the Perseus cluster of galaxies. Ap. J. **168**, 321 (1971).

CHRISTY, R. F.: Energy transport in the hydrogen ionization zone of giant stars. Ap. J. **136**, 887 (1962).

CHRISTY, R. F.: Pulsation theory. Ann. Rev. Astron. and Astrophys. **4**, 353 (1966).

CHRISTY, R. F.: A study of pulsation in RR Lyrae models. Ap. J. **144**, 108 (1966).

CLAIRAUT, A. C.: Théorie de la figure de la terre, tirée des principes de l'hydrostatique; par Clairaut, de l'Academie Royale des Sciences et de la Societe Royale de Londres (Theory of the figure of the earth, and the principles of hydrostatics by Clairaut for the Royal Academy of Sciences and the Royal Society of London). Paris, Du Mond 1743.

CLAYTON, D. D.: Cosmoradiogenic chronologies of nucleosynthesis. Ap. J. **139**, 637 (1964).

CLAYTON, D. D.: Isotropic composition of cosmic importance. Nature **224**, 56 (1969).

CLEMENCE, G. M.: The system of astronomical constants. Ann. Rev. Astron. Astrophys. **3**, 93 (1965).

CLERKE, A. M.: The system of the stars. London: Black 1905.

COLGATE, S. A., WHITE, R. H.: The hydrodynamic behavior of supernovae explosions. Ap. J. **143**, 626 (1966).

COLLINDER, P.: On structural properties of open galactic clusters and their spatial distribution. Lund Obs. Ann. No. 2 (1931).

CONKLIN, E. K., BRACEWELL, R. N.: Isotropy of cosmic background radiation at 10,690 MHz. Phys. Rev. Lett. **18**, 614 (1967).

CORWIN, H. G.: Notes on the Hercules cluster of galaxies. P.A.S.P. **83**, 320 (1971).

COX, A. N., STEWART, J. N.: Rosseland opacity tables for population II composition. Ap. J. Suppl. No. 174, **19**, 261 (1970).

COX, J. P., GIULI, R. T.: Principles of stellar structure. New York: Gordon & Breach 1968.

CRAWFORD, D. F., JAUNCEY, D. L., MURDOCH, H. S.: Maximum-likelihood estimation of the slope from number-flux-density counts of radio sources. Ap. J. **162**, 405 (1970).

DABBS, J. W. T., HARVEY, J. A., PAYA, D., HORSTMANN, H.: Gravitational acceleration of free neutrons. Phys. Rev. **139**, B 756 (1965).

DANZIGER, I. J.: The cosmic abundance of helium. Ann. Rev. Astron. Astrophys. **8**, 161 (1970).

DELHAYE, J.: Solar motion and velocity distribution of common stars. In: Galactic structure—Stars and stellar systems V (ed. A. BLAAUW and M. SCHMIDT). Chicago, Ill.: University of Chicago Press 1965.

DEMARQUE, P., MENGEL, J. G., AIZENMAN, M. L.: The early evolution of population II stars: II. Ap. J. **163**, 37 (1971).

DICKE, R. H.: Mach's principle and invariance under transformation of units. Phys. Rev. **125**, 2163 (1962).

DICKE, R. H.: The Sun's rotation and relativity. Nature **202**, 432 (1964).

DICKE, R. H.: The age of the Galaxy from the decay of uranium. Ap. J. **155**, 123 (1969).

DICKE, R. H., GOLDENBERG, H. M.: Solar oblateness and general relativity. Phys. Rev. Lett. **18**, 313 (1967).

DICKE, R. H., PEEBLES, P. J. E., ROLL, P. G., WILKINSON, D. T.: Cosmic black-body radiation. Ap. J. **142**, 414 (1965).

DICKSON, F. P.: The bowl of the night. Cambridge, Mass.: M.I.T. Press 1968.

DIRAC, P. A. M.: The cosmological constants. Nature **139**, 323 (1937).

DIRAC, P. A. M.: A new basis for cosmology. Proc. Roy. Soc. London **A 165**, 199 (1938).

DOLLFUS, A.: New optical measurements of the diameters of Jupiter, Saturn, Uranus, and Neptune. Icarus **12**, 101 (1970).

DOPPLER, C.: Über d. farbige Licht d. Dopplersterne u. s. w. (On the colored light of double stars, etc.). Abhandlungen d.k. Böhm. Gesell. Wiss. **2**, 467 (1842).

DOROSHKEVICH, A. G., ZEL'DOVICH, Ya. B., SUNYAEV, R. A.: In *The Formation of Stars and Galaxies* (Nauka Publ. House, 1976).

DUBE, R. R., WICKES, W. C., WILKINSON, W. T.: Ap. J. **215**, L51 (1977).

DUFAY, J.: Introduction to astrophysics—The stars. Engl. trans. O. Gingerich. London: G. Newnes 1964.

DUNCAN, J. C.: Three variable stars and a suspected novae in the spiral nebula M 33 Trianguli. P.A.S.P. **34**, 290 (1922).

DUNCAN, R. A.: Jupiters rotation. Planet Space Sci. **19**, 391 (1971).

DYCE, R. B., PETTENGILL, G. H., SHAPIRO, I. I.: Radar determinations of the rotations of Venus and Mercury. Astron. J. **72**, 351 (1967).

DYSON, F. W., EDDINGTON, A. S., DAVIDSON, C.: A determination of the deflection of light by the sun's gravitational field, from observations made at the total eclipse of May 29, 1919. Phil. Trans. **A 220**, 291 (1920).

EDDINGTON, A. S.: The systematic motions of stars of Professor Boss's 'Preliminary General Catalogue'. M.N.R.A.S. **71**, 4 (1910).

EDDINGTON, A. S.: The total eclipse of 1919 May 29 and the influence of gravitation on light. Observatory **42**, 119 (1919).

EDDINGTON, A. S.: The mathematical theory of relativity. Cambridge: Cambridge University Press 1923.

EDDINGTON, A. S.: A comparison of Whitehead's and Einstein's formulae. Nature **113**, 192 (1924).

EDDINGTON, A. S.: On the relation between the masses and luminosities of the stars. M.N.R.A.S. **84**, 308 (1924).

EDDINGTON, A. S.: The internal constitution of the stars. Cambridge: Cambridge University Press. New York: Dover 1926.

EDDINGTON, A. S.: On the instability of Einstein's spherical world. M.N.R.A.S. **90**, 668 (1930).

EDLÉN, B.: The dispersion of standard air. J. Opt. Soc. Amer. **43**, 339 (1953).

EHLERS, J., GEREN, P., SACHS, R. K.: Isotropic solutions to the Einstein-Liouville equations. J. Math. Phys. **9**, 1344 (1968).

EINSTEIN, A.: Zur Elektrodynamic bewegter Körper (On the electrodynamics of moving bodies). Ann. Physik **17**, 891 (1905). Eng. trans. in: The principle of relativity. New York: Dover 1952.

EINSTEIN, A.: Über den Einfluß der Schwerkraft auf die Ausbreitung des Lichtes (On the influence of gravitation on the propagation of light). Ann. Physik **35**, 898 (1911). Eng. trans. in: The principle of relativity. New York: Dover 1952.

EINSTEIN, A.: Die Grundlage der allgemeinen Relativitätstheorie (The foundation of the general theory of relativity). Ann. Physik **49**, 769 (1916). Eng. trans. in: The principle of relativity. New York: Dover 1952.

EINSTEIN, A.: Kosmologische Betrachtungen zur allgemeinen Relativitätstheorie (Cosmological considerations of the general theory of relativity). Acad. Wiss. **1**, 142 (1917). Eng. trans. in: The principle of relativity. New York: Dover 1952.

EINSTEIN, A.: Spielen Gravitationsfelder im Aufbau der materiellen Elementarteilchen eine wesentliche Rolle? (Do gravitational fields play an essential part in the structure of the elementary particles of matter?). Acad. Wiss. **1**, 349 (1919). Eng. trans. in: The principle of relativity. New York: Dover 1952.

EINSTEIN, A., DESITTER, W.: On the relation between the expansion and the mean density of the universe. Proc. Nat. Acad. Sci. **18**, 213 (1932).

ENCKE, J. F.: Die Entfernung der Sonne: Fortsetzung (The distance of the Sun: continuation): 1824; and Berlin Abhandlungen 295 (1835).

Eötvös, R.: Über die Anziehung der Erde auf verschiedene Substanzen (On the attraction by the earth on different materials). Math. Nat. Ber. Ungarn **8**, 65 (1889).
Essen, L.: The measurement of time. In: Vistas in astronomy XI (ed. A. Beer). New York: Pergamon Press 1969.
Estermann, I., Simpson, O. C., Stern, O.: The free fall of molecules. Phys. Rev. **53**, 947 (1938).
Etherington, I. M. H.: On the definition of distance in general relativity. Phil. Mag. **15**, 761 (1933).
Explanatory supplement to the american ephemeris and nautical almanac. London: H.M.S.O. 1961.
Field, G. B.: An attempt to observe neutral hydrogen between the galaxies. Ap. J. **129**, 525 (1959).
Field, G. B.: Absorption by intergalactic hydrogen. Ap. J. **135**, 684 (1962).
Field, G. B.: Intergalactic matter. Ann. Rev. Astron. Astrophys. **10**, 227 (1972).
Field, G. B., Henry, R. C.: Free-free emission by intergalactic hydrogen. Ap. J. **140**, 1002 (1964).
Field, G. B., Solomon, P. M., Wampler, E. J.: The density of intergalactic hydrogen molecules. Ap. J. **145**, 351 (1966).
Fish, R. A.: A mass-potential-energy relationship in elliptical galaxies and some inferences concerning the formation and evolution of galaxies. Ap. J. **139**, 284 (1964).
van Flandern, T. C.: The rate of change of G. M.N.R.A.S. **170**, 333 (1975).
Forbes, E. G.: The Einstein effect and the observed solar red shifts. Observatory **90**, 149 (1970).
Fowler, W. A.: The stability of supermassive stars. Ap. J. **144**, 180 (1966).
Fowler, W. A.: New observations and old nucleocosmochronologies. In: Cosmology, fusion, and other matter, a memorial to George Gamow (ed. F. Reines). Boulder, Colorado: Colorado Ass. University Press 1972.
Fowler, W. A., Hoyle, F.: Nuclear cosmochronology. Ann. Phys. (N. Y.) **10**, 280 (1960).
Fricke, W., Kopff, A.: The fourth fundamental catalogue (FK 4). Veroff./Astron. Rechen-Institut, Heidelberg, No. 10, 11. Karlsruhe: Braun 1963.
Friedmann, A.: Über die Krümmung des Raumes (On the curvature of space). Z. Physik **10**, 377 (1922).
Friedmann, A.: Über die Möglichkeit einer Welt mit konstanter negativer Krümmung des Raumes (On the possibility of a universe with a constant negative curvature). Z. Physik **21**, 326 (1924).
Gallouet, L., Heidmann, N.: Optical positions of bright galaxies. Astron. Astrophys. Suppl. **3**, 325 (1971).
Gamow, G.: Expanding universe and the origin of the elements. Phys. Rev. **70**, 572 (1946).
Gamow, G.: The origin of the elements and the separation of galaxies. Phys. Rev. **74**, 505 (1948).
Gamow, G.: The role of turbulence in the evolution of the universe. Phys. Rev. **86**, 251 (1952).
Gauss, C. F.: Theoria Motus ... (Theory of the motion of heavenly bodies moving about the Sun in conic sections). Trans. by C. H. Davis, Boston: Little Brown 1857.
Genkin, I. L., Genkina, L. M.: A new mass-luminosity relation for elliptical galaxies. Sov. Astron. **13**, 886 (1970).
Giacconi, R., Gursky, H., Paolini, F. R., Rossi, B. B.: Evidence for X-rays from sources outside the solar system. Phys. Rev. Lett. **9**, 439 (1962).
Glanfield, J. R., Cameron, M. J.: Optical positions of bright galaxies south of $+20°$ declination. Austr. J. Phys. **20**, 613 (1967).
Gliese, W.: Catalogue of nearby stars. Veröffentlichungen des Astronomischen Rechen-Instituts Heidelberg. Karlsruhe: Braun 1969.
Goldstein, H.: Classical mechanics. Reading, Mass.: Addison-Wesley 1950.
Gott, J. R.: Ann. Rev. Astron. Astrophys. **15**, 235 (1977); Proc. IAU Symp. No. 79 (Tallinn, 1977).
Greenberg, J. M.: Interstellar grains. In: Nebulae and interstellar matter—Stars and stellar systems, vol. VII (ed. B. M. Middlehurst and L. H. Aller). Chicago, Ill.: University of Chicago Press 1968.
Greenberg, J. M.: Discussion in: Interstellar gas dynamics. I.A.U. Symp. No. 39 (ed. H. J. Habing). Dordrecht, Holland: D. Reidel 1970.
Greenstein, J. L., Oke, J. B., Shipman, H. L.: Effective temperature, radius, and gravitational redshift of Sirius B. Ap. J. **169**, 563 (1971).
Greenstein, J. L., Trimble, V. L.: The Einstein redshift in white dwarfs. Ap. J. **149**, 283 (1967).
Greenstein, J. L., Trimble, V. L.: The gravitational redshift of 40 Eridani B. Ap. J. **175**, L 7 (1972).
Gunn, J. E., Peterson, B. E.: On the density of neutral hydrogen in intergalactic space. Ap. J. **142**, 1633 (1965).
Haddock, F. T., Sciama, D. W.: Proposal for the detection of dispersion of radio-wave propagation through intergalactic space. Phys. Rev. Lett. **14**, 1007 (1965).

HALLEY, E.: Considerations on the change in the latitudes of some of the principal fixed stars. Phil. Trans. **30**, 756 (1718).

HALLEY, E.: On the infinity of the sphere of fix'd stars. Phil. Trans. **31**, 22 (1720).

HALLEY, E.: Of the number, order, and light of the fix'd stars. Phil. Trans. **31**, 24 (1720). Reprod. in Jaki (1969).

HALM, J.: Further considerations relating to the systematic motions of the stars. M.N.R.A.S. **71**, 610 (1911).

HANBURY-BROWN, R., TWISS, R. Q.: Interferometry of the intensity fluctuations in light: III. Applications to astronomy. Proc. Roy. Soc. London **A 248**, 199 (1958).

HANBURY-BROWN, R., DAVIS, J., ALLEN, L. R., ROME, J. M.: The stellar interferometer at Narrabri Observatory—II. The angular diameters of 15 stars. M.N.R.A.S. **137**, 393 (1967).

HARGREAVES, R.: Integral forms and their connexion with physical equations. Trans. Camb. Phil. Soc. **21**, 107 (1908).

HARRIS, D. L.: Photometry and colorimetry of planets and satellites. In: Planets and satellites—The solar system II (ed. G. P. KUIPER and B. M. MIDDLEHURST). Chicago, Ill.: Univ. of Chicago Press 1961.

HARRIS, D. L.: The stellar temperature scale and bolometric corrections. In: Basic astronomical data—Stars and stellar systems III (ed. K. A. STRAND). Chicago, Ill.: University of Chicago Press 1963.

HARRIS, D. L., STRAND, K. A., WORLEY, C. E.: Empirical data on stellar masses, luminosities, and radii. In: Basic astronomical data—Stars and stellar systems III (ed. K. A. STRAND). Chicago, Ill.: University Chicago Press 1963.

HARRISON, E. R.: Matter, antimatter, and the origin of galaxies. Phys. Rev. Lett. **18**, 1011 (1967).

HARRISON, E. R.: Generation of magnetic fields in the radiation era. M.N.R.A.S. **147**, 279 (1970).

HAWKING, S. W.: Black holes in general relativity. Comm. Math. Phys. **25**, 152 (1972).

HAYASHI, C., HOSHI, R., SUGIMOTO, D.: Evolution of the stars. Prog. Theor. Phys. (Japan) Supp. **22** (1962).

HAYFORD, J. F.: The figure of the earth and isostasy from measurements in the United States. Coast and Geodetic Survey, U.S. Govt. Printing Office, Washington 1909.

HECKMANN, O.: Theorien der Kosmologie (Theory of cosmology). Berlin: Springer 1942.

HEESCHEN, D. S.: The absolute radio luminosity and surface brightness of extragalactic radio sources. Ap. J. **146**, 517 (1966).

HELMHOLTZ, H.: Popular Lectures (1854).

HENRY, R. C., FRITZ, G., MEEKINS, J. F., FRIEDMAN, H., BYRAM, E. T.: Possible detection of a dense intergalactic plasma. Ap. J. **153**, L 11 (1968).

HERSCHEL, SIR F. W.: On the proper motion of the Sun and solar system, with an account of several changes that have happened among the fixed stars since the time of Mr. Flamsteed. Phil. Trans. **73**, 247 (1753).

HERSCHEL, SIR F. W.: Account of the changes that have happened during the last 25 years, in the relative situation of double stars, with an investigation of the cause to which they are aiming. Phil. Trans. **93**, 340 (1803).

HERTZSPRUNG, E.: Zur Strahlung der Sterne (Giants and dwarfs). Z. Wiss. Photog. 3 (1905). Eng. trans. in: Source book in astronomy (ed. H. SHAPLEY). Cambridge, Mass.: Harvard University Press 1960.

HERTZSPRUNG, E.: Über die räumliche Verteilung der Veränderlichen vom δ Cephei-Typus (About the space distribution of the variables of the δ Cephei type). Astron. Nach. **196**, 201 (1913).

HERTZSPRUNG, E.: Bemerkungen zur Statistik der Sternparallaxen (Remarks on the statistics of stellar parallax). Astron. Nach. **208**, 89 (1919).

HEYL, P. R.: A redetermination of the constant of gravitation. J. Res. Nat. Bur. Stands. **5**, 1243 (1930).

HILL, E. R.: The component of the galactic gravitational field perpendicular to the galactic plane. B.A.N. **15**, 1 (1960).

HILL, J. M.: A measurement of the gravitational deflection of radio waves by the Sun. M.N.R.A.S. **153**, 7 (1971).

HIPPARCHUS—(125 B.C.) in: Ptolemy's Almagest. Cf. A history of astronomy from Thales to Kepler, by J. L. E. DREYER. New York: Dover 1953.

HOAG, A. A., JOHNSON, H. L., IRIARTE, B., MITCHELL, R. I., HALLAM, K. L., SHARPLESS, S.: Photometry of stars in galactic cluster fields. Publ. U.S. Nav. Obs. **17**, part VIII (1961).

HODGE, P. W.: Radii, orbital properties, and relaxation times of dwarf elliptical galaxies. Ap. J. **144**, 869 (1966).

HOERNER, S. VON: The internal structure of globular clusters. Ap. J. **128**, 451 (1957).

HOGG, H. B.: A bibliography of individual globular clusters. Publ. David Dunlap, Obs. Univ. Toronto Press 1963.

HOGG, H. S.: Star clusters. In: Handbuch der Physik, vol. LIII: Astrophysics IV—Stellar systems (ed. S. FLÜGGE), p. 129. Berlin-Göttingen-Heidelberg: Springer 1959.

HOLDEN, D. J.: An investigation of the clustering of radio sources. M.N.R.A.S. **133**, 225 (1966).

HOLMBERG, E.: A photometric study of nearby galaxies. Medd. Lund. Obs. Ser. 2 No. 128 (1950).

HOLMBERG, E.: A photographic photometry of extragalactic nebulae. Medd. Lund. Obs. Ser. 2 No. 136 (1958).

HOLMES, A., LAWSON, R. W.: Radioactivity of potassium and its geological significance. Phil. Mag. **2**, 1218 (1926).

HOYLE, F.: A new model for the expanding universe. M.N.R.A.S. **108**, 372 (1948).

HOYLE, F.: The ages of type I and type II subgiants. M.N.R.A.S. **119**, 124 (1959).

HOYLE, F.: Review of recent developments in cosmology. Proc. Roy. Soc. **A 308**, 1 (1968).

HOYLE, F., FOWLER, W. A.: Nucleosynthesis in supernovae. Ap. J. **132**, 565 (1960).

HOYLE, F., FOWLER, W. A.: On the nature of strong radio sources. M.N.R.A.S. **125**, 169 (1963).

HOYLE, F., FOWLER, W. A., BURBIDGE, G. R., BURBIDGE, E. M.: On relativistic astrophysics. Ap. J. **139**, 909 (1964).

HOYLE, F., NARLIKAR, J. V.: A conformal theory of gravitation. Proc. Roy. Soc. **A 294**, 138 (1966).

HOYLE, F., SANDAGE, A. R.: The second-order term in the redshift-magnitude relation. P.A.S.P. **68**, 301 (1956).

HOYLE, F., TAYLER, R. J.: The mystery of the cosmic helium abundance. Nature **203**, 1108 (1964).

HUBBLE, E. P.: Study of diffuse galactic nebulae. Ap. J. **56**, 162 (1922).

HUBBLE, E. P.: Extragalactic nebulae. Ap. J. **64**, 321 (1926).

HUBBLE, E. P.: A relation between distance and radial velocity among extra-galactic nebulae. Proc. Nat. Acad. Sci. (Wash.) **15**, 168 (1929).

HUBBLE, E. P.: The distribution of extra-galactic nebulae. Mt. Wilson Contr. No. 435, Ap. J. **79**, 8 (1934).

HUBBLE, E. P.: The realm of the nebulae. New Haven, Conn.: Yale University Press 1936. Republ. New York: Dover 1958.

HUBBLE, E. P., HUMASON, M. L.: The velocity-distance relation among extra-galactic nebulae. Ap. J. **74**, 43 (1931).

HUBBLE, E. P., HUMASON, M. L.: The velocity-distance relation for isolated extra-galactic nebulae. Proc. Nat. Acad. Sci. (Wash.) **20**, 264 (1934).

HUGGINS, SIR W.: On the spectra of some fixed stars. Phil. Mag. **28**, 152 (1864).

HUGHES, R. B., LONGAIR, M. S.: Evidence on the isotropy of faint radio sources. M.N.R.A.S. **135**, 131 (1967).

HUMASON, M. L.: The large radial velocity of N.G.C. 7619. Proc. Nat. Acad. Sci. (Wash.) **15**, 162 (1929).

HUMASON, M. L., MAYALL, N. U., SANDAGE, A. R.: Redshifts and magnitudes of extragalactic nebulae. Astron. J. **61**, 97 (1956).

IBEN, I. I.: Stellar evolution within and off the main sequence. Ann. Rev. Astron. Astrophys. **5**, 571 (1967).

IBEN, I. I., ROOD, R. T.: Metal-poor stars: III. On the evolution of horizontal-branch stars. Ap. J. **161**, 587 (1970).

IRWIN, J. B.: Orbit determination of eclipsing binaries. In: Astronomical techniques—Stars and stellar systems II (ed. W. A. HILTNER). Chicago, Ill.: Univ. of Chicago Press 1962.

IRWIN, J. B.: The differential limb darkening of S Cancri. Ap. J. **138**, 1104 (1963).

ISREAL, W.: Event horizons in static vacuum spacetimes. Phys. Rev. **164**, 1776 (1967).

ISREAL, W.: Event horizons in static electrovac spacetimes. Comm. Math. Phys. **8**, 245 (1968).

JAKI, S. L.: The paradox of Olbers paradox. New York: Herder and Herder 1969.

JEANS, SIR J.: The stability of a spherical nebula. Phil. Trans. Roy. Soc. London **A 199**, 1 (1902).

JEANS, SIR J.: Astronomy and cosmology. Cambridge: Cambridge Univ. Press 1929. Repr. New York: Dover 1961.

JOHNSON, H. L.: Photometric systems. In: Basic astronomical data-Stars and stellar systems III (ed. K. A. STRAND). Chicago, Ill.: University of Chicago Press 1963.

JOHNSON, H. L.: Infrared photometry of M-dwarf stars. Ap. J. **141**, 170 (1965).

JOHNSON, H. L., MORGAN, W. W.: Fundamental stellar photometry for standards of spectral type in the revised system of the Yerkes spectral atlas. Ap. J. **117**, 313 (1953).

JOHNSON, H. L., HOAG, A. A., IRIARTE, B., MITCHELL, R. I., HALLAM, K. L.: Galactic clusters as indicators of stellar evolution and galactic structure. Lowell Obs. Bull. No. 113 **V**, 133 (1963).

JONES, B. J.: Rev. Mod. Phys. **48**, 107 (1976).

JONES, H. S.: The rotation of the earth, and the secular accelerations of the sun, moon, and planets. M. N. R. A. S. **99**, 541 (1939).

VAN DE KAMP, P.: The nearby stars. Ann. Rev. Astron. Astrophys. **9**, 103 (1971).

KAPTEYN, J. C.: Discovery of the two stars streams. Brit. Assoc. Rept. (South Africa) **A**, 257 (1905). Reprod. in: Source book in astronomy (ed. H. SHAPLEY). Cambridge, Mass.: Harvard University Press 1960.

KAPTEYN, J. C.: On the derivation of the constants for the two star-streams. M. N. R. A. S. **72**, 743 (1912).

KARACHETSEV, I. D.: The virial mass-luminosity ratio and the instability of different galactic systems. Astrophys. **2**, 39 (1966).

KARDASHEV, N. S.: Nonstationariness of spectra of young sources of nonthermal radio emission. Sov. Astron. **6**, 317 (1962).

KARDASHEV, N. S.: Lemaître's universe and observations. Ap. J. **150**, L 135 (1967).

KAULA, W. M.: Theory of satellite geodesy. Waltham, Mass.: Blaisdell 1966.

KEENAN, P. C.: Classification of stellar spectra. In: Basic astronomical data—Stars and stellar systems III (ed. K. A. STRAND). Chicago, Ill.: University Chicago Press 1963.

KEENAN, P. C., MORGAN, W. W.: Classification of stellar spectra. In: Astrophysics (ed. J. A. HYNEK). New York: McGraw-Hill 1951.

KELLERMANN, K. I.: On the interpretation of radio-source spectra and the evolution of radio galaxies and quasi-stellar sources. Ap. J. **146**, 621 (1966).

KELLERMANN, K. I., DAVIS, M. M., PAULINY-TOTH, I. I. K.: Counts of radio sources at 6-centimeter wavelength. Ap. J. **170**, L 1 (1970).

KELLERMANN, K. I., PAULINY-TOTH, I. I. K.: The spectra of opaque radio sources. Ap. J. **155**, L 71 (1969).

KELVIN, LORD (W. THOMSON): Sur refroidissement séculaire du soleil. De la temperature actuelle du soleil. De l'origine et de la somme totale de la chaleur solaire (On the age of the Sun's heat, the actual temperature of the Sun, and the origin of the sum total of the Sun's heat). Les Mondes. **3**, 473 (1863).

KEPLER, J.: De harmonice monde (The harmony of the world). 1619.

KERR, F. J.: Galactic velocity models and the interpretation of 21-CM surveys. M. N. R. A. S. **123**, 327 (1962).

KERR, F. J.: The large-scale distribution of hydrogen in the galaxy. Ann. Rev. Astron. Astrophys. **7**, 39 (1969).

KERR, R. P.: Gravitational field of a spinning mass as an example of algebraically special metrics. Phys. Rev. Lett. **11**, 237 (1963).

KIANG, T.: The galaxian luminosity function. M. N. R. A. S. **122**, 263 (1961).

KING, E. S.: Standard velocity curves for spectroscopic binaries. Ann. Harv. Obs. **81**, 231 (1920).

KING-HELE, D. G.: The earth's gravitational potential, deduced from the orbits of artificial satellites. Geophys. J. **4**, 20 (1961).

KLEIN, O.: Arguments concerning relativity and cosmology. Science **171**, 339 (1971).

KOEHLER, J. A.: Intergalactic atomic neutral hydrogen II. Ap. J. **146**, 504 (1966).

KOMESAROFF, M. M.: Ionospheric refraction in radio astronomy. Austr. J. Phys. **13**, 353 (1960).

KOPYLOV, A. I., KARACHENTSEV, I. D.: Pis'ma Astron. Zh. No. 6 (1978).

KOZAI, Y.: The motion of a close earth satellite. Astron. J. **64**, 367 (1959).

KOZAI, Y.: The gravitational field of the earth derived from motions of three satellites. Astron. J. **66**, 8 (1961).

KOZAI, Y.: New determination of zonal harmonic coefficients of the earth's gravitational potential. In: Satellite geodesy 1958–1964. Nat. Aeronaut. and Space Admin. 1966.

KRAFT, R. P.: Color excesses for supergiants and classical Cepheids V. The period-color and period luminosity relations: A review. Ap. J. **134**, 616 (1961).

KRAUS, J. D.: Radio astronomy. New York: McGraw-Hill 1966.

KRAUSE, H. G. L.: Die sähulären und teriodèrchen Störungen der Bahn eines künstlichen Erdsatelliten (On the behavior of the orbit of an artificial earth satellite). Int. Astronaut. Kong. VII (1956).

KRON, G. E., MAYALL, N. U.: Photoelectric photometry of galactic and extragalactic star clusters. Ap. J. **65**, 581 (1960).
KRUSKAL, M. D.: Maximal extension of Schwarzschild metric. Phys. Rev. **119**, 1743 (1960).
KUIPER, G. P.: The magnitude of the Sun, the stellar temperature scale, and bolometric corrections. Ap. J. **88**, 429 (1938).
KWEE, K. K., MULLER, C. A., WESTERHOUT, G.: The rotation of the inner parts of the galactic system. B.A.N. **12**, 211 (1954).
VAN DER LAAN, H.: Radio galaxies: I. The interpretation of radio source data. M.N.R.A.S. **126**, 519 (1963).
LANCZOS, K.: Bemerkung zur de Sitterschen Welt (Notes on the de Sitter universe). Phys. Z. **23**, 539 (1922).
LANCZOS, K.: Über die Rotverschiebung in der de Sitterschen Welt (On the redshift in the de Sitter universe). Z. Physik **17**, 168 (1923).
LANDAU, L.: On the theory of stars. Phys. Z. Sowjetunion **1**, 285 (1932).
LANDAU, L., LIFSHITZ, E. M.: The classical theory of fields. Reading, Mass.: Addison-Wesley 1962.
LAPLACE, P. S. MARQUIS DE: Théorie des attractions des sphéroides et de la figure de planétes (Theory of the attraction of spheres and the figure of planets). Mem. de l'Acad. 113 (1782).
LAPLACE, P. S. MARQUIS DE: Le system du monde. Paris 1798. Eng. trans.: The system of the world. London: W. Flint 1809.
LAPLACE, P. S. MARQUIS DE: Sur la vitesse du son dans l'air et dans l'eau (On the velocity of sound in the air and the water). Ann. Chem. Phys. **3**, 238 (1816).
LAPLACE, P. S. MARQUIS DE: Mécanique céleste (Celestial mechanics). Paris 1817. Eng. trans. by N. Bowditch. Boston: Hilliard, Gray, Little, Wilkins 1829.
LEAVITT, H. S.: Periods of 25 variable stars in the small Magellanic cloud. Harv. Cir. No. **173**, 1912. Repr. in: Source book in astronomy (ed. H. SHAPLEY). Cambridge, Mass.: Harvard Univ. Press 1960.
LEBLANC, J. M., WILSON, J. R.: A numerical example of the collapse of a rotating magnetized star. Ap. J. **161**, 541 (1970).
LECAR, M., SORENSON, J., ECKELS, A.: A determination of the coefficient *J* of the second harmonic in the earth's gravitational potential from the orbit of satellite 1958 B. J. Geophys. Res. **64**, 209 (1959).
LEDOUX, P., WALRAVEN, T.: Variable stars. In: Handbuch der Physik, vol. LI, Astrophysics II. Stellar structure p. 353 (ed. S. FLÜGGE). Berlin-Göttingen-Heidelberg: Springer 1958.
LEGENDRE, A. M.: Suite des recherches sur la figure des planètes (Additional research on the figure of the planets). Mém. de l'Acad. Sci. (1789).
LEGENDRE, A. M.: Exercices de cacul intégral. Paris 1817.
LEMAÎTRE, G.: Un univers homogene de masse constante et de rayon croissant, redant compte de la vitesse radiale des nébuleuses extra-galactiques (A homogeneous universe of constant mass and curvature, understanding the radial velocity of extragalactic nebulae). Ann. Soc. Sci. Bruxelles **A 47**, 49 (1927). Eng. trans. in M.N.R.A.S. **91**, 483 (1931).
LEMAÎTRE, G.: The expanding universe. M.N.R.A.S. **91**, 490 (1931).
LEMAÎTRE, G.: L'Univers en expansion (The universe in expansion). Soc. Sci. Bruxelles Ann. **A 53**, 51 (1933).
LESLIE, P. R. R.: An investigation of the clustering of radio stars. M.N.R.A.S. **122**, 371 (1961).
LEVINE, J. L., GARWIN, R. L.: Absence of gravity-wave signals at a bar at 1695 Hz. Phys. Rev. Lett. **31**, 173 (1973).
LIFSHITZ, E. M.: On the gravitational stability of the expanding universe. J. Phys. USSR (Zurn Erksp. Theor. Fiz) **10**, 116 (1946).
LIMBER, D. N., MATHEWS, W. G.: The dynamical stability of Stephan's quintet. Ap. J. **132**, 286 (1960).
LIN, C. C., YUAN, C., SHU, F. H.: On the spiral structure of disk galaxies: III. Comparison with observations. Ap. J. **155**, 721 (1969).
LINDBLAD, B.: The small oscillations of a rotating stellar system and the development of spiral arms. Medd. Astron. Obs. Uppsala **19**, 1 (1927).
LINDBLAD, B.: A condensation theory for meteoric matter and its cosmological significance. Nature **135**, 133 (1935).
LONGAIR, M. S.: On the interpretation of radio source counts. M.N.R.A.S. **133**, 421 (1966).
LONGAIR, M. S.: Observational cosmology. Rep. Prog. Phys. **34**, 1125 (1971).
LONGAIR, M. S., SUNYAEV, R. A.: The origin of the X-ray background. Astrophys. Lett. **4**, 65 (1969).

LORENTZ, H. A.: Electromagnetic phenomena in a system moving with any velocity less than that of light. Proc. Ams. Acad. Sci. **6**, 809 (1904). Reprod. in: The principle of relativity. New York: Dover 1952.

Lund observatory tables for the conversion of equatorial and galactic coordinates. Ann. Obs. Lund **15**, **16**, **17** (1961).

LUYTEN, W.: The luminosity function. Ann. N. Y. Acad. Sci. **42**, 205 (1941).

MATTHEWS, T. A., SANDAGE, A. R.: Optical identification of 3C 48, 3C 196, and 3C 286 with stellar objects. Ap. J. **138**, 30 (1963).

MATTIG, W.: Über den Zusammenhang zwischen Rotverschiebung und scheinbarer Helligkeit (On the relationship between the redshift and the apparent light). Astron. Nach. **284**, 109 (1958).

MATTIG, W.: Über den Zusammenhang zwischen der Anzahl der extragalaktischen Objekte und der scheinbaren Helligkeit (About the relationship between the number of extragalactic objects and their apparent light). Astron. Nach. **285**, 1 (1959).

MAURY, A. C., PICKERING, E. C.: Spectra of bright stars. Ann. Harv. Coll. Obs. **28** (1897).

MAY, M. M., WHITE, R. H.: Hydrodynamic calculations of general-relativistic collapse. Phys. Rev. **141**, 1232 (1966).

MAYALL, N. U.: The radial velocities of fifty globular star clusters. Ap. J. **104**, 290 (1946).

MCCREA, W. H.: Observable relations in relativistic cosmology. Z. Ap. **9**, 290 (1935).

MCCUSKEY, S. W.: Variations in the stellar luminosity function VIII. A summary of the results. Ap. J. **123**, 458 (1956).

MELBOURNE, W. G., MUHLEMAN, D. O., O'HANDLEY, D. A.: Radar determination of the radius of Venus. Science **160**, 987 (1968).

MERKELIJN, J. K.: A determination of the luminosity function of radiogalaxies at 400 and 2700 MHz. Astron. Astrophys. **15**, 11 (1971).

MICHELSON, A. A.: On the application of interferance measurements to astronomical measurements. Phil. Mag. **30**, 1 (1890).

MICHELSON, A. A., PEASE, F. G.: Measurement of the diameter of α Orionis with the interferometer. Ap. J. **53**, 249 (1921).

MIHALAS, D.: Stellar atmospheres. San Francisco: W. H. Freeman 1970.

MILEY, G. K.: The radio structure of quasars—A statistical investigation. M.N.R.A.S. **152**, 477 (1971).

MILNE, E. A.: Relativity, gravitation and world structure. Oxford: Oxford at the Clarendon Press 1935.

MINKOWSKI, H.: Die Grundgleichungen für die elektromagnetischen Vorgänge in bewegten Körpern (The basic equations for the electromagnetic events in moving bodies). Gott. Nach. **1**, 53 (1908).

MINKOWSKI, R.: NGC 6166 and cluster Abell 2199. Astron. J. **66**, 558 (1961).

MISNER, C. W., THORNE, K. S., WHEELER, J. A.: Gravitation. New York: W. H. Freeman 1973.

MORGAN, W. W., KEENAN, P. C., KELLMAN, E.: An atlas of stellar spectra. Chicago, Ill.: Univ. of Chicago Press 1943.

MORRISON, P., PINES, S.: The reduction from geocentric to geodetic coordinates. Astron. J. **66**, 15 (1961).

MÖSSBAUER, R. L.: Kernresonanzfluoreszenz von Gammastrahlung in Ir^{191} (Nuclear resonance fluorescence by gamma radiation in Ir^{191}). Z. Physik **151**, 124 (1958).

MOWBRAY, A. G.: The diameters of globular clusters. Ap. J. **104**, 47 (1946).

MUHLEMAN, D. O.: On the radio method of determining the astronomical unit. M.N.R.A.S. **144**, 151 (1969).

MUHLEMAN, D. O., EKERS, R. D., FOMALONT, E. B.: Radio interferometric test of the general relativistic light bending near the Sun. Phys. Rev. Lett. **24**, 1377 (1970).

NEWCOMB, S.: On the dynamics of the earth's rotation with respect to periodic variations of latitude. M.N.R.A.S. **52**, 336 (1892).

NEWCOMB, S.: Tables of the Sun. A. P. Am. Ephem. **6**, 9 (1895).

NEWCOMB, S.: Tables of the motion of the earth on its axis and around the Sun. A. P. Am. Ephem. **6**, 7 (1898).

NEWCOMB, S.: A compendium of spherical astronomy. New York: Macmillan 1906.

NEWMAN, E. T., COUCH, E., CHINNAPARED, R., EXTON, H., PRAKASH, A., TORRENCE, R.: Metric of a rotating, charged mass. J. Math. Phys. **6**, 918 (1965).

NEWTON, I.: Philosophiae naturalis principia mathematica (Natural philosophy and the principles of mathematics). (1687). Eng. Trans. by A. Motte, revised and annotated by F. Cajori. University California Press 1966.

NEWTON, I.: The correspondence of Isaac Newton 1688–1694 (ed. H. W. TURNBILL). Cambridge: Cambridge University Press 1961.

NIER, A. O.: The isotopic constitution of radiogenic leads and the measurement of geological time II. Phys. Rev. **55**, 153 (1939).

NIETO, M. M.: The Titius-Bode law of planetary distances: Its history and theory. Oxford: Pergamon Press 1972.

NORDSTROM, G.: On the energy of the gravitation field in Einstein's theory. K. Ned. Akad. Wet. Amst. Proc. Sec. Sci. **20**, 1238 (1918).

OEPIK, E.: An estimate of the distance of the Andromeda nebula. Ap. J. **55**, 406 (1922).

OHLSSON, J.: Lund observatory tables for the conversion of equatorial coordinates into galactic coordinates. Ann. Lund Obs. **3** (1932).

OKE, J. B.: Redshifts and absolute spectral energy distributions of galaxies in distant clusters. Ap. J. **170**, 193 (1971).

OKE, J. B., SANDAGE, A. R.: Energy distributions, *K* corrections, and the Stebbins-Whitford effect for giant elliptical galaxies. Ap. J. **154**, 21 (1968).

O'KEEFE, J. A., ECKELS, A., SQUIRES, R. K.: The gravitational field of the earth. Astron. J. **64**, 245 (1959).

OLBERS, W.: Über die Durchsichtigkeit des Weltraums (About the transparency of the universe). Dode Jb. 15 (1826). Reprod. in Dickson (1968), Jaki (1969).

OMNÈS, R.: On the origin of matter and galaxies. Astron. Astrophys. **10**, 228 (1971).

OORT, J. H.: Observational evidence confirming Lindblad's hypothesis of a rotation of the galactic system. B. A. N. **3**, 275 (1927).

OORT, J. H.: Dynamics of the galactic system in the vicinity of the Sun. B. A. N. **4**, 269 (1928).

OORT, J. H.: The force exerted by the stellar system in the direction perpendicular to the galactic plane and some related problems. B. A. N. **6**, 249 (1932).

OORT, J. H.: Distribution of galaxies and the density of the universe. In: La structure et l'evolution de l'univers—Institut International de Physique Solvay (ed. R. STOOP). Bruxelles: Coodenberg 1958.

OORT, J. H.: Note on the determination of K_z and on the mass density near the Sun. B. A. N. **15**, 45 (1960).

OORT, J. H.: Infall of gas from intergalactic space. Nature **224**, 1158 (1969).

OPPENHEIMER, J. R., SNYDER, H.: On continued gravitational contraction. Phys. Rev. **56**, 455 (1939).

OPPENHEIMER, J. R., VOLKOFF, G. M.: On massive neutron cores. Phys. Rev. **55**, 374 (1939).

OSTIC, R. G., RUSSELL, R. D., REYNOLDS, P. H.: A new calculation for the age of the earth from abundances of lead isotopes. Nature **199**, 1150 (1963).

OSTRICKER, J. P., GUNN, J. E.: On the nature of pulsars: I. Theory. Ap. J. **157**, 1395 (1969).

OZERNOY, L. M., CHERNIN, A. D.: The fragmatation of matter in a turbulent matagalactic medium I. Sov. Astron. A. J. **11**, 907 (1968).

OZERNOY, L. M., CHIBISOV, G. V.: Galactic parameters as a consequence of cosmological turbulence. Astrophys. Lett. **7**, 201 (1971).

OZERNOY, L. M.: Proc. First Europ. Astron. Meeting **3**, 65 (1974); in *The Formation of Stars and Galaxies* (Nauka Publ. House, 1976) p. 133, and references therein (whirl perturbations).

PAGE, T.: Average masses and mass-luminosity ratios of the double galaxies. Ap. J. **132**, 910 (1960).

PAGE, T.: Average masses of the double galaxies. In: Fourth Berkeley symposium on mathematical statistics and probability **3**, 277 (1961).

PAGE, T.: *M/L* for double galaxies, a correction. Ap. J. **136**, 685 (1962).

PAGE, T. L.: Statistical evidence of the masses and evolution of galaxies. Smithsonian Astrophys. Obs. Rpt. No. 195 (1965).

PAULINY-TOTH, I. I. K., KELLERMANN, K. I., DAVIS, M. M.: Number counts and spectral distribution of radio sources. In: External galaxies and quasi-stellar objects—I. A. U. Symp. No. 44 (ed. D. S. EVANS). Dordrecht, Holland: D. Reidel 1972.

PARTRIDGE, R. B., PEEBLES, P. J. E.: Are young galaxies visible? II. The integrated background. Ap. J. **148**, 377 (1967).

PATTERSON, C.: Age of meteorites and the earth. Geochim. et. Cosmochim. Acta. **10**, 230 (1956).

PEACH, J. V.: The determination of the deceleration parameter and the cosmological constant from the redshift-magnitude relation. Ap. J. **159**, 753 (1970).

Peach, J. V.: Cosmological information from galaxies and radio galaxies. In: External galaxies and quasi-stellar objects—I. A. U. Symp. No. 44 (ed. D. S. Evans). Dordrecht, Holland: D. Reidel 1972.
Pease, F. G.: Interferometer methods in astronomy. Eigebn. Exactn. Naturwiss. **10**, 84 (1931).
Peebles, P. J. E.: Primordial helium abundance and the primordial fireball II. Ap. J. **146**, 542 (1966).
Peebles, P. J. E.: Recombination of the primeval plasma. Ap. J. **153**, 1 (1968).
Peebles, P. J. E.: Physical cosmology. Princeton, N. J.: Princeton University Press 1971.
Peebles, P. J. E.: Nonthermal primeval fireball? Astrophys. and Space Sci. **10**, 280 (1971).
Penrose, R.: Gravitational collapse: the role of general relativity. Revista del Nuovo Cimento **1**, 252 (1969).
Penzias, A. A., Scott, E. H.: Intergalactic H I absorption at 21 centimeters. Ap. J. **153**, L 7 (1968).
Penzias, A. A., Wilson, R. W.: A measurement of excess antenna temperature at 4080 MHz. Ap. J. **142**, 419 (1965).
Penzias, A. A., Wilson, R. W.: Intergalactic H I emission at 21 centimeters. Ap. J. **156**, 799 (1969).
Peters, P. C., Mathews, J.: Gravitational radiation from point masses in a Keplerian orbit. Phys. Rev. **131**, 435 (1963).
Petrosian, V., Salpeter, E. E., Szekeres, P.: Quasi-stellar objects in universes with non-zero cosmological constant. Ap. J. **147**, 1222 (1967).
Pickering, E. C.: On the spectrum of Ursae Majoris. Am. J. Sci. **39**, 46 (1889).
Pickering, E. C., Fleming, W. P.: Miscellaneous investigations of the Henry Draper memorial. Ann. Harv. Coll. Obs. **26** (1897).
Planck, M.: Über das Gesetz der Energieverteilung im Normalspektrum (On the theory of thermal radiation). Ann. Physik **4**, 553 (1901).
Pogson, N.: Magnitude of 36 of the minor planets. M. N. R. A. S. **17**, 12 (1856).
Poincaré, H.: L'état actual and l'avenir de la physique mathématica (The present and future state of mathematical physics). Bull. Sci. Math. **2**, 28 (1904).
Poincaré, H.: Lecons de mecanique céleste (Lessons of celestial mechanics). Gauthier-Violars, Imprimeur-Librarie (1905).
Popper, D. M.: Redshift in the spectrum of 40 Eridani B. Ap. J. **120**, 316 (1954).
Pound, R. V., Rebka, G. A.: Gravitational redshift in nuclear resonance. Phys. Rev. Lett. **3**, 439 (1959).
Pound, R. V., Rebka, G. A.: Apparent weight of photons. Phys. Rev. Lett. **4**, 337 (1960).
Pound, R. V., Snider, J. L.: Effect of gravity on nuclear resonance. Phys. Rev. Lett. **13**, 539 (1964).
Poynting, J. H.: Radiation in the solar system: Its effect on temperature and its pressure on small bodies. Phil. Tran. Roy. Soc. London **A 202**, 525 (1904).
Press, W. H., Thorne, K. S.: Gravitational wave astronomy. Ann. Rev. Astron. Astrophys. **10**, 335 (1972).
Raychaudhuri, A.: Relativistic cosmology I. Phys. Rev. **98**, 1123 (1955).
Raychaudhuri, A.: Relativistic and Newtonian cosmology. Z. Astrophys. **43**, 161 (1957).
Rees, M. J.: Scattering of background X-rays by metagalactic electrons. Astrophys. Lett. **4**, 113 (1969).
Reissner, H.: Über die Eigengravitation des elektrischen Feldes nach der Einsteinschen Theorie (On the special gravitation of the electric field given by Einstein's theory). Ann. Phys. **50**, 106 (1916).
van Rhijn, P. J.: On the frequency of the absolute magnitudes of the stars. Publ. Astr. Lab. Groningen **38** (1925).
van Rhijn, P. J.: The absorption of light in interstellar galactic space and the galactic density distribution. Publ. Astr. Lab. Groningen **47** (1936).
Roach, F. E., Smith, L. L.: An observational search for cosmic light. Geophys. J. R. A. S. **15**, 227 (1968).
Roberts, M. S.: Integral properties of spiral and irregular galaxies. Astron. J. **74**, 859 (1969).
Roberts, M. S., Rots, A. H.: Comparison of rotation curves of different galaxy types. Astron. and Astrophys. **26**, 483 (1973).
Robertson, H. P.: On relativistic cosmology. Phil. Mag. **5**, 835 (1928).
Robertson, H. P.: Relativistic cosmology. Rev. Mod. Phys. **5**, 62 (1933).
Robertson, H. P.: Kinematics and world-structure. Ap. J. **82**, 284 (1935).
Robertson, H. P.: Kinematics and world-structure II, III. Ap. J. **83**, 187, 257 (1936).
Robertson, H. P.: Dynamical effects of radiation in the solar system. M. N. R. A. S. **97**, 423 (1937).
Robertson, H. P.: The theoretical aspects of the nebular redshift. P. A. S. P. **67**, 82 (1955).
Robertson, H. P.: Relativity and cosmology. In: Space age astronomy (ed. A. J. Deutsch and W. B. Klemperer). New York: Academic Press 1962.

ROGSTAD, D. H., ROUGOOR, G. W., WHITEOAK, J. B.: Neutral hydrogen studies of galaxies with a single-spacing interferometer. Ap. J. **150**, 9 (1967).

ROLL, P. G., KROTOV, R., DICKE, R. H.: The equivalence of inertial and passive gravitational mass. Ann. Phys. (N. Y.) **26**, 442 (1964).

ROOD, H. J., ROTHMAN, V. C. A., TURNROSE, B. E.: Empirical properties of the mass discrepancy in groups and clusters of galaxies. Ap. J. **162**, 411 (1970).

ROSE, R. D., PARKER, H. M., LOWRY, R. A., KUHLTHAU, A. R., BEAMS, J. W.: Determination of the gravitational constant *G*. Phys. Rev. Lett. **23**, 655 (1969).

RUSSELL, H. N.: Relation between the spectra and other characteristics of the stars. Proc. Amer. Phil. Soc. **51**, 569 (1912).

RUSSELL, H. N.: Real brightness of variable stars. Science **37**, 651 (1913).

RUSSELL, H. N.: Relations between the spectra and other characteristics of the stars. Pop. Astron. **22**, 275, 331 (1914). Reprod. in: Source book in astronomy (ed. H. SHAPLEY). Cambridge, Mass.: Harvard University Press 1960.

RUSSELL, H. N., DUGAN, R. S., STEWART, J.: Astronomy. Boston, Mass.: Gunn 1926.

RUTHERFORD, E.: Origin of actinium and age of the earth. Nature **123**, 313 (1929).

RUTHERFURD, L. M.: Astronomical observations with the spectroscope. Sill. Amer. J. **35**, 71 (1863).

RYLE, M.: The nature of the cosmic radio sources. Proc. Roy. Soc. London **248**, 289 (1958).

RYLE, M.: The counts of radio sources. Ann. Rev. Astron. Astrophys. **6**, 249 (1968).

RYLE, M., CLARKE, R. W.: An examination of the steady-state model in the light of some recent observations of radio sources. M. N. R. A. S. **122**, 349 (1961).

SAHA, M. N.: On the physical theory of stellar spectra. Proc. Roy. Soc. London **A 99**, 135 (1921).

SALPETER, E. E.: The luminosity function and stellar evolution. Ap. J. **121**, 161 (1955).

SAMPSON, R. A.: On the rotation and mechanical state of the Sun. Mem. R. A. S. **51**, 123 (1894).

SANDAGE, A. R.: Observational approach to evolution: I. Luminosity functions; II. A computed luminosity function for K0–K2 stars from $M_v = +5$ to $M_v = -4.5$. Ap. J. **125**, 422, 435 (1957).

SANDAGE, A. R.: Current problems in the extragalactic distance scale. Ap. J. **127**, 513 (1958).

SANDAGE, A. R.: The ability of the 200-inch telescope to discriminate between selected world models. Ap. J. **133**, 335 (1961).

SANDAGE, A. R.: The distance scale. In: Problems in extragalactic research (ed. G. MCVITTIE). New York: Macmillan 1962.

SANDAGE, A. R.: The correlation of colors with redshifts for QSS leading to a smoothed mean energy distribution and new values for the *K*-correction. Ap. J. **146**, 13 (1966).

SANDAGE, A. R.: A new determination of the Hubble constant from globular clusters in M 87. Ap. J. **152**, L 149 (1968).

SANDAGE, A. R.: Main-sequence photometry, color-magnitude diagrams, and ages for the globular clusters M 3, M 13, M 15, and M 92. Ap. J. **162**, 841 (1970).

SANDAGE, A. R.: The age of the galaxies and globular clusters, problems of finding the Hubble constant and deceleration parameter. In: Nuclei of galaxies, pontifica academia scientiavm (ed. D. K. O'CONNELL). Amsterdam, Holland: North Holland 1971.

SANDAGE, A. R., SCHWARZSCHILD, M.: Inhomogeneous stellar models: II. Models with exhausted cores in gravitational contraction. Ap. J. **116**, 463 (1952).

SANDAGE, A. R., TAMMANN, G. A.: A composite period—luminosity relation for Cepheids at mean and maximum light. Ap. J. **151**, 531 (1968).

SANDAGE, A. R., TAMMANN, G. A.: Absolute magnitude of Cepheids: III. Amplitude as a function of position in the instability strip—A period-luminosity-amplitude relation. Ap. J. **167**, 293 (1971).

SATO, H., MATSUDA, T., TAKEDA, H.: Galaxy formation and the primordial turbulence in the expanding hot universe. Prog. Theor. Phys. **43**, 1115 (1970).

SCHARLEMANN, E. T., WAGONER, R. V.: Electromagnetic fields produced by relativistic rotating disks. Ap. J. **171**, 107 (1972).

SCHEUER, P. A. G.: A sensitive test for the presence of atomic hydrogen in intergalactic space. Nature **207**, 963 (1965).

SCHEUER, P. A. G.: Radio astronomy and cosmology. In: Stars and stellar systems, vol. IX—Galaxies and the universe (ed. A. R. SANDAGE). Chicago, Ill.: University of Chicago Press 1974.

SCHIFF, L. I.: Motion of a gyroscope according to Einstein's theory of gravitation. Proc. Nat. Acad. Sci. (Wash.) **46**, 871 (1960).

SCHILD, R., OKE, J. B.: Energy distributions and *K*-corrections for the total light from giant elliptical galaxies. Ap. J. **169**, 209 (1971).

SCHLESINGER, F.: Line of sight constants for the principal stars. Ap. J. **10**, 1 (1899).

SCHMIDT, M.: A model of the distribution of mass in the galactic system. B. A. N. **13**, 15 (1956).

SCHMIDT, M.: Spiral structure in the inner parts of the galactic system derived from the hydrogen emission at 21 cm wavelength. B. A. N. **13**, 247 (1957).

SCHMIDT, M.: The rate of star formation: II. The rate of formation of stars of different mass. Ap. J. **137**, 758 (1963).

SCHMIDT, M.: Rotation parameters and distribution of mass in the Galaxy. In: Galactic structure—Stars and stellar systems V (ed. A. BLAUUW and M. SCHMIDT). Chicago, Ill.: Univ. of Chicago Press 1965.

SCHMIDT, M.: Lifetimes of extragalactic radio sources. Ap. J. **146**, 7 (1966).

SCHMIDT, M.: Space distribution and luminosity functions of quasi-stellar radio sources. Ap. J. **151**, 393 (1968).

SCHMIDT, M.: Space distribution and luminosity functions of quasi-stellar objects. In: Nuclei of galaxies (ed. D. J. K. O'CONNELL). Amsterdam, Holland: North Holland 1971.

SCHÖNBERG, M., CHANDRASEKHAR, S.: On the evolution of main-sequence stars. Ap. J. **96**, 161 (1942).

SCHOTT, G. A.: Electromagnetic radiation. Cambridge: Cambridge Univ. Press 1912.

SCHRAMM, D. N.: Necleo-cosmochronology. Presented at Cosmochem. Symp., Cambridge, Mass. (1972).

SCHRAMM, N., WASSERBURG, G. J.: Nucleochronologies and the mean age of the elements. Ap. J. **162**, 57 (1970).

SCHWARTZ, D. A.: The isotropy of the diffuse cosmic X-rays determined by OSO III. Ap. J. **162**, 439 (1970).

SCHWARZSCHILD, K.: Über das Gravitationsfeld eines Massenpunktes nach der Einsteinschen Theorie (The gravitational field of a point mass according to Einstein's theory). Sitz. Acad. Wiss. **1**, 189 (1916).

SCHWARZSCHILD, M.: Mass distribution and mass-luminosity ratio in galaxies. Astron. J. **59**, 273 (1954).

SCHWARZSCHILD, M.: Stellar evolution in globular clusters. Q. J. R. A. S. **11**, 12 (1970).

SCOTT, E. L.: The brightest galaxy in a cluster as a distance indicator. Astron. J. **62**, 248 (1957).

SEARS, R. L.: Helium content and neutrino fluxes in solar models. Ap. J. **140**, 477 (1964).

SECCHI, P. A.: Spettri prismatici delle stelle fisse (Spectra of the fixed stars). Memoira (Memoirs) Roma, 1868, Firenze, 1869.

SEEGER, P. A., FOWLER, W. A., CLAYTON, D. D.: Nucleosynthesis of heavy elements by neutron capture. Ap. J. Suppl. No. 97, **11**, 121 (1965).

SEEGER, P. A., SCHRAMM, D. N.: *r*-process production ratios of chronologic importance. Ap. J. **160**, L 157 (1970).

SEIELSTAD, G. A., SRAMEK, R. A., WEILER, K. W.: Measurement of the deflection of 9.602 GHz radiation from 3C 279 in the solar gravitational field. Phys. Rev. Lett. **24**, 1373 (1970).

SHAKESHAFT, J. R., RYLE, M., BALDWIN, J. E., ELSMORE, B., THOMPSON, J. H.: A survey of radio sources between declinations $-38°$ and $+83°$. Mem. R. A. S. **67**, 106 (1955).

SHANE, W. W., BIEGER-SMITH, G. P.: The galactic rotation curve derived from observations of neutral hydrogen. B. A. N. **18**, 263 (1966).

SHAPIRO, I. I.: Fourth test of general relativity. Phys. Rev. Lett. **13**, 789 (1964).

SHAPIRO, I. I.: Spin and orbital motions of the planets. In: Radar astronomy (ed. J. V. EVANS and T. HAGFORS). New York: McGraw-Hill 1968.

SHAPIRO, I. I., SMITH, W. B., ASH, M. E., HERRICK, S.: General relativity and the orbit of Icarus. Astron. J. **76**, 588 (1971).

SHAPIRO, I. I., ASH, M. E., INGALLS, R. P., SMITH, W. B., CAMPBELL, D. B., DYCE, R. B., JURGENS, R. F., PETTENGILL, G. H.: Forth test of general relativity: New radar result. Phys. Rev. Lett. **26**, 1132 (1971).

SHAPIRO, I. I., SMITH, W. B., ASH, M. E., HERRICK, S.: General relativity and the orbit of Icarus. Astron. J. **76**, 588 (1971).

SHAPIRO, I. I., SMITH, W. B., ASH, M. E., INGALLS, R. P., PETTENGILL, G. H.: Gravitational constant: Experimental bound on its time variation. Phys. Rev. Lett. **26**, 27 (1971).

SHAPIRO, I. I., PETTENGILL, G. H., ASH, M. E., INGALLS, R. P., CAMPBELL, D. B., DYCE, R. B.: Mercury's perihelion advance: Determination by radar. Phys. Rev. Lett. **28**, 1594 (1972).

SHAPIRO, S. L.: The density of matter in the form of galaxies. Astron. J. **76**, 291 (1971).

SHAPLEY, H.: On the determination of the distances of globular clusters. Ap. J. **48**, 89 (1918).

SHAPLEY, H.: On the existence of external galaxies. P.A.S.P. **31**, 261 (1919).
SHAPLEY, H.: Luminosity distribution and average density in twenty five groups of galaxies. Proc. Nat. Acad. Sci. (Wash.) **19**, 591 (1933).
SHAPLEY, H.: Stellar clusters. In: Handbuch der Astrophysik VII, 534. Berlin: Springer 1933.
SHAPLEY, H., AMES, A.: A survey of the external galaxies brighter than the thirteenth magnitude. Ann. Harv. Coll. Obs. **88**, No. 2 (1932).
SHAPLEY, H., SAWYER, H. B.: A classification of globular clusters. Harv. Bull. No. 849 (1927).
SHAPLEY, H., SAYER, A. R.: The angular diameters of globular clusters. Harv. Obs. Rpt. No. 116 (1935).
SHECTMAN, S. A.: Clusters of galaxies and the cosmic light. Ap. J. **179**, 681 (1973).
SHKLOVSKII, I. S.: Physical conditions in the gaseous envelope of 3C 273. Astron. Zh. **41**, 408 (1964). Sov. Astron. **8**, 638 (1965).
SILK, J.: Cosmic black-body radiation and galaxy formation. Ap. J. **151**, 459 (1968).
SIMON, M., DRAKE, F. D.: An evolutionary sequence for strong radio sources. Nature **215**, 1457 (1967).
DESITTER, W.: On Einstein's theory of gravitation, and its astronomical consequences II. M.N.R.A.S. **77**, 155, 481 (1916).
DESITTER, W.: On the relativity of inertia. Remarks concerning Einstein's latest hypothesis. Proc. Akad. Wetensch. Amst. **19**, 1217 (1917).
DESITTER, W.: On Einstein's theory of gravitation and its astronomical consequences III. M.N.R.A.S. **78**, 3 (1917).
DESITTER, W.: On Einstein's term in the motion of the lunar perigee and node. M.N.R.A.S. **81**, 102 (1920).
SLIPHER, V. M.: The detection of nebular rotation. Lowell Obs. Bull. **2**, No. 62 (1914).
SLIPHER, V. M.: Nebulae. Proc. Amer. Phil. Soc. **56**, 403 (1917).
SLIPHER, V. M.: Analysis of radial velocities of globular clusters and non-galactic nebulae. Ap. J. **61**, 353 (1925).
SMART, W. M., GREEN, H. E.: The solar motion and galactic rotation from radial velocities. M.N.R.A.S. **96**, 471 (1936).
SMITH, S.: The mass of the Virgo cluster. Ap. J. **83**, 23 (1936).
Smithsonian astrophysical observatory star catalogue. Washington: Smithsonian Institute 1966.
SMOOT, G. F., GORENSTEIN, M. V., MULLEN, R. A.: Phys. Rev. Lett. **39**, 898 (1977).
SOLHEIM, J. E.: Relativistic world models and redshift-magnitude observations. M.N.R.A.S. **133**, 321 (1966).
SPITZER, L.: Dynamical evolution of dense spherical star systems. In: Nuclei of galaxies (ed. D. K. O'CONNELL). Amsterdam, Holland: North Holland 1971.
SPITZER, L., HÄRM, R.: Evaporation of stars from isolated clusters. Ap. J. **127**, 544 (1958).
SRAMEK, R. A.: A measurement of the gravitational deflection of microwave radiation near the Sun, 1970 October. Ap. J. **167**, L 55 (1971).
ST. JOHN, C. E.: Evidence for gravitational displacement of lines in the solar spectrum predicted by Einstein's theory. Ap. J. **67**, 195 (1928).
STEBBINS, J.: Absorption and space reddening in the galaxy as shown by the colors of globular clusters. Proc. Nat. Acad. Sci. **19**, 222 (1933).
STEBBINS, J., WHITFORD, A. E.: Six-color photometry of stars: I. The law of space reddening from the colors of O and B stars. Ap. J. **98**, 20 (1943).
STEBBINS, J., WHITFORD, A. E.: Six-color photometry of stars: VI. The colors of extragalactic nebulae. Ap. J. **108**, 413 (1948).
STECHER, T. P.: Interstellar extinction in the ultraviolet II. Ap. J. Lett. **157**, L 125 (1969).
STEFAN, A. J.: Beziehung zwischen Wärmestrahlung und Temperatur (Relation between thermal radiation and temperature). Wien. Ber. **79**, 397 (1879).
SUNYAEV, R. A., ZELDOVICH, Y. B.: Small-scale fluctuations of relic radiation. Astrophys. and Space Sci. **7**, 3 (1970).
SZEKERES, G.: On the singularities of a Riemannian manifold. Pub. Math. Debrecen **7**, 285 (1960).
TAMMANN, G. A.: Remarks on the radial velocities of galaxies in the Virgo cluster. Astron. Astrophys. **21**, 355 (1972).
TAYLER, R. J.: Stellar evolution. Rept. Prog. Phys. **31**, 167 (1968).
TEMPLE, G.: New systems of normal coordinates for relativistic optics. Proc. Roy. Soc. London **A 168**, 122 (1938).
TER HAAR, D., CAMERON, A. G. W.: Historical review of theories of the origin of the solar system. In: Origin of the solar system (ed. R. JASTROW and A. G. W. CAMERON). New York: Academic Press 1963.

THEKAEKARA, M. P., DRUMMOND, A. J.: Standard values for the solar constant and its spectral components. Nature-Phys. Sci. **229**, 6 (1971).

THORNE, K. S., WILL, C. M.: Theoretical frameworks for testing relativistic gravity: I. Foundations. Ap. J. **163**, 595 (1971).

TITIUS, J. D.: In contemplation de la nature. By C. BONNET. Amsterdam: Marc-Michel 1764.

TITIUS, J. D.: Betrachtung über die Natur, vom Herrn KARL BONNET (An overview of nature, by Mr. KARL BONNET). Leipzig: Johann Friedrich Junius 1766.

TOLMAN, R. C.: On the estimation of distance in a curved universe with a non-static line element. Proc. Nat. Acad. Sci. **16**, 511 (1930).

TOLMAN, R. C.: Relativity, thermodynamics, and cosmology. Oxford: Oxford at the Clarendon Press 1934.

TRIMBLE, V., GREENSTEIN, J. L.: The Einstein redshift in white dwarfs III. Ap. J. **177**, 441 (1972).

TRUMPLER, R. J.: Absorption of light in the galactic system. P. A. S. P. **42**, 214 (1930).

DE VAUCOULEURS, G.: Integrated colors of bright galaxies in the UBV system. Ap. J. Suppl. **5**, 233 (1961).

DE VAUCOULEURS, G.: The case for a hierarchical cosmology. Science **167**, 1203 (1970).

DE VAUCOULEURS, G., DE VAUCOULEURS, A.: Reference catalogue of bright galaxies. Austin, Texas: Univ. Texas Press 1964.

VOGT, H.: Die Beziehung zwischen den Massen und den absoluten Leuchtkräften der Sterne (The relationship between the mass and absolute luminosity of a star). Astron. Nach. **226**, 302 (1926).

WAGONER, R. V.: Physics of massive objects. Ann. Rev. Astron. Astrophys. **7**, 553 (1969).

WAGONER, R. V., FOWLER, W. A., HOYLE, F.: On the synthesis of elements at very high temperatures. Ap. J. **148**, 3 (1967).

WALKER, A. G.: Distance in an expanding universe. M. N. R. A. S. **94**, 159 (1933).

WALKER, A. G.: On Milne's theory of world-structure. Proc. Lon. Math. Soc. **42**, 90 (1936).

WEBER, J.: Detection and generation of gravitational waves. Phys. Rev. **117**, 306 (1960).

WEBER, J.: Anisotropy and polarization in the gravitational-radiation experiments. Phys. Rev. Lett. **25**, 180 (1970).

WEINBERG, S.: Entropy generation and the survival of proto-galaxies in an expanding universe. Ap. J. **168**, 175 (1971).

WEINBERG, S.: Gravitation and cosmology: Principles and applications of the general theory of relativity. New York: Wiley 1972.

WEIZSÄCKER, C. F.: The evolution of galaxies and stars. Ap. J. **114**, 165 (1951).

WESSELINK, A. J.: The observation of brightness, colour, and radial velocity of δ Cephei and the pulsation hypothesis. B. A. N. **10**, 91 (1949).

WESTERHOUT, G.: The distribution of atomic hydrogen in the outer parts of the galactic system. B. A. N. **13**, 201 (1957).

WEYL, H.: Zur allgemeinen Relativitätstheorie (On the general theory of relativity). Phys. Z. **24**, 230 (1923).

WEYL, H.: Redshift and relativistic cosmology. Phil. Mag. **9**, 936 (1930).

WEYMANN, R.: The energy spectrum of radiation in the expanding universe. Ap. J. **145**, 560 (1966).

WEYMANN, R.: Possible thermal histories of intergalactic gas. Ap. J. **147**, 887 (1967).

WHITFORD, A. E.: The law of interstellar reddening. Astron. J. **63**, 201 (1958).

WHITFORD, A. E.: Absolute energy curves and K-corrections for giant elliptical galaxies. Ap. J. **169**, 215 (1971).

WHITROW, G. J., MORDUCH, G. E.: Relativistic theories of gravitation. In: Vistas in astronomy, vol. 6 (ed. A. BEER). New York: Pergamon Press 1965.

WHITTACKER, E. T.: On the definition of distance in curved space, and the displacement of the spectral lines of distant sources. Proc. Roy. Soc. London **A 133**, 93 (1931).

WILKINSON, D. T., PARTRIDGE, R. B.: Large-scale density inhomogeneities in the universe. Nature **215**, 719 (1967).

WILSON, R. E.: General catalogue of stellar radial velocities. Carnegie Inst. Wash. 601 (1953).

WITTEBORN, F. C., FAIRBANK, W. M.: Experimental comparison of the gravitational force on freely falling electrons and metallic electrons. Phys. Rev. Lett. **19**, 1049 (1967).

WOOLARD, E. W.: A redevelopment of the theory of nutation. Astron. J. **58**, 1 (1953).

WOOLARD, E. W., CLEMENCE, G. M.: Spherical astronomy. New York: Academic Press 1966.

WRIGHT, M. C. H.: On the interpretation of observations of neutral hydrogen in external galaxies. Ap. J. **166**, 455 (1971).

Zeldovich, Y. B., Novikov, I. D.: Relativistic astrophysics, vol. 1: Stars and relativity. Chicago, Ill.: University of Chicago Press 1971.

Zwicky, F.: Die Rotverschiebung von extragalaktischen Nebeln (The redshift of extragalactic nebulae). Helv. Phys. Acta. **6**, 110 (1933).

Zwicky, F.: Morphological astronomy. Berlin-Göttingen-Heidelberg: Springer 1957.

Zwicky, F.: Multiple galaxies. In: Handbuch der Physik, vol. LIII: Astrophysics IV—Stellar systems (ed. S. Flügge), p. 373. Berlin-Göttingen-Heidelberg: Springer 1959.

Zwicky, F. Herzog, E., Wild, P.: Catalogue of galaxies and clusters of galaxies. Pasadena, Calif.: California Institute of Technology 1961.

Zwicky, F., Humason, M. L.: Spectra and other characteristics of interconnected galaxies and galaxies in groups and clusters III. Ap. J. **139**, 269 (1964).

Supplemental References to the Second Edition

General References

AUDOUZE, J. (ed.): CNO isotopes in astrophysics. Boston: Reidel 1977.

AVRETT, E. H. (ed.): Frontiers of astrophysics. Cambridge: Harvard University Press 1976.

BAITY, W. A., PETERSON, L. E.: X-Ray astronomy. New York: Pergamon Press 1979.

BASCHEK, B., KEGEL, W. H., TRAVING, G.: Problems in stellar atmospheres and envelopes. New York: Springer 1975.

BASINSKA-GRZESIK, E., MAYOR, M. (eds.): Chemical and dynamical evolution of our galaxy. Geneva: Geneva Observatory 1978.

BERGMAN, P. G., FENYVES, E. J., MOTZ, L. (eds.): Seventh texas symposium on astrophysics. New York: New York Academy of Sciences 1975.

BERKHUIJSEN, E. M., WIELEBINSKI, R. (eds.): Structure and properties of nearby galaxies. I.A.U. Symposium No. 77. Boston: Reidel 1978.

BRECHER, K., SETTI, G. (eds.): High energy astrophysics and its relation to elementary particle physics. Cambridge: MIT Press 1974.

BUMBA, V., KLECZEK, J. (eds.): Basic mechanics of solar activity. I.A.U. Symposium No. 71. Boston: Reidel 1976.

CAMERON, A. G. W. (ed.): Cosmochemistry. Boston: Reidel 1973.

CARSON, T. R., ROBERTS, M. J. (eds.): Atoms and molecules in astrophysics. New York: Academic Press 1972.

CHAMBERLAIN, J. W.: Theory of planetary atmospheres. New York: Academic Press 1978.

CHIU, H. Y., MURIEL, A.: Stellar evolution. Cambridge: MIT Press 1972.

CHUPP, E. L.: Gamma ray astronomy. Boston: Reidel 1976.

CLINE, T. L., RAMATY, R.: Gamma ray spectroscopy in astrophysics. N.A.S.A. Technical Memorandum 79619. Greenbelt: Goddard Space Flight Center 1978.

CONTI, P. S., LELOORE, C. W. H.: Mass loss and evolution of 0 type stars. I.A.U. Symposium No. 83. Boston: Reidel 1979.

DALGARNO, A., MASNOU-SEEUWS, F., MCWHIRTEN, R. W. P.: Atomic and molecular processes in astrophysics. Geneva: Geneva Observatory 1975.

DEJONG, T., MAEDER, A. (eds.): Star formation. I.A.U. Symposium No. 76. Boston: Reidel 1977.

DEWITT-MORETTE (ed.): Gravitational radiation and gravitational collapse. I.A.U. Symposium No. 64. Boston: Reidel 1974.

DUNCOMBE, R. L. (ed.): Dynamics of the solar system. I.A.U. Symposium No. 81. Boston: Reidel 1979.

GIACCONI, R., GURSKY, H.: X-ray astronomy. Boston: Reidel 1974.

GIACCONI, R., RUFFINI, R.: Physics and astrophysics of black holes. New York: North Holland 1978.

GIBSON, E. G.: The quiet sun. N.A.S.A. SP-303. Washington: N.A.S.A. 1973.

GREENBERG, J. M., VAN DE HULST, H. C. (eds.): Interstellar dust and related topics. I.A.U. Symposium No. 52. Boston: Reidel 1973.

GURSKY, H., RUFFINI, R.: Neutron stars, black holes and binary X-ray sources. Boston: Reidel 1975.

HANSEN, C. J. (ed.): Physics of dense matter. I.A.U. Symposium No. 53. Boston: Reidel 1974.

HAWKING, S. W., ELLIS, G. F. R.: The large scale structure of space-time. Cambridge: Cambridge University Press 1973.

HAWKING, S. W., ISRAEL, W.: General relativity an Einstein centenary survey. Cambridge: Cambridge University Press 1979.

HAYLI, A. (ed.): Dynamics of stellar systems. I.A.U. Symposium No. 69. Boston: Reidel 1975.
HEINTZ, W. D.: Double stars. Boston: Reidel 1978.
IBEN, I., RENZINI, A., SCHRAMM, D. N.: Advanced stages in stellar evolution. Geneva: Geneva Observatory 1977.
IRVINE, J. M.: Neutron stars. Oxford: Clarendon Press 1978.
ISRAEL, W. (ed.): Relativity astrophysics and cosmology. Boston: Reidel 1973.
JAUNCEY, D. L. (ed.): Radio astronomy and cosmology. I.A.U. Symposium No. 74. Boston: Reidel 1977.
KAPLAN, S. A., TSYTOVICH, V. N.: Plasma astrophysics. New York: Pergamon Press 1973.
KANE, S. R. (ed.): Solar gamma X and EUV radiation. I.A.U. Symposium No. 68. Boston: Reidel 1975.
KOPAL, I.: Dynamics of close binary systems. Boston: Reidel 1978.
KOZAI, Y. (ed.): The stability of the solar system and of small stellar systems. I.A.U. Symposium No. 62. Boston: Reidel 1974.
KUKARIN, B. V. (ed.): Pulsating stars. New York: John Wiley 1975.
LANG, K. R., GINGERICH, O.: Source book in astronomy and astrophysics 1900–1975. Cambridge: Harvard University Press 1979.
LANZEROTTI, L. J., KENNEL, C. F., PARKER, E. N. (eds.): Solar system plasma physics. New York: North Holland 1979.
LEBOVITZ, N. R., REID, W. H., VANDERVOORT, P. O. (eds.): Theoretical principles in astrophysics and relativity. Chicago: University of Chicago Press 1978.
LONGAIR, M. S. (ed.): Confrontation of cosmological theories with observational data. I.A.U. Symposium No. 63. Boston: Reidel 1974.
LONGAIR, M. S., EINASTO, J. (eds.): The large scale structure of the universe. I.A.U. Symposium No. 79. Boston: Reidel 1978.
MANCHESTER, R. N., TAYLOR, J. H.: Pulsars. San Fancisco: W. H. Freeman 1977.
MCCARTHY, D. D., PILKINGTON, J. D. H. (eds.): Time and the earth's rotation. I.A.U. Symposium No. 82. Boston: Reidel 1979.
MCDONALD, F. B., FICHTEL, C. E. (eds.): High energy particles and quanta in astrophysics. Cambridge: MIT Press 1974.
PAPAGIANNIS, M. D. (ed.): Eighth texas symposium on relativistic astrophysics. New York: New York Academy of Sciences 1977.
PINKAU, K. (ed.): The interstellar medium. Boston: Reidel 1974.
REES, M., RUFFINI, R., WHEELER, J. A.: Black holes, gravitational waves, and cosmology: An introduction to current research. New York: Gordon and Breach 1974.
RYAN, M. P., SHEPLEY, L. C.: Homogeneous relativistic cosmologies. Princeton: Princeton University Press 1975.
SAAKYAN, G. S.: Equilibrium configurations of degenerate gaseous masses. New York: John Wiley 1974.
SCHRAMM, D. N., ARNETT, W. D. (eds.): Explosive nucleosynthesis. Austin: University of Texas Press 1973.
SETTI, G. (ed.): Structure and evolution of galaxies. Boston: Reidel 1975.
SETTI, G. (ed.): The physics of non-thermal radio sources. Boston: Reidel 1976.
SHERWOOD, V. E., PLAUT, L. (eds.): Variable stars and stellar evolution. I.A.U. Symposium No. 67. Boston: Reidel 1975.
SOBOLEV, V. V.: Light scattering in planetary atmospheres. New York: Pergamon Press 1975.
SPITZER, L.: Physical processes in the interstellar medium. New York: John Wiley 1978.
TERZIAN, Y. (ed.): Planetary nebulae: I.A.U. Symposium No. 76. Boston: Reidel 1978.
TINSLEY, B. M., LARSON, R. B. (eds.): The evolution of galaxies and stellar populations. New Haven: Yale Observatory 1977.
TSESEVICH, V. P. (ed.): Eclipsing variable stars. New York: John Wiley 1973.
TUCKER, W. H.: Radiation processes in astrophysics. Cambridge: MIT Press 1975.
WEINBERG, S.: The first three minutes. New York: Basic Books 1977.
WESTERLUND, B. E. (ed.): Stars and star systems. Boston: Reidel 1979.
WILSON, T. L., DOWNES, D. (eds.): H II regions and related topics. New York: Springer 1975.

Chapter 1

Acquista, C., Anderson, J. L.: Radiative transfer of partially polarized light. Ap. J. **191**, 567 (1974).

Adams, R. C., et al.: Analytic pulsar models. Ap. J. **192**, 525 (1974).

Adams, T. F.: The mean photon path length in extremely opaque media. Ap. J. **201**, 350 (1975).

Anderson, J. L., Spiegel, E. A.: Radiative transfer through a flowing refractive medium. Ap. J. **202**, 454 (1975).

Arons, J., et al.: A multiple pulsar model for quasi-stellar objects and active galactic nuclei. Ap. J. **198**, 687 (1975).

Atlee Jackson, E.: A new pulsar atmospheric model I. aligned magnetic and rotational axes. Ap. J. **206**, 831 (1976).

Barkstrom, B. R.: An exact expression for the temperature structure of a homogeneous planetary atmosphere containing isotropic scatters. Ap. J. **190**, 225 (1974).

Blumenthal, G. R.: The Poynting-Robertson effect and Eddington limit for electrons scattering with hard photons. Ap. J. **188**, 121 (1974).

Burbidge, G. R., Jones, T. W., O'Dell, S. L.: Physics of compact nonthermal sources. III. energetic considerations. Ap. J. **193**, 43 (1974).

Cheng, A., Ruderman, M., Sutherland, P.: Current flow in pulsar magnetospheres. Ap. J. **203**, 209 (1976).

Cheng, A., Ruderman, M. A.: A crab pulsar model: X-ray optical, and radio energy. Ap. J. **216**, 865 (1977).

Christiansen, W. A., Scott, J. S.: Formation of double radio source structures and superluminal expansion. Ap. J. **216**, L1 (1977).

Cocke, W. J.: On the production of power-law spectra and the evolution of cosmic synchrotron sources: a model for the Crab Nebula. Ap. J. **202**, 773 (1975).

Cocke, W. J., Pacholczyk, A. G.: Coherent curvature radiation and low-frequency variable radio sources. Ap. J. **195**, 279 (1975).

Cocke, W. J., Pacholczyk, A. G.: Theory of the polarization of pulsar radio radiation. Ap. J. **204**, L13 (1976).

Condon, J. J., Backer, D. C.: Interstellar Scintillation of extragalactic radio sources. Ap. J. **197**, 31 (1975).

Dent, W. A., Aller, H. D., Olsen, E. T.: The evolution of the radio spectrum of Cassiopeia A. Ap. J. **188**, L11 (1974).

De Young, D. S.: Extended extragalactic radio sources. Ann. Rev. Astron. Ap. **14**, 447 (1976).

De Young, D. S.: The internal dynamics and brightness distributions of a class of extended radio source models. Ap. J. **211**, 329 (1977).

Earl, J. A.: The diffusive idealization of charged-particle transport in random magnetic fields. Ap. J. **193**, 231 (1974).

Elliot, J. L., Shapiro, S. L.: On the variability of the compact nonthermal sources. Ap. J. **192**, L3 (1974).

Gilman, R. C.: Free-free and free-bound emission in low-surface-gravity stars. Ap. J. **188**, 87 (1974).

Ginzburg, V. L., Zheleznyakov, V. V.: On the pulsar emission mechanisms. Ann. Rev. Astron. Ap. **13**, 511 (1975).

Hardee, P. E.: Production of pulsed emission from the crab and vela pulsars by the synchrotron mechanism. Ap. J. **227**, 958 (1979).

Hardee, P. E., Rose, W. K.: A mechanism for the production of pulsar radio radiation. Ap. J. **210**, 533 (1976).

Harris, D. E., Romanishin, W.: Inverse Compton radiation and the magnetic field in clusters of galaxies. Ap. J. **188**, 209 (1974).

Harrison, E. R., Tademaru, E.: Acceleration of pulsars by asymmetric radiation. Ap. J. **201**, 447 (1975).

Helfand, D. J., Taylor, J. H., Manchester, R. N.: Pulsar proper motions. Ap. J. **213**, L1 (1977).

Helfand, D. J., Tademaru, E.: Pulsar velocity observations: correlations, interpretations, and discussion. Ap. J. **216**, 842 (1977).

Henriksen, R. N., Norton, J. A.: Oblique rotating pulsar magnetospheres with wave zones. Ap. J. **201**, 719 (1975).

HINATA, S.: Relativistic plasma turbulence and its application to pulsar phenomena. Ap. J. **206**, 282 (1976).

HOUSE, L. L., STEMITZ, R.: The non LTE transport equation for polarized radiation in the presence of magnetic fields. Ap. J. **195**, 235 (1975).

JOKIPII, J. R.: Pitch-angle scattering of charged particles in a random field. Ap. J. **194**, 465 (1974).

JONES, P. B.: Pulsar magnetic alignment: the critical period and integrated pulse width. Ap. J. **209**, 602 (1976).

JONES, W., O'DELL, S. L.: Transfer of polarized radiation in self-absorbed synchrotron sources. Ap. J. **214**, 522 (1977).

JONES, T. W., O'DELL, S. L.: Transfer of polarized radiation in self-absorbed synchrotron sources. II. Treatment of inhomogeneous media and calculation of emergent polarization. Ap. J. **215**, 236 (1977).

JONES, T. W., TOBIN, W.: Restrictions on models for superlight flux variations in radio sources. Ap. J. **215**, 474 (1977).

JONES, T. W., O'DELL, S. L., STEIN, W. A.: Physics of compact nonthermal sources. I. Theory of radiation processes. Ap. J. **188**, 353 (1974).

JONES, T. W., O'DELL, S. L., STEIN, W. A.: Physics of compact nonthermal sources. II. Determination of physical parameters. Ap. J. **192**, 261 (1974).

KATZ, J. I.: Nonrelativistic Compton scattering and models of quasars. Ap. J. **206**, 910 (1976).

KEMP, J. C.: Magnetobremsstrahlung and optical polarization: an understanding and a correction. Ap. J. **213**, 794 (1977).

KO, H. C., CHUANG, C. W.: On the passage of radiation through moving astrophysical plasmas. Ap. J. **222**, 1012 (1978).

LEE, L. C., JOKIPII, J. R.: Strong scintillations in astrophysics. I. the Markov approximation, its validity and application to angular broadening. Ap. J. **196**, 695 (1975).

LEE, L. C., JOKIPII, J. R.: Strong scintillations in astrophysics. II. A theory of temporal broadening of pulses. Ap. J. **201**, 532 (1975).

LEE, L. C., JOKIPII, J. R.: Strong scintillations in astrophysics. III. The fluctuations in intensity. Ap. J. **202**, 439 (1975).

LERCHE, I.: On the passage of radiation through inhomogeneous, moving media. I. The plane, differentially sheared medium. Ap. J. **187**, 589 (1974).

LERCHE, I.: On the relativistic theory of electromagnetic dispersion relations and Poynting's theorem. Ap. J. Suppl. **29**, 113 (1975).

LERCHE, I.: On the passage of radiation through inhomogeneous, moving media. IV. Radiative transfer under single-particle Compton scattering. Ap. J. **191**, 191 (1974).

LERCHE, I.: On the passage of radiation through inhomogeneous, moving media. V. Line absorption and frequency variations of optical depth. Ap. J. **191**, 753 (1974).

LERCHE, I.: On the passage of radiation through inhomogeneous, moving media. VI. Dispersion effects on phase and ray paths in a plane, differentially shearing medium. Ap. J. **191**, 759 (1974).

LERCHE, I.: On the passage of radiation through inhomogeneous, moving media. X. Ray and phase paths in arbitrary velocity fields. Ap. J. **194**, 403 (1974).

LERCHE, I.: On the passage of radiation through inhomogeneous, moving media. II. The rotating, differentially shearing medium. Ap. J. **187**, 597 (1974).

MANCHESTER, R. N., TAYLOR, J. H., VAN, Y. Y.: Detection of pulsar proper motion. Ap. J. **189**, L119 (1974).

MARSCHER, A. P.: Effects of nonuniform structure on the derived physical parameters of compact synchrotron sources. Ap. J. **216**, 244 (1977).

MARTIN, P. G.: On the Kramers-Kronig relations for interstellar polarization. Ap. J. **202**, 389 (1975).

MICHEL, F. C.: Rotating magnetosphere: far-field solutions. Ap. J. **187**, 585 (1974).

MICHEL, F. C.: Rotating magnetosphere: acceleration of plasma from the surface. Ap. J. **192**, 713 (1974).

MICHEL, F. C.: Pulsar extinction. Ap. J. **195**, L69 (1975).

MICHEL, F. C.: Rotating magnetospheres: frozen-in-flux violations. Ap. J. **196**, 579 (1975).

MICHEL, F. C.: Self-consistent rotating magnetosphere. Ap. J. **197**, 193 (1975).

MICHEL, F. C.: Composition of the neutron star surface in pulsar models. Ap. J. **198**, 683 (1975).

OPHER, R.: The origin of the nonthermal radiation of quasistellar objects and galactic nuclei. Ap. J. **201**, 526 (1975).

PACHOLCZYK, A. G., SCOTT, J. S.: Physics of compact radio sources. I. particle acceleration and flux variations. Ap. J. **210**, 311 (1976).
PETERSON, F. W.: Spectral index dispersion in extragalactic radio sources. Ap. J. **202**, 603 (1975).
PETERSON, F. W., KING, C. III: A model for simultaneous synchrotron and inverse Compton fluxes. Ap. J. **195**, 753 (1975).
ROBERTS, D. H.: The luminosity distribution and total space density of pulsars. Ap. J. **205**, L29 (1976).
RUDERMAN, M. A., SUTHERLAND, P. G.: Theory of pulsars: polar gaps, sparks, and coherent microwave radiation. Ap. J. **196**, 51 (1975).
SALVATI, M.: Relativistic correlations in the theory of expanding synchrotron sources. Ap. J. **233**, 11 (1979).
STURROCK, P. A., BAKER, K., TURK, J. S.: Pulsar extinction. Ap. J. **206**, 273 (1976).
STURROCK, P. A., PETROSIAN, V., TURK, J. S.: Optical radiation from the crab pulsar. Ap. J. **196**, 73 (1975).
TADEMARU, E.: Acceleration of pulsars by asymmetric radiation. III. Observational evidence. Ap. J. **214**, 885 (1977).
VITELLO, P., PACINI, F.: The evolution of expanding nonthermal sources. I. Nonrelativistic expansion. Ap. J. **215**, 452 (1977).
VITELLO, P., PACINI, F.: The evolution of expanding nonthermal sources. II. Relativistic expansions. Ap. J. **220**, 756 (1978).
WEISHEIT, J. C., SHORE, B. W.: Plasma-screening effects upon atomic hydrogen photoabsorption. Ap. J. **194**, 519 (1974).
WITT, A. N.: Multiple scattering in reflection nebulae. I. A Monte Carlo approach. Ap. J. Suppl. **35**, 1 (1977).
WITT, A. N.: Multiple scattering in reflection nebulae. III. Nebulae with embedded illuminating stars. Ap. J. Suppl. **35**, 21 (1977).
YOUNG, A. T.: Seeing: Its cause and cure. Ap. J. **189**, 587 (1974).

Chapter 2

ADAMS, W. M., PETROSIAN, V.: Effect of inelastic electronatom collisions on the Balmer decrement. Ap. J. **192**, 199 (1974).
AHMAD, I. A.: Effects of a free-free radio continuum on the populations of high atomic levels at low temperatures and densities. Ap. J. **194**, 503 (1974).
ALEXANDER, D. R.: Low-temperature Rosseland opacity tables. Ap. J. Supp. **29**, 363 (1975).
ALLEN, M., ROBINSON, G. W.: Formation of molecules on small interstellar grains. Ap. J. **195**, 81 (1975).
ALLEN, M., ROBINSON, G. W.: Molecular hydrogen in interstellar dark clouds. Ap. J. **207**, 745 (1976).
ALLEN, M., ROBINSON, G. W.: Molecular hydrogen in interstellar dark clouds: erratum. Ap. J. **214**, 955 (1977).
ALLER, L. H., EPPS, H. W.: Electron densities in gaseous nebulae. Ap. J. **204**, 445 (1976).
ALLOIN, D., CRUZ-GONZALEZ, C., PEIMBERT, M.: On the number of planetary nebulae in our galaxy. Ap. J. **205**, 74 (1976).
ANDERS, E., HAYATSU, R., STUDIER, M. H.: Interstellar molecules: origin by catalytic reactions on grain surfaces? Ap. J. **192**, L101 (1974).
BALICK, B., SNEDEN, C.: The ionization structure of H II regions: the effects of stellar metal opacity. Ap. J. **208**, 336 (1976).
BARKER, T.: Spectrophotometry of planetary nebulae, II. chemical abundances. Ap. J. **220**, 193 (1978).
BARLOW, M. J., SILK, J.: H_2 recombination of interstellar grains. Ap. J. **207**, 131 (1976).
BECKWITH, S., PERSSON, S. E., GATLEY, I.: Detection of molecular hydrogen emission from five planetary nebulae. Ap. J. **219**, L33 (1978).
BHADURI, R. K., NOGAMI, Y., WARKE, C. S.: Hydrogen atom and hydrogen molecule ion in homogeneous magnetic fields of arbitrary strength. Ap. P. **217**, 324 (1977).
BLACK, J. H., DALGARNO, A., OPPENHEIMER, M.: The formation of CH+ in interstellar clouds. Ap. J. **199**, 633 (1975).

BLACK, J. H., DALGARNO, A.: Interstellar H_2: the population of excited rotational states and the infrared response to ultraviolet radiation. Ap. J. **203**, 132 (1976).
BOESGAARD, A. M.: Measurements of magnetic fields in young main-sequence stars. Ap. J. **188**, 567 (1974).
BOESHAAR, G. O.: Chemical abundances of planetary nebulae. Ap. J. **195**, 695 (1975).
BORRA, E. F.: Hα polarization and line profiles in white dwarfs with strong magnetic fields. Ap. J. **209**, 858 (1976).
BROWN, R. L., GOMEZ-GONZALEZ, J.: The transfer of radio recombination line radiation through a cold gas. I. Hydrogen and helium lines in compact H II regions. Ap. J. **200**, 598 (1975).
BUHL, D., et al.: Silicon monoxide: detection of maser emission from the second vibrationally excited state. Ap. J. **192**, L97 (1974).
BURBIDGE, E. M., BURBIDGE, G. R.: Empirical evidence concerning absorption lines and radiation pressure in quasi-stellar objects. Ap. J. **202**, 287 (1975).
BURBIDGE, G. R., CROWNE, A. H., SMITH, H. E.: An optical catalog of quasi-stellar objects. Ap. J. Suppl. **33**, 113 (1976).
BURBIDGE, G., CROWNE, A. H.: An optical catalog of radio galaxies. Ap. J. Suppl. **40**, 583 (1979).
BURBIDGE, G., et al.: On the origin of the absorption spectra of quasi-stellar and BL lacertae objects. Ap. J. **218**, 33 (1977).
BURTON, W. B., et al.: The overall distribution of carbon monoxide in the plane of the galaxy. Ap. J. **202**, 30 (1975).
CAHN, J. H., WYATT, S. P.: The birthrate of planetary nebulae. Ap. J. **210**, 508 (1976).
CANFIELD, R. C.: A simplified method for calculation of radiative energy loss due to spectral lines. Ap. J. **194**, 483 (1974).
CHAISSON, E. J.: High-frequency observations of possible "heavy-element" recombination lines. Ap. J. **191**, 411 (1974).
CHAN, Y. T., BURBIDGE, E. M.: Emission-line strengths and the chemical composition of quasi-stellar objects. Ap. J. **198**, 45 (1975).
CHANG, M.: Mean lives of some astrophysically important excited levels in carbon, nitrogen, oxygen. Ap. J. **211**, 300 (1977).
CHU, S. I., DALGARNO, A.: Angular distributions in the elastic scattering and rotational excitation of molecular hydrogen by atomic hydrogen. Ap. J. **199**, 637 (1975).
COCHRAN, W. D., OSTRIKER, J. P.: The development of compact dust-bounded H II regions. I. Their relation to infrared objects and maser sources. Ap. J. **211**, 392 (1977).
CODE, A. D., MEADE, M. R.: Ultraviolet photometry from the orbiting astronomical observatory XXXII. An atlas of ultraviolet stellar spectra. Ap. J. Suppl. **39**, 195 (1979).
CRUTCHER, R. M.: Excitation of OH toward interstellar dust clouds. Ap. J. **216**, 308 (1977).
CRUTCHER, R. M.: Detection and significance of carbon recombination lines in diffuse interstellar clouds. Ap. J. **217**, L109 (1977).
DANA, R. A., PETROSIAN, V.: Approximate solutions of radiative transfer in dusty nebulae. II. Hydrogen and helium. Ap. J. **208**, 354 (1976).
DAVIDSON, K., NETZER, H.: The emission lines of quasars and similar objects. Rev. Mod. Phys. **51**, 715 (1979).
DAUB, C. T.: Oxygen abundances in planetary nebulae. Ap. J. **200**, 82 (1975).
DELSEMME, A. H., COMBI, M. R.: Production rate and origin of H_2O+ in comet Bennett 1970 II. Ap. J. **209**, L153 (1976).
DICKINSON, D. F.: Water emisson from infrared stars. Ap. J. Suppl. **30**, 259 (1976).
DICKMAN, R. L.: The ratio of carbon monoxide to molecular hydrogen in interstellar dark clouds. Ap. J. Suppl. **37**, 407 (1978).
DUPREE, A. K.: Carbon recombination lines and interstellar hydrogen clouds. Ap. J. **187**, 25 (1974).
DUVAL, P., KARP, A. H.: The combined effects of expansion and rotation on spectral lines shapes. Ap. J. **222**, 220 (1978).
DWORETSKY, M. M.: Rotational velocities of A0 stars. Ap. J. Suppl. **28**, 101 (1974).
ELITZUR, M., GOLDREICH, P., SCOVILLE, N.: OH-IR stars. II. A model for the 1612 MHz masers. Ap. J. **205**, 384 (1976).
ERMAN, P.: Experimental oscillator strengths of molecular ions. Ap. J. **213**, L89 (1977).
FEHSENFELD, F. C.: Ion reactions with atomic oxygen and atomic nitrogen of astrophysical importance. Ap. J. **209**, 638 (1976).

FERCH, R. L., SALPETER, E. E.: Models of planetary nebulae with dust. Ap. J. **202**, 195 (1975).

FORD, A. L., DOCKEN, K. K., DALGARNO, A.: The photoionization and dissociative photoionization of H_2, HD, and D_2. Ap. J. **195**, 819 (1975).

FORD, A. L., DOCKEN, K. K., DALGARNO, A.: Cross sections for photoionization of vibrationally excited molecular hydrogen. Ap. J. **200**, 788 (1975).

FORD, H. C., JACOBY, G. H.: Planetary nebulae in local-group galaxies. VIII. A catalog of planetary nebulae in the Andromeda galaxy. Ap. J. Suppl. **38**, 351 (1978).

FUHR, J. R., WIESE, W. L.: Bibliography on atomic transition probabilities, NBSSP 320, Supplement I, Wash. (1971).

GAUTIER, T. N. III, et al.: Detection of molecular hydrogen quadrupole emission in the Orion nebula. Ap. J. **207**, L 129 (1976).

GEBALLE, T. R., TOWNES, C. H.: Infrared pumping processes for SiO masers. Ap. J. **191**, L 37 (1974).

GEROLA, H., GLASSGOLD, A. E.: Molecular evolution of contracting clouds: basic methods and initial results. Ap. J. Suppl. **37**, 1 (1978).

GLASSGOLD, A. E., LANGER, W. D.: Model calculations for diffuse molecular clouds. Ap. J. **193**, 73 (1974).

GLASSGOLD, A. E., LANGER, W. D.: Abundances of simple oxygenbearing molecules and ions in interstellar clouds. Ap. J. **206**, 85 (1976).

GLENNON, B. M., WIESE, W. L.: Bibliography on atomic transition probabilities, Nat. Bur. Stands. (Wash.), Mon. **50**: Addenda (1963), NBS Misc. Publ. 278 (1968).

GOLDREICH, P., KWAN, J.: Astrophysical masers, V. Pump mechanisms for H_2O masers. Ap. J. **191**, 93 (1974).

GOLDREICH, P., KWAN, J.: Molecular Clouds. Ap. J. **189**, 441 (1974).

GOLDREICH, P., SCOVILLE, N.: OH-IR stars, I. Physical properties of circumstellar envelopes. Ap. J. **205**, 144 (1976).

GOLDSMITH, P. F., LANGER, W. D.: Molecular cooling and thermal balance of dense interstellar clouds. Ap. J. **222**, 881 (1978).

GOODMAN, F. O.: Formation of hydrogen molecules on interstellar grain surfaces. Ap. J. **226**, 87 (1978).

GORDON, M. A., BURTON, W. B.: Carbon monoxide in the galaxy. I. The radial distribution of CO, H_2, and nucleons. Ap. J. **208**, 346 (1976).

GOULD, R. J.: Radiative recombination of complex ions. Ap. J. **219**, 250 (1978).

GOULD, R. J., LEVY, M.: Deviation from a Maxwellian velocity distribution in regions of interstellar molecular hydrogen. Ap. J. **206**, 435 (1976).

GREEN, S.: Rotational excitation of molecular ions in interstellar clouds. Ap. J. **201**, 366 (1975).

GREEN, S., THADDEUS, P.: Rotational excitation of CO by collisions with He, H, and H_2 under conditions in interstellar clouds. Ap. J. **205**, 766 (1976).

GREEN, S., CHAPMAN, S.: Collisional excitation of interstellar molecules: linear molecules of CO, CS, OCS, and HC_3N. Ap. J. Suppl. **37**, 167 (1978).

GREEN, S., et al.: Collisional excitation of interstellar formaldehyde. Ap. J. Suppl. **37**, 321 (1978).

GREEN, S., RAMASWAMY, R., RABITZ, H.: Collisional excitation of interstellar molecules: H_2. Ap. J. Suppl. **36**, 483 (1978).

GREENSTEIN, J. L., et al.: The rotation and gravitational redshift of white dwarfs. Ap. J. **212**, 186 (1977).

HABING, H. J., ISRAEL, F. P.: Compact H II regions and OB star formation. Ap. J. **17**, 345 (1979).

HAGAN, L.: Bibliography on atomic energy levels and spectra, NBS SP 363, Wash. (1977).

HARRISON, E. R.: Diffusion and sedimentation of dust in molecular clouds. Ap. J. **226**, L 95 (1978).

HARRISON, E. R., NOONAN, T. W.: Interpretation of extragalactic redshifts. Ap. J. **232**, 18 (1979).

HAYDEN SMITH, W., ZWEIBEL, E. G.: Radiative processes affecting the abundance of interstellar OH. Ap. J. **207**, 758 (1976).

HERBST, E.: Radiative association in dense, H_2-containing interstellar clouds. Ap. J. **205**, 94 (1976).

HERBST, E.: What are the products of polyatomic ion-electron dissociative recombination reactions? Ap. J. **222**, 508 (1978).

HILL, J. K., SILK, J.: Molecular hydrogen in H II region transition zones. Ap. J. **202**, L 97 (1975).

HILL, J. K., HOLLENBACH, D. J.: H_2 molecules and the intercloud medium. Ap. J. **209**, 445 (1976).

HJALMARSON, A., et al.: Radio observations of interstellar CH, II. Ap. J. Suppl. **35**, 263 (1977).

HODGE, P. W.: A second survey of H II regions in galaxies. Ap. J. Suppl. **27**, 113 (1974).

HOLLENBACH, D., CHU, S., MCCRAY, R.: H_2 in expanding circumstellar shells. Ap. J. **208**, 458 (1976).

HOLLENBACH, D., MCKEE, C. F.: Molecule formation and infrared emission in fast interstellar shocks. I. Physical processes. Ap. J. Suppl. **41**, 555 (1979).
HUNTER, D. A., WATSON, W. D.: The translational and rotational energy of hydrogen molecules after recombination on interstellar grains. Ap. J. **226**, 477 (1978).
HUNTER, J. H., Jr., NIGHTINGALE, S. L.: The influence of dust upon the dynamics and stability of planetary nebulae. Ap. J. **193**, 693 (1974).
HUNTRESS, W. T., Jr.: Laboratory studies of bimolecular reactions of positive ions in interstellar clouds, in comets, and in planetary atmospheres of reducing composition. Ap. J. Suppl. **33**, 495 (1977).
IGLESIAS, E.: The chemical evolution of molecular clouds. Ap. J. **218**, 697 (1977).
JACOBS, V. L., et al.: The influence of autoionization accompanied by excitation on dielectronic recombination and ionization equilibrium. Ap. J. **211**, 605 (1977).
JACOBS, V. L., et al.: Dielectric recombination rates, ionization equilibrium, and radiative energy-loss rates for neon, magnesium, and sulfur ions in low-density plasmas. Ap. J. **230**, 627 (1979).
JOHNSON, H. R., BEEBE, R. F., SNEDEN, C.: Molecular column densities in selected model atmospheres. Ap. J. Suppl. **29**, 123 (1975).
JURA, M.: Formation and destruction rates of interstellar H_2. Ap. J. **191**, 375 (1974).
JURA, M.: Interstellar clouds containing optically thick H_2. Ap. J. **197**, 581 (1975).
KALER, J. B.: The exciting stars of low-excitation planetary and diffuse nebulae. Ap. J. **210**, 843 (1976).
KALER, J. B.: A catalog of relative emission intensities observed in planetary and diffuse nebulae. Ap. J. Suppl. **31**, 517 (1976).
KALER, J. B.: The [0 III] lines as a quantitative indicator of nebular central-star temperature. Ap. J. **220**, 887 (1978).
KALER, J. B.: Galactic abundances of neon, argon, and chlorine derived from planetary nebulae. Ap. J. **225**, 527 (1978).
KALER, J. B., ALLER, L. H., CZYZAK, S. J.: A spectrographic survey of 21 planetary nebulae. Ap. J. **203**, 636 (1976).
KARP, A. H., et al.: The opacity of expanding media: the effect of spectral lines. Ap. J. **214**, 161 (1977).
KASABOV, G. A., ELISEYEV, V. V.: Spectroscopic tables for low temperature plasma, Atomizdat, Moscow (1973).
KATZ, J. I., MALONE, R. C., SALPETER, E. E.: Models for nuclei of planetary nebulae and ultraviolet stars. Ap. J. **190**, 359 (1974).
KOSLOFF, R., KAFRI, A., LEVINE, R. D.: Rotational excitation of interstellar OH molecules. Ap. J. **215**, 497 (1977).
KRAEMER, W. P., DIERCKSEN, G. H. F.: Identification of interstellar x-ogen as HCO+. Ap. J. **205**, L97 (1976).
KROLIK, J. H., MCKEE, C. F.: Hydrogen emission-line spectra in quasars and active galactic nuclei. Ap. J. Suppl. **37**, 459 (1978).
KUIPER, T. B. H., RODRIGUEZ KUIPER, E. N., ZUCKERMAN, B.: Spectral line shapes in spherically symmetric radially moving clouds. Ap. J. **219**, 129 (1978).
KUNTZ, P. J., MITCHELL, G. F., GINSBURG, J.: Fourier analysis of steady-state reaction schemes for interstellar molecules. Ap. J. **209**, 116 (1976).
KWAN, J., SCOVILLE, N.: Radiative trapping and population inversions of the SiO masers. Ap. J. **194**, L97 (1974).
LANGER, W.: The carbon monoxide abundance in interstellar clouds. Ap. J. **206**, 699 (1976).
LANGER, W. D.: The formation of molecules in interstellar clouds from singly and multiply ionized atoms. Ap. J. **225**, 860 (1978).
LEUNG, C. M.: Radiation transport in dense interstellar dust clouds. II. Infrared emission from molecular clouds associated with H II regions. Ap. J. **209**, 75 (1976).
LEUNG, C. M., LISZT, H. S.: Radiation transport and non-LTE analysis of interstellar molecular lines. I. Carbon monoxide. Ap. J. **208**, 732 (1976).
LISZT, H. S., LEUNG, C. M.: Radiation transport and non-LTE analysis of interstellar molecular lines. II. Carbon monosulfide. Ap. J. **218**, 396 (1977).
LOCKMAN, F. J.: A survey of ionized hydrogen in the plane of the galaxy. Ap. J. **209**, 429 (1976).
LOCKMAN, F. J., BROWN, R. L.: On the derivation of nebular electron temperatures from radio recombination line observations. Ap. J. **207**, 436 (1976).
LOVAS, F. J., SNYDER, L. E., JOHNSON, D. R.: Recommended rest frequencies for observed interstellar molecular transitions. Ap. J. Suppl. **41**, 451 (1979).

LYNDS, B. T.: An atlas of dust and H II regions in galaxies. Ap. J. Suppl. **28**, 391 (1974).

MARTIN, R. N., BARRETT, A. H.: Microwave spectral lines in galactic dust globules. Ap. J. Suppl. **36**, 1 (1978).

MERRILL, K. M., RUSSELL, R. W., SOIFER, B. T.: Infrared observations of ices and silicates in molecular clouds. Ap. J. **207**, 763 (1976).

MILES, B. M., WIESE, W. L.: Bibliography on atomic transition probabilities, Nat. Bur. Stands. Spec. Publ. 320, Wash. (1970).

MITCHELL, G. F., GINSBURG, J. L., KUNTZ, P. J.: The formation of molecules in diffuse interstellar clouds. Ap. J. **212**, 71 (1977).

MITCHELL, G. F., GINSBURG, J. L., KUNTZ, P. J.: A steady-state calculation of molecule abundances in interstellar clouds. Ap. J. Suppl. **38**, 39 (1978).

MONTES, C.: Variability of intensity of interstellar maser lines due to induced Compton scattering. Ap. J. **216**, 329 (1977).

MORAN, J. M., et al.: Very long baseline interferometric observations of OH masers associated with infrared stars. Ap. J. **211**, 160 (1977).

NICHOLLS, R. W.: Transition probability data for molecules of astrophysical interest. Ann. Rev. Astron. Ap. **15**, 197 (1977).

NOERDLINGER, P. D., RYBICKI, G. B.: Transfer of line radiation in differentially expanding atmospheres. IV. The two-level atom in plane-parallel geometry solved by the Feautrier method. Ap. J. **193**, 651 (1974).

O'CONNELL, R. F.: Highly excited states of atoms in a magnetic field. Ap. J. **187**, 275 (1974).

OLSON, E. C.: Hydrogen profiles, helium line strengths, and surface gravities of eclipsing binary stars. Ap. J. Suppl. **29**, 43 (1975).

OPPENHEIMER, M., DALGARNO, A.: The formation of carbon monoxide and the thermal balance in interstellar clouds. Ap. J. **200**, 419 (1975).

OSMER, P. S.: On the abundances of helium, nitrogen, and oxygen in the planetary nebulae of the Magellanic clouds. Ap. J. **203**, 352 (1976).

PARKS, A. D., SAMPSON, D. H.: Electron impact excitation cross sections for complex ions. III. Highly charged ions with three valence electrons. Ap. J. **209**, 312 (1976).

PEIMBERT, M., TORRES-PEIMBERT, S.: Chemical composition of H II regions in the large Magellanic cloud and its cosmological implications. Ap. J. **193**, 327 (1974).

PEIMBERT, M., TORRES-PEIMBERT, S.: Chemical composition of H II regions in the small Magellanic cloud and the pregalactic helium abundance. Ap. J. **203**, 581 (1976).

RAU, A. R. P., SPRUCH, L.: Energy levels of hydrogen in magnetic fields of arbitrary strength. Ap. J. **207**, 671 (1976).

RAYMOND, J. C.: On dielectric recombination and resonances in excitation cross sections. Ap. J. **222**, 1114 (1978).

REID, M. J., et al.: The structure of stellar hydroxyl masers. Ap. J. **214**, 60 (1977).

REILMAN, R. F., MANSON, S. T.: Photoabsorption cross sections for positive atomic ions with $Z \leqslant 30$. Ap. J. Suppl. **40**, 815 (1979).

RICKARD, L., et al.: Detection of extragalactic carbon monoxide at millimeter wavelengths. Ap. J. **199**, L75 (1975).

RICKARD, L. J., et al.: Extragalactic carbon monoxide. Ap. J. **213**, 673 (1977).

RICKARD, L. J., et al.: Observations of extragalactic molecules. II. HCN and CS. Ap. J. **214**, 390 (1977).

RICKARD, L. J., TURNER, B. E., PALMER, P.: Carbon monoxide in Maffei 2. Ap. J. **218**, L51 (1977).

ROBERTS, M. S., et al.: Detection at $z \approx 0.5$ of a 21-cm absorption line in AO 0235+164: the first coincidence of large radio and optical redshifts. Astron. J. **81**, 293 (1976).

ROBOUCH, B. V., RAGER, J. P.: Ionization equilibrium of the three highest stages of ionization of the elements carbon to argon at high temperatures. Ap. J. **208**, 609 (1976).

ROWAN-ROBINSON, M.: Clouds of dust and molecules in the galaxy. Ap. J. **234**, 111 (1979).

RYDBECK, O. E. H., et al.: Radio observations of interstellar CH. 1. Ap. J. Suppl. **31**, 333 (1976).

SALEM, M., BROCKLEHURST, M.: A table of departure coefficients from thermodynamic equilibrium (b_n factors) for hydrogenic ions. Ap. J. Suppl. **39**, 633 (1979).

SAMPSON, D. H.: Electron-impact excitation cross-sections for complex ions. I. Theory for ions with one and two valence electrons. Ap. J. Suppl. **28**, 309 (1974).

SAMPSON, D. H., PARKS, A. D.: Electron-impact excitation cross-sections for complex ions. II. Application to the isoelectronic series of helium and other light elements. Ap. J. Suppl. **28**, 323 (1974).
SARAZIN, C. L.: Abundance gradients in extragalactic H II regions and internal absorption by dust. Ap. J. **208**, 323 (1976).
SARAZIN, C. L.: Effects of dust on the structure of H II regions. Ap. J. **211**, 772 (1977).
SAVAGE, B. D., et al.: A survey of interstellar molecular hydrogen, I. Ap. J. **216**, 291 (1977).
SCHILD, R. E., CHAFFEE, F. H.: The Balmer discontinuities of 09-B2 supergiants. Ap. J. **196**, 503 (1975).
SCHNEIDER, S., ELMEGREEN, B. G.: A catalog of dark globular filaments. Ap. J. Suppl. **41**, 87 (1979).
SCHMIDT, M.: Optical spectra and redshifts of quasi-stellar radio sources in the NRAO 5 GHz and 4C radio catalogs. Ap. J. **217**, 358 (1977).
SCOVILLE, N. Z., SOLOMON, P. M.: Radiative transfer, excitation, and cooling of molecular emission lines (CO and CS). Ap. J. **187**, L67 (1974).
SCOVILLE, N. Z., SOLOMON, P. M., JEFFERTS, K. B.: Molecular clouds in the galactic nucleus. Ap. J. **187**, L63 (1974).
SCOVILLE, N. Z., SOLOMON, P. M.: Molecular clouds in the galaxy. Ap. J. **199**, L105 (1975).
SCOVILLE, N. Z., KWAN, J.: Infrared sources in molecular clouds. Ap. J. **206**, 718 (1976).
SHIPMAN, H. L., MEHAN, R. G.: The unimportance of pressure shifts in the measurement of gravitational redshifts in white dwarfs. Ap. J. **209**, 205 (1976).
SHULL, J. M., HOLLENBACH, D. J.: H_2 cooling, dissociation, and infrared emission in shocked molecular clouds. Ap. J. **220**, 525 (1978).
SOLOMON, P. M., DEZAFRA, R.: Carbon monoxide in external galaxies. Ap. J. **199**, L79 (1975).
SPITZER, L. Jr., COCHRAN, W. D., HIRSHFELD, A.: Column densities of interstellar molecular hydrogen. Ap. J. Suppl. **28**, 373 (1974).
SPITZER, L. Jr., ZWEIBEL, E. G.: On the theory of H_2 rotational excitation. Ap. J. **191**, L127 (1974).
SPITZER, L. Jr., MORTON, W. A.: Components in interstellar molecular hydrogen. Ap. J. **204**, 731 (1976).
STRY, P. E.: Envelope ejection to form planetary nebulae. Ap. J. **196**, 559 (1975).
SURMELIAN, G. L., O'CONNELL, R. F.: Energy spectrum of hydrogen-like atoms in a strong magnetic field. Ap. J. **190**, 741 (1974).
TROLAND, T. H., HEILES, C.: The Zeeman effect in radio frequency recombination lines. Ap. J. **214**, 703 (1977).
TURNER, B. E.: U93, 174: A new interstellar line with quadrupole hyperfine splitting. Ap. J. **193**, L83 (1974).
TURNER, B. E.: Microwave detection of interstellar ketene. Ap. J. **213**, L75 (1977).
TURNER, J., KIRBY-DOCKEN, K., DALGARNO, A.: The quadrupole vibration-rotation transition probabilities of molecular hydrogen. Ap. J. Suppl. **35**, 281 (1977).
ULRICH, M. H.: Redshifts of forty-three radio sources. Ap. J. **206**, 364 (1976).
VAINSTEIN, L. A., SOBELMAN, I. I., YUKOV, E. A.: Cross sections for excitation of atoms and ions by electron impact. NAUKA Publishers, Moscow (1973).
VAN VLECK, J. H., HUBER, D. L.: Absorption, emission and linebreadths: a semihistorical perspective. Rev. Mod. Phys. **49**, 939 (1977).
VINER, M. R., VALLÉE, J. P., HUGHES, V. A.: A theoretical study of the radio recombination line and continuum emission from compact inhomogeneous H II regions. Ap. J. Suppl. **39**, 405 (1979).
WADEHRA, J. M.: Transition probabilities and some expectation values for the hydrogen atom in intense magnetic fields. Ap. J. **226**, 372 (1978).
WANNIER, P. G., et al.: Isotope abundances in interstellar molecular clouds. Ap. J. **204**, 26 (1976).
WATSON, W. D.: Ion-molecule reactions, molecule formation, and hydrogen-isotope exchange in dense interstellar clouds. Ap. J. **188**, 35 (1974).
WATSON, W. D.: Molecular CH, CH^+, and H_2 in the interstellar gas. Ap. J. **189**, 221 (1974).
WATSON, W. D.: Interstellar molecule reactions. Rev. Mod. Phys. **49**, 513 (1976).
WATSON, W. D.: Gas phase reactions in astrophysics. Ann. Rev. Astron. Ap. **16**, 585 (1978).
WEHINGER, P. A., et al.: Identification of H_2O^+ in the tail of comet Kohoutek (1973f). Ap. J. **190**, L43 (1974).
WEHINGER, P., WYCKOFF, S.: H_2O^+ in spectra of comet Bradfield (1974b). Ap. J. **192**, L4 (1974).
WEISHEIT, J. C.: X-ray ionization cross-sections and ionization equilibrium equations modified by Auger transitions. Ap. J. **190**, 735 (1974).
WENTZEL, D. G.: Dynamics of envelopes of planetary nebulae. Ap. J. **204**, 452 (1976).

WICKRAMASINGHE, D. T., et al.: Spectra of white dwarfs. Ap. J. **202**, 191 (1975).
WIESE, W. L., SMITH, M. W., MILES, B. M.: Atomic transition probabilities: Na through Ca, Vol. 2, NSD DS-NBS, 22, Wash. (1969).
WILSON, L. W.: Analytical variational calculation of the ground-state binding energy of hydrogen in intermediate and intense magnetic fields. Ap. J. **188**, 349 (1974).
WILSON, W. J., et al.: Observations of galactic carbon monoxide emission at 2.6 millimeters. Ap. J. **191**, 357 (1974).
WYCKOFF, S., WEHINGER, P. A.: On the ionization and excitation of H_2O^+ in comet Kohoutek (1973f). Ap. J. **204**, 616 (1976).
WYNN-WILLIAMS, C. G., BECKLIN, E. E., NEUGEBAUER, G.: Infrared studies of H II regions and OH sources. Ap. J. **187**, 473 (1974).
YORK, D. G.: On the existence of molecular hydrogen along lines of sight with low reddening. Ap. J. **204**, 750 (1976).
ZAIDEL, A. N., et al.: Spectral lines tables, Fizmatgiz, Moscow (1962).
ZUCKERMAN, B., BALL, J. A.: On microwave recombination lines from H I regions. Ap. J. **190**, 35 (1974).
ZUCKERMAN, B., et al.: $^{12}C/^{13}C$ abundance ratios from observations of interstellar $H_2{}^{13}C^{16}($). Ap. J. **189**, 217 (1974).
ZUCKERMAN, B., EVANS, N. J. II: Models of massive molecular clouds. Ap. J. **192**, L 149 (1974).

Chapter 3

AANNESTAD, P. A., KENYON, S. J.: Temperature fluctuations and the size distribution of interstellar grains. Ap. J. **230**, 771 (1979).
ABRAMOWICZ, M. A., WAGONER, R. V.: Variational analysis of rotating neutron stars. Ap. J. **204**, 896 (1976).
AGGARWAL, H., OBERBECK, V. R.: Roche limit of a solid body. Ap. J. **191**, 577 (1974).
AHMAD, A., COHEN, L.: Dynamical friction in gravitational systems. Ap. J. **188**, 469 (1974).
AIZENMAN, M. L., COX, J. P.: Pulsational stability of stars in thermal imbalance. IV. Direct solution of differential equation. Ap. J. **194**, 663 (1974).
AIZENMAN, M. L., COX, J. P.: Pulsational stability of stars in thermal imbalance. VI. Physical mechanisms and extension to nonradial oscillations. Ap. J. **195**, 175 (1975).
AIZENMAN, M. L., COX, J. P.: Vibrational stability of differentially rotating stars. Ap. J. **202**, 137 (1975).
AIZENMAN, M. L., HANSEN, C. J., ROSS, R. R.: Pulsation properties of upper-main-sequence stars. Ap. J. **201**, 387 (1975).
ANGEL, J. R. P., LANDSTREET, J. D.: A determination by the Zeeman effect of the magnetic field strength in the white dwarf G 99–37. Ap. J. **191**, 457 (1974).
ANGEL, J. R. P.: Magnetism in white dwarfs. Ap. J. **216**, 1 (1977).
ANGEL, J. R. P.: Magnetic white dwarfs. Ann. Rev. Astron. Ap. **16**, 487 (1978).
APRUZESE, J. P.: Radiative transfer in spherical circumstellar dust envelopes. III. Dust envelope models of some well-known infrared stars. Ap. J. **196**, 761 (1975).
APRUZESE, J. P.: Radiative transfer in spherical circumstellar dust envelopes. V. Theoretical circumstellar graphite and silicate emission spectra. Ap. J. **207**, 799 (1976).
ARNETT, W. D.: Neutrino trapping during gravitational collapse of stars. Ap. J. **218**, 815 (1977).
ARNETT, W. D., BOWERS, R. L.: A microscopic interpretation of neutron star structure. Ap. J. Suppl. **33**, 415 (1977).
ARONS, J., LEA, S. M.: Accretion onto magnetized neutron stars: normal mode analysis of the interchange instability at the magnetopause. Ap. J. **210**, 792 (1976).
ASSOUSA, G. E., HERBST, W., TURNER, K. C.: Supernova-induced star formation in Cepheus OB3. Ap. J. **218**, L 13 (1977).
AHLUWALIA, D. V., WU, T. Y.: On the magnetic field of cosmological bodies. Nuovo Cimento Letters **23**, 406 (1978).
BARDEEN, J. M., et al.: A new criterion for secular instability of rapidly rotating stars. Ap. J. **217**, L 49 (1977).

BARKAT, Z.: Neutrino processes in stellar interiors. Ann. Rev. Astron. Ap. **13**, 45 (1975).

BARKAT, Z., REISS, Y., RAKAVY, G.: Stars in the mass range $7 \leqslant M/M_{\odot} \leqslant 10$ as candidates for pulsar progenitors. Ap. J. **193**, L 21 (1974).

BARLOW, M. J., SILK, J.: Graphite grain surface reactions in interstellar and protostellar environments. Ap. J. **215**, 800 (1977).

BARNES, A.: Acceleration of the solar wind by the interplanetary mganetic field. Ap. J. **188**, 645 (1974).

BAR-NUN, A.: Interstellar molecules: direct formation on graphite grains. Ap. J. **197**, 341 (1975).

BASH, F. N., GREEN, E., PETERS, W. L.: The galactic density wave, molecular clouds, and star formation. Ap. J. **217**, 464 (1977).

BAYM, G., LAMB, D. Q., LAMB, F. K.: Dynamical effects of possible solid cores in neutron stars and degenerate dwarfs. Ap. J. **208**, 829 (1976).

BAYM, G., PETHICK, C.: Physics of neutron stars. Ann. Rev. Astron. Ap. **17**, 415 (1979).

BERTIN, G., RADICATI, L. A.: The bifurcation from the Maclaurin to the Jacobi sequence as a second-order phase transition. Ap. J. **206**, 815 (1976).

BLACK, D. C., BODENHEIMER, P.: Evolution of rotating interstellar clouds. II. The collapse of protostars of 1, 2, and 5 $M_{\odot}$. Ap. J. **206**, 138 (1976).

BLACK, J. H., DALGARNO, A.: Models of interstellar clouds. I. The zeta ophiuchi cloud. Ap. J. Suppl. **34**, 405 (1977).

BLUDMAN, S. A., VAN RIPER, K. A.: Equation of state of an ideal Fermi gas. Ap. J. **212**, 859 (1977).

BÖHM, K.-H., et al.: Some properties of very low temperature, pure helium surface layers of degenerate dwarfs. Ap. J. **217**, 521 (1977).

BOWERS, R. L., et al.: A realistic lower bound for the maximum mass of neutron stars. Ap. J. **196**, 639 (1975).

BOWERS, R. L., GLEESON, A. M., PEDIGO, R. D.: A higher stability limit for neutron stars. Ap. J. **205**, 261 (1976).

BRECHER, K., CHANMUGAM, G.: Why do collapsed stars rotate so slowly? Ap. J. **221**, 969 (1978).

BREGER, M., BREGMAN, J. N.: Period-luminosity-color relations and pulsation modes of pulsating variable stars. Ap. J. **200**, 343 (1975).

BRUENN, S. W., MARROQUIN, A.: Structure and properties of detonation waves. I. Detonation waves in dense stellar material. Ap. J. **195**, 567 (1975).

BUCHLER, J. R., YUEH, W. R.: Compton scattering opacities in a partially degenerate electron plasma at high temperatures. Ap. J. **210**, 440 (1976).

BUCHLER, J. R., COON, S. A.: The interacting neutron gas at high density and temperature. Ap. J. **212**, 807 (1977).

BUCHLER, J. R.: On the vibrational stability of stars in thermal imbulance. Ap. J. **220**, 629 (1978).

BURKE, J. R., SILK, J.: Dust grains in a hot gas. I. Basic physics. Ap. J. **190**, 1 (1974).

BURKE, J. R., SILK, J.: The dynamical interaction of a newly formed protostar with infalling matter: the origin of interstellar grains. Ap. J. **210**, 341 (1976).

BURTON, W. B., LISZT, H. S.: The gas distribution in the central region of the galaxy. I. Atomic hydrogen. Ap. J. **225**, 815 (1978).

CANNON, C. J., THOMAS, R. N.: The origin of stellar winds: subatmospheric nonthermal storage modes versus radiation pressure. Ap. J. **211**, 910 (1977).

CANUTO, V.: Equation of state ultrahigh densities. Ann. Rev. Astron. Ap. **13**, 335 (1975).

CANUTO, V., DATTA, B., KALMAN, G.: Superdense neutron matter. Ap. J. **221**, 274 (1978).

CARSON, T. R.: Stellar opacity. Ann. Rev. Astron. Ap. **14**, 95 (1976).

CARTER, B., QUINTANA, H.: Stationary elastic rotational deformation of a relativistic neutron star model. Ap. J. **202**, 511 (1975).

CASSEN, P., PETTIBONE, D.: Steady accretion of a rotating fluid. Ap. J. **208**, 500 (1976).

CASSINELLI, J. P.: Stellar winds. Ann. Rev. Astron. Ap. **17**, 275 (1979).

CHANAN, G. A., MIDDLEDITCH, J., NELSON, J. E.: The geometry of the eclipse of a pointlike star by a Roche-lobefilling companion. Ap. J. **208**, 512 (1976).

CHANDRASEKHAR, S.: On a criterion for the onset of dynamical instability by a nonaxisymmetric mode of oscillation along a sequence of differentially rotating configurations. Ap. J. **187**, 169 (1974).

CHANDRASEKHAR, S., ELBERT, D. D.: The deformed figures of the Dedekind ellipsoids in the post-Newtonian approximation to general relativity. Ap. J. **192**, 731 (1974).

CHANG, M.-W., CHAMMUGAM, G.: Radial oscillations of zero-temperature white dwarfs and neutron stars below nuclear densities. Ap. J. **217**, 799 (1977).

CHEN, H.-H., RUDERMAN, M. A., SUTHERLAND, P. G.: Structure of solid iron in superstrong neutron-star magnetic fields. Ap. J. **191**, 473 (1974).
CHEVALIER, R. A.: The interaction of supernovae with the interstellar medium. Ann. Rev. Astron. Ao. **15**, 175 (1977).
CLARK, F. O., et al.: Upper limits to the ambient magnetic field in several dense molecular clouds. Ap. J. **226**, 824 (1978).
CLARK, J. P. A., EARDLEY, D. M.: Evolution of close neutron star binaries. Ap. J. **215**, 311 (1977).
CLAYTON, D. D., HOYLE, F.: Grains of anomalous isotopic composition from novae. Ap. J. **203**, 490 (1976).
CLAYTON, D. D., DWEK, E., WOOSLEY, S. E.: Isotopic anomalies and proton irradiation in the early solar system. Ap. J. **214**, 300 (1977).
COGAN, B. C.: The pulsation periods of stars with convection zones. Ap. J. **211**, 890 (1977).
CONTI, P. S.: Mass loss in early-type stars. Ann. Rev. Astron. Ap. **16**, 371 (1978).
COWAN, J. J., KAFATOS, M., ROSE, W. K.: Sources of excitation of the interstellar gas and galactic structure. Ap. J. **195**, 47 (1975).
COWIE, L. L., MCKEE, C. F.: The evaporation of spherical clouds in a hot gas. I. Classical and saturated mass loss rates. Ap. J. **211**, 135 (1977).
COX, A. N., TABOR, J. E.: Radiative opacity tables for 40 stellar mixtures. Ap. J. Suppl. **31**, 271 (1976).
COX, J. P.: Effects of thermal imbalance on the pulsational stability of stars undergoing thermal runaways. Ap. J. **192**, L 85 (1974).
COX, J. P., DAVEY, W. R., AIZENMAN, M. L.: Pulsational stability of stars in thermal imbalance. III. Analysis in terms of absolute variations. Ap. J. **191**, 439 (1974).
COX, J. P.: Nonradial oscillations of stars: theories and observations. Ann. Rev. Astron. Ap. **14**, 247 (1976).
CRAVENS, T. E., DALGARNO, A.: Ionization, dissociation, and heating efficiencies of cosmic rays in a gas of molecular hydrogen. Ap. J. **219**, 750 (1978).
CRUDDACE, R., et al.: On the opacity of the interstellar medium to ultrasoft X-rays and extreme-ultra-violet radiation. Ap. J. **187**, 497 (1974).
CRUTCHER, R. M., et al.: OH Zeeman observations of interstellar dust clouds. Ap. J. **198**, 91 (1975).
CZYZAK, S. J., MEESE, J. M., SANTIAGO, J. J.: Effects of stellar particle irradiation on interstellar grains. Ap. J. **207**, 425 (1976).

DALGARNO, A., OPPENHEIMER, M.: Chemical heating of interstellar clouds. Ap. J. **192**, 597 (1974).
DAVEY, W. R.: Pulsational stability of stars in thermal imbalance. V. Eigensolutions for quasi-adiabatic oscillations. Ap. J. **194**, 687 (1974).
DAVEY, W. R., COX, J. P.: Pulsational stability of stars in thermal imbalance. II. An energy approach. Ap. J. **189**, 113 (1974).
DEGREGORIA, A. J.: Linear radial and nonradial modes of oscillation of hot white dwarfs. Ap. J. **217**, 175 (1977).
DETWEILER, S. L., LINDBLOM, L.: On the evolution of the homogeneous ellipsoidal figures. Ap. J. **213**, 193 (1977).
DEUTSCHMAN, W. A., DAVIS, R. J., SCHILD, R. E.: The galactic distribution of interstellar absorption as determined from the celescope catalog of ultraviolet stellar observations and a new catalog of UBV, h-beta photoelectric observations. Ap. J. Suppl. **30**, 97 (1976).
DOPITA, M. A., MASON, D. J., ROBB, W. D.: Atomic nitrogen as a probe of physical conditions in the interstellar medium. Ap. J. **207**, 102 (1976).
DOPITA, M. A.: Optical emission from shocks. IV. The Herbig-Haro objects. Ap. J. Suppl. **37**, 117 (1978).
DRAINE, B. T.: Photoelectric heating of interstellar gas. Ap. J. Suppl. **36**, 595 (1978).
DRAINE, B. T., SALPETER, E. E.: On the physics of dust grains in hot gas. Ap. J. **231**, 77 (1979).
DRAINE, B. T., SALPETER, E. E.: Destruction mechanisms for interstellar dust. Ap. J. **231**, 438 (1979).
DURISEN, R. H.: Upper mass limits for stable rotating white dwarfs. Ap. J. **199**, 179 (1975).
DURISEN, R. H.: Viscous effects in rapidly rotating stars with application to white-dwarf models. III. Further numerical results. Ap. J. **195**, 483 (1975).

EASSON, I., PETHICK, C. J.: Magnetohydrodynamics of neutron star interiors. Ap. J. **227**, 995 (1979).
ELMEGREEN, B. G.: The ionization of a low-density intercloud medium by a single O star. Ap. J. Suppl. **32**, 147 (1976).

ELMEGREEN, B. G.: Gravitational collapse in dust lanes and the appearance of spiral structure in galaxies. Ap. J. **231**, 372 (1979).
ELMEGREEN, B. G., ELMEGREEN, D. M.: Star formation in shock-compressed layers. Ap. J. **220**, 1051 (1978).
ELMEGREEN, B. G., LADA, C. J.: Sequential formation of subgroups in OB associations. Ap. J. **214**, 725 (1977).
ELSNER, R. F., LAMB, F. K.: Accretion by magnetic neutron stars. I. Magnetospheric structure and stability. Ap. J. **215**, 897 (1977).
EPSTEIN, R. I., ARNETT, W. D.: Neutronization and thermal disintegration of dense stellar matter. Ap. J. **201**, 202 (1975).
EPSTEIN, R.: Neutrino angular momentum loss in rotating stars. Ap. J. **219**, L 39 (1978).
EWART, G. M., GUYER, R. A., GREENSTEIN, G.: Electrical conductivity and magnetic field decay in neutron stars. Ap. J. **202**, 238 (1975).
FAULKNER, D. J., FREEMAN, K. C.: Gas in globular clusters. I. Time-independent flow models. Ap. J. **211**, 77 (1977).
FIELD, G. B.: Interstellar abundances: gas and dust. Ap. J. **187**, 453 (1974).
FINN, G. D., SIMON, T.: Dust shell models for compact infrared sources. Ap. J. **212**, 472 (1977).
FISHBONE, L. G.: The relativistic Roche problem. II. Stability theory. Ap. J. **195**, 499 (1975).
FLOWERS, E., ITOH, N.: Transport properties of dense matter. Ap. J. **206**, 218 (1976).
FLOWERS, E., RUDERMAN, M., SUTHERLAND, P.: Neutrino pair emission from finite-temperature neutron superfluid and the cooling of young neutron stars. Ap. J. **205**, 541 (1976).
FLOWERS, E., RUDERMAN, M. A.: Evolution of pulsar magnetic fields. Ap. J. **215**, 302 (1977).
FLOWERS, E. G., et al.: Variational calculation of groundstate energy of iron atoms and condensed matter in strong magnetic fields. Ap. J. **215**, 291 (1977).
FLOWERS, E., ITOH, N.: Transport properties of dense matter. II. Ap. J. **230**, 847 (1979).
FONTAINE, G., et al.: The effects of differences in composition, equation of state, and mixing length upon the structure of white-dwarf conversion zones. Ap. J. **193**, 205 (1974).
FONTAINE, G., VAN HORN, H. M.: Analytic surface boundary conditions for white dwarf evolutionary calculations. Ap. J. **197**, 647 (1975).
FONTAINE, G., VAN HORN, H. M.: Convective white-dwarf envelope model grids for H-, He-, and C-rich compositions. Ap. J. Suppl. **31**, 467 (1976).
FONTAINE, G., GRABOSKE, H. C. Jr., VAN HORN, H. M.: Equations of state for stellar partial ionization zones. Ap. J. Suppl. **35**, 293 (1977).
FORREST, W. J., GILLETT, F. C., STEIN, W. A.: Circumstellar grains and the intrinsic polarization of starlight. Ap. J. **195**, 423 (1975).
FREEMAN, J., et al.: The local interstellar helium density. Ap. J. **215**, L 83 (1977).
FRIMAN, B. L., MAXWELL, O. V.: Neutrino emissivities of neutron stars. Ap. J. **232**, 541 (1979).
GEROLA, H., KAFATOS, M., MCCRAY, R.: Statistical time-dependent model for the interstellar gas. Ap. J. **189**, 55 (1974).
GHOSH, P., LAMB, F. K., PETHICK, C. J.: Accretion by rotating magnetic neutron stars. I. Flow of matter inside the magnetosphere and its implications for spin-up and spin-down of the star. Ap. J. **217**, 578 (1977).
GHOSH, P., LAMB, F. K.: Accretion by rotating magnetic neutron stars. II. Radial and vertical structure of the transition zone in disk accretion. Ap. J. **232**, 259 (1979).
GHOSH, P., LAMB, F. K.: Accretion by rotating magnetic neutron stars. III. Accretion torques and period changes in pulsating X-ray sources. Ap. J. **234**, 296 (1979).
GLASSER, M. L.: Ground state of electron matter in high magnetic fields. Ap. J. **199**, 206 (1975).
GLASSER, M. L., KAPLAN, J. I.: The surface of a neutron star in superstrong magnetic fields. Ap. J. **199**, 208 (1975).
GOUGH, D. O.: Mixing-length theory for pulsating stars. Ap. J. **214**, 196 (1977).
GRABOSKE, H. C. Jr., OLNESS, R. J., GROSSMAN, A. S.: Thermodynamics of dense hydrogen-helium fluids. Ap. J. **199**, 255 (1975).
GREENBERG, P. J.: The equations of hydrodynamics for a thermally conducting viscous compressible fluid in a special relativity. Ap. J. **195**, 761 (1975).
GREENSTEIN, G.: Superfluidity in neutron stars. I. Steadystate hydrodynamics and frictional heating. Ap. J. **200**, 281 (1975).
GREENSTEIN, G.: Superfluidity in neutron stars. II. After a period jump. Ap. J. **208**, 836 (1976).

GREENSTEIN, J. L.: A new list of 52 degenerate stars. VII. Ap. J. **189**, L 131 (1974).
GREENSTEIN, J. L.: A further list of degenerate stars. VIII. Ap. J. **196**, L 117 (1975).
GREENSTEIN, J. L.: Degenerate stars with helium atmospheres. Ap. J. **210**, 524 (1976).
GREENSTEIN, J. L.: Some further degenerate stars. IX. Ap. J. **207**, L 119 (1976).
GREISEN, E. W.: The small-scale structure of interstellar hydrogen. Ap. J. **203**, 371 (1976).
GROTH, E. J.: Timing of the crab pulsar. I. Arrival times. II. Method of analysis. III. The slowing down and the nature of random process. Ap. J. Suppl. **29**, 431 (1975).

HANSEN, C. J., AIZENMAN, M. L., ROSS, R. R.: The equilibrium and stability of uniformly rotating, isothermal gas cylinders. Ap. J. **207**, 736 (1976).
HANSEN, C. J., COX, J. P., CARROLL, B. W.: The quasi-adiabatic analysis of nonradial modes of stellar oscillation in the presence of slow rotation. Ap. J. **226**, 210 (1978).
HARDING, D., GUYER, R. A., GREENSTEIN, G.: Superfluidity in neutron stars. III. Relaxation processes between the superfluid and the crust. Ap. J. **222**, 991 (1978).
HARTLE, J. B., SAWYER, R. F., SCALAPINO, D. J.: Pion condensed matter at high densities: equation of state and stellar models. Ap. J. **199**, 471 (1975).
HARTLE, J. B., MUNN, M. W.: Slowly rotating relativistic stars. V. Static stability analysis of $n=3/2$ polytropes. Ap. J. **198**, 467 (1975).
HARTLE, J. B., SABBADINE, A. G.: The equation of state and bounds on the mass of nonrotating neutron stars. Ap. J. **213**, 831 (1977).
HARVEL, C. A.: Radiative transfer in circumstellar dust shells. Ap. J. **210**, 862 (1976).
HEGYI, D. J., LEE, T.-S. H., COHEN, J. M.: The maximum mass of nonrotating neutron stars. Ap. J. **201**, 462 (1975).
HEGYI, D. J.: The upper mass limit for neutron stars including differential rotation. Ap. J. **217**, 244 (1977).
HEILES, C.: The interstellar magnetic field. Ann. Rev. Astron. Ap. **14**, 1 (1976).
HEILES, C.: An almost complete survey of 21 cm line radiation for $|b| \geqslant 10^0$. III. The interdependence of H_I, galaxy counts, reddening, and galactic latitude. Ap. J. **204**, 379 (1976).
HEILES, C.: An almost complete survey of 21 centimeter line radiation for $|b| \geqslant 10^0$. VI. Energetic expanding H_I shells. Ap. J. **208**, L 137 (1976).
HENRIKSEN, R. N., CHAU, W. Y.: Neutrino angular momentum loss by the Poynting-Robertson effect. Ap. J. **225**, 712 (1978).
HERBST, W., ASSOUSA, G. E.: Observational evidence for supernovae-induced star formation: Canis Major R 1. Ap. J. **217**, 473 (1977).
HEWISH, A.: Pulsars and high density physics. Rev. Mod. Phys. **47**, 567 (1975).
HILL, H. A., CAUDELL, T. P., ROSENWALD, R. D.: On the use of spectral lines as a temperature indicator in a pulsating system. Ap. J. **213**, L 81 (1977).
HILL, J. K., SILK, J.: On the nature of the intercloud medium. Ap. J. **198**, 299 (1975).
HILL, J. K., HOLLENBACH, D. J.: Effects of expanding compact H II regions upon molecular clouds: molecular dissocation waves, shock waves, and carbon ionization. Ap. J. **225**, 390 (1978).
HILLEBRABDT, W., MÜLLER, E.: Matter in superstrong magnetic fields and the structure of a neutron star's surface. Ap. J. **207**, 589 (1976).
HILLS, J. G.: The rate of formation of white dwarfs in stellar systems. Ap. J. **219**, 550 (1978).
HILLS, J. G.: An upper limit to the rate of formation of neutron stars in the galaxy. Ap. J. **221**, 973 (1978).
HINTZEN, P., STRITTMATTER, P. A.: Problems in classifying cool degenerate stars. Ap. J. **201**, L 37 (1975).
HOBBS, L. M.: On ionization in H I regions. Ap. J. **188**, L 107 (1974).
HOBBS, L. M.: Statistical properties of interstellar clouds. Ap. J. **191**, 395 (1974).
HSIEH, S.-H., SPIEGEL, E. A.: The equations of photohydrodynamics. Ap. J. **207**, 244 (1976).
HUNTER, C.: On secular stability, secular instability, and points of bifurcation of rotating gaseous masses. Ap. J. **213**, 497 (1977).
HUNTER, C.: The collapse of unstable isothermal spheres. Ap. J. **218**, 834 (1977).
HUTCHINS, J. B.: The thermal effects of H_2 molecules in rotating and collapsing spheroidal gas clouds. Ap. J. **205**, 103 (1976).
HWANG, A. E., DYKLA, J. J.: Can a neutron star be compressed into a black hole? Ap. J. **192**, L 141 (1974).

IBEN, I. Jr., TRURAN, J. W.: On the surface composition of thermally pulsing stars of high luminosity and on the contribution of such stars to the element enrichment of the interstellar medium. Ap. J. **220**, 980 (1978).

INGHAM, W. H., BRECHER, K., WASSERMAN, I.: On the origin of continuum polarization in white dwarfs. Ap. J. **207**, 518 (1976).

IPSER, J. R.: On using entropy arguments to study the evolution and secular stability of spherical stellar-dynamical systems. Ap. J. **193**, 463 (1974).

JENKINS, E. B., MORTON, D. C., YORK, D. G.: Rocket-ultraviolet spectra of kappa, lambda, tau and upsilon scorpii. Ap. J. **194**, 77 (1974).

JURA, M.: Photoelectric heating of the interstellar gas. Ap. J. **204**, 12 (1976).

KARP, A. H.: Hydrodynamic models of a cepheid atmosphere. I. Deep envelope calculations. Ap. J. **199**, 448 (1975).

KARP, A. H.: Hydrodynamic models of a cepheid atmosphere. II. Continuous spectrum. Ap. J. **200**, 354 (1975).

KARP, A. H.: Hydrodynamic models of a cepheid atmosphere. III. Line spectrum and radius determinations. Ap. J. **201**, 641 (1975).

KATZ, J., HORWITZ, G., KLAPISCH, M.: Thermodynamic stability of relativistic stellar clusters. Ap. J. **199**, 307 (1975).

KAUFMAN, M.: Star formation and galactic evolution. I. General expressions and applications to our galaxy. Ap. J. **232**, 707 (1979).

KAZANAS, D., SCHRAMM, D. N.: Neutrino damping of nonradial pulsations in gravitational collapse. Ap. J. **214**, 819 (1977).

KIMMER, E.: Physical conditions in a hydrogen gas heated by suprathermal protons. Ap. J. **203**, 674 (1976).

KING, D., et al.: Applications of linear pulsation theory to the cepheid mass problem and the double-mode cepheids. Ap. J. **195**, 467 (1975).

KISLINGER, M. B., MORLEY, P. D.: Asymptotic freedom and dense stellar matter. II. The equation of state for neutron stars. Ap. J. **219**, 1017 (1978).

KRAFT, R. P.: On the nonhomogeneity of metal abundances in stars of globular clusters and satellite subsystems of the galaxy. Ann. Rev. Astron. Ap. **17**, 309 (1979).

KRUSKAL, M., SCHWARZSCHILD, M., HÄRM, R.: An instability due to the local mixing-length approximation. Ap. J. **214**, 498 (1977).

LAMB, D. Q., VAN HORN, H. M.: Evolution of crystallizing pure ^{12}C white dwarfs. Ap. J. **200**, 306 (1975).

LAMB, D. Q., et al.: Hot dense matter and stellar collapse. Phys. Rev. Lett. **41**, 1623 (1978).

LANDSTREET, J. D., ANGEL, J. R. P.: The wavelength dependence of circular polarization in GD 229. Ap. J. **190**, L 25 (1974).

LANDSTREET, J. D., ANGEL, J. R. P.: The polarization spectrum and magnetic field strength of the white dwarf Grw + 70°82–47. Ap. J. **196**, 819 (1975).

LANGER, W. D.: Interstellar cloud evolution and the abundance of formaldehyde. Ap. J. **210**, 328 (1976).

LANGER, W. D.: The stability of interstellar clouds containing magnetic fields. Ap. J. **225**, 95 (1978).

LATTIMER, J. M., et al.: The decompression of cold neutron star matter. Ap. J. **213**, 225 (1977).

LEBOVITZ, N. R.: The fission theory of binary stars. II. Stability to third-harmonics disturbances. Ap. J. **190**, 121 (1974).

LEE, L. C., JOKIPII, J. R.: The irregularity spectrum in interstellar space. Ap. J. **206**, 735 (1976).

LERCHE, I., SCHRAMM, D. N.: Magnetic fields greater than 10^{20} Gauss? Ap. J. **216**, 881 (1977).

LEVINE, R. H.: Acceleration of thermal particles in collapsing magnetic regions. Ap. J. **190**, 447 (1974).

LEVINE, R. H.: A new theory of control heating. Ap. J. **190**, 457 (1974).

LICHTENSTADT, I., et al.: Effects of neutrino degeneracy and of downscatter on neutrino radiation from dense stellar cores. Ap. J. **226**, 222 (1978).

LIEBERT, J., ANGEL, J. R. P., LANDSTREET, J. D.: The detection of an Hα Zeeman pattern in the cool magnetic white dwarf G 99–47. Ap. J. **202**, L 139 (1975).

LIEBERT, J., et al.: A hot rotating magnetic white dwarf. Ap. J. **214**, 457 (1977).

LIGHTMAN, A. P., PRESS, W. H., ODENWALD, S. F.: Present and past death rates for globular clusters. Ap. J. **219**, 629 (1978).

LIGHTMAN, A. P., SHAPIRO, S. L.: The dynamical evolution of globular clusters. Rev. Mod. Phys. **50**, 437 (1978).

LILLIE, C. F., WITT, A. N.: Ultraviolet photometry from the orbiting astronomical observatory. XXV. Diffuse galactic light in the 1500–4200 Å region and the scattering properties of interstellar dust grains. Ap. J. **208**, 64 (1976).

LINDBLOM, L., DETWEILER, S. L.: On the secular instabilities of the Maclaurin spheroids. Ap. J. **211**, 565 (1977).

LINDBLOM, L., DETWEILER, S.: The role of neutrino dissipation in gravitational collapse. Ap. J. **232**, L 101 (1979).

LISZT, H. S., BURTON, W. B.: The gas distribution in the central region of the galaxy. II. Carbon monoxide. Ap. J. **226**, 790 (1978).

LODENQUAI, J., et al.: Photon opacity in surfaces of magnetic neutron stars. Ap. J. **190**, 141 (1974).

LOREN, R. B.: Colliding clouds and star formation in NGC 1333. Ap. J. **209**, 466 (1976).

LOREN, R. B.: The Monoceros R 2 cloud: near-infrared and molecular observations of a rotating collapsing cloud. Ap. J. **215**, 129 (1977).

LOW, B. C.: Resistive diffusion of force-free magnetic fields in a passive medium. IV. The dynamical theory. Ap. J. **193**, 243 (1974).

MCCLINTOCK, W., et al.: Ultraviolet observations of cool stars. V. The local density of interstellar matter. Ap. J. **204**, L 103 (1976).

MCCRAY, R., SNOW, T. P. Jr.: The violet interstellar medium. Ann. Rev. Astron. Ap. **17**, 213 (1979).

MCKEE, C. F., COWIE, L. L.: The evaporation of spherical clouds in a hot gas. II. Effects of radiation. Ap. J. **215**, 213 (1977).

MCKEE, C. F., OSTRIKER, J. P.: A theory of the interstellar medium: three components regulated by supernova explosions in a inhomogeneous substrate. Ap. J. **218**, 148 (1977).

MADORE, B. F., VAN DEN BERGH, S., ROGSTAD, D. H.: Gas density and the rate of star formation in M 33. Ap. J. **191**, 317 (1974).

MALONE, R. C., JOHNSON, M. B., BETHE, H. A.: Neutron star models with realistic high-density equations of state. Ap. J. **199**, 741 (1975).

MANCHESTER, R. N.: Structure of the local galactic magnetic field. Ap. J. **188**, 637 (1974).

MARCUS, P. S., PRESS, W. H., TEUKOLSKY, S. A.: Stablest shapes for an axisymmetric body of gravitating, incompressible fluid. Ap. J. **214**, 584 (1977).

MARTIN, P. G.: Interstellar polarization from a medium with changing grain alignment. Ap. J. **187**, 461 (1974).

MARTIN, P. G., ANGEL, J. R. P.: Systematic variations in the wavelength dependence of interstellar circular polarization. Ap. J. **207**, 126 (1976).

MARTIN, P. G., CAMPBELL, B.: Circular polarization observations of the interstellar magnetic field. Ap. J. **208**, 727 (1976).

MASHHOON, B.: On tidal phenomena in a strong gravitational field. Ap. J. **197**, 705 (1975).

MATHEWS, W. G.: Stability of gas clouds near quasi-stellar objects. Ap. J. **207**, 351 (1976).

MATHEWS, W. G., BLUMENTHAL, G. R.: Rayleigh-Taylor stability of compressible and incompressible radiation-supported surfaces and slabs: application to QSO clouds. Ap. J. **214**, 10 (1977).

MATHIS, J. S., RUMPL, W., NORDSIECK, K. H.: The size distribution of interstellar grains. Ap. J. **217**, 425 (1977).

MAVKO, G. E., et al.: Observations of structure in the interstellar polarization curve: preliminary results. Ap. J. **187**, L 117 (1974).

MAXWELL, O., et al.: Beta decay of pion condensates as a cooling mechanism for neutron stars. Ap. J. **216**, 77 (1977).

MAZUREK, T. J., LATTIMER, J. M., BROWN, G. E.: Nuclear forces, partition functions, and dissociation in stellar collapse. Ap. J. **229**, 713 (1979).

MEIER, D. L., et al.: Magnetohydrodynamic phenomena in collapsing stellar cores. Ap. J. **204**, 869 (1976).

MÉSZÁROS, P.: Ionization mechanisms of the intercloud medium. Ap. J. **191**, 79 (1974).

MIHALAS, D., HUMMER, D. G.: Theory of extended stellar atmospheres. I. Computational method and first results for static spherical models. Ap. J. Suppl. **28**, 343 (1974).

MIKAELIAN, K. O.: New mechanism for slowing down the rotation of dense stars. Ap. J. **214**, L 23 (1977).

MILTON, R. L.: The effects of rapid, differential rotation on the spectra of white dwarfs. Ap. J. **189**, 543 (1974).
MORTON, D. C., SMITH, A. M., STECHER, T. P.: A new limit on the interstellar abundance of boron. Ap. J. **189**, L 109 (1974).
MOUSCHOVIAS, T. Ch.: Static equilibria of the interstellar gas in the presence of magnetic and gravitational fields. Ap. J. **192**, 37 (1974).
MOUSCHOVIAS, T. Ch.: Nonhomologous contraction and equilibria of self-gravitating, magnetic interstellar clouds embedded in an intercloud medium: star formation. I. formulation of the problem and method of solution. Ap. J. **206**, 753 (1976).
MOUSCHOVIAS, T. Ch.: Nonhomologous contraction and equilibria of self-gravitating, magnetic interstellar clouds embedded in an intercloud medium: star formation. II. Results. Ap. J. **207**, 141 (1976).
MOUSCHOVIAS, T. Ch., SPITZER, L. Jr.: Note on the collapse of magnetic interstellar clouds. Ap. J. **210**, 326 (1976).
MOUSCHOVIAS, T. Ch.: A connection between the rate of rotation of interstellar clouds, magnetic fields, ambipolar diffusion and the periods of binary stars. Ap. J. **211**, 147 (1977).
MOUSCHOVIAS, T. Ch.: Magnetic braking of self-gravitating, oblate interstellar clouds. Ap. J. **228**, 159 (1979).
MOUSCHOVIAS, T. Ch., PALEOLOGOU, E. V.: The angular momentum problem and magnetic braking: an exact, time-dependent solution. Ap. J. **230**, 204 (1979).
MUFSON, S. L.: The structure and stability of shock waves in a multiphase interstellar medium. Ap. J. **193**, 561 (1974).
MYERS, P. C.: A compilation of interstellar gas properties. Ap. J. **225**, 380 (1978).
NANDY, A.: Correlation effects on the energy shifts of excited nucleous in neutron-star matter. Ap. J. **190**, 385 (1974).
NI, W.-T.: Relativistic stellar stability: preferred-frame effects. Ap. J. **190**, 131 (1974).
O'CONNELL, R. F.: Internal magnetic fields of pulsars, white dwarfs, and other stars. Ap. J. **195**, 751 (1975).
OKE, J. B.: Absolute spectral energy distributions for white dwarfs. Ap. J. **188**, 443 (1974).
OKE, J. B.: Absolute spectral energy distributions for white dwarfs. Ap. J. Suppl. **27**, 21 (1974).
OPPENHEIMER, M., DALGARNO, A.: The chemistry of sulfur in interstellar clouds. Ap. J. **187**, 231 (1974).
OSAKI, J.: An excitation mechanism for pulsation in beta cephei stars. Ap. J. **189**, 469 (1974).
PANDHARIPANDE, V. R., PINES, D., SMITH, R. A.: Neutron star structure: theory, observation, and speculation. Ap. J. **208**, 550 (1976).
PARKER, E. N.: Hydraulic concentration of magnetic fields in the solar photosphere. I. Turbulent pumping. Ap. J. **189**, 563 (1974).
PARKER, E. N.: Hydraulic concentration of magnetic fields in the solar photosphere. II. Bernoulli effect. Ap. J. **190**, 429 (1974).
PARKER, E. N.: The relative diffusion of strong magnetic fields and tenuous gases. Ap. J. **215**, 374 (1977).
PIRAN, T.: The role of viscosity and cooling mechanisms in the stability of accretion disks. Ap. J. **221**, 652 (1978).
PRESS, W. H., WIITA, P. J., SMARR, L. L.: Mechanism for inducing synchronous rotation and small eccentricity in close binary systems. Ap. J. **202**, L 135 (1975).
PRESS, W. H., TEUKOLSKY, S. A.: On formation of close binaries by two-body tidal capture. Ap. J. **213**, 183 (1977).
PURCELL, E. M.: Temperature fluctuations in very small interstellar grains. Ap. J. **206**, 685 (1976).
PURCELL, E. M., SHAPIRO, P. R.: A model for the optical behavior of grains and resonant impurities. Ap. J. **214**, 92 (1977).
PURCELL, E. M.: Suprathermal rotation of interstellar grains. Ap. J. **231**, 404 (1979).
RASTALL, P.: The maximum mass of a neutron star. Ap. J. **213**, 234 (1977).
RAYMOND, J. C.: Shock waves in the interstellar medium. Ap. J. Suppl. **39**, 1 (1979).
RICHER, H. B., ULRYCH, T. J.: High-frequency optical variables. II. Luminosity-variable white dwarfs and maximum entropy spectral analysis. Ap. J. **192**, 719 (1974).
RICHSTONE, D. O.: The occurrence of a nonspherical thermal instability in red giant stars. Ap. J. **188**, 327 (1974).
ROBERTS, W. J.: Electromagnetic multipole fields of neutron stars. Ap. J. Suppl. **41**, 75 (1979).
ROSEN, J., ROSEN, N.: The maximum mass of a cold neutron star. Ap. J. **202**, 782 (1975).

ROSI, L. A., ZIMMERMAN, R. L., KEMP, J. C.: Polarized radiation in magnetic white dwarfs. Ap. J. **209**, 868 (1976).
RUDERMAN, M.: Crust-breaking by neutron superfluids and the vela pulsar glitches. Ap. J. **203**, 213 (1976).
RUDERMAN, M. A., SUTHERLAND, P. G.: Rotating superfluid in neutron stars. Ap. J. **190**, 137 (1974).
RUIZ, M. T., SCHWARZSCHILD, M.: An approximate dynamical model for spheroidal stellar systems. Ap. J. **207**, 376 (1976).
RUIZ, M. T.: A dynamical model for the central region of M 31. Ap. J. **207**, 382 (1976).
RYDGREN, A. E., STROM, S. E., STROM, K. M.: The nature of the objects of Joy: a study of the t tauri phenomenon. Ap. J. Suppl. **30**, 307 (1976).
RYTER, C., CESARSKY, C. J., AUDOUZE, J.: X-ray absorption, interstellar reddening and elemental abundances in the interstellar medium. Ap. J. **198**, 103 (1975).
SAENZ, R. A.: Maximum mass of neutron stars: dependence on the assumptions. Ap. J. **212**, 816 (1977).
SAENZ, R. A., SHAPIRO, S. L.: Gravitational and neutrino radiation from stellar core collapse: improved ellipsoidal model calculations. Ap. J. **229**, 1107 (1979).
SAGDEEV, R. Z.: The 1976 Oppenheimer lectures: critical problems in plasma astrophysics. I. Turbulence and nonlinear waves. Rev. Mod. Phys. **51**, 1 (1979).
SAGDEEV, R. Z.: The 1976 Oppenheimer lectures: critical problems in plasma astrophysics. II. Singular layers and reconnection. Rev. Mod. Phys. **51**, 11 (1979).
SALPETER, E. E.: Formation and flow of dust grains in cool stellar atmospheres. Ap. J. **193**, 585 (1974).
SALPETER, E. E.: Nucleation and growth of dust grains. Ap. J. **193**, 579 (1974).
SALPETER, E. E.: Dying stars and reborn dust. Rev. Mod. Phys. **46**, 433 (1974).
SALPETER, E. E.: Planetary nebulae, supernova remnants, and the interstellar medium. Ap. J. **206**, 673 (1976).
SALPETER, E. E.: Formation and destruction of dust grains. Ann. Rev. Astron. Ap. **15**, 267 (1977).
SASLAW, W. C.: Motion around a source whose luminosity changes. Ap. J. **226**, 240 (1978).
SASTRI, V. K., STOTHERS, R.: Influence of opacity on the pulsational stability of massive stars with uniform chemical composition. II. Modified Kramers opacity. Ap. J. **193**, 677 (1974).
SATO, H.: Slowly braked, rotating neutron stars. Ap. J. **195**, 743 (1975).
SAVAGE, B. D., MATHIS, J. S.: Observed properties of interstellar dust. Ann. Rev. Astron. Ap. **17**, 73 (1979).
SAWYER, R. F., SONI, A.: Neutrino transport in pion-condensed neutron stars. Ap. J. **216**, 73 (1977).
SAWYER, R. F., SONI, A.: Transport of neutrinos in hot neutron-star matter. Ap. J. **230**, 859 (1979).
SCALO, J. M.: On the limiting mass of carbon-oxygen white dwarfs. Ap. J. **206**, 215 (1976).
SCHARLEMAN, E. T.: The fate of matter and angular momentum in disk accretion onto a magnetized neutron star. Ap. J. **219**, 617 (1978).
SCHRAMM, D. N., ARNETT, W. D.: The weak interaction and gravitational collapse. Ap. J. **198**, 629 (1975).
SCHWARZT, R. A., STEIN, R. F.: Waves in the solar atmosphere. IV. Magneto-gravity and acoustic-gravity modes. Ap. J. **200**, 499 (1975).
SCHWARTZ, R. D.: Evidence of star formation triggered by expansion of the Gum nebula. Ap. J. **212**, L 25 (1977).
SERKOWSKI, K., MATHEWSON, D. S., FORD, V. L.: Wavelength dependence of interstellar polarization and ratio of total to selective extinction. Ap. J. **196**, 261 (1975).
SERVICE, A. T.: Concise approximation formulae for the Lane-Emden functions. Ap. J. **211**, 908 (1977).
SHAPIRO, P. R.: Interstellar polarization: magnetite dust. Ap. J. **201**, 151 (1975).
SHAPIRO, P. R., FIELD, G. B.: Consequences of a new hot component of the interstellar medium. Ap. J. **205**, 762 (1976).
SHAPIRO, S. L., TEUKOLSKY, S. A.: On the maximum gravitational redshift of white dwarfs. Ap. J. **203**, 697 (1976).
SHORE, S. N., ADELMAN, S. J.: Magnetic fields and diffusion processes in peculiar A stars. Ap. J. **191**, 165 (1974).
SHIPMAN, H. L.: Masses, radii, and model atmospheres for cool white-dwarf stars. Ap. J. **213**, 138 (1977).
SHIPMAN, H. L.: Masses and radii of white-dwarf stars. III. Results for 110 hydrogen-rich and 28 helium-rich stars. Ap. J. **228**, 240 (1979).

SHOUB, E. C.: Departures of the electron energy distribution from a Maxwellian in hydrogen. I. Formation and solution of the electron kinetic equation. Ap. J. Suppl. **34**, 259 (1977).

SHOUB, E. C.: Departures of the electron energy distribution from a Maxwellian in hydrogen. II. Consequences. Ap. J. Suppl. **34**, 277 (1977).

SHULL, J. M.: Grain disruption in interstellar hydromagnetic shocks. Ap. J. **215**, 805 (1977).

SHULL, J. M.: Disruption and sputtering of grains in intermediate-velocity interstellar clouds. Ap. J. **226**, 858 (1978).

SILK, J., BURKE, J. R.: Dust grains in a hot gas. II. Astrophysical applications. Ap. J. **190**, 11 (1974).

SILK, J.: Hydromagnetic waves and shock waves as an interstellar heat source. Ap. J. **198**, L 77 (1975).

SILK, J.: On the fragmentation of cosmic gas clouds. II. Opacity-limited star formation. Ap. J. **214**, 152 (1977).

SILK, J.: On the fragmentation of cosmic gas clouds. III. The initial stellar mass function. Ap. J. **214**, 918 (1977).

SILK, J., NORMAN, C.: Gas-rich dwarfs and accretion phenomena in early-type galaxies. Ap. J. **234**, 86 (1979).

SION, E. M., LIEBERT, J.: The space motions and luminosity function of white dwarfs. Ap. J. **213**, 468 (1977).

SION, E. M., ACIERNO, M. J., TOMCZYK, S.: Hydrogen shell flashes in massive accreting white dwarfs. Ap. J. **230**, 832 (1979).

SNOW, T. P. Jr.: The depletion of interstellar elements and the interaction between gas and dust in space. Ap. J. **202**, L 87 (1975).

SNOW, T. P. Jr., MORTON, D. C.: Copernicus ultraviolet observations of mass-low effects O and R stars. Ap. J. Suppl. **32**, 429 (1976).

SPITZER, L. Jr., JENKINS, E. B.: Ultraviolet studies of the interstellar gas. Ann. Rev. Astron. Ap. **13**, 133 (1975).

SPITZER, L. Jr., SHULL, J. M.: Random gravitational encounters and the evolution of spherical systems. VI. Plummer's model. Ap. J. **200**, 339 (1975).

SPITZER, L. Jr., SHULL, J. M.: Random gravitational encounters and the evolution of spherical systems. VII. Systems with several mass groups. Ap. J. **201**, 773 (1975).

SPITZER, L. Jr., MCGLYNN, T. A.: Disorientation of interstellar grains in suprathermal rotation. Ap. J. **231**, 417 (1979).

SRNKA, L. J., BIBHAS, R. De: Spin-related magnetism of interstellar grains. Ap. J. **225**, 422 (1978).

STEIGMAN, G.: Ion-atom charge-transfer reactions and a hot intercloud medium. Ap. J. **195**, L 39 (1975).

STEIGMAN, G.: Carbon-helium charge transfers and the ionization of carbon in the intercloud medium. Ap. J. **199**, 336 (1975).

STEIGMAN, G.: Charge transfer reactions in multiply charged ion-atom collisions. Ap. J. **199**, 642 (1975).

STELLINGWERF, R. F.: The calculation of periodic pulsations of stellar models. Ap. J. **192**, 139 (1974).

STEVENSON, D. J., SALPETER, E. E.: The phase diagram and transport properties for hydrogen-helium fluid planets. Ap. J. Suppl. **35**, 221 (1977).

STEVENSON, D. J., SALPETER, E. E.: The dynamics and helium distribution in hydrogen-helium fluid planets. Ap. J. Suppl. **35**, 239 (1977).

STOTHERS, R.: Influence of rotation on the maximum mass of pulsationally stable stars. Ap. J. **192**, 145 (1974).

SURMELIAN, G. I., O'CONNELL, R. F.: Quadratic Zeeman effect in the hydrogen Balmer lines from magnetic white dwarfs. Ap. J. **193**, 705 (1974).

SWEDLUND, J. B., et al.: Discovery of time-varying circular and linear polarization in the white-dwarf suspect GD 229. Ap. J. **187**, L 121 (1974).

TAAM, R. E., FAULKNER, J.: Ultrashort-period binaries. III. The accretion of hydrogen-rich matter onto a white dwarf of one solar mass. Ap. J. **198**, 435 (1975).

TAAM, R. E., SCHWARTZ, R. D.: Radiative transport in circumstellar dust shells. Ap. J. **204**, 842 (1976).

TAAM, R. E., PICKLUM, R. E.: Thermonuclear runaways on neutron stars. Ap. J. **233**, 327 (1979).

TALBOT, R. J. Jr.: Sensitivity of the star formation rate to the interstellar gas abundance of heavy elements. Ap. J. **188**, 209 (1974).

TALBOT, R. J. Jr., NEWMAN, M. J.: Encounters between stars and dense interstellar clouds. Ap. J. Suppl. **34**, 295 (1977).

TASSOUL, M.: On the stability of congruent Darwin ellipsoids. Ap. J. **202**, 803 (1975).
THORNE, K. S.: The relativistic equations of stellar structure and evolution. Ap. J. **212**, 825 (1977).
THORNE, K. S., ZYTKOW, A. N.: Stars with degenerate neutron cores. I. Structure of equilibrium models. Ap. J. **212**, 832 (1977).
THUAN, T. X.: On the ionization of the intercloud medium by runaway O-B stars. Ap. J. **198**, 307 (1975).
TORRES-PEIMBERT, S., LAZCANO-ARAUJO, A., PEIMBERT, M.: Ionization of the low-density interstellar medium. Ap. J. **191**, 401 (1974).
TUBBS, D. L.: Direct-simulation neutrino transport: aspects of equilibrium. Ap. J. Suppl. **37**, 287 (1978).
TUBBS, D. L.: Conservative scattering, electron scattering, and neutrino thermalization. Ap. J. **231**, 846 (1979).
TUCHMAN, Y., SACK, N., BARKAT, Z.: Mass loss from dynamically unstable stellar envelopes. Ap. J. **219**, 183 (1978).
ULRICH, R. K.: A nonlocal mixing-length theory of convection for use in numerical calculations. Ap. J. **207**, 564 (1976).
ULRICH, R. K., BURGER, H. L.: The accreting component of mass-exchange binaries. Ap. J. **206**, 509 (1976).
VAN DEN HEUVEL, E. P. J.: The upper mass limit for white dwarf formation as derived from the stellar content of the Hyades cluster. Ap. J. **196**, L 121 (1975).
VANDERVOORT, P. O.: New applications of the equations of stellar hydrodynamics. Ap. J. **195**, 333 (1975).
VAN RIPER, K. A., BLUDMAN, S. A.: Composition and equation of state of thermally dissociated matter. Ap. J. **213**, 239 (1977).
VAN RIPER, K. A.: The hydrodynamics of stellar collapse. Ap. J. **221**, 304 (1978).
VAN RIPER, K. A.: General relativistic hydrodynamics and the adiabatic collapse of stellar cores. Ap. J. **232**, 558 (1979).
VAUCLAIR, G., FONTAINE, G.: Convective mixing in helium white dwarfs. Ap. J. **230**, 563 (1979).
VILA, S. C., SION, E. M.: The pulsational properties of high-luminosity degenerate stars with hydrogen burning near the surface. Ap. J. **207**, 820 (1976).
VON HOERNER, S., SASLAW, W. C.: The evolution of massive collapsing gas clouds. Ap. J. **206**, 917 (1976).
WATSON, W. D., KUNZ, A. B.: Multiple electron ejection by X-ray photoionization and the abundance of interstellar ions. Ap. J. **201**, 165 (1975).
WATSON, W. D.: Multiple ionization by low-energy cosmic rays and the abundace of highly ionized interstellar atoms. Ap. J. **204**, 47 (1976).
WEBER, S. V.: Oscillation and collapse of interstellar clouds. Ap. J. **208**, 113 (1976).
WESTBROOK, C. K., TARTER, C. B.: On protostellar evolution. Ap. J. **200**, 48 (1975).
WHIPPLE, F. L., HUEBNER, W. F.: Physical processes in comets. Ann. Rev. Astron. Ap. **14**, 143 (1976).
WHITE, R. E.: Depletion of interstellar sodium and calcium. Ap. J. **187**, 449 (1974).
WIITA, P. J., PRESS, W. H.: Mass-angular-momentum regimes for certain instabilities of a compact, rotating stellar core. Ap. J. **208**, 525 (1976).
WILSON, J. R.: Coherent neutrino scattering and stellar collapse. Phys. Rev. Lett. **32**, 849 (1974).
WITHBROE, G. L., NOYES, R. W.: Mass and energy flow in the solar chromosphere and corona. Ann. Rev. Astron. Ap. **15**, 363 (1977).
WITTEN, T. A. Jr.: Compounds in neutron-star crusts. Ap. J. **188**, 615 (1974).
WOLFF, C. L.: Rigid and differential rotation driven by oscillations within the sun. Ap. J. **194**, 489 (1974).
WOLFF, C. L.: White-dwarf variability and the rotation of g-modes. Ap. J. **216**, 784 (1977).
WOLFSON, R.: Energy considerations in axisymmetric accretion. Ap. J. **213**, 208 (1977).
WOLFSON, R.: Axisymmetric accretion near compact objects. Ap. J. **213**, 200 (1977).
WONG, C.-Y.: Toroidal figures of equilibrium. Ap. J. **190**, 675 (1974).
WOOD, P. R., CAHN, J. H.: Mira variables, mass loss, and the fate of red giant stars. Ap. J. **211**, 499 (1977).
WOODWARD, P. R.: Shock-driven implosion of interstellar gas clouds and star formation. Ap. J. **207**, 484 (1976).

WOODWARD, P. R.: Theoretical models of star formation. Ann. Rev. Astron. Ap. **16**, 555 (1978).
YORK, D. G., ROGERSON, J. B. Jr.: The abundance of deuterium relative to hydrogen in interstellar space. Ap. J. **203**, 378 (1976).

Chapter 4, Part 1

ALASTUEY, A., JANCOVICI, B.: Nuclear reaction rate enhancement in dense stellar matter. Ap. J. **226**, 1034 (1978).
AUDOUZE, J., TRURAN, J. W.: p-Process nucleosynthesis in postshock supernova envelope environments. Ap. J. **202**, 204 (1975).
ARNETT, W. D.: Advanced evolution of massive stars. V. Neon burning. Ap. J. **193**, 169 (1974).
ARNETT, W. D.: Advanced evolution of massive stars. VI. Oxygen burning. Ap. J. **194**, 373 (1974).
ARNETT, W. D.: Iron production by ^{12}C-detonation supernovae. Ap. J. **191**, 727 (1974).
ARNETT, W. D.: Advanced evolution of massive stars. VII. Silicon burning. Ap. J. Suppl. **35**, 145 (1977).
ARNETT, W. D.: On the bulk yields of nucleosynthesis from massive stars. Ap. J. **219**, 1008 (1978).
BAHCALL, J. N.: Solar neutrino experiments. Rev. Mod. Phys. **50**, 881 (1978).
BAHCALL, J. N., et al.: Proposed solar-neutrino experiment using ^{71}Ga. Phys. Rev. Lett. **40**, 1351 (1978).
BARKAT, Z., et al.: On the collapse of iron stellar cores. Ap. J. **196**, 633 (1975).
BEAUDET, G., YAHIL, A.: More on big-bang nucleosynthesis with nonzero Lepton numbers. Ap. J. **218**, 253 (1977).
BECKER, S. A., IBEN, I. Jr.: The asymptotic giant branch evolution of intermediate-mass stars as a function of mass and composition. I. Through the second dredge-up phase. Ap. J. **232**, 831 (1979).
BHAVSAR, S. P., HÄRM, R.: The neutrino flux of inhomogeneous solar models. Ap. J. **216**, 138 (1977).
BIGNAMI, G. F., FICHTEL, C. E.: Galactic arm structure and gamma-ray astronomy. Ap. J. **189**, L 65 (1974).
BLAKE, J. B., SCHRAMM, D. N.: A consideration of the neutron capture time scale in the s-process. Ap. J. **197**, 615 (1975).
BLAKE, J. B., SCHRAMM, D. N.: A possible alternative to the r-process. Ap. J. **209**, 846 (1976).
BODANSKY, D., JACOBS, W. W., OBERG, D. L.: On the production of lithium, beryllium, and boron at low energies. Ap. J. **202**, 222 (1975).
BOESGAARD, A. M., CHESLEY, S. E.: Beryllium and post-main-sequence evolution. Ap. J. **210**, 475 (1976).
BUTCHER, H. R.: Studies of heavy-element synthesis in the galaxy. II. A survey of e-, r-, and s-process abundances. Ap. J. **199**, 710 (1975).
BUTCHER, H. R.: On s-process abundance evolution in the galactic disk. Ap. J. **210**, 489 (1976).
CAMERON, A. G. W.: The neutron-rich silicon-burning and equilibrium processes of nucleosynthesis. Ap. J. **230**, L 53 (1979).
CANAL, R., ISERN, J., SANAHUJA, B.: Low-energy nucleosynthesis of lithium, beryllium, and boron. Ap. J. **200**, 646 (1975).
CANAL, R., ISERN, J., SANAHUJA, B.: Synthesis of lithium by spallation reactions in red-giant stars. Ap. J. **214**, 189 (1977).
CARSON, T. R., EZER, D., STOTHERS, R.: Solar neutrinos and the influence of radiative opacities on solar models. Ap. J. **194**, 743 (1974).
CARSON, T. R., STOTHERS, R.: Evolutionary problems of Cepheids and other giants investigated with new radiative opacities. Ap. J. **204**, 461 (1976).
CLAYTON, D. D., NEWMAN, M. J.: s-Process studies: exact solution to a chain having two distinct cross-section values. Ap. J. **192**, 501 (1974).
CLAYTON, D. D., WARD, R. A.: s-process studies: exact evaluation of an exponential distribution of exposures. Ap. J. **193**, 397 (1974).
CLAYTON, D. D., WOOSLEY, S. E.: Thermonuclear astrophysics. Rev. Mod. Phys. **46**, 755 (1974).
CLAYTON, D. D., et al.: Solar models of low neutrino counting rate: the depleted Maxwellian tail. Ap. J. **199**, 494 (1975).
CLAYTON, D. D., et al.: Solar models of low neutrino-counting rate: the central black hole. Ap. J. **201**, 489 (1975).

COUCH, R. G., ARNETT, W. D.: On the thermal properties of the convective Urca process. Ap. J. **194**, 537 (1974).

COUCH, R. G., LOUMOS, G. L.: The Urca process in dense stellar interiors. Ap. J. **194**, 385 (1974).

COUCH, R. G., SCHMIEDEKAMP, A. B., ARNETT, W. D.: s-Process nucleosynthesis in massive stars: core helium burning. Ap. J. **190**, 95 (1974).

COUCH, R. G., ARNETT, W. D.: Carbon ignition and burning in degenerate stellar cores. Ap. J. **196**, 791 (1975).

COWAN, J. J., ROSE, W. K.: Production of ^{17}O and ^{18}O by means of the hot CNO tri-cycle. Ap. J. **201**, L 45 (1975).

COWAN, J. J., ROSE, W. K.: Production of ^{14}C and neutrons in red giants. Ap. J. **212**, 149 (1977).

CRAWFORD, J. P., HANSEN, C. J., MAHANTHAPPA, K. T.: Stellar neutrino pair emission from de-excitation of nuclear states via weak neutral currents. Ap. J. **206**, 208 (1976).

DEARBORN, D., SCHRAMM, D. N.: CNO tri-cycling as an ^{17}O enrichment mechanism. Ap. J. **194**, L 67 (1974).

DESPAIN, K. H.: Convective neutron and s-process element production in deeply mixed envelopes. Ap. J. **212**, 774 (1977).

DEUPREE, R. G.: On shallow convective envelopes. Ap. J. **201**, 183 (1975).

EDWARDS, T. W., HARRISON, T. G.: A photoneutron mechanism for the production of technetium-99 in the interior of evolved stars. Ap. J. **187**, 313 (1974).

ENDAL, A. S.: Carbon-burning nucleosynthesis with convection. Ap. J. **195**, 187 (1975).

ENDAL, A. S.: Theoretical studies of massive stars. I. Evolution of a 15 $M_\odot$ star from the zero-age main sequence to neon ignition. Ap. J. **197**, 405 (1975).

ENDAL, A. S., SPARKS, W. M.: On the lower mass limit for the carbon detonation scenario. Ap. J. **200**, L 77 (1975).

EPSTEIN, R. I., ARNETT, W. D., SCHRAMM, D. N.: Synthesis of the light elements in supernovae. Ap. J. Suppl. **31**, 111 (1976).

FINZI, A.: Solar neutrinos and the behavior of the Fermi coupling constant. Ap. J. **189**, 157 (1974).

FLOWERS, E.: Finite nuclear size effects on neutrino-pair bremsstrahlung in neutron stars. Ap. J. **190**, 381 (1974).

FOWLER, W. A., CAUGHLAN, G. R., ZIMMERMAN, B. A.: Thermonuclear reaction rates, II. Ann. Rev. Astron. Ap. **13**, 69 (1975).

FOWLER, W. A., ENGELBRECHT, C. A., WOOSLEY, S. E.: Nuclear partition functions. Ap. J. **226**, 984 (1978).

GINGOLD, R. A., FAULKNER, D. J.: Thermal pulses in helium shell-burning stars. III. Ap. J. **188**, 145 (1974).

GINGOLD, R. A.: Asymptotic giant-branch evolution of a 0.6 $M_\odot$ star. Ap. J. **193**, 177 (1974).

HÄRM, R., SCHWARZSCHILD, M.: Transition from a red giant to a blue nucleus after ejection of a planetary nebula. Ap. J. **200**, 324 (1975).

HAINEBACH, K. L., et al.: On the e-process: its components and their neutron excesses. Ap. J. **193**, 157 (1974).

HAINEBACH, K. L., SCHRAMM, D. N., BLAKE, J. B.: Cosmic-ray spallative origin of the rare odd-odd nuclei, consistent with light-element production. Ap. J. **205**, 920 (1976).

HANSEN, C. J., VAN HORN, H. M.: Steady-state nuclear fusion in accreting neutron-star envelopes. Ap. J. **195**, 735 (1975).

HARRISON, T. G., EDWARDS, T. W.: Low-temperature photoneutron sources for stellar nucleosynthesis. Ap. J. **187**, 303 (1974).

HARRISON, T. G.: The low-temperature photonuclear nucleosynthesis of the bypassed (p-) nuclei in degenerate hydrogen burning zones and its relationship to nova outburst. Ap. J. Suppl. **36**, 199 (1978).

HARWIT, M., PACINI, F.: Infrared galaxies: evolutionary stages of massive star formation. Ap. J. **200**, L 127 (1975).

HAUBOLD, H. J., JOHN, R. W.: On the evaluation of an integral connected with the thermonuclear reaction rate in closed form. Astron. Nach. **299**, 225 (1978).

HAUBOLD, H. J., JOHN, R. W.: On resonant thermonuclear reaction rate integrals—closed-form evaluation and approximation considerations. Astron. Nach. **300**, 63 (1979).

HAVAZELET, D., BARKAT, Z.: On core mass-luminosity relations for shell helium burning stages of stellar evolution. Ap. J. **233**, 589 (1979).

HOWARD, A. J., et al.: Measurement and theoretical analysis of some reaction rates of interest in silicon burning. Ap. J. **188**, 131 (1974).

HOYLE, F., CLAYTON, D. D.: Nucleosynthesis in white-dwarf atmospheres. Ap. J. **191**, 705 (1974).

IBEN, I. Jr.: Neon-22 as a neutron source, light elements as modulators, and s-process nucleosynthesis in a thermally pulsating star. Ap. J. **196**, 549 (1975).

IBEN, I. Jr.: Thermal pulses; p-capture, α-capture, s-process nucleosynthesis; and convective mixing in a star of intermediate mass. Ap. J. **196**, 549 (1975).

ICKO, I. Jr.: Further adventures of a thermally pulsing star. Ap. J. **208**, 165 (1976).

ICKO, I. Jr.: Thermal pulse and interpulse properties of intermediate-mass stellar models with carbon-oxygen cores of mass 0.96, 1.16, and 1.36 $M_\odot$. Ap. J. **217**, 788 (1977).

ICKO, I. Jr.: Urca neutrino-loss rates under conditions found in the carbon-oxygen cores of intermediate-mass stars. Ap. J. **219**, 213 (1978).

ICKO, I. Jr.: Thermal oscillations during carbon burning in an electron-degenerate stellar core. Ap. J. **226**, 996 (1978).

ITOH, N., TOTSUJI, H., ICHIMARU, S.: Enhancement of thermonuclear reaction rate due to strong screening. Ap. J. **218**, 477 (1977).

JOHNS, O., REEVES, H.: The r-process production ratios of long-lived radionuclides. Ap. J. **202**, 214 (1975).

JOSS, P. C.: Helium-burning flashes on an accreting neutron star: a model for X-ray burst sources. Ap. J. **225**, L 123 (1978).

KU, W., et al.: Energy dependence of the size of the X-ray source in the Crab Nebula. Ap. J. **204**, L 77 (1976).

KINAHAN, B. F., HÄRM, R.: Chemical composition and the Hertzsprung gap. Ap. J. **200**, 330 (1975).

KRISHNA, S. K. S., STECHER, T. P.: Non-LTE H_2^+ as the source of missing opacity in the solar atmosphere. Ap. J. **194**, L 153 (1974).

KUAN, P., KUHI, L. V. P.: Cygni stars and mass loss. Ap. J. **199**, 148 (1975).

KUTTER, G. S., SPARKS, W. M.: Studies of hydrodynamic events in stellar evolution. III. Ejection of planetary nebulae. Ap. J. **192**, 447 (1974).

KWOK, S.: Radiation pressure on grains as a mechanism for mass loss in red giants. Ap. J. **198**, 583 (1975).

LAMB, D. Q., LAMB, F. K.: Nuclear burning in accreting neutron stars and X-ray bursts. Ap. J. **220**, 291 (1978).

LAMB, S. A., IBEN, I. Jr., HOWARD, W. M.: On the evolution of massive stars through the core carbon-burning phase. Ap. J. **207**, 209 (1976).

LAMB, S. A., et al.: Neutron-capture nucleosynthesis in the helium-burning cores of massive stars. Ap. J. **217**, 213 (1977).

LANGER, G. E., KRAFT, R. P., ANDERSON, K. S.: FG Sagittae: the s-process episode. Ap. J. **189**, 509 (1974).

LATOUR, J., et al.: Stellar convection theory. I. The anelastic modal equations. Ap. J. **207**, 233 (1976).

LAUTERBORN, D., SIQUIG, R.: Multiple solutions and secular stability of a 7 $M_\odot$ star with core helium and shell hydrogen burning. Ap. J. **187**, 299 (1974).

LAUTERBORN, D., SIQUIG, R. A.: Island solutions in linear series of static stellar models with core helium and shell hydrogen burning for $M = 5.7$, and 9 $M_\odot$. Ap. J. **191**, 589 (1974).

LAZAREFF, B., et al.: Hot CNO—Ne cycle hydrogen burning: explosive hydrogen burning in novae. Ap. J. **228**, 875 (1979).

LEVY, E. H., ROSE, W. K.: Production of magnetic fields in the interiors of stars and several effects on stellar evolution. Ap. J. **193**, 419 (1974).

MACKLIN, R. L., HALPERIN, J., WINTERS, R. R.: Neutron capture by ^{208}Pb at stellar temperatures. Ap. J. **217**, 222 (1977).

MAHAFFY, J. H., HANSEN, C. J.: Carbon detonations in rapidly rotating stellar cores. Ap. J. **201**, 695 (1975).

MATHEWS, G. J., VIOLA, V. E. Jr.: On the light-element abundances, galactic evolution, and the universal baryon density. Ap. J. **228**, 375 (1979).

MICHAUD, G., et al.: Diffusion in main-sequence stars: radiation forces, time scales, anomalies. Ap. J. **210**, 447 (1976).

MITALAS, R.: Destruction of ^{14}N by ^{14}N$(e,n)^{14}$C$(\alpha,\gamma)^{18}$O in degenerate matter. Ap. J. **187**, 155 (1974).

MITLER, H. E.: Thermonuclear ion-electron screening at all densities. I. Static solution. Ap. J. **212**, 513 (1977).

MONTMERLE, T., MICHAUD, G.: Diffusion in stars: ionization and abundance effects. Ap. J. Suppl. **31**, 489 (1976).

MONTMERLE, T.: Light-element production by cosmological cosmic rays. Ap. J. **217**, 878 (1977).

NEWMAN, M. J., FOWLER, W. A.: Maximum rate for the proton-proton reaction compatible with conventional solar models. Phys. Rev. Lett. **36**, 895 (1976).

NEWMAN, M. J., FOWLER, W. A.: Solar models of low neutrino counting rate: energy transport by processes other than radiative transfer. Ap. J. **207**, 601 (1976).

NEWMAN, M. J.: s-Process studies: the exact solution. Ap. J. **219**, 676 (1978).

NORMAN, E. B., SCHRAMM, D. N.: On the conditions required for the r-process. Ap. J. **228**, 881 (1979).

OLSON, G. L., PEÑA, H. J.: Nucleosynthesis and star formation of the galaxy and Magellanic clouds. Ap. J. **205**, 527 (1976).

OSMER, P. S., PETERSON, D. M.: The composition and evolutionary status of the helium-rich stars. Ap. J. **187**, 117 (1974).

PACZYŃSKI, B.: Helium-shell flashes in population I stars. Ap. J. **192**, 483 (1974).

PACZYŃSKI, B.: Core mass-interflash period relation for double shell source stars. Ap. J. **202**, 558 (1975).

PACZYŃSKI, B.: Helium shell flashes. Ap. J. **214**, 812 (1977).

PACZYŃSKI, B., TREMAINE, S. D.: Core helium flash and the origing of CH and carbon stars. Ap. J. **216**, 57 (1977).

PACZYŃSKI, B., ZYTKOW, A. N.: Hydrogen shell flashes in a white dwarf with mass accretion. Ap. J. **222**, 604 (1978).

PARDO, R. C., COUCH, R. G., ARNETT, W. D.: A study of nucleosynthesis during explosive carbon burning. Ap. J. **191**, 711 (1974).

PENZIAS, A. A., et al.: Deuterium in the galaxy. Ap. J. **211**, 108 (1977).

PFEIFFER, L., et al.: Indium-loaded liquid scintillator for low-energy solar-neutrino spectroscopy. Phys. Rev. Lett. **41**, 63 (1978).

RAMATY, R., KOZLOVSKY, B.: Deuterium, tritium, and helium-3 production in solar flares. Ap. J. **193**, 729 (1974).

REEVES, H., JOHNS, O.: The long-lived radioisotopes as monitors of stellar, galactic, and cosmological phenomena. Ap. J. **206**, 958 (1976).

RIOS, M., SCHWEITZER, J. S., ANDERSON, B. D.: Stellar rates for some reactions of interest in astrophysics. Ap. J. **199**, 173 (1975).

ROBERTSON, J. W.: Core-helium-burning stars in young clusters in the large Magellanic cloud. Ap. J. **191**, 67 (1974).

ROLFS, C., RODNEY, W. S.: Experimental evidence for CNO tri-cycling. Ap. J. **194**, L 63 (1974).

ROOD, R. T., STEIGMAN, G., TINSLEY, B. M.: Stellar production as a source of ^{3}He in the interstellar medium. Ap. J. **207**, L 57 (1976).

ROUGHTON, N. A., et al.: Stellar reaction rates for proton capture on ^{28}Si, ^{50}CR, ^{54}Fe, ^{58}Ni, ^{60}Wi and ^{61}Ni. Ap. J. **193**, 187 (1974).

ROUGHTON, N. A., et al.: Thick-target measurement of the $(p.\gamma)$ stellar reaction rates of the nuclides ^{12}C, ^{29}Si, ^{46}Ti, ^{47}Ti, and ^{56}Fe. Ap. J. **188**, 595 (1974).

ROUGHTON, N. A., et al.: Thermonuclear reaction rates derived from thick target yields. Ap. J. **205**, 302 (1976).

SACKMANN, I.-J.: What quenches the helium shell flashes? Ap. J. **212**, 159 (1977).

SCALO, J. M., DESPAIN, K. H., ULRICH, R. K.: Studies of evolved stars. V. Nucleosynthesis in hot-bottom convective envelopes. Ap. J. **196**, 805 (1975).

SCALO, J. M., ULRICH, R. K.: The effect of composition changes on evolutionary tracks of double-shell models. Ap. J. **200**, 682 (1975).

SCHLESINGER, B. M.: Constraints on the evolutionary history of stars showing s-processed material. Ap. J. **188**, 141 (1974).

SCHLESINGER, B. M.: Patterns of convection in the evolution of massive stars. Ap. J. **199**, 166 (1975).

SCHLESINGER, B. M.: The hydrogen profile, previous mixing, and loops in the H—R diagram during core helium burning. Ap. J. **212**, 507 (1977).

SCHRAMM, D. N., TINSLEY, B. M.: On the origin and evolution of s-process elements. Ap. J. **193**, 151 (1974).

Schwarzschild, M.: On the scale of photospheric convection in red giants and supergiants. Ap. J. **195**, 137 (1975).

Share, G. H., Kinzer, R. L., Seeman, N.: Diffuse cosmic gamma radiation above 10 MeV. Ap. J. **187**, 511 (1974).

Share, G. H., Kinzer, R. L., Seeman, N.: Observation of gamma-radiation from the galactic center region. Ap. J. **187**, 45 (1974).

Silberberg, R., Tsao, C. H.: Composition of methods for calculating cross sections at high energies in astrophysics. Ap. J. Suppl. **35**, 137 (1977).

Silberberg, R., Tsao, C. H.: Cross sections for (p, xn) reactions, and astrophysical applications. Ap. J. Suppl. **35**, 129 (1977).

Sion, E. M., Vila, S. C.: The pulsational properties of high-luminosity degenerate stars with helium burning near the surface. Ap. J. **209**, 850 (1976).

Smith, R. L., Gonsiorowski, A.: Stellar mixing and s-process nucleosynthesis. Ap. J. **211**, 900 (1977).

Sparks, W. M.: Studies of hydrodynamic events in stellar evolution. III. Ejection of planetary nebulae. Ap. J. **192**, 447 (1974).

Sparks, W. M., Stecher, T. P.: The result of the death spiral of a white dwarf into a red giant. Ap. J. **188**, 149 (1974).

Starrfield, S., Sparks, W. M., Truran, J. W.: CNO abundances and hydrodynamic models of the nova outburst. II. 1.00 $M_\odot$ models with enhanced carbon and oxygen. Ap. J. Suppl. **28**, 247 (1974).

Starrfield, S., Sparks, W. M., Truran, J. W.: CNO abundances and hydrodynamic models of the nova outburst. III. 0.5 $M_\odot$ models with enhanced carbon, oxygen, and nitrogen. Ap. J. **192**, 647 (1974).

Stothers, R.: A comparison of homogeneous stellar models based on the Cox-Stewart and Carson opacities. Ap. J. **194**, 695 (1974).

Stothers, R.: Violation of the Vogt-Russell theorem for homogeneous nondegenerate stars. Ap. J. **194**, 699 (1974).

Stothers, R., Chin, C.-W.: Stellar evolution at high mass with semiconvective mixing according to the Ledoux criterion. Ap. J. **198**, 407 (1975).

Stothers, R.: Excitation of pulsations in the CNO ionization zone of luminous stars. Ap. J. **204**, 853 (1976).

Stothers, R., Chin, C.-W.: Stellar evolution at high mass with semiconvective mixing according to the Schwarzschild criterion. Ap. J. **204**, 472 (1976).

Stothers, R., Chin, C.-W.: Does the upper main sequence extend across the whole H—R diagram? Ap. J. **211**, 189 (1977).

Stothers, R., Chin, C.-W.: Evolution of helium stars. Ap. J. **216**, 61 (1977).

Stothers, R., Chin, C.-W.: Stellar evolution at high mass including the effect of a stellar wind. Ap. J. **233**, 267 (1979).

Straus, J. M., Blake, J. B., Schramm, D. N.: Effects of convective overshoot on lithium depletion in main-sequence stars. Ap. J. **204**, 481 (1976).

Sweigart, A. V.: Do helium-shell flashes cause extensive mixing in low-mass stars? Ap. J. **189**, 289 (1974).

Sweigart, A. V., Gross, P. G.: Horizontal-branch evolution with semiconvection. I. Interior evolution. Ap. J. **190**, 101 (1974).

Taam, R. E., Kraft, R. P., Suntzeff, N.: The origin and evolution of RR Lyrae stars of high metal abundance. Ap. J. **207**, 201 (1976).

Tarbell, T. D., Rood, R. T.: The triple-alpha rate, screening factors, and the helium flash. Ap. J. **199**, 443 (1975).

Tassoul, M.: Nuclear reactions in carbon-rich stars. Ap. J. **202**, 755 (1975).

Thorne, K. S., Zytkow, A. N.: Red giants and supergiants with degenerate neutron cores. Ap. J. **199**, L 19 (1975).

Tinsley, B. M., Gunn, J. E.: Luminosity functions and the evolution of low-mass population I giants. Ap. J. **206**, 525 (1976).

Trimble, V., Reines, F.: The solar neutrino problem—A progress (?) report. Rev. Mod. Phys. **45**, 1 (1973).

Trimble, V.: The origin and abundances of the chemical elements. Rev. Mod. Phys. **47**, 877 (1975).

Trivedi, B. M. P.: Mass range of supernovae for r-process nucleosynthesis. Ap. J. **225**, 209 (1978).

TRURAN, J. W., IBEN, I. Jr.: On s-process nucleosynthesis in thermally pulsing stars. Ap. J. **216**, 797 (1977).

TRURAN, J. W., CAMERON, A. G. W.: ^{26}Al production in explosive carbon burning. Ap. J. **219**, 226 (1978).

TRURAN, J. W., COWAN, J. J., CAMERON, A. G. W.: The helium-driven r-process in supernovae. Ap. J. **222**, L 63 (1978).

ULRICH, R. K.: Solar models with low neutrino fluxes. Ap. J. **188**, 369 (1974).

VILA, S. C.: Thermal stability of hydrogen-burning shells in white dwarfs. Ap. J. **217**, 171 (1977).

VLIEKS, E. A., MORGAN, J. F., BLATT, S. L.: Reaction rates of interest in late stages of stellar nucleosynthesis. Ap. J. **191**, 699 (1974).

WAGNER, R. L.: Theoretical evolution of extremely metal-poor stars. Ap. J. **191**, 173 (1974).

WARD, R. A., NEWMAN, M. J., CLAYTON, D. D.: s-Process studies: branching and the time scale. Ap. J. Suppl. **31**, 33 (1976).

WARD, R. A.: The importance of long-lived isomeric states in s-process branching. Ap. J. **216**, 540 (1977).

WARD, R. A., NEWMAN, M. J.: s-Process studies: the effects of a pulsed neutron flux. Ap. J. **219**, 195 (1978).

WEAVER, T. A., ZIMMERMAN, G. B., WOOSLEY, S. E.: Presupernova evolution of massive stars. Ap. J. **225**, 1021 (1978).

WEBBINK, R. F.: The evolution of low-mass close binary systems. II. $1.50\,M_\odot + 0.75\,M_\odot$: evolution into contact. Ap. J. Suppl. **32**, 583 (1976).

WHEELER, J. C., CAMERON, A. G. W.: The effect of primordial hydrogen/helium fractionation on the solar neutrino flux. Ap. J. **196**, 601 (1975).

WOOSLEY, S. E., HOWARD, W. M.: The p-process in supernovae. Ap. J. Suppl. **36**, 285 (1978).

YAHIL, A., BEAUDET, G.: Big-bang nucleosynthesis with nonzero Lepton numbers. Ap. J. **206**, 26 (1976).

YANG, J., et al.: Constraints on cosmology and neutrino physics from big bang nucleosynthesis. Ap. J. **227**, 697 (1979).

YORK, D. G., ROGERSON, J. B. Jr.: The abundance of deuterium relative to hydrogen in interstellar space. Ap. J. **203**, 378 (1976).

Chapter 4, Part 2

ADAIR, R. K., et al.: Determination of the neutron-proton ratio in primary cosmic rays. Phys. Rev. Lett. **39**, 112 (1977).

ALLKOFER, O. C., et al.: Sea-level muon charge ratio of cosmic rays at high energies. Phys. Rev. Lett. **41**, 832 (1978).

ALME, M. L., WILSON, J. R.: A possible mechanism for mass transfer in X-ray binary systems with OB supergiant companions. Ap. J. **210**, 233 (1976).

ARNETT, W. D., SCHRAMM, D. N.: Origin of cosmic rays, atomic nuclei, and pulsars in explosions of massive stars (Erratum). Ap. J. **187**, L 47 (1974).

ARNETT, W. D.: Supernova remnants and presupernova models. Ap. J. **195**, 727 (1975).

ARNETT, W. D.: On the theory of type I supernovae. Ap. J. **230**, L 37 (1979).

ARONS, J., LEA, S. M.: Accretion onto magnetized neutron stars: structure and interchange instability of a model magnetosphere. Ap. J. **207**, 914 (1976).

AVNI, Y., BAHCALL, J. N.: Masses for Vela X-1 and other X-ray binaries. Ap. J. **202**, L 131 (1975).

AXFORD, W. I., et al.: Cosmic-ray gradients from Pioneer-10 and Pioneer-11. Ap. J. **210**, 603 (1976).

BAAN, W. A.: Neutron stars as X-ray burst sources. Ap. J. **214**, 245 (1977).

BADHWAR, G. D., STEPHENS, S. A.: Hydrostatic equilibrium of gas, extent of cosmic ray confinement, and radio emission in the galaxy. Ap. J. **212**, 494 (1977).

BADHWAR, G. D., et al.: The cosmic-ray antiproton flux: an upper limit near that predicted for secondary production. Ap. J. **217**, L 135 (1977).

BAHCALL, J. N., et al.: Multiple star systems and X-ray sources. Ap. J. **189**, L 17 (1974).

BAHCALL, J. N., SARAZIN, C. L.: Parameters and predictions for the X-ray emitting gas of Coma, Perseus, and Virgo. Ap. J. **213**, L 99 (1977).

BAHCALL, J. N.: Masses of neutron stars and black holes in X-ray binaries. Ann. Rev. Astron. Ap. **16**, 241 (1978).
BAHCALL, N. A.: X-ray clusters of galaxies: correlations with optical morphology and galaxy density. Ap. J. **217**, L 77 (1977).
BAHCALL, N. A.: X-ray clusters of galaxies: correlations of X-ray luminosity with galactic content. Ap. J. **218**, L 93 (1977).
BAHCALL, N. A.: The X-ray luminosity function of clusters of galaxies: predictions from a thermal bremsstrahlung model. Ap. J. **232**, L 83 (1979).
BENVENUTI, A., et al.: Measurements of neutrino and antineutrino cross sections at high energies. Phys. Rev. Lett. **32**, 125 (1974).
BIGNAMI, G. F., et al.: High-energy galactic gamma radiation from cosmic rays concentrated in spiral arms. Ap. J. **199**, 54 (1975).
BLAKE, J. B., et al.: Ultraheavy cosmic rays: theoretical implications of recent observations. Ap. J. **221**, 694 (1978).
BLANDFORD, R. D., OSTRIKER, J. P.: Particle acceleration by astrophysical shocks. Ap. J. **221**, L 29 (1978).
BOWERS, R. L.: Gravitationally redshifted gamma rays and neutron star masses. Ap. J. **216**, L 63 (1977).
BRECHER, K., MORRISON, P.: Cosmic gamma-ray bursts from directed stellar flares. Ap. J. **187**, L 97 (1974).
BUFFINGTON, A., ORTH, C. D., SMOOT, G. F.: Measurement of the positron-electron ratio in the primary cosmic rays from 5 to 50 GeV. Phys. Rev. Lett. **33**, 34 (1974).
BUFFINGTON, A., ORTH, C. D., SMOOT, G. F.: Measurement of primary cosmic-ray electrons and positrons from 4 to 50 GeV. Ap. J. **199**, 669 (1975).
BUFFINGTON, A., ORTH, C. D., MAST, T. S.: Relativistic Be^7: a probe of cosmic-ray acceleration? Phys. Rev. Lett. **41**, 594 (1978).
CASSÉ, M., SOUTOUL, A.: Time delay between explosive nucleosynthesis and cosmic-ray acceleration. Ap. J. **200**, L 75 (1975).
CAVALLO, G., JELLEY, J. V.: Why are no radio pulses associated with the bursts of celestial gamma rays? Ap. J. **201**, L 113 (1975).
CHAN, J. H., PRICE, P. B.: Composition and energy spectra of heavy nuclei of unknown origin detected on Skylab. Phys. Rev. Lett. **35**, 539 (1975).
CHEVALIER, R. A.: The evolution of supernova remnants. I. Spherically symmetric models. Ap. J. **188**, 501 (1974).
CHEVALIER, R. A.: The evolution of supernova remnants. III. Thermal waves. Ap. J. **198**, 355 (1975).
CHEVALIER, R. A.: The evolution of supernova remnants. IV. The supernova ejecta. Ap. J. **200**, 698 (1975).
CHEVALIER, R. A.: The hydrodynamics of type II supernovae. Ap. J. **207**, 872 (1976).
CHEVALIER, R. A.: Cassiopeia A, faint supernovae, and heavy-element ejection by supernovae. Ap. J. **208**, 826 (1976).
CHEVALIER, R. A., ROBERTSON, J. W., SCOTT, J. S.: Cosmic-ray acceleration and the radio evolution of Cassiopeia A. Ap. J. **207**, 450 (1976).
CLARK, G. W.: X-ray binaries in globular clusters. Ap. J. **199**, L 143 (1975).
CLAYTON, D. D.: Line ^{57}Co gamma rays: new diagnostic of supernova structure. Ap. J. **188**, 155 (1974).
CLAYTON, D. D., HOYLE, F.: Gamma-ray lines from novae. Ap. J. **187**, L 101 (1974).
CLAYTON, D. D.: Gamma-ray lines: a ^{22}Na radioactive diagnostic of young supernovae. Ap. J. **198**, 151 (1975).
CLAYTON, D. D., DWEK, E.: Gamma-ray emission and nucleosynthesis of lithium by young pulsars. Ap. J. **206**, L 59 (1976).
CLINE, T. L., et al.: Helios 2-vela-ariel 5 gamma-ray burst source position. Ap. J. **229**, L 47 (1979).
CLINE, T. L., et al.: Gamma-ray burst observations from Helios 2. Ap. J. **232**, L 1 (1979).
COLGATE, S. A.: The formation of deuterium and the light elements by spallation in supernova shocks. Ap. J. **187**, 321 (1974).
COWIE, L. L., OSTRIKER, J. P., STARK, A. A.: Time-dependent spherically symmetric accretion onto compact X-ray sources. Ap. J. **226**, 1041 (1978).
COWSIK, R., LEE, M. A.: On the sources of cosmic ray electrons. Ap. J. **228**, 297 (1979).

COX, D. P., SMITH, B. W.: Large-scale effects of supernova remnants on the galaxy: generation and maintenance of a hot network of tunnels. Ap. J. **189**, L 105 (1974).

CRANNELL, C. J., et al.: Formation of the 0.511 MeV line in solar flares. Ap. J. **210**, 582 (1976).

DAUGHERTY, J. K., HARTMAN, R. C., SCHMIDT, P. J.: A measurement of cosmic-ray positron and negatron spectra between 50 and 800 MV. Ap. J. **198**, 493 (1975).

DEGREGORIA, A. J.: An investigation of accretion of matter onto white dwarfs as a possible X-ray mechanism. Ap. J. **189**, 555 (1974).

DENOYER, L. K.: Neutral hydrogen associated with supernova remnants. I. The Cygnus loop. Ap. J. **196**, 479 (1975).

DICKEL, J. R.: Do supernova remnants provide the cosmic-ray electrons? Ap. J. **193**, 755 (1974).

DICUS, D. A., et al.: Neutrino pair bremsstrahlung including neutral current effects. Ap. J. **210**, 481 (1976).

DWYER, R., MEYER, P.: Isotopic composition of cosmic-ray nitrogen at 1.5 GeV/amu. Phys. Rev. Lett. **35**, 601 (1975).

EARDLEY, D. M., LIGHTMAN, A. P.: Magnetic viscosity in relativistic accretion disks. Ap. J. **200**, 187 (1975).

EILEK, J. A., CAROFF, L. J.: Cloud acceleration by cosmic rays in the vicinity of compact luminous objects. Ap. J. **208**, 887 (1976).

EPSTEIN, R. I., ARNETT, W. D., SCHRAMM, D. N.: Can supernovae produce deuterium? Ap. J. **190**, L 13 (1974).

EPSTEIN, R. I., PETROSIAN, V.: Effects of primordial fluctuations on the abundances of the light elements. Ap. J. **197**, 281 (1975).

EPSTEIN, R. I.: Deuterium production by high-energy particles. Ap. J. **212**, 595 (1977).

EVANS, W. D., BELIAN, R. D., CONNER, J. P.: Observations of intense cosmic X-ray bursts. Ap. J. **207**, L 91 (1976).

FELTEN, J. E., GOULD, R. J.: The effect of repeated Compton scatterings on the diffuse X-ray background. Ap. J. **194**, L 39 (1974).

FICHTEL, C. E., et al.: Significance of medium-energy gamma-ray astronomy in the study of cosmic rays. Ap. J. **208**, 211 (1976).

FICHTEL, C. E., et al.: SAS-2 observations of the diffuse gamma radiation in the galactic latitude interval $10° < |b| \leqslant 90°$. Ap. J. **217**, L 9 (1977).

FISHER, A. J., et al.: The isotopic composition of cosmic rays with $5 \leqslant Z \leqslant 26$. Ap. J. **205**, 938 (1976).

FISK, L. A., KOZLOVSKY, B., RAMATY, R.: An interpretation of the observed oxygen and nitrogen enhancements in low-energy cosmic rays. Ap. J. **190**, L 35 (1974).

FLEISCHER, R. L., HART, H. R. Jr., RENSHAW, A.: Composition of heavy cosmic rays from 25 to 180 MeV per atomic mass unit. Ap. J. **193**, 575 (1974).

FLOWERS, E. G., SUTHERLAND, P. G.: Neutrino-neutrino scattering and supernovae. Ap. J. **208**, L 19 (1976).

FUCHS, B., SCHLICKEISER, R., THIELHEIM, K. O.: The structure of the galactic disk and its implications for gamma-ray astronomy. Ap. J. **206**, 589 (1976).

GAFFET, B.: Pulsar theory of supernova light curves. I. Dynamical effect and thermalization of the pulsar strong waves. Ap. J. **216**, 565 (1977).

GALLAGHER, J. S., STARRFIELD, S.: Theory and observations of classical novae. Ann. Rev. Astron. Ap. **16**, 171 (1978).

GARCIA-MUNOZ, M., MASON, G. M., SIMPSON, J. A.: The cosmic-ray age deduced from the ^{10}Be abundance. Ap. J. **201**, L 141 (1975).

GARCIA-MUNOZ, M., MASON, G. M., SIMPSON, J. A.: The isotopic composition of galactic cosmic-ray lithium, beryllium, and boron. Ap. J. **201**, L 145 (1975).

GARCIA-MUNOZ, M., MASON, G. M., SIMPSON, J. A.: The age of the galactic cosmic rays derived from the abundance of ^{10}Be. Ap. J. **217**, 859 (1977).

GIACCONI, R., et al.: The third UHURU catalog of X-ray sources. Ap. J. Suppl. **27**, 37 (1974).

GIACCONI, R., et al.: A high-sensitivity X-ray survey using the Einstein observatory and the discrete source contribution to the extragalactic X-ray background. Ap. J. **234**, L 1 (1979).

GINZBURG, V. L., PTUSKIN, V. S.: On the origin of cosmic rays: some problems in high-energy astrophysics. Rev. Mod. Phys. **48**, 161 (1976).

GOLDEN, R. L., et al.: Rigidity spectrum of $Z \geqslant 3$ cosmic-ray nuclei in the range 4 to 285 GV and a search for cosmic antimatter. Ap. J. **192**, 747 (1974).

GOLDEN, R. L., et al.: Evidence for the existence of cosmic-ray antiprotons. Phys. Rev. Lett. **43**, 1196 (1979).

GOODMAN, J. A., et al.: Composition of primary cosmic rays above 10^{13} eV from the study of time distributions of energetic Hadrons near air-shower cores. Phys. Rev. Lett. **42**, 854 (1979).

GOULD, R. J.: Energy loss of relativistic electrons and positrons traversing cosmic matter. Ap. J. **196**, 689 (1975).

GOULD, R. J., REPHAELI, Y.: The effective penetration distance of ultrahigh-energy electrons and photons traversing a cosmic blackbody photon gas. Ap. J. **225**, 318 (1978).

GRINDLAY, J. E., FAZIO, G. G.: Cosmic gamma-ray bursts from relativistic dust grains. Ap. J. **187**, L 93 (1974).

GRINDLAY, J. E.: Thermal limit for spherical accretion and X-ray bursts. Ap. J. **221**, 234 (1978).

HAGEN, F. A., FISHER, A. J., ORMES, J. F.: ^{10}Be abundance and the age of cosmic rays: a balloon measurement. Ap. J. **212**, 262 (1977).

HAINEBACH, K. L., NORMAN, E. B., SCHRAMM, D. N.: Consistency of cosmic-ray composition, acceleration mechanism, and supernova models. Ap. J. **203**, 245 (1976).

HALL, R. D., et al.: Detection of nuclear gamma rays from Centaurus A. Ap. J. **210**, 631 (1976).

HARTMAN, R. C., et al.: Galactic plane gamma-radiation. Ap. J. **230**, 597 (1979).

HATCHETT, S., WEAVER, R.: Structure of the iron fluorescence line in X-ray binaries. Ap. J. **215**, 285 (1977).

HAYAKAWA, S., et al.: Observation of the diffuse component of cosmic soft X-rays. Ap. J. **195**, 535 (1975).

HAYMES, R. C., et al.: Detection of nuclear gamma rays from the galactic center region. Ap. J. **201**, 593 (1975).

HEDRICK, D., COX, D. P.: Galactic infall and cosmic ray acceleration. Ap. J. **215**, 208 (1977).

HENRY, J. P., et al.: Detection of X-ray emission from distant clusters of galaxies. Ap. J. **234**, L 15 (1979).

HIGDON, J. C., LINGENFELTER, R. E.: The origin of cosmic rays and the Vela gamma-ray excess. Ap. J. **198**, L 17 (1975).

HIGDON, J. C., LINGENFELTER, R. E.: Nuclear γ-ray lines in accretion source spectra. Ap. J. **215**, L 53 (1977).

HIGDON, J. C.: Distribution of cosmic rays and magnetic field in the galaxy as deduced from synchrotron radio and gamma-ray observations. Ap. J. **232**, 113 (1979).

HOGAN, C., LAYZER, D.: Origin of the X-ray background. Ap. J. **212**, 360 (1977).

IPAVICH, F. M.: Galactic winds driven by cosmic rays. Ap. J. **196**, 107 (1975).

ISENBERG, P. A.: Adiabatic self-similar blast waves, their radial instabilities, and their application to supernova remnants. Ap. J. **217**, 597 (1977).

JENKINS, E. B., SILK, J., WALLERSTEIN, G.: Interaction of the Vela supernova remnant with the cloudy interstellar medium. Ap. J. **209**, L 87 (1976).

JOKIPII, J. R.: Consequences of a lifetime greater than 10^7 years for galactic cosmic rays. Ap. J. **208**, 900 (1976).

JONES, C.: Energy spectra of 43 galactic X-ray sources observed by Uhuru. Ap. J. **214**, 856 (1977).

KANE, S. R., ANDERSON, K. A.: Characteristics of cosmic X-ray bursts observed with the OGO-5 satellite. Ap. J. **210**, 875 (1976).

KATZ, J. I., SALPETER, E. E.: X-ray emission from vibrating white dwarfs. Ap. J. **193**, 429 (1974).

KATZ, J. I.: The origin of X-ray sources in clusters of galaxies. Ap. J. **207**, 25 (1976).

KATZ, J. I.: X-rays from spherical accretion onto degenerate dwarfs. Ap. J. **215**, 265 (1977).

KELLOGG, E., BALDWIN, J. R., KOCH, D.: Studies of cluster X-ray sources, energy spectra for the Perseus, Virgo, and Coma clusters. Ap. J. **199**, 299 (1975).

KELLOGG, E. M.: X-ray astronomy in the Uhuru epoch and beyond. Ap. J. **197**, 689 (1975).

KELLOGG, E. M.: Primeval gas clouds and the low-energy X-ray background. Ap. J. **218**, 582 (1977).

KESTENBAUM, H. L., et al.: Evidence for X-ray iron line emission in Cygnus X-3 obtained with a crystal spectrometer. Ap. J. **216**, L 19 (1977).

KIRSHNER, R. P.: Spectrophotometry of the Crab Nebula. Ap. J. **194**, 323 (1974).

KIRSHNER, R. P., KWAN, J.: The envelopes of Type II supernovae. Ap. J. **197**, 415 (1975).

KOZLOVSKY, B., RAMATY, R.: 478-keV and 431-keV line emissions from alpha-alpha reactions. Ap. J. **191**, L 43 (1974).

LAMB, D. Q., et al.: Neutron star wobble in binary X-ray sources. Ap. J. **198**, L 21 (1975).

Lamb, D. Q., Pethick, C. J.: Effects of neutrino degeneracy in supernova models. Ap. J. **209**, L 77 (1976).
Lamb, F. K., et al.: A model for bursting X-ray sources: time-dependent accretion by magnetic neutron stars and degenerate dwarfs. Ap. J. **217**, 197 (1977).
Langer, S. H., Petrosian, V.: Impulsive solar X-ray bursts. III. Polarization, directivity, and spectrum of the reflected and total bremsstrahlung radiation from a beam of electrons directed toward the photosphere. Ap. J. **215**, 666 (1977).
Lattimer, J. M., Schramm, D. N., Grossman, L.: Condensation in supernova ejecta and isotopic anomalies in meteorites. Ap. J. **219**, 230 (1978).
Lea, S. M.: Pulsating X-ray sources: slowly rotating neutron stars? Ap. J. **209**, L 69 (1976).
Leventhal, M., MacCallum, C. J., Stang, P. D.: Detection of 511 keV positron annihilation radiation from the galactic center direction. Ap. J. **225**, L 11 (1978).
Levine, A., et al.: On the ultrasoft X-ray background. Ap. J. **205**, 226 (1976).
Liang, E. P. T.: Convective accretion disks and X-ray bursters. Ap. J. **218**, 243 (1977).
Lingenfelter, R. E., Ramaty, R.: Gamma-ray lines from interstellar grains. Ap. J. **211**, L 19 (1977).
Lightman, A. P., Shapiro, S. L.: Spectrum and polarization of X-rays from accretion disks around black holes. Ap. J. **198**, L 73 (1975).
McCray, R., Hatchett, S.: Mass transfer in binary X-ray systems. Ap. J. **199**, 196 (1975).
McCray, R., Stein, R. F., Kafatos, M.: Thermal instability in supernova shells. Ap. J. **196**, 565 (1975).
McDonald, F. B., et al.: The anomalous abundance of cosmic-ray nitrogen and oxygen nuclei at low energies. Ap. J. **187**, L 105 (1974).
Maehl, R. C., et al.: Neutron-rich isotopes of cosmic rays with $9 \leqslant Z \leqslant 16$. Ap. J. **202**, L 119 (1975).
Malina, R., Lampton, M., Bowyer, S.: Soft X-ray morphology of the Virgo, Coma, and Perseus clusters of galaxies. Ap. J. **209**, 678 (1976).
Maraschi, L., Reina, C., Treves, A.: The effect of radiation pressure on accretion disks around black holes. Ap. J. **206**, 295 (1976).
Maraschi, L., Treves, A.: Temperature of an accretion disk around a black hole near the Eddington luminosity. Ap. J. **211**, 263 (1977).
Maraschi, L., Treves, A.: Gamma rays from accreting black holes. Ap. J. **218**, L 113 (1977).
Margolis, S. H., Schramm, D. N., Silberberg, R.: Ultrahigh-energy neutrino astronomy. Ap. J. **221**, 990 (1978).
Mathews, W. G.: The enormous mass of the elliptical galaxy M 87: a model for the extended X-ray source. Ap. J. **219**, 413 (1978).
Maza, J., van den Bergh, S.: Statistics of extragalactic supernovae. Ap. J. **204**, 519 (1976).
Mazurek, T. J.: Pauli constriction of the low-energy window in neutrino supernova models. Ap. J. **207**, L 87 (1976).
Meegan, C. A., Earl, J. A.: The spectrum of cosmic electrons with energies between 6 and 100 GeV. Ap. J. **197**, 219 (1975).
Metzger, A. E., et al.: Observation of a cosmic gamma-ray burst on Apollo 16. I. Temporal variability and energy spectrum. Ap. J. **194**, L 19 (1974).
Mewaldt, R. A., et al.: Isotopic and elemental composition of the anomalous low-energy cosmic-ray fluxes. Ap. J. **205**, 931 (1976).
Mewaldt, R. A., Stone, E. C., Vogt, R. E.: The isotopic composition of hydrogen and helium in low-energy cosmic rays. Ap. J. **206**, 616 (1976).
Michalsky, J. J., Swedlund, J. B., Stokes, R. A.: Cygnus X-1: discovery of variable circular polarization. Ap. J. **198**, L 101 (1975).
Michel, F. C.: Accretion magnetospheres: general solutions. Ap. J. **213**, 836 (1977).
Michel, F. C.: Accretion magnetosphere stability. I. Adiabatic gas model. Ap. J. **214**, 261 (1977).
Michel, F. C.: Accretion magnetosphere stability. II. Polar cap "drip". Ap. J. **216**, 838 (1977).
Mikaelian, K. O.: Supernova explosions, the new Leptons, and right-handed neutrinos. Phys. Rev. Lett. **36**, 1089 (1976).
Miller, J. S.: Spectrophotometry of filaments in the Crab Nebula. Ap. J. **220**, 490 (1978).
Montmerle, T.: On the ability of current experiments to test π^0 decay gamma-ray background theories. Ap. J. **197**, 285 (1975).
Montmerle, T.: On the possible existence of cosmological cosmic rays. I. The framework for light-element and gamma-ray production. Ap. J. **216**, 177 (1977).

Montmerle, T.: On the possible existence of cosmological cosmic rays. II. The observational constraints set by the γ-ray background spectrum and the lithium and deuterium abundances. Ap. J. **216**, 620 (1977).

Montmerle, T.: On the possible existence of cosmological cosmic rays. III. Nuclear γ-ray production. Ap. J. **218**, 263 (1977).

Moore, W. E., Garmire, G. P.: The X-ray stucture of the Vela supernova remnant. Ap. J. **199**, 680 (1975).

Mullan, D. J.: Flares on white dwarfs and gamma-ray bursts. Ap. J. **208**, 199 (1976).

Muraki, Y., et al.: Measurement of cosmic-ray muon spectrum and charge ratio at large zenith angles in the momentum range 100 GeV/c to 10 TeV/c using a magnet spectrograph. Phys. Rev. Lett. **43**, 974 (1979).

Noerdlinger, P. D.: Positrons in compact radio sources. Phys. Rev. Lett. **41**, 135 (1978).

Novaco, J. C., Brown, L. W.: Nonthermal galactic emission below 10 megahertz. Ap. J. **221**, 114 (1978).

Ögelman, H. B., Maran, S. P.: The origin of OB associations and extended regions of high-energy activity in the galaxy through supernova cascade processes. Ap. J. **209**, 124 (1976).

Ormes, J., Freier, P.: On the propagation of cosmic rays in the galaxy. Ap. J. **222**, 471 (1978).

Orth, C. D., Buffington, A.: Secondary cosmic-ray $e^{\pm}$ from 1 to 100 GeV in the upper atmosphere and interstellar space, and interpretation of a recent e^{+} flux measurement. Ap. J. **206**, 312 (1976).

Ostriker, J. P., et al.: A new luminosity limit for spherical accretion onto compact X-ray sources. Ap. J. **208**, L 61 (1976).

Otgonsuren, O., et al.: Abundances of $Z > 52$ nuclei in galactic cosmic rays: long-term averages based on studies of pallasites. Ap. J. **210**, 258 (1976).

Owens, A. J., Jokipii, J. R.: Cosmic rays in a dynamical halo. I. Age and matter traversal distributions and anisotropy for nuclei. Ap. J. **215**, 677 (1977).

Owens, A. J., Jokipii, J. R.: Cosmic rays in a dynamical halo. II. Electrons. Ap. J. **215**, 685 (1977).

Pacini, F., Salvati, M.: The active region in galactic nuclei: a spinar model. Ap. J. **225**, L 99 (1978).

Paczyński, B.: A model of accretion disks in close binaries. Ap. J. **216**, 822 (1977).

Paul, J., Cassé, M., Cerarsky, C. J.: Distribution of gas, magnetic fields, and cosmic rays in the galaxy. Ap. J. **207**, 62 (1976).

Petterson, J. A.: Twisted accretion disks. I. Derivation of the basic equations. Ap. J. **214**, 550 (1977).

Petterson, J. A.: Twisted accretion disks. II. Applications to X-ray binary systems. Ap. J. **216**, 827 (1977).

Podosek, F. A.: Isotopic structures in solar system materials. Ann. Rev. Astron. Ap. **16**, 293 (1978).

Puget, J. L., Stecker, F. W.: The distribution of cosmic rays in the galaxy and their dynamics as deduced from recent γ-ray observations. Ap. J. **191**, 323 (1974).

Puget, J. L., Stecker, F. W., Bredekamp, J. H.: Photonuclear interactions of ultrahigh energy cosmic rays and their astrophysical consequences. Ap. J. **205**, 638 (1976).

Ramaty, R., Kozlovsky, B., Lingenfelter, R. E.: Nuclear gamma-rays from energetic particle interactions. Ap. J. Suppl. **40**, 487 (1979).

Reeves, H., Meyer, J.-P.: Cosmic-ray nucleosynthesis and the infall rate of extragalactic matter in the solar neighborhood. Ap. J. **226**, 613 (1978).

Rephaeli, Y.: On Compton and thermal models for X-ray emission from clusters of galaxies. Ap. J. **218**, 323 (1977).

Ricker, G. R., et al.: High-energy X-ray observations of a lunar occultation of the Crab Nebula. Ap. J. **197**, L 83 (1975).

Rocchia, R., Ducros, R., Gaffet, B.: Spectrum and origin of X- and gamma-ray diffuse background. Ap. J. **209**, 350 (1976).

Rose, W. K., Scott, E. H.: Magnetic fields and the nova outburst. Ap. J. **204**, 516 (1976).

Rosner, R., Vaiana, G. S.: Cosmic flare transients: constraints upon models for energy storage and release derived from the event frequency distribution. Ap. J. **222**, 1104 (1978).

Rowan-Robinson, M.: On the unity of activity in galaxies. Ap. J. **213**, 635 (1977).

Sarazin, C. L., Bahcall, J. N.: X-ray line emission for clusters of galaxies. II. Numerical models. Ap. J. Suppl. **34**, 451 (1977).

Sarazin, C. L., Bahcall, J. N.: On the Zeeman splitting of X-ray lines by neutron-star magnetic fields. Ap. J. **216**, L 67 (1977).

SCHMIDT, G. D., ANGEL, J. R. P., BEAVER, E. A.: The small-scale polarization of the Crab Nebula. Ap. J. **227**, 106 (1979).

SCHWARTZ, D. A., MURRAY, S. S., GURSKY, H.: A measurement of fluctuations in the X-ray background by Uhuru. Ap. J. **204**, 315 (1976).

SERLEMITSOS, P. J., et al.: X-radiation from clusters of galaxies: spectral evidence for a hot evolved gas. Ap. J. **211**, L 63 (1977).

SHAPIRO, S. L., SALPETER, E. E.: Accretion onto neutron stars under adiabatic shock conditions. Ap. J. **198**, 671 (1975).

SHAPIRO, P. R., MOORE, R. T.: Time-dependent radiative cooling of a hot, diffuse cosmic gas, and the emergent X-ray spectrum. Ap. J. **207**, 460 (1976).

SHUKLA, P. G., PAUL, J.: Gamma-ray production by the inverse Compton process in interstellar space. Ap. J. **208**, 893 (1976).

SILK, J., ARONS, J.: On the nature of the globular cluster X-ray sources. Ap. J. **200**, L 131 (1975).

SILK, J.: Accretion by galaxy clusters and the relationship between X-ray luminosity and velocity dispersion. Ap. J. **208**, 646 (1976).

SMOOT, G. F., BUFFINGTON, A., ORTH, C. D.: Search for cosmic-ray antimatter. Phys. Rev. Lett. **35**, 258 (1975).

SOFIA, S., VAN HORN, H. M.: The origin of the cosmic gamma-ray bursts. Ap. J. **194**, 593 (1974).

SOFIA, S.: The bright stars associated with galactic X-ray sources. Ap. J. **188**, L 45 (1974).

SOLINGER, A., RAPPAPORT, S., BUFF, J.: Isothermal blast wave model of supernova remnants. Ap. J. **201**, 381 (1975).

SPARKS, W. M., STARRFIELD, S., TRURAN, J. W.: CNO abundances and hydrodynamic models of the nova outburst. IV. Comparison with observations. Ap. J. **208**, 819 (1976).

SPARKS, W. M., STARRFIELD, S., TRURAN, J. W.: A hydrodynamic study of a slow nova outburst. Ap. J. **220**, 1063 (1978).

STARRFIELD, S., TRURAN, J. W., SPARKS, W. M.: Novae, supernovae, and neutron sources. Ap. J. **198**, L 113 (1975).

STECKER, F. W., et al.: Molecular hydrogen in the galaxy and galactic gamma rays. Ap. J. **201**, 90 (1975).

STECKER, F. W.: Observations of galactic gamma-rays and their implications for galactic structure studies. Ap. J. **212**, 60 (1977).

STERN, R., BOWYER, S.: Apollo-Soyuz survey of the extreme-ultraviolet/soft X-ray background. Ap. J. **230**, 755 (1979).

STRAKA, W. C.: Numerical models of the evolution of supernova remnants: the shell-formation stage. Ap. J. **190**, 59 (1974).

STRAKA, W. C., LADA, C. J.: Emission from supernova remnants. I. Thermal bremsstrahlung in the Sedov-Taylor phase. Ap. J. **195**, 563 (1975).

STRONG, I. A., KLEBESADEL, R. W., OLSON, R. A.: A preliminary catalog of transient cosmic gamma-ray sources observed by the Vela satellites. Ap. J. **188**, L 1 (1974).

TARLÉ, G., AHLEN, S. P., CARTWRIGHT, B. G.: Measurement of the isotopic composition of the iron-group elements in the galactic cosmic radiation. Phys. Rev. Lett. **41**, 771 (1978).

THORNE, K. S., PRICE, R. H.: Cygnus X-1: an interpretation of the spectrum and its variability. Ap. J. **195**, L 101 (1975).

TUCHMAN, Y., SACK, N., BARKAT, Z.: A new high ($\geqslant 6\,M_\odot$) upper mass limit for planetary nebula formation, and a new high lower mass bound for carbon detonation supernova models. Ap. J. **225**, L 137 (1978).

VAN DEN BERGH, S.: A systematic search for galactic supernova remnants. Ap. J. Suppl. **38**, 119 (1978).

VAN DEN HEUVEL, E. P. J.: Modes of mass transfer and classes of binary X-ray sources. Ap. J. **198**, L 109 (1975).

VAN HORN, H. M., HANSEN, C. J.: A model for the transient X-ray sources. Ap. J. **191**, 479 (1974).

VAN RIPER, K. A., ARNETT, W. D.: Stellar collapse and explosion: hydrodynamics of the core. Ap. J. **225**, L 129 (1978).

WALKER, A. B. C. Jr., RUGGE, H. R., WEISS, K.: Relative coronal abundances derived from X-ray observations. I. Sodium, magnesium, aluminum, silicon, sulfur, and argon. Ap. J. **188**, 423 (1974).

WALKER, A. B. C. Jr., RUGGE, H. R., WEISS, K.: Relative coronal abundances derived from Y-ray observations. II. Nitrogen, oxygen, neon, magnesium, and iron. Ap. J. **192**, 169 (1974).

WATSON, W. D.: Production of galactic X-rays following charge exchange by cosmic-ray nuclei. Ap. J. **206**, 842 (1975).

WEAVER, T. A.: The structure of supernova shock waves. Ap. J. Suppl. **32**, 233 (1976).

WENTZEL, D. G., et al.: On the relationship between CO and gamma-ray observations, cosmic rays, and the thickness of the galactic disk. Ap. J. **201**, L 5 (1975).

WENTZEL, D. G.: Isotropy of cosmic rays caused by magnetic discontinuities. Ap. J. **216**, L 59 (1977).

WHEELER, J. C.: Type I supernovae. Ap. J. **187**, 337 (1974).

WHEELER, J. C., MCKEE, C. F., LECAR, M.: Neutron stars in close binary systems. Ap. J. **192**, L 71 (1974).

WHEELER, J. C., LECAR, M., MCKEE, C. F.: Supernovae in binary systems. Ap. J. **200**, 145 (1975).

WHEELER, J. C.: X-ray bursts from magnetized accretion disks. Ap. J. **214**, 560 (1977).

WHEELER, J. C.: Type I supernovae, R Coronae Borealis stars, and the Crab Nebula. Ap. J. **225**, 212 (1978).

WILLIAMSON, F. O., et al.: Observations of features in the soft X-ray background flux. Ap. J. **193**, L 133 (1974).

WOODS, R. T., HART, H. R. Jr., FLEISCHER, R. L.: Apollo 17 cosmic-ray experiment: interplanetary heavy nuclei of energies 0.05 to 5.0 MeV per atomic mass unit. Ap. J. **198**, 183 (1975).

YUEH, W. R., BUCHLER, J. R.: The effects of Fermi statistics on neutrino transport in supernova models. Ap. J. **211**, L 121 (1977).

YUEH, W. R., BUCHLER, J. R.: Neutrino transport in supernovae models: SN method. Ap. J. **217**, 565 (1977).

ZAUMEN, W. T.: Pair production in intense magnetic fields. Ap. J. **210**, 776 (1976).

Chapter 5, Part 1

ADAMS, T. F.: A survey of the Seyfert galaxies based on large-scale image-tube plates. Ap. J. Suppl. **33**, 19 (1977).

ANILE, A. M., et al.: Cosmological turbulence reexamined. Ap. J. **205**, L 59 (1976).

ANILE, A. M., MOTTA, S.: Perturbations of the general Robertson-Walker universes and angular variations of the cosmic blackbody radiation. Ap. J. **207**, 685 (1976).

AUDOUZE, J., TINSLEY, B. M.: Galactic evolution and the formation of the light elements. Ap. J. **192**, 487 (1974).

AUDOUZE, J., et al.: Implications of the presence of deuterium in the galactic center. Ap. J. **208**, L 51 (1976).

ADOUZE, J., TINSLEY, B. M.: Chemical evolution of galaxies. Ann. Rev. Astron. Ap. **14**, 43 (1976).

BAHCALL, J. N., SARAZIN, C. L.: X-ray line spectroscopy for clusters of galaxies. I. Ap. J. **219**, 781 (1978).

BAHCALL, N. A.: The Perseus cluster: galaxy distribution, anisotropy and the mass/luminosity ratio. Ap. J. **187**, 439 (1974).

BAHCALL, N. A.: Core radii and central densities of 15 rich clusters of galaxies. Ap. J. **198**, 249 (1975).

BAHCALL, N. A.: Clusters of galaxies. Ann. Rev. Astron. Ap. **15**, 505 (1977).

BAHCALL, N. A.: The luminosity function of galaxy systems: from single galaxies and small groups to rich clusters. Ap. J. **232**, 689 (1979).

BALICK, B., FABER, S. M., GALLAGHER, J. S.: H I in early-type galaxies. III. Observations of SO galaxies. Ap. J. **209**, 710 (1976).

BECKLIN, E. E., NEUGEBAUER, G.: High-resolution maps of the galactic center at 2.2 and 10 microns. Ap. J. **200**, L 71 (1975).

BERGERSON, J., GUNN, J. E.: The extended H I regions around spiral galaxies: a probe for galactic structure and the intergalactic medium. Ap. J. **217**, 892 (1977).

BLITZ, L.: The rotation curve of the galaxy to R = 16 kiloparsecs. Ap. J. **231**, L 115 (1979).

BÖHM-VITENSE, E., et al.: Masses and luminosities of population II Cepheids. Ap. J. **194**, 125 (1974).

BÖHM-VITENSE, E.: The luminosities of population II Cepheids. Ap. J. **188**, 571 (1974).

BONOMETTO, S. A., LUCCHIN, F.: Conditions for galaxy formation from adiabatic fluctuations. Ap. J. **206**, 391 (1976).

BURBIDGE, G.: On the masses and relative velocities of galaxies. Ap. J. **196**, L 7 (1975).

Burbidge, G. R., Narlikar, J. V.: The log N-log S curve for 3CR radio galaxies and the problem of identifying faint radio galaxies. Ap. J. **205**, 329 (1976).
Burbidge, G., Perry, J.: On the masses of the quasi-stellar objects. Ap. J. **205**, L 55 (1976).
Canuto, V.: Inhomogeneities in the early universe. Ap. J. **205**, 659 (1976).
Chincarini, G. L., Rood, H. J.: Empirical properties of the mass discrepancy in groups and clusters of galaxies. IV. Double compact galaxies. Ap. J. **194**, 21 (1974).
Contopoulos, G.: Inner Lindblad resonance in galaxies. Nonlinear theory. Ap. J. **201**, 566 (1975).
Cowie, L. L., McKee, C. F.: Extragalactic diffuse neutral hydrogen clouds: probes of a hot intergalactic medium. Ap. J. **209**, L 105 (1976).
Cowie, L. L., Perrenod, S. C.: The origin and distribution of gas within rich clusters of galaxies: the evolution of cluster X-ray sources over cosmological time scales. Ap. J. **219**, 354 (1978).
Cowsik, R., Lerche, I.: On galactic rotation and the intergalactic medium. Ap. J. **199**, 555 (1975).
Cox, D. P., Smith, B. W.: Accretion by the galaxy: effects of radiative cooling on the flow structure and infall rate. Ap. J. **203**, 361 (1976).
Crane, P.: Empirical evidence for galaxy evolution. Ap. J. **198**, L 9 (1975).
Davidson, K.: Dielectronic recombination and abundances near quasars. Ap. J. **195**, 285 (1975).
Davis, M., Wilkinson, D. T.: Search for primeval galaxies. Ap. J. **192**, 251 (1974).
Davis, M., Peebles, P. J. E.: On the integration of the BBGKY equations for the development of strongly nonlinear clustering in an expanding universe. Ap. J. Suppl. **34**, 425 (1977).
Defouw, R. J.: Dynamical response of the interstellar medium to explosions at the galactic center. Ap. J. **208**, 52 (1976).
de Vaucouleurs, G.: Supergalactic studies. III. The supergalactic distribution of nearby groups of galaxies. Ap. J. **202**, 610 (1975).
de Vaucouleurs, G., Corwin, H. G. Jr.: Supergalactic studies. II. Supergalactic distribution of the nearest intergalactic gas clouds. Ap. J. **202**, 327 (1975).
de Vaucouleurs, G., Bollinger, G.: Contributions to galaxy photometry. VI. Revised standard total magnitudes and colors of 228 multiply observed galaxies. Ap. J. Suppl. **34**, 469 (1977).
de Vaucouleurs, G., Corwin, H. G. Jr., Bollinger, G.: Contributions to galaxy photometry. III. Total magnitudes and colors of 296 galaxies in the B,V system derived from the Holmberg photographic photometry. Ap. J. Suppl. **33**, 229 (1977).
De Young, D. S., Roberts, M. S.: The stability of galaxy clusters: neutral hydrogen observations. Ap. J. **189**, 1 (1974).
Dressel, L. L., Condon, J. J.: Accurate optical positions of bright galaxies. Ap. J. Suppl. **31**, 187 (1976).
Dressel, L. L., Condon, J. J.: The Arecibo 2380 MHz survey of bright galaxies. Ap. J. Suppl. **36**, 53 (1978).
Duus, A., Newell, B.: A catalog of southern groups and clusters of galaxies. Ap. J. Suppl. **35**, 209 (1977).
Eichler, D.: A possible manifestation of unseen matter. Ap. J. **208**, L 5 (1976).
Eichler, D.: On the extragalactic interpretation of high-velocity H I clouds. Ap. J. **208**, 694 (1976).
Faber, S. M., Gallagher, J. S.: H I in early-type galaxies. II. Mass loss and galactic winds. Ap. J. **204**, 365 (1976).
Faber, S. M., Jackson, R. E.: Velocity dispersions and mass-to-light ratios for elliptical galaxies. Ap. J. **204**, 668 (1976).
Faber, S. M., et al.: The neutral hydrogen content, stellar rotation curve, and mass-to-light ratio of NGC 4594, the "Sombrero" galaxy. Ap. J. **214**, 383 (1977).
Faber, S. M., Gallagher, J. S.: Masses and mass-to-light ratios of galaxies. Ann. Rev. Astron. Ap. **17**, 135 (1979).
Falk, S. W., Arnett, W. D.: Radiation dynamics, envelope ejection, and supernova light curves. Ap. J. Suppl. **33**, 515 (1977).
Fall, S. M., Saslaw, W. C.: The growth of correlations in an expanding universe and the clustering of galaxies. Ap. J. **204**, 631 (1976).
Fall, S. M.: Galaxy correlations and cosmology. Rev. Mod. Phys. **51**, 21 (1979).
Felten, J. E.: On Schmidt's V_m estimator and other estimators of luminosity functions. Ap. J. **207**, 700 (1976).
Field, G. B., Perrenod, S. C.: Constraints on a dense hot intergalactic medium. Ap. J. **215**, 717 (1977).

GOLDREICH, P., TREMAINE, S.: The excitation and evolution of density waves. Ap. J. **222**, 850 (1978).
GORDON, M. A.: Determination of the spiral pattern speed of the galaxy. Ap. J. **222**, 100 (1978).
GOTT, J. R. III: On the formation of elliptical galaxies. Ap. J. **201**, 296 (1975).
GOTT, J. R. III, THUAN, T. X.: On the formation of spiral and elliptical galaxies. Ap. J. **204**, 649 (1976).
GOTT, J. R. III: Recent theories of galaxy formation. Ann. Rev. Astron. **15**, 235 (1977).
GOTT, J. R. III, TURNER, E. L.: Groups of galaxies. III. Mass-to-light ratios and crossing times. Ap. J. **213**, 309 (1977).
GOTT, J. R. III, TURNER, E. L.: Groups of galaxies. IV. The multiplicity function. Ap. J. **216**, 357 (1977).
GOTTLIEB, D. M.: Skymap: a new catalog of stellar data. Ap. J. Suppl. **38**, 287 (1978).
GRASDALEN, G. L.: (V-K) colors of galaxies: statistical differences between spirals and ellipticals and the color-diameter relation for elliptical galaxies. Ap. J. **195**, 605 (1975).
GREGORY, S. A.: Redshifts and morphology of galaxies in the Coma cluster. Ap. J. **199**, 1 (1975).
GRINDLAY, J. E.: The optical counterpart of GX 339-4, a possible black hole X-ray source. Ap. J. **232**, L 33 (1979).
GUNN, J. E.: Massive galactic halos. I. Formation and evolution. Ap. J. **218**, 592 (1977).
HAINEBACH, K. L., SCHRAMM, D. N.: Galactic evolution models and the rhenium-187-osmium-187 chronometer: a greater age for the galaxy. Ap. J. **207**, L 79 (1976).
HAINEBACH, K. L., SCHRAMM, D. N.: Comments on galactic evolution and nucleocosmochronology. Ap. J. **212**, 347 (1977).
HARRISON, E. R.: Interpretation of redshifts of galaxies in clusters. Ap. J. **191**, L 51 (1974).
HARTWICK, F. D. A., SARGENT, W. L. W.: The mass of M 31 as determined from the motions of its globular clusters. Ap. J. **190**, 283 (1974).
HARTWICK, F. D. A.: Upper limits to the mass of the Virgo clusters of galaxies. Ap. J. **208**, L 13 (1976).
HARTWICK, F. D. A.: The chemical evolution of the galactic halo. Ap. J. **209**, 418 (1976).
HARTWICK, F. D. A., SARGENT, W. L. W.: Radial velocities for outlying satellites and their implications for the mass of the galaxy. Ap. J. **221**, 512 (1978).
HUMPHREYS, R. M.: Studies of luminous stars in nearby galaxies. I. Supergiants and O stars in the Milky Way. Ap. J. Suppl. **38**, 309 (1978).
HUNTLEY, J. M., SASLAW, W. C.: The distribution of stars in galactic nuclei: loaded polytropes. Ap. J. **199**, 328 (1975).
ILLINGWORTH, G., ILLINGWORTH, W.: The masses of globular clusters. I. Surface brightness distributions and star counts. Ap. J. Suppl. **30**, 227 (1976).
JACKSON, P. D., KELLMAN, S. A.: A redetermination of the galactic H I half-thickness and a discussion of some dynamical consequences. Ap. J. **190**, 53 (1974).
JANES, K. A.: Evidence for an abundance gradient in the galactic disk. Ap. J. Suppl. **39**, 135 (1979).
JENSEN, E. B., STROM, K. M., STROM, S. E.: Composition gradients in spiral galaxies: a consistency check on the density-wave theory. Ap. J. **209**, 748 (1976).
JONES, B. J. T.: The origin of galaxies: a review of recent theoretical developments and their confrontation with observation. Rev. Mod. Phys. **48**, 107 (1976).
JURA, M.: Star formation in elliptical galaxies. Ap. J. **212**, 634 (1977).
KELLERMANN, K. I., et al.: Further observations of apparent changes in the structure of 3C 273 and 3C 279. Ap. J. **189**, L 19 (1974).
KELLERMANN, K. I., et al.: Observations of compact radio nuclei in Cygnus A, Centaurus A., and other extended radio sources. Ap. J. **197**, L 113 (1975).
KHACHIKIAN, E. Y., WEEDMAN, D. W.: An atlas of Seyfert galaxies. Ap. J. **192**, 581 (1974).
KNOBLOCH, E.: Tidal disruption of clusters. Ap. J. **209**, 411 (1976).
KNOBLOCH, E.: Stochastic tidal disruption of clusters. Ap. J. **218**, 406 (1977).
KNOBLOCH, E.: Tidal interactions of galaxies. Ap. J. Suppl. **38**, 253 (1978).
KRISTIAN, J., SANDAGE, A., KATEM, B.: On the systematic optical identification of the remaining 3C radio sources. I. A search in 47 fields. Ap. J. **191**, 43 (1974).
LARSON, R. B., TINSLEY, B. M.: Photometric properties of model spherical galaxies. Ap. J. **192**, 293 (1974).
LARSON, R. B., TINSLEY, B. M.: Star formation rates in normal and peculiar galaxies. Ap. J. **219**, 46 (1978).

LAU, Y. Y., BERTIN, G.: Discrete spiral modes, spiral waves, and the local dispersion relationship. Ap. J. **226**, 508 (1978).
LEA, S. M.: The dynamics of the intergalactic medium in the vicinity of clusters of galaxies. Ap. J. **203**, 569 (1976).
LEIR, A. A., VAN DEN BERGH, S.: A study of 1889 rich clusters of galaxies. Ap. J. Suppl. **34**, 381 (1977).
LIANG, E. P.: Dynamics of primordial inhomogeneities in model universes. Ap. J. **204**, 235 (1976).
LIANG, E. P. T.: Nonlinear periodic waves in a self-gravitating fluid and galaxy formation. Ap. J. **230**, 325 (1979).
LIGHT, E. S., DANIELSON, R. E., SCHWARZSCHILD, M.: The nucleus of M 31. Ap. J. **194**, 257 (1974).
LIGHTMAN, A. P., PRESS, W. H.: Time evolution of galaxy correlations in a model for gravitational instability. Ap. J. **225**, 677 (1978).
MCCLELLAND, J., SILK, J.: The correlation function for density perturbations in an expanding universe. Ap. J. **216**, 665 (1977).
MCCLELLAND, J., SILK, J.: The correlation function for density perturbations in an expanding universe. II. Nonlinear theory. Ap. J. **217**, 331 (1977).
MCCLELLAND, J., SILK, J.: The correlation function for density perturbations in an expanding universe. III. The three-point and predictions of the four-point and higher order correlation functions. Ap. J. Suppl. **36**, 389 (1978).
MCKEE, C. F., PETROSIAN, V.: Are quasars dusty? Ap. J. **189**, 17 (1974).
MCKEE, C. F., TARTER, C. B.: Radiation pressure in quasar clouds. Ap. J. **202**, 306 (1975).
MARCHANT, A. B., SHAPIRO, S. L.: The formation of elliptical galaxies by tidal interactions. Ap. J. **215**, 1 (1977).
MARGOLIS, S. H., SCHRAMM, D. N.: Dust in the universe? Ap. J. **214**, 339 (1977).
MARK, J. W.-K.: On density waves in galaxies. I. Source terms and action conservations. Ap. J. **193**, 539 (1974).
MARK, J. W.-K.: On density waves in galaxies. III. Wave amplification by stimulated emission. Ap. J. **205**, 363 (1976).
MARK, J. W.-K.: On density waves in galaxies. V. Maintenance of spiral structure and discrete spiral modes. Ap. J. **212**, 645 (1977).
MATHEWS, W. G.: Radiative acceleration of gas clouds near quasi-stellar objects and Seyfert galaxy nuclei. Ap. J. **189**, 23 (1974).
MEIER, D. L.: Have primeval galaxies been detected? Ap. J. **203**, L 103 (1976).
MEIER, D. L.: The optical appearance of model primeval galaxies. Ap. J. **207**, 343 (1976).
MILLER, R. H.: On the stability of a disk galaxy. Ap. J. **190**, 539 (1974).
MILLER, G. E., SCALO, J. M.: The initial mass function and stellar birthrate in the solar neighborhood. Ap. J. Suppl. **41**, 513 (1979).
MOTTMANN, J., ABELL, G. O.: The luminosity function of galaxies in cluster A 2670. Ap. J. **218**, 53 (1977).
MUELLER, M. W., ARNETT, W. D.: Propagating star formation and irregular structure in spiral galaxies. Ap. J. **210**, 670 (1976).
NOONAN, T. W.: Binary galaxies and the problem of masses of clusters of galaxies. Ap. J. **196**, 683 (1975).
O'CONNELL, R. W.: Galaxy spectral synthesis. I. Stellar populations in the nuclei of giant ellipticals. Ap. J. **206**, 370 (1976).
OLSON, D. W., SILK, J.: Primordial inhomogeneities in the expanding universe. II. General features of spherical models at late times. Ap. J. **233**, 395 (1979).
OORT, J. H.: The galactic center. Ann. Rev. Astron. Ap. **15**, 295 (1977).
OPHER, R.: Acceleration of QSO clouds by radiation pressure. Ap. J. **187**, 5 (1974).
OPHER, R.: A theory of galactic nuclei and quasi-stellar objects. Ap. J. **188**, 201 (1974).
OSTRIKER, J. P., PEEBLES, P. J. E., YAHIL, A.: The size and mass of galaxies, and the mass of the universe. Ap. J. **193**, L 1 (1974).
OSTRIKER, J. P., RICHSTONE, D. O., THUAN, T. X.: On the numbers, birthrates, and final states of moderate- and high-mass stars. Ap. J. **188**, L 87 (1974).
OSTRIKER, J. P., THUAN, T. X.: Galactic evolution. II. Disk galaxies with massive halos. Ap. J. **202**, 353 (1975).
OSTRIKER, J. P., TREMAINE, S. D.: Another evolutionary correction to the luminosity of giant galaxies. Ap. J. **202**, L 113 (1975).

Ostriker, J. P., Hausman, M. A.: Cannibalism among the galaxies: dynamically produced evolution of cluster luminosity functions. Ap. J. **217**, L 125 (1977).

Pacholczyk, A. G., Scott, J. S.: In situ particle acceleration and physical conditions in radio tail galaxies. Ap. J. **203**, 313 (1976).

Partridge, R. B.: A search for primeval galaxies at high redshifts. Ap. J. **192**, 241 (1974).

Peebles, P. J. E.: The gravitational-instability picture and the nature of the distribution of galaxies. Ap. J. **189**, L 51 (1974).

Peebles, P. J. E., Hauser, M. G.: Statistical analysis of catalogs of extragalactic objects. III. The Shane-Wirtanen and Zwicky catalogs. Ap. J. Suppl. **28**, 19 (1974).

Peebles, P. J. E.: A cosmic virial theorem. Ap. J. **205**, L 109 (1976).

Peimbert, M., Torres-Peimbert, S., Rayo, J. F.: Abundance gradients in the galaxy derived from H II regions. Ap. J. **220**, 516 (1978).

Peters, W. L.: Models for the inner regions of the galaxy. I. An elliptical streamline model. Ap. J. **195**, 617 (1975).

Peterson, C. J., Roberts, M. S.: Extended rotation curves of high-luminosity spiral galaxies. III. The spiral galaxy NGC 7217. Ap. J. **226**, 770 (1978).

Peterson, S. D.: Double galaxies. I. Observational data on a well-defined sample. Ap. J. Suppl. **40**, 527 (1979).

Petrosian, V.: Surface brightness and evolution of galaxies. Ap. J. **209**, L 1 (1976).

Poveda, A., Allen, C.: A new upper limit to the mass loss from the central region of the galaxy. Ap. J. **200**, 42 (1975).

Press, W. H., Schechter, P.: Formation of galaxies and clusters of galaxies by self-similar gravitational condensation. Ap. J. **187**, 425 (1974).

Press, W. H.: Exact evolution of photons in an anisotropic cosmology with scattering. Ap. J. **205**, 311 (1976).

Richstone, D. O.: Collisions of galaxies in dense clusters. I. Dynamics of collisions of two galaxies. Ap. J. **200**, 535 (1975).

Richstone, D. O.: Collisions of galaxies in dense clusters. II. Dynamical evolution of cluster galaxies. Ap. J. **204**, 642 (1976).

Roberts, M. S., Whitehurst, R. N.: The rotation curve and geometry of M 31 at large galactocentric distances. Ap. J. **201**, 327 (1975).

Roberts, M. S., Steigerwald, D. G.: A search for neutral hydrogen clouds in radio galaxies and in intergalactic space. Ap. J. **217**, 883 (1977).

Roberts, W. W. Jr., Roberts, M. S., Shu, F. H.: Density wave theory and the classification of spiral galaxies. Ap. J. **196**, 381 (1975).

Roberts, W. W. Jr., Huntley, J. M., van Albada, G. D.: Gas dynamics in barred spirals: gaseous density waves and galactic shocks. Ap. J. **233**, 67 (1979).

Rood, H. J.: Concerning two forces hypothesized to resolve the mass discrepancy of galaxy clusters. Ap. J. **193**, 15 (1974).

Rood, H. J.: Empirical properties of the mass discrepancy in groups and clusters of galaxies. II. Ap. J. **188**, 451 (1974).

Rood, H. J.: Empirical properties of the mass discrepancy in groups and clusters of galaxies. V. Nine samples. Ap. J. **194**, 27 (1974).

Rood, H. J.: Empirical properties of the mass discrepancy in groups and clusters of galaxies. VI. Projection effects. Ap. J. **199**, 549 (1975).

Rood, H. J.: Corrected ratios of average mass to average luminosity for double galaxies. Ap. J. **205**, 354 (1976).

Rood, H. J.: Nearby groups of galaxy clusters. Ap. J. **207**, 16 (1976).

Rood, H. J., Dickel, J. R.: Radial velocities and masses of galaxies in groups from 21-centimeter line observations. Ap. J. **205**, 346 (1976).

Rood, H. J.: The virial mass and mass-to-light ratio of the Andromeda (M 31) subgroup. Ap. J. **232**, 699 (1979).

Rood, H. J., Dickel, J. R.: Analytical examination of virial properties of groups of galaxies. Ap. J. **233**, 418 (1979).

Rose, W. K., Tinsley, B. M.: Late stages of stellar evolution in the light of elliptical galaxies. Ap. J. **190**, 243 (1974).

Rubin, V. C.: The scatter on the Hubble diagram and the motion of the local group. Ap. J. **211**, L 1 (1977).

Rubin, V. C., Ford, W. K. Jr., Thonnard, N.: Extended rotation curves of high-luminosity spiral galaxies. IV. Systematic dynamical properties, Sa→Sc. Ap. J. **225**, L 107 (1978).

Rudnick, L., Owen, F. N.: Head-tail radio sources in clusters of galaxies. Ap. J. **203**, L 107 (1976).

Saaf, A. F.: On gaseous flows in disk galaxies. Ap. J. **189**, 33 (1974).

Sanders, R. H.: Explosions in galactic nuclei and the formation of double radio sources. Ap. J. **205**, 335 (1976).

Sargent, W. L. W., Turner, E. L.: A statistical method for determining the cosmological density parameter from the redshifts of a complete sample of galaxies. Ap. J. **212**, L 3 (1977).

Saslaw, W. C.: The ejection of massive objects from galactic nuclei: gravitational scattering on the object by the nucleus. Ap. J. **195**, 773 (1975).

Saslaw, W. C.: Orbit segregation in evolving galaxies and clusters of galaxies. Ap. J. **216**, 690 (1977).

Schechter, P.: An analytic expression for the luminosity function for galaxies. Ap. J. **203**, 297 (1976).

Schmidt, M.: The mass of the galactic halo derived from the luminosity function of high-velocity stars. Ap. J. **202**, 22 (1975).

Schwarz, J., Ostriker, J. P., Yahil, A.: Explosive events in the early universe. Ap. J. **202**, 1 (1975).

Seldner, M., Peebles, P. J. E.: A new way to estimate the mean mass density associated with galaxies. Ap. J. **214**, L 1 (1977).

Shields, G. A.: Composition gradients across spiral galaxies. Ap. J. **193**, 335 (1974).

Shields, G. A., Tinsley, B. M.: Composition gradients across spiral galaxies. II. The stellar mass limit. Ap. J. **203**, 66 (1976).

Silk, J.: On the fragmentation of cosmic gas clouds. I. The formation of galaxies and the first generation of stars. Ap. J. **211**, 638 (1977).

Silk, J.: Protogalaxy interactions in newly formed clusters: galaxy luminosities, colors, and intergalactic gas. Ap. J. **220**, 390 (1978).

Soneira, R. M., Peebles, P. J. E.: Is there evidence for a spatially homogeneous population of field galaxies? Ap. J. **211**, 1 (1977).

Spinrad, H., et al.: Halos of spiral galaxies: photometry and mass-to-light ratios. Ap. J. **225**, 56 (1978).

Spinrad, H., Stone, R. P. S.: An upper limit to the extragalactic background light. Ap. J. **226**, 609 (1978).

Stoner, R., Ptak, R.: The effect of suprathermal protons on the physical conditions in Seyfert galaxy nuclei. II. Ap. J. **208**, 298 (1976).

Talbot, R. J. Jr., Arnett, W. D.: Some recent results from galactic and stellar evolution theory. Ap. J. **190**, 605 (1974).

Talbot, R. J. Jr., Arnett, W. D.: The evolution of galaxies. IV. Highly flattened disks. Ap. J. **197**, 551 (1975).

Terrell, J.: Size limits on fluctuating astronomical sources. Ap. J. **213**, L 93 (1977).

Tifft, W. G., Gregory, S. A.: Direct observations of the large-scale distribution of galaxies. Ap. J. **205**, 696 (1976).

Tinsley, B. M.: Constraints on models for chemical evolution in the solar neighborhood. Ap. J. **192**, 629 (1974).

Tinsley, B. M.: Interpretation of the stellar metallicity distribution. Ap. J. **197**, 159 (1975).

Tinsley, B. M.: Nucleochronology and chemical evolution. Ap. J. **198**, 145 (1975).

Tinsley, B. M.: Chemical evolution in the solar neighborhood. II. Statistical constraints, finite stellar lifetimes, and inhomogeneities. Ap. J. **208**, 797 (1976).

Tinsley, B. M.: Surface brightness parameters as tests of galactic evolution. Ap. J. **209**, L 7 (1976).

Tinsley, B. M.: Surface brightness parameters as tests of galactic evolution. Ap. J. **210**, L 49 (1976).

Tinsley, B. M., Gunn, J. E.: Evolutionary synthesis of the stellar population in elliptical galaxies. I. Ingredients, broad-band colors, and infrared features. Ap. J. **203**, 52 (1976).

Tinsley, B. M.: Chemical evolution in the solar neighborhood. III. Time scales and nucleochronology. Ap. J. **216**, 548 (1977).

Tinsley, B. M.: Evolutionary synthesis of the stellar population in elliptical galaxies. II. Late M giants and composition effects. Ap. J. **222**, 14 (1978).

Tinsley, B. M., Larson, R. B.: Chemical evolution and the formation of galactic disks. Ap. J. **221**, 554 (1978).

Tinsley, B. M.: Stellar lifetimes and abundance ratios in chemical evolution. Ap. J. **229**, 1046 (1979).

TOOMRE, A.: Theories of spiral structure. Ann. Rev. Astron. Ap. **15**, 437 (1977).
TREMAINE, S. D., OSTRIKER, J. P., SPITZER, L. Jr.: The formation of the nuclei of galaxies. I. M31. Ap. J. **196**, 407 (1975).
TREMAINE, S. D.: The formation of the nuclei of galaxies. II. The local group. Ap. J. **203**, 345 (1976).
THUAN, T. X., GOTT, J. R. III: The angular momentum of galaxies. Ap. J. **216**, 194 (1977).
TURNER, E. L., GOTT, J. R. III: Evidence for a spatially homogeneous component of the universe: single galaxies. Ap. J. **197**, L 89 (1975).
TURNER, E. L.: Binary galaxies. II. Dynamics and mass-to-light ratios. Ap. J. **208**, 304 (1976).
TURNER, E. L., GOTT, J. R. III: Groups of galaxies. I. A catalog. Ap. J. Suppl. **32**, 409 (1976).
TURNER, E. L., GOTT, J. R. III: Groups of galaxies. II. The luminosity function. Ap. J. **209**, 6 (1976).
VAN DEN BERGH, S.: Tentative identification of main-sequence stars in the nuclear bulge of the galaxy. Ap. J. **188**, L 9 (1974).
VAN DEN BERGH, S.: The nature of quasars. Ap. J. **198**, L 1 (1975).
VEEDER, G. J.: The local mass density. Ap. J. **191**, L 57 (1974).
VIDAL-MADJAR, A., et al.: The ratio of deuterium to hydrogen in interstellar space. III. The lines of sight to zeta Puppis and gamma Cassiopeiae. Ap. J. **211**, 91 (1977).
WEISHEIT, J. C., COLLINS, L. A.: Model galactic coronae: ionization structure and absorption-line spectra. Ap. J. **210**, 299 (1976).
WEYMANN, R.: Confinement and internal structure of radiatively accelerated quasar clouds. Ap. J. **208**, 286 (1976).
WITTELS, J. J., et al.: Apparent "superrelativistic" expansions of the extragalactic radio source 3C 345. Ap. J. **206**, L 75 (1976).
YAHIL, A.: The density profiles of rich clusters of galaxies. Ap. J. **191**, 623 (1974).
YAHIL, A.: The universe as a "non-ideal gas" of galaxies. Ap. J. **204**, L 59 (1976).
YAHIL, A.: On the dynamics of binary galaxies. Ap. J. **217**, 27 (1977).
YAHIL, A., TAMMANN, G. A., SANDAGE, A.: The local group: the solar motion relative to its centroid. Ap. J. **217**, 903 (1977).
YAHIL, A., VIDAL, N. V.: The velocity distribution of galaxies in clusters. Ap. J. **214**, 347 (1977).

Chapter 5, Part 2

AARONSON, M., HUCHRA, J., MOULD, J.: The infrared luminosity/H I velocity-width relation and its application to the distance scale. Ap. J. **229**, 1 (1979).
ABRAMOWICZ, M. A., WAGONER, R. V.: Analytic properties of relativistic, rotating bodies. Ap. J. **216**, 86 (1977).
ABRAMOWICZ, M. A., WAGONER, R. V.: Slowly rotating, relativistic stars. Ap. J. **226**, 1063 (1978).
ADAMS, R. C., COHEN, J. M.: Analytic stellar models in general relativity. Ap. J. **198**, 507 (1975).
ANDERSON, J. D., et al.: Experimental test of general relativity using time-delay data from Mariner 6 and Mariner 7. Ap. J. **200**, 221 (1975).
BAHCALL, J. N., WOLF, R. A.: Star distribution around a massive black hole in a globular cluster. Ap. J. **209**, 214 (1976).
BAHCALL, J. N., WOLF, R. A.: The star distribution around a massive black hole in a globular cluster. II. Unequal star masses. Ap. J. **216**, 883 (1977).
BALBUS, S. A., BRECHER, K.: Tidal friction in the binary pulsar system PSR 1913+16. Ap. J. **203**, 202 (1976).
BARDEEN, J. M., PETTERSON, J. A.: The Lense-Thirring effect and accretion disks around Kerr black holes. Ap. J. **195**, L 65 (1975).
BARKER, B. M., O'CONNELL, R. F.: Relativistic effects in the binary pulsar PSR 1913+16. Ap. J. **199**, L 25 (1975).
BARRY, G. W.: Charged cosmology. Ap. J. **190**, 279 (1974).
BAUM, W. A., FLORENTIN-NIELSEN, R.: Cosmological evidence against time variation of the fundamental atomic constants. Ap. J. **209**, 319 (1976).

BESSELL, M. A., et al.: 2U 1700-37; another black hole? Ap. J. **187**, 355 (1974).
BLANDFORD, R., TEUKOLSKY, S. A.: On the measurement of the mass of PSR 1913+16. Ap. J. **198**, L 27 (1975).
BLEYER, U., JOHN, R. W., LIEBSCHER, D.-E.: On a new method of determining the gravitational constant. Gerlands Beitr. Geophys. Leipzig **86**, 148 (1977).
BLUMENTHAL, G. R., MATHEWS, W. G.: Spherical winds and accretion in general relativity. Ap. J. **203**, 714 (1976).
BONAZZOLA, S., SCHNEIDER, J.: An exact study of rigidly and rapidly rotating stars in general relativity with application to the Crab pulsar. Ap. J. **191**, 273 (1974).
BRANS, C. H.: Propagation of electromagnetic polarization effects in anisotropic cosmologies. Ap. J. **197**, 1 (1975).
BRECHER, K.: Some implications of period changes in the first binary radio pulsar. Ap. J. **195**, L 113 (1975).
BRECHER, K.: Is the speed of light independent of the velocity of the source? Phys. Rev. Lett. **39**, 1051 (1977).
BRECHER, K.: Possible test of the strong principle of equivalence. Ap. J. **219**, L 117 (1978).
BROWN, G. S., TINSLEY, B. M.: Galaxy counts as a cosmological test. Ap. J. **194**, 555 (1974).
BROWN, R. W., STECKER, F. W.: Cosmological Baryon-number domain structure from symmetry breaking in grand unified field theories. Phys. Rev. Lett. **43**, 315 (1979).
BRUZUAL, G. A., SPINRAD, H.: The characteristic size of clusters of galaxies: a metric rod used for a determination of q_0. Ap. J. **220**, 1 (1978).
BURKE, W. L.: Large-scale random gravitational waves. Ap. J. **196**, 329 (1975).
BUTTERWORTH, E. M., IPSER, J. R.: Rapidly rotating fluid bodies in general relativity. Ap. J. **200**, L 103 (1975).
BUTTERWORTH, E. M.: On the structure and stability of rapidly rotating fluid bodies in general relativity. II. The structure of uniformly rotating pseudopolytropes. Ap. J. **204**, 561 (1976).
BUTTERWORTH, E. M., IPSER, J. R.: On the structure and stability of rapidly rotating fluid bodies in general relativity. I. The numerical method for computing structure and its application to uniformly rotating homogeneous bodies. Ap. J. **204**, 200 (1976).
CAMENZIND, M.: Theories of gravity with structure-dependent γ's. Phys. Rev. Lett. **35**, 1188 (1975).
CANDELAS, P., SCIAMA, D. W.: Irreversible thermodynamics of black holes. Phys. Rev. Lett. **38**, 1372 (1977).
CANUTO, V., HSIEH, S. H., ADAMS, P. J.: Scale-covariant theory of gravitation and astrophysical applications. Phys. Rev. Lett. **39**, 429 (1977).
CANUTO, V., LODENQUAI, J.: Dirac cosmology. Ap. J. **211**, 342 (1977).
CANUTO, V., TSIANG, E.: On the hydrodynamic expansion of a relativistic gas. Ap. J. **213**, 27 (1977).
CANUTO, V. M., HSIEH, S.-H.: Scale covariance and G-varying cosmology. II. Thermodynamics, radiation, and the 3K background. Ap. J. Suppl. **41**, 243 (1979).
CANUTO, V. M., HSIEH, S.-H., OWEN, J. R.: Scale covariance and G-varying cosmology. III. The (m,z), (θ_m, z), (θ_i, z) and $[N(m), m]$ tests. Ap. J. Suppl. **41**, 263 (1979).
CANUTO, V. M., OWEN, J. R.: Scale covariance and G-varying cosmology. IV. The log N-log S relation. Ap. J. Suppl. **41**, 301 (1979).
CARR, B. J.: Some cosmological consequences of primordial black-hole evaporations. Ap. J. **206**, 8 (1976).
CHAN, K. L., JONES, B. J. T.: Distortions of the microwave background radiation spectrum in the submillimeter wavelength region. Ap. J. **198**, 245 (1975).
CHAN, K. L., JONES, B. J. T.: Distortions of the 3°K background radiation spectrum: observational constraints on the early thermal history of the universe. Ap. J. **195**, 1 (1975).
CHAN, K. L., JONES, B. J. T.: The evolution of the cosmic radiation spectrum under the influence of turbulent heating. I. Theory. Ap. J. **200**, 454 (1975).
CHAN, K. L., JONES, B. J. T.: The evolution of the cosmic radiation spectrum under the influence of turbulent heating. II. Numerical calculation and application. Ap. J. **200**, 461 (1975).
CHAPMAN, G. A.: Recent measurements of the flux excess from solar faculae and the implication for the solar oblateness. Phys. Rev. Lett. **34**, 755 (1975).
CHENG, E. S., et al.: Large-scale anisotropy in the 2.7 K radiation. Ap. J. **232**, L 139 (1979).
CHIA, T. T., CHAU, W. Y., HENRIKSEN, R. N.: Gravitational radiation from a rotating collapsing gaseous ellipsoid. Ap. J. **214**, 576 (1977).

Chin, C.-W., Stothers, R.: Limit on the secular change of the gravitational constant based on studies of solar evolution. Phys. Rev. Lett. **36**, 833 (1976).

Chitre, D. M., Hartle, J. B.: Stationary configurations and the upper bound on the mass of non-rotating, causal neutron stars. Ap. J. **207**, 592 (1976).

Combley, F., et al.: g-2 experiments as a test of special relativity. Phys. Rev. Lett. **42**, 1383 (1979).

Contopoulos, G., Spyrou, N.: The center of mass in the post-Newtonian approximation of general relativity. Ap. J. **205**, 592 (1976).

Cooper, P. S., et al.: Experimental test of special relativity from a high-γ electron g-2 measurement. Phys. Rev. Lett. **42**, 1386 (1979).

Counselman, C. C. III, et al.: Solar gravitational deflection of radio waves measured by very-long-baseline interferometry. Phys. Rev. Lett. **33**, 1621 (1974).

Crowley, R. J., Thorne, K. S.: The generation of gravitational waves. II. The postlinear formalism revisited. Ap. J. **215**, 624 (1977).

Cunningham, C. T.: The effects of redshifts and focusing on the spectrum of an accretion disk around a Kerr black hole. Ap. J. **202**, 788 (1975).

Cunningham, C.: Returning radiation in accretion disks around black holes. Ap. J. **208**, 534 (1976).

Damour, T., Ruffini, R.: Quantum electrodynamical effects in Kerr-Newmann geometries. Phys. Rev. Lett. **35**, 463 (1975).

Davis, M., Geller, M. J., Huchra, J.: The local mean mass density of the universe: new methods for studying galaxy clustering. Ap. J. **221**, 1 (1978).

Detweiler, S. L.: A variational calculation of the fundamental frequencies of quadrupole pulsation of fluid spheres in general relativity. Ap. J. **197**, 203 (1975).

Detweiler, S. L.: A variational principle and a stability criterion for the dipole modes of pulsation of stellar models in general relativity. Ap. J. **201**, 440 (1975).

Detweiler, S. L.: Black holes and gravitational waves. I. Circular orbits about a rotating hole. Ap. J. **225**, 687 (1978).

de Vaucouleurs, G., Pence, W. D.: Type I supernovae as cosmological clocks. Ap. J. **209**, 687 (1976).

de Vaucouleurs, G., Bollinger, G.: The extragalactic distance scale. VII. The velocity-distance relations in different directions and the Hubble ratio within and without the local supercluster. Ap. J. **233**, 433 (1979).

de Vaucouleurs, G., Nieto, J.-L.: Luminosity distribution in the central regions of Messier 87: isothermal core, point source, or black hole? Ap. J. **230**, 697 (1979).

Dicke, R. H.: Faculae and the solar oblateness. II. Ap. J. **190**, 187 (1974).

Dicke, R. H., Goldenberg, H. M.: The oblateness of the sun. Ap. J. Suppl. **27**, 131 (1974).

Dicke, R. H.: New solar rotational period, the solar oblateness, and solar faculae. Phys. Rev. Lett. **37**, 1240 (1976).

Dicus, D. A., Kolb, E. W., Teplitz, V. L.: Cosmological implications of massive, unstable neutrinos. Ap. J. **221**, 327 (1978).

Douglass, D. H., et al.: Two-detector-coincidence search for bursts of gravitational radiation. Phys. Rev. Lett. **35**, 480 (1975).

Eardley, D. M.: Observable effects of a scalar gravitational field in a binary pulsar. Ap. J. **196**, L 59 (1975).

Eardley, D. M., Press, W. H.: Astrophysical processes near black holes. Ann. Rev. Astron. Ap. **13**, 381 (1975).

Ehlers, J., et al.: Comments on gravitational radiation damping and energy loss in binary systems. Ap. J. **208**, L 77 (1976).

Eichler, D., Solinger, A.: The electromagnetic background: limitations on models of unseen matter. Ap. J. **203**, 1 (1976).

Endal, A. S., Sofia, S.: Detectable gravitational radiation from stellar collapse. Phys. Rev. Lett. **39**, 1429 (1977).

Epstein, R., Wagoner, R. V.: Post-Newtonian generation of gravitational waves. Ap. J. **197**, 717 (1975).

Epstein, R.: The binary pulsar: Post-Newtonian timing effects. Ap. J. **216**, 92 (1977).

Esposito, L. W., Harrison, E. R.: Properties of the Hulse-Taylor binary pulsar system. Ap. J. **196**, L 1 (1975).

FISHBONE, L. G., MONCRIEF, V.: Relativistic fluid disks in orbit around Kerr black holes. Ap. J. **207**, 962 (1976).

FISHBONE, L. G.: Relativistic fluid disks in orbit around Kerr black holes. II. Equilibrium structure of disks with constant angular momentum per baryon. Ap. J. **215**, 323 (1977).

FOMALONT, E. B., SRAMEK, R. A.: A confirmation of Einstein's general theory of relativity by measuring the bending of microwave radiation in the gravitational field of the sun. Ap. J. **199**, 749 (1975).

FOMALONT, E. B., SRAMEK, R. A.: Measurements of the solar gravitational deflection of radio waves in agreement with general relativity. Phys. Rev. Lett. **36**, 1475 (1976).

FOX, J. G., et al.: On defining "rest" in Robertson-Walker cosmologies. Ap. J. **201**, 545 (1975).

FRIEDMAN, J. L., SCHUTZ, B. F.: Gravitational radiation instability in rotating stars. Ap. J. **199**, L 157 (1975).

FRIEDMAN, J. L., SCHUTZ, B. F.: On the stability of relativistic systems. Ap. J. **200**, 204 (1975).

GIBBONS, G. W., PERRY, M. J.: Black holes in thermal equilibrium. Phys. Rev. Lett. **36**, 985 (1976).

GOTT, J. R. III: A time-symmetric, matter, antimatter, tachyon cosmology. Ap. J. **187**, 1 (1974).

GOTT, J. R., et al.: An unbound universe? Ap. J. **194**, 543 (1974).

GRINDLAY, J., GURSKY, H.: Scattering model for X-ray bursts: massive black holes in globular clusters. Ap. J. **205**, L 131 (1976).

GUNN, J. E., OKE, J. B.: Spectrophotometry of faint cluster galaxies and the Hubble diagram: an approach to cosmology. Ap. J. **195**, 255 (1975).

GUNN, J. E., TINSLEY, B. M.: Dynamical friction: the Hubble diagram as a cosmological test. Ap. J. **210**, 1 (1976).

HARTLE, J. B.: Effective-potential approach to graviton production in the early universe. Phys. Rev. Lett. **39**, 1373 (1977).

HARTWICK, F. D. A.: Interstellar absorption and an apparent anisotropy in the Hubble expansion. Ap. J. **195**, L 7 (1975).

HAUGAN, M. P., WILL, C. M.: Weak interactions and Eötvös experiments. Phys. Rev. Lett. **37**, 1 (1976).

HICKSON, P.: The angular-size-redshift relation. II. A test for the decleration parameter. Ap. J. **217**, 964 (1977).

HIER, R. G., RASBAND, S. N.: Cylinders as gravitational radiation telescopes. Ap. J. **195**, 507 (1975).

HILL, H. A., et al.: Solar oblateness excess brightness, and relativity. Phys. Rev. Lett. **33**, 1497 (1974).

HILL, H. A., STEBBINS, R. T.: The intrinsic visual oblateness of the sun. Ap. J. **200**, 471 (1975).

HIRAKAWA, H., NARIHARA, K.: Search for gravitational radiation at 145 MHz. Phys. Rev. Lett. **35**, 330 (1975).

HISCOCK, W. A.: On tidal interactions with Kerr black holes. Ap. J. **216**, 908 (1977).

JAKOBSEN, H. P., KON, M., SEGAL, I. E.: Angular momentum of the cosmic background radiation. Phys. Rev. Lett. **42**, 1788 (1979).

JOHN, V. R. W.: The geodetic interval in a static spherically symmetric Riemannian space-time in the second post-Minkowskian approximation and the signal retardation effect of Einstein's gravitation theory. Experimentelle Technik der Physik **23**, 127 (1975).

JOHN, V. R. W.: Explizite Darstellungsformeln Greenscher Funktionen von kovarianten Wellengleichungen im schwachen Gravitationsfeld. Ann. Phys. **29**, 15 (1973).

KATZ, J., HORWITZ, G.: Thermodynamic stability of relativistic rotating stellar configurations and a maximum principle for the entropy. Ap. J. **194**, 439 (1974).

KENNICUTT, R. C. Jr.: H II regions as extragalactic distance indicators. I. Observations and structural considerations. Ap. J. **228**, 394 (1979).

KENNICUTT, R. C. Jr.: H II regions as extragalactic distance indicators. II. Application of isophotal diameters. Ap. J. **228**, 696 (1979).

KENNICUTT, R. C. Jr.: H II regions as extragalactic distance indicators. III. Application of H II region fluxes and galaxy diameters. Ap. J. **228**, 704 (1979).

KOVÁCS, S., THORNE, K. S.: The generation of gravitational waves. III. Derivation of bremsstrahlung formulae. Ap. J. **217**, 252 (1977).

KRISTIAN, J., SANDAGE, A., WESTPHAL, J. A.: The extension of the Hubble diagram. II. New redshifts and photometry of very distant galaxy clusters: first indication of a deviation of the Hubble diagram form a straight line. Ap. J. **221**, 383 (1978).

LANG, K. R., et al.: The composite Hubble diagram. Ap. J. **202**, 583 (1975).

LANSBERG, P. T., PATHRIA, R. K.: Cosmological parameters for a restricted class of closed big-bang universes. Ap. J. **192**, 577 (1974).

LATTIMER, J. M., SCHRAMM, D. N.: Black-hole-neutron-star collisions. Ap. J. **192**, L 145 (1974).

LATTIMER, J. M., SCHRAMM, D. N.: The tidal disruption of neutron stars by black holes in close binaries. Ap. J. **210**, 549 (1976).

LEE, D. L., et al.: Theoretical frameworks for testing relativistic gravity. V. Post-Newtonian limit of Rosen's theory. Ap. J. **206**, 555 (1976).

LEITER, D., KAFATOS, M.: Penrose pair production in massive, extreme Kerr black holes. Ap. J. **226**, 32 (1978).

LEVINE, J. L., GARWIN, R. L.: New negative result for gravitational wave detection, and comparison with reported detection. Phys. Rev. Lett. **33**, 794 (1974).

LIANG, R. P. T.: Relativistic simple waves: shock damping and entropy production. Ap. J. **211**, 361 (1977).

LIGHTMAN, A. P.: Time-dependent accretion disks around compact objects. I. Theory and basic equations. Ap. J. **194**, 419 (1974).

LIGHTMAN, A. P.: Time-dependent accretion disks around compact objects. II. Numerical models and instability of inner region. Ap. J. **194**, 429 (1974).

LIGHTMAN, A. P., EARDLEY, D. M.: Black holes in binary systems: instability of disk accretion. Ap. J. **187**, L 1 (1974).

LIGHTMAN, A. P., SHAPIRO, S. L.: The distribution and consumption rate of stars around a massive, collapsed object. Ap. J. **211**, 244 (1977).

LYTTLETON, R. A., FITCH, J. P.: Effect of a changing G on the moment of inertia of the earth. Ap. J. **221**, 412 (1978).

MANSFIELD, V. N., MALIN, S.: Some astrophysical effects of the time variation of all masses. Ap. J. **209**, 335 (1976).

MANSFIELD, V. N.: Dirac cosmologies and the microwave background. Ap. J. **210**, L 137 (1976).

MARKS, D. W.: On the spherical symmetry of static stars in general relativity. Ap. J. **211**, 266 (1977).

MASTERS, A. R., ROBERTS, D. H.: On the nature of the binary system containing the pulsar PSR 1913 + 16. Ap. J. **195**, L 107 (1975).

MATZNER, R. A., RYAN, M. P. Jr.: Scattering of gravitational radiation from vacuum black holes. Ap. J. Suppl. **36**, 451 (1978).

MEIER, D. L., et al.: A new determination of the luminosity function of radio galaxies and an investigation of the evolutionary properties of the radio galaxy population in the recent past. Ap. J. **229**, 25 (1979).

MISRA, R. M.: Interaction of neutron stars with black holes. Ap. J. **203**, 704 (1976).

NEWMAN, D., et al.: Precision experimental verification of special relativity. Phys. Rev. Lett. **40**, 1355 (1978).

NOERDLINGER, P. D.: An evaluation of parallax in Friedmann universes. Ap. J. **218**, 317 (1977).

NORDTVEDT, K. Jr.: Anisotropic gravity and the binary pulsar PSR 1913+16. Ap. J. **202**, 248 (1975).

NOWOTNY, E.: A matter and radiation filled universe: consequences of the astronomical observations. Ap. J. **206**, 402 (1976).

O'DELL, S. L., ROBERTS, D. H.: Comments on "the composite Hubble diagram". Ap. J. **210**, 294 (1976).

OLSON, D. W.: Helium production and limits on the anisotropy of the universe. Ap. J. **219**, 777 (1978).

OLSON, D. W., SILK, J.: Inhomogeneity in cosmological element formation. Ap. J. **226**, 50 (1978).

OSTRIKER, J. P., TINSLEY, B. M.: Is deuterium of cosmological or of galactic origin? Ap. J. **201**, L 51 (1975).

PAGE, D. N., THORNE, K. S.: Disk-accretion onto a black hole. I. Time-averaged structure of accretion disk. Ap. J. **191**, 499 (1974).

PAGE, D. N., HAWKING, S. W.: Gamma rays from primordial black holes. Ap. J. **206**, 1 (1976).

PARIJSKIJ, Y. N.: New limit on small-scale irregularities of "blackbody" radiation (erratum). Ap. J. **188**, L 113 (1974).

PASACHOFF, J. M., CESARSKY, D. A.: Further observations at the interstellar deuterium frequency. Ap. J. **193**, 65 (1974).

PENZIAS, A. A.: The origin of the elements. Rev. Mod. Phys. **51**, 425 (1979).

PETERSON, B. A., et al.: Number magnitude counts of faint galaxies. Ap. J. **233**, L 109 (1979).

PETROSIAN, V.: The Hubble relation for nonstandard candles and the origin of the redshift of quasars. Ap. J. **188**, 443 (1974).

QUINTANA, H.: The structure equations of a slowly rotating, fully relativistic solid star. Ap. J. **207**, 279 (1976).

RICHARDSON, M. B., VAN HORN, H. M., SAVEDOFF, M. P.: A code for general relativistic stellar evolution. Ap. J. Suppl. **39**, 29 (1979).

ROBERTS, D. H., MASTERS, A. R., ARNETT, W. D.: Determining the stellar masses in the binary system containing the pulsar PSR 1913+16: is the companion a helium main-sequence star? Ap. J. **203** 196 (1976).

ROEDER, R. C.: Apparent magnitudes, redshifts, and inhomogeneities in the universe. Ap. J. **196**, 671 (1975).

ROSEN, N.: Bimetric gravitation and cosmology. Ap. J. **211**, 357 (1977).

ROSEN, N.: Bimetric gravitation theory and PSR 1913+16. Ap. J. **221**, 284 (1978).

SAENZ, R. A., SHAPIRO, S. L.: Gravitational radiation from stellar collapse: ellipsoidal models. Ap. J. **221**, 286 (1978).

SANDAGE, A., TAMMANN, G. A.: Steps toward the Hubble constant. I. Calibration of the linear sizes of extragalactic H II regions. Ap. J. **190**, 525 (1974).

SANDAGE, A., TAMMANN, G. A.: Steps toward the Hubble constant. II. The brightest stars in late-type spiral galaxies. Ap. J. **191**, 603 (1974).

SANDAGE, A., TAMMANN, G. A.: Steps toward the Hubble constant. III. The distance and stellar content of the M 101 group of galaxies. Ap. J. **194**, 223 (1974).

SANDAGE, A., TAMMANN, G. A.: Steps toward the Hubble constant. IV. Distances to 39 galaxies in the general field leading to a calibration of the galaxy luminosity classes and a first hint of the value of H_0. Ap. J. **194**, 559 (1974).

SANDAGE, A.: The redshift-distance relation. VIII. Magnitudes and redshifts of southern galaxies in groups: a further mapping of the local velocity field and an estimate of q_0. Ap. J. **202**, 563 (1975).

SANDAGE, A., TAMMANN, G. A.: Steps toward the Hubble constant. V. The Hubble constant from nearby galaxies and the regularity of the local velocity field. Ap. J. **196**, 313 (1975).

SANDAGE, A., TAMMANN, G. A.: Steps toward the Hubble constant. VI. The Hubble constant determined from redshifts and magnitudes of remote Sc I galaxies: the value of q_0. Ap. J. **197**, 265 (1975).

SANDAGE, A., TAMMANN, G. A.: Steps toward the Hubble constant. VII. Distances to NGC 2403, M 101, and the Virgo cluster using 21 centimeter line widths compared with optical methods: the global value of H_0. Ap. J. **210**, 7 (1976).

SARGENT, W. L., et al.: Dynamical evidence for a central mass concentration in the galaxy M 87. Ap. J. **221**, 731 (1978).

SCHWARTZ, D. A.: A new cosmological test for q_0. Ap. J. **206**, L 95 (1976).

SEQUIN, F. H.: The stability of nonuniform rotation in relativistic stars. Ap. J. **197**, 745 (1975).

SHAHAM, J.: Free precession of neutron stars: role of possible vortex pinning. Ap. J. **214**, 251 (1977).

SHAPIRO, I. I., COUNSELMAN, C. C. III, KING, R. W.: Verification of the principle of equivalence for massive bodies. Phys. Rev. Lett. **36**, 555 (1976).

SHAPIRO, S. L.: Accretion onto black holes: the emergent radiation spectrum. III. Rotating (Kerr) black holes. Ap. J. **189**, 343 (1974).

SHAPIRO, S. L., LIGHTMAN, A. P.: Black holes in X-ray binaries: marginal existence and rotation reversals of accretion disks. Ap. J. **204**, 555 (1976).

SHAPIRO, S. L., LIGHTMAN, A. P.: Rapidly rotating, post-Newtonian neutron stars. Ap. J. **207**, 263 (1976).

SHAPIRO, S. L.: Gravitational radiation from stellar collapse: the initial burst. Ap. J. **214**, 566 (1977).

SHAPIRO, S. L.: The dissolution of globular clusters containing massive black holes. Ap. J. **217**, 281 (1977).

SHAPIRO, S. L., MARCHANT, A. B.: Star clusters containing massive, central black holes: Monte Carlo simulations in two-dimensional phase space. Ap. J. **225**, 603 (1978).

SHECTMAN, S. A.: The small-scale anisotropy of the cosmic light. Ap. J. **188**, 233 (1974).

SHIELDS, G. A., WHEELER, J. C.: Sustenance of a black hole in a galactic nucleus. Ap. J. **222**, 667 (1978).

SILK, J.: Large-scale inhomogeneity of the universe: implications for the deceleration parameter. Ap. J. **193**, 525 (1974).

SILK, J.: The primordial generation of random shear motions and small-scale angular anisotropy in the microwave background radiation. Ap. J. **194**, 215 (1974).

SILK, J., WHITE, S. D. M.: The determination of q_0 using X-ray and microwave observations of galaxy clusters. Ap. J. **226**, L 103 (1978).

SILK, J., WILSON, M. L.: Primordial inhomogeneities in the expanding universe. I. Density and velocity distributions of galaxies in the vicinities of rich clusters. Ap. J. **228**, 641 (1979).

SILK, J., WILSON, M. L.: Primordial inhomogeneities in the expanding universe. III. The density and velocity distributions in cluster halos as a cosmological probe. Ap. J. **233**, 769 (1979).

SMARR, L. L., BLANDFORD, R.: The binary pulsar: physical processes, possible companions, and evolutionary histories. Ap. J. **207**, 574 (1976).

SMOOT, G. F., GORENSTEIN, M. V., MULLER, R. A.: Detection of anisotropy in the cosmic blackbody radiation. Phys. Rev. Lett. **39**, 898 (1977).

SOLHEIM, J.-E., BARNES, T. G. III, SMITH, H. J.: Observational evidence against a time variation in Planck's constant. Ap. J. **209**, 330 (1976).

SPYROU, N.: The N-body problem in general relativity. Ap. J. **197**, 725 (1975).

STEIGMAN, G.: A crucial test of the Dirac cosmologies. Ap. J. **221**, 407 (1978).

TAYLOR, J. H., et al.: Further observations of the binary pulsar PSR 1913+16. Ap. J. **206**, L 53 (1976).

TAYLOR, J. H., FOWLER, L. A., MCCULLOCH, P. M.: Measurements of general relativistic effects in the binary pulsar PSR 1913+16. Nature **277**, 437 (1979).

TEUKOLSKY, S. A., PRESS, W. H.: Perturbations of a rotating black hole. III. Interaction of the hole with gravitational end electromagnetic radiation. Ap. J. **193**, 443 (1974).

THORNE, K. S.: Disk-accretion onto a black hole. II. Evolution of the hole. Ap. J. **191**, 507 (1975).

THORNE, K. S., KOVACS, S. J.: The generation of gravitational waves. I. Weak-field sources. Ap. J. **200**, 245 (1975).

THORNE, K. S., BRAGINSKY, V. B.: Gravitational-wave bursts from the nuclei of distant galaxies and quasars: proposal for detection using Doppler tracking of interplanetary spacecraft. Ap. J. **204**, L 1 (1976).

THUAN, T. X., OSTRIKER, J. P.: Gravitational radiation from stellar collapse. Ap. J. **191**, L 105 (1974).

TINSLEY, B. M.: Galaxy counts, color-redshift relations and related quantities as probes of cosmology and galactic evolution. Ap. J. **211**, 621 (1977).

TREMAINE, S., GUNN, J. E.: Dynamical role of light neutral leptons in cosmology. Phys. Rev. Lett. **42**, 407 (1979).

TURNER, E. L.: Statistics of the Hubble diagram. I. Determination of q_0 and luminosity evolution with application to quasars. Ap. J. **230**, 291 (1979).

TURNER, M.: Gravitational radiation from point-masses in unbound orbits: Newtonian results. Ap. J. **216**, 610 (1977).

TURNER, M.: Tidal generation of gravitational waves from orbiting Newtonian stars. I. General formalism. Ap. J. **214**, 914 (1977).

TURNER, M., WILL, C. M.: Post-Newtonian gravitational bremsstrahlung. Ap. J. **220**, 1107 (1978).

TYSON, J. A., GIFFARD, R. P.: Gravitational-wave astronomy. Ann. Rev. Astron. Ap. **16**, 521 (1978).

VAN PATTEN, R. A., EVERITT, C. W. F.: Possible experiment with two counter-orbiting drag-free satellites to obtain a new test of Einstein's general theory of relativity and improved measurements of geodesy. Phys. Rev. Lett. **36**, 629 (1976).

VILA, S. C.: Changing gravitational constant and white dwarfs. Ap. J. **206**, 213 (1976).

WAGONER, R. V., MALONE, R. C.: Post-Newtonian neutron stars. Ap. J. **189**, L 75 (1974).

WAGONER, R. V.: Test for the existence of gravitational radiation. Ap. J. **196**, L 63 (1975).

WAGONER, R. V., WILL, C. M.: Post-Newtonian gravitational radiation from orbiting point masses. Ap. J. **210**, 764 (1976).

WAGONER, R. V.: Determining q_0 from supernovae. Ap. J. **214**, L 5 (1977).

WAGONER, R. V., WILL, C. M.: Post-Newtonian gravitational radiation from orbiting point masses: erratum. Ap. J. **215**, 984 (1977).

WARBURTON, R. J., GOODKIND, J. M.: Search for evidence of a preferred reference frame. Ap. J. **208**, 881 (1976).

WASSERMAN, I., BRECHER, K.: Time invariance of Planck's constant. Phys. Rev. Lett. **41**, 920 (1978).

WEINBERG, S.: Apparent luminosities in a locally inhomogeneous universe. Ap. J. **208**, L 1 (1976).

WEINBERG, S.: Cosmological production of baryons. Phys. Rev. Lett. **42**, 850 (1979).

WHEELER, J. C.: The binary pulsar: preexplosion evolution. Ap. J. **205**, 578 (1976).

WHITMAN, P. G., PIZZO, J. F.: Rotating neutron stars in general relativity: exact solutions for the case of slow rotation. Ap. J. **230**, 893 (1979).

WICKRAMASINGHE, D. T., et al.: 2U 0900-40; A black hole? Ap. J. **88**, 167 (1974).

WILL, C. M.: On the stability of axisymmetric systems to axisymmetric perturbation in general relativity. V. Differentially rotating configurations. Ap. J. **190**, 403 (1974).

WILL, C. M.: Perturbations of a slowly rotating black hole by a stationary axisymmetric ring of matter. I. Equilibrium configurations. Ap. J. **191**, 521 (1974).

WILL, C. M.: Periastron shifts in the binary system PSR 1913+16: theoretical interpretations. Ap. J. **196**, L 3 (1975).

WILL, C. M.: Perturbation of a slowly rotating black hole by a stationary axisymmetric ring of matter. II. Penrose processes, circular orbits, and differential mass formulae. Ap. J. **196**, 41 (1975).

WILL, C. M.: Active mass in relativistic gravity: theoretical interpretation of the Kreuzer experiment. Ap. J. **204**, 224 (1976).

WILL, C. M.: A test of post-Newtonian conservation laws in the binary system PSR 1913+16. Ap. J. **205**, 861 (1976).

WILL, C. M.: Gravitational radiation from binary systems in alternative metric theories of gravity: dipole radiation and the binary pulsar. Ap. J. **214**, 826 (1977).

WILLIAMS, J. G., et al.: New test of the equivalence principle from lunar ranging. Phys. Rev. Lett. **36**, 551 (1976).

WILSON, R. W.: The cosmic microwave background radiation. Rev. Mod. Phys. **51**, 433 (1979).

WOLFE, A. M., BROWN, R. L., ROBERTS, M. S.: Limits on the variation of fundamental atomic quantities over cosmic time scales. Phys. Rev. Lett. **37**, 179 (1976).

WOLFF, S. C., BEICHMAN, C. A.: The physical properties and orbital parameters of the BO Ia star HD 152667 = V 861 Scorpii: a supergiant with a black hole companion? Ap. J. **230**, 519 (1979).

WOODY, D. P., et al.: Measurement of the spectrum of the submillimeter cosmic background. Phys. Rev. Lett. **34**, 1036 (1975).

WOODY, D. P., RICHARDS, P. L.: Spectrum of the cosmic background radiation. Phys. Rev. Lett. **42**, 925 (1979).

YOUNG, P. J.: Angular momentum of a black hole in a dense stella system. Ap. J. **212**, 227 (1977).

YOUNG, P. J., et al.: Evidence for a supermassive object in the nucleus of the galaxy M 87 from SIT and CCD area photometry. Ap. J. **221**, 721 (1978).

Author Index

Subject Index

Astronomy and Astrophysics

A European Journal

Recognized as a "Europhysics Journal" by the European Physical Society

ISSN 0004-6361 Title No. 230

Astronomy & Astrophysics occupies an eminent position among journals in its field. Established in 1969, it is the result of the merging of six renowned European journals in astronomy and astrophysics.

Astronomy & Astrophysics presents papers on all aspects of astronomy and astrophysics – theoretical, observational, and instrumental – regardless of the techniques employed – optical, radio, particles, space vehicles, numerical analysis, etc. Letters to the editor, research notes and occasional review papers are also included.

Astronomy & Astrophysics is edited by an international staff of scientists.

Springer-Verlag
Berlin
Heidelberg
New York

Published by Springer-Verlag Berlin Heidelberg New York on behalf of the Board of Directors European Southern Observatory (ESO)

A. Unsöld

The New Cosmos

Translated from the German by R. C. Smith
Based on the translation by W. H. McCrea of the 1st edition

2nd revised and enlarged edition
1977. 166 figures, 21 tables. XII, 451 pages
(Heidelberger Science Library)
ISBN 3-540-90223-6

The book is the authorized English translation of Professor A. Unsöld's „Der Neue Kosmos", first published in 1967. It provides a comprehensive introductory text to present-day astronomy. It takes the reader through the entire subject, from apparent motions on the celestial sphere through studies of the solar system, the sun and stars, to stellar atmospheres, stellar evolution, radio-astronomy, high energy astrophysics and cosmology. It concludes with some considerations regarding the origin of life on the earth.

At the present time many graduates in mathematics and physics are turning to research in optical or radioastronomy, in astrophysics, in space science, or in cosmology. This book provides exactly the concise, but comprehensive introduction to modern astronomy that all such students need at the outset of their work.

Springer-Verlag
Berlin
Heidelberg
New York

Contents: Classical Astronomy. – Sun and Stars: Astrophysics of Individual Stars. – Stellar Systems: Milky Way and Galaxies; Cosmogony and Cosmology. – Physical Constants and Astronomical Quantities.